Life on Earth

Life on Earth

Third Edition

Teresa Audesirk

Gerald Audesirk

Bruce E. Byers

Pearson Custom Publishing

Prentice Hall

Taken from:

Life on Earth, Third Edition
by Teresa Audesirk, Gerald Audesirk and Bruce E. Byers

Published by Prentice Hall
Upper Saddle River, New Jersey 07458

This special edition published in cooperation with Pearson Custom Publishing.

Published in the United States of America

10 9 8 7 6 5 4 3 2

Please visit our web site at www.pearsoncustom.com

ISBN 0–536–72074–6

BA 996354

PEARSON CUSTOM PUBLISHING
75 Arlington Street, Suite 300, Boston, MA 02116
A Pearson Education Company

About the Authors

Both **Terry and Gerry Audesirk** grew up in New Jersey, where they met as undergraduates. After marrying in 1970, they moved to California, where Terry earned her doctorate in marine ecology at the University of Southern California and Gerry earned his doctorate in neurobiology at the California Institute of Technology. As postdoctoral students at the University of Washington's marine laboratories, they worked together on the neural bases of behavior, using a marine mollusk as a model system.

Terry and Gerry are now professors of biology at the University of Colorado at Denver, where they have taught introductory biology and neurobiology since 1982. In their research lab, funded by the National Institutes of Health, they investigate the mechanisms by which neurons are harmed by low levels of environmental pollutants.

Terry and Gerry share a deep appreciation of nature and of the outdoors. They enjoy hiking in the Rockies, running near their home in the foothills west of Denver, and attempting to garden at 7000 feet in the presence of hungry deer and elk. They are long-time members of many conservation organizations. Their daughter, Heather, has added another focus to their lives.

Bruce E. Byers, a midwesterner transplanted to the hills of western Massachusetts, is a professor in the biology department at the University of Massachusetts, Amherst. He's been a member of the faculty at UMass (where he also completed his doctoral degree) since 1993. Bruce teaches introductory biology courses for both nonmajors and majors; he also teaches courses in ornithology and animal behavior.

A lifelong fascination with birds ultimately led Bruce to scientific exploration of avian biology. His current research focuses on the behavioral ecology of birds, especially on the function and evolution of the vocal signals that birds use to communicate. The pursuit of vocalizations often takes Bruce outdoors, where he can be found before dawn, tape recorder in hand, awaiting the first songs of a new day.

Panel of Biology Educators
Life on Earth, Third Edition

We express sincere gratitude to the contributors who worked closely with the authors to ensure *Life on Earth's* continuing tradition of readability, accuracy, and relevance.

Rita Farrar
Louisiana State University

Research Areas: Immunoparasitology and immunopathology

I enjoy watching "non-science" majors lose their fear of science and find that they really can understand and learn biology. It is very gratifying to see them discover that they truly are a part of a fragile "web of life" on this planet and that even though they are not scientists, this web needs their understanding and diligent care.

James A. Hewlett
Finger Lakes Community College

Research Areas: Biotechnology and molecular ecology

My true love is in the teaching of a mixed-majors' undergraduate biology class. The mention of recombinant DNA brings out clamors for detailed protocols from majors, and concerns over the ethical and social aspects of such a practice from a liberal arts student. Art students become scientific illustrators as part of a class project while a business major is enthused over the marriage of biology and the NASDAQ. These combinations make teaching undergraduates enjoyable.

Edward Levri
Pennsylvania State University at Altoona

Research Areas: Evolutionary ecology: host–parasite and predator–prey relationships

Teaching nonmajors' biology is one of the most rewarding aspects of my career. Nonmajors bring many different views about the world into the course. These different perspectives make each class unique and make the discovery process more enjoyable. I learn something new every time I teach it. It's always special to find some students who realize for the first time that they actually like science.

Kenneth A. Mason
Purdue University

Research Area: Genetics of pigmentation in lower vertebrates

I enjoy teaching nonmajors' biology because I feel like I am doing something of benefit to a larger community than just the scientific community. If I can do my small part to increase science literacy and make biology more accessible to non-scientists, then I have made a genuine contribution to society.

Timothy Metz
Campbell University

Research Area: Plant biotechnology

I enjoy the challenge of helping students, especially those with limited science backgrounds, develop an appreciation of scientific investigation and an understanding of how biological knowledge is relevant to our individual and corporate lives. It is exciting to see students turned on to the critical-thinking process involved in science; to know that when the course is over, they leave better equipped to evaluate issues, ideas, and sources of information encountered throughout a lifetime.

Rhoda E. Perozzi
Virginia Commonwealth University

Research Areas: Plant ecology and biology education

A former student recently told me that my nonmajors' biology class was the only class in which she learned things that she uses every day. It is enabling students to grasp this relevance of biology to daily life that makes teaching a joy.

Susan M. Wadkowski
Lakeland Community College

Research Areas: Physiological plant biology and ecology

I enjoy teaching nonmajors' biology courses because they are a challenge. Students come into the course having "failed biology in high school" or "hated biology in high school," or "didn't do well." By the end of my course, most realize that biology is "not all that bad," "interesting," or even "fun." I love it when I know the students have learned, I've learned from them, and we had fun doing it.

Robin Wright
University of Washington

Research Areas: Cell biology and genetics

Teaching undergraduates is exciting to me because they bring fresh ideas and questions to problems that I frequently take for granted. These new perspectives are often valuable in helping to reveal preconceived notions in my own research approaches. In addition, since most of us AREN'T biologists, by bringing the excitement and wonder of biology to a more general audience, I hope to make a larger impact on how we, as a society, view science and research.

Brief Contents

Essays

Earth Watch

Health Watch

Scientific Inquiry

Evolutionary Connections

Contents

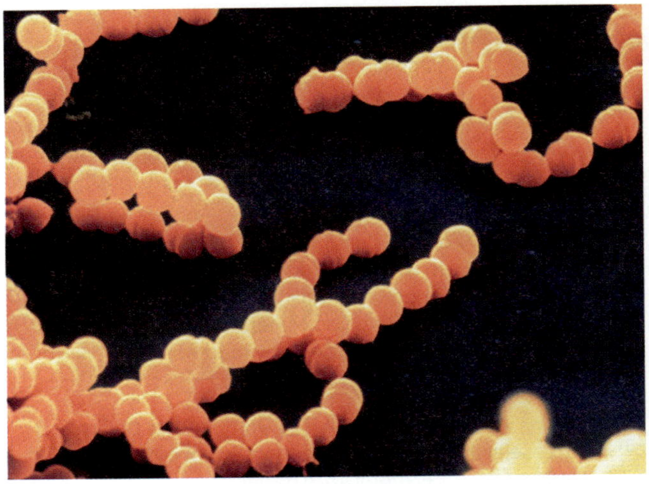

USA
Barcelona'92
USA
NGR

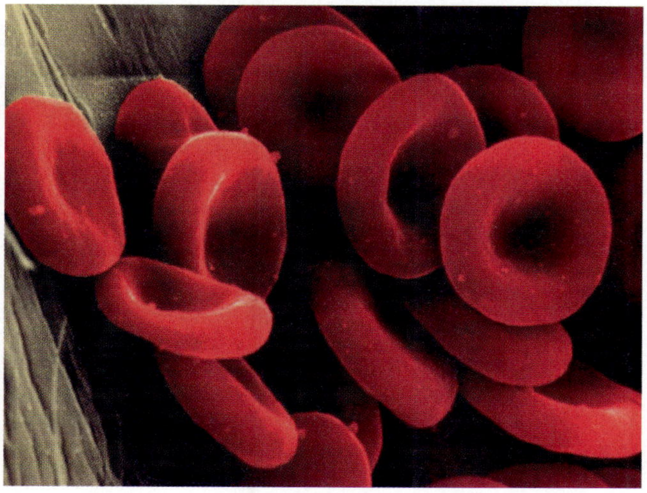

UNIT THREE

EVOLUTION 207

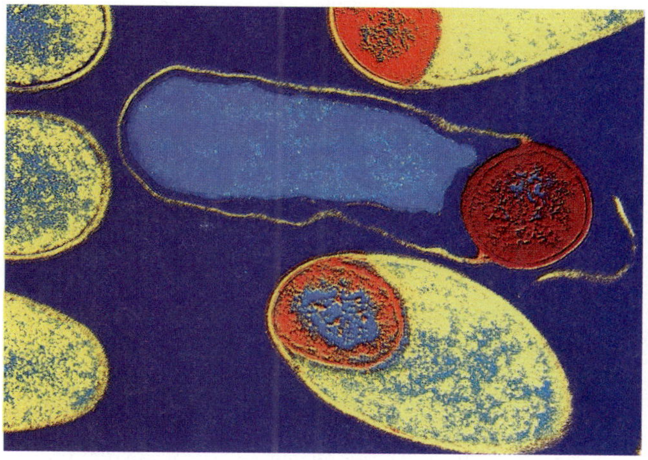

UNIT FOUR

PLANT ANATOMY AND PHYSIOLOGY 311

Chapter 17 Plant Form and Function 312

Chapter 18 The Plant Life Cycle 334

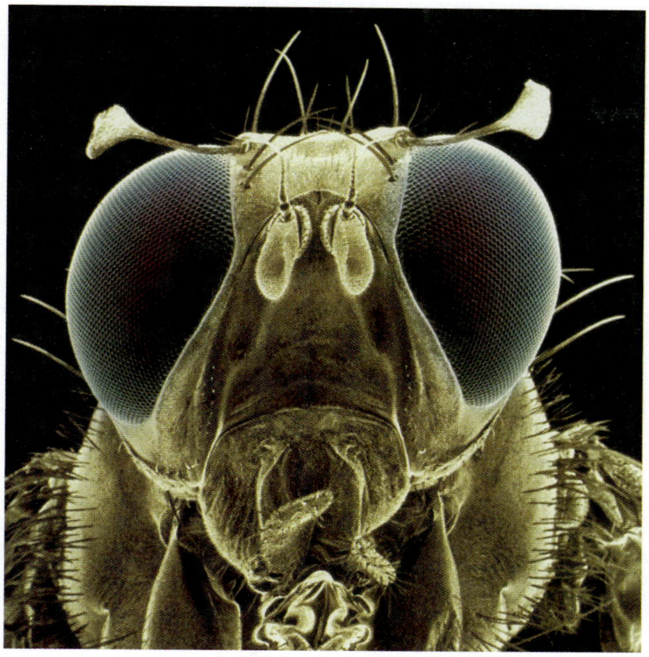

UNIT FIVE

ANIMAL ANATOMY AND PHYSIOLOGY 357

Chapter 28 Community Interactions 560

Chapter 29 How Do Ecosystems Work? 580

Chapter 30 Earth's Diverse Ecosystems 602

Preface

Will scientists clone a person in the foreseeable future? Are genetically engineered crops safe? Are people causing climate change? Is AIDS still spreading? Will physicians soon be transplanting pig hearts into people? The need for citizens to understand the basic concepts and issues of biology has never been more urgent. We have met with biology educators throughout the United States to discuss everything from the "big picture" to the details of how specific topics should be presented. These people combine exceptional teaching skills with expertise in diverse areas of biology. Such collaboration has been at the heart of this revision and of our effort to create a text that is responsive to the needs of today's students. Many of you are teaching a class that will be your students' final exposure to biology before they go out into the world. You told us that they should emerge from your course able to scrutinize science articles in the popular press with an educated and critical eye. Likewise, the course should prepare students to ask intelligent questions and make informed choices, both as voters and as consumers. You want your students to understand and appreciate the working of their own bodies. They should know about the other organisms with which we share Earth, the evolutionary forces that molded all life-forms, and how the complex interactions within ecosystems sustain us and all other life on Earth. Finally, you told us that you want a text with a relevancy that will leave students with a fascination for life that will inspire them to keep learning. This revision is our response. Now in its third edition, *Life on Earth* . . .

. . . Actively Engages Students

Each chapter opens with a strikingly illustrated Case Study. Our case studies are based on recent news items, on situations in which students might find themselves, or on particularly fascinating biological topics. For example, your students will investigate blood doping by elite athletes (p. 95); contemplate the controversy over genetically modified food (p. 197); see cloned, genetically engineered pigs that are being raised as potential organ donors for humans (p. 373); and follow along as a scientist estimates the number of species on Earth (p. 271). Each Case Study is revisited at the end of the chapter, allowing students to explore the topic a bit further in light of what they have learned and, often, to find answers to questions raised in the initial study. Finally, at the end of each chapter are Web Investigations, which take the students to the internet to further explore and apply the concepts addressed in the Case Study.

Throughout each chapter, our major headings pose important questions that encourage students to seek answers as they read. The full-sentence, conceptual subheadings both suggest answers to these questions and help students focus on the key points in each subsection.

We have placed icons next to key illustrations, tables, and concepts. These icons direct the student to a new section called the "MediaTutor" at the end of each chapter. Here, students will find descriptions of related topics that they can pursue (1) on our Web site, *Audesirk Live!*, or (2) on the MediaTutor student CD-ROM, which contains interactive exercises and animations that engage students in learning and help them visualize biological concepts.

We have continued to provide the popular group activities in each chapter. These exercises, used successfully by our Panel of Biology Educators, encourage students to work out interesting problems in small groups and thereby become active participants in the learning process. The *Instructor's Guide to Print and Media* includes additional Group Activities and provides suggestions about how instructors can incorporate these exercises into the classroom.

Finally, particularly where processes are illustrated, we have added annotations to figures. These annotations place descriptions of each process at the points where they are most needed for clarity and reduce the need for lengthy, multi-part captions.

. . . Conveys That Biology Is Everywhere

Throughout the text, we relate key biological concepts both to everyday experiences and to important issues facing society. These references are woven into the text, introduced in end-of-chapter critical-thinking questions ("Applying the Concepts"), and highlighted in updated boxed essays. These essays cover a wide range of current topics in biology, from environmental issues ("Earth Watch"), to clinical discussions ("Health Watch"), to the procedures biologists use in doing their work ("Scientific Inquiry"). For example:

Earth Watch: These environmental essays explore pressing issues, such as the loss of biodiversity, the growing ozone hole, and invasions of exotic species.

Health Watch: These clinical essays investigate topics such as sexually transmitted diseases, the dangers of artificial steroids, and how smoking damages the lungs.

. . . Is Flexible and Easy to Use

The chapters in *Life on Earth* have been written to allow instructors flexibility in the order of use. Chapter 1 sets the stage for the coverage of topics in different sequence by providing an overview of the diversity of life. Cross-referencing among chapters allows students to seek additional information on specific topics in other parts of the book.

At the start of each chapter, the conceptual questions are brought together in an effective summary, "At a Glance." A consistent numbering system identifies the major conceptual headings—all the way from At a Glance, through the body of the chapter in major headings, and into the "Summary of Key Concepts." A "Key Terms" list identifies all of the boldface terms that appear in the chapter, along with the page on which each term is introduced; key terms (and many others) are defined in the Glossary at the back of the book.

. . . Has Supporting Technology That Enhances Learning and Is Integrated into the Textbook

Life on Earth is fully supported by an array of both electronic and print ancillary materials. The Prentice Hall team has worked closely with biology educators and revised the text's supplements to ensure that they meet the needs both of students and instructors.

Instructional support in the classroom is the focus of the ancillary program. The Instructor Resource CD-ROM acts as a warehouse of presentation materials that allows the instructor to access all the available tools in one place at one time. Driven by a powerful database that allows instructors to browse by chapter or to search easily by table of contents, term, subject, media type, or figure number, it contains all the line art from the book (in both labeled and unlabeled styles) and many of the photos. For each chapter, there are also deluxe PowerPoint files that include animated art and editable labels. The entire Student MediaTutor CD-ROM, along with guidelines for integrating the animations and exercises into lectures, is fully accessible from the Instructor Resource CD-ROM. As an instructor, you will have the *Instructor's Guide to Print and Media Resources* on the CD-ROM and all of the questions from the Test Item File. The instructor's package includes 220 full-color acetates. Prentice Hall's *TestGen EQ* and *QuizMaster* testing software provides a selection of more than 2400 sample test questions in a variety of styles, allowing you to create and edit exams electronically.

An entirely new student CD-ROM accompanies each text and provides interactive exercises and animations. For easy reference and tight integration with our multimedia supplements, icons have been placed next to key illustrations, tables, and concepts throughout each chapter that direct students to the MediaTutor at the end of the chapter. Each MediaTutor is a resource for students that describes the specific content and objectives on the student CD-ROM and the "Web Investigation" for that particular chapter. By informing students about what to expect, MediaTutor allows them to follow their interests and needs, thus to make the best use of these resources.

The companion Web site, *Audesirk Live!*, has been completely updated with new questions, activities, and links to provide students with extensive opportunities for study and to provide instructors with additional sources for assignments. Our chapter Web sites are updated frequently, allowing students to explore topics relevant to each chapter on the World Wide Web. In *Audesirk Live!*, students will discover that biology is everywhere by exploring our "Issues in Biology" and "Bizarre Facts" sections. *The New York Times Themes of the Times*, our free newspaper-format supplement, provides a collection of recent science articles written by world-renowned science writers for this outstanding newspaper. These effective teaching enhancements connect biology in the classroom to what's happening in the world today.

The Fundamental Philosophy of *Life on Earth*

Although our textbook continues to evolve in response to the changing needs of our audience, there are fundamental issues in the teaching of biology that don't change. *Life on Earth* still does the following:

Focuses on Concepts

Our conceptual questions and subheadings cast as sentences, At a Glance chapter-opening outlines, and end-of-chapter Summary of Key Concepts sections keep students focused on the important themes in each chapter. Figure captions include caption titles, which give the theme of each image and then provide more specific information. Because it is easy for students to lose sight of the underlying concepts in a welter of technical detail, we provide a general overview of complex subjects without overwhelming the reader with unnecessary details and terminology.

Communicates the Scientific Process

Biology is not just a compendium of facts and ideas; rather, it is the outgrowth of a dynamic process of inquiry and human endeavor. In many instances, we describe how scientists discovered specific facts. The scientific process is further highlighted in Scientific Inquiry essays:

Scientific Inquiry: With these essays, students will learn how fossils are dated and how PET scans are performed. They follow the development of the science of genetics from the laboratory of Watson and Crick to the Roslin Center in Scotland, where Ian Wilmut cloned a sheep from adult DNA, and even to Lake Maracaibo in Venezuela, where Nancy Wexler began to unlock the secrets of Huntington disease.

Emphasizes Unifying Themes

To help students appreciate *Life on Earth's* "take home messages" further in the third edition, most chapters contain passages designated as "The Big Picture." Each Big Picture passage briefly summarizes a broad, key idea and provides a conceptual overview and a sense of context for chapter material. In addition, the unifying theme of evolution permeates the text. Throughout, students will find examples of how natural selection has produced organisms that are adapted to specific environments. In addition, many chapters end with an "Evolutionary Connections" section:

Evolutionary Connections: These lively discussions link chapter concepts into the broader perspective of evolution.

Our own concern for the environment can be found interwoven throughout this text and highlighted in Earth Watch essays as well. Whenever appropriate, we have tried to present students with the biological rationale for making sound environmental decisions in their daily lives.

Strives for Accuracy

A text is useless if it does not convey accurate information. To this end, we utilize multiple, high-level sources for our basic facts. Several talented educators who are experts in the unit subject areas carefully scrutinize each unit of each new edition. They help us in our quest to present the material in a way that is both accurate and comprehensible. In between editions, our users serve as informal reviewers. We never fail to follow up on your queries, and the book is better as a result. As always, however, the buck stops with the authors. We take responsibility for the accuracy of the text material, and we take this responsibility seriously.

A Final Word

A course in introductory biology may be a student's first—and sometimes last—in-depth exposure to the fascinating complexity of life. As teachers, we recognize how easily a student can become mired in a plethora of new facts and unfamiliar terms while losing sight of the underlying concepts of biology. We have carefully revised *Life on Earth* to reduce unnecessary detail and excess terminology and to emphasize the ways in which an understanding of biology can enrich and enlighten day-to-day living. This text can serve the student in many ways—from an "owner's manual" of the human body to a "user's guide" for the environment. Why study biology? Maybe we're biased, but what can be more fascinating than learning about Life on Earth?

Acknowledgments

Life on Earth is truly a group effort. We are fortunate that Robin Wright, a member of our second edition Panel of Biology Educators, agreed to participate in the revision of Unit Two. In addition to her teaching skills, she brings specialized knowledge of the rapidly changing applications of molecular biology to this task. Her careful revision has made the text more accurate and timely.

To meet the dauntingly complex challenge of putting together a text and supplement package of this magnitude, Prentice Hall has assembled an experienced and skilled development team. The text benefited considerably from the thoughtful suggestions of Developmental Editor Elmarie Hutchinson. She not only helped us keep the text clear, consistent, and student-friendly, she has also contributed substantially to the clarity of the revised art. Tim Flem, our Production Editor, coordinated the efforts of the photo researcher, copy editor, art studio, and authors. He skillfully brought the art, photos, and manuscript together into a seamless whole while dealing good-naturedly with last-minute improvements. Formatting this book is not an easy endeavor, but Tim applied his expertise with great attention to detail. Photo Researcher Jerry Marshall tracked down excellent photos. Roberta Dempsey tackled the job of copyediting with exceptional skill. Proofreader Margaret Buresch read the final manuscript with scrupulous care to help ensure accuracy.

We also wish to thank Art Director Wanda Espana and Director of Design Carole Anson for guiding the text and cover design with flair and talent, and Art Editor Adam Velthaus for coordinating such an immense art program.

Travis Moses-Westphal, our Project Manager, played multiple roles from the beginning with much creativity and gusto. He deftly carried the vision for the media program and oversaw the seamless integration of the text, media, and print supplements. Editorial Assistant Colleen Lee was always around just when we needed her with good cheer and brought exceptional skills to traffic the manuscript.

Colleagues at other institutions have helped us enormously. Many, listed in the following pages, have stimulated us to rethink our presentation with careful, thoughtful reviews. Our new Panel of Biology Educators, listed on page vi, has met with us personally and made a special contribution to the third edition.

The expansive and crucial supplements package has many valued contributors. Joseph Chinnici and Susan Wadkowski prepared the *Study Guide*; Gail Gasparich, Charles Good, and Kate Lajtha, prepared the *Test Item File*; Timothy Metz was responsible for the *Instructor's Guide to Print and Media Resources*; Joseph Coehlo, David Demers, Jim Hewlett, Kerry Kilburn, Ken Mason, Darcy Medica, Paul Ramp, Diana Wheat, and

Cal Young spent countless hours readying the content for the Student MediaTutor CD-ROM; and Kelly Johnson, Jeff Kenton, Chris Romero, Ina Pour-el, William Hayes, Lydia Daniels, and John Braun were responsible for updating the Web site.

Shari Meffert, our Marketing Manager, oversees a large and dedicated sales force with energy, talent, and enthusiasm. Shari provides inspired marketing concepts, shares success stories, and makes sure that user's comments always get through to the authors. We thank Paul Corey, now president of the Engineering, Science, and Mathematics Division of Prentice Hall, for his confidence and support through this and the past two editions. Finally, but most importantly, our editors: Editor in Chief Sheri Snavely has supported us now through three editions. Executive Editor Teresa Ryu Chung has assumed leadership of the team with talent and zeal, combined with a clear sense of where the project should be going and how to get it there without killing the authors. Her total commitment to the project, her organizational ability, and her sensitivity to all the people involved have been crucial to its success.

So here we acknowledge, with deep appreciation, our "coach" and all our teammates!

Terry and Gerry Audesirk
Bruce E. Byers

Panel of Multimedia Advisors and Reviewers

Media Contributors
Joseph Chinnici, *Virginia Commonwealth University*
Joseph Coelho, *Culver Stockton College*
Lydia Daniels, *University of Pittsburgh*
Lewis Deaton, *University of Southwestern Louisiana*
Gail Gasparich, *Towson State University*
William Hayes, *Delta State University*
James Hewlett, *Finger Lakes Community College*
David Huffman, *Southwest Texas State University*
J. Kelly Johnson, *University of Kansas*
Jeff Kenton, *Iowa State University*
Kate Lajtha, *Oregon State University*
Timothy Metz, *Campbell University*
Ina Pour-el, *DMACC–Boone Campus*
Paul Ramp, *Pellissippi State Technical College*
Chris Romero, *Front Range Community College*
Cal Young, *Fullerton College*

Media Reviewers
J. Gregory Burg, *University of Kansas*
Jerry Button, *Portland Community College*
Walter J. Conley, *State University of New York at Potsdam*
Jerry Cook, *Sam Houston State University*
David M. Demers, *University of Hartford*
Susannah Feldman, *Towson University*
Timothy L. Henry, *University of Texas Arlington*
James Hewlett, *Finger Lakes Community College*
Kelly Johnson, *University of Kansas*
Jeffrey Kiggins, *Blue Ridge Community College*
Harry Kurtz, *Sam Houston State University*
Kenneth A. Mason, *University of Kansas*
Timothy Metz, *Campbell University*
Marvin Price, *Cedar Valley College*
Chris Romero, *Front Range Community College*
Patricia Shields, *George Mason University*
Susan M. Wadkowski, *Lakeland Community College*
Stacy Wolfe, *Art Institutes International*
Robin Wright, *University of Washington*
Cal Young, *Fullerton College*

Third Edition Reviewers

Michelle Baker, *Utah State University*
Neil Baker, *Ohio State University*
Robert Boyd, *Auburn University*
Neil Buckley, *State University of New York, Plattsburgh*
Martin Cohen, *University of Hartford*
Deborah A. Fahey, *Wheaton College*
Matthew Fountain, *State University of New York, Fredonia*
Carole Griffiths, *Long Island University*
Elizabeth E. LeClair, *DePaul University*
Douglas Lyng, *Indiana University Purdue University, Fort Wayne*
James Murphy, *Monroe Community College*
Lance Myler, *State University of New York, Canton*
Gary Pettibone, *Buffalo State College*
Karen Raines, *Colorado State University*
Todd Rimkus, *Marymount University*
William D. Rogers, *Ball State University*
Sonya J. Williams, *Langston University*

Previous Editions Reviewers

W. Sylvester Allred, *Northern Arizona University*
Judith Keller Amand, *Delaware County Community College*
William Anderson, *Abraham Baldwin Agriculture College*
Steve Arch, *Reed College*
Kerri Lynn Armstrong, *Community College of Philadelphia*
G. D. Aumann, *University of Houston*
Vernon Avila, *San Diego State University*
J. Wesley Bahorik, *Kutztown University of Pennsylvania*
Bill Barstow, *University of Georgia, Athens*
Colleen Belk, *University of Minnesota, Duluth*

Michael C. Bell, *Richland College*
Gerald Bergtrom, *University of Wisconsin*
Arlene Billock, *University of Southwestern Louisiana*
Brenda C. Blackwelder, *Central Piedmont Community College*
Raymond Bower, *University of Arkansas*
Marilyn Brady, *Centennial College of Applied Arts & Technology*
Virginia Buckner, *Johnson County Community College*
Arthur L. Buikema, Jr., *Virginia Polytechnic Institute*
William F. Burke, *University of Hawaii*
Robert Burkholter, *Louisiana State University*
Kathleen Burt-Utley, *University of New Orleans*
Linda Butler, *University of Texas, Austin*
W. Barkley Butler, *Indiana University of Pennsylvania*
Bruce E. Byers, *University of Massachusetts, Amherst*
Sara Chambers, *Long Island University*
Nora L. Chee, *Chaminade University*
Joseph P. Chinnici, *Virginia Commonwealth University*
Dan Chiras, *University of Colorado, Denver*
Bob Coburn, *Middlesex Community College*
Martin Cohen, *University of Hartford*
Mary U. Connell, *Appalachian State University*
Joyce Corban, *Wright State University*
Ethel Cornforth, *San Jacinto College, South*
David J. Cotter, *Georgia College*
Lee Couch, *Albuquerque Technical Vocational Institute*
Donald C. Cox, *Miami University of Ohio*
Patricia B. Cox, *University of Tennessee*
Peter Crowcroft, *University of Texas, Austin*
Carol Crowder, *North Harris Montgomery College*
Donald E. Culwell, *University of Central Arkansas*
Robert A. Cunningham, *Erie Community College, North*
Karen Dalton, *Community College of Baltimore County–Catonsville Campus*
David H. Davis, *Asheville-Buncombe Technical Community College*
Jerry Davis, *University of Wisconsin, LaCrosse*
Douglas M. Deardon, *University of Minnesota*
Lewis Deaton, *University of Southwestern Louisiana*
Fred Delcomyn, *University of Illinois, Urbana*
Lorren Denney, *Southwest Missouri State University*
Katherine J. Denniston, *Towson State University*
Charles F. Denny, *University of South Carolina, Sumter*
Jean DeSaix, *University of North Carolina, Chapel Hill*
Ed DeWalt, *Louisiana State University*
Daniel F. Doak, *University of California, Santa Cruz*
Matthew M. Douglas, *University of Kansas*
Ronald J. Downey, *Ohio University*
Ernest Dubrul, *University of Toledo*
Michael Dufresne, *University of Windsor*
Susan A. Dunford, *University of Cincinnati*
Mary Durant, *North Harris College*
Ronald Edwards, *University of Florida*
Rosemarie Elizondo, *Reedley College*
George Ellmore, *Tufts University*
Joanne T. Ellzey, *University of Texas, El Paso*
Wayne Elmore, *Marshall University*
Carl Estrella, *Merced College*
Nancy Eyster-Smith, *Bentley College*
Gerald Farr, *Southwest Texas State University*
Rita Farrar, *Louisiana State University*
Marianne Feaver, *North Carolina State University*
Linnea Fletcher, *Austin Community College, Northridge*
Charles V. Foltz, *Rhode Island College*
Douglas Fratianne, *Ohio State University*
Scott Freeman, *University of Washington*
Donald P. French, *Oklahoma State University*
Don Fritsch, *Virginia Commonwealth University*
Teresa Lane Fulcher, *Pellissippi State Technical Community College*
Michael Gaines, *University of Kansas*
Irja Galvan, *Western Oregon University*
Gail E. Gasparich, *Towson University*
Farooka Gauhari, *University of Nebraska, Omaha*
George W. Gilchrist, *University of Washington*
David Glenn-Lewin, *Iowa State University*
Elmer Gless, *Montana College of Mineral Sciences*
Charles W. Good, *Ohio State University, Lima*
Margaret Green, *Broward Community College*
Lonnie J. Guralnick, *Western Oregon University*
Martin E. Hahn, *William Paterson College*
Madeline Hall, *Cleveland State University*
Georgia Ann Hammond, *Radford University*
Blanche C. Haning, *North Carolina State University*
Helen B. Hanten, *University of Minnesota*
John P. Harley, *Eastern Kentucky University*
Stephen Hedman, *University of Minnesota*
Jean Helgeson, *Collins County Community College*
Alexander Henderson, *Millersville University*
James Hewlett, *Finger Lakes Community College*
Alison G. Hoffman, *University of Tennessee, Chattanooga*
Leland N. Holland, *Paso-Hernando Community College*
Laura Mays Hoopes, *Occidental College*
Michael D. Hudgins, *Alabama State University*
Donald A. Ingold, *East Texas State University*
Jon W. Jacklet, *State University of New York, Albany*
Rebecca M. Jessen, *Bowling Green State University*
Florence Juillerat, *Indiana University–Purdue University at Indianapolis*
Thomas W. Jurik, *Iowa State University*
Arnold Karpoff, *University of Louisville*
L. Kavaljian, *California State University*
Hendrick J. Ketellapper, *University of California, Davis*
Kate Lajtha, *Oregon State University*
Patricia Lee-Robinson, *Chaminade University of Honolulu*
William H. Leonard, *Clemson University*
Edward Levri, *Indiana University of Pennsylvania*
Graeme Lindbeck, *University of Central Florida*
Jerri K. Lindsey, *Tarrant County Junior College, Northeast*
John Logue, *University of South Carolina, Sumter*
William Lowen, *Suffolk Community College*
Ann S. Lumsden, *Florida State University*
Steele R. Lunt, *University of Nebraska, Omaha*
Daniel D. Magoulick, *The University of Central Arkansas*
Paul Mangum, *Midland College*
Michael Martin, *University of Michigan*
Linda Martin-Morris, *University of Washington*
Kenneth A. Mason, *University of Kansas*
Margaret May, *Virginia Commonwealth University*
D. J. McWhinnie, *De Paul University*
Gary L. Meeker, *California State University, Sacramento*
Thoyd Melton, *North Carolina State University*
Joseph R. Mendelson III, *Utah State University*
Karen E. Messley, *Rockvalley College*
Timothy Metz, *Campbell University*
Glendon R. Miller, *Wichita State University*
Neil Miller, *Memphis State University*

Jack E. Mobley, *University of Central Arkansas*
John W. Moon, *Harding University*
Richard Mortenson, *Albion College*
Gisele Muller-Parker, *Western Washington University*
Kathleen Murray, *University of Maine*
Robert Neill, *University of Texas*
Harry Nickla, *Creighton University*
Daniel Nickrent, *Southern Illinois University*
Jane Noble-Harvey, *University of Delaware*
David J. O'Neill, *Community College of Baltimore County, Dundalk Campus*
James T. Oris, *Miami University, Ohio*
Marcy Osgood, *University of Michigan*
C. O. Patterson, *Texas A & M University*
Fred Peabody, *University of South Dakota*
Harry Peery, *Tompkins–Cortland Community College*
Rhoda E. Perozzi, *Virginia Commonwealth University*
Bill Pfitsch, *Hamilton College*
Ronald Pfohl, *Miami University, Ohio*
Bernard Possident, *Skidmore College*
Elsa C. Price, *Wallace State Community College*
James A. Raines, *North Harris College*
Mark Richter, *University of Kansas*
Robert Robbins, *Michigan State University*
Paul Rosenbloom, *Southwest Texas State University*
K. Ross, *University of Delaware*
Mary Lou Rottman, *University of Colorado, Denver*
Albert Ruesink, *Indiana University*
Christopher F. Sacchi, *Kutztown University*
Alan Schoenherr, *Fullerton College*
Edna Seaman, *University of Massachusetts, Boston*
Linda Simpson, *University of North Carolina, Charlotte*
Anu Singh-Cundy, *Western Washington University*
Russel V. Skavaril, *Ohio State University*
John Smarelli, *Loyola University*
Shari Snitovsky, *Skyline College*
Jim Sorenson, *Radford University*
Mary Spratt, *University of Missouri, Kansas City*
Benjamin Stark, *Illinois Institute of Technology*
William Stark, *Saint Louis University*
Kathleen M. Steinert, *Bellevue Community College*
Barbara Stotler, *Southern Illinois University*
Gerald Summers, *University of Missouri, Columbia*
Marshall Sundberg, *Louisiana State University*
Bill Surver, *Clemson University*
Eldon Sutton, *University of Texas, Austin*
Dan Tallman, *Northern State University*
David Thorndill, *Essex Community College*
William Thwaites, *San Diego State University*
Professor Tobiessen, *Union College*
Richard Tolman, *Brigham Young University*
Dennis Trelka, *Washington & Jefferson College*
Sharon Tucker, *University of Delaware*
Gail Turner, *Virginia Commonwealth University*
Glyn Turnipseed, *Arkansas Technical University*
Lloyd W. Turtinen, *University of Wisconsin, Eau Claire*
Robert Tyser, *University of Wisconsin, La Crosse*
Robin W. Tyser, *University of Wisconsin, LaCrosse*
Kristin Uthus, *Virginia Commonwealth University*
F. Daniel Vogt, *State University of New York, Plattsburgh*
Nancy Wade, *Old Dominion University*
Susan M. Wadkowski, *Lakeland Community College*
Jyoti R. Wagle, *Houston Community College, Central*
Michael Weis, *University of Windsor*
DeLoris Wenzel, *University of Georgia*
Jerry Wermuth, *Purdue University, Calumet*
Jacob Wiebers, *Purdue University*
Carolyn Wilczynski, *Binghamton University*
P. Kelly Williams, *University of Dayton*
Roberta Williams, *University of Nevada, Las Vegas*
Sandra Winicur, *Indiana University, South Bend*
Bill Wischusen, *Louisiana State University*
Chris Wolfe, *North Virginia Community College*
Colleen Wong, *Wilbur Wright College*
Wade Worthen, *Furman University*
Robin Wright, *University of Washington*
Brenda L. Young, *Daemen College*
Cal Young, *Fullerton College*
Tim Young, *Mercer University*

To Heather, Jack, and Lori and in memory of Eve and Joe
T. A. & G. A.

To Maija, Varis, and Ivars
B. E. B.

"Viewed from the distance of the moon, the astonishing thing about the earth, catching the breath, is that it is alive. The photographs show the dry, pounded surface of the moon in the foreground, dead as an old bone. Aloft, floating free beneath the moist, gleaming surface of bright blue sky, is the rising earth, the only exuberant thing in this part of the cosmos."

Lewis Thomas in* The Lives of a Cell *(1974)

Earth, seen from the Moon, is an oasis of life in our solar system.

1 An Introduction to Life on Earth

AT A GLANCE

1) What Is Life?

What is life? This short, simple question does not have a short, simple answer. Although each of us has an intuitive understanding of what it means to say that something is alive, that intuition cannot be easily translated to a precise verbal definition. Life is so diverse and complex that it has proved impossible to devise a definition that neatly divides the living from the nonliving. Standard dictionary definitions are of little help, as they typically fall back on phrases like "the quality that distinguishes living organisms from dead organisms," without providing much insight as to what that mysterious "quality" might be.

Because we cannot define life precisely, we must instead build our definition bit by bit, by describing a series of different features of living things. In fact, this entire textbook is really an extended attempt to define life. As you read the book and attend your biology class sessions, you will learn about many different aspects of the living world. Our hope is that as you proceed through this learning process, a picture of life will emerge, in much the same way that a painted image gradually takes shape from the patches of color an artist applies to a canvas. As a prelude to that process, we begin with a brief discussion of some properties that are shared by living things. Taken together, these properties form a combination of characteristics that is not found in nonliving objects.

Living Things Are Both Complex and Organized

Compared with nonliving matter of similar size, living things are highly complex and organized. A crystal of table salt, for example, consists of just two chemical elements, sodium and chlorine, arranged in a precise way: The salt crystal is organized but simple (Fig. 1-1a). The oceans contain some atoms of all the naturally occurring elements, but these atoms are randomly distributed: The oceans are complex but not organized (Fig. 1-1b). In contrast, even the simplest organism contains dozens of different elements linked together in thousands of specific combinations. In many organisms, these combinations are further organized into ever larger and more complex assemblies to form structures such as the eyes, legs, digestive tract, and brain of a waterflea (Fig. 1-1c).

MediaTutor 1.1 Defining Life

Living Things Grow and Reproduce

At some time in its life cycle, every organism becomes larger—that is, it *grows*. This characteristic is most obvious in plants and animals, which tend to start out very small and undergo tremendous growth during their lives. Even single-celled bacteria, however, grow to about double their original size before they divide.

Organisms are also able to reproduce, giving rise to offspring of the same type (Fig. 1-2). Methods for producing offspring vary quite a bit among different kinds of organisms, but the result—the production of new individuals—is always the same.

Living Things Respond to Stimuli

Organisms detect and respond to stimuli in their environments. For example, plants bend to grow toward a source of light. Some kinds of bacteria can move to get away from a poisonous chemical. Animals have sensory organs and muscular systems that allow them to detect and respond to light, sound, chemicals, and many other stimuli in their surroundings. Animals can also respond to stimuli inside their bodies. For example, when you feel hungry, you sense the contractions of your empty stomach and the low levels of sugars and fats in your blood. You then respond by acquiring food and eating.

Figure 1-1 Life is both complex and organized
(a) Each crystal of table salt, sodium chloride, is a cube, showing great organization but minimal complexity. ***(b)*** The water and dissolved materials in the ocean have complexity but very little organization. ***(c)*** The waterflea, *Daphnia pulex*, is only 1 millimeter long (1/1000 meter; smaller than the letter *i*), yet it has legs, a mouth, a digestive tract, reproductive organs, light-sensing eyes, and even a rather impressive brain in relation to its size. Living things have both complexity and organization.

Living Things Acquire and Use Materials and Energy

To grow, reproduce, and maintain their high level of organized complexity, organisms must obtain materials and energy from the environment (Fig. 1-3). The materials and energy power an organism's **metabolism**, the sum total of all the chemical reactions needed to sustain life.

Organisms obtain **energy**—the ability to do work, such as carrying out chemical reactions, growing leaves in the spring, or contracting a muscle—in one of two ways. Plants and some single-celled organisms capture the energy of sunlight and store it in sugar molecules, a process called **photosynthesis**. The remaining organisms, which are not capable of photosynthesis, must consume energy-rich molecules contained in the bodies of other organisms.

Figure 1-2 Living things reproduce
As they grow, these polar bear cubs will resemble, but not be identical to, their parents. The similarity and variability of offspring are crucial to the evolution of life.

Living Things Use DNA to Store Information

All known forms of life use a molecule called **deoxyribonucleic acid**, or **DNA**, to store information (Fig. 1-4). Each organism's DNA acts as an instruction manual, a blueprint that guides the construction and operation of the organism's body. Each segment of DNA that contains a set of instructions is a **gene**; when an organism reproduces, it passes a copy of its DNA to its offspring. This transmission of genes to offspring is the basis of heredity, the means by which offspring inherit the characteristics of parents.

2 Why Is Life So Diverse?

Of the characteristics of living things, one of the most striking is the universal presence of DNA. The DNA molecule is a rather complex structure, consisting of a distinctive arrangement of a specific set of elements. How could this particular unique structure have ended up in the bodies of every living thing from the smallest bacterium to the largest whale? The answer to this question lies in the common ancestry of all life. That is, every

Figure 1-3 Living things acquire energy and materials from the environment
The green plants seen here capture energy from the sun and materials from the air, water, and soil. The insect gains both energy and materials from the plants, while the toad extracts energy and materials from its insect prey.

organism has descended from the same ancestor, and that common ancestor used DNA to store information. DNA has been passed down from generation to generation (remember that when an organism reproduces, it passes a copy of its DNA to its offspring). Every organism living today carries the legacy of a shared ancient heritage.

If modern organisms descended from preexisting life-forms, and if all life therefore shares such fundamental qualities as the presence of DNA, why are living things so diverse? If an oak tree is related to a butterfly, why are they so different?

Evolution Accounts for Both Life's Unity and Its Diversity

Over long periods of time, organisms change. The kind of change under discussion here is not the kind that occurs within an individual, the kind that makes you look different now than you did when you were a small child. Instead, we're talking about the change that takes place within groups of organisms, from generation to generation. Each generation is slightly different than the one before it so that, after many generations have passed, the characteristics of a typical member of the group may bear little resemblance to those of typical group members of earlier times. This kind of accumulating generation-to-generation change within a group is known as **evolution**.

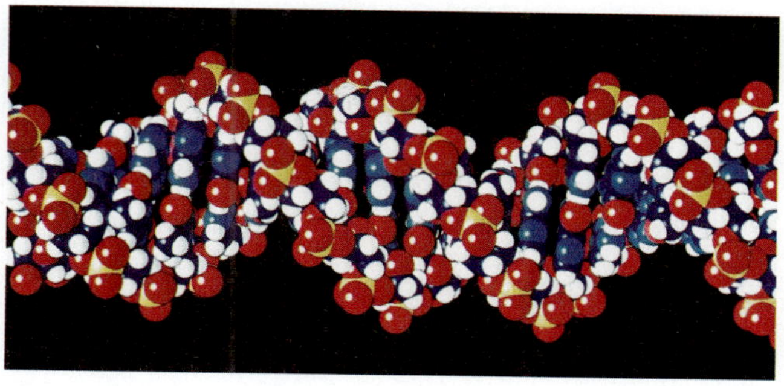

Figure 1-4 DNA
A computer-generated model of DNA, the molecule of heredity.

Evolutionary change has led to the amazing variety of life-forms on Earth. At the same time, evolution accounts for the presence of features that demonstrate life's unity, because evolutionary change has taken place along the branches of a single, gigantic "family tree" that formed as ancestors gave rise to new groups of modified descendants. All the biological structures, mechanisms, systems, and interactions that we will discuss in this book arose through evolution by natural selection. The idea that evolution gave rise to all of life's key features and characteristics is the guiding principle that unifies the study of biology.

Natural Selection Is the Primary Mechanism of Evolution

The most important process by which evolution occurs is **natural selection.** Natural selection occurs because the characteristics of the different individuals in a group vary, and some individuals possess characteristics that help them survive and reproduce more successfully than do others that lack those traits. The individuals with these favorable traits tend to have a greater number of offspring and those offspring tend to inherit the favorable traits from their parents. Those traits thus become more common in the group. For example, beavers with larger teeth might be able to chew down trees more efficiently, build bigger dams and lodges, and eat more bark than "ordinary" beavers. Because these big-toothed beavers obtain more food and better shelter than their smaller-toothed relatives, they will probably raise more offspring. The offspring will inherit their parents' genes for larger teeth. Over time, less-successful,

Earth Watch
Why Preserve Biodiversity?

"The loss of species is the folly our descendants are least likely to forgive us."
E. O. Wilson, Professor, Harvard University

Ever since the United Nations' 1992 "Earth Summit" in Rio de Janeiro, Brazil, the word "biodiversity" has jumped out at us from magazines and news articles. What is biodiversity, and why should we be concerned with preserving it? Biodiversity refers to the total number of species within a community and to the interactions among them; in short, it defines the "richness" of a community.

Scientists have estimated that there are 8 to 10 million species on Earth today. Of these, scientists have named only about 1.4 million, and only a tiny fraction of this number have been studied. Evolution has not, however, merely been churning out millions of independent species. Over thousands of years, organisms in a given area have been molded by forces of natural selection exerted by other living species as well as by the nonliving environment in which they live. The outcome is the community, a highly complex web of interdependent life-forms whose interactions sustain one another. By participating in the natural cycling of water, oxygen, and other substances, and by producing rich soil and purifying wastes, these communities contribute to the sustenance of human life as well. The concept of biodiversity has emerged as a result of our increasing concern over the loss of countless forms of life and the habitat that sustains them.

The Tropics are home to the vast majority of all the species on Earth, perhaps 7 million to 8 million of them, living in complex communities. The rapid destruction of habitats in the Tropics, from rain forests to coral reefs, as a result of human activities is producing high rates of extinction of many species (Fig. E1-1). Most of these species have never been named, and others never even discovered. Aside from ethical concerns over eradicating irreplaceable forms of life, as we drive unknown organisms to extinction, we lose potential sources of medicine, food, and raw materials for industry. Many scientists are also concerned that as species are eliminated, the communities of which they were a part might change and become less stable and more vulnerable to damage by diseases or adverse environmental conditions.

Figure E1-1 Biodiversity threatened
Destruction of tropical rain forests by indiscriminant logging threatens Earth's greatest storehouse of biological diversity. Diverse communities are sustained by interrelationships such as those that have evolved between this *Heliconia* flower and its hummingbird pollinator.

smaller-toothed beavers will become increasingly scarce as natural selection operates against them. After many generations, all beavers will have large teeth.

Life's Diversity Is Currently Threatened

One outcome of evolution is that Earth is now home to a tremendous variety of **species**, or types of organisms. In many places, these species have evolved complex interrelationships with one another and with their surroundings. The word **biodiversity** is often used to describe this wealth of species and the complex interrelationships that sustain them. In recent decades, a single species, *Homo sapiens* (modern humans), has drastically increased the rate at which the environment is changing. Unfortunately, many other species have been unable to cope with this rapid change. In the habitats most affected by humans, many species are being driven to extinction. This problem is further explored in "Earth Watch: Why Preserve Biodiversity?"

3 What Is Biology?

Biology is the science of life, and biologists try to answer questions about life. What causes cancer? Why are frog populations shrinking worldwide? What happens when a sperm and egg meet? How does HIV cause AIDS? When did the earliest mammal appear? How do bees fly? The list of questions is both endless and endlessly fascinating.

The Scientific Method Is the Basis for Scientific Inquiry

How do biologists study the workings of life? How do they explain what they observe? Ideally, biologists use the **scientific method**, which consists of four interrelated operations: (1) *observation*; (2) *hypothesis*; (3) *experiment*; and (4) *conclusion*. All scientific inquiry begins with an **observation** of a specific phenomenon. The observation, in turn, leads to questions, such as "How did this come about?" Then in a flash of insight, or more typically after long, hard thought, a **hypothesis** is formulated. A hypothesis is a proposed explanation for the observed phenomenon. To be useful, the hypothesis must lead to predictions that can be tested by additional controlled observations, or **experiments**. These experiments produce results that either support or refute the hypothesis, and a **conclusion** is drawn about its validity. A single experiment is never an adequate basis for a conclusion; the experiment must be repeated not only by the original researcher but also by others.

Well-Designed Experiments Incorporate Controls

Simple experiments test the hypothesis that a single factor, or **variable**, is the cause of an observed phenomenon. The most straightforward test is usually an experiment in which only the variable being tested is changed. To be scientifically valid, however, the experiment must also rule out other possible variables that might have caused the phenomenon. So experiments usually also include **controls**, sections of the experiment in which no variable is changed. The results from the control sections can then be compared with those from the experimental sections. For a classic example of a well-designed experiment, see "Scientific Inquiry: Does Spoiled Meat Produce Maggots?" which describes Francesco Redi's demonstration that flies do not arise spontaneously from rotting meat.

The Scientific Method Is Useful in Everyday Life

If you think about it for a moment, you will probably realize that you use the scientific method regularly in everyday life. For example, imagine this scenario: Late to an appointment, you rush to your car and make the *observation* that it won't start. Immediately, you form a *hypothesis*: The battery is dead. Quickly, you design an *experiment*: You replace your battery with the battery from your roommate's new car and try to start your car again. The result seems to confirm your hypothesis, because your car starts immediately.

But wait! You haven't provided controls for several variables. Perhaps your battery was fine all along, and you just needed to try to start the car again. Or perhaps the battery cable was loose and simply needed to be tightened. Realizing the need for a good *control*, you replace your old battery, making sure the cables are secured tightly, and attempt to restart the car. If your car repeatedly refuses to start with the old battery but starts immediately when you put in your roommate's new battery, you have isolated a single *variable*, the battery. And (although you may have missed your appointment) you can now safely draw the *conclusion* that your old battery is dead.

The scientific method is powerful, but it is important to recognize its limitations. In particular, scientists can seldom be sure that they have controlled *all* the variables other than the one they are trying to study. Therefore, scientific conclusions remain tentative and are subject to revision if new observations or experiments demand it.

A final important element of science is *communication*. No matter how well designed an experiment is, it is useless if it is not communicated. Redi's experimental design and conclusions survive today because he recorded his methods and observations and made them available to other scientists. If experiments are not communicated to other scientists in enough detail to be repeated, their conclusions cannot be verified. Without verification, scientific findings cannot be safely used as the basis for new hypotheses and further experiments. Scientific effort is wasted if it is not reported in clear and accurate detail.

A wonderful aspect of scientific inquiry is that nearly every conclusion immediately raises further questions that lead to further hypotheses and more experiments (why did your battery die?). Science is a never-ending quest for knowledge.

Life Can Be Studied at Different Levels of Organization

Biologists often view the living world as a series of *levels of organization*, with each level providing the building blocks for the next level (Fig. 1-5). For example, the **cell** is the smallest unit of life (Fig. 1-6), but in many life-forms, cells of similar type combine to form structures known as **tissues**. Different tissues, in turn, can combine to form **organs** (for example, a heart or a kidney) and a collection of different organs forms the basis of an **organism**. A collection of organisms of the same type constitutes a **population**, and a collection of different populations constitutes a **community**. Note that each level of organization incorporates multiple members of the previous level; a community contains many populations, a population contains many organisms, and so on. (Also note that we could specify many more levels of organization than we have mentioned here.)

Biologists choose to work at an appropriate level of organization. One of the first decisions a biologist must make when designing an experiment is to choose a level of organization at which to study a problem. This decision is ordinarily based on the question to be answered. For example, if you wanted to know how frogs make croaking sounds, you would study frog organs

Scientific Inquiry
Does Spoiled Meat Produce Maggots?

An experiment of Italian physician Francesco Redi (1621–1697) beautifully demonstrates the scientific method. Redi investigated why maggots appear on spoiled meat. When he began this work, the appearance of maggots was considered to be evidence of **spontaneous generation**, the production of living things from nonliving matter.

Redi *observed* that flies swarm around fresh meat and that maggots appear on meat left out for a few days. He formed a testable *hypothesis*: The flies produce the maggots. In his *experiment*, Redi wanted to test just one variable: the access of flies to the meat. Therefore, he took two clean jars and filled them with similar pieces of meat. He left one jar open (the *control* jar) and covered the other with gauze to keep out flies (the *experimental* jar). He did his best to keep all the other variables the same (for example, the type of jar, the type of meat, and the temperature). After a few days, he observed maggots on the meat in the open jar, but saw none on the meat in the covered jar. Redi *concluded* that his hypothesis was correct and that maggots are produced by flies, not by the nonliving meat (Fig. E1-2). Only through controlled experiments could the age-old hypothesis of spontaneous generation be laid to rest.

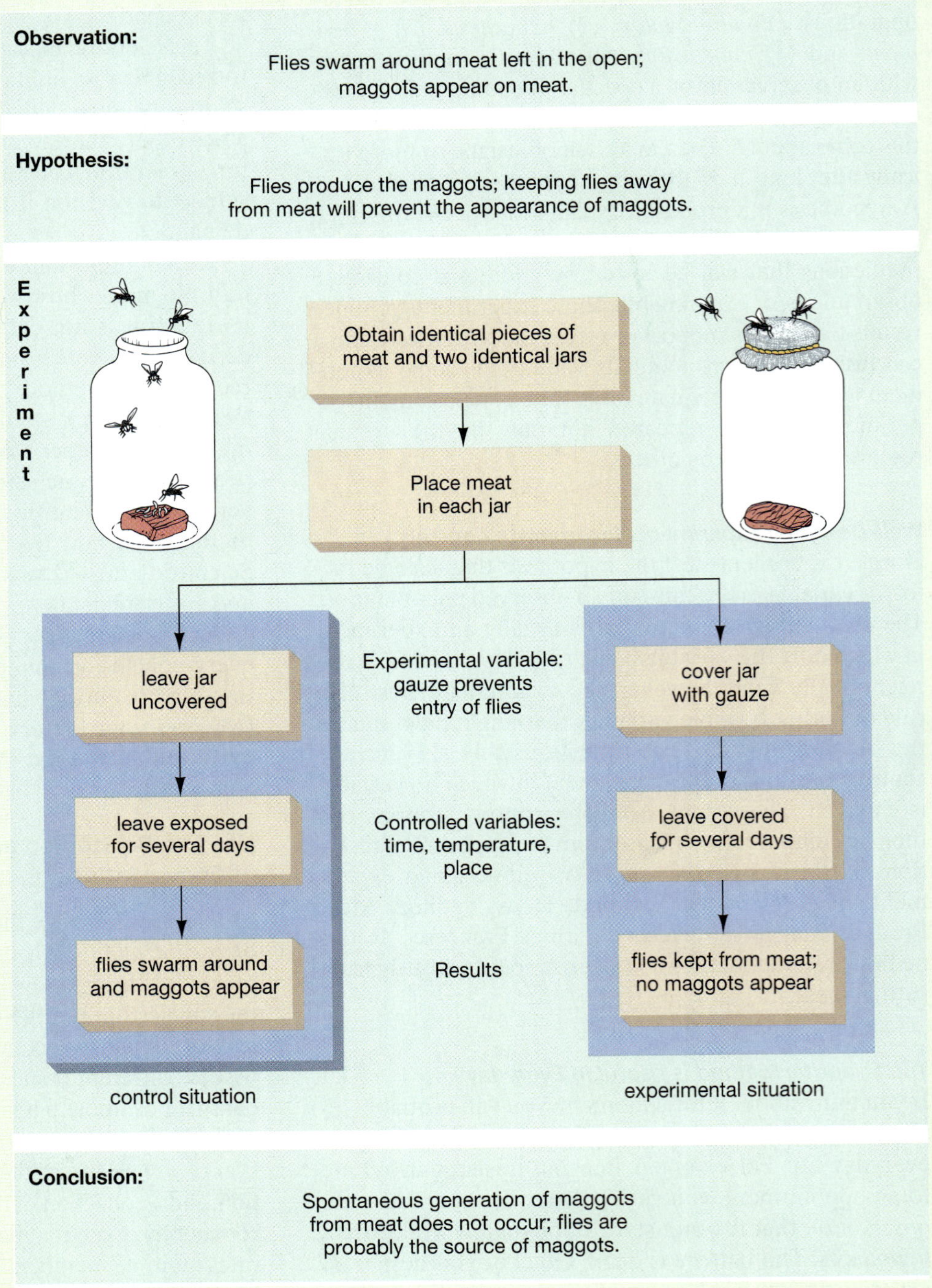

Figure E1-2 The experiments of Francesco Redi

(structures within the frog body). The question of how frogs croak would be impossible to answer if you focused on frog cells or frog communities. On the other hand, if you want to know whether global warming is reducing the number of frogs in the world, it would do you no good to study frog organs. To answer that question, you would have to study frog populations. It is important for scientists to recognize and choose the level of organization that is most appropriate to the question at hand.

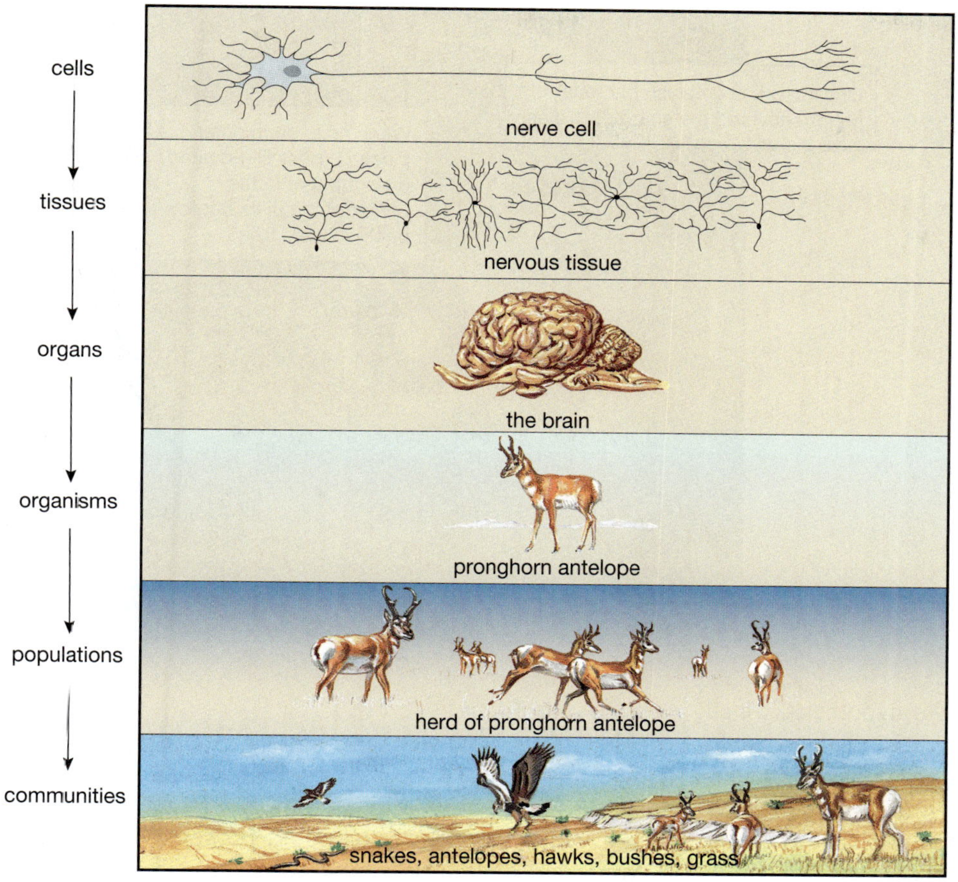

Figure 1-5 Levels of organization of life
Life is organized in levels of organization, some of which are represented here. Items at each level are the building blocks of the next level.

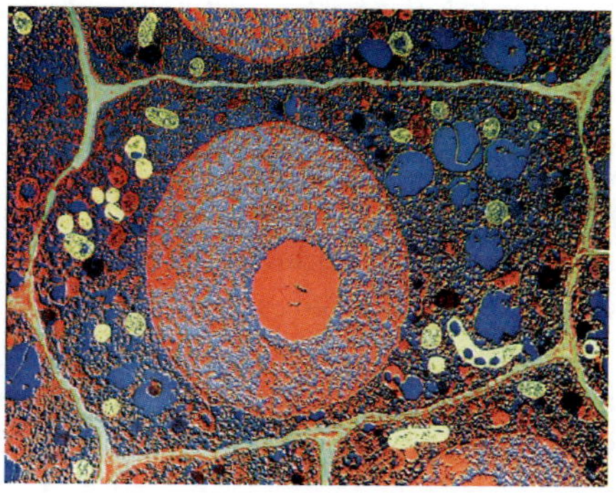

Figure 1-6 The smallest unit of life
The boundary of this plant cell is formed by a cell wall, which is green in this colorized micrograph.

Science Is a Human Endeavor

Scientists are real people. They are driven by the same ambitions, pride, and fears as other people, and they sometimes make mistakes. As you will read in Chapter 8, ambition played an important role in the discovery by James Watson and Francis Crick of the structure of DNA. Accidents, lucky guesses, controversies with competing scientists, and, of course, the intellectual powers of individual scientists contribute greatly to scientific advances. To illustrate what we might call "real science," let's consider an actual case.

To study bacteria, microbiologists must use pure *cultures*—that is, plates containing colonies of a single type of bacteria free from contamination by other bacteria, molds, and so on. Only by studying a single type at a time can we learn about the properties of that particular bacterium. Consequently, at the first sign of contamination, a culture is normally thrown out, often with mutterings about sloppy technique. On one occasion, however, Scottish bacteriologist Alexander Fleming turned a ruined culture into one of the greatest medical advances in history.

In the 1920s, one of Fleming's bacterial cultures became contaminated with a patch of a mold called *Penicillium*. As he was about to throw out the culture dish, Fleming observed that no bacteria were growing near the mold. Why not? Fleming hypothesized that perhaps *Penicillium* had released a substance that killed off the bacteria growing nearby. To test this hypothesis, Fleming grew some pure *Penicillium* in a liquid nutrient broth. He then filtered out the *Penicillium* mold and applied the liquid in which the mold had grown to an uncontaminated bacterial culture. Sure enough, something in the liquid killed the bacteria. Further research into these mold extracts resulted in the production of the first *antibiotic*: penicillin, a bacteria-killing substance that has since saved millions of lives (Fig. 1-7).

Fleming used the scientific method. His experiment began with an observation and proceeded to a hypothesis,

Figure 1-7 Penicillin kills bacteria

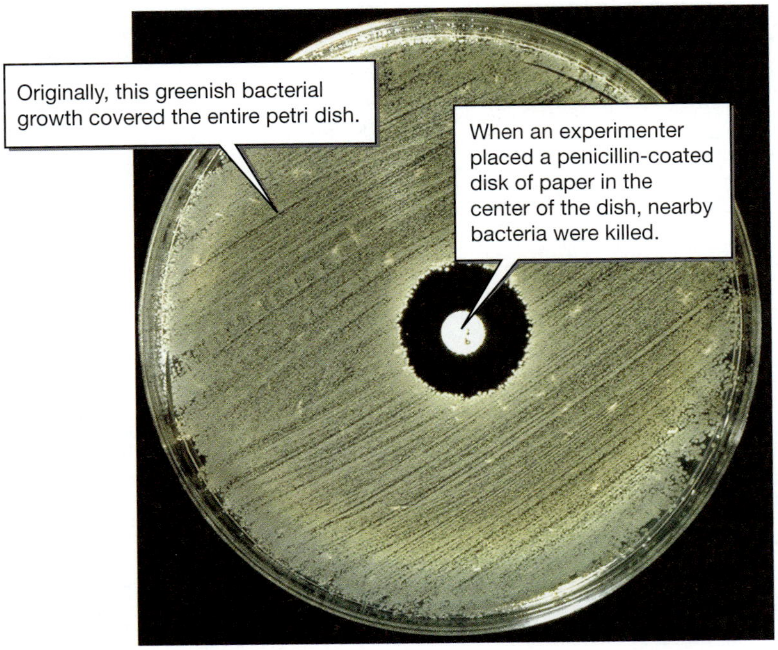

followed by experimental tests of the hypothesis, which led to a conclusion. But the scientific method alone would have been useless without the lucky combination of accident and a brilliant scientific mind. Had Fleming been a "perfect" microbiologist, he wouldn't have had any contaminated cultures. Had he been less observant, the contamination would have been just another spoiled culture dish. Instead, it was the beginning of antibiotic therapy for bacterial diseases. As French microbiologist Louis Pasteur said, "Chance favors the prepared mind."

Scientific Theories Have Been Thoroughly Tested

Scientists use the word *theory* in a way that is different from its everyday usage. If Dr. Watson were to ask Sherlock Holmes, "Do you have a theory as to the perpetrator of this foul deed?" he would be asking for what a scientist would call a hypothesis—an educated guess based on observable evidence, or clues. A **scientific theory** is far more general and more reliable than a hypothesis. Far from being an educated guess, a scientific theory is a general explanation of important natural phenomena, developed through extensive and reproducible tests. For example, the theory of gravity, which states that objects exert attraction for one another, is fundamental to the science of physics. Similarly, the *cell theory*, which states that all living things are composed of cells, is fundamental to the study of biology. Perhaps the most important theory in biology is the theory of evolution. Since its formulation in the mid-1800s by Charles Darwin and Alfred Russel Wallace, the theory of evolution has been supported by an overwhelming accumulation of evidence, including fossil finds, geological studies, radioactive dating of rocks, breeding experiments, and research results in genetics, molecular biology, and biochemistry. People who say that evolution is "just a theory" profoundly misunderstand what scientists mean by the word *theory*.

4 Why Should You Study Biology?

As you begin your study of biology, you may find yourself wondering why you should bother. We hope that you've chosen to study biology because you are already fascinated by the study of life, but in case you have not yet reached that point we offer for your consideration a few reasons to pursue this subject.

Biology Helps You Understand Your Body

A large part of biology is devoted to understanding how bodies work and understanding the disruptions and diseases that can prevent bodies from working properly. As you learn more biology, you will develop a greater understanding of the inner workings of your own body, and you will learn more about the manner in which different diseases do their damage. This kind of knowledge can be empowering; when you discuss your health with your physician, you will have a much better understanding of the subject matter. Biological knowledge helps you become an informed consumer of health care, and helps you take charge of your own health.

Biology Helps You Become an Informed Citizen

Our society is currently confronted with a host of complex social and ethical issues, many of which require that difficult decisions be made. As a member of society, you will have a voice in those decisions and a responsibility to make your views known (or to elect representatives who share your views). Many of the most controversial problems involve biology: Should we do anything to halt the extinction of endangered species? Do we need to worry about the impact of pollution on our health? Are genetically modified foods beneficial or detrimental?

Should cloning of humans be legal? Should scientists be allowed to use cells from human embryos to develop cures for diseases? Is population control a good idea? The more you know about the basic biology that underlies these and other issues, the more likely you will be to have an informed opinion and to be an effective advocate for your point of view.

Biology Can Enrich Your Appreciation of the World

Some people regard science as a "dehumanizing" activity, feeling that too deep an understanding of the world robs us of vision and awe. Nothing could be further from the truth, as we repeatedly discover anew in our own lives. Years ago, we watched a bee foraging at lupine flowers. Two fused petals on the lower half of these flowers enclose the pollen-laden male reproductive structures (*stamens*) and sticky pollen-capturing female reproductive structure (*stigma*). We had recently learned that in young lupine flowers (Fig. 1-8a), the weight of a bee pushing on these petals compresses the stamens, pushing pollen onto the bee's abdomen (Fig. 1-8b). In older flowers, the stigma (not shown in Fig. 1-8) protrudes through the lower petals; when a pollen-dusted bee visits an older flower, it usually leaves behind a few grains of pollen. These pollen grains carry sperm that can fertilize the eggs that lie below the stigma, and thus begin the production of new lupine seeds.

Did our newfound insights into the functioning of lupine flowers detract from our appreciation of them? Far from it. Rather, we now looked on lupines with new delight, understanding something of the interplay of form and function, and of bee and flower. A few months later we ventured atop Hurricane Ridge in Olympic National Park in Washington State, where the alpine meadows burst with color in August (Fig. 1-9). As we crouched beside a wild lupine, an elderly man stopped to ask what we were looking at so intently. He listened with interest as we explained the structure to him; he then went off to another patch of lupines to watch the bees foraging. He too felt the increased sense of wonder that comes with understanding.

We hope that you will find this dual sense of understanding and wonder as you read this text. Biology is

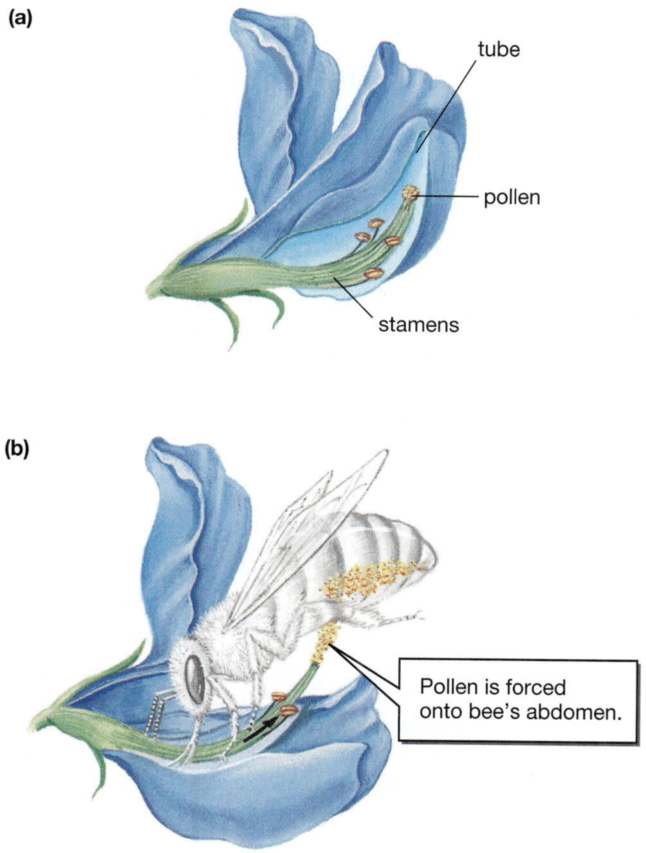

Figure 1-8 The structure of lupine flowers helps ensure pollen transfer
(a) The lower petals of lupine flowers form a tube within which the stamens fit snugly. The stamens shed pollen within the tube. ***(b)*** When the weight of a foraging bee pushes on the lower petals, the stamens are thrust forward, and pollen is ejected onto the bee's abdomen.

Figure 1-9 Wild lupines and fir trees
Thousands of people visit Hurricane Ridge in Washington State's Olympic National Park each summer to gaze in awe at Mt. Olympus, but few bother to investigate the wonders at their feet.

not a completed work but an exploration that we have really just begun. As Lewis Thomas, a physician and natural philosopher, eloquently stated: "The only solid piece of scientific truth about which I feel totally confident is that we are profoundly ignorant about nature. Indeed, I regard this as the major discovery of the past hundred years of biology . . . but we are making a beginning."

We cannot urge you strongly enough, even if you are not contemplating a career in biology, to join in the journey of biological discovery throughout your life. Don't think of biology as just another course to take, just another set of facts to memorize. Biology can be much more than that. It can be a pathway to a new understanding of yourself and of the life on Earth around you.

Summary of Key Concepts

1) What Is Life?

Organisms possess the following characteristics: Their structure is complex and organized; they grow; they reproduce; they acquire energy and materials from the environment; they respond to stimuli; and they use DNA to store information.

2) Why Is Life So Diverse?

Evolution by natural selection has brought about both life's unity (shared features) and its diversity (huge number of incredibly different forms). Evolution is the theory that modern organisms descended, with modification, from preexisting life-forms. Species evolve as a consequence of (1) variation among members of a population, (2) inheritance of those variations by offspring, and (3) natural selection of the variations that best help an organism to survive and reproduce in its environment.

3) What Is Biology?

Biology is the science of life. Knowledge in biology is acquired through the application of the scientific method. First, an observation is made. Then a hypothesis is formulated that suggests a natural cause for the observation. The hypothesis is used to predict the outcome of further observations or experiments. A conclusion is then drawn about the validity of the hypothesis. A scientific theory is a general explanation of natural phenomena, developed through extensive and reproducible experiments and observations.

4) Why Should You Study Biology?

Practical benefits include greater understanding of health and disease, and more informed opinions on environmental and bioethical controversies. Most importantly, the more you know about living things, the more fascinating they become!

Key Terms

biodiversity *p. 4*
cell *p. 5*
community *p. 5*
conclusion *p. 5*
control *p. 5*
deoxyribonucleic acid (DNA) *p. 2*
energy *p. 2*
evolution *p. 3*
experiment *p. 5*
gene *p. 2*
hypothesis *p. 5*
metabolism *p. 2*
natural selection *p. 3*
observation *p. 5*
organ *p. 5*
organism *p. 5*
photosynthesis *p. 2*
population *p. 5*
scientific method *p. 5*
scientific theory *p. 8*
species *p. 4*
spontaneous generation *p. 6*
tissue *p. 5*
variable *p. 5*

Thinking Through the Concepts

Multiple Choice

1. *Which of the following is paired incorrectly?*
   a. organ—a structure formed of tissues
   b. cell—the smallest unit of life
   c. genes—units of heredity
   d. population—a group of tissues of the same type
   e. tissue—a structure formed of cells

2. *Fleming's discovery of penicillin demonstrated that*
   a. a methodical approach is the best way to get new ideas
   b. ambition motivates the best scientists
   c. accidents and chance occurrences can be important in science
   d. a scientific theory is similar to an educated guess
   e. molds generally kill bacteria

3. *Choose the answer that best describes the scientific method.*
   a. observation, hypothesis, experiment, absolute proof
   b. guess, hypothesis, experiment, conclusion
   c. observation, hypothesis, experiment, conclusion
   d. hypothesis, experiment, observation, conclusion
   e. experiment, observation, hypothesis, conclusion
4. *Which of the following are characteristics of living things?*
   a. They reproduce.
   b. They respond to stimuli.
   c. They are complex and organized.
   d. They acquire energy.
   e. all of the above
5. *All organisms contain DNA because*
   a. organisms with favorable traits leave more offspring
   b. the characteristics of individuals within a population vary
   c. DNA is abundant in the environment and all organisms absorb it
   d. DNA is a simple molecule that has formed by chance in many different organisms
   e. all life has descended from a common ancestor

? Review Questions

1. What are the differences between a salt crystal and a tree? Which is living? How do you know? How would you test your "knowledge"? What controls would you use?
2. What is the difference between a scientific theory and a hypothesis? Explain how each is used by scientists.
3. Define and explain the terms *natural selection*, *evolution*, and *population*.
4. Starting with the cell, list the levels of organization of life, briefly explaining each level.
5. Describe the scientific method. In what ways do you use the scientific method in everyday life?
6. What is evolution? Briefly describe how evolution occurs.

Applying the Concepts

1. In the heavily populated state of California, natural occurrences, including earthquakes, heavy rains, and grass and forest wildfires, have reduced the quality of life for many Californians. Look at it another way for a moment. What habitats are found in this state? What impacts have humans had on these natural habitats? Are changes in these areas the reason for fires and floods and mudslides? What needs to be done to alter the balance of "humans and nature"?
2. Design an experiment to test the effects of a new dog food, "Super Dog," on the thickness and water-shedding properties of the coats of golden retrievers. Include all the parts of a scientific experiment. Design objective methods to assess coat thickness and water-shedding ability.

For More Information

Attenborough, D. *Life on Earth.* Boston: Little, Brown, 1979. Gorgeously illustrated and beautifully written introduction to the diversity of life on Earth; the inspiration for our title.

Dawkins, R. *The Blind Watchmaker.* New York: W. W. Norton & Co., 1986. An engagingly written description of the process of evolution, which Dawkins compares to a blind watchmaker.

Ehrlich, P. R. *The Machinery of Nature.* New York: Simon & Schuster, 1986. Using layperson's terms, a foremost ecologist and author explains the science of ecology and the biological rationale for environmental concern.

Leopold, A. *A Sand County Almanac.* New York: Oxford University Press, 1949 (reprinted in 1989). A classic by a natural philosopher; provides an eloquent foundation for the conservation ethic.

Swain, R. B. *Earthly Pleasures.* New York: Charles Scribner's Sons, 1981. Insightful essays stress the interrelatedness and diversity of life.

Thomas, L. *The Medusa and the Snail.* New York: Bantam Books, 1980. The late physician, researcher, and philosopher Lewis Thomas shares his awe of the living world in a series of delightful essays.

Wilson, E. O. *The Diversity of Life.* New York: W. W. Norton & Co., 1992. A celebration of the diversity of life, how it evolved, and how humans are impacting it. Wilson's writings have won two Pulitzer prizes.

Answers to Multiple-Choice Questions

1.d 2.c 3.c 4.e 5.e

MediaTutor

CD Activities

Activity 1.1: Defining Life

Estimated time: 15 minutes

In this simulation you will explore the characteristics that define life and the way that we use the scientific method to ask questions about life. You will play the part of the scientist provided with some material that may contain living organisms—or not. It is up to you to make the determination.

Start the MediaTutor Student CD-ROM and enter the activity number in the Quick Search box to be taken directly to that activity.

Web Investigation

Properties of Life

Estimated time: 15 minutes

Through several centuries of scientific investigation, biologists have developed a list of properties that define life. Some recent findings, however, have blurred the boundary between the living and the non-living. This Web Investigation will lead you through an assessment of the properites of life, and show you why some scientists have come to question the criteria by which we answer the question "what is life?"

Go to http://www.prenhall.com/audesirk3, the Audesirk Companion Web site. Select Chapter 1 and the Web Investigation to begin.

UNIT ONE

The Life of a Cell

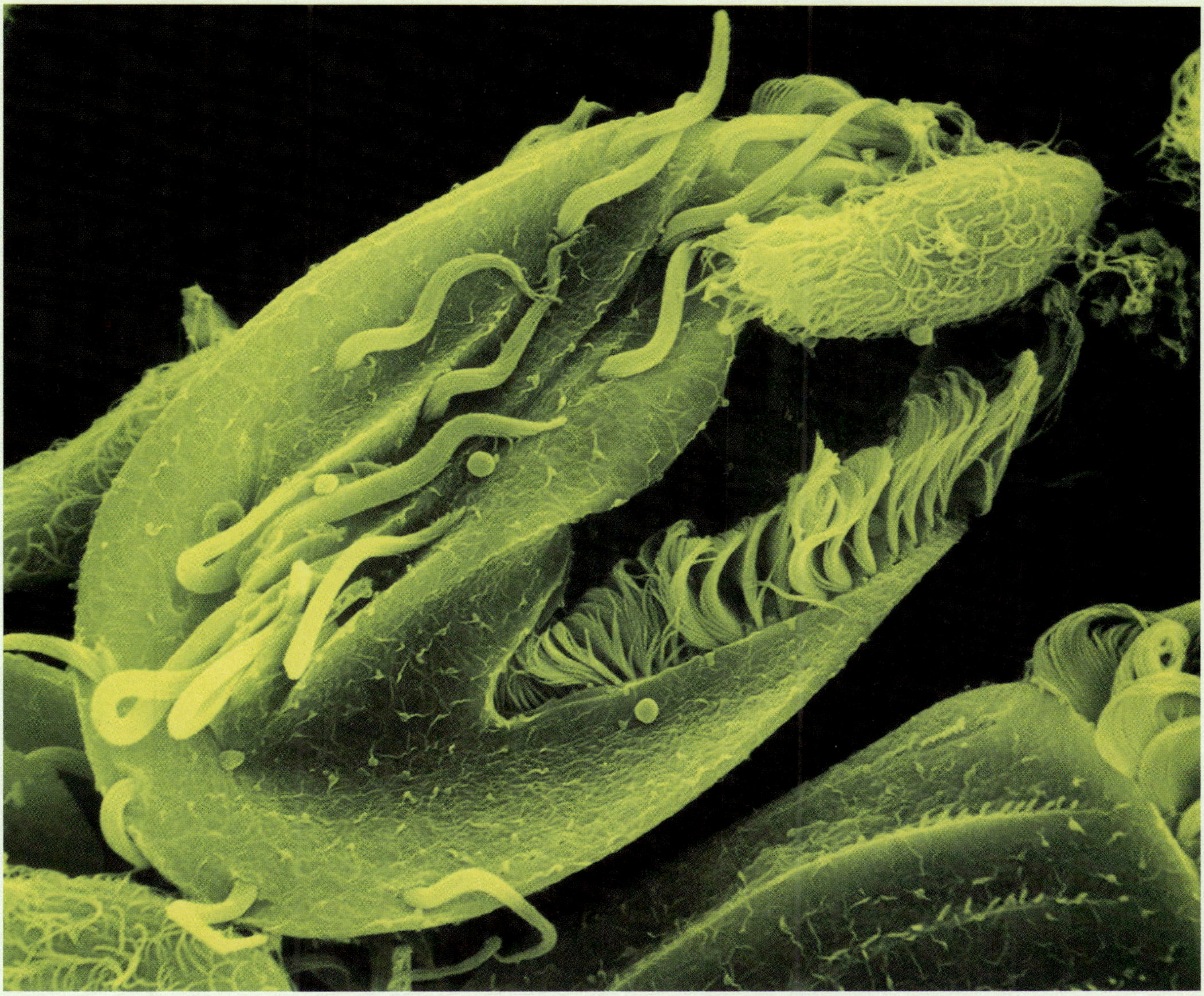

Single cells can be complex, independent organisms. Here, a large Euplotes *prepares to eat a much smaller* Paramecium.

"If there is magic on this planet, it is contained in water."

Loren Eiseley in* The Immense Journey *(1957)

Many of today's most popular foods contain artificially modified molecules. Olestra, a new molecule that mimics fat, has proven to be an especially controversial addition to food.

2 Atoms, Molecules, and Life

AT A GLANCE

CASESTUDY

Improving on Nature?

Protein, carbohydrate, fat, sugar. These words have become part of our everyday vocabulary, in large part because they appear on the "Nutrition Facts" panels that adorn packaged foods. You probably know that proteins, carbohydrates, fats, and sugars are substances found in food, but you may not know that each of these words describes a particular kind of molecule. These special biological molecules are the building blocks of all living things (including our bodies), and are also the focus of scientific efforts to improve on nature's handiwork. As our understanding of biological molecules increases, so do efforts to modify them.

Strangely, a major goal of scientists who seek to modify biological molecules is to make them nonnutritious. For example, food scientists at Procter & Gamble spent 25 years and $200 million to invent an artificial fat known as olestra. Like real fat, olestra adds a pleasing texture and taste to foods. Our bodies, however, are not able to extract the energy stored in the manufactured molecules. Olestra is completely indigestible and passes through the body unchanged. This indigestibility ensures that potato chips made with olestra have far fewer calories than normal chips, but the indigestible substance irritates some people's digestive systems. Thus, when the U.S. Food and Drug Administration (FDA) approved olestra for use in snack foods, it also required that foods containing the fake fat be identified with a warning label that reads, "This product contains olestra. Olestra may cause abdominal cramping and loose stools. Olestra inhibits the absorption of some vitamins and nutrients."

Even after olestra's approval, some scientists and consumer advocates argued that olestra should be removed from the market, on the grounds that the health effects of olestra consumption had not been sufficiently studied. For its part, Procter & Gamble petitioned the FDA for permission to remove the warning label, on the grounds that olestra had been shown to be perfectly safe. In response to the continuing dispute, the FDA agreed to revisit its position after foods with olestra had been on the market for a couple of years. An advisory panel convened in June 1998 to examine consumer complaints and new scientific evidence of olestra's health effects.

When different experts present conflicting interpretations of the same scientific evidence, how can an educated consumer decide whether it's safe to add "non-fat" potato chips to the shopping cart? Perhaps you will find it easier to make your own decision after learning about the chemistry of biological molecules. ■

1 What Are Atoms?

An **element** is a substance that can neither be broken down nor converted to other substances by ordinary chemical means. For example, carbon is an element, so if you took a diamond (a form of carbon) and cut it into pieces, each piece would still be carbon. If you could make finer and finer divisions, you would eventually produce a pile of carbon **atoms**. Atoms are the basic structural units of matter.

Atoms Are Composed of Even Smaller Particles

Each atom contains a central **atomic nucleus** that is composed of two types of subatomic particles of equal weight: positively charged **protons** and uncharged **neutrons**. Other subatomic particles, called **electrons**, orbit the atomic nucleus (Fig. 2-1). Electrons are lighter and negatively charged. An atom by itself has an equal number of electrons and protons and is therefore electrically neutral.

There are 92 different types of atoms, and therefore 92 different elements in nature. Each type of atom is the basic unit of a different element. The number of protons in the nucleus, called the **atomic number**, is unique for each element. For example, a hydrogen atom has one proton in its nucleus, a carbon atom has six protons, and an oxygen atom has eight. Each element also has other chemical properties. For example, some elements, such as oxygen and hydrogen, are gases at room temperature; others, such as lead and iron, are solids. Atoms of the same element may have different numbers of neutrons; when they do, the differing atoms are called **isotopes** of the element. Some, but not all, isotopes are **radioactive**; that is, they spontaneously break apart, forming different types of atoms and releasing energy in the process.

MediaTutor 2.1 Interactive Atoms

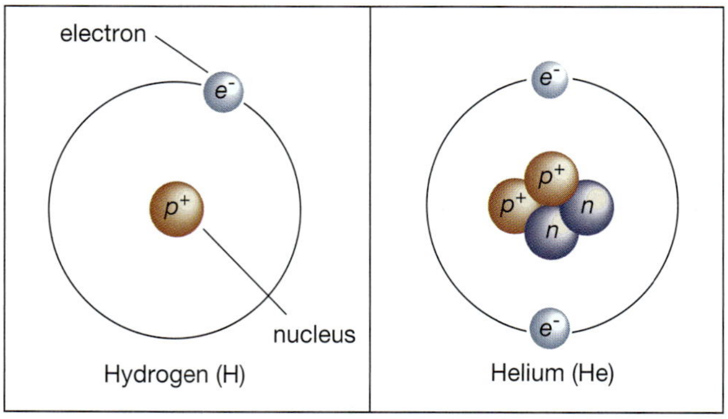

Figure 2-1 Atomic models
The two smallest atoms, hydrogen and helium. In these simplified models, the electrons are represented as miniature planets, circling around a nucleus. The nucleus of hydrogen consists of a proton; the nucleus of helium consists of two protons and two neutrons.

Electrons Orbit the Nucleus, Forming Electron Shells

As you may know from experimenting with a magnet, like poles repel each other and opposite poles attract each other. In a similar way, negatively charged electrons repel one another, but are drawn to the positively charged protons of the nucleus. This attraction holds an atom together as its electrons orbit around the nucleus. Because of their mutual repulsion, only limited numbers of electrons can be at the same distance from the nucleus. Electrons orbit at only a few different distances from the nucleus; the different distances are called **electron shells** (Fig. 2-2). The electrons in an atom normally first fill the shell closest to the nucleus and then begin to occupy the next shell. The electron shell closest to the atomic nucleus can hold only two electrons. The second shell can hold up to eight electrons. Thus, a carbon atom, with six electrons, has two electrons in the first shell, closest to the nucleus, and four electrons in its second shell.

The Big Picture

All matter in the universe is composed of atoms, which are themselves composed of still smaller particles. The incredibly diverse array of different structures in the universe is based on only a small number (92) of different kinds of atoms.

2 How Do Atoms Form Molecules?

A **molecule** consists of two or more atoms, which can be of the same or of different elements. The atoms are held together by interactions among their outermost electron shells.

Atoms Interact When There Are Vacancies in Their Outermost Electron Shells

Atoms interact with one another according to two basic principles:

- An atom will not react with other atoms when its outermost electron shell is completely full or empty. Such an atom is described as being *inert.*
- An atom will react with other atoms when its outermost electron shell is only partially full. Such atoms are described as *reactive.*

To demonstrate these principles, consider three atoms: helium, hydrogen, and oxygen (see Figs. 2-1 and 2-2). Helium has two protons in its nucleus, and two electrons fill its single electron shell. With its full outer shell, helium is inert. In contrast, hydrogen and oxygen have partially empty outer shells. Hydrogen has one proton in its nucleus and one electron in its single (and therefore outermost) electron shell, which can hold up to two electrons. The oxygen atom has six electrons in its outer

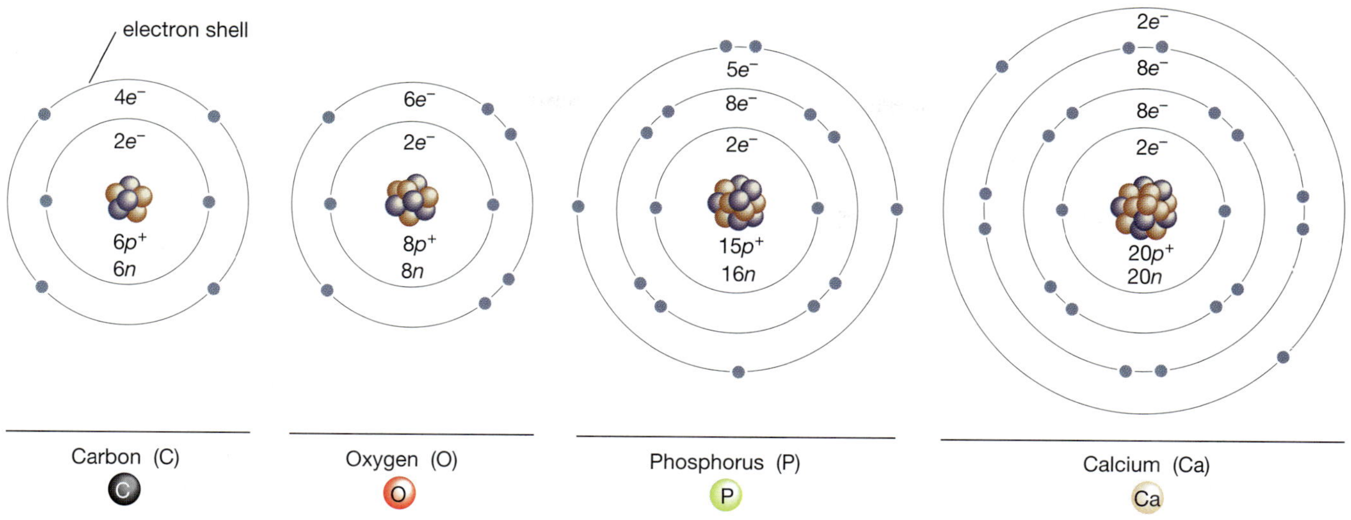

Figure 2-2 Electron shells in atoms
Most biologically important atoms have at least two shells of electrons. The first shell, closest to the nucleus, can hold two electrons; the next shell can hold a maximum of eight electrons. More distant shells can also hold eight electrons each.

shell, which can hold eight. Therefore, we might predict that hydrogen and oxygen atoms will be reactive. We might further predict that oxygen atoms could fill their outer shells by reacting with hydrogen atoms. And, in fact, oxygen reacts readily with hydrogen to form water (H_2O). In this reaction, the outer shell of an oxygen atom is filled by the electrons from two hydrogen atoms. The resulting water molecule is comparatively less likely to undergo further reactions; that is, the water molecule is more stable than the hydrogen and oxygen atoms that gave rise to it.

An atom with an outermost electron shell that is partially full can gain stability by losing electrons (emptying the shell), by gaining electrons (filling the shell), or by sharing electrons with another atom (with both atoms behaving as though they had full outer shells). The results of losing, gaining, and sharing electrons are **chemical bonds**, attractive forces that hold atoms together in molecules. Each element has chemical bonding properties that arise from the configuration of electrons in its outer shell. **Chemical reactions** make and break chemical bonds to form new substances. Such reactions are essential for the maintenance of life and for the working of modern society. Whether chemical reactions occur in a plant cell as it captures solar energy, in your brain as it forms new memories, or in your car's engine as it guzzles gas, the reactions make new chemical bonds, or break existing ones, or both.

Charged Atoms Interact to Form Ionic Bonds

The stability that atoms gain by emptying or filling their outermost shells is demonstrated by the formation of table salt (sodium chloride). Sodium (Na) has only one electron in its outermost electron shell, and chlorine (Cl) has seven electrons in its outer shell—one electron short of being full (Fig. 2-3a). Sodium, therefore, can become stable by losing an electron to chlorine (leaving its outer shell empty), and chlorine can fill its outer shell by gaining that electron. Once sodium loses an electron, the protons in the atom outnumber the electrons, and the sodium atom becomes positively charged (Na^+). Similarly, when chlorine picks up an electron, it becomes negatively charged (Cl^-). Positively or negatively charged atoms are called **ions** (Fig. 2-3b).

Opposite charges attract; therefore, sodium ions and chloride ions tend to stay near one another. They form crystals that contain repeating orderly arrangements of the two ions (Fig. 2-3c). The electrical attraction between oppositely charged ions that holds them together in crystals is called an **ionic bond**. Ionic bonds are weak and easily broken, as occurs when salt is dissolved in water (Table 2-1).

Uncharged Atoms Share Electrons to Form Covalent Bonds

An atom with a partially full outermost electron shell can also become stable by sharing electrons with another atom, forming a **covalent bond**. Consider the hydrogen atom, which has one electron in a shell built for two. A hydrogen atom can become stable if it shares its single electron with another hydrogen atom, forming a molecule of hydrogen gas, H_2. Because the two hydrogen atoms are identical, neither nucleus can exert more attraction and capture the other's electron. So the two electrons orbit around both nuclei for equal amounts of time, forming a single covalent bond; each hydrogen atom behaves almost as if it had two electrons in its shell. Covalent bonds are stronger than ionic bonds.

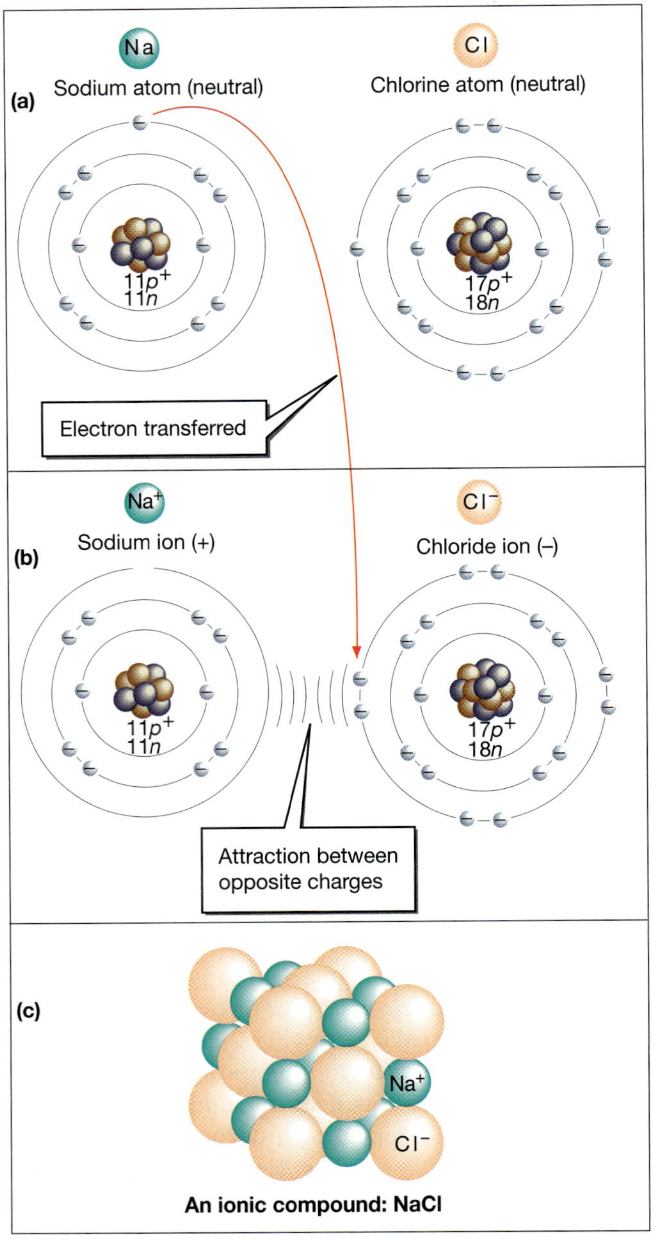

Figure 2-3 **The formation of ions and ionic bonds**

Table 2-1 Chemical Bonds

Type of Bond	Bond Forms:
Weak Bonds: allow interactions between individual atoms or molecules	
Ionic bonds	Between positive and negative ions
Hydrogen bonds	Between a hydrogen atom in a polar covalent bond and another atom in a polar covalent bond
Hydrophobic interactions	Because interactions between water molecules exclude hydrophobic molecules
Strong Bonds: hold atoms together within molecules	
Covalent bonds	By the sharing of electron pairs; equal sharing produces nonpolar covalent bonds; unequal sharing produces polar covalent bonds

Most Biological Molecules Use Covalent Bonding

The molecules in proteins, sugars, carbohydrates, fats, and virtually every other biological molecule are formed of atoms held together by covalent bonds. The most common atoms found in biological molecules (hydrogen, carbon, oxygen, nitrogen, phosphorus, and sulfur) all need at least two electrons to fill their outermost electron shells and can share electrons with two or more other atoms. Hydrogen can form a covalent bond with one other atom; oxygen and sulfur with two other atoms; nitrogen with three; and phosphorus and carbon with up to four. This diversity of bonding arrangements permits biological molecules to be constructed in almost infinite variety and complexity.

Covalent Bonds Are Either Nonpolar or Polar

In hydrogen gas the two nuclei are identical, and the shared electrons spend equal time near each nucleus. Therefore, not only is the molecule as a whole electrically neutral, but each end, or *pole*, of the molecule is also electrically neutral. Such an electrically symmetrical bond is called a nonpolar covalent bond and a molecule (such as H_2) formed with such bonds is a **nonpolar molecule** (Fig. 2-4a). Electron sharing in covalent bonds, however, is not always equal. In many molecules, one nucleus may have a larger positive charge, and therefore attract the electrons more strongly than does the other nucleus. This situation produces a polar covalent bond. Although the **polar molecule** as a whole is electrically neutral, it has charged parts: The atom that attracts the electrons more strongly picks up a slightly negative charge, and is thus the negative pole of the molecule. The other atom has a slightly positive charge and is the positive pole. In water, for example, oxygen attracts electrons more strongly than does hydrogen, so the oxygen end of a water molecule is negative and each hydrogen is positive (Fig. 2-4b). Water, with its charged poles, is a polar molecule.

Hydrogen Bonds Form Between Molecules with Polar Covalent Bonds

Because their covalent bonds are polar, water molecules attract one another. The partially negatively charged oxygen atoms of water molecules attract the partially positively charged hydrogen atoms of other nearby water molecules. This electrical attraction is called a **hydrogen bond** and is weaker than ionic and covalent bonds (Fig. 2-5; see Table 2-1). As we shall see shortly, hydrogen bonds give water several unusual properties that are essential to life on Earth. Hydrogen bonds are also important in many biological molecules.

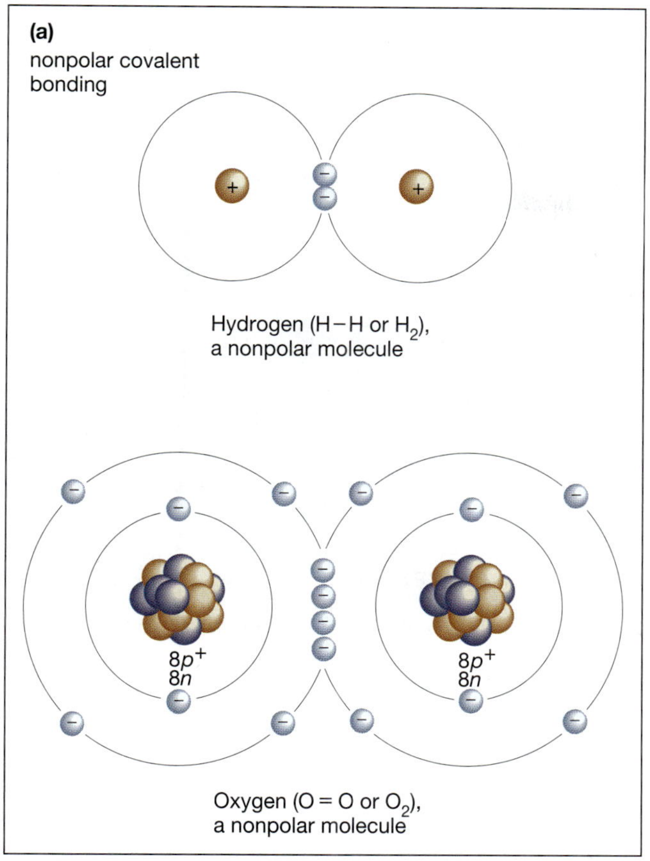

Figure 2-4 Covalent bonds
Shared electrons form covalent bonds. ***(a)*** In hydrogen gas, one electron from each hydrogen atom is shared, forming a single covalent bond. In oxygen gas, two oxygen atoms share four electrons, forming a double covalent bond. ***(b)*** Oxygen, lacking two electrons to fill its outer shell, makes one bond with each of two hydrogen atoms to form water.

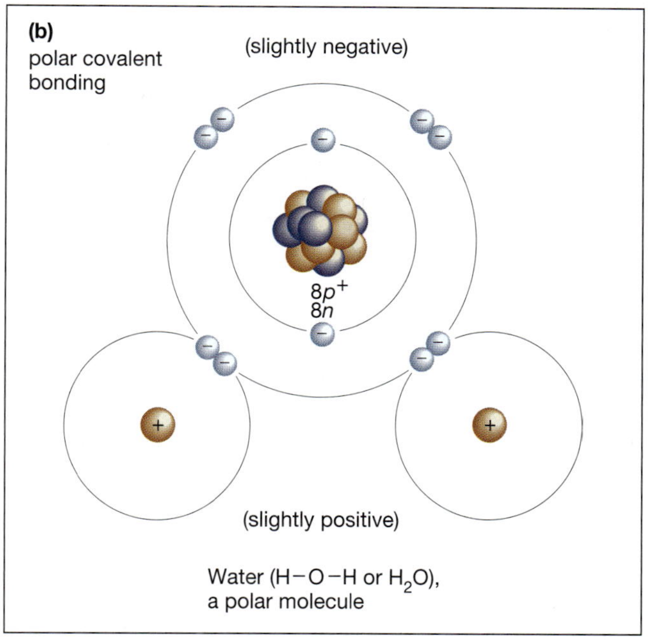

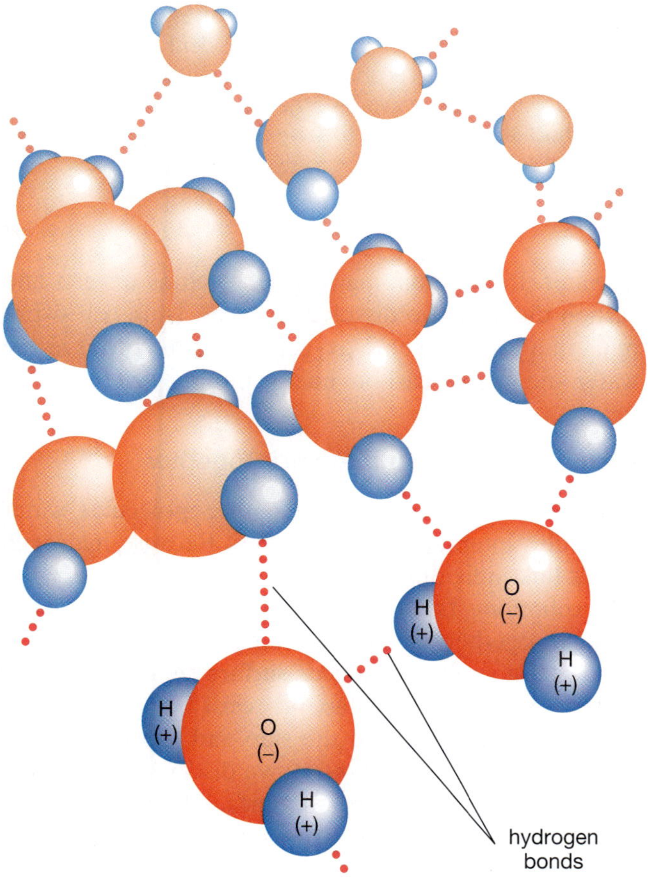

Figure 2-5 Hydrogen bonds
The partial charges on different parts of water molecules produce weak hydrogen bonds, with each bond joining the hydrogen of one water molecule to the oxygen of another water molecule.

3 Why Is Water So Important to Life?

Water is one of the most abundant substances on Earth and plays a crucial role in sustaining life. In fact, most scientists believe that life would be impossible without water. Life on Earth probably originated in water, all living things require water, and 60% to 90% of an organism consists of water. The importance of water derives from some properties of the water molecule that allow it to perform a unique role in support of life. Let's consider some of these properties.

MEDIATUTOR
2.2 Water and Life

Water Interacts with Many Other Molecules

Water enters into many of the chemical reactions in living cells. For example, water supplies the oxygen that green plants release into the air during photosynthesis. Water is also used when animals digest the molecules in food; conversely, water is produced in the reactions that produce proteins, fats, and sugars. Whenever a chemical reaction occurs in a living cell, water is very likely to be involved.

Many Molecules Dissolve Easily in Water

Why is water so important in biological chemical reactions? One reason is that water is an extremely good **solvent**—that is, it is capable of dissolving a wide range of substances, including proteins, salts, and sugars. Because water can dissolve so many molecules, the watery environment inside a cell provides an excellent setting for the countless chemical reactions essential to life.

Earth Watch
Deadly Elements

Combinations of elements form our bodies, our surroundings, and everything we eat or drink. Unfortunately, however, certain elements can harm us if we ingest them. Lead and mercury are two well-known examples of elements whose atoms have the capacity to harm us, and recognition of this danger has led to unleaded gasoline, programs to remove lead-based paint from homes, and the gradual disappearance of thermometers filled with mercury. Nonetheless, significant amounts of these dangerous substances remain in our air, water, and food supplies.

Another hazardous element that is widespread in the environment is arsenic. In high doses, arsenic is a strong poison that can kill within hours. It is also quite dangerous in lower doses. In particular, steady consumption of smaller amounts of arsenic can cause cancer in humans, as studies of some unfortunate populations in Chile and Taiwan have shown. The ability of arsenic to cause cancer is bad news indeed, because arsenic is found in the drinking water of a huge number of people around the world. An estimated 35 million people in Bangladesh alone drink water that is heavily contaminated with arsenic. In the United States, levels of arsenic are high in many public water supplies, especially in the western states.

How does arsenic get into drinking water? In many cases, it seems to have dissolved from naturally occurring deposits in rocks. Industrial waste, however, is an important source of arsenic pollution. The waste from mining operations is especially likely to introduce previously inaccessible arsenic to a water supply. Ultimately, however, the question of how arsenic got into the water is less important than the question of just how much arsenic it takes to endanger human health. The U.S. government first formally recognized the danger of arsenic in drinking water in 1962, when it agreed to set a standard for safe levels of arsenic in water. Finally, after years of scientific study, the U.S. Environmental Protection Agency (EPA) ordered in January 2001 that the concentration of arsenic in drinking water should not exceed 10 micrograms per liter (sometimes also expressed as 10 parts per billion). Just 58 days later—following the inauguration of a new president who was concerned about the high economic cost of reducing arsenic—the new administration cancelled the EPA order, saying, "We pulled back the decision, so that we can make a decision based on sound science."

Figure E2-1 Arsenic pollution from mining waste endangers drinking water
Much of the arsenic in our water supply comes from mining and industrial waste, but naturally occurring deposits of arsenic are also major contributors to the problem.

Water is such an excellent solvent because it is a polar molecule, with positive and negative poles. Thus, if a crystal of table salt (sodium chloride) is dropped into water, the positively charged ends of water molecules will be attracted to and will surround the negatively charged chloride ions in the salt crystal. At the same time, the negatively charged poles of water molecules will surround the positively charged sodium ions. As water molecules enclose the sodium and chloride ions and shield them from interacting with each other, the ions separate from the crystal and drift away in the water—the ionic bonds are broken and the salt dissolves (Fig. 2-6).

Water also dissolves molecules that are held together by polar covalent bonds. Its positive and negative poles are attracted to oppositely charged regions of dissolving molecules. Ions and polar molecules that dissolve readily in water are termed *hydrophilic* (Greek for "water-loving") because of their electrical attraction for water molecules. (Molecules, such as fats and oils, that are uncharged and nonpolar and do not dissolve in water are called *hydrophobic.*) Many biological molecules, including sugars and amino acids, are hydrophilic and dissolve in water. In addition, many gases, such as oxygen and carbon dioxide, also dissolve in water. Thus, fish swimming in a lake consume oxygen that is dissolved in the water and release carbon dioxide into the water.

Water Molecules Tend to Stick Together

In addition to interacting with other molecules, water molecules interact with each other. Because hydrogen bonds link individual water molecules, liquid water has high **cohesion**—that is, water molecules have a tendency to stick together. Cohesion among water molecules at the water's surface produces **surface tension**, the tendency for the water surface to resist being broken. If you've ever experienced the slap and sting of a belly flop into a swimming pool, you've discovered firsthand the power of surface tension. Surface tension can support fallen leaves, some spiders and water insects, and even a running lizard (Fig. 2-7a).

Cohesion plays a more important role in land plants. A plant absorbs water through its roots, but is then faced with the big problem of moving the water to the aboveground parts, which can be more than 300 feet up

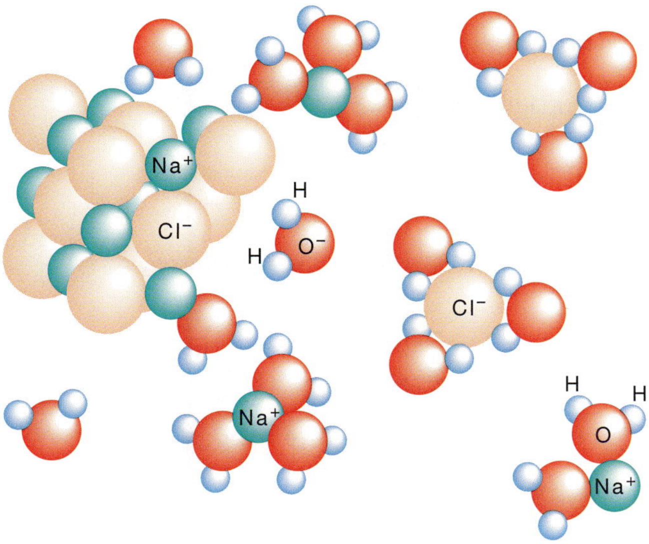

Figure 2-6 Water as a solvent
The polarity of water molecules allows water to dissolve both charged molecules and polar molecules. When a salt crystal is dropped into water, the oppositely charged poles of the water molecules surround the sodium and chloride ions. Thus insulated from the attractiveness of other molecules of salt, the ions disperse, and the whole crystal gradually dissolves.

in a tall redwood (Fig. 2-7b). The problem is solved by cohesion. The tiny tubes that connect the leaves, stem, and roots are filled with water, and when water molecules evaporate from the leaves atop the plant, water is pulled up the tubes from below to fill the empty space. The system works because the hydrogen bonds between water molecules are stronger than the weight of the water in the tubes. Even in a 300-foot-tall tree, the water "chain" doesn't break.

Water exhibits another property, *adhesion*, a word that describes its tendency to stick to surfaces that have a slight charge that attracts polar water molecules. Adhesion helps water move within small spaces, such as the thin tubes in plants that carry water from roots to leaves. If you stick the end of a narrow glass tube into water, the water will move a short distance up the tube. Put some water in a narrow glass bud vase or test tube and you'll see that the upper surface is curved; the water is using both adhesion and cohesion to pull itself up the sides of the glass—by adhesion to the surface of the glass and cohesion among water molecules.

Water Can Form Ions

Although water is generally regarded as a stable compound, individual water molecules constantly gain, lose, and swap hydrogen atoms. As a result, at any given time about two of every billion water molecules are ionized—that is, broken apart into hydrogen ions (H^+) and hydroxide ions (OH^-):

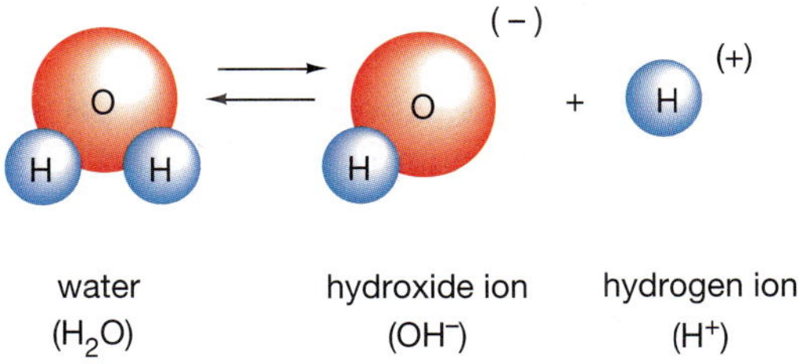

(a)

(b)

Figure 2-7 Cohesion among water molecules
(a) With webbed feet bearing specialized scales, the basilisk lizard of South America makes use of surface tension, caused by cohesion, to support its weight as it races across the surface of a pond. ***(b)*** Within a redwood tree, cohesion holds water molecules together in continuous strands from the roots to the topmost leaves, which may be more than 300 feet above the ground.

A hydroxide ion has gained an electron from the hydrogen atom in a water molecule, and it has a negative charge. The hydrogen ion, which has lost its electron, now has a positive charge. Pure water contains equal concentrations of hydrogen ions and hydroxide ions.

In many solutions, however, the concentrations of H^+ and OH^- are not the same. If the concentration of H^+ exceeds the concentration of OH^-, the solution is *acidic*. An **acid** is a substance that releases hydrogen ions when it is dissolved in water. When hydrochloric acid (HCl), for example, is added to pure water, almost all of the HCl molecules separate into H^+ and Cl^-. Therefore, the concentration of H^+ greatly exceeds the concentration of OH^-, and the resulting solution is acidic. (Many acidic substances, such as lemon juice and vinegar, have a sour taste. This is because the sour-taste receptors on your tongue are specialized to respond to the excess of H^+.)

If the concentration of OH^- is greater than that of H^+, the solution is *basic*. A **base** is a substance that combines with hydrogen ions, reducing their number. If, for instance, sodium hydroxide (NaOH) is added to water, the NaOH molecules separate into Na^+ and OH^-. The OH^- combine with H^+, reducing the number of H^+ ions. The solution then contains an excess of OH^- and is basic.

pH Measures Acidity

The degree of acidity is expressed on the **pH scale**, in which neutrality (equal numbers of H^+ and OH^-) is assigned the number 7. Acids have a pH below 7; bases have a pH above 7 (Fig. 2-8). Pure water, with equal concentrations of H^+ and OH^-, has a pH of 7. Each unit on the pH scale represents a tenfold change in the concentration of H^+. Thus, a carbonated soft drink (pH = 3) has a concentration of H^+ 10,000 times higher than water (pH = 7)—no wonder it is bad for your teeth!

A Buffer Maintains a Solution at a Constant pH

In most mammals, including humans, both the cell interior and the fluids that bathe the cells are nearly neutral (pH about 7.3 to 7.4). Small increases or decreases in pH may cause drastic changes in both the structure and function of biological molecules, leading to the death of cells or entire organisms. Nevertheless, living cells seethe with chemical reactions that take up or give off H^+. How, then, does the pH remain constant overall? The answer lies in the many buffers found in organisms. A **buffer** is a substance that tends to maintain a solution at a constant pH by accepting or releasing H^+ in response to small changes in H^+ concentration. If the H^+ concentration rises, buffers combine with them; if the H^+ concentration falls, buffers release H^+. The result in both cases is that the concentration of H^+ is restored to its original level.

The Big Picture

Water is absolutely essential to life on Earth. Most scientists believe that if life exists on any other planet, it is present only on planets on which liquid water is present.

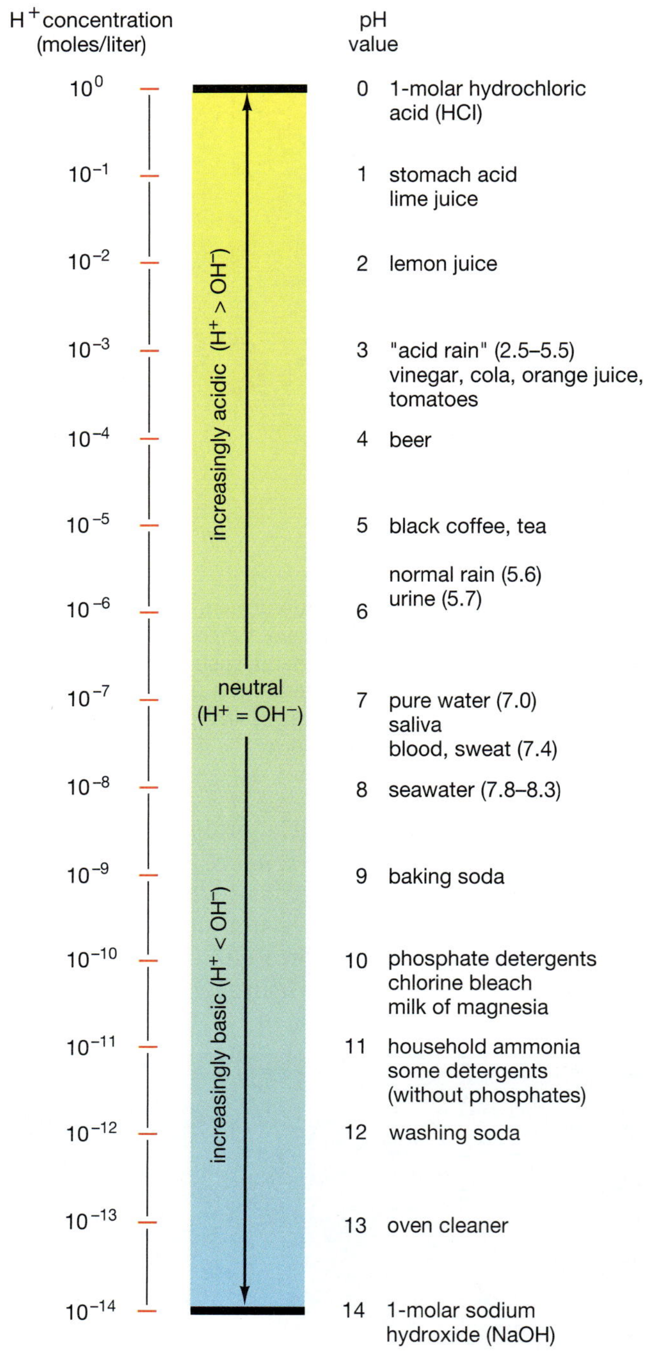

Figure 2-8 The pH scale
The pH scale expresses the concentration of hydrogen ions in a solution on a scale of 0 (very acidic) to 14 (very basic).

4 Why Is Carbon So Important to Life?

Organisms contain a tremendous variety of different substances, but a great many of these diverse molecules have one property in common: they contain carbon. The pervasiveness of carbon in living things is related to the versatility of the carbon atom. Carbon can combine with other atoms in many different ways to form a huge number of different molecules. This vast array of different combinations is possible because a carbon atom has four electrons in its outermost shell, leaving room for four more. Therefore, carbon atoms are able to form many bonds. They become stable by sharing four electrons with other atoms, forming up to four covalent bonds. Carbon-containing molecules can contain many carbon atoms and can assume a variety of complex shapes, including chains, branches, and rings. Carbon thus forms the basis for an amazing diversity of molecules, and makes it possible for living things to construct the many different substances required to sustain life from a limited pool of different molecules.

In chemistry, molecules that have a carbon skeleton and also contain some hydrogen atoms are known as **organic molecules.** The carbon skeletons of organic molecules can be quite complex, but carbon alone does not account for the diversity of organic molecules. Instead, groups of atoms, called **functional groups**, attach to the carbon backbone and determine the characteristics and chemical reactivity of the molecules. These functional groups are far less stable than the carbon backbone and are more likely to participate in chemical reactions. The common functional groups found in organic molecules are shown in Table 2-2.

5 How Are Organic Molecules Constructed?

In living organisms, large, complex molecules are constructed by first assembling smaller molecules and then hooking them together. Small organic molecules (for example, *sugars*) are used as subunits to construct longer molecules (for example, *starches*), like cars in a train.

Biological Molecules Are Joined Together or Broken Apart by Adding or Removing Water

In organisms, biological molecules must be both built and broken down. Water plays a central role in both types of reaction. When complex biological molecules are built in an organism, water is often produced as a by-product. When molecules, such as those consumed as food, are broken down to liberate subunits that an organism can use, water is usually consumed.

The chemical reactions that link subunits together to make large biological molecules are called **dehydration synthesis** reactions. In a dehydration synthesis, a hydrogen atom (–H) is removed from one subunit and

Table 2-2 Important Functional Groups in Biological Molecules

Group	Structure	Properties	Types of Molecules
Hydrogen (–H)	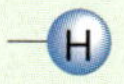	Polar or nonpolar, depending on which atom hydrogen is bonded to; involved in condensation and hydrolysis	Almost all organic molecules
Hydroxyl (–OH)	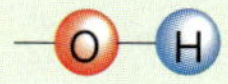	Polar; involved in condensation and hydrolysis	Carbohydrates, nucleic acids, alcohols, some acids, and steroids
Carboxyl (–COOH)	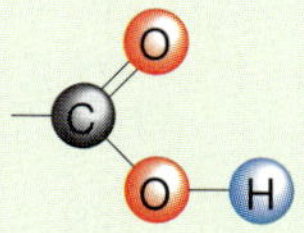	Acidic; negatively charged when H^+ dissociates; involved in peptide bonds	Amino acids, fatty acids
Amino ($-NH_2$)	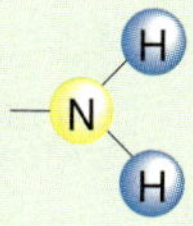	Basic; may bond an additional H^+, becoming positively charged; involved in peptide bonds	Amino acids, nucleic acids
Phosphate ($-H_2PO_4$)	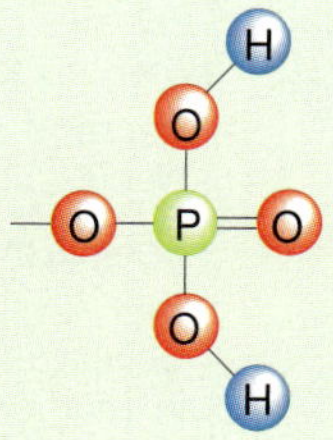	Acidic; up to two negative charges when H^+ dissociates; links nucleotides in nucleic acids; energy-carrier group in ATP	Nucleic acids, phospholipids
Methyl ($-CH_3$)	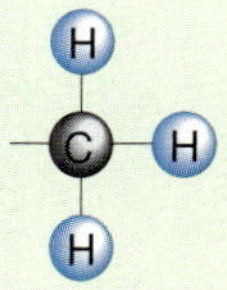	Nonpolar; tends to make molecules hydrophobic	Many organic molecules; especially common in lipids

MEDIATUTOR
2.3 Structure of Biological Molecules

a hydroxyl group (–OH) is removed from a second subunit, creating openings in the outer electron shells of the two subunits. These openings are then filled by electrons shared between the subunits, creating a covalent bond that links them. The free hydrogen and hydroxyl ions combine to form a molecule of water (H_2O):

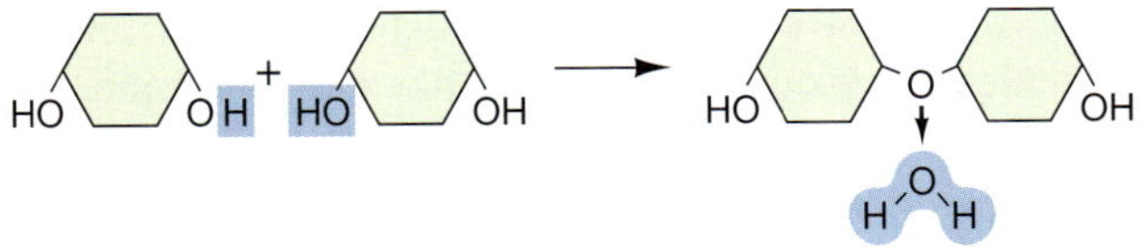

Dehydration synthesis

The reverse reaction, **hydrolysis** ("to break apart with water"), can split the molecule into individual subunits:

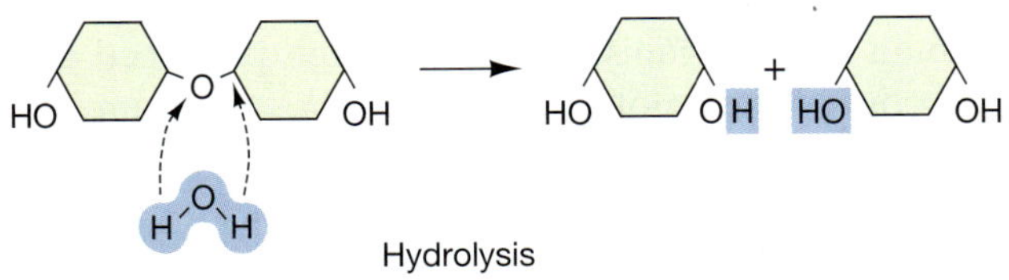

Hydrolysis

Considering how complicated living things are, it might surprise you to learn that nearly all biological molecules fall into one of only four general categories: carbohydrates, lipids, proteins, or nucleic acids (Table 2-3).

6 What Are Carbohydrates?

Carbohydrates are molecules composed of carbon, hydrogen, and oxygen in the approximate ratio of 1:2:1. Carbohydrates can be small, water-soluble sugars or long chains that are made by stringing sugar subunits together. A carbohydrate consisting of just one sugar molecule is called a simple sugar, or *monosaccharide*. When two monosaccharides are linked, they form a *disaccharide*; three or more form a *polysaccharide*.

Carbohydrates are important energy sources for most organisms. Consider a breakfast that includes blueberry pancakes, syrup, and orange juice. The pan-

Table 2-3 The Principal Biological Molecules

Class of Molecule	Principal Subtypes (subunits in parentheses)	Example	Function
Carbohydrate: Normally contains carbon, oxygen, and hydrogen, in the approximate formula $(CH_2O)_n$	(*Monosaccharide*: Simple sugar)	Glucose	Important energy source for cells; subunit of which most polysaccharides are made
	Disaccharide: Two monosaccharides bonded together	Sucrose	Principal sugar transported throughout bodies of land plants
	Polysaccharide: Many monosaccharides (usually glucose) bonded together	Starch Glycogen Cellulose	Energy storage in plants Energy storage in animals Structural material in plants
Lipid: Contains high proportion of carbon and hydrogen; usually nonpolar and insoluble in water	*Oil, fat*: Three fatty acids bonded to glycerol	Peanut oil Beef fat	Energy storage in animals, some plants
	Wax: Variable numbers of fatty acids bonded to long-chain alcohol	Waxes in plant cuticle	Waterproof covering on leaves and stems of land plants
	Phospholipid: Polar phosphate group and two fatty acids bonded to glycerol	Phosphatidylcholine	Part of membranes in cells
	Steroid: Four fused rings of carbon atoms with functional groups attached	Cholesterol	Part of membranes of eukaryotic cells; precursor for other steroids such as testosterone, bile salts
Protein: Chains of amino acids; contain carbon, hydrogen, oxygen, nitrogen, and sulfur	(*amino acids*)	Keratin	Principal component of hair
		Silk	Principal component of silk moth cocoons and spider webs
		Hemoglobin	Transport of oxygen in vertebrate blood
Nucleic acid: Made of nucleotide subunits; may consist of a single nucleotide or long chain of nucleotides	*Long-chain nucleic acids*	Deoxyribonucleic acid (DNA)	Genetic material of all living cells
		Ribonucleic acid (RNA)	Genetic material of some viruses; in living cells, essential in transfer of genetic information from DNA to protein
	(*Single nucleotides*)	Adenosine triphosphate (ATP)	Principal short-term energy-carrier molecule in cells
		Cyclic adenosine monophosphate (cyclic AMP)	Intracellular messenger

cakes consist mainly of carbohydrates that were originally stored in the seeds of wheat or other grains. The sugar that sweetens the syrup, blueberries, and orange juice was also stored by plants as an energy source. The carbohydrate molecules that once served the plants that manufactured them now provide energy to the humans who consume them. Other carbohydrates, such as cellulose and similar molecules, provide structural support for individual cells or even for the entire bodies of organisms as diverse as plants, fungi, bacteria, and insects.

Most simple sugars have a backbone of three to seven carbon atoms. Each carbon atom in the backbone generally has both a hydrogen (–H) and a hydroxyl group (–OH) attached to it, so carbohydrates have the general chemical formula $(CH_2O)_n$, where n is the number of carbons in the backbone. This formula explains the origin of the name "carbohydrate," which literally means "carbon plus water." When dissolved in water, the carbon backbone of a simple sugar usually "circles up" to form a ring (Fig. 2-9); simple sugars also assume the ring form when they link together to form disaccharides and polysaccharides.

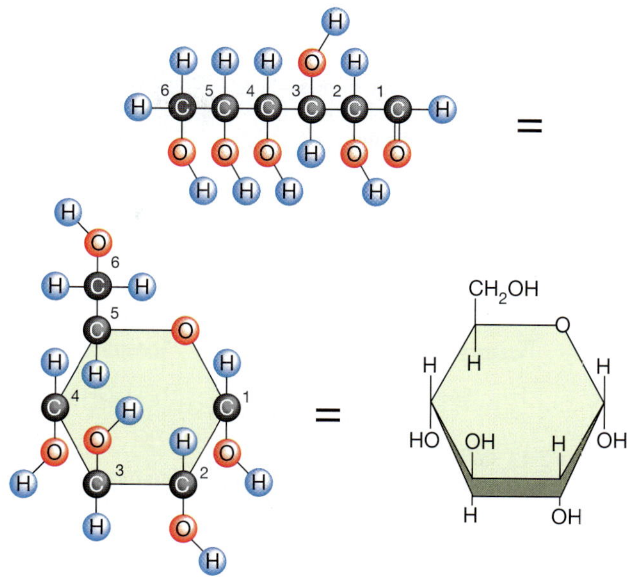

Figure 2-9 A simple sugar
Three diagrams of glucose, a six-carbon monosaccharide. The top diagram represents the molecule in its linear (straight) form. The two diagrams at the bottom show two different ways to represent the ring form of glucose. In the ring diagram on the right, the carbon at each corner of the ring is not labeled; this is a convention always used on this type of diagram.

A Variety of Simple Sugars Occur in Organisms

Glucose is the most common simple sugar in organisms and is the subunit of which most polysaccharides are made. Glucose has six carbons, and hence its chemical formula is $C_6H_{12}O_6$. Many organisms synthesize other simple sugars that have the same chemical formula as glucose but have slightly different structures. These include fructose (the "corn sugar" found in corn syrup; also the molecule that makes fruits taste sweet). Some other common simple sugars, such as ribose and deoxyribose, have five carbons.

Disaccharides Store Energy and Serve as Building Blocks

Simple sugars, especially glucose and its relatives, have a short life span in a cell. Most either quickly break down, freeing their chemical energy to fuel various cellular activities, or become linked to form disaccharides or polysaccharides (Fig. 2-10). Disaccharides are often used for short-term energy storage, especially in plants. Common disaccharides include sucrose (table sugar: glucose plus fructose) and lactose (milk sugar: glucose plus galactose). When cells require energy, disaccharides are broken apart into their monosaccharide subunits to release energy.

Polysaccharides Store Energy and Provide Support

Try chewing a cracker for a long time. Does it taste sweeter the longer you chew? It should, because the cracker contains starch, a polysaccharide that breaks down into its sweet-tasting glucose subunits as you chew. Certain polysaccharides like starch (in plants) and glycogen (in animals) are used mainly for long-term energy storage. Starch is commonly formed in roots and seeds, most commonly as huge, branched chains of up to half a million glucose subunits. Glycogen molecules, which are stored as an energy source in the liver and muscles of animals such as humans, are generally much smaller than starch molecules.

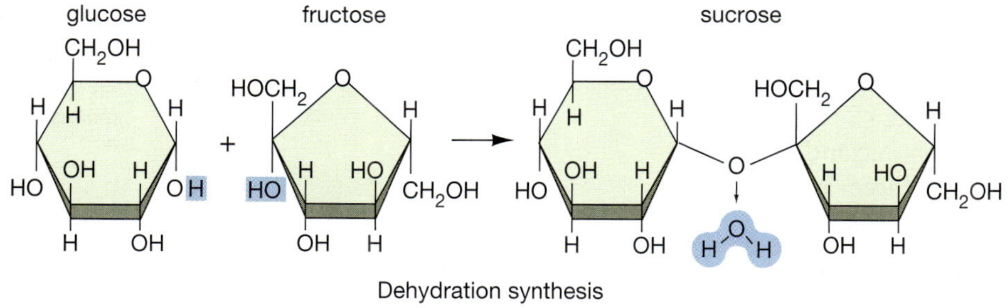

Figure 2-10 Manufacture of a disaccharide
The disaccharide sucrose is made by a dehydration synthesis reaction in which a hydrogen (–H) is removed from glucose and a hydroxyl group (–OH) is removed from fructose, another simple sugar. The two simple sugars join by a covalent bond, and a water molecule forms in the process.

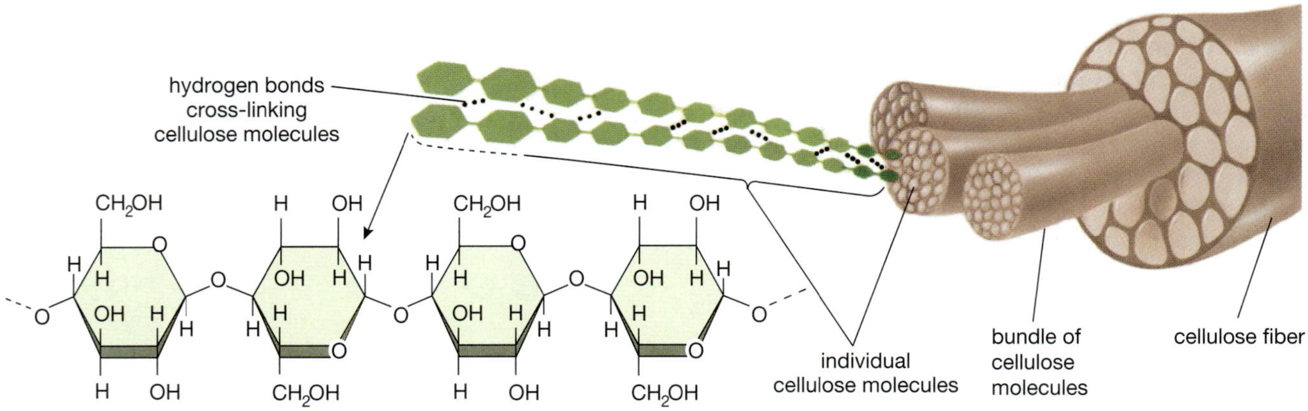

Figure 2-11 Cellulose structure
Cellulose, like starch, is composed of glucose subunits. Unlike starch, however, cellulose has great structural strength, due partly to the arrangement of parallel molecules of cellulose into long, cross-linked fibers.

MediaTutor 2.4 Functions of Macromolecules

Many organisms also use polysaccharides as structural materials. One of the most important structural polysaccharides is cellulose, which provides support for plant cells and makes up about half the bulk of a typical tree trunk (Fig. 2-11). Cellulose is by far the most abundant organic molecule on Earth (not that surprising when you consider the vast fields and forests that blanket much of our planet). Ecologists estimate that about a trillion tons of cellulose are made each year.

Like starch, cellulose consists of long chains of glucose subunits. Cellulose's digestibility, however, differs sharply from that of starch. Whereas most animals can readily digest starch, they cannot digest cellulose at all. For most animals, cellulose is roughage or fiber, material that passes undigested through the digestive tract. The only exceptions are animals such as cows or termites, which have special one-celled organisms living in their digestive tracts. These microscopic organisms are among the few that are able to break down cellulose, so the animals in whose guts they live can eat leaves, wood, and other foods that contain a lot of cellulose. The microbes digest the cellulose and the animal hosts can absorb the nutrient molecules that are thereby freed.

7 What Are Lipids?

Lipids are molecules that share two important features. First, they contain large regions composed almost entirely of hydrogen and carbon, with nonpolar carbon–carbon or carbon–hydrogen bonds. Second, these nonpolar regions make lipids hydrophobic and insoluble in water. Lipids are classified into three major groups: (1) oils, fats, and waxes; (2) phospholipids; and (3) steroids.

Oils, Fats, and Waxes Contain Only Carbon, Hydrogen, and Oxygen

Oils, fats, and waxes have several features in common: (1) They contain only carbon, hydrogen, and oxygen; (2) they contain one or more *fatty acid* subunits, which are long chains of carbon and hydrogen with a *carboxyl group* (–COOH) at one end; and (3) they usually do not have ring structures. Fats and oils form by dehydration synthesis from three fatty acid subunits and one molecule of glycerol, a short, three-carbon molecule with one hydroxyl group (–OH) per carbon:

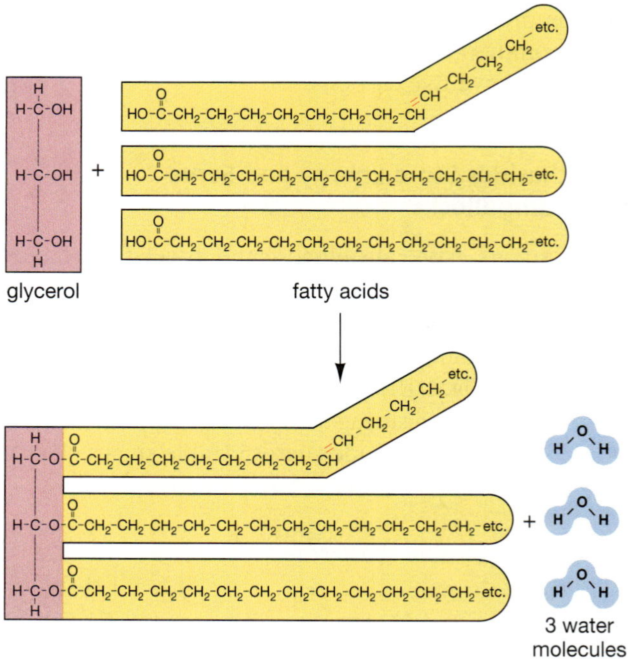

Fats and oils have a high concentration of chemical energy, about 9.3 Calories per gram, compared with 4.1

for sugars and proteins (the Calorie is a unit that measures the energy content of foods). Because fats are so high in calories, fat substitutes such as olestra may be especially appealing to dieters.

Fats and oils are used for long-term energy storage in both plants and animals. For example, during summer and fall, bears consume more energy than they spend and store fat on their bodies, which tides them over during their winter hibernation. Because fats store the same energy with less weight than do carbohydrates, fat is an efficient way for animals to store energy (Fig. 2-12a).

The difference between a fat (such as beef fat), which is a solid at room temperature, and an oil (such as peanut oil), which is liquid at room temperature, lies in their fatty acids. In fats, all carbons in the fatty acid chains are joined with single bonds. The remaining bond positions on the carbons are occupied by hydrogens. Such fatty acids are said to be saturated, because they are "saturated" with hydrogens—they have as many hydrogens as possible. Saturated fatty acids tend to be very straight, and they nestle closely together, forming solid lumps at room temperature:

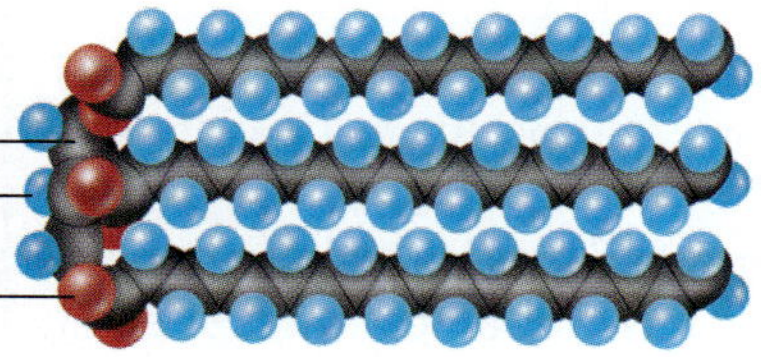

Beef fat (saturated)

If there are double bonds between some of the carbons, and consequently fewer attached hydrogens, the fatty acid is said to be unsaturated. Unsaturated fatty acids, which predominate in oils, tend to have bends and kinks in the fatty acid chains:

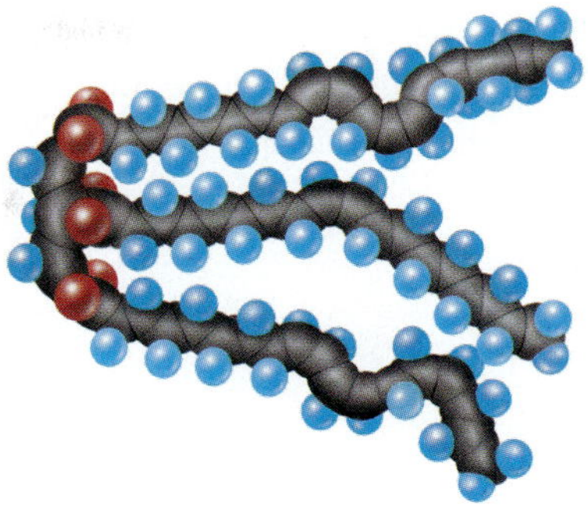

Peanut oil (unsaturated)

The kinks keep oil molecules apart; as a result, oil is liquid at room temperature. An oil can be converted to a fat by breaking the double bonds between carbons, replacing them with single bonds, and adding hydrogens to the remaining bond positions. This is the "hydrogenated oil" listed in the ingredients on a box of margarine, which makes the margarine solid at room temperature.

Waxes are chemically similar to fats. They are highly saturated, making them solid at normal outdoor temperatures. Waxes form a waterproof coating over the leaves and stems of land plants. Some animals synthesize waxes as waterproofing (for mammalian fur, for example) and a few others use it to build elaborate structures such as beehives (Fig. 2-12b).

Phospholipids Have Water-Soluble Heads and Water-Insoluble Tails

The membrane that separates the inside of a cell from the outside world contains several types of *phospholipids*. A phospholipid is similar to an oil, except that

(a) (b)

Figure 2-12 Lipids
(a) Fat is an efficient way to store energy. If this bear stored the same amount of energy in carbohydrates instead of fat, she probably would be unable to walk! ***(b)*** Wax is a highly saturated lipid that remains very firm at normal outdoor temperatures. Its rigidity serves well in the strong but thin-walled hexagons of this honeycomb.

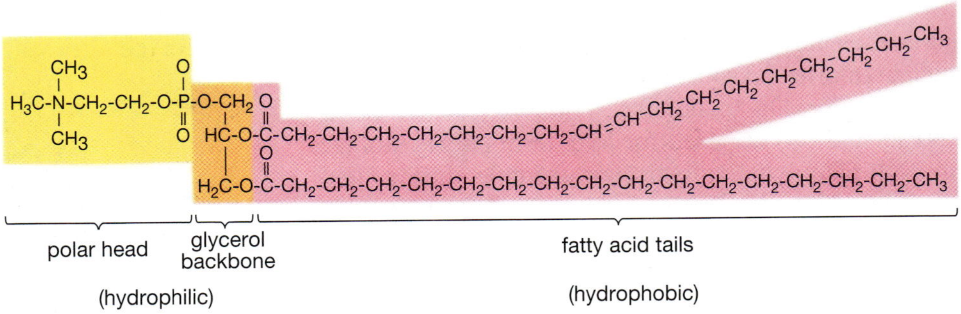

Figure 2-13 Phospholipids
Phospholipids are similar to fats and oils, except that only two fatty acid tails are attached to the glycerol backbone. The backbone is bonded to a polar head composed of a phosphate group plus another functional group.

one of the three fatty acids is replaced by a smaller, phosphate-containing subunit. This subunit sits at one end of the phospholipid molecule, where it forms a polar head that is water soluble. The two fatty acids form tails that are not soluble in water. Thus, a phospholipid has two dissimilar ends: a hydrophilic head attached to hydrophobic tails (Fig. 2-13). As you will see in Chapter 3, this dual nature of phospholipids is crucial to the structure and function of cell membranes.

Steroids Consist of Four Carbon Rings Fused Together

Steroids are structurally different from the other lipids. All steroids are composed of four rings of carbon fused together with various functional groups protruding from them (Fig. 2-14; note the basic steroid "skeleton" in color). One type of steroid is cholesterol. Cholesterol is a vital component of the membranes of animal cells and is also used by cells to synthesize other steroids. Other steroids made from cholesterol include male and female sex hormones, the hormones that regulate salt levels, and bile, which assists in fat digestion. Why, then, has cholesterol gotten so much bad publicity? Find out in "Health Watch: Cholesterol—Friend and Foe."

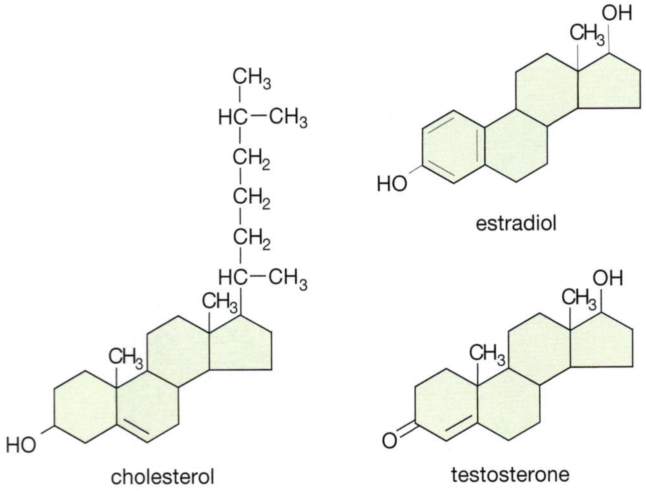

Figure 2-14 Steroids
Other steroids are synthesized from cholesterol. All steroids have almost the same molecular structure (colored rings). Great differences in steroid function result from the differences in functional groups attached to the rings. Notice the similarity in structure between the male sex hormone testosterone and the female sex hormone estradiol (a type of estrogen).

8 What Are Proteins?

Proteins perform many different functions in organisms. An especially important role is played by *enzymes*, proteins that guide almost all the chemical reactions that occur inside cells. Because each enzyme assists only one or a few specific reactions, most cells contain hundreds of different enzymes. Other types of proteins are used for structure (Fig. 2-15) or energy storage. Proteins may also function in transport or movement (a protein carries oxygen in the blood, and others help muscle cells move).

Proteins Are Formed from Chains of Amino Acids

A molecule of **protein** consists one or more chains of amino acids. All **amino acids** have the same fundamental structure, in which a central carbon is bonded to four different functional groups: an amino group ($-NH_2$); a carboxyl group ($-COOH$); a hydrogen; and a variable group (represented by the letter R):

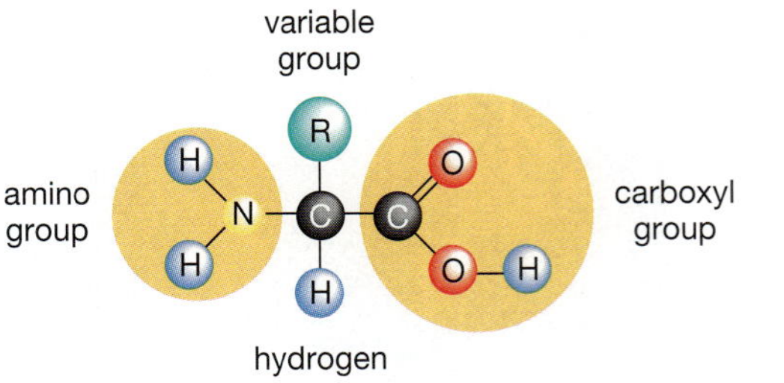

Each of the twenty different amino acids that are found in proteins has a different R group, so amino acids differ in their chemical and physical properties—size, water solubility, electrical charge, and so on. Therefore, the sequence in which the amino acids of a protein are arranged dictates the properties of the protein and

Figure 2-15 **Structural proteins**
Common structural proteins include those of hair, horn, and spider web silk.

determines whether it is an enzyme or a hormone or a structural protein.

Amino Acids Join to Form Chains by Dehydration Synthesis

Like lipids and polysaccharides, proteins form by dehydration synthesis. The nitrogen of the amino group ($-NH_2$) of one amino acid is joined to the carbon of the carboxyl group ($-COOH$) of another amino acid by a single covalent bond. This bond is called a peptide bond, and the resulting chain of two amino acids is called a *peptide* (Fig. 2-16). More amino acids are added, one by one, until the protein is complete. Amino acid chains in living cells vary in length from three to thousands of amino acids. Often, the word "protein," or "polypeptide," is reserved for long chains—say, 50 or more amino acids in length—and "peptide" is used for shorter chains.

Three-Dimensional Structures Give Proteins Their Functions

The phrase "amino acid chains" may evoke images of proteins as floppy, featureless structures, but they are not. Proteins are, instead, highly organized molecules that fold themselves into complex, three-dimensional shapes. In many cases, two or more peptides join together to form even more complicated structures (Fig. 2-17).

The shape of a protein is determined by the exact type, position, and number of its amino acids. If the shape of a protein is disrupted, the protein may no longer be able to function correctly, even if the peptide bonds between amino acids remain intact. Proteins whose shape has been disrupted are said to be **denatured**. There are many ways to denature a protein. For example, when an egg is fried, the heat of the frying pan denatures the albumin protein in the egg white, causing its appearance to change from clear to white and its texture to change from liquid to solid. Sterilization using heat or ultraviolet rays kills bacteria and viruses by denaturing the proteins that they need to live. Salty or acidic solutions also denature proteins—dill pickles are preserved in this way.

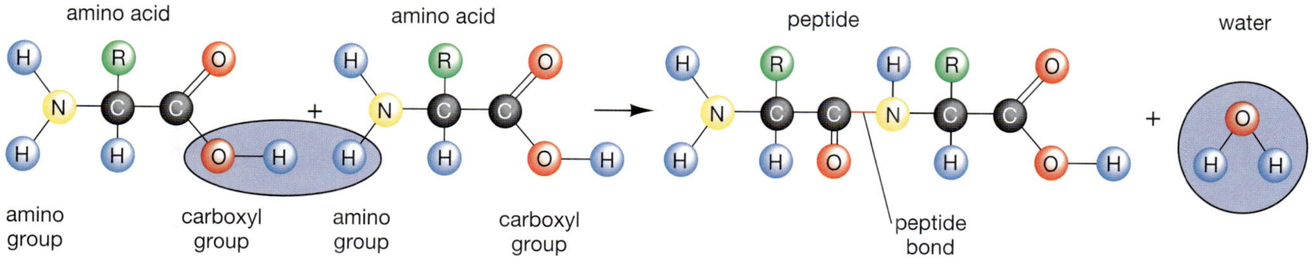

Figure 2-16 **Protein synthesis**
In protein synthesis, a dehydration synthesis joins the carbon of the carboxyl acid group of one amino acid to the nitrogen of the amino group of a second amino acid.

Health Watch
Cholesterol—Friend and Foe

Cholesterol is a steroid with a bad reputation. Why are so many products now advertising themselves as "cholesterol free" or "low in cholesterol"? After all, cholesterol is a crucial component of cell membranes. It is also the raw material for the production of bile (which helps us digest fats), vitamin D, and both male and female sex hormones.

Although cholesterol is crucial to life, medical researchers have found that people with excessively high levels of cholesterol in their blood are at increased risk for heart attacks and strokes. Unfortunately, the cholesterol builds up "silently" and gives no warning signs. A person may not know that anything is wrong until he or she actually suffers a heart attack. Cholesterol contributes to the formation of obstructions in arteries, called *plaques,* which in turn can promote the formation of blood clots. These clots can break loose and block an artery carrying blood to the heart, causing a heart attack, or to the brain, causing a stroke.

Have you ever heard of "good cholesterol" and "bad cholesterol"? Because cholesterol molecules are nonpolar, they do not dissolve in blood (which is mostly water). Thus, the cholesterol molecules are transported through blood in packets surrounded by special carrier molecules called *lipoproteins* (phospholipids plus proteins). Cholesterol in high-density lipoprotein packets ("HDL cholesterol," which has more protein and less lipid) is the good kind; these packets transport cholesterol to the liver, where it is removed from circulation. Cholesterol in low-density lipoprotein packets ("LDL cholesterol" with less protein and more lipid) is the bad kind; this is the form in which cholesterol circulates to cells throughout the body and can be deposited on artery walls. A high ratio of HDL (good) to LDL (bad) is correlated with reduced risk of heart disease. A complete cholesterol screening test will distinguish between these two forms in your blood.

Where does cholesterol come from? Cholesterol comes from animal-derived foods; it is essentially nonexistent in plants. Egg yolks are a particularly rich source; meat, whole milk, and butter contain it as well. Another source of cholesterol is your own body, which synthesizes cholesterol from other lipids. Because of genetic differences, some people's bodies manufacture more than others. People with high cholesterol (about 25% of all adults in the United States) can often reduce their levels by eating a diet low in both cholesterol and saturated fats. For people with excessively high cholesterol who are unable to reduce it adequately by changing their diets, doctors often prescribe cholesterol-reducing drugs.

9 What Are Nucleic Acids?

Nucleic acids are long chains of similar but not identical subunits called nucleotides. All nucleotides have a three-part structure: (1) a five-carbon sugar (ribose or deoxyribose), (2) a phosphate group, and (3) a nitrogen-containing molecule called a *base* that differs among nucleotides:

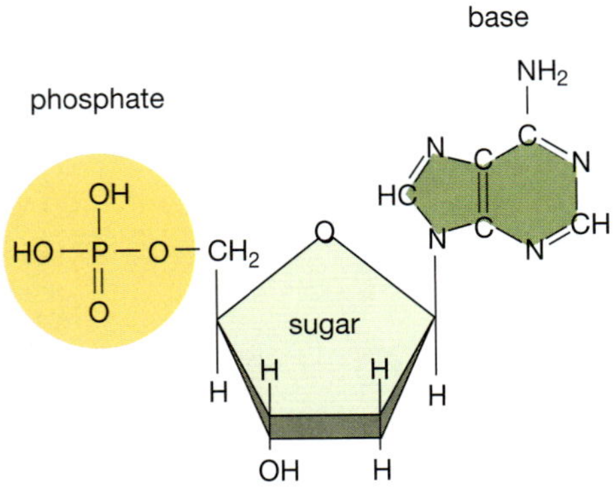

Deoxyribose nucleotide

There are two types of nucleotides, the ribose nucleotides (containing the sugar ribose) and the deoxyribose nucleotides (containing the sugar deoxyribose). Nucleotides string together in long chains as nucleic acids, with the phosphate group of one nucleotide covalently bonded to the sugar of another:

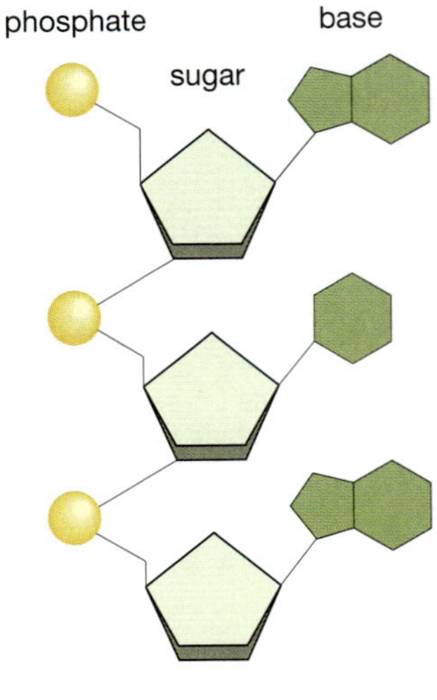

Nucleotide chain

DNA and RNA, the Molecules of Heredity, Are Nucleic Acids

There are two types of nucleic acids: (1) *deoxyribonucleic acid*, or *DNA*, and (2) *ribonucleic acid*, or *RNA*. DNA consists of chains of deoxyribose nucleotides millions of units long. DNA is found in all living things, and

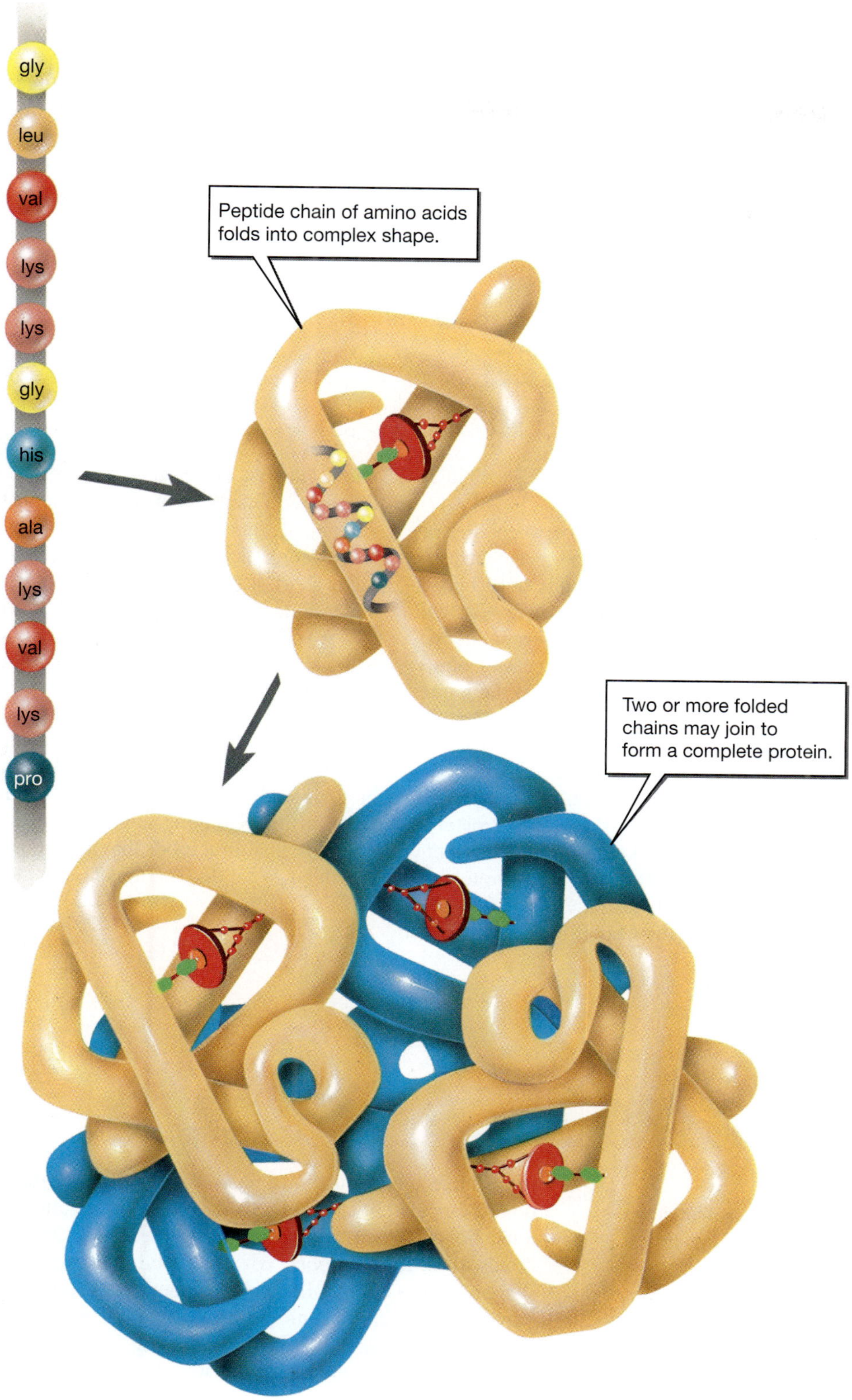

Figure 2-17 Protein chains fold into complex shapes
Hemoglobin, the oxygen-carrying protein in red blood cells, is a combination of four peptide subunits (two each of two different types). Even minor alterations to its shape can prevent hemoglobin from doing its job.

its sequence of nucleotides, like the dots and dashes of a biological Morse code, spells out the genetic information needed to construct the proteins of each organism. Chains of RNA are copied from the central repository of DNA in the nucleus of each cell. These RNA copies carry the message of DNA's genetic code and direct the manufacture of proteins.

Other Nucleotides Perform Other Functions

Not all nucleotides are part of nucleic acids. Some exist singly in the cell or as parts of other molecules. Cyclic nucleotides, such as *cyclic adenosine monophosphate* (*cyclic AMP*), act as messengers within cells, and carry information to other molecules in the cell. Nucleotides with extra phosphate groups, such as *adenosine triphosphate* (*ATP*), carry energy from place to place within a cell. They pick up energy where it is produced (during photosynthesis, for example) and deliver it to drive energy-demanding reactions elsewhere (say, to manufacture a protein).

The Big Picture

Life is largely based on just a dozen or so different chemical elements. These few elements, however, combine and recombine in a multitude of different ways to yield the intricate, organized complexity of growing, reproducing organisms.

CASESTUDYREVISITED

Improving on Nature?

Now that you know more about biological molecules, let's revisit the topic of olestra and see how scientists have modified fat molecules to make them indigestible. Earlier fat substitutes used protein, carbohydrate, and fat molecules to build new molecules that imitated the molecular shape of fat—one molecule of glycerol attached to three molecules of fatty acids. In olestra, however, the glycerol molecule is replaced with sucrose and has six, seven, or eight fatty acids attached (the fatty acids come from vegetable oil). Apparently, the large number of fatty acid chains prevents digestive enzymes from reaching the digestible sucrose at the center of the olestra molecule. The molecule thus never breaks down into fragments that can be absorbed by the body, and is excreted unchanged.

The latest face-off between olestra's critics and its promoters was inconclusive. After reviewing the results of recent scientific studies, the panel of experts convened in 1998 by the FDA recommended that the status quo should prevail: olestra may remain on the market in the United States, but it must continue to carry the warning label. The panel's report did little to sway partisans on either side of the dispute. "Frankly, I think it's completely immoral for corporations like Procter & Gamble to market products they know will cause gastrointestinal symptoms in many of their customers," said Michael Jacobsen of the Center for Science in the Public Interest. But Suzette Middleton, a Procter & Gamble nutritional scientist, responded that "Olestra snacks don't make people sick." Still, olestra sales have fallen short of Procter & Gamble's hopes, and in June 2001 the company announced that it will cut production of olestra and will make no new investments in marketing the product.

Some experts argue that fake fats help people by allowing them to eat appealing food while limiting fat consumption. Others contend that people should shun these artificial dieting aids and instead eat more healthful foods that are already low in fat. What do you think?

Summary of Key Concepts

1) What Are Atoms?

An element is a substance that can neither be broken down nor converted to different substances by ordinary chemical means. The smallest possible particle of an element is the atom, which is itself composed of a central nucleus, containing protons and neutrons, and electrons outside the nucleus. All atoms of a given element have the same number of protons, which is different from the number of protons in the atoms of every other element. Electrons orbit the nucleus in electron shells, at specific distances from the nucleus. Each shell can contain a fixed maximum number of electrons.

2) How Do Atoms Form Molecules?

The chemical reactivity of an atom depends on the number of electrons in its outermost electron shell: An atom is most stable, and therefore least reactive, when its outermost shell is either completely full or empty. Atoms combine to form molecules, which are held together by chemical bonds. Oppositely charged ions may be held together by ionic bonds. When two atoms share electrons, covalent bonds form. In a nonpolar covalent bond, the two atoms share electrons equally. In a polar covalent bond, one atom may attract the electron more strongly than the other atom does; in this case, the strongly attracting atom bears a slightly negative charge, and the weakly attracting atom bears a slightly positive charge. Some polar covalent bonds give rise to hydrogen bonding, the attraction between charged regions of individual polar molecules.

3) Why Is Water So Important to Life?

For organisms, important properties of the water molecule include its ability to interact with many other molecules and to dissolve many polar and charged substances, to participate in chemical reactions, and to cohere to itself.

4) Why Is Carbon So Important to Life?

Organisms depend on carbon's ability to form a huge variety of different molecules. Carbon-containing molecules are so diverse because the carbon atom is able to form many types of bonds. This ability, in turn, allows organic molecules (molecules with a backbone of carbon and hydrogen atoms) to form many complex shapes, including chains, branches, and rings.

5) How Are Organic Molecules Constructed?

Most large biological molecules are synthesized by linking together many smaller subunits. Chains of subunits are connected by covalent bonds through dehydration synthesis; the chains may be broken apart by hydrolysis reactions. The most important organic molecules fall into one of four classes: carbohydrates, lipids, proteins, and nucleic acids. Their major characteristics are summarized in Table 2-3.

6) What Are Carbohydrates?

Carbohydrate molecules are generally composed of carbon, hydrogen, and oxygen in the ratio 1 carbon : 2 hydrogens : 1 oxygen. Carbohydrates include sugars, starches, and cellulose. Sugars (monosaccharides and disaccharides) are used for temporary storage of energy and for the construction of other molecules. Starches and glycogen are polysaccharides that serve for longer-term energy storage in plants and animals, respectively. Cellulose and related polysaccharides form the cell walls of bacteria, fungi, plants, and some microorganisms.

7) What Are Lipids?

Lipids are nonpolar, water-insoluble molecules of diverse chemical structure that include oils, fats, waxes, phospholipids, and steroids. Lipids are used for energy storage (oils and fats), as waterproofing for the outside of plants and animals (waxes), as the principal component of cellular membranes (phospholipids), and as hormones (steroids).

8) What Are Proteins?

Proteins are chains of amino acids. The structure of a protein is determined by the sequence of amino acids in the chain. A protein is functional when folded into its characteristic three-dimensional shape. Proteins include enzymes (which guide chemical reactions), structural molecules (hair, horn), hormones (insulin), or transport molecules (hemoglobin).

9) What Are Nucleic Acids?

Nucleic acid molecules are chains of nucleotides. Each nucleotide is composed of a phosphate group, a sugar group, and a nitrogen-containing base. The two types of nucleic acids are deoxyribonucleic acid (DNA) and ribonucleic acid (RNA). Nucleotides that function singly include intracellular messengers (cyclic AMP) and energy-carrier molecules (ATP).

Key Terms

acid *p. 22*
amino acid *p. 28*
atom *p. 16*
atomic nucleus *p. 16*
atomic number *p. 16*
base *p. 22*
buffer *p. 22*
carbohydrate *p. 24*
chemical bond *p. 17*
chemical reaction *p. 17*
cohesion *p. 20*
covalent bond *p. 17*
dehydration synthesis *p. 23*
denatured *p. 29*
electron *p. 16*
electron shell *p. 16*
element *p. 16*
functional group *p. 23*
hydrogen bond *p. 18*
hydrolysis *p. 24*
ion *p. 17*
ionic bond *p. 17*
isotope *p. 16*
lipid *p. 26*
molecule *p. 16*
neutron *p. 16*
nonpolar molecule *p. 18*
nucleic acid *p. 30*
organic molecule *p. 23*
pH scale *p. 22*
polar molecule *p. 18*
protein *p. 28*
proton *p. 16*
radioactive *p. 16*
solvent *p. 19*
surface tension *p. 20*

Thinking Through the Concepts

Multiple Choice

1. *When an atom ionizes, what happens?*
a. It shares one or more electrons with another atom.
b. It emits energy as it loses extra neutrons.
c. It gives up or takes up one or more electrons.
d. It shares a hydrogen atom with another atom.
e. none of the above

2. *A covalent bond forms*
a. when two ions are attracted to one another
b. between adjacent water molecules, producing surface tension
c. when one atom gives up its electron to another atom
d. when two atoms share electrons
e. between water molecules and fat globules

3. *What is the defining characteristic of an acid?*
   a. It donates hydrogen ions.
   b. It accepts hydrogen ions.
   c. It will donate or accept hydrogen ions, depending on the pH.
   d. It has an excess of hydroxide ions.
   e. It has a pH greater than 7.
4. *Which of the following is not a function of polysaccharides in organisms?*
   a. energy storage in plants
   b. storage of hereditary information
   c. formation of cell walls
   d. structural support
   e. energy storage in animals
5. *Proteins differ from one another because*
   a. the peptide bonds linking amino acids differ from protein to protein
   b. the sequence of amino acids in the polypeptide chain differs from protein to protein
   c. each protein molecule contains its own unique sequence of sugar molecules
   d. the number of nucleotides in each protein varies from molecule to molecule
   e. the number of nitrogen atoms in each amino acid differs from the number in all others
6. *Which of the following statements about lipids is false?*
   a. A wax is a lipid.
   b. Unsaturated fats are liquid at room temperature.
   c. The body doesn't need any cholesterol.
   d. Both male and female sex hormones are steroids.
   e. Beef fat is highly saturated.

? Review Questions

1. Distinguish among atoms and molecules; elements and compounds; and protons, neutrons, and electrons.
2. Compare and contrast covalent bonds and ionic bonds.
3. Describe how water dissolves a salt.
4. Define *acid*, *base*, and *buffer*. How do buffers reduce changes in pH when hydrogen ions or hydroxide ions are added to a solution? Why is this phenomenon important in organisms?
5. Which elements are common components of biological molecules?
6. List the four principal types of biological molecules, and give an example of each.
7. What roles do nucleotides play in organisms?
8. Distinguish among the following: monosaccharide, disaccharide, and polysaccharide. Give two examples of each and their functions.
9. Describe the synthesis of a protein from amino acids.

Applying the Concepts

1. Many "over-the-counter" substances are sold to bring relief from "acid stomach" or "heartburn." What is the chemical basis for these compounds? Why do they work?
2. Fats and oils do not dissolve in water; polar and ionic molecules dissolve easily in water. Detergents and soaps help clean by dispersing fats and oils in water so that they can be rinsed away. From your knowledge of the structure of water and the hydrophobic nature of fats, what general chemical structures (for example, polar or nonpolar parts) must a soap or detergent have, and why?
3. Many birds must store large amounts of energy to power flight during migration. Which type of organic molecule would be the most advantageous for energy storage? Why?
4. Remember the nuclear accident at Chernobyl in 1986? A scientist suspects that the food in a nearby ecosystem may have been contaminated with radioactive nitrogen over a period of months. Which substances in plants and animals could be examined for radioactivity to test his hypothesis?
5. Fat contains twice as many calories per unit weight as carbohydrate does, so fat is an efficient way for animals, who must move about, to store energy. Compare the way fat and carbohydrates interact with water, and explain why this interaction also gives fat an advantage for weight-efficient energy storage.

Group Activity

Work in groups of three or four to play "molecular 20 questions." Each member of the group should select a molecule important for life (water, oxygen, carbohydrate, sugar, protein, amino acid, nucleic acid, DNA, RNA, lipid, phospholipid, cholesterol, and so on) without telling the others in the group. Then, the others in the group can ask "yes-or-no" type questions to try to identify each molecule.

For More Information

Atkins, P. W. *Molecules*. New York: Scientific American Library, 1987. A layperson's introduction to atoms and molecules, with superb illustrations.

Doolittle, R. F. "Proteins." *Scientific American*, October 1985. Proteins are the molecular tools that, directly or indirectly, carry out almost all the functions of a cell.

Goodsell, D. S. *The Machinery of Life*. New York: Springer, 1993. Goodsell depicts the molecules of a cell in all their three-dimensional, interactive glory. A great way to gain a feel for the beauty and intricacy of the organic molecules of life.

Hill, J. W., and Kolb, D. K. *Chemistry for Changing Times*. 9th ed. Englewood Cliffs, NJ: Prentice Hall, 2001. A chemistry textbook for nonscience majors that is both clearly readable and thoroughly enjoyable.

Morrison, P., and Morrison, P. *Powers of Ten*. New York: W. H. Freeman, 1982. A fascinating journey from the universe to the nucleus of an atom.

Answers to Multiple-Choice Questions

1. c 2. d 3. a 4. b 5. b 6. c

MediaTutor

CD Activities

Activity 2.1: Interactive Atoms

Estimated time: 5 minutes

An understanding of simple chemistry is critical to understanding biology. This tutorial provides the background you need to learn about more complex biological molecules. The bonds that hold together large, complex, biologically important molecules like DNA are the same as the bonds you will learn about here.

Activity 2.2: Water and Life

Estimated time: 5 minutes

Water is essential for life. In fact, water may be the only absolute essential for living systems. This tutorial explores the properties of water and how they relate to living systems.

Activity 2.3: Structure of Biological Molecules

Estimated time: 5 minutes

In this tutorial you will explore the various classes of important biological molecules.

Activity 2.4: Functions of Macromolecules

Estimated time: 5 minutes

The important biological molecules in cells are necessary for most cellular functions. This tutorial introduces the major functions of the four main types of biological molecules: carbohydrates, lipids, proteins, and nucleic acids.

Start the MediaTutor Student CD-ROM and enter the activity number in the Quick Search box to be taken directly to that activity.

Web Investigation

Improving on Nature?

Estimated time: 10 minutes

Let's take a closer look at olestra. While most artificial sweeteners, food colors, and fat replacers can be used for any food product, olestra is currently approved by the U.S. Food and Drug Administration (FDA) for use only in "savory snacks," such as potato chips. Some experts say that olestra is safe. Others think the side effects outweigh the benefits. In this exercise we will examine the pros and cons of the issue.

Go to http://www.prenhall.com/audesirk3, the Audesirk Companion Web site. Select Chapter 2 and the Web Investigation to begin.

"To stay alive, you have to be able to hold out against equilibrium, maintain imbalance, bank against entropy, and you can only transact this business with membranes in our kind of world."

Lewis Thomas in* The Lives of a Cell *(1974)

Our bodies cannot function without cholesterol, yet too much cholesterol can threaten our health by contributing to hard deposits (colored gray in this photo) that restrict the flow of blood through arteries. Inside our bodies, the cell membrane shoulders much of the task of keeping blood cholesterol levels within bounds.

3 Cell Membrane Structure and Function

AT A GLANCE

CASESTUDY

A Thin Line Between Health and Disease

A few years ago, a 30-year-old seamstress and part-time bank teller from Quebec stood before a group of reporters at a press conference. "Marie" (she wishes to keep her real name private) was appearing before the press to report her recovery from a case of high cholesterol. Why, you may be wondering, would busy news professionals take the time to report on recovery from such a commonplace condition? After all, millions of people suffer from blood cholesterol levels that are high enough to increase their risk of heart disease. High cholesterol is a serious problem for many people, but hardly news. Marie, however, was not a typical victim of high cholesterol. For one thing, her cholesterol problem was not new to her; she'd been living with it since birth. And her blood cholesterol level had been extraordinarily high, many times the maximum safe level. In fact, Marie suffered from a disorder known as familial hypercholesterolemia.

Like others with her disease, Marie was born with extremely high levels of circulating cholesterol. For most victims, the extreme levels persist throughout their entire lives. Unfortunately, many of those lives are tragically short. It is not unusual for a person with familial hypercholesterolemia to die of a heart attack while still in his or her teens or twenties. Oddly, the profound and disastrous physiological changes wrought by this disorder stem from a minute defect in an incredibly small structure: the cell membrane. That membrane, the microscopically thin barrier that separates the contents of a cell from its surroundings, is a structure of sublime complexity, with marvelous special properties that make cellular life possible. But when something goes awry in that complex structure, the consequences can be life-threatening.

Marie was very fortunate to be alive at age 30. The reason for her good fortune was also the reason that the reporters attended her news conference. Thanks to an innovative new treatment, Marie's blood contained less cholesterol than it ever had before. Her health had improved dramatically, and her prospects for a long life were increased. She owed her health to a team of scientists and physicians that had managed to repair the crucial defect in her cell membranes. What was that defect, and what was done to repair it? Before answering that question, let's look more closely at the nature of cell membranes. ■

1 What Does the Plasma Membrane Do?

The cell is the smallest unit of life. Each cell is surrounded by a thin **plasma membrane**, which isolates the cell's contents from the external environment. The membrane also acts as a gatekeeper, controlling which substances are allowed to pass in or out and transferring chemical messages from the external environment to the cell's interior. These are formidable tasks for a structure so thin that 10,000 plasma membranes stacked atop one another would scarcely equal the thickness of this page.

At first glance, a plasma membrane might appear to be a simple film surrounding a cell, similar to a soap bubble. Membranes, however, are complex structures that contain many different components, each of which performs a particular function.

2 What Is the Structure of the Plasma Membrane?

The ability of the plasma membrane to accomplish its jobs is closely tied to the manner in which membranes are structured. The overall organization of this structure can be described as proteins floating in a double layer of lipids. The lipids are responsible for the isolating function of membranes; the proteins regulate the exchange of substances across the membrane and communication with the cell's environment.

Membranes Are "Fluid Mosaics"

One way to picture the structure of a membrane is to think of it as a **fluid mosaic**. A fluid is a liquid or a gas—that is, any substance that can move or change shape in response to external forces without breaking apart. The word "mosaic" refers to anything that resembles a mosaic artwork made by setting small colored pieces, such as tile, in mortar. A membrane, then, when viewed at high magnification, looks something like a lumpy, constantly shifting mosaic of tiles (Fig. 3-1). A double layer of phospholipids forms a thick but still liquid "mortar" for the mosaic; an assortment of proteins are the "tiles," which can move about within the phospholipid layers. This movement gives a dynamic, ever-changing quality to membranes (even though the actual ingredients of the membrane remain relatively constant).

The Phospholipid Bilayer Is the Fluid Portion of the Membrane

As you learned in Chapter 2, a phospholipid consists of two very different parts: a polar, hydrophilic head and a pair of nonpolar, hydrophobic tails:

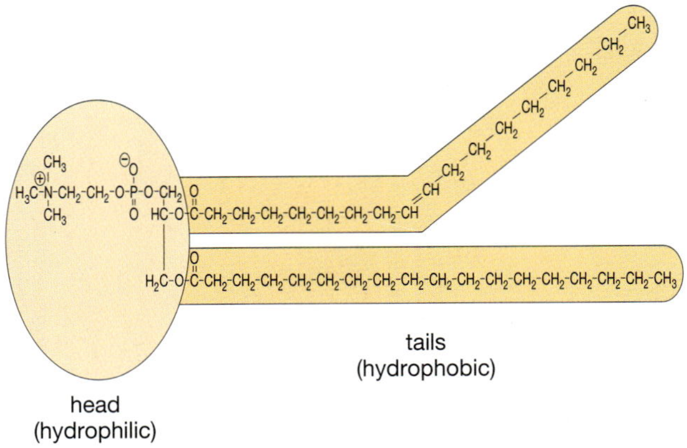

Both the internal and external environments of a cell are watery. Inside the cell, the cytoplasm, which consists of all of a cell's internal contents (except the nucleus), is mostly water. Outside the cell, there is more water. Single-celled organisms may live in fresh water, the ocean, or in the moisture that clings to soil particles and other surfaces. In multicelled organisms, cells are bathed in a fluid that is mostly water. So, plasma membranes separate a cell's watery cytoplasm from its watery external environment. Under these conditions, phospholipids spontaneously arrange themselves into a double layer called a **phospholipid bilayer**. In this arrangement, the phospholipid heads form the outer surface, because they are **hydrophilic**, which means that they are attracted to water molecules and tend to dissolve in water. In contrast, the phospholipid tails are **hydrophobic**; they repel water and "hide" inside the membrane:

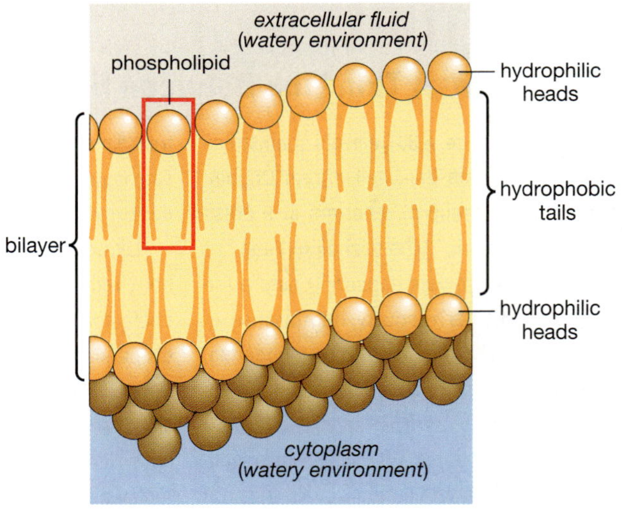

Hydrogen bonds can form between water and the phospholipid heads, so each hydrophilic head faces either the cytoplasm or the extracellular fluid. Because individual phospholipid molecules are not bonded to one another, the double layer is quite fluid; individual phospholipids move about easily within each layer.

In most cells, the phospholipid bilayer of membranes also contains cholesterol. Some cellular membranes have just a few cholesterol molecules; others have as many cholesterol molecules as they do phospholipids. Cholesterol affects membrane structure and function in several ways:

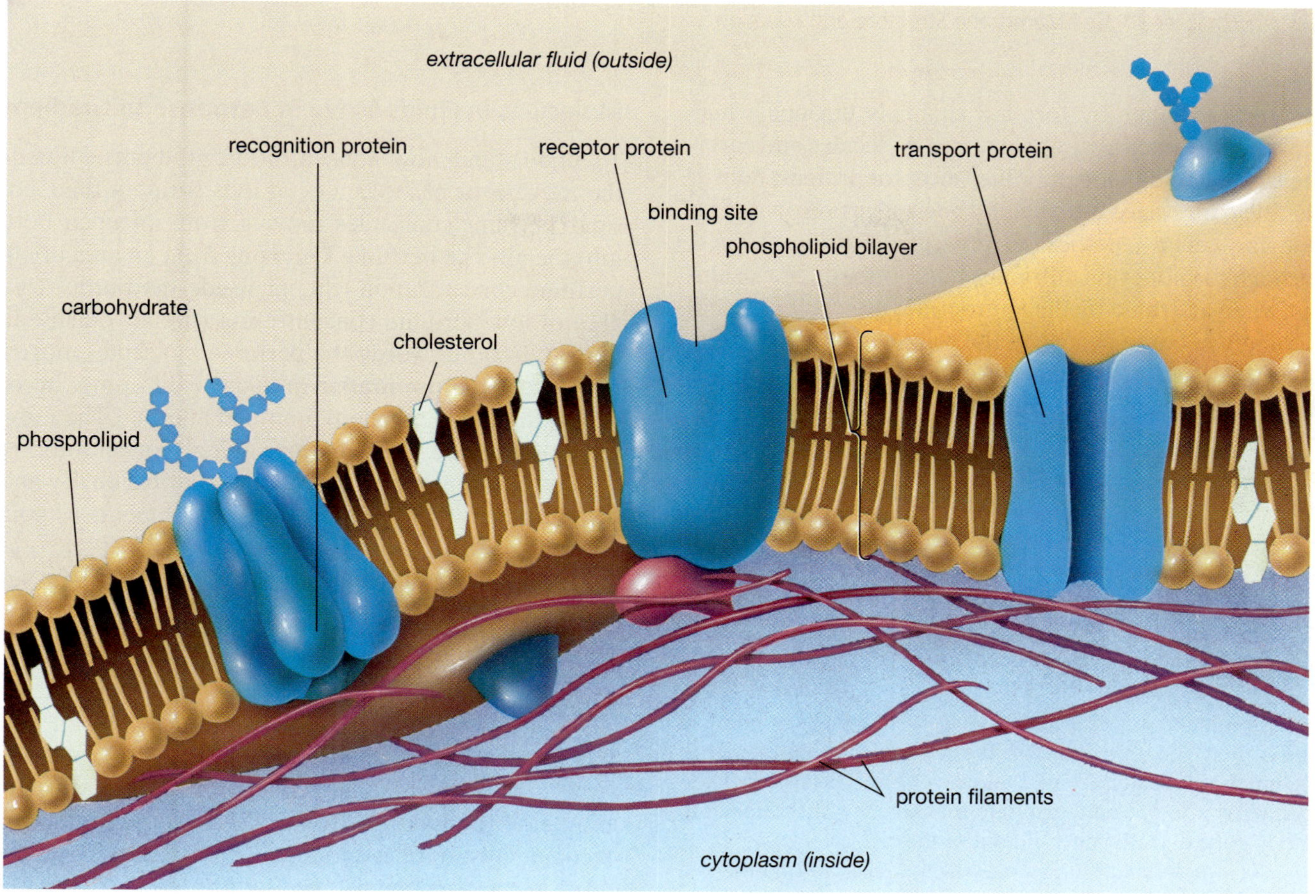

Figure 3-1 The plasma membrane is a fluid mosaic
The plasma membrane is a bilayer of phospholipids in which various proteins are embedded. Many proteins have carbohydrates attached to them. The wide variety of membrane proteins fall mostly into three categories: recognition proteins, receptor proteins, and transport proteins.

MediaTutor 3.1 Membrane Structure and Transport

It makes the bilayer stronger, more flexible, and less permeable to water-soluble substances such as ions or simple sugars. (You may want to review Fig. 2-14 and "Health Watch: Cholesterol—Friend and Foe" in Chapter 2.)

A Mosaic of Proteins Is Embedded in the Membrane

Thousands of proteins are embedded within or attached to the surface of a membrane's phospholipid bilayer. Some of them can move about within the relatively fluid phospholipid bilayer, but others are anchored to a network of protein filaments within the cytoplasm. Many of the proteins have carbohydrate groups attached to them, especially to the parts that stick outside the cell. Together, this collection of proteins is responsible for moving substances across the membrane and for communicating with other cells.

3 How Does the Plasma Membrane Play Its Gatekeeper Role?

One of the main functions of the plasma membrane is to control the movement of substances into and out of cells. Both the phospholipid bilayer and the embedded proteins play roles in this gatekeeper job.

The Phospholipid Bilayer Blocks the Passage of Most Molecules

Most biological molecules, including salts, amino acids, and sugars, are hydrophilic: polar and water soluble. These substances cannot easily pass through the nonpolar, hydrophobic fatty acid tails of the phospholipid bilayer. The phospholipid bilayer thus plays the main role in isolating the cell's contents from the external environment. The isolation is not complete, however. Very small molecules, such as water, and uncharged, lipid-soluble molecules can pass relatively freely through the lipid bilayer.

The Embedded Proteins Selectively Recognize, Respond to, and Transport Molecules

Most molecules that cross the membrane do so with the assistance of membrane proteins. In addition to providing transportation services, membrane proteins also play an important role in a cell's responses to the substances in its environment. Each protein is specialized to do a specific job but, in general, membrane proteins fall into three major categories, each of which serves a different function: (1) transport proteins, (2) receptor proteins, and (3) recognition proteins (see Fig. 3-1).

Transport proteins allow the movement of hydrophilic (water-soluble) molecules through the plasma membrane, either by forming channels through which molecules pass or by grabbing onto molecules and carrying them across the membrane. **Receptor proteins** deliver chemical messages to the cell. They trigger responses inside the cell when specific molecules outside the cell, such as hormones or nutrients, bind to them. For example, when activated by the appropriate molecule, a receptor might set off a change in the cell, such as cell division or secretion of a hormone. **Recognition proteins** serve as identification tags and cell-surface attachment sites. For example, the cells of your immune system use recognition proteins to distinguish your own cells from those of disease-causing invaders. Thus, a bacterium is recognized as a foreign invader to be destroyed while your blood cells are ignored, because the two types of cells have different recognition proteins on their surfaces.

The Big Picture

A cell's contents are separated from its surroundings by nothing more than the thin film of the plasma membrane. This seemingly insubstantial barrier, however, has special properties that allow it to not only maintain the integrity of the cell, but also to control traffic into and out of the cell.

4 What Is Diffusion?

Before continuing our discussion of transport across membranes, we must first review some characteristics of molecules in fluids. We begin with some definitions:

- The *concentration* of molecules in a fluid is the number of molecules in a given unit of volume.
- A *gradient* is a physical difference between two regions of space that causes molecules to move from one region to another. Molecules frequently encounter gradients of concentration, pressure, and electrical charge.

Molecules in Fluids Move in Response to Gradients

To understand how concentration gradients influence the movement of molecules or ions within a fluid, consider perfume molecules moving from an open bottle into the air. The perfume is moving from an area of high perfume concentration (the air inside the bottle) to an area of low perfume concentration (the air outside the bottle). In other words, the perfume molecules move in response to a **concentration gradient**, a difference in concentration between one region and another. The movement of molecules from regions of high concentration to regions of low concentration is called **diffusion**. By analogy with gravity, we refer to such movements as going "down" the concentration gradient. If there are no other factors opposing diffusion, it will continue until the substance is evenly dispersed throughout the fluid.

A Drop of Dye in Water Illustrates Diffusion

To watch diffusion in action, place a drop of food coloring in a glass of water, and check its progress every few minutes. With time, the drop will seem to spread out and become paler, until eventually, even without stirring, the entire glass of water will be uniformly faintly colored. Molecules of dye move from the region of high dye concentration into the surrounding water where the dye concentration is low (Fig. 3-2). Simultaneously, some water molecules enter the dye droplet, and the net movement of water is from the high water concentration outside the drop into the lower water concentration inside the drop.

At first, there is a very steep concentration gradient, and the dye diffuses rapidly. As the concentration differences lessen, the dye diffuses more and more slowly. In other words, the greater the concentration gradient, the faster the rate of diffusion. However, as long as the concentration of dye within the expanding drop is greater than the concentration of dye in the rest of the glass, the net movement of dye will continue until it

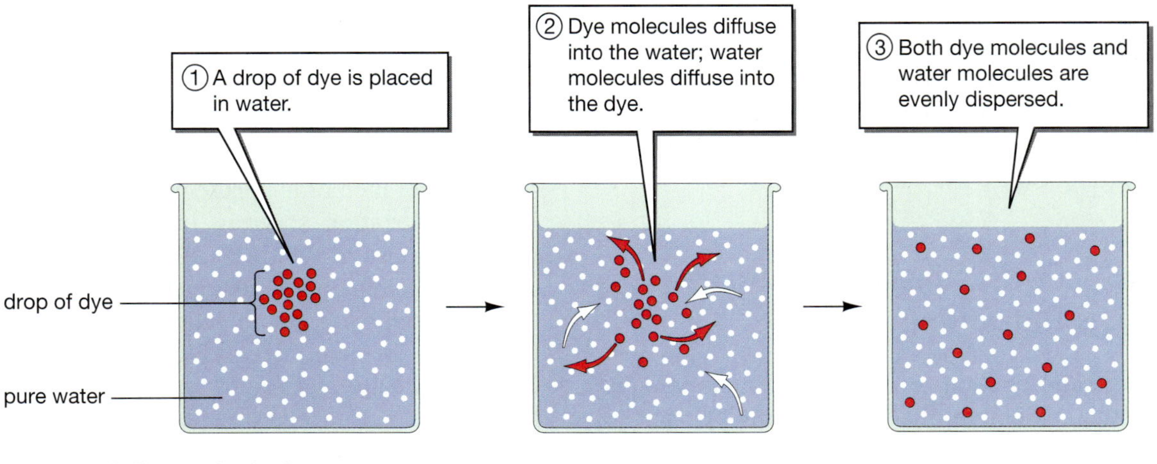

Figure 3-2 Diffusion of a dye in water

becomes uniformly dispersed in the water. Then, with no concentration gradient of either dye or water, diffusion stops.

Summary: The Principles of Diffusion

1. Diffusion is the movement of molecules down a gradient from high concentration to low concentration.
2. The greater the concentration gradient, the faster the rate of diffusion.
3. If no other processes intervene, diffusion will continue until the concentration gradient is eliminated.

5 What Is Osmosis?

Like other molecules, water diffuses from regions of high water concentration to regions of low water concentration. What do we mean when we describe a solution as having a "high water concentration" or a "low water concentration"? The answer is simple: Pure water has the highest water concentration. Any substance added to pure water displaces some of the water molecules; the resulting solution will have a lower water content than pure water. The higher the concentration of dissolved substances, the lower the concentration of water.

The movement of water across membranes from areas of high water concentration to areas of low water concentration has such dramatic and important effects that we refer to it by a special name: **osmosis.** Osmosis moves water across **selectively permeable** membranes, which are so called because they allow the passage of some molecules but prevent the passage of other molecules. A very simple selectively permeable membrane might have pores just large enough for water to pass through but small enough to be impermeable to sugar molecules (Fig. 3-3a). Consider a bag made of a special plastic that is permeable to water but not to sugar. What will happen if we place a sugar solution in the bag and then immerse the sealed bag in pure water? The principles of osmosis tell us that the bag will swell. If it is weak enough, it will eventually burst (Fig. 3-3b).

Summary: The Principles of Osmosis

1. Osmosis is the diffusion of water across a selectively permeable membrane.
2. Dissolved substances reduce the concentration of free water molecules in a solution.
3. Water moves across a membrane down its concentration gradient from a high concentration of free water molecules to a low concentration of free water molecules.

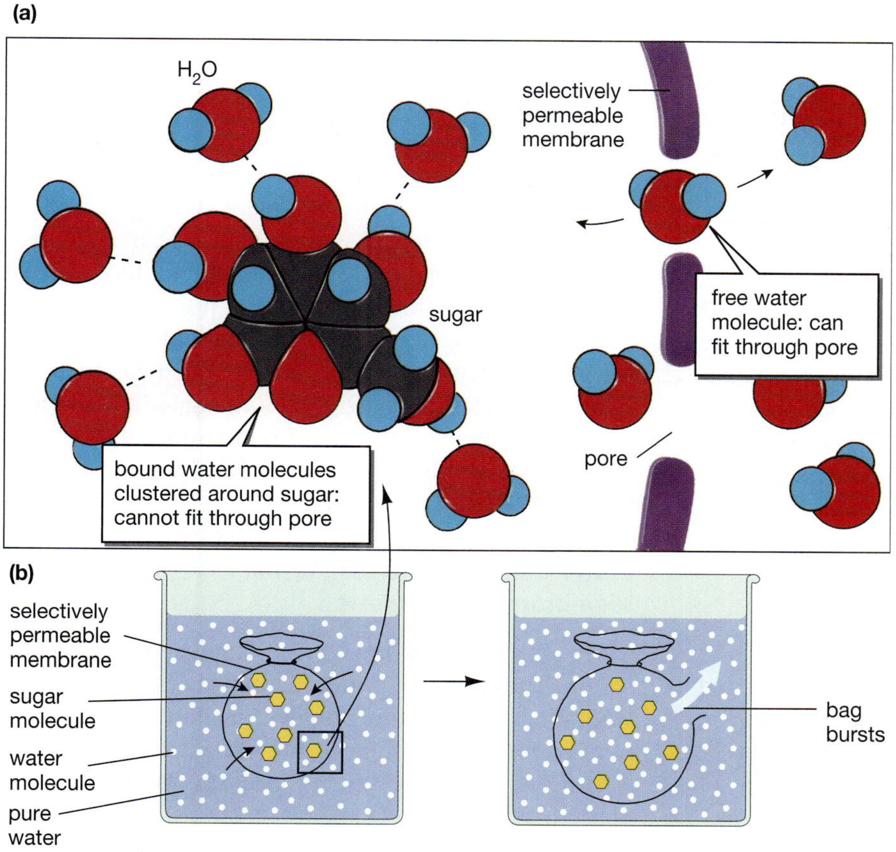

Figure 3-3 Osmosis
Free water molecules diffuse down their concentration gradient across a selectively permeable membrane, from a region of high water concentration to a region of lower water concentration.

MediaTutor 3.2 Osmosis

Table 3-1 Transport Across Membranes

Passive transport	Movement of substances across a membrane, going down a gradient of concentration, pressure, or electrical charge. Does not require the cell to expend energy.
Simple diffusion	Diffusion of water, dissolved gases, or lipid-soluble molecules through the phospholipid bilayer of a membrane.
Facilitated diffusion	Diffusion of (normally water-soluble) molecules through a channel or carrier protein.
Osmosis	Diffusion of water across a differentially permeable membrane—that is, a membrane that is more permeable to water than to dissolved molecules.
Energy-requiring transport	Movement of substances across a membrane, usually against a concentration gradient, using cellular energy.
Active transport	Movement of individual small molecules or ions through membrane-spanning proteins, using cellular energy, normally ATP.
Endocytosis	Movement of large particles, including large molecules or entire microorganisms, into a cell by engulfing extracellular material, as the plasma membrane forms membrane-bound sacs that enter the cytoplasm.
Exocytosis	Movement of materials out of a cell by enclosing the material in a membranous sac that moves to the cell surface, fuses with the plasma membrane, and opens to the outside, allowing its contents to diffuse away.

6 How Do Diffusion and Osmosis Affect Transport Across the Plasma Membrane?

As shown in Table 3-1, there are two main types of transport across membranes. One of these types is **passive transport**. During passive transport, substances move into or out of cells down concentration gradients. Passive transport requires no expenditure of energy, because the concentration gradients drive the movement. Concentration gradients of ions and molecules across the plasma membranes of cells are present because the composition of the cytoplasm of a cell is very different from that of the extracellular fluid. The lipids and protein pores of the plasma membrane regulate which molecules can cross by passive transport, but they do not influence the direction of movement. Passive transport includes simple diffusion, facilitated diffusion, and osmosis.

Plasma Membranes Are Selectively Permeable

Molecules are able to diffuse across membranes, and many molecules do in fact cross plasma membranes by diffusion, driven by differences between their concentration in the cytoplasm and in the external environment. Not all molecules, however, are equally likely to be able to diffuse across a membrane, because plasma membranes are selectively permeable; some molecules can pass through but other molecules cannot.

Some Molecules Move Across Membranes by Simple Diffusion

Molecules that dissolve in lipids, such as ethyl alcohol and vitamin A, easily diffuse across the phospholipid bilayer, as do very small molecules, including water and dissolved gases such as oxygen and carbon dioxide. This type of passive transport is called **simple diffusion** (Fig. 3-4a). Generally, the rate of simple diffusion depends on the concentration gradient across the membrane, the size of the molecule, and how easily it dissolves in lipids. Large concentration gradients, small molecule size, and high lipid solubility all increase the rate of simple diffusion.

Other Molecules Cross the Membrane by Facilitated Diffusion

Most water-soluble molecules, such as ions (K^+, Na^+, Ca^{2+}), amino acids, and simple sugars, cannot move through the phospholipid bilayer on their own. These molecules can diffuse across only by **facilitated diffusion**, a type of passive transport in which diffusion is aided by either of two types of transport proteins: channel proteins and carrier proteins.

Channel proteins form pores, or channels, in the lipid bilayer through which certain ions can cross the membrane (Fig. 3-4b). In general, each type of channel protein is specialized and allows only particular ions to pass through. Nerve cells, for example, have separate channels for sodium ions, potassium ions, and calcium ions.

A **carrier protein** grabs onto a specific molecule on one side of a membrane and carries it to the other side. Each different type of carrier protein is able to bind to a specific molecule (typically an amino acid, sugar, or small protein). Binding triggers a change in the shape of the carrier that allows the bound molecule to pass through the protein and across the plasma membrane (Fig. 3-4c).

Water Can Move Across Plasma Membranes by Osmosis

Most plasma membranes are highly permeable to water. The flow of water across them depends on whether water is more highly concentrated inside or outside the cell. Ordinarily, the extracellular fluid outside the cell has the same concentration as the cytoplasm inside the cell. In that case, the extracellular fluid is said to be isotonic to the cytoplasm, and there will be

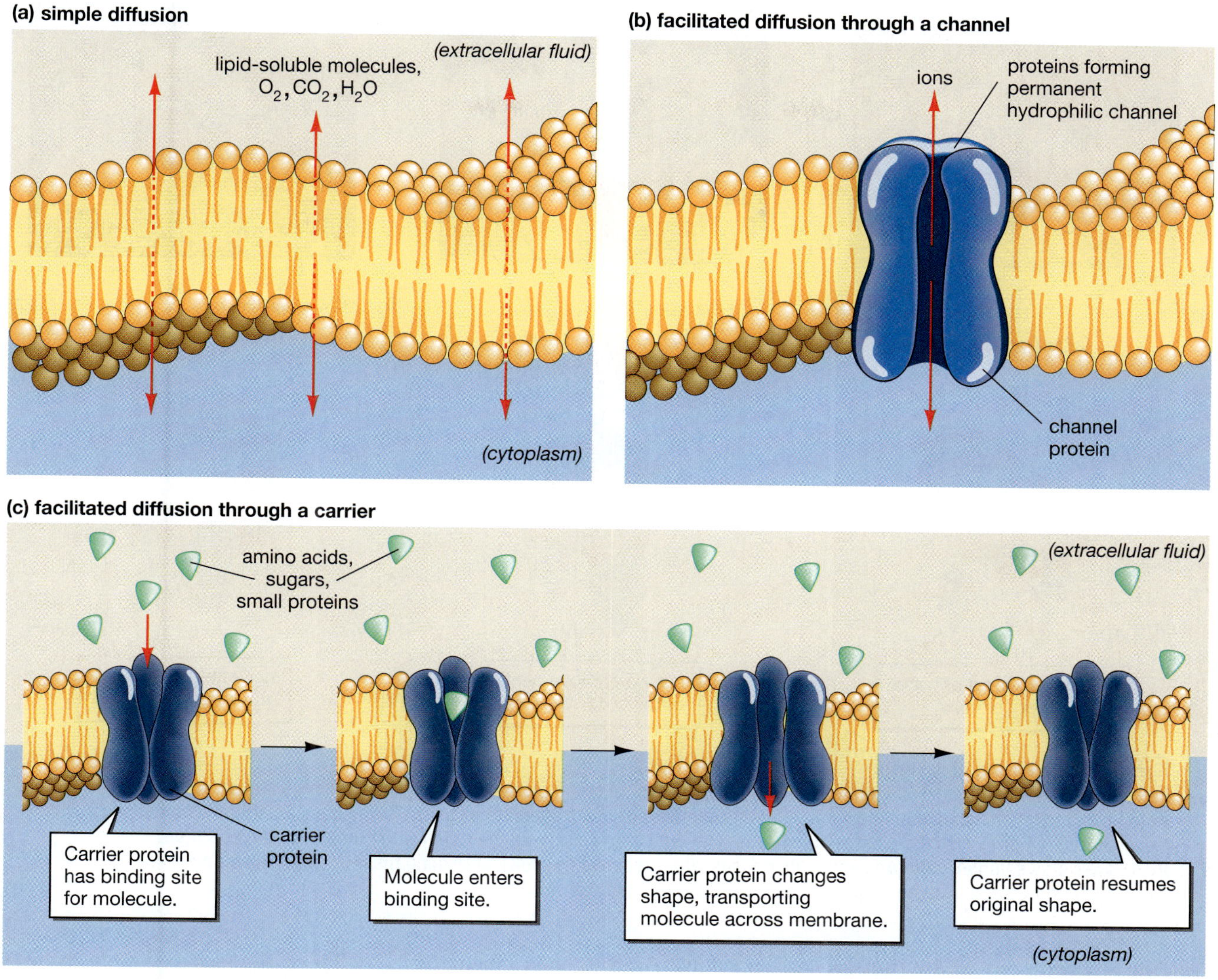

Figure 3-4 Diffusion through the plasma membrane
(a) Lipid-soluble molecules and gases such as oxygen and carbon dioxide can pass by simple diffusion directly through the phospholipids. ***(b)*** Some water-soluble molecules enter or exit the cell by facilitated diffusion through a channel protein. ***(c)*** Certain molecules cross a membrane by facilitated diffusion through a carrier protein that changes shape to allow the passage.

no net movement of water into or out of the cell (Fig. 3-5a). Note that the *types* of dissolved particles are seldom the same inside and outside the cells, but the *total concentration* of all dissolved particles is equal, with the result that the water concentration inside is equal to that outside the cells.

If red blood cells are taken out of the body and immersed in salt solutions of varying concentrations, the effects of osmosis become dramatically apparent. If the solution has a higher salt concentration than the cytoplasm of the red blood cell (that is, if the solution has a lower water concentration), water will leave the cells by osmosis. The cells in such a hypertonic solution will shrivel up until the concentrations of water inside and outside become equal (Fig. 3-5b). If, on the other hand, the solution has little salt (is hypotonic), water will enter the cells, causing them to swell (Fig. 3-5c). This process is why your fingers wrinkle up after a long bath. It may seem like your fingers are shrinking, but they're not. Instead, water is moving into the outer skin cells of your fingers, swelling them more rapidly than the cells underneath and causing the wrinkling.

7 How Do Molecules Move Against a Concentration Gradient?

Passive transport, while undeniably efficient, could never by itself meet all of a cell's needs for moving molecules across membranes. Passive transport can carry substances only down concentration gradients, but all cells also need to move some materials "uphill" across their plasma membranes, against concentration gradients. For example, it is often necessary to move nutrients

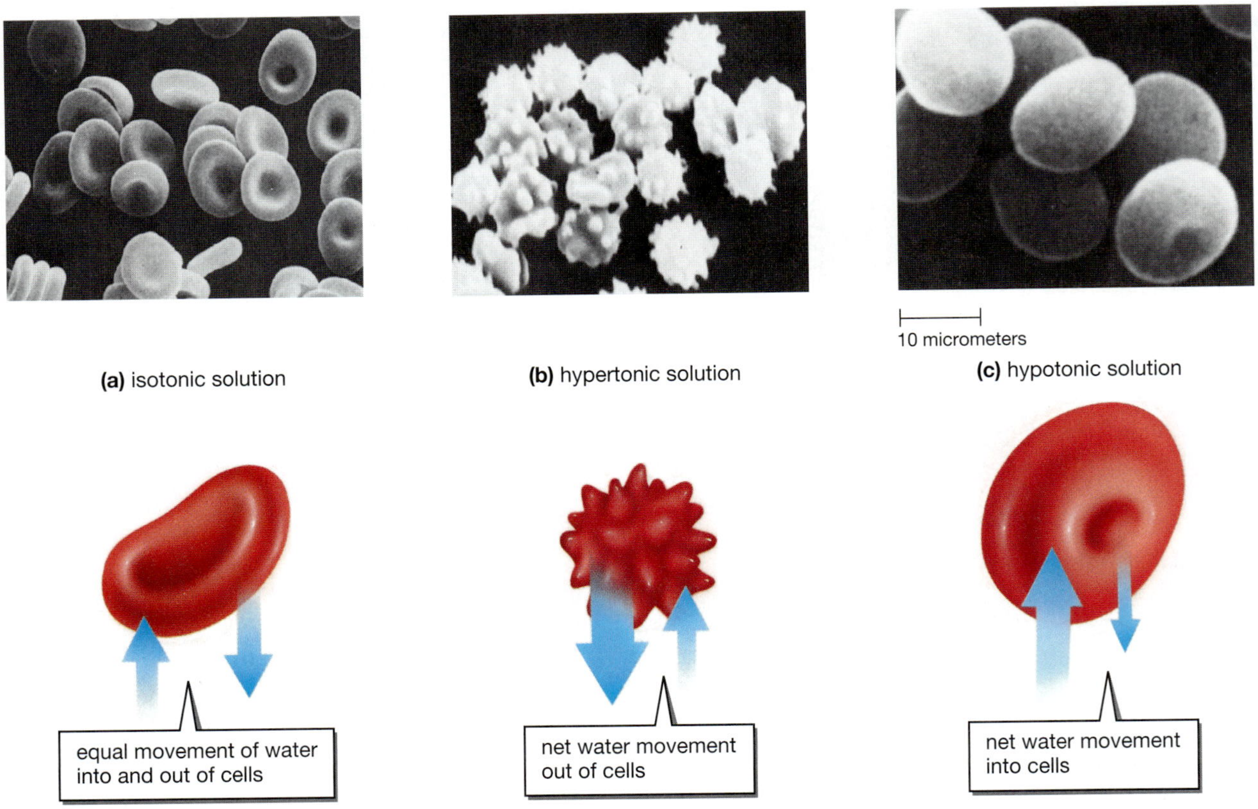

Figure 3-5 The effects of osmosis
Red blood cells are normally suspended in the fluid environment of the blood. ***(a)*** If red blood cells are immersed in an isotonic salt solution, which has the same concentration of dissolved substances as the blood cells do, there is no net movement of water across the plasma membrane. The red blood cells keep their characteristic dimpled disk shape. ***(b)*** A hypertonic solution, with too much salt, causes water to leave the cells, shriveling them up. ***(c)*** A hypotonic solution, with less salt than is in the cells, causes water to enter, and the cells swell.

from the environment, where they are less concentrated, into cell's cytoplasm, where they are more concentrated. Passive transport, however, would move nutrient molecules out of the cell, in the direction opposite the one required by the cell. A different mode of transport is necessary.

Active Transport Uses Energy to Move Molecules Against Their Concentration Gradients

During **active transport**, the other main type of transport across membranes, the cell uses energy to move substances *against* a concentration gradient (see Table 3-1). A helpful analogy for understanding the difference between passive and active transports is to consider what happens when you ride a bike. If you don't pedal, you can go only downhill, as in passive transport. However, if you put enough energy into pedaling, you can go uphill as well, as in active transport.

Membrane Proteins Regulate Active Transport

In active transport, membrane proteins use cellular energy to move individual molecules across the plasma membrane (Fig. 3-6). Active-transport proteins span the width of the membrane and have two active sites. One active site (which may be either on the face of the plasma membrane in contact with the cytoplasm or on the face in contact with the extracellular fluid, depending on the protein) binds a particular molecule, say a calcium ion. The second site (always on the inside of the membrane) binds an energy-carrier molecule, normally adenosine triphosphate (ATP). The ATP donates energy to the active-transport protein, causing it to change shape and move the calcium ion across the membrane. Active-transport proteins are often called *pumps*, in analogy to water pumps, because they use energy to move molecules "uphill" against a concentration gradient.

Cells Engulf Particles or Fluids by Endocytosis

Sometimes cells must acquire particles that are too large to move across a membrane by either passive transport or active transport. One way in which cells can accomplish this task is by an energy-consuming process called **endocytosis**. During endocytosis, the plasma membrane engulfs an extracellular particle or a droplet of fluid and pinches off a membranous sac called a **vesicle**, with the particle or fluid inside, into the cytoplasm.

Endocytosis can proceed in several different ways, depending on size and nature of the material to be captured. When a drop of liquid must be moved into a cell,

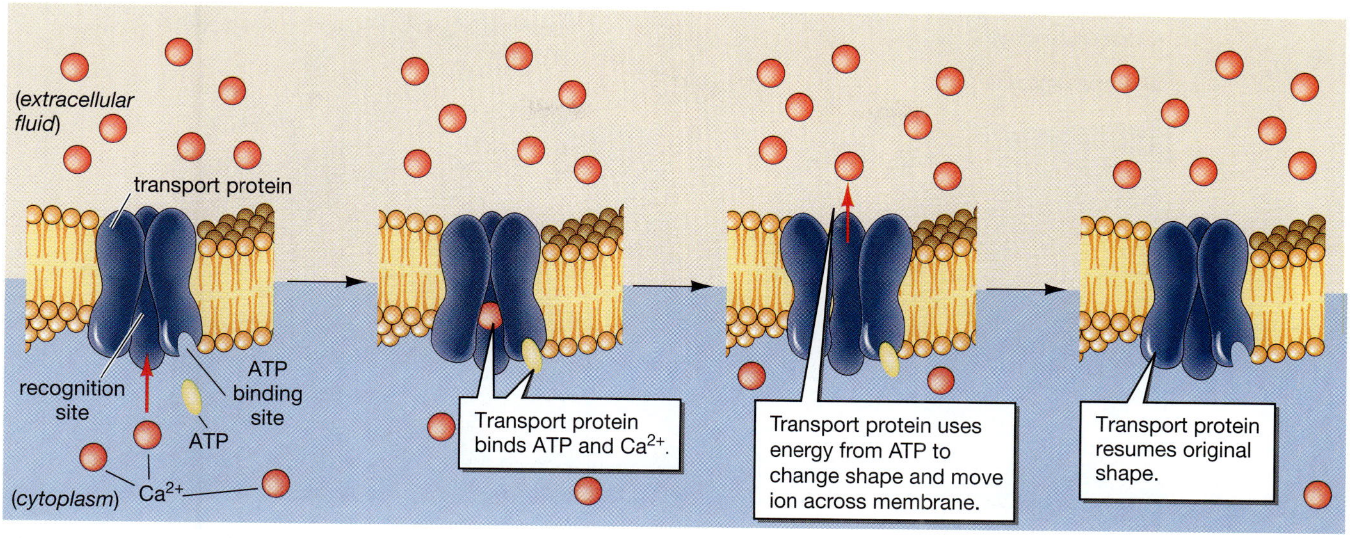

Figure 3-6 Active transport
Active transport uses cellular energy to move molecules across the plasma membrane, often against a concentration gradient. An active-transport protein binds ATP and the molecule to be transported, and then changes shape to move the ion across the membrane.

a very small patch of plasma membrane dimples inward as it surrounds extracellular fluid and buds off into the cytoplasm as a tiny vesicle (Fig. 3-7a), a process called pinocytosis. Another process, phagocytosis, is used to pick up larger particles, such as when an amoeba engulfs food or when a white blood cell destroys bacteria that have invaded a human body. In phagocytosis, the cell first extends parts of its surface membrane. These extensions are called *pseudopodia* (singular, pseudopod). The tips of the pseudopodia fuse around the particle and carry it to the interior of the cell inside a vesicle (Fig 3-7b).

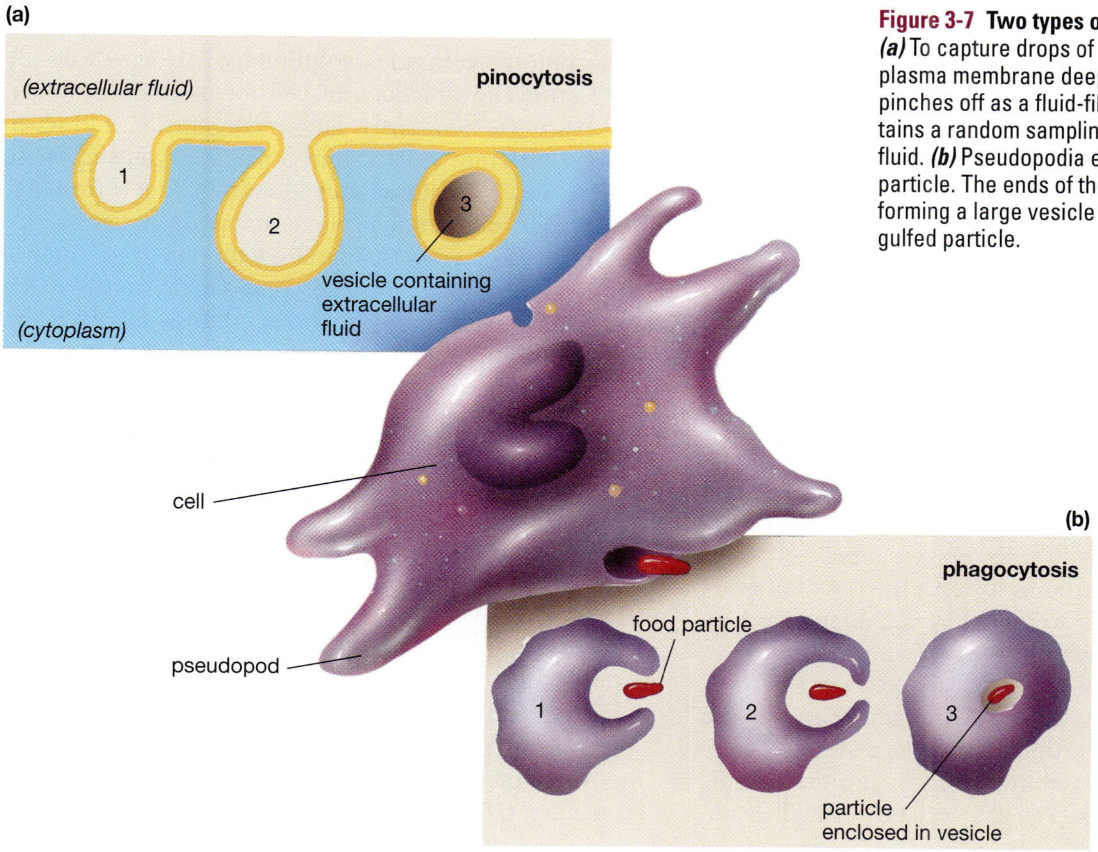

Figure 3-7 Two types of endocytosis
(a) To capture drops of liquid, a dimple in the plasma membrane deepens and eventually pinches off as a fluid-filled vesicle, which contains a random sampling of the extracellular fluid. ***(b)*** Pseudopodia encircle an extracellular particle. The ends of the pseudopodia fuse, forming a large vesicle that contains the engulfed particle.

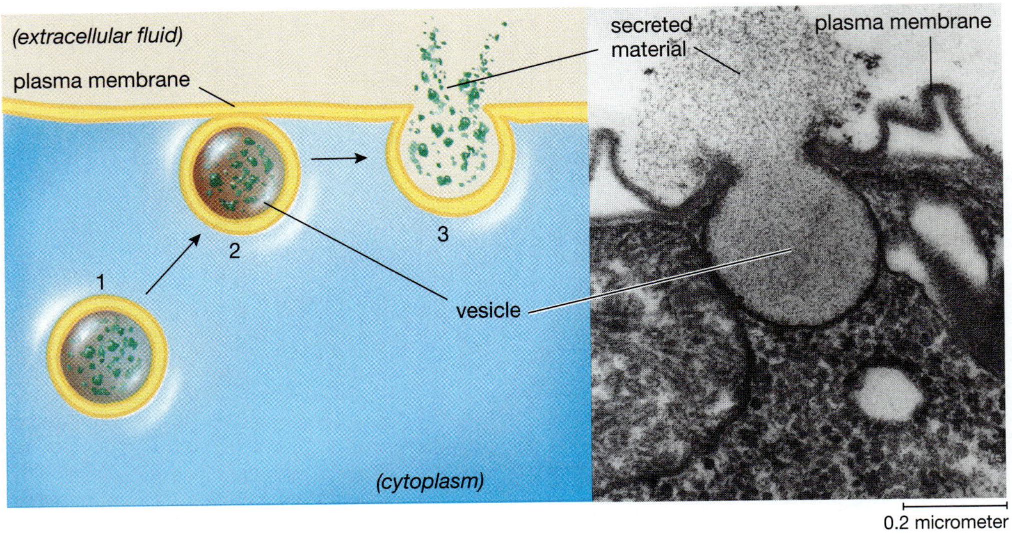

Figure 3-8 Exocytosis
Exocytosis is functionally the reverse of endocytosis. The material to be ejected from the cell is encapsulated into a membrane-bound vesicle that moves to the plasma membrane and fuses with it. After the vesicle opens to the outside, its contents leave by diffusion.

Exocytosis Moves Material Out of the Cell

Cells often use the reverse of endocytosis, a process called **exocytosis**, to dispose of unwanted materials, such as the waste products of digestion, or to secrete materials, such as hormones (Fig. 3-8). During exocytosis, a membrane-enclosed vesicle carrying material to be expelled moves to the cell surface, where the vesicle's membrane fuses with the cell's plasma membrane. The vesicle then opens to the extracellular fluid, and its contents diffuse out.

The Big Picture

A multitude of molecules moves across the plasma membrane in a constant flow of two-way traffic. This continuous exchange between a cell and its surroundings ensures that the chemical conditions inside the cell are quite different from those on the outside.

Some Plasma Membranes Are Surrounded by Cell Walls

The plasma membranes of bacteria, plants, and fungi lie inside stiff coatings called **cell walls**. Cell walls, which are produced by the cells they surround, support and protect otherwise fragile cells. For example, cell walls allow plants and mushrooms to resist the forces of gravity and blowing winds so that they can stand erect on land. Tree trunks, which can support enormous weight, demonstrate just how great the collective strength of cell walls can be. Despite their strength, cell walls are normally porous, permitting easy passage of small molecules such as minerals, water, oxygen, carbon dioxide, amino acids, and sugars. The cell wall, however, does not control the interaction between a plant, fungal, or bacterial cell and its external environment; this function is instead filled by the plasma membrane that lies inside the cell wall.

Evolutionary Connections

Caribou Legs and Membrane Diversity

The basic structure of membranes is similar in all cells, reflecting the common evolutionary heritage of all life on Earth. The functions of cell membranes, however, can vary tremendously among organisms and even among different cells within a single organism. These differences in membrane function stem largely from differences in membrane proteins and phospholipids.

To see how environmental conditions can influence the evolution of membrane components, consider the caribou, animals that live in very cold regions of North America (Fig. 3-9). During the long arctic winters of these regions, temperatures plummet far below freezing. These conditions have favored the evolution of anatomical specializations that permit the temperature of the lower legs to drop almost to freezing (thus allowing caribou to conserve body heat and avoid using precious energy to keep their lower legs warm). In contrast, the upper legs and main trunk of the body remain at about 99 °F (37 °C).

The radically different temperatures in different body regions place different demands on the plasma membranes of cells. Phospholipids in the cell membranes of the upper leg must be very different from those near the hooves, if membranes in both places are to remain fluid. Remember, the membrane of a cell needs to be some-

Figure 3-9 Caribou browse on the frozen Alaskan tundra
How do they conserve body heat?

what fluid to allow the membrane proteins to move to sites where they are needed. The fluidity of a membrane is a function of the fatty acid tails of its phospholipids: Unsaturated fatty acids remain fluid at lower temperatures than do saturated fatty acids (see Chapter 2). Caribou have a range of fatty acids in the phospholipids of the cells in their legs. The membranes of cells near the chilly hoof have lots of unsaturated fatty acids, whereas the membranes of cells near the warmer trunk have more saturated ones. This arrangement gives the plasma membranes throughout the leg the proper fluidity despite great differences in temperature.

CASE STUDY REVISITED

A Thin Line Between Health and Disease

As you may have guessed while reading this chapter, familial hypercholesterolemia occurs when circulating cholesterol is unable to cross cell membranes. Much of the cholesterol in the blood travels in the form of low-density lipoprotein (LDL), particles consisting of thousands of cholesterol molecules (and other lipids) bound to a protein molecule. In a healthy person, cells in the liver remove LDL from the blood. This removal is accomplished when circulating LDL particles bind to receptor molecules on a liver cell's surface and are engulfed by the cell (an example of endocytosis). In a person with familial hypercholesterolemia, however, the LDL receptor molecules are defective. Circulating LDL cannot bind to cell membranes, and therefore cannot be transported into cells. Most LDL remains in the blood, where it accumulates to dangerous levels.

What was the treatment that so changed the life of Marie, the seamstress with familial hypercholesterolemia? Her doctors had removed some cells from her liver and then used the techniques of biotechnology to replace her defective LDL-receptor genes with normal genes that produced a functional receptor protein. When surgeons restored the newly engineered liver cells to the woman's body, they proliferated there. In time, the woman had enough normal, cholesterol-removing cells that her blood cholesterol levels began to decline.

Summary of Key Concepts

1) What Does the Plasma Membrane Do?

The plasma membrane isolates the cytoplasm from the external environment and regulates the flow of materials into and out of the cell and the flow of chemical messages into the cell.

2) What Is the Structure of the Plasma Membrane?

The membrane consists of a bilayer of phospholipids in which proteins are embedded.

3) How Does the Plasma Membrane Play Its Gatekeeper Role?

The phospholipid bilayer blocks most molecules (though some small molecules and lipid-soluble molecules do pass through). Larger molecules can cross the membrane only with the help of membrane proteins. Transport proteins regulate the movement of water-soluble substances through the membrane; receptor proteins bind molecules in the external environment, triggering changes in the cell; and recognition proteins serve as identification tags and attachment sites.

4) What Is Diffusion?

Diffusion is the movement of particles from regions of higher concentration to regions of lower concentration.

5) What Is Osmosis?

Osmosis is the diffusion of water across a selectively permeable membrane and down its concentration gradient. Dissolved substances decrease the concentration of free water molecules.

6) How Do Diffusion and Osmosis Affect Transport Across the Plasma Membrane?

In simple diffusion, water, dissolved gases, and lipid-soluble molecules diffuse through the phospholipid bilayer. In facilitated diffusion, water-soluble molecules cross the membrane through protein channels or with the assistance of carrier proteins. In all these types of passive transport, molecules move down their concentration gradients, and cellular energy is not required. Water often moves across the cell membranes by osmosis. Osmosis does not require cellular energy.

7) How Do Molecules Move Against a Concentration Gradient?

In active transport, proteins spanning the membrane use cellular energy (ATP) to drive the movement of molecules across the plasma membrane, usually against concentration gradients. Large molecules (for example, proteins), particles of food, microorganisms, and extracellular fluid may be acquired by endocytosis. The secretion of substances such as hormones and the excretion of wastes from a cell are accomplished by exocytosis.

Key Terms

active transport *p. 44*
carrier protein *p. 42*
cell wall *p. 46*
channel protein *p. 42*
concentration gradient *p. 40*
diffusion *p. 40*
endocytosis *p. 44*
exocytosis *p. 46*
facilitated diffusion *p. 42*
fluid mosaic *p. 38*
hydrophilic *p. 38*
hydrophobic *p. 38*
osmosis *p. 41*
passive transport *p. 42*
phospholipid bilayer *p. 38*
plasma membrane *p. 38*
receptor protein *p. 40*
recognition protein *p. 40*
selectively permeable *p. 41*
simple diffusion *p. 42*
transport protein *p. 39*
vesicle *p. 44*

Thinking Through the Concepts

Multiple Choice

1. *Active transport through the plasma membrane occurs through the action of*
   a. diffusion
   b. membrane proteins
   c. DNA
   d. water
   e. osmosis

2. *The following is characteristic of a plasma membrane:*
   a. It separates the cell contents from its environment.
   b. It is permeable to certain substances.
   c. It is a lipid bilayer with embedded proteins.
   d. It contains pumps for moving molecules against their concentration gradient.
   e. all of the above

3. *If an animal cell is placed into a solution whose concentration of dissolved substances is higher than that inside the cell,*
   a. the cell will swell
   b. the cell will shrivel
   c. the cell will remain the same size
   d. the solution is described as hypertonic
   e. both b and d are correct

4. *Small, nonpolar hydrophobic molecules such as fatty acids*
   a. pass readily through a membrane's lipid bilayer
   b. diffuse very slowly through the lipid bilayer
   c. require special channels to enter a cell
   d. are actively transported across cell membranes
   e. must enter the cell via endocytosis

5. *Which of the following would be least likely to diffuse through a lipid bilayer?*
   a. water
   b. oxygen
   c. carbon dioxide
   d. sodium ions
   e. the small, nonpolar molecule butane

6. *Which of the following processes causes substances to move across membranes without the expenditure of cellular energy?*
   a. endocytosis
   b. exocytosis
   c. active transport
   d. diffusion
   e. pseudopodia

? Review Questions

1. Describe and diagram the structure of a plasma membrane. What are the two principal types of molecules in plasma membranes? What are the four principal functions of plasma membranes?
2. What are the three main categories of proteins commonly found in plasma membranes, and what is the function of each?
3. Define *diffusion,* and compare that process to osmosis. How do these two processes help plant leaves remain firm?
4. Define *hypotonic, hypertonic,* and *isotonic.* What would be the fate of an animal cell immersed in each of the three types of solution?
5. Describe the following types of transport processes: simple diffusion, facilitated diffusion, active transport, pinocytosis, phagocytosis, and exocytosis.

Applying the Concepts

1. Different cells have somewhat different plasma membranes. The plasma membrane of a *Paramecium*, for example, is only about 1% as permeable to water as the plasma membrane of a human red blood cell. Referring back to our discussion of the effects of osmosis on red blood cells and the role of contractile vacuoles in *Paramecium,* what do you think is the function of the low water permeability of *Paramecium*? What molecular differences do you think might account for this low water permeability?
2. A preview question for Chapter 22: The integrity of the plasma membrane is essential for cellular survival. Could the immune system utilize this fact to destroy foreign cells that have invaded the body? How might cells of the immune system disrupt membranes of foreign cells? (Two hints: Virtually all cells can secrete proteins, and some proteins form pores in membranes.)
3. A preview question for Chapter 17: Plant roots take up minerals (inorganic ions such as potassium) that are dissolved in the water of the soil. The concentration of such ions is usually much lower in the soil water than in the cytoplasm of root cells. Design the plasma membrane of a hypothetical mineral-absorbing cell, with special reference to mineral-permeable channel proteins and mineral-transporting active-transport proteins. Justify your choice of channel proteins and active-transport proteins.
4. Red blood cells will swell up and burst when placed in a hypotonic solution such as pure water. Why don't we swell up and burst when we swim in water that is hypotonic to our cells and body fluids?

For More Information

McNeil, P. L. "Cell Wounding and Healing." *American Scientist*, May–June 1991. The fluidity of plasma membrane phospholipids makes cells able to withstand minor damage.

Rothman, J. E., and Orci, L. "Budding Vesicles in Living Cells." *Scientific American*, March 1996. Membranes within cells form vesicles that transport materials inside the cell. Researchers are discovering the mechanisms by which these vesicles form.

Answers to Multiple-Choice Questions

1. b 2. e 3. e 4. a 5. d 6. d

MediaTutor

CD Activities

Activity 3.1: Membrane Structure and Transport

Estimated time: 10 minutes

The plasma membrane enclosing a cell controls both what is inside and what is outside the cell. In this tutorial you will observe how the cell controls the transport of various molecules into and out of the cell.

Activity 3.2: Osmosis

Estimated time: 5 minutes

Cells are both filled with and surrounded by water. In this tutorial you will explore how osmosis, the movement of water in response to a concentration gradient, affects the cell.

Start the MediaTutor Student CD-ROM and enter the activity number in the Quick Search box to be taken directly to that activity.

Web Investigation

A Thin Line Between Health and Disease

Estimated time: 15 minutes

Over the past few decades, health care professionals have become increasingly concerned as the number of patients with high blood cholesterol has grown. The increasing incidence of high cholesterol and the associated high rates of heart disease have spurred a major effort to develop new cholesterol-lowering drugs. These drugs, however, have little or no effect on patients with familial hypercholesterolemia (FH). This exercise examines the properties and functions of cholesterol and describes some strategies for treating patients with FH.

Go to http://www.prenhall.com/audesirk3, the Audesirk Companion Web site. Select Chapter 3 and the Web Investigation to begin.

"It is all too easy now to underestimate cells. We have known about them for such large fractions of our lives that, for the most part, we cease being aware of how remarkable they really are."

Bruce Alberts, Dennis Bray, Julian Lewis, Martin Raff, Keith Roberts, and James Watson from the Prologue to* Molecular Biology of the Cell *(1983)

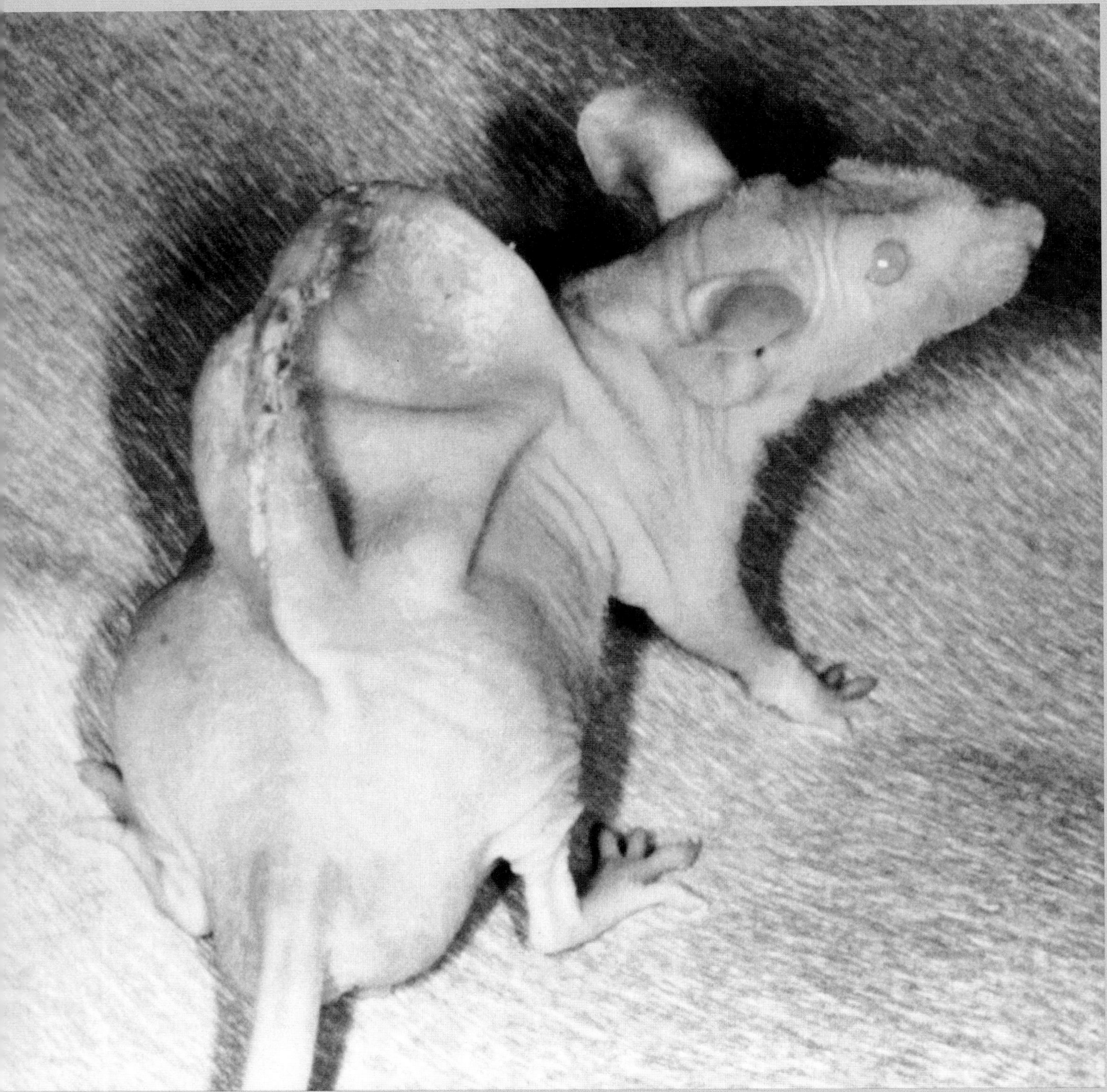

This artificial ear, grafted onto the back of a mouse, demonstrates our newly acquired power to manipulate cells, the building blocks of life.

4 Cell Structure and Function

AT A GLANCE

CASE STUDY

Spare Parts For Human Bodies

Raul Mercia lost part of his thumb in an industrial accident, but doctors are growing a new one for him. After the accident, doctors harvested some of his bone cells and kept them alive in the laboratory. Later, surgeons grafted a piece of sterile coral, shaped like a thumb bone, onto Mr. Mercia's damaged digit. His cultured bone cells were then seeded onto the coral's surface. In time, the bone cells will increase in number and form a real bone that will replace the biodegradable coral support.

Mr. Mercia was one of the first humans to benefit from a new technology known as tissue engineering, in which new body parts are grown in the laboratory. Researchers have been working for years on techniques to create lab-grown tissues and organs, but the field took a great leap forward in 1995, when scientists produced a mouse with a human ear growing on its back. Actually, it wasn't a real ear, but rather an artificial ear made out of cartilage. To make it, researchers used biodegradable plastic to construct an ear-shaped "scaffold." Then they painted cartilage cells onto the scaffold, and the resulting artificial "ear" was grafted onto the back of a mouse (so that the cartilage cells could be nourished by the mouse's circulatory system). As the cartilage cells reproduced, they spread over the surface of the scaffold and developed into living cartilage tissue in the shape of an ear. Eventually, the biodegradable plastic scaffold disappeared, and the cartilage "ear" could be surgically removed from the mouse. In theory, an ear grown in this manner could be used to replace a damaged or missing human ear.

The success of the rather bizarre ear-on-a-mouse experiment demonstrated our newfound power to manipulate cells, the fundamental units of life. All living things are constructed of cells, including organs and tissues that can be damaged by disease or injury. If scientists can shape cells into a living ear, might they not someday be able to sculpt cells into working livers, kidneys, and lungs? If so, perhaps fewer lives will be lost to failed organs. ■

1) What Features Are Shared by All Cells?

Cells are the smallest units that retain the properties of life. All living things, from the tiniest bacterium to the largest whale, are composed of cells. The smallest organisms consist of only a single cell, but larger organisms can contain trillions of cells, each specialized to perform a specific function. Because cells can be specialized for many different tasks, and because there are many different species of single-celled organisms, there are many different types of cells. Despite their diversity, cells share certain features, which are described below.

Cells Are Enclosed by a Plasma Membrane

As we saw in Chapter 3, all cells are surrounded by a **plasma membrane**. The plasma membrane consists of a phospholipid bilayer in which proteins are embedded, and it controls the movement of substances into and out of the cell. A description of plasma membrane structure and function can be found in Chapter 3.

Cells Use DNA as a Hereditary Blueprint

All cells contain **deoxyribonucleic acid (DNA)**, genetic material that stores the instructions for making all the other parts of the cell and for producing new cells. Each cell inherits its genetic material from the cells that give rise to it, so all cells in a single body generally contain identical DNA molecules.

Cells Contain Cytoplasm

The **cytoplasm** consists of all the material inside the plasma membrane and outside the DNA-containing region. The fluid portion of the cytoplasm contains water, salts, and an assortment of organic molecules; it forms a thick soup of proteins, lipids, carbohydrates, salts, sugars, amino acids, and nucleotides (see Chapter 2). Most of the cell's *metabolic activities*—that is, the sum of all the biochemical reactions that underlie life—occur in the cell cytoplasm. For example, the manufacture of proteins takes place in the cytoplasm on special structures called ribosomes.

Cells Obtain Energy and Nutrients from Their Environment

To maintain themselves, all cells must continuously acquire and expend energy. The ultimate source of this energy is sunlight. Cells in some organisms, such as plants, can harness the sun's energy directly and store it in high-energy molecules. Once solar energy is captured in this way, some of the energy-containing molecules become available to cells that are not capable of capturing solar energy on their own. For example, an animal may consume plant tissue, thereby making the energy captured by the plant cells available to its own cells.

In addition to energy, cells also need nutrients. Carbon, nitrogen, oxygen, minerals, and other building blocks of biological molecules are found in the environment—the air, water, rocks, and living things. As with energy, cells must obtain these materials from their living and nonliving environment.

Table 4-1 summarizes the features of cells we have just discussed.

Cell Function Limits Cell Size

Most cells are small, ranging from about 1 to 100 micrometers (millionths of a meter) in diameter (Fig. 4-1). Why are cells so small? The answer lies in the method by which cells move nutrients and wastes across the plasma membrane. As you learned in Chapter 3, many nutrients and wastes move into and out of cells by *diffusion*, the movement of molecules from places of high concentration to places of low concentration. Diffusion, however, is a slow process; oxygen molecules take more than 200 days to diffuse over a distance of 4 inches in cytoplasm. At speeds like that, diffusion is effective only over short distances. If cells were any larger than they are, vital materials could not diffuse from the plasma membrane to the innermost portion of the cell (or vice versa) quickly enough to be useful.

Table 4-1 Features Common to All Cells

Molecular components	Protein, amino acids, lipids, carbohydrates, sugars, nucleotides, DNA, RNA
Structural components	Plasma membrane, cytoplasm, ribosomes
Metabolism	Extracts energy and nutrients from the environment; uses energy and nutrients to build, repair, and replace cellular parts

Further, as a cell enlarges, its volume increases more rapidly than its surface area does. For example, a cell that doubles in radius becomes eight times greater in volume but only four times greater in surface area.

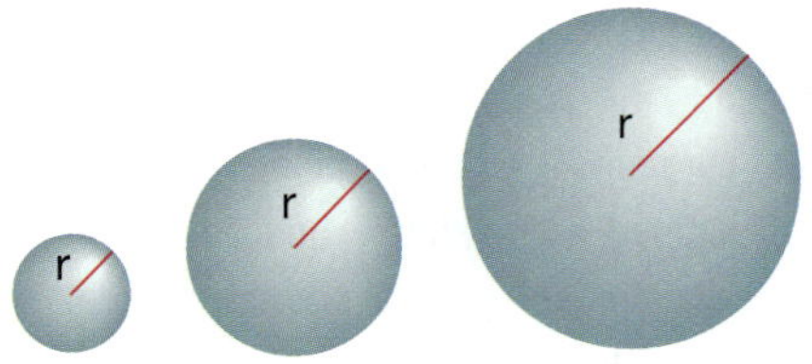

distance to center (r)	1.0	2.0	3.0
surface area ($4\pi r^2$)	12.6	50.3	113.1
volume ($4/3\ \pi r^3$)	4.2	33.5	113.1
area/volume	3.0	1.5	1.0

So, as a cell grows, its volume of cytoplasm (and thus the amount of nutrients and wastes that must be exchanged with the environment) grows much faster than the expanse of plasma membrane through which the exchanges are made. If cells get too big, metabolic needs will overwhelm available membrane surface area.

2 How Do Prokaryotic and Eukaryotic Cells Differ?

Despite the features that are shared by all cells, cells can be grouped into two fundamentally different types. The first type, the **prokaryotic cell**, is found only in two groups of single-celled organisms, the bacteria and the archaea. Biologists often refer to bacteria and

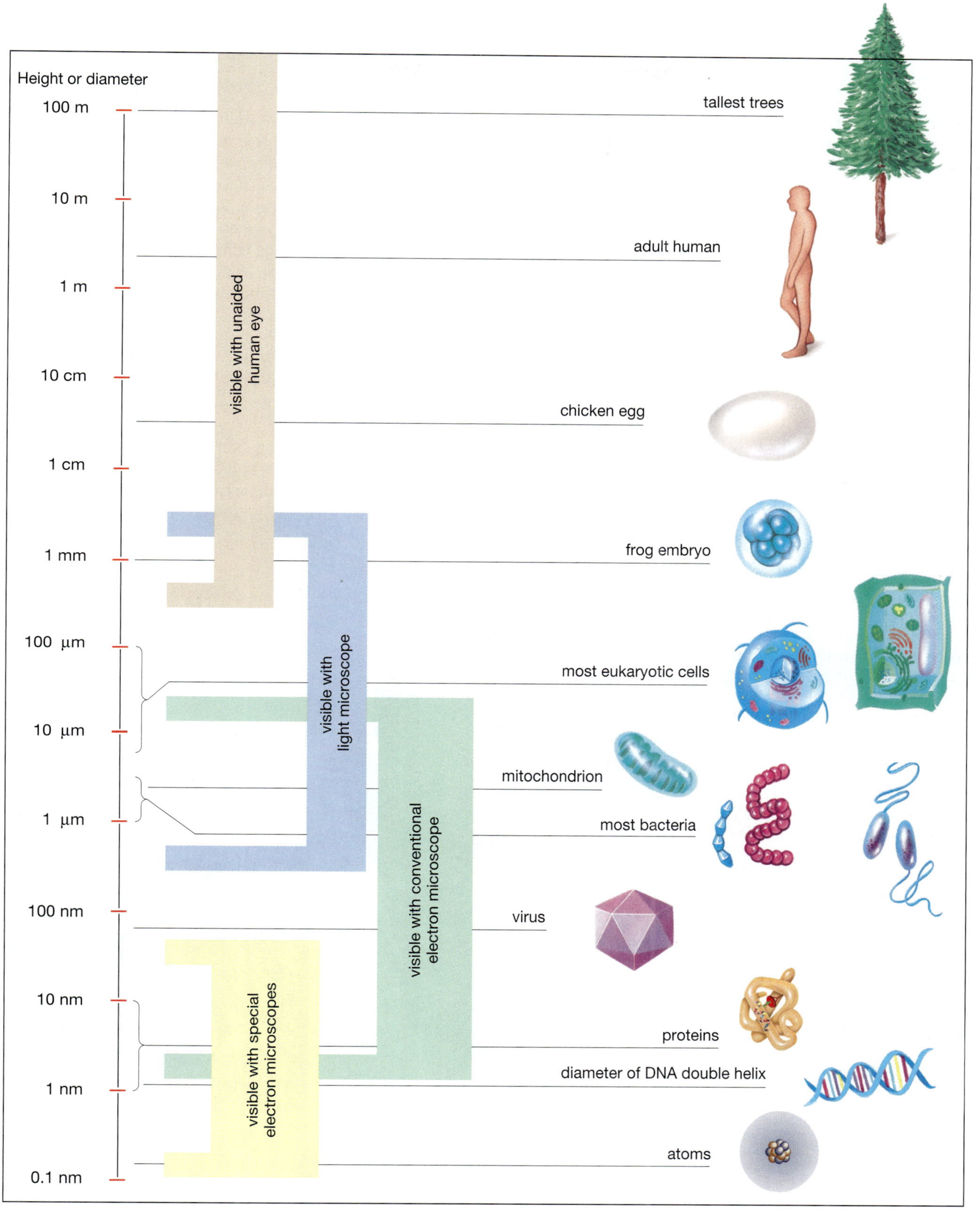

Units of measurement:
1 meter (m) = 39.37 inches
1 centimeter (cm) = 1/100 m
1 millimeter (mm) = 1/1000 m
1 micrometer (µm) = 1/1,000,000 m
1 nanometer (nm) = 1/1,000,000,000 m

Figure 4-1 Relative sizes
Dimensions commonly encountered in biology range from about 100 meters (the height of the tallest redwoods) through a few micrometers (the diameter of most cells) to a few nanometers (the diameter of many large molecules).

Scientific Inquiry
The Search for the Cell

Human understanding of the cellular nature of life came slowly. In 1665 English scientist and inventor Robert Hooke reported observations that he made through a primitive microscope. He aimed his instrument at a thin piece of cork (which comes from the bark of an oak tree) and saw "a great many little Boxes" (Fig. E4-1a). Hooke called the boxes "cells," because he thought they resembled the tiny rooms, or cells, occupied by monks. He wrote that in the living oak and other plants, "These cells [are] fill'd with juices."

By the 1670s, Dutch microscopist Anton van Leeuwenhoek was observing a previously unknown world through simple microscopes of his own construction. His descriptions of myriad "animalcules" (his term for single-celled organisms) in rain, pond water, and well water caused quite an uproar, because water was consumed without treatment in those days. Leeuwenhoek made careful observations of an enormous range of microscopic specimens, including red blood cells, sperm, and the eggs of small insects.

More than a century passed before biologists began to understand more about the role of cells in life on Earth. Microscopists first noted that many plants consist entirely of cells. The thick wall surrounding all plant cells, first observed by Hooke, made their observations easier. Animal cells, however, escaped notice until the 1830s, when German zoologist Theodor Schwann saw that cartilage contains cells that "exactly resemble [the cells of] plants." In 1839, after studying cells for years, Schwann was confident enough to publish his *cell theory*, calling cells the elementary particles of both plants and animals. By the mid-1800s, German botanist Matthias Schleiden had further refined science's view of cells when he wrote: "It is . . . easy to perceive that the vital process of the individual cells must form the first, absolutely indispensable fundamental basis" of life.

Ever since the pioneering efforts of Hooke and van Leeuwenhoek, biologists, physicists, and engineers have collaborated to improve the capabilities of microscopes. Today's microscopes

(a) (b)

Figure E4-1 Microscopes yesterday and today
(a) Robert Hooke's drawings of the cells of cork, as he viewed them with an early light microscope similar to the one shown here. The cells of cork are not living, and only the cell walls remain to outline the cell. ***(b)*** A scanning electron microscope. This allows a three-dimensional image of the surfaces of specimens. The specimen is prepared by coating it with a thin layer of heavy metal (such as platinum), which is vaporized and deposited on the surface of the object to be viewed. The microscope directs an electron beam over the coated surface, and the metal ions on the surface release electrons in response. The number of electrons released depends on the angle of the electron beam relative to the surface of the specimen, which in turn is determined by the contours of the specimen. Thus, the final image, formed by detection of the emitted electrons, has a three-dimensional appearance.

fall into two basic categories: *light microscopes* and *electron microscopes*.

Light microscopes use lenses, usually made of glass, to focus and magnify light rays that either pass through or bounce off a specimen. Light microscopes provide a wide range of images, depending on how the specimen is illuminated and whether it has been stained (Fig. E4-2a). The resolving power of light microscopes—that is, the smallest structure that can be seen—is about 1 micrometer (a millionth of a meter).

Electron microscopes use beams of electrons instead of light. The electrons are focused by magnetic fields rather than by lenses. Some types of electron microscopes can resolve structures as small as a few nanometers (billionths of a meter). *Transmission electron microscopes* (TEMs) pass electrons through a thin specimen and can reveal the details of interior cell structure, including organelles and plasma membranes (Fig. E4-2b). *Scanning electron microscopes* (SEMs; see Fig. E4-1b) bounce electrons off specimens that have been coated with metals and provide three-dimensional images. SEMs can be used to view the surface details of structures that range in size from entire insects down to cells and even organelles (Fig. E4-2c,d). Photographs made through electron microscopes are often called **electron micrographs**.

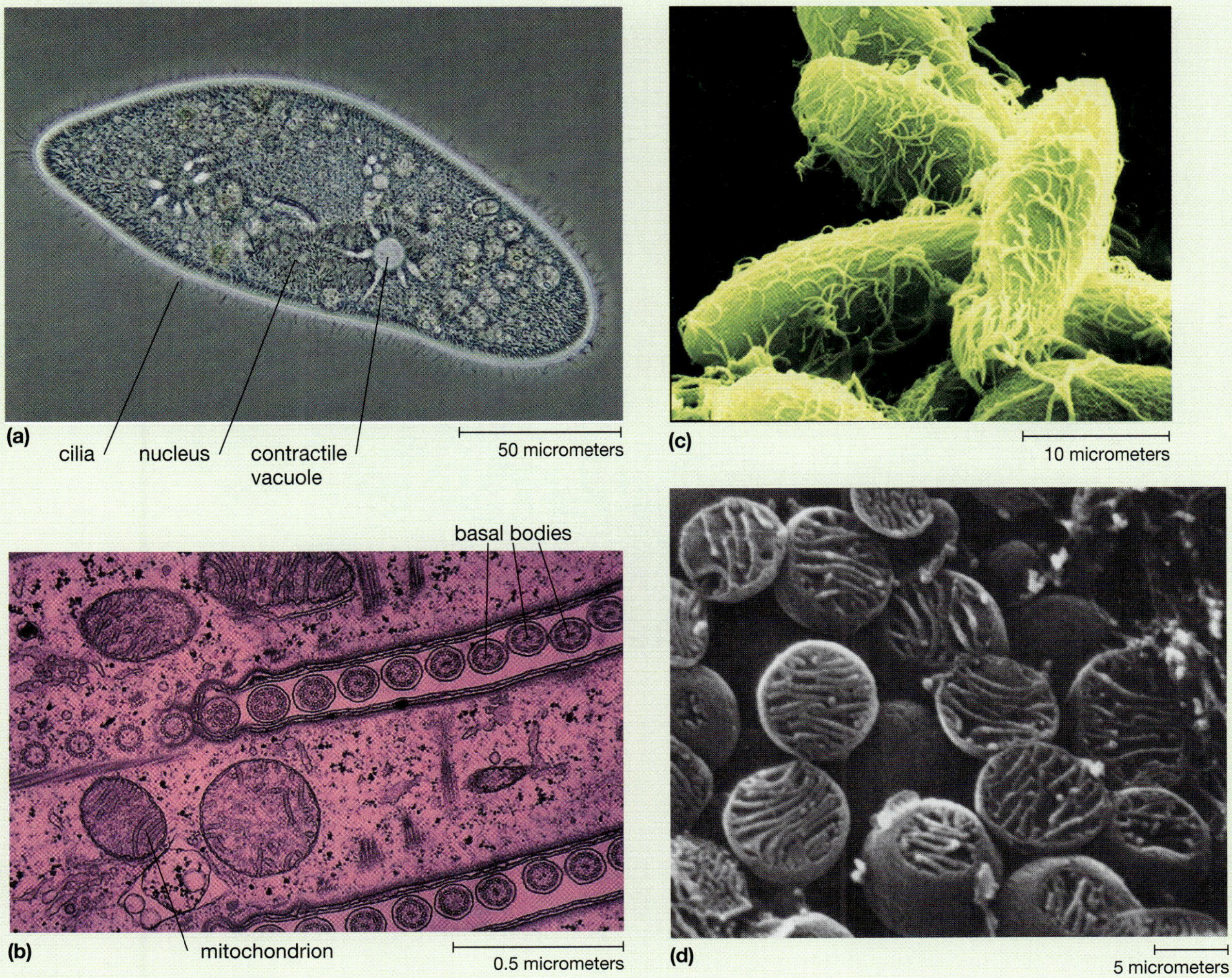

Figure E4-2 A comparison of microscope images
(a) A living *Paramecium* (a single-celled freshwater organism) photographed through a light microscope. ***(b)*** A false-color TEM photo showing the basal bodies at the bases of the cilia that cover *Paramecium*. Mitochondria are also visible. ***(c)*** A false-color SEM photo of *Paramecium*. ***(d)*** An SEM photo at much higher magnification, showing mitochondria, many of which are sliced open.

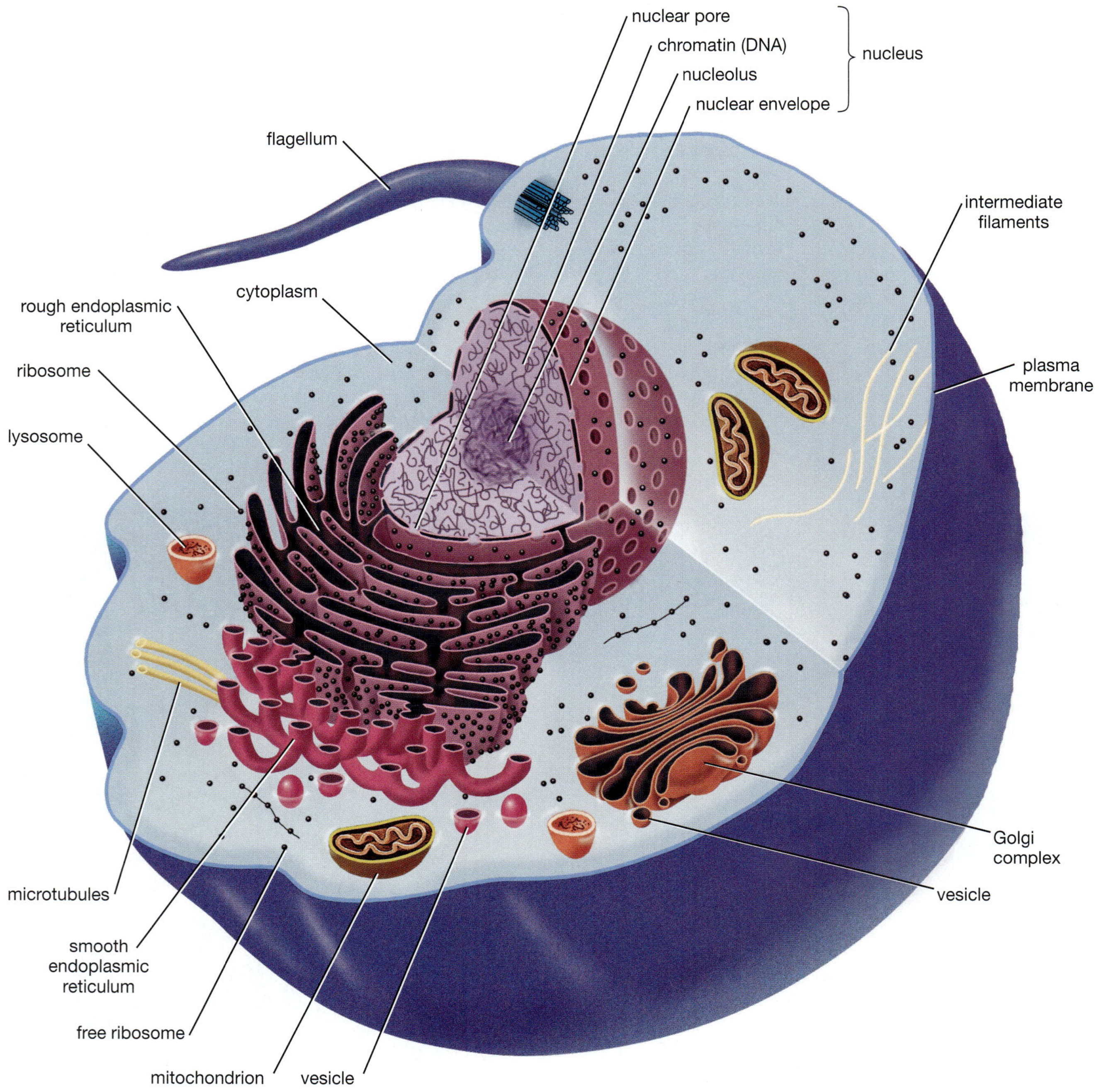

MEDIATUTOR 4.1 Cell Structure

Figure 4-2 A generalized animal cell

archaea as prokaryotes. The second type of cell, which makes up the bodies of all other living things, is called the **eukaryotic cell**, and organisms with these cells are known as eukaryotes. One striking difference between the two cell types is that the DNA of eukaryotic cells is contained within a separate, membrane-bound structure called the **nucleus**. In contrast, the genetic material of prokaryotic cells is not enclosed within a membrane.

The earliest cells on Earth were probably prokaryotic. Eukaryotic cells arose much later, perhaps as the result of a partnership among different kinds of prokaryotic cells. This scenario for the origin of eukaryotic cells is described in more detail in Chapter 15.

The Big Picture

Cells are the building blocks of life. All cells share certain fundamental features inherited from the ancient single-celled organisms that were Earth's first living things.

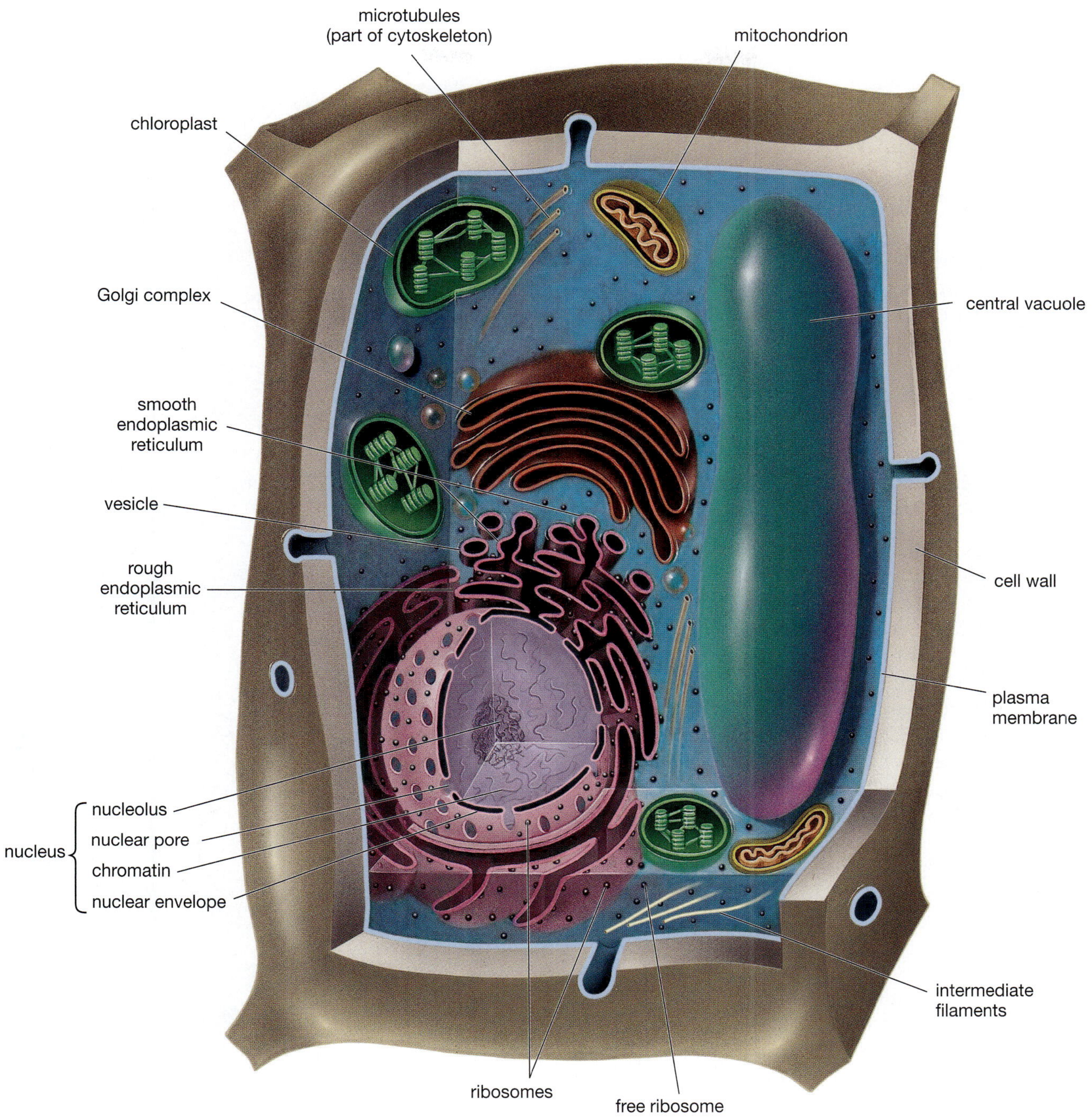

Figure 4-3 A generalized plant cell

MEDIATUTOR
4.1 Cell Structure

3 What Are the Main Features of Eukaryotic Cells?

Eukaryotic cells differ from prokaryotic cells in many ways. For one thing, they are usually larger than prokaryotic cells—typically more than 10 micrometers in diameter (see Fig. 4-1). Also, the cytoplasm of eukaryotic cells houses membrane-enclosed structures called **organelles** that perform specific functions within the cell. A network of protein fibers, the cytoskeleton, gives shape and organization to the cytoplasm of eukaryotic cells. Many of the organelles are attached to the cytoskeleton.

Different types of eukaryotic cells may contain different kinds of organelles. For example, compare the structures that are found in animal and plant cells (Figs. 4-2 and 4-3). Each type of cell has a few unique organelles not found in the other. As you read the following sections, which describe the structures of the cell in more detail, you may wish to refer to the illustrations in Figures 4-2 and 4-3.

4 What Role Does the Nucleus Play?

The nucleus is usually the largest organelle in the cell. It consists of three readily distinguishable parts (Fig. 4-4). The nuclear envelope separates the nuclear material from the cytoplasm. Inside the nuclear envelope, the nucleus contains a granular-looking material called chromatin and a darker region called the nucleolus.

The Nuclear Envelope Controls Passage of Materials

The nucleus is isolated from the rest of the cell by a **nuclear envelope** that consists of a double membrane. The membrane is perforated with tiny membrane-lined channels called *pores*. Water, ions, and small molecules can pass freely through the pores, but the passage of large molecules, such as proteins, pieces of ribosomes, and RNA, is regulated by special "gatekeeper proteins" that line each nuclear pore. These gatekeepers permit the passage of certain molecules and prevent the passage of others. DNA does not cross the nuclear membrane; it remains within the nucleus for the life of the cell.

The Nucleus Contains Chromosomes

The DNA molecules contained inside the nucleus are closely associated with certain kinds of protein molecules, and this DNA-protein complex is known as **chromatin**. A eukaryotic cell's chromatin is arranged in a set of long, thread-shaped structures called **chromosomes**. When cells divide, each chromosome coils upon itself, becoming thicker and shorter. The resulting "condensed" chromosomes are easily visible under even relatively low-power microscopes (Fig. 4-5).

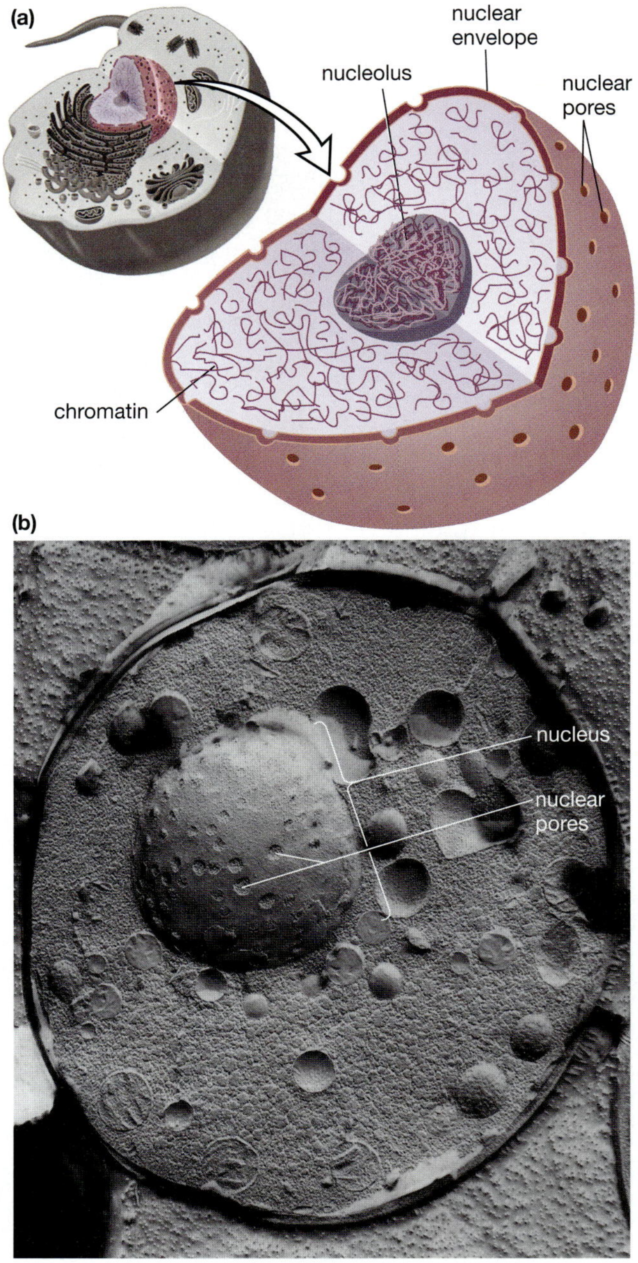

Figure 4-4 The nucleus
(a) The nucleus is bounded by a nuclear envelope. Inside are chromatin (DNA and associated proteins) and a nucleolus. ***(b)*** An electron micrograph of a yeast cell that was frozen and broken open to reveal its internal structures. The large nucleus, with nuclear pores penetrating its nuclear envelope, is clearly visible.

Ribosome Components Are Made at the Nucleolus

Most eukaryotic nuclei have one or more darkly staining regions called *nucleoli* (a **nucleolus** is shown in Fig. 4-4a). The nucleolus consists of DNA, RNA, proteins, and ribosomes in various stages of construction.

Nucleoli are the sites where the components of **ribosomes** are made. The components leave the nucleus and move to the cytoplasm, where they come together to form complete ribosomes. A ribosome is composed of RNA and proteins and serves as a kind of "workbench" for the manufacture of proteins. Just as a workbench can be used to construct many different objects, any ribosome can be used to synthesize any of the thousands of proteins made by a cell. In electron micrographs, ribosomes appear as dark granules, either distributed in the cytoplasm (Fig. 4-6) or clustered along membranes (see Figs. 4-2 and 4-3).

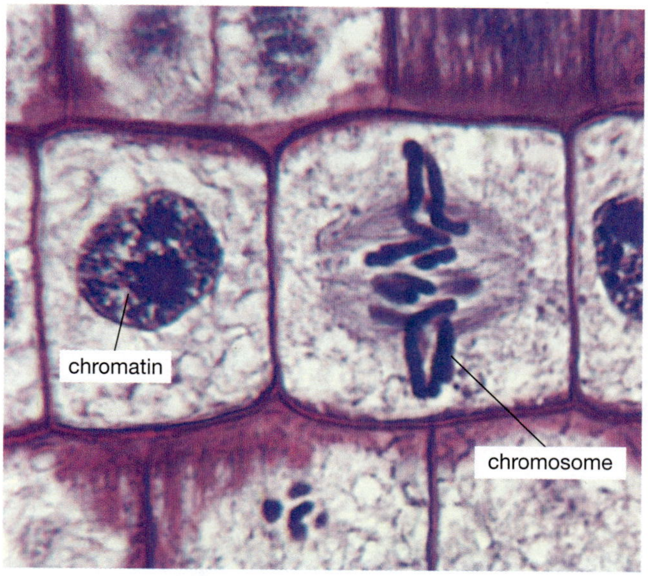

Figure 4-5 Chromosomes
Chromosomes, seen here in a light micrograph of a dividing cell (on the right) in an onion root tip, are the same material (DNA and proteins) as the chromatin seen in nondividing cells adjacent to it, but in a more compact state.

5 What Roles Do Membranes Play in Eukaryotic Cells?

Eukaryotic cells have an elaborate system of membranes that not only encloses the cell but also creates internal compartments within the cytoplasm. This membrane system is composed of the plasma membrane and several organelles, including the endoplasmic reticulum, nuclear envelope, Golgi complex, and membrane-enclosed sacs such as lysosomes (see Figs. 4-2 and 4-3).

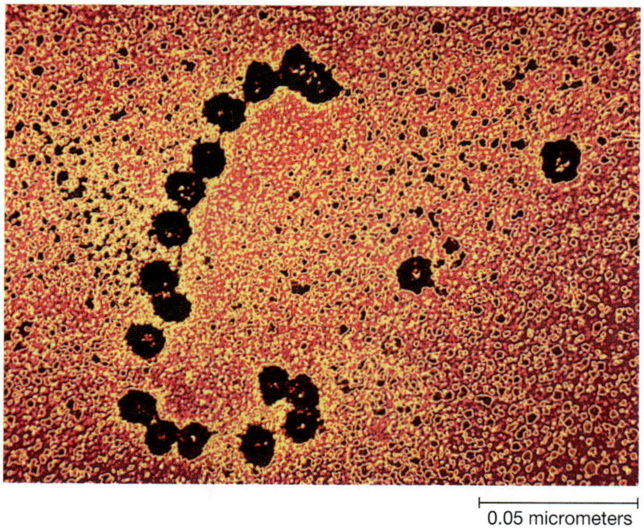

Figure 4-6 Ribosomes

The Plasma Membrane Isolates the Cell and Helps It Interact with Its Environment

The plasma membrane forms the outer boundary of a cell, enclosing the cytoplasm. It is a complex structure that performs the seemingly contradictory functions of separating the cytoplasm of the cell from the outside environment while providing for the transport of selected substances into or out of the cell. In addition, plants, fungi, and some single-celled organisms have a stiff *cell wall* that forms an outer protective coating.

The Endoplasmic Reticulum Forms Channels Within the Cytoplasm

The endoplasmic reticulum (ER) is a series of interconnected, membrane-enclosed channels in the cytoplasm (Fig. 4-7); the ER membrane is continuous with the nuclear membrane. Eukaryotic cells have two forms of ER: rough and smooth. Numerous ribosomes stud the outside of the **rough endoplasmic reticulum**; in contrast, **smooth endoplasmic reticulum** lacks ribosomes.

The rough ER and smooth ER have different functions. Rough ER, with its embedded ribosomes, provides sites where proteins are manufactured. As they are made, many of the proteins move through the ER membrane into the ER interior. The proteins then move through channels in the ER and accumulate in pockets. The pockets bud off, forming membrane-bound sacs called **vesicles**, that carry their protein cargo to its destination.

Smooth ER provides sites for the manufacture of other molecules that, like the proteins made in the rough ER, can be packaged for shipment elsewhere in the cell. Among the molecules made at the smooth ER are the phospholipids and cholesterol that form membranes. Because smooth ER can manufacture phospholipids and rough ER can make membrane proteins, ER is the cell's site for membrane construction. Most of this new membrane is used for new or replacement ER, but some of it moves toward the nucleus to replace nuclear membrane or into the cytoplasm to maintain other membranes of the cell. Smooth ER also performs other functions in certain types of cells. For example, enzymes bound to the smooth ER in liver cells help detoxify substances such as drugs and alcohol.

The Golgi Complex Sorts, Chemically Alters, and Packages Important Molecules

The **Golgi complex** is a specialized set of membranes derived from the endoplasmic reticulum (Fig. 4-8). It looks very much like a stack of flattened sacs and functions as a kind of transfer station for the substances that are transported around the cell. Vesicles arrive from the ER and fuse with one face of the Golgi complex, adding

MediaTutor 4.2 Membrane Traffic

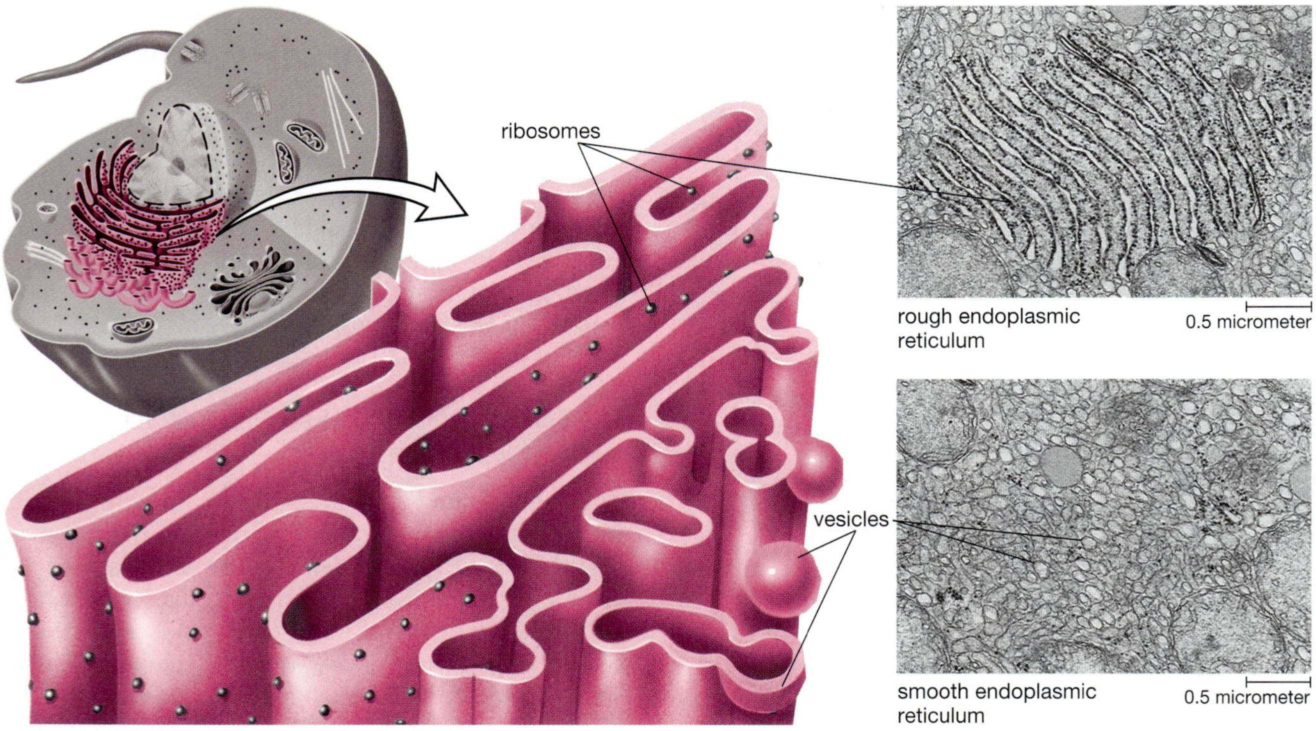

Figure 4-7 Endoplasmic reticulum
There are two types of endoplasmic reticulum: rough ER, coated with ribosomes, and smooth ER, without ribosomes. Although in electron micrographs the ER looks like a series of tubes and sacs, it is actually a maze of folded sheets and interlocking channels.

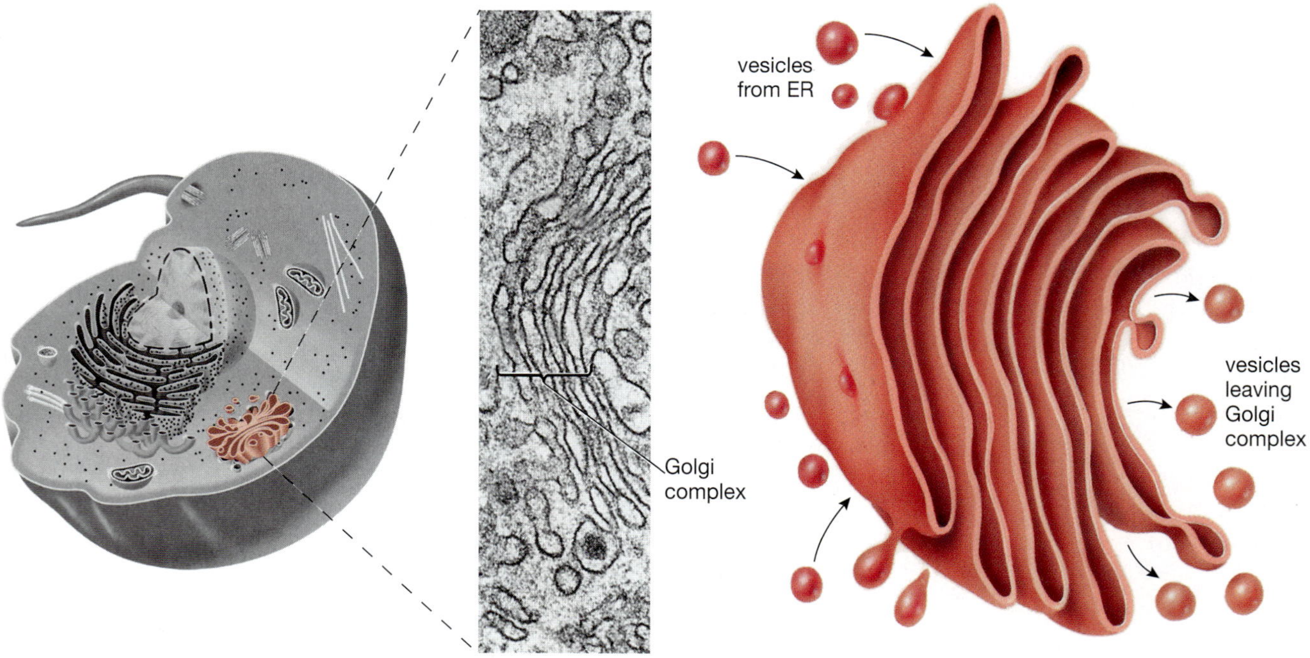

Figure 4-8 The Golgi complex
The Golgi complex is a stack of flat membranous sacs. Vesicles transport material from the ER to the Golgi (and vice versa) and from the Golgi to plasma membrane, lysosomes, and vesicles. Departing vesicles bud off from the Golgi on one face; arriving vesicles join it on the opposite face.

their membrane to the Golgi complex and emptying their contents into the Golgi sacs. At the same time, other vesicles bud off the Golgi complex on the opposite face of the stack, carrying away proteins, lipids, and other complex molecules. These vesicles then move to other parts of the cell or to the plasma membrane for export.

As molecules pass through the Golgi, they are sorted and sometimes undergo further processing. Proteins and lipids received from the ER are sorted according to their destinations. For example, proteins that will remain inside the cell are separated from those that the cell will secrete. The Golgi modifies some molecules. For instance, it adds sugars to proteins to make glycoproteins.

We Can Follow the Travels of a Secreted Protein

To understand how the membranous organelles work together, let's look at the secretion of an antibody within your body. An antibody is a protein, secreted by a type of white blood cell, that binds to foreign invaders (such as bacteria) and helps destroy them. The antibody protein is manufactured on ribosomes of the rough ER and then packaged into vesicles formed from the ER membrane. These vesicles travel to the Golgi, where the membranes fuse, releasing the protein into the Golgi complex. Here, carbohydrates are attached to the protein, which is then repackaged into vesicles formed from Golgi membrane. The vesicle containing the completed antibody travels to the plasma membrane and fuses with it, releasing the antibody outside the cell, where it will make its way into the bloodstream to help defend your body against infection.

Lysosomes Serve as the Cell's Digestive System

Some of the proteins manufactured in the ER and sent to the Golgi complex are digestive enzymes that can break down proteins, fats, and carbohydrates into their component subunits. In the Golgi these enzymes are packaged in vesicles called **lysosomes**. One function of lysosomes is to digest excess cellular membranes or defective organelles. After identifying these organelles, the cell encloses them in vesicles made of membrane from the ER. These vesicles fuse with lysosomes, and digestive enzymes within the lysosome enable the cell to recycle valuable materials from the defunct organelles.

6 Which Other Structures Play Key Roles in Eukaryotic Cells?

In addition to the organelles described above, eukaryotic cells include a number of other important structures. Among these are vacuoles, mitochondria, chloroplasts, and the elements of the cytoskeleton. Some eukaryotic cells also have structures called cilia and flagella that help with locomotion.

Vacuoles Regulate Water and Store Substances

Most cells contain one or more vacuoles, which are fluid-filled sacs surrounded by a single membrane. For example, many plant cells contain a large **central vacuole** (see Fig. 4-3). Filled mostly with water, the central vacuole helps keep the right amount of water in the cell. It also provides a dump site for hazardous wastes, which plant cells often cannot excrete. Some plant cells store extremely poisonous substances, such as sulfuric acid, in their vacuoles, which deters animals from feeding on the otherwise nutritious leaves. Vacuoles may also store sugars and amino acids not immediately needed by the cell.

Mitochondria Extract Energy from Food Molecules

The energy that cells need to survive, grow, and reproduce is harvested in the **mitochondria** (singular, **mitochondrion**). The mitochondrion is sometimes called the "powerhouse of the cell" because it is the organelle in which energy is extracted from sugar molecules and stored in the high-energy bonds of ATP. Once the energy is stored in ATP molecules, it is available to power the multitude of energy-consuming reactions that occur in every cell. Mitochondria are round, oval, or tubular sacs made of a pair of membranes. The outer mitochondrial membrane is smooth, but the inner membrane loops back and forth to form deep folds (see Figs. 4-2 and 4-3). The structure of the mitochondrion is described in greater detail in Chapter 7, as is the mitochondrion's role in energy production.

Chloroplasts Capture Solar Energy

Plant cells contain **chloroplasts**, organelles that capture energy directly from sunlight and store it in sugar molecules. Chloroplasts are the site of photosynthesis, the energy-capturing process on which all life ultimately depends. Each chloroplast is surrounded by a double membrane, and contains stacks of hollow membranous sacs (see Fig. 4-3). The green pigment chlorophyll gives chloroplasts a green color (plants are green because they contain chloroplasts). The structure of chloroplasts and the reactions of photosynthesis are described in more detail in Chapter 6.

The Cytoskeleton Provides Shape, Support, and Movement

Organelles do not drift haphazardly about the cytoplasm; most are attached to a network of protein fibers called the **cytoskeleton** (Fig. 4-9). Several types of protein fibers, including thin **microfilaments**, medium-sized **intermediate filaments**, and thick **microtubules**, make up

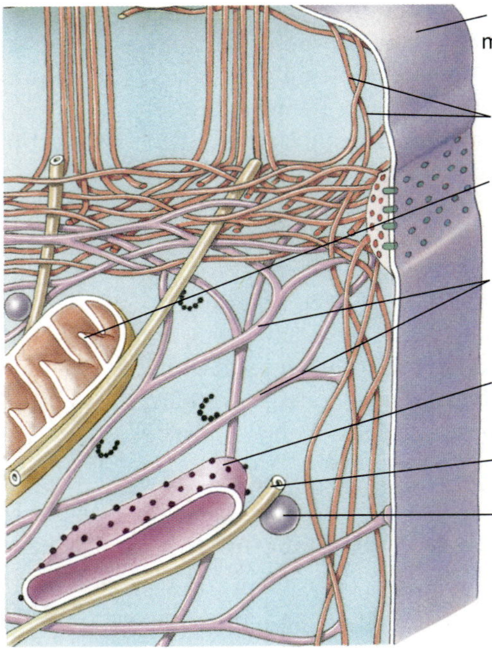

Figure 4-9 The cytoskeleton
Eukaryotic cells are given shape and organization by the cytoskeleton, which consists of three types of proteins: microtubules, intermediate filaments, and microfilaments.

the cytoskeleton. Table 4-2 provides more detail about the cytoskeleton and its components.

The cytoskeleton performs the following important functions:

- *Cell shape*. In cells without cell walls, the cytoskeleton determines the shape of the cell.
- *Cell movement*. Many types of cells move about. Much of this movement is generated when microfilaments and microtubules assemble, disassemble, and slide past one another.
- *Organelle movement*. Microtubules and microfilaments move organelles from place to place within a cell. For example, the vesicles that bud off of the ER and Golgi complex are guided to their destinations by the cytoskeleton.
- *Cell division*. Microtubules and microfilaments are essential to cell division in eukaryotic cells. When eukaryotic nuclei divide, microtubules move the chromosomes into the daughter nuclei. When cells of animals divide, the cytoplasm is divided by a ring of microfilaments that pinches the "waist" of the parent cell around its middle.

Cilia and Flagella Move the Cell or Move Fluid Past the Cell

Both **cilia** and **flagella** (singular, **cilium** and **flagellum**) are slender extensions of the plasma membrane. They typically contain microtubules that extend along their length. Some single-celled eukaryotic organisms use cilia or flagella to move about. Most animal sperm rely on flagella for movement. Some small animals use cilia for movement, swimming by the coordinated beating of rows of cilia, like the oars on a Roman galley ship. More commonly, however, cilia are used to move fluids and suspended particles past a surface. Cells with cilia line such diverse structures as the gills of oysters (moving food- and oxygen-rich water), the oviducts of female mammals (moving the eggs along from the ovary to the uterus), and the respiratory tracts of most land vertebrates (clearing mucus that carries debris and microorganisms from the windpipe and lungs).

The main differences between cilia and flagella lie in their length, number, and the direction of the force they generate. In general, cilia are shorter (about 10 to 25 micrometers long) and more numerous than flagella. They provide force in a direction parallel to the plasma membrane, like the oars in a canoe. This is accomplished through a "rowing" motion. Flagella are longer (50 to 75 micrometers), usually fewer in number, and provide force perpendicular to the plasma membrane, like the engine on a motorboat.

7 What Are the Features of Prokaryotic Cells?

Most prokaryotic cells are very small (less than 5 micrometers long), with a relatively simple internal structure (compare Fig. 4-10 with Figs. 4-2 and 4-3). In general, they are surrounded by a relatively stiff cell wall, which

Table 4-2 Components of the Cytoskeleton

	Structure	Protein Structure	Function
Microfilaments	Twisted double strands, each consisting of a string of protein subunits; about 7 nm in diameter and up to several centimeters long (in muscle cells)	Actin 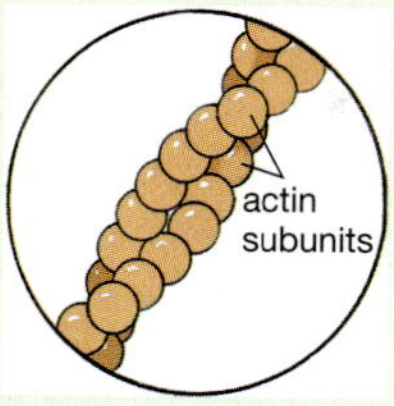	Muscle contraction; changes in cell shape, including cytoplasmic division in animal cells; cytoplasmic movement; movement of pseudopodia
Intermediate filaments	Consist of eight subunits composed of ropelike protein strands; 8–12 nm in diameter and 10–100 μm in length	Protein varies with tissue type 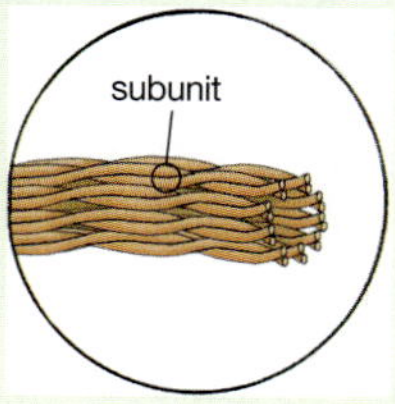	Maintenance of cell shape; attachments of microfilaments in muscle cells; support of nerve cell extensions; attach cells together
Microtubules	Tubes consisting of spiraling two-part protein subunits; about 25 nm in diameter and can be 50 μm in length	Tubuli 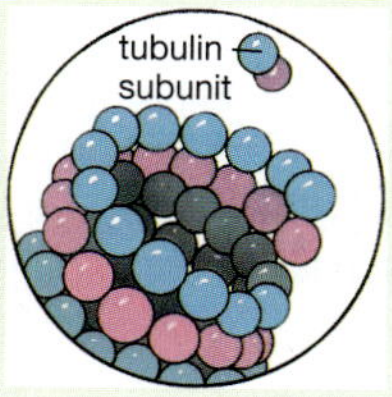	Movement of chromosomes during cell division; movement of organelles within cytoplasm; movement of cilia and flagella

provides shape and protects the cell. Some prokaryotes can move, propelled by whiplike protrusions that undulate or spin. However, these "flagella" of prokaryotic cells do not contain microtubules and have no evolutionary relationship to eukaryotic flagella or cilia.

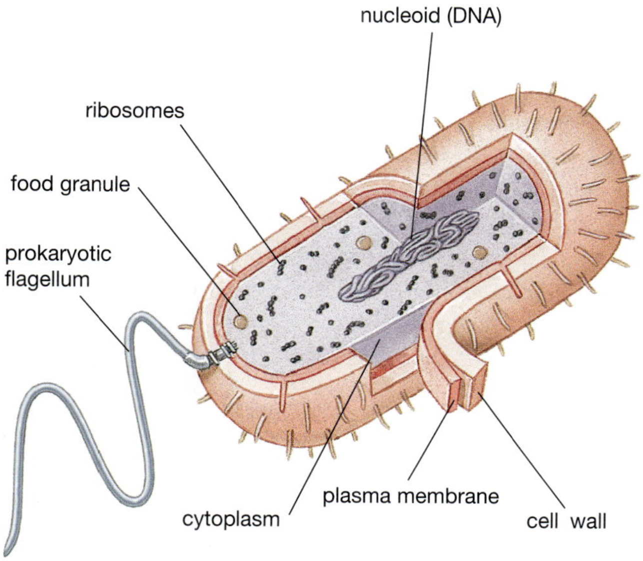

Figure 4-10 A generalized prokaryotic cell

The cytoplasm of most prokaryotic cells is relatively uniform in appearance. Each prokaryotic cell has a single, circular strand of DNA. It is usually coiled, attached to the plasma membrane, and concentrated in a region of the cell called the **nucleoid**. The nucleoid is not, however, separated from the rest of the cytoplasm by a membrane.

Prokaryotic cells lack nuclei as well as most of the other membrane-enclosed organelles that characterize eukaryotic cells. Prokaryote cytoplasm does, however, contain ribosomes, the workbenches on which proteins are manufactured in both prokaryotes and eukaryotes. The cytoplasm also may contain food granules that store energy-rich materials, such as glycogen.

Prokaryotic cells are compared with the eukaryotic cells of plants and animals in Table 4-3.

The Big Picture

Cells are stunningly complex. Each one is like a tiny, automated factory, containing an extensive assortment of parts and subsystems that work together in intricate ways. The sum of this coordinated activity allows the cell to survive, carry out its functions, and eventually produce additional cells.

Table 4-3 Cell Structures, Their Functions, and Their Distribution in Living Cells

Structure	Function	Prokaryotes	Plants	Animals
Cell surface				
Cell wall	Protects, supports cell	Present	Present	Absent
Plasma membrane	Isolates cell contents from environment; regulates movement of materials into and out of cell; communicates with other cells	Present	Present	Present
Organization of genetic material				
Genetic material	Encodes information needed to construct cell and control cellular activity	DNA	DNA	DNA
Chromosomes	Contain and control use of DNA	Single, circular, no proteins	Many, linear, with proteins	Many, linear, with proteins
Nucleus	Membrane-bound container for chromosomes	Absent	Present	Present
Nuclear envelope	Encloses nucleus; regulates movement of materials into and out of nucleus	Absent	Present	Present
Nucleolus	Synthesizes components of ribosomes	Absent	Present	Present
Cytoplasmic structures				
Mitochondria	Produce energy by aerobic metabolism	Absent	Present	Present
Chloroplasts	Perform photosynthesis	Absent	Present	Absent
Ribosomes	Provide site of protein synthesis	Present	Present	Present
Endoplasmic reticulum	Synthesizes membrane components and lipids	Absent	Present	Present
Golgi complex	Modifies and packages proteins and lipids; synthesizes carbohydrates	Absent	Present	Present
Lysosomes	Contain digestive enzymes	Absent	Present	Present
Central vacuole	Contains water and wastes	Absent	Present	Absent
Cytoskeleton	Gives shape and support to cell; positions and moves cell parts	Absent	Present	Present
Centrioles	Synthesize microtubules of cilia and flagella	Absent	Absent (in most)	Present
Cilia and flagella	Move cell through fluid or move fluid past cell surface	Present[a]	Absent (in most)	Present

[a]Many prokaryotes have structures called flagella, but these are not made of microtubules and move in a fundamentally different way than eukaryotic cilia or flagella do.

CASE STUDY REVISITED

Spare Parts for Human Bodies

By the middle of 2001, Raul Mercia's thumb was developing nicely. His own bone cells were filling the pores in the coral framework, and the new structure retained the shape of a normal thumb bone. Still, Mercia remains one of only a few human recipients of bioengineered tissues. In the lab, however, scientists have managed to grow liver, skin, cartilage, bone, heart valves, tendons, intestines, blood vessels, and breast tissue on plastic scaffolds, and to implant some of these tissues into animals. Researchers continue to refine tissue construction methods and to develop better scaffolding materials. With each passing day, our ability to control the behavior of cells gets a little better. And Raul Mercia will not have to live without a thumb.

It seems that our newfound ability to rearrange the building blocks of life will produce at least some medical benefits. These benefits (and in fact a large proportion of modern medical technology) flow directly from a longstanding scientific endeavor: trying to understand how cells work. Not everyone agrees, however, that we ought to use the fruits of this project to alter the makeup of human bodies. What do you think?

Summary of Key Concepts

1) What Features Are Shared by All Cells?

The smallest units of life are single cells. Cells are the functional units of multicellular organisms; every living organism is made up of one or more cells. Cells are limited in size because they must exchange materials with their surroundings by diffusion. Because diffusion is relatively slow, the interior of the cell must never be too far from the plasma membrane, and the plasma membrane must have a large surface area (relative to the volume of its cytoplasm) through which materials can diffuse. Both of these constraints limit the size of cells.

2) How Do Prokaryotic and Eukaryotic Cells Differ?

All cells are either prokaryotic or eukaryotic. Prokaryotic cells are small and relatively simple in structure and are found only in bacteria and archaea. More complex eukaryotic cells make up all other forms of life.

3) What Are the Main Features of Eukaryotic Cells?

Eukaryotic cells are generally much larger than prokaryotic cells. Eukaryotic cells also have organelles and a cytoskeleton.

4) What Role Does the Nucleus Play?

The nucleus is bounded by the double membrane of the nuclear envelope and contains the cell's genetic material (DNA). Pores in the nuclear envelope regulate the movement of molecules between nucleus and cytoplasm. The genetic material of eukaryotic cells is organized into chromosomes, which consist of DNA and proteins. Ribosomes are particles that are the sites of protein synthesis. The nucleolus is the site at which the components of ribosomes are manufactured.

5) What Roles Do Membranes Play in Eukaryotic Cells?

The membrane system of a cell consists of the plasma membrane, endoplasmic reticulum (ER), Golgi complex, and vesicles derived from these membranes. Rough ER manufactures many cellular proteins. Smooth ER manufactures phospholipids. The ER is the site of all membrane synthesis within the cell. The Golgi complex is a series of membranous sacs derived from the ER. The Golgi complex processes and modifies materials synthesized in the rough and smooth ER. Some substances in the Golgi are packaged into vesicles for transport elsewhere in the cell. Lysosomes are vesicles that contain enzymes that digest food particles and defective organelles.

6) Which Other Structures Play Key Roles in Eukaryotic Cells?

All eukaryotic cells contain mitochondria, organelles that use oxygen to complete the breakdown of food molecules, capturing much of their energy as ATP. Plant cells contain chloroplasts, which capture the energy of sunlight during photosynthesis, enabling the cells to manufacture organic molecules, particularly sugars, from simple inorganic molecules. Many eukaryotic cells contain vacuoles that are bounded by a single membrane and that store food or wastes.

The cytoskeleton organizes and gives shape to eukaryotic cells and moves and anchors organelles. The cytoskeleton is composed of microfilaments, intermediate filaments, and microtubules (see Table 4-3). Cilia and flagella are whiplike extensions of the plasma membrane that contain microtubules. These structures move fluids past the cell or move the cell through its fluid environment.

7) What Are the Features of Prokaryotic Cells?

Prokaryotic cells are generally very small with a relatively simple internal structure. Most are surrounded by a relatively stiff cell wall. The cytoplasm of prokaryotic cells lacks membrane-enclosed organelles and contains a single, circular strand of DNA.

Study Note

Figures 4-2 and 4-3 illustrate the overall structure of animal and plant cells, respectively. Table 4-3 lists the principal organelles, their functions, and their occurrence in prokaryotic cells, animal cells, and plant cells.

Key Terms

cell *p. 51*
central vacuole *p. 61*
chloroplast *p. 61*
chromatin *p. 58*
chromosome *p. 58*
cilium *p. 62*
cytoplasm *p. 52*
cytoskeleton *p. 61*
deoxyribonucleic acid (DNA) *p. 52*
electron micrograph *p. 55*
eukaryotic cell *p. 56*
flagellum *p. 62*
Golgi complex *p. 59*
intermediate filament *p. 61*
lysosome *p. 61*
microfilament *p. 61*
microtubule *p. 61*
mitochondrion *p. 61*
nuclear envelope *p. 58*
nucleoid *p. 63*
nucleolus *p. 58*
nucleus *p. 56*
organelle *p. 57*
plasma membrane *p. 51*
prokaryotic cell *p. 52*
ribosome *p. 58*
rough endoplasmic reticulum *p. 59*
smooth endoplasmic reticulum *p. 59*
vesicle *p. 59*

Thinking Through the Concepts

Multiple Choice

1. *The outermost boundary of an animal cell is the*
   a. plasma membrane b. nucleus
   c. cytoplasm d. cytoskeleton
   e. cell wall
2. *Which organelle contains a eukaryotic cell's chromosomes?*
   a. Golgi complex b. ribosomes
   c. nucleus d. mitochondria
   e. chloroplast
3. *Most of the cell's ATP is synthesized in the*
   a. Golgi complex b. ribosomes
   c. nucleus d. mitochondria
   e. chloroplast
4. *Which organelle sorts, chemically modifies, and packages newly synthesized protein?*
   a. Golgi complex b. ribosomes
   c. nucleus d. mitochondria
   e. chloroplast
5. *Membrane-enclosed organelles that contain digestive enzymes are called*
   a. lysosomes
   b. smooth endoplasmic reticulum
   c. cilia
   d. Golgi complex
   e. mitochondria
6. *A series of membrane-enclosed channels studded with ribosomes are called*
   a. lysosomes
   b. Golgi complex
   c. rough endoplasmic reticulum
   d. mitochondria
   e. smooth endoplasmic reticulum

? Review Questions

1. Diagram "typical" prokaryotic and eukaryotic cells, and describe their important similarities and differences.
2. Which organelles are common to both plant and animal cells, and which are unique to each?
3. Describe the nucleus, including the nuclear envelope, chromatin, chromosomes, DNA, and the nucleolus.
4. What are the functions of mitochondria and chloroplasts?
5. What is the function of ribosomes? Where in the cell are they typically found?
6. Describe the structure and function of the endoplasmic reticulum and Golgi complex.
7. How are lysosomes formed? What is their function?

Applying the Concepts

1. If muscle biopsies (samples of tissue) were taken from the legs of a world-class marathon runner and a typical couch potato, which would you expect to have a higher density of mitochondria? Why? What about a muscle biopsy from the biceps of a weight lifter?
2. One of the functions of the cytoskeleton in animal cells is to give shape to the cell. Plant cells have a fairly rigid cell wall surrounding the plasma membrane. Does this mean that a cytoskeleton is superfluous for a plant cell? Defend your answer in terms of other functions of the cytoskeleton.
3. Most cells are very small. What physical and metabolic constraints limit cell size? What problems would an enormous cell encounter? What adaptations might help a very large cell survive?

Group Activity

Work in groups of two to estimate the number of cells in your little finger. Assume that your finger is a cylinder, for which volume ≈ 3(radius)^2(height). Assume also that the average size of one of your body cells is 1000 cubic micrometers.

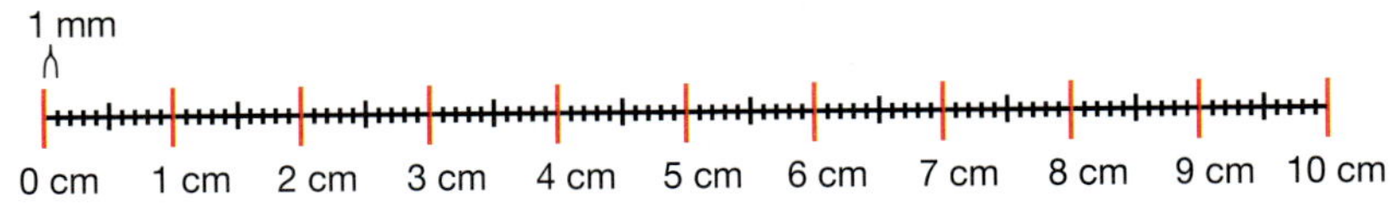

1 centimeter (cm) = 10 millimeters (mm)
1 millimeter = 1000 micrometers (μm)

For More Information

de Duve, C. "The Birth of Complex Cells." *Scientific American*, April 1996. Describes the mechanisms by which the first eukaryotic cells were produced from prokaryotic ancestors.

Ford, B. J. "The Earliest Views." *Scientific American*, April 1998. The author used the original microscopes of Anton van Leeuwenhoek to see the microscopic world as van Leeuwenhoek saw it. Photographic images taken through these early and very primitive instruments reveal remarkable detail.

Goodsell, D. S. *The Machinery of Life*. New York: Springer, 1993. Wonderful, drawn-to-scale images of the organelles and molecules of the cell.

Ingber, D. E. "The Architecture of Life." *Scientific American*, January 1998. Counteracting forces stabilize the design of organic structures, from carbon compounds to the cytoskeleton-reinforced architecture of the cell.

Kiester, E., Jr. "A Bug in the System." *Discover*, February 1991. Mitochondria may be the descendants of bacteria. Because mitochondria are essential for human life, defects in mitochondrial genes can cause some devastating diseases.

Murray, M. "Life on the Move." *Discover*, March 1991. Many cells, including *Amoeba* and white blood cells, crawl about with the help of tiny protein "motors" that manipulate their cytoskeleton.

Rothman, J. E., and Orci, L. "Budding Vesicles in Living Cells." *Scientific American*, March 1996. Researchers are uncovering the mechanisms by which cells form vesicles.

Answers to Multiple-Choice Questions

1. a 2. c 3. d 4. a 5. a 6. c

MediaTutor

CD Activities

Activity 4.1: Cell Structure

Estimated time: 5 minutes

The basic unit of biological structure is the cell. All living things consist of cells. In this tutorial you will explore the structure of the basic cell types.

Activity 4.2: Membrane Traffic

Estimated time: 5 minutes

This tutorial presents the details of membrane traffic in the cell. This includes where new membrane is synthesized and how proteins that are synthesized on the rough endoplasmic reticulum are sent to their proper locations. The complementary processes of phagocytosis and exocytosis are also explored. These allow the import and export of large molecules and larger complexes that could never pass through even the largest transport proteins in the cell.

Start the MediaTutor Student CD-ROM and enter the activity number in the Quick Search box to be taken directly to that activity.

Web Investigation

Spare Parts for Human Bodies

Estimated time: 15 minutes

As you have just read, the fundamental unit of life is the cell. Given the great diversity of life that exists on our planet, it is not surprising that there is also great diversity in the shape and morphology of cells that make up all living organisms. Despite this variety, you have seen that all cells share many common properties. This exercise will examine the properties common to all cells and explore the many uses that man has devised for cells.

Go to http://www.prenhall.com/audesirk3, the Audesirk Companion Web site. Select Chapter 4 and the Web Investigation to begin.

"Disorder spreads through the universe, and life alone battles against it."

G. Evelyn Hutchinson

The bodies of these runners are efficiently converting energy stored in fats and carbohydrates to the energy of movement and heat.

5 Energy Flow in the Life of a Cell

AT A GLANCE

CASESTUDY

A Missing Molecule Makes Mischief

Imagine that you could never eat pizza, hamburgers, ice cream, or bagels with cream cheese. For one out of every 10,000 or so children born in the United States, the simple act of eating such foods can lead directly to neurological disorders and severe mental retardation. These children, who have a syndrome known as phenylketonuria (PKU), cannot fully digest proteins. In particular, their bodies are unable to break down the amino acid phenylalanine (recall from Chapter 2 that proteins consist of long chains of amino acid molecules). As a result, phenylalanine builds up in the blood of PKU victims. For reasons that remain poorly understood, high levels of phenylalanine are toxic, and they have especially devastating effects on developing nervous systems.

The only known effective treatment for PKU is to avoid eating phenylalanine. If phenylalanine is absent from the diet, excess levels of the amino acid cannot accumulate. In practice, restricting dietary phenylalanine means restricting proteins (almost all of which contain at least some phenylalanine). So people with PKU must avoid many of the foods that the rest of us take for granted, including meat, cheese, and fish. Even bread, pasta, and crackers must be replaced by special low-protein versions. PKU sufferers must also be on guard against the artificial sweetener aspartame, which contains phenylalanine. Unfortunately, aspartame is widespread. It is found not only in a host of "sugarless" foods and diet soft drinks, but also in many other consumer products such as toothpaste, cough syrup, and children's aspirin.

Why is it that PKU victims cannot process phenylalanine? After all, their digestive organs and tissues are normal in every respect, and they are able to process all other amino acids in the normal fashion. It turns out that the only difference between PKU sufferers and everyone else is found in a single type of molecule. The victims of PKU have had their lives turned upside-down due to a small shift in molecular structure. What kind of molecule could have such a dramatic effect, turning ordinary foods into poisons? ■

1 What Is Energy?

Energy is the capacity to do work. In this definition, "work" refers to any of a wide range of different actions, including synthesizing molecules, moving objects, and generating heat and light. Energy can be either kinetic or potential, depending on whether it is in use or stored. **Kinetic energy**, or energy of movement, includes light (movement of photons), heat (movement of molecules), electricity (movement of electrically charged particles), and movement of large objects. **Potential energy**, or stored energy, includes chemical energy stored in the bonds that hold atoms together in molecules, electrical energy stored in a battery, and positional energy stored in a diver poised to spring. Under the right conditions, kinetic energy can be transformed into potential energy, and vice versa. For example, when a diver climbs up to a platform, he converts kinetic energy of movement into potential energy. When he jumps off, the potential energy is converted back into kinetic energy (Fig. 5-1).

Figure 5-1 From potential energy to kinetic energy
The body of a diver on the platform has potential energy, because the heights of the platform and the pool are different. As he dives, the potential energy is converted to the kinetic energy of motion of his body. Finally, some of this kinetic energy is transferred to the water, which itself is set in motion.

MediaTutor 5.1 Energy and Chemical Reactions

Energy Cannot Be Created or Destroyed

The laws of thermodynamics describe the basic properties and behavior of energy. The **first law of thermodynamics**, often called the law of conservation of energy, states that energy can neither be created nor destroyed, so the total amount of energy remains constant. Energy can, however, change its form. For example, a car converts the potential chemical energy of gasoline into the kinetic energy of movement and heat. Similarly, a runner converts the potential chemical energy of food into the kinetic energy of movement plus heat. In both of these cases, potential energy changes form, but the total amount of energy remains unchanged.

Energy Tends to Become Distributed Evenly

The **second law of thermodynamics** states that whenever energy is converted from one form to another, the amount of energy that is in useful forms decreases. All spontaneous changes result in a more even distribution of energy, which reduces the energy differences that are required to do work. In other words, energy tends to be converted from more useful into less useful forms.

To illustrate the second law, let's examine a car engine burning gasoline. In a moving vehicle, the kinetic energy of movement is much smaller than the amount of chemical energy in the gasoline that was burned to generate the movement. The first law tells us that energy cannot be destroyed, so where is the "missing" energy? It has been transferred to the car's materials and surroundings. The burning gas not only moved the car but also heated up the engine, the exhaust system, and the air around the car. In addition, the friction of tires on the pavement slightly heated the road. No energy is missing. However, the energy that was released as heat has been converted to a less useful form; it merely heated the engine, the road, and the air. The second law tells us that all conversions of energy entail a loss of usefulness; no process is 100% efficient.

Matter Tends to Become Less Organized

Regions in which energy is concentrated tend to be more organized than regions with less energy. For example, the eight carbon atoms in a single molecule of gasoline have a much more orderly arrangement than do the eight carbon atoms in the eight separate, randomly moving molecules of carbon dioxide that are formed when the gasoline molecule burns. Therefore, we can also phrase the second law in terms of the organization of matter: processes that proceed spontaneously lead to increasing randomness and disorder. This randomness and disorder is called **entropy**. Entropy always tends to increase, a tendency that can be overcome only by adding energy.

We all experience the tendency toward entropy in our homes. Frequent inputs of energy are required to keep debris confined to the trash can, newspapers to folded stacks, books to their shelves, and clothes to drawers and closets. Without our energetic cleaning and organizing efforts, these items tend to end up in their lowest-energy state—a state of disorder. When the ecologist G. Evelyn Hutchinson said, "Disorder spreads through the universe, and life alone battles against it," he was making an eloquent reference to entropy and the second law of thermodynamics.

Living Things Use the Energy of Sunlight to Create Low-Entropy Conditions

If you think about the second law of thermodynamics, you may wonder how life can exist at all. All chemical reactions, including those inside living cells, cause the amount of usable energy to decrease. Matter tends toward increasing randomness and disorder. Given these facts, how can organisms accumulate the concentrated energy and precisely ordered molecules that characterize living things? The answer is that living things use a continuous input of solar energy to construct complex molecules and maintain orderly structures—to "battle against disorder." This solar energy arrives on Earth in the form of sunlight, which is produced by nuclear reactions in the sun. These reactions produce huge increases in entropy, so the highly organized, low-entropy systems of life do not violate the second law; the increased order of living things is achieved at the expense of an enormous loss of order in the sun. The entropy of the solar system as a whole constantly increases.

The Big Picture

Life operates within constraints imposed by the laws of physics. None of life's processes may violate these laws, and knowledge of them can help us understand why organisms operate as they do.

2 How Does Energy Flow in Chemical Reactions?

A **chemical reaction** converts one set of chemical substances, the **reactants**, into another set, the **products**. Some chemical reactions release energy; others consume it. A reaction that releases energy is classified as an **exergonic reaction**, and its products contain less energy than the reactants:

Exergonic reaction

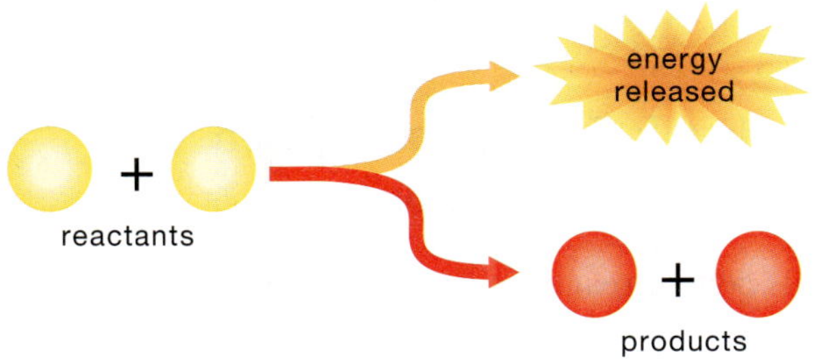

Conversely, an **endergonic reaction** requires an input of energy from some outside source. The products of an endergonic reaction contain more energy than the reactants:

Endergonic reaction

Let's look at two processes that illustrate these types of reactions: burning sugar and photosynthesis.

Exergonic Reactions Release Energy

When sugar is burned by a flame, it reacts with oxygen (O_2) to produce carbon dioxide (CO_2) and water (H_2O) and energy, as described by this equation:

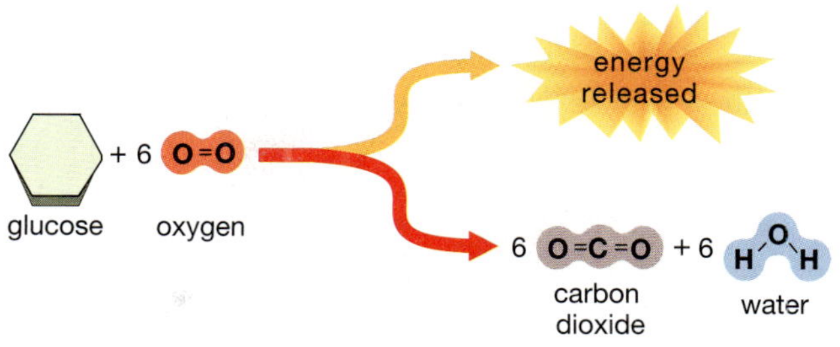

The combustion of sugar is an exergonic reaction, in which the reactants contain more energy than the products. That is, the molecules of sugar and oxygen contain much more energy than the molecules of carbon dioxide and water that are produced when the sugar breaks down. The extra energy is released as heat.

Once started, exergonic reactions proceed without an input of energy. Once ignited, a spoonful of sugar continues to burn spontaneously until all sugar molecules are consumed. It's as if exergonic reactions run "downhill," from high energy to low energy, just as a rock pushed from atop a hill rolls to the bottom.

Endergonic Reactions Require an Input of Energy

Unlike the energy-releasing reaction that takes place when sugar burns, many reactions in living things create products that contain *more* energy than the reactants. For example, the sugar produced by photosynthesis contains far more energy than the carbon dioxide and water from which it is formed:

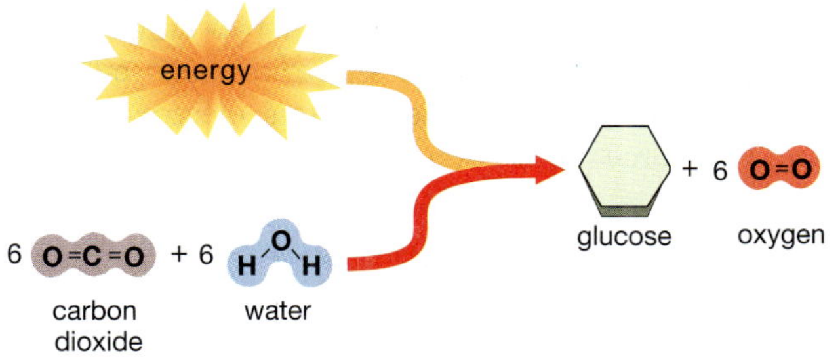

Similarly, a protein molecule contains more energy than the individual amino acids that were joined together to build it.

Completing reactions that form complex biological molecules requires an overall input of energy. The energy for photosynthesis, for example, comes from sunlight. We can view endergonic reactions such as photosynthesis as "uphill" reactions, going from low energy to high energy, like pushing a rock to the top of a hill.

All Reactions Require an Initial Input of Energy

All chemical reactions, even exergonic ones that release energy overall, require an initial input of energy to get started. For example, even though burning sugar releases energy, a spoonful of sugar can't burst into flames by itself; the fire doesn't begin until some energy is added.

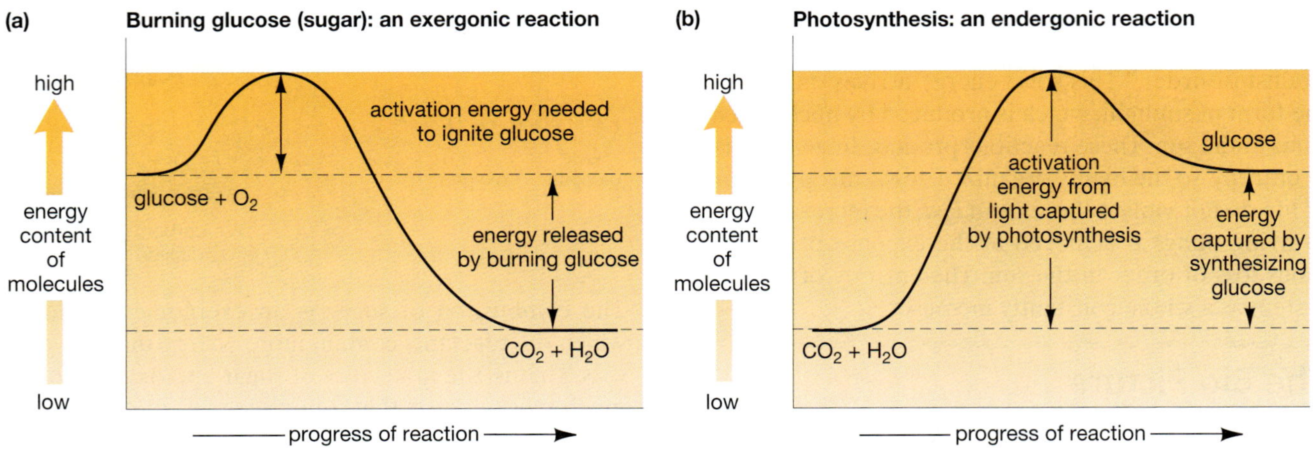

Figure 5-2 Energy relations in exergonic and endergonic reactions
(a) An exergonic ("downhill") reaction, such as the burning of sugar, proceeds from high-energy reactants (here, glucose) to low-energy products (CO_2 and H_2O). The energy difference between the chemical bonds of the reactants and products is released as heat. To start the reaction, however, an initial input of energy—the activation energy—is required. ***(b)*** An endergonic ("uphill") reaction, such as photosynthesis, proceeds from low-energy reactants (CO_2 and H_2O) to high-energy products (glucose) and therefore requires a large input of energy, in this case from sunlight.

This initial energy input to a chemical reaction is called the **activation energy**, and is something like the push one gives a rock poised at the top of a hill to start it rolling down (Fig. 5-2).

Chemical reactions require activation energy to get started because a shell of negatively charged electrons surrounds atoms and molecules. For two molecules to react with each other, their electron shells must be forced together, despite their mutual electrical repulsion. That force requires energy. The usual source of activation energy is the kinetic energy of movement. Molecules moving with sufficient speed collide hard enough to force their electron shells to mingle and react. Because molecules move faster as the temperature increases, most chemical reactions occur more readily at high temperatures.

Exergonic Reactions May Be Linked with Endergonic Reactions

Endergonic reactions require energy from other sources; they obtain this energy from exergonic, energy-releasing reactions. In a **coupled reaction**, an exergonic reaction provides the energy needed to drive an endergonic reaction. For example, when you drive a car, the exergonic reaction of burning gasoline provides the energy for the endergonic reaction of starting a stationary car into motion and keeping it moving. Photosynthesis is another coupled reaction. In photosynthesis, the exergonic reaction occurs in the sun, and the endergonic reaction occurs in the plant. Because some energy is always lost as heat during the transfer (remember the second law of thermodynamics), the energy provided by an exergonic reaction must exceed that needed to drive the endergonic reaction.

3 How Is Energy Carried Between Coupled Reactions?

Cells couple reactions so that the energy released by exergonic reactions is used to drive endergonic reactions. The exergonic and endergonic parts of coupled reactions often occur in different parts of the cell, so there must be some way to transfer energy from the exergonic reaction that releases energy to the endergonic reaction that consumes it. The job of transferring energy from place to place is done by **energy-carrier molecules**.

Energy carriers work something like rechargeable batteries, picking up an energy charge at an exergonic reaction, moving to another location within the cell, and releasing the energy to drive an endergonic reaction. Because energy-carrier molecules are unstable, they are used only for temporary energy transfer within cells. They are not used to transfer energy from cell to cell, nor are they used for long-term energy storage.

ATP Is the Principal Energy Carrier in Cells

The most common energy-carrier molecule in cells is **adenosine triphosphate**, or **ATP**. ATP provides energy for a wide variety of endergonic reactions, serving as a common currency of energy transfer. ATP is made from **adenosine diphosphate (ADP)** and phosphate, using energy released in cells through glucose breakdown:

ATP synthesis: Energy is stored in ATP

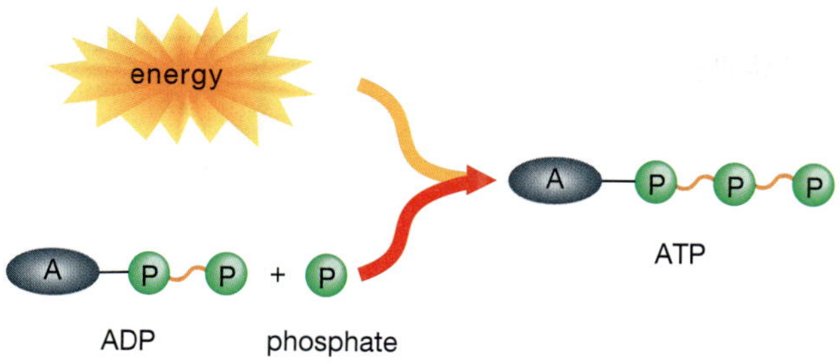

ATP stores energy in its chemical bonds and can carry the energy to sites in the cell that perform energy-requiring reactions. Then, the bond to one of ATP's phosphate groups breaks (yielding ADP and phosphate, which are later recycled to make more ATP). The energy contained in the broken bond is transferred to the energy-requiring reaction:

ATP breakdown: Energy of ATP is released

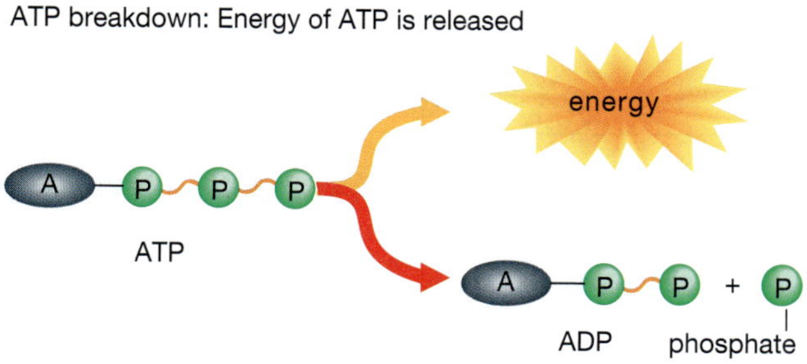

In summary, exergonic reactions (such as glucose breakdown) drive the endergonic reaction of ATP synthesis from ADP. The ATP molecule moves to a different part of the cell, where the ATP breaks down and liberates some of its energy to drive an endergonic reaction (such as muscle cell contraction, Fig. 5-3). During these energy transfers, heat is given off and there is an overall loss of usable energy (Fig. 5-4).

The life span of an ATP molecule in a living cell is very short, because this energy carrier is continuously formed, broken down, and remade. If the molecules of ATP that you use just sitting at your desk all day could be captured (instead of recycled), they would weigh nearly 90 pounds. A marathon runner may use a pound of ATP every minute. (ADP must be quickly converted back to ATP, or it would be a very brief run.) As you can see, ATP is not a long-term energy-storage molecule. Long-term storage is the job of more stable molecules, such as sucrose, glycogen, starch, or fat, that can store energy in your body for hours, days, or months.

MediaTutor
5.2 Energy and Life

Electron Carriers Also Transport Energy Within Cells

In addition to ATP, other carrier molecules also transport energy within a cell. **Electron carriers** capture the energetic electrons to which energy is transferred in some exergonic reactions, and the loaded electron carriers then donate the electrons, along with their energy, to other molecules (Fig. 5-5). Common electron carriers include nicotinamide adenine dinucleotide (NAD^+) and its relative flavin adenine dinucleotide (FAD). You will learn more about electron carriers and their role in cells in Chapters 6 and 7.

The Big Picture

Much of the activity of organisms and the cells that compose them is devoted to obtaining energy and converting it to forms in which it can be stored or used for work. Acquiring useable energy is a prerequisite for all of the other activities of life.

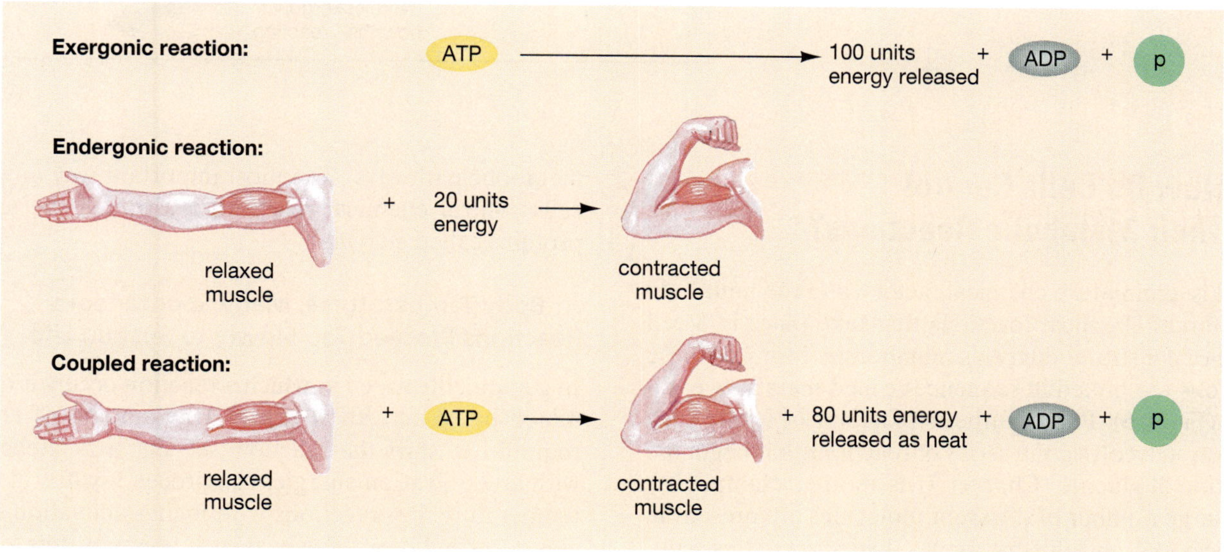

Figure 5-3 A coupled reaction
The exergonic reaction of ATP breakdown must precede the endergonic reaction of muscle movement.

Figure 5-4 Coupled reactions give off heat
The overall reaction is "downhill": More energy is produced by the exergonic reaction than is needed to drive the endergonic reaction. The extra energy is released as heat.

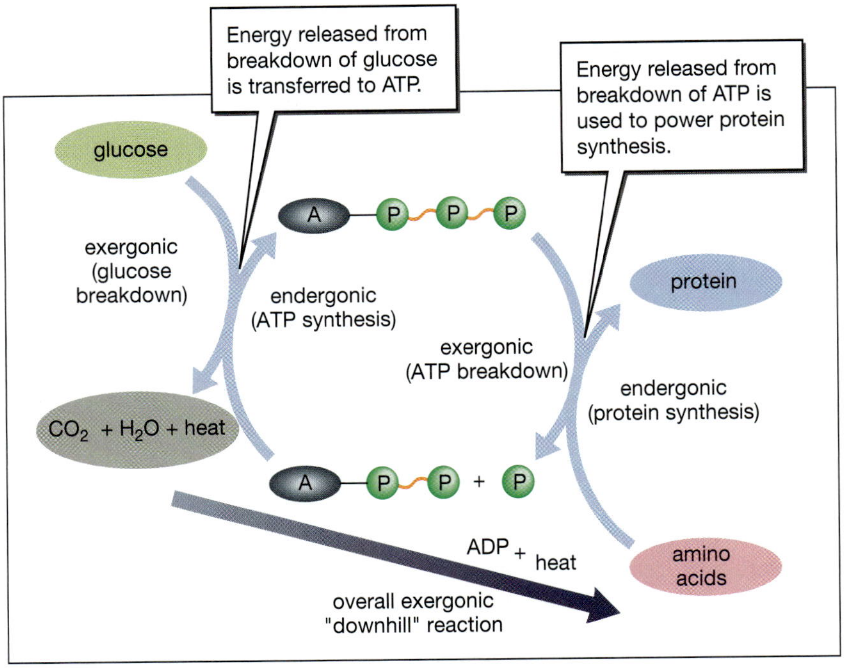

Figure 5-5 Electron carriers
An electron-carrier molecule such as NAD^+ picks up an electron generated by an exergonic reaction and holds it in a high-energy outer electron shell. The electron is then deposited, energy and all, with another molecule to drive an endergonic reaction, typically the synthesis of ATP.

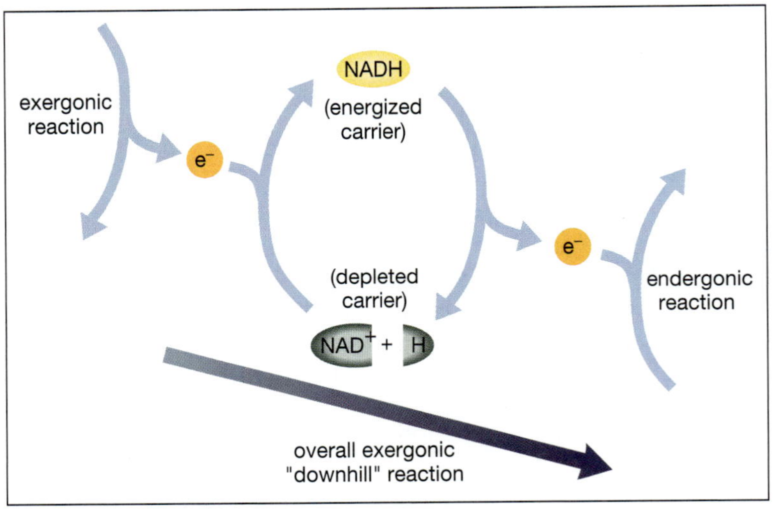

4 How Do Cells Control Their Metabolic Reactions?

A cell is a miniature chemical factory, and the multitude of different chemical reactions that take place in a cell together constitute the cell's **metabolism**. Many of these reactions are linked in sequences called **metabolic pathways** (Fig. 5-6). Photosynthesis (Chapter 6) is one such pathway. Glycolysis, the series of reactions that begin the digestion of glucose (Chapter 7), is another example.

A large number of different molecules are present in a typical cell, and this molecular diversity creates a potential problem for the cell's metabolism. With so many potentially reactive molecules present, how does the cell control reactions so that they proceed in orderly metabolic pathways? The most important way in which cells control chemical reactions is through the use of proteins called *enzymes*.

At Body Temperatures, Many Spontaneous Reactions Proceed Too Slowly to Sustain Life

In general, the speed at which a reaction occurs is determined by its activation energy (the amount of energy required to start the reaction; see Fig. 5-2). Reactions with low activation energies can proceed swiftly at body temperature, but reactions with higher activation energies occur only very slowly at such temperatures.

Many crucial metabolic reactions have activation energies high enough that they do not occur spontaneously in the body. For example, at the temperatures found in

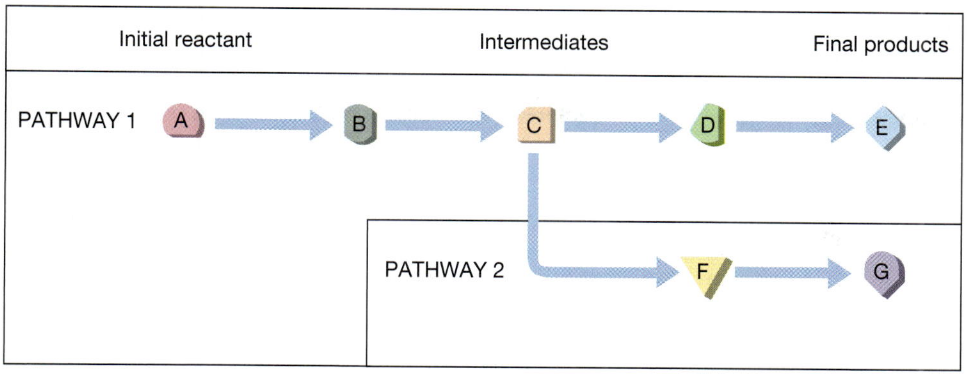

Figure 5-6 Simplified view of metabolic pathways
The original reactant molecule, A, undergoes a series of reactions. The product of each reaction serves as the reactant for the next reaction in the pathway or for a reaction in another pathway.

organisms, sugar molecules almost never spontaneously break down and give up their energy. Nonetheless, the breakdown of sugar is one of the most important energy sources for living cells. Cells are able to take advantage of the energy stored in sugar molecules because of the action of *catalysts*, especially enzymes.

Catalysts Reduce Activation Energy

Catalysts are molecules that speed up a reaction without themselves being used up or permanently altered. They speed up reactions by reducing the activation energy (Fig. 5-7). In the absence of a catalyst, a reaction with high activation energy will proceed only slowly at low temperatures, because few molecules collide hard enough to react. But when the activation energy is lowered by a catalyst, a much higher proportion of molecules move fast enough to react when they collide. Therefore, the reaction proceeds much more rapidly.

As an example of catalytic action, let's consider the catalytic converters in many automobile engines. When gasoline is burned completely, the final products are carbon dioxide and water. In engines, however, flaws in the combustion process generate other substances, including poisonous carbon monoxide (CO). In the air, carbon monoxide reacts spontaneously but slowly with oxygen to form less-harmful carbon dioxide (CO_2). This reaction eventually removes CO from the air we breathe, but many vehicles emit so much CO that the spontaneous reaction of CO with O_2 can't keep pace, and unhealthy levels of carbon monoxide accumulate. Enter the catalytic converter. Catalysts in the converter speed up the conversion of CO to CO_2, thereby reducing air pollution.

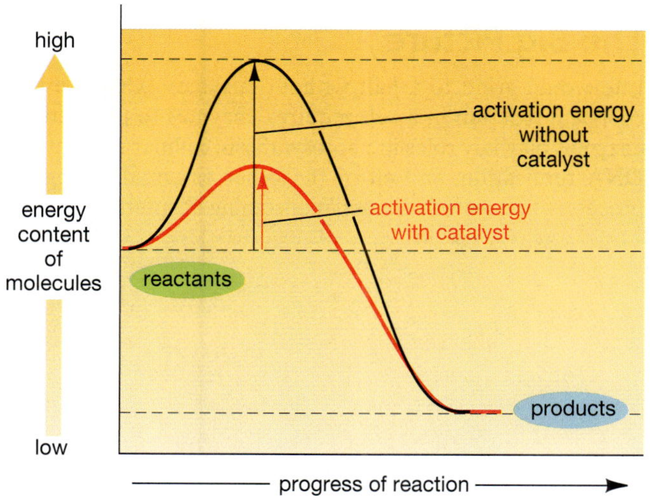

Figure 5-7 Catalysts reduce activation energy

Note three important principles about all catalysts:

- Catalysts speed up reactions.
- Catalysts can speed up only those reactions that would occur spontaneously anyway (at a much slower rate).
- Catalysts are not consumed in the reactions they promote. No matter how many reactions they participate in, the catalysts themselves are not permanently changed.

Enzymes Are Biological Catalysts

Enzymes are catalysts, usually proteins, that are made by living organisms. In addition to the three characteristics of catalysts that we just described, enzymes have additional attributes that set them apart from nonbiological catalysts. One of the most important of these is that each enzyme is usually very specific, catalyzing a single reaction or at most only a few reactions. In metabolic pathways such as those in Figure 5-6, each reaction is catalyzed by a different enzyme.

The Structure of Enzymes Allows Them to Catalyze Specific Reactions

The function of enzymes is intimately related to their structure. Each enzyme has a complex three-dimensional shape that includes a "pocket," called the **active site**, into which reactant molecules, called **substrates**, can enter.

Why are enzymes so specific? The active site of each enzyme has a distinctive shape and distribution of electrical charge. Because the enzyme and its substrate must fit together, only certain substrate molecules can enter the active site. As a result, a given enzyme can interact with only one or a small number of substrates. For example, several enzymes are required to completely digest all the proteins we eat, because each enzyme breaks apart only a specific sequence of amino acids.

How does an enzyme catalyze a reaction? When substrates enter an enzyme's active site, both the substrates and the active site change shape (Fig. 5-8). Chemical interactions between the active site and the substrate molecules may alter the chemical bonds within the substrates. These changes in the substrates promote the particular chemical reaction catalyzed by the enzyme. When the reaction between the substrates is finished, the products no longer fit properly into the active site and are expelled. The enzyme is ready to accept another set of substrates.

The breakdown or synthesis of a molecule within a cell normally occurs in many separate steps. Just as carving a staircase into a cliff allows the cliff to be climbed one small step at a time, breaking a reaction into small steps (each with a low activation energy), allows the overall reaction to surmount its high total activation energy. Each step is catalyzed by a different enzyme, which lowers the activation energy for that particular step (see Fig. 5-7). For example, our cells break down sugar in many small steps, with each step liberating a small amount of energy. This step-by-step procedure is the reason that the reaction can proceed at body temperature, and we are not roasted by the sugar that is "burning" in our bodies.

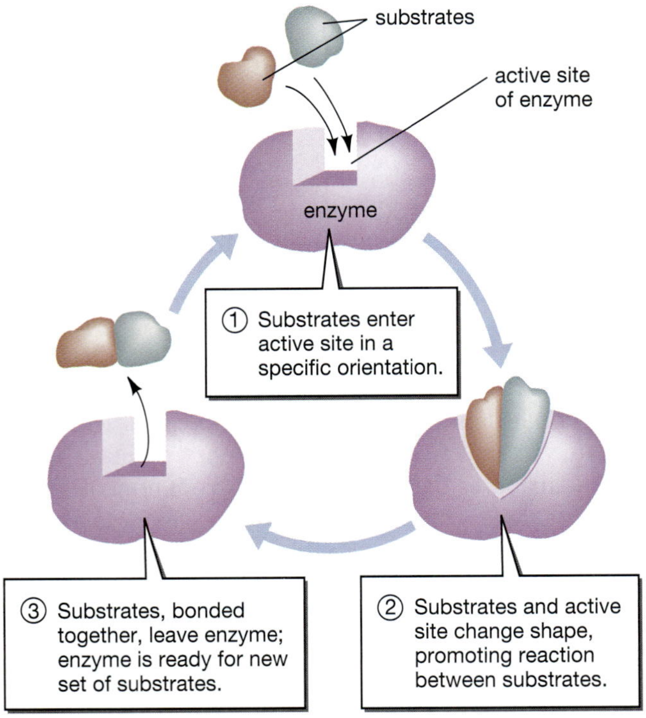

Figure 5-8 The cycle of enzyme-substrate interactions

The Activity of Enzymes Is Influenced by Their Environment

The complex three-dimensional structure of an enzyme is required for proper function, but the shape is also sensitive to changes in environmental conditions. The hydrogen bonds that play a key role in determining enzyme structure can be altered by their chemical and physical surroundings. Each enzyme functions optimally at a particular pH, salt concentration, and temperature, and deviations from these optimal conditions can change an enzyme's shape and destroy its effectiveness. For example, one reason that dill pickles stay well preserved in a vinegar-salt solution is that that salty, acidic conditions in the solution distort the shape of enzymes used in bacterial metabolism. With their enzymes inoperative, the bacteria that ordinarily cause food to spoil cannot survive.

When temperatures rise too high, the hydrogen bonds that determine enzyme shape may be broken apart and the enzyme rendered useless. Smaller changes in temperature, however, may simply change the rate at which enzyme-catalyzed reactions proceed. At higher temperatures, substrate molecules move more rapidly and their random movements are more likely to bring them into contact with the active site of an appropriate enzyme. Thus, many reactions are accelerated by moderately higher temperatures and slowed by lower temperatures.

The Big Picture

Life would grind to a halt without enzymes. Almost every chemical reaction in a cell requires enzymes to proceed, so enzymes play key roles in photosynthesis, cellular respiration, DNA replication, and all of the other essential biological processes that we will discuss in upcoming chapters.

CASE STUDY REVISITED

A Missing Molecule Makes Mischief

The molecule missing from people with PKU is an enzyme. PKU is caused by the absence of the enzyme phenylalanine hydroxylase. This enzyme normally catalyzes a key reaction in the metabolic pathway of phenylalanine, namely its conversion to another amino acid, tyrosine. In the absence of the hydroxylase enzyme, the activation energy of the conversion reaction remains too high for the reaction to proceed spontaneously. The reactant phenylalanine is not converted to the product tyrosine, no matter how high the concentration of reactant rises. The reactant instead accumulates, reaching toxic levels.

Newborns in virtually all industrialized countries are now routinely tested for PKU, so those born with the disease can be placed on a special low-protein diet right away. As a result, few children in these countries suffer the devastating damage that they would have experienced in earlier times. Still, the special diet can be a struggle to maintain, and some PKU patients fail to stick to the dietary strictures after they reach adolescence or adulthood (with sometimes unfortunate health consequences). A more efficient, less disruptive treatment would be welcome. A good first step in developing such a treatment would be to increase our understanding of the pathways that link the failure to produce the functional enzyme to the developmental damage. Scientists at several labs around the world are currently working to understand this connection.

Summary of Key Concepts

1) What Is Energy?

Energy is the capacity to do work. Kinetic energy is the energy of movement (light, heat, electricity, movement of large particles). Potential energy is stored energy (chemical energy, positional energy). The first law of thermodynamics states that energy can be neither created nor destroyed, so the total amount of energy remains constant, although it may change in form. The second law of thermodynamics states that any use of energy causes a decrease in the quantity of concentrated, useful energy and an increase in the randomness and disorder of matter. Entropy is a measure of disorder within a system.

2) How Does Energy Flow in Chemical Reactions?

Chemical reactions fall into two categories. In exergonic reactions, the product molecules have less energy than do the reactant molecules, so the reaction releases energy. In endergonic reactions, the products have more energy than do the reactants, so the reaction requires an input of energy. Exergonic reactions can occur spontaneously, but all reactions, including exergonic ones, require an input of activation energy to overcome electrical repulsions between reactant molecules. In a coupled reaction, the energy liberated by an exergonic reaction drives the endergonic reaction. Organisms couple exergonic reactions, such as light-energy capture or sugar metabolism, with endergonic reactions, such as the synthesis of organic molecules.

3) How Is Energy Carried Between Coupled Reactions?

Energy released by chemical reactions within a cell is captured and transported within the cell by energy-carrier molecules, such as ATP and electron carriers. These molecules are the major means by which cells couple exergonic and endergonic reactions that occur at different places in the cell.

4) How Do Cells Control Their Metabolic Reactions?

Cellular reactions are linked in metabolic pathways, and are regulated in large measure by enzymes. High activation energies cause many reactions, even exergonic ones, to proceed very slowly under normal environmental conditions. Catalysts lower the activation energy and thereby speed up chemical reactions without being permanently changed themselves. Organisms synthesize enzymes that catalyze one or a few reactions. The reactants temporarily bind to the active site of the enzyme, making it easier to form the new chemical bonds of the products. Environmental conditions including pH, salt concentration, and temperature can promote or inhibit the function of enzymes, by altering their three-dimensional structure.

Key Terms

activation energy *p. 72*
active site *p. 75*
adenosine diphosphate (ADP) *p. 73*
adenosine triphosphate (ATP) *p. 72*
catalyst *p. 75*
chemical reaction *p. 71*
coupled reaction *p. 72*
electron carrier *p. 73*
endergonic reaction *p. 71*
energy *p. 69*
energy-carrier molecule *p. 72*
entropy *p. 70*
enzyme *p. 75*
exergonic reaction *p. 71*
first law of thermodynamics *p. 70*
kinetic energy *p. 69*
metabolic pathway *p. 74*
metabolism *p. 74*
potential energy *p. 69*
product *p. 71*
reactant *p. 71*
second law of thermodynamics *p. 70*
substrate *p. 75*

Thinking Through the Concepts

Multiple Choice

1. *According to the first law of thermodynamics, the total amount of energy in the universe*
   a. is always increasing
   b. is always decreasing
   c. varies up and down
   d. is constant
   e. cannot be determined
2. *What is predicted by the second law of thermodynamics?*
   a. Energy is always decreasing.
   b. Disorder cannot be created or destroyed.
   c. Systems always tend toward greater states of disorder.
   d. All potential energy exists as chemical energy.
   e. all of the above
3. *Which statement about exergonic reactions is true?*
   a. The products have more energy than do the reactants.
   b. The reactants have more energy than do the products.
   c. They will not proceed spontaneously.
   d. Energy input reverses entropy.
   e. none of the above
4. *ATP is important in cells because*
   a. it transfers energy from exergonic reactions to endergonic reactions
   b. it is assembled into long chains that make up cell membranes
   c. it acts as an enzyme
   d. it accelerates diffusion
   e. all of the above
5. *How does an enzyme increase the speed of a reaction?*
   a. by changing an endergonic to an exergonic reaction
   b. by providing activation energy
   c. by lowering the activation energy
   d. by decreasing the concentration of reactants
   e. by increasing the concentration of products
6. *Which of the following statements about enzymes is (are) true?*
   a. They interact with specific reactants (substrates).
   b. Their three-dimensional shapes are closely related to their activities.
   c. They change the shape of the reactants.
   d. They have active sites.
   e. all of the above

? Review Questions

1. Explain why organisms do not violate the second law of thermodynamics. What is the ultimate energy source for most forms of life on Earth?
2. Define *metabolism*, and explain how reactions can be coupled to one another.
3. What is activation energy? How do catalysts affect activation energy? How does this change the rate of reactions?
4. Describe some exergonic and endergonic reactions that occur in plants and animals very regularly.
5. Describe the structure and function of enzymes.

Applying the Concepts

1. A preview question for ecology: When a brown bear eats a salmon, does the bear acquire all the energy contained in the body of the fish? Why or why not? What implications do you think this answer would have for the relative abundance (by weight) of predators and their prey? Does the second law of thermodynamics help explain the title of the book *Why Big, Fierce Animals Are Rare*?
2. Many people in sub-Saharan Africa have experienced the effects of malnutrition and starvation, but the very young are most severely affected. Some individuals suffer permanent disability even if food is provided. How could a lack of food intake interfere with functions of individual cells and tissues? Which tissues are likely to suffer the most irreversible damage?
3. As you learned in Chapter 3, the subunits of virtually all organic molecules are joined by condensation reactions and can be broken apart by hydrolysis reactions. Why, then, does your digestive system produce separate enzymes to digest proteins, fats, and carbohydrates—in fact, several of each type?
4. Suppose someone tried to disprove the existence of evolution with the following argument: "According to evolutionary theory, organisms have increased in complexity through time. However, evolution of increased biological complexity contradicts the second law of thermodynamics. Therefore, evolution is impossible." How would you respond to this argument in support of evolution?

Group Activity

Shake hands with the student next to you, noting the relative temperature of his or her hand. Now, rub your own hands together and clasp and unclasp them for 30 seconds. Shake hands again; you should detect a warmer handshake this time. Discuss how the additional heat was generated.

For More Information

Baker, J. J. W., and Allen, G. E. *Matter, Energy, and Life*. 4th ed. Reading, MA: Addison-Wesley, 1981. An excellent introduction to chemical-energy principles for those interested in biology.

Farid, R. S. "Enzymes Heat Up." *Science News*, May 9, 1998. Scientists explore new ways to synthesize enzymes that will function at high temperatures.

Fenn, J. *Engines, Energy, and Entropy*. New York: W. H. Freeman, 1982. Elegantly simple introduction to the laws of thermodynamics and their relationship to everyday life.

Answers to Multiple-Choice Questions

1. d 2. c 3. b 4. a 5. c 6. e

MediaTutor

CD Activities

Activity 5.1: Energy and Chemical Reactions

Estimated time: 5 minutes

The cell is a chemical factory. To understand it, we must understand its basic chemistry, including the concepts of energy, energy conversions, and the laws of thermodynamics.

Activity 5.2: Energy and Life

Estimated time: 10 minutes

The concepts of thermodynamics introduced in the last tutorial are applied to living systems in this tutorial. The central importance of ATP as an energy carrier is detailed, electron carriers are introduced, and the importance of enzymes is described.

Start the MediaTutor Student CD-ROM and enter the activity number in the Quick Search box to be taken directly to that activity.

Web Investigation

A Missing Molecule Makes Mischief

Estimated time: 15 minutes

Cells are like tiny factories that use energy to manufacture chemicals. In this analogy, enzymes act as essential tools in the cellular assembly lines. When one of these key tools is missing, the effects can be devastating, as illustrated in this chapter's Case Study. This activity offers further exploration of the role of enzymes.

Go to http://www.prenhall.com/audesirk3, the Audesirk Companion Web site. Select Chapter 5 and the Web Investigation to begin.

". . . the cells of life bound themselves into clans, a multitude of cells
To make one being—as the molecules before
Had made of many one cell. Meanwhile they had invented
Chlorophyll and ate sunlight, cradled in peace
On the warm waves."

Robinson Jeffers in* The Beginning and the End *(1963)

As *Tyrannosaurus* threatens *Triceratops*, a giant meteor streaks toward Earth. The meteor's impact will drastically reduce photosynthesis, with disastrous consequences for the dinosaurs.

6 Capturing Solar Energy: Photosynthesis

AT A GLANCE

CASE STUDY

Did the Dinosaurs Die from Lack of Sunlight?

About 65 million years ago, life on Earth suffered a catastrophic blow. Within a short period, most of the species on the planet were destroyed. This devastating mass extinction eliminated more than 70% of the species then existing, including the dinosaurs. *Triceratops*, *Tyrannosaurus*, and all the other dinosaur species disappeared forever. Land and sea were nearly emptied of life, and it would be many millions of years before new species arose to take the places of those that had disappeared.

Most scientists believe that this devastation began when a gigantic meteorite, 6 miles in diameter, entered the atmosphere and crashed into Earth. As the meteorite plowed into the ocean at the tip of the Yucatan Peninsula, it dug a crater a mile deep and 120 miles wide. Any organism in the immediate area was, of course, immediately killed by the blast wave from the impact. This kind of direct destruction, however, must have been limited to a relatively small area. How, then, did the meteorite impact eliminate thousands of species in all corners of the globe? In all likelihood, the most serious damage was done not by the meteorite itself, but by the lingering aftereffects of its sudden arrival. In particular, the meteorite's most damaging effect was its disruption of the most important chemical reaction on Earth: photosynthesis. ■

1 What Is Photosynthesis?

More than 2 billion years ago, some cells acquired the ability to harness the energy in sunlight, using a process known as photosynthesis. These cells captured some of sunlight's energy, and stored it as chemical energy. Exploiting this new source of energy with little competition, early photosynthetic cells filled the seas. The form of photosynthesis that eventually became most common (and that is the dominant form today) uses carbon dioxide and water as its raw materials and releases oxygen gas as a by-product.

Before the advent of photosynthesis, oxygen gas was more or less absent from the atmosphere. As photosynthetic organisms grew in number, however, oxygen began to accumulate in the atmosphere. Today, most forms of life on Earth depend on the oxygen produced by photosynthesis. Equally important to life is the energy captured by photosynthetic organisms and stored in sugar and other organic molecules. Virtually every living thing depends on this energy, which provides nourishment for photosynthetic organisms and for the other organisms that consume them.

MEDIATUTOR 6.1 Photosynthesis

Photosynthesis Converts Carbon Dioxide and Water to Glucose

Starting with the simple molecules of carbon dioxide (CO_2) and water, **photosynthesis** converts the energy of sunlight into chemical energy stored in the bonds of glucose ($C_6H_{12}O_6$); it also releases oxygen. When written in its simplest form, the overall chemical reaction for photosynthesis is:

$$6\ CO_2 + 6\ H_2O + \text{light energy} \rightarrow C_6H_{12}O_6 + 6\ O_2$$

This basic photosynthetic reaction takes place in almost all photosynthetic organisms, which include plants, seaweeds, and various types of single-celled organisms. Each group of photosynthetic organisms, however, exhibits distinctive variations on the basic process. In this chapter, we will focus on photosynthesis in plants.

Plant Photosynthesis Takes Place in Leaves

Leaves have several features that make them especially well suited for their role as the main location of photosynthesis. For example, the leaves of most land plants are only a few cells thick, which ensures that sunlight can easily penetrate them (Fig. 6-1a). In addition, the flattened shape of leaves exposes a large surface area to the sun.

A leaf obtains CO_2 for photosynthesis from the air, through adjustable pores in the leaf surface called **stomata** (singular, **stoma**). Stomata can open and close, and they open at appropriate times to admit CO_2. Inside the

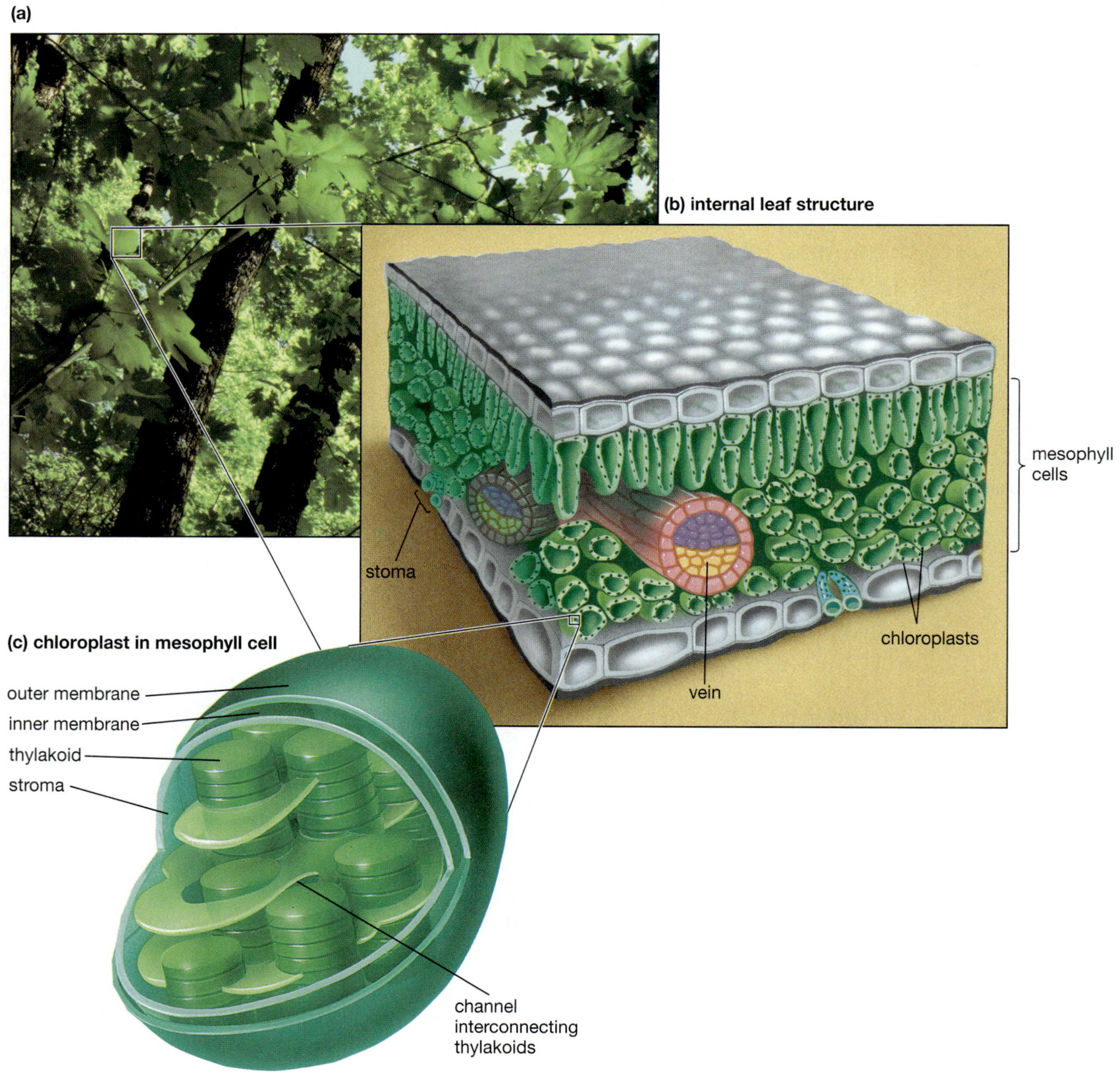

Figure 6-1 An overview of photosynthetic structures

leaf are a few layers of cells collectively called **mesophyll**; photosynthesis occurs principally in these cells (Fig. 6-1b). A system of veins supplies water and minerals to the mesophyll cells and carries the sugars they produce to other parts of the plant.

Leaf Cells Contain Chloroplasts

Leaf cells, especially mesophyll cells, contain large numbers of **chloroplasts**, the organelles in which photosynthesis occurs. Chloroplasts have a double outer membrane that encloses a semifluid medium, the **stroma** (Fig. 6-1c). Embedded in the stroma are disk-shaped, interconnected membranous sacs called **thylakoids**. The light-dependent reactions of photosynthesis occur within the membranes of the thylakoids.

Photosynthesis Consists of Light-Dependent and Light-Independent Reactions

The simple equation that summarizes the overall reaction of photosynthesis obscures the fact that photosynthesis actually requires dozens of steps, each catalyzed by a different enzyme. The steps can be divided into two groups, the light-dependent reactions and the light-independent reactions. The two groups of reactions occur at different locations within the chloroplast, but are linked by energy-carrier molecules.

In the light-dependent reactions, molecules in the membranes of the thylakoids capture sunlight energy and convert some of it into chemical energy stored, for the short term, in two different energy-carrier molecules. The molecules are the familiar energy carrier **adenosine triphosphate (ATP)** and the electron carrier **nicotinamide adenine dinucleotide phosphate (NADPH)**. Oxygen gas is released as a by-product. In the light-independent reactions, enzymes in the stroma use the chemical energy of the carrier molecules in the synthesis of glucose or other organic molecules.

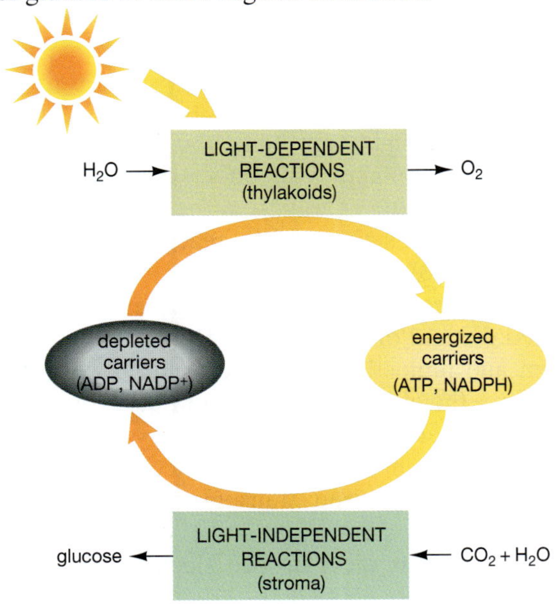

The Big Picture

Photosynthesis is absolutely essential to life on Earth. Virtually all the energy used by living things is captured by photosynthesis, and all the oxygen in Earth's atmosphere was produced by photosynthesis.

2 How Is Light Energy Converted to Chemical Energy?

In the first stage of photosynthesis, the light-dependent reactions convert the energy of sunlight into chemical energy stored in carrier molecules. The light-dependent reactions, however, do not use the full range of electromagnetic radiation emitted by the sun. The sun emits energy as an array of electromagnetic waves of various sizes (the size of a wave is measured by its wavelength). That is, the energy that Earth receives from the sun is a portion of the electromagnetic spectrum (the full spectrum ranges from short-wavelength gamma rays, through ultraviolet, visible, and infrared light, to very long wavelength radio waves). Of the range of wavelengths emitted by the sun, photosynthesis uses only those that fall into the range of visible light (Fig. 6-2a). Note that, within the range of visible light, we perceive different wavelengths as different colors.

Light Energy Is First Captured by Pigments in Chloroplasts

The job of absorbing light energy for use in photosynthesis falls to certain molecules in the thylakoid membranes of chloroplasts. Thylakoid membranes contain several types of light-absorbing molecules (pigments), each of which absorbs a different range of wavelengths. **Chlorophyll**, the key light-capturing molecule, strongly absorbs violet, blue, and red light but reflects green. The reflected wavelengths reach our eyes, and we see chlorophyll as green (Fig. 6-2b). Most leaves appear green because they are rich in chlorophyll.

Thylakoids also contain other molecules, called accessory pigments, that capture light energy and transfer it to chlorophyll. Carotenoids absorb blue and green light; phycocyanins absorb green and yellow. All wavelengths of light are thus absorbed to some degree by either chlorophyll, carotenoids, or phycocyanins, so all wavelengths of visible light can drive photosynthesis in plants with those pigments (Fig. 6-2c).

The Light-Dependent Reactions Generate Energy-Carrier Molecules

The light-dependent reactions take place in special locations called photosystems, which are located in the

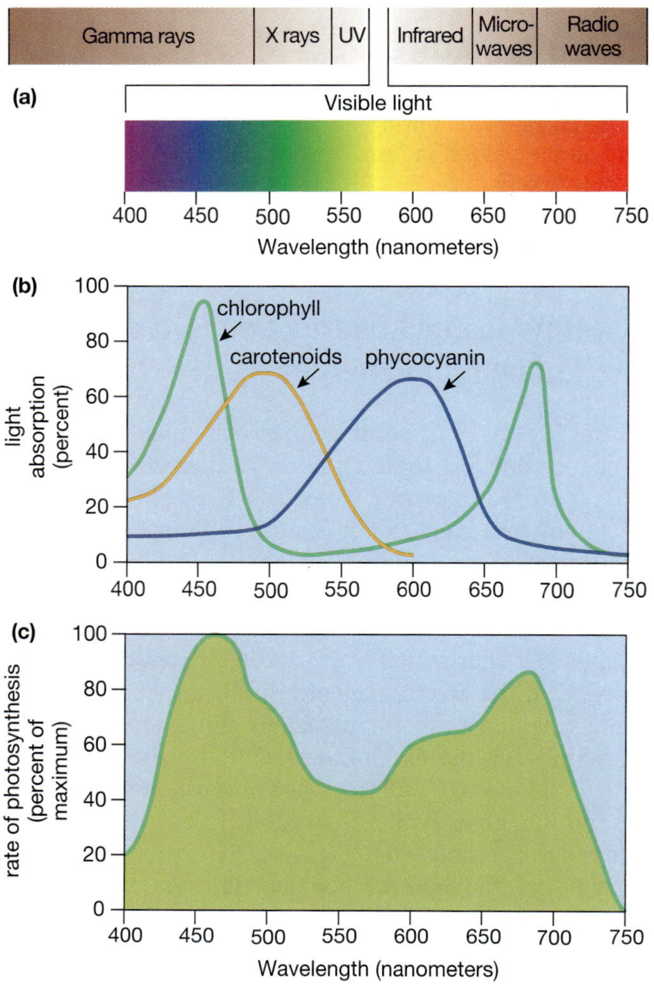

Figure 6-2 Light, chloroplast pigments, and photosynthesis
(a) Visible light, a small part of the electromagnetic spectrum (top line), consists of wavelengths that correspond to the colors of the rainbow. ***(b)*** Different pigments selectively absorb certain colors; the height of the curves represents the ability to absorb light of the particular color. Chlorophyll strongly absorbs violet, blue, and red light. The other pigments in chloroplasts absorb mostly other colors of light. ***(c)*** Photosynthesis is driven to some extent by all colors of light, because collectively, the pigments in chloroplasts absorb some of each visible wavelength.

thylakoid membranes. Each **photosystem** consists of a highly organized assemblage of proteins, chlorophyll, accessory pigment molecules, and electron-carrier molecules. In each thylakoid, there are thousands of photosystems of two types: photosystem I and photosystem II.

Each photosystem consists of two major parts. The first part, the **light-harvesting complex**, contains about 300 chlorophyll and accessory pigment molecules. These molecules absorb light and pass the energy to a specific chlorophyll molecule called the **reaction center**. The reaction-center chlorophyll is located next to the second part of the photosystem: the **electron transport system**, a series of electron carrier molecules also embedded in the thylakoid membrane.

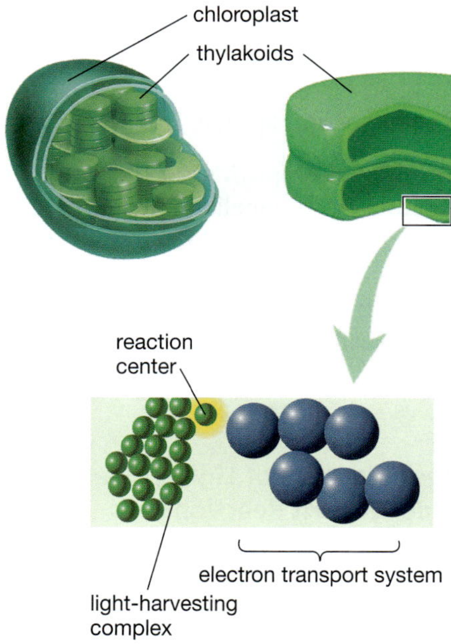

Photosystem II Generates ATP

As you read the following sections about the photosystems, you may wish consult Figure 6-3 to help you follow the steps. ① The light-dependent reactions begin when the light-harvesting complex of photosystem II absorbs energy from light. The energy passes from molecule to molecule until it reaches the reaction center. ② When the reaction-center chlorophyll molecule receives the energy, electrons absorb the energy and jump from the chlorophyll molecule over to ③ the adjacent electron transport system. After arriving at the electron transport system, these energetic electrons keep jumping, moving from one carrier molecule to the next. During some of the jumps, the electrons release energy. ④ This released energy powers reactions that synthesize the energy carrier ATP.

Photosystem I Generates NADPH

⑤ Meanwhile, light rays have also been striking the light-harvesting complex of photosystem I. Just as in photosystem II, the energy from the light striking photosystem I is transmitted to the reaction-center chlorophyll, which ⑥ ejects electrons. Almost immediately, these lost electrons are replaced by electrons that have reached the end of photosystem II's electron transport system and jumped over to photosystem I. ⑦ At the same time, the ejected photosystem I electrons jump to the photosystem I electron transport system. The electrons move from molecule to molecule through the photosystem I electron transport system until they reach the electron carrier molecule $NADP^+$. ⑧ Each $NADP^+$ molecule picks up two electrons and one hydrogen ion, forming NADPH.

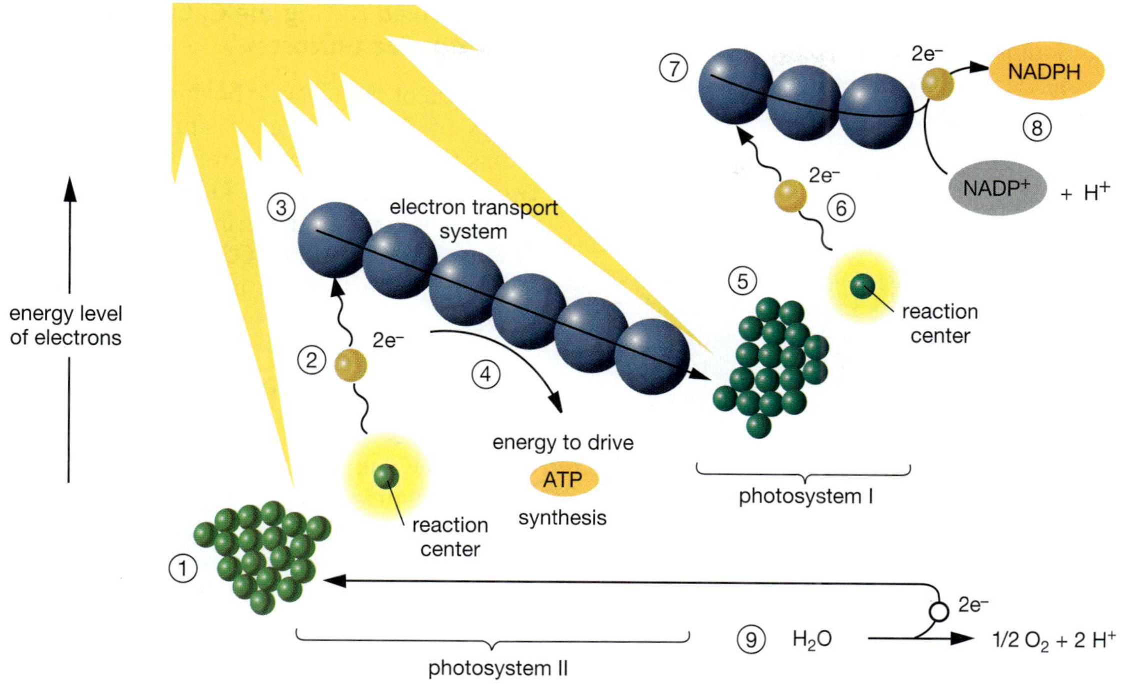

Figure 6-3 The light-dependent reactions of photosynthesis

Splitting Water Maintains the Flow of Electrons Through the Photosystems

Overall, electrons flow from the reaction center of photosystem II, through the photosystem II electron transport system, to the reaction center of photosystem I, through the photosystem I electron transport system, and on to form NADPH. To sustain this one-way flow of electrons, photosystem II's reaction center must be continuously supplied with new electrons to replace the ones it gives up. These replacement electrons come from water (step ⑨ in Fig. 6-3) by the reaction:

$$H_2O \rightarrow \frac{1}{2} O_2 + 2\,H^+ + 2e^-$$

As the water molecules are split, their oxygen atoms combine to form molecules of oxygen gas, O_2. The oxygen may be used directly by the plant in its own respiration or given off to the atmosphere.

Summary: Light-Dependent Reactions

① The light-dependent reactions begin when light is absorbed by the light-harvesting complex of photosystem II. ② The absorbed energy causes electrons from the reaction center of the complex to be ejected. ③ The ejected electrons are transferred to photosystem II's electron transport system. As the electrons pass through the transport system, they release energy. ④ Some of the energy is used to drive ATP synthesis. ⑤ Meanwhile, light is absorbed by the light-harvesting complex of photosystem I. ⑥ The absorbed energy ejects electrons from the reaction center. ⑦ The ejected photosystem I electrons are picked up by photosystem I's electron transport system. ⑧ Some of the energy released as electrons move through the electron transport system is captured as NADPH. ⑨ A water molecule splits, donating electrons to the photosystem II chlorophyll and generating oxygen as a by-product. The products of the light-dependent reactions are NADPH, ATP, and O_2.

3 How Is Chemical Energy Stored in Glucose Molecules?

The ATP and NADPH molecules generated in the light-dependent reactions are used in the light-independent reactions to synthesize molecules for the long-term storage of energy. The light-independent reactions, so called because they can occur independently of light as long as ATP and NADPH are available, take place in the fluid stroma that surrounds the thylakoids. There, the reactions synthesize glucose from carbon dioxide and water, a process that requires enzymes that are present in the stroma.

The C_3 Cycle Captures Carbon Dioxide

The first step on the path to glucose is to capture carbon dioxide. It is captured in a set of reactions known as the **C_3 cycle** (three-carbon cycle) because some of the important molecules in the cycle have three carbon atoms in them; the reactions are also sometimes called the Calvin-Benson cycle. The C_3 cycle requires the five-carbon sugar **ribulose biphosphate (RuBP)**, CO_2 (normally from the air), enzymes to catalyze all the reactions, and energy in the form of ATP and NADPH (normally from the light-dependent reactions).

The C_3 cycle begins (and ends) with RuBP. Six molecules of RuBP combine with CO_2 in a series of reactions that yields 12 three-carbon molecules of **phosphoglyceric acid (PGA)**. In effect, the CO_2 molecules are captured from the atmosphere, a process known as **carbon fixation**, because it "fixes" gaseous CO_2 into a relatively stable organic molecule, PGA (step ① in Fig. 6-4).

As the cycle proceeds, the 12 molecules of PGA are converted to 12 molecules of **glyceraldehyde-3-phosphate (G3P)**; step ② in Fig. 6-4), using energy donated by ATP and NADPH (recall that these energy carriers were generated during the light-dependent reactions). Then, through a series of reactions requiring ATP energy, 10 of the G3P molecules are used to regenerate the 6 molecules of RuBP needed to re-start the cycle (step ③ in Fig. 6-4).

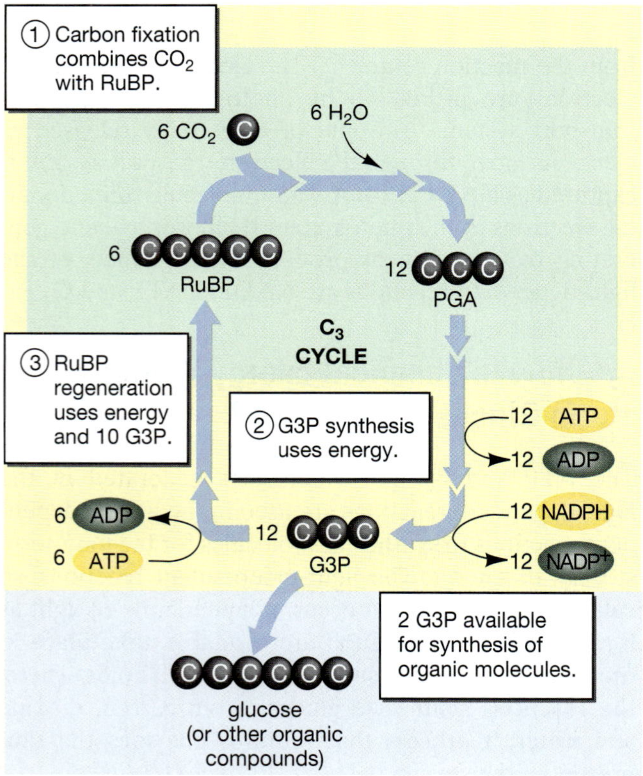

Figure 6-4 The C_3 cycle of carbon fixation
Keep track of the carbons as you follow the illustration around. ① Six molecules of RuBP react with 6 molecules of CO_2 and 6 molecules of H_2O to form 12 molecules of PGA. ② The energy of 12 ATPs and the electrons and hydrogens of 12 NADPHs are used to convert the 12 PGA molecules to 12 G3Ps. ③ Energy from 6 ATPs is used to rearrange 10 G3Ps into 6 RuBPs, completing one turn of the C_3 cycle. The remaining 2 G3P molecules are further processed into glucose.

Carbon Fixed During the C_3 Cycle Is Used to Synthesize Glucose

Each turn of the C_3 cycle begins and ends with six molecules of RuBP but also captures additional carbon atoms from CO_2. These extra carbons are found in the two "extra" molecules of G3P that are left over after RuBP is regenerated. As indicated in the lower panel of Figure 6-4, these two G3P molecules (three carbons each) combine to form one molecule of glucose (six carbons).

Summary: Light-Independent Reactions

Each passage through the C_3 cycle yields one molecule of glucose and depleted energy carriers (ADP and $NADP^+$) that will be recharged during light-dependent reactions. In the cycle, 6 molecules of RuBP capture 6 molecules of CO_2 to produce 12 molecules of PGA. A series of reactions driven by energy from ATP and NADPH uses the PGA to produce 12 molecules of G3P. Ten of these G3P molecules are used to regenerate the 6 RuBP molecules. Thus, 2 molecules of G3P are left over, and they join to form 1 molecule of glucose.

4 What Is the Relationship Between Light-Dependent and Light-Independent Reactions?

As you learn some of the details of the light-dependent and light-independent reactions of photosynthesis, it is important to not lose sight of the close interdependence of the two sets of reactions. The connection between the light-dependent and light-independent reactions is illustrated in Figure 6-5; you might also wish to review the figure on page 83. Stated simply, the "photo" part of photosynthesis refers to the capture of light energy by the light-dependent reactions. The "synthesis" refers to the light-independent reactions' use of this captured energy to synthesize glucose. Stated in more detail, the light-dependent reactions in the membranes of the thylakoids use light energy to "charge up" the energy-carrier molecules ADP and $NADP^+$ to form ATP and NADPH. These energized carriers move to the stroma, where their energy is used to drive the light-independent reactions that synthesize glucose. The depleted carriers, ADP and $NADP^+$, then return to the light-dependent reactions for recharging to ATP and NADPH.

The Big Picture

Photosynthesis converts the energy of electromagnetic radiation to chemical energy in the bonds of energy-carrier molecules. Unlike electromagnetic energy, the chemical energy can be used to power the metabolic reactions of cells.

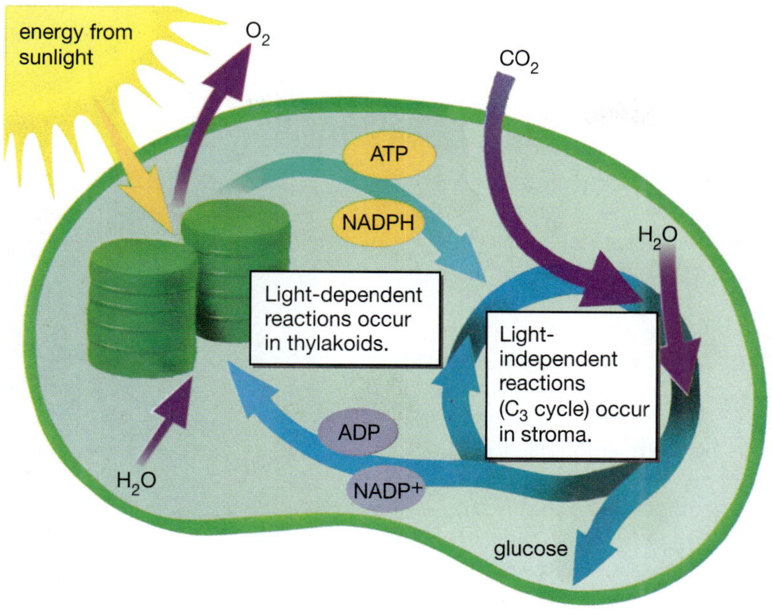

Figure 6-5 Two sets of reactions are connected in photosynthesis

5 How Does the Need to Conserve Water Affect Photosynthesis?

Photosynthesis requires carbon dioxide. Therefore, you might think that an ideal leaf would be very porous to allow lots of CO_2 to enter the leaf from the air. For land plants, however, being porous to CO_2 also allows water to evaporate easily from the leaf. Loss of water from leaves can be stressful for land plants and may even be fatal. Leaf structure, therefore, must balance the need to obtain CO_2 with the need to retain water.

One solution to the problem of water loss has been the evolution of stomata, pores in leaves that can be opened and closed. When water supplies are adequate, the stomata open, letting in CO_2. If the plant is in danger of drying out, the stomata close. Closing the stomata reduces evaporation but, unfortunately, it also reduces CO_2 intake and restricts the release of O_2, which is produced during photosynthesis.

When Stomata Are Closed to Conserve Water, Wasteful Photorespiration Occurs

When the stomata close, CO_2 levels drop, O_2 levels rise, and photosynthesis is hampered. The main problem is that RuBP can combine with *either* CO_2 *or* O_2 (Fig. 6-6a). When RuBP combines with O_2, rather than CO_2, a wasteful process called **photorespiration** takes place. Photorespiration does not produce any useful cellular energy, and it prevents the C_3 pathway from synthesizing glucose. Thus, photorespiration undermines the plant's ability to fix carbon. During hot, dry weather, when the stomata seldom open, photorespiration dominates and plants may die because they are unable to capture enough energy.

An Alternative Pathway Reduces Photorespiration in C_4 Plants

Some plants, described as *C_4 plants*, have evolved a way to reduce photorespiration and boost photosynthesis during dry weather. These plants, which include species such as corn and crabgrass that thrive in relatively hot and dry conditions, use a two-stage carbon-fixation pathway, the **C_4 pathway**.

In C_4 plants, the mesophyll cells contain a three-carbon molecule called **phosphoenolpyruvate (PEP)** instead of RuBP. Carbon dioxide reacts with PEP to form the four-carbon molecule **oxaloacetate** (for which C_4 plants are named). This reaction is highly specific for CO_2 and is not hindered by high O_2 concentrations. Oxaloacetate, with its payload of carbon, is then shuttled from mesophyll cells to another group of cells known as **bundle-sheath cells** (Fig. 6-6b). In the bundle-sheath cells, oxaloacetate breaks down, releasing CO_2 again. This creates a high CO_2 concentration in the bundle-sheath cells, where the regular C_3 cycle can now proceed with less competition from oxygen. The remnant of the shuttle molecule (pyruvic acid) returns to the mesophyll cells, where ATP energy is used to regenerate the PEP molecule.

Earth Watch
Can Forests Prevent Global Warming?

The earth's climate is growing warmer. Although no one can say with certainty what the consequences of this warming trend will be, many biologists and climate scientists fear that a hotter future climate will place extraordinary stress on Earth's inhabitants, including humans. The main cause of global warming appears to be a buildup in the atmosphere of *greenhouse gases*, which cause Earth to retain more heat (a phenomenon described in more detail in Chapter 29). The increase of greenhouse gases stems from the activities of our industrial society, especially from the tremendous amounts of carbon dioxide (a greenhouse gas) that we are adding to the atmosphere from the combustion of wood, gasoline, coal, and other carbon-containing fuels.

In recent years, the looming threat of global warming has prompted governments to seek a worldwide strategy for reducing carbon dioxide emissions. Negotiations have frequently been contentious, as each nation seeks to avoid commitments that would put it at an economic disadvantage. One controversial proposal would have forests counted as offsetting carbon dioxide emissions. Under this idea, a country that agreed to reduce emissions by a certain amount would be able to count its forested area toward that total. After all, the reasoning goes, forests are the site of a great deal of photosynthesis, and photosynthesis consumes carbon dioxide. Consuming carbon dioxide could be just as helpful as reducing the amount produced.

The question of how forests should be counted in carbon dioxide budgets depends on whether forests really do absorb more carbon dioxide than they give off. Although trees undeniably photosynthesize, they also respire—that is, they liberate cellular energy in a process that consumes oxygen and emits carbon dioxide (the subject of the next chapter). And we must also consider respiration by the animals that live in a forest, as well as respiration by the huge number of microorganisms that inhabit forest soils (Fig. E6-1).

Researchers have shown that mature forests do indeed absorb more carbon dioxide than they give off. But, as is so often the case in biology, the situation is not as simple as it first appeared. For example, some research indicates that, as temperature increases, both plants and soil microorganisms respire more rapidly. In a warmer future world, forests might produce more carbon dioxide than they consume. Some scientists have argued, however, that increased respiration might be offset by increased photosynthesis. A recent experiment raises doubts about this scenario. In the experiment, researchers increased the carbon dioxide level in a patch of forest. The trees in the patch experienced a burst of photosynthesis and growth, but only for a short period. The growth spurt quickly used up nutrients in the soil; with nutrients in short supply, the trees could no longer take advantage of the extra carbon dioxide.

Can forests prevent global warming? Scientists are working hard to uncover the answer to this question. For now, though, the only real answer is that it's too soon to tell.

Figure E6-1 Forests both consume and emit carbon dioxide Forest plants photosynthesize, but they also respire. Other forest organisms, such as the mushrooms visible here, also respire.

C_3 and C_4 Plants Are Each Adapted to Different Environmental Conditions

Plants that use the C_4 process to fix carbon are locked into this pathway, which uses up more energy to produce glucose than the C_3 pathway. Nonetheless, C_4 plants have an advantage when light energy is abundant but water is scarce. Consequently, C_4 plants thrive in deserts and in midsummer in temperate climates. However, in places where light levels are low or where water is plentiful and stomata can stay open, the C_3 carbon fixation pathway is more efficient. C_3 plants thus have the advantage in cool, wet, cloudy climates.

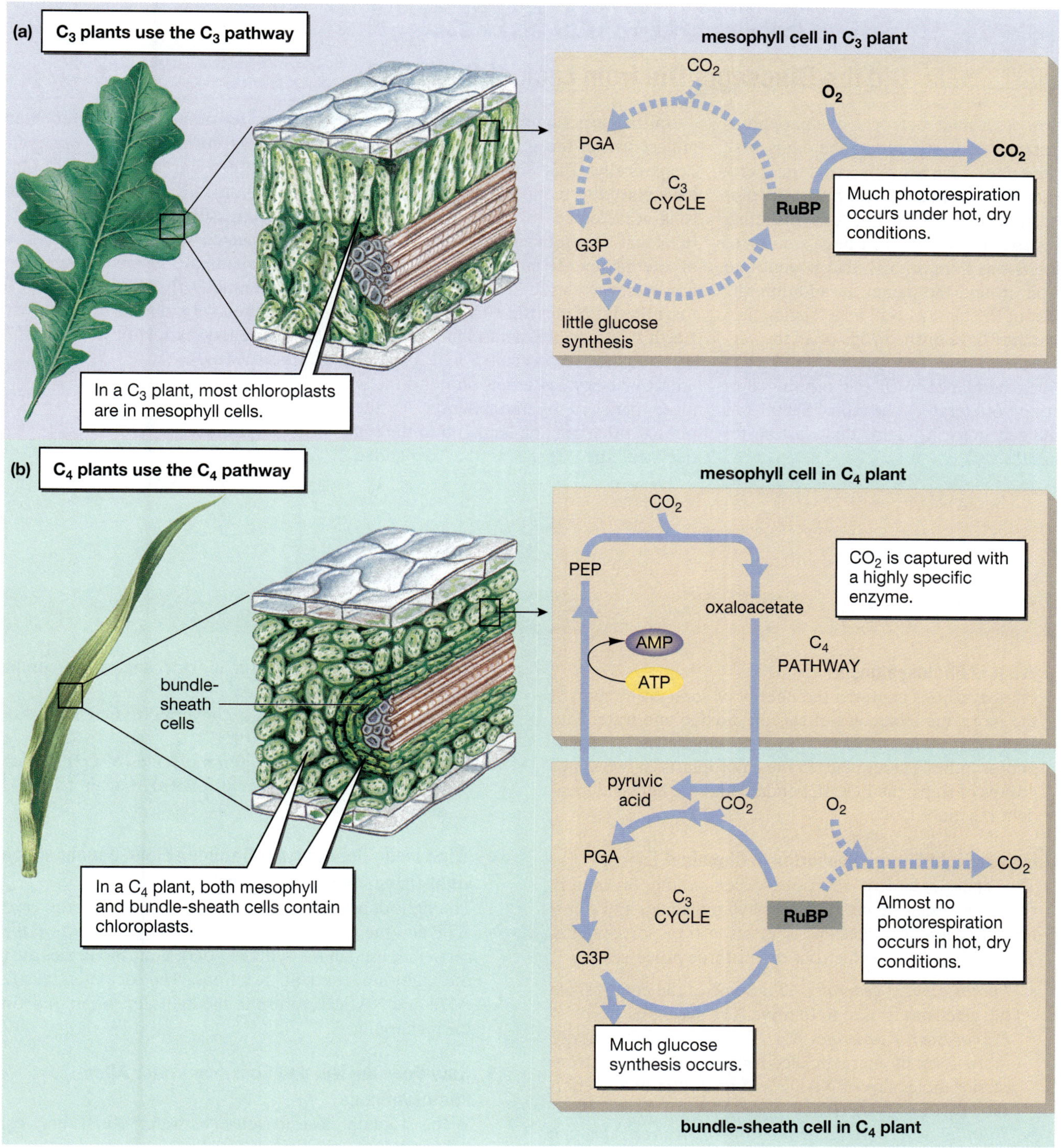

Figure 6-6 Comparison of C_3 and C_4 plants
(a) All carbon fixation in C_3 plants is by the C_3 pathway. ***(b)*** The initial carbon-fixation step in the mesophyll cells of C_4 plants is a reaction between PEP and CO_2, with which O_2 does not compete. The resulting four-carbon molecule of oxaloacetate moves to the bundle-sheath cells, where it releases CO_2. The bundle-sheath cells thus have higher CO_2 levels, which allow efficient carbon fixation via the C_3 pathway. Note that the regeneration of PEP requires energy: Two phosphates are removed from ATP to produce AMP (adenosine monophosphate).

CASE STUDY REVISITED

Did the Dinosaurs Die from Lack of Sunlight?

How did a giant meteorite hamper photosynthesis and exterminate the dinosaurs? The force of impact pulverized the meteorite and smashed the Earth's crust, sending trillions of tons of debris rocketing into the atmosphere. This huge amount of debris was flung so high that it went into orbit around the planet. In addition, the heat of the blast caused huge, raging fires that may have burned 25% of all the vegetation on land. The resulting smoke and ashes, combined with the orbiting dust cloud, obliterated the sun. Earth was plunged into a night that lasted for months.

In the perpetual darkness of the post-impact world, the temperature dropped dramatically. This sudden winter must have placed tremendous stress on organisms accustomed to the tropical conditions that then prevailed over large areas of the planet. Even more devastating, however, was the sudden decline in photosynthesis. With little sunlight reaching Earth's surface, the land plants that had survived the fires could not capture enough energy. In the sea, photosynthetic algae perished by the billions. As the photosynthesizers on land and in the sea declined, the organisms that depended on them for food also suffered. Large plant-eaters such as *Triceratops,* which needed to consume hundreds of pounds of vegetation daily, must have been especially vulnerable. Predators such as *Tyrannosaurus,* which fed on the disappearing plant-eaters, were soon without prey. Eventually, the severe stress of life on a planet with diminished photosynthesis spelled extinction for most of Earth's species.

Summary of Key Concepts

1) What Is Photosynthesis?

Photosynthesis captures the energy of sunlight to convert the inorganic molecules of carbon dioxide and water into high-energy organic molecules such as glucose. In plants, photosynthesis takes place in the chloroplasts, in two major groups of steps: the light-dependent and the light-independent reactions.

2) How Is Light Energy Converted to Chemical Energy?

The light-dependent reactions occur in the thylakoids. Light excites electrons in chlorophyll molecules and transfers the energetic electrons to electron transport systems. The energy of these electrons drives three processes:

- *Photosystem II generates ATP.* Some of the energy from the electrons is used to to drive ATP synthesis.
- *Photosystem I generates NADPH.* Some of the energy, in the form of energetic electrons, is added to electron-carrier molecules of $NADP^+$ to make the highly energetic carrier NADPH.
- *Splitting water maintains the flow of electrons through the photosystems.* Some of the energy is used to split water, generating the electrons needed by photosystem II and producing hydrogen ions and oxygen as by-products.

3) How Is Chemical Energy Stored in Glucose Molecules?

In the stroma of the chloroplasts, both ATP and NADPH provide the energy that drives the light-independent reactions that synthesize glucose from CO_2 and H_2O. The light-independent reactions constitute the C_3 cycle, which has three major parts:

- *Carbon fixation.* Carbon dioxide and water combine with RuBP to form PGA.
- *Synthesis of G3P.* PGA is converted to G3P, using energy from ATP and NADPH.
- *Regeneration of RuBP.* Ten molecules of G3P are used to regenerate six molecules of RuBP, again using ATP energy.

4) What Is the Relationship Between Light-Dependent and Light-Independent Reactions?

The light-dependent reactions produce the energy carrier ATP and the electron carrier NADPH. Energy from these carriers is used in the synthesis of organic molecules during the light-independent reactions. The depleted carriers, ADP and $NADP^+$, return to the light-dependent reactions for recharging.

5) How Does the Need to Conserve Water Affect Photosynthesis?

When stomata close to conserve water, RuBP may combine with O_2 (rather than CO_2). Such photorespiration prevents carbon fixation and does not generate ATP. Some plants have evolved an additional step for carbon fixation that minimizes photorespiration. In the mesophyll cells of these C_4 plants, CO_2 combines with PEP to form oxaloacetate. Oxaloacetate is transported into adjacent bundle-sheath cells, where it releases CO_2, thereby maintaining a high CO_2 concentration in those cells. This CO_2 is then fixed in the C_3 cycle.

Key Terms

adenosine triphosphate (ATP) *p. 83*
bundle-sheath cell *p. 87*
C_3 cycle *p. 85*
C_4 pathway *p. 87*
carbon fixation *p. 86*
chlorophyll *p. 83*
chloroplasts *p. 83*
electron transport system *p. 84*
glyceraldehyde-3-phosphate (G3P) *p. 86*
light-harvesting complex *p. 84*
mesophyll *p. 83*
nicotinamide adenine dinucleotide phosphate (NADPH) *p. 83*
oxaloacetate *p. 87*
phosphoenolpyruvate *p. 87*
phosphoglyceric acid (PGA) *p. 86*
photorespiration *p. 87*
photosynthesis *p. 81*
photosystem *p. 84*
reaction center *p. 84*
ribulose biphosphate (RuBP) *p. 86*
stoma *p. 82*
stroma *p. 83*
thylakoid *p. 83*

Thinking Through the Concepts

Multiple Choice

1. *Photosynthesis is measured in the leaf of a green plant exposed to different wavelengths of light. Photosynthesis is*
   a. highest in green light
   b. highest in red light
   c. highest in blue light
   d. highest in red and blue light
   e. the same at all wavelengths

2. *Where do the light-dependent reactions of photosynthesis occur?*
   a. in the stomata
   b. in the chloroplast stroma
   c. within the thylakoid membranes of the chloroplast
   d. in the leaf cell cytoplasm
   e. in leaf cell mitochondria

3. *The oxygen produced during photosynthesis comes from*
   a. the breakdown of CO_2
   b. the breakdown of H_2O
   c. the breakdown of both CO_2 and H_2O
   d. the breakdown of oxaloacetate
   e. photorespiration

4. *The role of accessory pigments is to*
   a. provide an additional photosystem to generate more ATP
   b. allow photosynthesis to occur in the dark
   c. prevent photorespiration
   d. donate electrons to chlorophyll reaction centers
   e. capture additional light energy and transfer it to the chlorophyll reaction centers

5. *Where do the light-independent, carbon-fixing reactions occur?*
   a. in the guard cell cytoplasm
   b. in the chloroplast stroma
   c. within the thylakoid membranes
   d. at night in the thylakoids
   e. in mitochondria

? Review Questions

1. Write the overall equation for photosynthesis. Does the overall equation differ between C_3 and C_4 plants?
2. Draw a diagram of a chloroplast and label it. Explain specifically how chloroplast structure is related to its function.
3. Briefly describe the light-dependent and light-independent reactions. In what part of the chloroplast does each occur?
4. What is the difference between carbon fixation in C_3 and in C_4 plants? Under what conditions does each mechanism of carbon fixation work most effectively?

Applying the Concepts

1. Many lawns and golf courses are planted with bluegrass, a C_3 plant. In the spring, the bluegrass grows luxuriously. In the summer, crabgrass, a weed and a C_4 plant, often appears and spreads rapidly. Explain this sequence of events, given the normal weather conditions of spring and summer and the characteristics of C_3 versus C_4 plants.
2. Suppose an experiment is performed in which plant I is supplied with normal carbon dioxide but with water that contains radioactive oxygen atoms. Plant II is supplied with normal water but with carbon dioxide that contains radioactive oxygen atoms. Each plant is allowed to perform photosynthesis, and the oxygen gas and sugars produced are tested for radioactivity. Which plant would you expect to produce radioactive sugars, and which plant would you expect to produce radioactive oxygen gas? Why?
3. You continuously monitor the photosynthetic oxygen production from the leaf of a plant illuminated by white light. Explain what will happen (and why) if you place (a) red, (b) blue, and (c) green filters between the light source and the leaf.
4. A plant is placed in a CO_2-free atmosphere in bright light. Will the light-dependent reactions continue to generate ATP and NADPH indefinitely? Explain how you reached your conclusion.
5. You are called before the Ways and Means Committee of the U.S. House of Representatives to explain why the U.S. Department of Agriculture should continue to fund photosynthesis research. How would you justify the expense of producing, by genetic engineering, an enzyme that catalyzes the reaction of RuBP with CO_2 but prevents RuBP from reacting with oxygen as well as CO_2? What are the potential applied benefits of this research?

Group Activity

Work in groups of two. Each of you should write down all components of the last meal you ate. Then, for each food item, your partner should explain how that food contains energy derived originally from the sun. Do the same with your partner's list.

For More Information

Bazzazz, F. A., and Fajer, E. D. "Plant Life in a CO_2-Rich World." *Scientific American*, January 1992. Burning fossil fuels is increasing CO_2 levels in the atmosphere (see Chapter 40). This increase could tip the balance between C_3 and C_4 plants.

Govindjee, and Coleman, W. J. "How Plants Make Oxygen." *Scientific American*, February 1990. The generation of oxygen during photosynthesis is just beginning to be understood.

Grodzinski, B. "Plant Nutrition and Growth Regulation by CO_2 Enrichment." *BioScience*, 1992. How higher CO_2 levels influence plant metabolism.

Hall, D. O., and Rao, K. K. *Photosynthesis*. 5th ed. New York: Cambridge University Press, 1994. An excellent short book recommended to any student interested in finding out more about photosynthesis.

Hinkle, P. C., and McCarthy, R. E. "How Cells Make ATP." *Scientific American*, March 1978. A good explanation of chemiosmosis, which is a difficult concept for many students.

Monastersky, R. "Children of the C_4 World." *Science News*, January 3, 1998. What role did a shift in global vegetation toward C_4 photosynthesis play in the evolution of humans?

Mooney, H. A., Drake, B. G., Luxmoore, R. J., Oechel, W. C., and Pitelka, L. F. "Predicting Ecosystems' Responses to Elevated CO_2 Concentrations." *BioScience*, 1994. What effects will CO_2 enrichment of the atmosphere due to human activities have on ecosystems?

Zimmer, C. "The Processing Plant." *Discover*, September 1995. Describes organisms inhabiting the watery digestive chamber of the pitcher plant, which is both photosynthetic and carnivorous.

Answers to Multiple-Choice Questions

1.d 2.c 3.b 4.e 5.b

MediaTutor

CD Activities

Activity 6.1: Photosynthesis

Estimated Time: 15 minutes

The energy that powers the biosphere comes from the sun. The process that converts radiant energy into useful energy for living systems is photosynthesis. This process also emits oxygen, which animals breathe and use during cellular respiration. The process of photosynthesis is explored in depth, and related to the overall energy flow in the biosphere and to the process of respiration.

Start the MediaTutor Student CD-ROM and enter the activity number in the Quick Search box to be taken directly to that activity.

Web Investigation

Did the Dinosaurs Die from Lack of Sunlight?

Estimated time: 10 minutes

While most people view the death of the dinosaurs as a unique biological disaster, geological records suggest that global mass extinctions have occurred several times. In the past two hundred years more species have become extinct than in many previous millennia. This exercise will examine the causes of some modern extinctions in the hope that they might shed some light on the other mass extinctions, including the one that ended the age of dinosaurs.

Go to http://www.prenhall.com/audesirk3, the Audesirk Companion Web site. Select Chapter 6 and the Web Investigation to begin.

As these bicyclists race, the cells of their leg muscles use energy to do work. To meet the muscle cells' demand for energy, the body's energy-storage molecules undergo a complex transformation that makes energy available for cellular work.

7 Harvesting Energy: Glycolysis and Cellular Respiration

AT A GLANCE

CASESTUDY

Unfair Advantage

On a sunny July morning in 1998, the members of the Festina bicycle racing team, the top-rated team in the world, sat in a French café. Nearby, dozens of other professional bicyclists made their final preparations for the impending start of the day's segment of the grueling Tour de France race. The Festina riders, however, would not be joining the race. Hours earlier, the entire team had been expelled from the Tour de France for the offense of blood doping.

The dramatic expulsion of top athletes from the world's premier bicycle race focused attention on the arcane practice of blood doping. By using blood-doping techniques, some athletes try to gain a competitive edge. But what is blood doping, and how does it enhance athletic performance?

Blood doping increases a person's physical endurance by increasing the capacity of the blood to carry oxygen. One crude method for accomplishing this goal is to simply inject extra red blood cells into the bloodstream. Red blood cells transport oxygen to the body's tissues, so simply adding more of them is a straightforward way of increasing the amount of oxygen that reaches the tissues. In recent years, however, blood-doping athletes have increasingly turned to injections of erythropoietin (Epo) as a more effective approach to increasing blood oxygen.

Epo is a protein molecule that is present in a normal human body, where it functions as a chemical messenger that stimulates bone marrow to produce more red blood cells. Under normal circumstances, the body produces just enough Epo to ensure that red blood cells are replaced as they age and die. An injection of extra Epo, however, can stimulate the production of a huge number of extra red blood cells. The extra cells greatly increase the oxygen-carrying capacity of the blood. Unfortunately, the excess blood cells also thicken the blood and make it harder to move through blood vessels, so those who inject Epo suffer increased risk of heart failure.

Why would professional athletes take such a risk to get more oxygen molecules into their bloodstreams? How does extra oxygen increase endurance? The answers to those questions lie in the role of oxygen in supplying energy to muscle cells. ■

1) What Is the Source of a Cell's Energy?

Living cells need energy to power the multitude of metabolic reactions that are necessary to sustain life. To power a reaction, however, the energy must be in a useable form. To be useable, energy must, in most cases, be stored in the bonds of energy-carrier molecules, especially **adenosine triphosphate** (**ATP**; see chapters 5 and 6). Thus, some of the most important reactions in cells are the ones that transfer energy from energy-storage molecules to energy-carrier molecules.

Glucose Is a Key Energy-Storage Molecule

Although most cells can harvest energy from a variety of organic molecules, this chapter's description of the harvesting process will focus on glucose. Almost all cells metabolize glucose for energy at least part of the time. Even when cells draw energy from other types of molecule, those molecules are typically first converted to glucose.

Photosynthesis Is the Ultimate Source of Cellular Energy

All the energy released during glucose metabolism was originally acquired by photosynthesis. Photosynthetic cells capture and store the energy of sunlight, and this stored energy is later used by cells. Those cells may be in the photosynthetic organisms themselves, or in other organisms that directly or indirectly consume photosynthesizers. For example, caterpillars eat plants, bluebirds eat caterpillars, and hawks eat bluebirds. The cells that make up the plants, caterpillars, bluebirds, and hawks are all powered by energy stored in glucose that was originally captured by photosynthesis in the plant's leaves. The captured energy is passed along a chain of consumption.

Glucose Metabolism and Photosynthesis Are Complementary Processes

As photosynthesis captures solar energy for storage in cells, it consumes water and carbon dioxide and yields glucose and oxygen gas. When glucose is broken down in cells, oxygen is consumed and water and carbon dioxide are released. Thus, the two processes are almost perfectly complementary; the products of each process provide reactants for the other. This symmetry is apparent in the chemical equations for glucose formation by photosynthesis and for the complete metabolism of glucose back to CO_2 and H_2O:

Photosynthesis:

$$6\ CO_2 + 6\ H_2O + \text{sunlight energy} \rightarrow C_6H_{12}O_6 + 6\ O_2$$

Complete Glucose Metabolism:

$$C_6H_{12}O_6 + 6O_2 \rightarrow 6\ CO_2 + 6\ H_2O + \text{chemical energy and heat energy}$$

This symmetry might lead you to believe that a cell can convert all the chemical energy in a glucose molecule to high-energy bonds of ATP. But remember that the second law of thermodynamics tells us that conversion of energy into different forms always decreases the amount of useful energy. Thus, most of the energy released during the breakdown of glucose is lost as heat rather than being converted to the chemical energy of ATP. Nevertheless, a cell can extract a great deal of energy, in the form of ATP, when a glucose molecule is completely broken down.

The Big Picture

Life depends on the reciprocal relationship between the reactions of photosynthesis and those of glucose metabolism. Photosynthesis produces the glucose and oxygen that feed glucose metabolism; glucose metabolism in turn gives off the carbon dioxide that photosynthesis requires. A balance between the two processes keeps levels of oxygen and carbon dioxide in the atmosphere within bounds conducive to life.

2 How Do Cells Harvest Energy from Glucose?

Glucose metabolism proceeds in stages (Fig. 7-1). The first stage, *glycolysis*, does not require oxygen and proceeds in exactly the same way under both aerobic (with oxygen) and anaerobic (without oxygen) conditions. The products of glycolysis are further processed in the second stage, which is *cellular respiration* if oxygen is available or *fermentation* in the absence of oxygen.

Glycolysis, which occurs in the cytoplasm, splits apart a single glucose molecule (a six-carbon sugar) into two three-carbon molecules of **pyruvate**. The split releases a small amount of chemical energy, which is used to generate two ATP molecules.

The pyruvate produced by glycolysis may enter the **mitochondria**, which are organelles specialized for the aerobic breakdown of pyruvate. There, if oxygen is available, cellular respiration uses oxygen to break pyruvate down completely to carbon dioxide and water, generating an additional 34 or 36 ATP molecules (the amount differs from cell to cell). The extra ATP produced by cellular respiration is so important to most organisms that anything that interferes with its production, such as a sustained lack of oxygen, quickly results in death.

Under anaerobic conditions, pyruvate remains in the cytoplasm and is usually converted by fermentation into lactate or ethanol. Fermentation does not produce additional ATP energy.

3 What Happens During Glycolysis?

The reactions that split glucose are collectively called **glycolysis**. Glycolysis, which requires no oxygen, cleaves a molecule of glucose into two molecules of pyruvate. The reactions produce relatively little energy, just two molecules of ATP and two molecules of the electron carrier **nicotinamide adenine dinucleotide (NADH)**. Nonetheless, glycolysis is an essential part of glucose metabolism; without it, life would rapidly be extinguished. Glycolysis consists of two major parts (each with several steps): glucose activation and energy harvest (Fig. 7-2).

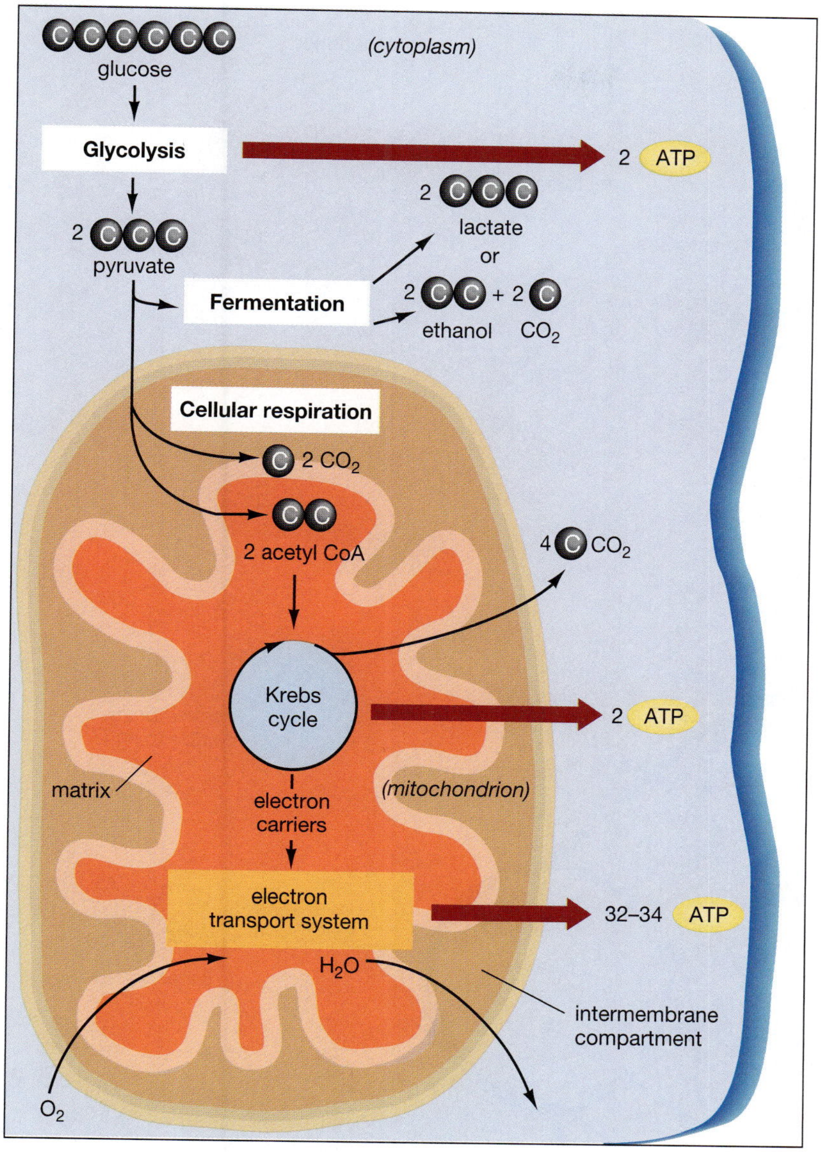

Figure 7-1 A summary of glucose metabolism Refer to this diagram as we progress through the reactions of glycolysis (in the fluid portion of the cytoplasm) and cellular respiration (in the mitochondrion). Glucose is broken down in stages, with various amounts of energy harvested as ATP along the way. The vast majority of the ATP is produced in mitochondria, justifying their nickname, "powerhouse of the cell."

MEDIATUTOR 7.1 Glucose Metabolism

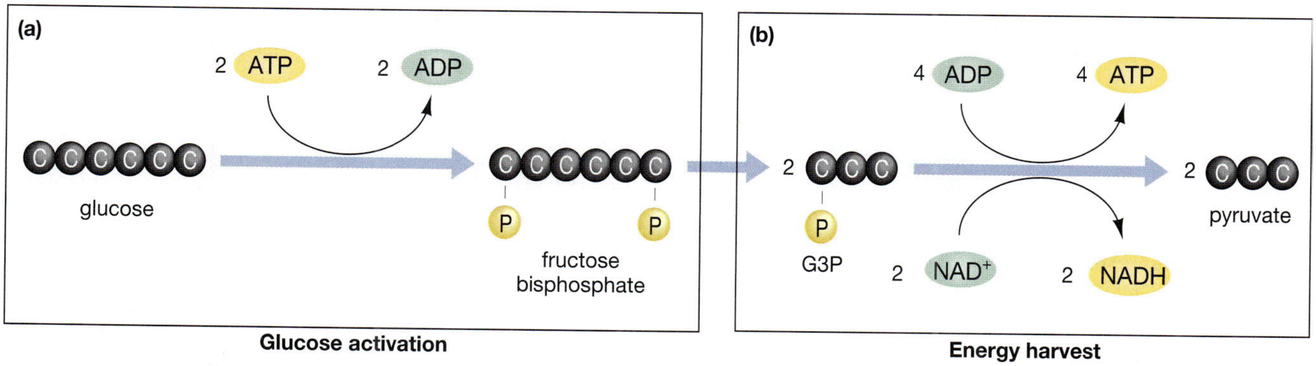

Figure 7-2 The essentials of glycolysis

Activation Consumes Energy

Before glucose is broken down to release its energy, it must be activated—a process that actually uses up energy. During glucose activation, a molecule of glucose undergoes two enzyme-catalyzed reactions, each of which uses ATP energy. These reactions use two phosphates from ATP to convert a relatively stable glucose molecule into a highly unstable, activated molecule of fructose biphosphate (Fig. 7-2a). Forming fructose bisphosphate costs the cell two ATP molecules, a cost that is necessary to produce much greater energy returns in the long run.

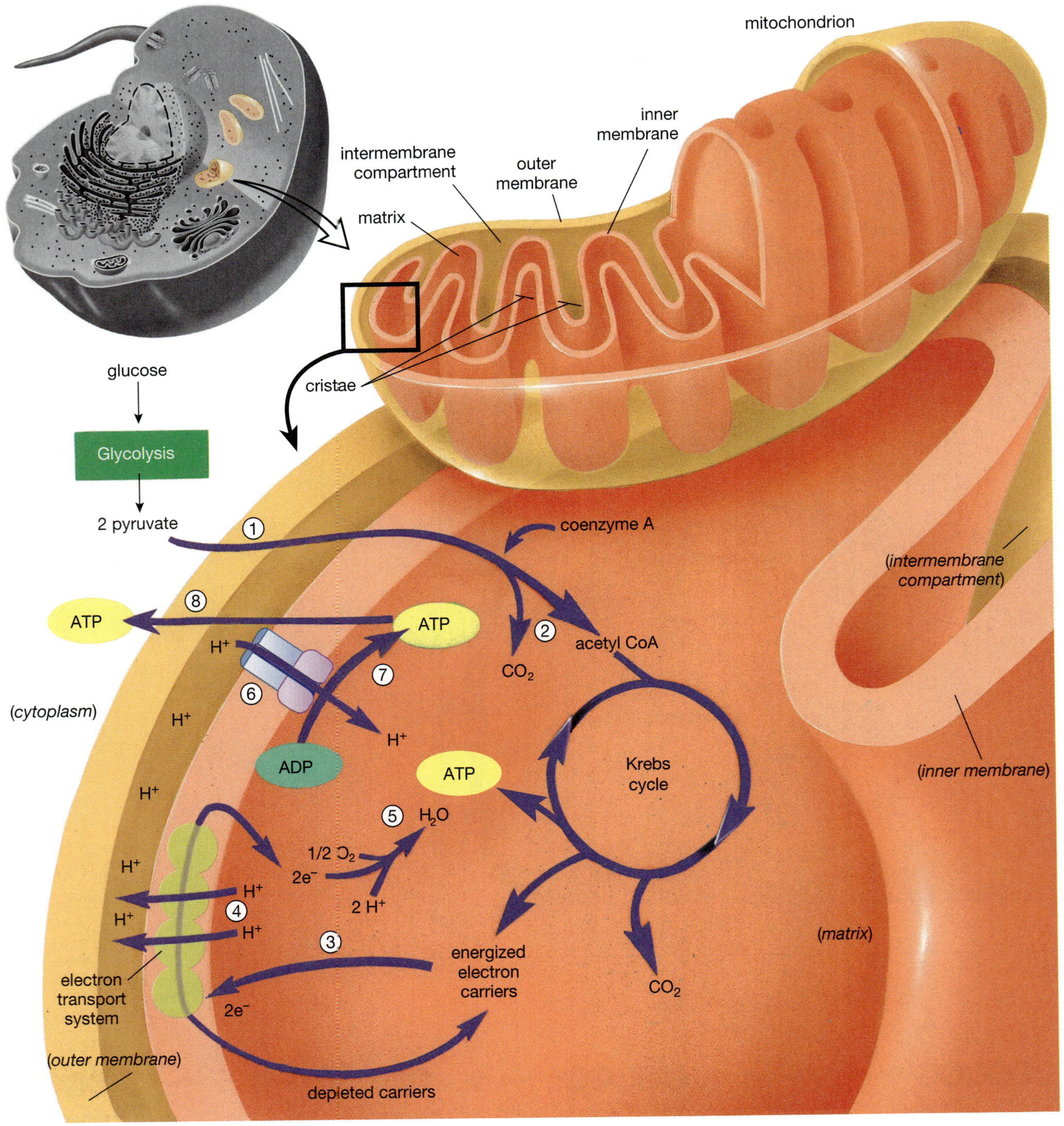

Figure 7-3 Cellular respiration
Cellular respiration takes place in mitochondria. A mitochondrion's inner membrane separates the matrix (inner compartment) from the intermembrane compartment (between the inner and outer membranes).

Energy Harvest Yields Energy-Carrier Molecules

In the energy-harvest steps, fructose bisphosphate splits apart into two three-carbon molecules of glyceraldehyde 3-phosphate (G3P). (In Chapter 6 we encountered G3P in the C_3 cycle of photosynthesis.) Each G3P molecule then goes through a series of reactions that converts it to pyruvate (Fig. 7-2b). During these reactions, two ATP are generated for each G3P, for a total of four ATPs. Because two ATPs were used to activate the glucose molecule in the first place, there is a net gain of only two ATPs per glucose molecule.

At another step along the way from G3P to pyruvate, the energized electron carrier NADH is produced (Fig. 7-2b). Each molecule of G3P yields one molecule of NADH, so two NADH carrier molecules are formed for each glucose molecule that is converted to pyruvate.

Summary: Glycolysis

Each molecule of glucose is broken down to two molecules of pyruvate. During this process, ATP is both consumed and produced, but the process as a whole yields a net increase of 2 ATP molecules. In addition, two molecules of the electron carrier NADH are formed.

4 What Happens During Cellular Respiration?

Cellular respiration is a series of reactions, occurring under aerobic conditions, in which the pyruvate produced by glycolysis is broken down to carbon dioxide and water. Over the course of the reactions, large amounts of ATP are produced. In eukaryotic cells, cellular respiration takes place in the mitochondria.

For an overview of cellular respiration, review Figure 7-3 with the help of the summary below. A more detailed description follows.

① The two molecules of pyruvate produced by glycolysis are transported into a mitochondrion. A mitochondrion is surrounded by two membranes, the inner membrane and the outer membrane; the pyruvate molecules cross both of them and pass into the fluid **matrix** that lies inside the inner membrane.

② Each three-carbon pyruvate is split into CO_2 and a two-carbon molecule, which enters a series of reactions known as the Krebs cycle. The Krebs cycle produces one ATP from each pyruvate and donates energetic electrons to several electron-carrier molecules.

③ The electron carriers donate their energetic electrons to the electron transport system of the inner membrane.

④ In the electron transport system, the energy of the electrons is used to transport H^+ from the matrix to the **intermembrane compartment** that lies between the inner and outer membranes.

⑤ At the end of the electron transport system, the electrons combine with O_2 and H^+ to form H_2O.

⑥ The hydrogen ions that were moved to the intermembrane compartment by the electron transport system move back across the inner membrane, along their concentration gradient.

⑦ As the ions pass back into the matrix, their flow provides the energy to produce ATP.

⑧ ATP moves out of the mitochondrion into the fluid of the cytoplasm, where it provides energy for cellular activities.

The Krebs Cycle Breaks Down Pyruvate in the Mitochondrial Matrix

After being produced by glycolysis in the cell cytoplasm, pyruvate diffuses down its concentration gradient through pores in the two mitochondrial membranes. After the pyruvate reaches the mitochondrial matrix, it reacts with a molecule called **coenzyme A** (Fig. 7-4, step 1). Each pyruvate is split into CO_2 and a two-carbon molecule called an acetyl group, which immediately attaches to coenzyme A, forming an **acetyl-coenzyme A complex (acetyl CoA)**. During this reaction, two energetic electrons and a hydrogen ion are transferred to NAD^+, forming NADH.

The next group of reactions forms the cyclic pathway of the **Krebs cycle** (Fig. 7-4, step 2), which is sometimes called the citric-acid cycle because citrate (the ionized form of citric acid) is the first molecule produced in the cycle. Citrate is formed when acetyl CoA combines with the four-carbon molecule oxaloacetate (which, as you may recall from Chapter 6, also plays a role in C_4 photosynthesis) to form the six-carbon citrate. As citrate forms, coenzyme A is released for reuse in further turns of the cycle. Each citrate then undergoes a series of enzyme-catalyzed rearrangements that regenerate oxaloacetate, give off two more CO_2 molecules, and capture most of the energy of the original acetyl group in one ATP and four electron carriers—one **flavin adenine dinucleotide ($FADH_2$)** and three NADH.

Summary: The Mitochondrial Matrix Reactions

Each pyruvate that enters the matrix reactions produces one CO_2 and one NADH during the synthesis of acetyl CoA. Each acetyl CoA in turn yields two more

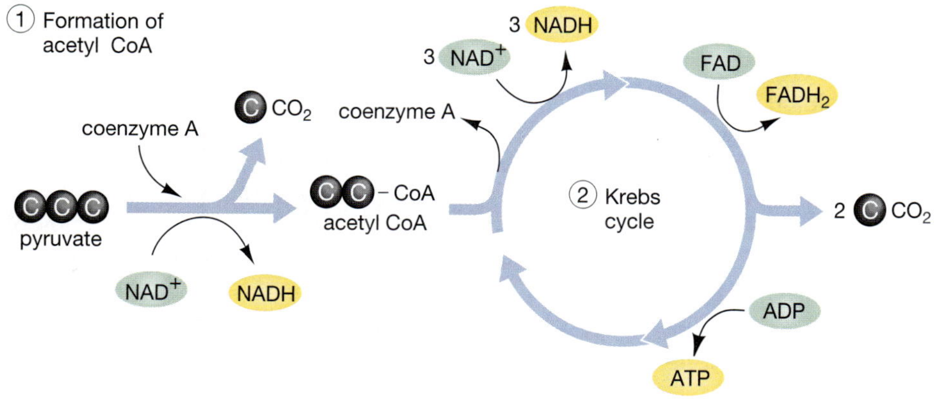

Figure 7-4 The essential reactions in the mitochondrial matrix

Table 7-1 Summary of Glycolysis and Cellular Respiration

Process	Location	Reactions	Electron Carriers Formed	ATP Yield (per glucose molecule)
Glycolysis	Fluid cytoplasm	Glucose breaks down to two pyruvates	2 NADH	2 ATP
Cellular Respiration				
Acetyl CoA formation	Matrix of mitochondrion	Each pyruvate combines with coenzyme A to form acetyl CoA and CO_2	2 NADH	
Krebs cycle	Matrix of mitochondrion	Acetyl group of acetyl CoA metabolized to two CO_2	6 NADH, 2 $FADH_2$	2 ATP
Electron transport	Inner membrane, intermembrane compartment	Energy of electrons from NADH and $FADH_2$ used to pump H^+ into intermembrane compartment, H^+ gradient used to synthesize ATP		32–34 ATP

CO_2, one ATP, three more NADH, and one $FADH_2$ via the Krebs cycle. Because each glucose feeds two pyruvates to the matrix reactions, the reactions yield a total of six CO_2 molecules, two ATPs, eight NADH electron carriers, and two $FADH_2$ electron carriers per glucose molecule.

Energetic Electrons Are Carried to Electron Transport Systems

At the end of the matrix reactions, only a small portion of the energy of glucose has been captured in ATP. The cell has thus far gained only four ATP molecules from each original glucose molecule: two during glycolysis and two during the Krebs cycle. The cell has, however, captured many energetic electrons in carrier molecules: 2 NADH during glycolysis plus 8 more NADH and 2 $FADH_2$ from the matrix reactions, for a total of 10 NADH and 2 $FADH_2$ (Table 7-1).

The carriers deposit their electrons in **electron transport systems** located in the inner mitochondrial membrane (step 1 in Fig. 7-5). Within the transport systems, the energetic electrons move systematically from molecule to molecule. During these transfers, energy is released and used to pump hydrogen ions out of the matrix, across the inner mitochondrial membrane, and into the intermembrane compartment (step 2 in Fig. 7-5).

At the end of the electron transport system, oxygen finally plays its part. An oxygen atom combines with two hydrogen ions and two energetically depleted electrons to form water (step 3 in Fig. 7-5). Oxygen thus accepts electrons after they run their course through the electron transport system. If oxygen were not present to pull spent electrons from the end of the transport chain, the electrons would "pile up" in the system, and the process of pumping hydrogen ions out of the matrix would come to a halt.

Energy from a Hydrogen-Ion Gradient Is Used to Produce ATP

The hydrogen ions that are pumped out of the matrix become a source of energy for the production of more ATP. The process of pumping hydrogen ions across the inner mitochondrial membrane generates a large H^+ concentration gradient—that is, the concentration of hydrogen ions in the intermembrane compartment becomes high and the concentration in the matrix becomes low. From the second law of thermodynamics (see Chapter 5), we know that creating this nonuniform distribution of hydrogen ions requires an input of energy; it's the chemical equivalent of pumping water upward into an elevated storage tank. Energy is released when the hydrogen ions are allowed to move down their concentration gradient—like opening the valves of the storage tank and allowing the water to rush out.

In mitochondria, the energy of the hydrogen-ion concentration gradient can be captured because the inner

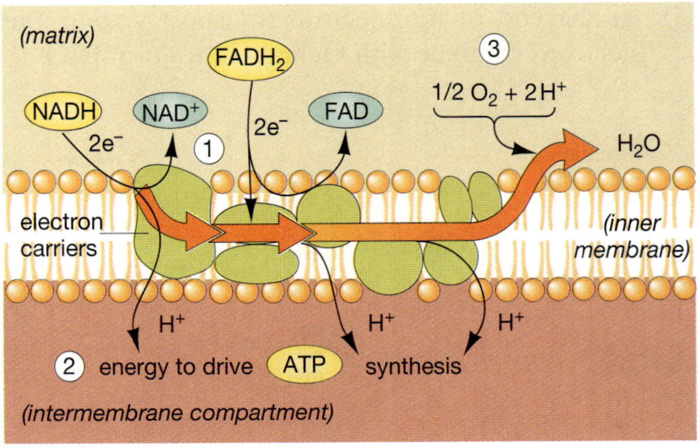

Figure 7-5 The electron transport system of mitochondria

Health Watch

Why Can You Get Fat by Eating Sugar?

As you know, humans do not live on glucose alone. We consume a wide variety of nutrients, but even the best-planned diet cannot contain the exact mix of molecules required by the body's cells. The body must be able to both cope with diverse inputs and take advantage of them to meet its needs. Consequently, the cells of the human body seethe with biochemical reactions that transform molecules to meet metabolic needs. Let's look at two examples of such transformations: the production of ATP from fats and proteins and the synthesis of fats from sugars.

How Are Fats and Proteins Metabolized? Even the leanest people have some fat in their bodies. During fasting or starvation, the body mobilizes these fat reserves for ATP synthesis, because even the bare maintenance of life requires a continuous supply of ATP. Fat metabolism flows directly into the pathways of glucose metabolism.

A fat molecule consists of three fatty acids connected to a glycerol backbone (see Chapter 2). In fat metabolism, the bonds between the fatty acids and glycerol break. The glycerol portion, after activation by ATP, feeds directly into the middle of the glycolysis pathway (Fig. E7-1). The fatty acids are transported into the mitochondria, where enzymes in the inner membrane and matrix chop them up into acetyl groups. These groups attach to coenzyme A to form acetyl CoA, which enters the Krebs cycle.

Severe starvation (or a diet consisting almost exclusively of protein) can cause amino acids to be used as a source of energy. Amino acids used in this fashion are first converted to pyruvate, acetyl CoA, or one of the compounds of the Krebs cycle. These molecules then proceed through the remaining stages of cellular respiration, yielding amounts of ATP that vary with their point of entry into the pathway.

How Is Fat Synthesized from Sugar? The body, in addition to having developed ways of coping with fasting or starvation, has also evolved strategies for coping with situations in which food intake exceeds current energy needs. Sugars and starches can be converted into fats for energy storage. Complex sugars, such as starches and sucrose, are first broken down to simple sugars (see Chapter 2), which split to yield pyruvate. The pyruvate is converted to acetyl CoA and, if the cell needs ATP, the acetyl CoA will enter the Krebs cycle. If the cell has plenty of ATP, acetyl CoA will instead be used to make fatty acids. Thus, eating sugar can generate fat. In humans, fat is stored in specialized fat cells, which tend to accumulate in certain areas of the body, especially around the waist and hips.

Energy use, fat storage, and nutrient intake are usually precisely balanced. Where the balance point lies, however, varies from person to person. Some people seem able to eat nearly continuously without ever storing much fat; other people crave high-calorie foods even when they have a lot of stored fat. From an evolutionary perspective, overeating during times of easy food availability is highly beneficial behavior. If hard times come, heavier people may survive while the more trim succumb to starvation. Only recently (in evolutionary terms) have people in societies such as ours had continuous access to inexpensive, high-calorie food. Under these conditions, the drive to eat and the adaptation of storing excess food as fat leads to obesity, a growing health problem in the United States.

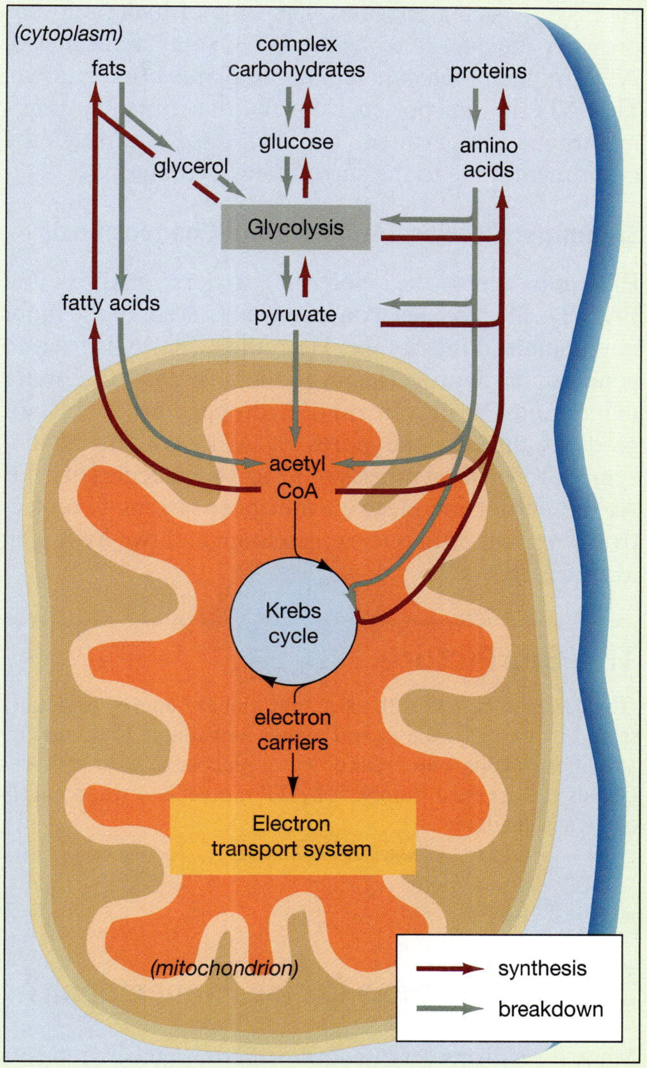

Figure E7-1 How various nutrients yield energy
Fats, carbohydrates such as starches, and proteins can all be broken down into molecules that enter glycolysis or the Krebs cycle, where they are used to generate ATP.

membrane is impermeable to hydrogen ions except at protein channels that are part of ATP-synthesizing enzymes. In the process of **chemiosmosis**, hydrogen ions move down their concentration gradient from the intermembrane compartment to the matrix through these ATP-synthesizing enzymes. The flow of hydrogen ions provides the energy to synthesize 32 to 34 molecules of ATP for each molecule of glucose that is broken down. This ATP is transported from the matrix to the intermembrane compartment, and then diffuses through the outer membrane to the surrounding cytoplasm.

Summary: Electron Transport and Chemiosmosis

Electrons from the electron carriers NADH and $FADH_2$ enter the electron transport system of the inner mitochondrial membrane. Here their energy is used to generate a hydrogen-ion gradient across the inner membrane. The movement of hydrogen ions down their gradient through the pores of ATP-synthesizing enzymes drives the synthesis of 32 to 34 molecules of ATP. At the end of the electron transport system, two electrons combine with one oxygen atom and two hydrogen ions to form water.

The Big Picture

Glucose metabolism is the key that unlocks energy originally stored by photosynthesis, and makes it available to power the activities of cells. The breakdown of glucose is most efficient (yields the greatest number of ATP molecules) when ample oxygen is available.

5 What Happens During Fermentation?

Many organisms (particularly microorganisms) thrive in places where oxygen is scarce or absent. Such organisms can be found in animal intestines, in sediments beneath lakes and oceans, deep beneath the surface of Earth, in bogs and marshes. Even some of our own body cells must sometimes cope without oxygen for brief periods. The anaerobic conditions in all these places rule out cellular respiration, because oxygen is required as the final electron acceptor. Instead, the products of glycolysis may be metabolized by **fermentation**.

Unlike the reactions of cellular respiration, the reactions of fermentation generate no ATP. Instead, they serve to regenerate the NAD^+ molecules that are needed for glycolysis. In fermentation, the pyruvate produced by glycolysis acts as an electron acceptor for the NADH produced by glycolysis, such that the pyruvate is converted to ethanol or lactate, and the NADH to NAD^+. Thus, in anaerobic conditions glycolysis becomes the main source of ATP, and NADH production serves not as a method of capturing energy for subsequent production of ATP, but as a way to soak up the hydrogen ions and electrons produced during glycolysis. This disposal method, however, poses a problem for the cell, because NAD^+ is used up as it accepts electrons and hydrogen ions to become NADH. Fermentation provides a way to regenerate NAD^+ so that glycolysis can continue.

Some Cells Ferment Pyruvate to Form Alcohol

There are two main types of fermentation: one converts pyruvate to ethanol and carbon dioxide and the other converts pyruvate to lactate. The form of fermentation that yields ethanol and CO_2 is known as alcoholic fermentation and is the primary mode of metabolism in many microorganisms. The reactions of alcoholic fermentation use hydrogen ions and electrons from NADH to regenerate NAD^+ under anaerobic conditions:

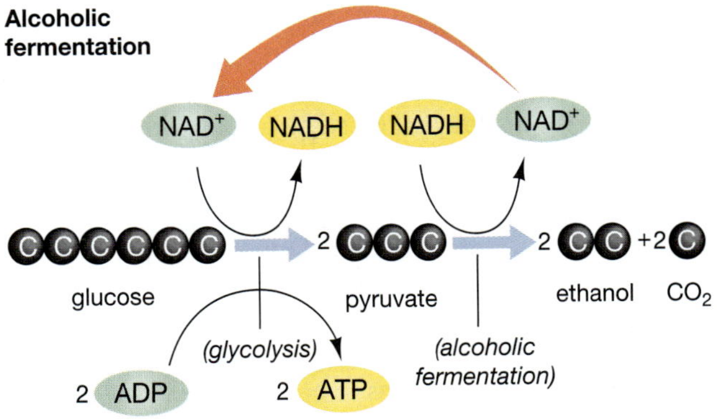

Alcoholic fermentation by single-celled fungi called yeasts is responsible for some gastronomic staples: wine, beer, and bread. Wine retains the ethanol product of the yeast's fermentation of grape sugar, while beer benefits from both the alcohol and carbon dioxide (retained as bubbles) products of fermentation. The yeast in bread dough likewise produces both alcohol and CO_2; the carbon dioxide makes the bread rise, but the alcohol evaporates during baking (Fig. 7-6a).

Other Cells Ferment Pyruvate to Lactate

Various microorganisms ferment pyruvate to lactate, including the bacteria that give yogurt, sour cream, and cheese their distinctive flavors. Lactate fermentation also occurs in muscles during vigorous exercise (Fig. 7-6b). Although working muscles need lots of ATP and cellular respiration generates much more ATP than does glycolysis, cellular respiration is limited by the organism's ability to provide oxygen (by breathing, for example). When you exercise vigorously, you may not be able to supply your muscles with sufficient oxygen to allow cellular respiration to meet all their energy needs.

When deprived of adequate oxygen, your muscles do not immediately stop working. Instead, glycolysis con-

(a)

(b)

Figure 7-6 Fermentation
(a) Bread dough rises as CO_2 is liberated by fermenting yeast. ***(b)*** A runner's lungs and blood vessels cannot supply oxygen to her leg muscles fast enough to keep up with the demand for energy, so glycolysis and lactate fermentation must provide the ATP.

tinues for a while, providing its meager two ATP molecules per glucose and generating both pyruvate and NADH. Then, to regenerate NAD^+, muscle cells ferment pyruvate molecules to lactate, using electrons and hydrogen ions from NADH:

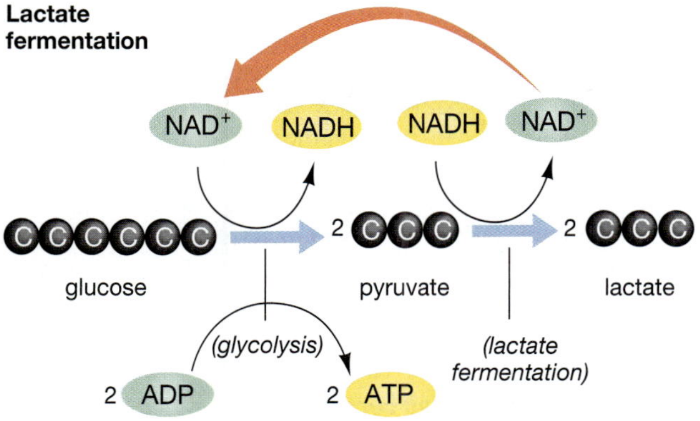

In high concentrations, however, lactate is toxic to your cells. Soon, it causes intense discomfort and fatigue, forcing you to stop or at least slow down. As you rest, breathing rapidly after your sprint, oxygen once more becomes available and the lactate is converted back to pyruvate.

Fermentation's Effects Limit Human Muscle Performance

Why is the average speed of the 5000-meter run in the Olympics slower than that of the 100-meter dash? During a sprint, runners' leg muscles use more ATP than cellular respiration can supply, because their bodies cannot deliver enough oxygen to keep up with the demand. With oxygen in short supply, glycolysis and lactate fermentation keep the muscles supplied with ATP. This system can work for a short time, long enough for a sprinter to complete a 100-meter dash. Soon, however, the toxic effects of lactate buildup cause fatigue and cramps. To avoid these effects, distance runners must pace themselves, so that cellular respiration can power their muscles for most of the race. An anaerobic sprint is reserved for the finish.

The Big Picture

Although it yields far fewer ATP molecules than cellular respiration, fermentation is nonetheless a widespread and important method of glucose metabolism. Fermenters can colonize oxygen-free environments that are off-limits to organisms that require oxygen for cellular respiration.

CASE STUDY REVISITED

Unfair Advantage

As you have seen, human cells most efficiently extract energy from glucose when an ample supply of oxygen is available to them. The aim of blood-doping cyclists, then, is to extend as long as possible the period in which their muscle cells have access to oxygen. During a difficult hill-climb, a cyclist who has doped his blood with erythropoietin may be able to pedal efficiently, his muscle cells using the Krebs cycle to turn out abundant ATP. At the same time, his "clean" competitor may labor painfully, leg muscles laden with lactate from fermentation.

The extra endurance that Epo supplies can tempt elite athletes to risk their health for an extra edge, despite the penalties that await those who are caught using a banned substance. The temptation may be especially large in the case of Epo. Because this substance forms naturally in the human body, its abuse is hard to detect; standard drug-screening procedures cannot distinguish between natural Epo and that injected in blood doping. In fact, the Festina team that was banned from the 1998 Tour de France was caught not by any blood or urine test, but only because a large supply of Epo was found in the team's car at a border crossing.

Since the Festina fiasco, however, researchers have worked hard at developing better tests for blood doping with Epo. A new urine test for Epo was introduced at the 2000 summer Olympics. This test detects a subtle chemical difference between the breakdown products of natural Epo and those of the manufactured version that is used in doping injections. Olympic officials, however, are not totally confident in the accuracy of the new urine test, and they use it only in conjunction with blood tests that screen for unusually high density of blood cells. Only athletes who fail both tests are disqualified. Meanwhile, researchers continue to explore the chemistry of Epo metabolism in hope of discovering a definitive test for blood doping.

Summary of Key Concepts

1) What Is the Source of a Cell's Energy?

Cells require energy to survive and function. To be useable by a cell, energy must be in the form of chemical energy in the bonds of an energy-carrier molecule, usually ATP. A key component of cellular metabolism consists of reactions that harvest energy from food molecules such as glucose and convert it to ATP energy. The ultimate source of all cellular energy is solar energy that is captured by photosynthesis.

2) How Do Cells Harvest Energy from Glucose?

Cells produce usable energy by breaking down glucose to lower-energy compounds and capturing some of the released energy as ATP. In glycolysis, glucose is metabolized in the fluid portion of the cytoplasm to two molecules of pyruvate, generating two ATP molecules. If oxygen is available, the pyruvates are metabolized to CO_2 and H_2O through cellular respiration in the mitochondria, generating much additional ATP.

3) What Happens During Glycolysis?

During glycolysis, a molecule of glucose is activated to form fructose bisphosphate. In a series of reactions, the fructose bisphosphate is broken down into two molecules of pyruvate. These reactions produce four ATP molecules and two NADH electron carriers. Because two ATPs were used in the activation steps, the net yield from glycolysis is two ATPs and two NADHs.

4) What Happens During Cellular Respiration?

If oxygen is available, cellular respiration can occur. The pyruvates are transported into the matrix of the mitochondria. In the matrix, each pyruvate reacts with coenzyme A to form acetyl CoA plus CO_2. One NADH is also formed at this step. The two-carbon acetyl group of acetyl CoA enters the Krebs cycle, which releases the two carbons as CO_2. One ATP, three NADHs, and one $FADH_2$ are also formed for each acetyl group that goes through the cycle.

The NADHs and $FADH_2$s from glycolysis and the matrix reactions deliver their energetic electrons to the electron transport system. The energy of the electrons is used to pump hydrogen ions across the inner membrane from the matrix to the intermembrane compartment. At the end of the electron transport system, the depleted electrons combine with hydrogen ions and oxygen to form water. This is the oxygen-requiring step of cellular respiration. During chemiosmosis, the hydrogen-ion gradient created by the electron transport system is used to produce ATP, as the hydrogen ions diffuse back across the inner membrane through channels in ATP-synthesizing enzymes. Electron transport and chemiosmosis yield 32 to 34 additional ATPs, for a net yield of 36 to 38 ATPs per glucose molecule.

Figure 7-7 and Table 7-1 summarize the locations, major mechanisms, and overall energy harvest for the complete metabolism of glucose from glycolysis through cellular respiration.

5) What Happens During Fermentation?

In the absence of oxygen, the pyruvate produced by glycolysis cannot enter the reactions of cellular respiration. Instead, it is converted by fermentation to lactate or ethanol and CO_2. Fermentation regenerates NAD^+ so glycolysis may continue. However, no additional ATP is gained by fermentation.

Key Terms

Thinking Through the Concepts

Multiple Choice

1. *Where does glycolysis occur?*
   a. cytoplasm
   b. matrix of mitochondria
   c. inner membrane of mitochondria
   d. outer membrane of mitochondria
   e. stroma of chloroplast

2. *Where does respiratory electron transport occur?*
   a. cytoplasm
   b. matrix of mitochondria
   c. inner membrane of mitochondria
   d. outer membrane of mitochondria
   e. stroma of chloroplast

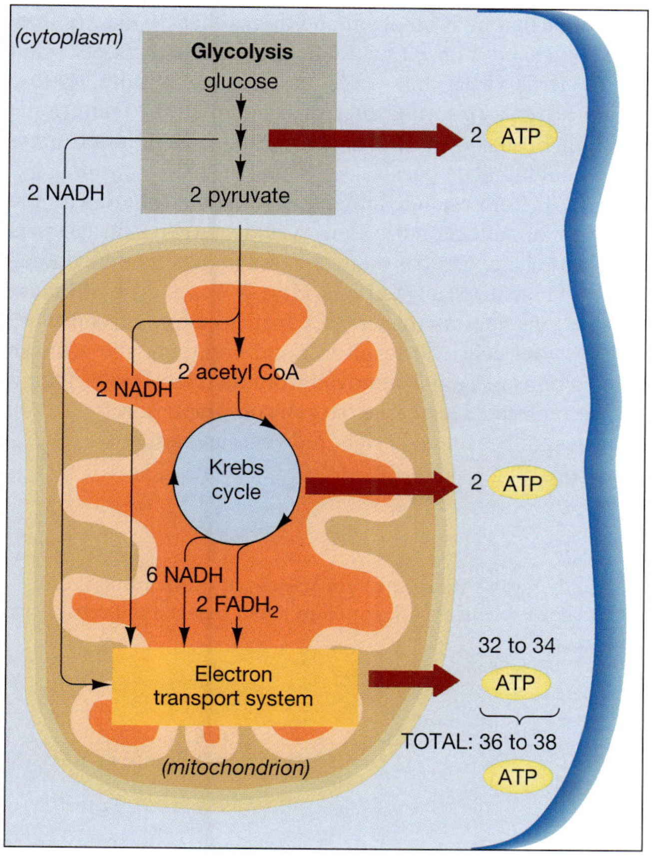

Figure 7-7 The energy harvest from the complete metabolism of one glucose molecule

3. *What is the product of the fermentation of sugar by yeast in bread dough that is essential for the rising of the dough?*
   a. lactate
   b. ATP
   c. ethanol
   d. CO_2
   e. O_2

4. *The majority of ATP produced in aerobic respiration comes from*
   a. glycolysis
   b. the Krebs cycle
   c. chemiosmosis
   d. fermentation
   e. photosynthesis

5. *The process that converts glucose into two molecules of pyruvate is*
   a. glycolysis
   b. fermentation
   c. the Krebs cycle
   d. electron transport
   e. the Calvin-Benson cycle

6. *The process that causes lactate buildup in muscles during strenuous exercise is*
   a. glycolysis
   b. fermentation
   c. the Krebs cycle
   d. electron transport
   e. the Calvin-Benson cycle

? Review Questions

1. Starting with glucose ($C_6H_{12}O_6$), write the overall reactions for (a) cellular respiration and (b) fermentation in yeast.
2. Draw a labeled diagram of a mitochondrion, and explain how its structure is related to its function.
3. What role do the following play in the complete metabolism of glucose: (a) glycolysis, (b) mitochondrial matrix, (c) inner membrane of mitochondria, (d) fermentation, and (e) NAD^+?
4. Outline the major steps in (a) cellular respiration and (b) fermentation, indicating the sites of ATP production. What is the overall energy harvest (in terms of ATP molecules generated per glucose molecule) for each?
5. Describe the Krebs cycle. In what form is most of the energy harvested?
6. Describe the mitochondrial electron transport system and the process of chemiosmosis.
7. Why is oxygen necessary for cellular respiration?

Applying the Concepts

1. Some years ago a freight train overturned, spilling a load of grain. Because the grain was unusable, it was buried in the embankment. Yeasts are common in the soil. Although there is no shortage of other food locally, the local bear population has created a nuisance by continually uncovering the grain. What do you think has happened to the grain to make them do this, and how is it related to human cultural evolution?
2. In detective novels, "the odor of bitter almonds" is the telltale clue to murder by cyanide poisoning. Cyanide works by attacking the enzyme that transfers electrons from the electron transport system to O_2. Why is it not possible for the victim to survive by using lactate fermentation? Why is cyanide poisoning almost immediately fatal?
3. More than a century ago, French biochemist Louis Pasteur described a phenomenon, now called "the Pasteur effect," in the wine-making process. He observed that in a sealed container of grape juice containing yeast, the yeast will consume the sugar very slowly as long as oxygen remains in the container. As soon as the oxygen is gone, however, the rate of sugar consumption by the yeast increases greatly and the alcohol content in the container rises. Discuss the Pasteur effect on the basis of what you know about cellular respiration and fermentation.
4. The dumping of large amounts of raw sewage into rivers or lakes typically leads to massive fish kills, although sewage itself is not toxic to fish. Similar fish kills also occur in shallow lakes that become covered in ice during the winter. What kills the fish? How might you reduce fish mortality after raw sewage is accidentally released into a small pond containing large bass?
5. Although respiration occurs in all living cells, different cells respire at different rates. Explain why. How could you predict from the microscopic examination of cells the relative rates of cellular respiration in different tissues of a fish?
6. Imagine that a starving cell reached the stage where every bit of its ATP was depleted and converted to ADP plus phosphate. If that cell were placed in fresh nutrient broth at this point, will it recover and survive? Explain your answer based on what you know of glucose breakdown.

Group Activity

Bo and Luke are competitive body builders who train at the same Florida gym but are bitter enemies. At a workout a week before an upcoming national competition, Bo begins to feel very hot, sweats profusely, and then collapses. When the paramedics arrive, they note that Bo is feverish, has a rapid heartbeat and a rapid breathing rate, and appears blue around the mouth. On the way to the hospital, Bo goes into a coma; he dies a day later.

During the subsequent investigation, crystals of some substance are discovered on the top of Bo's water bottle. In addition, the police discover that Luke has an illegal tropical fish business in which he uses cyanide to stun fish that swim near the reefs around the Florida Keys. When they float to the top, he scoops them up in a net. Those fish that recover are sent to aquarium stores throughout the United States.

The police suspect that, to increase his chances of winning the competition, Luke has poisoned Bo with cyanide. Luke maintains that he is innocent and thinks that Bo has accidentally poisoned himself with dinitrophenol, an illegal and highly toxic drug that some body builders take to lose weight. Break up into groups of three or four to determine whether Bo has died of cyanide poisoning or of dinitrophenol poisoning.

HINT 1: Both cyanide and dinitrophenol interfere with the function of mitochondria, but cyanide targets the electron transport chain whereas dinitrophenol targets the functioning of an ATP-synthesizing enzyme. HINT 2: Which drug will cause oxygen consumption to increase? Increased oxygen consumption by your cells will result in rapid breathing and a bluish skin color as oxygen is removed from the blood faster than it can be replaced. Decreased oxygen consumption will result in shortness of breath and bright red veins and arteries.

For More Information

McCarty, R. E. "H^+-ATPases in Oxidative and Photosynthetic Phosphorylation." *BioScience*, January 1985. A description of the structure and function of the ATP-synthesizing enzymes in mitochondria and chloroplasts.

Nelson, M., Burgess, T. L., Alling, A., Alverez-Romo, N., Dempster, W. F., Walford, R. L., and Allen, J. P. "Using a Closed Ecological System to Study Earth's Biosphere." *BioScience*, 1993. An artificial ecosystem allows scientists to learn more about how natural ecosystems function.

Answers to Multiple-Choice Questions

1. a 2. c 3. d 4. c 5. a 6. b

MEDIATUTOR

CD Activities

Activity 7.1: Glucose Metabolism

Estimated Time: 15 minutes

The mechanisms that cells use to derive energy from the sugar glucose are explored. The passage of glucose through glycolysis and the Krebs cycle are detailed, along with the fates of products of glycolysis under both anaerobic and aerobic conditions. The oxidative phosphorylation of ATP is also presented.

Start the MediaTutor Student CD-ROM and enter the activity number in the Quick Search box to be taken directly to that activity.

Web Investigation

Unfair Advantgage?

Estimated time: 15 minutes

Living organisms catabolize organic molecules and use the energy released to manufacture ATP, the energy molecule of life. For aerobic organisms, this process occurs via the mechanism of cellular respiration. Oxygen use and ATP production are mediated in eukaryotic cells through the mitochondria. This important organelle is central to a series of complex biochemical reactions that provide the cell with the energy it needs to carry out its normal functions. This investigation will further explore the role oxygen plays in this process and how certain compounds can assist or interfere with the production of ATP in the cells of our body.

Go to http://www.prenhall.com/audesirk3, the Audesirk Companion Web site. Select Chapter 7 and the Web Investigation to begin.

UNIT TWO

Inheritance

These apples share the same general shape and structure because their genes are nearly identical. Tiny differences in some genes, however, give the apples different colors, shapes, and sizes.

"A structure this pretty just had to exist."

James Watson in* The Double Helix *(1968)

Sunbathers increase their chances of developing skin cancer because sunlight can damage the double helix structure of DNA.

8 DNA: The Molecule of Heredity

AT A GLANCE

CASESTUDY

Sunshine Perils

Do you plan to spend a lot of time at the beach or pool next summer? If so, be careful, because the consequences of exposing unprotected skin to the sun can be severe. Consider the case of Rachel, a college student with a summer job as a lifeguard.

Preoccupied with her job and her busy summer social life, Rachel paid little attention to the bumpy, black mole on her back. Her mother, however, noticed the mole and urged Rachel to schedule an appointment with the family doctor. The doctor advised Rachel to have the mole removed, and told her that he could perform the procedure right there in his office.

Once the mole was gone, Rachel gave it little thought. A few days later, however, the doctor phoned her. As a precaution, he had sent the tissue he removed to a laboratory for examination. The lab report was sobering: Rachel had cancer.

Rachel's cancer was diagnosed as melanoma, a relatively common type of skin cancer. The cancer can spread to other parts of the body, including internal organs, with frequently deadly results. This year, about 7000 people in the United States will die as a result of melanoma, and more than 44,000 new cases will be diagnosed. The incidence of melanoma appears to be rising, and many physicians suspect that the increase is the result of people spending more time in the sun.

How can sunlight, the carrier of life-giving energy for the planet, cause cancer? To answer this question, we need to understand the basic structure of DNA, the molecule of heredity. ■

1) What Are Genes Made of?

Just 60 years ago, no one knew that deoxyribonucleic acid, or **DNA**, is the molecule that carries the blueprints for all forms of life on Earth. We now know that the molecular recipe in DNA directs the life of each cell in an organism, and confers on each cell its special characteristics. DNA also enables organisms to transmit characteristics from one generation to the next. A parent passes copies of its DNA to its offspring, and those offspring thus inherit some of the characteristics of the parent.

Even before the discovery of DNA's role in inheritance, scientists understood that characteristics were transmitted from parents to offspring in discrete units called **genes**, which determine many of the inherited differences among individuals. For example, a gene for flower color determines whether an individual pea plant has white flowers or purple flowers. Today, we know that genes are made of DNA; each gene is a particular segment of DNA. Genes are located within cells in threadlike structures called **chromosomes**, which contain protein as well as DNA.

2 What Is the Structure of DNA?

Knowing that genes are made of DNA does not by itself tell us much about how inheritance works. But before we can describe the methods by which DNA copies itself and transmits information, we must first describe the structure of the DNA molecule.

DNA Is Composed of Four Different Subunits

The DNA of every organism on Earth is composed of four small subunits called **nucleotides**. Each nucleotide in DNA consists of three parts: (1) a phosphate group, (2) a sugar called deoxyribose, and (3) one of four different nitrogen-containing **bases**: **adenine (A)**, **thymine (T)**, **guanine (G)**, or **cytosine (C)**. The nucleotides join together in long chains to form a DNA molecule.

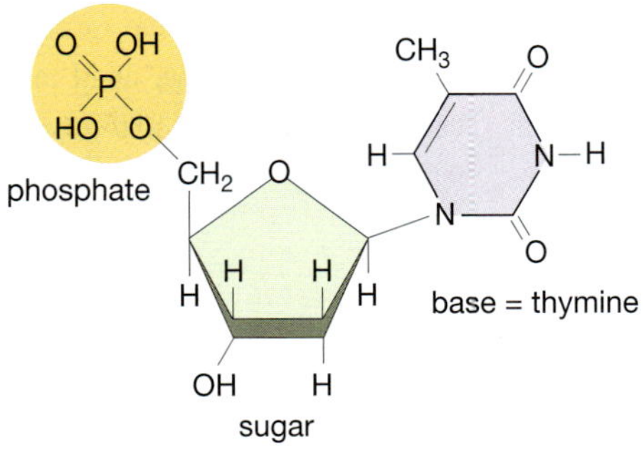

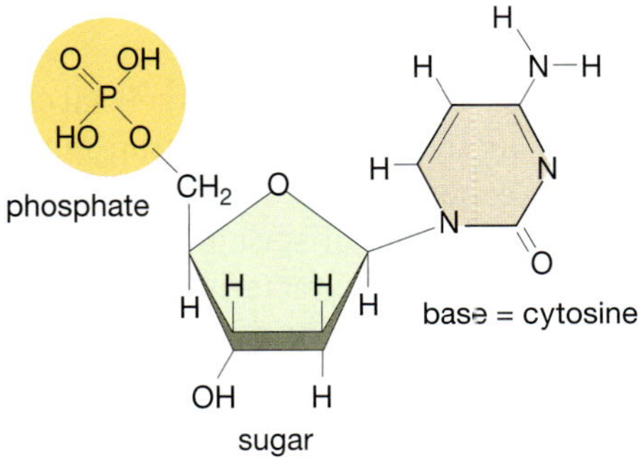

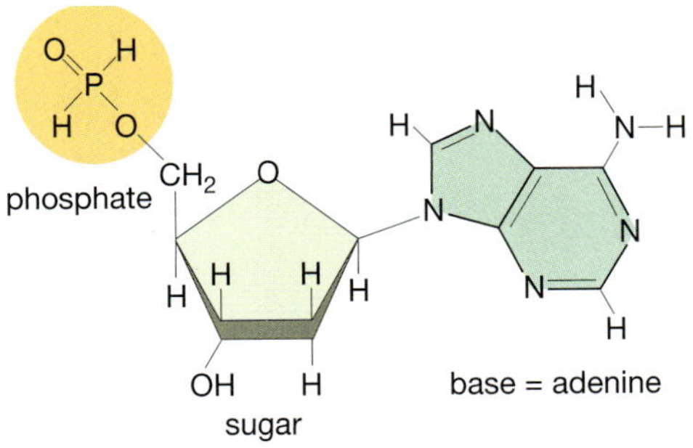

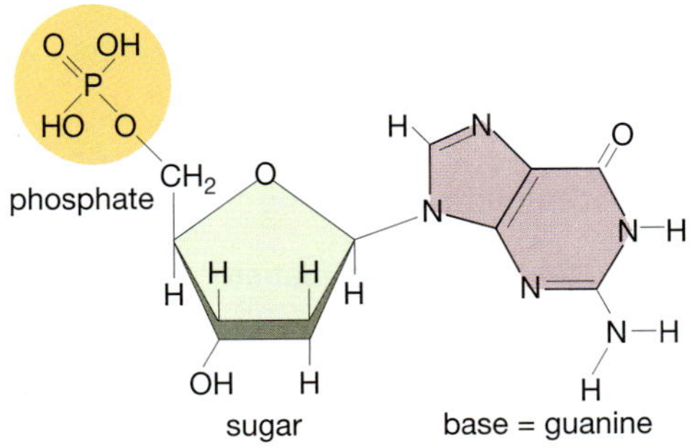

DNA Is a Double Helix of Two Nucleotide Strands

After earlier researchers discovered the basic elements of DNA's composition and structure, the final secrets of DNA's structure were uncovered in the early 1950s by James Watson and Francis Crick (see "Scientific Inquiry: The Discovery of the Double Helix"). Combining a knowledge of how complex organic molecules bond together with an intuition that "important biological objects come in pairs," Watson and Crick found that the DNA molecule consists of two separate DNA **strands** of linked nucleotides. Within each strand, the phosphate group of one nucleotide bonds to the sugar of the next nucleotide in the strand. This bonding pattern produces a "backbone" of alternating, covalently bonded sugars and phosphates. The nucleotide bases protrude from this **sugar–phosphate backbone**.

All of the nucleotides within a single DNA strand are oriented in the same direction. This consistent orientation makes the two ends of a single DNA strand different: One end of the DNA strand has an unbonded sugar and the other end of the DNA strand has an unbonded phosphate (Fig. 8-1a).

Hydrogen Bonds Hold the Two DNA Strands Together

Watson and Crick discovered that the two DNA strands are held together by hydrogen bonds that form between the protruding bases of the two separate strands. These bonds give DNA a ladder-like structure, with the sugar-phosphate backbones forming the sides and the nucleotide bases forming the rungs (Fig. 8-1b). However, the DNA strands are not straight. Instead, they are

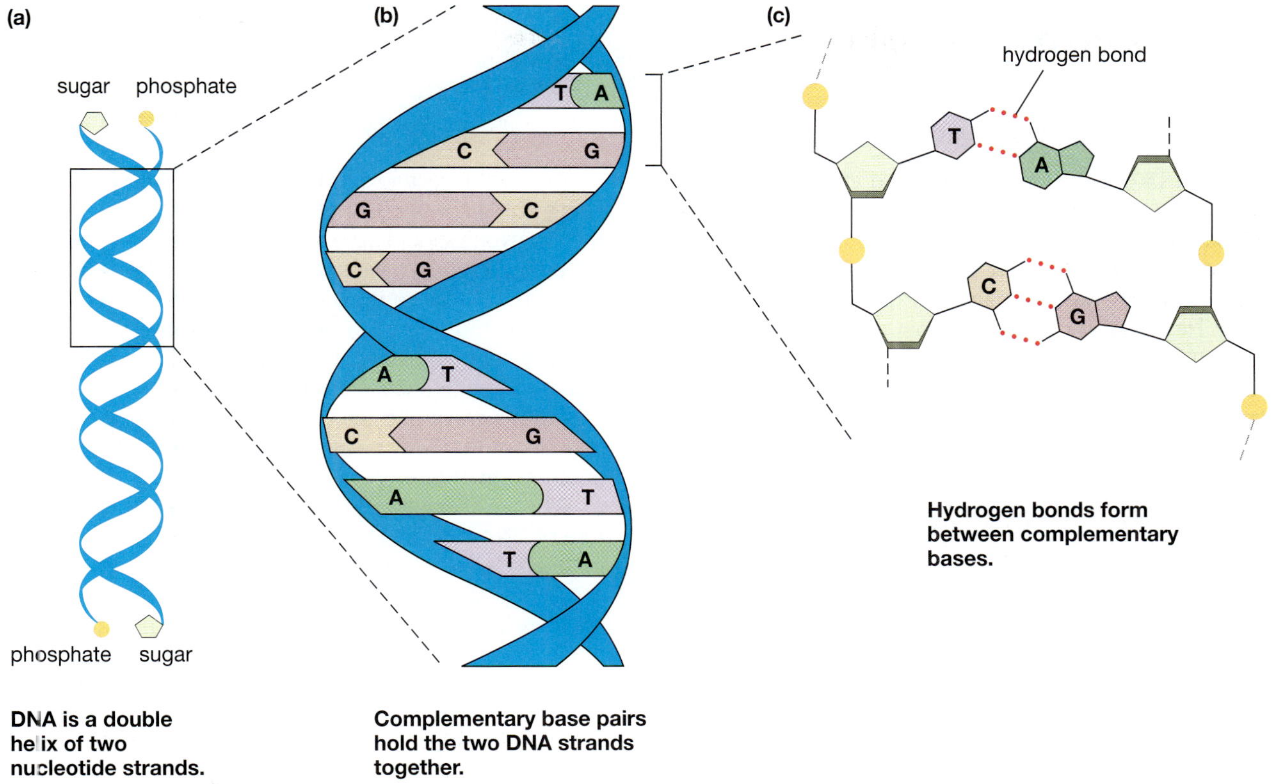

Figure 8-1 The Watson-Crick model of DNA structure
(a) Two strands of DNA, running in opposite directions, wind about each other in a double helix, like a twisted ladder.
(b) Complementary base pairs (adenine and thymine, guanine and cytosine) hold the two DNA strands together.
(c) In DNA, hydrogen bonds form only between T and A and between C and G.

MEDIATUTOR 8.1 DNA Structure

twisted about each other to form a **double helix**, much like a ladder twisted into a circular staircase shape. The two strands in a DNA double helix are oriented in opposite directions so that each end of the helix has one strand with an unbonded phosphate and one with an unbonded sugar (see Fig. 8-1a).

Take a closer look at the pairs of hydrogen-bonded bases that form each rung of the double helix ladder. Notice that adenine forms hydrogen bonds only with thymine and that guanine forms hydrogen bonds only with cytosine (Fig. 8-1c). These A–T and G–C pairs are called **complementary base pairs**. Because an A in one DNA strand always pairs with a T in the other strand, the amount of A in DNA always equals the amount of T. Similarly, because a G in one DNA strand always pairs with a C in the other DNA strand, the amount of G always equals the amount of C.

3 How Does DNA Encode Information?

Look again at the elegant structure of DNA shown in Figure 8-1. The relative simplicity of the structure and the small number of different bases incorporated in it raise the question of how DNA could possibly encode the vast amount of information required to build and operate an organism. How, for example, can the color of a bird's feathers, the size and shape of its beak, its ability to make a nest, and its urge to migrate all be determined by a molecule with just four simple parts?

Although it took many experiments to explain DNA's tremendous capacity for information storage, scientists eventually found evidence to support a simple explanation. It's not the *number* of different subunits that is important, but rather their *order*. Within a DNA strand, the four types of bases can be arranged in any order. Each sequence of bases represents a unique set of genetic instructions, like a biological Morse code. A stretch of DNA that is just 10 nucleotides long can have more than a million possible sequences of the four bases. Because an organism has millions (in bacteria) to billions (in plants or animals) of nucleotides, DNA molecules can encode a staggering amount of information.

The Big Picture

DNA is a marvel of efficiency and miniaturization. Composed of only a handful of chemical building blocks and confined to the microscopically small spaces inside of cells, it nonetheless stores the coded instructions for every living thing.

Scientific Inquiry
The Discovery of the Double Helix

In the early 1950s many biologists realized that the key to understanding inheritance lay in the structure of DNA. They also knew that whoever deduced the correct structure of DNA would receive recognition, very possibly the Nobel Prize. Linus Pauling of Caltech was the person most likely to solve DNA structure. He probably knew more about the chemistry of large organic molecules than did any person alive. Pauling, however, had two main handicaps. First, for years he had concentrated on protein research, and therefore he had little data about DNA. Second, he was active in the peace movement. At that time some government officials, including Senator Joseph McCarthy, considered such activity to be potentially subversive and possibly dangerous to national security. This latter handicap may have proved decisive.

The second most likely competitors were the British scientists Rosalind Franklin and Maurice Wilkins, who were experts in using X rays to determine the structures of large molecules. In fact, Wilkins and Franklin's X-ray studies made them the only scientists who had very good data about the general shape of the DNA molecule. Unfortunately for them, their methodical approach was also slow.

The door was open for the eventual discoverers of the double helix, James Watson and Francis Crick, two scientists with neither Pauling's tremendous understanding of chemical bonds nor Franklin and Wilkins's expertise in X-ray analysis. Watson and Crick did no experiments in the ordinary sense of the word; instead, they spent their time thinking about DNA, trying to construct a molecular model that made sense and fit the data. Because they were working in England and because Wilkins was very open about his and Franklin's data, Watson and Crick were familiar with all the X-ray information relating to DNA.

This X-ray information was just what Pauling lacked. Because of Pauling's presumed subversive tendencies, the U.S. State Department refused to issue him a passport to leave the United States, so he could neither attend meetings at which Wilkins presented the X-ray data nor visit England to talk with Franklin and Wilkins directly. Watson and Crick knew that Pauling was working on DNA structure and were driven by the fear that he would beat them to it. In his book *The Double Helix*, Watson recounts his belief that, if Pauling could have seen the X-ray pictures, "in a week at most, Linus would have the structure."

You might be thinking by now, "But wait just a minute! That's not fair. If the goal of science is to advance knowledge, then everyone should have access to all the data. If Pauling was the best, he should have discovered the double helix first." Perhaps so. But science is an activity of scientists, who, after all, are people too. Although virtually all scientists want to see the advancement and benefit of humanity, each individual also wants to be the one responsible for that advancement and to receive the credit and the glory. Linus Pauling remained in the dark about the X-ray pictures of DNA and the clues they gave about its structure.

When Watson and Crick discovered the double helix structure of DNA (Fig. E8-1), Watson described it in a letter to Max Delbruck, a friend and adviser at Caltech. He asked Delbruck not to reveal the contents of the letter to Pauling until their structure was formally published. Delbruck, perhaps more of a model scientist, firmly believed that scientific discoveries belong in the public domain and promptly told Pauling all about it. Showing himself to be a great scientist and a great person, Pauling graciously congratulated Watson and Crick on their brilliant solution to the DNA structure. The race was over.

Figure E8-1 The discovery of DNA
James Watson and Francis Crick with a model of DNA structure.

4 How Does DNA Copy Itself?

Cells reproduce by a complex process of cell division that produces two daughter cells from a single parental cell. To produce healthy daughter cells, cell division must ensure that each daughter cell receives a copy of all the parental cell's genetic information.

DNA Is Copied During Cell Division

At an early stage of cell division, the parental cell must synthesize two exact copies of its DNA, a process known as **DNA replication**. DNA replication produces two identical DNA double helices, each of which will be passed to one of the new daughter cells. Each double helix makes up the DNA component of a single chromosome, so each daughter cell receives its DNA as a set of chromosomes.

Although many cells in an adult organism never divide, most of the thousands of cells that do divide are irreversibly committed to do so once DNA replication begins. If a cell tries to replicate its DNA without stockpiling enough raw materials and energy to complete the process of cell division, it can die. Consequently, the timing of replication is tightly controlled. This control ensures that DNA replication does not begin unless and until a cell is ready to divide.

DNA Replication Produces Two DNA Double Helices

DNA replication begins when enzymes pull apart the parental DNA double helix, so that the bases of the two parental strands no longer form base pairs with one another. Other enzymes move along each separated parental strand, selecting individual free nucleotides from the surrounding fluid. The selected free nucleotides have bases that are complementary to the bases in the parental strand, so they form hydrogen bonds with them. As free nucleotides line up along the two parental strands, enzymes join the free nucleotides together to form two new strands, each complementary to the parental strand it is joining with (Fig. 8-2).

When replication is complete, one parental strand and its newly synthesized complementary strand wind together into one double helix. At the same time, the other parental strand and its new strand wind together into a second double helix.

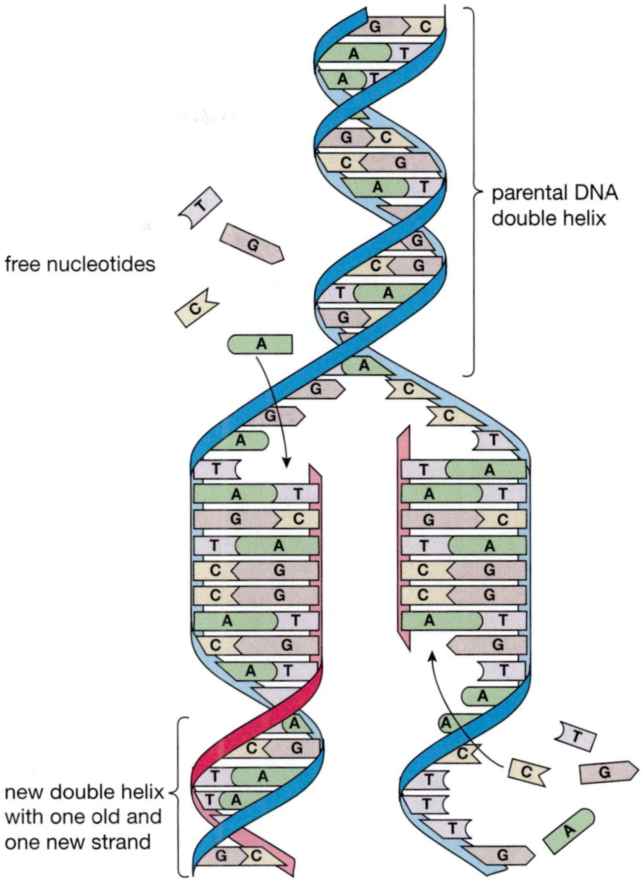

Figure 8-2 DNA replication
The two strands of a DNA molecule separate during replication. Each strand becomes part of a new DNA molecule.

Each New Double Helix Has One Old Strand and One New Strand

Note that, when DNA replicates, each new molecule contains one parental strand and one newly synthesized strand (see Figure 8-2). The process thus conserves one parental strand in each new molecule and is therefore called **semiconservative replication**.

5 Which Enzymes Are Important in DNA Replication?

DNA replication requires the action of dozens of enzymes. We describe the roles of a few of the most important ones below.

DNA Helicase Separates Parental Strands

One key enzyme is **DNA helicase**. Together with related enzymes, DNA helicase uses energy from ATP to break the hydrogen bonds between complementary base pairs that hold the two parental strands of a DNA molecule together. This activity separates and unwinds the parental double helix, forming a gap called a **replication bubble** (Fig. 8-3). Within the replication bubble, the nucleotide bases of the parental strands are no longer paired with one another, and replication can proceed.

A replication bubble is bounded on each end by a spot where the two parental strands have not yet been unwound. These locations are called replication forks, perhaps because they resemble a fork in a road. Imagine that you are traveling along a DNA double helix, moving toward a replication bubble. When you reach the replication bubble, the two strands separate and you must choose whether to follow one strand or the other. DNA replication continues from the replication forks until it is completed.

DNA Polymerase Synthesizes New DNA Strands

The enzyme **DNA polymerase** plays a critical role in the synthesis of new DNA strands. Working at each replication fork, DNA polymerase recognizes unpaired nucleotide bases in the parental strands and matches them

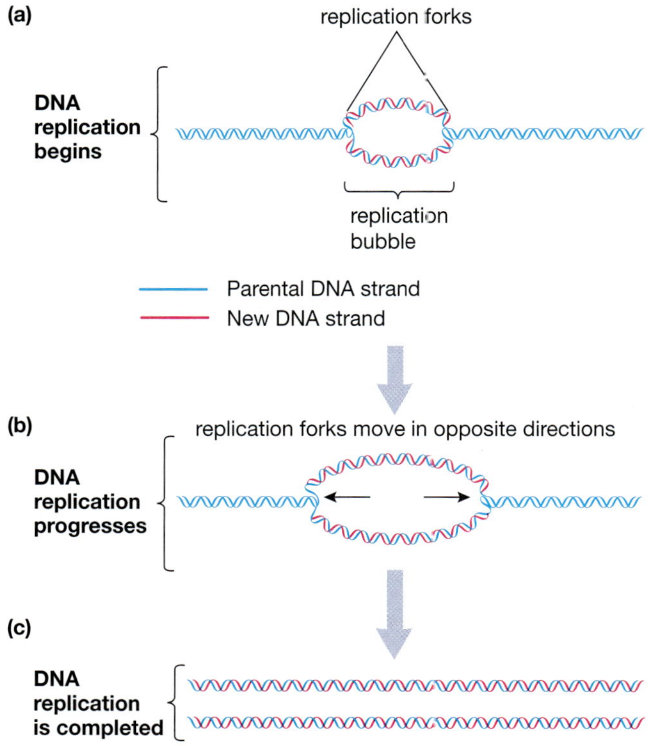

Figure 8-3 Replication bubble
For DNA to replicate, DNA helicase and related enzymes must unwind portions of the parental double helix, creating a replication bubble.

up with free nucleotides that have the correct complementary base. For example, DNA polymerase pairs up an exposed adenine base with a thymine base. Then DNA polymerase catalyzes the formation of new covalent bonds that link the phosphate of the incoming free nucleotide to the sugar of the previously added nucleotide in the growing new strand. In this way, DNA polymerase helps build the sugar-phosphate backbone of the new strand.

Like other enzymes, DNA polymerase is extremely specific. It can add new nucleotides only onto the unbonded-sugar end of the new strand (not onto the unbonded-phosphate end). Thus, a new strand can grow in only one direction (toward the unbonded-sugar end of the parental strand). Because the two parental strands are oriented in opposite directions, the two new strands grow in opposite directions. Consequently, as DNA helicase moves along the parental double helix, separating the strands, the DNA polymerases on the two parental strands behave differently. On one strand, the DNA polymerase moves toward the replication fork, adding nucleotides to form a long, continuous new strand. But, to attach new nucleotides onto the unbonded-sugar end of the opposite strand, the DNA polymerase must move in the opposite direction, away from the replication fork. As this new strand grows away from the fork, the fork continues to open behind the growing strand. For this reason, the strand growing away from the fork must continually begin anew at the fork, synthesizing a new strand in short segments (Fig. 8-4).

The Strand Synthesized in Short Segments Is United by DNA Ligase

The short segments that grow during DNA replication on one strand of parental DNA are subsequently connected together by the enzyme **DNA ligase**. This joining process is repeated as many as 10 million times during replication of a single human chromosome. DNA ligase also plays an important role in repairing DNA that has been damaged.

The Big Picture

DNA's ability to replicate itself is essential for the continuity of life. The capacity of life's information-carrying molecules to accurately replicate themselves enables organisms to make genetically identical copies of existing cells. Such copies are required for organisms to grow and reproduce.

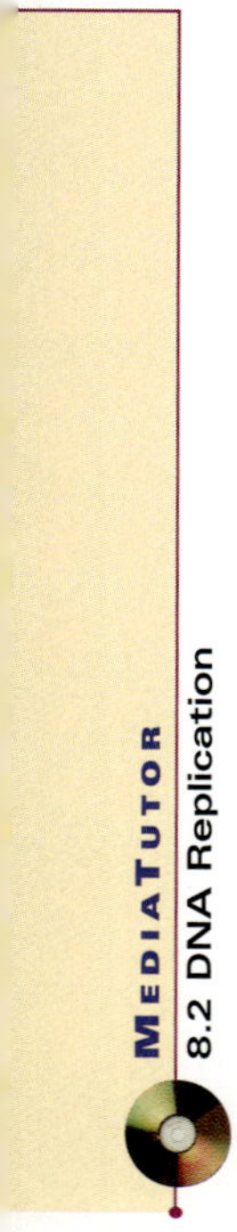

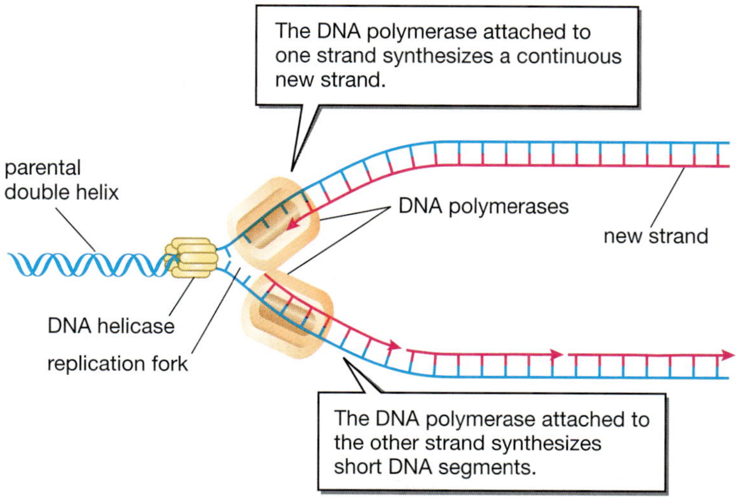

Figure 8-4 One new DNA strand grows continuously, the other grows in segments
The short segments of the lower new strand will be joined together by DNA ligase.

Health Watch
Genes on a Chip

A new device called the DNA chip promises to foster dramatic improvements in medical diagnosis, genetic screening, and drug development. A DNA chip is constructed on a thin wafer of silicon, just like a computer microchip. But instead of circuits, a DNA chip holds millions of strands of DNA, each strand firmly attached to the chip's surface.

As many as a million short (8 to 20 bases), single-stranded fragments of DNA can be placed on a single chip. Chip manufacturers specify the precise point of attachment for each DNA strand and the exact sequence of nucleotides in each strand. Each strand in the array will, under appropriate conditions, bind to any other strand that has a complementary sequence of bases (just as complementary bases bind to the template strand of a replicating segment of DNA in a cell).

To use one of these chips, an investigator secures a sample of DNA to be tested, perhaps from a drop of blood. The DNA is chopped into short, single-stranded fragments. Each fragment is labeled by attaching molecules of a fluorescent chemical to it. The labeled DNA fragments are then spread over the surface of a chip. Any fragments in the sample that are complementary to any of those embedded in the chip will bind at the appropriate location. Fragments that find no mate are washed away. The investigator can then view the chip under a microscope; the locations on the chip at which labeled sample fragments have bound will glow. The pattern of fluorescence on the chip tells the investigator the degree to which the DNA in the sample matches the DNA in the chip (Fig. E8-2).

Why would we want to know if the DNA in a sample matched the DNA on a chip? One reason might be to diagnose infections. For example, we could manufacture a chip loaded with the gene sequences of many different disease-causing bacteria and viruses. A single drop of blood from a sick patient would then be sufficient to screen for a wide range of diseases. This approach is already used to screen AIDS patients to see if they harbor the strains of HIV that are resistant to the most common anti-AIDS drugs.

Gene chips could also be used to reveal genetic defects that cause diseases. Many diseases, including diabetes, cystic fibrosis, and breast cancer, are associated with known genetic defects. Doctors can now, in theory, detect such genetic defects before a patient actually gets sick. In practice, however, there is no comprehensive genetic screening of patients, because accomplishing such screening using standard laboratory techniques would require prohibitive amounts of time and money. DNA chips, however, with their ability to screen vast stretches of DNA rapidly and cheaply, could solve the problem. Some boosters of DNA technology believe that DNA chips will soon make large-scale screening easy and affordable. From there, it will be just a short step to "designer drugs," medicines tailor-made to treat each individual's particular combination of genetic propensities to disease.

Of course, not everyone is pleased with the prospect of universal genetic screening. Will insurance companies use such information to deny coverage to some people? How will new parents cope with information about the potential disabilities of their children? Is it really beneficial for everyone to be informed of his or her risks for acne, obesity, and alcoholism? The ethical dilemmas raised by genetic screening are profound and complex. But the technology has already arrived.

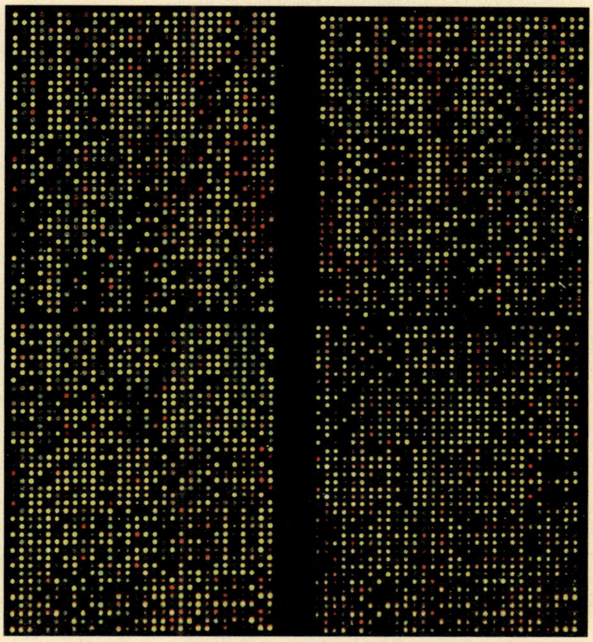

Figure E8-2 A DNA chip
The location, color, and brightness of the spots on the chip reveal sequences present in a DNA sample.

6 Do Errors Occur During DNA Replication?

Bonding between complementary base pairs makes DNA replication highly accurate. Nevertheless, no process in cells, not even DNA replication, is perfect.

Proofreading Produces Almost Error-Free Replication of DNA

Partly because replication is so fast (up to about 700 nucleotides per second), DNA polymerase matches bases incorrectly about once in every 10,000 base pairs. Completed DNA strands, however, contain only about one mistake in every billion base pairs. This improvement is due to various DNA repair enzymes that "proofread" each new strand. Working during and after construction of a new strand, these proofreading enzymes also make any necessary repairs.

But a Few Mistakes Do Remain

Despite the amazing accuracy of DNA replication, no organism is completely free of DNA damage. For example, in addition to mistakes made during DNA replication, the DNA in each cell in a human body loses about 10,000 bases every day due to spontaneous chemical breakdown of DNA. Certain environmental conditions can also damage DNA. For example, whenever you go out in the sunshine, DNA in some of your skin cells is damaged by ultraviolet light. Although special DNA repair enzymes are continuously on call to repair this damage, inevitably some errors are not repaired. A cell with DNA damage may function normally (if, for example, the damage is in a stretch of DNA whose function is not important); or such a cell may survive but not function as efficiently as before; or it may die.

CASE STUDY REVISITED

Sunshine Perils

Rachel had frequently been sunburned as a child, and all that sun may have caused Rachel's cancer. Studies indicate that getting sunburned when we are kids increases our risk of developing melanoma to about three times higher than that of those who were not sunburned.

Why is sunburn connected to cancer? Because sunlight can damage DNA. Even on a cloudy day, ultraviolet rays in sunlight penetrate uncovered skin and sometimes directly hit nucleotides in DNA. Thymine and cytosine are particularly prone to UV damage. For example, if the UV rays hit a DNA strand with two adjacent thymines, the two thymines can become improperly linked together. Unless the damage is repaired, DNA polymerase may be unable to properly replicate this region of DNA and may insert incorrect nucleotides into the new strand. If the damage is to one of the many genes that help control cell division, the damaged cell may begin the out-of-control reproduction that causes cancerous tumors to form.

Because her melanoma was caught at a very early stage, Rachel's prognosis for total recovery is good. You, too, can learn to spot potential problems before they become deadly, since recognizing a possible melanoma is as easy as ABCD: Look at your moles for *A*symmetry, irregular *B*order, unusual *C*olor, or a *D*iameter larger than the eraser on the end of a pencil. Be sure to visit your doctor promptly to have any suspicious spots checked out.

Summary of Key Concepts

1) What Are Genes Made of?

DNA, located in each cell's chromosomes, is the molecule of heredity. Genes, which transmit characteristics of organisms to succeeding generations, are made of DNA.

2) What Is the Structure of DNA?

DNA is composed of nucleotides that are linked together into long strands. Each nucleotide consists of a phosphate group, the five-carbon sugar deoxyribose, and a nitrogen-containing base. There are four types of base in DNA: adenine, guanine, thymine, and cytosine. Two nucleotide strands wind about one another to form a DNA double helix. Within each strand, the sugar of one nucleotide is linked to the phosphate of the next nucleotide, forming a sugar-phosphate "backbone" on each side of the double helix. The nucleotide bases of each strand pair up in the middle of the helix, held together by hydrogen bonds. Only complementary base pairs can bond together in the helix: adenine bonds with thymine, and guanine bonds with cytosine.

3) How Does DNA Encode Information?

Even though DNA is composed of only four different nucleotides, it is able to encode huge amounts of information. This ability stems from the huge number of different sequences in which the four bases can be arranged.

4) How Does DNA Copy Itself?

When cells reproduce, they must replicate their DNA so that each daughter cell receives all of the original genetic information. During DNA replication, the two parental DNA strands unwind. Then, free nucleotides bond with complementary bases on the parental strands, and the free nucleotides link together to form new DNA strands. The sequence of nucleotides in each newly formed strand is complementary to the sequence of a parental strand. Replication is semiconservative because, when DNA replication is complete, both new DNA double helices consist of one parental DNA strand and one newly synthesized, complementary strand.

5) Which Enzymes Are Important in DNA Replication?

DNA helicase breaks the bonds between complementary bases so that the parental DNA strands can unwind and separate to form a replication bubble. DNA polymerase synthesizes new DNA strands by selecting and attaching to the parental strand complementary free nucleotides, and then linking the free nucleotides together. DNA ligase links the short fragments of daughter DNA that form on one of the parental strands during replication.

6) Do Errors Occur During DNA Replication?

DNA polymerase mismatches some base pairs during replication, but proofreading enzymes detect and repair most of these errors. Nonetheless, a few errors do remain even after the activity of the proofreading and repair systems. In addition, environmental factors such as ultraviolet light can damage DNA.

Key Terms

adenine *p. 112*
base *p. 112*
chromosome *p. 112*
complementary base pair *p. 113*
cytosine *p. 112*
DNA *p. 111*
DNA helicase *p. 115*
DNA ligase *p. 116*
DNA polymerase *p. 115*
DNA replication *p. 114*
double helix *p. 113*
gene *p. 112*
guanine *p. 112*
nucleotide *p. 112*
replication bubble *p. 115*
semiconservative replication *p. 115*
strand *p. 112*
sugar–phosphate backbone *p. 112*
thymine *p. 112*

Thinking Through the Concepts

Multiple Choice

1. *How many different possible base sequences are there in a nucleotide chain three nucleotides in length?*
   a. 1
   b. 3
   c. 9
   d. 64
   e. more than 64

2. *Because each base pairs with a complementary base, in every DNA molecule the amount of*
   a. cytosine equals that of guanine
   b. cytosine equals that of thymine
   c. cytosine equals that of adenine
   d. each nucleotide is unrelated to all others
   e. each nucleotide is equal to all others

3. *Semiconservative replication refers to the fact that*
   a. each new DNA molecule contains two new single DNA strands
   b. DNA polymerase uses free nucleotides to synthesize new DNA molecules
   c. certain bases pair with specific bases
   d. each parental DNA strand is joined with a new strand containing complementary base pairs
   e. mistakes are made during DNA replication

4. *DNA helicase*
   a. cleaves hydrogen bonds that join the two strands of DNA
   b. converts two single strands of DNA into a double helix
   c. is a unique form of DNA
   d. adds nucleotides to newly forming DNA molecules
   e. proofreads the newly formed DNA strand

5. *DNA polymerase*
   a. can advance in either direction along a single strand of DNA
   b. cleaves hydrogen bonds that join the two strands of DNA
   c. creates a polymer that consists of many molecules of DNA
   d. adds appropriate nucleotides to a newly forming DNA strand

6. *Which of the following are incorrectly matched?*
   a. complementary base pairs ↔ adenine and cytosine
   b. bases ↔ adenine, thymine, cytosine, and guanine
   c. nucleotide ↔ phosphate and sugar and base
   d. eukaryotic chromosome ↔ DNA and protein
   e. enzymes for DNA replication ↔ DNA polymerase and DNA helicase

? Review Questions

1. Draw the general structure of a nucleotide. Which parts are identical in all nucleotides, and which can vary?
2. Name the four types of nitrogen-containing bases found in DNA.
3. Which bases are complementary to one another? How are they held together in the double helix of DNA?
4. Describe the structure of DNA. Where are the bases, sugars, and phosphates in the structure?
5. Describe the process of DNA replication.

Applying the Concepts

1. As you learned in "Scientific Inquiry: The Discovery of the Double Helix," scientists in different laboratories often compete with one another to make new discoveries. Do you think this competition helps promote scientific discoveries? Sometimes, researchers in different laboratories collaborate with one another. What advantages does collaboration offer over competition? What factors might provide barriers to collaboration and lead to competition?
2. Today, scientific advances are being made at an astounding rate, and nowhere is this more evident than in our understanding of the biology of heredity. Using DNA as a starting point, do you believe there are limits to the knowledge people should acquire? Defend your answer.

Group Activity

Make a class model of DNA: Form two lines of students, with the students in each line facing a different direction. Each of you in the line should put your left hand on the shoulder of the person in front of you and extend your right hand out toward the other line, palm facing forward. Each of you is a nucleotide. The hand on the shoulder represents the sugar–phosphate link. The hands extended, almost touching, represent the base pairs. It is easy to see the antiparallel nature of DNA in this model.

As an additional exercise, have one student play the role of DNA polymerase by matching "complementary bases"—that is, by pairing up students in opposite lines on the basis of eye color or some other basis. For example, pair brown eyes with green eyes, and pair blue eyes with gray or black eyes.

For More Information

Crick, F. *What Mad Pursuit: A Personal View of Scientific Discovery*. New York: Basic Books, 1998. A view of the race to determine the structure of DNA, by Francis Crick himself.

Gibbs, W. W. "Peeking and Poking at DNA." *Scientific American (Explorations)*, March 31, 1997. An update on new techniques for studying the DNA molecule.

Judson, H. F. *The Eighth Day of Creation*. Cold Spring Harbor, NY: Cold Spring Harbor Laboratory Press, 1993. A very readable historical perspective on the development of genetics.

Leutwyler, K. "Turning Back the Strands of Time." *Scientific American (Explorations)*, February 2, 1998. A brief discussion of telomeres, the repeating DNA regions at the ends of chromosomes.

Mirsky, A. E. "The Discovery of DNA." *Scientific American*, June 1968. The early history of DNA research.

Olby, R. *The Path to the Double Helix*. Seattle: University of Washington Press, 1975. A historical perspective on the development of genetics in the twentieth century.

Radman, M., and Wagner, R. "The High Fidelity of DNA Duplication." *Scientific American*, August 1988. Faithful duplication of chromosomes requires both reasonably accurate initial replication of DNA sequences and final proofreading.

Rennie, J. "DNA's New Twists." *Scientific American*, March 1993. A reprise of new information on DNA structure and function.

Watson, J. D. *The Double Helix*. New York, Atheneum, 1968. If you still believe the Hollywood images that scientists are either maniacs or cold-blooded, logical machines, be sure to read this book. Although hardly models for the behavior of future scientists, Watson and Crick are certainly human enough!

Weinberg, R. "How Cancer Arises." *Scientific American*, September 1996. An overview of the molecular basis of cancer: mutations in DNA.

Answers to Multiple-Choice Questions

1. d 2. a 3. d 4. a 5. d 6. a

MediaTutor

CD Activities

Activity 8.1: DNA Structure

Estimated time: 5 minutes

The structure of the DNA helix is key to its ability to store and transmit genetic information. In this tutorial, you will explore the parts of the DNA helix.

Activity 8.2: DNA Replication

Estimated time: 10 minutes

Making exact copies of our genetic material (chromosomes) is crucial to life. In this tutorial, you will explore the mechanism of DNA replication.

Start the MediaTutor Student CD-ROM and enter the activity number in the Quick Search box to be taken directly to that activity.

Web Investigation

Sunshine Peril

Estimated time: 10 minutes

Skin cancer is now the most common cancer in the United States. Who is most at risk? What are the treatment alternatives? How can it be prevented? This exercise explores the biochemical, medical, and psychological features of skin cancers.

Go to http://www.prenhall.com/audesirk3, the Audesirk Companion Web site. Select Chapter 8 and Web Investigation to begin.

People with hemophilia, whose cells cannot produce all the proteins necessary for blood clotting, currently depend on expensive injections of clotting proteins to prevent uncontrolled bleeding. Gene therapy may soon produce a cure for this genetic disease.

9 Gene Expression and Regulation

AT A GLANCE

CASESTUDY

Fixing Broken Genes

To repair a malfunctioning automobile, a mechanic searches for the defective parts causing the problem and then replaces them with new, properly functioning parts. This approach may also be used in medicine, as when a transplant surgeon removes a diseased kidney and replaces it with a healthy one. The strategy of fixing problems by replacing defective parts has a straightforward appeal and makes intuitive sense to almost everyone. Only in recent years, however, has it become possible to apply this approach to diseases such as cystic fibrosis and cancer. A growing understanding of how genetic defects cause disease, combined with the rapid advance of techniques for manipulating cells and DNA, has led to the advent of gene therapy. Practitioners of gene therapy attempt to treat diseases by "fixing" the damaged genes that cause disease.

Hemophilia, for example, is caused by a defect in a single gene. As you will discover later in this chapter, the instructions encoded in genes are used to manufacture proteins. In people with hemophilia, the failure of a defective gene to specify a functional protein has grave consequences. In people who have this genetic defect, the body is unable to produce one or more of the proteins required for blood clotting. As a result, hemophiliacs can bleed to death from even the smallest wounds. Gene therapy for hemophilia attempts to replace some of the patient's defective cells with cells whose nucleus contains a normal version of the affected gene, cells that can produce normal blood-clotting proteins.

Any of the many diseases with a genetic component could potentially be treated by gene therapy, and the National Institutes of Health (which oversees all government-funded gene therapy research in the United States) has authorized more than 200 different gene therapy procedures. None of these treatments has yet proven to be an unequivocal success, but researchers continue to make steady progress toward successful gene therapy.

Advances in techniques to correct genetic defects in body cells suggest that we will soon see similar techniques for modifying genes in sperm and egg cells. Such modifications, known as germline genetic engineering, would alter the genetic makeup of embryos, and the changes would be inherited by offspring. What implications could germline genetic engineering have for future generations? We'll look at the possibilities at the end of this chapter. ■

1 How Are Genes and Proteins Related?

Information by itself cannot accomplish work. For example, a recipe may contain detailed information describing how to make a cake, but unless that information is converted into action, no cake will ever be made. Similarly, the information encoded in a DNA molecule has no effect until it is used to guide the construction of other molecules that directly influence the characteristics of organisms. In particular, the information encoded in DNA makes its impact by providing instructions for the manufacture of proteins.

Most Genes Contain Information for the Synthesis of a Single Protein

Most chromosomes contain hundreds or thousands of genes, each of which occupies a particular position in the DNA of the chromosome. In general, each **gene** is a stretch of DNA that encodes the instructions for the manufacture of a single protein. Proteins, in turn, are the molecular workhorses of the cell, forming both cellular structures and enzymes that catalyze the chemical reactions required by cells.

Technically, the "one gene, one protein" relationship should really be expressed as "one gene, one polypeptide." A polypeptide is a chain of amino acids. Although many proteins consist of a single polypeptide, others are composed of more than one polypeptide subunit. For example, DNA polymerase is composed of more than a dozen polypeptides, each one encoded by a different gene. Nonetheless, many biochemists informally use "polypeptide" and "protein" interchangeably, and this text uses "protein" to refer to all polypeptides.

Genetic Information for Protein Synthesis Is Carried by RNA Intermediaries

DNA does not directly guide protein synthesis, but instead relies on an intermediary that carries information from the nucleus to the cytoplasm. This intermediary is the molecule **ribonucleic acid**, or **RNA**. RNA is structurally similar to DNA but differs in three respects: RNA is normally single-stranded; RNA has the sugar ribose instead of deoxyribose in its backbone; and RNA contains the base uracil instead of the base thymine (Table 9-1).

Table 9-1 A Comparison of DNA and RNA

	DNA	RNA	
Strands	2	1	
Sugar	deoxyribose	ribose	
Types of Bases	adenine (A), thymine (T) cytosine (C), guanine (G)	adenine (A), uracil (U) cytosine (C), guanine (G)	
Base Pairs	DNA:DNA A–T T–A C–G G–C	RNA:DNA A–T U–A C–G G–C	RNA:RNA A–U U–A C–G G–C
Function	Contains genes; sequence of bases in most genes determines the amino acid sequence of a protein	**Messenger RNA (mRNA):** carries the code for a protein-coding gene from DNA to ribosomes **Ribosomal RNA (rRNA):** combines with proteins to form ribosomes, the structures that link amino acids to form a protein **Transfer RNA (tRNA):** carries the proper amino acid to the ribosome, per the code in mRNA	

2 What Is the Genetic Code?

The information in DNA specifies the amino acid sequence of proteins, but the information is encoded. This **genetic code** is conceptually similar to Morse code, which is a set of symbols that can be translated into a different set of symbols. In Morse code, a sequence of dots and dashes is translated into a sequence of letters of the alphabet; in the genetic code, a sequence of nucleotide bases in DNA is translated into a sequence of amino acids in a protein.

A Sequence of Three Bases Codes for an Amino Acid

In the simplest possible genetic code, each different base would code for a particular amino acid. We can see, however, that such a code is not possible because proteins contain 20 different amino acids but DNA contains only four different bases (A, T, G, and C; see Table 9-1). Similarly, a system in which a sequence of two bases codes for an amino acid is also impossible, as there are only 16 possible combinations of two bases, still not enough to account for all 20 amino acids. A three-base sequence, however, gives 64 possible combinations, more than enough. If nature operates as economically as possible, the genetic code would be a triplet code in which a sequence of three bases specifies a single amino acid. In fact, this hypothesis is correct. (The method by which scientists confirmed the hypothesis is explained later in this chapter in "Scientific Inquiry: Cracking the Genetic Code.") Each three-base triplet in a gene specifies one amino acid in the corresponding protein. These triplets are called **codons**.

Some Codons Provide Punctuation

As shown in Table 9-2, almost all of the 64 codons code for particular amino acids. (Note that the table follows the common practice of showing the genetic code in terms of the codons of the RNA intermediaries, rather than of DNA.) A few codons, however, instead mark

Scientific Inquiry
Cracking the Genetic Code

The hypothesis that three bases in DNA code for one amino acid in a protein is an attractive and logical possibility. But how could you prove that nature actually uses a triplet code?

Francis Crick and his co-workers exposed bacteriophages, viruses that multiply inside bacteria, to a chemical called acridine. Acridine causes one, two, or three nucleotides to be inserted into the virus's DNA molecule at a random location near the beginning of a particular gene. The researchers found that inserting one or two nucleotides into the DNA caused the synthesis of defective enzymes and prevented the viruses from multiplying inside their host bacteria. Inserting three nucleotides, however, sometimes produced phages that synthesized normal or nearly normal enzymes.

Crick concluded that during RNA synthesis, a gene's DNA is "read" in a linear order, in which each sequence of three bases makes up a "word"; that is, three bases in DNA encode a single amino acid. The code must somehow specify the beginning and end of a gene, but genes have no spaces or punctuation between words. How did he arrive at these conclusions?

To understand Crick's reasoning, let's suppose that we have a gene with multiple repeats of the sequence of bases TAG. If DNA always has three-letter words, then this base sequence spells out the word "TAG" over and over (Fig. E9-1a). You don't need spaces to separate the words if they all have three and only three letters.

Now suppose that the acridine inserts another nucleotide with the base cytosine near the beginning of the gene (Fig. E9-1b). Most of the gene now reads GTA . . ., which is the code for a different amino acid and produces a nonfunctional enzyme. Inserting an adenine next to the cytosine again changes the amino acid specified (Fig. E9-1c), as most of the gene now reads AGT . . ., and the product is another nonfunctional enzyme. A third insertion next to the first two, however, causes most of the gene once again to be TAG . . . (Fig. E9-1d). In this case, most of the enzyme would be synthesized correctly and might work well enough to allow the viruses to reproduce.

Crick concluded that only a triplet code could account for the fact that three insertions, but not one or two, restore near-normal enzyme function. Further, only a code without punctuation or spaces between words would be confused by insertions. Finally, a code without punctuation between words could work only if the start points and stop points of protein synthesis are clearly marked. All three conclusions have been found to be correct, as we describe in the text.

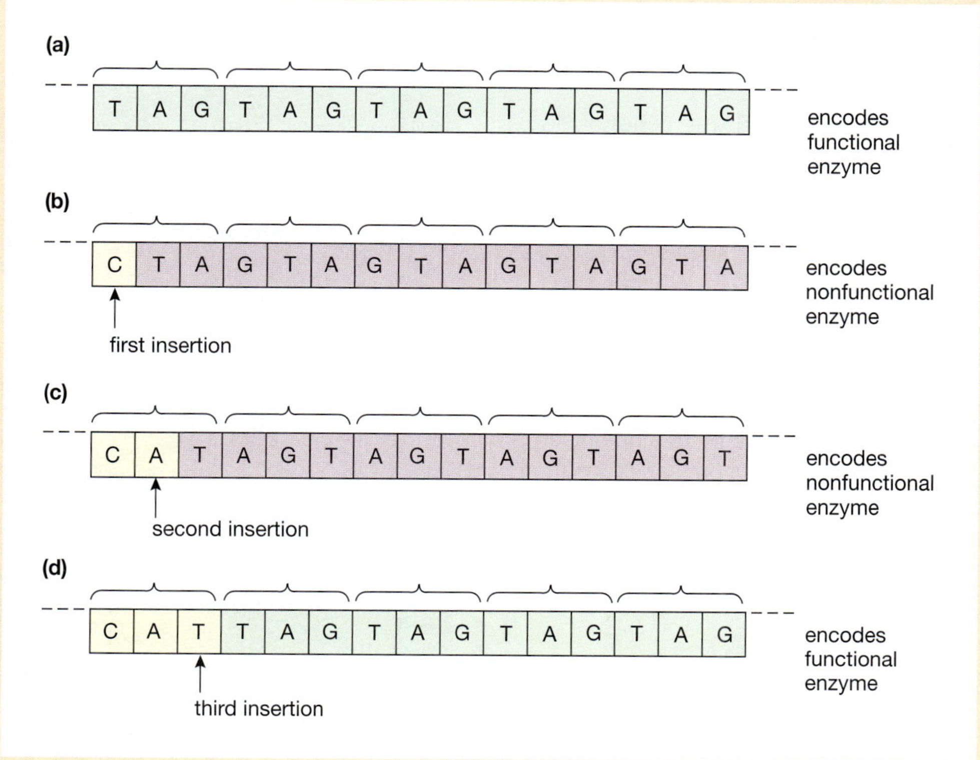

Figure E9-1 How insertion mutations helped crack the genetic code

Table 9-2 The Genetic Code Written as Codons of RNA Intermediaries (mRNA)

First Base	Second Base: U		Second Base: C		Second Base: A		Second Base: G		Third Base
U	UUU	Phenylalanine (Phe)	UCU	Serine (Ser)	UAU	Tyrosine (Tyr)	UGU	Cysteine (Cys)	U
	UUC	Phenylalanine	UUC	Serine	UAC	Tyrosine	UGC	Cysteine	C
	UUA	Leucine	UCA	Serine	UAA	Stop	UGA	Stop	A
	UUG	Leucine	UCG	Serine	UAG	Stop	UGG	Tryptophan (Trp)	G
C	CUU	Leucine (Leu)	CCU	Proline (Pro)	CAU	Histidine (His)	CGU	Arginine (Arg)	U
	CUC	Leucine	CCC	Proline	CAC	Histidine	CGC	Arginine	C
	CUA	Leucine	CCA	Proline	CAA	Glutamine (Gln)	CGA	Arginine	A
	CUG	Leucine	CCG	Proline	CAG	Glutamine	CGG	Arginine	G
A	AUU	Isoleucine (Ile)	ACU	Threonine (Thr)	AAU	Asparagine (Asp)	AGU	Serine (Ser)	U
	AUC	Isoleucine	ACC	Threonine	AAC	Asparagine	AGC	Serine	C
	AUA	Isoleucine	ACA	Threonine	AAA	Lysine (Lys)	AGA	Arginine (Arg)	A
	AUG	Methionine (Met)	ACG	Threonine	AAG	Lysine	AGG	Arginine	G
G	GUU	Valine (Val)	GCU	Alanine (Ala)	GAU	Aspartic acid (Asp)	GGU	Glycine (Gly)	U
	GUC	Valine	GCC	Alanine	GAC	Aspartic acid	GGC	Glycine	C
	GUA	Valine	GCA	Alanine	GAA	Glutamic acid (Glu)	GGA	Glycine	A
	GUG	Valine	GCG	Alanine	GAG	Glutamic acid	GGG	Glycine	G

the boundaries of a protein-coding sequence. These special codons act much like the punctuation marks that allow a reader to determine the beginning and end of a sentence; they allow the cell to recognize where the code for a given protein starts and stops. The location at which protein synthesis begins is indicated by the codon AUG, which codes for the amino acid methionine. All proteins thus begin with methionine. The end of a protein-coding sequence is marked by a **stop codon**. When the protein-making machinery reaches one of the three stop codons (UAG, UAA, and UGA), it releases the newly synthesized protein.

All 64 possible codons are used in the genetic code, even though only 20 amino acids are encoded. Thus, almost all amino acids are specified by more than one codon. For example, six different codons code for leucine (see Table 9-2). Each codon, however, specifies one, and only one, amino acid.

The Big Picture

All the instructions for building and operating the amazingly complex and diverse array of living things are contained in molecules of DNA. Incredibly, this vast amount of information is stored in a simple code that uses only four symbols, and is converted to living form by a system that can perform only a single task: manufacturing proteins.

3 How Is the Information in DNA Used to Synthesize Proteins?

The nucleotide sequences of genes are used to specify the amino acid sequences of proteins in a process that incorporates two major steps, which are known as transcription and translation. In **transcription**, the information contained in the DNA of a gene is transcribed into RNA. In everyday language, the word "transcribe" means "to transfer information from one recording and storing system to another," as when a typist transcribes spoken words into written text. The biological process of transcription is similar; the DNA that has been transcribed remains unchanged but a copy of its information is transferred to a newly created RNA molecule. DNA is transcribed into RNA in the nucleus (Fig. 9-1a),

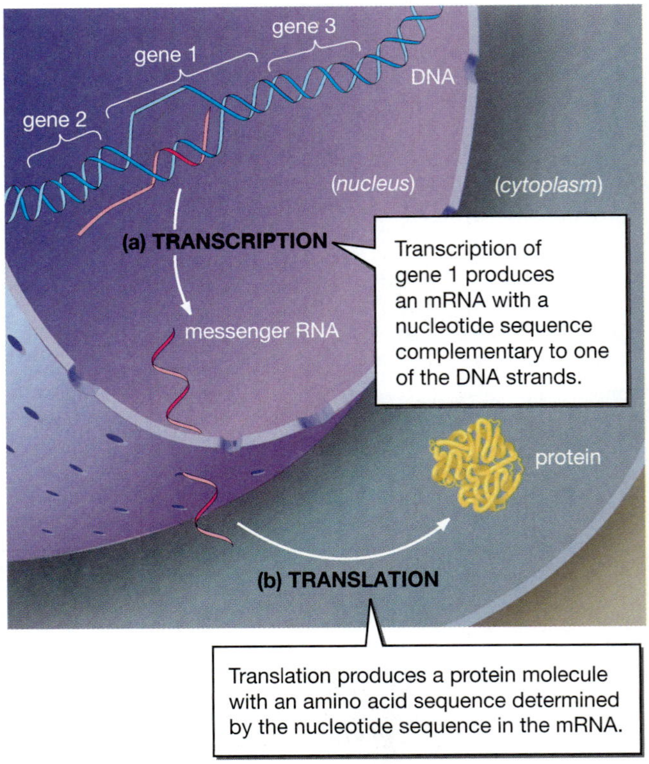

Figure 9-1 Genetic information flows from DNA to RNA to protein After being transcribed, the mRNA molecule moves from the nucleus to the cytoplasm, where it is translated.

and the new RNA can be any one of three types: **messenger RNA (mRNA)**, **transfer RNA (tRNA)**, or **ribosomal RNA (rRNA)**.

Protein is synthesized during **translation** (Fig. 9-1b), a process analogous to the translation of a letter from English to French. In the biological process of translation, a message is converted from the "language" of RNA to the "language" of protein. The message is a sequence of codons in messenger RNA. Transfer RNA and ribosomal RNA, working together, use the message to link amino acids in a specific sequence.

Messenger RNA Carries the Code for a Protein to the Cytoplasm

Among the various types of RNA produced by transcription of DNA, only messenger RNA carries the code for the amino acid sequence of a protein (Fig. 9-2a). Messenger RNA carries this information from the nucleus to the cytoplasm, while the gene, like a valuable document in a library, remains safely stored in the nucleus. Messenger RNA molecules move to the cytoplasm through pores in the nuclear envelope.

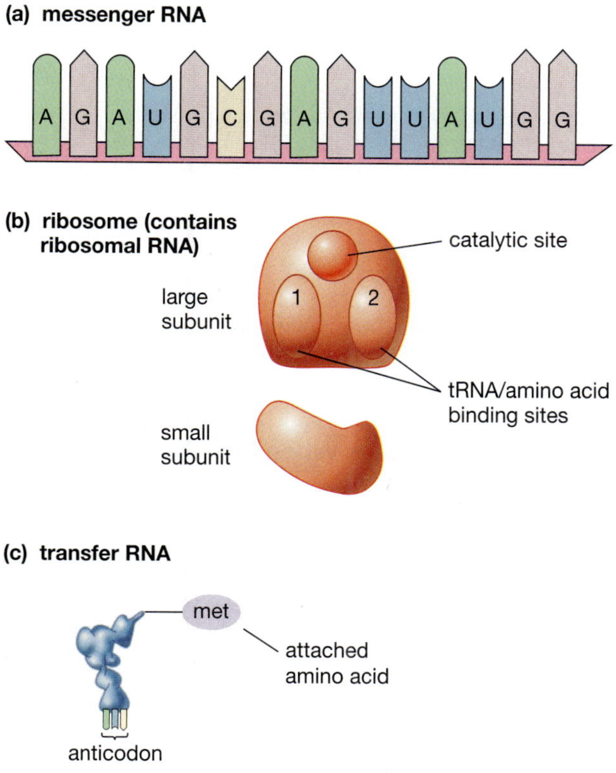

Figure 9-2 Cells synthesize three major types of RNA
RNA consists of a single strand of nucleotides whose sequence is complementary to that of the DNA from which it was transcribed. ***(a)*** mRNA is a sequence of nucleotides that encodes a protein. ***(b)*** rRNA is a major component of the two subunits that join together to form a ribosome. ***(c)*** One side of tRNA contains an anticodon that is complementary to an mRNA codon; the opposite side binds an amino acid.

Ribosomal RNA Is a Component of the Structures That Construct Proteins

Translation, the process of constructing the proteins encoded in mRNA base sequences, is carried out by **ribosomes**. Ribosomes contain ribosomal RNA and many different proteins, and each one is composed of two subunits—one small and one large. The large subunit has binding sites for two tRNA molecules and a catalytic site where the amino acids are joined together in a growing protein chain. The two subunits remain separate unless they are actively synthesizing proteins (Fig. 9-2b). For protein synthesis, the two subunits come together, clasping an mRNA molecule between them.

Transfer RNA Molecules Carry Amino Acids to the Site of Protein Synthesis

Each cell synthesizes 61 different types of transfer RNA, one type for each different amino acid–encoding codon (recall that 3 of the 64 codons are stop codons). These tRNA molecules deliver the appropriate amino acids to the ribosome, where the amino acids are incorporated into the growing protein chain. The tRNA molecules are recognized in the cytoplasm by 20 different enzymes, one for each amino acid. Each of these enzymes recognizes particular tRNA molecules and attaches the correct amino acid to one end. For example, one enzyme recognizes the six different leucine tRNAs and attaches leucine molecules to them.

The ability of tRNA to deliver the proper amino acid to the growing protein depends on base-pairing between tRNA and mRNA. Each tRNA has three exposed bases, called the **anticodon**, which form complementary base pairs with an mRNA codon (Fig. 9-2c). For example, the mRNA codon AUG bonds to the anticodon UAC of a tRNA that has an attached methionine amino acid.

The following sections describe transcription and translation in greater detail.

4 What Happens During Transcription?

We can view transcription as a process consisting of (1) initiation, (2) elongation, and (3) termination. These three steps correspond to the three major parts of most genes: (1) a promoter region at the beginning of the gene, where transcription begins; (2) the main body of the gene, where the RNA strand is elongated; and (3) a termination signal at the end of the gene, where RNA synthesis stops.

Transcription Begins When RNA Polymerase Binds to the Promoter of a Gene

The enzyme **RNA polymerase** catalyzes the transcription of DNA to RNA. To initiate transcription, RNA polymerase must first locate the **promoter** region. The promoter is a nontranscribed sequence of DNA bases

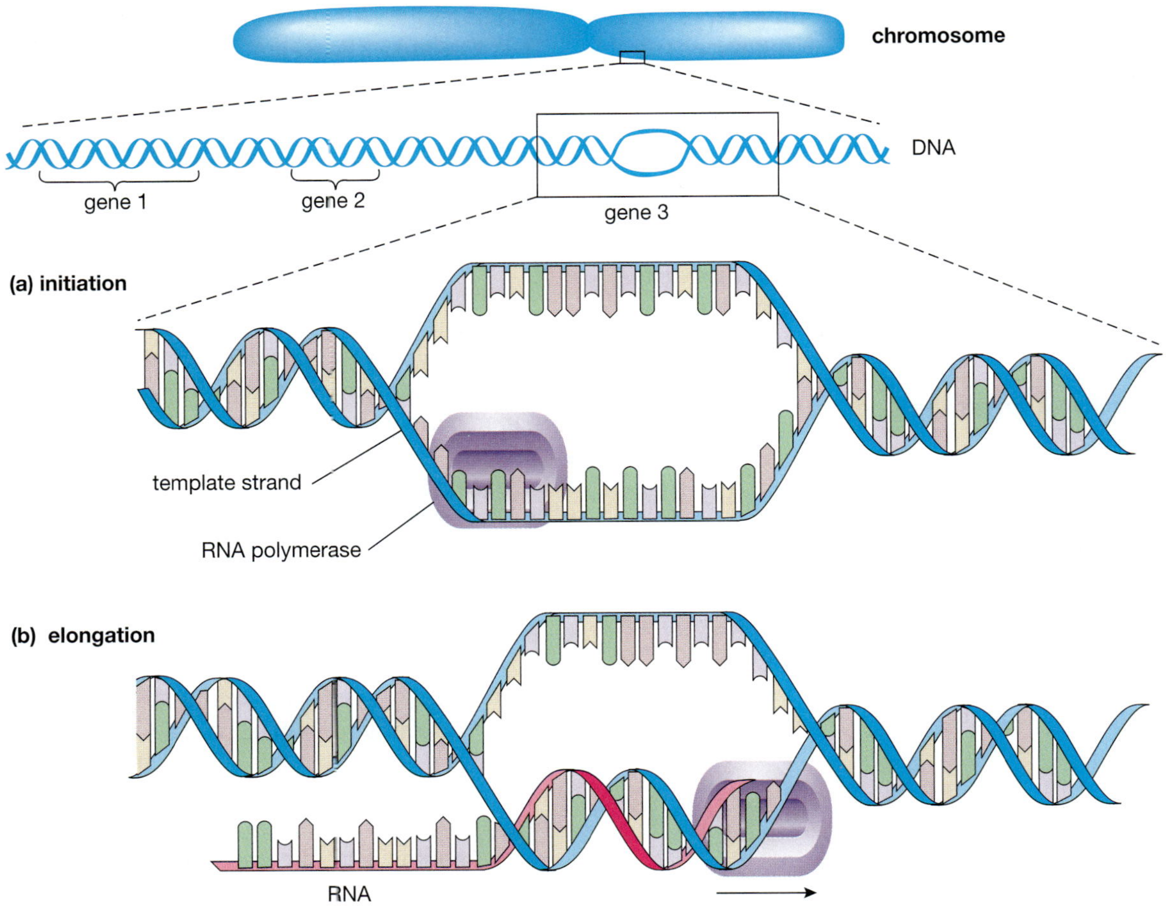

Figure 9-3 Initiation and elongation are the first two steps of transcription
Within a gene, one of the DNA strands serves as the template for RNA synthesis. ***(a)*** Initiation: The enzyme RNA polymerase binds to the promoter region of DNA near the beginning of a gene, forcing the DNA double helix to separate in front of it. ***(b)*** Elongation: RNA polymerase travels along the DNA template strand, catalyzing the formation of a complementary RNA molecule from free RNA nucleotides. As RNA polymerase moves along, the already transcribed portion of the DNA double helix begins to rewind.

MEDIATUTOR 9.1 Transcription

that marks the beginning of the gene. When RNA polymerase binds to a promoter, transcription of that gene begins (Fig. 9-3a).

Elongation Generates a Growing Strand of RNA

After it binds to the promoter of a gene, RNA polymerase changes shape, forcing the DNA double helix at the beginning of the gene to partially unwind. RNA polymerase then travels in one direction along one of the DNA strands, designated the **template strand**, and synthesizes a single strand of RNA that is complementary to the template strand (Fig. 9-3b). Because the newly created complementary strand is RNA rather than DNA, uracil, rather than thymine, pairs with any adenines on the template strand (see Table 9-1).

After about 10 nucleotides have been added to the growing RNA chain, the first nucleotides in the RNA molecule separate from the DNA template strand (Fig. 9-3b). As the RNA strand separates, the two DNA strands re-form the base pairs of the original double helix. This process of RNA growth followed by separation continues, and one end of the growing RNA drifts away from the DNA. The other end remains attached to the DNA template strand by RNA polymerase (Fig. 9-4).

Elongation Terminates When RNA Polymerase Reaches a Signal

RNA polymerase continues along the template strand until it reaches a sequence of DNA bases known as the termination signal. At this point, RNA polymerase releases the completed RNA molecule and detaches from the DNA. The RNA polymerase is then free to bind to another promoter and synthesize another RNA molecule (Fig. 9-5).

Transcription Is Selective

The transcription of genes into RNA is selective. In any given cell, only a portion of the collection of genes is transcribed into RNA. Although some genes, because they encode proteins that are essential for the life of any cell, are transcribed in all cells, many genes are transcribed only in certain types of cells. For example, even

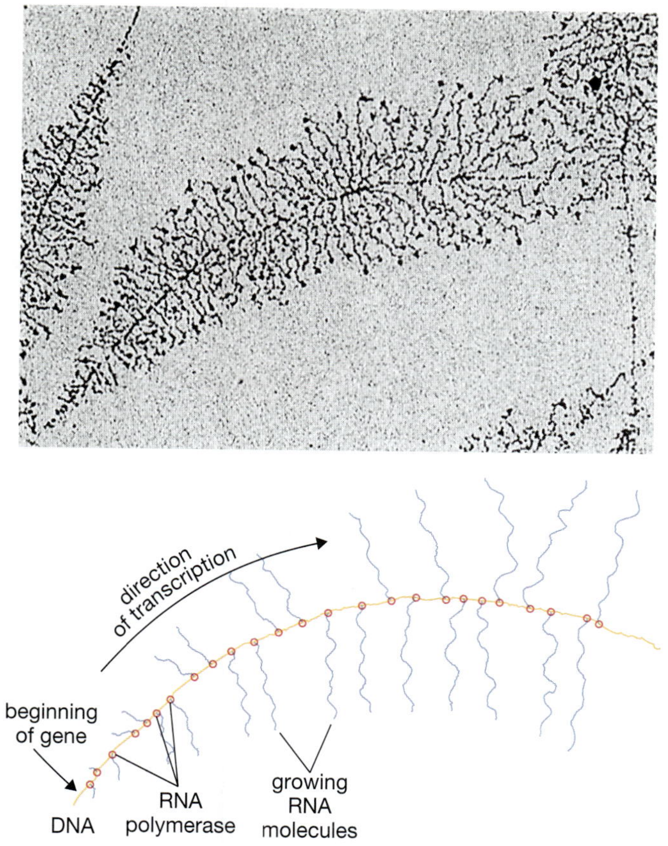

Figure 9-4 RNA transcription in action
The images in this electron micrograph show RNA transcription. In each treelike structure, the central "trunk" is DNA and the "branches" are RNA molecules. RNA polymerase molecules are traveling along the DNA, synthesizing RNA as they go. The beginning of the gene is on the left, and the short RNA molecules on the left have just begun to be synthesized; the long RNA molecules on the right are almost finished.

though every cell in your body contains the gene for the protein hormone insulin, that gene is transcribed only in certain cells in your pancreas. In general, transcription is restricted to the genes needed by a particular type of cell at a particular time.

5 What Happens During Translation?

Like transcription, translation has three steps: (1) initiation of protein synthesis, (2) elongation of the protein chain, and (3) termination of translation.

Translation Begins When tRNA and mRNA Bind to a Ribosome

The first amino acid in any newly synthesized protein is methionine. Methionine is specified only by the AUG codon (see Table 9-2), so the first AUG in an mRNA sequence is where translation begins. The process gets under way with the formation of an **initiation complex** composed of a small ribosomal subunit, a methionine tRNA with a methionine amino acid attached to it, and several proteins (Fig. 9-6a). The initiation complex binds to an mRNA molecule and is joined by a large ribosomal subunit, which completes the assembly of the ribosome (Fig. 9-6b,c). The ribosome is now ready to begin translation.

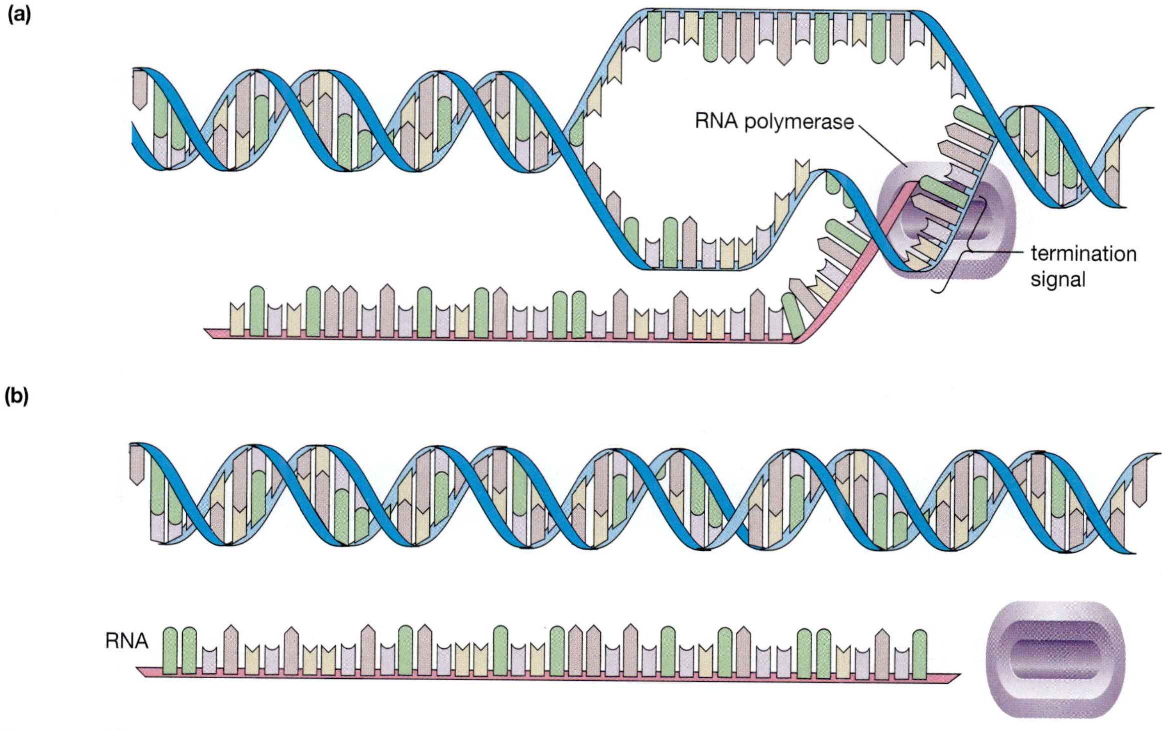

Figure 9-5 Termination is the third step of transcription
Elongation of the RNA molecule continues until *(a)* RNA polymerase encounters, at the end of gene, a termination signal. *(b)* RNA polymerase detaches from the DNA and releases the RNA molecule, allowing the DNA double helix to completely rewind.

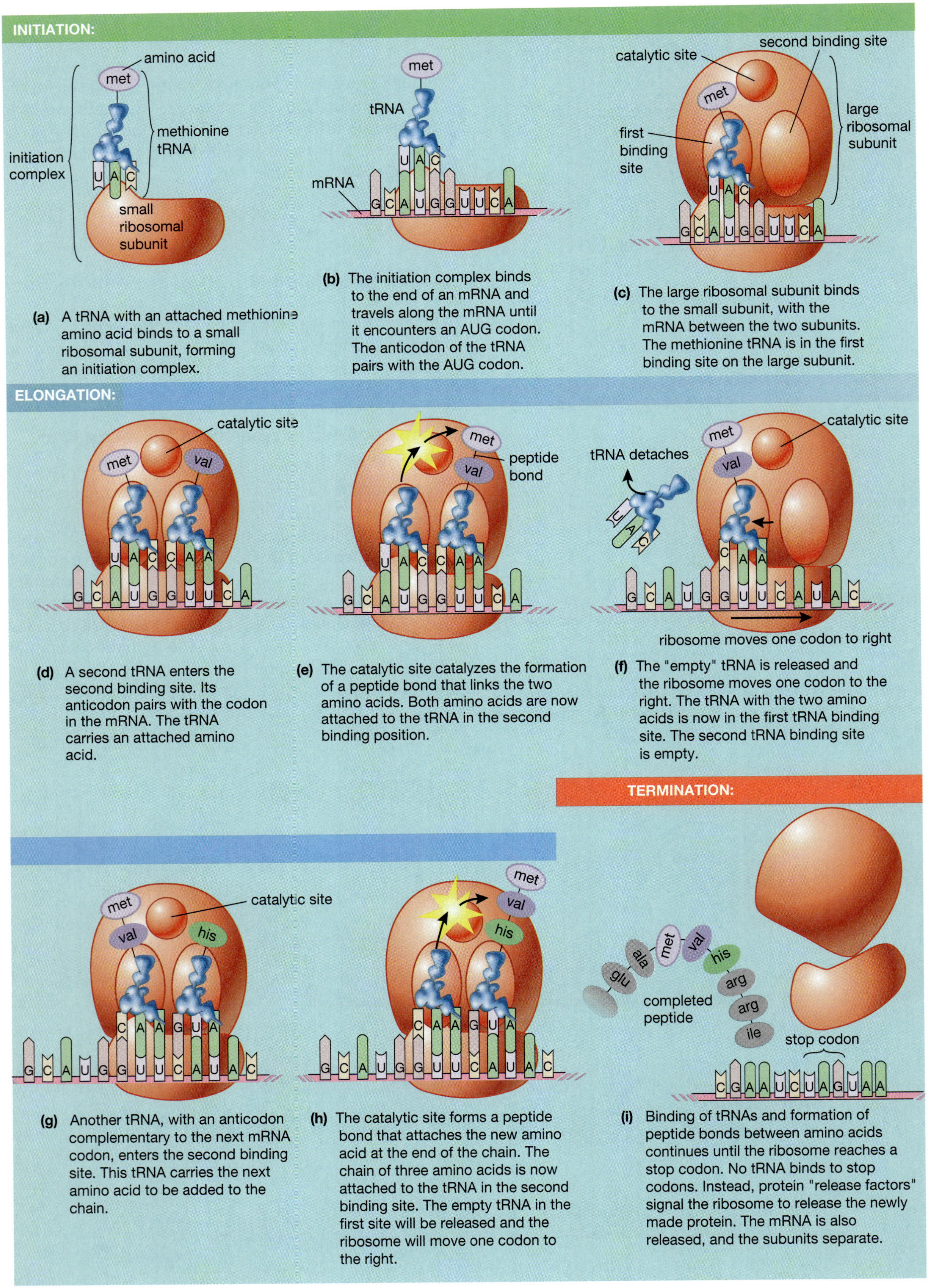

(a) A tRNA with an attached methionine amino acid binds to a small ribosomal subunit, forming an initiation complex.

(b) The initiation complex binds to the end of an mRNA and travels along the mRNA until it encounters an AUG codon. The anticodon of the tRNA pairs with the AUG codon.

(c) The large ribosomal subunit binds to the small subunit, with the mRNA between the two subunits. The methionine tRNA is in the first binding site on the large subunit.

(d) A second tRNA enters the second binding site. Its anticodon pairs with the codon in the mRNA. The tRNA carries an attached amino acid.

(e) The catalytic site catalyzes the formation of a peptide bond that links the two amino acids. Both amino acids are now attached to the tRNA in the second binding position.

(f) The "empty" tRNA is released and the ribosome moves one codon to the right. The tRNA with the two amino acids is now in the first tRNA binding site. The second tRNA binding site is empty.

(g) Another tRNA, with an anticodon complementary to the next mRNA codon, enters the second binding site. This tRNA carries the next amino acid to be added to the chain.

(h) The catalytic site forms a peptide bond that attaches the new amino acid at the end of the chain. The chain of three amino acids is now attached to the tRNA in the second binding site. The empty tRNA in the first site will be released and the ribosome will move one codon to the right.

(i) Binding of tRNAs and formation of peptide bonds between amino acids continues until the ribosome reaches a stop codon. No tRNA binds to stop codons. Instead, protein "release factors" signal the ribosome to release the newly made protein. The mRNA is also released, and the subunits separate.

MEDIATUTOR 9.2 Translation

Figure 9-6 Translation: the base sequence of an mRNA is translated into the amino acid sequence of a protein

Elongation Generates a Growing Chain of Amino Acids

The assembled ribosome has two binding sites and a catalytic site. When the first binding site is occupied by a tRNA and its attached amino acid, a second tRNA, with an anticodon complementary to next mRNA codon, moves into the second binding site. The catalytic site forms a peptide bond between the two amino acids (Fig. 9-6d,e). The ribosome then moves to the next codon, shifting the growing chain from the second to the first binding site, and the pattern repeats, joining a third amino acid to the chain (Fig. 9-6f,g,h). The whole process repeats over and over again as the ribosome moves along the mRNA one codon at a time, elongating the chain.

A Stop Codon Signals Termination

Encountering a stop codon in the mRNA molecule signals the ribosome that synthesis of the protein is complete. Stop codons, unlike the other 61 codons, do not bind to a tRNA. Instead, special proteins bind to the ribosome when it encounters a stop codon, causing the ribosome to release the finished protein chain and the mRNA (Fig. 9-6i). The ribosome disassembles into large and small subunits, which can then be used to translate another mRNA.

Summary: Transcription and Translation

We can now summarize how a cell decodes the genetic information stored in its DNA to synthesize a protein (Fig. 9-7).

(a) With few exceptions, each gene codes for a single protein.

(b) Transcription of a protein-coding gene produces a messenger RNA molecule that is complementary to one strand of the gene. The mRNA molecule is composed of codons, which are three-base sequences that specify either a single amino acid or a "stop."

(c) An amino acid is attached to each transfer RNA molecule; the particular amino acid that is attached depends on the tRNA's anticodon.

(d) During translation, tRNAs carry their attached amino acids to the ribosome. There, the bases in tRNA anticodons bind to the complementary bases in mRNA codons, so that the amino acids attached to the tRNAs line up in the sequence specified by the codons. The ribosome links the amino acids together to form a protein.

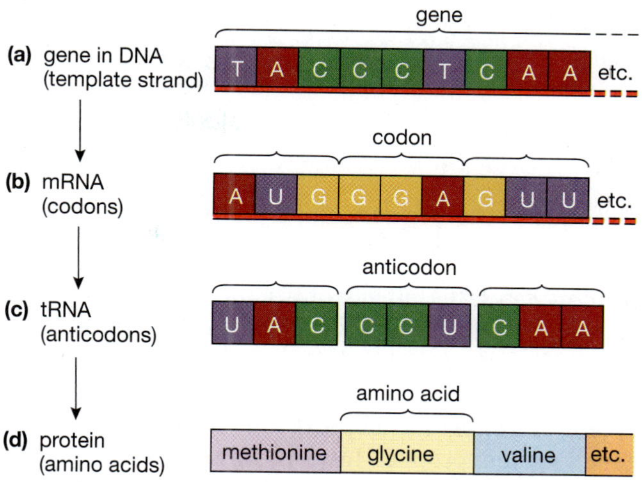

Figure 9-7 Complementary base pairing is critical at each step in decoding genetic information
(a) DNA contains two strands. One of them, the template strand, is used to synthesize an RNA molecule. *(b)* Bases in the template strand are transcribed into complementary mRNA codons. *(c)* Unless it is a stop codon, each mRNA codon forms base pairs with the anticodon of a tRNA molecule that carries a specific amino acid. *(d)* The amino acids are linked together, forming a protein.

6 What Is a Mutation?

Despite the best efforts of proofreading enzymes, DNA replication generates a few mistakes. These errors can cause **mutations**, which are changes in the sequence of bases in a DNA molecule.

Chemicals and Radiation Can Increase the Frequency of Mutations

The frequency of replication errors, and thus of mutations, may increase when cells are exposed to certain chemicals. Chemicals that have been shown to have this mutation-increasing effect include aflatoxins, which are produced by molds that live on grain and peanuts, and several dozen different substances that are found in tobacco smoke. The mutation rate can also be increased by exposure to some types of radiation, especially X rays and the ultraviolet rays in sunlight. Chemicals and radiation can also cause mutations in DNA that is not replicating.

Mutations Result from Nucleotide Substitutions, Insertions, or Deletions

Mutations can be classified into several categories. One type is a **nucleotide substitution**, in which a single base pair in a gene is replaced with a different, incorrect base pair. Because nucleotide substitutions change individual nucleotides in the DNA sequence, they are also sometimes called **point mutations**. An **insertion mutation** occurs when one or more new nucleotide pairs are inserted into a gene. A **deletion mutation** occurs when one or more nucleotide pairs are removed from a gene.

7 What Are the Effects of Mutations?

Like DNA replication, transcription and translation are subject to an occasional error. The effects of errors in transcription and translation are usually not serious, though, because such errors result in only a few faulty mRNA or protein molecules in a cell that also contains many correct copies of these molecules. Errors in DNA replication, however, can have more severe effects, because a mutation in a gene affects every protein molecule produced by the flawed gene.

Changes in the Sequence of a Gene Usually Have Harmful or Neutral Effects

In most cases, the effects of a mutation in a gene are harmful. Just as randomly changing a word in the middle of *Hamlet* would be unlikely to improve the quality of Shakespeare's writing, changing the base sequence of a gene is unlikely to improve the performance of the gene's protein product. Instead, a mutation in a gene is likely to disrupt the proper functioning of the gene, and is also likely to have a negative effect on the organism in which the mutation occurs. (For examples in which single mutations have far-reaching consequences, see "Health Watch: Sex, Aging, and Mutations.") Many mutations, however, have no effect on the organism. In rare cases, a mutation may even be beneficial.

Mutations Can Affect Proteins in Different Ways

The effect of a gene mutation depends on the type and extent of the mutation. For example, deletions and insertions of one or two nucleotides usually have catastrophic effects on a gene, because all the codons that follow the deletion or insertion are altered. The protein synthesized from an mRNA containing such changed codons is almost certain to be nonfunctional (see "Scientific Inquiry: Cracking the Genetic Code"). Point mutations (nucleotide substitutions) within a protein-coding gene, however, will not always render the code useless. To see why, let's consider some mutations in the gene that encodes the protein beta-globin. At least four different outcomes are possible.

- *The protein may be unchanged:* Because most amino acids are encoded by several different codons, some point mutations do not change the protein. Consider the sixth codon in the beta-globin mRNA, GAG, which specifies glutamic acid. If a point mutation changes this codon from GAG to GAA, the new codon still codes for glutamic acid. The protein synthesized from the mutated gene remains the same even though the DNA sequence is different.
- *The new protein may be equivalent to the original one:* Many proteins have regions whose exact amino acid sequence is relatively unimportant. For example, in beta-globin, the amino acids on the outside of the protein must be hydrophilic to keep the protein dissolved in the cytoplasm of red blood cells. Exactly *which* hydrophilic amino acids are on the outside doesn't matter much. For example, a mutation from GAG to CAG, which replaces glutamic acid (hydrophilic) with glutamine (also hydrophilic), does not affect the ability of beta-globin to function as it should.
- *The new protein may be significantly different:* A mutation that changes the sixth codon in beta-globin mRNA from GAG to GUG replaces glutamic acid (hydrophilic) with valine (hydrophobic). This substitution prevents beta-globin from functioning properly and causes the genetic disease sickle-cell anemia (see Chapter 12).
- *Protein function may be destroyed by a premature stop codon:* The 17th codon in the beta-globin gene, AAG, specifies lysine. A mutation that changes the codon from AAG to UAG changes the lysine codon to a stop codon. This premature stop codon halts translation of beta-globin mRNA before the protein is completed. As a result, no beta-globin protein is made in cells with this mutant gene. Individuals with this mutation in all of their beta-globin genes have a condition called beta-thalassemia, which can be fatal unless treated with blood transfusions.

Mutations Are the Raw Material for Evolution

Mutations are the ultimate source of all genetic differences among individuals. If there were no mutations, all individuals would share the same DNA sequence. Although most mutations are neutral or harmful to the individuals that carry them, a mutation may occasionally improve an individual's ability to survive and reproduce. If such a beneficial mutation is in a cell that gives rise to gametes (sperm or eggs), it may be passed on to individuals in the next generation. If possessors of the mutant gene produce more offspring than individuals that lack the mutation, the mutant gene and the characteristics it bestows will become more common over time. This process, known as natural selection, is a major cause of evolutionary change, which is described in detail in Unit III.

The Big Picture

The genetic machinery of cells performs almost flawlessly, but errors are inevitable in a system that performs such complex tasks on such a huge scale. Mutations resulting from genetic mistakes are the source of many diseases and physical defects, but without such mistakes there would be no genetic variability among individuals and therefore no evolution.

8 Are All Genes Expressed?

Each cell contains many genes, but only a portion of them is active at any given time. For example, all of the 30,000 or more genes in the human genome are present in each body cell, but individual cells express (transcribe and translate) only a small fraction of these genes. The particular set of genes that is expressed depends on the type of cell and the needs of the organism at a particular time. Cells must control, or regulate,

Health Watch
Sex, Aging, and Mutations

Imagine that you are a girl in her mid-teens. Your breasts have filled out, but you have not begun to menstruate. Your concerned parents take you to visit your family doctor, who takes a small blood sample for a chromosome test. When the results come in, you and your parents are called back to the office, where a genetic counselor gives you amazing news: Your sex chromosomes are XY, a combination that would normally give rise to a male. The reason you have not begun to menstruate is that you lack ovaries and a uterus. Instead, you have testes that have remained inside your abdominal cavity. Your blood contains levels of male hormones (collectively called androgens) that would be normal for a male. In fact, these male hormones are produced by your testes and have been present since early in your development. The problem is that your cells cannot respond to them—a rare condition called androgen insensitivity. This condition was a problem for Maria Jose Martinez Patino, an outstanding Spanish athlete who reached the Olympics, only to be barred from the women's hurdles competition because her chromosomes indicated that she was male. After three years of struggle, the fact that she had developed as a female was finally recognized, and she was allowed to compete against others of her gender.

Androgen insensitivity is caused by a mutation. The affected gene codes for a protein known as an androgen receptor. In normal males, androgen receptor proteins are present in the cytoplasm of many body cells. Male hormones such as testosterone bind to the receptor molecules, and the hormone–receptor combination enters the cell nucleus and binds to DNA, stimulating the transcription of genes into mRNA. The mRNA molecules serve as templates for the synthesis of specific proteins that contribute to maleness. These proteins help generate many male features, including the formation of a penis, the descent of the testes into sacs outside the body cavity, and sexual characteristics that develop at puberty, such as a beard and increased muscle mass.

A person with XY chromosomes who inherits a mutant androgen receptor gene from both parents will be unable to make a functional androgen receptor protein. Her cells will be unable to respond to the testosterone that the testes produce, and male characteristics will not develop. Thus, a change in the nucleotide sequence of a single gene, causing a single type of defective protein to be produced, can cause a person who is genetically male (XY) to look like and perceive herself to be a woman (Fig. E9-2).

A second type of mutation provides clues to the mystery of why people age. Why will your hair whiten, skin wrinkle, joints ache, and eyes cloud as you become elderly? A few people carry a defective gene that causes Werner syndrome, a type of premature aging. This condition, most common among people of Japanese ancestry, is the result of a mutation that interferes with proper DNA replication, increasing the incidence of mutations throughout the body (Fig. E9-3). People with this disorder die of age-related conditions, typically by age 50. Recent research has identified the gene responsible for Werner syndrome. As you have seen, accurate replication of DNA is crucial to the production of normal cells. A mutation that interferes with accurate DNA replication, proofreading, and repair allows mutations to accumulate in cells throughout the body.

The observation that an overall increase in mutations produces symptoms of old age provides support for one hypothesis about the symptoms of normal aging. During a long (say, 80-year) life span, mutations gradually accumulate because of mistakes in DNA replication and environmentally induced DNA damage. Eventually, these mutations interfere with nearly every aspect of body functioning and contribute to death from "old age."

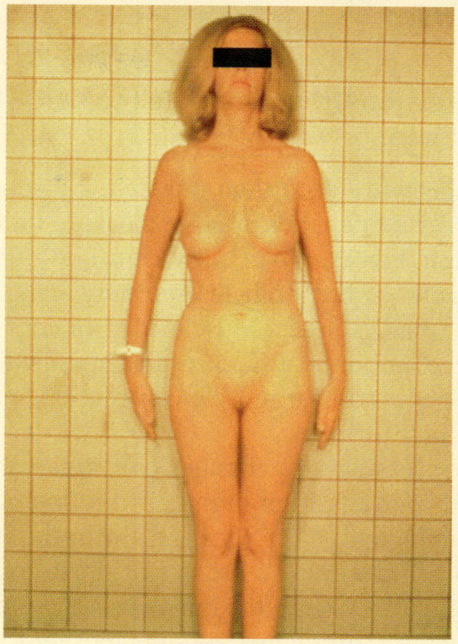

Figure E9-2 Androgen insensitivity leads to female features This individual is genetically male; she has an X and a Y chromosome. As a result, she has testes (located in her abdomen) that produce testosterone, and she lacks ovaries and a uterus. A mutation in her androgen receptor genes prevents her cells from responding to the testosterone produced by her testes, and causes her obviously female appearance.

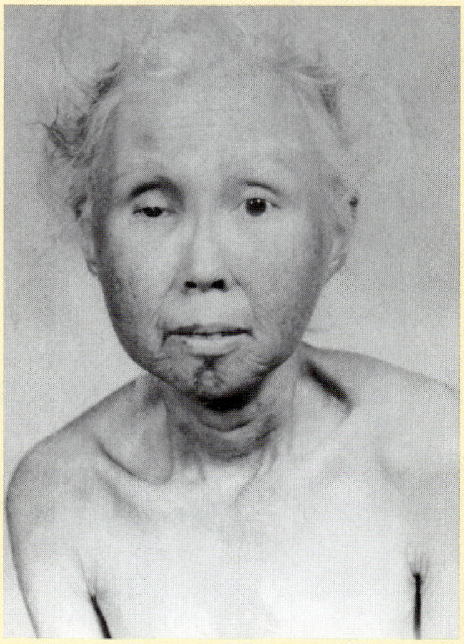

Figure E9-3 A 48-year-old woman with Werner syndrome

which genes are expressed and when they are expressed. This regulation of gene expression is crucial for proper cell function.

Gene Expression Differs from Cell to Cell and over Time

In organisms with more than one cell, the set of genes that is expressed depends in large measure on the function of a cell. For example, in humans and other mammals, the cells of hair follicles synthesize the proteins from which hair is made. Muscle cells, in contrast, synthesize large amounts of the proteins actin and myosin, which are necessary for muscle contraction, but do not synthesize hair proteins.

Gene expression also changes over time, depending on the body's needs from moment to moment. For example, immediately after a baby's birth, the milk-producing cells in a woman's breasts begin to express the gene that encodes casein, the major protein in milk. This change in gene expression allows the mother to produce large amounts of protein-laden milk to feed her new baby. Circumstances may even dictate that some genes are never expressed. A human male, for example, does not express the casein gene. However, he will pass a copy of this gene to his daughters, who will express it if they give birth.

Environmental Cues Influence Gene Expression

Changes in an organism's environment also help determine which genes are transcribed. For example, in temperate climates the length of the day increases as winter turns into spring. Among birds that live in such climates, increasing day length stimulates sharp increases in the size of sex organs (testes and ovaries) and in the production of hormones by the sex organs. Under the influence of the sex hormones, birds begin to produce eggs and sperm, to sing, to mate, and to build nests. The proliferation of cells in the sex organs, the production of hormones by these cells, and the effects of those hormones on other cells throughout the body all result, directly or indirectly, from changes in gene expression.

9 How Is Gene Expression Regulated?

The mechanisms by which cells regulate gene expression are many and varied. We briefly describe a few of them here.

Regulatory Proteins That Bind to Promoters Alter the Transcription of Genes

Transcription is often regulated by molecules that act directly on the promoter region of a gene. Many steroid hormones accomplish their functions by acting in this manner. In birds, for example, the sex hormone estrogen regulates expression of the gene for the egg-white protein albumin. The albumin gene is not transcribed in the winter when birds are not breeding and sex hormone levels are low. During the breeding season, however, estrogen enters cells in the female reproductive system and binds to a receptor protein in the cytoplasm. The estrogen–receptor combination then enters the nucleus and attaches to the DNA in a region near the promoter of the albumin gene. This attachment makes it easier for RNA polymerase to bind to the promoter. As a result, the cells begin to transcribe large amounts of albumin mRNA that is translated into the large amounts of albumin protein needed to make eggs.

Some Regions of Chromosomes Are Condensed and Not Normally Transcribed

Certain parts of eukaryotic chromosomes are in a highly condensed, compact state in which most of the DNA seems to be inaccessible to RNA polymerase. Some of these tightly condensed regions contain genes that are not currently being transcribed. When the product of a gene is needed, the portion of the chromosome containing that gene becomes "decondensed"—loosened so that the nucleotide sequence is accessible to RNA polymerase and transcription can occur.

Entire Chromosomes May Be Inactivated and Not Transcribed

In some cases an entire chromosome may be condensed, making it largely inaccessible to RNA polymerase. The sex chromosomes of mammals provide an example of such inactivation. Male mammals usually have an X and a Y chromosome (XY) and females usually have two X chromosomes (XX). As a consequence, females have the capacity to synthesize twice as much protein from the genes on their two X chromosomes as males with only one X chromosome. Such imbalance in protein synthesis might be detrimental, but there is a mechanism to ensure that the expression of genes on the female's X chromosomes is balanced with that of the male. One of her X chromosomes is condensed into a tight mass. Under a light microscope, the inactivated, condensed X chromosome shows up in the nucleus as a dark spot called a Barr body.

CASE STUDY REVISITED

Fixing Broken Genes

In some ways, germline engineering will be easier than current gene therapy techniques, because it will require only a single successfully transformed cell (the fertilized egg). The person who develops from that egg will have the altered genome in every cell of his or her body. By instructing a clinician to insert new genes or even an entire artificial chromosome into a fertilized egg, a prospective parent could ensure that a child was free from inherited disorders. The parents' inherited vulnerabilities to cancer, heart disease, cystic fibrosis, or any other heritable malady could be repaired before the fertilized egg was allowed to develop.

Many commentators predict that the chance to protect prospective offspring from health hazards will produce strong demand for germline engineering. And advocates see nothing wrong with that. Nobel prize winner James Watson asks, "If we can make better human beings by knowing how to add genes, why shouldn't we do it?" W. French Anderson, who performed the very first gene therapy treatment in 1990, comments, "In principle, I'm for it on the most fundamental of grounds: human nature. None of us want to pass on to our children lethal genes if we can prevent it." Molecular biologist John Campbell adds, "We could probably engineer people to be totally resistant to AIDS, or to certain kinds of cancers. We might engineer people to live much longer. I would say all these are good qualities."

The future prospects of germline engineering, however, are far from clear. The field has its share of skeptics who note that even conventional gene therapy for body cells has yet to prove itself safe and effective. Will we ever really understand the incredibly complex workings of the human genome well enough to safely engineer the DNA of an embryo? Other observers worry that if germline engineering becomes available to treat diseases and defects, it will inevitably be used to treat genetically based cosmetic problems such as baldness. From there, in the view of the skeptics, it is a short conceptual step to actively engineered children, constructed to be tall, athletic, and intelligent. Anyone who has seen the movie *Gattaca*, which depicts a future society in which social and economic status is determined by the quality of one's genes, will understand the potential pitfalls that accompany our increased prowess in genetic manipulation. "Evolution is being superseded by technology, and the time scale will be far more rapid," says biophysicist Gregory Stock. "Humans are becoming the objects of conscious design."

Summary of Key Concepts

1) How Are Genes and Proteins Related?

Genes are segments of DNA that can be transcribed into RNA and, for most genes, translated into protein. Each gene codes for a single protein.

2) What Is the Genetic Code?

The genetic code consists of codons, sequences of three bases, that specify either an amino acid in the protein chain or the end of protein synthesis (stop codons).

3) How Is the Information in DNA Used to Synthesize Proteins?

Transcription produces mRNA, tRNA, and rRNA, all needed for translation. During translation, tRNA and rRNA collaborate with enzymes and other proteins to decode the sequence of bases in mRNA and produce a protein with the amino acid sequence specified by the gene.

4) What Happens During Transcription?

Within an individual cell, only certain genes are transcribed. When the cell requires the product of a gene, RNA polymerase binds to the promoter region of the gene and synthesizes a single strand of RNA. This RNA is complementary to the template strand in the gene's DNA double helix.

5) What Happens During Translation?

mRNA carries the genetic information from the nucleus to the cytoplasm, where ribosomes use the information to synthesize a protein. Ribosomes contain rRNA and proteins organized into large and small subunits. These subunits come together at the first AUG codon of the mRNA molecule to form the complete protein-synthesizing machine. tRNAs deliver the appropriate amino acids to the ribosome for incorporation into the growing protein. Base pairing between the anticodons of the tRNAs and the codons of the mRNA puts the amino acids in the order coded in the mRNA. Two tRNAs, each carrying an amino acid, bind simultaneously to the ribosome's large subunit, which then catalyzes the formation of peptide bonds between the amino acids. As each new amino acid is attached, one tRNA detaches, and the ribosome moves down the mRNA one codon and binds to another tRNA that carries the amino acid specified by that codon. Addition of amino acids to the growing protein continues until a stop codon is reached, signaling the ribosome to disassemble and to release both the mRNA and the newly formed protein.

6) What Is a Mutation?

A mutation is a change in the nucleotide sequence of a gene. Mutations can be caused by mistakes in base pairing during replication, by chemical agents, and by environmental factors such as radiation. Common types of mutations include changes in a nucleotide base pair (point mutations) and insertions or deletions of nucleotide base pairs.

7) What Are the Effects of Mutations?

Mutations are generally neutral or harmful, but in rare cases a beneficial mutation will encode features that will be favored by natural selection.

8) Are All Genes Expressed?

To be expressed, a gene is transcribed and translated, and the resulting protein performs some action within the cell. Which genes are expressed in a cell at any given time depends on the function of the cell, the developmental stage of the organism, and the environment.

9) How Is Gene Expression Regulated?

The amount of mRNA synthesized from a particular gene may be regulated by molecules that increase or decrease the rate of transcription. For example, entire chromosomes or parts of chromosomes may be condensed and inaccessible to RNA polymerase, whereas other portions are expanded, allowing transcription to occur.

Key Terms

anticodon *p. 127*
codon *p. 124*
deletion mutation *p. 131*
gene *p. 124*
genetic code *p. 124*
initiation complex *p. 129*
insertion mutation *p. 131*
messenger RNA (mRNA) *p. 127*
mutation *p. 131*
nucleotide substitution *p. 131*
point mutation *p. 131*
promoter *p. 127*
ribonucleic acid (RNA) *p. 124*
ribosomal RNA (rRNA) *p. 127*
ribosome *p. 127*
RNA polymerase *p. 127*
stop codon *p. 126*
template strand *p. 128*
transcription *p. 126*
transfer RNA (tRNA) *p. 127*
translation *p. 127*

Thinking Through the Concepts

Multiple Choice

1. *A gene*
   a. is synonymous with a chromosome
   b. is composed of mRNA
   c. is a specific segment of nucleotides in DNA
   d. contains only those nucleotides required to synthesize a protein
   e. specifies the sequence of nutrients required by the body
2. *Which of the following is a single-stranded molecule that contains the information for assembly of a specific protein?*
   a. transfer RNA
   b. messenger RNA
   c. exon DNA
   d. intron DNA
   e. ribosomal RNA
3. *Anticodon is the term applied to*
   a. the list of amino acids that corresponds to the genetic code
   b. the concept that multiple codons sometimes code for a single amino acid
   c. the part of the tRNA that interacts with the codon
   d. the several three-nucleotide stretches that code for "stop"
   e. the part of the tRNA that binds to an amino acid
4. *DNA*
   a. takes part directly in protein synthesis by leaving the nucleus and being translated on the ribosome
   b. takes part indirectly in protein synthesis; the DNA itself stays in the nucleus
   c. has nothing to do with protein synthesis; it is involved only in cell division
   d. is involved in protein synthesis within the nucleus
   e. codes for mRNA but not for tRNA or rRNA
5. *Synthesis of a protein based on the sequence of messenger RNA*
   a. is catalyzed by DNA polymerase
   b. is catalyzed by RNA polymerase
   c. is called translation
   d. is called transcription
   e. occurs in the nucleus
6. *A Barr body is*
   a. a condensed X chromosome
   b. an organelle involved in protein synthesis
   c. another term for chromosomes as they are seen during cell division
   d. visible in cells of both males and females
   e. present only in males, because their female chromosomes are inactivated

? Review Questions

1. How does RNA differ from DNA?
2. What are the three types of RNA? What is the function of each?
3. Define the following terms: genetic code; codon; anticodon. What is the relationship among the bases in DNA, the codons of mRNA, and the anticodons of tRNA?
4. How is mRNA formed from a eukaryotic gene?
5. Diagram and describe protein synthesis.
6. Explain how both transcription and translation require complementary base pairing.
7. Describe some mechanisms of gene regulation.
8. Define mutation, and give one example of how a gene might mutate. Would you expect most mutations to be beneficial or harmful? Explain your answer.

Applying the Concepts

1. Some antibiotics, such as erythromycin and streptomycin, bind to the small ribosomal subunit in bacteria and inhibit translation. Why do these drugs kill bacteria? How might bacteria evolve to become resistant to such antibiotics?
2. As you have learned in this chapter, many factors influence gene expression, including hormones. The use of anabolic steroids and growth hormones among athletes has created controversy in recent years. Hormones certainly affect gene expression, but, in the broadest sense, so do vitamins and foods. What do you think are appropriate guidelines for the use of hormones? Should athletes take steroids and growth hormones? Should children at risk of being unusually short be given growth hormones? Should parents be allowed to request growth hormones for their children of normal height in the hope of producing a future basketball player?

Group Activity

Form groups of three or four students. Discuss and answer the following question: Will a change in a single DNA nucleotide (a nucleotide substitution mutation) always result in a changed protein? Why or why not? Next, design a random 15-nucleotide DNA sequence, starting with the sequence TAC. Transcribe and translate this sequence to determine the amino acid sequence for which it codes. Use the genetic code, Table 9-2. Now "mutate" your DNA sequence by changing the first nucleotide; repeat transcription and translation of this new sequence. Note whether the resulting amino acid sequence is different. Repeat the mutation process by mutating every other nucleotide (the third, fifth, seventh, and so on) of your original sequence one at a time, and repeating transcription and translation. Each time, your original DNA nucleotide sequence should differ from the original DNA sequence by only one nucleotide. Note any differences in the resulting protein each time. What conclusions can you derive? Do your results support your answer to the original question?

For More Information

Grunstein, M. "Histones as Regulators of Genes." *Scientific American*, October 1992. Histones are proteins associated with DNA in eukaryotic chromosomes. Once thought to be merely a scaffold for DNA, they are actually important in gene regulation.

Nirenberg, M. W. "The Genetic Code: II." *Scientific American*, March 1963. Nirenberg describes some of the experiments in which he deciphered much of the genetic code.

Tjian, R. "Molecular Machines That Control Genes." *Scientific American*, February 1995. Complexes of proteins regulate which genes are transcribed in a cell and therefore help determine the cell's structure and function.

Travis, J. "Biology's Periodic Table." *Science News*, March 22, 1997. A brief synopsis of the Human Genome Project.

Answers to Multiple-Choice Questions

1. c 2. b 3. c 4. b 5. c 6. a

MediaTutor

CD Activities

Activity 9.1: Transcription

Estimated time: 10 minutes

The genetic information that is stored in the DNA found in the nucleus of a cell must be conveyed to the cytoplasm, where it can be translated into a specific sequence of amino acids. The intermediate molecule is RNA, synthesized by the process of transcription. This tutorial investigates the steps involved in transcription.

Activity 9.2: Translation

Estimated time: 10 minutes

Transcription provides a messenger RNA transcript of the gene. Now it is the task of the cell to take the information encoded in this messenger RNA and translate it into a specific sequence of amino acids. This process occurs on the ribosomes and involves a number of steps illustrated in this tutorial.

Start the MediaTutor Student CD-ROM and enter the activity number in the Quick Search box to be taken directly to that activity.

Web Investigation

Fixing Broken Genes

Estimated time: 20 minutes

Recent advances in genetics and molecular biology have provided an unprecedented opportunity to combat human genetic diseases. As discussed in Case Study: Fixing Broken Genes, gene therapy is one approach that holds such promise. In order to develop "genetic" treatments for such disorders, scientists must first understand the role the gene plays within the cell and, if possible, how the gene is regulated. Much of this can be determined by deciphering the nucleotide sequence, or code, of the gene. This investigation will explore our understanding of the genetic code and how our current knowledge can potentially lead us to develop cures for various diseases.

Go to http://www.prenhall.com/audesirk3, the Audesirk Companion Web site. Select Chapter 9 and Web Investigation to begin.

"All cells come from cells."

Rudolf Virchow, German physician and early proponent of the cell theory (1858)

These calves are genetically identical to one another and to the adult cow from which they were cloned. The technique of cloning is a by-product of advances in our understanding of how cells reproduce.

10 The Continuity of Life: How Cells Reproduce

AT A GLANCE

CASE STUDY

Call in the Clones

Though scientists have already cloned cows, sheep, goats, pigs, and mice, nearly everyone agrees that cloning should never be used to re-create a human. Nonetheless, the technology of cloning has a role to play in human affairs. For example, cloning may lead to dramatic improvements in drug production. In particular, cloning can improve our ability to use genetically engineered animals to produce proteins of medicinal value. Such animals include sheep whose DNA has been altered to include a copy of the human gene for the hormone insulin, which is used in the treatment of diabetes. The engineered sheep secrete the hormone in their milk, becoming living insulin factories and producing a valuable product for as long as they live.

Our ability to produce drugs in this manner, however, has been hampered by the difficulty of reliably inserting a foreign gene into an animal cell. Only a small fraction of attempted gene insertions actually yield a correctly transformed adult animal that produces the desired protein. But cloning can solve this problem. The new techniques for duplicating adult animals will enable drug companies to make many genetically identical copies of their successfully altered animals. With cloning, it becomes much easier to create large numbers of drug-producing animals.

Another potential application of cloning technology has generated even more excitement and far more controversy. The techniques used to clone animals can also be used to create laboratory cultures of stem cells. In living animals, stem cells form in the hours immediately after an egg is fertilized and give rise to all the specialized cell types that eventually compose the adult body (bone cells, nerve cells, muscle cells, and so on). Cultures of stem cells could, in theory, be used to grow any kind of tissue; these tissues would then provide an inexhaustible supply for transplants to treat a host of different diseases and injuries.

We'll revisit the connection between cloning and stem cells at the end of this chapter, after you've had a chance to familiarize yourself with the ways in which cells reproduce themselves (cellular reproduction) and with the methods that were used to produce Dolly, the cloned sheep. ■

1) Why Do Cells Divide?

One of the essential qualities of living things is that they grow and reproduce. Both growth and reproduction depend heavily on **cell division**, the process by which one cell gives rise to two or more cells. The process is organized such that a parent cell accurately distributes both genes and other cell components to its daughter cells. In eukaryotes, there are two types of cell division: mitotic and meiotic. **Mitotic cell division** yields daughter cells that are genetically identical to the original cell. In **meiotic cell division** the daughter cells are not genetically identical to each other or to the original cell.

Some Organisms Reproduce Asexually by Mitotic Cell Division

Many eukaryotic organisms are single-celled, and such organisms usually reproduce by mitotic cell division that produces two daughter cells that are genetically identical to the original cell. This method of producing new single-celled eukaryotes is a form of asexual reproduction (Fig. 10-1a,b), so-called because offspring are formed from a single parent without the uniting of sperm and egg (that is, without sex).

Asexual reproduction is not limited to single-celled organisms; many multicellular organisms can also reproduce asexually. In this process, small replicas of the parent grow by means of repeated mitotic cell division. For example, a *Hydra* can reproduce by growing a miniature replica of itself as a bud (Fig. 10-1c). Eventually the bud separates from its parent, going off to live independently. Because mitotic cell division produces genetically identical cells, these offspring are genetically identical to their parents and each other. A group of such identical offspring is called a **clone**.

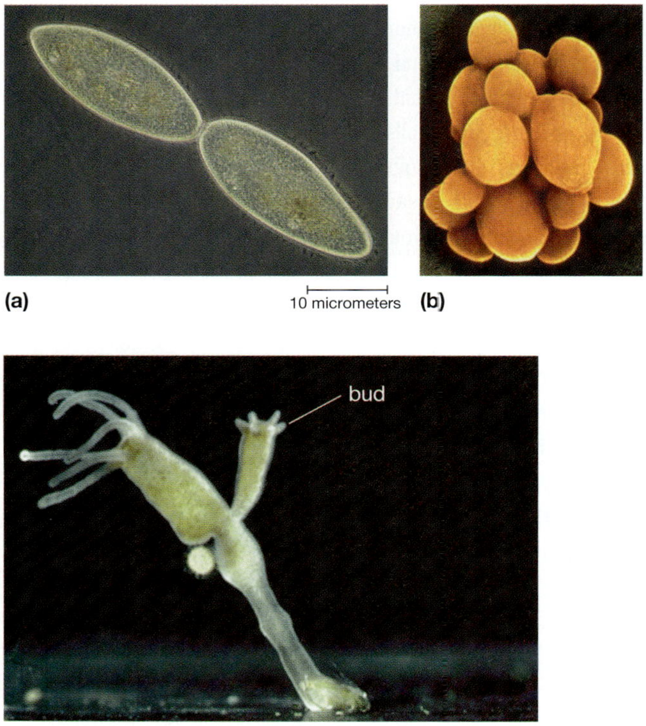

(c)

Figure 10-1 Asexual reproduction by mitotic cell division
(a) A single-celled eukaryotic microorganism called *Tetrahymena* has just divided mitotically and is about to separate as two new, independent organisms. ***(b)*** These yeast cells, each one a separate organism, were produced asexually by mitotic cell division. ***(c)*** This *Hydra*, a freshwater relative of jellyfish, is growing a miniature replica of itself (a bud) protruding from its side. When fully developed, the bud will break off and assume independent life.

Organisms Develop, Grow, and Maintain and Repair Tissues by Mitotic Cell Division

The role of mitotic cell division is not limited to asexual reproduction. For example, the fertilized eggs of multicellular organisms divide mitotically to develop—eventually into an adult (Fig. 10-2a–c). Your own body began as a single fertilized egg cell. This cell underwent a mitotic cell division to produce two genetically identical daughter cells; these two daughter cells underwent mitotic cell division to produce four genetically identical daughter cells. The divisions continued, ultimately producing the trillions of genetically identical cells that make up your body.

Mitotic cell division also allows an organism to maintain its tissues, many of which require frequent replacement (Fig. 10-2d). For example, your skin cells live for only about 2 weeks, and your red blood cells become worn out after about 4 months. The cells of your stomach lining, exposed to acid and digestive enzymes, survive only about 3 days. Without mitotic cell division to replace these short-lived cells, your body would soon be unable to function properly.

The ability of organisms to repair their bodies following injury also depends on mitotic cell division. For example, mitotic cell division is partly responsible for the ability of the human liver to regenerate—the only organ in our bodies with this capability. Even if half of the liver is lost, mitotic cell divisions will replace the lost part in just a few weeks, restoring normal liver function. The regenerative property of the liver makes it possible for a living person to donate a part of his or her liver to someone who needs a liver transplant: Only the right lobe of the donor's liver is removed. In the donor, cells in the remaining portion of the liver undergo mitotic cell division until it is the same size as before. In the recipient, cells in the donated portion of the liver undergo mitotic cell division to form a full-sized liver.

Sexual Reproduction Is Based on Meiotic Cell Division

In contrast to mitotic cell division, meiotic cell division does not simply generate additional copies of the original cell. Instead, each new cell produced by meiotic cell division contains only a portion of the genetic material of the original cell. This special feature of meiotic cell division is the foundation of sexual reproduction, a process that produces offspring with a mixture of genetic material from two parents.

Sexual reproduction is based on specialized cells called **gametes** (for example, eggs and sperm) that are produced by meiotic cell division. Each gamete carries half the genetic material of the parent's body cells. Fusion of two gametes, one from each parent, reconstitutes a full complement of genetic material, giving rise to an

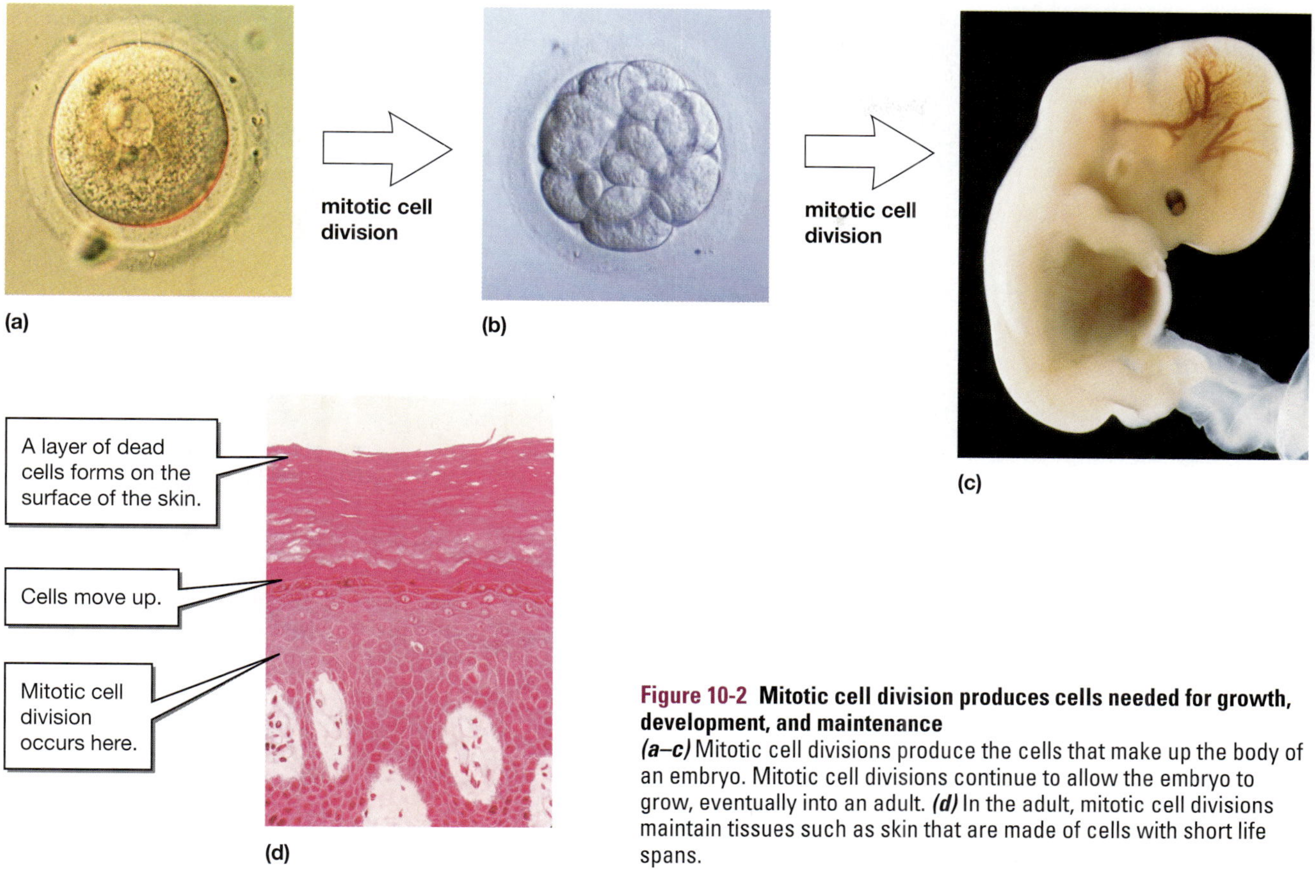

Figure 10-2 Mitotic cell division produces cells needed for growth, development, and maintenance
(a–c) Mitotic cell divisions produce the cells that make up the body of an embryo. Mitotic cell divisions continue to allow the embryo to grow, eventually into an adult. ***(d)*** In the adult, mitotic cell divisions maintain tissues such as skin that are made of cells with short life spans.

offspring that is similar to both parents, but identical to neither. In contrast to asexual reproduction by mitotic cell division (which can produce only clones), sexual reproduction reshuffles genes among individuals to produce genetically varied offspring.

The Life Cycles of Many Organisms Include Both Meiotic and Mitotic Cell Division

The life of an organism can be described in terms of its life cycle, the progression of events in which new individuals arise, grow and develop, and ultimately reproduce to create new individuals that begin the cycle anew. Life cycles can assume many different patterns, reflecting the great diversity of life forms on Earth. Despite this variability, however, the life cycles of many organisms share a common overall pattern (Fig. 10-3). A new life begins when two gametes fuse, bringing together genes from two parents. The new individual uses mitotic cell division to grow and develop a multicellular body. At some point in the individual's life, meiotic cell division creates new gametes that unite with other gametes to produce the next generation. This basic pattern recurs, with great variation in its details, throughout the tree of life.

The Big Picture

We are all familiar with the human progression from fetus in the womb to helpless infant, through stages of growth and development, to adulthood and, in most cases, parenthood. Organisms of all types grow, develop, and, eventually, reproduce. Beneath the surface, all of these changes depend entirely on countless repetitions of the beautifully choreographed processes of cell division.

2 What Happens During Cell Division?

Given the crucial role that cell division plays in the maintenance of life, it is not surprising that uncovering the mechanisms by which cells divide has long been one of the main goals of cell biologists. The main elements of cell division are now understood, but great gaps remain in our understanding of the details.

The Nucleus Divides Before the Cytoplasm

Cell division has two major components: division of the nucleus and division of the cytoplasm. In mitotic cell division, the nucleus divides by a process known as **mitosis**,

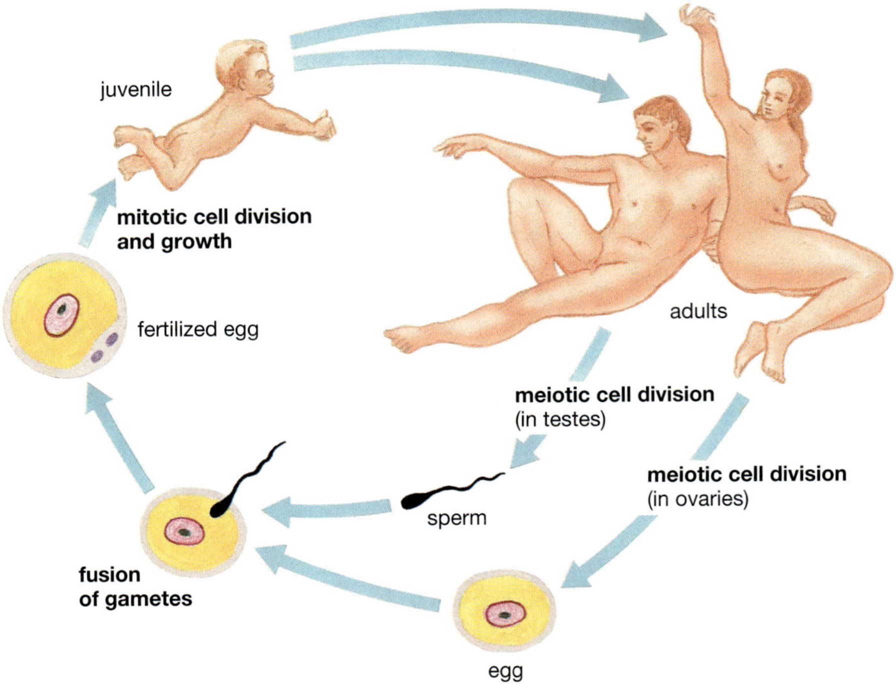

Figure 10-3 Mitotic and meiotic cell division in humans

which yields two nuclei, each containing a copy of every chromosome that was present in the original nucleus. In meiotic cell division, the nucleus divides by **meiosis**, which yields four nuclei, each containing half the number of chromosomes that was present in the original nucleus. The process that divides the cytoplasm is known as **cytokinesis**. When cytokinesis is complete, there are two daughter cells, each containing one nucleus and the organelles, nutrients, enzymes, and other molecules needed to maintain life.

Mitotic Cell Division Is One Part of the Cell Cycle

Mitotic cell division is one part of a cycle of eukaryotic cell growth and reproduction called the **cell cycle** (Fig. 10-4). The other part of the cell cycle is **interphase**. During interphase, the cell acquires nutrients from its environment, grows, and duplicates its chromosomes. After a period whose length depends on the species, the type of cell, and the nutrients available, the cell divides. Each daughter cell may then enter another cell cycle, producing additional cells. Alternatively, the newly formed cells may exit the cell cycle and never divide again. In our bodies, cells in our bone marrow and skin divide as frequently as once a day. Cells in our liver divide only once or twice a year. Most of our nerve cells and our muscle cells never divide after they mature; if one of these cells dies, it is not replaced.

3 How Is the Eukaryotic Chromosome Organized?

In both mitosis and meiosis, accurate distribution of genes to the new daughter cells is a major challenge. Chromosomes play a large role in the methods by which this challenge is met, so before moving on to describe the details of cell division, we first describe some key features of eukaryotic chromosomes.

Chromosomes Consist of DNA Bound to Proteins

DNA does not float free within the nucleus. Instead, each double helix is packaged into a chromosome (see Chapter 4). The DNA in a chromosome is bound to proteins that organize and compress the DNA so that it can fit into the nucleus. Fitting the DNA into this tiny space is no small task. If the total DNA in a single human body cell were laid end to end, it would be about 6 feet long, or about a million times as long as the nucleus.

Chromosomes Condense During Cell Division

The degree to which the DNA is compressed varies over time. When the cell is not dividing, the DNA is dispersed, so that the enzymes needed for transcription and replication can gain access. During cell division, however, the chromosomes must be sorted and moved into two daughter nuclei. Just as thread is easier to organize when it is wound tightly onto spools, chromosomes

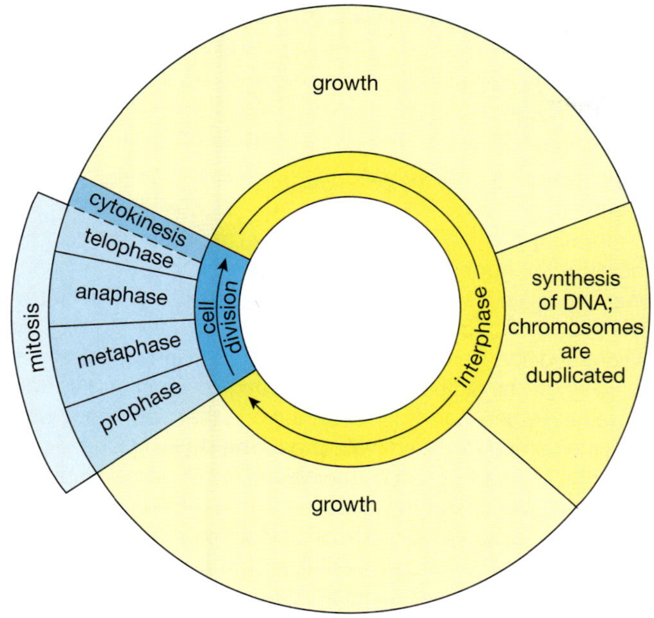

Figure 10-4 The eukaryotic cell cycle
Interphase and cell division are the two major phases of the cell cycle.

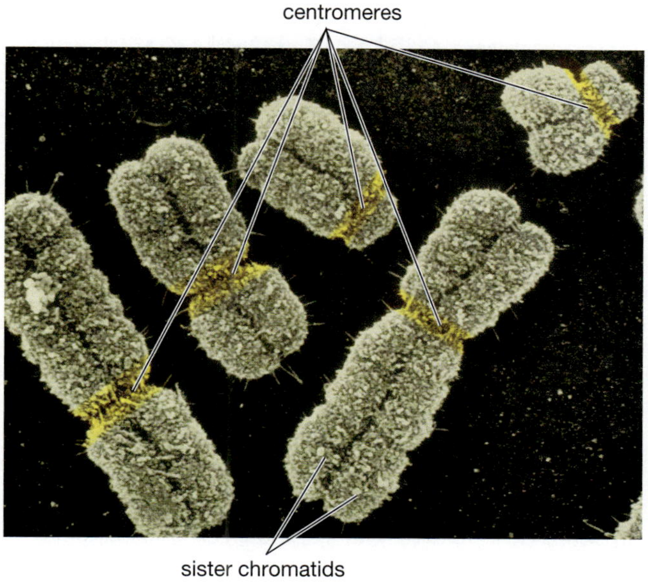

Figure 10-5 Human chromosomes during mitosis
The DNA and associated proteins in these duplicated human chromosomes have coiled up into the thick, short sister chromatids attached at the centromere. During cell division, the condensed chromosomes are about 5 to 20 micrometers long. At other times, the chromosomes uncoil until they are about 10,000 to 40,000 micrometers long.

are easier to sort and transport when they are compressed and shortened. Prior to cell division, the DNA of each chromosome folds up into a compact structure, and is then said to be condensed.

At First, Duplicated Chromosomes Remain Attached

By the time cell division begins, the DNA within each chromosome has already replicated, and the two resulting chromosomes remain attached at a specialized region of DNA called the **centromere**. While the two chromosomes are attached at their centromeres, we refer to each one as a **chromatid**, and to the pair as a set of sister chromatids. Thus, DNA replication produces a **duplicated chromosome** with two identical sister chromatids (Fig. 10-5).

Chromosomes Usually Occur in Pairs

The body cells of many organisms, including humans, contain pairs of chromosomes, as can be seen when we view an entire set of stained chromosomes from a single cell such as the one shown in Figure 10-6. With a single exception that we discuss in the next paragraph, each pair consists of two chromosomes that carry the same genes, arranged in the same order. As a result, the members of a pair, known as **homologues**, are similar in size, shape, and staining pattern. (A pair of homologues is often referred to as "homologous chromosomes.")

An exception to the rule that homologues contain the same genes is found in organisms that have **sex chromosomes** that carry the genes for sex determination. In

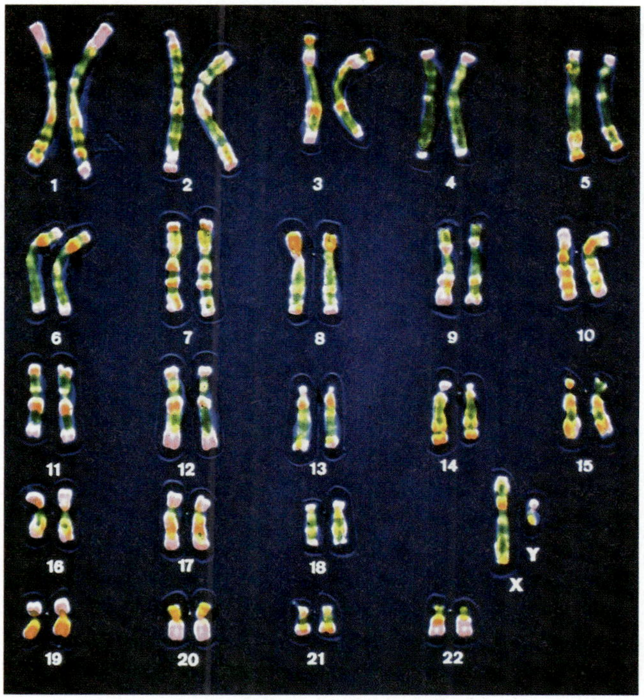

Figure 10-6 A human male's chromsomes
To make this picture, one male's chromosomes were stained and photographed. Individual chromosomes were then cut out of the photo and arranged in descending order of size. The two chromosomes above each number are a pair that is similar in size and staining patterns and that has similar genetic material. In a set like this from a female, there would be two X chromosomes and no Y chromosome. Notice that the Y chromosome is much smaller than the X chromosome.

Scientific Inquiry
Much Ado About Dolly

"And what is the Scientific Community doing? THEY'RE CLONING SHEEP. Great! Just what we need! Sheep that look MORE ALIKE than they already do!"

Dave Barry, 1997

Early in 1997 a short article in the journal *Nature* caused headlines throughout the world. Ian Wilmut and his associates at the Roslin Institute in Edinburgh, Scotland, had done what many scientists believed to be impossible: They cloned a mammal, using a nucleus taken from adult tissue. This type of **cloning** is the process of artificially creating a clone—a new individual that is genetically identical to an existing individual.

At the time of Wilmut's breakthrough, researchers had already discovered that they could replace the nucleus of an egg cell with a nucleus taken from a cell in an extremely early stage of embryonic development. After such a nuclear transfer, the introduced embryonic cell nucleus directs the development of a new individual, with no need for a fusion of sperm and egg. Until 1997, however, attempts to make clones using nuclei taken from older tissues (from embryos that had more than 16 cells) had failed. Researchers hypothesized that these attempts failed because it is impossible to undo the changes in gene expression that cause cells to specialize, to become, say, nerve cells instead of skin cells or liver cells or fat cells. Wilmut's work, however, showed that cell specialization is reversible under the right conditions. The researchers took cells from the udder of a 6-year-old ewe and grew them in culture dishes. The cells were deprived of nutrients, which forced them into the nondividing stage of the cell cycle. Apparently, this change made possible the expression of all necessary genes when the nucleus from one of the cells was transplanted into an egg cell (Fig. E10-1).

Cloning is a highly artificial form of asexual reproduction, because the nucleus that directs development of the offspring was produced by mitotic division of a diploid cell and not by the fusion of haploid gametes produced by meiosis. Like the hydra budding from its parent, the cloned mammal is a genetic "chip off the old block." The advantage of cloning cells from an adult rather than from an embryo is that you know what you are getting. A prized wool-producing sheep could theoretically be cloned to produce a whole herd. The same might be done with a cow that has exceptional milk production. Cloning an adult allows scientists to use asexual reproduction to take advantage of the natural variability provided by sexual reproduction. By observing the adults, the most advantageous gene combinations produced by meiosis and the union of gametes can be selected and then perpetuated in clones.

Some of the potential uses of cloning, and the difficult issues raised by those uses, are discussed in this chapter's Case Study and Case Study Revisited. Dolly and other cloned mammals also promise to provide insights into aging. Do the genes of each species dictate a maximum life span? If so, Dolly may have been born middle-aged, because her genes came from the cell of a 6-year-old ewe. Recent results support this idea. The ends of eukaryotic chromosomes have special repetitive nucleotide sequences called telomeres. In most cells in an animal, each cell division leaves the telomeres a tiny bit shorter. Old animals, then, have shorter telomeres than young animals. Dolly's telomeres are shorter than usually seen in a 3-year-old sheep. Will her life be shorter as a result? It may take some time to know, since a sheep's average life span is 13 years.

mammals, females have two copies of the X chromosome, but males have one X and one Y. Males thus have one pair of chromosomes (XY) in which the members have very different sizes, staining patterns, and genes. The X and Y chromosomes, however, behave like homologues during the cell division so we treat them as a pair.

Cells May Contain One or Two Sets of Chromosomes

Each different species has a chracteristic number of chromosomes. Human body cells, for example, have 46 chromosomes. These are not, however, 46 *different* chromosomes. Instead, each body cell contains 23 different types of chromosomes, each present as a pair of homologues. The cell has two copies of chromosome 1, two copies of chromosome 2, and so on, up through chromosome 22. Each cell also has two sex chromosomes. Thus, a human body cell has two sets of 23 chromosomes.

Cells that contain two homologous chromosomes of each type are called **diploid**. Most cells in our bodies are diploid. Our gametes, however, are produced by meiotic cell division, which leaves them with only one set of 23 chromosomes. Cells that contain only one of each type of chromosome are called **haploid**. If two haploid cells fuse, as happens during sexual reproduction, a diploid cell is formed.

The number of different chromosomes in a species' cells is called its haploid number and is designated n. Similarly, the total number of chromosomes in a species' diploid cells is designated $2n$. Thus, for humans, $n = 23$ (because we have 23 types of chromosomes) and $2n = 46$.

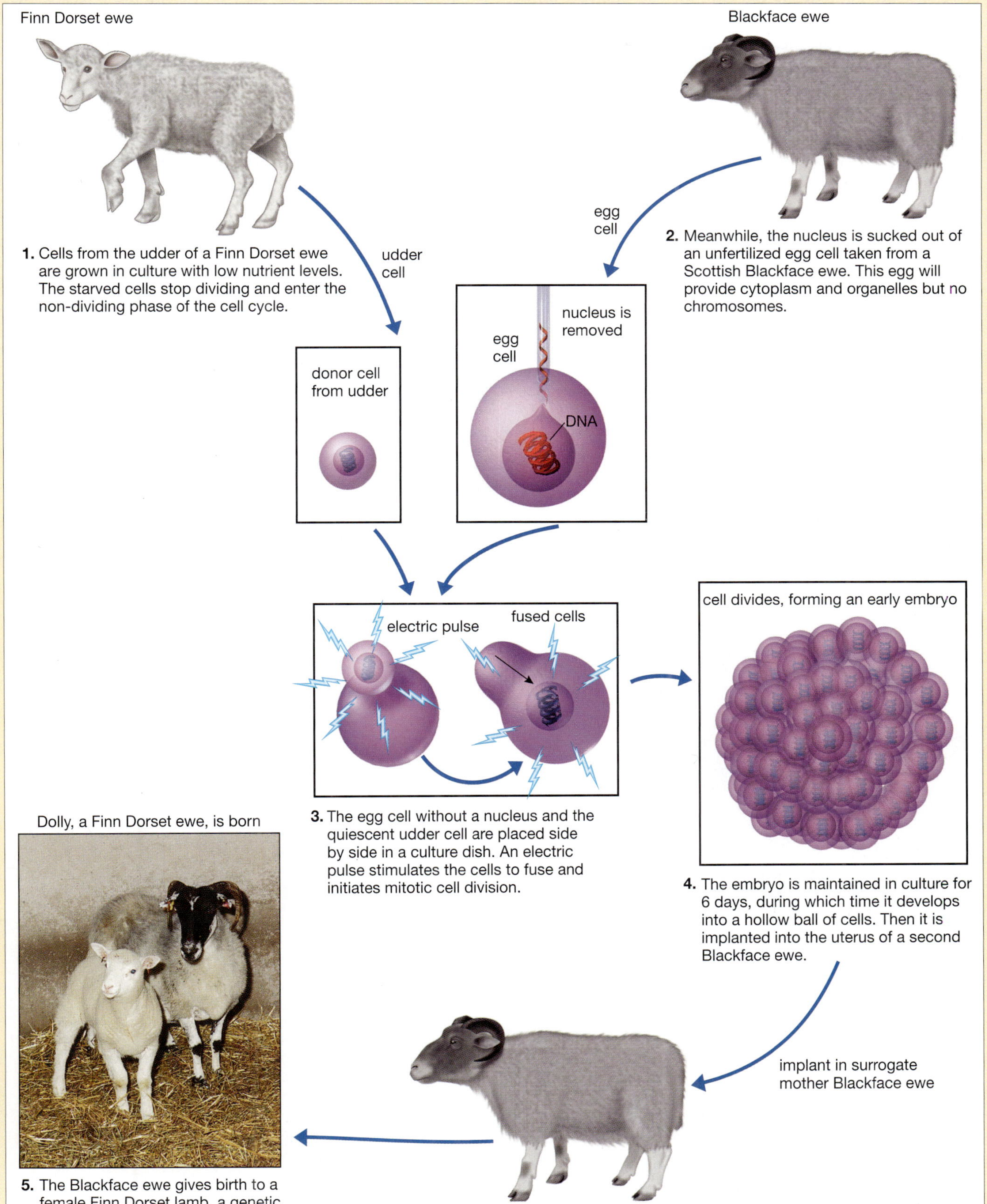

Figure E10-1 The making of Dolly

4 What Happens During Mitosis?

Mitosis begins after interphase (when chromosomes are duplicated), so at the onset of mitosis each chromosome already consists of two sister chromatids attached to one another at the centromere (Figs. 10-7a and 10-8a). For convenience, we divide mitosis into four phases, based on the appearance and behavior of the chromosomes: prophase, metaphase, anaphase, and telophase. These phases are not separate events. Rather, they form a continuum, with each phase merging into the next.

During Prophase, the Chromosomes Condense and Attach to Microtubules

As **prophase** begins, the duplicated chromosomes coil up and condense (Figs. 10-7b and 10-8b,c). After the duplicated chromosomes condense, the **spindle microtubules** that will later orchestrate chromosome movement begin to assemble in the cytoplasm. The spindle microtubules are rod-like structures that extend both inward toward the nucleus and outward toward the plasma membrane from two points called spindle poles, one on each side of the nucleus. In animal cells (but not in plant cells) each spindle pole is centered on a pair of microtubule-containing organelles called **centrioles**.

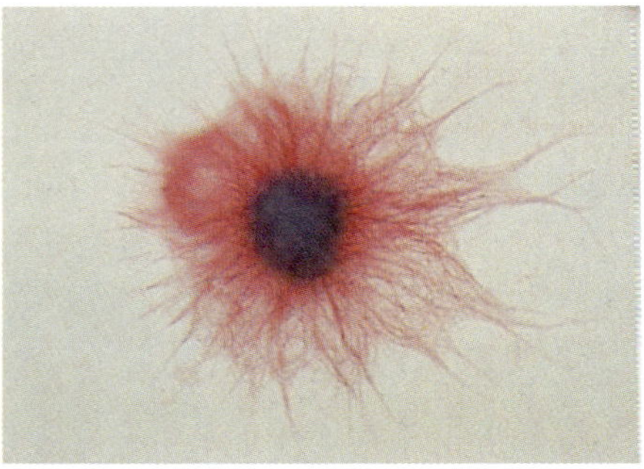

(a) Interphase in a seed cell: The chromosomes (blue) are in the thin, extended state and appear as a mass in the center of the cell. The spindle microtubules (red) extend outward from the nucleus to all parts of the cell.

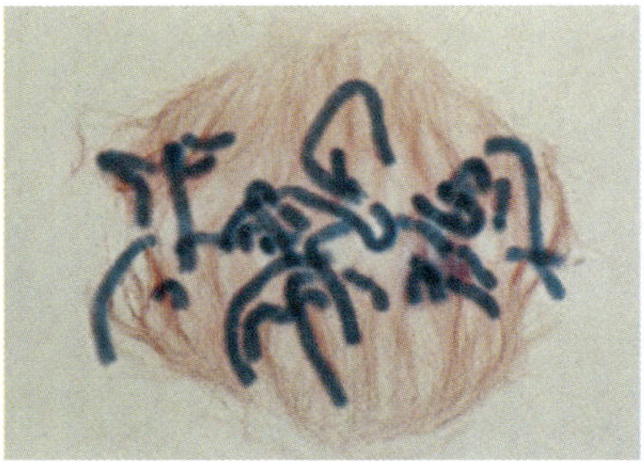

(b) Late prophase: The chromosomes (blue) have condensed and attached to the spindle microtubules (red).

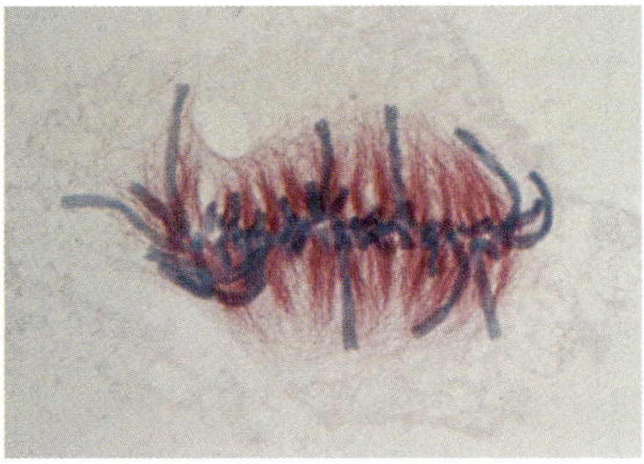

(c) Metaphase: The chromosomes have moved to the equator of the cell.

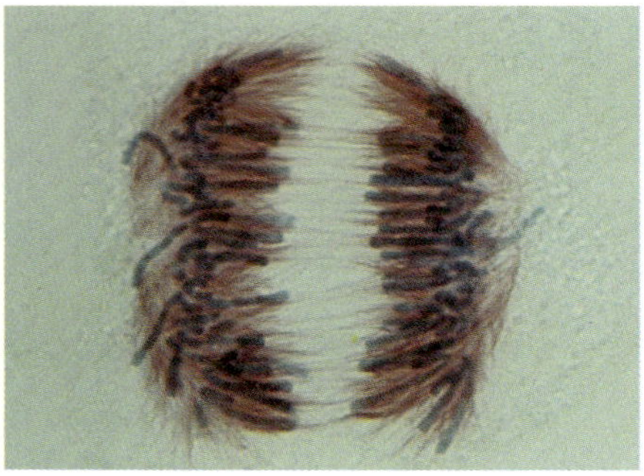

(d) Anaphase: Sister chromatids have separated, and one set has moved toward each pole.

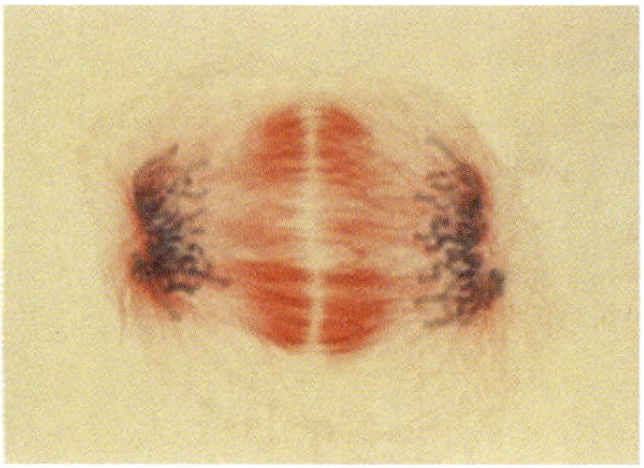

(e) Telophase: The chromosomes have gathered into two clusters, one at the site of each future nucleus.

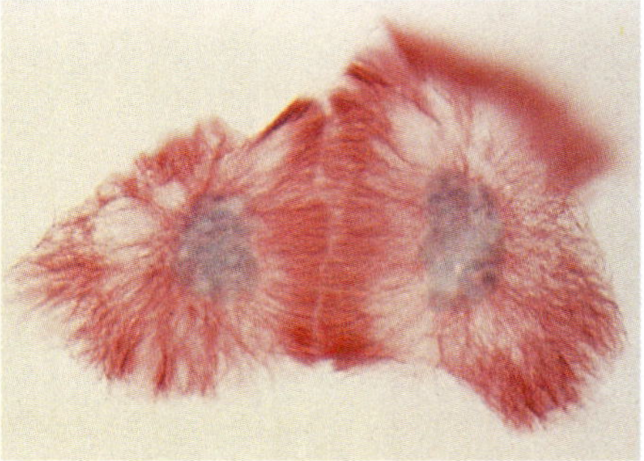

(f) Resumption of interphase: The chromosomes are relaxing again into their extended state. The spindle microtubules are disappearing, and the microtubules of the two daughter cells are rearranging into the interphase pattern.

Figure 10-7 The cell cycle in a plant cell
Interphase and mitosis in the African blood lily. The chromosomes are stained bluish-purple, and the spindle microtubules are stained pink to red. Compare these micrographs with the drawings of mitosis in an animal cell shown in Figure 10-8. (Each micrograph is of a different single cell that has been fixed and stained at a particular stage of mitosis. Only chromosomes and microtubules have been stained.)

Health Watch
Cancer: Mitotic Cell Division Runs Wild

How does a cell know when to divide and when not to? What keeps adult nerve cells and muscle cells from dividing while skin cells divide continuously? The answers lie in the same amazing stretches of DNA that determine all the specialized structures and functions of the different cell types—the genes. The cell cycle itself is under genetic control. Some genes, when activated, suppress cell division, whereas other genes promote it.

If the precise, tissue-specific regulation of the cell cycle is disrupted, a cell may become cancerous, capable of starting the growth of a large tissue mass called a tumor. For example, if a gene that normally represses cell division suffers a mutation that renders its protein product ineffective, the cell may divide out of control. In the fastest-growing cancers, a mass the size of a very small pea will contain a billion cells and produce about a million new cells each hour.

Because mitosis preserves the genetic makeup of each parent cell, all the offspring of a cancerous cell will be cancerous as well. If a cancerous cell is carried by the circulatory system to another part of the body, it may start a new cancerous growth there.

One gene, known as *p53*, may be defective in as many as 50 percent of all cancers. The *p53* gene codes for a protein that suppresses cell division during interphase. Normally the *p53* gene is activated in a cell in which the DNA has been damaged, so that the cell has time to repair its DNA before it is replicated. But cells in which the *p53* protein is defective cannot fix their DNA errors. If, in one of these cells, a gene that regulates the cell cycle mutates, the damage is less likely to be repaired. The result: unregulated cell division and cancer.

Damage to *p53* and other genes can result from environmental assaults on DNA, such as those caused by exposure to sunlight, certain viruses, and certain chemicals that contaminate food, water, and air. For example, when cells grown in the laboratory are exposed to benzo(*a*)pyrene, one of the many *carcinogens* (cancer-causing agents) in cigarette smoke, the *p53* gene usually mutates. This same type of mutation is found in a common form of lung cancer seen in heavy smokers. Carcinogens such as those in cigarette smoke exert a "double whammy" effect—increasing the numbers of errors in DNA replication and simultaneously damaging the *p53* gene that would normally allow many of these errors to be repaired. We shall discuss cancer in more detail in Chapter 22.

Eventually, the spindle microtubules form a complete basket around the nucleus, and the nuclear envelope disintegrates. Each duplicated chromosome attaches to two sets of spindle microtubules, one set from each pole of the cell (Fig. 10-8c). The point of attachment is at the chromosome's centromere, where each sister chromatid has a protein-containing structure called a **kinetochore**. One sister chromatid of the duplicated chromosome is tethered by its kinetochore to a spindle microtubule that leads to one pole, and the other sister chromatid is tethered to a spindle microtubule that leads to the other pole.

During Metaphase, the Chromosomes Line Up at the Cell's Equator

During **metaphase**, the spindle microtubules lengthen and shorten, adjusting their lengths to tug the duplicated chromosomes and change their positions. Eventually, the chromosomes become lined up along the equator of the cell, with one kinetochore facing each pole (Figs. 10-7c and 10-8d).

During Anaphase, Sister Chromatids Move to Opposite Poles

At the beginning of **anaphase**, the sister chromatids separate, and the kinetochores (with their attached chromosomes) move along the spindle microtubules toward the poles (Figs. 10-7d and 10-8e). As the kinetochores tow their chromosomes toward the poles, unattached spindle microtubules lengthen to push the poles of the cell farther apart, forcing the cell into an oval shape. The chromosomes keep moving until each pole of the cell contains a cluster of chromosomes. Note that each cluster contains one copy of each chromosome that was in the original cell (because each sister chromatid in a duplicated chromosome that splits during anaphase is an identical copy of one of the original chromosomes that was present before mitosis began).

During Telophase, Nuclear Envelopes Form Around Chromosomes

When the chromosomes reach the poles, **telophase** begins (Figs. 10-7e and 10-8f). The spindle microtubules disintegrate, and a nuclear envelope forms around each group of chromosomes. The chromosomes revert to their extended, non-condensed state. Cytokinesis separates each daughter nucleus into a separate cell (Fig. 10-8g). The daughter cells enter interphase (Figs. 10-7f and 10-8h).

5 What Happens During Cytokinesis?

Cytokinesis in animal cells depends on a ring of microfilaments that is attached to the plasma membrane and extends around the equator of the cell. The ring contracts

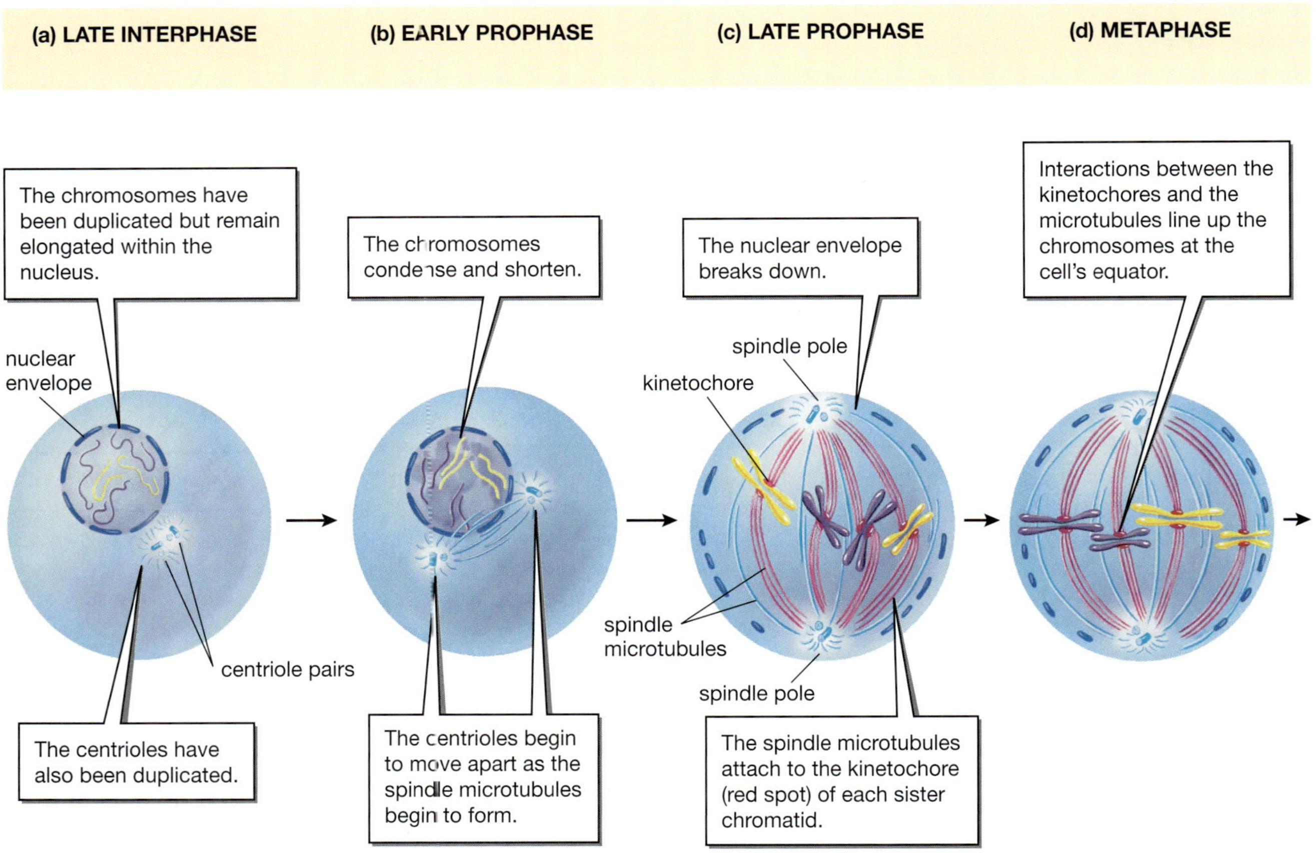

MEDIATUTOR 10.1 Mitosis

Figure 10-8 The cell cycle in an animal cell

during cytokinesis and constricts the cell's equator, much as pulling the drawstring of pair of sweatpants constricts the pants' waist. Eventually the cell's "waist" contracts down to nothing, dividing the cytoplasm into two new daughter cells (Fig. 10-9).

Cytokinesis in plant cells proceeds differently, perhaps because a plant cell's stiff cell wall makes it impossible to divide the cell by pinching at the waist. Instead, carbohydrate-filled vesicles bud off from the Golgi complex and line up along the cell's equator (Fig. 10-10). The vesicles fuse, producing a structure called the cell plate, which is a flattened sac, surrounded by plasma membrane and filled with sticky carbohydrates. The edges of the cell plate merge with the plasma membranes surrounding the daughter cells, and the carbohydrate contained in the cell plate becomes part of the cell walls.

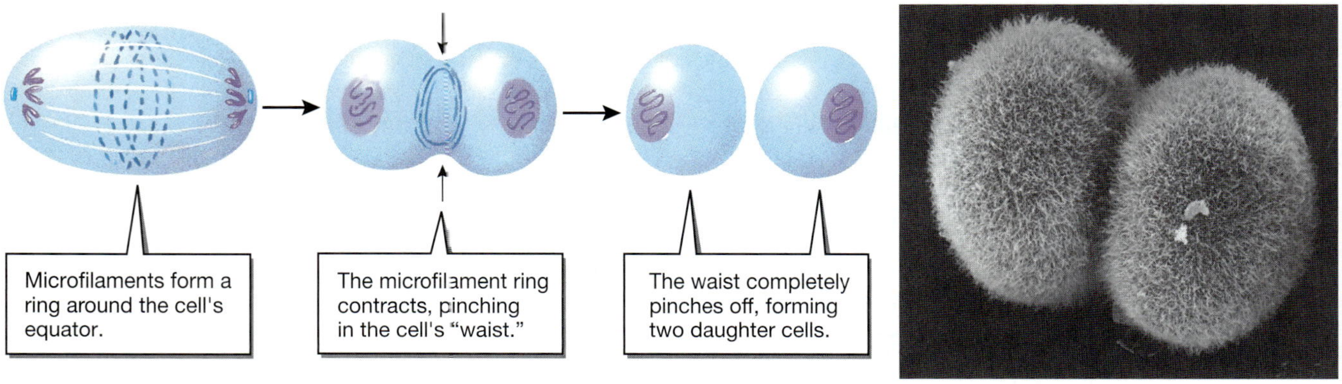

Figure 10-9 Cytokinesis in an animal cell

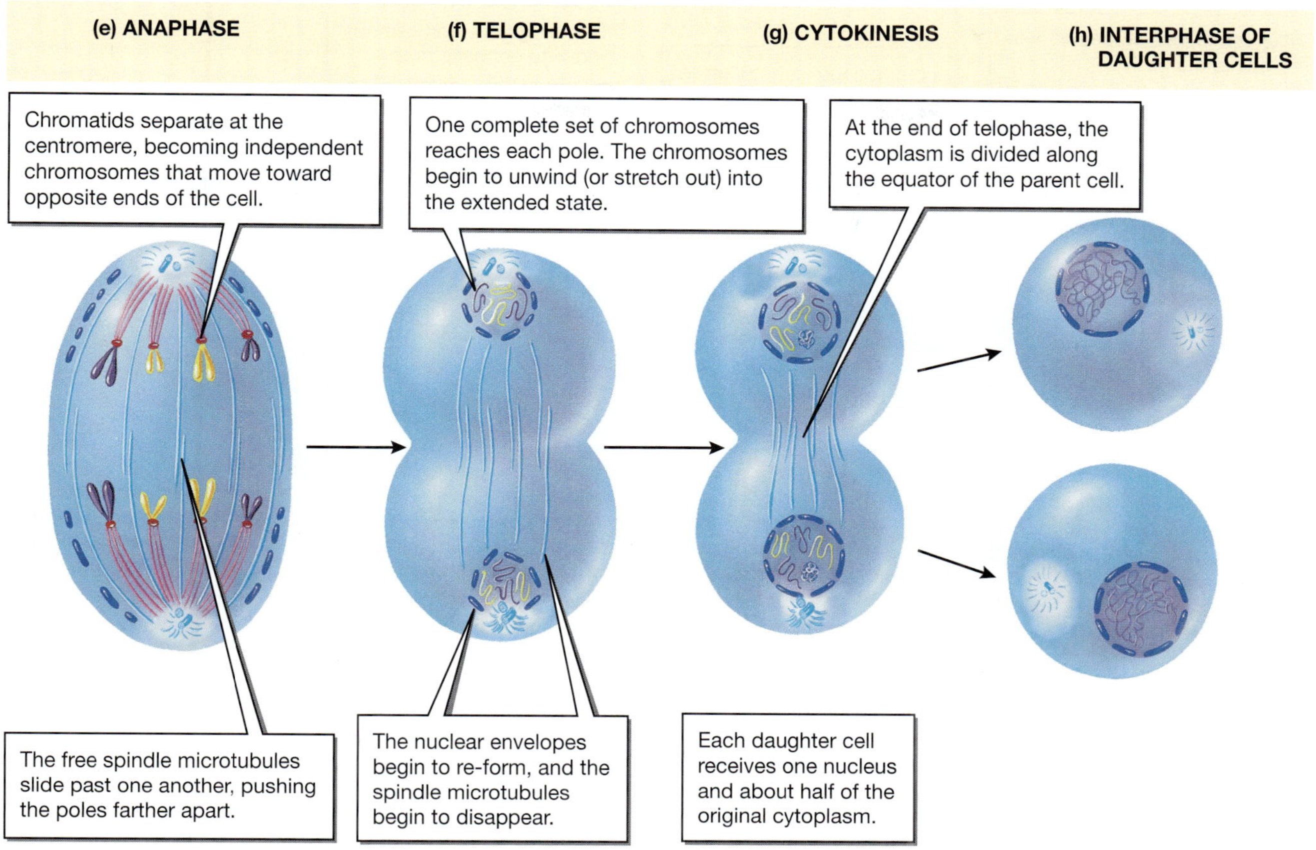

Figure 10-8 ***(continued)***

6 What Happens During Meiosis?

The key to sexual reproduction is meiosis, the production of haploid nuclei with unpaired chromosomes from diploid parent nuclei with paired chromosomes. Meiosis reduces the number of chromosomes in the nucleus by half; each daughter nucleus receives one member from each pair of homologous chromosomes. For example, because each diploid cell in a human body contains 23 pairs of chromosomes, meiotic cell division produces gametes with 23 chromosomes, one from each pair.

Meiosis Produces Four Haploid Daughter Nuclei

Many of the events in meiosis are similar or identical to those of mitosis. However, meiosis includes *two* nuclear divisions, so each cell that undergoes meiotic cell division

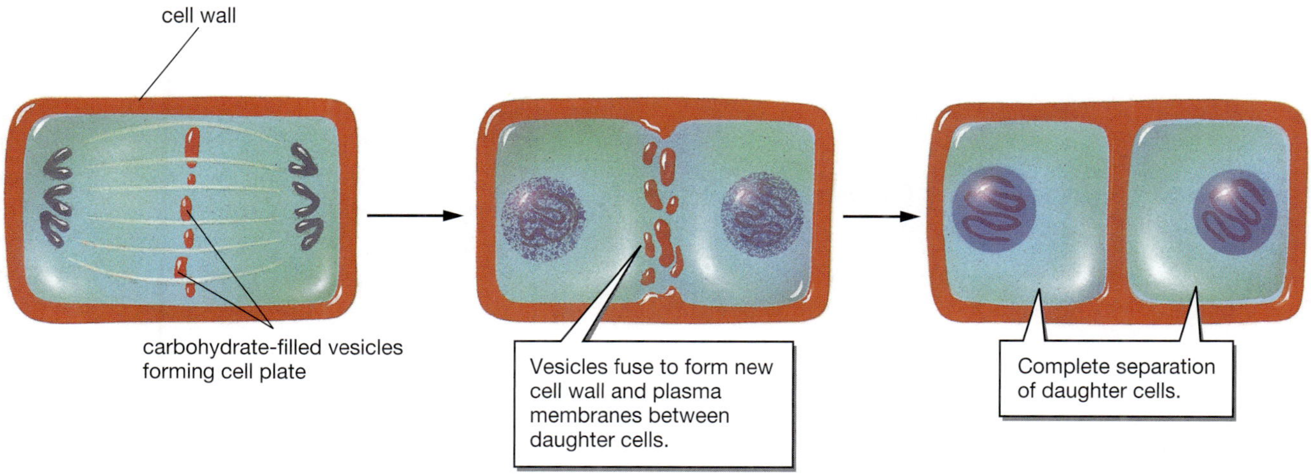

Figure 10-10 Cytokinesis in a plant cell

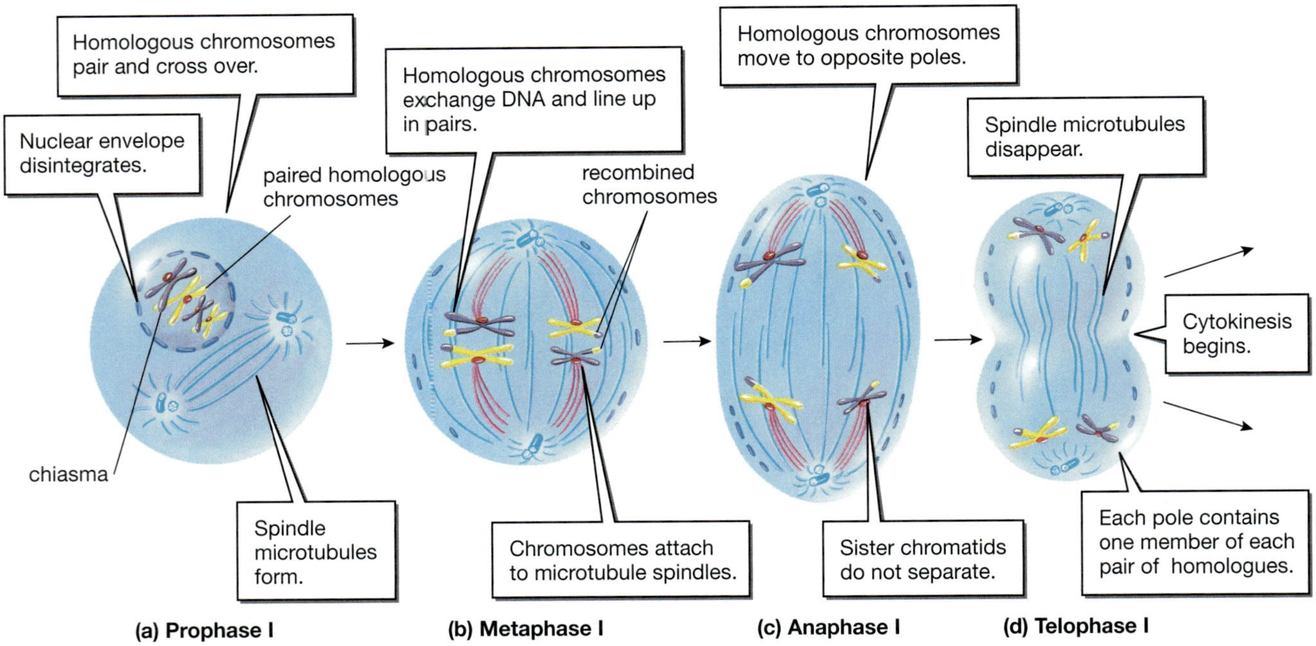

Figure 10-11 The details of meiotic cell division
In these diagrams, two pairs of homologous chromosomes are shown, large and small. The yellow chromosomes are from one parent and the violet chromosomes are from the other parent. Note that each daughter cell contains one member of each pair of parental homologous chromosomes.

yields four daughter cells. The two nuclear divisions are known as meiosis I and meiosis II. In meiosis I, homologous chromosomes pair up but sister chromatids remain connected to each other (differing from mitosis). In meiosis II, chromosomes behave as they do in mitosis; sister chromatids separate and are pulled to opposite ends of the cell.

Each of the four daughter nuclei produced by meiosis is haploid. Recall that DNA replication in a diploid cell produces four chromatids for each type of chromosome (that is, two chromatids for each member of each pair of homologous chromosomes). Meiosis separates each of these four chromatids into a different nucleus, thus reducing the number of chromosomes in each nucleus from diploid to haploid.

The phases of meiosis have the same names as the roughly equivalent phases in mitosis, followed by a I or II to distinguish the two nuclear divisions.

Meiosis I Separates Homologous Chromosomes into Two Daughter Nuclei

Like mitosis, meiosis I begins after DNA replication when duplicated chromosomes condense. Just as in mitosis, the sister chromatids of each duplicated chromosome are attached to one another at the centromere.

During Prophase I, Homologues Pair Up

After chromosome replication, meiosis I differs considerably from mitosis. In mitosis, each chromosome moves independently, and the members of a pair of homologous chromosomes do not interact. In contrast, during *prophase I* of meiosis, pairs of homologues line up side by side (Fig. 10-11a). Proteins bind the homologues together so that they match up exactly along their entire length. The two homologues in a bound pair often intertwine, forming crosses, or **chiasmata** (singular, **chiasma**; Fig. 10-12). Eventually, the proteins that hold the homologues together disappear, but the homologues remain associated with each other via the chiasmata.

As prophase I continues, spindle microtubules begin to assemble outside the nucleus. Near the end of prophase I, the nuclear envelope breaks down and the chromosomes attach to spindle microtubules by their kinetochores.

Paired Homologues Exchange DNA

Homologous chromosomes remain in contact at their chiasmata for some time, which allows them to exchange DNA at the points where they cross. This mutual exchange of DNA is known as **crossing over**. After crossing over, neither chromosome is quite the same as it was before the chiasmata formed. The result of crossing over, then, is genetic **recombination**: the formation of new combinations of genes on a chromosome.

During Metaphase I, Paired Homologues Line Up at the Equator of the Cell

During *metaphase I*, interactions between the kinetochores and the spindle microtubules move the paired homologues to the equator of the cell (see Fig. 10-11b).

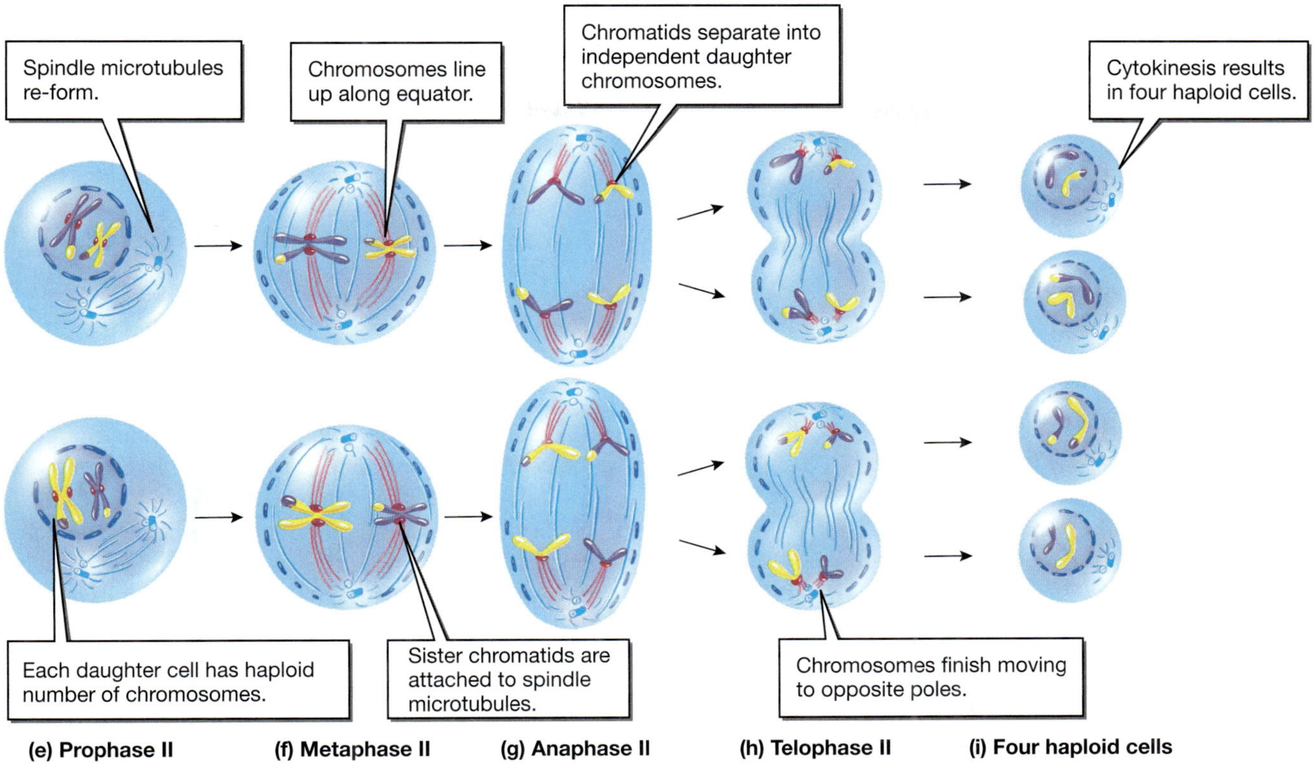

Figure 10-11 ***(continued)***

Note that the pairing of homologues in meiosis I makes the arrangement of chromosomes along the equator quite different from the arrangement in mitosis. In mitosis, *individual* duplicated chromosomes line up along the equator, but during metaphase I of meiosis *homologous pairs* of duplicated chromosomes line up along the equator.

Attachment of Chromosomes to Spindles Differs from Mitosis

The behavior of chromosomes during meiosis I is determined by how they attach to the microtubule spindles. It might be helpful, then, to compare chromosome attachment in meiosis I to that in mitosis (Fig. 10-13). In mitosis, each chromosome attaches independently to the spindle. But in meiosis I, homologues remain associated with each other via chiasmata and attach to the spindle as a unit containing both homologues. Also, in mitosis a duplicated chromosome has two functional kinetochores, each of which links one sister chromatid to spindle microtubules, so that the two sister chromatids are attached to microtubules that pull toward opposite poles (Fig. 10-13a). In contrast, a duplicated chromosome in meiosis I has only one functional kinetochore, and the two members of a homologous pair attach to spindle microtubules that pull toward opposite poles.

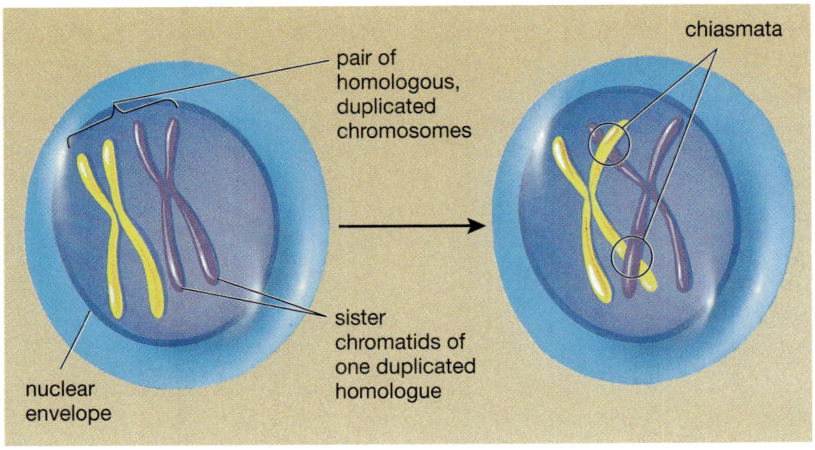

Figure 10-12 Chiasmata
Chiasmata are locations where nonsister chromatids cross and connect with each other, helping to hold homologues together.

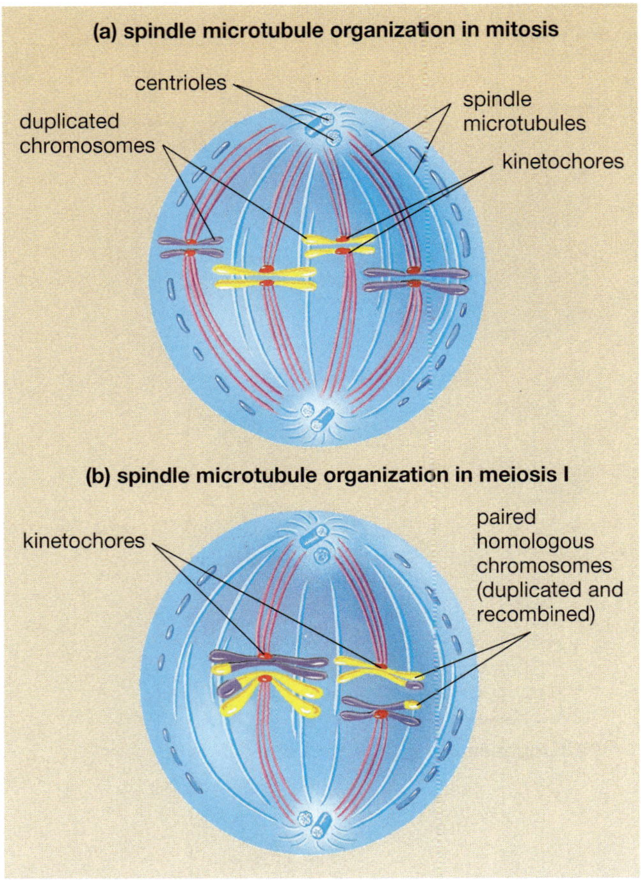

Figure 10-13 A comparison of the spindles formed during mitosis and meiosis I
(a) In mitosis, homologous chromosomes are not paired. The kinetochores of sister chromatids are attached to spindle microtubules that lead to opposite poles. When the sister chromatids separate during anaphase, the newly independent daughter chromosomes move to opposite poles of the cell. ***(b)*** In meiosis I, homologous chromosomes are paired. Each duplicated chromosome has one functional kinetochore. Homologous chromosomes (not sister chromatids) attach to microtubules from opposite poles. Sister chromatids of each chromosome remain together, moving to the same pole, but homologous chromosomes separate and move to opposite poles.

Thus, in mitosis, the *sister chromatids* separate and move to opposite poles, but in meiosis I the sister chromatids of each duplicated chromosome remain attached to each other and the *paired homologues* separate and move to opposite poles (Fig. 10-13b).

During Anaphase I, Homologous Chromosomes Separate

In *anaphase I*, the homologues separate from one another and are towed by their kinetochores to opposite poles of the cell (see Fig. 10-11c). One duplicated chromosome (still consisting of two sister chromatids) from each homologous pair moves to each pole of the dividing cell. Thus, at the end of anaphase I, the cluster of chromosomes at each pole contains one member of each pair of homologous chromosomes.

After Telophase I and Cytokinesis, There Are Two Daughter Cells

In *telophase I*, the spindle microtubules disappear and the nuclear envelope may reappear (see Fig. 10-11d). In most cases, cytokinesis takes place and divides the cell into two daughter cells. Each daughter cell has only one of each type of chromosome. Each chromosome, however, is a duplicated chromosome with two sister chromatids.

Meiosis II Separates Sister Chromatids into Four Daughter Nuclei

Meiosis II usually begins immediately after meiosis I ends; there is no intervening interphase and the chromosomes are not replicated again. Typically, the chromosomes remain condensed between meiosis I and meiosis II. Meiosis II is virtually identical to mitosis; the sister chromatids of each duplicated chromosome separate and are encapsulated in the nuclei of new daughter cells. Because meiosis I yielded two daughter cells, division of these cells by meiosis II yields a total of four daughter cells for each original cell.

During *prophase II*, the spindle microtubules reform (see Fig. 10-11e). The duplicated chromosomes attach individually to spindle microtubules as they did in mitosis. Each sister chromatid now has a functional kinetochore by which it attaches to spindle microtubules that extend to one or the other pole of the cell. During *metaphase II*, the duplicated chromosomes line up at the cell's equator (see Fig. 10-11f). During *anaphase II*, the sister chromatids separate and are towed to opposite poles (see Fig. 10-11g). As *telophase II* and cytokinesis conclude meiosis II, nuclear envelopes form, the chromosomes relax into their extended, non-condensed state, and the cytoplasm divides (see Fig. 10-11h, i).

Now that we have covered cell division in some detail, you may wish to examine Table 10-1 for a review and to compare mitotic and meiotic cell division.

The Big Picture

Without meiosis, there could be no sex. Sexual reproduction depends on the union of two gametes, each of which must contain an exact subset of the chromosomes necessary for a new organism. Only the intricate chromosomal maneuvers of meiosis, which divide a cell's genetic contents neatly in half, can provide gametes suitable for sexual reproduction.

7 How Do Meiosis and Sexual Reproduction Produce Genetic Variability?

Life has been shaped by evolution, and a crucial prerequisite for evolutionary change is genetic variability among organisms (we will explore this theme at length

Table 10-1 A Comparison of Mitotic and Meiotic Cell Divisions in Animal Cells

Feature	Mitotic Cell Division	Meiotic Cell Division
Cells in which it occurs	Body cells	Gamete-producing cells
Final chromosome number	Diploid—2*n* (46 for humans); two copies of each chromosome (homologous pairs)	Haploid—1*n* (23 for humans); one member of each homologous pair
Number of daughter cells	Two, identical to the parent cell and to each other	Four, containing recombined chromosomes due to crossing over
Number of cell divisions per DNA replication	One	Two
Function in animals	Development, growth, repair and maintenance of tissues, asexual reproduction	Gamete production for sexual reproduction

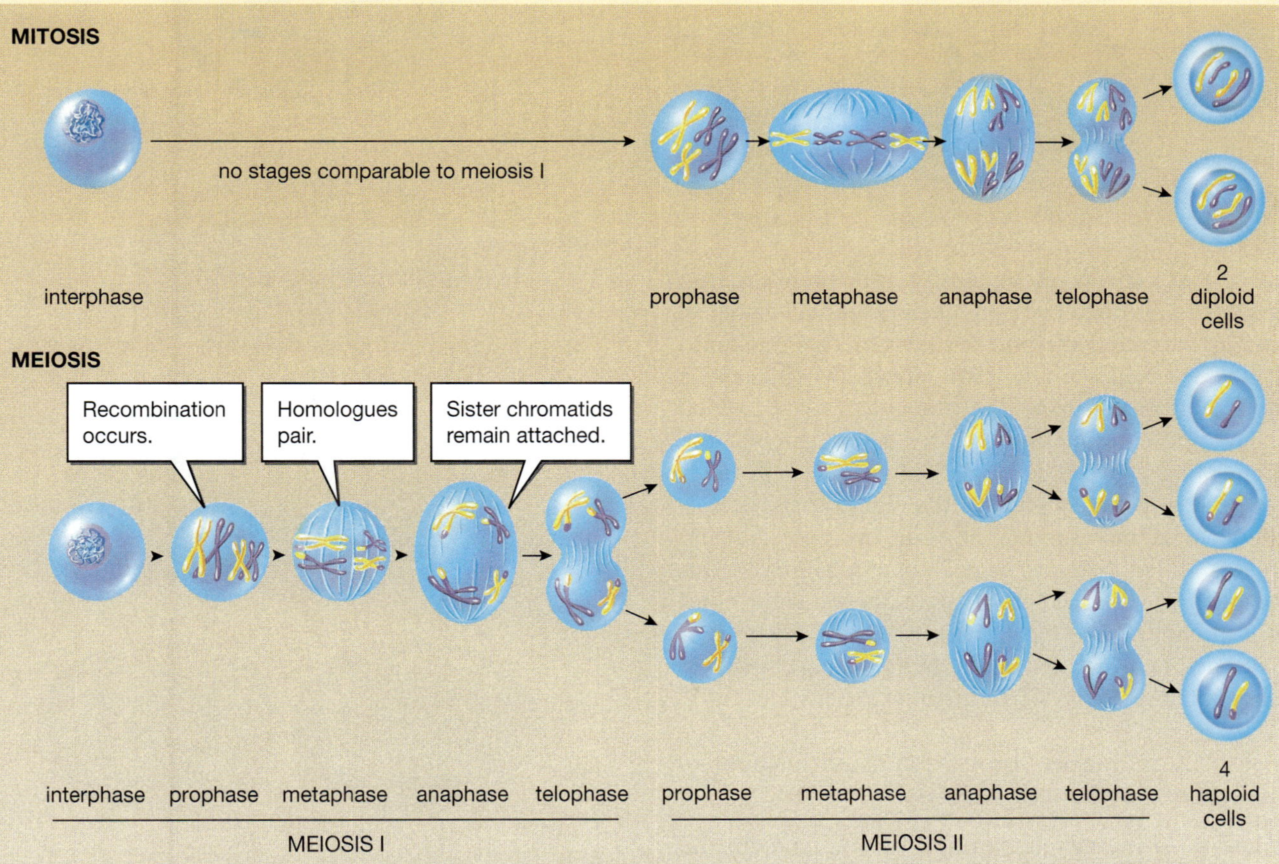

In these diagrams, comparable phases are aligned. In both mitosis and meiosis, chromosomes are replicated during interphase. Meiosis I, with the pairing of homologous chromosomes, formation of chiasmata, exchange of chromosome parts, and separation of homologues, has no counterpart in mitosis. Meiosis II, however, is similar to mitosis.

in Unit III). Mutations are the ultimate source of new genes, but meiosis and sexual reproduction play key roles in ensuring that individuals differ in the particular combinations of genes that they carry.

Shuffling of Homologues Creates Novel Combinations of Chromosomes

One important mechanism by which meiosis produces genetic diversity is the random assortment of homologues to daughter cells at meiosis I. Recall that at metaphase I, paired homologues line up at the cell's equator. For each pair of homologues, one chromosome faces one pole and the other chromosome faces the opposite pole. The determination of which chromosome faces which pole, however, is a matter of random chance that follows no particular pattern.

To see the effects of this random arrangement of homologues, let us consider meiosis in mosquitoes, which have three pairs of homologous chromosomes ($n = 3, 2n = 6$). For the sake of simplicity, we'll represent these chromosomes as large, medium, and small. To keep track of the homologues, let's color-code one set of chromosomes yellow and the other set violet. At metaphase I, the chromosomes can align in four configurations:

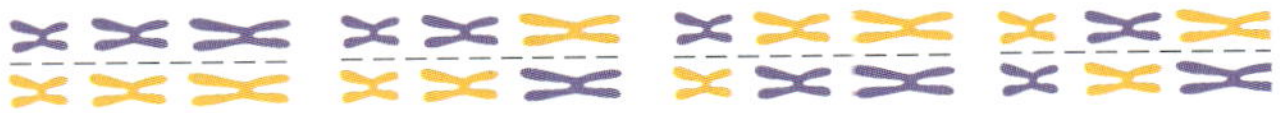

These four possible metaphase configurations yield eight possible combinations of chromosomes ($2^3 = 8$) when they separate during anaphase I:

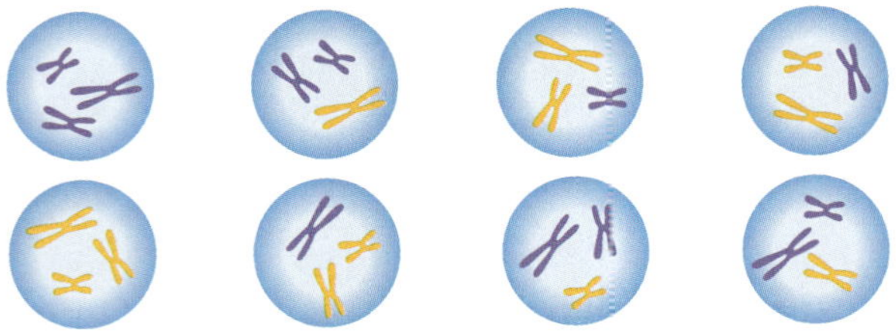

Each of these chromosome clusters will then undergo meiosis II to produce two genetically identical gametes. Thus, any particular meiotic cell division in a mosquito will produce four gametes, two of which will have a combination from the top row above and two of which will have the combination directly below it. Over the millions of meiotic cell divisions that produce a mosquito's output of gametes, each of the eight possible chromosome combinations will be on average equally represented. For humans, with 23 pairs of homologous chromosomes, meiosis produces gametes with more than 8 million (2^{23}) different combinations of chromosomes, each of which is equally likely to participate in fertilization and be passed to the next generation.

Crossing Over Creates Chromosomes with Novel Combinations of Genetic Material

In addition to the genetic variation resulting from the random combination of chromosomes, crossing over during meiosis produces chromosomes with combinations of genetic material that differ from those of either parent. In fact, these new combinations may have never before existed. Because homologous chromosomes cross over in new and different places at each meiotic division, a parent is unlikely to produce two gametes that carry exactly the same combinations of genetic material. In essence, every egg and every sperm is genetically unique.

Fusion of Gametes Creates Genetically Variable Offspring

The gametes generated by meiosis pass their genetic payload to the next generation at fertilization. Two haploid gametes, each containing a unique combination of

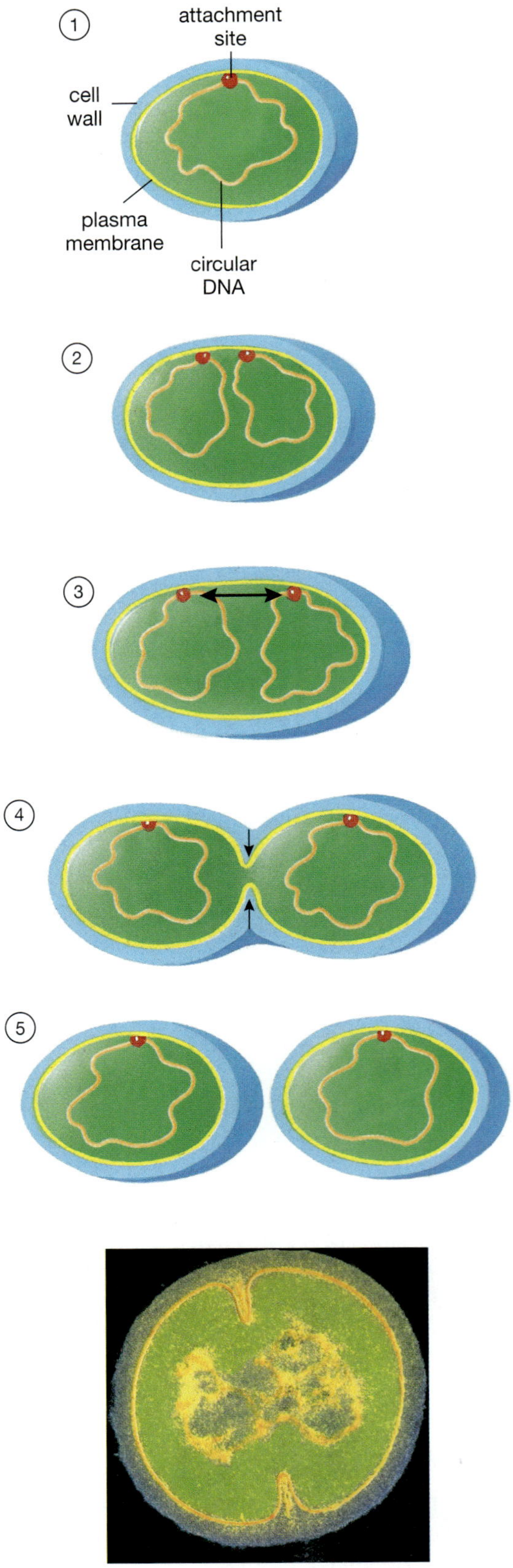

Figure 10-14 Prokaryotic cells divide by binary fission

genes, fuse to form a diploid offspring. The number of possible genetic outcomes is large. As we saw, one person can produce 2^{23} (about 8 million) different gametes, based solely on the random separation of the homologues. Fusion of gametes from just two people could produce 8 million × 8 million = more than 6 trillion genetically different children. This variation takes into account only random assortment and fertilization. When you also consider the almost endless additional variability produced by crossing over, is it any wonder that (unless you are an identical twin) there is truly no one just like you?

8 How Do Prokaryotes Reproduce?

Only eukaryotic cells use mitosis and meiosis in reproduction. Prokaryotic organisms, such as bacteria, reproduce by a type of cell division called **binary fission**, which begins with one cell and yields two genetically identical cells. Under suitable conditions, binary fission can proceed rapidly, allowing prokaryote populations to increase rapidly. For example, the number of *Salmonella* bacteria living on a piece of raw hamburger can double every 15 minutes when conditions are right. As *Salmonella* can cause food poisoning, it's a good idea to store meat under conditions, such as low temperature, that do not favor rapid bacterial reproduction.

At the beginning of binary fission, a prokaryotic cell duplicates its genetic material, which typically consists of a single, circular DNA molecule. Each of the two resulting DNA double helices attaches to the plasma membrane at a different point (Fig. 10-14, ②). The size of the cell begins to increase and, as the plasma membrane grows, the two DNA double helices are pulled apart (Fig. 10-14, ③). When the cell has approximately doubled in size, the plasma membrane begins to pinch inward between the two DNA attachment sites (Fig. 10-14, ④). Ultimately, the plasma membrane pinches off to form two daughter cells (Fig. 10-14, ⑤).

CASE STUDY REVISITED

Call in the Clones

Using techniques developed in the late 1990s, researchers are now able to keep cultures of human stem cells alive in the laboratory. The stem cells that have the most potential to become the basis for medical treatments are found in embryos. These embryonic stem cells represent a very early stage in development, when each cell in the embryo is still capable of differentiating into any type of cell. So a scientist with a human embryonic stem cell culture and the right knowledge could potentially induce stem cells to grow into nerve tissue to repair brains damaged by Alzheimer's disease, into bone tissue to repair bones damaged by osteoporosis, or into muscle tissue to repair hearts damaged by heart attacks.

Scientists' ability to extract and culture embryonic stem cells has created an ethical dilemma. The process of extracting embryonic stem cells destroys embryos. This destruction is troublesome to many people, and the U.S. government strictly limits the conditions under which it will fund research on embryonic stem cells. Yet stem cell research has the potential to save lives and reduce suffering. The debate between those who wish to push ahead with research and those who wish to prevent the destruction of embryos defies easy resolution.

This is where cloning comes into the picture. Cloning is based on the transfer of a cell nucleus from a body cell to an egg from which the nucleus has been removed. Some scientists have indicated that they expect to soon be able to create human stem cells by transferring the nucleus of an adult's body cell to an egg. Would such a cell still be considered an embryo? Would the procedure constitute human cloning? The scientists who are developing the technique argue that, because the egg is not fertilized by a sperm, the procedure does not generate an embryo and does not constitute human cloning (no new individual is made, just a stem cell). Would the new procedure make stem cell research acceptable to those who currently object on ethical grounds? If so, the rewards would be tremendous, because such cells could be created using donor cells from any human, and the resulting cell culture would therefore be genetically identical to the donor. Any new tissues grown from the cells would be ideal for transplants to the original nucleus donor because they would not be rejected by the patient's immune system.

Summary of Key Concepts

1) Why Do Cells Divide?

Meiotic cell division produces gametes, which have only half of the parent's DNA. Fusion of gametes creates a fertilized egg that has a genetic makeup different from that of either parent. Mitotic cell division of a fertilized egg produces genetically identical cells that develop into an embryo. Additional mitotic cell divisions allow the embryo to grow, eventually, into an adult. Mitotic cell division also maintains body tissues, such as skin and blood, and repairs damage to some organs. Asexual reproduction is based on mitotic cell divisions and forms clones that are genetically identical to the parent.

2) What Happens During Cell Division?

Cell division consists of nuclear division and cytokinesis. Mitosis parcels out one copy of every chromosome into each of two nuclei, producing two genetically identical daughter nuclei. Meiosis separates homologous chromosomes and produces four daughter nuclei with only one homologue from each pair. Both mitosis and meiosis are followed by cytokinesis, which encloses each daughter nucleus in a separate cell. Mitotic cell division is part of the eukaryotic cell cycle, which consists of interphase and cell division. During interphase, the cell grows and duplicates its chromosomes.

3) How Is the Eukaryotic Chromosome Organized?

Each eukaryotic chromosome consists of a single DNA double helix and proteins that organize the DNA. During cell growth, the chromosomes are extended and accessible for the reading of their genetic instructions. During cell division, the chromosomes condense into short, thick structures. Eukaryotic cells typically contain pairs of homologous chromosomes. Homologues have virtually identical appearance because they carry the same genes with similar nucleotide sequences. Cells with pairs of homologous chromosomes are diploid. Cells with only a single member of each chromosome pair are haploid.

4) What Happens During Mitosis?

The chromosomes are duplicated during interphase, prior to mitosis. The two identical chromatids remain attached to one another at the centromere during the early stages of mitosis. Mitosis consists of four phases (see Fig. 10-8):

- **Prophase:** The chromosomes condense and their kinetochores attach to the spindle microtubules that form at this time.
- **Metaphase:** The chromosomes move along their attached spindle microtubules to the equator of the cell.
- **Anaphase:** The two chromatids of each duplicated chromosome separate and move along the spindle microtubules to opposite poles of the cell.
- **Telophase:** The chromosomes relax into their extended state, and nuclear envelopes re-form around each new daughter nucleus.

5) What Happens During Cytokinesis?

At the end of telophase, cytokinesis divides the cytoplasm into approximately equal halves, each containing a nucleus. In animal cells, the plasma membrane is pinched in along the equator by a ring of microfilaments. In plant cells, new plasma membrane forms along the equator by the fusion of vesicles produced by the Golgi complex.

6) What Happens During Meiosis?

During interphase before meiosis, chromosomes are duplicated. The cell then undergoes two specialized cell divisions—meiosis I and meiosis II—to produce four haploid daughter cells.

- **Meiosis I:** During prophase I, homologous duplicated chromosomes, each consisting of two chromatids, pair up and exchange parts by crossing over. During metaphase I, homologues move together as a pair to the cell's equator, one member of each pair facing opposite poles of the cell. Homologous chromosomes separate during anaphase I, and two nuclei form during telophase I. Each daughter nucleus receives only one member of each pair of homologues. The sister chromatids remain attached to each other throughout meiosis I.
- **Meiosis II:** Both daughter nuclei go through meiosis II, which resembles mitosis. The duplicated chromosomes move to the cell's equator during metaphase II. The two chromatids of each chromosome separate and move to opposite poles of the cell during anaphase II. This second division produces four haploid nuclei. Telophase II and cytokinesis complete the production of four haploid cells.

7) How Do Meiosis and Sexual Reproduction Produce Genetic Variability?

The random shuffling of homologous maternal and paternal chromosomes creates new chromosome combinations. Crossing over creates chromosomes with combinations of genetic material that may never before have been together on the same chromosome. Because of crossing over, a parent probably never produces any two gametes that are completely identical. The fusion of two such genetically unique gametes creates genetically differing offspring.

8) How Do Prokaryotes Reproduce?

Prokaryotes reproduce by binary fission in which the cell grows, replicates its DNA, and divides.

Key Terms

anaphase *p. 147*
binary fission *p. 155*
cell cycle *p. 142*
cell division *p. 139*
centriole *p. 146*
centromere *p. 143*
chiasma (chiasmata) *p. 150*
chromatid *p. 143*
clone *p. 140*
cloning *p. 144*
crossing over *p. 150*
cytokinesis *p. 142*
diploid *p. 144*
duplicated chromosome *p. 143*
gamete *p. 140*
haploid *p. 144*
homologue *p. 143*
interphase *p. 142*
kinetochore *p. 147*
meiosis *p. 142*
meiotic cell division *p. 139*
metaphase *p. 147*
mitosis *p. 141*
mitotic cell division *p. 139*
prophase *p. 146*
recombination *p. 150*
sex chromosome *p. 143*
spindle microtubule *p. 146*
telophase *p. 147*

Thinking Through the Concepts

Multiple Choice

1. *At which stage of mitosis are chromosomes arranged along a plane at the midline of the cell?*
a. anaphase
b. telophase
c. metaphase
d. prophase
e. interphase

2. *A diploid cell contains in its nucleus*
a. an even number of chromosomes
b. an odd number of chromosomes
c. one copy of each homologue
d. either an even or an odd number of chromosomes
e. four sister chromatids of each chromosome

3. *Synthesis of new DNA occurs during*
a. prophase
b. interphase
c. mitosis
d. cytokinesis
e. formation of the cell plate

4. *When do homologous chromosomes pair up?*
a. only in mitosis
b. only in meiosis I
c. only in meiosis II
d. in both mitosis and meiosis
e. in neither mitosis nor meiosis

5. *Curiously, there is no crossing over of any chromosome in the male fruit fly* Drosophila, *which has four pairs of chromosomes. How many different combinations of chromosomes are possible in a male fruit fly's sperm?*
a. 2
b. 4
c. 8
d. 16
e. many more than the above

? Review Questions

1. Define mitosis and cytokinesis. What changes in cell structure result when cytokinesis does not occur after mitosis?

2. Diagram the stages of mitosis. How does mitosis ensure that each daughter nucleus receives a full set of chromosomes?

3. Define the following terms: homologue, centromere, kinetochore, chromatid, diploid, haploid.

4. Describe and compare cytokinesis in animal cells and in plant cells.

5. Diagram the events of meiosis. At which stage do homologous chromosomes separate?

6. In what ways are mitosis and meiosis similar? In what ways are they different?

7. Describe how meiosis provides for genetic variability. If an animal had a haploid number of 2 (no sex chromosomes), how many genetically different types of gametes could it produce? (Assume no crossing over.) If it had a haploid number of 5?

Applying the Concepts

1. Nerve cells in the adult human central nervous system, as well as heart muscle cells, remain in interphase and do not divide. In contrast, cells lining the inside of the small intestine divide frequently. Discuss this difference in terms of why damage to the nervous system and heart muscle cells (such as caused by a stroke or heart attack) is so dangerous. What do you think might happen to tissues such as the intestinal lining if some disorder or drug blocked mitoses in all cells of the body?
2. Cancer cells divide out of control. Side effects of chemotherapy and radiation therapy that fight cancers include loss of hair and of the gastrointestinal lining, producing severe nausea. Note that cells in hair follicles and intestinal lining divide frequently. What can you infer about the mechanisms of these treatments? What would you look for in an improved cancer therapy?
3. Some animal species can reproduce either asexually or sexually, depending on the state of the environment. Asexual reproduction tends to occur in stable, favorable environments; sexual reproduction is more common in unstable or unfavorable circumstances. Discuss the advantages or disadvantages this pattern might have on survival of the species in an evolutionary sense or on survival of individuals.
4. Would you predict that a clone produced as Dolly was would be a carbon copy of the animal from which the nucleus was obtained? What factors might cause the two sheep to differ? What variables might cause a human clone to differ from the individual who donated the genetic material?

Group Activity

Your instructor will choose several traits (such as a widow's peak, the ability to curl your tongue, and gender) that you can easily evaluate and will show you the appropriate symbols for those traits and which traits are dominant or recessive. If you have the dominant phenotype for a given trait and can't determine whether you are heterozygous dominant or homozygous dominant, assume that you are heterozygous dominant for that trait. On one side of a strip of paper, label one allele (for example, *T*); on the other side, label the other allele (say, *t*, if you are heterozygous). For simplicity's sake, assume that each allele is on a separate chromosome, so label each trait on separate paper strips; you'll have as many strips of paper (chromosomes) as traits chosen. Put your initials on each paper strip so that you'll be able to identify it as yours. Label each strip as chromosome 1, 2, 3, and so on according to the traits assigned. (In other words, each student should show the same trait on the same numbered chromosome.)

Drop your "chromosome 1" strip onto the floor so that the allele that faces up is determined at random. Do the same thing for each trait. The alleles that face up are the ones you'll pass on to your offspring—one per trait. Record those alleles, indicating the trait for which each applies. Then "mate" with—that is, share chromosomes with—another student. Record your partner's alleles for each trait. You'll get pairs of alleles that result from random mating. Then determine the genotype and phenotype of your offspring. If you have time, try to get the results of a few different random matings, and compare them.

For More Information

Grant, M. C. "The Trembling Giant." *Discover*, October 1993. Aspen groves are really single individuals: huge, slowly spreading from the roots of the original parent tree, and potentially almost immortal.

Lanza, R. P., Dresser, B. L., and Damiani, P. "Cloning Noah's Ark." *Scientific American,* November 2000. Cloning rare and endangered species may offer hope to prevent extinction.

Nash, M. "The Age of Cloning." *Time*, March 10, 1997. Well-written and well-illustrated description of the technique and implications of using an adult cell to clone a sheep.

Travis, J. "A Fantastical Experiment." *Science News*, April 5, 1997. A clear description of the cloning of Dolly the sheep and some of its implications.

Wilmut, I. "Cloning for Medicine." *Scientific American,* December 1998. Explanation of why cloning experiments might have medical applications.

Answers to Multiple-Choice Questions

1. c 2. a 3. b 4. b 5. d

MediaTutor

CD Activities

Activity 10.1: Mitosis

Estimated time: 8 minutes

Each of us starts as a single cell—the fusion between sperm and egg. Yet by the time you are old enough to read this, you are made of trillions of cells, all descended from that original single cell. This is the result of repeated cell divisions. Every time one cell divides, it must accurately distribute copies of the chromosomes of the parental cell, producing two genetically identical cells. This process of distributing chromosomes occurs during mitosis. In this tutorial you will explore the steps involved in mitosis.

Activity 10.2: Meiosis

Estimated time: 8 minutes

"Variety is the spice of life." Sexual reproduction generates genetic variety, or diversity. In order for the fusion of sperm and egg to produce a cell with the correct number of chromosomes, the sperm and egg must each have one-half the normal number of chromosomes. To accomplish this, certain cells in the body undergo chromosome reduction division, meiosis. This tutorial will lead you through the steps of meiosis and ask you to determine how meiosis compares with mitosis.

Start the MediaTutor Student CD-ROM and enter the activity number in the Quick Search box to be taken directly to that activity.

Web Investigation

Call in the Clones

Estimated time: 20 minutes

Understanding the process of cell division is extremely important for addressing future approaches to combating human disease. For example, cancer occurs when normal tissue cells lose their ability to control the process of cell division. In addition, our understanding of reproduction at the cellular level has led to advancements in areas such as cloning and stem cell research—biotechnologies that hold promise to combat a multitude of ailments. This Web Investigation will explore how our current knowledge of cell division plays a role in these new medical advancements.

Go to http://www.prenhall.com/audesirk3, the Audesirk Companion Web site. Select Chapter 10 and Web Investigation to begin.

Each year, more than a million Americans seek treatment for alcoholism. Many of these alcoholics may have inherited a predisposition to dependence on alcohol.

11 Patterns of Inheritance

AT A GLANCE

CASESTUDY

Born to Drink?

"Hello. My name is Mike and I'm an alcoholic." So begins a meeting of a typical "12-step program" for people trying to overcome the heavy drinking and dependence on alcohol that characterize alcoholism. Mike, like many of the estimated 14 million alcoholics in the United States, is relying on group support, psychological therapy, and his own willpower to defeat his craving for alcohol. Attendance at monthly meetings sponsored by his self-help program is a big part of his effort. At one such meeting, however, Mike is shaken up when a long-standing group member rises and says, "My name is Al and I just read in the paper that alcoholism is controlled by genes. I guess I inherited my alcoholism from my parents. I'm programmed by my DNA to drink, and I can't do a thing to stop it. I might as well stop trying."

Is Al right? Is alcoholism a purely biological, inherited malady with a genetic cause, ultimately treatable only by drugs and other medical treatments? Are psychological approaches to alcoholism doomed to failure, best rejected along with theories that alcoholism's origin is rooted in childhood experiences or a deficient social environment? In fact, a sweeping rejection of any approach to treating alcoholism is premature. Alcoholism's ultimate cause remains unknown, but it is clearly far too complex to be explained by any single factor. As with most complex behavioral patterns, alcoholism is influenced by a range of both biological and environmental factors.

The complexity of alcoholism has not prevented researchers from trying to understand it. Biologists have been especially intrigued by the observation that alcoholism tends to run in families, which suggests that a tendency to succumb to alcoholism can be inherited. Inheritance, of course, means that genes may be involved, and one of the major threads of the alcoholism research effort has been a search for genes connected to alcoholism.

The search for the genetic underpinning of alcoholism has included both experiments on nonhuman animals and direct investigation of human genes. But before we turn our attention to some of the results of these efforts, let's take a look at the processes that govern inheritance of traits as simple as the color of a flower and as complex as a compulsion to consume too much alcohol. ■

1 What Is the Physical Basis of Inheritance?

The modern science of genetics got its start as the study of **inheritance**, the process by which the characteristics of individuals are passed to their offspring. The scope of genetics has grown to encompass far more than the mechanisms of inheritance, but even today geneticists remain fascinated by inheritance and many of them continue to explore its mysteries. In this chapter, we will explore some of the early findings about inheritance, as well as some more recent discoveries about how inheritance affects human health. Before we can begin our exploration, however, we must review a bit about the physical structures that underlie inheritance, and introduce a few terms that will make it easier to discuss the process of inheritance. (Note that the discussion in this chapter applies to organisms, such as humans, that have diploid body cells and that reproduce sexually by the fusion of haploid gametes.)

A Gene May Have Variant Forms

Many traits of individuals are influenced by genes that encode the instructions for proteins; each gene is a particular segment of DNA. A gene's specific physical location on a chromosome is called its **locus** (plural, *loci*). Each member of a pair of homologous chromosomes carries the same genes, located at the same loci. The nucleotide sequence at a given gene locus, however, may not be precisely identical on the two homologues. Instead, the nucleotide sequences at a locus may differ slightly from one homologous chromosome to the other, producing alternate forms of the same gene, called **alleles**. For example, the human blood types A, B, and O are produced by the three different alleles of the blood-type gene.

An Individual's Two Alleles May Be the Same or Different

At any given gene locus, an individual may carry either two copies of the same allele or one copy of each of two different alleles. If both homologous chromosomes have the *same* allele at a locus, the organism is said to be **homozygous** at that locus. For example, the chromosomes in Figure 11-1 are homozygous at two loci. If two homologous chromosomes have *different* alleles at a locus, the organism is **heterozygous** at that locus. The chromosomes in Figure 11-1 are heterozygous at one locus.

Whether an individual is homozygous or heterozygous at a locus affects the genetic content of the gametes produced by that individual. Because homologous chromosomes separate during meiosis (see Chapter 10), each gamete receives one member of each pair of homologous chromosomes. As a result, each gamete has only one allele for each gene. If an individual is homozygous at a particular locus, all the gametes it produces contain the same allele at that locus. In contrast, an individual that is heterozygous at a locus produces two kinds of gametes: half of them contain one allele, and half contain the other allele.

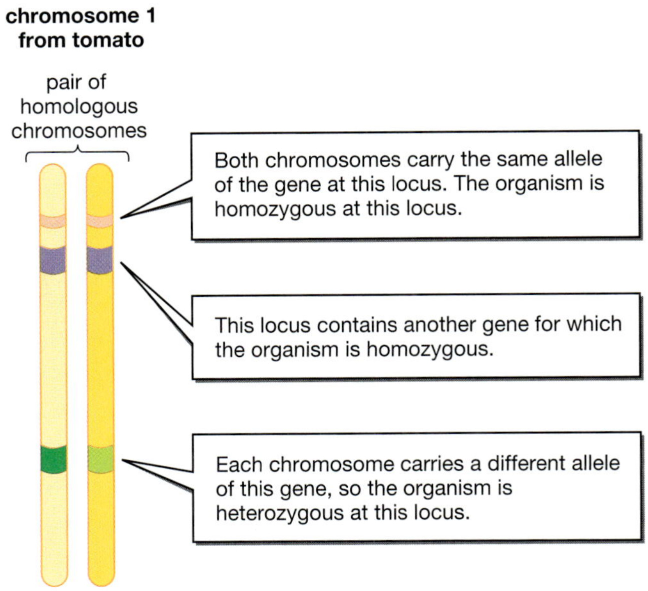

Figure 11-1 The relationships among genes, alleles, and chromosomes
Homologues carry the same gene loci.

2 What Was Mendel's Contribution?

Modern genetics is built on a foundation laid down in the mid-1800s by Gregor Mendel (Fig. 11-2) during his time as a monk in the monastery of St. Thomas in Brünn (now Brno, in the Czech Republic). At the St. Thomas monastery, Mendel carried out both his monastic duties and a groundbreaking series of experiments on inheritance in the common edible pea. Even today, reviewing the results of Mendel's work is one of the best ways to learn the basic principles of inheritance.

Mendel Took a New Approach

Mendel was able to make his breakthroughs largely because his approach to the study of inheritance was innovative. In contrast to earlier scientists, who generally tried to understand inheritance by comparing entire organisms to their offspring, Mendel studied different characteristics one at a time. He chose traits that had unmistakably different forms, such as white flowers versus purple flowers. In addition, Mendel followed the in-

Figure 11-2 Gregor Mendel
A portrait of Mendel, painted in about 1888, after he had completed his pioneering genetics experiments.

Figure 11-3 The seeds and flowers of the edible pea
In the intact pea flower, the lower petals form a container enclosing the reproductive structures. Pollen normally cannot enter the flower from outside, so peas normally self-fertilize.

heritance of those traits for several generations, and counted the numbers of offspring with each type of trait. Mendel's quantitative analysis of sharply defined traits over many generations enabled him to uncover the basic patterns of inheritance.

Mendel's Work Went Unrecognized in His Lifetime

Mendel presented his ideas on inheritance in 1865, and they were published the following year. Oddly, his revolutionary paper, which eventually spawned one of the most important theories in biology, made little immediate impression on the study of biology. Apparently, few contemporary biologists read his paper, and those who did failed to recognize its significance. It was not until 1900 that three biologists—Carl Correns, Hugo de Vries, and Erich Tschermak—working independently and knowing nothing of Mendel's work, rediscovered the principles of inheritance. When they searched the scientific literature before publishing their results, they found (no doubt to their intense disappointment) that Mendel had scooped them more than 30 years earlier. To their credit, they graciously acknowledged the important work of the Augustinian monk, who had died in 1884.

Mendel's Choice of the Edible Pea Was Especially Fortunate

The success of Mendel's experiments owes a great deal to his choice of the edible pea as an experimental subject (Fig. 11-3). One key feature of peas is that their reproduction can be easily controlled and manipulated. If no human intervenes, peas reproduce by **self-fertilization**, in which the egg cells in a flower are fertilized by sperm from the same flower. Pea plants can easily self-fertilize because each pea flower contains both the structures that produce eggs and the structures that produce pollen (which contains the sperm). Thus, each pea flower can supply pollen to fertilize its own eggs and, because the petals of a pea flower enclose both the male and female structures, pollen from other flowers normally cannot enter.

Although peas left to their own devices self-fertilize, a plant breeder can move pollen from a different plant onto the female part of a flower (the tip of the carpel; Fig. 11-3). If sperm from this "foreign" pollen fertilizes an egg, that egg is said to be **cross-fertilized**. To accomplish cross-fertilization, a breeder first prevents self-fertilization by removing the pollen-producing structures (stamens) from a flower. Next, the breeder dusts the tip of the carpel with pollen from another plant. In this way, two different plants can be mated, or crossed, to see what types of offspring they produce. If the crossed plants have different traits, the offspring are said to be **hybrids**.

3 What Were the Results of Mendel's Experiments?

MEDIATUTOR 11.1 Monohybrid Crosses

Mendel started simply. He first raised varieties of pea plants that were true-breeding for different forms of a single trait. The term "true-breeding" refers to plants (or other organisms) with a trait, such as purple flowers, that is always inherited by all offspring that result from self-fertilization. In his earliest experiments, Mendel cross-fertilized plants with different true-breeding traits, saved the resulting hybrid seeds, and then grew the seeds to observe the characteristics of the resulting plants.

Crossing White-Flowered and Purple-Flowered Plants Yields Only Purple-Flowered Offspring

In one of his experiments, Mendel cross-fertilized white-flowered plants with purple-flowered ones. The plants he crossed were the parental generation, denoted by the letter *P*. When he grew the resulting seeds, he found that all the first-generation offspring (the "first filial," or F_1, generation) produced purple flowers:

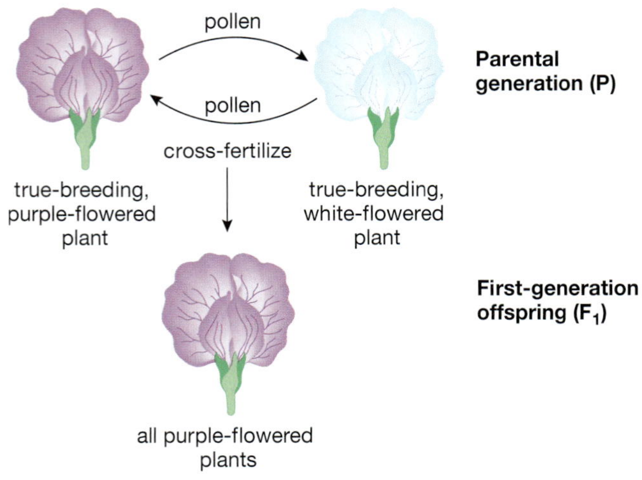

What had happened to the white color? The flowers of the F_1 hybrids were just as purple as the flowers of their purple parents. The white color seemed to have disappeared in these offspring.

Self-Fertilization of F_1 Yields a 3:1 Ratio of Purple to White Flowers

Mendel allowed the F_1 flowers to self-fertilize, collected the seeds, and planted them the next spring. In the second generation (F_2), about three-fourths of the plants had purple flowers and one-fourth had white flowers:

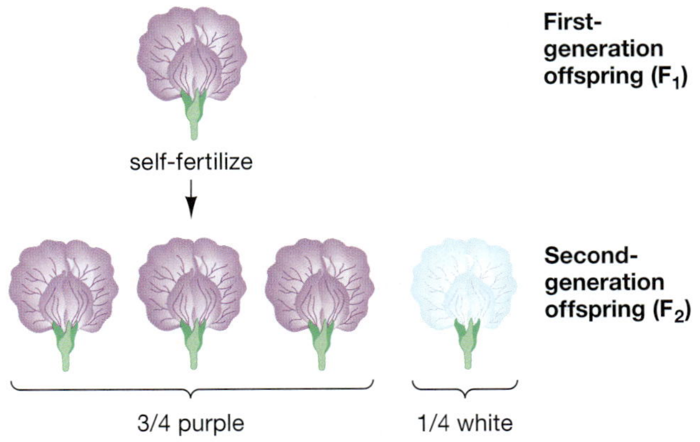

The exact numbers were 705 purple and 224 white, or a ratio of about 3 purple to 1 white. This result showed that the gene that produced white flowers had not disappeared but had only been "hidden."

Self-Fertilized F_2 Either Breed True or Yield a 3:1 Ratio

Mendel allowed the F_2 plants to self-fertilize and produce a third (F_3) generation. He found that all the white-flowered F_2 plants were true-breeding; that is, they produced only white-flowered offspring. For as many generations as he had time and patience to raise, white-flowered parents always gave rise to white-flowered offspring.

The purple-flowered F_2 plants were of two types: About $\frac{1}{3}$ were true-breeding for purple. The other $\frac{2}{3}$ produced both purple- and white-flowered offspring, again in the ratio of 3 to 1. Overall, therefore, the F_2 generation included $\frac{1}{4}$ true-breeding purple plants, $\frac{1}{2}$ hybrid purple, and $\frac{1}{4}$ true-breeding white.

4 What Did Mendel's Findings Reveal About Inheritance?

Mendel developed a five-part hypothesis to explain his results. (Our descriptions here of Mendel's ideas make use of the modern understanding of genes and chromosomes, entities that had not yet been discovered in Mendel's time.)

1. Traits are determined by distinct physical units (genes) that come in pairs. Each individual has two alleles for a given gene, such as the gene that determines flower color. One allele of the gene is present on each homologous chromosome.
2. The pairs of alleles on homologous chromosomes separate from each other during gamete formation, so each gamete receives only one allele from each pair. This conclusion is known as Mendel's *law of segregation*, because it results from the segregation of the two alleles of a gene during meiosis. When a sperm fertilizes an egg, the resulting offspring receives one allele from the father and one from the mother.
3. The particular allele that ends up in any given gamete is determined by chance. Because homologous chromosomes separate at random during meiosis, the assignment of alleles to gametes is random.
4. When two different alleles are present in an individual, one—the *dominant* allele—may mask the expression of the other—the *recessive* allele. The recessive allele, however, is still present. Both the dominant and the recessive alleles are passed into the individual's gametes. In Mendel's experiments with flower color, the allele for purple flowers is dominant, and the allele for white flowers is recessive.
5. True-breeding organisms have two copies of the same allele; they are homozygous at the gene locus for the true-breeding trait. All the gametes from a homozygous individual have the same allele at that locus:

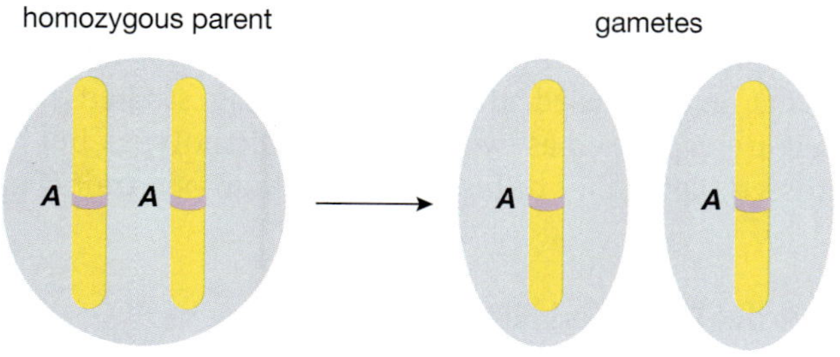

Individuals with two different alleles at a locus are heterozygous. Half of a heterozygote's gametes contain one allele for that gene; half contain the other allele.

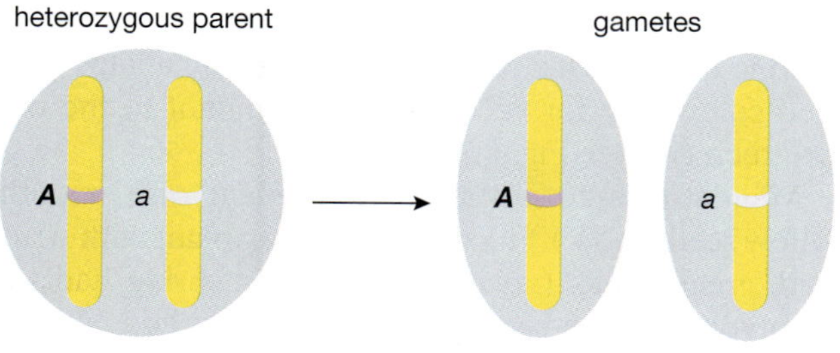

Mendel's Hypothesis Was Consistent with His Results

Let's see how Mendel's hypothesis explained the results of his experiments with flower color. Using letters to represent the different alleles, we assign the uppercase letter *P* to the dominant allele for purple and the lowercase letter *p* to the recessive allele for white. (By Mendel's convention, the dominant allele is represented by a capital letter.) A true-breeding (homozygous) purple-flowered plant has two alleles for purple flowers (*PP*), whereas a white-flowered plant has two alleles for white flowers (*pp*). All the sperm and eggs produced by a *PP* plant carry the *P* allele; all the sperm and eggs of a *pp* plant carry the *p* allele:

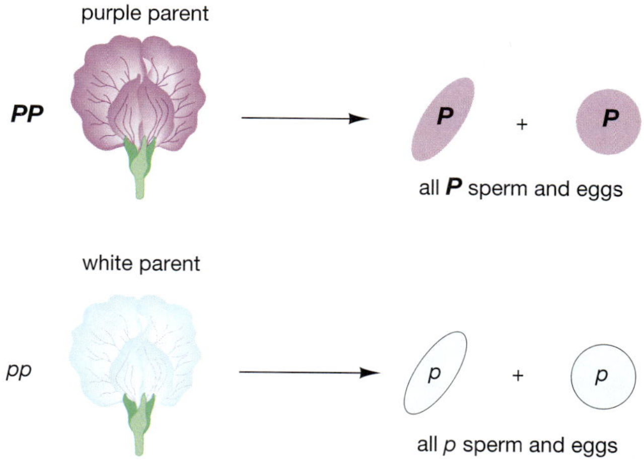

Mendel's F_1 hybrid offspring were produced when *P* sperm fertilized *p* eggs or when *p* sperm fertilized *P* eggs. In both cases, the F_1 offspring were *Pp*. Because *P* is dominant to *p*, all the offspring were purple:

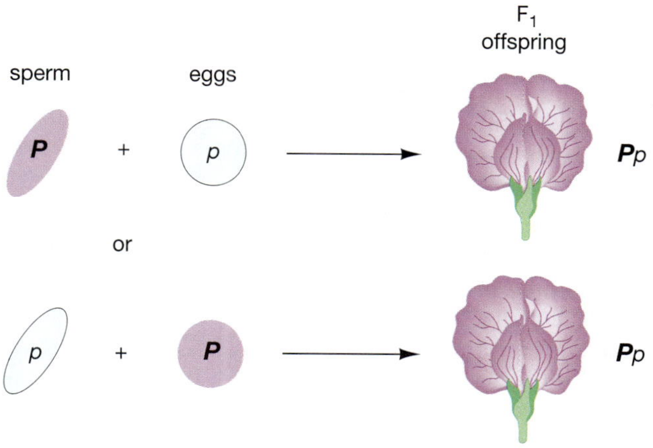

Mendel's F_2 offspring resulted from self-fertilization of the heterozygous F_1 plants. Each gamete produced by a heterozygous *Pp* plant has an equal chance of receiving either the *P* allele or the *p* allele. That is, the heterozygous plant produces equal numbers of *P* and *p* sperm and equal numbers of *P* and *p* eggs. When a *Pp*

plant self-fertilizes, each type of sperm has an equal chance of fertilizing each type of egg:

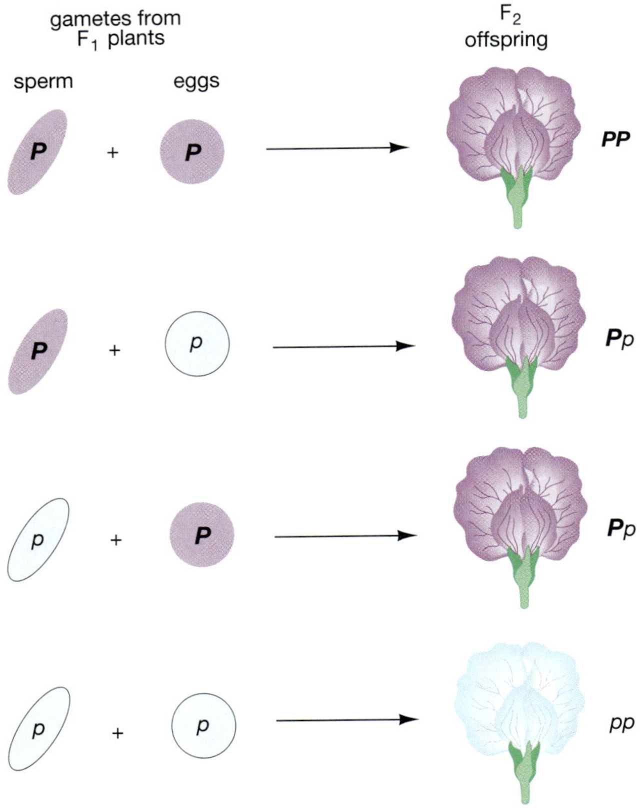

Therefore, Mendel's F_2 contained three types of offspring (*PP*, *Pp*, and *pp*), and the three types were in the approximate proportions of ¼ *PP*, ½ *Pp*, and ¼ *pp*. Because both *PP* and *Pp* plants have purple flowers, ¾ of the plants in the F_2 generation were purple-flowered, and ¼ were white-flowered.

Observable Traits Do Not Always Reveal Underlying Alleles

Mendel's hypotheses imply that two plants that look alike may actually carry different combinations of alleles. The combination of alleles carried by an organism (for example, *PP* or *Pp*) is its **genotype**. The organism's traits, including its outward appearance, behavior, digestive enzymes, blood type, or any other observable or measurable feature, make up its **phenotype**. As we have seen, plants with either the *PP* or the *Pp* genotype make purple flowers. Thus, even though the two types of plant have different genotypes, they have the same phenotype for flower color. So, even though the F_2 generation of Mendel's peas contained three genotypes (¼ *PP*, ½ *Pp*, and ¼ *pp*), it contained only two phenotypes (¾ purple and ¼ white).

Mendel's Hypothesis Can Predict the Outcome of Single-Trait Crosses

You have probably recognized that Mendel used the scientific method, observing results and formulating a hypothesis based on them. The scientific method has a third step: to see if the hypothesis accurately predicts the results of further experiments. For example, based on his hypothesis that the heterozygous F_1 plants had one allele for purple flowers and one for white (*Pp*), Mendel predicted the outcome of cross-fertilizing these *Pp* plants with homozygous recessive white plants (*pp*). In particular, he predicted that there would be equal numbers of *Pp* (purple) and *pp* (white) offspring, and this is indeed what he found. Can you see why?

A Punnett Square Helps Make Predictions about Inheritance

The **Punnett square** is a useful tool for predicting the genotypes and phenotypes of offspring. Figure 11-4 shows how to use a Punnett square to predict the offspring expected from self-fertilization of a flower that is heterozygous for color (or from any combination of gametes from individuals that are heterozygous for a single trait).

The Big Picture

The principles of inheritance rank among the great scientific discoveries of all time. Progress in our understanding of inheritance set the stage for dramatic advances in fields ranging from medicine to agriculture to evolutionary biology.

5 Are Multiple Traits Inherited Together?

Having determined how single traits are inherited, Mendel turned to the more complex question of how multiple traits are inherited. In his next series of experiments, he again used edible peas and again studied traits that are expressed as two alternative phenotypes, but he crossed plants that differed in more than one trait (Fig. 11-5).

Mendel Extended His Experiments with More Complex Crosses

Mendel began his exploration of multiple traits by crossing plants that differed in two traits—for example, seed color (yellow or green) and seed shape (smooth or wrinkled). From earlier crosses of plants with these traits, he already knew that the allele of the seed shape gene that causes seeds to be smooth (*S*) is dominant to the one for wrinkled seeds (*s*), and that the allele of the seed color gene for yellow (*Y*) is dominant to the one for green (*y*) (see Fig. 11-5).

Mendel crossed a true-breeding plant with smooth, yellow seeds (*SSYY*) to a true-breeding plant with wrinkled, green seeds (*ssyy*). All the F_1 offspring had the genotype *SsYy* and had smooth, yellow seeds. Mendel

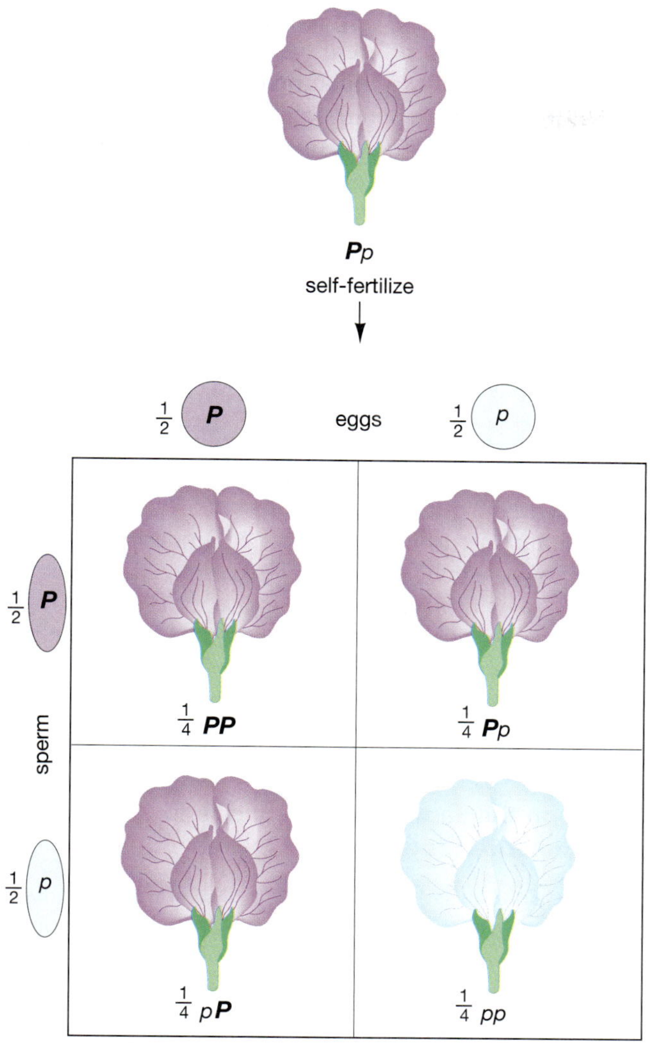

Figure 11-4 The Punnett square method
The Punnett square method allows you to predict both genotypes and phenotypes of specific crosses; here we use it for a cross between plants that are heterozygous for a single trait, flower color.

① Assign letters to the different alleles; use uppercase for dominant and lowercase for recessive.

② Determine all genetically different gametes that can be produced by the male and female parents.

③ Draw the Punnett square, with the columns labeled with the egg genotypes and the rows labeled with the sperm genotypes. (We have included the fractions of these genotypes with each label.)

④ Fill in the genotype of the offspring in each box by combining the genotype of sperm in its row with the genotype of the egg in its column. (We have placed the fractions in each box.)

⑤ Count the number of offspring with each genotype. (Note that *Pp* is the same as *pP*.)

⑥ Convert the number of offspring of each genotype to a fraction of the total number of offspring. In this example, out of four fertilizations, only one is predicted to produce the *pp* genotype, so ¼ of the total number of offspring produced by this cross is predicted to be white. To determine phenotypic fractions, add the fractions of genotypes that would produce a given phenotype. For example, purple flowers are produced by ¼ *PP* + ¼ *Pp* + ¼ *pP*, for a total of ¾ of the offspring.

Figure 11-5 Traits of pea plants that Mendel studied

MEDIATUTOR 11.2 Dihybrid Crosses

allowed these F_1 plants to self-fertilize, and observed that the F_2 generation consisted of 315 plants with smooth yellow seeds, 101 with wrinkled yellow seeds, 108 with smooth green seeds, and 32 with wrinkled green seeds, a ratio of about 9:3:3:1.

Mendel Found That Multiple Traits Are Inherited Independently

Mendel reasoned that this F_2 was what would be expected if the genes for seed color and seed shape were inherited independently. For each trait, ¾ of the offspring would exhibit the dominant phenotype (encoded by *SS* and *Ss* genotypes) and ¼ would show the recessive phenotype (encoded by *ss*), producing a 3:1 ratio. Combining two independent 3:1 ratios yields a 9:3:3:1 ratio of four two-trait phenotypes (as shown in the Punnett square in Fig. 11-6). This result is just what Mendel observed in his two-trait experiment. If we count single traits within the 9:3:3:1 ratio of phenotypes in the F_2, we see 423 plants with smooth seeds (of either color) versus 133 with wrinkled ones (a ratio of about 3:1), and 416 plants with yellow seeds (of either shape) versus 140 with green ones (also about 3:1).

The observation that two or more distinct traits are inherited independently gave rise to a principle known

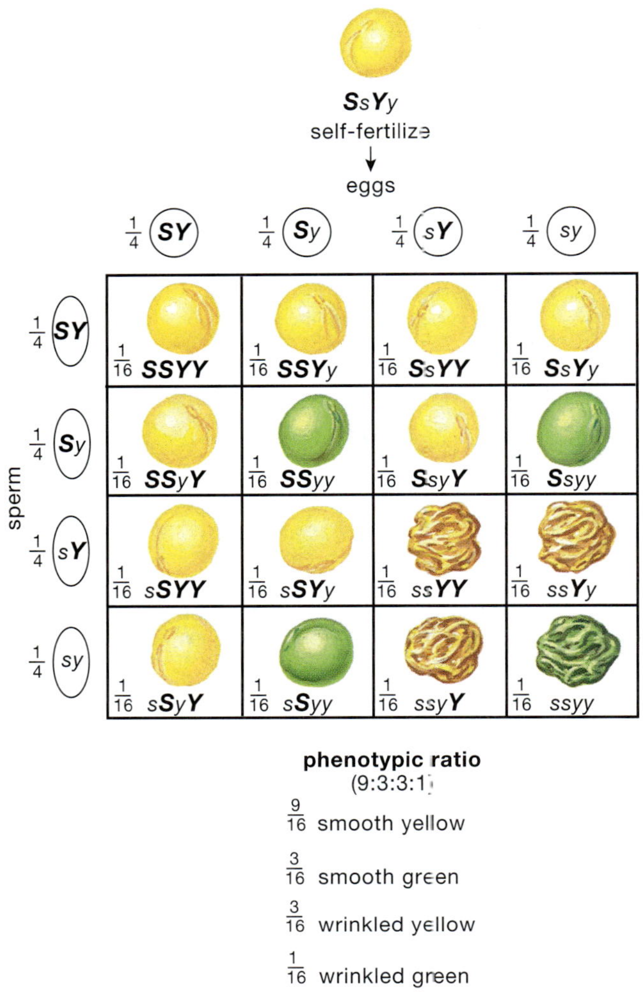

Figure 11-6 Predicting genotypes and phenotypes for a cross between parents that are heterozygous for two traits
Here we are working with both seed color and shape, with yellow (*Y*) dominant to green (*y*), and smooth (*S*) dominant to wrinkled (*s*). In this cross, both parents are heterozygous for each trait (or a single individual heterozygous for both traits self-fertilizes). There are now 16 boxes in the Punnett square. In addition to predicting all the genotypic combinations and the 9:3:3:1 overall phenotypic ratio, the Punnett square predicts ¾ yellow seeds, ¼ green seeds, ¾ smooth seeds, and ¼ wrinkled seeds, just as we would expect from crosses made of each trait separately.

as the **law of independent assortment**. This principle states that multiple traits are inherited independently because the alleles of one gene are distributed to gametes independently of the alleles of other genes. The physical basis for this independence is found in the events of meiosis (see Chapter 10). When paired homologous chromosomes line up during metaphase I, homologues face the poles of the cell at random, and the orientation of any one homologous pair does not influence the orientation of other pairs. Therefore, when the homologues separate during anaphase I, the alleles of genes on different chromosomes are distributed, or "assorted," independently (Fig. 11-7). Note that only genes that are located on different chromosomes assort independently. Mendel's formulation of the law of independent assortment depended on the rather amazing coincidence that all of the several traits he chose to study happened to be controlled by genes located on different chromosomes.

6 How Are Genes Located on the Same Chromosome Inherited?

Mendel, of course, knew nothing about chromosomes or the physical basis of genes, and his work provided only a partial explanation of the inheritance of multiple traits. Later, as scientists began to understand more about how the genetic material is organized, it became obvious that an organism has many more genes than chromosomes, so each chromosome must contain many genes. For two or more traits that are encoded by different genes on a single chromosome, the law of independent assortment does not apply.

Genes on the Same Chromosome Tend to Be Inherited Together

Genetic **linkage** is the inheritance of certain genes as a group because they are on the same chromosome. Linked genes tend to be inherited together and do not assort independently. One of the first pairs of linked genes to be discovered was found in the sweet pea, a different species from Mendel's garden pea. In sweet peas, the gene for flower color and the gene for pollen grain shape are located on the same chromosome. Thus, alleles of these genes normally assort together into gametes during meiosis and are inherited together as a result.

Consider a sweet pea plant with red flowers and long pollen that has the following pair of homologous chromosomes:

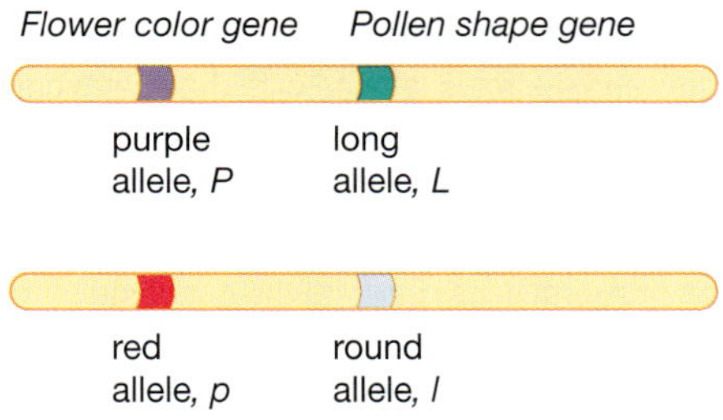

Note that the allele for purple flower color and the allele for long pollen shape are located on one homologue. The allele for red flower color and the one for round pollen shape are located on the other homologue. The gametes produced by this sweet pea plant are likely to have *either* alleles for purple and long *or*

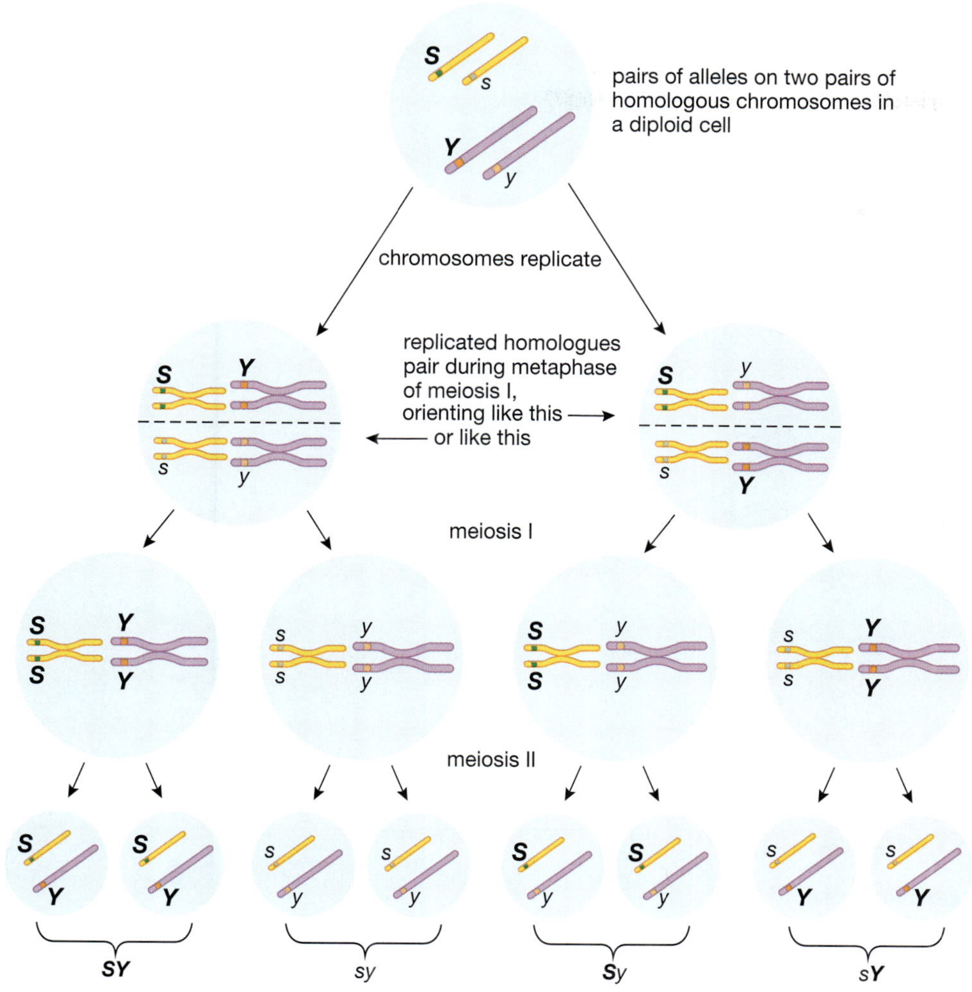

Figure 11-7 Independent assortment of alleles
Chromosome movements during meiosis produce independent assortment of alleles, as shown here for two genes. Each possible combination is equally likely, producing gametes in the proportions ¼ *SY*, ¼ *sy*, ¼ *sY*, and ¼ *Sy*.

alleles for red and round. This pattern of inheritance breaks the law of independent assortment, which states that the alleles for flower color and pollen shape should segregate independently. Instead, they tend to stick together.

Crossing Over Can Create New Combinations of Linked Alleles

Although they tend to be inherited together, genes on the same chromosome do not always stay together. In the sweet pea cross just described, for example, an F_2 generation commonly includes a few plants in which the genes for flower color and pollen shape have been inherited as if they were not linked. What accounts for these exceptions?

As you learned in Chapter 10, during prophase I of meiosis, the nonsister chromatids of homologous chromosomes intertwine at sites called "chiasmata." At chiasmata, homologous chromosomes exchange segments, a process called **crossing over**. Crossing over is common; there is at least one exchange between each homologous chromosome pair during each meiosis. The exchange of corresponding segments of DNA during crossing over produces new allele combinations on both homologous chromosomes. Then, when recombined homologous chromosomes separate at anaphase I, the chromosomes that each haploid daughter cell receives have different combinations of alleles from those of the parent cell.

7 How Is Sex Determined?

In animals of many species, an individual's sex is determined by a special pair of chromosomes, the **sex chromosomes**. In mammals, for example, males and females have the same total number of chromosomes, but females

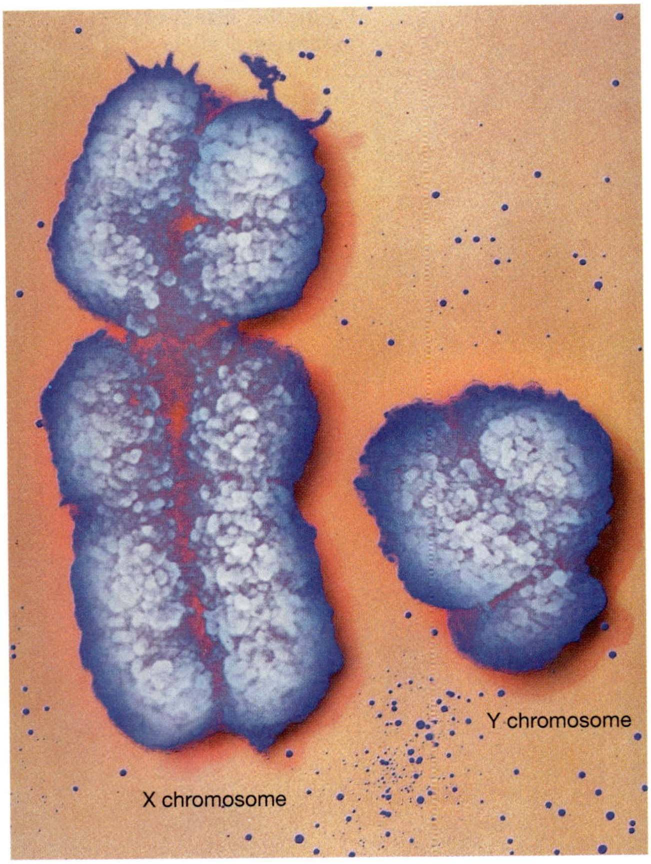

Figure 11-8 Photomicrograph of human sex chromosomes
Notice the small size of the Y chromosome, which carries relatively few genes.

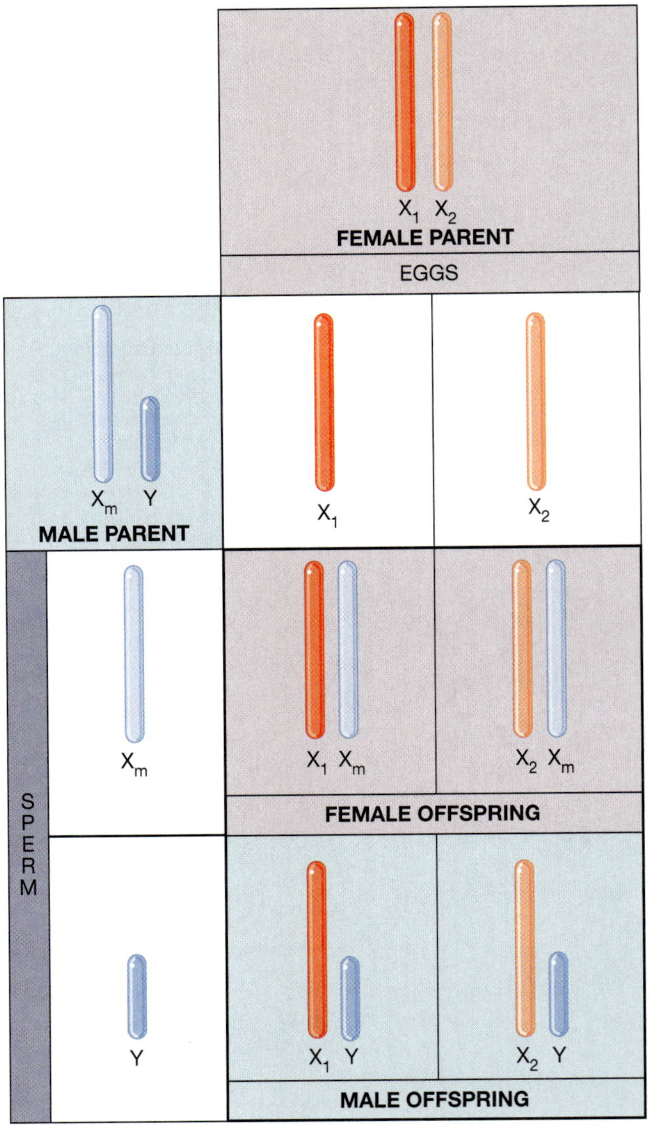

Figure 11-9 Sex determination in mammals
Male offspring receive their Y chromosome from the father; female offspring receive the father's X chromosome (labeled X_m). Both male and female offspring receive an X chromosome (either X_1 or X_2) from the mother.

have two identical sex chromosomes, called X chromosomes, whereas males have one X chromosome and one Y chromosome (Fig. 11-8). Even though the Y is much smaller than the X, the X and Y chromosomes act as homologues, pairing up during prophase of meiosis I and separating during anaphase I. During gamete formation, the sex chromosomes segregate. Each egg receives one X chromosome (because females are XX) and each sperm receives either an X or a Y chromosome (because males are XY). Thus, an egg that is fertilized by a Y-bearing sperm yields a male offspring, and an egg fertilized by an X-bearing sperm yields a female (Fig. 11-9).

The number of chromosomes varies tremendously among species. In animals, the chromosome number includes a single pair of sex chromosomes and additional pairs of **autosomes** (the name for all the chromosomes that are not sex chromosomes). It is the number of autosomes that varies among species. For example, the fruit fly *Drosophila* has four pairs of chromosomes (three pairs of autosomes and one pair of sex chromosomes), humans have 23 pairs (22 pairs of autosomes and one pair of sex chromosomes), and dogs have 39 pairs (38 pairs of autosomes and one pair of sex chromosomes).

8 How Are Sex-Linked Genes Inherited?

Genes that are on one sex chromosome but not on the other are said to be **sex-linked**. In many animal species, the Y chromosome carries only a few genes. The human Y chromosome contains about 20 genes, most of which determine maleness or male fertility. In contrast, the X chromosome contains many genes that have nothing to do with specifically female traits. The human X chromosome contains about 1500 genes that have no counterpart on the Y chromosome, including genes for color vision, blood clotting, and certain structural proteins in muscles.

Because they have two X chromosomes, females can be either homozygous or heterozygous for genes on the X chromosome, and normal dominant versus recessive relationships among alleles apply. In contrast, males have only a single X chromosome, and thus express *all* alleles they have on their X, whether those alleles are dominant or recessive. For this reason, recessive traits encoded by genes on the X chromosome show up in males. In humans such traits include color blindness, hemophilia, and certain types of muscular dystrophy. We will return to this concept later in the chapter.

9 What Are Some Variations on the Mendelian Theme?

In our discussion of patterns of inheritance thus far, we looked at only the simplest kinds of traits: those that are completely controlled by a single gene, those for which there are only two possible alleles of each gene, and those in which one allele is completely dominant to the other, recessive, allele. Most traits, however, are influenced in more varied and subtle ways.

Alleles May Exhibit Incomplete Dominance

In snapdragons, crossing homozygous red-flowered plants (*RR*) with homozygous white-flowered ones (*R′R′*) does not, as we might expect based on our understanding of Mendel's pea flower experiments, produce F_1 hybrids with red flowers. Instead, the F_1 (*RR′*) flowers are pink. This pattern of inheritance, in which the heterozygous phenotype is intermediate between the two homozygous phenotypes, is called **incomplete dominance.**

How can an allele be incompletely dominant? For some traits, *two* copies of a particular allele are required for expression of the dominant phenotype. For example, the *R* allele in snapdragons codes for an enzyme that catalyzes the formation of red pigment; the *R′* allele codes for a defective, nonfunctional enzyme. When only one copy of the *R* allele is present, less red pigment is produced and the resulting flower appears pink. Plants with the *RR* genotype produce lots of red pigment and have red flowers; those with the *R′R′* genotype produce no pigment and have white flowers (Fig. 11-10). The *R* allele is incompletely dominant because heterozygotes exhibit a phenotype that is intermediate between the phenotypes of homozygous dominants and homozygous recessives.

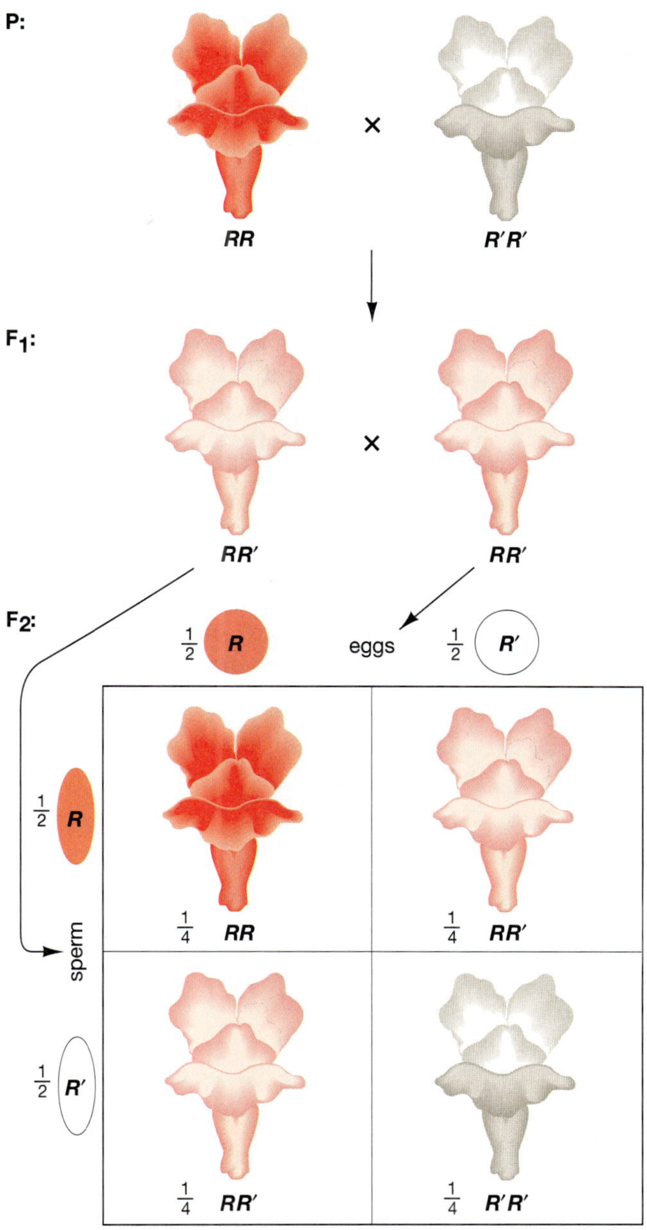

Figure 11-10 Incomplete dominance
The inheritance of flower color in snapdragons is an example of incomplete dominance. (In such cases, we use capital letters for both alleles, here *R* and *R′*.) Heterozygtes (*RR′*) have pink flowers, whereas the homozygotes are red (*RR*) or white (*R′R′*).

A Single Gene May Have Multiple Alleles

If we could sample all the individuals of a species, we would typically find dozens of alleles for every gene. Thus, rather than each gene having only two alleles, most genes have multiple alleles. One eye-color gene in fruit flies, for example, has more than a thousand alleles. Depending on how the alleles are combined, the eyes of fruit flies can be white, yellow, orange, pink, brown, or red. Remember, however, that an *individual* diploid organism can have only two alleles for a given gene.

A Single Trait May Be Influenced by Several Genes

If you look around your class, you are likely to see people of varied heights, skin colors, and body builds, to consider just a few obvious traits. These are examples of traits that exhibit **polygenic inheritance**; they are governed not by single genes but by interactions among two or more genes.

Eye color is another good example of polygenic inheritance. The color of the human iris varies from very pale blue through green to almost black, depending on the amount of pigment in the outer layer of the iris. If there is little or no pigment here, the iris appears blue, owing to the same type of light scattering that makes the sky appear blue. People with more pigment in the outer iris may have eyes that appear green, brown, or almost black. At least two (and probably more) genes direct the synthesis of pigment in the front of the iris, with each gene having two alleles that show incomplete dominance (Fig. 11-11).

As you might imagine, the more genes that contribute to a single trait, the greater the number of phenotypes and the finer the distinctions among them. When more than three genes contribute to a trait, differences between phenotypes are small, and it is extremely difficult to classify the phenotypes reliably. For example, at least three or four genes control skin color, producing smooth variation from very dark to very light coloration.

A Single Gene May Have Multiple Effects on Phenotype

We have just seen that several different genes may influence a single trait. The reverse is also true; a single gene may influence several different traits, a phenomenon called **pleiotropy**. A good example is the *SRY* gene, discovered in 1990 on the Y chromosome. The *SRY* gene (short for "sex-determining region of the Y chromosome") initiates a sequence of events that affects many different body structures. *SRY* codes for a protein that activates other genes, which in turn code for proteins that cause an embryo to develop as a male. Testes form and then secrete sex hormones that stimulate the development of male reproductive structures, such as the epididymis, seminal vesicles, prostate gland, penis, and scrotum.

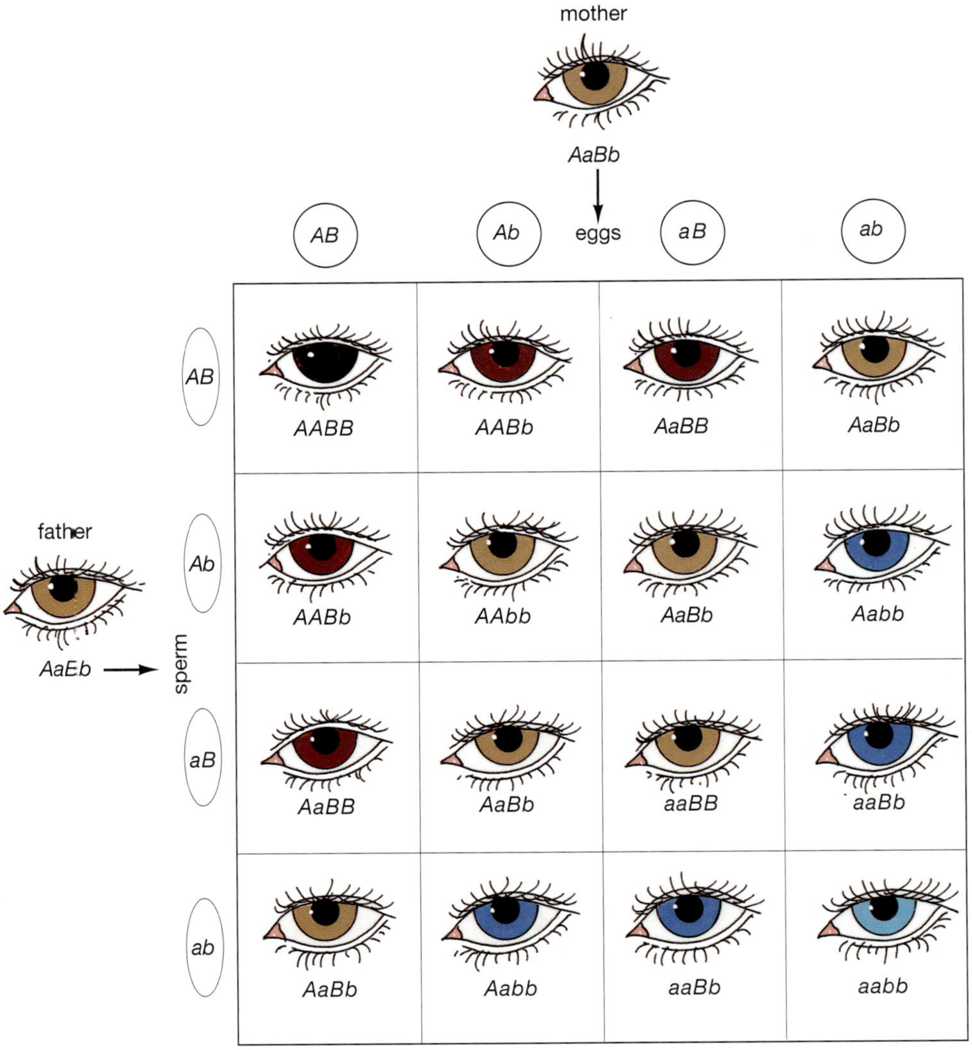

Figure 11-11 Human eye color At least two genes, each with two incompletely dominant alleles, determine human eye color. A brown-eyed man and a brown-eyed woman, each heterozygous for both genes, could have children with five different eye colors, ranging from light blue (no dominant alleles) through light brown (two dominant alleles) to almost black (all four dominant alleles).

The Environment Influences the Expression of Genes

An organism's phenotype is not entirely determined by its genes. The environment in which an organism lives also plays a large role in determining phenotype. A striking example of environmental effects on gene action is provided by the Himalayan rabbit, which, like the Siamese cat, has pale body fur but black ears, nose, tail, and feet. The Himalayan rabbit's genotype contains alleles that code for black fur and, as the genotype is the same in each body cell, one might expect the rabbit to have black fur all over its body. The enzyme that produces the black pigment, however, is inactive at temperatures above about 93 °F (34 °C). Because the temperature of a rabbit's body surface is warmer than 93 °F, most of the rabbit's fur is pale. A rabbit's body extremities, however, are typically cooler than the rest of the body. As a result, black pigment can be produced at the extremities, and the fur on ears, nose, tail, and feet is black (Fig. 11-12).

All traits that are influenced by genes are also influenced by environmental factors. For example, the polygenic trait of human skin color is modified by the environmental effects of sun exposure. Height, another polygenic trait, is affected by nutrition, an environmental factor. On the whole, it is impossible to conceive of a trait that has a purely genetic basis or one that is purely environmental. Even traits that seem to be under rigid genetic control, such as the color of pea flowers, depend on environmental conditions in the sense that pea plants will not produce flowers unless the environment provides the correct mix of water, nutrients, sunlight, and so on. By the same token, even the most complex human behaviors depend on physical structures (such as the brain) whose construction is influenced by genes. In fact, genetic differences have been shown to influence differences in human traits as diverse as intelligence, sexuality, and susceptibility to cancer. Complex traits stem from a complex interplay between genetic and environmental influences, and efforts to understand the genetic basis of such traits will always be difficult and subject to uncertainty.

Figure 11-12 Environmental influence on phenotype
Expression of the gene for black fur in the Himalayan rabbit depends on an interaction between genotype and environment. The gene is expressed only on the cooler parts of the rabbit's body.

The Big Picture

Debates about the degree to which human behavior is influenced by genes are often informed more by emotion than by evidence, but ethical limits on the means by which human genetics can be studied mean that firm conclusions will remain elusive.

10 How Are Human Genetic Disorders Investigated?

Among human traits, those in which genes influence susceptibility to disease are of special interest. Because experimental crosses with humans are out of the question, human geneticists study the inheritance of disease-causing alleles by searching medical, historical, and family records to study past crosses. Records extending across several generations can be arranged in the form of family **pedigrees**, diagrams that show the genetic relationships among a set of related individuals (Fig. 11-13). Careful analysis of pedigrees can reveal whether a particular trait is inherited in a dominant, recessive, or sex-linked pattern. Since the mid-1960s, analysis of human pedigrees, combined with molecular genetic technology, has produced great strides in understanding human genetic diseases. For instance, geneticists now know the genes responsible for dozens of inherited diseases, such as sickle-cell anemia, prostate cancer, and cystic fibrosis. Research in molecular genetics promises to increase our ability to predict genetic diseases and perhaps even to cure them, a topic we explore further in Chapter 12.

11 How Are Single-Gene Disorders Inherited?

Many common human traits, such as freckles, long eyelashes, cleft chin, and widow's peak hairline, are inherited in a simple Mendelian fashion; that is, each trait is controlled by a single gene with a dominant and a recessive allele. Here, we will focus on a few examples of medically important disorders that also exhibit simple Mendelian inheritance.

Recessive Alleles Cause Many Disorders

Disorders can be caused by single-gene mutations that damage or destroy key enzymes. In some disorders, however, the effects of such a mutation are masked when one normal allele is present and can generate a sufficient amount of functional enzyme. In such cases, the person who is heterozygous (with one normal and

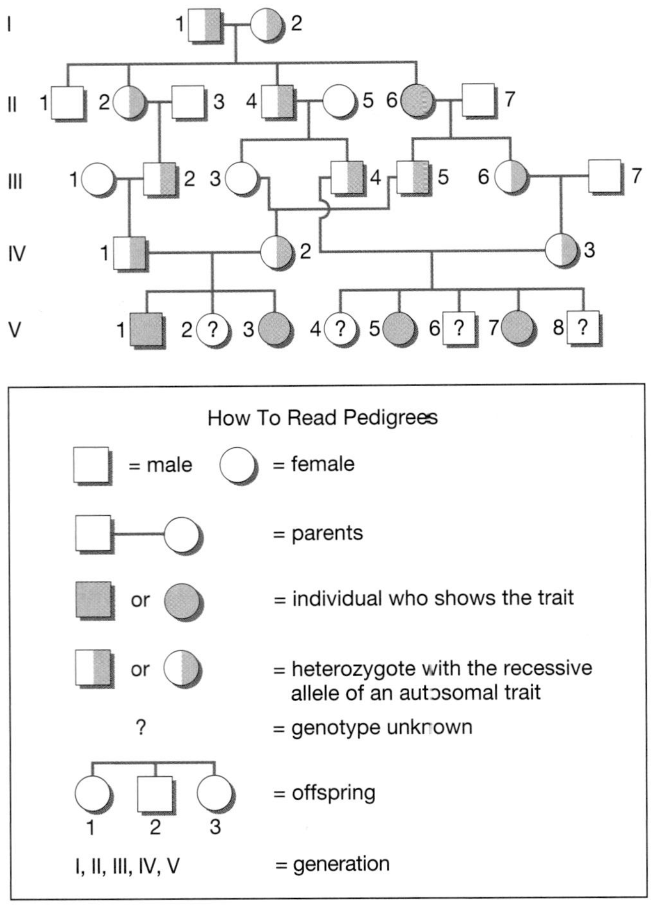

Figure 11-13 A family pedigree
This pedigree is for a rare recessive trait. Both of the original parents are heterozygotes who carry the recessive allele. Because the recessive allele is rare, pairing between heterozygotes is an unlikely event. However, the chance that each of two *related* people carry a rare recessive allele (inherited from a common ancestor) is much higher than that of two unrelated people. As a result, a disproportionate number of cases of recessive diseases are among offspring of pairings between cousins or other relatives. In this family, there are three pairings between cousins—between III 3 and III 5, III 4 and IV 3, and IV 1 and IV 2.

one mutant allele) will remain healthy and have a phenotype that is indistinguishable from that of homozygotes with two copies of the normal allele. In other words, the mutant allele is recessive, and only people who inherit two copies of the mutant allele have the disorder.

Individuals who are heterozygous for a harmful recessive allele have a normal phenotype but can pass the defective allele to their offspring. Geneticists estimate that each of us carries 5 to 15 potentially harmful recessive alleles, each of which would cause a serious genetic defect if we were homozygous for it. When a couple has a child, each of the defective alleles in each parent has a 50:50 chance of being passed on, so the probability is very high that the child will inherit some harmful recessive alleles. Fortunately, however, the child is unlikely to be homozygous for a recessive allele for a genetic disease because it is unlikely that both parents carry defective alleles of the same gene.

Recessive disorders in offspring are more common if the parents are related (especially if they are first cousins or closer). Relatives share a recent common ancestor and are therefore much more likely to share common alleles, including defective recessive alleles. Each child borne by a couple that shares a harmful recessive allele has a 1 in 4 chance of being affected by the genetic disorder (see Fig. 11-13).

A Defective Allele for Melanin Production Causes Albinism

Albinism is a recessive trait caused by an absence of melanin, the dark pigment in skin cells. Human and other mammalian albinos have white skin and hair and pink eyes (the absence of melanin pigment in the iris makes blood vessels at the back of the eye visible). The genetic basis of albinism is a recessive allele of a gene called *TYR* that codes for the enzyme tyrosinase. Tyrosinase is needed to produce melanin, and an individual who is homozygous for a mutant *TYR* allele that produces defective tyrosinase has the phenotype of albinism (Fig. 11-14).

A Defective Allele for Hemoglobin Synthesis Causes Sickle-Cell Anemia

Sickle-cell anemia, an inherited disease that damages red blood cells, is caused by a particular mutation in the gene that encodes the protein hemoglobin. Hemoglobin is concentrated in red blood cells, where it binds to oxygen molecules so that they can be transported in the blood. In people with sickle-cell anemia, a single nucleotide substitution in the hemoglobin gene places an incorrect amino acid at a crucial position in hemoglobin protein molecule, disrupting its function. The defective hemoglobin molecules tend to stick together in clumps when oxygen levels are low (such as in muscles during exercise). The clumps force the red blood cell out of its normal disk shape (Fig. 11-15a) into a longer, sickle shape (Fig. 11-15b). The sickled cells are fragile, making them likely to break; they also tend to aggregate, clogging blood vessels. The resulting lack of bloodflow can cause pain, especially in joints. If blood vessels in the brain become blocked, a person with this disorder may suffer paralyzing strokes. Other symptoms include anemia (insufficient number of red blood cells) because so many red blood cells are destroyed, and reduced immunity to disease.

The symptoms of sickle-cell anemia are generally expressed only in those who are homozygous for the recessive sickle-cell allele. Heterozygotes have about half normal and half abnormal hemoglobin, but they usually have few sickled cells and are not disabled by the dis-

(a) (b) (c)

Figure 11-14 Albinism
Albinism occurs throughout the animal kingdom, including *(a)* humans, *(b)* rattlesnakes, and *(c)* wallabies. This female wallaby, having mated with a normally pigmented male, carries a normally colored offspring in her pouch.

ease. About 8% of the African-American population is heterozygous for sickle-cell anemia, reflecting a genetic legacy of African origins. In some regions of Africa, 15% to 20% of the population carries the allele.

Dominant Alleles Cause Many Disorders

Many serious genetic diseases are caused by dominant alleles. How can a harmful mutant allele be dominant to the normal allele? Why aren't heterozygotes rescued by the presence of one normal allele of these genes? Disease-causing dominant alleles may do their damage by producing an abnormal protein that interferes with the function of the normal one. Alternatively, dominant alleles may encode proteins that carry out destructive, toxic reactions. A dominant allele may also encode a protein that is overactive, performing its function at inappropriate times and places.

Dominant genetic diseases tend to be non-fatal or to have symptoms that are not expressed until late

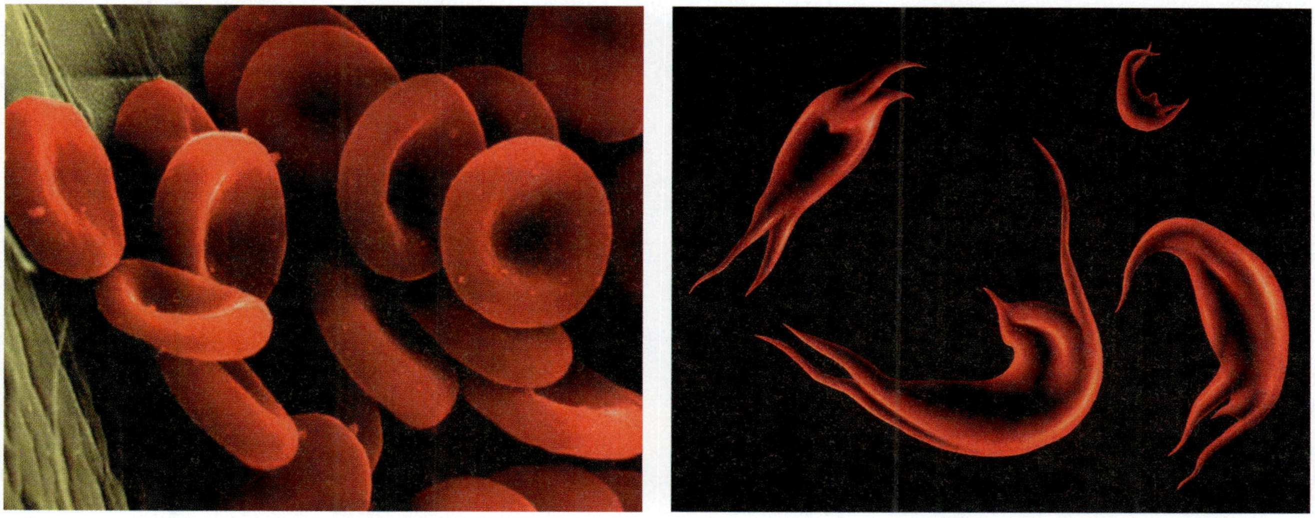

(a) (b)

Figure 11-15 Sickle-cell anemia
(a) Normal red blood cells are disc-shaped with indented centers. *(b)* The red blood cells of a person with sickle-cell anemia become sickled.

Scientific Inquiry
In Search of the Huntington Gene

In 1968 the clinical psychologist Milton Wexler founded the Hereditary Disease Foundation, in hope of finding a cure or treatment for Huntington disease (HD), the inherited neurological disorder that had killed his wife. His daughter Nancy Wexler, then a young woman of 22, was profoundly affected by her mother's suffering, and decided to devote her life to finding a cure for HD. Though trained as a psychologist like her father, Dr. Wexler taught herself genetics by attending lectures, holding informal discussions with geneticists, and reading extensively. Today, she serves as president of the Hereditary Disease Foundation.

Dr. Wexler's initial research focused on several villages near Lake Maracaibo in Venezuela, where an unusually high percentage of the villagers died of HD. When she visited these villages, she found that about 1 in 5 people there were stricken with the disease (Fig. E11-1). All of the victims were descended from a Portuguese sailor who married a local woman seven generations earlier, in the 1800s. Dr. Wexler began assembling a pedigree showing the relationships among the members of this genetically interconnected population.

Eventually, more than 10,000 individuals were added to this extensive pedigree (families are large in this area of Venezuela). Wexler also collected blood samples from the villagers with HD and shipped the samples to Dr. James Gusella at Massachusetts General Hospital in Boston. In 1993 Gusella's lab identified a region of DNA near the tip of human chromosome 4 as the location of the gene for HD. Comparing the genetic information in the blood samples with the relationships depicted in the pedigree, Gusella's group devised a screening test. This test very accurately determines whether a person has inherited the gene for HD from an affected parent or is free of this "genetic time bomb," as Dr. Wexler describes the Huntington allele. Unfortunately, even though we know the specific gene responsible, no treatment has yet been discovered to slow or halt the progression of Huntington disease.

This lack of treatments may partly explain why Nancy Wexler has decided not to be tested to determine whether she has inherited her mother's Huntington allele. As Dr. Wexler explains, each person at high risk of having the allele must make a decision as to whether the potential joy of learning that the allele is absent makes it worth risking the pain of finding out the allele is present. For some, not knowing is preferable. Perhaps the test is of greatest importance to people who may have inherited the Huntington allele but who want to have children unaffected by the disease. Genetic counseling can help such individuals make informed, personal decisions about their reproduction options.

Figure E11-1 Nancy Wexler tracks the gene for Huntington disease
Nancy Wexler, right, with one of the many inhabitants of the Lake Maracaibo region of Venezuela who have Huntington disease.

adulthood. Any dominant allele that caused its bearers to die before reproducing would never be passed to offspring, and would quickly disappear from the population.

One dominant genetic disease in humans is Huntington disease, which causes a slow, progressive deterioration of parts of the brain. This unstoppable deterioration results in the loss of motor coordination, flailing movements, personality disturbances, and eventual death. The symptoms of Huntington disease typically do not appear until 30 to 50 years of age. Therefore, many people pass the allele to their children before they suffer the first symptoms. In 1983 both molecular genetic technology and painstaking pedigree analysis were used to localize the Huntington gene to a small part of chromosome 4 (see "Scientific Inquiry: In Search of the Huntington Gene"). Geneticists finally isolated the Huntington gene in 1993 and a few years later identified the gene's product, a protein dubbed "huntingtin." The function of the protein encoded by the normal allele of the gene remains unknown. The mutant protein forms large aggregates in nerve cells that appear to result, ultimately, in nerve cell death.

Some Disorders Are Sex-Linked

If a gene is present on one sex chromosome but not on the other, its pattern of inheritance is sex linked. For example, the X chromosome contains many genes that have no counterpart on the Y chromosome. Because males have only one X chromosome, they have only one allele for each of these genes. This single allele is expressed, even if it is recessive.

A son receives his X chromosome from his mother and passes it only to his daughters. Thus, sex-linked disorders caused by a recessive allele have a distinctive pattern of inheritance. Such disorders appear far more frequently in males and typically skip generations: An affected male passes the allele to a phenotypically normal daughter, who in turn bears affected sons. The most familiar genetic defects caused by recessive alleles of X-chromosome genes are red-green color blindness and hemophilia. Hemophilia is caused by a recessive allele on the X chromosome that results in a deficiency of one of the proteins needed for blood clotting. People with hemophilia bleed excessively from a wound or from mild damage to internal structures, and they bruise easily. Hemophiliacs often have anemia due to blood loss. Today, modern medical treatment allows many hemophiliacs to live long lives, but even in earlier times some hemophiliac males survived long enough to pass their defective allele to their daughters, who carried the allele and could pass it to their sons.

The Big Picture

Our increased ability to detect the mutations responsible for inherited diseases and defects has opened new avenues for potential treatments and cures. But these same advances also create new responsibilities and ethical dilemmas, as we struggle to find the most compassionate and effective way to use our newfound power to screen a person's genes.

12 How Do Errors in Chromosome Number Affect Humans?

In Chapter 10 we examined the intricate mechanisms of meiosis, which ensure that each sperm and egg receive only one chromosome from each homologous pair. Not surprisingly, this elaborate dance of the chromosomes occasionally misses a step, resulting in gametes that have too many or too few chromosomes. Such errors in meiosis, called **nondisjunction**, can affect the number of either sex chromosomes or autosomes. Most embryos that arise from gametes with abnormal chromosome numbers spontaneously abort, accounting for 20% to 50% of all miscarriages. Some embryos with abnormal chromosome numbers survive to birth or beyond.

Abnormal Numbers of Sex Chromosomes Cause Some Disorders

Because the X and Y chromosomes pair up during meiosis, sperm normally carry either an X or a Y chromosome. Nondisjunction of sex chromosomes in a male, however, produces sperm that have two sex chromosomes (XX, YY, or XY) or that have no sex chromosomes at all (these are designated O). Similarly, nondisjunction in a female can produce XX or O eggs instead of normal eggs with a single X. When these defective sperm or eggs fuse with normal gametes, the resulting offspring have abnormal numbers of sex chromosomes (Table 11-1). The most common abnormalities are XO, XXX, XXY, and XYY. (Genes on the X chromosome are essential to survival, and any embryo without at least one X chromosome spontaneously aborts very early in development.)

Turner Syndrome (XO)

About one in every 3000 phenotypically female babies has only one X chromosome, a condition known as Turner syndrome. At puberty, hormone deficiencies prevent XO females from menstruating or developing secondary sexual characteristics, such as enlarged breasts. Treatment with hormones can promote physical development but, because most women with Turner syndrome lack mature eggs, hormone treatment does

Table 11-1 Effects of Nondisjunction of the Sex Chromosomes During Meiosis

Nondisjunction in Father

Sex Chromosomes of Defective Sperm	Sex Chromosomes of Normal Egg	Sex Chromosomes of Offspring	Phenotype
O (none)	X	XO	Female—Turner syndrome
XX	X	XXX	Female—Trisomy X
YY	X	XYY	Male—"XYY male"
XY	X	XXY	Male—Klinefelter syndrome

Nondisjunction in Mother

Sex Chromosomes of Normal Sperm	Sex Chromosomes of Defective Egg	Sex Chromosomes of Offspring	Phenotype
X	O (none)	XO	Female—Turner syndrome
Y	O (none)	YO	Dies as embryo
X	XX	XXX	Female—Trisomy X
Y	XX	XXY	Male—Klinefelter syndrome

(a) (b)

Figure 11-16 Trisomy 21, or Down syndrome
(a) Chromosomes of a Down syndrome child reveals three copies of chromosome 21. *(b)* These girls have the relaxed mouth and distinctively shaped eyes typical of Down syndrome.

not induce menstruation or reverse the infertility typically seen in XO women. Other symptoms of Turner syndrome include short stature, folds of skin around the neck, and increased risk of cardiovascular disease, kidney defects, and hearing loss. Because women with Turner syndrome have only one X chromosome, they have more X-linked recessive disorders, such as hemophilia and color blindness, than do XX women.

Trisomy X (XXX)

About one in every 1000 women has three X chromosomes, a condition known as trisomy X. Most such women have no detectable defects, except for a tendency to be tall. There is, however, a higher incidence of below-normal intelligence among women with trisomy X. Unlike women with Turner syndrome, most trisomy X women are fertile and, interestingly enough, almost always bear normal XX and XY children. Some unknown mechanism must operate during meiosis to prevent the extra X chromosome from being included in eggs.

Klinefelter Syndrome (XXY)

About one male in every 1000 is born with two X chromosomes and one Y chromosome. Most of these men go through life never realizing that they have an extra X chromosome. However, at puberty, some of these men show mixed secondary sexual characteristics, including partial breast development, broadening of the hips, and small testes. These symptoms are known as Klinefelter syndrome. XXY men are usually infertile because of low sperm count but are not impotent. They are usually diagnosed when an affected man and his partner seek medical help for their inability to conceive a baby.

XYY Males

Another common type of sex chromosome abnormality is XYY, present in about one male in every 1000. You might expect that having an extra Y chromosome, which presumably has few genes, would not make very much difference, and this seems to be true in most cases. However, XYY males usually have high levels of testosterone, often have severe acne, and are tall (about two-thirds of XYY males are more than 6 feet tall, compared with the average male height of 5 feet 9 inches). Some appear to have slightly lower scores on IQ tests than their brothers who are XY.

Abnormal Numbers of Autosomes Cause Some Disorders

Nondisjunction of autosomes produces eggs or sperm that are missing an autosome or that have two copies of an autosome. Fusion of one of these abnormal gametes with a normal one (that bears one copy of each autosome) leads to an embryo with either one or three copies of the affected autosome. Embryos that have only one copy of any autosome abort so early in development that the woman never knows she was pregnant. Embryos with three copies of an autosome (trisomy) also usually abort spontaneously. However, a small fraction of embryos with three copies of chromosomes 13, 18, or 21 can develop sufficiently to be born. A baby with trisomy 21 can live into adulthood.

Trisomy 21, or Down Syndrome

In about one of every 900 births, the child inherits an extra copy of chromosome 21, a condition called trisomy 21 or Down syndrome. Children with Down syndrome have several distinctive physical characteristics, including weak muscle tone, a small mouth held partially open because it cannot accommodate the tongue, and distinctively shaped eyelids (Fig. 11-16). More serious defects include low resistance to infectious diseases, heart malformations, and mental retardation so severe that only about one in 25 ever learns to read and only one in 50 learns to write.

The frequency of nondisjunction in gametes increases with age. There is a small increase in defective sperm with increasing age of the father, and nondisjunction in sperm accounts for about 25% of Down syndrome cases. The mother's age, however, is a more significant factor in the probability of Down syndrome (Fig. 11-17). Since the 1970s, it has become more common for couples to delay having children, increasing the probability of trisomy 21. Even today, however, older women as a group account for far fewer babies than younger women, so the majority of Down syndrome babies are born to young women.

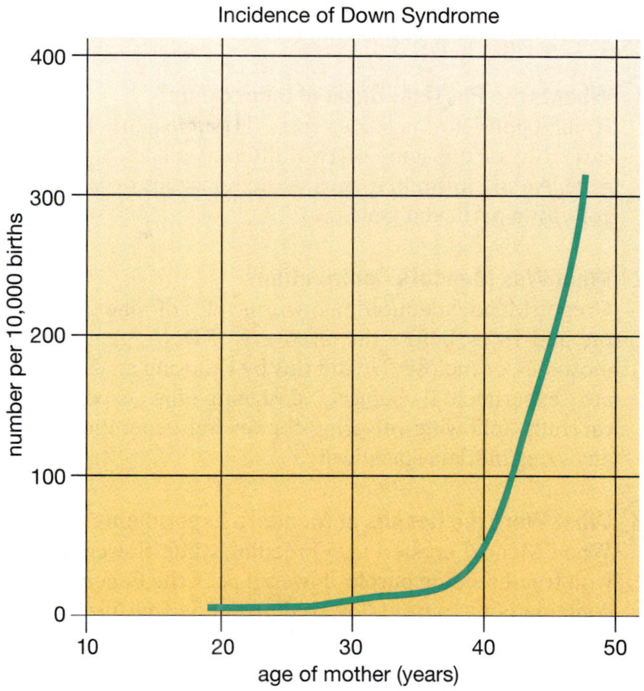

Figure 11-17 Down syndrome frequency increases with maternal age
The shape of the blue line shows that a low percentage of 20-year-old mothers give birth to a child with Down syndrome, but the percentage begins to grow before age 30 and increases dramatically after age 35.

CASESTUDYREVISITED

Born to Drink?

Researchers working with fruit flies, mice, and other animals have been able to identify particular mutations that are associated with increased susceptibility to the effects of alcohol. Similar mutations may well be at work in human alcoholics, but it is difficult to know the degree to which results in animals can help us understand humans. So it is important to study directly the genetics of alcoholism in humans. Human geneticists cannot, of course, perform the type of experimental manipulations that researchers use with animals. They can, however, identify families that contain an especially large number of alcoholics and can screen members of these families for mutations. Mutations that are present in many alcoholics and their families may be associated with alcoholism.

A study sponsored by the National Institutes of Health (NIH), recently finished screening almost 1000 people from more than 100 alcoholism-prone families. The study identified four chromosomal regions, on chromosomes 1, 2, 4, and 7, that are linked to vulnerability to alcoholism. Dr. Enoch Gordis, director of the NIH unit that oversees research on alcoholism, believes that these findings represent the first step in a process that will inevitably discover the genetic basis of alcoholism. He stated that "when, over the next few years, we pinpoint the actual genes that confer and mitigate alcoholism risk, we can target preventive interventions and devise pharmacologic treatments based on that knowledge. Understanding genetic factors also is the first critical step in unraveling the more complex contribution of environmental factors."

Dr. Gordis's optimism notwithstanding, full understanding of the genetics of alcoholism remains elusive. Even our current, rather rudimentary understanding indicates that susceptibility to alcohol dependency is a polygenic trait, influenced by loci on different chromosomes. The particular genetic underpinnings of the malady probably differ from family to family and even from person to person within a family. In any case, we need not be prisoners of our genes. Even people who are born with a special vulnerability to alcoholism are not necessarily destined to become alcoholics. Many people in alcoholism-prone families never become dependent on alcohol. Alcoholics can and do recover, many with the help of 12-step programs. Still, even though it is impossible to predict the precise outcome of the stunningly complex interaction between the myriad genetic and environmental factors that influence behavior, it makes a great deal of sense to gain what advantage we can by exploring the genetic bases of destructive behavior patterns such as alcoholism.

Summary of Key Concepts

1) What Is the Physical Basis of Inheritance?

Homologous chromosomes carry the same genes but may carry two of the same or two different alleles of a given gene. An organism may thus be homozygous or heterozygous for a particular gene.

2) What Was Mendel's Contribution?

Gregor Mendel deduced many principles of inheritance in the mid-1800s, before the discovery of DNA, genes, chromosomes, or meiosis. He did this by choosing an appropriate experimental subject, designing his experiments carefully, following offspring for several generations, and analyzing his data statistically.

3) What Were the Results of Mendel's Experiments?

When Mendel crossed true-breeding white-flowered peas with true-breeding purple-flowered peas, the F_1 generation contained only purple-flowered plants. Self-fertilization of the F_1 yielded an F_2 in which white flowers reappeared in a 3:1 ratio of purple-flowered to white-flowered plants.

4) What Did Mendel's Findings Reveal About Inheritance?

Traits are inherited in patterns that depend on alleles that parents pass on to their offspring. Each parent provides its offspring with one allele of every gene, so that the offspring inherits a pair of alleles for every gene. The combination of alleles in an offspring determines whether it displays a particular phenotype for a trait. Dominant alleles mask the expression of recessive alleles. Because recessive alleles can be masked, organisms with different genotypes may have the same phenotype. That is, organisms that are homozygous dominant have the same phenotype as do heterozygous organisms. Because each allele segregates randomly during meiosis, we can predict the relative proportions of offspring with a particular trait.

5) Are Multiple Traits Inherited Together?

If the genes for two traits are located on different chromosome pairs, they assort independently of one another into the egg or sperm. Thus, crossing two organisms that are heterozygous at two loci on different chromosomes produces offspring with 16 different genotypes. If the alleles are typical dominant and recessive alleles, these progeny will display only 4 different phenotypes.

6) How Are Genes Located on the Same Chromosome Inherited?

Genes located on the same chromosome are linked and tend to be inherited together. Unless two linked alleles are separated by crossing over during meiosis, they pass together to the offspring.

7) How Is Sex Determined?

In many animals, sex is determined by sex chromosomes, often designated X and Y. Female mammals have two X chromosomes, whereas males have one X and one Y chromosome. The Y chromosome has many fewer genes than the X chromosome.

8) How Are Sex-Linked Genes Inherited?

Because males have only one copy of most X chromosome genes, recessive alleles on the X chromosome are more likely to be phenotypically expressed in males.

9) What Are Some Variations on the Mendelian Theme?

Not all inheritance follows the simple dominant–recessive pattern:

- In incomplete dominance, heterozygotes have a phenotype that is intermediate between the two homozygous phenotypes.
- Many traits are determined by polygenic inheritance, in which several different genes at different loci contribute to the phenotype.
- Many genes have multiple effects on the phenotype (pleiotropy).
- The environment influences the phenotypic expression of all traits.

10) How Are Human Genetic Disorders Investigated?

Studying the genetics of humans is similar to studying the genetics of other animals, except that experimental crosses are not feasible. Analysis of family pedigrees and, more recently, molecular genetic techniques help to determine the mode of inheritance of human traits.

11) How Are Single-Gene Disorders Inherited?

Many genetic disorders are inherited as recessive traits, with only homozygous recessive persons showing symptoms of the disease. Heterozygotes carry the recessive allele but do not express the trait. Many other diseases are inherited as simple dominant traits. In such cases, only one copy of the dominant allele is needed to cause the disease symptoms. Some disorders stem from recessive alleles on the X chromosome; these disorders appear more frequently in males, who may pass some of them to their grandsons through their daughters.

12) How Do Errors in Chromosome Number Affect Humans?

Errors in meiosis can result in gametes with abnormal numbers of sex chromosomes or autosomes. Many people with abnormal numbers of sex chromosomes have distinguishing physical characteristics. Abnormal numbers of autosomes typically lead to spontaneous abortion early in pregnancy. In rare instances, babies trisomic for chromosomes 13, 18, or 21 are born. Of these, only some of the Down syndrome babies survive to adulthood; those who do survive have mental and physical deficiencies.

Key Terms

allele *p. 162*
autosome *p. 170*
cross-fertilization *p. 164*
crossing over *p. 169*
dominant *p. 165*
genotype *p. 166*
heterozygous *p. 162*
homozygous *p. 162*
hybrid *p. 164*
incomplete dominance *p. 171*
inheritance *p. 162*
law of independent assortment *p. 168*
law of segregation *p. 165*
linkage *p. 168*
locus *p. 162*
nondisjunction *p. 177*
pedigree *p. 173*
phenotype *p. 166*
pleiotropy *p. 172*
polygenic inheritance *p. 172*
Punnett square *p. 166*
recessive *p. 165*
self-fertilization *p. 163*
sex chromosome *p. 169*
sex-linked *p. 170*

Thinking Through the Concepts

Multiple Choice

1. *An organism is described as* Rr*:red. The* Rr *is the organism's [A]; red is the organism's [B]; and the organism is [C].*
   a. [A] phenotype; [B] genotype; [C] degenerate
   b. [A] karyotype; [B] hybrid; [C] recessive
   c. [A] genotype; [B] phenotype; [C] heterozygous
   d. [A] gamete; [B] linkage; [C] pleiotropic
   e. [A] zygote; [B] phenotype; [C] homozygous

2. *The 9:3:3:1 ratio is a ratio of*
   a. phenotypes and genotypes in the offspring from a cross of individuals that differ in one trait.
   b. phenotypes in the offspring from a cross of individuals that differ in one trait
   c. phenotypes in the offspring from cross of individuals that differ in two traits
   d. genotypes in the offspring from a cross of individuals that differ in one trait
   e. genotypes in the offspring from a cross of individuals that differ in two traits

3. *A heterozygous pink-flowered snapdragon mated with a white-flowered snapdragon would produce*
   a. all white-flowered plants in the F_1
   b. all pink-flowered plants in the F_1
   c. ½ red-flowered and ½ white-flowered plants in the F_1
   d. ¾ pink-flowered and ¼ white-flowered plants in the F_1
   e. ½ pink-flowered and ½ white-flowered plants in the F_1

4. *Which is NOT true of sickle-cell anemia?*
   a. It is most common in African Americans.
   b. It involves a one-amino-acid change in hemoglobin.
   c. It involves red blood cells.
   d. It is lethal in heterozygotes because it is dominant.
   e. It affects blood circulation.

5. *Sex-linked disorders such as color blindness and hemophilia are*
   a. caused by genes on the X chromosome
   b. caused by genes on the autosome
   c. caused by genes on the Y chromosome
   d. expressed only in men
   e. expressed only when two chromosomes are homozygous recessive

? Review Questions

1. Define the following terms: gene, allele, dominant, recessive, true breeding, homozygous, heterozygous, cross-fertilization, self-fertilization.
2. Explain why genes located on the same chromosome are said to be linked. Why do alleles of linked genes sometimes separate during meiosis?
3. Define polygenic inheritance. Why is it possible for a couple to have offspring that are notably different in eye or skin color than either parent?
4. What is sex linkage? In mammals, which sex would be most likely to show recessive sex-linked traits?
5. What is the difference between a phenotype and a genotype? Does knowledge of an organism's phenotype always allow you to determine the genotype? What type of experiment would you perform to determine the genotype of a phenotypically dominant individual?
6. If one (heterozygous) parent of a couple has Huntington disease, calculate the fraction of that couple's children that would be expected to develop the disease. What if both parents were heterozygous?
7. Why are most genetic diseases inherited as recessives rather than dominants?
8. Define nondisjunction, and describe the common syndromes caused by nondisjunction of sex chromosomes and autosomes.

Applying the Concepts

1. Sometimes the term "gene" is used rather casually. Compare and contrast use of the terms "allele" and "locus" as alternatives to "gene."
2. Mendel's numbers seem almost too perfect to be real; some believe he may have cheated a bit on his data. Perhaps he continued to collect data until the numbers matched his predicted ratios, then stopped. Recently, there has been much publicity over violations of scientific ethics, including researchers' plagiarizing others' work, using other scientists' methods to develop lucrative patents, or just plain fabricating data. How important an issue is this for society? What are the boundary lines of ethical scientific behavior? How should the scientific community or society "police" scientists? What punishments would be appropriate for violations of scientific ethics?
3. Although American society has been described as a "melting pot," people often engage in "assortative mating," in which they marry others of similar height, socioeconomic status, race, and IQ. Discuss the consequences to society of assortative mating among humans. Would society be better off if people mated more randomly? Discuss why or why not.
4. "Eugenics" is the term applied to the notion that the human condition might be improved by improving human genes. Do you think there are both good and bad sides to eugenics? What examples can you think of to back up your stand? What would a eugenicist think of the medical advances that have lessened the problems of hemophilia?
5. Think about some of the personal, religious, and economic issues related to prenatal counseling and diagnosis. Would you avoid having children if you knew that both you and your spouse were heterozygous for a recessive disorder that is fatal at an early age and may involve considerable suffering? What would you do if you or your spouse were pregnant and learned that your offspring, if born, would be homozygous for such a disorder? Is the situation qualitatively different for Down syndrome? (Down syndrome children have a life expectancy of 20 to 30 years with mental retardation but can have productive lives.) If you were heterozygous for Huntington disease, would you want to avail yourself of the medical diagnostic tests now available to find out? Would your answer be different if you were planning to have a baby?

Genetics Problems

(Note: An extensive group of genetics problems, with answers, can be found in the Study Guide.)

1. In certain cattle, hair color can be red (homozygous *RR*), white (homozygous *R′R′*), or roan (a mixture of red and white hairs, heterozygous *RR′*).
   a. When a red bull is mated to a white cow, what genotypes and phenotypes of offspring could be obtained?
   b. If one of the offspring in (a) were mated to a white cow, what genotypes and phenotypes of offspring could be produced? In what proportion?
2. The palomino horse is golden in color. Unfortunately for horse fanciers, palominos do not breed true. In a series of matings between palominos, the following offspring were obtained:

   65 palominos, 32 cream-colored, 34 chestnut (reddish brown)

   What is the probable mode of inheritance of palomino coloration?
3. In the edible pea, tall (*T*) is dominant to short (*t*), and green pods (*G*) are dominant to yellow pods (*g*). List the gametes and offspring phenotypes that would be produced in the following crosses:
   a. *TtGg* × *TtGg*
   b. *TtGg* × *TTGG*
   c. *TtGg* × *Ttgg*
4. In tomatoes, round fruit (*R*) is dominant to long fruit (*r*), and smooth skin (*S*) is dominant to fuzzy skin (*s*). A true-breeding round, smooth tomato (*RRSS*) is crossed with a true-breeding long, fuzzy tomato (*rrss*). All the F_1 offspring are round and smooth (*RrSs*). When these F_1 plants are crossed with each other, the following F_2 generation is obtained:

   Round, smooth: 43 Long, fuzzy: 13

   Are the genes for skin texture and fruit shape likely to be on the same chromosome or on different chromosomes? Explain your answer.
5. In the tomatoes of problem 4, an F_1 offspring (*RrSs*) is mated with a homozygous recessive plant (*rrss*). The following offspring are obtained:

   Round, smooth: 583 Round, fuzzy: 21
   Long, fuzzy: 602 Long, smooth: 16

What is the most likely explanation for this distribution of phenotypes?

6. In humans, hair color is controlled by two interacting genes. The same pigment, melanin, is present in both brown-haired and blond-haired people, but brown hair has much more of it. The allele for brown hair (*B*) is dominant to the allele for blond (*b*). Whether any melanin can be synthesized depends on another gene. The dominant allele (*M*) allows melanin synthesis; the recessive allele (*m*) prevents melanin synthesis. Homozygous recessives (*mm*) are albino. What will be the expected proportions of phenotypes in the children of the following parents?

a. *BBMM* × *BbMm*

b. *BbMm* × *BbMm*

c. *BbMm* × *bbmm*

7. In humans, one of the genes determining color vision is located on the X chromosome. The dominant allele (*C*) produces normal color vision; red-green color blindness is caused by a recessive allele (*c*). If a man with normal color vision marries a color-blind woman, what is the probability of their having a color-blind son? A color-blind daughter?

8. In the couple described in problem 7, the woman gives birth to a color-blind but otherwise normal daughter. The husband sues for a divorce on the grounds of adultery. Will his case stand up in court? Explain your answer.

Answers to Genetics Problems

1. a. A red bull (*RR*) is mated to a white cow (*R′R′*). The bull will produce all *R* sperm; the cow will produce all *R′* eggs. All the offspring will be *RR′* and will have roan hair.

b. A roan bull (*RR′*) is mated to a white cow (*R′R′*). The bull produces half *R* and half *R′* sperm; the cow produces *R′* eggs. Using the Punnett square method:

sperm \ eggs	*R′*
R	*R R′*
R′	*R′ R′*

Using probabilities:

sperm	egg	offspring
1/2 *R*	*R′*	1/2 *RR′*
1/2 *R′*	*R′*	1/2 *R′R′*

The predicted offspring will be 1/2 *RR′* (roan) and 1/2 *R′R′* (white).

2. The offspring are of three types, dark (chestnut), light (cream), and intermediate (palomino). This distribution suggests incomplete dominance, with the allele for chestnut (*C*) combining with the allele for cream (*C′*) to produce palomino heterozygotes (*CC′*). We can test this hypothesis by examining the offspring numbers. There are approximately 1/4 chestnut (*CC*), 1/2 palomino (*CC′*), and 1/4 cream (*C′C′*). If palominos are heterozygotes, we would expect the cross *CC′* × *CC′* to yield 1/4 *CC*, 1/2 *CC′*, and 1/4 *C′C′*. Our hypothesis is supported.

3. a. *TtGg* × *TtGg*. This is a "standard" cross of F_1 heterozygotes whose parents were true breeding for different phenotypes of two traits. Both parents produce *TG, Tg, tG*, and *tg* gametes. The expected proportions of offspring are 9/16 tall green, 3/16 tall yellow, 3/16 short green, 1/16 short yellow.

b. *TtGg* × *TTGG*. In this cross, the heterozygous parent produces *TG, Tg, tG*, and *tg* gametes. However, the homozygous dominant parent produces only *TG* gametes. Therefore, all offspring receive at least one *T* allele for tallness and one *G* allele for green pods and are tall with green pods.

c. *TtGg* × *Ttgg*. The second parent produces two types of gametes, *Tg* and *tg*. Using a Punnett square:

sperm \ eggs	*Tg*	*tg*
TG	*TTGg*	*TtGg*
Tg	*TTgg*	*Ttgg*
tG	*TtGg*	*ttGg*
tg	*Ttgg*	*ttgg*

The expected proportions of offspring are 3/8 tall green, 3/8 tall yellow, 1/8 short green, 1/8 short yellow.

4. If the genes were on separate chromosomes, then this would be a cross of individuals heterozygous for two independently assorting traits with expected offspring in the proportions 9/16 round smooth, 3/16 round fuzzy, 3/16 long smooth, and 1/16 long fuzzy. However, only the parental combinations show up in the offspring, indicating that the genes are on the same chromosome.

5. The genes are on the same chromosome and are quite close together. On rare occasions, crossing over occurs between the two genes, recombining the alleles.

6. a. *BBMM* (brown) × *BbMm* (brown). The first parent can produce only *BM* gametes, so all offspring will receive at least one dominant allele for each gene. Therefore, all offspring will have brown hair.

b. *BbMm* (brown) × *BbMm* (brown). Both parents can produce four types of gametes: *BM, Bm, bM,* and *bm*. Filling in the Punnett square:

eggs

sperm	*BM*	*Bm*	*bM*	*bm*
BM	*BBMM*	*BBMm*	*BbMM*	*BbMm*
Bm	*BBMm*	*BBmm*	*BbMm*	*Bbmm*
bM	*BbMM*	*BbMm*	*bbMM*	*bbMm*
bm	*BbMm*	*Bbmm*	*bbMm*	*bbmm*

All *mm* offspring are albino, so we get the expected proportions 9⁄16 brown-haired, 3⁄16 blond-haired, 4⁄16 albino.

c. *BbMm* (brown) × *bbmm* (albino):

eggs

sperm	*bm*
BM	*BbMm*
Bm	*Bbmm*
bM	*bbMm*
bm	*bbmm*

The expected proportions of offspring are 1⁄4 brown-haired, 1⁄4 blond-haired, 1⁄2 albino.

7. A man with normal color vision is *CY* (remember, the Y chromosome does not have the gene for color vision). His color-blind wife is *cc*. Their expected offspring will be:

eggs

sperm	*c*
C	*Cc*
Y	*cY*

We therefore expect that all the sons will be color-blind and all the daughters will have normal color vision.

8. The husband should win his case. All his daughters must receive one X chromosome, with the *C* allele, from him and therefore should have normal color vision. If his wife gives birth to a color-blind daughter, her husband cannot be the father (unless there was a new mutation for color blindness in his sperm line, which is very unlikely).

Group Activity

Genetic counseling is a field that has been growing with our increasing knowledge of human genes. Individuals and prospective parents can now obtain information that helps them make decisions. Counselors provide information and discuss the consequences of various decision scenarios but do not tell their clients what they should do. Your instructor will provide your group with a case study for you to discuss to determine how you, as genetic counselors, might proceed.

For More Information

Horgan, J. "Eugenics Revisited." *Scientific American*, June 1993. Determining patterns of inheritance, let alone finding suspect genes, for behavioral disorders is notoriously difficult. What's more, it isn't at all clear what one should do with such data, if they become available.

Hoyt, E. "Wild Relatives." *Living Wilderness*, Summer 1990. Undisturbed ecosystems are a valuable gene bank with great potential for improving crop plants.

Kahn, P. "Gene Hunters Close in on Elusive Prey." *Science*, March 8, 1996. Using pedigrees and biotechnology, researchers are uncovering the causes of diseases that involve complex interactions between genes and the environment.

McGue, M. "The Democracy of the Genes." *Nature*. July 1997. The environment has a more important role in the development of intelligence than previously thought.

Pines, M. "In the Shadow of Huntington's." *Science 84,* May 1984; Grady, D. "The Ticking of a Time Bomb in the Genes." *Discover,* June 1987; Roberts, L. "Huntington's Gene: So Near, Yet So Far." *Science,* February 1990; and Morell, V. "Huntington's Gene Finally Found." *Science*, April 1993. These articles focus on the story of the scientific detective work of diagnosing Huntington disease by using recombinant DNA techniques, the frustrations of searching for the gene, and, finally, finding it. Particularly poignant because one of the lead investigators may herself be a victim.

Sapienza, C. "Parental Imprinting of Genes." *Scientific American*, October 1990. It is not quite true that all genes are equal, regardless of whether they have been inherited from mother or father. In some cases, which parent a gene comes from greatly alters its expression in the offspring.

Stern, C., and Sherwood, E. R. *The Origin of Genetics: A Mendel Source Book*. San Francisco: W. H. Freeman, 1966. There is no substitute for the real thing, in this case a translation of Mendel's original paper to the Brünn Society.

Wivel, N. A., and Walters, L. "Germ-line Gene Modification and Disease Prevention: Some Medical and Ethical Perspectives." *Science*, 1993. If we can fix a defective allele in an egg or zygote, should we?

Answers to Multiple-Choice Questions

1. c 2. c 3. e 4. d 5. a

MEDIATUTOR

CD Activities

Activity 11.1: Monohybrid Crosses

Estimated time: 10 minutes

When Gregor Mendel first examined crosses of pea plants, he analyzed single-trait differences and found some very interesting results. This tutorial will recreate one of Mendel's early experiments and show that a Punnett square can be used to determine how probability figures into genetic analysis.

Activity 11.2: Dihybrid Crosses

Estimated time: 10 minutes

If a Punnett square can be used to demonstrate the probability of genetics with monohybrid crosses, can it also be used for a dihybrid cross, tracking two separate traits at once? In this tutorial, we will examine a set of crosses, focusing on two separate genetic traits. We will see how probability can help us understand this situation.

Start the MediaTutor Student CD-ROM and enter the activity number in the Quick Search box to be taken directly to that activity.

Web Investigation

Born to Drink

Estimated time: 20 minutes

The principles of inheritance deduced by Gregor Mendel hold true for many human characteristics, including diseases. However, the expression of most human characteristics (including diseases) is determined by the interaction of several genes (polygenic) and possibly the environment, as discussed in the chapter. This investigation will further explore how the inheritance of certain alleles can contribute to the development of certain disorders.

Go to http://www.prenhall.com/audesirk3, the Audesirk Companion Web site. Select Chapter 11 and Web Investigation to begin.

Does this field of soybeans represent our best chance to feed a hungry world? Or is it a symptom of humanity's dangerous compulsion to tinker with the essence of life? It depends on whom you ask.

12 Biotechnology

AT A GLANCE

CASESTUDY

Food Fight

The spring sun shines brightly on the Mississippi countryside as a tractor crawls slowly across a field. A farmer is planting his soybean crop. But some critics see the farmer's seemingly harmless activity as extremely dangerous. In this view, the farmer's activity threatens our health, our food supply, and the environment. According to writer Jeremy Rifkin, the farmer and others like him are "spreading chaos throughout the biological world, drowning out the ancient language of creation." Around the world, impassioned protesters demand protection from the farmer's crops.

Why does something as ordinary as farming inspire such intense opposition? Because more and more farmers are growing genetically altered, bioengineered crops. These genetically modified foods have been singled out for especially harsh attack by critics of biotechnology, in part because, unlike most nonagricultural applications of biotechnology, genetic engineering in agriculture is already widespread and common; its products are used by farmers everywhere. You've almost certainly consumed a bioengineered food; such foods are widely sold in the United States and no law requires that they be specially labeled.

Agricultural biotechnology is based on the introduction of foreign genes into crop plants. The purpose of such gene transfers is often to confer resistance to a disease, pest, or herbicide. For example, farmers are now planting corn, cotton, and potatoes that contain a gene that helps the plants repel insect pests, reducing the farmers' need to apply pesticides. New, bioengineered strains of squash, potatoes, and tomatoes resist infection by viral diseases. Cotton, corn, soybeans, and sugar beets have been altered so that they are resistant to particular herbicides; farmers who grow these altered plants can use the herbicides to control weeds without worrying that the herbicides will kill the crops. Other fruits, vegetables, and even farm animals have been genetically engineered for improved yields, better nutritive value, or better resistance to the stress of shipping and handling.

The genes that are used to confer desirable qualities on crops are typically drawn from fungi or bacteria. Later in this chapter, we'll explain more about these genes that make the big leap across species boundaries and the techniques that are used to induce the leaps. At the chapter's close, we'll discuss some of the reasons why these gene-jumps concern antibiotechnology activists. First, however, let's examine the processes that make it possible for humans to move genes from one species to another. ■

1 What Is Biotechnology?

In its broadest sense, **biotechnology** is any use or alteration of organisms, cells, or biological molecules to achieve specific practical goals. By this definition, biotechnology is nothing new. For example, people have been using yeast cells to produce bread, beer, and wine for the past 10,000 years, and selective breeding of plants and animals has an equally long history. Evidence of early selective breeding includes fragments of squash, 8000 to 10,000 years old, that were found in a dry cave in Mexico. These ancient squash have larger seeds and thicker rinds than wild varieties, suggesting that humans had selectively bred them to preserve desirable traits. Similarly, prehistoric art and animal remains suggest that dogs, sheep, goats, pigs, and camels were domesticated and selectively bred beginning about 10,000 years ago. Because selective breeding is in essence a technique for changing the frequency of particular alleles in populations of domestic organisms, it was in effect an early method of genetic manipulation.

Even today, selective breeding remains an important part of biotechnology. Modern biotechnology, however, also commonly uses **genetic engineering**, a term that refers to more direct methods for modifying genetic material. Genetically engineered cells or organisms may have genes deleted, added, or changed. Genetic engineering can be used to learn more about how cells and genes work, to develop better treatments for diseases, to produce valuable biological molecules, and to improve plants and animals for agriculture.

A key tool in genetic engineering is **recombinant DNA**, which is DNA that has been altered to contain genes or portions of genes from different organisms, often from different species. Recombinant DNA is often multiplied in bacteria, viruses, or yeast and then transferred into other species, including animals and plants. These **transgenic** plants and animals then express genes that have been transferred from other species.

Since its development in the 1970s, recombinant DNA technology has grown explosively, transforming both research and commerce. Today, researchers in almost every area of biology routinely use recombinant DNA technology in their experiments. In the pharmaceutical and other industries, genetic engineering has become the preferred way to manufacture many products, including human insulin and the enzymes used for making cheese. Some applications of genetic engineering, however, have raised concerns about the safety and ethical implications of biotechnology.

2 How Does DNA Recombine in Nature?

DNA can be transferred from one organism to another by natural processes, and many recombinant DNA technologies are based on these naturally occurring transfers. Our exploration of biotechnology will therefore be aided by a brief review of natural DNA recombination.

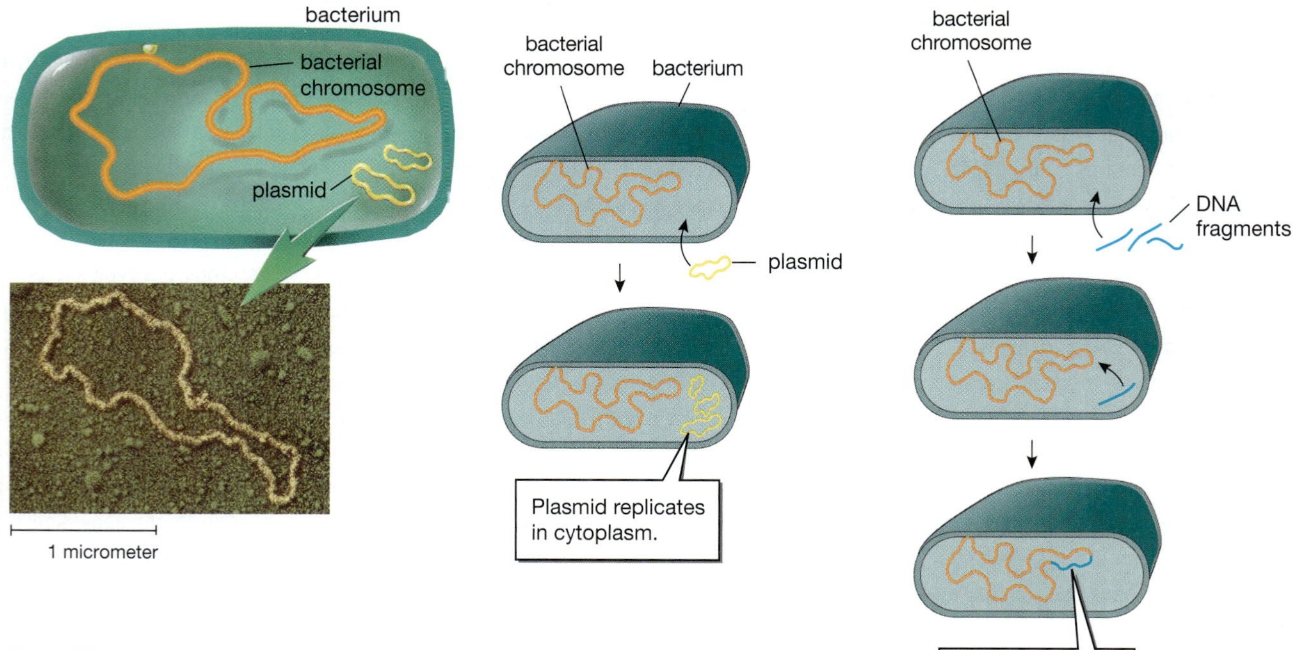

Figure 12-1 Recombination in bacteria
(a) In addition to their large circular chromosome, bacteria commonly possess small plasmids. Bacterial transformation occurs when living bacteria take up ***(b)*** these plasmids or ***(c)*** fragments of chromosomes.

MediaTutor 12.1 Genetic Recombination

Sexual Reproduction Recombines DNA

During sexual reproduction, the genetic makeup of chromosomes can change. In particular, crossing over during meiosis I fosters an exchange of DNA between the chromosomes in a homologous pair. Thus, the chromosomes in gametes tend to differ from the parental chromosomes; each chromosome contains a mixture of alleles from two different parental chromosomes. In this sense, each egg and sperm contains recombinant DNA, as does the organism that arises from fusion of a sperm and egg.

Bacterial Transformation May Combine DNA from Different Species

To some people, recombination of DNA within a species during sexual reproduction is "natural" and therefore a good thing, but such recombination performed in the laboratory between different species is "unnatural" and therefore bad. Recombination of DNA between species, however, is not confined to the lab; it also occurs in nature.

Bacteria can undergo several types of recombination that allow gene transfer between species. In a process called **transformation**, for example, bacteria pick up pieces of DNA from the environment. The DNA that is picked up may be part of the chromosome of another bacterium, including DNA from another bacterial species, or may be in the form of tiny circular DNA molecules called **plasmids** (Fig. 12-1). A single bacterium may contain dozens or even hundreds of copies of a plasmid and, when the bacterium dies, it releases its plasmids into the environment, where they can be picked up by bacteria of the same or different species. In addition, living bacteria can often pass their plasmids directly to other living bacteria, or even to yeast cells.

Viruses May Transfer DNA Between Species

Viruses, which are little more than genetic material encased in a protein coat, transfer their genetic material to host cells that they infect. (A **host cell** is a cell susceptible to infection by a particular virus.) Inside the infected host cell, viral genes replicate and direct the synthesis of viral proteins. The replicated genes and new viral proteins assemble inside the cell, forming new viruses that are released to infect other cells (Fig. 12-2).

In some cases, the cycle of viral reproduction may cause genes to be transferred from one organism to another. Such transfers can occur because certain types of viruses can insert their DNA into a host cell's chromosome. The viral DNA can remain there for days, months, or even years. Every time the cell divides, it replicates the incorporated viral DNA along with its own DNA. When new viruses are finally produced, they can carry away pieces of the host's DNA along with the viral DNA. If such recombinant viruses infect other cells and insert their DNA into the new host cells' chromosomes, pieces of the previous host cell's DNA will also be inserted. Most of the time, host DNA that is transferred in that manner will move between different individuals of a single species, but some viruses can infect more than one species; such viruses may transfer genes from one species to another.

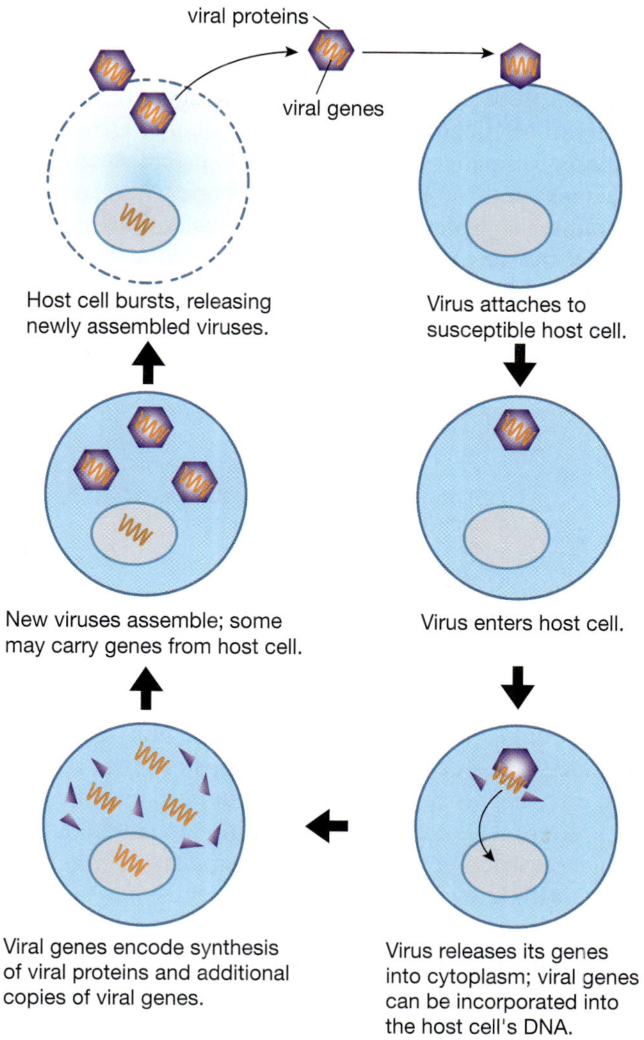

Figure 12-2 Viruses infect host cells and use them to make more viruses

3 How Is Recombinant DNA Made in the Lab?

Suppose that you want to develop a genetically engineered rice plant that would produce rice grains highly enriched in vitamin A. One approach would be to transfer the genes needed to make vitamin A from another organism to a rice plant. To find such genes, you would probably look in an organism that naturally manufactures high levels of vitamin A, such as a daffodil. But before you can use daffodil genes in your genetic engineering project, you must convert the daffodil genome to a useful form. (A **genome** is the entire set of genetic material carried by an organism.)

Restriction Enzymes Cut DNA at Specific Nucleotide Sequences

The first steps in the process will take advantage of the ability of restriction enzymes to cut up large DNA molecules into smaller, more manageable pieces. **Restriction enzymes** cut through the DNA backbone wherever they encounter a particular sequence of nucleotides. For example, the EcoRI restriction enzyme cuts through a DNA double helix wherever it encounters the following sequence:

...AATTGCTTAGAATTCGATTTG...
...TTAACGAATCTTAAGCTAAAC...

EcoRI binds to the GAATTC sequence and cuts the DNA, creating DNA fragments.

...AATTGCTTAG AATTCGATTTG...
...TTAACGAATCTTAA GCTAAAC...

Note that EcoRI produces a staggered cut through the DNA double helix, leaving a small region of single-stranded DNA at the cut ends. Dozens of other restriction enzymes work the same way, but each enzyme cuts at a different sequence. Thus, treating DNA with a particular restriction enzyme yields smaller DNA fragments with known sequences at their ends.

In nature, restriction enzymes are produced by bacteria, and serve to defend the bacteria against viral infections by cutting apart viral DNA. The bacteria's own DNA is protected by methyl groups ($-CH_3$) that are attached to some of the nucleotides. This modified DNA is not recognized by restriction enzymes, and so remains undamaged.

Foreign DNA Can Be Inserted into a Plasmid or Virus

As you continue working to create your transgenic rice plant, you will need a convenient way to store the daffodil DNA fragments. One possibility is to store the fragments inside bacterial cells. But if you're going to use bacteria for storage, you must first get daffodil DNA into the bacteria. Fortunately, plasmids and viruses can be used to transfer DNA between species. Such plasmids and viruses are examples of vectors. A **vector** is a vehicle of biological transfer.

Inserting foreign DNA into a vector is straightforward. For example, DNA can be isolated from a daffodil, or any other organism, and cut with EcoRI, producing hundreds of thousands of small DNA fragments. Each fragment of daffodil DNA will have the same nucleotide sequence at its ends: TTAA on one end of each fragment, and AATT on the other end. If the DNA of a plasmid (or other vector) is also cut with EcoRI, the cut ends will be complementary to those of the daffodil DNA. Next, the plasmid fragments and the daffodil fragments are combined. DNA ligase (an enzyme that joins DNA strands together; see Chapter 8) is added to the mixture, and it joins the daffodil DNA to the plasmid DNA, forming recombinant molecules.

Recombinant DNA Moves into Bacteria, Building a DNA Library

The daffodil–bacteria recombinant DNA molecules are then added to a culture of bacteria. The bacteria pick up the recombinant molecules by bacterial transformation. Each bacterium now contains a particular portion of the daffodil genome, carried in a plasmid vector. Whenever the bacterium divides, the recombinant plasmid is replicated, so a growing culture of these bacteria produces billions of copies of the various recombinant DNA molecules.

The bacterial culture now contains a **DNA library**, a readily accessible, easily duplicated collection of all the DNA from a particular organism (Fig. 12-3). Each genetically modified bacterium in the culture contains just a small portion of the daffodil genome, but the collection of bacteria contains the entire daffodil genome. Although the name "library" implies that the genome is nicely organized, in fact a DNA library is typically an unorganized collection of billions of genetically modified bacteria in a single test tube. Further steps are necessary to identify and isolate the particular daffodil genes needed to produce vitamin A.

4 How Can Researchers Identify Specific Genes?

After developing a library of recombinant DNA molecules, the next step is to identify and isolate the specific genes of interest. Researchers have devised a number of techniques for identifying a particular gene from among the recombinant molecules in a DNA library. One of the most commonly used methods takes advantage of DNA probes.

If an investigator already knows at least a portion of the target gene's sequence, he or she can find the gene by using a **DNA probe**, a short sequence of single-stranded DNA that is complementary to the gene being sought. When a DNA library is exposed to a probe, the only bacteria to which the probe binds are the ones that contain the target gene. This procedure is, of course, useful only if the researcher can easily detect the presence of the probe, so a probe usually includes a chemical marker of some kind. In many cases, the marker is a radioactive atom that has been attached to the probe.

To use a probe, a researcher first spreads the bacteria from a DNA library on a plate of solid growth medium. On the plate, each individual bacterium divides many times to form a colony of genetically identical cells. A

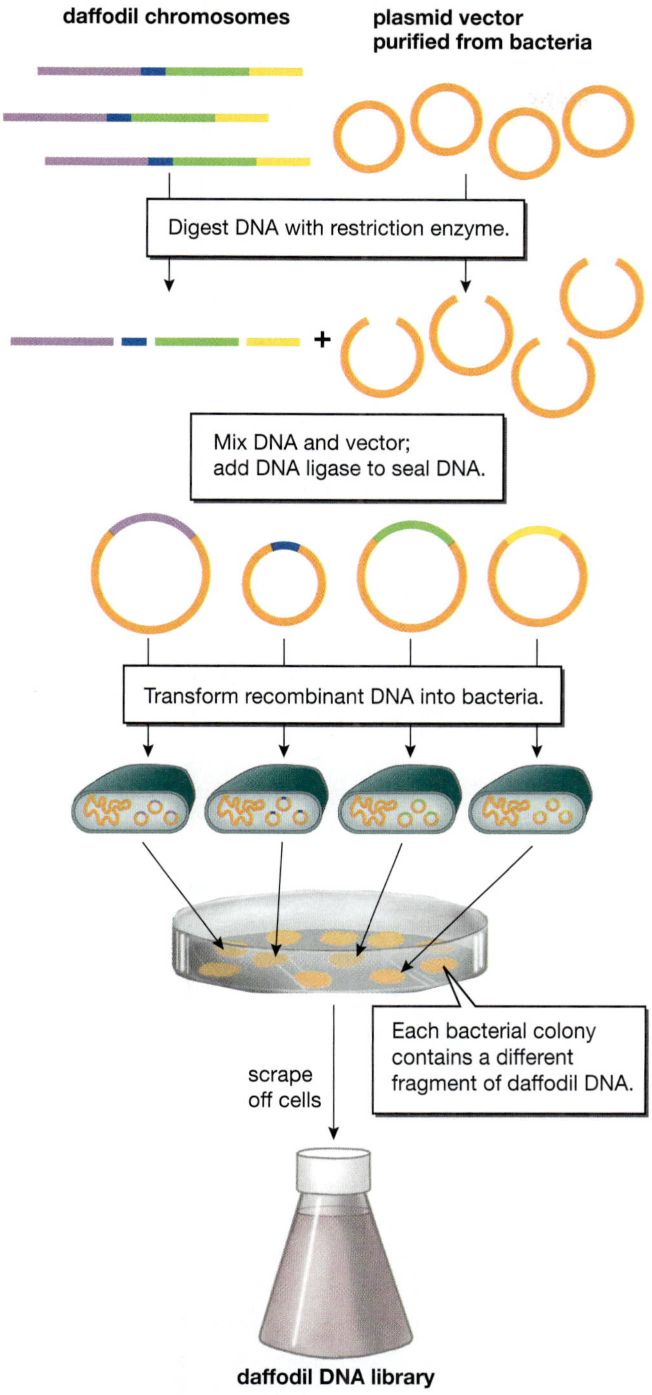

Figure 12-3 Building a DNA library

piece of filter paper is placed on top of the colonies, and some of the cells from each colony stick to the paper. Next, the cells on the paper are broken open and bathed with the DNA probe. The probe binds only to the cells that contain DNA complementary to the probe. Because the probe is radioactive, its location can be detected by laying a piece of X-ray film on top of the filter paper. The radioactivity in the probe exposes the film, producing dark spots on top of the bacterial colonies that contain the gene of interest. The researcher now knows which colonies on the original plate contain the target gene.

5 How Can Researchers Make Many Copies of a Gene?

Once a gene has been isolated, it can be put to a variety of uses. In most cases, however, those uses require many copies of the gene. For some purposes, making copies of the gene may require only growing large colonies of bacteria that contain the appropriate recombinant plasmids. For other purposes, however, a different approach is required.

Perhaps the most commonly used method of copying genes is the **polymerase chain reaction (PCR)**. Developed in 1986 by Kary B. Mullis, PCR allows billions of copies of selected genes to be made faster and more cheaply than they can be grown in bacteria. PCR is so elegant and so crucial to continued advances in molecular biology and biotechnology that it earned Mullis a share of the Nobel Prize for Chemistry in 1993.

PCR relies on a special type of DNA polymerase (the enzyme that synthesizes new DNA strands) to produce large quantities of specific DNA segments. This distinctive polymerase synthesizes DNA in species of bacteria that live in the near-boiling water of hot springs. Unlike the polymerase of more conventional organisms, this version works perfectly well at high temperatures, and the PCR technique takes advantage of this resistance to heat (see "Scientific Inquiry: Biotech Tools of the Trade").

MEDIATUTOR 12.2 Building a DNA Library

6 Why Is It Useful to Know the Sequence of a Gene?

After generating copies of a gene, researchers often determine the exact sequence of nucleotides in the gene. The techniques for discovering gene sequences are varied and often ingenious, but today the process has been almost completely automated. With the right machine, an instruction booklet, and a bit of training, just about anyone can determine the sequence of a DNA sample.

Why is knowing the nucleotide sequence of a gene useful? One reason is that knowing the nucleotide sequence can help determine how the gene exerts its effect. If one knows the gene's sequence, after all, one also knows the amino acid sequence of the encoded protein. Once the protein is known, its mechanism of action and its function can often be deduced.

DNA sequences are also essential for diagnosing genetic diseases. For example, the scientists who sequenced the cystic fibrosis gene went on to determine how the sequence of the alleles found in cystic fibrosis patients differed from the normal allele. The defective allele of most patients was missing three nucleotides. Using this knowledge, researchers developed methods for rapidly determining whether a person is a heterozygote carrying this allele. Such genetic testing can also tell parents whether an unborn child has inherited two copies of the defective

Scientific Inquiry
Biotech Tools of the Trade

Although the techniques of biotechnology are many and varied, a few basic procedures are especially important. Among these, the polymerase chain reaction (PCR) and restriction fragment length polymorphism (RFLP) analysis are used so widely and for so many purposes that we single them out for description here.

The Polymerase Chain Reaction The polymerase chain reaction is used to amplify (make many copies of) a particular piece of DNA. PCR begins with a mixture of four main components: a sample of DNA that contains the sequence to be copied, heat-resistant DNA polymerase (isolated from bacteria that thrive in hot springs or hot ocean vents), a combination of the four different types of individual nucleotides, and two short sequences of single-stranded DNA called primers. The primers are chosen (or synthesized) such that their sequences are complementary to sequences at each end of the target DNA sequence. Thus, during PCR the primers bind to each end of the segment of DNA to be copied, and DNA polymerase synthesizes the DNA that lies between the primers' positions.

Once the PCR mixture is prepared, the reaction is controlled by the temperature as follows:

1. The PCR mixture is heated to about 90 °C, which causes the DNA sample to separate into single strands.
2. The temperature is lowered to about 50 °C, which allows the two primers to bind to the DNA strands. The primers establish the locations, one on each strand, at which DNA synthesis will begin.
3. The temperature is raised to about 72 °C, the temperature at which the heat-resistant DNA polymerase works best. The polymerase catalyzes the synthesis of new strands that are complementary to those in the DNA sample.

These three temperature-changing steps are repeated as many times as necessary to generate the desired amount of amplified DNA (Fig. E12-1). Typically, PCR is run in a machine that automatically generates the required cycles of heating and cooling. Each cycle takes only a few minutes, so PCR can produce billions of copies of a gene or other DNA segment in a few hours.

allele and will thus develop the disease. Genetic tests are available for other inherited diseases, including sickle-cell anemia (see "Health Watch: Prenatal Genetic Screening"). In the United States, state legislation mandates that all newborns be tested for several genetic diseases, including phenylketonuria and hypothyroidism. Most states also mandate testing for sickle-cell anemia. Diagnosis of infections ranging from tuberculosis to HIV also increasingly relies on genetic tests; these tests determine whether DNA from the disease-causing bacterium or virus is present in the patient.

7 What Is DNA Fingerprinting?

The DNA sequences of different individuals of the same species are similar but not identical, and the differences between individuals can be used as a means of identification. When DNA from different individuals is digested with restriction enzymes, the sizes of the resulting fragments differ among those individuals, because small sequence differences cause the restriction enzymes to cut in different places (see "Scientific Inquiry: Biotech Tools of the Trade"). Each individual's DNA thus produces a unique set of restriction fragments, and the pattern of these fragments is a unique **DNA fingerprint** that can be used to identify an individual (Fig. 12-4).

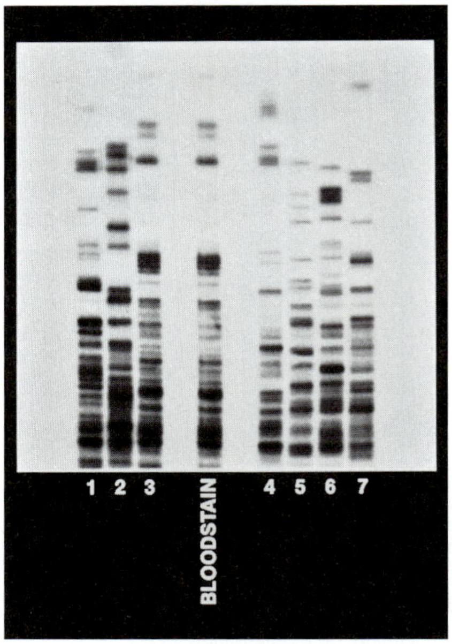

Figure 12-4 DNA fingerprinting in forensics
This set of DNA fingerprints clearly shows that only one suspect's DNA matches that found in a bloodstain at the crime scene.

DNA Fingerprinting Has Revolutionized Forensics

DNA fingerprinting has proved useful in many fields, but its impact has been especially dramatic in forensics, the collection of information to be used as evidence in court proceedings. DNA tests were first used in a court

Restriction Fragment Length Polymorphisms Analysis of **restriction fragment length polymorphisms** is best known for its role in DNA fingerprinting, but RFLP analysis also has a host of other uses in molecular biology and biotechnology. The procedure relies on the observation that when different DNA molecules are digested with a restriction enzyme, they yield different collections of fragments. To see why, imagine two segments of DNA. One segment happens to contain the sequence GAATTC at three different locations. The second segment is identical to the first, except that a mutation has changed one of the GAATTC sequences to GAACTC. Now imagine that both DNA segments are digested with the restriction enzyme EcoRI, which cuts wherever it encounters GAATTC. (In the drawings below, the locations at which EcoRI will cut are marked by arrows.)

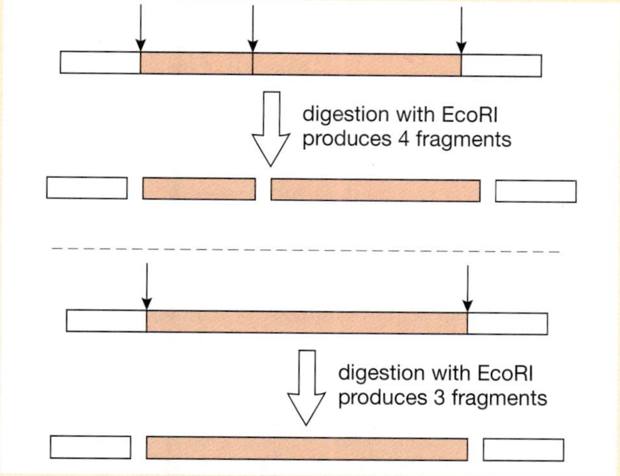

Even though the difference between the two DNA segments is small, EcoRI cuts at three places in the top segment, but at only two places in the bottom segment. Thus, the top molecule yields four fragments, while the bottom one yields only three fragments. Also note that the fragments generated from the two molecules differ not only in number, but also in size.

An investigator conducting an RFLP analysis typically digests a number of different DNA samples with one or more restriction enzymes, and then compares the fragments from the different samples. To make such a comparison, the investigator needs some way to organize and view the fragments. The most common way to accomplish this is by **gel electrophoresis**. In this procedure, the mixture of fragments is placed on one end of a gel, which is slab of gelatinous material (often agar, a carbohydrate produced by seaweed). When an electrical current is applied to the gel, the DNA fragments (which are negatively charged) move toward the positively charged electrode. Smaller fragments can slip through the spaces between the gel's molecules more easily than larger fragments, so they move more rapidly, leaving the larger fragments behind. Eventually, the fragments are separated by size, forming distinct bands on the gel (see Fig. 12-4).

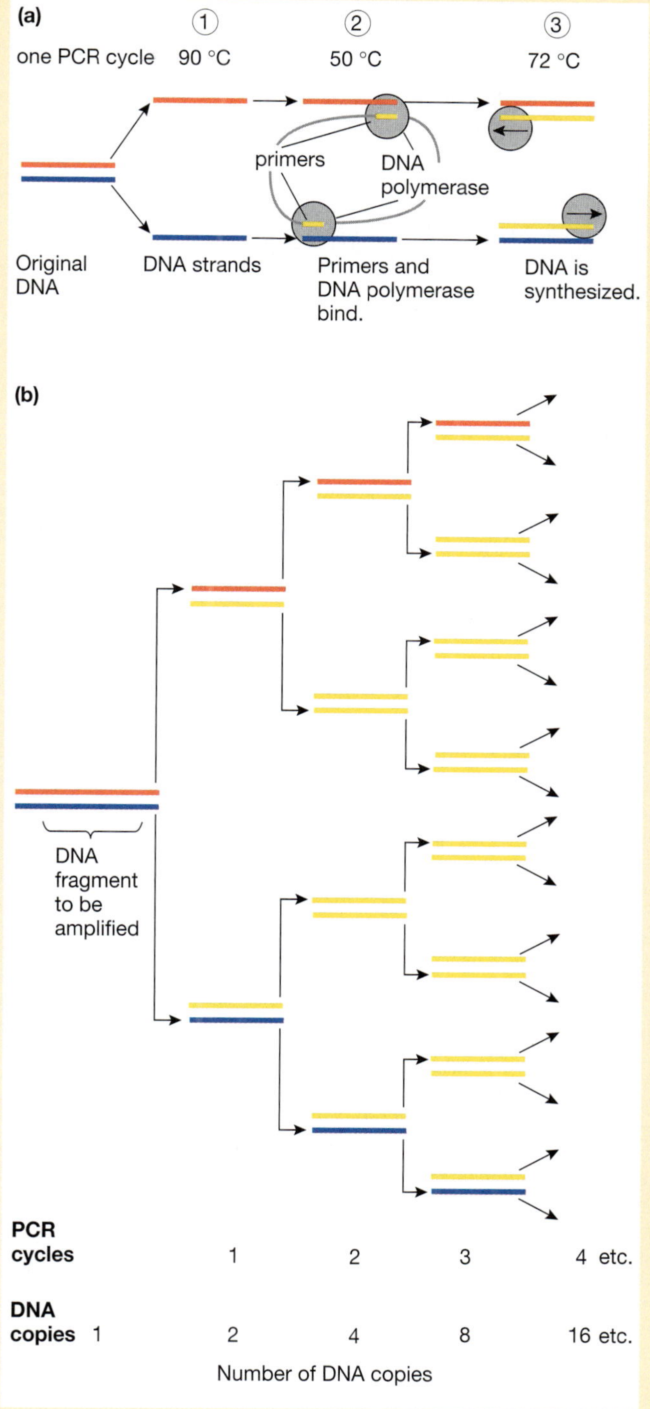

Figure E12-1 PCR copies a specific DNA sequence
(a) During each cycle of PCR, one double-stranded DNA molecule is used as a template to yield two double-stranded molecules. ***(b)*** With each cycle, the amount of DNA doubles.

Health Watch
Prenatal Genetic Screening

Prenatal (before-birth) diagnosis of many genetic disorders, including cystic fibrosis, sickle-cell anemia, Tay-Sachs disease, and Down syndrome, requires samples of fetal cells (cells from the fetus) or chemicals produced by the developing fetus. Presently, there are two main techniques used to obtain these samples—*amniocentesis* and *chorionic villus sampling*. In addition, new techniques for isolating fetal cells from the mother's blood are currently under development. Once samples are collected, they can be used for tests that allow prenatal diagnosis of genetic disorders.

Amniocentesis A human fetus is surrounded by a watery liquid, which is enclosed in a waterproof membrane called the amnion. As the fetus grows, it sheds some cells into the surrounding fluid. When a fetus is 16 weeks or older, this amniotic fluid can be collected safely by a procedure called amniocentesis. In amniocentesis, a physician, who has determined the position of the fetus by ultrasound scanning, inserts a sterilized needle through the abdominal wall, the uterus, and the amnion, and withdraws 10 to 20 milliliters of fluid (Fig. E12-2). Biochemical analysis may be performed on the fluid immediately, but other types of analysis must wait, because there are very few cells in the fluid sample. For any analyses requiring cells, such as testing for Down syndrome, the cells must first be allowed to multiply in a laboratory culture for a week or two.

Chorionic Villus Sampling The chorion, a membrane produced by the fetus, has many small projections called villi. In chorionic villus sampling (CVS), a physician inserts a small tube into the uterus through the mother's vagina and, by suction, collects a small sample of tissue from the villi for analysis (see Fig. E12-2). CVS has two major advantages over amniocentesis. First, it can be done much earlier in pregnancy—as early as the 8th week. This factor is especially important if the woman is contemplating a therapeutic abortion in the event that the fetus has a major defect. Second, the sample contains enough fetal cells that analyses can be performed immediately rather than a week or two later.

Fetal Cells from Maternal Blood Fetal cells become detectable in a mother's blood between the 6th and 12th weeks of pregnancy, and a new procedure seeks to harvest these cells from a small sample of maternal blood. If successful, this procedure might replace both amniocentesis and CVS as a means of obtaining fetal cells. By eliminating the need to penetrate the uterus, blood sampling would greatly reduce the risk of injury and infection to both mother and fetus. But only about 1 of every 100,000 cells in maternal blood comes from the fetus, so learning how to efficiently separate fetal cells from maternal ones is a major challenge.

Analyzing the Samples Several types of analyses can be performed on the fetal cells or amniotic fluid. For amniotic fluid, analyses that determine the concentration of chemicals are most common. For example, Tay-Sachs disease and many other metabolic disorders can be detected by abnormally low concentrations of key enzymes, or by abnormally high concentrations of the reactants that those enzymes are supposed to break down. Other tests require whole cells. For example, analysis of the chromosomes of the fetal cells can show if the correct number of chromosomes is present, or if any chromosomes show structural abnormalities.

Defective alleles, such as those that cause cystic fibrosis or sickle-cell anemia, can be detected by recombinant DNA techniques. In some cases, these techniques have eliminated the need to grow cells in culture before testing. Now, for example, one can extract DNA from just a few cells, use PCR to amplify the hemoglobin gene, and then use RFLP analysis to detect the allele that causes sickle-cell anemia. (RFLP analysis can detect the sickle-cell allele because the mutation that causes the disease also destroys a cutting site for a particular restriction enzyme; see "Scientific Inquiry: Biotech Tools of the Trade.") If prenatal screening reveals that a fetus is homozygous for the sickle-cell allele, some therapeutic measures can be taken. In particular, regular doses of penicillin after the infant is born will greatly reduce the bacterial infections that otherwise kill about 15% of homozygous children.

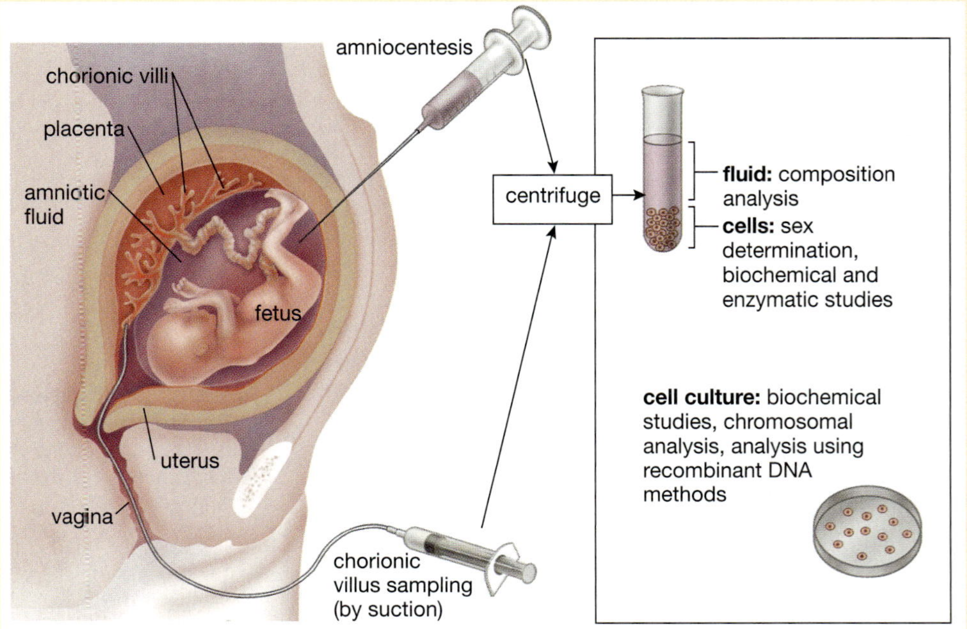

Figure E12-2 Prenatal cell-sampling techniques
Two methods of obtaining fetal cell samples—amniocentesis and chorionic villus sampling—and some of the tests performed on the fetal cells.

case in 1986, when blood samples from all the men in a British village were tested to solve the rape and murder of two women. Since then, DNA fingerprinting has entered the mainstream of forensic science and become a common form of evidence in criminal cases.

Thanks to the ability of PCR to make many copies from a small amount of DNA, investigators can produce a DNA fingerprint from even tiny samples. All that's required is a semen stain, a few cells from the base of a hair shaft, tiny skin fragments found under a victim's fingernails after a struggle, or a speck of dried blood. Even the back of a stamp that was licked to attach it to an envelope has enough DNA for analysis. A match between a suspect's DNA and DNA found on a victim or at a crime scene can provide powerful evidence for conviction.

DNA fingerprinting evidence can, of course, also be used to establish the innocence of suspected criminals. In some cases, DNA fingerprinting has even been used to overturn the convictions of innocent people. For example, in the last 10 years, DNA fingerprinting has shown that at least 10 death row inmates were actually innocent of the crimes for which they had received the death penalty.

DNA Fingerprinting Is Also Used for Other Purposes

Many areas of biology and medicine have found uses for DNA fingerprinting methods. For example, DNA fingerprinting can be used to determine paternity. Paternity testing has proved useful in studies of animal mating behavior, in legal proceedings that turn on determining a child's biological father, and even in the study of human history. In 1998 investigators tested DNA samples from living descendants of Thomas Jefferson, the third president of the United States, and of his slave Sally Hemings, and showed that Hemings' youngest son was fathered by Jefferson.

DNA fingerprinting has also become an important tool for identifying human remains. For example, DNA tests have been performed on the remains of unidentified war casualties who are interred in the "Tomb of the Unknowns." As individuals are identified, their remains are removed from the tomb and buried in accordance with the wishes of surviving family members. Today, the armed forces collect DNA samples from all members, so that no remains of casualties of future wars will remain unidentified.

DNA fingerprinting has dozens of other uses. It enables us to test potential organ donors for compatibility with a particular patient, to monitor pedigree claims for livestock, to examine relationships among human populations, and to determine compliance with endangered species laws. For example, DNA fingerprinting of whale meat purchased in a Japanese market showed that some of the meat came from endangered whale species, whose harvest was prohibited by international law.

8 What Are Some Agricultural Uses of Biotechnology?

Genetic engineering is rapidly replacing traditional breeding techniques as the method of choice for producing improved crops. Transgenic plants can be produced in several ways. One commonly used technique inserts recombinant DNA via a plasmid found in a soil bacterium that infects plants. The plasmid is introduced to plant cells growing in a laboratory culture; these genetically altered cells must then be induced to grow into individual plants. Alternately, genes can be inserted into the cells of plant seedlings by a "gene gun" that blasts DNA-coated particles directly into cells. This method is well suited to plant species whose cells cannot be easily cultured in the lab.

In many cases, the inserted gene is from a species that is only distantly related to plants. For example, the transgenic tobacco plant in Figure 12-5 contains a gene from a firefly. The firefly gene encodes the enzyme luciferase, which catalyzes a reaction that produces visible light. The plant's eerie glow clearly demonstrates that the firefly gene has been incorporated successfully into the plant genome.

Before a transgenic crop (or animal) can be marketed, it must be approved by one or more government agencies. More than 50 transgenic varieties have been deemed safe for commercial production in the United States; these crops can now be grown and crossed with no special government control or oversight. In Europe

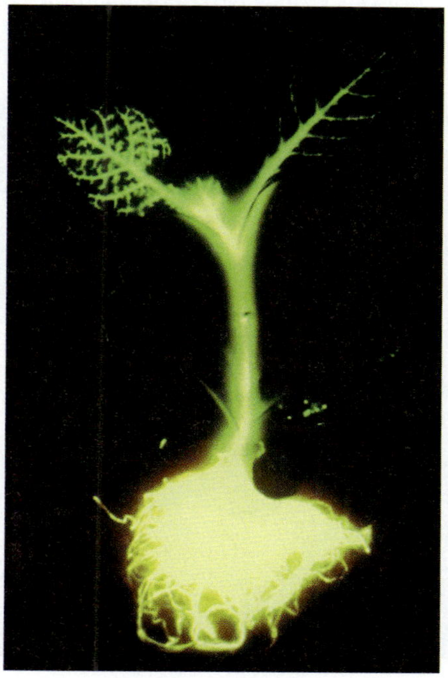

Figure 12-5 Inserting genes into plant cells
Genetic engineering has inserted a firefly gene into this tobacco plant, and the plant has taken up water containing the chemical luciferin. The firefly gene codes for an enzyme that breaks down the chemical luciferin, releasing light in the process.

Table 12-1 Genetically Engineered Crops with USDA Approval

Genetically Engineered Trait	Potential Advantage	Examples of Bioengineered Crops Approved by USDA
Resistance to herbicide	Herbicide kills weeds, but not crop plants.	beet, canola, corn, cotton, flax, potato, rice, soybean, tomato
Resistance to pests	Crop plants suffer less damage from insects, producing higher crop yields. Pesticides need not be used.	corn, cotton, potato
Resistance to disease	Plants are less prone to infection by viruses, bacteria, or fungi, producing higher crop yields.	papaya, potato, squash
Sterile	Transgenic plants cannot cross with wild varieties, making them safer for the environment and more economically valuable to the seed companies that produce them.	chicory, corn
Altered ripening	Fruits can be more easily shipped with less damage, producing higher returns for farmer.	tomato

and many other parts of the world, however, transgenic crops are more strictly regulated. Some of the crops that have been approved for commercial use in the United States are listed in Table 12-1.

Genetic Engineering Can Make Plants Resistant to Diseases, Insects, and Weed Killers

The most economically important agricultural application of genetic engineering is the production of crops that have been engineered to be more resistant to diseases, pests, and poisons (Fig. 12-6). For example, one widespread strategy is to insert genes that make crop plants resistant to the effects of herbicides. A farmer who plants an herbicide-resistant crop can apply herbicide more aggressively, killing weeds without fear of simultaneously killing crop plants. Herbicide-resistant varieties are increasingly popular; the U.S. Department of Agriculture estimates that they account for 8% of the corn, 57% of the soybeans, and 46% of the cotton grown in the United States.

Crop plants are under constant attack by viruses, bacteria, fungi, and insects, and genetic engineers have modified many crops by inserting genes that help

(a)

(b)

Figure 12-6 Transgenic crops
(a) This squash has been genetically engineered to resist two common plant viruses. ***(b)*** NewLeaf® potatoes (middle row) have been genetically engineered to resist the Colorado potato beetle. The genetically engineered strain has a clear advantage over the nonengineered strains in the rows on either side.

plants resist these threats. The most common strategy for creating resistance to insects is to incorporate genes that encode a chemical that is toxic to insects. The inserted genes typically come from a bacterium (*Bacillus thuringiensis*, or Bt) that produces a natural insecticide called Bt toxin. A crop plant that expresses the Bt toxin gene poisons the insects that try to eat it, reducing the need for farmers to apply pesticide sprays. A number of crop plants, including corn, soybeans, and potatoes, have been engineered to produce Bt toxin.

Genetic Engineering May Improve Domestic Animals

Unlike plants, individual animals cannot be grown in the lab from cultured cells, so methods for creating transgenic animals differ from those that are used with plants. In mammals, the desired foreign gene is injected into a fertilized egg, which is then implanted into a surrogate mother and allowed to develop. The resulting offspring are tested to see if they express the foreign gene, and those that do are bred with each other to produce animals homozygous for the foreign gene.

Progress in genetic engineering of livestock has been slower than in crop plants, and successes fewer. For example, when researchers tried to produce faster-growing, leaner hogs by transferring human and cow growth-hormone genes into pigs, the pigs did grow faster, but they developed serious health problems and died prematurely. Results from work with fish have been more promising. Growth-hormone genes from rainbow trout have been introduced into salmon, carp, and catfish, causing them to grow 40% faster than normal (Fig. 12-7). These fish, however, are not yet available for human consumption.

9 What Are Some Medical Uses of Biotechnology?

Biotechnology applied to the human genome promises to have a profound and growing impact on human health. One increasingly available application is the use of this technology to screen for genetic defects. Potential parents can learn if they carry an allele that causes a genetic disorder; whether a fetus will develop a genetic disorder can be diagnosed early in a pregnancy (see "Health Watch: Prenatal Genetic Screening"). In the following sections we discuss some other important medical applications of biotechnology for today and tomorrow.

Genetically Engineered Microorganisms Produce Therapeutic Proteins

Organisms can be genetically engineered to produce human proteins that have medicinal uses. Single-celled organisms are especially useful for this purpose. Large quantities of the medicinal human proteins are generated in giant vats of recombinant bacteria or yeast that express human genes. The first protein to be manufactured using this approach was insulin. Before 1982, when recombinant human insulin was first licensed for use, diabetics used insulin extracted from the pancreas of cattle or pigs slaughtered for meat. Although these insulin proteins are very similar to human insulin, slight differences caused an allergic reaction in about 5% of diabetics. This problem is avoided with the recombinant human insulin.

Animals Can Also Be Genetically Modified to Produce Human Proteins

Cows, goats, pigs, rabbits, rats, and sheep have also been used to produce therapeutic human proteins. For example, the human gene for the protein alpha-1-antitrypsin

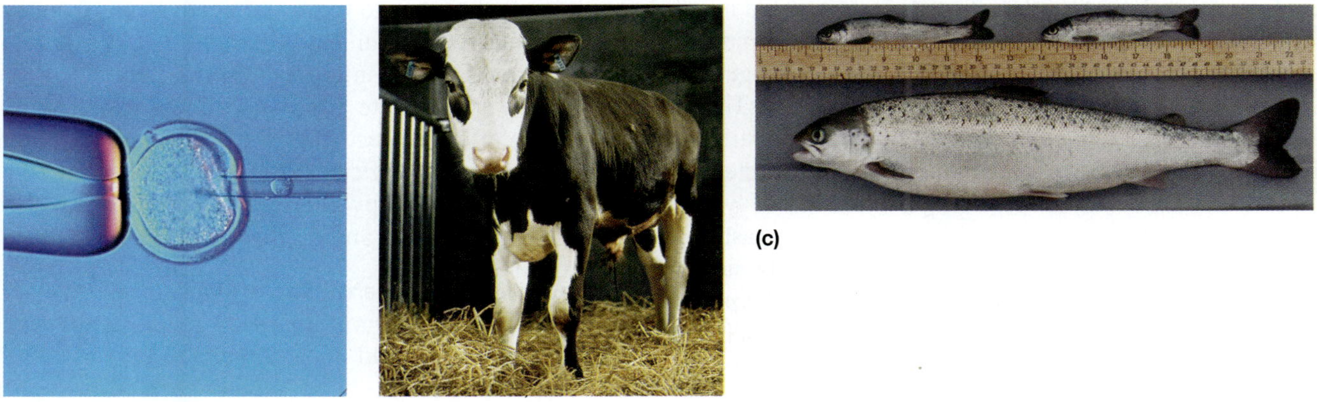

(a) (b) (c)

Figure 12-7 Transgenic animals
(a) A fertilized egg is injected with cloned DNA. Afterward, the egg will be transferred into a surrogate mother (if it is a mammal egg) or allowed to develop (if it is a fish egg). ***(b)*** This transgenic calf has been genetically engineered to secrete lactoferrin in its milk. Lactoferrin is a protein normally found in human breast milk, making this cow's milk better for human babies. ***(c)*** A gene for growth hormone has been genetically engineered into this Atlantic salmon (shown with two normal siblings).

Table 12-2 Examples of Currently Available Products Produced by Recombinant DNA Methods

Type of Product	Purpose	Typical Method of Genetic Engineering
Human hormones	Used in treatment of diabetes, growth deficiency	Human gene inserted into bacteria
Human cytokines (regulators of immune system function)	Used in bone marrow transplants and to treat cancers and viral infections, including hepatitis and genital warts	Human gene inserted into yeast
Antibodies (immune system proteins)	Used to fight infections, cancers, diabetes, organ rejection, and multiple sclerosis	Recombinant antibody genes inserted into cultured hamster cells
Viral proteins	Used to generate vaccines against viral diseases and for diagnosing viral infections	Viral gene inserted into yeast
Enzymes	Used in treatment of heart attacks, cystic fibrosis, and other diseases, and in production of cheeses and detergents	Human gene inserted into cultured hamster cells

(AAT) has been introduced to sheep, who then secrete AAT into their milk—as much as 1 ounce of protein per quart of milk. AAT is an important treatment for people who have inherited a defective allele that prevents normal production of the protein; these individuals are at high risk for a range of serious lung and liver diseases. AAT from transgenic sheep has not yet been approved for medical use but, as of early 2002, testing for safety and effectiveness had entered its final phase. Some of the other human proteins produced by recombinant DNA technology are listed in Table 12-2.

Transgenic Plants Can Deliver Vaccines

Transgenic plants can be ready-made vehicles for medicinal proteins, because many plants produce ready-to-eat tissues. This feature of plants has stimulated research into edible vaccines. For example, scientists have inserted into potatoes genes for proteins that stimulate an immune response in humans. When eaten raw, these transgenic potatoes confer resistance to a particular disease. Thus far, potato vaccines have been developed for hepatitis, cholera, and a strain of *Escherichia coli* (*E. coli*) bacteria that causes serious diarrhea. Attempts are under way to use bananas instead of potatoes, as bananas are tastier than raw potatoes and can be fed to infants. A banana vaccine against the dangerous *E. coli* strain would be particularly valuable in developing nations, where millions of infants and children die each year of diarrhea that results from *E. coli* infections.

Human Gene Therapy Is Just Beginning

Defective alleles of different single genes cause more than 4000 diseases in humans, including cystic fibrosis, sickle-cell anemia, and Huntington disease. Many more diseases, such as cancer, heart disease, arthritis, and asthma, are caused or worsened by impaired functioning of multiple genes. Gene therapy, which seeks to treat ailments by replacing defective genes with functional ones, offers potential treatments for these diseases, as well as for infectious diseases such as AIDS. Thus, the promise of human gene therapy is enormous. The challenges are equally large.

Gene Insertion Is a Difficult Task

One major challenge to effective gene therapy is to deliver functional genes into a large number of the patient's cells in a way that ensures that the genes will be expressed, producing the functional protein when and where it is needed. Two main approaches are used. One approach is to introduce gene-carrying vectors directly into appropriate sites in the body. The second and more common approach is to remove defective cells from the patient, culture them in test tubes with vectors that deliver the normal gene into the cells, and then return the cells to an appropriate site in the body. In either approach, an effective vector is essential. A variety of different viruses and artificial structures have been used as vectors for gene therapy treatments. Each has proved to have both advantages and disadvantages.

Despite Intensive Research, Success Remains Elusive

In the United States, more than 400 gene therapy trials were under way or had been completed by early 2002. But only a few patients have actually been permanently cured by gene therapy. Perhaps the greatest gene-therapy success to date has been in treating children with a rare genetic disorder called SCID (severe combined immunodeficiency). Children with SCID cannot produce an enzyme called adenosine deaminase (ADA). Without ADA, individuals cannot develop a functioning im-

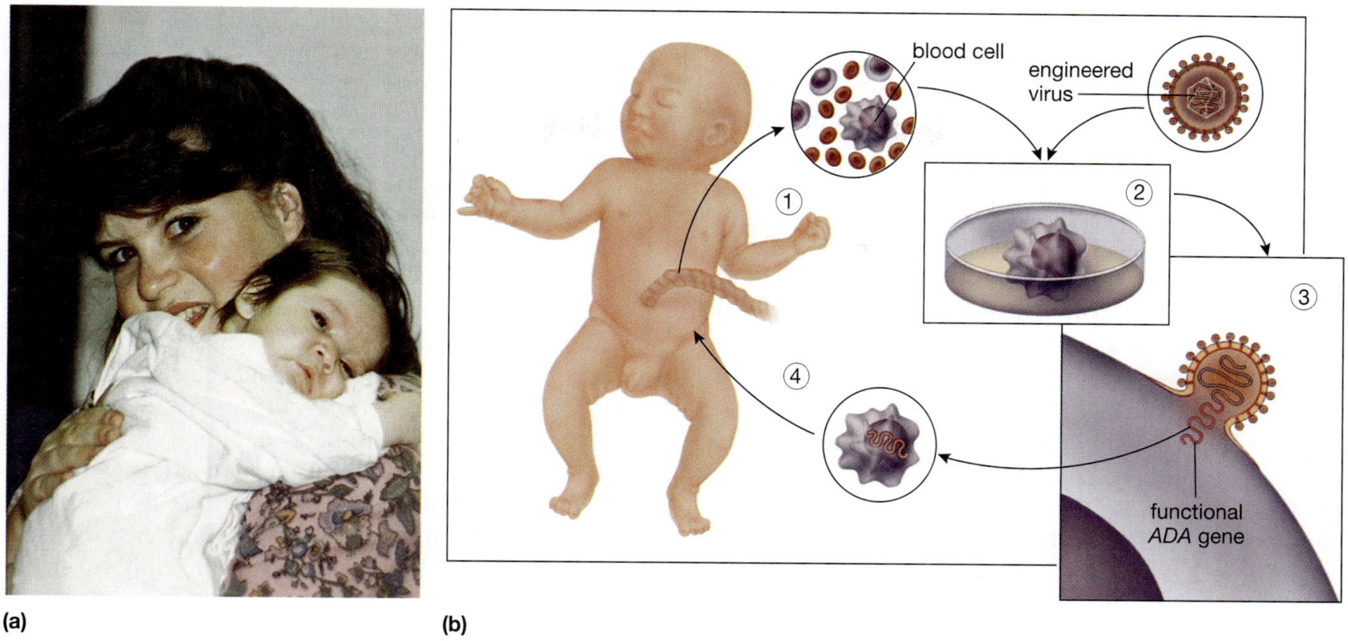

Figure 12-8 Hope through gene therapy
(a) This infant received gene therapy to treat the genetic disease SCID (severe combined immunodeficiency). Eventually, he may produce adequate numbers of normal immune cells on his own. ***(b)*** Genetic engineering of stem cells may permanently replace the defective *ADA* allele. ① Cells are harvested from umbilical cord blood. ② A virus engineered to contain normal human *ADA* alleles is mixed with the stem cells in culture. ③ The virus transmits its DNA, including the normal *ADA* allele, into the cells. ④ The engineered cells are injected into the same newborn to take up residence in bone marrow and produce normal immune cells.

mune system, are unable to fight infections, and die in infancy or early childhood unless treated with regular injections of ADA.

Since 1990, at least 18 children have received gene therapy treatments for SCID. In the treatment, some white blood cells are removed, genetically altered with a virus vector containing a functional ADA gene, and returned to the bloodstream (Fig. 12-8). Doctors believe that two of the treated children have been permanently cured by this therapy.

The limited success of gene therapy has been disappointing, and misgivings about its value grew when a patient died as a direct result of his participation in a gene therapy trial. In 1999, 18-year-old Jesse Gelsinger suffered a fatal immune reaction to a virus used in a gene therapy trial. This fatality disheartened researchers and caused the National Institutes of Health to increase scrutiny of all gene therapy trials. Nevertheless, most researchers in the field remain optimistic that gene therapy will ultimately provide cures for currently incurable diseases.

A Working Draft of the Entire Human Genome Has Been Completed

The Human Genome Project was launched in 1990 by the National Institutes of Health (NIH) and the Department of Energy (DOE). The project is an international effort to determine the sequence of the three billion or so nucleotides of the human genome, and to identify every human gene (current estimates suggest that humans have between 30,000 and 60,000 genes). Initially, the time, effort, and cost (an estimated $3 billion) of sequencing the human genome seemed daunting, but technological improvements made the goal attainable much more rapidly than researchers originally imagined possible. In February 2001 two groups of scientists involved in the Human Genome Project published independent working drafts of the human genome sequence. The final, confirmed listing of all three billion base pairs should be completed in 2003.

The Human Genome Project is probably science's most ambitious response to the command of the ancient Oracle at Delphi: "Know thyself." The answers to a host of questions about humankind reside in the nucleotide sequences of our genes. Knowing the nucleotide sequence of the entire human genome will eventually allow rapid screening and prenatal diagnosis of genetic disorders and better therapies for many diseases.

10 Will Biotechnology Create a Real Jurassic Park?

The 1990s saw the publication of some astonishing reports by scientists who claimed to have recovered DNA from ancient fossilized organisms. Among the more spectacular pronouncements were claims that DNA had been isolated from a 15-million-year-old magnolia leaf, a 25-million-year-old termite, a 40-million-year-old bee, and a 120-million-year-old weevil. Since the publication of those claims, however, other researchers have repeatedly tried and failed to extract DNA from similar specimens, and the original findings have become controversial. Many scientists suspect that original researchers' samples simply became contaminated with modern DNA (when samples are amplified by PCR, even a single stray DNA molecule can be copied millions of times). For now, it seems that recovery of DNA from truly ancient organisms such as dinosaurs must remain an intriguing but unproved possibility.

Although recovery of DNA that is millions of years old may be unlikely, researchers do seem to have successfully extracted DNA from organisms that have been dead for as long as tens of thousands of years. For example, recent analyses of DNA from 30,000-year-old fossils of Neanderthals (extinct relatives of modern humans) have been replicated in several different labs and have withstood scientific scrutiny.

Given that some fairly old DNA has been recovered, should we expect that scientists will soon be able to recreate long-extinct species? Probably not. Even if the DNA of a wooly mammoth (or a dinosaur, for that matter) were recovered, it is likely that such old DNA would be too damaged and fragmented for a complete reconstruction of the organism's genes. Some scientists, however, do plan to use biotechnology to revive some more recently extinct organisms (see "Earth Watch: Cloning Endangered Species").

11 What Are Some Ethical Implications of Biotechnology?

The rapid advance of biotechnology is delivering to humans new powers and abilities for which society is, in many ways, ill prepared. The debate over genetically engineered foods is one instance that illustrates the complexity and contentiousness of the issues that biotechnology has forced upon us. Another dilemma stems from our newfound ability to screen individuals for genetic diseases and defects. Physicians will soon have access to dozens of easily administered tests that will reveal whether a patient carries a specific gene. These new powers will require that new decisions, both ethical and economic, be made by patients, physicians, and society.

A Genetic Test for Cystic Fibrosis Illustrates Potential Problems

Cystic fibrosis (CF) is the most common fatal genetic disease in the United States. Approximately 850 babies are born in the United States each year with cystic fibrosis, most to families with no known history of this disease. To reduce the number of babies afflicted with CF, an advisory panel of the National Institutes of Health recommended that testing for CF be offered to all couples who consider having a baby. If this policy is adopted, many couples who have never heard of CF will learn that they have a 1 in 4 chance of producing a child with the disease because both partners carry a CF allele. Will these couples get the medical information and counseling they need to deal with the results? They will need to be educated about the symptoms, treatment, and monetary cost of CF, and about the emotional difficulty of the "genetic lottery" they are about to enter. Should they remain childless or adopt? Should they conceive and then use prenatal genetic diagnosis to determine whether their child will develop CF? If tests are positive, then what? Is abortion acceptable to the couple? If not, they will give birth to a child who will struggle for breath and be hospitalized repeatedly for respiratory infections despite daily doses of antibiotics. The child will need to lie down and be pounded thoroughly on the chest and back twice a day to dislodge the thick buildup of mucus and will require a special diet fortified with digestive enzymes. Despite these treatments, their child will have a 50% chance of dying before he or she reaches age 30. The costs will be staggering. Should society bear these costs for a family who knowingly takes this risk? Are insurance companies justified in denying coverage for CF treatment to the children of parents who knew before having children that they both carry the CF allele?

The increasing availability of tests for genetic diseases also raises major concerns about genetic discrimination by insurance companies and employers. Many people worry that, under pressure to hold down rising costs, insurance companies will demand information about an applicant's genetic makeup and will use this information as a basis for denying or restricting the availability of life insurance or health insurance. Insurance companies have already terminated or denied health insurance based on required genetic tests. As more genetic information becomes available and treatments become more costly, this trend is sure to intensify.

The Potential to Clone Humans Raises Further Ethical Issues

The cloning of Dolly the sheep (see "Scientific Inquiry: Much Ado About Dolly" in Chapter 10) led to a flurry of debate and proposed legislation regarding human cloning. There are no known technical barriers to the eventual development of human cloning, but many people believe that ethical barriers make human clones unthinkable. One argument against cloning humans is that there is no valid reason to do so. Infertile couples now have many other options for conceiving a

Earth Watch
Cloning Endangered Species

Environmentalists tend to be skeptical of claims that biotechnology can help solve environmental problems, but one group of scientists is determined to apply bioengineering to conservation problems. Researchers at the biotech company Advanced Cell Technologies (ACT) have embarked on an ambitious plan to clone species that are in danger of extinction. In early 2001, they had their first success with the birth of a gaur, a wild ox native to India and southeast Asia (Fig. E12-3). Gaurs are very rare and in danger of extinction, as their habitat is disappearing rapidly.

The gaur was cloned using a preserved skin cell from an individual that had died 8 years earlier. The genetic material from this cell was inserted into a cow egg from which the nucleus had been removed (see "Scientific Inquiry: Much Ado About Dolly" in Chapter 10). The resulting cloned embryo was implanted in the uterus of a surrogate mother cow. (Cows were chosen to be the egg donor and surrogate mother because of easy availability and to avoid posing undue risk to any gaur.) The embryo developed properly and the cow gave birth to the gaur calf. Unfortunately, the calf died 2 days later of dysentery, an infection that the researchers say was unrelated to the cloning procedure.

Buoyed by their first success, the cloning team plans to move its project forward. They recently received permission from the Spanish government to clone the bucardo, a mountain goat species of Europe that is already extinct (the last captive individual recently died). The scientists from ACT have spoken of being especially inspired by the idea of resurrecting an extinct species (they will use preserved bucardo tissues that are in storage in Spain). The researchers have also focused on the endangered and charismatic giant panda, as well as endangered big cats such as tigers, cheetahs, and leopards. The techniques for cloning such species, however, have not yet been developed. Unlike gaurs and bucardos, these species are not closely related to the few types of domestic animals that have already been cloned.

The idea of cloning endangered species is highly controversial among conservationists. Many feel that producing new animals one at a time can do little to truly help endangered species, and that in any case the ability to artificially reproduce members of a species does little good if the species' habitat has been destroyed. Advocates reply that cloning may be a way to keep a species alive until efforts to restore its habitat are successful, and to preserve genetic diversity by saving the genotypes of the most-threatened populations of a species. Many of those who wish to continue efforts to clone endangered species argue that it doesn't make sense to abandon any approach to conserving species, as we will need more than one strategy to solve such a difficult problem.

Figure E12-3 The gaur, an endangered Asian ox, was recently cloned

child, so where's the pressing need for clones? Cloning also seems to raise the possibility that unscrupulous individuals in positions of power would use the technology to perpetuate themselves and their grip on power. Perhaps the most widespread argument against human cloning is a deeply felt sense that biological uniqueness is a fundamental human right and intrinsic to human dignity.

Countering these arguments are those who say that we should not hamper efforts to develop human cloning technology, because medical advances of great benefit to humanity could arise from efforts to clone human tissues and organs. Cloning proponents also point out that making precise copies of humans is impossible, even with cloning technology. After all, identical twins are more similar to one another than clones would ever be, because twins share not only the same genes, but the same original egg cell, the same uterus, and the same home environment and upbringing. An individual trying to produce a copy of him- or herself might be sorely disappointed at the differences produced by differences in these environmental factors.

As governments struggle to decide whether and how much to restrict cloning and related techniques, one fact is painfully clear: Advances in our ability to manipulate the genes of humans and other species threaten to outpace the ability of society to understand and use the resulting technologies effectively. We can expect continuing upheaval as society struggles to come to grips with this new knowledge and the power that accompanies it.

CASE STUDY REVISITED

Food Fight

To many people, the improvements gained through agricultural biotechnology represent advances that are crucial to human well-being. Robert Shapiro, a former executive of Monsanto, a company that sells bioengineered seeds, believes that biotechnology is necessary to feed a growing population. He asks, "How are you going to feed 10 to 12 billion people a decent diet when in fact there will not only be no more acreage available to produce that food, there may very well be less?" To Shapiro, the answer is obvious. But critics of transgenic foods are equally certain that the potential benefits of biotechnology are dwarfed by the risks.

The critics worry that introducing new genes to food may also inadvertently introduce new substances that are toxic or that cause allergic reactions. They are concerned that herbicide-resistant crops encourage farmers to use excessive amounts of dangerous herbicides. They believe that farmers will come to regret their increasing dependence on the large corporations that produce bioengineered seeds. They also fear that bioengineered organisms could escape from farm fields, establish a foothold in the wild, and disrupt ecosystems.

Another concern is that bioengineered crops will foster the evolution of resistant weeds and pests. If, for example, farmers who grow bioengineered, herbicide-resistant crops do indeed apply larger amounts of herbicide, any weeds with mutations for herbicide-resistance will survive and prosper. Eventually, herbicide-resistant "super weeds" will predominate and herbicides will ultimately become much less effective. Similarly, widespread cultivation of plants containing the Bt toxin could favor the evolution of insects that tolerate or inactivate the toxin, rendering it useless.

Are the critics right to worry? Some scientists disparage the critics as doomsayers and alarmists who needlessly inflame the public. The U.S. Food and Drug Administration has announced that genetically modified foods "are safe for introduction into the marketplace." The critics' warnings, however, do resonate with the public, especially in Europe, where many stores and restaurants refuse to sell genetically modified foods and where public concern has limited farmers' willingness to plant bioengineered seeds.

In the future, you will be increasingly called on to add your voice to the debate about biotechnology. We hope that this chapter has helped ensure that your views, be they pro or con, are well informed.

Summary of Key Concepts

1) What Is Biotechnology?

Biotechnology is any use or alteration of organisms, cells, or biological molecules to achieve specific practical goals. Modern biotechnology alters genetic material by means of genetic engineering. In many applications of genetic engineering, recombinant DNA is produced by combining DNA from different organisms. DNA is transferred between organisms by vectors, such as bacterial plasmids, to create transgenic organisms. Some major goals of genetic engineering are to increase our understanding of gene function, to treat disease, and to improve agriculture.

2) How Does DNA Recombine in Nature?

DNA recombination occurs naturally through processes such as bacterial transformation, viral infection, and crossing over during sexual reproduction.

3) How Is Recombinant DNA Made in the Lab?

DNA is cut into specific fragments by restriction enzymes. The fragments are inserted into a vector, which allows the DNA to be replicated within a host cell. The joining of a DNA fragment and a vector produces recombinant DNA molecules.

4) How Can Researchers Identify Specific Genes?

Researchers often identify specific genes within a library of recombinant DNA molecules by using DNA probes, which are complementary to the sequence being sought. Probes usually include chemical or radioactive labels, so that their presence can be detected.

5) How Can Researchers Make Many Copies of a Gene?

Once a gene has been identified, most applications of biotechnology require that the gene be amplified. The most popular way to make many copies of a DNA segment is the polymerase chain reaction (PCR), which uses a heat-resistant DNA polymerase to synthesize DNA in a test tube.

6) Why Is It Useful to Know the Sequence of a Gene?

Among the benefits that can stem from knowing the nucleotide sequence of a gene are the ability to synthesize and study the encoded protein and the ability to devise tests for defective alleles.

7) What Is DNA Fingerprinting?

DNA fingerprinting can establish the relatedness of individuals. The procedure is based on analysis of restriction fragment length polymorphisms; different individuals are characterized by different fingerprints (fragment patterns). DNA fingerprinting is used in forensics to analyze biological material at a crime scene and link it to suspects.

8) What Are Some Agricultural Uses of Biotechnology?

Biotechnology is revolutionizing agriculture by allowing scientists to engineer plants that resist pests and herbicides and that have superior qualities for shipping and eating.

9) What Are Some Medical Uses of Biotechnology?

Genetically engineered therapeutic proteins, such as human insulin, are produced by growing huge quantities of bacteria, yeast, or mammalian cells that contain recombinant genes and then extracting the protein. Farm animals can also be genetically engineered to secrete proteins of interest in milk. Human gene therapy is in its infancy. Vectors such as viruses have been used to deliver functional genes to specific cells that lack them. This technique has been used successfully in treating children with SCID. An ambitious collaborative project to sequence the entire human genome is nearing completion.

10) Will Technology Create a Real Jurassic Park?

Although DNA has been extracted from preserved organisms (including humans) that are tens of thousands of years old, the DNA is so degraded that cloning whole organisms from it will most likely be impossible, even if we do manage to invent the technology for doing so. Most scientists regard with great skepticism past claims that DNA was extracted from fossils that are tens of millions of years old.

11) What Are Some Ethical Implications of Biotechnology?

To administer tests for genetic diseases appropriately, the physician or genetic counselor must decide who should receive the tests, and then educate the patients about the disorder and assist them in making a series of difficult decisions about their own health or that of potential offspring. Questions include whether to have prenatal testing, whether to conceive a child, and whether to abort a fetus with a fatal or permanently debilitating disorder. Should insurance companies pay the lifelong costs to treat children with genetic diseases born to parents who knew they carry the defective allele? How can society protect individuals from various types of discrimination on the basis of their genetic makeup? Should research on human cloning techniques be permitted? Both the rapid acquisition of knowledge about genetics and the power to manipulate the genome of humans and other organisms threaten to outpace our ability to deal with the consequences.

Key Terms

biotechnology *p. 188*
DNA fingerprint *p. 192*
DNA library *p. 190*
DNA probe *p. 193*
gel electrophoresis *p. 193*
genetic engineering *p. 188*
genome *p. 189*
host cell *p. 189*
plasmid *p. 189*
polymerase chain reaction (PCR) *p. 191*
recombinant DNA *p. 188*
restriction enzyme *p. 190*
restriction fragment length polymorphism (RFLP) *p. 193*
transformation *p. 189*
transgenic *p. 188*
vector *p. 190*
virus *p. 189*

Thinking Through the Concepts

Multiple Choice

1. *Restriction enzymes are*
   a. isolated from bacterial cells
   b. used to produce DNA fingerprints
   c. used to create recombinant plasmids
   d. used to create a DNA library
   e. all of the above
2. *Restriction fragment length polymorphisms*
   a. can be used to detect differences in DNA among individuals
   b. have been used to identify human genes
   c. can be used to analyze fetal cells to detect disorders prenatally
   d. are produced through the use of restriction enzymes
   e. all of the above
3. *The polymerase chain reaction*
   a. is a method of synthesizing human protein from human DNA
   b. takes place naturally in bacteria
   c. can produce billions of copies of a DNA fragment in several hours
   d. uses restriction enzymes
   e. is relatively slow and expensive compared with other types of DNA purification
4. *DNA fingerprinting*
   a. requires large amounts of DNA
   b. is useful only for forensic analysis
   c. may include analysis of RFLPs
   d. can only prove guilt, never innocence
   e. is a constitutional right
5. *Genetic engineers often use ________ to insert foreign DNA into a genome.*
   a. bacteria
   b. unfertilized egg cells
   c. *Bacillus thuringiensis*
   d. viruses
   e. all of the above

? Review Questions

1. Describe three natural forms of genetic recombination, and discuss the similarities and differences between recombinant DNA technology and these natural forms of genetic recombination.
2. What is a plasmid? How are plasmids involved in bacterial transformation?
3. What is a restriction enzyme? How can restriction enzymes be used to splice a piece of human DNA within a plasmid?

4. What is a DNA library? Briefly describe the steps for creating a DNA library of a mouse genome.
5. What is a restriction fragment length polymorphism (RFLP)? Describe how it might be useful in proving the innocence of a murder suspect or determining relatedness among people.
6. Describe the polymerase chain reaction.
7. Describe several uses of genetic engineering in agriculture.
8. Describe several uses of genetic engineering in human medicine.
9. Describe amniocentesis and chorionic villus sampling, including the advantages and disadvantages of each. What are their medical uses?

Applying the Concepts

1. Discuss the ethical issues that surround the release of bioengineered organisms (plants, animals, or bacteria) into the environment. What could go wrong? What precautions might prevent the problems you listed from occurring? What benefits do you think would justify the risks? Do you think some agricultural applications of biotechnology are cause for concern but not others? How would you assess the risk that bioengineered organisms might harm the environment or harm humans who consumed them as food?
2. Do you think that using recombinant DNA technologies to change the genetic composition of a human egg cell is ever justified? If so, what restrictions should be placed on such a use? What about human cloning?
3. Are there any conditions under which insurance companies or employers should be allowed access to the results of tests for genetic defects? Discuss this issue from the standpoint of the company or employer, the individual who was tested, and society as a whole.
4. If you were contemplating having a child, would you want both yourself and your spouse tested for the cystic fibrosis allele? If both of you were found to carry this allele, how would you deal with this decision?

Group Activity

DNA evidence is often used to identify individuals in criminal investigations and paternity suits. The uniqueness of each person's DNA results in each person having a unique DNA fingerprint. The diagram of a gel on the right shows DNA evidence from a paternity suit. A child's DNA fingerprint is a mixture of the father's and mother's DNA fingerprints. In groups of three, answer the questions concerning the DNA evidence in this paternity suit.

(a) Would you expect the child's DNA fingerprint to contain all the fragments from either parent? Why or why not?
(b) Which male(s) cannot be the father of this child?
(c) Which male(s) might be the father?
(d) Does this evidence prove that one of these men is the child's father? Explain.
(e) Does this evidence prove that one or more of these men could not be the father? Explain.

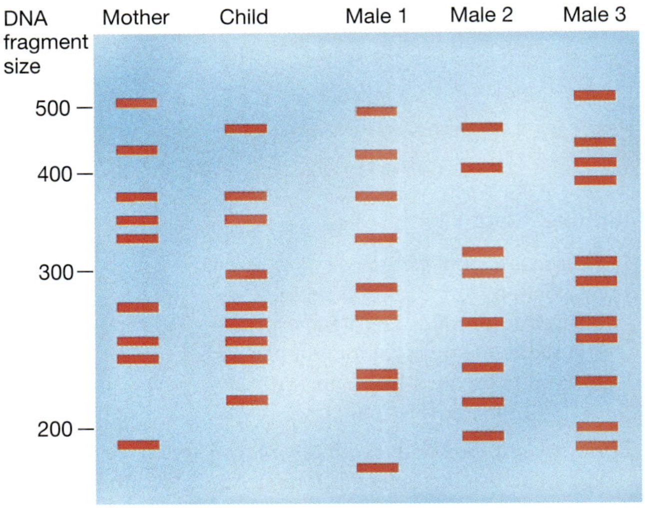

For More Information

Beardsley, T. "Vital Data." *Scientific American*, March 1996. Discusses the Human Genome Project and the problems and prospects of the tests for genetic diseases that are arising as a spin-off of this project.

Collins, F. S., and others. "New Goals for the U.S. Human Genome Project: 1998–2003." *Science*, October 1998. Explains goals of human genome project, including those involving bioethics.

Fackelmann, K. A. "DNA Dilemmas." *Science News*, December 17, 1994. Uses case histories to explore the ethical controversies surrounding human applications of genetic engineering.

Gibbs, N. "Baby, It's You! And You, and You. . . ." *Time*, February 19, 2001. How close are we to cloning a human?

Gibbs, W. W. "Plantibodies." *Scientific American*, November 1997. Clinical trials using human antibodies engineered into corn began in 1998; the target was tooth decay!

Gura, T. "New Genes Boost Rice Nutrients," *Science*, August 1999. Explanation of how rice was genetically engineered to produce vitamin A precursors.

Karapelou, J. "Gene Therapy: Special Delivery." *Discover*, January 1996. Describes several methods on trial for delivering therapeutic genes into human cells.

Langridge, W. H. R. "Edible Vaccines." *Scientific American*, September 2000. Current state of attempts to develop plants that produce vaccines or treatments for diseases.

Lanza, R. P., Dresser, B.L., and Damiani, P. "Cloning Noah's Ark." *Scientific American*, November 2000. Using domestic animals as surrogate mothers for endangered species provides some hope for staving off extinction.

Marvier, M. "Ecology of Transgenic Crops." *American Scientist*, March/April 2001. A thoughtful article that weighs the benefits, risks, and uncertainties of bioengineered crops.

Miller, R. V. "Bacterial Gene Swapping in Nature." *Scientific American*, January 1998. How likely is it that genes introduced into bioengineered organisms might be inadvertently transferred to wild organisms?

Mirsky, S., and Rennie, J. "What Cloning Means for Gene Therapy." *Scientific American*, June 1997. Could human cloning technology help permanently repair genetic defects?

Roberts, L. "To Test or Not to Test." *Science*, January 1990. Should the entire childbearing population be screened for carriers of cystic fibrosis?

Ronald, P. C. "Making Rice Disease-Resistant." *Scientific American*, November 1997. Rice, a staple crop for millions of people, has finally been engineered for disease resistance.

Scientific American, June 1997. A special issue devoted to the prospects of human gene therapy.

Travis, J. "Cystic Fibrosis Controversy." *Science News*, May 10, 1997. Might gene therapy in the womb help CF patients?

Answers to Multiple-Choice Questions

1. e 2. e 3. c 4. c 5. d

MediaTutor

CD Activities

Activity 12.1: Genetic Recombination

Estimated time: 10 minutes

We often think of DNA as being static and unchanging. In fact, it is anything but static. This tutorial will demonstrate some of the ways that genetic recombination can occur in bacteria. As you will see, there are many opportunities for bacteria to exchange genetic material with other bacteria. What are the consequences of these exchanges? Does this happen in other organisms too?

Activity 12.2: Building a DNA Library

Estimated time: 10 minutes

Biotechnology uses modern technology to modify biological molecules. For example, biotechnology enables scientists to produce certain molecules that are needed by humans. Often, this technique can be used to compensate for an individual's inability to produce these molecules. An important step in this technique is to find the particular gene in question. The traditional way to find the gene is to create a DNA library containing all of the DNA of an organism, but in small pieces. This tutorial will lead you through the steps involved in producing a DNA library from a daffodil.

Start the MediaTutor Student CD-ROM and enter the activity number in the Quick Search box to be taken directly to that activity.

Web Investigation

Case Study: Food Fight

Estimated time: 20 minutes

The application of biotechnology in the development of medicine, agriculture, and even forensics is discussed in the chapter. Case Study: Food Fight highlights some of the issues, including consumer concerns associated with the development and sale of bioengineered food products. This activity will provide additional information on these topics and reveal different perspectives on the implementation of these new technologies.

Go to http://www.prenhall.com/audesirk3, the Audesirk Companion Web site. Select Chapter 12 and Web Investigation to begin.

UNIT THREE

Evolution

The ghostly grandeur of ancient bones evokes images of a lost world. Fossil remnants of extinct creatures, such as this Triceratops *dinosaur skeleton, provide clues to the biologists who attempt to reconstruct the history of life.*

"Nothing in biology makes sense except in the light of evolution."

Theodosius Dobzhansky

Newly discovered fossils of feathered dinosaurs such as *Caudipteryx* (shown here in an artist's reconstruction) provide powerful evidence that today's birds descended from dinosaur ancestors.

13 Principles of Evolution

AT A GLANCE

CASESTUDY

A Missing Link Unearthed

As the twentieth century drew to an end, some of history's most breathtaking fossil discoveries were made in the little village of Sihetun in northeastern China. Working in dusty quarries around the village, Chinese fossil hunters extracted the exquisitely preserved remains of some previously undiscovered types of dinosaurs. New dinosaur discoveries are always cause for celebration, but the new Chinese specimens displayed a distinctive feature that propelled them to superstar status: They had feathers. Plainly visible along the margin of these clearly dinosaurian fossil skeletons were impressions of what were unmistakably feathers. For the first time, scientists had hard evidence of the existence of feathered dinosaurs.

As descriptions of the new fossils were published in 1998 and 1999, surprise gave way to delight among many paleontologists (the scientists who study fossils). A flurry of news stories alerted the public to the exciting new evidence. Why such a big fuss over feathers? Because the new discoveries seem to confirm the controversial theory that dinosaurs were the ancestors of the birds that today flit about the trees in our backyards and flock to our feeders. The fossils are evidence that modern birds arose from a branch of the dinosaur family tree—a branch whose members had feathers. That is, the fossils show that today's birds arose through **evolution**, the process by which the characteristics of a population change over time. ■

1) How Did Evolutionary Thought Evolve?

Modern biology is based on our understanding that life has evolved, but this fundamental principle was not recognized by early scientists. The main ideas of evolutionary biology developed over a long period and eventually received a large boost from the work of Charles Darwin.

Early Biological Thought Did Not Include the Concept of Evolution

Pre-Darwinian science, heavily influenced by theology, held that all organisms were created simultaneously by God and that each distinct life-form remained fixed and unchanging from the moment of its creation. This explanation of how life's diversity arose was elegantly expressed by the ancient Greek philosophers, especially Plato and Aristotle. Plato (427–347 B.C.) proposed that each object on Earth was merely a temporary reflection of its divinely inspired "ideal form." Plato's student Aristotle (384–322 B.C.) categorized all organisms into a linear hierarchy that he called the "ladder of Nature."

These ideas formed the intellectual basis of the view that the form of each type of organism is permanently fixed. This view reigned unchallenged for nearly 2000 years. By the eighteenth century, however, several lines of newly emerging evidence began to erode the dominance of this static view of creation.

Exploration of New Lands Revealed a Staggering Diversity of Life

As early European naturalists explored the newly discovered lands of Africa, Asia, and America, they found that the number of species, or different types of organisms, was much greater than anyone had suspected. Naturalists also observed that some species closely resembled one another, yet also differed in some characteristics. These observations led some naturalists to consider that perhaps species could change after all. Perhaps some of the similar species might have developed from a common ancestor.

Fossil Discoveries Showed That Life Has Changed over Time

At the same time, excavations for roads, mines, and canals revealed that many rocks occur in layers (Fig. 13-1). In some cases, impressions or fragments that resembled parts of living organisms were found embedded within one of these layers. At first, these **fossils** were thought to be ordinary rocks that wind, water, or people had worked into lifelike forms. As more and more fossils were discovered, however, it became obvious that they were the remains or impressions of plants or animals that had died long ago and been changed into or in some way preserved in rock (Fig. 13-2).

Figure 13-1 The Grand Canyon of the Colorado River
Layer upon layer of sedimentary rock forms the walls of the Grand Canyon. The layers formed from the accumulation of sediments over more than a billion years.

How the fossils were distributed in the rock was also revealing. After studying fossils carefully, British surveyor William Smith (1769–1839) realized that certain fossils were always found in the same layers of rock. Further, the organization of fossils and rock layers was consistent: Fossil type A could always be found in a rock layer resting atop an older layer containing fossil type B, which in turn rested atop a still-older layer containing fossil type C, and so on. (Because the layers are formed from sediments deposited on top of one another, the lowest layers are the oldest, and the ones on top are progressively younger.)

Fossil remains also showed a remarkable progression. Most fossils found in the oldest layers were very different from modern organisms; the resemblance to modern organisms gradually increased upward in younger rocks. Many of the fossils were from plant or animal species that had gone *extinct*—that is, no members of the species still lived on Earth (Fig. 13-3). Putting these facts together, scientists came to the inescapable conclusion that different types of organisms had lived at various times in the past.

Some Scientists Devised Nonevolutionary Explanations for Fossils

What did this newfound richness of organisms mean? To account for the multitude of species, both living and extinct, while preserving the notion of creation by God, Georges Cuvier (1769–1832) proposed the theory of **catastrophism**. Cuvier, a French paleontologist, hypothesized that a vast supply of species was created initially. Successive catastrophes (such as the Great Flood described in the Bible) produced layers of rock and destroyed many species, fossilizing some of their remains in the process. The organisms of the modern world, he theorized, are the species that survived the catastrophes.

If Cuvier were correct in believing that modern species have survived from an original creation, then many individuals of the surviving species should have died in the ancient catastrophes. Surely some of these would have been fossilized, and even the oldest layers should contain fossils of present-day species. Contrary to Cuvier's hypothesis, however, the vast majority of fossils are of extinct species. To account for this observation, French geologist Louis Agassiz (1807–1873) proposed that there was a new creation after each catastrophe and that modern species result from the most recent creation. The fossils known at that time forced Agassiz to hypothesize at least 50 separate catastrophes and creations!

Figure 13-2 Types of fossils
Any part or trace of an organism that is preserved in rock or sediments is a fossil. Examples include ① eggs, ② impressions of skin, ③ bones, ④ feces, and ⑤ footprints.

A Few Scientists Speculated That Life Had Evolved

Some scientists could not readily accept that the creator would create so many types of organisms, only to let thousands go extinct. One alternative explanation was proposed by French naturalist Georges Louis LeClerc (1707–1788), known by the title Comte de Buffon, who suggested that perhaps the original creation provided a relatively small number of founding species and that some modern species had been "conceived by Nature and produced by Time"—that is, they had evolved through natural processes. Most people were not convinced. First, Buffon could not provide a mechanism whereby Nature could "conceive" new species. Second, no one thought that Earth was old enough to allow sufficient time for the "production" of new species.

Geology Provided Evidence That Earth Is Exceedingly Old

In the early 1700s, few scientists suspected that Earth could be more than a few thousand years old. Counting generations in the Old Testament, for example, yields a maximum age of 4000 to 6000 years. An Earth this young poses problems for the idea that life has evolved. For example, ancient writers such as Aristotle describe wolves, deer, lions, and other organisms that are identical to those present in Europe more than 2000 years later. If organisms had changed so little over that time, how could whole new species possibly have arisen if Earth was created only a couple of thousand years before Aristotle's time?

One explanation, of course, is that perhaps Earth *is* old enough to allow for the production of new species. Such was the view of geologists James Hutton (1726–1797) and Charles Lyell (1797–1875), who contemplated the forces of wind, water, earthquakes, and volcanism. Hutton and Lyell concluded that there was no need to invoke catastrophes to explain the findings of geology. Don't flooding rivers lay down layers of sediment? Don't lava flows produce layers of basalt? Why, then, should we assume that layers of rock are evidence of anything but ordinary natural processes, occurring repeatedly over long periods of time? This concept, called **uniformitarianism**, had profound implications. If slow, natural processes alone are able to produce layers of rock thousands of feet thick, then Earth must be old indeed, many millions of years old. Hutton and Lyell, in fact, concluded that Earth was eternal: "No Vestige of a Beginning, no Prospect of an End," in Hutton's words. (Modern geologists estimate that Earth is about 4.5 billion years old; see "Scientific Inquiry: How Do We Know How Old a Fossil Is?" in

Figure 13-3 Fossils of extinct organisms
Fossils provide strong support for the idea that today's organisms were not created all at once but arose over time by the process of evolution. If all species were created simultaneously, we would not expect ***(a)*** trilobites to be found in older rock layers than ***(b)*** seed ferns, which in turn would not be expected in older layers than ***(c)*** dinosaurs, such as *Allosaurus.* Trilobites had gone extinct by about 230 million years ago, seed ferns by about 150 million years ago, and dinosaurs by 65 million years ago.

Chapter 15.) Thus, Hutton and Lyell provided the time for evolution to occur. But there was still no convincing mechanism.

Some Pre-Darwin Biologists Proposed Mechanisms for Evolution

One of the first scientists to propose a mechanism for evolution was French biologist Jean Baptiste Lamarck (1744–1829). Lamarck was impressed by the sequences of organisms in the rock layers. He observed that older fossils tend to be simpler, whereas younger fossils tend to be more complex and more like existing organisms. In 1801 Lamarck hypothesized that organisms evolved through the inheritance of acquired characteristics, a process in which the bodies of living organisms are modified through the use or disuse of parts, and these modifications are inherited by offspring. Why would bodies be modified? Lamarck proposed that all organisms possess an innate drive for perfection. For example, if ancestral giraffes tried to improve their lot by stretching upward to feed on leaves that grow high up in trees, their necks became slightly longer as a result. Their offspring would inherit these longer necks, and then stretch even farther to reach still higher leaves. Eventually, this process would produce modern giraffes with very long necks indeed.

Today, we understand how inheritance works and can see that Lamarck's proposed evolutionary mechanism could not work as he described it. Acquired characteristics are not inherited; the fact that a prospective father pumps iron doesn't mean that his children will look like Arnold Schwarzenegger. Remember, though, that in Lamarck's time the principles of inheritance had not yet been discovered (Mendel was born a few years before Lamarck's death). In any case, Lamarck's insight that inheritance plays an important role in evolution was an important influence on the later biologists who discovered the key mechanism of evolution.

Darwin and Wallace Proposed a Mechanism of Evolution

By the mid-1800s, a growing number of biologists had concluded that present-day species had evolved from earlier ones. But how? In 1858 Charles Darwin and Alfred Russel Wallace, working separately, provided convincing evidence that evolution was driven by a simple yet powerful process.

Although their social and educational backgrounds were very different, Darwin and Wallace were quite similar in some respects. Both had traveled extensively in the Tropics and had studied the plants and animals living there. Both found that some species differed only in a few features (Fig. 13-4). Both were familiar with the fossils that had been discovered, which showed a trend of increasing complexity through time. Finally, both were aware of the studies of Hutton and Lyell, who had proposed that Earth is extremely ancient. These facts suggested to both Darwin and Wallace that species change over time. Both men sought a mechanism that might cause such evolutionary change.

In 1858 Darwin and Wallace each described the same mechanism for evolution in remarkably similar papers that were presented to the Linnaean Society in London. Like Gregor Mendel's manuscript on the principles of genetics, their papers had little impact. The secretary of the society, in fact, wrote in his annual report that nothing very interesting happened that year. Fortunately, the next year Darwin published his monumental *On the Origin of Species by Means of Natural Selection*, which attracted a great deal of attention to the new theory.

(a) large ground finch, beak suited to large seeds

(b) small ground finch, beak suited to small seeds

(c) warbler finch, beak suited to insects

(d) vegetarian tree finch, beak suited to leaves

Figure 13-4 Darwin's finches, residents of the Galapagos Islands
The Galapagos Islands are home to a group of closely related species of finches that Darwin studied. Each species specializes in eating a different type of food and has a beak of characteristic size and shape. The finches are otherwise quite similar.

Scientific Inquiry
Charles Darwin—Nature Was His Laboratory

Charles Darwin's Voyage on the *Beagle* Sowed the Seeds for His Theory of Evolution

Like many students, Charles Darwin excelled only in subjects that intrigued him. Although his father was a physician, Darwin was uninterested in medicine and unable to stand the sight of surgery. He eventually obtained a degree in theology from Cambridge University, although theology too was of minor interest to him. What he really liked to do was to tramp over the hills, observing plants and animals, collecting new specimens, scrutinizing their structures, and categorizing them.

In 1831, when Darwin was 22 years old (Fig. E13-1), the British government was readying Her Majesty's Ship *Beagle* for a 5-year surveying expedition along the coastline of South America and then around the world. The *Beagle* was to carry a naturalist to observe and collect geological and biological specimens. Thanks to a professor's recommendation, Darwin was offered the position. The *Beagle* sailed to South America, and made many stops along its coast. There Darwin observed the plants and animals of the Tropics and was stunned by the greater diversity of species compared with that of Europe. Although he had boarded the *Beagle* convinced of the permanence of species, his experiences soon led him to doubt this. He discovered a snake with rudimentary hind limbs, calling it "the passage by which Nature joins the lizards to the snakes." Another snake he encountered vibrated its tail like a rattlesnake but had no rattles and therefore made no noise. Similarly, Darwin noticed that penguins used their wings to paddle through the water rather than fly through the air. If a creator had individually created each animal in its present form, to suit its present environment, what could be the purpose behind these makeshift arrangements?

Perhaps the most significant stopover of the voyage was the month spent on the Galapagos Islands, off the northwestern coast of South America. There Darwin found huge tortoises; in fact, *galapagos* means "tortoise" in Spanish. Different islands were home to distinctively different types of tortoises (Fig. E13-2). Darwin also found several types of finches; as with the tortoises, different islands had slightly different finches. Could the differences in these organisms have arisen after they became isolated from one another on separate islands? The diversity of tortoises and finches haunted him for years afterward.

Figure E13-1 A painting of Charles Darwin as a young man

The Big Picture

The concept of evolution is perhaps the most powerful and important idea in biology, a theme that unites and informs all aspects of the study of life. Development of this key idea, however, proceeded slowly until the work of Wallace and Darwin provided the foundation of modern evolutionary biology.

2 How Does Natural Selection Work?

Darwin and Wallace concluded that life's huge variety of excellent designs arose by a process of descent with modification, in which the members of each generation differ slightly from the members of the preceding generation, and these small changes accumulate over long stretches of time to produce major transformations. They called the process natural selection. The chain of logic leading to their powerful conclusion turns out to be surprisingly simple and straightforward. Darwin and Wallace based their theory of evolution by natural selection on four observations and the conclusions that followed naturally from those observations.

Observation 1: A natural **population**, which consists of all the individuals of one species in a particular area, has the potential to grow rapidly, because organisms can produce far more offspring than are required merely to replace the parents.

Observation 2: Nevertheless, the number of individuals in a natural population tends to remain relatively constant over time.

Figure E13-2 A Galapagos tortoise

In 1836 Darwin returned to England after 5 years on the *Beagle* and became established as one of the foremost naturalists of his time. But constantly gnawing on his mind was the problem of how isolated populations come to differ from each other. Part of the solution came to him from an unlikely source: the writings of an English economist and clergyman, Thomas Malthus. In his *Essay on Population,* Malthus wrote, "It may safely be pronounced, therefore, that [human] population, when unchecked, goes on doubling itself every 25 years, or increases in a geometrical ratio." Darwin realized that a similar principle holds true for plant and animal populations. In fact, most organisms can reproduce much more rapidly than can humans (consider rabbits, dandelions, and houseflies) and consequently could produce overwhelming populations in short order. Nonetheless, the world is not chest-deep in rabbits, dandelions, or flies: Natural populations do not grow "unchecked" but tend to remain approximately constant in size. Clearly, vast numbers of individuals must die in each generation, and most must not reproduce.

From his experience as a naturalist, Darwin realized that members of a species typically differ from one another. Further, which individuals die without reproducing in each generation is not arbitrary but depends to some extent on the structures and abilities of the organisms. This observation was the source of the theory of evolution by natural selection. As Darwin's colleague Alfred Wallace put it, "Those which, year by year, survived this terrible destruction must be, on the whole, those which have some little superiority enabling them to escape each special form of death to which the great majority succumbed." Here you see the origin of the expression "survival of the fittest." That "little superiority" that confers greater fitness might be better resistance to cold, more efficient digestion, or any of hundreds of other advantages, some very subtle. Everything now fell into place. Darwin wrote, "It at once struck me that under these circumstances favorable variations would tend to be preserved, and unfavorable ones to be destroyed." If a favorable variation were inheritable, then the entire species would eventually consist of individuals possessing the favorable trait. With the continual appearance of new variations (due, as we now know, to mutations), which in turn are subject to further natural selection, "the result . . . would be the formation of new species. Here, then, I had at last got a theory by which to work."

When Darwin finally published *On the Origin of Species* in 1859, his evidence had become truly overwhelming. Although its full implications would not be realized for decades, Darwin's theory of evolution by natural selection has become a unifying concept for virtually all of biology.

Conclusion 1: Therefore, it must be the case that more organisms are born than survive and reproduce. If some individuals fail to survive, it must also be true that organisms compete to survive and reproduce. In each generation, many individuals must die young. Even among those that survive, many must fail to reproduce, or produce few offspring, or produce less-fit offspring that fail to survive and reproduce in their turn.

Observation 3: Individual members of a population differ from one another in their ability to obtain resources, withstand environmental extremes, escape predators, and so on.

Conclusion 2: These differences among individuals help determine which individuals survive and reproduce most successfully, thereby leaving the most offspring. This process, in which those individuals whose traits are most advantageous leave a larger number of offspring, is known as **natural selection**.

Observation 4: At least some of the variation among individuals in traits that affect survival and reproduction is due to differences in characteristics that may be passed from parent to offspring.

Conclusion 3: Because the individuals that are best suited to their environment leave more offspring, the traits (and underlying genes) of these individuals are passed to a larger proportion of the individuals in subsequent generations. Over many generations, this differential, or unequal, reproduction among individuals with different genetic makeup changes the overall genetic composition of the population. This process is evolution by natural selection.

Figure 13-5 charts these observations and conclusions, and "Scientific Inquiry: Charles Darwin—Nature Was His Laboratory" tells how Darwin was able to make the observations that led to his understanding of how organisms evolve.

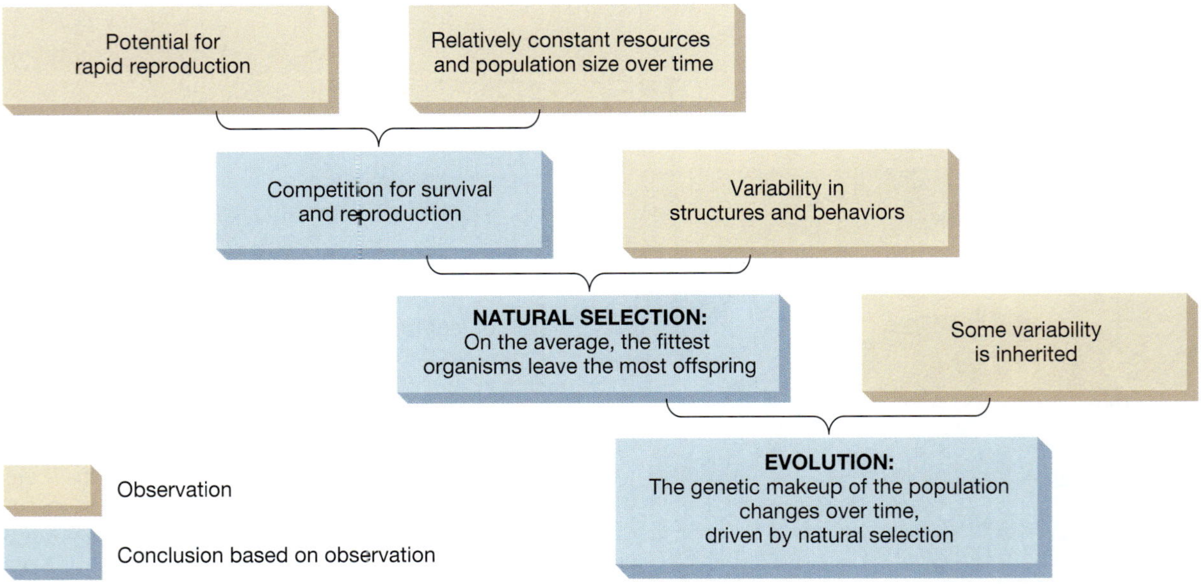

Figure 13-5 A flowchart of evolutionary reasoning
This chart is based on the hypotheses of Darwin and Wallace but incorporates ideas from modern genetics.

Modern Genetics Confirmed Darwin's Assumption of Inheritance

As you know, the principles of genetics had not yet been discovered when Darwin published *On the Origin of Species*. Observation 4, therefore, was an untested assumption for Darwin and Wallace—and thus a weakness in their theory at the time of its publication. Mendel's later work, of course, showed that Darwin was correct in assuming that particular traits can be passed to offspring. In addition, we now also know that the variations in natural populations arise purely by chance, as a result of random mutations in DNA (see Chapters 8 and 9). New variations can be good, bad, or indifferent; there is no mechanism to ensure that favorable variations arise.

Natural Selection Modifies Populations over Time

How might natural selection among chance variations change the makeup of a species? In *On the Origin of Species*, Darwin proposed the following example. "Let us take the case of a wolf, which preys on various animals, securing [them] by . . . fleetness. . . . The swiftest and slimmest wolves would have the best chance of surviving, and so be preserved or selected. . . . Now if any slight innate change of habit or structure benefited an individual wolf, it would have the best chance of surviving and of leaving offspring. Some of its young would probably inherit the same habits or structure, and by the repetition of this process, a new variety might be formed." The same argument would apply to the wolf's prey, in which the fastest or most alert would be the most likely to avoid predation and would pass on these traits to its offspring. Notice that natural selection is acting on individuals within a population. Over generations, the population would change as the percentage of its individuals inheriting favorable traits increased. An individual cannot evolve, but a population can.

Although it is easiest to understand how natural selection would cause changes *within* a species, under the right circumstances, the same principles might produce entirely *new* species. In Chapter 14 we will discuss the circumstances that give rise to new species.

3 How Do We Know That Evolution Has Occurred?

Virtually all biologists consider evolution to be a fact. Why? Because an overwhelming body of evidence permits no other conclusion. The key lines of evidence come from fossils, comparative anatomy (the study of how body structures differ among species), embryology, and biochemistry and genetics.

Fossils Provide Evidence of Evolutionary Change over Time

If it is true that many fossils are the remains of species ancestral to modern species, we might expect to find progressive series of fossils leading from an ancient, primitive organism, through several intermediate stages, and culminating in the modern species. Such series have indeed been found, and probably the best known one is of the fossil horses (Fig. 13-6). Series of fossil giraffes, elephants, and mollusks also show the evolution of body structures over time. These fossil se-

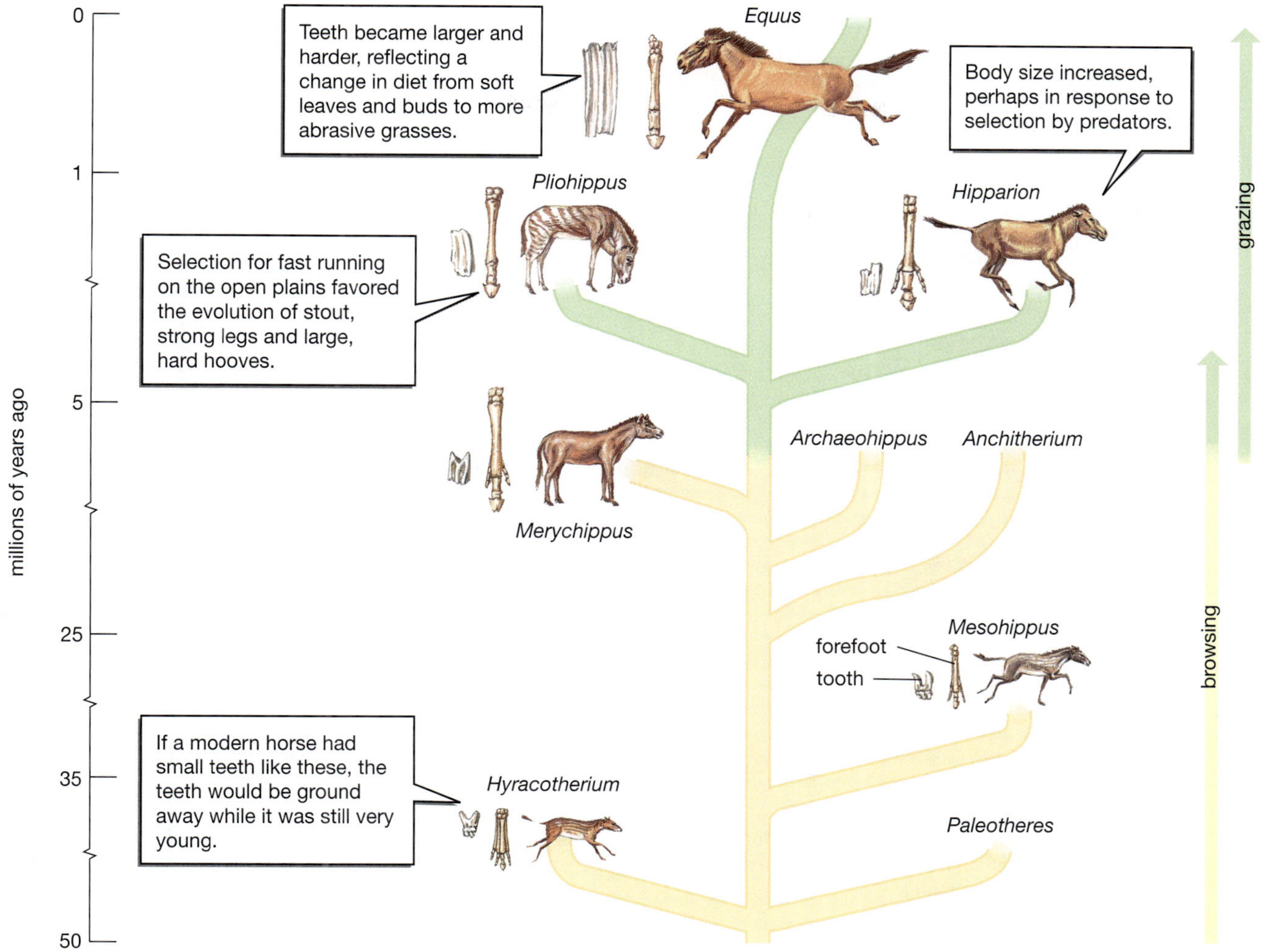

Figure 13-6 The evolution of the horse
Over the past 50 million years, horses evolved from small woodland browsers to large plains-dwelling grazers. Three major changes include size, leg anatomy, and tooth anatomy.

ries suggest that new species evolved from, and replaced, previous species. Certain sequences of fossil snails have such slight gradations in body structures between successive rock layers that paleontologists cannot easily decide where one species leaves off and the next one begins. Fossils also document larger-scale evolutionary transitions, such as the link between dinosaurs and birds that was confirmed by fossils of dinosaurs with feathers.

Comparative Anatomy Gives Evidence of Descent with Modification

Fossils provide snapshots of the past that allow biologists to trace evolutionary changes, but careful examination of today's organisms can also uncover evolution's story. In particular, comparing the bodies of organisms of different species can reveal similarities that can be explained only by shared ancestry, and differences that could result only from evolutionary change during descent from a common ancestor. In this way, the study of comparative anatomy has supplied strong evidence that different species are linked by a common evolutionary heritage.

Homologous Structures Provide Evidence of Common Ancestry

In different species, the same body structure may be modified by evolution to serve different functions. The forelimbs of birds and mammals, for example, are variously used for flying, swimming, running over several types of terrain, and grasping objects, such as branches and tools. Despite this enormous diversity of function, the internal anatomy of all bird and mammal forelimbs is remarkably similar (Fig. 13-7). It seems inconceivable that the same bone arrangements would be used to serve such diverse functions if each animal had been created separately. Such similarity is exactly what we would expect, however, if bird and mammal forelimbs were derived from a common ancestor. Through natural selection, each has been modified and now performs a particular function. Such internally similar structures

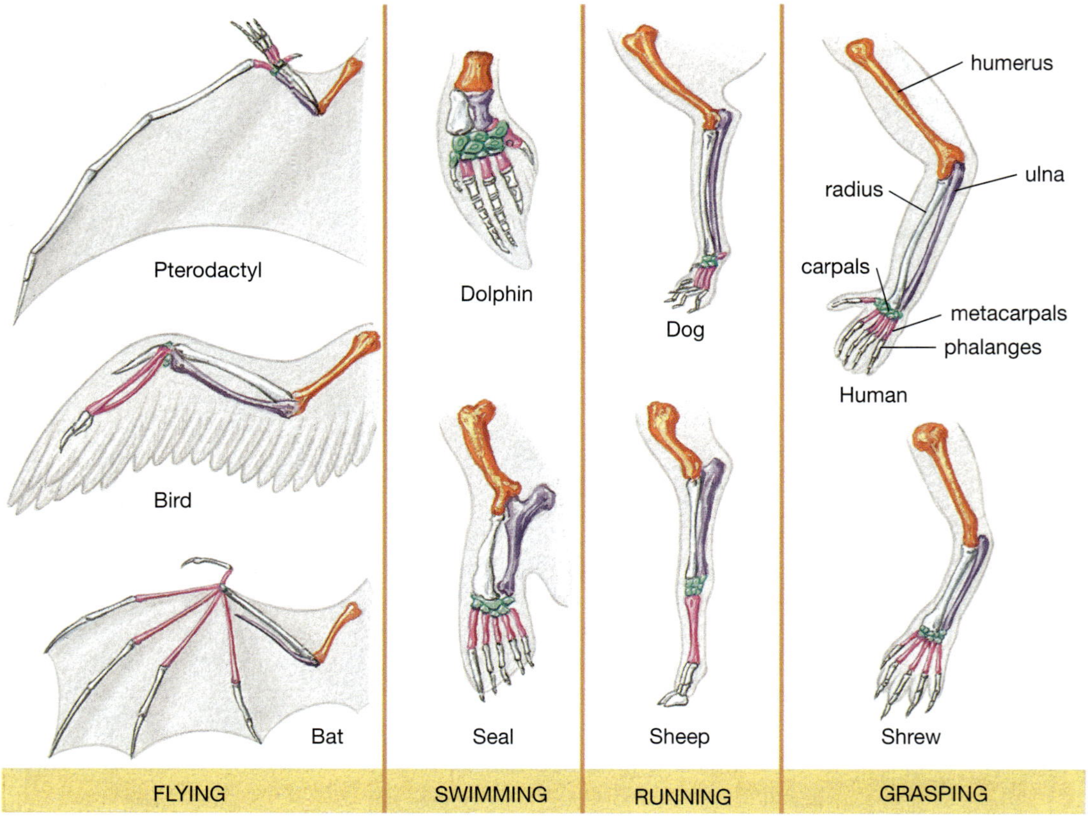

Figure 13-7 Homologous structures
Despite wide differences in function, the forelimbs of all of these animals contain the same set of bones, inherited through evolution from a common ancestor. The different colors of the bones highlight the correspondences among the various species.

are called **homologous structures**, meaning that they have the same evolutionary origin despite any differences in current function or appearance.

Functionless Structures Are Inherited from Ancestors

Evolution by natural selection also helps explain the curious circumstance of **vestigial structures** that serve no apparent purpose. Examples include such things as molar teeth in vampire bats (which live on a diet of blood and therefore don't chew their food) and pelvic bones in whales and certain snakes (Fig. 13-8). Both of these vestigial structures are clearly homologous to structures that are found in—and used by—other vertebrates (animals with a backbone). Their continued existence in animals that have no use for them is best explained as a sort of "evolutionary baggage." For example, the ancestral mammals from which whales evolved had four legs and a well-developed set of pelvic bones. Whales do not have hind legs, yet they have small pelvic and leg bones embedded in their sides. During whale evolution, losing the hind legs provided an advantage, better streamlining the body for movement through water. The result is the modern whale with small, useless, and unused pelvic bones.

Some Anatomical Similarities Result from Evolution in Similar Environments

Although the study of comparative anatomy has demonstrated the shared ancestry of life by identifying a host of homologous structures that different species have inherited from common ancestors, comparative anatomists have also identified many anatomical similarities do not stem from common ancestry. Instead, these similarities stem from **convergent evolution**, in which natural selection causes non-homologous structures that serve similar functions to resemble one another. For example, both birds and insects have wings, but this similarity arose not from evolutionary modification of a structure that both birds and insects inherited from a common ancestor, but rather from modification of two different, non-homologous structures that eventually gave rise to superficially similar structures. Because natural selection favored flight in both birds and insects, the two groups evolved superficially similar structures that are useful for flight. Such outwardly similar but non-homologous structures are called **analogous structures** (Fig. 13-9). Analogous structures are typically very different in internal anatomy, because the parts are not derived from common ancestral structures.

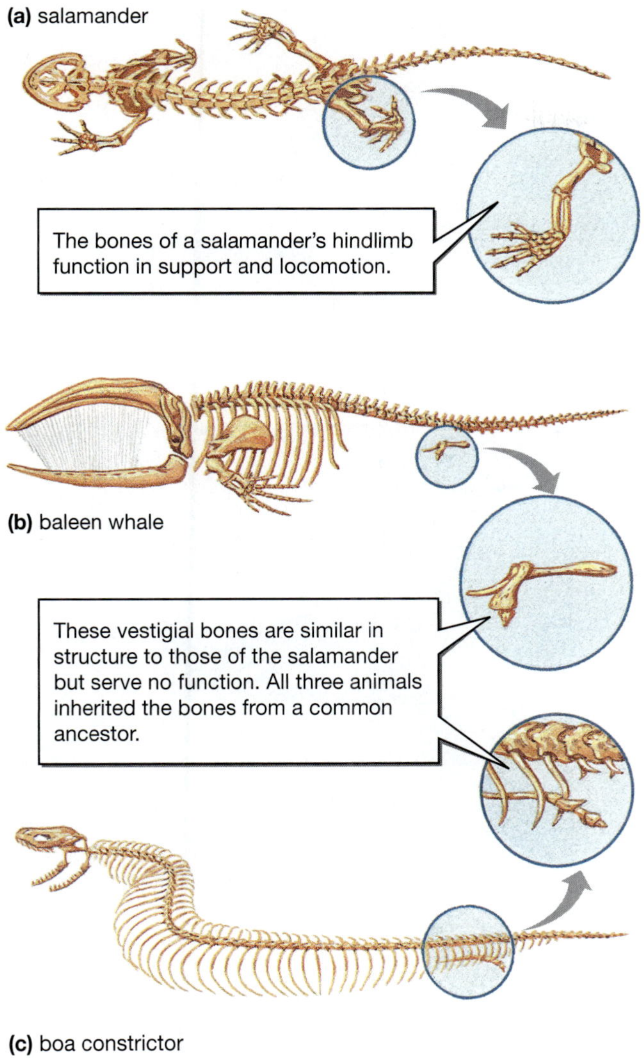

Figure 13-8 Vestigial structures
Many organisms have vestigial structures that serve no apparent function. The *(a)* salamander, *(b)* whale, and *(c)* snake all inherited hindlimb bones from a common ancestor; the bones remain functional in the salamander but are vestigial in the whale and snake.

Embryological Similarity Suggests Common Ancestry

In the early 1800s German embryologist Karl von Baer noted that all vertebrate embryos (developing organisms in the period from fertilization to birth or hatching) look quite similar to one another early in their development (Fig. 13-10). In their early embryonic stages, fish, turtles, chickens, mice, and humans all develop tails and gill slits. Only fish go on to develop gills, and only fish, turtles, and mice retain substantial tails. Why do vertebrates that are so different have similar developmental stages? The only plausible explanation is that ancestral vertebrates possessed genes that direct the development of gills and tails. All of their descendants still have those genes. In fish these genes are active throughout development, resulting in gill-bearing and tail-bearing adults. In humans and chickens these genes are active only during early developmental stages, and the structures are lost or inconspicuous in adults.

Modern Biochemical and Genetic Analyses Reveal Relatedness Among Diverse Organisms

Biochemistry and molecular biology provide striking evidence of the evolutionary relatedness of all organisms. At the most fundamental biochemical levels, all living cells are very similar. For example, all cells have DNA as the carrier of genetic information; all use RNA, ribosomes, and approximately the same genetic code to translate that genetic information into proteins; all use roughly the same set of 20 amino acids to build proteins; and all use ATP as an intracellular energy carrier. The most plausible explanation for such widespread sharing of such complex and specific biochemical traits is that the traits arose in the common ancestor of all living things, from which all of today's organisms inherited them.

The Big Picture

The evidence for life's common ancestry and its evolution by descent with modification is widespread, diverse, and abundant. No evidence that contradicts these findings has been found.

4 What Is the Evidence That Populations Evolve by Natural Selection?

We have seen that evidence of evolution comes from several sources. But what is the evidence that evolution occurs by the mechanism of natural selection?

Controlled Breeding Modifies Organisms

One line of evidence supporting evolution by natural selection that particularly impressed Charles Darwin was **artificial selection**: the breeding of domestic plants and animals to produce specific desirable features. The various breeds of dogs provide a striking example of artificial selection (Fig. 13-11). Dogs descended from wolves, and even today the two will readily cross-breed. With rare exceptions, however, few modern dogs resemble wolves. Some breeds are so different from one another that they would be considered separate species if they were found in the wild. If humans could breed such radically different dogs in a few thousand years, Darwin reasoned, it is quite plausible that natural selection could produce the spectrum of living organisms in hundreds of millions of years.

MEDIATUTOR 13.1 Analogous and Homologous Structures

Figure 13-9 Analogous structures
Convergent evolution can produce outwardly similar structures that differ anatomically. The wings of ***(a)*** insects and ***(b)*** birds and the sleek, streamlined shapes of ***(c)*** seals and ***(d)*** penguins are examples of such analogous structures.

(a)

(b)

(c)

(d)

Evolution by Natural Selection Occurs Today

The logic of natural selection gives us no reason to believe that evolutionary change is limited to the past. After all, inherited variation and competition for access to resources are certainly not limited to the past. If Darwin and Wallace were correct that those conditions lead inevitably to evolution by natural selection, then scientific observers and experimenters ought to be able to detect evolutionary change as it occurs. And they have. We consider next some examples that give us a glimpse of natural selection at work.

Brighter Coloration Can Evolve When Fewer Predators Are Present

On the island of Trinidad, guppies live in streams that are also inhabited by several species of larger, predatory fish that frequently dine on guppies. In upstream portions of

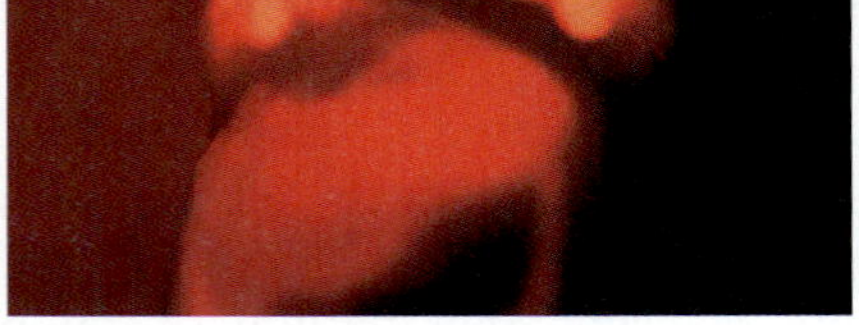

(a)

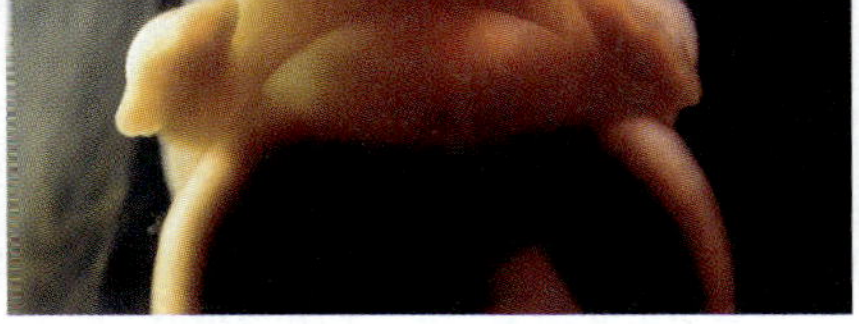

(b)

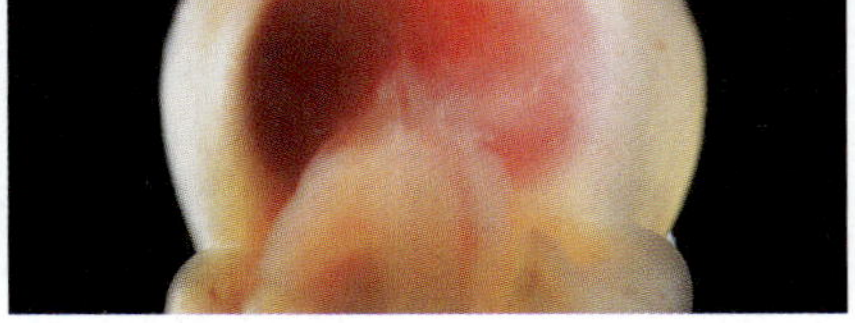

(c)

Figure 13-10 Embryological stages reveal evolutionary relationships
Early embryonic stages of a ***(a)*** lemur, ***(b)*** pig, and ***(c)*** human, showing strikingly similar anatomical features.

(a) (b)

Figure 13-11 Dog diversity illustrates artificial selection
A comparison of ***(a)*** the ancestral dog (the gray wolf, *Canis lupus*) and ***(b)*** various breeds of dog. Artificial selection by humans has caused great divergence in the size and shape of dogs in only a few thousand years.

these streams, however, the water is too shallow for the predators, and guppies are free of danger from predators. When scientists compared male guppies in an upstream area with ones in a downstream area, they found that the upstream guppies were much more brightly colored than the downstream guppies. The scientists knew that the source of the upstream population was guppies that had found their way up into the shallower waters many generations earlier. The explanation for the difference in coloration between the two populations stems from the sexual preferences of female guppies. The females prefer to mate with the most brightly colored males, so the brightest males have a large advantage when it comes to reproduction. In predator-free areas, male guppies with the bright colors that females prefer have more offspring than duller males. Bright color, however, makes guppies more conspicuous to predators, and therefore more likely to be eaten. Thus, where predators are common, they act as agents of natural selection by eliminating the bright-colored males before they can reproduce. In these areas, the duller males have the advantage and produce more offspring. The color difference between the upstream and downstream guppy populations is a direct result of natural selection.

Natural Selection Can Lead to Pesticide Resistance

Natural selection is also evident in numerous instances of insect pests evolving resistance to the pesticides with which we try to control them. For example, Florida homeowners were dismayed to realize that roaches were ignoring a formerly effective poison bait called Combat®. Researchers discovered that the bait had acted as an agent of natural selection. Roaches that liked it were consistently killed; those that survived had inherited a rare mutation that caused them to dislike glucose, a type of sugar found in the corn syrup used as bait in Combat. By the time researchers identified the problem in the early 1990s, the formerly rare mutation had become common in Florida's urban roach population.

Experiments Can Demonstrate Natural Selection

In addition to observing natural selection in the wild, scientists have also devised numerous experiments that confirm the action of natural selection. For example, one group of evolutionary biologists released small groups of *Anolis sagrei* lizards onto 14 small Bahamian islands that were previously uninhabited by lizards. The original lizards came from a population on Staniel Cay, an island with tall vegetation, including plenty of trees. In contrast, the islands to which the small colonial groups were introduced had few or no trees and were covered mainly with small shrubs and other low-growing plants.

The biologists returned to those islands 14 years after releasing the colonists and found that the original small groups of lizards had given rise to thriving populations of hundreds of individuals. On all 14 of the experimental islands, lizards had legs that were shorter and thinner than lizards from the original source population on Staniel Cay. In just over a decade, it appeared, the lizard populations had changed in response to new environments. Why had the new lizard populations evolved shorter, thinner legs? Long legs allow greater speed for escaping predators, but shorter legs allow for more

MEDIATUTOR
13.2 Natural Selection for Antibiotic Resistance

agility and maneuverability on narrow surfaces. So, natural selection favors legs that are as long and thick as possible while still allowing sufficient maneuverability. When the lizards were moved from an environment with thick-branched trees to an environment with only thin-branched bushes, the individuals with formerly favorable long legs were at a disadvantage. In the new environment, more agile, shorter-legged individuals were better able to escape predators and survive to produce a greater number of offspring. Thus, members of subsequent generations would have shorter legs on average.

Selection Acts on Random Variation to Favor the Phenotypes That Work Best in Particular Environments

Two important points underlie the evolutionary changes just described:

- *The variations on which natural selection works are produced by chance mutations.* The bright coloration in Trinidadian guppies, distaste for glucose in Florida cockroaches, and shorter legs in Bahamian lizards were not *produced* by the female mating preferences, poisoned corn syrup, or the thinner branches. The mutations that produced each of these beneficial traits had arisen spontaneously and were later made more common by natural selection.
- *Natural selection selects for organisms that are best adapted to a particular environment.* Natural selection is not a mechanism for producing ever-greater degrees of perfection. Natural selection does not select for the "best" in any absolute sense, but only in the context of a particular environment, which varies from place to place and which may change over time. A trait that is advantageous under one set of conditions may become disadvantageous if conditions change. In the presence of poisoned corn syrup, a distaste for glucose yields an advantage to a cockroach, but under natural conditions avoiding glucose would cause the insect to bypass good sources of food.

5 A Postscript by Charles Darwin

"It is interesting to contemplate an entangled bank, clothed with many plants of many kinds, with birds singing on the bushes, with various insects flitting about, and with worms crawling through the damp earth, and to reflect that these elaborately constructed forms . . . have all been produced by laws acting around us. These laws, taken in the highest sense, being Growth with Reproduction; Inheritance [and] Variability; a Ratio of Increase so high as to lead to a Struggle for Life, and as a consequence to Natural Selection, entailing Divergence of Character and Extinction of less-improved forms. . . . There is grandeur in this view of life, with its several powers, having been originally breathed into a few forms or into one; and that, whilst this planet has gone cycling on according to the fixed law of gravity, from so simple a beginning endless forms most beautiful and most wonderful have been, and are being, evolved."

These are the concluding sentences of Darwin's *On the Origin of Species*.

CASESTUDYREVISITED

A Missing Link Unearthed

The feathered dinosaur bandwagon hit a bit of a bump in early 2000 when a much-publicized new fossil of a feathered dinosaur species turned out to be a fake. The fraudulent fossil had been purchased from a dealer in Utah, and someone had apparently doctored it by gluing a dinosaur's tail onto a fossil bird. The embarrassed scientists who fell for the hoax, however, maintain that the fraud in no way affects the validity of previously discovered feathered dinosaur fossils.

Fossil fakery notwithstanding, even well-accepted feathered dinosaurs have failed to completely resolve the controversy over the evolutionary origin of birds. A small group of skeptical paleontologists questions whether the new fossils really provide truly conclusive evidence that dinosaurs gave rise to birds. For one thing, say the skeptics, the fossils of the feathered dinosaurs are much younger (by about 30 million years) than the fossils of *Archaeopteryx*, the oldest known bird, so these feathered dinosaur species could not have been the ancestors of birds. To the skeptics, this anomaly suggests that the fossils represent not dinosaurs, but rather birds that had lost the power of flight (as have some modern birds, such as the ostrich and penguin).

Most paleontologists, however, shrug off the question of the new fossils' age. To them, the comparatively young age of the fossils means nothing more than that some species of feathered dinosaurs persisted after birds first evolved from earlier dinosaurs (much as apes and humans both persist today, millions of years after a common ancestor gave rise to both groups). To proponents of the bird–dinosaur connection, feathered dinosaurs of any age demonstrate conclusively that pre-bird dinosaurs evolved feathers and that these feathered dinosaurs were the ancestors of birds.

Summary of Key Concepts

1) How Did Evolutionary Thought Evolve?

Historically, the most common explanation for the origin of species was the divine creation of each species in its present form; species were believed to remain unchanged after their creation. This view was challenged by evidence from fossils, geology, and biological exploration of the Tropics. Since the middle of the nineteenth century, scientists have realized that species originate and evolve by the operation of natural processes that change the genetic makeup of populations.

2) How Does Natural Selection Work?

Charles Darwin and Alfred Russel Wallace independently proposed the theory of evolution by natural selection. Their theory can be concisely expressed as three conclusions based on four observations. These are summarized in modern biological terms in Figure 13-5.

3) How Do We Know That Evolution Has Occurred?

Many lines of evidence indicate that evolution has occurred, including the following:

- Fossils of ancient species tend to be simpler in form than modern species. Sequences of fossils have been discovered that show a graded series of changes in form. Both of these observations would be expected if modern species evolved from older species.
- Species thought to be related through evolution from a common ancestor show many similar anatomical structures. An example is the forelimbs of amphibians, reptiles, birds, and mammals.
- Stages in early embryological development are quite similar among very different types of vertebrates.
- Similarities in such biochemical traits as the use of DNA as the carrier of genetic information support the notion of descent of related species through evolution from common ancestors.

4) What Is the Evidence That Populations Evolve by Natural Selection?

Similarly, many lines of evidence indicate that natural selection is the chief mechanism driving changes in the characteristics of species over time, including the following:

- Inheritable traits have been changed rapidly in populations of domestic animals and plants by selectively breeding organisms with desired features (artificial selection). The immense variations in species produced in a few thousand years of artificial selection by humans makes it seem likely that much larger changes could be wrought by hundreds of millions of years of natural selection.
- Evolution can be observed today. Both natural and human activities drastically change the environment over short periods of time. Characteristics of species have been observed to change significantly in response to such environmental changes.

Key Terms

analogous structures *p. 218*
artificial selection *p. 219*
catastrophism *p. 210*
convergent evolution *p. 218*
evolution *p. 209*
fossil *p. 210*
homologous structures *p. 218*
natural selection *p. 215*
population *p. 214*
uniformitarianism *p. 211*
vestigial structure *p. 218*

Thinking Through the Concepts

Multiple Choice

1. *Your arm is homologous with*
   a. a seal flipper
   b. an octopus tentacle
   c. a bird wing
   d. a sea star arm
   e. both a and c

2. *All organisms share the same genetic code. This commonality is evidence that*
   a. evolution is occurring now
   b. convergent evolution has occurred
   c. evolution occurs gradually
   d. all organisms are descended from a common ancestor
   e. life began a long time ago

3. *Which of the following are fossils?*
   a. pollen grains buried in the bottom of a peat bog
   b. the petrified cast of a clam's burrow
   c. the impression a clam shell made in mud, preserved in mudstone
   d. an insect leg sealed in plant resin
   e. all of the above

4. *In Africa, there is a species of bird called the yellow-throated longclaw. It looks almost exactly like the meadowlark found in North America, but they are not closely related. This is an example of*
   a. uniformitarianism
   b. artificial selection
   c. gradualism
   d. vestigial structures
   e. convergent evolution

5. *Which of the following are examples of vestigial structures?*
   a. your tailbone
   b. your nostrils
   c. sixth fingers found in some humans
   d. your kneecap
   e. none of the above

6. *Which of the following would stop evolution by natural selection from occurring?*
   a. if humans became extinct because of a disease epidemic
   b. if a thermonuclear war killed most living organisms and changed the environment drastically
   c. if ozone depletion led to increased ultraviolet radiation, which caused many new mutations
   d. if all individuals in a population were genetically identical, and there were no genetic recombination, sexual reproduction, or mutation
   e. all of the above

? Review Questions

1. Selection acts on individuals, but only populations evolve. Explain why this is true.
2. Distinguish between catastrophism and uniformitarianism. How did these hypotheses contribute to the development of evolutionary theory?
3. Describe Lamarck's theory of inheritance of acquired characteristics. Why is it invalid?
4. What is natural selection? Describe how natural selection might have caused differential reproduction among the ancestors of a fast-swimming predatory fish, such as the barracuda.
5. Describe how evolution occurs through the interactions among the reproductive potential of a species, the normally constant size of natural populations, variation among individuals of a species, natural selection, and inheritance.
6. What is convergent evolution? Give an example.
7. How do biochemistry and molecular genetics contribute to the evidence that evolution occurred?

Applying the Concepts

1. Does evolution through natural selection produce "better" organisms in an absolute sense? Are we climbing the "ladder of Nature"? Defend your answer.
2. Both the study of fossils and the idea of special creation have had an impact on evolutionary thought. Discuss why one is considered scientific endeavor and the other not scientific.
3. In evolutionary terms, "success" can be defined in many different ways. What are the most successful organisms you can think of in terms of (a) persistence over time, (b) sheer numbers of individuals alive now, (c) numbers of species, and (d) geographical range?
4. In what sense are humans currently acting as agents of selection on other species? Name some organisms that are *favored* by the environmental changes humans cause.
5. Darwin and Wallace's discovery of natural selection is one of the great revolutions in scientific thought. Some scientific revolutions spill over and affect the development of philosophy and religion. Is this true of evolution? Does (or should) the idea of evolution by natural selection affect the way humans view their place in the world?
6. In your mind, what scientific question currently represents the "mystery of mysteries"?

Group Activity

Form a group of four students. Your group has been asked to analyze some data from a long-term study in the Galapagos Islands. Another group of researchers spent 8 years measuring beaks, observing feeding behavior, and measuring rainfall in a population of a particular species of Darwin's finch. They found that this species eats only small seeds and that its beak enables the bird to crack and eat such seeds quickly and efficiently. The beak-size data look like this:

Year	1	2	3	4	5	6	7	8
Rainfall	normal	normal	normal	very low	very low	normal	normal	normal
Average beak size (in mm)	10.1	10.3	10.2	12.1	12.5	10.1	10.0	10.1

As a group, develop a hypothesis that explains what caused beak size to increase in dry years. Your hypothesis should specify how the fates of individuals determine changes in the population.

For More Information

Altman, S. A. "The Monkey and the Fig." *American Scientist,* May–June 1989. An amusing, informative discussion of many evolutionary themes, cast as a Socratic dialogue.

Darwin, C. *On the Origin of Species by Means of Natural Selection.* Garden City, NY: Doubleday, 1960 (originally published in 1859). An impressive array of evidence amassed to convince a skeptical world.

Dawkins, R. "God's Utility Function." *Scientific American,* November 1995. One evolutionary biologist's view of the nature of evolutionary change and how it relates to the meaning of life. Dawkins is an articulate advocate of the "selfish gene" view of evolution, in which genes use organisms as tools in a contest for superiority in self-replication.

Dennet, D. *Darwin's Dangerous Idea.* New York: Simon & Schuster, 1995. A philosopher's view of Darwinian ideas and their application to the world outside biology. A thought-provoking book that seems to have inspired admiration and condemnation in roughly equal proportions.

Eiseley, L. C. "Charles Darwin." *Scientific American,* February 1956. An essay on the life of Darwin by one of his foremost American biographers. Even if you need no introduction to Darwin, read this as an introduction to Eiseley, author of many marvelous essays.

Gould, S. J. *Ever Since Darwin,* 1977; *The Panda's Thumb,* 1980; and *The Flamingo's Smile,* 1985. New York: W. W. Norton. A series of witty, imaginative, and informative essays, mostly from *Natural History* magazine. Many deal with various aspects of evolution.

Grant, P. R. "Natural Selection and Darwin's Finches." *Scientific American,* October 1991. A drought in the Galapagos Islands provides dramatic evidence of natural selection as an agent of evolutionary change.

Weiner, J. "Evolution Made Visible." *Science,* January 6, 1995. A clear summary of modern evidence for evolution in action.

Answers to Multiple-Choice Questions

1. e 2. d 3. e 4. e 5. a 6. d

MediaTutor

CD Activities

Activity 13.1: Analogous and Homologous Structures

Estimated time: 5 minutes

Different animal species commonly share structures but use them in strikingly different ways. Such differences were part of the evidence that led Charles Darwin to conclude that all animals were not created in their current form. This exercise will help you to learn the difference between structures that are shared as a result of inheritance from a common ancestor and those that are similar as a result of common function.

Activity 13.2: Natural Selection for Antibiotic Resistance

Estimated time: 15 minutes

Rifampin is a drug that has been used to kill the bacterium that causes tuberculosis (TB) in humans. Many populations of this bacterium, *Mycobacterium tuberculosis*, have evolved resistance to Rifampin. After completing the tutorial, you will be able to describe the process of evolution by natural selection using the case study of *Mycobacterium tuberculosis*.

Start the MediaTutor Student CD-ROM and enter the activity number in the Quick Search box to be taken directly to that activity.

Web Investigation

A Missing Link Unearthed

Estimated time: 10 minutes

Evolutionary biologists and paleontologists probably attract more attention (and sometimes hostility) from non-scientists than any other scientific group. Finding and analyzing the fossils is hard work. Sometimes the results are quite unexpected. This exercise will explore the fossil record of birds.

Go to http://www.prenhall.com/audesirk3, the Audesirk Companion Web site. Select Chapter 13 and Web Investigation to begin.

"What but the wolf's tooth whittled so fine
The fleet limbs of the antelope?
What but fear winged the birds, and hunger
Jewelled with such eyes the great goshawk's head?"

Robinson Jeffers in **The Bloody Sire** ***(1941)***

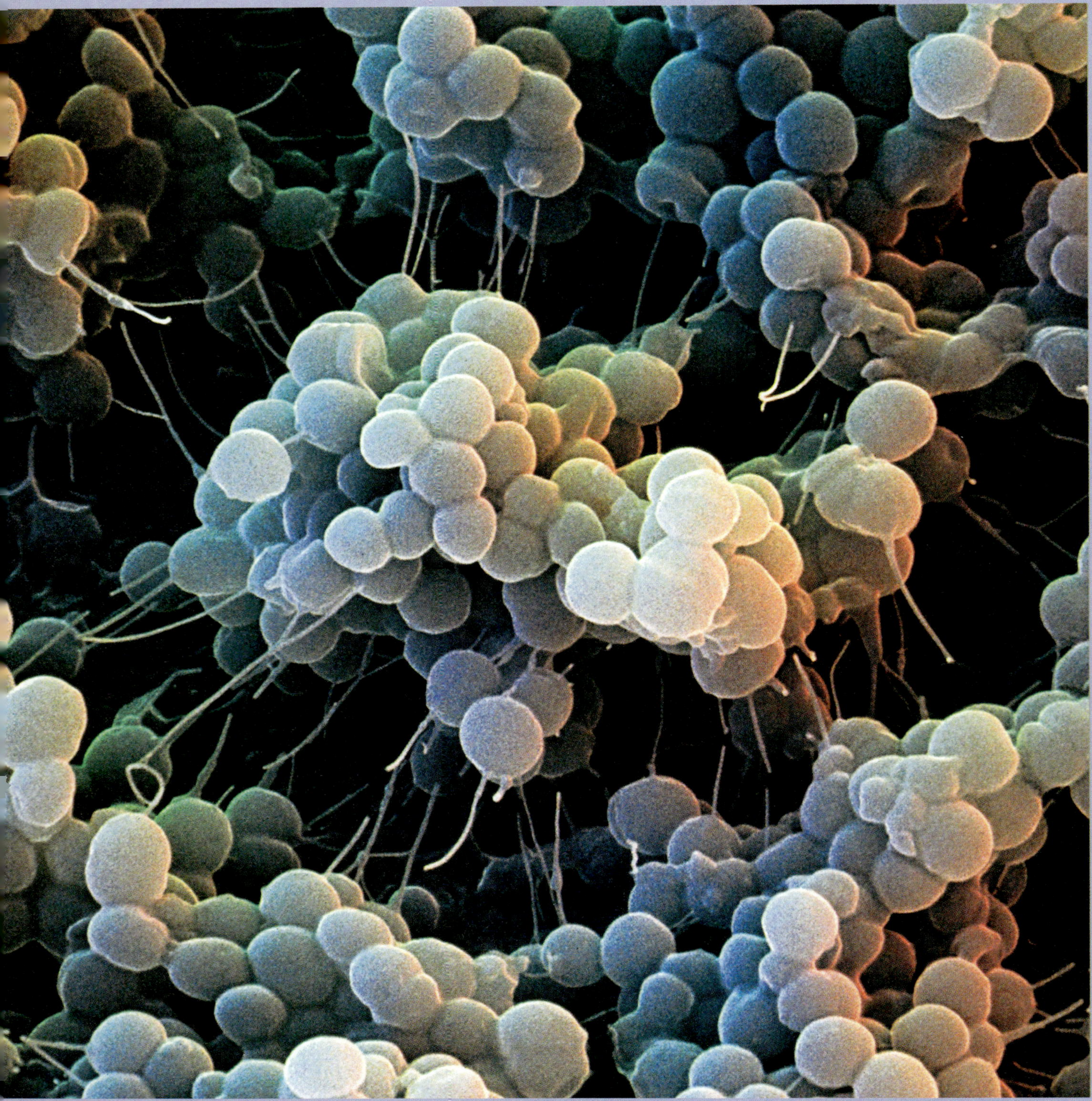

Staphylococcus aureus is among the many bacterial species that have evolved resistance to antibiotics. The evolution of such supergerms threatens to reverse our success in the battle against infectious disease.

14 How Organisms Evolve

AT A GLANCE

CASESTUDY

Cause of Death: Evolution

On a cold winter night in 1999, an 11-month-old infant arrived at a North Dakota emergency room. The child was ill with pneumonia and a blood infection, and doctors immediately began treatment with antibiotics. Despite this prompt treatment, however, the baby died within hours. The death of any child is a terrible tragedy, but this incident was especially chilling because it represented the escalation of a frightening trend: the rise of "supergerms," strains of disease-causing bacteria that defy destruction by antibiotic medicines. The young victim was the fourth child in the area to die of infection by antibiotic-resistant *Staphylococcus aureus* (commonly known as "staph"). The four deaths marked a disturbing new development in the story of drug-resistant staph, because the fatal infections were acquired not in a hospital (the usual source of drug-resistant infections) but in the children's homes and communities. As has long been feared, resistant staph bacteria are spreading beyond the hospital door.

The spread of drug-resistant staph is the latest sign of a trend that has been developing since antibiotics first came into widespread use in the 1940s. Antibiotics are losing their effectiveness against a host of diseases caused by bacteria, including food poisoning, tuberculosis, gonorrhea, meningitis, and urinary and respiratory tract infections. The rapid spread of resistance in bacterial populations raises the specter of incurable diseases, resistant to all known treatments.

In retrospect, this disturbing state of affairs is an unsurprising outcome of our unrelenting attack on disease-causing bacteria. The spread of a new characteristic (such as antibiotic resistance) through a population in response to changing environmental conditions (such as the introduction of antibiotics) is precisely the sort of outcome that a biologist might predict, in accordance with our understanding of the mechanics of evolutionary change. ■

1 How Are Populations, Genes, and Evolution Related?

The changes that we see in an individual organism as it grows and develops are not evolutionary changes. Instead, evolutionary changes are those that occur from generation to generation, the changes that cause descendants to be different from their ancestors. Furthermore, we can't detect evolutionary change across generations by looking at a single set of parents and offspring. For example, if you observed that a six-foot-tall man had an adult son who stood five feet tall, could you conclude that humans were evolving to become shorter? Obviously not. Rather, if you wanted to learn about evolutionary change in human height, you would begin by measuring many humans of many generations to see if the average height is changing over time. Clearly, evolution is a property not of individuals but of populations; a **population** is a group that includes all the members of a species living in a given area.

The recognition that evolution is a population-level phenomenon was one of Darwin's key insights. But populations are composed of individuals, and it is the actions and fates of individuals that determine which characteristics will be passed to descendant populations. In this fashion, inheritance provides the link between the lives of individual organisms and the evolution of populations. We will therefore begin our discussion of the processes of evolution by reviewing some principles of genetics as they apply to individuals, and we will then extend those principles to the genetics of populations.

Genes and the Environment Interact to Determine Traits

Each cell of every organism contains genetic information encoded in the DNA of its chromosomes. Recall that a *gene* is a segment of DNA located at a particular place on a chromosome. The sequence of nucleotides in a gene encodes the sequence of amino acids in a protein, usually an enzyme that catalyzes a particular reaction in the cell. At a given gene's location, different members of a species may have slightly different nucleotide sequences, called *alleles*, as we saw in Chapter 11. Different alleles generate different forms of the same enzyme. In this way, various alleles of the gene that influences eye color in humans, for example, help produce eyes that are brown, or blue, or green, and so on. In any population of organisms, there are usually two or more alleles of each gene. An individual of a diploid species whose alleles of a particular gene are both the same is *homozygous* for that gene, and an individual with different alleles for that gene is *heterozygous*. The specific alleles borne on an organism's chromosomes (its *genotype*) interact with the environment to influence development of its physical and behavioral traits (its *phenotype*).

Let's illustrate these principles with an example familiar to you from Unit II. A pea flower is colored purple because a chemical reaction in its petals converts a colorless molecule to a purple pigment. When we say that a pea plant has the allele for purple flowers, we mean that a particular stretch of DNA on one of its chromosomes contains a sequence of nucleotides that codes for the enzyme that catalyzes this reaction. A pea with the allele for white flowers has a different sequence of nucleotides at the corresponding chromosomal position. The enzyme for which that different sequence codes cannot produce purple pigment. If a pea is homozygous for the white allele, its flowers produce no pigment and thus are white.

The Gene Pool Is the Sum of the Genes in a Population

We can often deepen our understanding of a subject by looking at it from more than one perspective. In studying evolution, looking at the process from the point of view of a gene has proven to be an enormously effective tool. In particular, evolutionary biologists have made excellent use of the tools of a branch of genetics called population genetics, which deals with the frequency, distribution, and inheritance of alleles in populations. To take advantage of this powerful aid to understanding evolution, you will need to learn a few of the basic concepts of population genetics.

Population genetics defines the **gene pool** as the sum of all the genes in a population. In other words, the gene pool consists of all the alleles of all the genes in all the individuals of a population. Each particular gene can also be considered to have its own gene pool, which consists of all the alleles of that specific gene in a population. If we added up all the copies of each allele of that gene in all the individuals in a population, we could determine the relative proportion of each allele, a number called the **allele frequency**. For example, a population of 100 pea plants would contain 200 alleles of the gene that controls flower color (because pea plants are diploid). If 50 of those 200 alleles were of the type that codes for white flowers, then we would say that the frequency of that allele in the population is 0.25 (or 25%), because $50/200 = 0.25$.

Evolution Is the Change of Allele Frequencies Within a Population

A casual observer might define evolution on the basis of changes in the outward appearance or behaviors of the members of a population. The population geneticist, however, looks at a population and sees a gene pool that just happens to be divided into the packages that we call individual organisms. So any outward changes that we observe in the individuals that make up the population can also be viewed as the visible expression of underlying changes to the gene pool. A population geneticist therefore defines evolution as the changes in allele frequencies that occur in a gene pool over time. Evolution is a change in the genetic makeup of populations over generations.

The Equilibrium Population Is a Hypothetical Population in Which Evolution Does Not Occur

It is easier to understand what causes populations to evolve if the characteristics of a population that would *not* evolve are considered first. In 1908 English mathematician Godfrey H. Hardy and German physician Wilhelm Weinberg independently developed a simple mathematical model now known as the **Hardy-Weinberg principle**. This model showed that under certain conditions allele frequencies and genotype frequencies in a population will remain constant no matter how many generations pass. In other words, this population will not evolve. Population geneticists use the term **equilibrium population** for this idealized, evolution-free population in which allele frequencies will not change as long as the following conditions are met:

- There must be no mutation.
- There must be no **gene flow** between populations; that is, there must be no movement of alleles into or out of the population (as would be caused, for example, by the movement of organisms into or out of the population).
- The population must be very large.
- All mating must be random, with no tendency for certain genotypes to mate with specific other genotypes.
- There must be no natural selection; that is, all genotypes must reproduce with equal success.

Under these conditions, allele frequencies within a population will remain the same indefinitely. If one or more of these conditions are violated, then allele frequencies may change: The population will evolve.

As you might expect, few if any natural populations are truly in equilibrium. What, then, is the importance of the Hardy-Weinberg principle? The Hardy-Weinberg conditions are useful starting points for studying the mechanisms of evolution. In the following sections, we will examine some of the conditions, show that natural populations often fail to meet them, and illustrate the consequences of such failures. In this way, we can better understand both the inevitability of evolution and the forces that drive evolutionary change.

2 What Causes Evolution?

Population genetics theory predicts that the Hardy-Weinberg equilibrium can be disturbed by deviations from any of its five conditions. Most evolutionary biologists agree, however, that the most important sources of evolutionary change are mutation, small population size, and natural selection.

Mutations Are the Original Source of Genetic Variability

A population remains in genetic equilibrium only if there are no **mutations** (changes in DNA sequence), but mutations are inevitable. Although cells have efficient mechanisms that protect the integrity of their genes, some changes in nucleotide sequence slip past the checking and repair systems. An unrepaired mutation in a cell that produces gametes may be passed to an offspring and enter the gene pool of a population.

Mutations Are Rare But Important

How significant is mutation in changing the gene pool of a population? Mutations are rare; for example, only one out of every 100,000 to 1,000,000 human gametes carries a mutation for a given gene. Therefore, mutation by itself is not a major force in evolution. However, mutations are the source of new alleles, new variations on which other evolutionary processes can work. As such, they are the foundation of evolutionary change. Without mutations there would be no evolution and no diversity among organisms.

Mutations Are Not Goal-Directed

A mutation does not arise as a result of, or in anticipation of, environmental necessities (Fig. 14-1). A mutation simply happens and may in turn produce a change in a structure or function of the organism. Whether that change is helpful or harmful or neutral, now or in the future, depends on environmental conditions over which the organism has little or no control. The mutation provides a potential for evolutionary change. Other forces, especially natural selection, may act on that potential to spread the mutation through the population or to eliminate it from the population.

Allele Frequencies May Drift in Small Populations

To remain in equilibrium, a population must be so large that chance events have no impact on its overall genetic makeup. Disaster may befall even the fittest organism. The maple seed that falls into a pond never sprouts; the yearling deer and elk blasted away by the eruption of Mount St. Helens left no descendants. If a population is sufficiently large, chance events are unlikely to alter its genetic composition because such events would interfere equally with the reproduction of organisms of all genotypes. In a small population, however, certain alleles may be carried by only a few organisms. Chance events could eliminate some or all of such alleles from the population, altering its genetic makeup. This process by which chance events change allele frequencies in a small population is called **genetic drift**.

Population Size Matters

To see how genetic drift works, imagine two populations of amoebas in which each amoeba is either red or blue, and color is controlled by two alleles (*A* and *a*) of a gene. Half of the amoebas in each of our two populations are red and half are blue. One population, however, has only four individuals in it, while the other has 10,000.

Now let's picture reproduction in our imaginary populations. Let's select at random half of the individuals in each population and allow them to reproduce by binary fission. To do so, each reproducing amoeba splits in half to yield two amoebas, each of which is the same color as the parent. In the large population, 5000 amoebas reproduce,

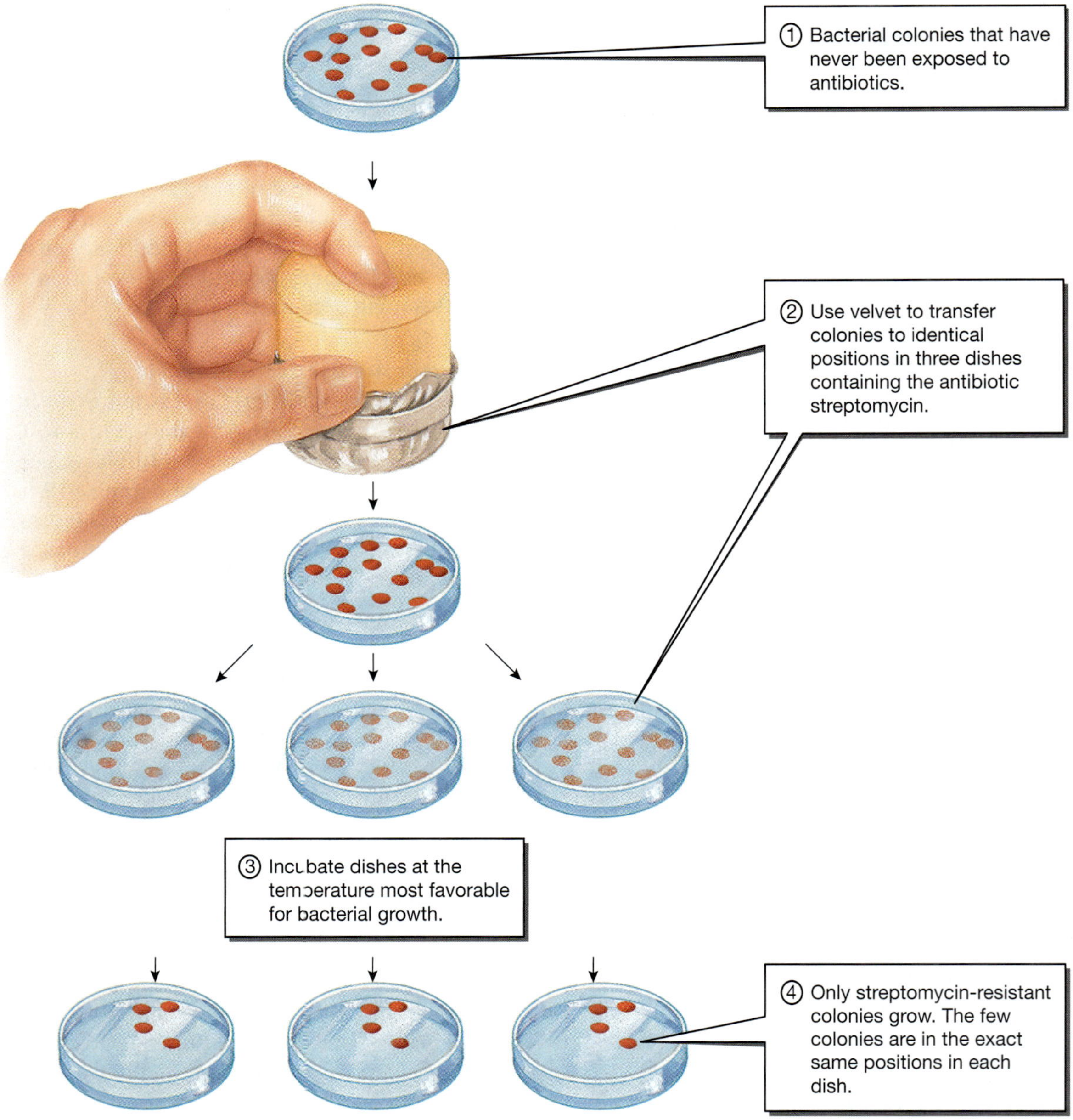

Figure 14-1 Mutations occur spontaneously
This experiment demonstrates that mutations occur spontaneously and not in response to environmental pressures. When bacterial colonies that have never been exposed to antibiotics are exposed to the antibiotic streptomycin, only a few colonies grow. The observation that these surviving colonies grow in the exact same positions in all dishes shows that the mutations for resistance to streptomycin were present in the original dish before exposure to the environmental pressure, streptomycin.

yielding a new generation of 10,000. What are the chances that all 10,000 members of the new generation will be red? Just about nil. In fact, it would be extremely unlikely for even 3000 amoebas to be red or for 7000 to be red. The most likely outcome is that about half will be red and half blue, just as in the original population. In this large population, then, we would not expect a major change in allele frequencies from generation to generation. One way to test this expectation is to write a computer program that simulates how the allele frequencies of the alleles could change over generations. Figure 14-2a shows, from four runs of such a simulation, that the frequency of allele *A*, encoding red color, remains close to 0.5, consistent with the expectation that half of the amoebas would be red.

In the small population, the situation is different. Only two amoebas reproduce, and there is a 25% chance that both reproducers will be red. (This outcome is as likely as flipping two coins and having both come up heads.) If only red amoebas reproduce, then the next generation will consist entirely of red amoebas—a relatively likely outcome. It is thus possible,

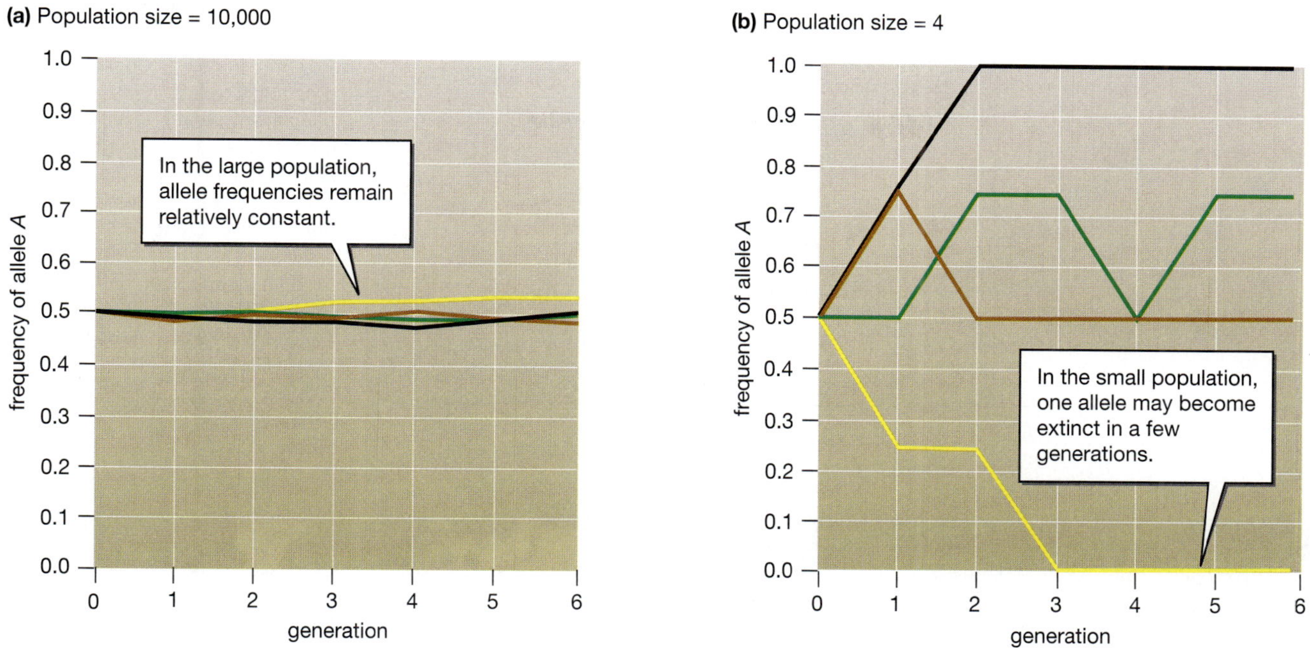

Figure 14-2 The effect of population size on genetic drift
Each colored line represents one computer simulation of the change over time in the frequency of allele *A* in a ***(a)*** small or ***(b)*** large population in which half the alleles were *A* (0.5), and in which randomly chosen individuals reproduced.

within a single generation, for the allele for blue color to disappear from the population. Figure 14-2b shows the fate of allele *A* in four runs of a simulation of our small population. In one of the four runs, allele *A* reaches a frequency of 1.0 (100 percent) in the second generation, meaning that all the amoebas in the third and following generations are red. In another run, the frequency of *A* drifts to 0.0 in the fourth generation, and the population subsequently is all blue. Thus, one of the two amoeba phenotypes disappeared in half of the simulations.

A Population Bottleneck Is an Example of Genetic Drift

Two causes of genetic drift, called the population bottleneck and the founder effect, further illustrate the consequences that small population size may have on the allele frequencies of a species. In a **population bottleneck**, a population is drastically reduced—as a result of a natural catastrophe or overhunting, for example. Then only a few individuals are available to contribute genes to the next generation. As our amoeba example showed, population bottlenecks can both change allele frequencies and reduce genetic variability (Fig. 14-3a). Even if the population later increases, the genetic effects of the bottleneck may remain for hundreds or thousands of generations (Fig. 14-3b,c).

A special case of a population bottleneck is the **founder effect**, which occurs when isolated colonies are founded by a small number of organisms. A flock of birds, for instance, may become lost during migration or may be blown off course by a storm, and may settle on an isolated island. Following both founder events and other bottlenecks, only a few individuals contribute genes to the future population. Such a small group may have allele frequencies that are very different from the frequencies of the parent population.

All Genotypes Are Not Equally Beneficial

In a hypothetical equilibrium population, individuals of all genotypes survive and reproduce equally well—that is, no genotype has any advantage over the others. This condition, however, is probably met only rarely, if ever, in real populations. Even though some alleles are neutral, in the sense that organisms possessing any of several alleles are equally likely to survive and reproduce, this is clearly not true of all alleles in all environments. Any time an allele confers, in Alfred Russel Wallace's words, "some little superiority," natural selection favors the individuals who possess it; that is, those individuals have higher reproductive success. This phenomenon is well illustrated by an example concerning an antibiotic drug.

Antibiotic Resistance Evolves by Natural Selection

The antibiotic penicillin first came into widespread use during World War II, when it was used to combat infections in wounded soldiers. Suppose that an infantryman, brought to a field hospital after suffering a gunshot wound in his arm, develops a staph infection in that arm. A medic sizes up the situation and resolves to treat the wounded soldier with an intravenous drip of penicillin. As the antibiotic courses through the soldier's blood vessels, millions of staph bacteria die before they can reproduce. A few bacteria, however, carry a rare allele that codes for an enzyme that destroys any penicillin that

(a)

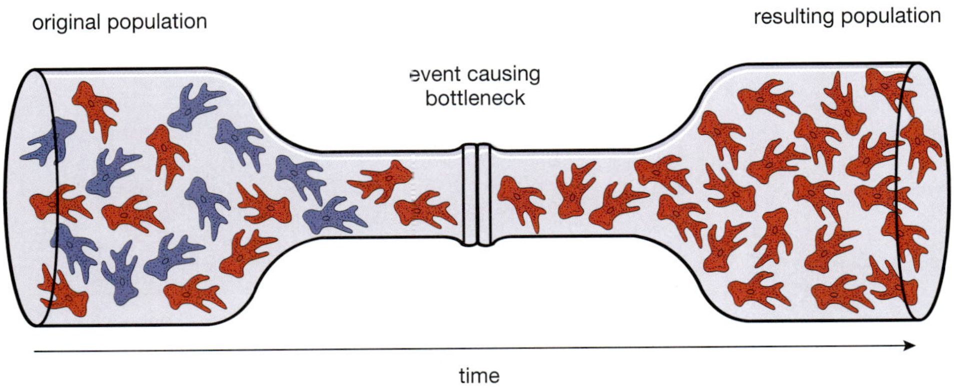

(b)

(c)

Figure 14-3 Population bottlenecks reduce variation
(a) A population bottleneck may drastically reduce genetic and phenotypic variation because the few organisms that survive may all carry similar sets of alleles. Both ***(b)*** the northern elephant seal, which was hunted almost to extinction in the 1800s, and ***(c)*** the cheetah passed through a population bottleneck in the recent past, resulting in an almost total loss of genetic diversity.

MediaTutor 14.1 The Bottleneck Effect

comes in contact with the bacterial cell. (This allele is a variant of a gene that normally codes for an enzyme that breaks down the bacterium's waste products.) The bacteria carrying this rare allele are able to survive and reproduce, and their offspring inherit the penicillin-destroying allele. After a few generations, the frequency of the penicillin-destroying allele has soared to nearly 100%, and the frequency of the normal, waste-processing allele has declined to near zero. As a result of natural selection imposed by the antibiotic's killing power, the population of staph bacteria within the soldier's body has evolved. The gene pool of the staph population has changed, and natural selection, in the form of bacterial destruction by penicillin, has caused the change.

Penicillin Resistance Illustrates Key Points About Evolution

The example of penicillin resistance highlights some important features of natural selection and evolution.

Natural selection does not cause genetic changes in individuals. The allele for penicillin resistance arose spontaneously, long before penicillin was dripped into the soldier's vein. Penicillin did not cause resistance to appear; its presence merely favored the survival of bacteria with penicillin-destroying alleles over that of bacteria with waste-processing alleles.

Natural selection acts on individuals, but it is populations that are changed by evolution. The agent of natural selection, penicillin, acted on individual staph bacteria. As a result, some individuals reproduced and some did not. However, it was the population as a whole that evolved as its allele frequencies changed.

Evolution is change in allele frequencies of a population, owing to unequal success at reproduction among organisms bearing different alleles. In evolutionary terminology, the **fitness** of an organism is measured by its reproductive success: In our example, the penicillin-resistant bacteria had greater fitness than the normal bacteria did, because the resistant bacteria produced greater numbers of viable (able to survive) offspring.

Evolutionary changes are not "good" or "progressive" in any absolute sense. The resistant bacteria were favored only because of the presence of penicillin in the soldier's body. At a later time, when the environment of the soldier's body no longer contained penicillin, the resistant bacteria may have been at a disadvantage relative to other bacteria that could process waste more effectively. Similarly, the long legs and necks of giraffes are helpful when the animals are feeding on leaves high up in trees, but they are a hindrance to drinking (Fig. 14-4). The lengths of giraffe legs and necks represent an evolutionary compromise between the advantage of reaching higher leaves for food and the disadvantage of vulnerability while drinking water.

Figure 14-4 A compromise between opposing environmental pressures ***(a)*** The long neck and legs of a giraffe are a definite advantage in feeding on acacia leaves high up in trees. ***(b)*** But a giraffe has to get into an extremely awkward and vulnerable position to drink. Feeding and drinking thus place opposing environmental pressures on the length of neck and legs.

The Big Picture

Evolution is change in the composition of gene pools over time. The changes are driven by the differing fates of individual organisms, whose success or failure in contributing genes to future gene pools depends on both the traits they possess and on chance events.

3 How Does Natural Selection Work?

Natural selection is not the *only* evolutionary force. As we have seen, mutation provides initial variability in heritable traits. The chance effects of genetic drift may change allele frequencies, even spawning new species. Further, evolutionary biologists are now beginning to appreciate the power of random catastrophe in shaping the history of life on Earth—massively destructive events that may exterminate thriving and failing species alike. Nevertheless, it is natural selection that shapes the evolution of populations as they adapt to their changing environment. For this reason, we will examine natural selection in more detail.

Natural Selection Stems from Unequal Reproduction

To most people, the words **natural selection** are synonymous with the phrase "survival of the fittest." Natural selection evokes images of wolves chasing caribou, of lions snarling angrily in competition over a zebra carcass. Natural selection, however, is not about survival alone; it is also about reproduction. It is certainly true that if an organism is to reproduce, it must survive long enough to do so. In some cases, it is also true that a longer-lived organism has more chances to reproduce. But no organism lives forever, and the only way that its genes continue into the future is through successful reproduction. When an organism that fails to reproduce dies, its genes die with it. The organism that reproduces lives on, in a sense, through the genes that it has passed to its offspring. Therefore, although evolutionary biologists often discuss survival, partly because survival is usually easier to observe than reproduction, the main issue of natural selection is *differences in reproduction*: Individuals bearing certain alleles leave more offspring (who inherit those alleles) than do other individuals with different alleles.

MEDIATUTOR
14.2 Natural Selection at Work: Alpine Skypilots

Natural Selection Acts on Phenotypes

Although we have defined evolution as changes in the genetic composition of a population, it is important to recognize that natural selection cannot act directly on the genotypes of individual organisms. Rather, natural selection acts on phenotypes, the actual structures and behaviors displayed by the members of a population. Selection acting on phenotypes, however, inevitably affects the genotypes present in a population, because phenotypes and genotypes are closely tied. For example, we know that a pea plant's height is strongly influenced by the plant's alleles of certain genes. If a population of pea plants were to encounter environmental conditions that favored taller plants, then taller plants would leave more offspring. These offspring would carry the alleles that contributed to their parents' height. Thus, if natural selection favors a particular phenotype, it will necessarily also favor the underlying genotype.

Some Phenotypes Reproduce More Successfully Than Others

As we have seen, natural selection consists of nothing more than the plain fact that some phenotypes reproduce more successfully than others do. What makes this simple process such a powerful agent of change is that only the "best" phenotypes pass traits to subsequent generations. But what makes a phenotype "best"? Successful phenotypes are those that have the best adaptations to their particular environment. **Adaptations** are characteristics that help an individual survive and reproduce.

An Environment Has Nonliving and Living Components

Individual organisms must cope with an environment that includes not only physical factors but also the other organisms with which the individual interacts. The nonliving (abiotic) component of the environment includes such factors as climate, availability of water, and minerals in the soil. The abiotic environment establishes the "bottom line" requirements that an organism must meet to survive and reproduce. However, many of the adaptations that we see in modern organisms have arisen because of interactions with other organisms—the living (biotic) component of the environment. As Darwin wrote, "The structure of every organic being is related . . . to that of all other organic beings, with which it comes into competition for food or residence, or from which it has to escape, or on which it preys." A simple example illustrates this concept.

A buffalo grass plant sprouts in a small patch of soil in the eastern Wyoming plains. Its roots must be able to take up enough water and minerals for growth and reproduction, and to that extent it must be adapted to its abiotic environment. But even in the dry prairies of Wyoming, this requirement is relatively trivial, provided that the plant is alone and protected in its square yard of soil. In reality, however, many other plants—including other buffalo grass plants but also other grasses, sagebrush bushes, and annual wildflowers—also sprout in that same patch of soil. If our buffalo grass is to survive, it must compete for resources with the other plants. Its long, deep roots and efficient methods of mineral uptake have evolved not so much because the plains are dry as because the buffalo grass must share the dry prairies with other plants. Further, buffalo grass must also coexist with animals that wish to eat it, such as the cattle that graze the prairie (and the bison that grazed it in the past). As a result, buffalo grass is extremely tough, with silica (glass) compounds reinforcing its leaves, an adaptation that discourages grazing. Over millennia, tougher plants were harder to eat and so survived better and reproduced more than did less-tough plants—another adaptation to the biotic environment.

Competition Acts as an Agent of Selection

As the buffalo grass example shows, one of the major agents of natural selection in the biotic environment is **competition** with other organisms for scarce resources. Competition for resources is most intense among members of the same species. As Darwin wrote in *On the Origin of Species*, "The struggle almost invariably will be most severe between the individuals of the same species, for they frequent the same districts, require the same food, and are exposed to the same dangers." In other words, no competing organism has such similar requirements for survival as does another member of the same species. Different species may also compete for the same resources, although generally to a lesser extent than do individuals within a species.

Both Predator and Prey Act as Agents of Selection

When two species interact extensively, each exerts strong selection pressures on the other. When one evolves a new feature or modifies an old one, the other typically evolves new adaptations in response. This constant, mutual feedback between two species is called **coevolution**. Perhaps the most familiar form of coevolution is found in predator–prey relationships.

Predation includes any situation in which one organism eats another. In some instances, coevolution between predators (those who do the eating) and prey (those who are eaten) is a sort of "biological arms race," with each side evolving new adaptations in response to "escalations" by the other. Darwin used the example of wolves and deer: Wolf predation selects against slow or careless deer, thus leaving faster, more-alert deer to reproduce and continue the species. In their turn, alert, swift deer select against slow, clumsy wolves, because such predators cannot acquire enough food.

Sexual Selection Favors Traits That Help an Organism Mate

In many animal species, males have conspicuous features such as bright colors, long feathers or fins, or elaborate antlers. Males may also exhibit bizarre courtship behaviors or sing loud, complex songs. While these extravagant features typically play a role in mating, they also seem to be at odds with efficient survival and reproduction. Exaggerated ornaments and displays may help males gain access to females, but also make the males more vulnerable to predators. Darwin was intrigued by this apparent contradiction. He

Figure 14-5 Competition between males favors the evolution, through sexual selection, of structures for ritual combat
Two male bighorn sheep spar during the fall mating season. In many species, the losers of such contests are unlikely to mate, while winners enjoy tremendous reproductive success.

coined the term **sexual selection** to describe the special kind of selection that acts on traits that help an animal acquire a mate.

Darwin recognized that sexual selection could be driven either by sexual contests among males or by female preference for particular male phenotypes. Male-male competition for access to females can favor the evolution of features that provide an advantage in fights or ritual displays of aggression (Fig. 14-5). Female mate choice provides a second source of sexual selection. In animal species in which females actively choose their mates from among males, females often seem to prefer males with the most elaborate ornaments or most extravagant displays (Fig. 14-6). Why? A popular hypothesis is that male structures, colors, and displays that do not enhance survival might instead provide a female with an outward sign of a male's condition. Only a vigorous, energetic male can survive when burdened with conspicuous coloration or a large tail that might make him more vulnerable to predators. Conversely, males that are sick or under parasitic attack are dull and frumpy compared with healthy males. A female that chooses the brightest, most ornamented male is also choosing the healthiest, most vigorous male. By doing so, she gains fitness if, for example, the most vigorous male provides superior parental care to offspring or if he carries alleles for disease resistance that will be inherited by offspring and help ensure their survival. Females thus gain a reproductive advantage by choosing the most highly ornamented males, and the traits (including the exaggerated male ornament) of these flashy males are more likely to be passed to subsequent generations.

Figure 14-6 The peacock's showy tail has evolved through sexual selection
The ancestors of today's peahens were apparently picky when deciding on a male with which to mate, favoring males with longer and more colorful tails. Their sons inherited alleles that made their tails larger and more colorful than average. Females in the next generation also preferred long, colorful tails, and the cycle continued. From countless cycles of female selectivity, the scene pictured above became the species' norm.

The Big Picture

Natural selection is evolution's editor. From life's vast sea of variability, natural selection determines which traits will be preserved and which will be eliminated.

4 What Is a Species?

Although Darwin brilliantly explained how evolution shapes complex, amazingly well-designed organisms, his ideas did not fully explain life's diversity. In particular, the mechanism of natural selection cannot by itself explain how living things came to be divided into groups, with each group distinctly different from all other groups. When we look at big cats, we don't see a continuous array of different tiger phenotypes that gradually grades into a lion phenotype. We see lions and tigers as separate, distinct entities with no overlap. Each distinct group is known as a species.

Biologists Need a Clear Definition of Species

Before we can study the origin of species, we must first clarify our definition of the term. Throughout most of human history, "species" was a poorly defined concept. To most Europeans, the word species simply referred to one of the "kinds" produced by the biblical creation. In this view, humans could not possibly know the criteria of the creator, but could only attempt to distinguish among species on the basis of visible differences in structure. In fact, species is Latin for "appearance."

On a coarse scale, it is easy to use quick visual comparisons to distinguish species. For example, warblers are clearly different from eagles, which are obviously different from ducks. But it is far more difficult to distinguish among different species of warblers or eagles or ducks. By what standard do scientists make these finer distinctions?

Species Are Groups of Interbreeding Populations

Today, biologists define **species** as "groups of actually or potentially interbreeding natural populations, which are reproductively isolated from other such groups." This definition, known as the biological-species concept, has at least one major limitation. It does not help us discern species boundaries among asexually reproducing organisms that do not interbreed. Despite this limitation, most biologists accept the biological-species concept for identifying species of sexually reproducing organisms. But scientists who study bacteria and other organisms whose reproduction is mainly asexual must use alternate species definitions.

Appearance Can Be Misleading

Biologists have found that some organisms with very similar appearances belong to different species. Conversely, differences in appearance do not always mean that two populations belong to different species. For example, bird field guides published in the 1970s list the myrtle warbler and Audubon's warbler (Fig. 14-7) as distinct species; these birds differ in geographic range and in the color of their throat feathers. More recently, scientists decided that these birds are local varieties of the same species. The reason: Where their ranges overlap, these warblers interbreed, and the offspring are just as vigorous and fertile as the parents.

The Big Picture

One of life's most distinctive features is that it comes in so many different forms. Explaining how evolution divides life into species is one of the great challenges of evolutionary biology.

5 How Do New Species Form?

Despite his exhaustive exploration of the process of natural selection, Charles Darwin never proposed a complete mechanism of **speciation**, the process by which new species form. One scientist who did play a large role in describing the process of speciation was Ernst Mayr of Harvard University, an ornithologist (expert on birds) and a pivotal figure in the history of evolutionary biology. Mayr developed the species definition given above. He was also among the first to recognize that speciation depends on two factors acting on a pair of populations: isolation and genetic divergence.

- *Isolation of populations.* If individuals can move freely between two populations, interbreeding will not become difficult or impossible. Gene flow will cause changes in one population to soon become wide-

(a)

(b)

Figure 14-7 Members of a species may differ in appearance
(a) The myrtle warbler and *(b)* Audubon's warbler are members of the same species.

spread in the other as well, and the two populations will not grow increasingly different. Speciation depends on isolation.

- *Genetic divergence of populations.* It is not sufficient for two populations simply to be isolated. They will become separate species only if, during the period of isolation, they evolve sufficiently large genetic differences. The differences must be large enough that, if the isolated populations are reunited, they can no longer interbreed and produce vigorous, fertile offspring. If isolated populations are small, chance events may generate significant genetic differences by genetic drift. In both small and large populations, different environmental pressures in separate environments may favor the evolution of large genetic differences.

Geographic Separation of a Population Can Lead to Speciation

Speciation can occur when an impassible barrier physically separates different parts of a population. Physical separation could occur if, for example, some members of a population of land-dwelling organisms drifted, swam, or flew to a remote oceanic island. Populations of water-dwelling organisms might be split when geological processes such as volcanism or continental drift create new land barriers that subdivide previously continuous seas or lakes. Geological change can also divide terrestrial populations (Fig. 14-8). Portions of populations can become stranded in patches of suitable habitat that become isolated by climate shifts. You can probably imagine myriad other scenarios that could lead to the geographic subdivision of a population.

If two populations become geographically isolated for any reason, there will be no gene flow between them. If the pressures of natural selection differ in the separate locations, then the populations may accumulate genetic differences. Alternatively, genetic differences may arise if one or more of the separated populations is small enough for genetic drift to occur, which may be especially likely in the aftermath of a founder event (in which a few individuals become isolated from the main body of the species). In either case, genetic differences between the separated populations may eventually become large enough to make interbreeding impossible. At that point, the two populations will have become separate species. Most evolutionary biologists believe that geographic isolation followed by speciation has been the most common source of new species, especially among animals.

Under Some Conditions, Many New Species May Arise

In some cases, many new species arise in a relatively short time. This process, called **adaptive radiation**, can occur when populations of one species invade a variety of new habitats and evolve in response to the differing environmental pressures in those habitats. Adaptive radiation has occurred many times and in many groups of organisms, typically when species encounter a wide variety of unoccupied habitats. For example, episodes of adaptive radiation took place when the ancestors of Darwin's finches colonized the Galapagos Islands, when marsupial mammals first invaded Australia, and when an ancestral cichlid fish species arrived at Lake Malawi (Fig. 14-9). In these examples, the invading species faced no competitors except other members of their own species, all the available habitats and food sources were rapidly exploited by new species that evolved from the original invaders.

MediaTutor
14.3 Speciation

6 How Is Reproductive Isolation Between Species Maintained?

The process of speciation is primarily a story about the evolution of mechanisms that prevent interbreeding. Genetic divergence during the period of isolation is a necessary condition for the origin of a new species. It is not, however, a sufficient condition unless the genetic

(a) (b)

Figure 14-8 Geographical isolation To determine if these two squirrels are members of different species, we must know if they're "actually or potentially interbreeding." Unfortunately, it is hard to tell, because ***(a)*** the Kaibab squirrel lives only on the north rim of the Grand Canyon and ***(b)*** the Abert squirrel lives only on the south rim. The two populations are geographically isolated but still quite similar. Have they diverged enough since their separation to be considered separate species? On the basis of our current knowledge, it is impossible to say.

Figure 14-9 Adaptive radiation
More than 300 species of cichlid fishes inhabit Lake Malawi in East Africa. These species are found nowhere else, and all of them descended from a single ancestral population within a million years. This dramatic adaptive radiation has led to a collection of closely related species with an array of adaptations for exploiting the many different food sources in the lake.

divergence includes changes that would block reproduction between the two separated groups were they to come into contact again. If members of a population are unable to interbreed with members of other populations, the population is said to be in **reproductive isolation**.

The structural and behavioral modifications that prevent interbreeding and maintain reproductive isolation are called **isolating mechanisms**. Isolating mechanisms give a clear benefit to individuals. Any individual that mates with a member of another species will probably produce unfit or sterile offspring, thereby wasting its reproductive effort and contributing nothing to future generations. Thus, there is evolutionary pressure to avoid mating across species boundaries. Mechanisms that prevent mating between species are called **pre-mating isolating mechanisms**.

When pre-mating isolating mechanisms fail or have not yet evolved, members of different species may mate. If, however, all resulting hybrid offspring die during development, then the two species are still reproductively isolated from one another. In some cases, however, viable hybrid offspring are produced. Even so, if these hybrids are less fit than their parents or are themselves infertile, the two species may still remain separate, with little or no gene flow between them. Mechanisms that prevent the formation of vigorous, fertile hybrids between species are called **post-mating isolating mechanisms**.

Pre-mating Isolating Mechanisms Prevent Mating Between Species

Reproductive isolation can be maintained by a variety of different mechanisms; those that prevent mating attempts are especially effective. We next describe the most important types of such pre-mating isolating mechanisms.

Members of Different Species May Be Prevented from Meeting

Members of different species cannot mate if they never get near one another. As we have already seen, geographical isolation typically provides the conditions for speciation in the first place. However, we cannot determine if geographically separated populations are actually distinct species. Should the physical barrier separating the two populations disappear (an intervening river might change course, for example), the reunited populations might interbreed freely and not be separate species. If they cannot interbreed, then other mechanisms, such as those considered below, must have developed during their isolation. Geographical isolation, therefore, is usually considered to be a mechanism that allows new species to form rather than a mechanism that maintains reproductive isolation between species.

Different Species May Occupy Different Habitats

Two populations that use different resources may spend time in different habitats within the same general area and thus exhibit ecological isolation. White-crowned and white-throated sparrows, for example, have extensively overlapping ranges. The white-throated sparrow, however, frequents dense thickets, whereas the white-crowned sparrow inhabits fields and meadows, seldom penetrating far into dense growth. The two species may coexist within a few hundred yards of one another and yet seldom meet during the breeding season. A more dramatic example is provided by the more than 750 species of fig wasp (Fig. 14-10). Each species of fig wasp breeds in (and pollinates) the fruits of a particular species of fig, and each fig species hosts one and only one species of pollinating wasp. Although ecological isolation may slow down interbreeding, it seems unlikely that it could prevent gene flow entirely. Other mechanisms normally also contribute to reproductive isolation.

Figure 14-10 Ecological isolation
This female fig wasp is carrying fertilized eggs from a mating that took place within a fig. She will find another fig of the same species, enter it through a pore, lay eggs, and die. Her offspring will hatch, develop, and mate within the fig. Because each species of fig wasp reproduces only in its own particular fig species, each wasp species is reproductively isolated.

Figure 14-11 Temporal isolation
Bishop pines, such as these, and Monterey pines live together in nature. In the laboratory they produce fertile hybrids. In the wild, however, they do not interbreed, because they release pollen at different times of the year.

Different Species May Breed at Different Times

Even if two species occupy similar habitats, they cannot mate if they have different breeding seasons, a phenomenon called temporal (time-related) isolation. Bishop pines and Monterey pines grow together near Monterey on the California coast (Fig. 14-11). Viable hybrids have been produced between these two species in the laboratory. In the wild, however, the two species release their sperm-containing pollen and have eggs ready to receive the pollen at different times: The Monterey pine releases pollen in early spring, the bishop pine in summer. Therefore, the two species never interbreed under natural conditions.

Different Species May Have Different Courtship Rituals

Among animals, the elaborate courtship colors and behaviors that so enthrall human observers serve as recognition and evaluation signals between male and female, but they also prevent mating with members of other species. Signals and behaviors that differ from species to species create behavioral isolation. The striking colors and calls of male songbirds, for example, may attract females of their own species, but females of other species treat them with the utmost indifference. Among frogs, males are often impressively indiscriminate, jumping on every female in sight, regardless of the species, when the spirit moves them. Females, however, approach only male frogs that croak the "ribbet" appropriate to their species. If they do find themselves in an unwanted embrace, they utter the "release call," which causes the male to let go. As a result, few hybrids are produced.

Species' Differing Sexual Organs May Foil Mating Attempts

In rare instances, a male and a female of different species attempt to mate. Their attempt is likely to fail. Among animal species with internal fertilization (in which the sperm is deposited inside the female's reproductive tract), the male's and female's sexual organs simply may not fit together. Among plants, differences in flower size or structure may prevent pollen transfer between species because the differing flowers may attract different pollinators. Isolating mechanisms of this type are called mechanical incompatibilities.

Post-mating Isolating Mechanisms Limit Hybrid Offspring

In some cases pre-mating isolation fails, members of different species mate, and the sperm of one species may reach the egg of another species. Such matings, however, often fail to produce vigorous, fertile hybrid offspring, due to post-mating isolating mechanisms.

One Species' Sperm May Fail to Fertilize Another Species' Eggs

Even if a male inseminates a female of a different species, his sperm may not be able to fertilize her eggs, an isolating mechanism called gametic incompatibility. For example, the fluids of the female reproductive tract may weaken or kill sperm of other species. Among plants, chemical incompatibility may prevent the germination of pollen from one species that lands on the stigma (pollen-catching structure) of the flower of another species.

Hybrid Offspring May Survive Poorly

If cross-species fertilization does occur, the resulting hybrid may be unable to survive, a situation called hybrid inviability. The genetic instructions directing development

Table 14-1 Mechanisms of Reproductive Isolation

Pre-mating Isolating Mechanisms: factors that prevent organisms of two populations from mating

- **Geographical isolation:** The populations cannot interbreed because a physical barrier separates them.
- **Ecological isolation:** The populations do not interbreed even if they are within the same area because they occupy different habitats.
- **Temporal isolation:** The populations cannot interbreed because they breed at different times.
- **Behavioral isolation:** The populations do not interbreed because they have different courtship and mating rituals.
- **Mechanical incompatibility:** Females cannot receive sperm from males of other species because their reproductive structures are incompatible.

Post-mating Isolating Mechanisms: factors that prevent organisms of two populations from producing vigorous, fertile offspring after mating

- **Gametic incompatibility:** Sperm from one population cannot fertilize eggs of another population.
- **Hybrid inviability:** Hybrid offspring fail to survive to maturity.
- **Hybrid infertility:** Hybrid offspring are sterile or have reduced fertility.

of the two species may be so different that hybrids abort early in development. If the hybrid does survive, it may display behaviors that are mixtures of the two parental types. In attempting to do some things the way species *A* does them and other things the way species *B* does them, the hybrid may be hopelessly uncoordinated and therefore unable to reproduce. Hybrids between certain species of lovebirds, for example, have great difficulty learning to carry nest materials during flight and probably could not reproduce in the wild.

Hybrid Offspring May Be Infertile

Most animal hybrids, such as the mule (a cross between a horse and a donkey) or the liger (a zoo-based cross between a male lion and a female tiger), are sterile. Hybrid infertility prevents hybrids from passing on their genetic material to offspring. A common reason for such infertility is the failure of chromosomes to pair properly during meiosis, so eggs and sperm never develop.

Table 14-1 reviews the different types of isolating mechanisms.

7 What Causes Extinction?

Every living organism must eventually die, and the same is true of species. Just like individuals, species are "born" (through the process of speciation), persist for some period of time, and then perish. The ultimate fate of any species is **extinction**, the death of the last of its members. In fact, at least 99.9% of all the species that have ever existed are now extinct. The natural course of evolution, as revealed by fossils, is continual turnover of species as new ones arise and old ones go extinct.

The immediate cause of extinction is probably always environmental change, either in the living or the nonliving parts of the environment. Two major environmental factors that may drive a species to extinction are competition among species and habitat destruction.

Interactions with Other Species May Drive a Species to Extinction

As described earlier, interactions such as competition and predation serve as agents of natural selection. In some cases, these same interactions can lead to extinction rather than to adaptation.

Organisms compete for limited resources in all environments. If a species' competitors evolve superior adaptations and the species doesn't evolve fast enough to keep up, it may become extinct. A particularly striking example of extinction through competition occurred in South America, beginning about 2.5 million years ago. At that time, the isthmus of Panama rose above sea level and formed a land bridge between North America and South America. After the previously separated continents were connected, the mammal species that had evolved in isolation on each continent were able to mix. Many species did indeed expand their ranges, as North American mammals moved southward and South American mammals moved northward. As they moved, each species encountered resident species that occupied the same kinds of habitats and exploited the same kinds of resources. The ultimate result of the ensuing competition was that the North American species diversified and underwent an adaptive radiation that displaced the vast majority of the South American species, many of which went extinct. Clearly, evolution had bestowed on the North American species some (as yet unknown) set of adaptations that enabled their descendants to exploit resources more efficiently and effectively than their South American counterparts could.

Habitat Change and Destruction Are the Leading Causes of Extinction

Habitat change, both contemporary and prehistoric, is the single greatest cause of extinctions. Present-day habitat destruction due to human activities is proceeding at a frightening pace. Many biologists believe that we are presently in the midst of the fastest-paced and most widespread episode of species extinction in the history of life. Loss of tropical forests is especially devastating to species diversity. As many as half the species

Earth Watch

Endangered Species: From Gene Pools to Gene Puddles

Ever since the Endangered Species Act was passed in 1973, the United States has had an official policy of protecting rare species. In fact, the real goal of the act is not protection but recovery; as one U.S. Fish and Wildlife Service official said, "The goal is to get species *off* the list." A species that has been officially listed as endangered can be removed from the list if wildlife biologists determine that its population has grown sufficiently large that it is no longer in danger of extinction from unpredictable events, such as a couple of years of drought or an epidemic of parasites. If a species reaches this critical population size, it is no longer legally "endangered."

Does a "large enough" population (which usually is still very small by historical standards) really ensure a species' survival? From our discussion of genetic drift and population bottlenecks, you probably realize that the answer is no. If the population of a species has been reduced to the point at which it is declared an endangered species, then it probably has lost much of its genetic diversity. Loss of habitat and the consequent reduction in population size mean that, as ecologist Thomas Foose aptly put it, "gene pools are being converted into gene puddles." Even if the species recovers its numbers, the variety of its original gene pool has been lost. When the forces of natural selection change at some future time, the species may not have the necessary genetic variability to produce individuals well suited to survival in the new environment, and it may then go extinct.

What can be done? The best solution, of course, is to save enough habitat of diverse types so that species never become endangered in the first place. The human population, however, has grown so large and appropriated so large a share of Earth's resources that this solution is not possible in many places (Fig. E14-1). For many species, the only solution is to preserve enough habitat so that the remaining population is large enough to retain all or most of the total genetic diversity of the species. If we are to be the caretakers of the planet and not merely its ultimate consumers, then protection of other life-forms and their genetic heritage will be our continuing responsibility.

Figure E14-1 Endangered by habitat destruction
This orangutan and its young, who live in the tropical rain forests of Borneo, are among the innumerable species whose continued survival is threatened by habitat destruction.

presently on Earth may be lost over the next 50 years as the tropical forests that contain them are cut for timber and to clear land for cattle and crops (see "Earth Watch: Endangered Species: From Gene Pools to Gene Puddles."). We will discuss extinctions due to prehistoric habitat change in Chapter 15.

Evolutionary Connections

Scientists Don't Doubt Evolution

In the popular press, conflicts among evolutionary biologists are sometimes seen as conflicts about evolution itself. We occasionally read statements to the effect that new theories are overthrowing Darwin's and casting doubt on the reality of evolution. Nothing could be further from the truth. Despite some disagreements about the details of the evolutionary process, biologists unanimously agree that evolution occurred in the past and is still occurring today. The only argument is over the relative importance of the various mechanisms of evolutionary change in the history of life on Earth, their pace, and which forces were most important in shaping the evolution of particular species. Meanwhile, wolves still tend to catch the slowest caribou, small populations still undergo genetic drift, and habitats still change or disappear. Evolution continues, still generating, in Darwin's words, "endless forms most beautiful."

CASE STUDY REVISITED

Cause of Death: Evolution

The evolution of antibiotic resistance in bacterial populations is a direct consequence of natural selection applied by widespread use of antibiotic drugs. When a new antibiotic is first introduced, it kills the vast majority of bacteria exposed to it. The surviving bacteria, however, may include individuals whose genomes happen to include a mutant allele that confers resistance. Bacteria carrying the resistance allele leave behind a disproportionately large share of offspring, which inherit the allele. If the environment consistently contains an antibiotic, bacteria carrying the resistance allele will eventually come to predominate. Because bacteria reproduce so rapidly and have comparatively high rates of mutation, evolutionary change leading to resistant populations is often rapid.

We have accelerated the pace of the evolution of antibiotic resistance by introducing massive quantities of antibiotics into the bacteria's environment. Each year, U.S. physicians prescribe more than 100 million courses of antibiotics; the Centers for Disease Control estimates that about half of these prescriptions are unnecessary. An additional 20 million pounds of antibiotics are fed to farm animals annually. The use of antibacterial soaps and cleansers has become routine in many households. As a result of this massive alteration of the bacterial environment, resistant bacteria are now found not only in hospitals and the bodies of sick people but are also widespread in our food supply and in the environment. Our heavy use (many would say overuse) of antibiotics means that susceptible bacteria are under constant attack and that resistant strains have little competition. In our fight against disease, we rashly overlooked some basic principles of evolutionary biology and are now paying a heavy price.

Summary of Key Concepts

1) How Are Populations, Genes, and Evolution Related?

Evolution is change in frequencies of alleles in a population's gene pool. Allele frequencies in a population will remain constant over generations only if the following conditions are met: (1) There is no mutation; (2) there is no gene flow; (3) the population is very large; (4) all mating is random; and (5) all genotypes reproduce equally well (that is, there is no natural selection). These conditions are rarely, if ever, met in nature. Understanding what happens when they are not met helps reveal the mechanisms of evolution.

2) What Causes Evolution?

- Mutations are random, undirected changes in DNA composition. Although most mutations are neutral or harmful to the organism, some prove advantageous in certain environments. Mutations are rare and do not change allele frequencies very much, but they provide the raw material for evolution.
- In any population, chance events kill or prevent reproduction by some of the individuals. If the population is small, chance events may eliminate a disproportionate number of individuals who bear a particular allele, thereby greatly changing the allele frequency in the population: This is genetic drift.
- The survival and reproduction of organisms are influenced by their phenotypes. Because phenotype depends at least partly on genotype, natural selection tends to favor the reproduction of certain alleles at the expense of others.

3) How Does Natural Selection Work?

Natural selection is driven by differences in reproductive success among different genotypes. Natural selection proceeds from the interactions of organisms with both the biotic and abiotic parts of their environments. When two or more species exert mutual environmental pressures on each other for long periods of time, both of them evolve in response. Such coevolution can result from any type of relationship between organisms, including competition and predation. Phenotypes that help organisms mate can evolve by sexual selection.

4) What Is a Species?

According to the biological-species concept, a species consists of all the populations of organisms that are potentially capable of interbreeding under natural conditions and that are reproductively isolated from other populations.

5) How Do New Species Form?

Speciation, the formation of new species, takes place when gene flow between two populations is reduced or eliminated and the populations diverge genetically. Most commonly, speciation follows geographical isolation and subsequent genetic divergence of the separated populations through genetic drift or natural selection.

6) How Is Reproductive Isolation Between Species Maintained?

Reproductive isolation between species may be maintained by one or more of several mechanisms, collectively known as pre-mating isolating mechanisms and post-mating isolating mechanisms. Pre-mating isolating mechanisms include geographical isolation, ecological isolation, temporal isolation, behavioral isolation, and mechanical incompatibility. Post-mating isolating mechanisms include gametic incompatibility, hybrid inviability, and hybrid infertility.

7) What Causes Extinction?

Factors that cause extinctions include competition among species and habitat destruction.

Key Terms

adaptation *p. 234*
adaptive radiation *p. 237*
allele frequency *p. 228*
coevolution *p. 234*
competition *p. 234*
equilibrium population *p. 229*
extinction *p. 240*
fitness *p. 232*
founder effect *p. 231*
gene flow *p. 229*
gene pool *p. 228*
genetic drift *p. 229*
Hardy-Weinberg principle *p. 228*
isolating mechanism *p. 238*
mutation *p. 229*
natural selection *p. 235*
population *p. 228*
population bottleneck *p. 231*
post-mating isolating mechanism *p. 238*
predation *p. 234*
pre-mating isolating mechanism *p. 238*
reproductive isolation *p. 238*
sexual selection *p. 235*
speciation *p. 236*
species *p. 236*

Thinking Through the Concepts

Multiple Choice

1. *Genetic drift is a _______________ process.*
   a. random
   b. directed
   c. selection-driven
   d. coevolutionary
   e. uniformitarian
2. *Of the following possibilities, the best way to estimate an organism's evolutionary fitness is to measure the*
   a. size of its offspring
   b. number of eggs it produces
   c. number of eggs it produces over its lifetime
   d. number of offspring it produces over its lifetime
   e. number of offspring it produces over its lifetime that survive to breed
3. *In many species of fireflies, males flash to attract females. Each species has a different flashing pattern. This is probably an example of*
   a. ecological isolation
   b. temporal isolation
   c. geographical isolation
   d. pre-mating isolation
   e. post-mating isolation
4. *Under the biological-species concept, the main criterion for identifying a species is*
   a. anatomical distinctiveness
   b. behavioral distinctiveness
   c. geographic isolation
   d. reproductive isolation
   e. gametic incompatibility
5. *In terms of changes in allele frequencies, founder events result in*
   a. gradual accumulation of many small changes
   b. large, rapid changes
   c. coevolution
   d. hybridization
   e. mechanical incompatibility
6. *After the demise of the dinosaurs, mammals evolved rapidly into many new forms because of*
   a. the founder effect
   b. a genetic bottleneck
   c. adaptive radiation
   d. geological time
   e. genetic drift

? Review Questions

1. What is a gene pool? How would you determine the allele frequencies in a gene pool?
2. Define *equilibrium population*. Outline the conditions that must be met for a population to stay in genetic equilibrium.
3. How does population size affect the likelihood of changes in allele frequencies by chance alone? Can significant changes in allele frequencies (that is, evolution) occur as a result of genetic drift?
4. If you measured the allele frequencies of a gene and found large differences from the proportions predicted by the Hardy-Weinberg principle, would that prove that natural selection is occurring in the population you are studying? Review the conditions that lead to an equilibrium population, and explain your answer.
5. People like to say that "you can't prove a negative." Study the experiment in Figure 14-1 again, and comment on what it demonstrates.
6. What is sexual selection? How is sexual selection similar to and different from other forms of natural selection?
7. Many of the oak tree species in central and eastern North America hybridize (interbreed). Are they "true species"?
8. What are the two major types of reproductive isolating mechanisms? Give examples of each type and describe how they work.

Applying the Concepts

1. By the 1940s the whooping crane population had been reduced to fewer than 50 individuals. Thanks to conservation measures, their numbers are now increasing. What special evolutionary problems do whooping cranes have now that they have passed through a population bottleneck?
2. In many countries, conservationists are trying to design national park systems so that "islands" of natural area

(the big parks) are connected by thin "corridors" of undisturbed habitat. The idea is that this arrangement will allow animals and plants to migrate between refuges. Why would such migration be important?

3. A species is all the populations of organisms that potentially interbreed with one another but that are reproductively isolated from other populations. Using the five conditions of the Hardy-Weinberg principle as a starting point, what factors do you think would be important in the splitting of a single ancestral species into two modern species?
4. The biological-species concept has no meaning with regard to asexual organisms, and it is difficult to apply to extinct organisms that we know only as fossils. Can you devise a meaningful, useful species definition that would apply in all situations?
5. Why do you suppose there are so many *endemic* species—that is, species found nowhere else—on islands? And why have the overwhelming majority of recent extinctions occurred on islands?
6. A contrarian biologist you've met claims that the fact that humans are pushing other species into small, isolated populations is good for biodiversity because these are the conditions that lead to new speciation events. Comment.
7. Southern Wisconsin is home to several populations of gray squirrels (*Sciurus carolinensis*) with black fur. Design a study to determine if they are actually separate species.

Group Activity

Form a group of four students. Your group has been asked to study two recently discovered fish species, both found in waters off a small island. The two species are very similar in general structure, but there are two key differences between them. In species *A*, males are brightly patterned with blue, red, and purple scales, whereas the females are drab, and males are much larger than females. In species *B*, males and females are the same size, and both are a dull brown color that blends in with the sandy bottom. You've decided to do an in-depth study of these two species, but to get your research grant, you must first develop some predictions about what you'll find. What differences in behavior do you predict (list as many as you can think of)? How might the evolutionary histories of the two species differ? Discuss these questions, and submit a brief report of your answers.

For More Information

Dawkins, R. *Climbing Mount Improbable.* New York: Norton, 1996. An eloquent book-length tribute to the power of natural selection to design intricate adaptations. The chapter on the evolution of the eye is an instant classic.

Eldredge, N. *Fossils: The Evolution and Extinction of Species.* New York: Harry N. Abrams, 1991. A nicely illustrated exploration of a paleontologist's approach to examining and interpreting the past, including speciation events.

Gould, S. J. "The Evolution of Life on the Earth." *Scientific American,* October 1994. The importance of chance and catastrophe in shaping modern life.

Grant, P. R. "Natural Selection and Darwin's Finches." *Scientific American,* October 1991. A temporary climate change altering the food supply created selection pressures on finches that altered their average beak shape within several generations.

Levy, S. B. "The Challenge of Antibiotic Resistance." *Scientific American,* March 1998. An excellent summary of the public health implications of antibiotic resistance. Also discusses some strategies for ameliorating the problem.

O'Brien, S. J., Wildt, D. E., and Bush, M. "The Cheetah in Peril." *Scientific American,* May 1986. According to molecular and immunological studies, a population bottleneck has reduced the genetic variation of the world's cheetahs almost to zero.

Quammen, D. *The Song of the Dodo.* New York: Scribner, 1996. Beautifully written exposition of the biology of islands. Read this book to understand why islands are known as "natural laboratories of speciation."

Ryan, M. J. "Signals, Species, and Sexual Selection." *American Scientist,* January–February 1990. Ryan explores a variety of experiments on sexual selection, including the investigation of genetic basis of male characteristics and female choice.

Wilson, E. O. *The Diversity of Life.* New York: W. W. Norton, 1992. Elegant description of how species arise, how they disappear, and why we should preserve them.

Answers to Multiple-Choice Questions

1. a 2. e 3. d 4. d 5. b 6. c

MediaTutor

CD Activities

Activity 14.1: The Bottleneck Effect

Estimated time: 5 minutes

Genetic drift is one of the major mechanisms of evolutionary change and has significant implications for human interactions with natural populations of organisms on Earth. This simple animation illustrates the bottleneck effect, a cause of genetic drift, in a hypothetical population of beetles.

Activity 14.2: Natural Selection at Work: Alpine Skypilots

Estimated time: 15 minutes

The size of alpine skypilot flowers differs between the bare alpine and forested sub-alpine environments. What factors contribute to this difference? Is it natural selection or some other factor at work? This tutorial will allow you to walk through Candace Galen's experiments that solved this mystery.

Activity 14.3: Speciation

Estimated time: 10 minutes

Geographic separation is the primary means by which new species evolve. This animation demonstrates how one species splits to form two that are separated in space. You will be challenged at various stages along the way to predict what will happen next.

Start the MediaTutor Student CD-ROM and enter the activity number in the Quick Search box to be taken directly to that activity.

Web Investigation

Cause of Death: Evolution

Estimated time: 10 minutes

Antibiotic resistance is reaching crisis proportions. Diseases that were all but eradicated only a few years ago are back and resistant to all known treatments. This exercise will take a brief look at the basic biology, medical repercussions, and social implications of bacterial antibiotic resistance.

Go to http://www.prenhall.com/audesirk3, the Audesirk Companion Web site. Select Chapter 14 and Web Investigation to begin.

"[T]hence life was born,
Its nitrogen from ammonia, carbon from methane,
Water from the cloud and salts from the young seas . . . the cells of life
Bound themselves together into clans, a multitude of cells
To make one being—as the molecules before
Had made of many one cell."

Robinson Jeffers in* The Beginning and the End *(1963)

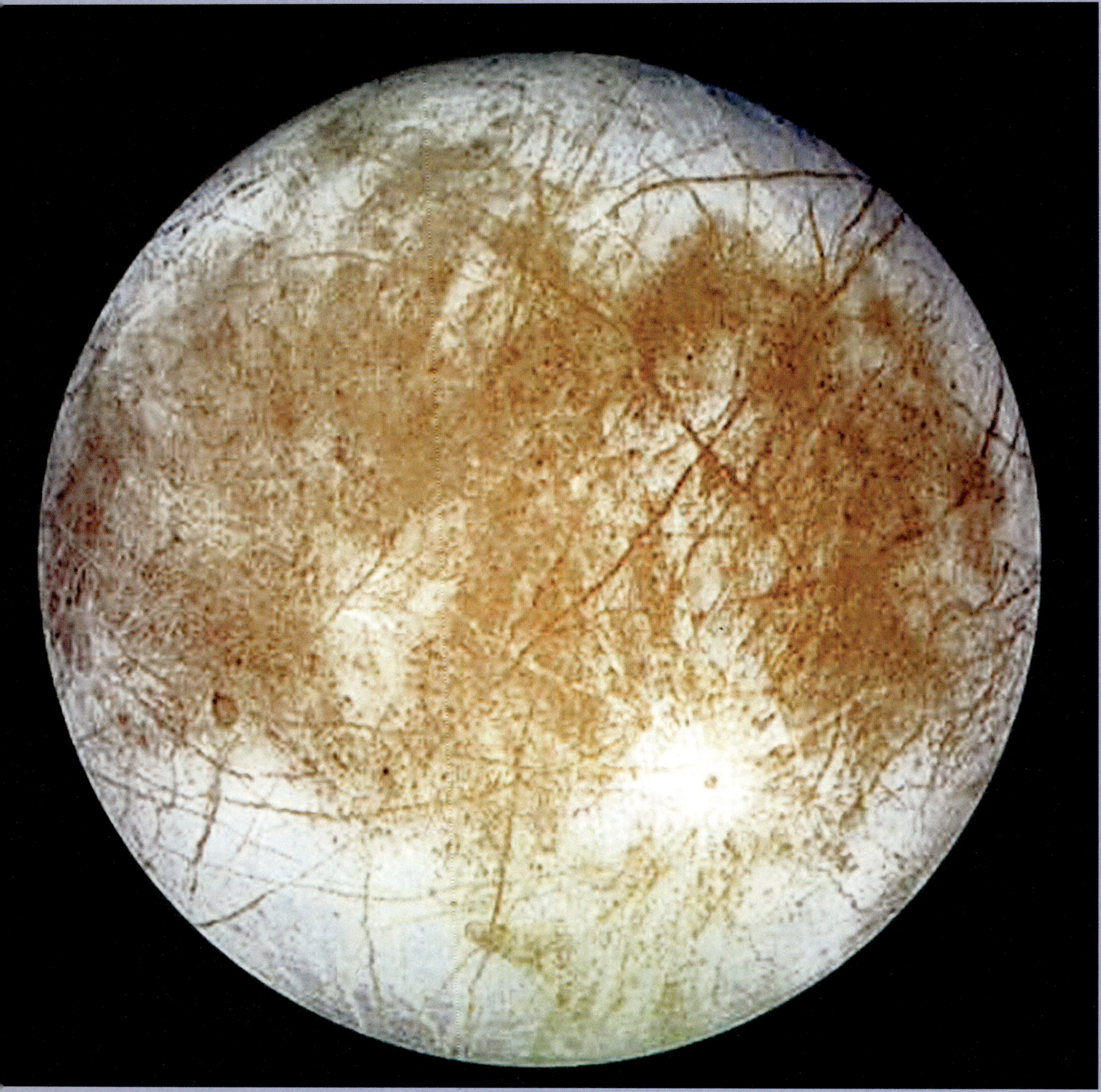

The icy surface of Jupiter's moon Europa may conceal a liquid ocean. Could this extraterrestrial ocean harbor life?

15 The History of Life on Earth

AT A GLANCE

CASESTUDY

Life on a Frozen Moon?

On the cold, dark surface of Europa, all is silent. On the horizon, the huge orange and purple mass of Jupiter looms. High overhead, a spidery metal spacecraft speeds across the sky, its instruments whirring. Within moments it is gone, hurtling back into space.

Back on Earth, a group of NASA scientists gathers around a computer screen. They gaze intently at the screen, examining images of Europa that have just been beamed home by *Galileo*, an unmanned spacecraft. Why so much interest in pictures from one of Jupiter's 16 moons? Because Europa has been deemed to be among the celestial objects most likely to hold extraterrestrial life.

Europa's status as a possible home to extraterrestrial life is new. Data gathered in the 1970s showed the moon's surface to be entirely covered with frozen water, and this icy world did not appear to be particularly hospitable to life. More recently, however, the *Galileo* spacecraft has flown quite close to Europa's surface and has produced a wealth of detailed pictures. These images, together with other data gathered by *Galileo*, have fostered a new outlook on the frozen moon.

The most recent images of Europa show that its frozen surface is littered with huge cracks, humps, and crevices. These deformities indicate that the ice sheet is moving and shifting, and careful analyses by scientists suggest that the observed pattern of cracks is what would be expected if the ice is resting atop liquid water. This seemingly cold and lifeless world may actually conceal a huge, liquid ocean, warmed by heat from the moon's rocky core and hidden deep beneath a vast sheet of ice.

What is the significance of this finding? Water is essential to life; all forms of life on Earth require water to survive. According to our current understanding of the origin of life on Earth, water is a key component of the conditions under which life can arise. The discovery of Europa's shifting ice sheets is our first indication that liquid water may exist somewhere in the solar system other than Earth. If life could arise in the watery environment of early Earth, why not in the watery environment of Europa? ■

1 How Did Life Begin?

How and when did life first appear on Earth? Just a few centuries ago, this question would have been considered trivial. Although no one knew how life *first* arose, people thought that new living things appeared all the time, through **spontaneous generation** from both nonliving matter and other, unrelated forms of life. In 1609 a French botanist wrote, "There is a tree . . . frequently observed in Scotland. From this tree leaves are falling; upon one side they strike the water and slowly turn into fishes, upon the other they strike the land and turn into birds." Medieval writings abound with similar observations and delightful recipes for creating life—even human beings. Microorganisms were thought to arise spontaneously from broth, maggots from meat, and mice from mixtures of sweaty shirts and wheat.

Experiments Refuted Spontaneous Generation

In 1668, Italian physician Francesco Redi disproved the maggots-from-meat hypothesis simply by keeping flies (whose eggs hatch into maggots) away from uncontaminated meat (see Chapter 1). In the mid-1800s Louis Pasteur in France and John Tyndall in England disproved the broth-to-microorganism idea (Fig. 15-1). Although their work effectively demolished the notion of spontaneous generation, it did not address the question of how life on Earth originated in the first place. Or, as biochemist Stanley Miller put it, "Pasteur never proved it didn't happen once, he only showed that it doesn't happen all the time."

The First Living Things Arose from Nonliving Ones

For almost half a century, the subject lay dormant. Eventually, biologists returned to the question of the origin of life. In the 1920s and 1930s, Alexander Oparin in Russia and John B. S. Haldane in England noted that the oxygen-rich atmosphere that we know would not have permitted the spontaneous formation of the complex organic molecules necessary for life. Oxygen reacts readily with other molecules, disrupting chemical bonds and thus tending to keep molecules simple. Oparin and Haldane speculated that the atmosphere of the young Earth was very low in oxygen and rich in hydrogen and that, under these atmospheric conditions, life could have arisen from nonliving matter through ordinary chemical reactions. This process of chemical evolution is called **prebiotic evolution**, or evolution before life existed.

Prebiotic Evolution Was Controlled by the Early Atmosphere and Climate

The primordial Earth differed greatly from the planet we now enjoy. When Earth first formed about 4.5 billion years ago, it was quite hot. A multitude of meteorites smashed into the forming planet, and the kinetic energy of these extraterrestrial rocks was converted into heat on impact. Still more heat was released as radioactive atoms decayed. The rock composing Earth melted, and heavier elements such as iron and nickel sank to the center of the planet, where they remain molten today. Gradually Earth cooled, and elements combined to form compounds of many sorts. Virtually all of the oxygen combined with hydrogen to form water, with carbon to form carbon dioxide, or with heavier elements to form minerals. After millions of years, Earth cooled enough to allow water to exist as a liquid; it must have rained for thousands of years as water vapor condensed out of the cooling atmosphere. As the water struck the surface, it dissolved many minerals, forming a weakly salty ocean. Lightning from storms, heat from volcanoes, and intense ultraviolet light from the sun all poured energy into the young seas.

Judging from the chemical composition of the rocks that formed at this time, geochemists have deduced that the early atmosphere probably contained the gases carbon dioxide, methane, ammonia, hydrogen, nitrogen,

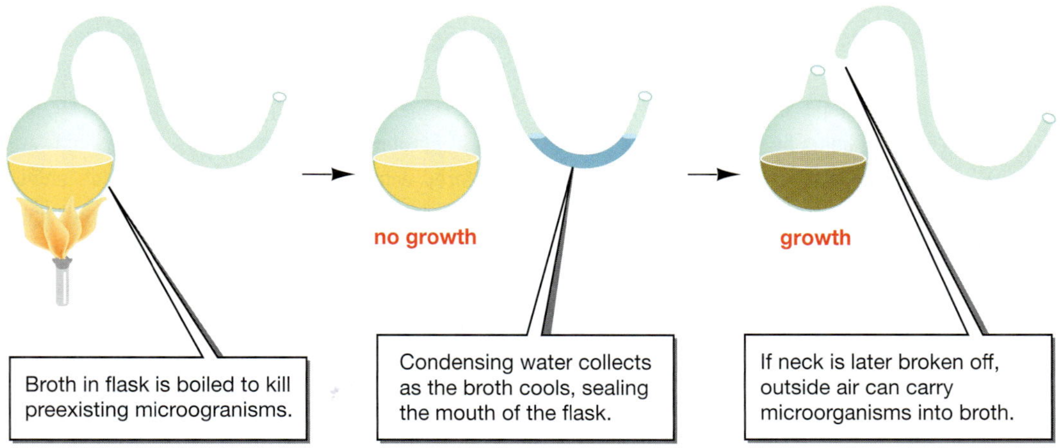

Figure 15-1 Spontaneous generation refuted
Louis Pasteur's experiment disproving the spontaneous generation of microorganisms in broth.

hydrochloric acid, hydrogen sulfide, and water vapor. But there was virtually no oxygen gas in the early atmosphere, because oxygen atoms were bound up in water, carbon dioxide, and minerals.

Organic Molecules Can Form Spontaneously Under Prebiotic Conditions

Inspired by the ideas of Oparin and Haldane, Stanley Miller, then a graduate student, and his advisor, Harold Urey of The University of Chicago, set out in 1953 to simulate prebiotic evolution in the laboratory. They tried to reproduce the atmosphere of early Earth by mixing water vapor, ammonia, hydrogen, and methane in a flask. An electrical discharge mimicked the intense energy of early Earth's lightning storms. In this experimental microcosm, Miller and Urey found that simple organic molecules appeared after just a few days (Fig. 15-2). Similar experiments by Miller and others have produced amino acids, short proteins, nucleotides, adenosine triphosphate (ATP), and other molecules characteristic of living things. Even though we may never know exactly what the earliest atmosphere was like, we can be confident that organic molecules formed spontaneously on early Earth. Additional organic molecules probably arrived from space when meteorites and comets crashed into the surface of Earth. (Analysis of present-day meteorites recovered from impact craters on Earth has revealed that some meteorites contain relatively high concentrations of amino acids and other simple organic molecules similar to those generated in the Miller-Urey experiment.)

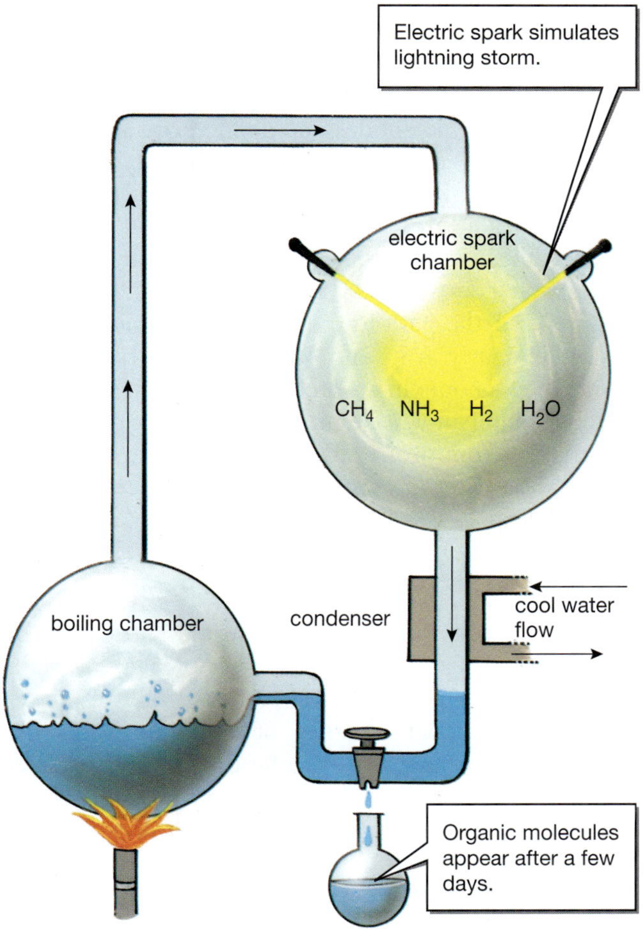

Figure 15-2 The experimental apparatus of Stanley Miller and Harold Urey
Life's very earliest stages left no fossils, so evolutionary historians have pursued a strategy of re-creating in the laboratory the conditions that may have prevailed on early Earth. The mixture of gases in the spark chamber simulates Earth's early atmosphere.

Organic Molecules Can Accumulate Under Prebiotic Conditions

Prebiotic synthesis would not have been very efficient or very fast. Nonetheless, in a few hundred million years, large quantities of organic molecules would have accumulated in the oceans, because they didn't break down very quickly back then. On Earth today, most organic molecules have a short life; they are either digested by living organisms or they react with atmospheric oxygen. Early Earth, however, lacked both life and free oxygen. But the early Earth also lacked an ozone layer, a region high in today's atmosphere that is enriched with ozone (O_3) molecules, which absorb some of the sun's high-energy ultraviolet (UV) light before it reaches Earth. During the time of prebiotic evolution before the ozone layer formed, UV bombardment, which can break apart organic molecules, must have been fierce. Some places, however, such as those beneath rock ledges or at the bottoms of even fairly shallow seas, would have been protected from UV radiation. In these locations, organic molecules may have accumulated.

Organic Molecules May Have Become Concentrated in Tidal Pools

In the next stage of prebiotic evolution, simple molecules combined to form larger molecules. The chemical reactions that formed the larger molecules required that the reacting molecules be packed closely together. Scientists have proposed several mechanisms by which the requisite high concentrations might have been achieved on early Earth. One possibility is that shallow pools at the ocean's edge were filled with water by waves crashing onto the shore during high tides. Afterward, during low tides, some of the water in the pools might have evaporated, concentrating the dissolved substances. Given enough cycles of refilling and evaporation, the molecules in these pools could have become a concentrated "primordial soup" in which spontaneous chemical reactions could generate complex organic molecules. These molecules could then have become the building blocks of the first living organisms.

RNA May Have Been the First Self-Reproducing Molecule

One of the key features of living things is their capacity to reproduce. This capacity depends on the special ability of the DNA molecule to replicate itself. In modern cells, DNA encodes the information the cells need to synthesize the proteins that carry out most cellular functions. Self-replication of DNA ensures that this information is transferred to subsequent generations of cells. But how did this system of self-copying arise? It is far from obvious how any of the components of a primordial soup could have become endowed with the ability to make copies of themselves.

In the 1980s Thomas Cech of the University of Colorado and Sidney Altman of Yale University offered an intriguing solution to this question. They discovered that certain small RNA molecules, dubbed **ribozymes**, act as enzymes that catalyze cellular reactions, including the synthesis of more RNA molecules. During hundreds of millions of years of prebiotic chemical synthesis, RNA nucleotides would have been among the molecules synthesized, and it seems inevitable that some of them may have bonded together to form short RNA chains. Let us suppose that, purely by chance, one of these RNA chains was a ribozyme that could catalyze the synthesis of copies of itself from free ribonucleotides in the surrounding waters. This first ribozyme probably wasn't very good at its job and made lots of mistakes. These mistakes were the first mutations. Like modern mutations, most of the mistakes undoubtedly ruined the catalytic abilities of the "daughter molecules," but a few may have been improvements. Molecular evolution could begin, as ribozymes with increased speed and accuracy of replication reproduced faster, making more and more copies of themselves.

Membrane-Like Microspheres May Have Enclosed Ribozymes

Self-replicating molecules alone do not constitute life; some kind of enclosing membrane is also required. The precursors of the earliest biological membranes may have been simple structures that formed spontaneously from purely physical, mechanical processes. For example, chemists have shown that if water that contains proteins and lipids is agitated to simulate waves beating against ancient shores, the proteins and lipids combine to form hollow structures called **microspheres** (Fig. 15-3). These hollow balls resemble living cells in several respects. They have a well-defined outer boundary that separates internal contents from the external solution. If the composition of the microsphere is right, a "membrane" forms that is remarkably similar in appearance to a real cell membrane. Under certain conditions microspheres can absorb material from the solution ("feed"), grow, and even divide.

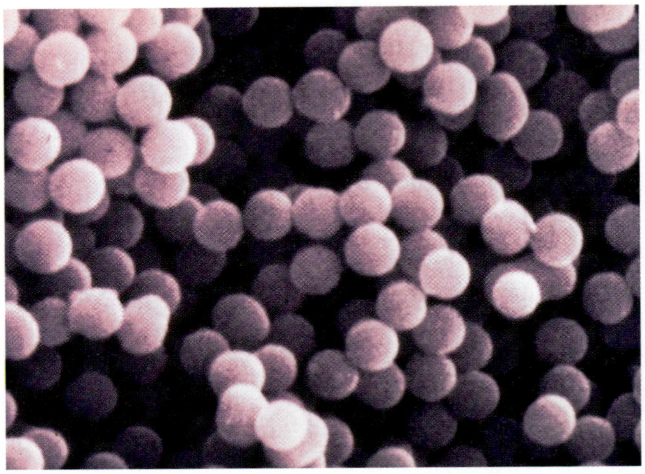

Figure 15-3 Did microspheres enclose the earliest cells? Cell-like microspheres can be formed by agitating proteins and lipids in a liquid medium. Each microsphere is this photo is about 5 μm in diameter.

If a microsphere happened to surround the right ribozymes, something a bit like a living cell would have been formed. We could call it a **protocell**, structurally similar to a cell but not a living thing. In the protocell, the ribozymes and their protein products would have been protected from free-roaming ribozymes in the primordial soup. Nucleotides and amino acids might have diffused across the membrane and might have been used to synthesize new RNA and protein molecules. After sufficient growth, the microsphere may have divided, with a few copies of both ribozymes and proteins becoming incorporated into each daughter microsphere. If so, the path to the evolution of the first cells would be nearly at its end. Was there a particular moment when a nonliving protocell gave rise to something alive? Probably not. Like most evolutionary transitions, the change from protocell to living cell was a continuous process, with no sharp boundary between one state and the next.

But Did All This Happen?

The above scenario, although plausible and consistent with many research findings, is by no means certain. One of the most striking aspects of origin-of-life research is the great diversity of assumptions, experiments, and contradictory hypotheses (see the suggested reading by Iris Fry, *The Emergence of Life on Earth*, for a taste of the controversies). Researchers disagree as to whether life arose in quiet pools, in the sea, in moist films on the surfaces of clay or iron pyrite (fool's gold), or in furiously hot deep-sea vents. (Proponents of a deep-sea vent origin of life suggest that similar vents might keep Europa's seas thawed and provide the conditions for an independent origin of life there.) A few researchers even argue that life arrived on Earth from space. Can we draw any conclusions from the research done so far? No one really knows the answer to that question, but we can offer a few observations.

First, the experiments of Miller and others show that amino acids, nucleotides, and other organic molecules, along with simple membrane-like structures, would have formed in abundance on the early Earth. Second, chemical evolution had long periods of time and huge areas of the Earth available to it. Given sufficient time and a sufficiently large pool of reactant molecules, even extremely rare events can occur many times. So, even if prebiotic evolution yielded only simple molecules, and the earliest catalysts were not very efficient, and the earliest membranes were simple, vast expanses of available time and space would have increased the likelihood of each small step on the path from primordial soup to living cell.

Most biologists accept that the origin of life was probably an inevitable consequence of the working of natural laws. We should emphasize, however, that this proposition is not proved and never will be. Biologists investigating the origin of life have neither millions of years nor trillions of gallons of reaction solutions with which to work!

The Big Picture

One of the grandest and most profound questions that a human can ask is: How did life begin? The transition from nonlife to life is shrouded in mystery, and the nature of the transition will never be known with any certainty. Still, laboratory experiments have yielded results that form the basis for some scientifically informed speculation about life's origin.

2 What Were the Earliest Organisms Like?

Life began during the *Precambrian era*, an interval designated by geologists and paleontologists, who have devised a hierarchical naming system of eras, periods, and epochs to delineate the immense span of geologic time (Table 15-1). The oldest fossil organisms found so far are in Precambrian rocks that are about 3.5 billion years old, based on radiometric dating techniques (see "Scientific Inquiry: How Do We Know How Old a Fossil Is?"). Chemical traces in older rocks have led some paleontologists to believe that life is even older, perhaps as old as 3.9 billion years.

The First Organisms Were Anaerobic Prokaryotes

The first cells to arise in Earth's oceans were **prokaryotes**; that is, their genetic material was not contained within a nucleus, separate from the rest of the cell. These cells probably obtained nutrients and energy by absorbing organic molecules from their environment. There was no oxygen gas in the atmosphere, so the cells must have metabolized the organic molecules anaerobically. You will recall from Chapter 7 that anaerobic metabolism yields only small amounts of energy.

The earliest cells were thus primitive anaerobic bacteria. As these bacteria multiplied, they must have eventually used up the organic molecules produced by prebiotic chemical reactions. Simpler molecules, such as carbon dioxide and water, would still have been very abundant, as was energy, in the form of sunlight. What was lacking, then, was not materials or energy itself, but energetic molecules—molecules in which energy is stored in chemical bonds.

Some Organisms Evolved the Ability to Capture the Sun's Energy

Eventually, some cells evolved the ability to use the energy of sunlight to drive the synthesis of complex, high-energy molecules from simpler molecules: In other words, photosynthesis appeared. Photosynthesis requires a source of hydrogen, and the very earliest photosynthetic bacteria probably used hydrogen sulfide gas dissolved in water for this purpose (much as today's purple photosynthetic bacteria do). Eventually, however, Earth's supply of hydrogen sulfide (which is produced mainly by volcanoes) must have run low. The shortage of hydrogen sulfide set the stage for the evolution of photosynthetic bacteria that were able to use the planet's most abundant source of hydrogen: water (H_2O).

Photosynthesis Increased the Amount of Oxygen in the Atmosphere

Water-based photosynthesis converts water and carbon dioxide to energetic molecules of sugar, releasing oxygen as a by-product. The emergence of this new method for capturing energy introduced significant amounts of free oxygen to the atmosphere for the first time. At first, the newly liberated oxygen was quickly consumed by reactions with other molecules in the atmosphere and in Earth's crust, or surface layer. One especially common reactive atom in the crust was iron, and much of the new oxygen combined with iron atoms to form huge deposits of iron oxide (also known as rust).

Once all the accessible iron had turned to rust, the concentration of free oxygen in the atmosphere began to increase. Chemical analysis of rocks suggests that significant amounts of free oxygen first appeared in the atmosphere about 2.2 billion years ago, produced by bacteria that were probably very similar to modern cyanobacteria (you will undoubtedly breathe in some oxygen molecules today that were expelled 2 billion years ago by one of those cyanobacteria). Atmospheric oxygen levels increased steadily until reaching a stable level about 1.5 billion years ago. Since that time, the proportion of oxygen in the atmosphere has been nearly constant, as

Table 15-1 The History of Life on Earth

Era	Period	Epoch	Millions of Years Ago*	Major Events
Precambrian			4600	Origin of solar system and Earth.
			4000–3900	Appearance of first rocks on Earth.
			3900–3500	First living cells (prokaryotes).
			3500	Origin of photosynthesis (in cyanobacteria).
			2200	Accumulation of free oxygen in atmosphere.
			2000–1700	First eukaryotes.
			By 1000	First multicellular organisms.
			About 1000	First animals (soft-bodied marine invertebrates).
Paleozoic	Cambrian		544–505	Primitive marine algae flourish; origin of most marine invertebrate types; first fishes.
	Ordovician		505–440	Invertebrates, especially arthropods and mollusks, dominant in sea; first fungi.
	Silurian		440–410	Many fishes, trilobites, mollusks in sea; first vascular plants; invasion of land by plants; invasion of land by arthropods.
	Devonian		410–360	Fishes and trilobites flourish in sea; first amphibians and insects; first seeds and pollen.
	Carboniferous		360–286	Swamp forests of tree ferns and club mosses; first conifers; dominance of amphibians; numerous insects; first reptiles.
	Permian		286–245	Massive marine extinctions, including last of trilobites; flourishing of reptiles and decline of amphibians; continents aggregated into one land mass, Pangaea.
Mesozoic	Triassic		245–208	First mammals and dinosaurs; forests of gymnosperms and tree ferns; breakup of Pangaea begins.
	Jurassic		208–146	Dominance of dinosaurs and conifers; first birds; continents partially separated.
	Cretaceous		146–65	Flowering plants appear and become dominant; mass extinctions of marine life and some terrestrial life, including last dinosaurs; modern continents well separated.
Cenozoic	Tertiary	Paleocene Eocene Oligocene Miocene Pliocene	65–54 54–38 38–23 23–5 5–1.8	Widespread flourishing of birds, mammals, insects, and flowering plants; shifting of continents into modern positions; mild climate at beginning of period, with extensive mountain building and cooling toward end.
	Quaternary	Pleistocene Recent	1.8–0.01 0.01–present	Evolution of genus *Homo*; repeated glaciations in Northern Hemisphere; extinction of many giant mammals.

* From University of California Museum of Paleontology, April 2000.

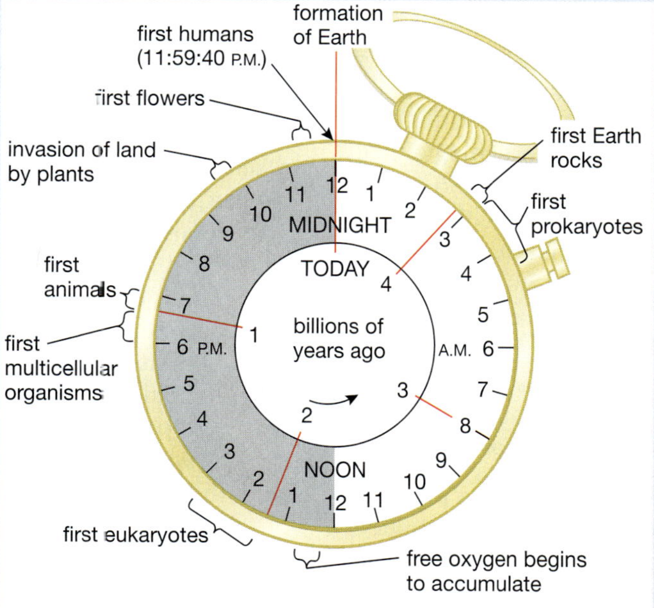

Earth's history projected on a 24-hour day

Scientific Inquiry
How Do We Know How Old a Fossil Is?

Early geologists could date rock layers and their accompanying fossils only in a *relative* way: Fossils found in deeper layers of rock were generally older than those found in shallower layers. With the discovery of radioactivity, it became possible to determine *absolute* dates, within certain limits of uncertainty. The nuclei of radioactive elements spontaneously break down, or decay, into other elements. For example, carbon-14 (usually written ^{14}C) emits an electron to become nitrogen-14 (^{14}N). Each radioactive element decays at a rate that is independent of temperature, pressure, or the chemical compound of which the element is a part. The time it takes for half the radioactive nuclei to decay at this characteristic rate is called the *half-life*. The half-life of ^{14}C, for example, is 5730 years.

How are radioactive elements used in determining the age of rocks? If we know the rate of decay and measure the proportion of decayed nuclei to undecayed nuclei, we can estimate how much time has passed since these radioactive elements were trapped in rock. This process is called *radiometric dating*. A particularly straightforward dating technique uses the decay of potassium-40 (^{40}K), which has a half-life of about 1.25 billion years, into argon-40 (^{40}Ar). Potassium is a very reactive element commonly found in volcanic rocks such as granite and basalt. Argon, however, is an unreactive gas. Let us suppose that a volcano erupts with a massive lava flow, covering the countryside. All the ^{40}Ar, being a gas, will bubble out of the molten lava, so when the lava cools and solidifies into rock, it will start out with no ^{40}Ar. Any ^{40}K present in the hardened lava will decay to ^{40}Ar, with half of the ^{40}K decaying every 1.25 billion years. The ^{40}Ar gas will be trapped in the rock. A geologist could take a sample of the rock and determine the proportion of ^{40}K to ^{40}Ar (Fig. E15-1). If the analysis finds equal amounts of the two elements, the geologist will conclude that the lava hardened 1.25 billion years ago. With appropriate care, such age estimates are quite reliable. If a fossil is found beneath a lava flow dated at, say, 500 million years, then we know that the fossil is at least that old.

Some radioactive elements, as they decay, can even give an estimate of the age of the solar system. Analysis of uranium, which decays to lead, has shown that the oldest meteorites and moon rocks collected by astronauts are about 4.6 billion years old.

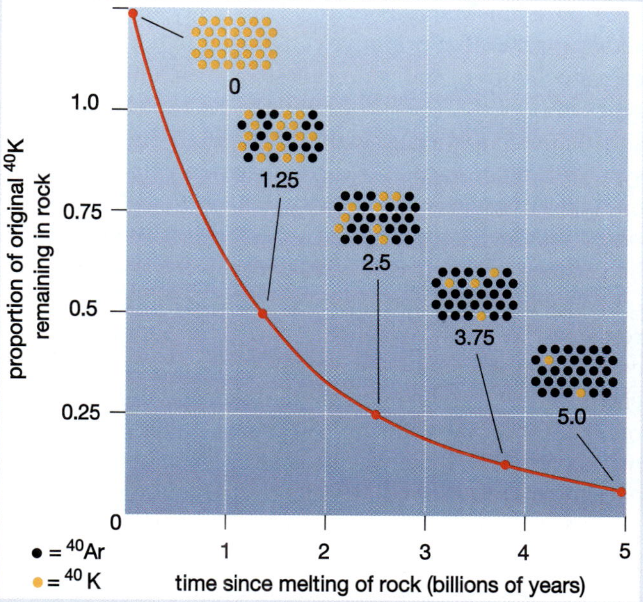

Figure E15-1 The relationship between time and the decay of radioactive ^{40}K to ^{40}Ar

the amount of oxygen released by photosynthesis worldwide is neatly balanced by the amount that is consumed by aerobic respiration.

Aerobic Metabolism Arose in Response to the Oxygen Crisis

Oxygen is potentially very dangerous to living things, because it reacts with organic molecules, destroying them. Many of today's anaerobic bacteria perish when exposed to what is for them a deadly poison, oxygen. The accumulation of oxygen in the atmosphere of early Earth probably exterminated many organisms and fostered the evolution of cellular mechanisms for detoxifying oxygen. This crisis for evolving life also provided the environmental pressure for the next great advance in the Age of Microbes: the ability to use oxygen in metabolism. This ability not only provides a defense against the chemical action of oxygen, but actually channels oxygen's destructive power through aerobic respiration to generate useful energy for the cell. Because the amount of energy available to a cell is vastly increased when oxygen is used to metabolize food molecules, aerobic cells had a significant selective advantage.

Some Organisms Acquired Membrane-Enclosed Organelles

Hordes of bacteria would offer a rich food supply to any organism that could eat them. There are no fossils of the first predatory cells to roam the seas, but paleobiologists speculate that once a suitable prey population (such as these bacteria) appeared, predation would have evolved quickly. These early predators would have been specialized prokaryotic cells that were able to engulf whole bacteria as prey. According to the most widely accepted hypothesis, these predators were otherwise quite primitive, being capable of neither photosynthesis nor aerobic metabolism. Although they could capture large food particles, namely bacteria, they metabolized them inefficiently. By about 1.7 billion years ago, however, one predator probably gave rise to the first eukaryotic cell.

As you know, the cells of **eukaryotes** differ from prokaryotic cells not only in having a nucleus that contains the cell's genetic material but in other ways as well. Other key eukaryotic structures are the organelles used for energy metabolism: mitochondria and (in plants only) chloroplasts. How did these organelles evolve?

Mitochondria and Chloroplasts May Have Arisen from Engulfed Bacteria

The **endosymbiont hypothesis**, championed most forcefully by Lynn Margulis of the University of Massachusetts, proposes that primitive cells acquired the precursors of mitochondria and chloroplasts by engulfing certain types of bacteria. These cells and the bacteria trapped inside them (*endo* means "within") gradually entered into a *symbiotic* relationship, a close association between different types of organisms over an extended time. Let us suppose that an anaerobic predatory cell captured an aerobic bacterium for food, as it often did, but for some reason failed to digest this particular prey. The aerobic bacterium remained alive and well. In fact, it was better off than ever, because the cytoplasm of its predator–host was chock-full of half-digested food molecules, the remnants of anaerobic metabolism. The aerobe absorbed these molecules and used oxygen to metabolize them, thereby gaining enormous amounts of energy. So abundant were the aerobe's food resources, and so bountiful its energy production, that the aerobe must have leaked energy, probably as ATP or similar molecules, back into its host's cytoplasm. The anaerobic predatory cell with its symbiotic bacteria could now metabolize food aerobically, gaining a great advantage over other anaerobic cells and leaving a greater number of offspring. Eventually, the endosymbiotic bacterium lost its ability to live independently of its host, and the mitochondrion was born (Fig. 15-4, ① and ②).

MediaTutor 15.1 Endosymbiosis

One of these successful new cellular partnerships must have managed a second feat: It captured a photosynthetic cyanobacterium and similarly failed to digest its prey. The cyanobacterium flourished in its new host and gradually evolved into the first chloroplast (Fig. 15-4, ③ and ④). Other eukaryotic organelles may have also originated through endosymbiosis. Many biologists believe that cilia, flagella, centrioles, and microtubules may all have evolved from a symbiosis between a spirilla-like bacterium (a form of bacterium with an elongated corkscrew shape) and a primitive eukaryotic cell.

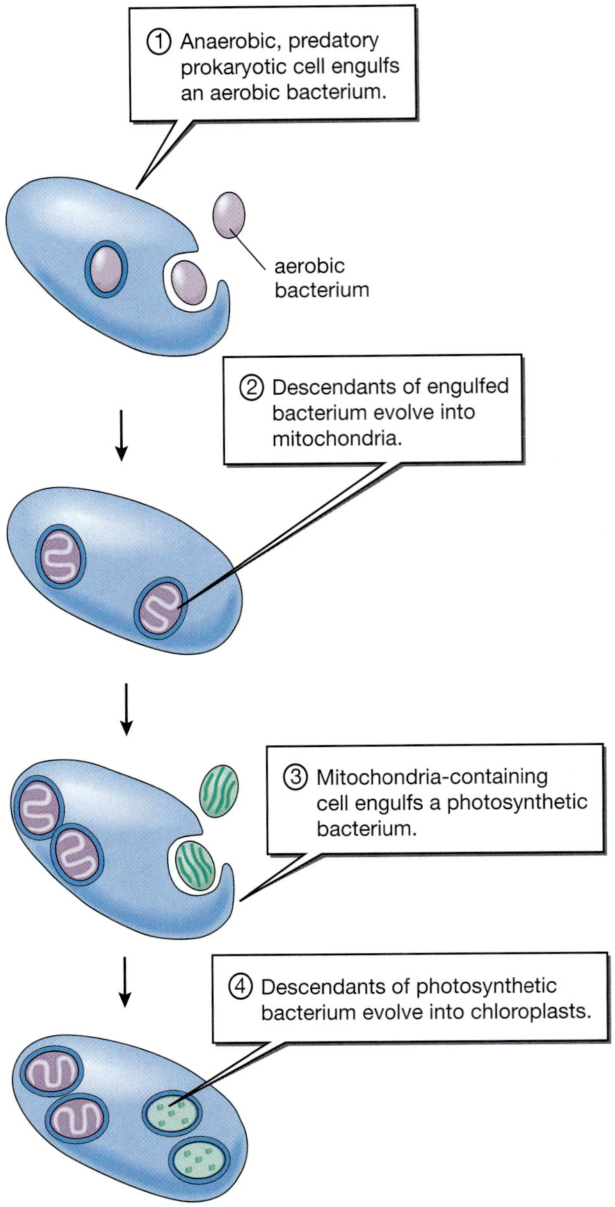

Figure 15-4 The probable origin of mitochondria and chloroplasts in eukaryotic cells

Evidence for the Endosymbiont Hypothesis Is Strong

Several types of evidence support the endosymbiont hypothesis. A particularly compelling line of evidence is the many distinctive biochemical features shared by eukaryotic organelles and living bacteria. In addition, mitochondria, chloroplasts, and centrioles each contain their own minute supply of DNA, which many researchers interpret as remnants of the DNA originally contained within the engulfed bacteria. Another kind of support comes from *living intermediates*, organisms alive today that are similar to hypothetical ancestors and thus help show that a proposed evolutionary pathway is plausible. For example, the amoeba *Pelomyxa palustris* lacks mitochondria but hosts a permanent population of aerobic bacteria that carry out much the same role. Similarly, a variety of corals, some clams, a few snails, and at least one species of *Paramecium* harbor a permanent collection of photosynthetic algae in their cells (Fig. 15-5). These examples of modern cells that host bacterial endosymbionts suggest that we have no

Figure 15-5 A modern symbiosis within a cell
The ancestors of the chloroplasts in today's plant cells may have resembled *Chlorella*, the green, photosynthetic, single-celled algae living symbiotically within the cytoplasm of the *Paramecium* pictured here.

reason to doubt that similar symbiotic associations could have occurred almost 2 billion years ago and led to the first eukaryotic cells.

The Big Picture

Some biologists are historians whose task is to reconstruct the 3.5-billion-year history of life on Earth. These scientists cannot, of course, directly observe the evolutionary changes that occurred in the past. Instead, evolutionary historians must infer the events of the past from whatever evidence is available in the present.

3 What Were the Earliest Multicellular Organisms Like?

Once predation had evolved, increased size became an advantage. In the marine environments to which life was restricted, a larger cell could more easily engulf a smaller cell while also being more difficult for other predatory cells to ingest. Most larger organisms can also move faster than small ones, making successful predation and escape more likely. But enormous single cells have problems. Oxygen and nutrients going into the cell and waste products going out must diffuse through the plasma membrane. As we pointed out in Chapter 4, the larger a cell becomes, the less surface membrane is available per unit volume of cytoplasm. There are only two ways that an organism larger than a millimeter or so in diameter can survive. First, it can have a low metabolic rate so that it doesn't need much oxygen or produce much carbon dioxide. This seems to work for certain very large single-celled algae. Alternatively, an organism may be multicellular—that is, it may consist of many small cells packaged into a larger, unified body.

Some Algae Became Multicellular

The oldest fossils of multicellular organisms are about one billion years old and include impressions of the first multicellular algae, which arose from single-celled eukaryotic cells containing chloroplasts. Multicellularity would have provided at least two advantages for these seaweeds. First, large, many-celled algae would have been difficult for single-celled predators to engulf. Second, specialization of cells would have conferred the potential for staying in one place in the brightly lit waters of the shoreline, as rootlike structures burrowed in sand or clutched onto rocks, while leaflike structures floated above in the sunlight. The green, brown, and red algae lining our shores today—some, such as the brown kelp, more than 200 feet long—are the descendants of these early multicellular algae.

Animal Diversity Arose in the Precambrian

In addition to fossil algae, billion-year-old rocks have yielded fossil traces of animal tracks and burrows. This evidence of early animal life notwithstanding, fossils of animal bodies first appear in rocks laid down between 650 million and 544 million years ago. Many of these ancient invertebrate animals (animals lacking backbones) are quite different in appearance from any animals that appear in later fossil layers. Some paleontologists believe that they represent a kind of failed evolutionary experiment in body plans that failed to survive the test of time. Most of these Precambrian animals appear rather suddenly in the fossil record and disappear just as suddenly, so it is difficult to determine either their ancestry or the fate of their descendants. (The phrase "fossil record" is a shorthand reference to the entire collection of all fossil evidence that has been found to date.) Fossil ancestors of today's sponges, however, have been found in Precambrian rocks about 570 million years old, so at least some representatives of modern animal groups had arisen by that time. For the most part, though, the familiar forms of modern invertebrate animals do not appear as fossils until the Cambrian period, marking the beginning of the Paleozoic era, about 544 million years ago. Like the Precambrian animals, the first fossils of Cambrian animals reveal an adaptive radiation that had already yielded a diverse array of complex body plans. The evolutionary history that produced such an impressive range of different animal forms remains unknown.

For animals, one of the advantages of multicellularity is the potential for eating larger prey. Coevolution of predator and prey led to increased sophistication in many kinds of animals. By the Silurian period (440 million to 410 million years ago), mud-skimming, armored

Figure 15-6 Diversity of ocean life during the Silurian period
(a) Artist's reconstruction of characteristic life of the oceans during the Silurian period, 440 million to 410 million years ago. Among the most common fossils from that time are ***(b)*** the trilobites and their predators, the nautiloids, and ***(c)*** the ammonites. ***(d)*** This living *Nautilus* is very similar in structure to the Silurian nautiloids, showing that a successful body plan may exist virtually unchanged for hundreds of millions of years.

trilobites were preyed on by (now-extinct) ammonites and the chambered nautilus, which still survives almost unchanged in deep Pacific waters (Fig. 15-6). Many animals of this era were more mobile than their evolutionary predecessors. Predators gain an advantage by being able to travel over wide areas in search of suitable prey, and speedy escape is also an advantage for prey. The evolution of an ability to move more efficiently was often associated with the evolution of greater sensory capabilities and more-sophisticated nervous systems. Senses for detecting touch, chemicals, and light became highly developed, along with nervous systems capable of handling the sensory information and directing appropriate behaviors.

About 500 million years ago, one group of animals—the fishes—developed a new form of support for the body: an internal skeleton. These early fishes were inconspicuous members of the ocean community, but by 400 million years ago fishes were a diverse and prominent group. By and large, the fishes proved to be faster than the invertebrates, with more-acute senses and larger brains. Eventually, they became the dominant predators of the open seas.

The Big Picture

For most of its history, life was exclusively single-celled and restricted to watery environments. On evolution's time scale, multicelled, land-dwelling organisms are recent developments, newcomers whose time on Earth has been brief compared to the nearly 4-billion-year history of marine microbes.

4 How Did Life Invade the Land?

A compelling subplot in the long tale of life's history is the story of life's invasion of land after more than 3 billion years of a strictly watery existence. In moving to solid ground, organisms had many obstacles to overcome. Life in the sea provides buoyant support against gravity, but on land an organism must bear its weight against the crushing force of gravity. The sea provides ready access to life-sustaining water, but a terrestrial organism must find adequate water. Sea-dwelling plants and animals can reproduce by means of mobile sperm and/or eggs that swim to each other through the water, but the gametes of land-dwellers must be protected from drying out.

Despite the obstacles to life on land, the vast empty spaces of the Paleozoic landmass represented a tremendous evolutionary opportunity. The potential rewards of terrestrial life were especially great for plants. Water strongly absorbs light, so even in the clearest water, photosynthesis is limited to the upper few hundred meters, and usually much less. Out of the water, the sun is dazzlingly bright, permitting rapid photosynthesis. Furthermore, terrestrial soils are rich storehouses of nutrients, whereas seawater tends to be low in certain nutrients, particularly nitrogen and phosphorus. Finally, the Paleozoic sea swarmed with plant-eating animals, but the land was devoid of animal life. The plants that first colonized the land would have had ample sunlight, untouched nutrient sources, and no predators.

Some Plants Became Adapted to Life on Dry Land

In moist soils at water's edge, a few small green algae began to grow, taking advantage of the sunlight and nutrients. They didn't have large bodies to support against the force of gravity, and, living right in the film of water on the soil, they could easily obtain water. About 400 million years ago, some of these algae gave rise to the first multicellular land plants. Initially simple, low-growing forms, land plants rapidly evolved solutions to two of the main difficulties of plant life on land: (1) obtaining and conserving water and (2) staying upright despite gravity and winds. Waterproof coatings on aboveground parts reduced water loss by evaporation, and rootlike structures delved into the soil, mining water and minerals. Specialized cells formed tubes called vascular tissues to conduct water from roots to leaves. Extra-thick walls surrounding certain cells enabled stems to stand erect.

Primitive Land Plants Retained Swimming Sperm and Required Water to Reproduce

Reproduction out of water presented challenges. Like animals, plants produce sperm and eggs, which must be able to meet to produce the next generation. The first land plants had swimming sperm, presumably much like those of some of today's marine algae that have swimming sperm (and in some cases swimming eggs as well). Consequently, the earliest plants were restricted to swamps and marshes, where the sperm and eggs could be released into the water, or to areas with abundant rainfall, where the ground would occasionally be covered with water. Later, plants with swimming sperm prospered during periods in which the climate was warm and moist. For example, the Carboniferous period (360 million to 286 million years ago) was characterized by vast forests of giant tree ferns and club mosses (Fig. 15-7). The coal we mine today is derived from the fossilized remains of those forests.

Seed Plants Encased Sperm in Pollen Grains

Meanwhile, some plants inhabiting drier regions had evolved a means of reproduction that no longer depended on water. The eggs of these plants were retained on the parent plant, and the sperm were encased in

Figure 15-7 The swamp forest of the Carboniferous period
The treelike plants in this artist's reconstruction are tree ferns and giant club mosses, most species of which are now extinct.

drought-resistant pollen grains that blew on the wind from plant to plant. When the pollen grains landed near an egg, they released sperm cells directly into living tissue, eliminating the need for a surface film of water. The fertilized egg remained on the parent plant, where it developed inside a seed, which provided protection and nutrients for the developing embryo within.

The earliest seed-bearing plants appeared in the late Devonian period (375 million years ago) and produced their seeds along branches, without any specialized structures to hold them. By the middle of the Carboniferous period, however, a new kind of seed-bearing plant had arisen. These plants, called **conifers**, protected their developing seeds inside cones. Conifers, which did not depend on water for reproduction, flourished and spread during the Permian period (286 to 245 million years ago), when mountains rose, swamps drained, and the climate became much drier. The good fortune of the conifers, however, was not shared by the tree ferns and giant club mosses, which, with their swimming sperm, largely went extinct.

Flowering Plants Enticed Animals to Carry Pollen

About 140 million years ago, during the Cretaceous period, the flowering plants appeared, having evolved from a group of conifer-like plants. Many flowering plants are pollinated by insects and other animals, and this mode of pollination seems to have conferred an evolutionary advantage. Flower pollination by animals can be far more efficient than pollination by wind; wind-pollinated plants must produce an enormous amount of pollen because the vast majority of pollen grains fail to reach their target. Flowering plants also evolved other advantages, including more-rapid reproduction and, in some cases, much more rapid growth. Today, flowering plants dominate the land, except in cold northern regions, where conifers still prevail.

Some Animals Became Adapted to Life on Dry Land

Soon after land plants evolved, providing potential food sources for other organisms, animals emerged from the sea. The first animals to move onto land were **arthropods** (the group that today includes insects, spiders, scorpions, centipedes, and crabs). Why arthropods? The answer seems to be that they already possessed certain structures that, purely by chance, were suited to life on land. Foremost among these structures was the external skeleton, or **exoskeleton**, a hard covering surrounding the body, such as the shell of a lobster or crab. Exoskeletons are both waterproof and strong enough to support a small animal against the force of gravity.

For millions of years, arthropods had the land and its plants to themselves, and for tens of millions of years more, they were the dominant land animals. Dragonflies with a wingspan of 28 inches (70 centimeters) flew among the Carboniferous tree ferns, while millipedes 6.5 feet (2 meters) long munched their way across the swampy forest floor. Eventually, however, the arthropods' splendid isolation came to an end.

Amphibians Evolved from Lobefin Fishes

About 400 million years ago, a group of Silurian fishes called the lobefins appeared, probably in freshwater. **Lobefins** had two important features that would later enable their descendants to colonize land: (1) stout, fleshy fins with which they crawled about on the bottoms of shallow, quiet waters, and (2) an outpouching of the digestive tract that could be filled with air, like a primitive lung. One group of lobefins colonized very shallow ponds and streams, which shrank during droughts and whose water often became oxygen-poor. By taking air into their lungs, these lobefins could obtain oxygen anyway. Some began to use their fins to crawl from pond to pond in search of prey or water, as some modern fish do today (Fig. 15-8).

The benefits of feeding on land and moving from pool to pool favored the evolution of a group of animals that could stay out of water for longer periods and that could move about more effectively on land. With improvements in lungs and legs, **amphibians** evolved from lobefins, first appearing in the fossil record about 350 million years ago. To an amphibian, the Carboniferous swamp forests were a kind of paradise: no predators to speak of, abundant prey, and a warm, moist climate. As with the insects and millipedes, some amphibians evolved gigantic size, including salamanders more than 10 feet (3 meters) long.

Despite their success, the early amphibians were not fully adapted to life on land. Their lungs were simple sacs without very much surface area, so they had to obtain some of their oxygen through their skin. Therefore, their skin had to be kept moist, which restricted them to swampy habitats where they wouldn't dry out. Further, amphibian sperm and eggs could not survive in dry surroundings and had to be deposited in watery environ-

Figure 15-8 A fish that walks on land
Some modern fishes, such as this mudskipper, walk on land. Like the ancient lobefin fishes that gave rise to amphibians, mudskippers use their strong pectoral fins to move across dry areas in their swampy habitats.

ments. So, although amphibians could move about on land, they could not stray too far from the water's edge. As with the tree ferns and club mosses, when the climate turned dry at the beginning of the Permian period about 286 million years ago, amphibians declined.

Reptiles Evolved from Amphibians

Just as the conifers were evolving on the fringes of the swamp forests, so too was a group of amphibians evolving, developing adaptations to drier conditions. These amphibians ultimately gave rise to the **reptiles**, which were characterized by four major adaptations to life on land. (1) They evolved internal fertilization, in which the reproductive tract of the female provided a watery environment for the sperm. Thus, sperm could be transferred on land, without the reptiles' having to venture back to the dangerous swamps full of fish and amphibian predators. (2) Reptiles evolved shelled, waterproof eggs that enclosed a supply of water for the developing embryo, again providing freedom from the swamps. (3) The ancestral reptiles evolved scaly, waterproof skin that helped prevent the loss of body water to the dry air. (4) Reptiles evolved improved lungs that were able to provide the entire oxygen supply for an active animal. As the climate dried during the Permian period, reptiles became the dominant land vertebrates, relegating amphibians to swampy backwaters, where most remain today.

A few tens of millions of years later, the climate returned to more moist and equable conditions. This period saw the evolution of some very large reptiles, in particular the dinosaurs. The variety of dinosaur forms was enormous—from predators (Fig. 15-9) to plant eaters; from those that dominated the land, to others that took to the air, to still others that returned to the sea. Dinosaurs were among the most successful animals ever, if we consider persistence as a measure of success. They flourished for more than a hundred million years, until about 65 million years ago, when the last dinosaurs went extinct. No one is certain why they died out, but the aftereffects of a gigantic meteorite's impact with Earth seems to have been the final blow (see the discussion later in this chapter on page 261).

Even during the age of dinosaurs, many reptiles remained quite small. One major difficulty faced by small reptiles is maintaining a high body temperature. Being active on land is helped by a warm body that maximizes the efficiency of the nervous system and muscles. But a warm body loses heat to the environment unless the air is also warm. Heat loss is a bigger problem for smaller animals, which have a larger surface area per unit of weight than do larger animals. Many species of small reptiles have retained slow metabolisms and have coped with the heat-loss problem by developing lifestyles in which they remain active only when the air is sufficiently warm. Two groups of small reptiles, however, independently followed a different evolutionary pathway: they developed insulation. One group evolved feathers, and another evolved hair.

Reptiles Gave Rise to Both Birds and Mammals

In the ancestral birds, insulating feathers helped retain body heat. Consequently, these animals could be active in cool habitats and during the night, when their scaly relatives became sluggish. Later, some ancestral birds evolved longer, stronger feathers on their forelimbs, perhaps under selection for better ability to glide from trees or to jump after insect prey. Ultimately, feathers evolved into structures capable of supporting powered

Figure 15-9 A reconstruction of a Cretaceous forest
By the Cretaceous period, flowering plants dominated terrestrial vegetation. Dinosaurs, such as the predatory pack of 6-foot-long *Velociraptors* shown here, were the preeminent land animals. Although small by dinosaur standards, *Velociraptors* were formidable predators with great running speed, sharp teeth, and deadly, sickle-like claws on their hind feet.

flight. Fully developed, flight-capable feathers are present in 150-million-year-old fossils, so the earlier insulating structures that eventually developed into flight feathers must have been present well before that time.

The earliest mammals coexisted with the dinosaurs but were small creatures, probably living in trees and being active mostly at night. When the dinosaurs went extinct, mammals colonized the habitats left empty by the extinctions. Mammal species prospered, diversifying into the array of modern forms.

Unlike birds, which retained the reptilian habit of laying eggs, mammals evolved live birth and the ability to feed their young with secretions of the mammary (milk-producing) glands. Ancestral mammals also developed hair, which provided insulation. Because the uterus, mammary glands, and hair do not fossilize, we may never know when these structures first appeared, or what their intermediate forms looked like. Recently, however, a team of paleontologists found bits of fossil hair preserved in coprolites, which are fossilized animal feces. These coprolites, found in the Gobi Desert of China, were deposited by an anonymous predator 55 million years ago, so mammals have presumably had hair at least that long.

5 What Role Has Extinction Played in the History of Life?

If there is a moral to the great tale of life's history, it is that nothing lasts forever. The story of life can be read as a long series of evolutionary dynasties, with each new dominant group rising, ruling the land or the seas for a time and, inevitably, falling into decline and extinction. Dinosaurs are the most famous of these fallen dynasties, but the list of extinct groups known only from fossils is impressively long. Despite the inevitability of extinction, however, the overall trend has been for species to arise at a faster rate than they disappear, so the number of different species on Earth has tended to increase over time.

Evolutionary History Has Been Marked by Periodic Mass Extinctions

Over much of life's history, dynastic succession has proceeded in a steady, relentless manner. This slow and steady turnover of species, however, has been interrupted by episodes of **mass extinction** (Fig. 15-10). These mass extinctions are characterized by the relatively sudden disappearance of a wide variety of species over a large part of Earth. In the most catastrophic episodes of extinction, more than half of the planet's species disappeared. The worst episode of all, which occurred 245 million years ago at the end of the Permian period, wiped out more than 90% of the world's species, and life came perilously close to disappearing altogether.

Climate Change Contributed to Mass Extinctions

Mass extinctions have had a profound impact on the course of life's history, repeatedly redrawing the picture of life's diversity. What could have caused such dramatic changes in the fortunes of so many species? Many evolutionary biologists believe that changes in climate must have played an important role. When the climate changes, as it has done many times over the course of Earth's history, organisms that were adapted for survival in one climate may be unable to survive under in a drastically different climate. In particular, at times when warm climates gave way to drier, colder climates with more variable temperatures, species may have gone extinct after failing to adapt to the harsh new conditions.

One cause of climate change is the changing positions of continents. Earth's surface is divided into portions called plates, which include the continents and the seafloor. The solid plates slowly move above a viscous

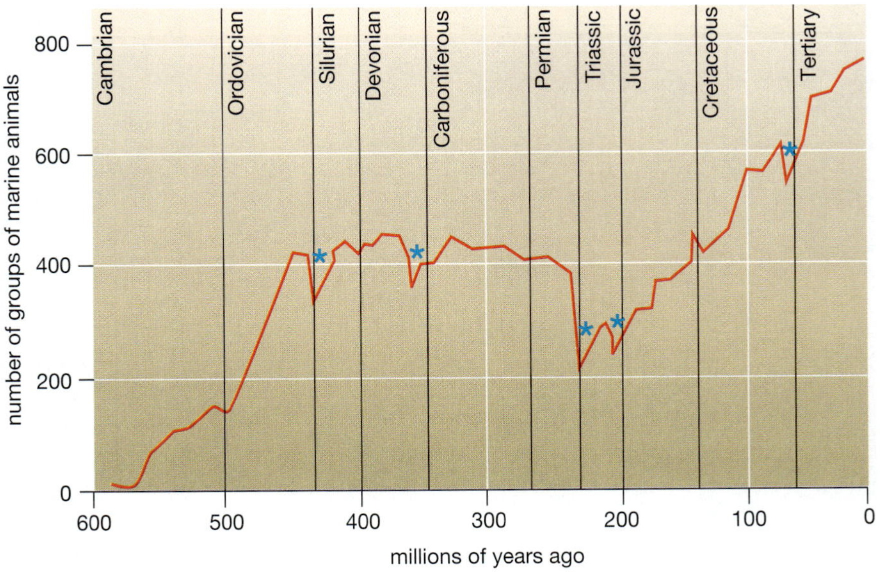

Figure 15-10 Mass extinctions This graph plots number of marine-animal groups against time, as reconstructed from the fossil record. Note the general trend toward an increasing number of groups, punctuated by periods of sometimes rapid extinction. Five of these declines, marked by asterisks, are so steep that they qualify as catastrophic mass extinctions.

but fluid layer; this movement is called **plate tectonics.** As the plates wander, their positions may change in latitude (Fig. 15-11). For example, 350 million years ago much of North America was located at or near the equator, an area characterized by consistently warm and wet tropical weather. But plate tectonics carried the continent up into temperate and arctic regions. As a result, the tropical climate was replaced by a regime of seasonal changes, cooler temperatures, and less rainfall. Plate tectonics continues today; the Atlantic Ocean, for example, widens by a few centimeters each year.

Catastrophic Events May Have Caused the Worst Mass Extinctions

Geological data indicates that most mass extinction events coincided with periods of climatic change. To many scientists, however, the rapidity of mass extinctions suggests that the slow process of climate change could not, by itself, be responsible for such large-scale disappearances of species. Perhaps more sudden events also play a role. For example, catastrophic geological events, such as massive volcanic eruptions, could have had devastating effects. Geologists have found evidence of past volcanic eruptions so huge that they make the 1980 Mount St. Helens explosion look like a firecracker by comparison. Even such gigantic eruptions, however, would directly affect only a relatively small portion of Earth's surface.

The search for the causes of mass extinctions took a fascinating turn in the early 1980s when Luis and Walter Alvarez proposed that the extinction event of 65 million years ago, which wiped out the dinosaurs and many other species, was caused by the impact of a huge meteorite. The Alvarezes' idea was met with great skepticism when it was first introduced, but geological research since that time has generated a great deal of evidence that a massive impact did indeed occur 65 million years ago. In fact, researchers have identified the Chicxulub crater, a 100-mile-wide crater buried beneath the Yucatan Peninsula of Mexico, as the impact site of a giant meteorite, 10 miles in diameter, that collided with Earth just at the time that dinosaurs disappeared.

Could this immense meteorite strike have caused the mass extinction that coincided with it? No one knows for sure, but scientists suggest that such a massive impact would have thrown so much debris into the atmosphere that the entire planet would have been plunged into darkness for a period of years. With little light reaching the planet, temperatures would have dropped precipitously and the photosynthetic capture of energy (upon which all life ultimately depends) would have declined drastically. The worldwide "impact winter" would have spelled doom for the dinosaurs and a host of other species.

6 How Did Humans Evolve?

Scientists are intensely interested in the origin and evolution of humans, especially in the evolution of the gigantic human brain. The outline of human evolution that we present in this section is a synthesis of current thought on the subject, but we note that it is speculative, as the fossil evidence of human evolution is comparatively scarce. Paleontologists disagree about the interpretation of the fossil evidence, and many ideas may have to be revised as new fossils are found.

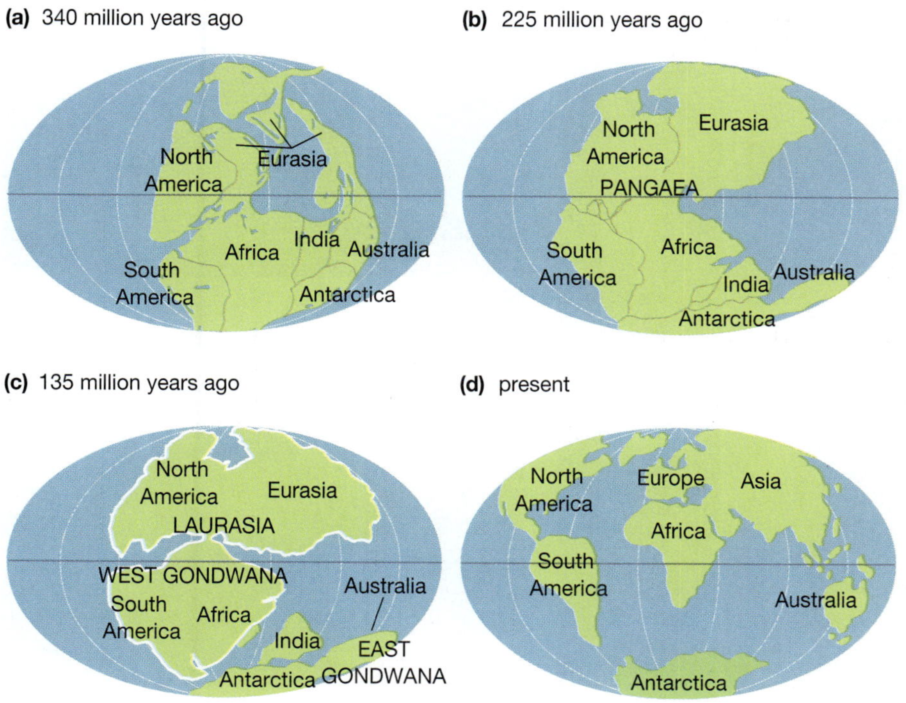

Figure 15-11 Continental drift from plate tectonics
The continents are passengers on plates moving on Earth's surface as a result of plate tectonics. ***(a)*** About 340 million years ago, much of what is now North America was positioned at the equator. ***(b)*** All the plates eventually fused together into one gigantic landmass, which geologists call Pangaea. ***(c)*** Gradually Pangaea broke up into Laurasia and Gondwanaland, which itself eventually broke up into West and East Gondwana. ***(d)*** Further plate motion eventually resulted in the modern positions of the continents.

MediaTutor 15.2 Plate Tectonics

Early Primates Lived in Trees

Fossils of **primates** (lemurs, monkeys, apes, and humans) are relatively rare compared to those of many other animals. The primate fossil evidence does show, however, that the most likely primate ancestors were insect-eating tree shrews, whose fossils are found in rocks about 80 million years old. Nimble, probably nocturnal animals, tree shrews were smaller than all but the tiniest modern primates.

Over the next 50 million years, the descendants of the tree shrews evolved forms that are similar to modern tarsiers, lemurs, and monkeys (Fig. 15-12). Early primates, which probably fed on fruits and leaves, evolved several adaptations for life in the trees. Many modern primates retain the tree-dwelling lifestyle of their ancestors. Some species are nocturnal, but many others are active during the day.

Human Ancestors Evolved from Monkey-like Primates

Between 20 million and 30 million years ago, in the moist tropical forests of Africa, a group of primates called the *dryopithecines* diverged from the monkey line and became land dwellers rather than tree dwellers. The dryopithecines appear to have been ancestral to the **hominids** (humans and their fossil relatives) and pongids (the great apes). About 18 million years ago, global climatic cooling began to shrink the vast expanses of forest, splitting up the woodlands into isolated islands dotted on a sea of grasslands. Diversification of habitat and isolation of small populations led to the diversification of dryopithecines. One of these primate groups gave rise to the later hominids and pongids.

The Oldest Hominid Fossils Are from Africa

The hominid line is believed to have diverged from the ape lineage sometime between 5 million and 8 million years ago, but direct evidence of that transition has not been found in the fossil record. The first trace of our hominid ancestors appears in 4.4-million-year-old rocks in Ethiopia, where excavations have yielded fragments of *Ardipithecus ramidus*. *Ardipithecus* was clearly a hominid, sharing several anatomical features with better-known, later members of the group. But this oldest known member of our family also exhibits other features (such as thin tooth enamel and thick arm bones) that are more characteristic of apes, so *Ardipithecus* may represent a point on our family tree that is fairly near the split from the apes.

Ardipithecus is known from only a few bones and teeth. A more extensive record of early hominid evolution does not really begin until about 4 million years ago. That date marks the beginning of the fossil record of the genus *Australopithecus* ("southern ape," after their original discovery site in southern Africa; Fig. 15-13). The brains of australopithecines (as the various species of *Australopithecus* are collectively known) were fairly large but still much smaller than those of modern humans.

Figure 15-12 Representative primates
(a) Tarsier, ***(b)*** lemur, and ***(c)*** lion-tail macaque monkey. Note that all have relatively flat faces, with forward-looking eyes providing binocular vision. All also have color vision and grasping hands. These features, retained from the earliest primates, are shared by humans.

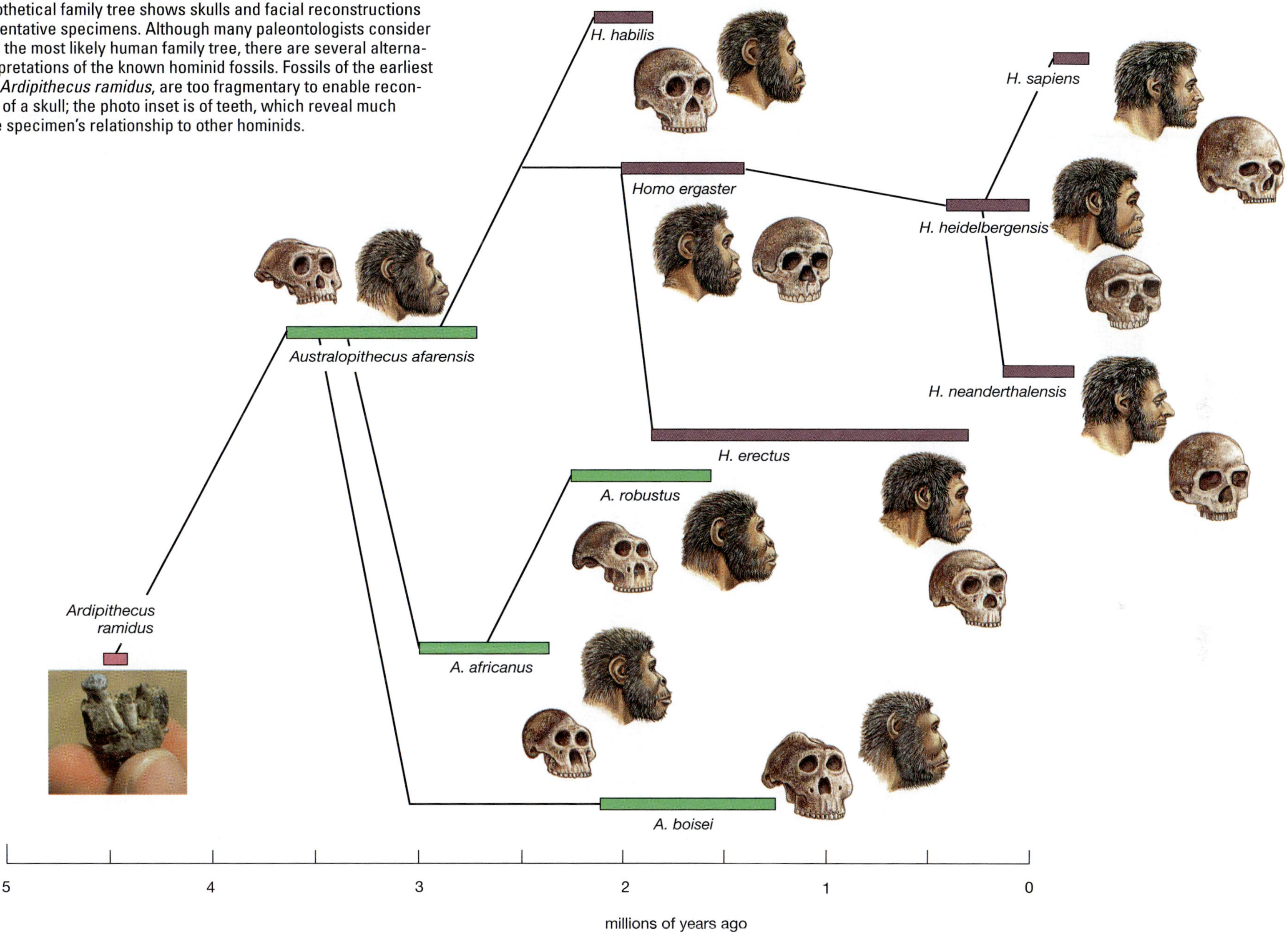

Figure 15-13 A possible evolutionary tree for humans
This hypothetical family tree shows skulls and facial reconstructions of representative specimens. Although many paleontologists consider this to be the most likely human family tree, there are several alternative interpretations of the known hominid fossils. Fossils of the earliest hominid, *Ardipithecus ramidus*, are too fragmentary to enable reconstruction of a skull; the photo inset is of teeth, which reveal much about the specimen's relationship to other hominids.

The Earliest Hominids Could Stand and Walk Upright

Although the earliest australopithecines had legs that were shorter, for their height, than those of modern humans, their knee joints allowed them to straighten their legs fully, permitting efficient bipedal (upright, two-legged) locomotion. Footprints almost 4 million years old, discovered in Tanzania by anthropologist Mary Leakey, show that even the earliest australopithecines could, and at least sometimes did, walk upright. Upright posture is an extremely important characteristic of hominids, because it freed their hands from use in walking. Later hominids thus were able to carry weapons, manipulate tools, and eventually achieve the cultural revolutions produced by modern *Homo sapiens*.

Several Species of *Australopithecus* Emerged in Africa

The oldest australopithecine species, represented by fossilized teeth, skull fragments, and arm bones, was unearthed near an ancient lake bed in Kenya from sediments that were dated, by means of radioactive isotopes, as between 3.9 million and 4.1 million years old (see "Scientific Inquiry: How Do We Know How Old a Fossil Is?"). It was named *Australopithecus anamensis* by its discoverers (*anam* means "lake" in the local Ethiopian language). The second most ancient australopithecine, called *Australopithecus afarensis*, was discovered in the Afar region of Ethiopia in 1974. Fossil remains of this species as old as 3.9 million years have been unearthed. Later, the *A. afarensis* line apparently gave rise to other species. All of the australopithecine species had apparently gone extinct by 1.2 million years ago, but one of them (*A. afarensis* in the interpretation shown in Fig. 15-13) first gave rise to a new branch of the hominid family tree, the genus *Homo*.

The Genus *Homo* Diverged from the Australopithecines 2.5 Million Years Ago

Hominids that are sufficiently similar to modern humans to be placed in the genus *Homo* first appear in African fossils that are about 2.5 million years old. Among the earliest African *Homo* fossils are *H. habilis* (see Fig. 15-13), a species whose body and brain were larger than those of the australopithecines but which retained the apelike long arms and short legs of their australopithecine forebears. In contrast, the skeletal anatomy of *H. ergaster*, a species whose fossils first appear 2 million years ago, has limb proportions more like those of modern humans. This species is believed by many paleoanthropologists (scientists who study human origins) to be on the evolutionary branch that led ultimately to our own species, *H. sapiens*. In this view, *H. ergaster* was the common ancestor of two distinct branches of hominids. The first branch led to *H. erectus*, which was the first hominid species to leave Africa. The second branch from *H. ergaster* ultimately led both to the species commonly known as the Neanderthals and to *H. sapiens*, modern humans.

The Evolution of *Homo* Was Accompanied by Advances in Tool Technology

Hominid evolution is closely tied to the development of tools, a hallmark of hominid behavior. The oldest tools discovered so far were found in 2.5-million-year-old East African rocks, concurrent with the early emergence of the genus *Homo*. Early *Homo*, whose cheek teeth were much smaller than those of its australopithecine ancestors, might have first used stone tools to break and crush tough foods that were hard to chew. Hominids constructed their earliest tools by striking one rock with another to chip off fragments and leave a sharp edge behind. Over the next several hundred thousand years, tool-making techniques in Africa gradually became more advanced. By 1.7 million years ago, tools were more sophisticated, with flakes chipped symmetrically from both sides of a rock to form double-edged tools ranging from hand axes, used for cutting and chopping, to points, probably used on spears (Fig. 15-14a,b). *Homo ergaster* and other bearers of these weapons presumably ate meat, probably acquired from both hunting and scavenging for the remains of prey killed by other predators. Double-edged tools were carried to Europe at least 600,000 years ago by migrating hominid populations, and the Neanderthal descendants of these emigrants took stone-tool construction to new heights of skill and delicacy (Fig. 15-14c).

Neanderthals Had Large Brains and Ritualistic Behaviors

Neanderthals first appear in the European fossil record about 150,000 years ago; by about 70,000 years ago, they had spread throughout Europe and western Asia. By 30,000 years ago, however, Neanderthals were extinct. Contrary to the popular image of a hulking, stoop-shouldered "cave man," Neanderthals were quite similar to modern humans in many ways. Although more heavily muscled, Neanderthals walked fully erect, were dexterous enough to manufacture finely crafted stone tools, and had brains, on the average, slightly larger than those of modern humans. Although many of the European fossils show heavy brow ridges and a broad, flat skull, others, particularly from areas around the eastern shores of the Mediterranean Sea, were remarkably like ourselves. Neanderthal remains also show evidence of modern behaviors, particularly ritualistic burial ceremonies (Fig. 15-15). Neanderthal skeletons have been discovered in clearly marked burial sites that were surrounded with stones and typically included offerings of flowers, bear skulls, and food. Altars found at other Neanderthal sites were probably used in religious rites associated with a bear cult.

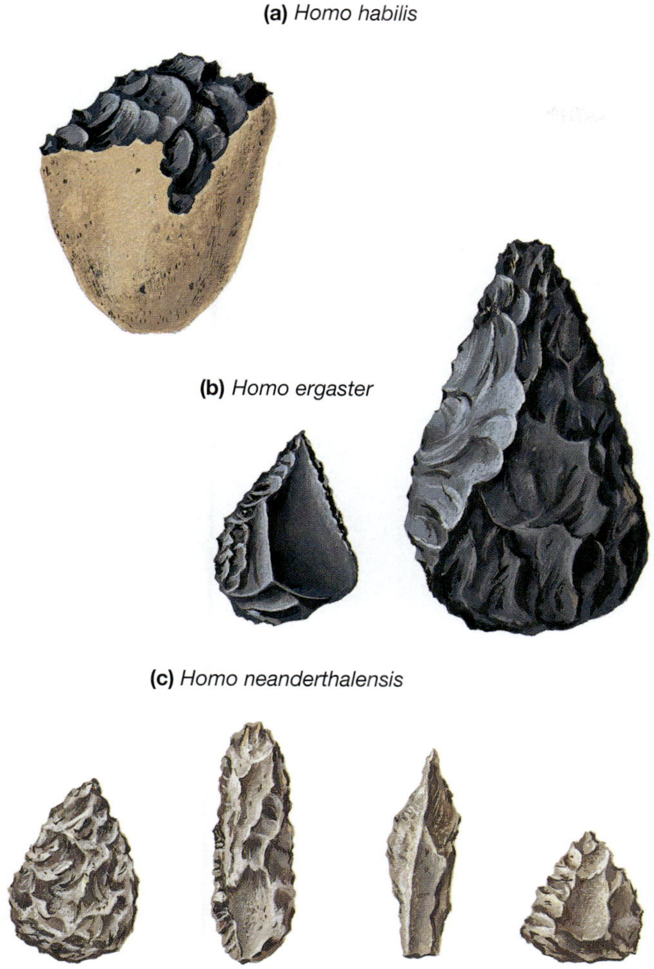

Figure 15-14 Representative hominid tools
(a) *Homo habilis* produced only fairly crude chopping tools called *hand axes*, usually unchipped on one end to hold in the hand. ***(b)*** *Homo ergaster* manufactured much finer tools. The tools were typically sharp all the way around the stone, so at least some of these blades were probably tied to spears rather than held in the hand. ***(c)*** Neanderthal tools were works of art, with extremely sharp edges made by flaking off tiny bits of stone. In comparing these weapons, note the progressive increase in the number of flakes taken off the blades and the corresponding decrease in flake size. Smaller, more numerous flakes produce a sharper blade and suggest more insight into toolmaking, more patience, finer control of hand movements, or perhaps all three.

Figure 15-15 Neanderthal burial
Museum reconstruction of a Neanderthal burial site. Some Neanderthals were buried with elaborate ceremonies, with bear skulls placed on the grave of the deceased.

Until recently, most anthropologists believed that Neanderthals were simply a variety of *H. sapiens*. It is becoming increasingly clear, however, that Neanderthals were a separate species. Dramatic evidence in support of this hypothesis has come from two groups of researchers who were able to isolate and analyze ancient DNA from two different 30,000-year-old Neanderthal skeletons. The researchers determined the nucleotide sequence of a Neanderthal gene and compared it to the same gene from a large number of different modern humans from various parts of the world. The Neanderthal gene sequence was very different from that of the modern humans and indicated that the evolutionary branch leading to Neanderthals diverged from the ancestral human line hundreds of thousands of years prior to the emergence of modern *H. sapiens*.

Modern Humans Emerged Only 150,000 Years Ago

The fossil record shows that anatomically modern humans appeared in Africa about 150,000 years ago. The location of these fossils suggests that *Homo sapiens* originated in Africa, but most of our knowledge about our own early history comes from European and Middle Eastern fossil *H. sapiens* collectively known as Cro-Magnons (after the district in France in which their remains were first discovered). Cro-Magnons appeared about 90,000 years ago and had domed heads, smooth brows, and prominent chins (just like us!). Their tools were precision instruments not very different from the stone tools used until recently in many parts of the world. Behaviorally, Cro-Magnons seem to have been similar to, but more sophisticated than, Neanderthals. Perhaps the most remarkable accomplishment of Cro-Magnons is the magnificent art left in caves such as Altamira in Spain and Lascaux in France (Fig. 15-16). No one knows exactly why these paintings were made, but they attest to minds fully as human as our own.

Cro-Magnons and Neanderthals Lived Side by Side

Cro-Magnons coexisted with Neanderthals in Europe and the Middle East for perhaps as many as 50,000 years before the Neanderthals disappeared. Some researchers believe that Cro-Magnons interbred extensively with Neanderthals, so Neanderthals were essentially absorbed into the human genetic mainstream. Other scientists disagree, citing mounting evidence such as the fossil DNA described earlier, and suggest that later-arriving Cro-Magnons simply overran and displaced the less-well-adapted Neanderthals.

Figure 15-16 The sophistication of Cro-Magnon people Cave paintings by Cro-Magnons have been remarkably preserved by the relatively constant underground conditions of a cave in Lascaux, France.

CASESTUDYREVISITED

Life on a Frozen Moon?

Is there really liquid water on Europa? Calculations that estimate the amount of heat produced by the gravitational push and pull exerted on Europa by Jupiter's other moons indicate that the moon's interior should be warm enough to sustain a liquid ocean. A more definitive answer, however, appears to be within reach. In 2003 (according to current plans) the *Europa Orbiter* spacecraft will be launched from Earth and, after a few years of travel to the vicinity of Jupiter, will enter into orbit around Europa. Its mission will include subsurface radar measurements that should definitively determine whether liquid is present.

If liquid water *is* present on Europa, has life arisen in it? If so what form has European life assumed? Answers to these questions do not appear to be close at hand, but some tantalizing clues might be found right here on Earth. The potential source of such clues is Lake Vostok, a huge lake (similar in size to Lake Ontario) that is buried more than two miles beneath the Antarctic ice pack. The cold waters of Lake Vostok are under extremely high pressure from the immense weight of the ice above it and have been isolated from Earth's atmosphere for perhaps hundreds of thousands of years; the environment of these waters may thus be similar to the environment beneath Europa's ice sheet.

Biologists would very much like to know if Lake Vostok harbors any living organisms and, if so, how they survive. If, for example, bacteria are alive in the lake, what is their source of energy? How do they cope with a cold environment? Answers to questions such as these could help set the stage for investigation of Europa's subsurface seas. Recent drilling to within a few hundred feet of Lake Vostok's surface revealed some preserved bacteria (apparently not still alive), raising hopes that the lake itself will harbor life. Scientists have held off drilling all the way to the lake for fear of contaminating the lake with microbes from the surface or of inadvertently releasing unknown organisms from the lake into the surface environment. Efforts are under way to devise methods to explore the lake without risking cross-contamination. Once such methods are devised, they could be adapted for use in future exploration of similar environments on Europa.

Summary of Key Concepts

1) How Did Life Begin?

Before life arose, lightning, ultraviolet light, and heat formed organic molecules from water and the components of primordial Earth's atmosphere. These molecules probably included nucleic acids, amino acids, short proteins, and lipids. By chance, some molecules of RNA may have had enzymatic properties, catalyzing the assembly of copies of themselves from nucleotides in Earth's waters. These may have been the forerunners of life. Protein-lipid microspheres enclosing these ribozymes may have formed the first protocells.

2) What Were the Earliest Organisms Like?

The oldest fossils, about 3.5 billion years old, are of prokaryotic cells that fed by absorbing organic molecules that had been synthesized in the environment. Because there was no free oxygen in the atmosphere, their energy metabolism must have been anaerobic. As the cells multiplied, they depleted the organic molecules that had been formed by prebiotic synthesis. Some cells developed the ability to synthesize their own food molecules by using simple inorganic molecules and the energy of sunlight. These earliest photosynthetic cells were probably ancestors of today's cyanobacteria.

Photosynthesis releases oxygen as a by-product, and by about 2.2 billion years ago significant amounts of free oxygen were accumulating in the atmosphere. Aerobic metabolism, which generates more cellular energy than does anaerobic metabolism, probably arose about this time.

Eukaryotic cells had evolved by about 1.7 billion years ago. The first eukaryotic cells probably arose as symbiotic associations between predatory prokaryotic cells and other bacteria. Mitochondria may have evolved from aerobic bacteria engulfed by predatory cells. Similarly, chloroplasts may have evolved from photosynthetic cyanobacteria.

3) What Were the Earliest Multicellular Organisms Like?

Multicellular organisms evolved from eukaryotic cells and first appeared, in the seas, about 1 billion years ago. Multicellularity offers several advantages, including greater size. In plants, increased size offered some protection from predation. Specialization of cells allowed plants to anchor themselves in the nutrient-rich, well-lit waters of the shore. For animals, multicellularity allowed more-efficient predation and more-effective escape from predators. These in turn provided environmental pressures for faster locomotion, improved senses, and greater intelligence.

4) How Did Life Invade the Land?

The first land organisms were probably algae. The first multicellular land plants appeared about 400 million years ago. Although life on land required special adaptations for support of the body, reproduction, and the acquisition, distribution, and retention of water, the land also offered abundant sunlight and protection from aquatic herbivores. Soon after land plants evolved, arthropods invaded the land. Absence of predators and abundant land plants for food probably facilitated the invasion of the land by animals.

The earliest land vertebrates evolved from lobefin fishes, which had leglike fins and a primitive lung. A group of lobefins evolved into the amphibians about 350 million years ago. Reptiles evolved from amphibians, with several further adaptations for life on land: internal fertilization, waterproof eggs that could be laid on land, waterproof skin, and better lungs. Birds and mammals evolved independently from separate groups of reptiles. A major advance in the evolution of both birds and mmmals was insulation over the body surface.

5) What Role Has Extinction Played in the History of Life?

The history of life has been characterized by constant turnover of species as species go extinct and are replaced by new ones. Mass extinctions, in which large numbers of species disappear within a relatively short time, have occurred periodically. Mass extinctions were probably caused by some combination of climate change and catastrophic events, such as volcanic eruptions and meteorite impacts.

6) How Did Humans Evolve?

One group of mammals evolved into the tree-dwelling primates. Between 20 million and 30 million years ago, some primates descended from the trees; these were the ancestors of apes and humans. The australopithecines arose in Africa about 4 million years ago. These hominids walked erect, had larger brains than did their forebears, and fashioned primitive tools. One group of australopithecines gave rise to a line of hominids in the genus *Homo*, which in turn gave rise to modern humans.

Key Terms

amphibian *p. 258*
arthropod *p. 258*
conifer *p. 258*
endosymbiont hypothesis *p. 254*
eukaryotes *p. 254*
exoskeleton *p. 258*
hominid *p. 262*
lobefin *p. 258*
mass extinction *p. 260*
microsphere *p. 250*
plate tectonics *p. 261*
prebiotic evolution *p. 248*
primate *p. 262*
prokaryote *p. 251*
protocell *p. 250*
reptile *p. 259*
ribozyme *p. 250*
spontaneous generation *p. 248*

Thinking Through the Concepts

Multiple Choice

1. *There was no free oxygen in the early atmosphere because most of it was tied up in*
   a. water
   b. ammonia
   c. methane
   d. rock
   e. radioactive isotopes

2. *RNA became a candidate for the first information-carrying molecule when Tom Cech and Sidney Altman discovered that some RNA molecules can act as enzymes that*
   a. degrade proteins
   b. turn light into chemical energy
   c. split water and release oxygen gas
   d. synthesize copies of themselves
   e. synthesize amino acids

3. *The earliest living organisms were*
   a. multicellular
   b. eukaryotes
   c. prokaryotes
   d. photosynthesizers
   e. aerobic
4. *Which two of the following observations support Lynn Margulis's endosymbiont hypothesis for the origin of chloroplasts and mitochondria from ingested bacteria?*
   a. Aerobic respiration takes place in mitochondria.
   b. Mitochondria have their own DNA.
   c. Photosynthesis takes place in chloroplasts.
   d. Chloroplasts have their own DNA.
   e. Some bacteria use anaerobic respiration.
5. *Early marine-dwelling arthropods were well suited for life on land because an exoskeleton*
   a. can support an animal's weight against the pull of gravity
   b. allows a wide diversity of body types
   c. resists drying
   d. absorbs light
   e. both a and c
6. *The evolution of the shelled, waterproof egg was an important event in vertebrate evolution because it*
   a. led to the Cambrian explosion
   b. was the first example of parents caring for their young
   c. allowed the colonization of freshwater environments
   d. freed organisms from having to lay their eggs in water
   e. allowed internal fertilization of eggs

? Review Questions

1. What is the evidence that life might have originated from nonliving matter on early Earth? What kind of evidence would you like to see before you would accept this hypothesis?
2. If the first cells with aerobic metabolism were so much more efficient at producing energy, why didn't they drive to extinction cells with only anaerobic metabolism?
3. Explain the endosymbiont hypothesis for the origin of chloroplasts and mitochondria.
4. Name two advantages of multicellularity for plants and two for animals.
5. What advantages and disadvantages would terrestrial existence have had for the first plants to invade the land? For the first land animals?
6. Outline the major adaptations that emerged during the evolution of vertebrates, from fish to amphibians to reptiles to birds and mammals. Explain how these adaptations increased the fitness of the various groups for life on land.
7. Outline the evolution of humans from early primates. Include in your discussion such features as binocular vision, grasping hands, bipedal locomotion, social living, tool making, and brain expansion.

Applying the Concepts

1. Do you think that studying our ancestors can shed light on the behavior of modern humans? Why or why not?
2. A biologist would probably answer the age-old question "What is life?" by saying "the ability to self-replicate." Do you agree with this definition? If so, why? If not, how would you define life in biological terms?
3. Traditional definitions of humans have emphasized "the uniqueness of humans" because we possess language and use tools. But most animals can communicate with other individuals in sophisticated ways, and many vertebrates use tools to accomplish tasks. Pretend that you are a biologist from Mars, and write a taxonomic description of the species *Homo sapiens*.
4. Extinctions have occurred throughout the history of life on Earth. Why should we care if humans are causing a mass extinction event now?
5. In biological terms, what do you think was the most significant event in the history of life? Explain your answer.

Group Activity

Like a grand mosaic, our picture of the evolutionary history of life has been pieced together from a multitude of small clues pulled from the fossil record by paleontologists. To understand a bit about the process of adding a piece to the mosaic, form a group of three students, and examine the accompanying drawing of a skull that paleontologists unearthed in South Dakota. Use the observable features of the skull to develop a hypothesis that describes what this creature's life was like. To which vertebrate group does it belong? What can you predict about this species' feeding habits, posture, sensory abilities, and social behavior? For each prediction that you make, cite the specific anatomical evidence that supports your claim. Finally, list five additional pieces of fossil evidence that, if found, would help confirm your hypothesis.

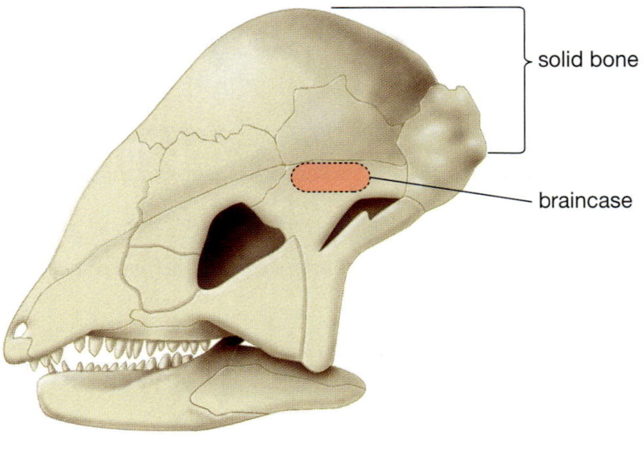

For More Information

de Duve, C. "The Birth of Complex Cells." *Scientific American*, April 1996. Narrative describing the origin of complex eukaryotic cells by repeated instances of endosymbiosis.

Diamond, J. "How to Speak Neanderthal." *Discover*, January 1990. Could the Neanderthals speak? How could we tell? Jared Diamond lucidly explains the arguments and methodology behind a scientific dispute.

Eigen, M. *Steps Towards Life*. New York: Oxford University Press, 1992. A Nobel laureate lays out a scenario for the origin of life and describes supporting experimental evidence. Challenging material, concisely and precisely explained.

Fenchel, T., and Finlay, B. J. "The Evolution of Life Without Oxygen." *American Scientist*, January–February 1994. Clues to the origin of the first eukaryotic cells are provided by symbiotic relationships of organisms in oxygen-free environments.

Fry, I. *The Emergence of Life on Earth: A Historical and Scientific Overview*. Brunswick, NJ: Rutgers University Press, 2000. A thorough review of research and hypotheses concerning the origin of life.

Hay, R. L., and Leakey, M. D. "The Fossil Footprints of Laetoli." *Scientific American*, February 1982. The actual footprints of a hominid family were discovered by Hay and Leakey in volcanic ash 3.5 million years old.

Maynard Smith, J., and Szathmary, E. *The Origins of Life: From the Birth of Life to the Origin of Language*. New York: Oxford University Press, 1999. A thought-provoking review of the major shifts that have occurred over the 3.5-billion-year history of life.

Monastersky, R. "The Rise of Life on Earth." *National Geographic*, March 1998. A beautifully illustrated and engaging description of current ideas and evidence of how life arose.

Morell, V. "Announcing the Birth of a Heresy." *Discover*, March 1987. Paleontologists Robert Bakker and Jack Horner speculate that dinosaurs were not the plodding beasts of monster movies but warm-blooded, advanced animals that may even have cared for their young.

Pappalardo, R., Head, J., and Greeley, R. "The Hidden Ocean of Europa." *Scientific American*, October 1999. A summary of the evidence that Europa's icy surface conceals a liquid ocean; illustrated with photos from the *Galileo* mission.

Tattersall, I. "Out of Africa . . . Again and Again?" *Scientific American*, April 1997. A review of the fossil evidence of human evolutionary history.

Ward, P. D. *The End of Evolution: On Mass Extinctions and the Preservation of Biodiversity*. New York: Bantam Books, 1994. An engaging first-person account of a paleontologist's investigation of the causes of mass extinction.

Answers to Multiple-Choice Questions

1. a 2. d 3. c 4. b, d 5. e 6. d

MediaTutor

CD Activities

Activity 15.1: Endosymbiosis

Estimated time: 5 minutes

Endosymbiosis is thought to have generated one of the most important events in evolutionary history: the origin of eukaryotic organisms. Recent advances have uncovered a great deal of evidence in support of the endosymbiont theory. This animation demonstrates how symbiosis may have given rise to such organelles as mitochondria and chloroplasts.

Activity 15.2: Plate Tectonics

Estimated time: 5 minutes

The idea of continental drift was largely scoffed at when it was first proposed. Now, the theory of plate tectonics forms a major foundation of geophysics. Like the theory of evolution, plate tectonics represents a profound change in our understanding of the world. This animation demonstrates how the continents have moved over the course of time and points out some effects of continental drift on the distributions and evolutionary histories of organisms.

Start the MediaTutor Student CD-ROM and enter the activity number in the Quick Search box to be taken directly to that activity.

Web Investigation

Life on a Frozen Moon?

Estimated time: 10 minutes

What are the origins of life? Since water is one of the basic elements of life as we know it, many astrobiologists believe that the best way to detect unknown life in space is to look for liquid water. Recent observations suggest that there may be liquid water on Europa, one of the moons of Jupiter. This exercise will take a closer look at the fourth biggest satellite of Jupiter.

Go to http://www.prenhall.com/audesirk3, the Audesirk Companion Web site. Select Chapter 15 and Web Investigation to begin.

"The life of this planet, in short, is largely unexplored. The study of biodiversity is one of science's greatest challenges, and humanity's greatest needs."

E. O. Wilson

More than half of all known species are insects. Beetles, such as the harlequin beetle shown here, are by far the largest group of insects.

16 The Diversity of Life

AT A GLANCE

CASESTUDY

A Biologist "Bombs" Beetles to Estimate Diversity

One of life's most conspicuous features is its diversity; life encompasses a seemingly infinite array of shapes, sizes, and ways of perpetuating itself. Despite centuries of scientific effort, the tremendous diversity of life overwhelms our ability to describe it. Scientists cannot even say with certainty how many species inhabit Earth. About 1.4 million species have been discovered, described, and named, but that total represents only a fraction of the actual number of species. Most species remain undiscovered, mainly because relatively few scientists are engaged in the search for new species. What's more, the search for species has historically focused on temperate regions, even though species diversity is highest in the Tropics. And a great many species remain undiscovered because they are small, inconspicuous, or live in inaccessible habitats such as the floor of the ocean or the topmost branches of the rain forest.

Because we will probably never be able to count all the species on Earth, some scientists have tried to estimate the number. One particularly fascinating attempt was made by Terry L. Erwin of the Smithsonian Institution. Working in a Panamanian rain forest, Erwin began by arranging jars of alcohol around the base of a tree. He placed a funnel into the top of each jar. Then, at night, when the breeze had died down, he fogged the tree's canopy with insecticide (in the rain forest, living things tend to be concentrated in the treetops, high above the forest floor). As this "bug bomb" took effect, dead and dying insects rained down from above and were funneled into the waiting jars.

Over the course of three years, Erwin repeated his procedure on dozens of trees, all of which were of the same species, *Luehea seemannii*. The preserved insects in the jars were studied in the lab, with special emphasis on counting the number of different beetle species. The focus was on beetles because the vast majority of known animal species are insects, and beetles are the largest group of insects. A total of 1200 different species of beetles, most of them previously unknown, were found in the sample. From this result, Erwin calculated that the world's tropical rain forests must hold about 30 million species of insects. How did Erwin work his way from his sample of Panamanian beetles to a worldwide estimate of species numbers? We'll reveal his methods at the end of this chapter, following a brief appreciation of the diversity of life. ■

1 How Are Organisms Named and Classified?

Organisms are placed into categories on the basis of their evolutionary relationships. There are eight major categories: (1) domain, (2) kingdom, (3) phylum, (4) class, (5) order, (6) family, (7) genus, and (8) species. These categories form a nested hierarchy in which each level includes all of the other levels below it. Each domain contains many kingdoms, each kingdom contains many phyla, each phylum includes many classes, each class includes many orders, and so on. As we move down the hierarchy, smaller and smaller groups are included. Each category is increasingly narrow and specifies a group whose common ancestor is increasingly recent. Table 16-1 includes some examples of classifications of specific organisms.

Each Species Has a Unique, Two-Part Name

The scientific name of an organism is formed from the two smallest categories, the genus and the species. Each genus includes a group of very closely related species, and each species within a genus includes populations of organisms that can potentially interbreed under natural conditions. Thus, the genus *Sialia* (bluebirds) includes the eastern bluebird (*Sialia sialis*), the western bluebird (*Sialia mexicana*), and the mountain bluebird (*Sialia currucoides*)—very similar birds that normally do not interbreed (Fig. 16-1).

Each two-part scientific name is unique, so referring to an organism by its scientific name rules out any chance of ambiguity or confusion. For example, the bird *Gavia immer* is commonly known in North America as the common loon, in Great Britain as the northern diver, and by still other names in non-English-speaking countries. But the Latin scientific name *Gavia immer* is recognized by biologists worldwide, overcoming language barriers and allowing precise communication.

Note that, by convention, scientific names are always underlined or *italicized*. The first letter of the genus name is always capitalized, and the first letter of the species name is always lowercase. The species name is never used alone but is always paired with its genus name.

Classification Originated as a Hierarchy of Categories

Aristotle (384–322 B.C.) was among the first to attempt to formulate a logical, standardized language for naming living things. Based on characteristics such as structural complexity, behavior, and degree of development at birth, he classified about 500 organisms into 11 categories. Aristotle's categories formed a hierarchical structure, with each category more inclusive than the one beneath it, a concept that is still used today.

Building on this foundation more than 2000 years later, Swedish naturalist Carl von Linné (1707–1778)—who called himself Carolus Linnaeus, a latinized version—laid the groundwork for the modern classification system. He placed each organism into a series of hierarchically arranged categories on the basis of its resemblance to other life-forms, and he also introduced the scientific name composed of genus and species. Nearly 100 years later, Charles Darwin (1809–1882) published *On the Origin of Species*, which demonstrated that all organisms are connected by common ancestry. Biologists then began to recognize that the categories ought to reflect the pattern of evolutionary relatedness among organisms. The more categories two organisms share, the closer their evolutionary relationship.

Biologists Try to Identify Features That Reveal Evolutionary Relationships

Biologists seek to reconstruct the tree of life, but they must do so without much direct knowledge of evolutionary history. Because they can't see into the past, they

Table 16-1 Classification of Selected Organisms, Reflecting Their Degree of Relatedness*

	Human	Chimpanzee	Wolf	Fruit Fly	Sequoia Tree	Sunflower
Domain	**Eukarya**	**Eukarya**	**Eukarya**	**Eukarya**	**Eukarya**	**Eukarya**
Kingdom	**Animalia**	**Animalia**	**Animalia**	**Animalia**	**Plantae**	**Plantae**
Phylum	**Chordata**	**Chordata**	**Chordata**	Arthropoda	Coniferophyta	Anthophyta
Class	**Mammalia**	**Mammalia**	**Mammalia**	Insecta	Coniferosida	Dicotyledoneae
Order	**Primates**	**Primates**	Carnivora	Diptera	Coniferales	Asterales
Family	Hominidae	Pongidae	Canidae	Drosophilidae	Taxodiaceae	Asteraceae
Genus	*Homo*	*Pan*	*Canis*	*Drosophila*	*Sequoiadendron*	*Helianthus*
Species	*sapiens*	*troglodytes*	*lupus*	*melanogaster*	*giganteum*	*annuus*

*Boldface categories are those that are shared by more than one of the organisms classified. Genus and species names are always italicized or underlined.

MediaTutor 16.1 Taxonomic Classification

Figure 16-1 Three species of bluebird
Despite their obvious similarity, these three species of bluebird remain distinct because they do not interbreed. The three species shown are (from left to right) the eastern bluebird (*Sialia sialis*), the western bluebird (*Sialia mexicana*), and the mountain bluebird (*Sialia currucoides*).

must infer the past as best they can, based on similarities among living organisms. Not just any similarity will do, however. Some observed similarities stem from convergent evolution in organisms that are not closely related; such similarities are not useful for classification. Instead, biologists value the similarities that exist because two kinds of organisms both inherited the characteristic from a common ancestor. Therefore, an important task is to distinguish informative similarities caused by common ancestry from similarities that result from convergent evolution. In the search for informative similarities, biologists look at many different kinds of characteristics.

Anatomy Plays a Key Role in Classification

Historically, the most important and useful distinguishing characteristics have been anatomical. Biologists look carefully at similarities in external body structure (see Fig. 16-1) and in internal structures, such as skeletons and muscles. For example, homologous structures, such as the finger bones of dolphins, bats, seals, and humans (see Fig. 13-7), provide evidence of a common ancestor. To detect relationships between more closely related species, biologists may use microscopes to discern finer details—the number and shape of the "teeth" on the tonguelike radula of a snail, the shape and position of the bristles on a marine worm, or the external structure of pollen grains of a flowering plant (Fig. 16-2).

Molecular Similarities Are Also Useful for Classification

The anatomical characteristics shared by related organisms are expressions of underlying genetic similarities, so it stands to reason that evolutionary relationships among species must also be reflected in genetic similarities. Of course, direct genetic comparisons were not possible for most of the history of biology. In the past two decades, however, advances in the techniques of molecular genetics have revolutionized studies of evolutionary relationships. For the first time, the nucleotide sequence of DNA (that is, genotype), rather than phenotypical features such as appearance or behavior, can be used to investigate relatedness among different types of organisms. Genetic relatedness among organisms can also be evaluated by examining the structure of their chromosomes. Among the findings derived from this technique is that the chromosomes of chimpanzees and humans are extremely similar, showing that these two species are very closely related.

2 What Are the Domains of Life?

Before 1970, all forms of life were classified into two kingdoms: Animalia and Plantae. All bacteria, fungi, and single-celled photosynthetic eukaryotes were considered to be plants, and the single-celled eukaryotes were classified as animals. As scientists learned more about fungi and microorganisms, however, it became apparent that the two-kingdom system oversimplified evolutionary history. To help rectify this problem, in 1969 Robert H. Whittaker proposed a five-kingdom classification that was eventually adopted by most biologists.

The Five-Kingdom System Improved Classification

Whittaker's five-kingdom system divides single-celled microorganisms into two kingdoms, based on whether the cells are prokaryotic or eukaryotic. Single-celled

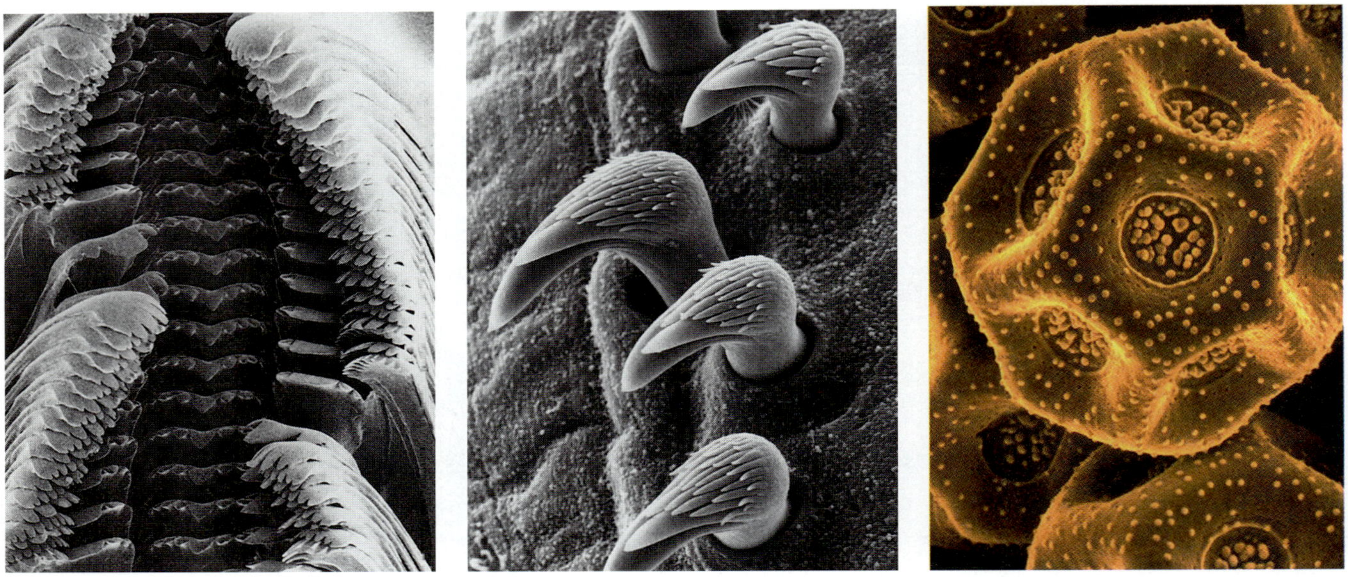

(a) (b) (c)

Figure 16-2 Microscopic structures may be used to classify organisms
(a) The "teeth" on a snail's tonguelike radula (a structure used in feeding), ***(b)*** the bristles on a marine worm, and ***(c)*** the shape and surface features of pollen grains are characteristics potentially useful in classification. Such finely detailed structures can reveal similarities between species that are not apparent in larger and more obvious structures.

prokaryotic organisms are in the kingdom Monera; single-celled eukaryotic organisms are in the kingdom Protista. All organisms in the remaining three kingdoms (Plantae, Fungi, and Animalia) are eukaryotic, and most are multicellular. These three kingdoms of multicellular eukaryotes can be distinguished on the basis of their methods of acquiring nutrients. Members of the kingdom Plantae photosynthesize, and members of the kingdom Fungi secrete enzymes outside their bodies and then absorb the externally digested nutrients. Members of the kingdom Animalia ingest their food and then digest it, either within an internal cavity or within individual cells.

Because it more accurately reflected our understanding of evolutionary history, the five-kingdom system was an improvement over the old two-kingdom system. As our understanding has continued to grow, however, it has proved necessary to change our view of life's most fundamental categories. The pioneering work of microbiologist Carl Woese has shown that biologists had overlooked a fundamental event in the early history of life, one that demands a new and more-accurate classification of life.

A Three-Domain System More Accurately Reflects Life's History

Since the 1970s Woese and other biologists interested in the evolutionary history of microorganisms have studied the biochemistry of prokaryotic organisms. These researchers, focusing on nucleotide sequences of the RNA that is found in the organisms' ribosomes, determined that the kingdom Monera encompassed two very different kinds of organisms. Woese has dubbed these two groups the Bacteria and the Archaea (Fig. 16-3).

Despite superficial similarities in their appearance under the microscope, the Bacteria and the Archaea are radically different. These two groups are no more closely related to one another than either one is to any

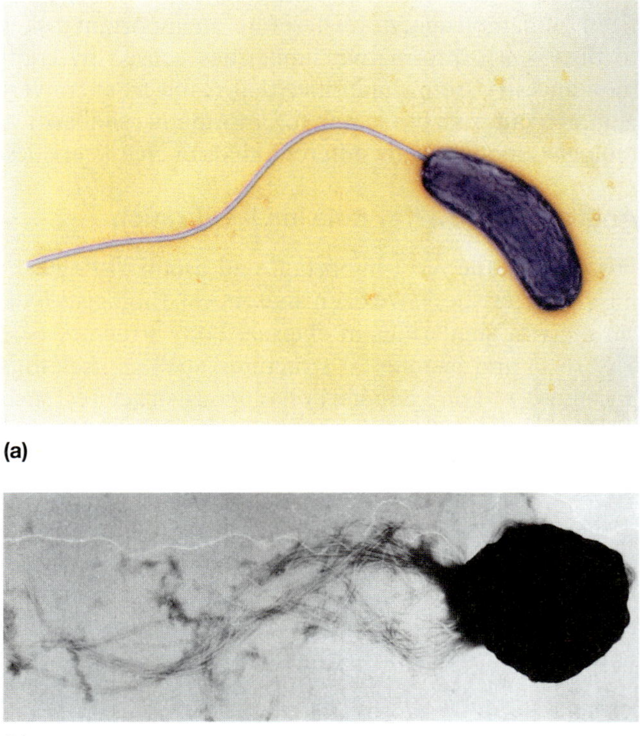

(a)

(b)

Figure 16-3 Two domains of prokaryotic organisms
Although superficially similar in appearance, ***(a)*** *Vibrio cholerae* and ***(b)*** *Methanococcus jannaschi* are less closely related than a mushroom and an elephant. *Vibrio* is in the domain Bacteria, and *Methanococcus* is in the Archaea.

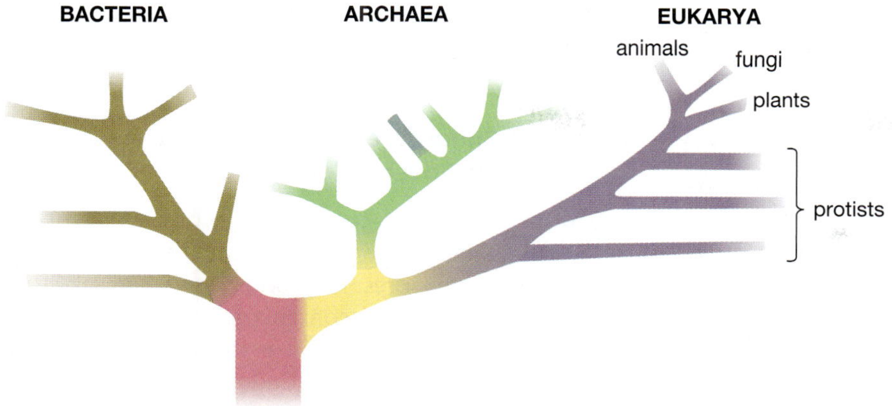

Figure 16-4 The tree of life
The three domains of life represent the earliest branches in evolutionary history.

eukaryote. The tree of life split into three parts very early in the history of life, long before the appearance of plants, animals, and fungi. So Woese has proposed, and many biologists agree, that we ought to classify life into three domains: the **Bacteria**, **Archaea**, and **Eukarya** (Fig. 16-4).

Kingdom-level Classification Remains Unsettled

If we accept the three-domain classification system, we're left with some dilemmas with regard to the status of our five kingdoms. It does not make sense to lump all prokaryotes into the single kingdom Monera, because the Bacteria and the Archaea are only very distantly related. If life has three main branches, we can't justify lumping two of them (Archaea and Bacteria) into one kingdom while splitting the third branch (Eukarya) into four kingdoms. A related problem is that if the domain Eukarya can contain four kingdoms, then the other two domains may also contain more than one kingdom. In fact, there were a number of very early splits both within the Archaea and within the Bacteria that long predated, for example, the split between plants and animals. If the plant–animal split is ancient enough to warrant kingdom status for each group, then each of the even-older splits within the Bacteria and Archaea should also merit kingdom status. Thus, we would need a lot more than five kingdoms to construct an accurate classification of life. Some biologists have even suggested that plants, animals, and fungi don't deserve kingdom status. After all, they represent only tiny twigs way down at the end of the eukaryote branch of the tree of life.

Overall, we're now at a rather awkward moment in the evolution of high-level classification. The molecular data strongly support the three-domain system, but the thorny issues of kingdom-level classification have yet to be definitively worked out. At the same time, it is unsatisfying to downplay the seemingly fundamental differences among protists, plants, animals, and fungi. In this text, our imperfect solution will be to reserve judgment on kingdom-level classification within the two prokaryotic domains and to retain the four kingdoms within the Eukarya. This arrangement is useful, but it is important for you to understand that it is derived from a five-kingdom system that may be replaced some time in the future by a more accurate system.

3 What Are Some Characteristics of Bacteria and Archaea?

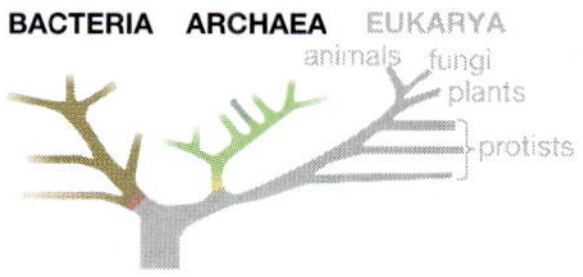

Two of life's three domains, Bacteria and Archaea, consist entirely of prokaryotes, single-celled microbes that lack organelles such as the nucleus, chloroplasts, and mitochondria. (See Chapter 4 for a comparison of prokaryotic and eukaryotic cells.) Both bacteria and archaea are normally very small, ranging from about 0.2 to 10 micrometers in diameter, compared with eukaryotic cells, whose diameters range from about 10 to 100 micrometers. About 250,000 average-sized bacteria or archaea could congregate on the period at the end of this sentence.

Bacteria and Archaea Are Fundamentally Different

Bacteria and archaea are superficially similar in appearance under the microscope, but the ancient evolutionary separation between them is revealed by striking differences in their structural and biochemical features. For example, the structure and composition of cell walls, plasma membranes, ribosomes, and RNA polymerases differ between the bacteria and archaea, as do key features of basic processes such as transcription and translation.

Within-Domain Classification of Prokaryotes Is Difficult

The sharp biochemical differences between archaea and bacteria make distinguishing the two domains a straightforward matter, but classification within each domain poses challenges. Prokaryotes are tiny and structurally simple and simply do not exhibit the huge array of anatomical and developmental differences that are used to infer the evolutionary history of plants, animals, and other eukaryotes. Consequently, prokaryotes have been classified on the basis of such features as shape, means of locomotion, pigments, nutrient requirements, the appearance of colonies (groups of individuals that descended from a single cell), and staining properties.

In recent years our understanding of the evolutionary history of the prokaryotic domains has been greatly expanded by comparisons of DNA and RNA nucleotide sequences. On the basis of this new information, some biologists now classify bacteria into about 13 to 15 kingdoms and archaea into about 3 kingdoms. Prokaryote classification, however, is a rapidly changing field, and consensus on kingdom-level classification has thus far proved elusive. With new DNA sequence data being generated at a furious pace, and with new and distinctive types of bacteria and archaea being discovered and described on a regular basis, prokaryote classification schemes will likely continue to be dynamic and frequently revised for some time to come.

Prokaryotes Differ in Shape and Structure

The cell walls that surround prokaryotic cells give characteristic shapes to different types of bacteria and archaea. The most common shapes are rodlike, spherical, and corkscrew-shaped (Fig. 16-5). Some bacteria and archaea have **flagella** (singular, **flagellum**). These are simpler in structure than eukaryotic flagella, as discussed in Chapter 4. Prokaryote flagella may appear singly at one end of a cell, in pairs (one at each end of the cell), as a tuft at one end of the cell (Fig. 16-6), or scattered over the entire cell surface. Flagella can rotate rapidly, propelling the organism through its liquid environment.

Protective Endospores Allow Some Bacteria to Withstand Adverse Conditions

When environmental conditions become inhospitable, many rod-shaped bacteria form protective resting structures called **endospores.** The endospore forms inside the bacterium (Fig. 16-7). It contains genetic material and a few enzymes encased within a thick protective coat. Metabolic activity ceases. Endospores are resistant structures that can survive extremely unfavorable conditions. Some can withstand boiling for an hour or more. Others can survive for extraordinarily long periods. In the most extreme example of such longevity, scientists recently discovered bacterial spores that had been sealed inside rock for 250 million years. After being carefully extracted from their rocky "tomb," the spores were incubated in test tubes. Amazingly, live bacteria developed from the ancient spores, which were older than the oldest dinosaur fossils.

Endospores are one of the main reasons that the bacterial disease anthrax has become an agent of biological terrorism. The bacterium that causes anthrax forms endospores, and the spores provide the means by which terrorists (or governments) can disperse the bacteria. The spores can be stored indefinitely and can survive any harsh conditions that they might en-

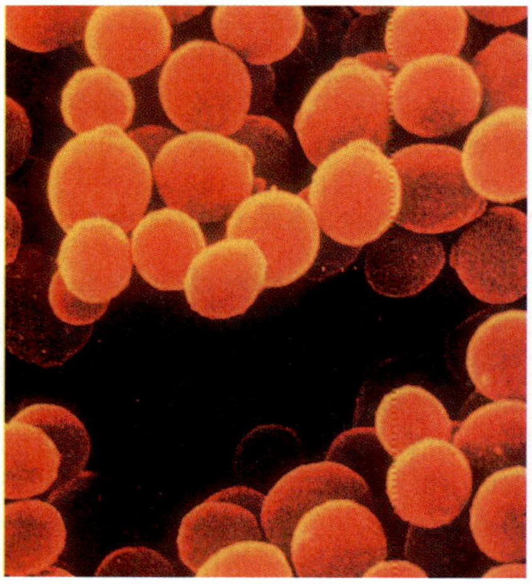

(a) 0.10 μm

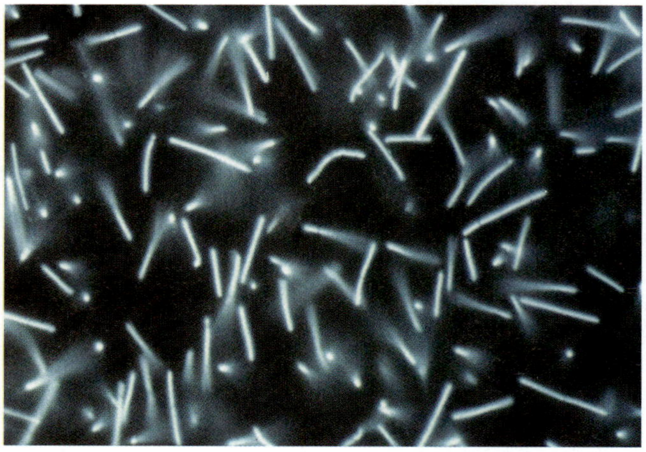

(b)

Figure 16-5 Three common prokaryote shapes ***(a)*** Spherical bacteria of the genus *Micrococcus*, ***(b)*** rod-shaped archaea of the genus *Methanopyrus*, and ***(c)*** corkscrew-shaped bacteria of the species *Leptospirosis interrogans*.

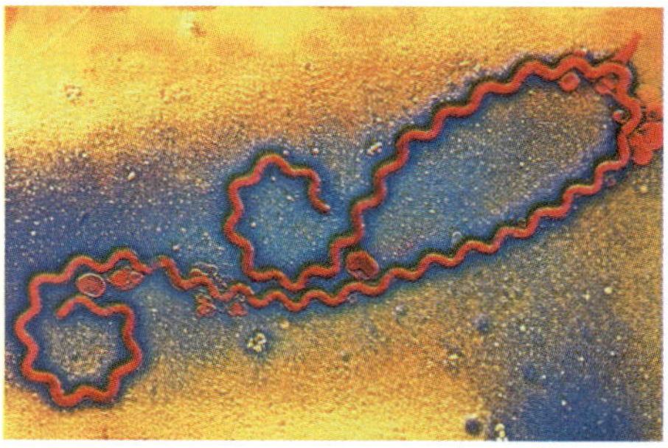

(c) 0.25 μm

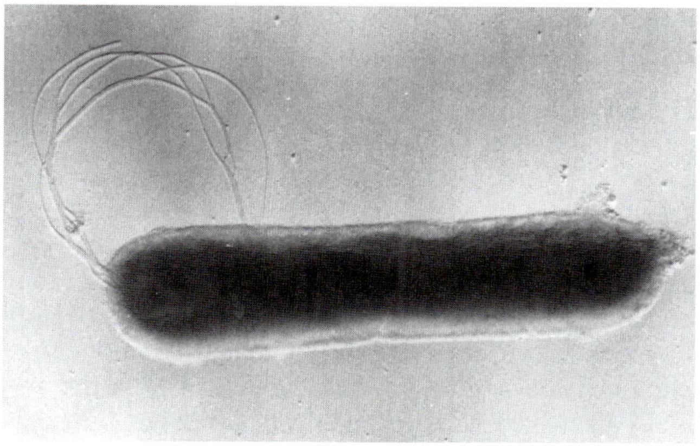

Figure 16-6 The prokaryote flagellum
An archaean of the genus *Aquifex* uses its flagella to move toward favorable environments.

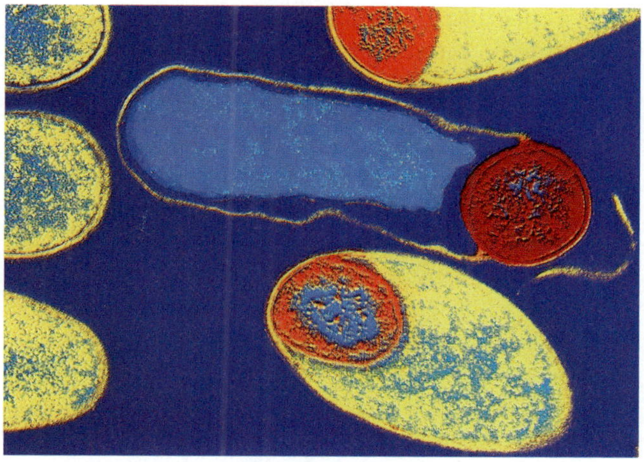

Figure 16-7 Spores protect some bacteria
Resistant endospores, here colored red, have formed inside bacteria of the genus *Clostridium*, which causes the potentially fatal food poisoning called botulism.

counter en route to their destination, even the stress of a missile launch and high-altitude travel. At their target, the spores can survive dispersal into the atmosphere and can remain viable until inhaled by a potential victim.

Prokaryotes Reproduce by Binary Fission

Most prokaryotes reproduce asexually by a simple form of cell division called binary fission (see Chapter 10), which produces genetically identical copies of the original cell (Fig. 16-8). Under ideal conditions, a prokaryotic cell can divide about once every 20 minutes, potentially giving rise to sextillions (10^{21}) of offspring in a single day! This rapid reproduction allows bacteria to exploit temporary habitats, such as a mud puddle or warm pudding. Recall that many mutations, the source of genetic variability, occur as a result of mistakes in DNA replication during cell division (see Chapter 9). Thus, the rapid reproductive rate of bacteria provides ample opportunity for new forms to arise and also allows mutations that enhance survival to spread quickly.

Prokaryotes Are Specialized for Specific Habitats

Prokaryotes occupy virtually every habitat, including those where extreme conditions prevent occupation by other forms of life. For example, some bacteria thrive in near-boiling environments, such as the hot springs of Yellowstone National Park (Fig. 16-9). Many archaea live in even hotter environments, including springs where the water actually boils or deep-ocean vents, where superheated water is spewed through cracks in

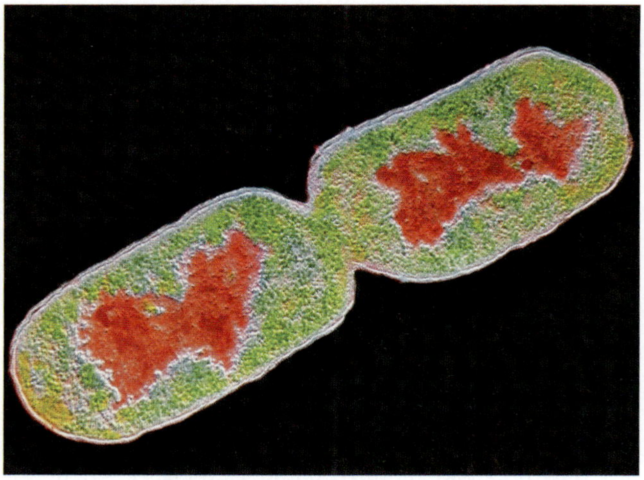

Figure 16-8 Reproduction in prokaryotes
Prokaryotic cells reproduce by binary fission. In this color-enhanced electron micrograph, *Escherichia coli*, a normal component of the human intestine, is dividing. Red areas are genetic material.

Figure 16-9 Some prokaryotes thrive in extreme conditions
Hot springs harbor bacteria and archaea that are both heat-tolerant and mineral-tolerant. Several species of bacteria paint these hot springs in Yellowstone National Park with vivid colors, and each is confined to a specific area determined by temperature range. The bacterial pigments function in photosynthesis.

Earth's crust at temperatures of up to 230 °F (110 °C). It is also pretty warm 1.7 miles (2.8 kilometers) below Earth's surface, which is where scientists recently discovered a new bacterial species. Bacteria and archaea are also found in very cold environments, such as Antarctic sea ice that remains frozen nearly year-round. Even extreme chemical conditions fail to impede invasion by prokaryotes. Thriving colonies of bacteria and archaea live in the Dead Sea, where a salt concentration seven times that of the oceans precludes all other life, and in waters that are as acidic as vinegar or as alkaline as household ammonia. Of course, rich bacterial communities also reside in a full range of more-moderate habitats, including in and on the healthy human body. An animal need not remain healthy to harbor bacteria, though. Recently, a colony of bacteria was found dormant within the intestinal contents of a mammoth that had lain in a peat bog for 11,000 years.

No single species of prokaryote, however, is as versatile as these examples may suggest. In fact, most prokaryotes are specialists. One species of archaea that inhabits deep-sea vents, for example, grows optimally at 223 °F (106 °C) and stops growing altogether at temperatures below 194 °F (90 °C). Clearly, this species could not survive in a less extreme habitat. Bacteria that live on the human body are also specialized; different species colonize the skin, the mouth, the respiratory tract, the large intestine, and the urogenital tract.

Prokaryotes Exhibit Diverse Metabolisms

Prokaryotes are able to colonize diverse habitats partly because they are able to use many different nutrient sources. Some species of bacteria engage in plantlike photosynthesis. Like green plants, these bacteria possess chlorophyll, produce oxygen as a byproduct of photosynthesis, and exist only where light and oxygen are available. Other photosynthetic bacteria, known as the sulfur bacteria, use hydrogen sulfide (H_2S) instead of water (H_2O) in photosynthesis, releasing sulfur instead of oxygen. No photosynthetic archaea are known.

Many species of bacteria and archaea can derive energy through reactions that combine oxygen with inorganic molecules such as sulfur, ammonia, or nitrite. In the process they may release sulfates or nitrates, crucial plant nutrients, into the soil. Some archaea convert carbon dioxide to methane (sometimes called "swamp gas"). Many prokaryotes are anaerobes, which means that they do not require oxygen to extract energy. Some anaerobes, such as many of the archaea found in hot springs and the bacterium that causes tetanus, are actually poisoned by oxygen. Others are opportunists, engaging in fermentation when oxygen is lacking and switching to aerobic respiration (a more efficient process) when oxygen becomes available.

Prokaryotes Perform Functions Important to Other Organisms

Did you know that if it weren't for bacteria, we would not be eating steaks and hamburgers? Cows, along with sheep and some nondomestic mammals such as deer, are members of a group known as ruminants. Ruminants subsist by eating leaves, but they can't actually digest that plant material themselves. Instead, they depend on certain bacteria that have the unusual ability to break down cellulose, the principal component of plant cell walls. Some of these bacteria live in the animals' digestive tracts, where they help liberate nutrients from plant fodder that the animals are unable to break down themselves. Without the bacteria, cows could not survive. No cows, no hamburgers.

Prokaryotes Capture the Nitrogen Needed by Plants

Humans could not live without plants, and plants are entirely dependent on bacteria. In particular, plants are unable to capture nitrogen from that element's most abundant reservoir, the atmosphere. Plants need nitrogen to grow. To acquire it, they depend on **nitrogen-fixing bacteria**, which live both in soil and in specialized nodules, small, rounded lumps on the roots of certain plants (legumes, which include alfalfa, soybeans, lupines, and clover). The nitrogen-fixing bacteria capture nitrogen gas (N_2) from air trapped in the soil and combine it with hydrogen to produce ammonium (NH_4^+), a nitrogen-containing nutrient that plants can use directly.

Prokaryotes Are Nature's Recyclers

Prokaryotes play a crucial role in recycling waste. Many substances that we consider to be waste can serve as food for archaea and bacteria, most species of which obtain energy by breaking down complex organic (carbon-containing) molecules. The range of compounds attacked by prokaryotes is staggering. Nearly anything that human beings can synthesize, including detergents and the poisonous solvent benzene, some prokaryote can destroy. The term "biodegradable" (meaning "broken down by living things") refers largely to the work of prokaryotes. Even oil is biodegradable. Soon after the tanker *Exxon Valdez* dumped 11 million gallons of crude oil into Prince William Sound, Alaska, researchers from Exxon sprayed oil-soaked beaches with a fertilizer that encouraged the growth of natural populations of oil-eating bacteria. Within 15 days the oil deposits were noticeably reduced compared with unsprayed areas. Prokaryotes are much less successful, however, at degrading most kinds of plastic. The search for useful, economical plastics that are also biodegradable is therefore a major focus of industrial research.

The appetite of some prokaryotes for nearly any organic compound is the key to their important role as decomposers in ecosystems. While feeding themselves, bacteria break down the waste products and dead bodies of plants and animals, freeing nutrients for reuse. The recycling of nutrients provides the basis for continued life on Earth, as we shall see in Chapter 29.

Some Bacteria Pose a Threat to Human Health

Despite the benefits some bacteria provide, the feeding habits of certain bacteria threaten our health and well-being. These **pathogenic** (disease-producing) bacteria synthesize toxic substances that cause disease symptoms. (So far, no pathogenic archaea have been identified.)

Some Anaerobic Bacteria Produce Dangerous Poisons

Some bacteria produce toxins that attack the nervous system. Examples of such pathogens include *Clostridium tetani*, which causes tetanus, and *Clostridium botulinum*, which causes botulism (a sometimes lethal food poisoning). Both of these bacterial species are anaerobes that survive as spores until introduced into a favorable, oxygen-free environment. For example, a deep puncture wound may provide a means through which tetanus bacteria can penetrate a human body and reach a place where they will be protected from contact with oxygen. As they multiply, the bacteria release their paralyzing poison into the bloodstream. For botulism bacteria, a sealed container of canned food that has been improperly sterilized may provide a haven. Thriving on the nutrients in the can, these anaerobes produce a toxin so potent that a single gram could kill 15 million people. Perhaps inevitably, this potent poison has caught the attention of biological weapon designers, who are presumed to have added it to their arsenals.

Humans Battle Bacterial Diseases Old and New

Bacterial diseases have had a significant impact on human history. Perhaps the most infamous example is bubonic plague, or "Black Death," which killed 100 million people during the mid-fourteenth century. In many parts of the world, one-third or more of the population died. Plague is caused by the highly infectious bacteria *Yersinia pestis*, which is spread by fleas that feed on infected rats and then move to human hosts. Although plague has not reemerged as a large-scale epidemic, about 2000 to 3000 people worldwide are diagnosed with the disease each year.

Some bacterial pathogens seem to emerge suddenly. Lyme disease, for example, was unknown until 1975. This disease, named after the town of Old Lyme, Connecticut, where it was first described, is caused by the spiral-shaped bacterium *Borrelia burgdorferi*. The bacterium is carried by the deer tick and transmitted to humans who are bitten by the tick. At first, the symptoms resemble flu, with chills, fever, and body aches. If untreated, weeks or months later the victim may experience rashes, bouts of arthritis, and in some cases abnormalities of the heart and nervous system. Both physicians and the general public are becoming more familiar with the disease, so more victims are receiving treatment before serious symptoms develop.

Perhaps the most frustrating pathogens are those that come back to haunt us long after we believed that we had them under control. Tuberculosis, a bacterial disease once almost vanquished in developed countries, is again on the rise in the United States and elsewhere. Two sexually transmitted bacterial diseases, gonorrhea and syphilis, have reached epidemic proportions around the globe. Cholera, a water-transmitted bacterial disease that flourishes when raw sewage contaminates drinking water or fishing areas, is under control in developed countries but remains a major killer in poorer parts of the world.

Some Common Bacterial Species Can Be Harmful

Some pathogenic bacteria are so widespread and ubiquitous that we cannot expect to ever be totally free of their damaging effects. For example, different species of the abundant streptococcus bacterium produce several diseases. One streptococcus causes strep throat. Another, *Streptococcus pneumoniae*, causes pneumonia by stimulating an allergic reaction that clogs the lungs with fluid. Yet another streptococcus has gained fame as the "flesh-eating bacterium." A small percentage of people who become infected with this one experience severe symptoms, described luridly in tabloid newspapers with such headlines as "Killer Bug Ate My Face." About 800 Americans each year are victims of necrotizing fasciitis (as the "flesh-eating" infection is more properly known), and about 15% of these victims die. The streptococci enter through broken skin and spew out toxins that either destroy flesh directly or stimulate an overwhelming and misdirected attack by the immune system against the body's own cells. A limb can be destroyed in hours, and in some cases only amputation can halt the rapid tissue destruction. In other cases, these rare strep infections sweep through the body, causing death within a matter of days; Jim Henson, creator of the Muppets, died as a result of such an infection.

But Most Bacteria Do No Harm

Although some bacteria assault the human body, most of the bacteria with which we share our bodies are harmless and many are beneficial. For example, the normal bacterial community in the female vagina creates an environment that is hostile to infections by parasites such as yeasts. Bacteria harmlessly inhabiting our intestines are an important source of vitamin K. As the late physician, researcher, and author Lewis Thomas so aptly put it, "Pathogenicity is, in a sense, a highly skilled trade, and only a tiny minority of all the numberless tons of microbes on the Earth has ever been involved in it; most bacteria are busy with their own business, browsing and recycling the rest of life."

The Big Picture

The vast majority of Earth's organisms are so tiny that that they are invisible to us, and their lives escape our notice. But microscopic creatures occupy virtually every corner of Earth, and their number is so immense as to be beyond comprehension.

4 Which Organisms Are in the Kingdom Protista?

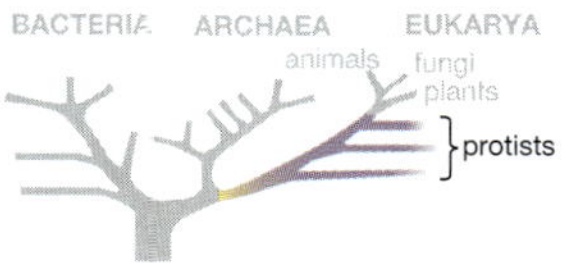

The third domain, Eukarya, includes one kingdom of microbes, the Protista. To appreciate the single-celled eukaryotes of the kingdom Protista, we have to overcome our size bias. These **protists** are largely invisible to us as we go about our daily lives. If we could somehow inhabit their microscopic scale, however, we might be more impressed with their spectacular and beautiful forms, their varied and active lifestyles, their astonishingly diverse modes of reproduction, and the structural and physiological innovations that are possible within the limits of a single cell.

Since Anton van Leeuwenhoek first observed protists through his simple homemade microscope in 1674, at least 50,000 species have been described. Although some protists form colonies, most live singly. Most protists reproduce asexually by mitotic cell division, but many are also capable of a form of sexual reproduction called conjugation (Fig. 16-10). All three major modes of nutrition are represented in this kingdom: For example, algae capture solar energy through photosynthesis; predatory protists ingest their food; and parasitic forms can absorb nutrients from their surroundings.

Protists Are a Diverse Group Including Funguslike, Plantlike, and Animal-like Forms

The kingdom Protista is extremely diverse. Until recently, the mostly single-celled organisms that compose this kingdom were grouped on the basis of apparent similarities to the cells of fungi, plants, or animals. We now recognize that these similarities are for the most part superficial and do not generally reflect shared evolutionary history with members of the other eukaryotic kingdoms. Nonetheless, categorizing protists as funguslike, plantlike, or animal-like is both traditional and convenient, and we will follow that convention in the descriptions that follow. Keep in mind, however, that most of the protist groups in each of these categories of convenience represent branches of the tree of life that split off long before the branches that led to fungi, plants, and animals. Thus, the funguslike, plantlike, and animal-like protists are not closely related to, respectively, fungi, plants, or animals, and are not necessarily closely related to one another.

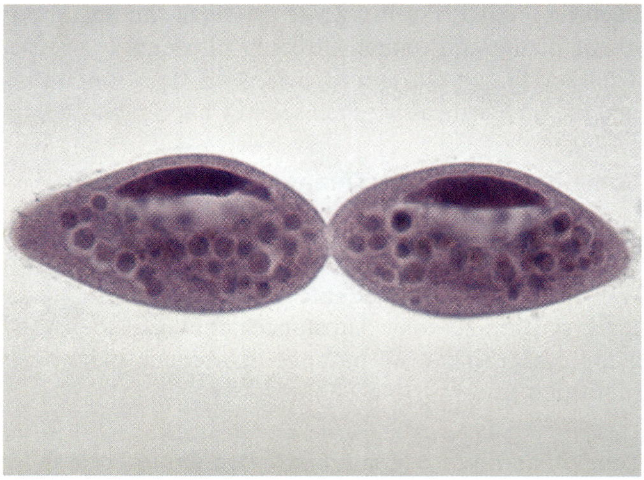

(a)

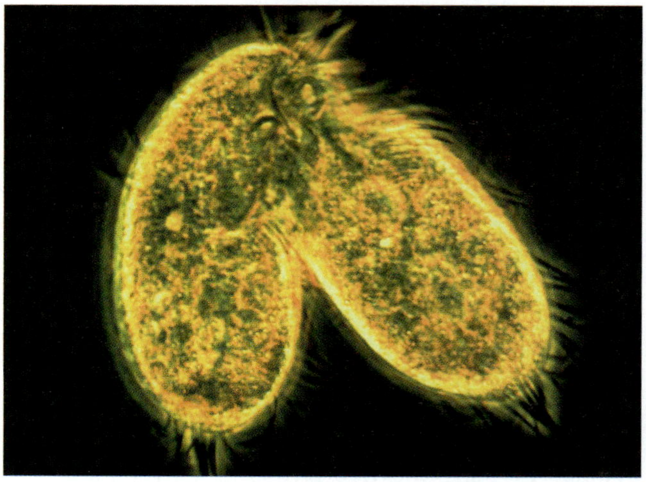

(b)

Figure 16-10 Two modes of protistan reproduction
(a) *Paramecium*, a ciliate, reproduces asexually by cell division that results in two daughter cells identical to the original parental cell. ***(b)*** Mating in a ciliate, *Euplotes*. Genetic material is exchanged across a cytoplasmic bridge. New individuals formed by cell division after the exchange will have gene combinations different from those of either parent cell.

The Slime Molds Are Funguslike Protists

Some protists are characterized by physical similarity to fungi and have a funguslike mode of nutrition. Like fungi, they absorb nutrients from the soil, water, or tissues of other organisms and typically help decompose dead organisms.

One group of funguslike protists includes the acellular slime molds. These protists consist of a mass of cytoplasm that may spread thinly over an area of several square meters. Although the mass, called a plasmodium, contains thousands of diploid nuclei, the nuclei are not confined in separate cells surrounded by plasma membranes, as in most multicellular organisms. This explains why these protists are described as "acellular" (without cells). The plasmodium oozes through decaying leaves and rotting logs, engulfing food such as bacteria and

(a)

(b)

Figure 16-11 The acellular slime mold *Physarum*
(a) *Physarum* oozes over a stone on the damp forest floor. ***(b)*** When food becomes scarce, the mass differentiates into black fruiting bodies in which spores are formed.

particles of organic material. The mass may be bright yellow or orange—a large plasmodium can be rather startling (Fig. 16-11a). Dry conditions or starvation stimulate the plasmodium to form reproductive structures called fruiting bodies in which haploid spores are produced (Fig. 16-11b). The spores are dispersed and germinate under favorable conditions, eventually giving rise to new plasmodia.

The Algae Are Plantlike Protists

Photosynthetic protists, collectively known as **algae** (singular, **alga**), are widely distributed in oceans and lakes. Most species are single-celled, but some form multicellular aggregations that are commonly known as seaweeds. The single-celled algal species are collectively known as **phytoplankton**. Although they are microscopic, phytoplankton are immensely important. Marine phytoplankton account for nearly 70% of all the photosynthetic activity on Earth, absorbing carbon dioxide, recharging the atmosphere with oxygen, and supporting the complex web of aquatic life.

There are several phyla of these plantlike protists. Three of them, the dinoflagellates, the brown algae, and the green algae, are described below.

Dinoflagellates Swim by Means of Two Whiplike Flagella

Dinoflagellates are named for the motion created by their two whiplike flagella (*dino* is Greek for "whirlpool"). One flagellum encircles the cell, and the second projects behind it. Some dinoflagellates are enclosed only by a cell membrane; others have cellulose walls that resemble armor plates (Fig. 16-12). Although some species live in fresh water, dinoflagellates are especially abundant in the ocean, where they are an important food source for larger organisms. Many dinoflagellates are bioluminescent, producing a brilliant blue-green light when disturbed. Specialized dinoflagellates live within the tissues of corals, some clams, and even other protists, where the algae provide photosynthetic nutrients and remove carbon dioxide. Reef-building corals live only in the shallow, well-lit waters in which their embedded dinoflagellates can survive.

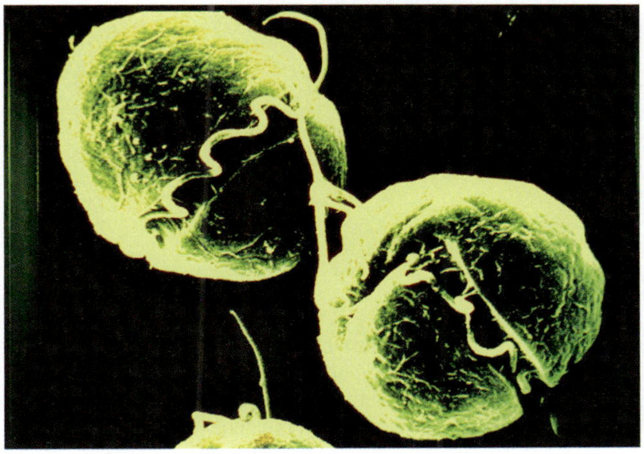

Figure 16-12 Dinoflagellates
Two dinoflagellates covered with protective cellulose armor. Visible on each is a flagellum in a groove that encircles the body.

Warm water that is rich in nutrients may bring on a dinoflagellate population explosion. Dinoflagellates can become so numerous that the water is dyed red by the color of their bodies, causing a "red tide" (Fig. 16-13). During red tides, fish die by the thousands, suffocated by clogged gills or by the oxygen depletion that results from the decay of billions of dinoflagellates. But oysters, mussels, and clams have a feast, filtering millions from the water for food. In the process, however, they concentrate a nerve poison the dinoflagellates produce in their bodies. Humans who eat these mollusks may be stricken with potentially lethal paralytic shellfish poisoning.

Figure 16-13 A red tide
The explosive reproductive rate of certain dinoflagellates under the right environmental conditions can produce dinoflagellate concentrations so great that their microscopic bodies dye the seawater red or brown, as in this bay in Mexico.

The Brown Algae Dominate in Cool Coastal Waters

The brown algae increase their light-gathering ability by means of brownish yellow pigments that (in combination with green chlorophyll) produce the brown to olive-green color of these algae. Almost all brown algae are marine and all are multicellular. This group includes the dominant seaweed species that dwell along rocky shores in the temperate (cooler) oceans of the world, including the eastern and western coasts of the United States. Brown algae live in habitats ranging from nearshore, where they cling to rocks that are exposed at low tide, to far offshore. Several species use gas-filled floats to support their bodies (Fig. 16-14a). Some of the giant kelp found along the Pacific coast reach heights of 325 feet (100 meters) and may grow more than 6 inches (15 centimeters) in a single day. With their dense growth and towering height (Fig. 16-14b), kelp form undersea forests that provide food, shelter, and breeding areas for marine animals.

(a)

(b)

Figure 16-14 Brown algae
(a) *Fucus*, a genus found near shores, is shown here exposed at low tide. Notice the gas-filled floats, which provide buoyancy in water. ***(b)*** The giant kelp *Macrocystis* forms underwater forests off southern California.

The Green Algae Live Mostly in Ponds and Lakes

The green algae, a large and diverse group, include both multicellular and unicellular species. Most species live in freshwater ponds and lakes, but some live in the seas. Some green algae, such as *Spirogyra*, form thin filaments from long chains of cells (Fig. 16-15). Other species of green algae form colonies containing clusters of cells that are somewhat interdependent and that con-

Figure 16-15 A green alga
Spirogyra is a filamentous green alga composed of strands only one cell thick.

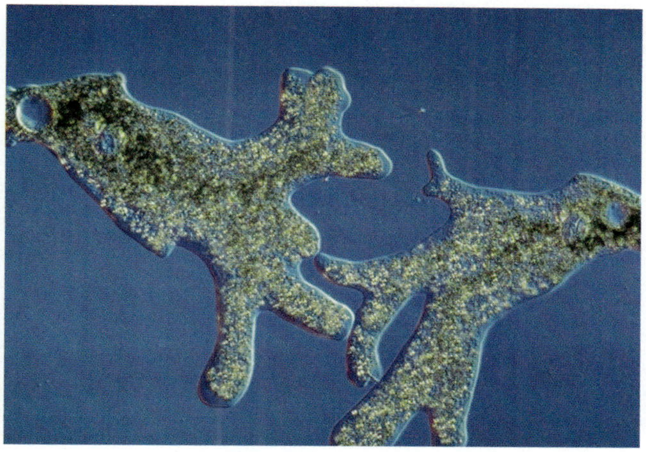

Figure 16-16 Amoebas
Amoebas like these use their pseudopods to move and capture prey.

stitute a structure intermediate between unicellular and multicellular forms. These colonies range from a few cells to a few thousand cells, as in species of *Volvox*. Most green algae are small, but some marine species are large. For example, the green alga *Ulva*, or sea lettuce, is similar in size to the leaves of its namesake.

The green algae are of special interest because, unlike the other groups of plantlike protists, green algae are actually closely related to plants. Plants share a common ancestor with some types of green algae, and many researchers believe that the very earliest plants were similar to today's multicellular green algae.

The Protozoa Are Animal-like Protists

The **protozoa** (singular, **protozoan**) are described as animal-like because they can move and they obtain their food from other organisms. All are unicellular and obtain food from their surroundings; they differ in methods of locomotion. We describe two representative phyla, the sarcodines and the ciliates.

Sarcodines Move by Means of Pseudopods

Sarcodines possess flexible plasma membranes that can extend the cytoplasm in any direction to form pseudopods, which are used for locomotion and for engulfing food (Fig. 16-16). The group includes amoebas, foraminiferans, and radiolarians.

Amoebas are common in freshwater lakes and ponds. Many amoebas are predators that stalk and engulf prey, but some species are parasites. One parasitic form causes amoebic dysentery, a disease that is prevalent in warm climates. The dysentery-causing amoeba multiplies in the intestinal wall, triggering severe diarrhea.

The foraminiferans and radiolarians are primarily marine sarcodines that produce beautiful shells. The shells of foraminiferans are constructed mostly of calcium carbonate (chalk; Fig. 16-17a); those of radiolarians are of silica (Fig. 16-17b). These elaborate shells are pierced by myriad openings through which pseudopods extend. The chalky shells of foraminiferans, accumulating over millions of years, have resulted in immense deposits of limestone such as those that form the famous White Cliffs of Dover, England.

Ciliates Are the Most Complex of the Protozoa

Ciliates, which inhabit fresh and salt water, represent the peak of unicellular complexity. They possess many specialized organelles, including **cilia** (singular, **cilium**), the short hairlike outgrowths after which they are named. The cilia may cover the cell, or they may be localized. In the well-known freshwater genus *Paramecium*, rows of cilia cover the entire body surface. Their coordinated beating propels the cell through the water at a rate of a millimeter per second—a protistan speed record. Although only a single cell, *Paramecium* responds to its environment as if it had a well-developed nervous system. Confronted with some noxious chemical or with a physical barrier, the cell immediately backs up by reversing the beating of its cilia and then proceeds in a new direction. Some ciliates, such as *Didinium*, are accomplished predators (Fig. 16-18).

MediaTutor 16.2 The Fungi

5 What Are the Key Features of Fungi?

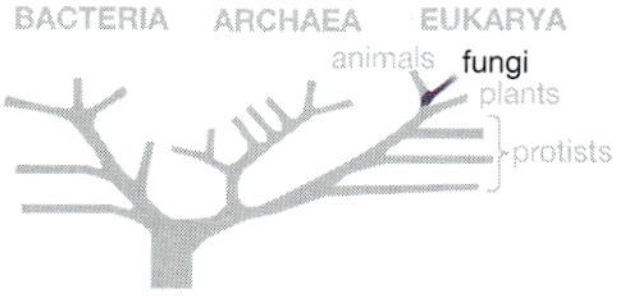

When you think of a fungus, you probably picture a mushroom. Mushrooms, however, are just temporary reproductive structures that extend from the main bodies of certain kinds of fungus. The body of almost all fungi is a **mycelium** (Fig. 16-19a), which is an interwoven mass of one-cell-thick, threadlike filaments called **hyphae** (singular, **hypha**; Fig. 16-19b). Fungi are not able to move from place to place. They compensate for this lack of mobility with hyphae that can grow rapidly in any direction within a suitable environment. The fungal mycelium quickly infuses itself into aging bread or

(a)

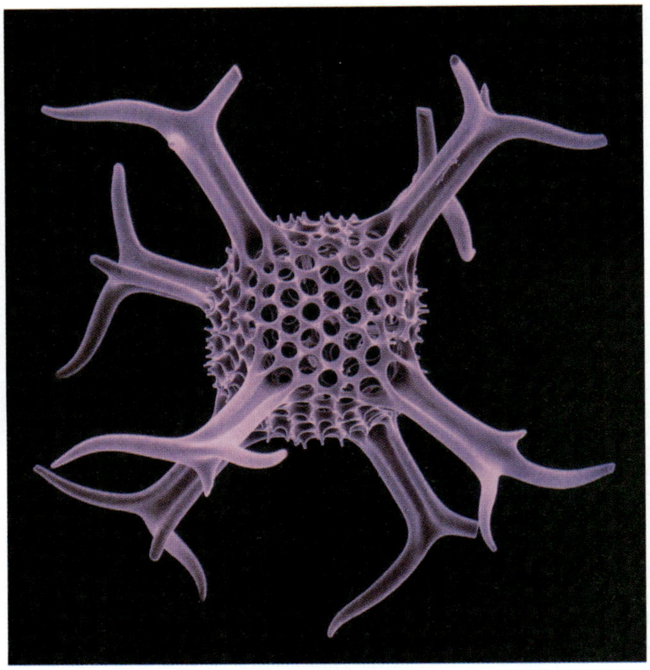

(b)

Figure 16-17 Foraminiferans and radiolarians
(a) The chalky shells of foraminiferans show numerous interior chambers. *(b)* The delicate, glassy shell of a radiolarian. Pseudopods, which sense the environment and capture food, extend out through the openings in the shell.

Figure 16-18 A microscopic predator
In this scanning electron micrograph, the predatory ciliate *Didinium* attacks a *Paramecium*. Note that the cilia of *Didinium* are confined to two bands, whereas *Paramecium* has cilia over its entire body. Ultimately, the predator will engulf and consume its prey. This microscopic drama could take place on a pinpoint, with room to spare.

cheese, beneath the bark of decaying logs, or into the soil. Periodically, the hyphae grow together and differentiate into reproductive structures that project above the surface beneath which the mycelium grows. These structures, including mushrooms, puffballs, and the powdery molds on unrefrigerated food, represent only a fraction of the complete fungal body, but are typically the only part of the fungus that we can easily see.

Like plant cells, fungal cells are surrounded by cell walls. Unlike plant cells, however, fungal cell walls are strengthened by chitin, the same substance found in the exoskeletons of arthropods.

Fungi Obtain Their Nutrients from Other Organisms

Like animals, fungi survive by breaking down nutrients stored in the bodies or wastes of other organisms. Some fungi digest the bodies of dead organisms. Others are parasitic, feeding on living organisms and causing disease. There are even a few predatory fungi, which attack tiny worms in soil.

Unlike animals, fungi do not ingest food. Instead, they secrete enzymes that digest complex molecules outside their bodies, breaking down the molecules into smaller subunits that can be absorbed. Fungal filaments can penetrate deeply into a source of nutrients and are only one cell thick, presenting an enormous surface area through which to secrete enzymes and absorb nutrients. This mode of securing nutrition has served the fungi well. Almost every biological material can be consumed by one or another fungal species, so nutritional support for fungi is likely to be at hand in nearly every terrestrial habitat.

Most Fungi Can Reproduce Both Sexually and Asexually

Unlike plants and animals, fungi form no embryos. Instead, fungi propagate by means of tiny, lightweight reproductive packages known as **spores**, which are extraordinarily mobile, even though most lack a means

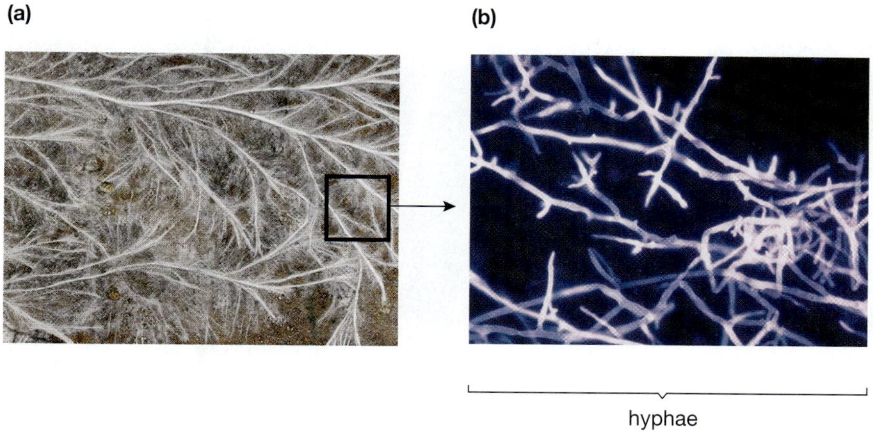

Figure 16-19 The filamentous body of a fungus
(a) A fungal mycelium spreads over decaying vegetation. The mycelium is composed of *(b)* a tangle of microscopic hyphae, only one cell thick.

MEDIATUTOR 16.3 Fungi Structure

for self-propulsion. Spores are distributed far and wide as hitchhikers on the outside of animal bodies, as passengers inside the digestive systems of animals that have eaten them, or as airborne drifters, cast aloft by chance or shot into the atmosphere by elaborate reproductive structures (Fig. 16-20). In addition, spores are often produced in great numbers (a single giant puffball may contain 5 trillion sexual spores). The fungal combination of prodigious reproductive capacity and highly mobile spores ensures that fungi are ubiquitous in terrestrial environments and accounts for the inevitable growth of fungi on every uneaten sandwich and container of leftovers.

In general, fungi are capable of both asexual and sexual reproduction. For the most part, asexual reproduction is the default mode under stable conditions, with sexual reproduction occurring mainly under conditions of environmental change or stress. Both asexual and sexual reproduction ordinarily involve the production of spores within special fruiting bodies that project above the mycelium.

Figure 16-20 Some fungi can eject spores
A ripe earthstar mushroom, struck by a drop of water, releases a cloud of spores that will be dispersed by air currents.

6 How Do Fungi Affect Humans?

The average person gives little thought to fungi, except perhaps for an occasional, momentary appreciation for the mushrooms on a pizza. We might do well, however, to pay them more attention, as fungi play a large part in human affairs.

Fungi Attack Plants That Are Important to People

Fungi cause the majority of plant diseases, and some of the plants that they infect are important to humans. For example, fungal pathogens have a devastating effect on the world's food supply. Especially damaging are the plant pests descriptively called rusts and smuts, which cause billions of dollars' worth of damage to grain crops annually (Fig. 16-21). Fungal diseases also affect the appearance of our landscape. The American elm and the American chestnut, two tree species that were once prominent in many of America's parks, yards, and forests, were destroyed on a massive scale by the fungi that cause Dutch elm disease and chestnut blight. Today, few people can recall the graceful forms of large elms and chestnuts, as these are almost entirely absent from the landscape.

The fungal impact on agriculture and forestry is not entirely negative, however. Fungal parasites that attack insects and other arthropod pests can be an important

Figure 16-21 Corn smut
This fungal pathogen destroys millions of dollars' worth of corn each year. Even a pest like corn smut has its admirers, though. In Mexico this fungus is known as *huitlacoche* and is considered to be a great delicacy.

Figure 16-22 A helpful fungal parasite
Some parasitic fungi are used by farmers to control insect pests. Here, a *Cordyceps* species has killed a grasshopper.

ally in pest control (Fig. 16-22). Farmers who wish to reduce their dependence on toxic and expensive chemical pesticides are increasingly turning to biological methods of pest control, including the application of "fungal pesticides." Fungal pathogens are currently used to control termites, rice weevils, tent caterpillars, aphids, citrus mites, and other pests.

Fungi Cause Human Diseases

The kingdom Fungi includes parasitic species that attack humans directly. Among the most familiar fungal diseases are those caused by fungi that attack the skin: athlete's foot, jock itch, and ringworm. These diseases, though unpleasant, are not life-threatening and can usually be treated with antifungal ointments. Prompt treatment can also usually control another common fungal disease, vaginal infections caused by the yeast *Candida albicans*. Fungi can also infect the lungs when victims inhale spores of the fungi that cause diseases such as valley fever and histoplasmosis. Like other fungal infections, these diseases can, if promptly diagnosed, be controlled with antifungal drugs. If untreated, however, they can develop into serious, systemic infections. Singer Bob Dylan, for instance, became gravely ill with histoplasmosis when the fungus infected the pericardial membrane surrounding his heart.

Fungi Can Produce Toxins

In addition to their role as agents of infectious disease, some fungi produce toxins that are dangerous to humans. Of particular concern are toxins produced by fungi that grow on grains and other foodstuffs that have been stored in too-moist conditions. For example, molds of the genus *Aspergillus* produce highly toxic, carcinogenic compounds known as aflatoxins. Some foods, such as peanuts, seem especially susceptible to attack by *Aspergillus*. Since aflatoxins were discovered in the 1960s, food growers and processors have developed methods for reducing the growth of *Aspergillus* in stored crops, so the amount of aflatoxin in the nation's peanut butter supply has been greatly reduced.

Many Antibiotics Are Derived from Fungi

Fungi have also had positive impacts on human health. The modern era of life-saving antibiotic medicines was ushered in by the discovery of penicillin, which is produced by a mold. Penicillin was the first antibiotic and is still used, along with other fungus-derived antibiotics, to combat bacterial diseases. Other important drugs are also derived from fungi, among them cyclosporin, which is used to suppress the immune response during organ transplants so that the body is less likely to reject the foreign organ.

Fungi Make Important Contributions to Gastronomy

Fungi make an important contribution to human nutrition. The most obvious components of this contribution are the fungi that we consume directly: wild and cultivated mushrooms and other fungi such as morels (Fig. 16-23) and the rare and prized truffle. The role of fungi in cuisine, however, also has less visible manifestations. For example, some of the world's most famous cheeses, including Roquefort, Camembert, Stilton, and Gorgonzola, gain their distinctive flavors from molds that grow on them as they ripen. Perhaps the most important and pervasive fungal contributors to our food supply, however, are the single-celled fungi known as yeasts.

Wine and Beer Are Made Using Yeasts

The discovery that we could harness yeasts to enliven our culinary experience is surely a key event in human history. Among the many foods and beverages that de-

Figure 16-23 A delicious fungus
The morel, an edible delicacy. (Consult an expert before sampling any wild fungus—some are deadly!)

pend on yeasts for their production are bread, wine, and beer, which are consumed so widely that it is difficult to imagine a world without them. All derive their special qualities from fermentation by yeasts. Fermentation occurs when yeasts extract energy from sugar and, as by-products of the metabolic process, emit carbon dioxide and ethyl alcohol. So, as yeasts consume the fruit sugars in grape juice, the sugars are replaced by alcohol, and wine is the result. Eventually, the increasing concentration of alcohol kills the yeasts, ending fermentation. If the yeast die before all available grape sugar is consumed, the wine will be sweet; if the sugar is exhausted, the wine will be dry.

Beer is brewed from grain (usually barley), but yeasts are not very effective at consuming the carbohydrates that compose grain kernels. For the yeasts to do their work, the barley grains must first be sprouted (recall that grains are actually seeds). Germination converts carbohydrates to sugar, so the sprouted barley provides an excellent food source for the yeasts. As with wine, fermentation converts sugars to alcohol, but beer brewers capture the carbon dioxide by-product as well, giving the beer its characteristic bubbly carbonation.

Yeasts Make Bread Rise

In bread making, carbon dioxide is the crucial fermentation product. The yeasts added to bread dough do produce alcohol as well as carbon dioxide, but the alcohol evaporates during baking. In contrast, the carbon dioxide is trapped in the dough, where it forms the bubbles that give bread its light, airy texture (and saves us from a life of eating sandwiches made of crackers). So the next time you're enjoying a slice of French bread with Camembert cheese and a nice glass of Chardonnay, or a slice of pizza and a cold bottle of your favorite brew, you might want to quietly give thanks to the yeasts. Our diets would certainly be a lot duller without the help we get from fungal partners.

Fungi Play a Crucial Ecological Role

No account of the fungi would be complete without mention of their fundamental ecological importance. The fungi are Earth's undertakers, consuming the dead of all kingdoms and returning their component substances to the ecosystems from which they came. The extracellular digestive activities of many fungi liberate nutrients such as carbon, nitrogen, and phosphorus compounds and minerals that can be used by plants. If fungi and bacteria were suddenly to disappear, the consequences would be disastrous. Nutrients would remain locked in the bodies of dead plants and animals, the recycling of nutrients would grind to a halt, soil fertility would rapidly decline, and waste and organic debris would accumulate. In short, ecosystems would collapse.

7 What Are the Major Groups of Plants?

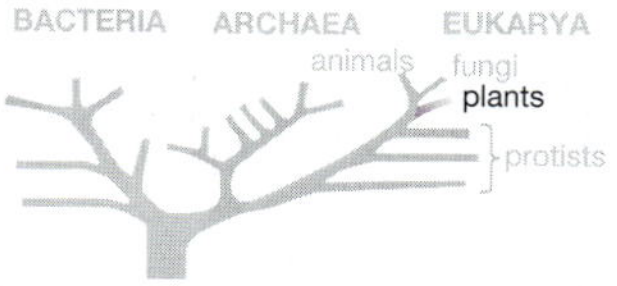

MEDIATUTOR
16.4 Evolution of Plant Structure

Two major groups of land plants arose from ancient algal ancestors. One group, the **bryophytes** (also called nonvascular plants), requires a moist environment to reproduce and thus straddles the boundary between aquatic and terrestrial life, much like the amphibians of the animal kingdom. The other group, the **vascular plants** (also called tracheophytes), has been able to colonize drier habitats.

Bryophytes Lack Conducting Structures

Bryophytes retain some characteristics of their algal ancestors. They lack true roots, leaves, and stems. They do possess rootlike anchoring structures called rhizoids that bring water and nutrients into the plant body, but bryophytes are nonvascular—they lack well-developed structures for conducting water and nutrients. They must instead rely on slow diffusion or poorly developed conducting tissues to distribute water and other nutrients. As a result, their body size is limited. Size is also limited by the absence of any stiffening agent in their bodies; without such material, they cannot achieve much upward growth. Most bryophytes are less than 1 inch (2.5 centimeters) tall.

The major bryophyte representatives are the liverworts and mosses (Fig. 16-24). Liverworts and mosses are generally most abundant in areas where moisture is plentiful. Some mosses, however, have a waterproof covering that retains moisture, preventing water loss. These mosses can survive in deserts, on bare rock, and in far northern and southern latitudes where humidity is low and liquid water is scarce for much of the year.

The Reproductive Structures of Bryophytes Are Protected

Among the bryophyte features that do represent adaptations to terrestrial existence are their enclosed reproductive structures, which prevent the gametes from

(a)

(b)

Figure 16-24 Bryophytes
Both plants shown here are less than a half inch (about 1 centimeter) in height. ***(a)*** Liverworts grow in moist, shaded areas. This is the female plant, bearing umbrella-like archegonia, which hold the eggs. Sperm must swim up the stalks through a film of water to fertilize the eggs. ***(b)*** Moss plants, showing the stalks that carry spore-bearing capsules.

drying out. There are two types of structure, one in which eggs develop and one in which sperm are formed. In some bryophyte species both egg-producing and sperm-producing structures are located on the same plant; in other species each individual plant is either male or female. In all bryophytes the sperm must swim to the egg (which emits a chemical attractant) through a film of water. Bryophytes that live in drier areas must time their reproduction to coincide with rains.

The Vascular Plants Have Conducting Vessels That Also Provide Support

Vascular plants are distinguished by specialized groups of conducting cells called vessels. The vessels are impregnated with the stiffening substance lignin and serve both supportive and conducting functions. Vessels allow vascular plants to grow taller than nonvascular plants, both because of the extra support provided by lignin and because the conducting cells allow water and nutrients absorbed by the roots to move to the upper portions of the plant. The vascular plants can be divided into two groups: the seedless vascular plants and the seed plants.

The Seedless Vascular Plants Include the Club Mosses, Horsetails, and Ferns

Like the bryophytes, seedless vascular plants have swimming sperm and require water for reproduction. As their name implies, they do not produce seeds but rather propagate by spores. The present-day seedless vascular plants—the club mosses, horsetails, and ferns—are much diminished in size compared to their ancestors, which dominated the landscape hundreds of millions of years ago. The bodies of ancient seedless vascular plants—transformed by heat, pressure, and time—are burned today as coal. The formerly dominant role of the seedless vascular plants has been assumed by the more-versatile seed plants.

The club mosses are now limited to representatives a few inches in height. Their leaves are small and scalelike (Fig. 16-25a). Modern horsetails form a single genus, *Equisetum*, that contains only 15 species, most less than 1 meter tall. The bushy branches of some species lend them the common name horsetails; the leaves are reduced to tiny scales on the branches (Fig. 16-25b). The ferns, with 12,000 species, are the most diverse of the seedless vascular plants. In the Tropics, tree ferns still reach heights reminiscent of their ancestors in the Carboniferous period. Ferns are the only seedless vascular plants that have broad leaves (Fig. 16-25c).

The Seed Plants Dominate the Land, Aided by Two Important Adaptations: Pollen and Seeds

In seed plants, sperm-producing cells are carried within tiny grains of **pollen**. Pollen grains are dispersed by wind or by animal pollinators, such as bees. In this way, sperm move through the air to fertilize egg cells. Thus, the distribution of seed plants is not limited by the need for

(a) (c) (b)

Figure 16-25 Seedless vascular plants
Seedless vascular plants are found in moist woodland habitats. ***(a)*** The club mosses (sometimes called ground pines) grow in temperate forests. This specimen is releasing spores. ***(b)*** The giant horsetail (genus *Equisetum*) extends long, narrow branches in a series of rosettes. Its leaves are insignificant scales. ***(c)*** The leaves of this deer fern are emerging from coiled fiddleheads.

water in which sperm can swim to the egg; seed plants are fully adapted to life on dry land.

The second reproductive adaptation is the seed itself. Analogous to the eggs of birds and reptiles, **seeds** consist of an embryonic plant, a supply of food for the embryo, and a protective outer coat. The seed coat maintains the embryo in a state of suspended animation or dormancy until conditions are proper for growth. The stored food helps sustain the emerging plant until it develops roots and leaves and can make its own food by photosynthesis. Some seeds possess elaborate adaptations that make possible dispersal by wind, water, and animals. These adaptations have helped them invade nearly every habitat on Earth.

Seed plants are grouped into two general types: gymnosperms, which lack flowers, and angiosperms, the flowering plants.

Gymnosperms Are Nonflowering Seed Plants

Gymnosperms evolved earlier than the flowering plants. One group, the conifers, with 500 species, still dominates large areas of our planet. They are most abundant in the cold latitudes of the far north and at high elevations where conditions are dry. Not only is rainfall limited in these areas, but soil water remains frozen and unavailable during the long winters. Conifers, including pines, firs, spruce, hemlocks, and cypresses, are adapted to dry, cold conditions in several ways. First, conifers retain green leaves throughout the year, enabling these plants to continue photosynthesizing and growing slowly during times when most other plants become dormant. Conifers are often called "evergreens" for this reason. Second, conifer leaves are actually thin needles covered with a thick, waterproof surface that minimizes evaporation. Finally, conifers produce an "antifreeze" in their sap that enables them to continue transporting nutrients in subfreezing temperatures. This substance gives them their fragrant "piney" scent. Other gymnosperms, such as the ginkgos and cycads (Fig. 16-26), have declined to a small remnant of their former range and abundance.

Angiosperms Are Flowering Seed Plants

Modern flowering plants, or **angiosperms**, have dominated Earth for more than 100 million years. The group is incredibly diverse, with more than 230,000 species. Angiosperms range in size from the diminutive duckweed (Fig. 16-27a) to the towering eucalyptus tree (Fig. 16-27b). From desert cactus to tropical orchids to grasses to parasitic mistletoe, angiosperms rule over the plant kingdom.

Flowers Attract Pollinators

Three major adaptations have contributed to the enormous success of angiosperms: flowers, fruits, and broad leaves. **Flowers**, the structures in which both male and female gametes are formed, may have evolved when gymnosperm ancestors formed an association with animals (most likely insects) that carried their pollen from plant to plant. According to this scenario, the relationship between these ancient gymnosperms and their animal pollinators was so beneficial that natural selection favored the evolution of showy flowers that advertised the presence of pollen to insects and other animals (Fig. 16-27b,e). The animals benefited by eating some of the protein-rich pollen, whereas the plant benefited from the animals' unwitting transportation of pollen from plant to plant. With this animal assistance, many flowering plants no longer needed to produce prodigious quantities of pollen and send it flying on the fickle winds to ensure fertilization. But there are also many wind-pollinated angiosperms (Fig. 16-27c,d).

Fruits Encourage Seed Dispersal

The ovary surrounding the seed of an angiosperm matures into a **fruit**, the second adaptation that has contributed to the success of angiosperms. Just as flowers encourage animals to transport pollen, so, too, many fruits entice animals to disperse seeds. If an animal eats a fruit, many of the enclosed seeds may pass through the animal's digestive tract unharmed, perhaps to fall at a suitable location for germination. Not all fruits, however, depend on edibility for dispersal. Dog owners are well aware, for example, that some fruits (called burs) disperse by clinging to animal fur. Other fruits, such as those of maples, form wings that carry the seed through the air. The variety of dispersal mechanisms made possible by fruits has helped the angiosperms invade nearly all possible terrestrial habitats.

Broad Leaves Capture More Sunlight

A third feature that gives angiosperms an adaptive advantage in warmer, wetter climates is broad leaves. When water is plentiful, as it is during the warm growing

(a)

(b)

Figure 16-26 Two uncommon gymnosperms
(a) The ginkgo, or maidenhair tree, has been kept from extinction by cultivation in China and Japan. Relatively resistant to pollution, these trees have become popular in U.S. cities. This ginkgo is female and bears fleshy seeds the size of large cherries, which are noted for their foul smell when ripe. ***(b)*** A cycad. Common in the age of dinosaurs, cycads are now limited to about 160 species living in warm, moist climates. Like ginkgos, cycads have separate sexes.

Figure 16-27 Angiosperms
(a) The smallest angiosperm is the duckweed, found floating on ponds. These specimens are about 1/8 inch (3 millimeters) in diameter. *(b)* The largest angiosperms are eucalyptus trees, which can reach 325 feet (100 meters) in height. Both *(c)* grasses and many trees, such as *(d)* this birch, in which flowers are shown as buds (green) and blossom (brown), have inconspicuous flowers and rely on wind for pollination. *(e)* Flowers, such as those shown on this butterfly weed and eucalyptus tree (inset in part b), entice insects and other animals that carry pollen between individual plants.

season of temperate and tropical climates, broad leaves give trees an advantage by collecting more sunlight for photosynthesis. However, broad leaves also increase the area from which water is lost, so many trees and shrubs drop their leaves during periods when water is in short supply. In regions with cold winters, water is much less available during the cold months and most angiosperm trees and shrubs drop their leaves during that period.

8 What Characteristics Define an Animal?

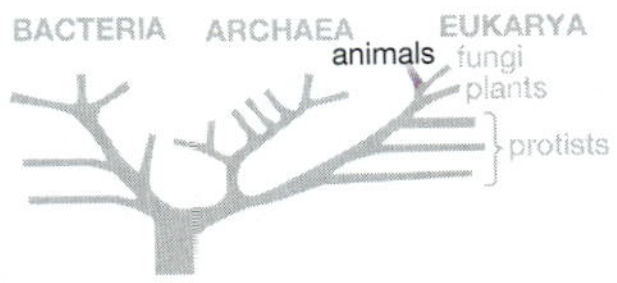

It is difficult to devise a concise definition of the term "animal." No single feature fully characterizes animals, so we define the group with a list of characteristics. None of these characteristics is unique to animals, but, taken together, they distinguish animals from members of other kingdoms:

- Animals are multicellular.
- Animals obtain their energy by consuming the bodies of other organisms.
- Animals typically reproduce sexually. Although animal species exhibit a tremendous diversity of reproductive styles, most are capable of sexual reproduction.
- Animal cells lack a cell wall.
- Animals are motile (able to move about) during some stage of their life. Even the stationary sponges have a free-swimming larval stage (a juvenile form).
- Most animals are able to respond rapidly to external stimuli as a result of the activity of nerve cells, muscle tissue, or both.

The Big Picture

When we think of animals, we tend to picture the ones that are most like us: large, active, land-dwelling creatures with bones inside them. These animals—mainly birds and mammals—are indeed conspicuous, but they make up only an insignificant portion of the animal world. In fact, the animal world is dominated by small, inconspicuous, boneless creatures, many of which live in habitats that we rarely contemplate and cannot easily visit.

(a) (b) (c)

Figure 16-28 Sponges
Some sponges, such as ***(a)*** this fire sponge, grow in free-form pattern over undersea rocks. ***(b)*** Tiny appendages attach this tubular sponge to rocks, whereas ***(c)*** this reef sponge with flared tubular openings attaches to a coral reef.

9 What Are the Major Animal Phyla?

It's easy to overlook the differences among the multitude of small, boneless animals in the world. Even Carolus Linnaeus, the originator of modern biological classification, recognized only two phyla of animals without backbones (insects and "worms"). Today, however, biologists recognize about 27 phyla of animals.

For convenience, biologists often place animals in one of two major categories: (1) **vertebrates**, those with a backbone, or vertebral column, and (2) **invertebrates**, those lacking a backbone. The vertebrates—fish, amphibians, reptiles, birds, and mammals—are perhaps the most conspicuous animals from a human point of view, but less than 3% of all known animal species on Earth are vertebrates. The vast majority of animals are invertebrates.

The earliest animals probably originated from colonies of protists whose members had become specialized to perform distinct roles within the colonial body. In our survey of the kingdom Animalia, we will begin with the sponges, whose body plan most closely resembles the probable ancestral protozoan colonies. Our discussion will proceed through a few representative invertebrate phyla and will end with the vertebrates in the chordate phylum.

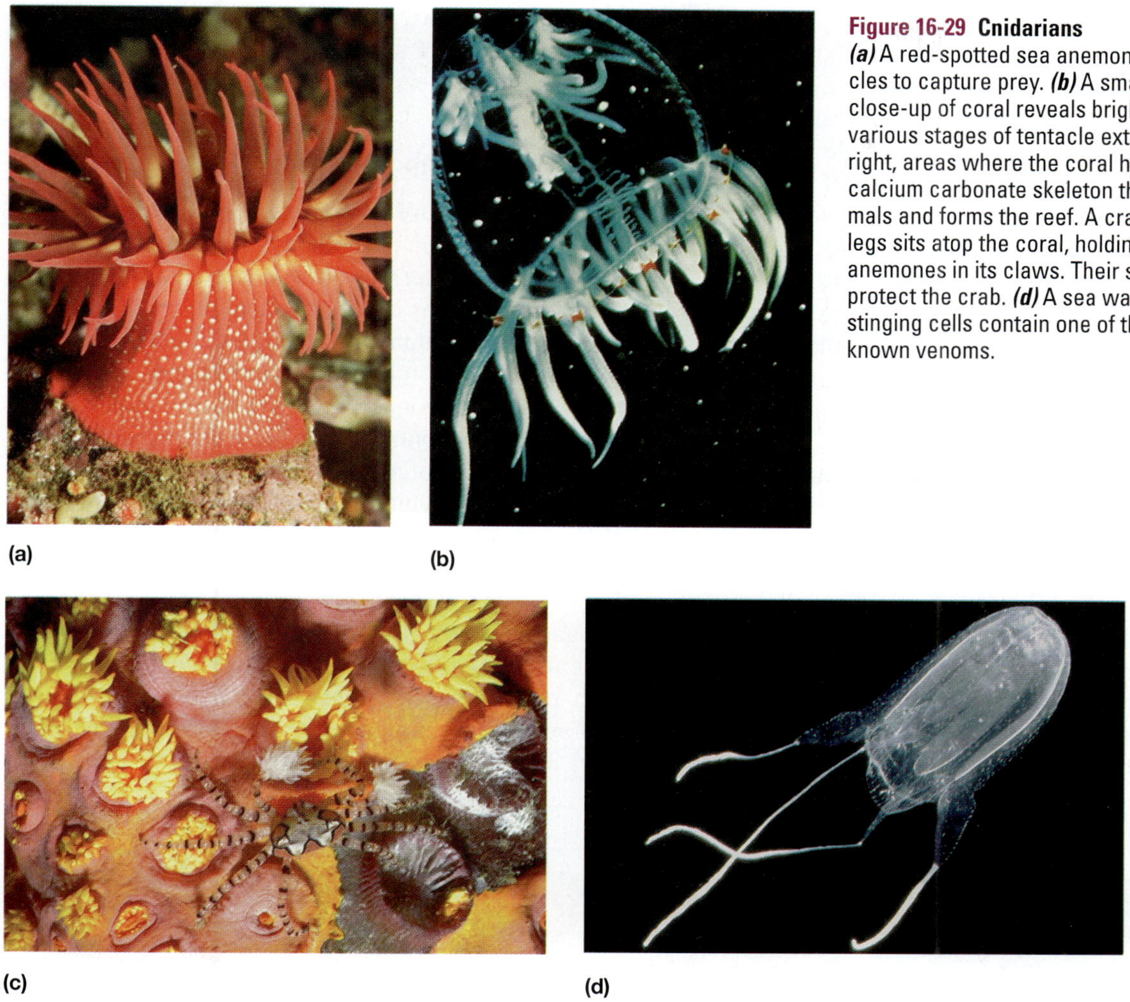

Figure 16-29 Cnidarians
(a) A red-spotted sea anemone spreads its tentacles to capture prey. ***(b)*** A small jellyfish. ***(c)*** A close-up of coral reveals bright yellow individuals in various stages of tentacle extension. At the lower right, areas where the coral has died expose the calcium carbonate skeleton that supports the animals and forms the reef. A crab with long banded legs sits atop the coral, holding tiny white sea anemones in its claws. Their stinging tentacles help protect the crab. ***(d)*** A sea wasp, a cnidarian whose stinging cells contain one of the most toxic of all known venoms.

Sponges Have a Simple Body Plan

Sponges lack true tissues and organs. In some ways, a sponge resembles a colony of single-celled organisms, and a few biologists believe that sponges should be classified as the protist group most closely related to animals, rather than within the animal kingdom.

All sponges have a similar body plan. The body is perforated by numerous tiny pores, through which water enters, and by fewer, large openings, through which it is expelled. Within the sponge, water travels through canals. During its passage, oxygen is extracted, microorganisms are filtered out and taken into individual cells where they are digested, and wastes are released.

Sponges come in a variety of shapes and sizes. Some species have a well-defined shape, but others grow free-form over underwater rocks (Fig. 16-28). The largest sponges can grow to more than 3 feet (1 meter) in height. An internal skeleton composed of calcium carbonate (chalk), silica (glass), or protein spines provides support for the sponge's body. The natural bath sponge, now largely replaced by factory-made cellulose imitations, is actually a proteinaceous sponge skeleton.

All sponges live in water, mostly in saltwater environments. Adult sponges live attached to rocks or other underwater surfaces and do not move about. Sponges may reproduce asexually by budding, in which the adult produces miniature versions of itself that drop off and assume an independent existence. Alternatively, they may reproduce sexually through the fusion of sperm and eggs. Fertilized eggs develop inside the adult into active larvae that escape through the openings in the sponge body. Water currents disperse the larvae to new areas, where they settle and develop into adult sponges.

Cnidarians Are Well-armed Predators

Cnidarians—jellyfish, sea anemones, corals, and hydrozoans—come in a bewildering and beautiful variety of forms (Fig. 16-29). In cnidarians, body parts are arranged in a circle around the mouth and digestive cavity. This arrangement of parts is well suited to these animals, which are either fixed in place or carried randomly by water currents, because they are prepared to respond to prey or threats from any direction.

Cnidarian tentacles are armed with cells containing structures that, when stimulated by contact, explosively inject poisonous or sticky filaments into prey. The venom of some cnidarians can cause painful stings in humans unfortunate enough to come into contact with them, and the stings of a few jellyfish species can even be life-threatening. The most deadly of these species is a "sea wasp," *Chironex fleckeri*, which is found in the waters off northern Australia and southeast Asia. The amount of venom in a single sea wasp could kill up to 60 people, and the victim of a serious stinging may die within minutes of being stung. The function of stinging cells, of course, is not to sting human swimmers but to capture prey. Although all cnidarians are predatory, none hunt actively. Instead, they rely on their victims' blundering by chance into the grasp of their enveloping tentacles. Stung and firmly grasped, the prey is forced through an expansible mouth into a digestive sac.

Like sponges, cnidarians are confined to watery habitats, and most species are marine. One group of cnidarians, the corals, is of particular ecological importance (see Fig. 16-29c). Coral individuals form large colonies, and each member of the colony secretes a hard skeleton of calcium carbonate. The skeletons persist long after the individuals die, serving as a base to which other individuals may attach themselves. The cycle continues until, after thousands of years, massive coral reefs are formed. Corals are restricted to the warm, clear waters of the Tropics, where their reefs form undersea habitats that are the basis of an ecosystem of stunning diversity and unparalleled beauty (see Chapter 30).

Annelids Are Composed of Identical Segments

A prominent feature of the segmented worms, or annelids, is the division of the body into a series of repeating segments. Externally, these segments appear as ringlike depressions on the surface. Internally, most of the segments contain identical copies of nerves, excretory structures, and muscles. Segmentation is advantageous for locomotion, because the body compartments, each of which is controlled by separate muscles, collectively are capable of more complex movement.

The annelid phylum includes three main subgroups, the oligochaetes, the polychaetes, and the leeches. The oligochaetes include the familiar earthworm and its relatives. Polychaetes live primarily in the ocean. Some polychaetes have paired fleshy paddles on most of their segments, used in locomotion. Others live in tubes from which they project feathery gills that both exchange gases and sift the water for microscopic food (Fig. 16-30a,b). Leeches (Fig. 16-30c) live in freshwater or moist

(a)

(b)

(c)

Figure 16-30 Annelids
(a) A polychaete annelid projects brightly spiraling gills from a tube, made by the worm and attached to rock. When the gills retract, the tube is covered by a "trap door" visible as a reddish collar encircling the top of the tube. ***(b)*** The "fireworm" polychaete swims by using paddles on each segment. The bristles on each paddle can deliver a fiery sting. ***(c)*** This leech, a freshwater annelid, shows numerous segments. The whitish disk is a sucker encircling its mouth, allowing it to attach to its prey.

terrestrial habitats, and are either carnivorous or parasitic—some preying on smaller invertebrates, others sucking the blood of larger animals.

Arthropods Are the Dominant Animals on Earth

In terms of both number of individuals and number of species, no other phylum comes close to the arthropods, which include insects, arachnids, and crustaceans. About 1 million arthropod species have been discovered, and scientists estimate that millions more remain undescribed.

All arthropods have an **exoskeleton**, an external skeleton that encloses the arthropod body like a suit of armor. This external skeleton protects against predators and is responsible for arthropods' greatly increased agility relative to their wormlike ancestors. In places the exoskeleton is thin and flexible, allowing movement of the paired, jointed appendages. By providing stiff but flexible appendages and rigid attachment sites for muscles, the exoskeleton makes possible the flight of the bumblebee and the intricate, delicate manipulations of the spider as it weaves its web. The exoskeleton also contributed enormously to the arthropod invasion of dry terrestrial habitats (arthropods were the earliest terrestrial animals; see Chapter 15) by providing a watertight covering for delicate, moist tissues such as those used for gas exchange.

Like a suit of armor, the arthropod exoskeleton poses some problems. First, because it cannot expand as the animal grows, the exoskeleton must be shed, or **molted**, periodically and replaced with a larger one. Molting uses energy and leaves the animal temporarily vulnerable until the new skeleton hardens. ("Soft-shelled" crabs are simply regular "hard-shelled" crabs that are caught just after molting.) The exoskeleton is also heavy; its weight increases exponentially as the animal grows. It is no coincidence that the largest arthropods are crustaceans (crabs and lobsters), whose watery habitat supports much of their weight.

Segmentation in arthropods is evidence that they share a common ancestor with annelids. Arthropod segments, however, tend to be fewer, less distinct from one another, and specialized for different functions such as sensing the environment, feeding, and movement. For example, in insects, sensory and feeding structures are concentrated on the front segment, known as the **head**, and digestive structures are largely confined to the **abdomen**, which is the rear segment. Between the head and the abdomen is the **thorax**, the segment to which structures used in locomotion, such as wings and walking legs, are attached.

Insects Are the Only Flying Invertebrates

The number of described insect species is about 850,000, roughly three times the total number of known species in all other classes of animals combined (Fig. 16-31). Insects have three pairs of legs, normally supplemented by two pairs of wings. Insects' capacity for flight distinguishes them from all other invertebrates and has contributed to their enormous success. As anyone who has pursued a fly can testify, flight helps in escaping from predators. It also allows the insect to find widely dispersed food. Swarms of locusts (Fig. 16-31d) have been traced all the way from Saskatchewan, Canada, to Texas on the trail of food.

During their development, insects undergo **metamorphosis**, a radical change from a juvenile body form to an adult body form. In insects with complete metamorphosis, the immature stage, called a **larva**, is worm-shaped (for example, the maggot of a housefly or the caterpillar of a moth or butterfly; see Fig. 16-31e). The larva hatches from an egg, grows by eating voraciously and shedding its exoskeleton several times, and then forms a nonfeeding stage called a **pupa**. Encased in an outer covering, the pupa undergoes a radical change in body form, emerging in its adult winged form. The adults mate and lay eggs, continuing the cycle. Metamorphosis may include a change in diet as well as in shape, thereby eliminating competition for food between adults and juveniles and in some cases allowing the insect to exploit different foods when they are most available. For instance, a caterpillar that feeds on new green shoots in springtime metamorphoses into a butterfly that drinks nectar from the summer's blooming flowers. Some insects undergo a more gradual metamorphosis (called incomplete metamorphosis), hatching as young that bear some resemblance to the adult, then gradually acquiring more-adult features as they grow and molt.

Most Arachnids Are Predatory Meat Eaters

The arachnids include spiders, mites, ticks, and scorpions (Fig. 16-32). All members of the class Arachnida have eight walking legs, and most are carnivorous; many subsist on a liquid diet of blood or predigested prey. For example, spiders, the most numerous arachnids, first immobilize their prey with a paralyzing venom. They then inject digestive enzymes into the helpless victim (typically an insect) and suck in the resulting soup. Arachnid eyes are particularly sensitive to movement, and in some species they probably can form images. Most spiders have eight eyes placed in such a way as to give them a panoramic view of predators and prey.

Most Crustaceans Are Aquatic

The crustaceans, including crabs, crayfish, lobster, shrimp, and barnacles, make up the only class of arthropods whose members live primarily in the water (Fig. 16-33). Crustaceans range in size from microscopic maxillopods that live in the spaces between grains of sand to the largest of all arthropods, the Japanese crab, with legs spanning nearly 12 feet (4 meters). Crustaceans have two pairs of sensory antennae, but the rest of their appendages are highly variable in form and number, depending on the habitat and lifestyle of the species.

Figure 16-31 Insects
(a) The rose aphid sucks sugar-rich juice from plants. ***(b)*** A mating pair of Hercules beetles. Only the male has the large "horns." ***(c)*** A June beetle displays its two pairs of wings as it comes in for a landing. The outer wings protect the abdomen and the inner wings, which are relatively thin and fragile. ***(d)*** Insects such as this locust can cause devastation of both crops and natural vegetation. ***(e)*** Caterpillars are larval forms of moths or butterflies. This caterpillar larva of the Australian fruit-sucking moth displays large eyespot patterns that may frighten potential predators, who mistake them for eyes of a large animal.

MEDIATUTOR 16.5 Insect Metamorphosis

Most Mollusks Have Shells

In terms of number of species, the mollusks are second (albeit a distant second) only to the arthropods. Some mollusks protect their bodies with a shell of calcium carbonate; others escape predation by moving swiftly or, if caught, by tasting terrible. With the exception of some snails and slugs, mollusks are water dwellers. Among the many classes of mollusks, we will discuss three in more detail: snails and their relatives, clams and their relatives, and octopuses and their relatives.

Gastropods Are One-Footed Crawlers

Snails and slugs, collectively known as gastropods, crawl on a muscular foot, and many are protected by shells that vary widely in form and color (Fig. 16-34a). Not all gastropods are shelled, however. Sea slugs, for example, lack shells, but their brilliant colors warn potential predators that they are poisonous, or at least taste bad (Fig. 16-34b).

Gastropods feed with a radula, a flexible ribbon of tissue studded with spines that is used to scrape algae from rocks or to grasp larger plants or prey. Most snails use gills, typically enclosed in a cavity beneath the shell, for respiration. Gases can also diffuse readily through the skin of most gastropods, and most sea slugs rely on this mode of gas exchange. The few gastropod species (including the destructive garden snails and slugs) that live in terrestrial habitats use a simple lung for breathing.

Bivalves Are Filter Feeders

Included among the bivalves are scallops, oysters, mussels, and clams (Fig. 16-35). Not only do members of the class lend exotic variety to the human diet, they are also important members of the nearshore marine community. Bivalves possess two shells connected by a flexible hinge. A strong muscle clamps the shells closed in response to danger; this muscle is what you are served when you order scallops in a restaurant.

(a) (b) (c)

Figure 16-32 Arachnids
(a) The tarantula is among the largest spiders but is relatively harmless. ***(b)*** Scorpions, found in warm climates, including deserts of the southwestern United States, paralyze their prey with venom from a stinger at the tip of the abdomen. A few species can harm humans. ***(c)*** Ticks before (left) and after feeding on blood. The uninflated exoskeleton is flexible and folded, allowing the animal to enlarge while feeding.

(a) (b) (c) (d)

Figure 16-33 Crustaceans
(a) The microscopic waterflea *Daphnia* is common in freshwater ponds. Notice the eggs developing within the body. ***(b)*** The sowbug, found in dark, moist places such as under rocks, leaves, and decaying logs, is one of the few crustaceans whose ancestors successfully invaded the land. ***(c)*** The hermit crab protects its soft abdomen by inhabiting an abandoned snail shell. ***(d)*** The goose-neck barnacle uses a tough, flexible stalk to anchor itself to rocks, boats, or even animals such as whales. Other types of barnacles attach with shells that resemble miniature volcanoes (see Fig. 16-35b). Early naturalists thought barnacles were mollusks until the jointed legs, seen extending into the water, were observed.

(a) (b)

Figure 16-34 Gastropod mollusks
(a) A Florida tree snail displays a brightly striped shell and eyes at the tip of stalks that are retracted instantly if touched. *(b)* Spanish shawl sea slugs prepare to mate.

(a) (b)

Figure 16-35 Bivalve mollusks
(a) This swimming scallop parts its hinged shells. The upper shell is covered with an encrusting sponge. *(b)* Mussels attach to rocks in dense aggregations exposed at low tide. White barnacles are attached to the mussel shells and surrounding rock.

Clams use a muscular foot for burrowing in sand or mud. In mussels, which live attached to rocks, the foot is smaller and is used to help secrete threads that anchor the animal to the rocks. Scallops lack a foot and move by a sort of whimsical jet propulsion achieved by flapping their shells together.

Bivalves are filter feeders, using their gills as both respiratory and feeding structures. Water is circulated over the gills, which are covered with a thin layer of mucus that traps microscopic food particles. Food is conveyed to the mouth by the beating of cilia on the gills. Probably because they filter-feed and do not move extensively, bivalves have "lost their heads" during the course of their evolution.

Cephalopods Are Marine Predators

The cephalopods are a fascinating group that contains octopuses, nautiluses, cuttlefish, and squids (Fig. 16-36), including the largest invertebrate, the giant squid, which can be up to 60 feet (18 meters) long. All cephalopods are predatory carnivores, and all are marine. In these mollusks, the foot has evolved into tentacles with well-developed sensory abilities and suction disks for detecting and grasping prey. Prey grasped by tentacles may be immobilized by a paralyzing venom in the saliva before being torn apart by beaklike jaws.

Cephalopods have highly developed brains and sensory systems. The cephalopod eye rivals our own in

(a)

(b)

Figure 16-36 Cephalopod mollusks
(a) An octopus can crawl rapidly by using its eight suckered tentacles. It can alter its color and skin texture to blend with its surroundings. In emergencies this mollusk can jet backward by vigorously contracting its mantle. ***(b)*** The squid moves by contracting its mantle to generate jet propulsion, which pushes the animal backward through the water. ***(c)*** The chambered nautilus secretes a shell with internal, gas-filled chambers that provide buoyancy. Note the well-developed eye and the tentacles used to capture prey.

(c)

complexity and exceeds it in efficiency of design. The cephalopod brain, especially that of the octopus, is exceptionally large and complex for an invertebrate brain. It is enclosed in a skull-like case of cartilage and endows the octopus with highly developed capabilities to learn and remember. In the laboratory, octopuses can rapidly learn to associate certain symbols with food and to open a screw-cap jar to obtain food.

The Chordates Include Both Invertebrates and Vertebrates

It's hard to believe that humans have anything in common with a sea squirt. Nonetheless, we share the chordate phylum not only with the birds and the apes but also with the tunicates (sea squirts) and little fishy-looking creatures called lancelets. What characteristics unite us with these creatures that seem so different? All chordates are united by four features that all possess at some stage of their lives:

- A **notochord.** A stiff but flexible rod that extends the length of the body and provides an attachment site for muscles.
- A dorsal, hollow **nerve cord.** This hollow structure develops a thickening at one end that becomes a brain.
- **Pharyngeal gill slits.** Located in the pharynx (the cavity behind the mouth), these may form functional openings for gills (organs for gas exchange in water) or may appear only as grooves during an early stage of development.
- A **post-anal tail.** An extension of the body past the anus.

This list may seem puzzling because, although humans are chordates, at first glance we seem to lack every feature except the nerve cord. But evolutionary relationships are sometimes seen most clearly during early stages of development; it is during our embryonic life that we develop, and lose, our notochord, our gill slits, and our tails. Humans share these chordate features with all other vertebrates and with two invertebrate chordate groups, the lancelets and the tunicates.

The Invertebrate Chordates Live in the Seas

The invertebrate chordates lack the backbone that is the defining feature of vertebrates. These chordates comprise two groups, the lancelets and the tunicates. The small (2 inches, or about 5 centimeters, long) fishlike lancelet spends most of its time half-buried in the sandy sea bottom, filtering tiny food particles from the water. All four chordate features are present in the adult lancelet.

The tunicates form a larger group of marine invertebrate chordates that includes the sea squirts. It is difficult to imagine a less likely relative of humans than the immobile, filter-feeding, vaselike sea squirt (Fig. 16-37). Its ability to move is limited to forceful contractions of

Figure 16-37 An invertebrate chordate
This adult sea squirt (a type of tunicate) has lost its tail and notochord and has assumed a sedentary life.

its saclike body, which can send a jet of seawater into the face of anyone who plucks it from its undersea home; hence the name sea squirt. Although adult sea squirts are immobile, their larvae swim actively and possess the four chordate features.

Vertebrates Have a Backbone

In vertebrates the embryonic notochord is normally replaced during development by a backbone, or **vertebral column**. The vertebral column is composed of bone or cartilage, a tissue that resembles bone but is less brittle and more flexible. This column supports the body, offers attachment sites for muscles, and protects the delicate nerve cord and brain. It is also part of a living internal skeleton that can grow and repair itself.

Vertebrates show other adaptations that have contributed to their successful invasion of most habitats. One is paired appendages. These first appeared as fins in fish and served as stabilizers for swimming. Over millions of years, some fins were modified by natural selection into legs that allowed animals to crawl onto dry land, and later into wings that allowed some to take to the air. Another adaptation that has contributed to the success of vertebrates is an increase in the size and complexity of their brains and sensory structures, which allow vertebrates to perceive their environment in detail and to respond to it in a great variety of ways.

The sequence of evolutionary events that led from invertebrate chordates to the first, fishlike vertebrates remains shrouded in mystery, because fossils of intermediate forms have never been discovered. Today, ver-

(a) (b) (c) (d)

Figure 16-38 Bony fishes
Bony fishes have colonized nearly every aquatic habitat. ***(a)*** This female deep-sea angler fish attracts prey with a living lure that projects just above her mouth. The fish is ghostly white because in the 2000-meter depth where anglers live, no light penetrates and thus colors are superfluous. Male deep-sea angler fish are extremely small and remain as permanent parasites attached to the female, always available to fertilize her eggs. Two parasitic males can be seen attached to this female. ***(b)*** This tropical green moray eel lives in rocky crevices. A small fish (a banded cleaner goby) on its lower jaw eats parasites that cling to the moray's skin. ***(c)*** The tropical seahorse may anchor itself with its tail, which is adapted for grasping, while the animal feeds on small crustaceans. ***(d)*** A rare photo of a coelacanth, a fish once believed to be long extinct, in its natural habitat off the coast of South Africa.

(a) (b) (c)

Figure 16-39 Amphibians
In many amphibian species, individuals make the transition from ***(a)*** the completely aquatic larval tadpole to ***(b)*** the adult animal, here a bullfrog, leading a semiterrestrial life. ***(c)*** The red salamander is restricted to moist habitats in the eastern United States. Salamanders hatch in a form that closely resembles the adult.

tebrates are represented by several classes of fishes, and by amphibians, reptiles, birds, and mammals.

Bony Fishes Just as our size bias makes us overlook the most-diverse invertebrate groups, our habitat bias makes us overlook the most-diverse vertebrates. The most diverse and abundant vertebrates are not the birds or predominantly terrestrial mammals. The vertebrate diversity crown belongs instead to the lords of the oceans and freshwater, the bony fishes. From the snakelike moray eel to the bizarre, luminescent deep-sea forms, to the streamlined tuna, this enormously successful group has spread to nearly every possible watery habitat, both freshwater and marine (Fig. 16-38). Although about 17,000 species have been identified, scientists estimate that perhaps twice this number may exist, including many in deep waters and remote areas. The ocean's potential to conceal species was illustrated in dramatic fashion by the case of the coelacanth, a bony fish that had been known only from 75-million-year-old fossils. In 1939 coelacanths were found to be alive and well in deep water off the coast of South Africa (Fig. 16-38d). Since that initial discovery, coelacanths have also been found in other locations, including waters near Madagascar and Indonesia.

Amphibians The amphibians straddle the boundary between aquatic and terrestrial existence (Fig. 16-39). The limbs of amphibians show varying degrees of adaptation to movement on land, from the belly-dragging crawl of salamanders to the long leaps of frogs and toads. A three-chambered heart (in contrast to the two-chambered heart of fishes) circulates blood more efficiently, and lungs replace gills in most adult forms. Amphibian lungs, however, are poorly developed and must be supplemented by the skin, which serves as an additional respiratory organ. This respiratory function requires that the skin remains moist, a constraint that greatly restricts the range of amphibian habitats on land.

Amphibians are also tied to moist habitats by their breeding behavior, which requires water. Their fertilization is external and takes place in water, where the sperm can swim to the eggs. The eggs must remain moist, as they are protected only by a jellylike coating that leaves them vulnerable to loss of water by evaporation. The mechanics of keeping the eggs moist varies considerably among different amphibian species, but in many species the requisite moisture is secured by the simple expedient of laying the eggs in water. In some amphibian species, fertilized eggs develop into aquatic larvae such as the tadpoles of some frogs and toads. The dramatic transformation of these aquatic larvae into semiterrestrial adults gives the amphibians their name, which means "double life." Their double life and their thin, permeable skin have made amphibians particularly vulnerable to pollutants and to environmental degradation, as described in "Earth Watch: Frogs in Peril."

Reptiles and Birds The reptiles include the lizards and snakes (by far the most successful of the modern groups) and the turtles, alligators, and crocodiles (Fig. 16-40). Some

(a) (b) (c)

Figure 16-40 Reptiles
(a) The mountain king snake has a color pattern very similar to that of the poisonous coral snake, which potential predators avoid. This mimicry helps the harmless king snake elude predation. ***(b)*** The outward appearance of the American alligator, found in swampy areas of the South, is almost identical to reconstructions of 150-million-year-old fossil alligators. ***(c)*** The tortoises of the Galapagos Islands, Ecuador, may live to be more than 100 years old.

reptiles, particularly desert dwellers such as tortoises and lizards, are completely independent of their aquatic origins. This independence was achieved through a series of adaptations, of which three are outstanding: (1) Reptiles evolved a tough, scaly skin that resists water loss and protects the body. (2) Reptiles evolved internal fertilization, in which the male deposits sperm within the female's body. (3) Reptiles evolved a shelled egg, which can be buried in sand or dirt, far from water with its hungry predators. The shell prevents the egg from drying out on land. An internal membrane encloses the embryo in the watery environment that all developing animals require.

One very distinctive group of "reptiles" is the birds (Fig. 16-41). Although birds have traditionally been classified as a group separate from reptiles, biologists have shown that birds are really a subset of an evolutionary group that includes both birds and the groups that have been traditionally designated as reptiles. The first birds appear in the fossil record roughly 150 million years ago and are distinguished from other reptiles by the presence of feathers, which are essentially a highly specialized version of reptilian body scales. Modern birds retain scales on their legs—a testimony to the ancestry they share with the rest of the reptiles.

Bird anatomy and physiology are dominated by adaptations that help meet the rigorous demands of flight. In particular, birds are exceptionally light for their size. Hollow bones reduce the weight of the bird skeleton to a fraction of that of other vertebrates, and many bones present in other reptiles have been lost in the course of evolution or fused with other bones. Reproductive organs are considerably reduced in size during nonbreeding periods, and female birds possess only a single ovary, further minimizing weight. The shelled egg that contributed to the reptiles' success on land frees the mother bird from carrying her developing offspring internally. Feathers form lightweight extensions to the wings and the tail for the lift and control required

(a) (b) (c)

Figure 16-41 Birds
(a) The delicate hummingbird beats its wings about 60 times per second and weighs about 0.15 ounce (4 grams). ***(b)*** This young frigate bird, a fish-eater from the Galapagos Islands, has nearly outgrown its nest. ***(c)*** The ostrich is the largest of all birds, weighing more than 300 pounds (136 kilograms); its eggs weigh more than 3 pounds (1500 grams).

Earth Watch

Frogs in Peril

Frogs and toads have frequented Earth's ponds and swamps for nearly 150 million years and somehow survived the Cretaceous catastrophe that extinguished the dinosaurs and so many other species about 65 million years ago. Their evolutionary longevity, however, appears to offer inadequate defense against the environmental changes wrought by human activities. Over the past decade, herpetologists (biologists who study reptiles and amphibians) from around the world have documented an alarming decline in amphibian populations. Thousands of species of frogs, toads, and salamanders are experiencing a dramatic decrease in numbers, and many have apparently gone extinct.

The phenomenon is worldwide; population crashes have been reported from every part of the globe. Yosemite toads and yellow-legged frogs are disappearing from the mountains of California; tiger salamanders have been nearly wiped out in the Colorado Rockies. Leopard frogs, eagerly chased by rural children, are becoming rare in the United States. Logging destroys the habitats of amphibians from the Pacific Northwest to the Tropics, but even amphibians in protected areas are dying. In the Monteverde Cloud Forest Preserve in Costa Rica, the golden toad was common in the early 1980s but has not been seen since 1989. The gastric brooding frog of Australia fascinated biologists by swallowing its eggs, brooding them in its stomach, and later regurgitating fully formed offspring. This species was abundant and seemed safe in a national park. Suddenly, in 1980, the gastric brooding frog disappeared and hasn't been seen since.

The causes of the worldwide decline in amphibian diversity are poorly understood, but researchers have recently discovered that frogs and toads in many places are succumbing to infection by a pathogenic fungus. The fungus has been found in the skin of dead and dying frogs in widespread locations, including Australia, Central America, and the western United States. In those places, discovery of the fungus has coincided with massive frog and toad die-offs, and most herpetologists agree that the fungus is causing the deaths. It seems unlikely, however, that the fungus alone is responsible for the worldwide decline of amphibians. For one thing, die-offs have occurred in many places where the fungus has not been found. In addition, many herpetologists believe that the fungal epidemic would not have arisen if the frogs and toads had not first been weakened by other stresses. So, if the fungus is not doing all of the damage on its own, what are the other possible causes of amphibian decline? All of the most likely causes stem from human modification of the biosphere—the portion of Earth that sustains life.

Habitat destruction, especially the draining of wetland habitats that are hospitable to amphibian life, is one major cause of the decline. The unique biology of amphibians also makes them especially vulnerable to toxic substances in the environment. In particular, amphibian bodies at all stages of life are protected only by a thin, permeable skin through which pollutants can easily penetrate (Fig. E16-1). To make matters worse, the double life of many amphibians exposes their permeable skin to a wide range of aquatic and terrestrial habitats and to a correspondingly wide range of environmental toxins. For example, many temperate amphibians' eggs and larvae develop in ponds and streams during the spring, a time when the melting of snow and ice formed from acid rain causes a pulse of intense acidity in freshwater ecosystems. In addition, many amphibians are herbivorous as larvae and insectivorous as adults, so they're exposed to both herbicides and insecticides in their food.

Amphibian eggs can also be damaged by ultraviolet (UV) light, according to recent research by Andrew Blaustein, an ecologist at Oregon State University. Blaustein demonstrated that the eggs of some species of frogs in the Pacific Northwest are sensitive to damage from UV light and that the most sensitive species are experiencing the most drastic declines. Unfortunately, many parts of Earth are subject to increasingly intense UV radiation levels, because atmospheric pollutants have caused a thinning of the protective ozone layer (as we shall see in Chapter 30).

Many scientists believe that the troubles of amphibians signal an overall deterioration of Earth's ability to support life. According to this line of reasoning, the highly sensitive amphibians are providing an early warning of environmental degradation that will eventually affect more-resistant organisms as well. Equally worrisome is the observation that amphibians are not just sensitive indicators of the health of the biosphere; they are also crucial components of many ecosystems. They may keep insect populations in check, and, in turn, serve as food for larger carnivores. Their decline will disrupt the balance of these ecosystems. Margaret Stewart, an ecologist at the State University of New York, Albany, aptly summarized the problem: "There's a famous saying among ecologists and environmentalists: 'Everything is related to everything else.' . . . You can't wipe out one large component of the system and not see dramatic changes in other parts of the system."

Figure E16-1 Amphibians in danger
The corroboree toad, shown here with its eggs, is rapidly declining in its native Australia. Tadpoles are developing within the eggs. The thin water-permeable and gas-permeable skin of the adult and the jellylike coating around the eggs provide little, if any protection against air and water pollutants.

for flight, and they also provide lightweight protection and insulation for the body.

Birds are also able to maintain body temperatures high enough to allow their muscles and metabolic processes to operate at peak efficiency, supplying the power necessary to fly regardless of the outside temperature. This physiological ability to maintain an internal temperature that is usually higher than that of the surrounding environment is characteristic of both birds and mammals, which are therefore sometimes described as warm-blooded. In contrast, the body temperature of invertebrates, fish, amphibians, and reptiles varies with the temperature of their environment, though these animals may exert some control of their body temperature by behavioral means (such as basking in the sun or seeking shade).

Mammals One branch of the reptile evolutionary tree gave rise to a group that evolved hair and diverged to form the mammals. The mammals first appeared approximately 250 million years ago but did not diversify and assume terrestrial prominence until after the dinosaurs went extinct roughly 65 million years ago. In most mammals, fur protects and insulates the warm body. Legs designed for running rather than crawling make many mammals fast and agile. In contrast to birds, whose bodies are almost uniformly molded to the requirements of flight, mammals have evolved a remarkable diversity of form. The bat, mole, impala, whale, seal, monkey, and cheetah exemplify the radiation of mammals into nearly all habitats, with bodies finely adapted to their varied lifestyles (Fig. 16-42).

Mammals are named for the milk-producing mammary glands used by all female members of this class to suckle their young (Fig. 16-42c). In addition to these unique glands, the mammalian body is arrayed with sweat, scent, and sebaceous (oil-producing) glands, none of which are found in reptiles.

The mammalian nervous system has contributed significantly to the success of the mammals by making pos-

(a) (b) (c) (d)

Figure 16-42 Mammals
(a) A humpback whale gives its offspring a boost. ***(b)*** A bat, the only mammal capable of true flight, navigates at night by using a kind of sonar. Large ears aid in detecting echoes as its high-pitched cries bounce off nearby objects. ***(c)*** Mammals are named after the mammary glands with which females nurse their young, as illustrated by this mother cheetah. ***(d)*** The male orangutan can reach 165 pounds (75 kilograms). These gentle, intelligent apes occupy swamp forests in limited areas of the Tropics. They are endangered by hunting and habitat destruction.

sible behavioral adaptation to changing and varied environments. The brain is more highly developed than in any other class, endowing mammals with unparalleled curiosity and learning ability. The highly developed brain allows mammals to alter their behavior on the basis of experience and helps them survive in a changing environment. Relatively long periods of parental care after birth allow some mammals to learn extensively under parental guidance; humans and other primates are good examples. In fact, the large brains of humans have been the major factor leading to human domination of Earth.

CASESTUDYREVISITED

A Biologist "Bombs" Beetles to Estimate Diversity

How did Terry Erwin, the author of the bug bomb estimate of tropical insect diversity, arrive at his prediction that 30 million species of insects inhabit tropical rain forests? Recall that Erwin found 1200 species of beetle in the canopy of a single tree species, *Luehea seemannii*. As a beetle expert, he knew that many beetle species are highly specialized and live exclusively on a single species of tree. He estimated that about 13% of beetles have such specializations, so about 160 (13% of 1200) of the beetle species he collected were unique to *L. seemannii*. Recall also that Erwin collected only beetles. About 40% of all known insect species are beetles, and if this proportion also applies to the insects inhabiting *L. seemannii*, then the 160 beetle species would represent 40% of a total of 400 insect species specialized to inhabit the tree. Finally, recall that Erwin sampled only from the tree canopy. He assumed that the canopy holds about two-thirds of the insect species on the tree, with the remaining one-third living on the trunk and in the roots. That one-third would represent an additional 200 species, giving a total of 600 insect species specialized to live on *L. seemannii*. If *L. seemannii* is a typical tropical tree, then each of the other rainforest tree species ought to host about 600 specialized insect species. As about 50,000 species of rainforest trees are known, there must be about $600 \times 50{,}000 = 30$ million species of insects altogether.

Each step in Erwin's chain of logic required an assumption, and each assumption added a bit of uncertainty to the final estimate. Other scientists have repeated the calculation using more conservative assumptions and have arrived at lower estimates of the number of insect species. Nonetheless, each of Erwin's assumptions is at least defensible, and his estimate established the upper bound of the range of different estimates that have appeared in the scientific literature. His work shows us the extent to which we may have underestimated life's true diversity and underscores our current lack of certainty.

Summary of Key Concepts

1) How Are Organisms Named and Classified?

Organisms are classified and placed into hierarchical categories that reflect their evolutionary relationships. The eight major categories, in order of decreasing inclusiveness, are (1) domain, (2) kingdom, (3) phylum, (4) class, (5) order, (6) family, (7) genus, and (8) species. The scientific name of an organism is its genus name and species name. Anatomical and molecular similarities are used to classify organisms. Molecular similarities among organisms are a measure of evolutionary relatedness.

2) What Are the Domains of Life?

A five-kingdom classification system consisting of the kingdoms Monera, Protista, Animalia, Fungi, and Plantae is widely used, but gaining acceptance is a new three-domain system, representing the three main branches of life, Bacteria, Archaea, and Eukarya.

3) What Are Some Characteristics of Bacteria and Archaea?

Members of the domains Bacteria and Archaea—the bacteria and archaea—are unicellular and prokaryotic. Archaea and bacteria are not closely related and differ in fundamental features, including cell wall composition and membrane structure. A cell wall determines the characteristic shapes of prokaryotes: round, rodlike, or spiral. Certain types of bacteria can form spores that disperse widely and withstand inhospitable environmental conditions. Prokaryotes obtain energy in a variety of ways. Some rely on photosynthesis. Others break down inorganic molecules to obtain energy. Many are anaerobic, able to obtain energy from fermentation when oxygen is not available.

Some bacteria are pathogenic, causing disorders such as pneumonia, tetanus, botulism, and the venereal diseases gonorrhea and syphilis. Most bacteria, however, are harmless to humans and play important roles in natural ecosystems. Bacteria and archaea have colonized nearly every habitat on Earth, including hot, acidic, very salty, and anaerobic environments. Some live in the digestive tracts of ruminants and break down cellulose. Nitrogen-fixing bacteria enrich the soil and aid in plant growth; many others live off the dead bodies and wastes of other organisms, liberating nutrients for reuse.

4) Which Organisms Are in the Kingdom Protista?

The kingdom Protista consists of organisms composed of single, highly complex eukaryotic cells. They are classified as funguslike, plantlike, and animal-like.

The funguslike acellular slime molds form a multinucleate plasmodium that crawls in amoeboid fashion, ingesting decaying organic matter. Drought or starvation stimulates the formation of a fruiting body on which spores are formed.

The plantlike algae are important photosynthetic organisms in marine and freshwater ecosystems. They include dinoflagellates, brown algae, and green algae.

Protozoa are nonphotosynthetic protists that absorb or ingest their food. They are widely distributed in soil and water; some are parasitic. They include the amoeboid sarcodines and the predatory ciliates.

5) What Are the Key Features of Fungi?

Fungal bodies generally consist of filamentous hyphae that form large, intertwined mycelia. A cell wall of chitin surrounds fungal cells. Fungi obtain energy by secreting digestive enzymes outside their bodies and absorbing the liberated nutrients. Fungal reproduction depends on spores, and most species reproduce both sexually and asexually.

6) How Do Fungi Affect Humans?

The majority of plant diseases are caused by parasitic fungi. Some parasitic fungi can help control insect crop pests. Others can cause human diseases, including ringworm, athlete's foot, and common vaginal infections. Some fungi produce toxins that can harm humans. Nonetheless, fungi add variety to the human food supply, and fermentation by fungi helps make wine, beer, and bread.

Fungi are extremely important decomposers in ecosystems. Their filamentous bodies penetrate rich soil and decaying organic material, liberating nutrients through extracellular digestion.

7) What Are the Major Groups of Plants?

Two groups of plants, bryophytes and vascular plants, arose from ancient algal ancestors. Bryophytes, including the liverworts and mosses, are small land plants that lack conducting vessels. Although some have adapted to dry areas, most live in moist habitats. Reproduction in bryophytes requires water through which the sperm swims to the egg.

In vascular plants a system of vessels, stiffened by lignin, conducts water and nutrients absorbed by the roots into the upper portions of the plant and supports the body as well. Owing to this support system, seedless vascular plants, including the club mosses, horsetails, and ferns, can grow larger than bryophytes.

Vascular plants with seeds have two major additional adaptive features: pollen and seeds. Seed plants are often classified into two categories: gymnosperms and angiosperms. Gymnosperms include ginkgos, cycads, and the highly successful conifers. Angiosperms, the flowering plants, dominate much of the land today. In addition to pollen and seeds, angiosperms also produce flowers and fruits. The flower allows angiosperms to utilize animals as pollinators. In contrast to wind, animals can in some cases carry pollen farther and with greater accuracy and less waste. Fruits may attract animal consumers, which incidentally disperse the seeds in their feces.

8) What Characteristics Define an Animal?

Animals are multicellular, sexually reproducing, heterotrophic organisms. Most can perceive and react rapidly to environmental stimuli and are motile at some stage in their lives. Their cells lack cell walls.

9) What Are the Major Animal Phyla?

The bodies of sponges are typically free-form in shape and cannot move. Sponges have relatively few types of cells. Digestion occurs exclusively within the individual cells.

Cnidarians (jellyfish, sea anemones, corals, and hydrozoans) use armed tentacles to capture prey.

The annelids, the most complex of the worms, include earthworms, marine tubeworms, and leaches.

Arthropods, which include the insects, arachnids, and crustaceans, are the most diverse and abundant organisms on Earth. They have invaded nearly every available terrestrial and aquatic habitat. Jointed appendages and well-developed nervous systems make possible complex, finely coordinated behavior. The exoskeleton (which conserves water and provides support) helps enable the insects and arachnids to inhabit dry land. The diversification of insects has been enhanced by their ability to fly. Crustaceans, which include the largest arthropods, are restricted to moist, usually aquatic habitats and respire by using gills.

The mollusks (gastropods, bivalves, and cephalopods) lack a skeleton; some forms protect the soft, moist, muscular body with a single shell (many gastropods and a few cephalopods) or a pair of hinged shells (the bivalves). The lack of a waterproof external covering limits this phylum to aquatic and moist terrestrial habitats. Although the body plan of gastropods and bivalves limits the complexity of their behavior, the cephalopod's tentacles are capable of precisely controlled movements. The octopus has the most complex brain and the best-developed learning capacity of any invertebrate.

The chordate phylum includes two invertebrate groups, the lancelets and tunicates, as well as the familiar vertebrates. All chordates possess a notochord, a nerve cord, pharyngeal gill slits, and a post-anal tail at some stage in their development. Vertebrates have a backbone that is part of a living internal skeleton.

During the transitions from fishes to amphibians to reptiles, some vertebrates evolved adaptations that helped them colonize dry land. All amphibians have legs, and most have simple lungs for breathing in air rather than in water. Most are confined to relatively damp terrestrial habitats by their need to keep their skin moist, use of external fertilization, and requirement that their eggs and larvae develop in water. Reptiles, with well-developed lungs, dry skin covered with relatively waterproof scales, internal fertilization, and the shelled egg with its own water supply, are well adapted to the driest terrestrial habitats. Birds and mammals are also fully terrestrial and have additional adaptations. The bird body is molded for flight, with feathers and hollow bones. Mammals have insulating hair and give birth to live young that are nourished with milk from the mothers' mammary glands. The mammalian nervous system is the most complex in the animal kingdom, providing mammals with enhanced learning ability that helps them adapt to changing environments.

Key Terms

abdomen *p. 295*
alga *p. 281*
angiosperm *p. 290*
Archaea *p. 275*
Bacteria *p. 275*
bryophyte *p. 287*
cilium *p. 283*
endospore *p. 276*
Eukarya *p. 275*
exoskeleton *p. 295*
flagellum *p. 276*
flower *p. 290*
fruit *p. 290*
gymnosperm *p. 289*
head *p. 295*
hypha *p. 283*
invertebrate *p. 292*
larva *p. 295*
metamorphosis *p. 295*
molt *p. 295*
mycelium *p. 283*
nerve cord *p. 299*
nitrogen-fixing bacterium *p. 278*
notochord *p. 299*
pathogenic *p. 279*
pharyngeal gill slit *p. 299*
phytoplankton *p. 281*
pollen *p. 288*
post-anal tail *p. 299*
protist *p. 280*
protozoa *p. 283*
pupa *p. 295*
seed *p. 289*
spore *p. 284*
thorax *p. 295*
vascular plant *p. 287*
vertebral column *p. 300*
vertebrate *p. 292*

Thinking Through the Concepts

Multiple Choice

1. *Which of the following pairs is the most distantly related?*
   a. archaea and bacteria
   b. protists and fungi
   c. plants and animals
   d. fish and starfish
   e. fungi and plants
2. *Most pathogenic bacteria cause disease by*
   a. directly destroying individual cells of the host
   b. fixing nitrogen and depriving the host of this nutrient
   c. producing toxins that disrupt normal functions
   d. depleting the energy supply of the host
   e. depriving the host of oxygen
3. *Which of the following statements is true both of fungi and of animals?*
   a. Both photosynthesize.
   b. Both form embryos.
   c. They interact to form lichens.
   d. Both fix nitrogen.
   e. Both gain nutrition by consuming the bodies of other organisms.
4. *What is the function of a fruit?*
   a. It attracts pollinators.
   b. It provides food for the developing embryo.
   c. It stores excess food produced by photosynthesis.
   d. It helps ensure seed dispersal from the parent plant.
   e. It evolved so that people would cultivate the plant, ensuring its survival.
5. *Which of the following plants produces sperm that swim to the egg?*
   a. walnut tree
   b. Douglas fir tree, a conifer
   c. rattlesnake fern
   d. common dandelion
   e. none of the above
6. *Which of the following animals molts its exoskeleton, enabling it to grow larger?*
   a. the blue crab
   b. the earthworm
   c. the scallop
   d. the octopus
   e. the Venus clam

? Review Questions

1. What features would you study to determine whether a dolphin is more closely related to a fish or to a bear?
2. Only a small fraction of the total number of species on Earth has been scientifically described. Why?
3. Describe some of the ways in which bacteria obtain energy and nutrients.
4. What are nitrogen-fixing bacteria, and what role do they play in ecosystems?
5. What is the major ecological role played by unicellular algae?
6. What portion of the fungal body is represented by mushrooms, puffballs, and similar structures? Why are these structures elevated above the ground?
7. List the adaptations that helped plants become successful terrestrial organisms. Which of these adaptations are possessed by bryophytes? By ferns? By gymnosperms and angiosperms?
8. The number of species of flowering plants is greater than the number of species in the rest of the plant kingdom. What feature(s) are responsible for the enormous success of angiosperms? Explain why.
9. Distinguish between vertebrates and invertebrates.
10. List four distinguishing features of chordates.
11. Describe the ways in which amphibians are adapted to life on land. In what ways are amphibians still restricted to a watery or moist environment?
12. List the adaptations that distinguish reptiles from amphibians and helped reptiles adapt to life in dry terrestrial environments.

Applying the Concepts

1. There are many areas of disagreement about the classification of organisms. For example, there is no consensus about whether the red wolf is a distinct species or about how many kingdoms are within the domain Bacteria. What difference does it make whether biologists consider the red wolf a species, or into which kingdom a bacterial species falls? As Shakespeare put it, "What's in a name?"
2. The pressures created by human population growth and economic expansion place storehouses of biological diversity such as the Tropics in peril. The seriousness of the situation is clear when we consider that probably only 1 out of every 20 tropical species is known to science at present. What arguments can you make for preserving biological diversity in poor and developing countries? Does such preservation require that these countries sacrifice economic development? Suggest some solutions to the conflict between the growing demand for resources and the importance of conserving biodiversity.
3. Recent research shows that ocean water off southern California has warmed by 2–3 °F (1–1.5 °C) over the past four decades, possibly due to the greenhouse effect. This warming has led indirectly to a depletion of nutrients in the water and thus a decline in photosynthetic protists such as diatoms. What effects is this warming likely to have for the life in the oceans?
4. The internal structure of many protists is much more complex than that of cells of multicellular organisms. Does this mean that the protist is engaged in more complex activities than the multicellular organism? If not, why should the protistan cell be much more complicated?
5. The discovery of penicillin was the result of a chance observation by an observant microbiologist, Alexander Fleming. How would you search systematically for new antibiotics produced by fungi? Where would you look for these fungi?
6. The class of insects is the largest group of animals on Earth. Its greatest diversity is in the Tropics, where habitats are being destroyed and species are going extinct at alarming rates. What biological, economic, and ethical arguments can you advance to persuade people and governments to preserve the biological diversity of tropical insects?
7. Discuss and defend what attributes you would use to define biological success among animals. Are humans a biological success by these standards? Why or why not?

Group Activity

Form a group of four students. Imagine that you've been working in a museum collection and you discover a dusty old cabinet that hasn't been opened in 150 years. Inside are some mysterious specimens, previously unknown to science. Your job is to classify these organisms. Develop a classification, draw an evolutionary tree, and explain your reasoning. The specimens are shown below.

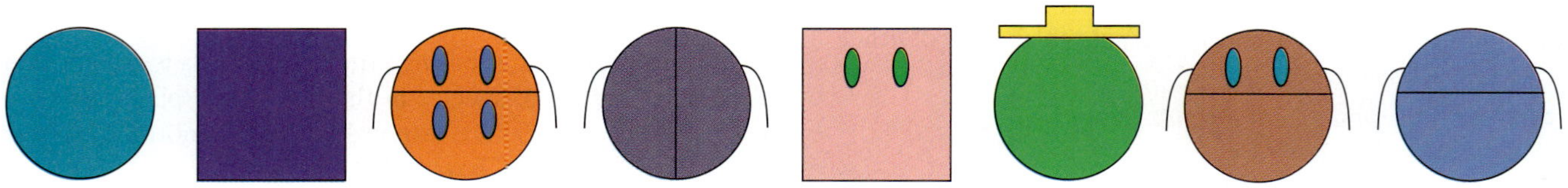

For More Information

Avise, J. C. "Nature's Family Archives." *Natural History*, March 1989. Shows how evolutionary relationships can be determined by analyzing differences in DNA contained in mitochondria.

Blaustein, A. "Amphibians in a Bad Light." *Natural History*, October 1994. Recent declines in amphibian population size and overall diversity are linked to possible harm from ultraviolet light that is penetrating a depleted ozone layer.

Brusca, R. C., and Brusca, G. J. *Invertebrates*. Sunderland, MA: Sinauer, 1990. A thorough survey of the invertebrate animals in textbook format but readable and filled with beautiful and informative drawings.

Chadwick, D. H. "Planet of the Beetles." *National Geographic*, March 1998. The beauty and diversity of beetles, which comprise one-third of the world's insects, are described in text and photographs.

Cole, L. A. "The Specter of Biological Weapons." *Scientific American*, December 1996. Assesses the likelihood of germ warfare, the politics of preventing it, and the biological agents that make it possible.

Gould, S. J. "What Is a Species?" *Discover*, December 1992. Discusses the difficulties of distinguishing separate species.

Hamner, W. "A Killer Down Under." *National Geographic*, August 1994. Among the most poisonous animals in the world is the box jellyfish, which lives off the coast of northern Australia.

Hudler, G. W. *Magical Mushrooms, Mischievous Molds*. Princeton, NJ: Princeton University Press, 1998. Engaging treatment of the fungi, focusing on their importance in human affairs.

Kaufman, P. B. *Plants—Their Biology and Importance.* New York: Harper & Row, 1989. Complete, readable coverage of all aspects of plant classification, physiology, and evolution.

Luoma, J. R. "Vanishing Frogs." *Audubon*, May–June 1997. Brief summary of declining amphibian populations, accompanied by striking photographs of some of the affected frog species.

Madigan, M., and Marrs, B. "Extremophiles." *Scientific American*, April 1997. Prokaryotes that prosper under extreme conditions, and potential industrial uses of the enzymes that they use to do so.

Mann, C., and Plummer, M. *Noah's Choice: The Future of Endangered Species*. New York: Knopf, 1995. A thought-provoking look at the hard choices we must make with regard to protecting biodiversity. Which species will we choose to preserve? What price are we willing to pay?

Margulis, L., and Sagan, D. *What Is Life?* London: Weidenfeld & Nicolson, 1995. A lavishly illustrated survey of life's diversity. Also includes an account of life's history and a meditation on the question posed by the title.

May, R. M. "How Many Species Inhabit the Earth?" *Scientific American*, October 1992. Although no one knows the precise answer to this question, an effective estimate is crucial to our effort to manage our biological resources.

Moffett, M. W. *The High Frontier: Exploring the Tropical Rainforest Canopy*. Cambridge, MA: Harvard University Press, 1994. The tremendous diversity of life in the rainforest treetops is only now becoming known to us. This book documents the unexpected and spectacular diversity of animals in the upper reaches of this endangered habitat.

Morell, V. "Life on a Grain of Sand." *Discover*, April 1995. The sand beneath shallow waters is home to an incredible range of microscopic creatures.

Schaechter, E. *In the Company of Mushrooms*. Cambridge: Harvard University Press, 1997. An accessible account of the world of mushrooms, written in a warm, personal style.

Wilson, E. O. *The Diversity of Life*. Cambridge, MA: Harvard University Press, 1992. An outline of the processes that created the diversity of life and a discussion of the threats to that diversity and the steps required to preserve it.

Answers to Multiple-Choice Questions

1. a 2. c 3. e 4. d 5. c 6. a

MediaTutor

CD Activities

Activity 16.1: Taxonomic Classification

Estimated time: 5 minutes

Taxonomic classification is a means of categorizing all species of organisms on Earth. This exercise will help you understand the hierarchy of taxonomic groups used by biologists.

Activity 16.2: The Fungi

Estimated time: 5 minutes

In this tutorial you are shown examples of each of the major phyla of fungi and their distinguishing characteristics. Afterward you will be quizzed on your understanding of those characteristics.

Activity 16.3: Fungi Structure

Estimated time: 5 minutes

In this tutorial you will study the basic structure of fungi and reproduction in two common groups of fungi: basidiomycetes and zygomycetes.

Activity 16.4: Evolution of Plant Structure

Estimated time: 10 minutes

One important theme in land plant evolution is selection favoring traits that allowed plants to colonize and thrive in increasingly terrestrial environments. In this tutorial you will explore some of these adaptations, with particular attention to the evolution of new tissue types and modifications in reproductive structures and function. Trends will be illustrated by examining four groups of land plants: Bryophytes, ferns, gymnosperms (conifers and their allies), and angiosperms (flowering plants).

Activity 16.5: Insect Metamorphosis

Estimated time: 5 minutes

Insects are perhaps the most successful group of organisms on Earth in terms of species diversity and sheer numbers. One key to the success of insects may be their ability to undergo metamorphosis, a profound change in body form from the juvenile to the adult stage. This simple animation demonstrates the metamorphosis of a periodic cicada. A nymph, having lived underground for 17 years feeding on tree roots, changes into a winged adult that will fly, feed on tree sap, mate, and die within a few weeks.

Start the MediaTutor Student CD-ROM and enter the activity number in the Quick Search box to be taken directly to that activity.

Web Investigation

A Biologist "Bombs" Beetles to Estimate Diversity

Estimated time: 15 minutes

As discussed in the Case Study: A Biologist "Bombs" Beetles to Estimate Diversity, scientists have begun to revise earlier estimates of the number of species living on earth. Until recently, we assumed that life could not exist in the harshest environments on Earth. That assumption changed with the discovery that archaea live in some of the planet's most extreme environments. This activity will further explore the diversity of life on our planet and some of the modern techniques that are being used to expand our understanding of the relationships between groups of species.

Go to http://www.prenhall.com/audesirk3, the Audesirk Companion Web site. Select Chapter 16 and Web Investigation to begin.

UNIT FOUR

Plant Anatomy and Physiology

Flowers, such as these Texas bluebonnets, provide beauty and delight to human admirers. The real function of flowers, however, is to attract animal pollinators.

The bark of the Pacific yew is the source of a potent anticancer drug. Plants hold many such medicinal treasures, most of them still undiscovered.

17 Plant Form and Function

AT A GLANCE

CASESTUDY

Hunting for Medical Treasures

Nineteenth-century prospectors panned for gold in mountain streams, but today's prospectors are more likely to find treasure in a plant. For example, the bark of the American yew tree (*Taxus brevifolia*) is the source of the anti-cancer drug Taxol, which racks up annual sales of more than one billion dollars. The financial and medical success of Taxol and other plant-derived medicines inspires a continuing search for the additional treasures that are no doubt hidden in plants. This search for biological riches is sometimes known as bioprospecting.

Bioprospectors must overcome a big hurdle: there are more than 250,000 known plant species. In such a vast sea of potential targets, how can someone find the treasures? One approach is to simply test every species, one by one. But many bioprospectors try to increase their chances of success by taking advantage of local wisdom and experience. In many parts of the world, especially in the tropical countries that hold most of the world's plant diversity, traditional healers—medicine men, herbalists, and midwives—can point out species with medicinal properties. This local knowledge can be a boon to a bioprospector looking for the best place to begin a search.

With millions of dollars to be made and millions of human lives to be saved, the planet's forests ought to be filled with armies of bioprospectors. There is, in fact, a fair amount of bioprospecting activity, but nothing like the frenzy of activity that many observers predicted. What's the holdup? One problem is disagreement over ownership of the products and profits derived from bioprospecting. Most of the world's plant species are located in economically poor tropical countries in Asia, Latin America, and Africa, but most of the world's large drug companies are located in Europe and North America. Governments in the poorer countries wonder if any of the wealth derived from plants found within their borders will be returned to their countries. Some countries have passed laws restricting outsiders' access to search for and exploit biological riches. For their part, the drug companies wonder if they ought to invest large sums in the search for plant-derived drugs if they will be denied the opportunity to profit from their investments. At times, mutual mistrust threatens to strangle the growth of bioprospecting and to short-circuit a potentially important path to improved human health. ■

1 How Are Plant Bodies Organized?

As we discovered in Chapters 15 and 16, the evolutionary history of the plant kingdom has produced many distinctly different types of plant bodies. In the current chapter, which discusses how plant bodies survive and grow, we focus mainly on the flowering plants, or angiosperms, which have become the most diverse and widespread of plant groups.

Flowering Plants Consist of a Root System and a Shoot System

Flowering plant bodies consist of two major parts, the root system and the shoot system (Fig. 17-1). **Roots** are branched portions of the plant body that are usually embedded in soil. They anchor the plant in the ground, absorb and transport water and nutrients from the soil, and store surplus sugars manufactured during photosynthesis.

The rest of the plant is the shoot system, usually located above ground. The shoot system consists of leaves, buds, and (in season) flowers and fruits, all borne on **stems**, which are typically branched. The functions of shoots include photosynthesis, mainly in leaves and young green stems; transport of materials among leaves, flowers, fruits, and roots; and reproduction.

Flowering Plants Can Be Divided into Two Categories

Flowering plants comprise two main groups: the **monocots**, which include grasses, lilies, palms, and orchids, and the **dicots**, which include deciduous trees (those that drop their leaves in winter), bushes, and many garden flowers. Monocots and dicots differ in the structure of their flowers, leaves, vascular tissue, root pattern, and seeds. These differences are illustrated in Figure 17-2. For now, don't worry about unfamiliar terms in the figure; look it over and refer to it later as we examine the parts of the flowering plants in more detail.

MEDIATUTOR 17.1 Plant Anatomy

Figure 17-1 The structure of a typical flowering plant

2 How Do Plants Grow?

Animals and plants develop in dramatically different ways. One difference is the timing and distribution of growth. As you grew from a baby to an adult, all parts of your body became larger. When you reached your adult height, you stopped growing (upward, at least!). In contrast, flowering plants grow throughout their lives, never reaching a stable "adult" body form. Moreover, most plants grow longer only at the tips of their branches and roots, and structures that developed earlier remain in exactly the same place; a swing tied to a tree branch does not move farther from the ground each year. Why do plants grow this way?

During Growth, Meristem Cells Give Rise to Differentiated Cells

Plant bodies are composites of two fundamentally different categories of cells: meristem cells and differentiated cells. **Meristem cells** are capable of mitotic cell division. Some of the offspring, or daughter cells, of dividing meristem cells lose the ability to divide and

Figure 17-2 Monocots and dicots compared
Distinguishing traits of the two major classes of flowering plants, the monocots and the dicots.

become **differentiated cells** that make up the nongrowing portions of the plant body. Thus, continued divisions of meristem cells keep a plant growing throughout its life, and their differentiated daughter cells form more stable or permanent parts of the plant, such as mature leaves or the trunks of trees.

Plants grow through the division and differentiation of two major types of meristem cells: apical meristems and lateral meristems. **Apical meristems** ("tip meristems") are located at the tips of roots and shoots, including main stems and branches (see Figs. 17-1, 17-8). **Lateral meristems** ("side meristems"), also called **cambia** (singular, **cambium**), form cylinders that run lengthwise inside roots and stems (see Fig. 17-12).

Different Processes Are Responsible for Growth in Length and Width

Plant growth takes two forms: primary growth and secondary growth. **Primary growth** occurs by the mitotic division of *apical* meristem cells. This type of growth takes place in the growing tips of roots and shoots. Primary growth is responsible both for increase in length and for the development of the specialized plant structures. The elongation of roots and shoots through primary growth allows them to enter new space from which to collect light, nutrients, and water.

Secondary growth, which is responsible for increases in diameter, occurs by the mitotic division of *lateral* meristem cells. Secondary growth causes the stems and roots of most conifers (cone-bearing trees) and dicots to become thicker and woodier as they age. Although we discuss only stem secondary growth in this chapter, keep in mind that secondary growth also occurs in roots.

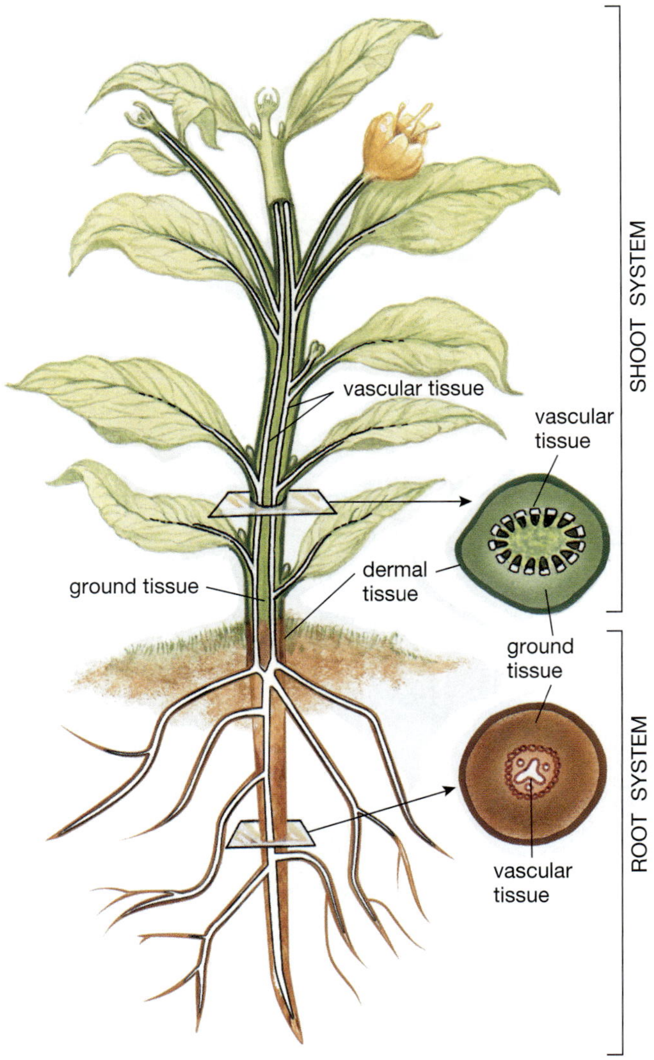

Figure 17-3 The structure of the root and shoot
Both the root and shoot of a flowering plant consist of dermal, ground, and vascular tissues. (The structure of the leaves, vascular tissue system, and roots indicate that our sample plant is a dicot.)

3 What Are the Tissues and Cell Types of Plants?

The structures of plants, including roots, stems, and leaves, consist of three types of tissue—dermal, ground, and vascular. Within each type there are subtypes composed of specialized cells (Fig. 17-3). **Dermal tissue** covers the outer surfaces of the plant body. **Ground tissue**, which consists of all nondermal and nonvascular tissues, makes up most of the body of young plants. Its functions include photosynthesis, support, and storage. **Vascular tissue** transports water, nutrients, sugars, and hormones throughout the plant.

Some flowering plants, described as herbaceous, are soft-bodied with flexible stems; herbaceous plants include lettuce, beans, and grasses. Such plants usually live only one year. Other flowering plants, such as trees and bushes, are described as woody; most are perennial (living many years) and develop hard, thickened, woody stems as a result of secondary growth. As you will see, different types of tissue are present in herbaceous and woody plants.

Dermal Tissue Covers the Plant Body

An important part of the dermal tissue that covers a plant's outer surface is the **epidermis**, the outermost cell layer that covers the leaves, stems, and roots of all young plants (Fig. 17-4a). The epidermis of the aboveground parts of a plant is generally composed of tightly packed thin-walled cells, covered with a waterproof, waxy **cuticle**. The cuticle reduces the evaporation of water from the plant and helps protect it from the invasion of disease microorganisms. In contrast, the epider-

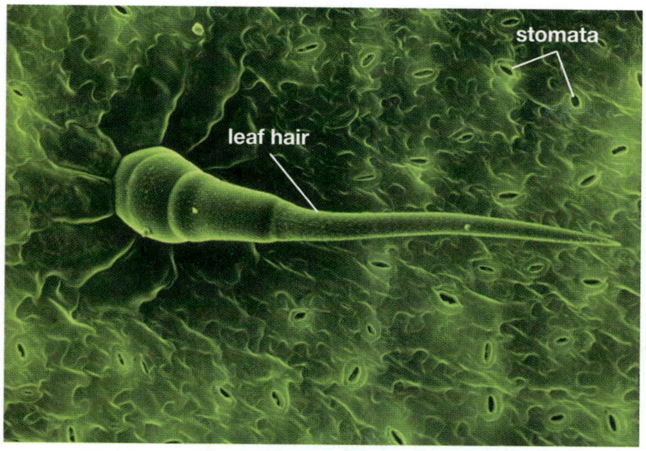

(a)

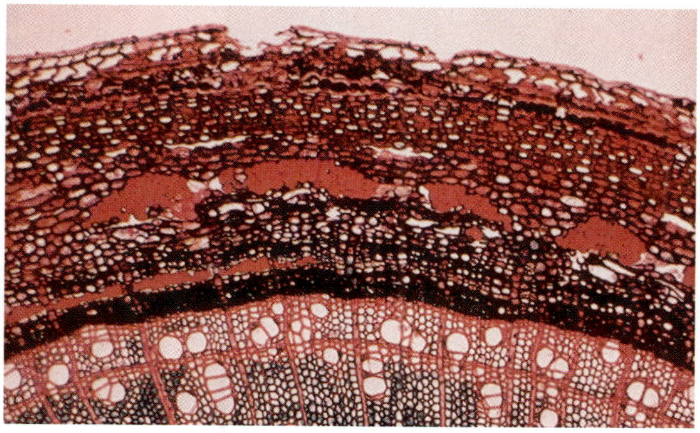

(b)

Figure 17-4 Dermal tissues cover plant surfaces
(a) The outer surface of shoot epidermis, such as the epidermis of this zinnia leaf, is covered with cuticle that reduces the evaporation of water. The leaf hair protruding from the epidermis also reduces evaporation by slowing the movement of air across the leaf surface. ***(b)*** Woody stems and roots develop a dense, tough outer layer, which consists mostly of thick-walled cork cells.

mis of roots is not covered with cuticle, because a cuticle would prevent the absorption of water and nutrients.

In herbaceous plants, the epidermis is retained as the outer covering of the entire body of the plant throughout its life. In woody plants, however, the epidermis of stems and roots is replaced as the plants age. The replacement tissue is composed primarily of **cork cells**, which have thick, waterproof walls and are dead at maturity (Fig. 17-4b). Cork cells form the protective outer layers of the bark of trees and woody shrubs and the tough covering of their roots.

Ground Tissue Makes Up Most of the Young Plant Body

The bulk of a young plant is made up of ground tissue, which consists of all nondermal and nonvascular tissues. Ground tissue cells typically carry out most of the metabolic activities of the plant. Depending on their location within the plant body, ground tissue cells have such diverse functions as photosynthesis, storage of sugars and starches, and secretion of hormones. Roots that are adapted for storage, including carrots and sweet potatoes, are packed with ground tissue cells that store carbohydrates, such as starch and sugar. In herbaceous plants and in the leaf stalks and young growing stems of all plants, ground tissue is an important source of support. The strings in celery stalks, for example, are mostly ground tissue cells in association with vascular tissue.

Vascular Tissue Consists of Xylem and Phloem

Vascular tissue consists of two complex conducting tissues—xylem and phloem. The major role of each of these is the transport of materials.

Xylem Conducts Water and Dissolved Nutrients from the Roots to the Rest of the Plant

Xylem conducts water and nutrients from roots to shoots in tubes that are made from cells called tracheids and vessel elements (Fig. 17-5). Tracheids are thin, tubelike cells with thick cell walls; their slanted ends resemble the tips of hypodermic needles. Tracheids are stacked atop one another with the slanted ends overlapping; porous sections in the overlapping portions allow water and nutrients to pass from one tracheid to the next. Vessel elements also meet end to end, but they are larger in diameter than tracheids. Perforations connect adjoining vessel elements. In some cases, the ends of vessel elements completely disintegrate, leaving an open tube (see Fig. 17-5). Thus, vessel elements form wide-diameter, relatively unobstructed pipelines from root to leaf. The development of both tracheids and vessel elements ends in cell death: The cytoplasm and plasma membrane of each cell disintegrate, leaving behind a hollow tube of cell wall.

Phloem Conducts Substances Throughout the Plant

Phloem carries water containing dissolved substances synthesized by the plant, including sugars, amino acids, and hormones. Scattered among the supportive ground tissue cells of phloem are sieve tubes, constructed of a single strand of cells called sieve-tube elements (Fig. 17-6). As sieve-tube elements mature, most of their internal contents disintegrate, leaving behind only a thin layer of cytoplasm that lines the plasma membrane. Sieve-tube elements thus lack most organelles and must be nourished by smaller, adjacent companion cells.

Where adjacent sieve-tube elements meet, end to end, holes form in the cell walls, creating structures called sieve plates. The interiors of adjacent sieve-tube elements are connected through the openings in the sieve plate. A continuous conducting tube is forged by the linking of many sieve-tube elements end to end in this way.

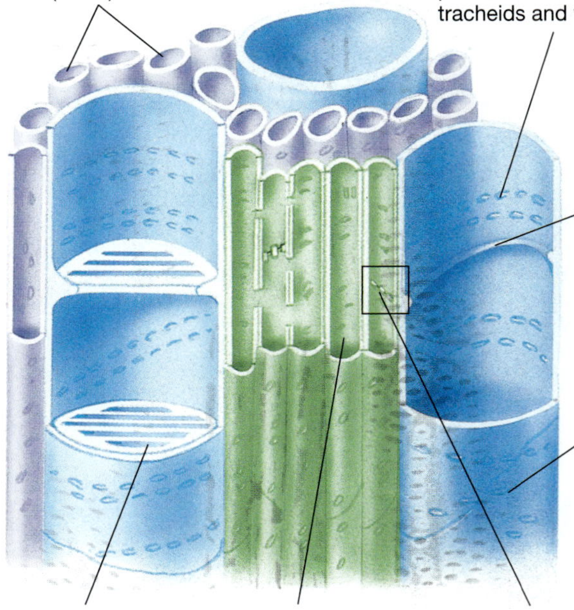

Figure 17-5 The structure of xylem
Xylem is a mixture of ground tissue and two types of conducting cells, tracheids and vessel elements. Both tracheids and vessel elements have porous side walls, allowing water and dissolved minerals to move sideways between adjacent conducting tubes.

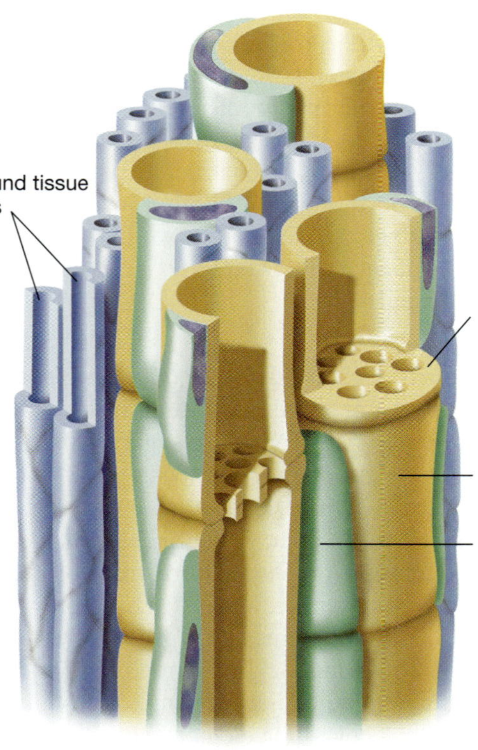

Figure 17-6 The structure of phloem
Phloem is a mixture of ground tissue, sieve-tube elements, and companion cells. Sieve-tube elements, stacked end to end, form the conducting tubes of phloem.

4 How Do Roots Grow and What Do They Do?

As a seed sprouts, the primary root—the first root to develop—grows down into the soil. Many dicots, such as carrots and dandelions, develop a taproot. A **taproot** consists of the primary root, which usually becomes longer and stouter with time, and many smaller roots that grow out from the sides (Fig. 17-7a). In contrast, in monocots such as grasses and palms, the primary root soon dies off, replaced by many new roots that emerge from the base of the stem. These new roots, nearly equal in size, are called **fibrous roots** (Fig. 17-7b).

Roots Elongate by Primary Growth

In young taproots and fibrous roots, divisions of the apical meristem give rise to four distinct regions: an outer envelope of epidermis; a vascular cylinder at the core of the root; a cortex between the two, and a **root cap** (Fig. 17-8). The root cap lies at the very tip of the root, and forms from daughter cells produced on the "soil side" of the apical meristem. The root cap protects the apical meristem from being scraped off as the root pushes down between particles of soil. Root-cap cells have thick cell walls and secrete a slimy lubricant that helps

(a) (b)

Figure 17-7 The roots of dicots and monocots
(a) Dicots typically have a long central taproot with many smaller roots branching from it. ***(b)*** Monocots typically have many fibrous roots of equal size.

ease the way between soil particles. Nevertheless, root-cap cells wear away and must be continuously replaced by new cells from the meristem.

The Epidermis of the Root Is Very Permeable to Water

The root's outermost covering of cells is the epidermis, which is in contact with the soil and any air or water trapped among the soil particles. The cell walls of the epidermal cells are highly permeable to water, allowing water to penetrate into the interior of the root. Many epidermal cells have **root hairs**, which are long, thin extensions that grow into the surrounding soil (Fig. 17-9). By increasing the surface area, root hairs increase the root's ability to absorb water and nutrients. Root hairs may add dozens of square yards of surface area to the roots of even small plants.

The Cortex Controls the Absorption of Water and Nutrients

Cortex is a type of ground tissue that makes up most of the inside of a young root. The cortex consists of an

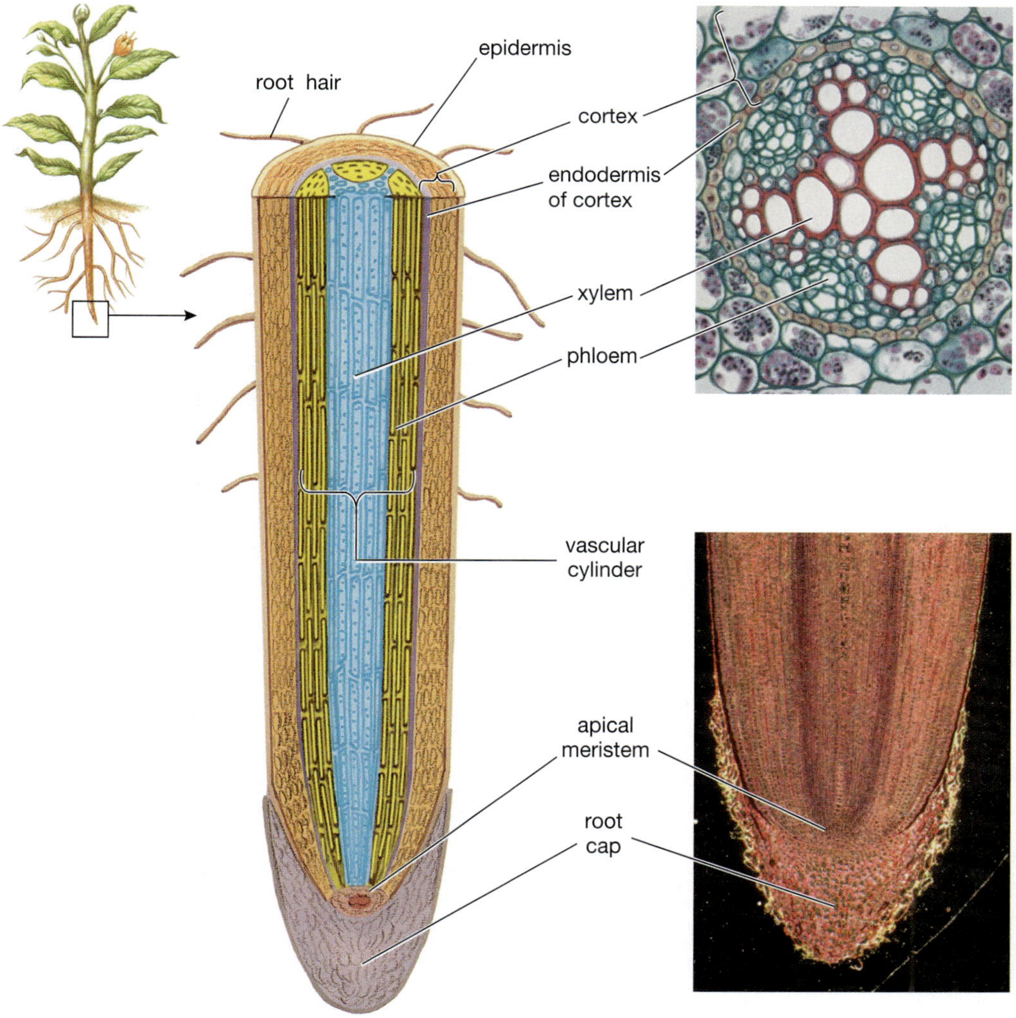

Figure 17-8 Primary growth in roots
Primary growth in roots results from cell divisions in the apical meristem, located near the tip of the root. Four regions are formed: the root cap, at the very tip of the root, and the epidermis, cortex, and vascular cylinder.

Figure 17-9 Root hairs
Root hairs, shown here in a sprouting radish, greatly increase a root's surface area for the absorption of water and nutrients from the soil.

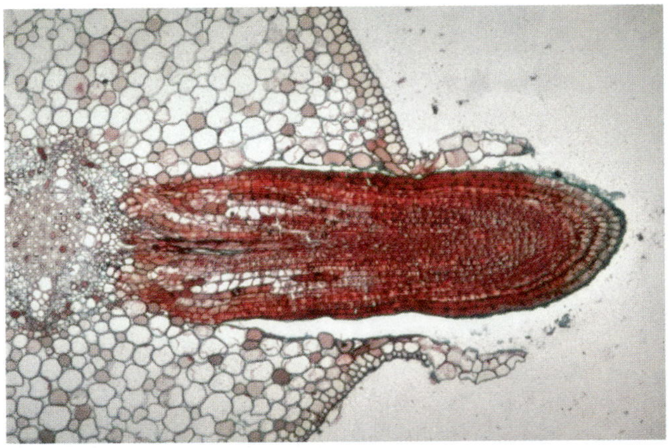

Figure 17-10 Branch roots
A branch root emerges from a primary root. The central axis of this branch root is already differentiating into vascular tissue.

outer mass of large, loosely packed cells just beneath the epidermis, and an inner layer of smaller, close-fitting cells called the **endodermis** that encircles the vascular cylinder (see Fig. 17-8). The cells of the endodermis have highly specialized cell walls. Where endodermal cells contact one another, their cell walls have a waxy, waterproof surface. But where they contact the cortex or vascular cylinder, their cell walls are permeable. Water and dissolved nutrients thus cannot travel between the endodermal cells, and can enter the vascular cylinder only by passing through the living membranes of the endodermal cells. These membranes regulate the types and amounts of materials that the roots can absorb.

The Vascular Cylinder Contains Xylem and Phloem and Meristem for Branch Roots

The **vascular cylinder** contains the conducting tissues of xylem and phloem, which transport water and dissolved materials within the plant. The outermost layer of the vascular cylinder retains the ability to divide and form the apical meristems of new roots that grow as branches from existing roots (Fig. 17-10). Branch root development is similar to primary root development except that the branch must first break out through the cortex and epidermis of the primary root. It does so by both crushing the cells that lie in its path and by secreting enzymes that digest them away. The vascular tissues of the branch root connect with the vascular tissues of the primary root.

The Big Picture

Natural selection has ensured that the structure of plant bodies is well matched to the distinctive needs of plants. Plants cannot move about, but their body structures excel at gathering from the immediate surroundings the required resources (sunlight, water, nutrients), and at moving resources from the point at which they are acquired or synthesized to the body locations where they are needed.

5 How Do Stems Grow and What Do They Do?

Like roots, stems develop from a small group of actively dividing cells. These cells lie at the tip of the young shoot within the **terminal bud** and make up the apical meristem. The daughter cells of the apical meristem differentiate into the specialized cell types of stem, buds, leaves, and flowers.

Most young stems are composed of four tissues: (1) epidermis (dermal tissue), (2) cortex (ground tissue), (3) pith (ground tissue), and (4) vascular tissue (Fig. 17-11). As Figure 17-2 illustrates, monocots and dicots differ somewhat in the arrangement of vascular tissues. We will discuss only dicot stems here.

The Epidermis of the Stem Retards Water Loss While Allowing Carbon Dioxide to Enter

In the stem (and leaves), the epidermis is exposed to dry air, making it a potential pathway for water loss. Epidermal cells of the stem secrete a waxy cuticle that reduces evaporation of water. The cuticle also, however, reduces the diffusion of carbon dioxide and oxygen into and out of the plant. Hence the epidermis is perforated with adjustable pores called **stomata** (singular, **stoma**) that regulate this exchange. Stomata are discussed in more detail later in this chapter.

The Cortex and Pith Support the Stem, Store Food, and Photosynthesize

The two types of ground tissue in stems, cortex (located between the epidermis and vascular tissues) and pith (inside the vascular tissues at the center of the stem), are similar in most respects; in fact, in some stems it is difficult to tell where cortex ends and pith begins. Cortex and pith perform three major functions: support, storage, and, in some cases, photosynthesis.

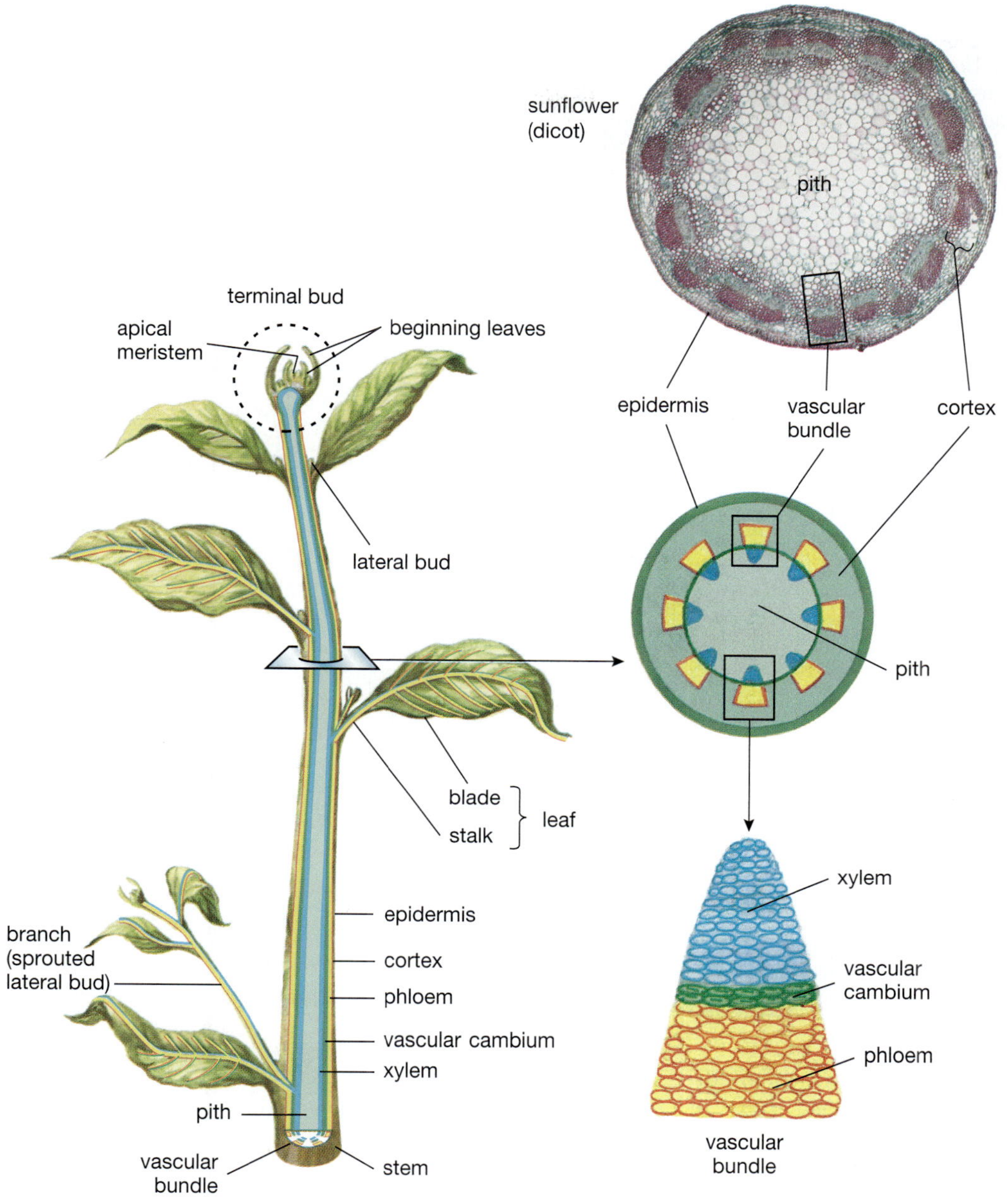

Figure 17-11 The structure of a young dicot stem
In cross section, vascular tissue forms a ring of vascular bundles in dicots such as the sunflower shown in the photomicrograph.

MEDIATUTOR 17.2 Primary and Secondary Growth

- *Support*. In very young stems, water fills the central vacuoles of cortex and pith cells, causing pressure that stiffens the cells, much as air inflates a tire. Somewhat older stems also have other ground cells with thickened cell walls, which don't depend on water pressure for strength.
- *Storage*. Cells in both cortex and pith convert the sugar produced by photosynthesis into starch and store the starch as a food reserve.
- *Photosynthesis*. In many stems, the outer layers of cortex cells contain chloroplasts and carry out photosynthesis. In some desert plants, such as cacti, the leaves are reduced or absent, and the cortex of the stem is the only green photosynthetic part of the plant.

Vascular Tissues in Stems Transport Water, Dissolved Nutrients, and Hormones

The xylem and phloem of stems, like those of roots, transport water, nutrients, sugars, and hormones. These vascular tissues are continuous in root, stem, and leaf, interconnecting all the parts of the plant. The xylem and phloem found in young stems arise from the apical meristem and are called primary xylem and primary phloem.

Branches Form from Lateral Buds Consisting of Meristem Cells

As the shoot grows, small clusters of meristem cells are "left behind" at the surface of the stem. These meristem cells form the leaf primordia that develop into mature

Figure 17-12 How branches form Stem branches grow from lateral buds located at the outer surface of a stem. The lateral bud's apical meristem generates an outward-growing branch, replicating the structure of the main stem.

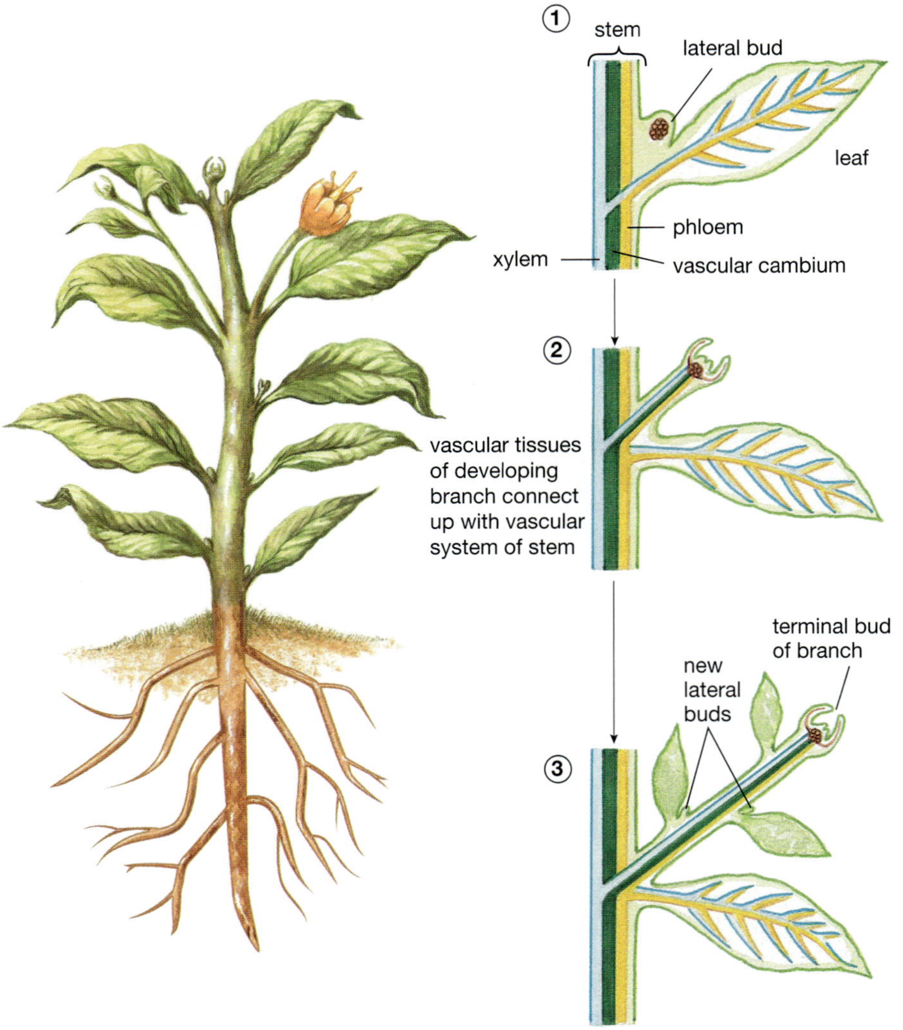

leaves, and they also produce **lateral buds**, which, under appropriate conditions, grow into branches. Lateral buds are located just above the attachment points of the leaves. When stimulated by the appropriate hormones, the meristem cells of a lateral bud are activated and the bud sprouts, growing into a branch (Fig. 17-12). As the branch grows, it duplicates the development of the stem: It has an apical meristem at its tip and produces its own leaves and lateral buds as it grows.

Secondary Growth Produces Thicker, Stronger Stems

In conifers and perennial dicots, stems may survive up to hundreds of years, becoming thicker and stronger each year. This secondary growth in stem thickness results from cell division in two lateral meristems: the vascular cambium and cork cambium.

Vascular Cambium Produces Secondary Xylem and Phloem

The **vascular cambium** is a cylinder of meristem cells located between the primary xylem and primary phloem. Daughter cells of the vascular cambium produced toward the inside of the stem differentiate into secondary xylem; those produced toward the outside of the stem differentiate into secondary phloem (Fig. 17-13). Because the center of the stem is already filled with pith and primary xylem, newly formed secondary xylem pushes the vascular cambium and all outer tissues farther out, increasing the diameter of the stem. This secondary xylem, with its thick cell walls, forms the wood that makes up most of the trunk of a tree. Young xylem, called sapwood (located just inside the vascular cambium), transports water and minerals; older xylem, the heartwood, contributes only to the strength of the trunk and no longer carries water and solutes. Heartwood serves as a collection site for metabolic wastes of the tree, such as gums, resins, and oils. These wastes increase the density of the heartwood, contribute to its darker color, and help the heartwood resist rotting.

Phloem cells are much weaker than xylem cells. As they die over time, the phloem cells are crushed between the hard xylem on the inside of the trunk and the tough cork on the outside (described below). Only a thin strip of recently formed phloem remains alive and functioning.

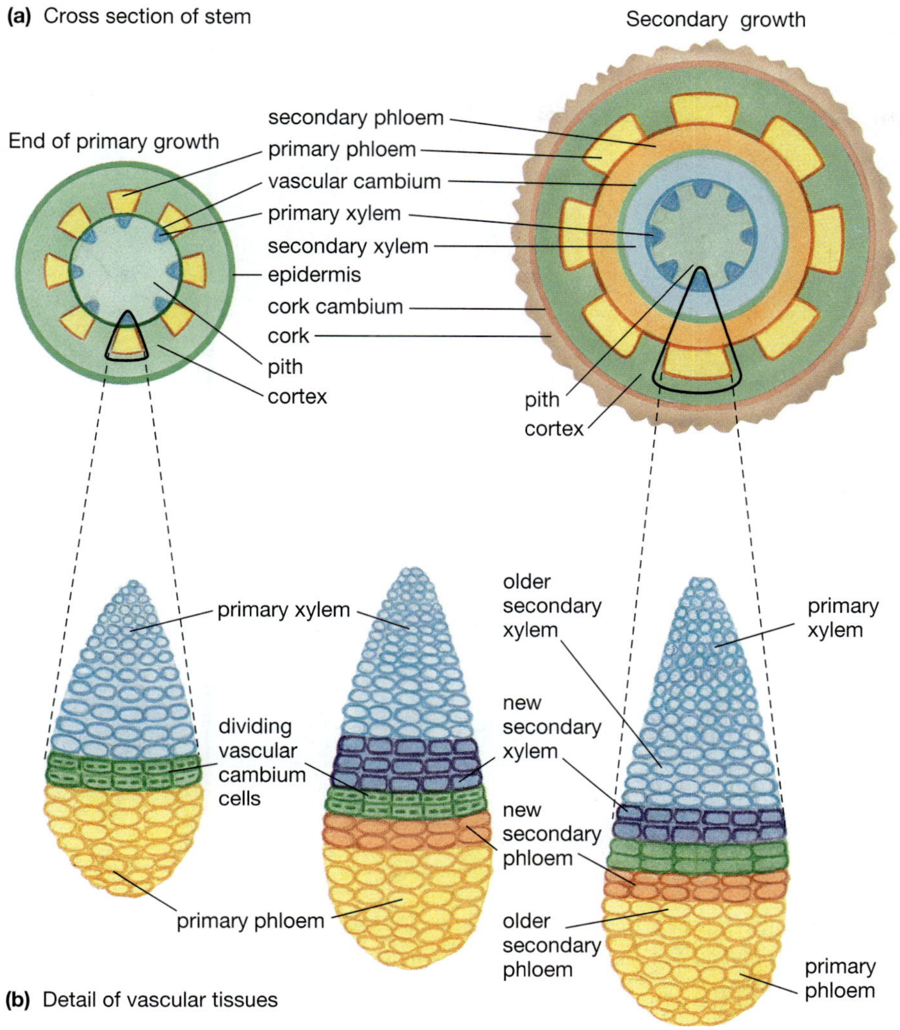

Figure 17-13 Secondary growth in a dicot stem
(a) Cross section of a dicot stem at the end of primary growth (left) and during early secondary growth (right). ***(b)*** Cells of the vascular bundle. When vascular cambium cells divide, daughter cells formed on the inside of the vascular cambium differentiate into secondary xylem; cells formed on the outside of the cambium differentiate into secondary phloem.

MEDIATUTOR 17.2 Primary and Secondary Growth

Tree Rings Form from Seasonal Changes in Vascular Cambium Growth

In trees adapted to temperate latitudes, such as oaks and pines, cell division in the vascular cambium ceases during the cold of winter. In spring the cambium cells divide, forming new xylem and phloem. The young cells grow by absorbing water and swelling while their cell walls are still soft. As the cells mature, the cell walls thicken and harden, preventing further growth. Water is readily available in spring; therefore, young xylem cells swell considerably and are large when mature. As summer progresses and water becomes scarcer, new xylem cells absorb less water and consequently are smaller when they mature. As a result, cross sections of tree trunks show a pattern of alternating pale regions (large cells formed in spring) and dark regions (small cells formed in summer). This pattern is the source of the annual growth rings of temperate trees. You can determine the approximate age of a tree that has been cut by counting the dark regions.

Secondary Growth Causes the Epidermis to Be Replaced by Woody Cork

Recall that epidermal cells are mature, differentiated cells that can no longer divide. Therefore, as new secondary xylem and phloem are added each year, enlarging the stem, the epidermis can't expand to keep up with the increasing circumference. The epidermis splits off and dies. Apparently prodded by hormones, some cells in the cortex become rejuvenated and form a new lateral meristem, the **cork cambium** (see Fig. 17-13). These cells divide, forming daughter cells toward the outside of the stem. These daughter cells, called cork cells, or simply cork, develop tough, waterproof walls that protect the trunk both from drying out and from physical damage. Cork cells die as they mature and may form a protective layer two feet thick in some tree species. As the trunk expands from year to year, the outermost layers of cork split apart or peel off, accommodating the growth.

The common term "bark" includes all the tissues outside the vascular cambium: phloem, cork cambium, and

(a) (b)

Figure 17-14 Some adaptations of stems
(a) The beach strawberry can reproduce via runners, horizontal stems that spread out over the surface of the sand. If a lateral bud of a runner touches the soil, it will sprout roots and develop into a complete plant. *(b)* The baobab tree develops an enormous water-storing trunk. The baobab grows in dry regions; when it rains, it is to the tree's advantage to store all the water it can get.

cork. The complete removal of a strip of bark all the way around a tree, called girdling, is invariably fatal to a tree because it severs the phloem. With the phloem gone, sugars synthesized in the leaves cannot reach the roots. Deprived of energy, the roots no longer take up water and nutrients, causing the tree to die.

Some Specialized Stems Produce New Plants or Store Water or Food

Many plants have modified stems that perform functions that are very different from the original one of raising leaves up to the light. Strawberries, for example, grow horizontal runners that snake out over the soil, sprouting new strawberry plants where lateral buds touch the soil (Fig. 17-14a). Some plants, such as the baobab tree (Fig. 17-14b), store water in aboveground stems. Many other plants store carbohydrates in underground stems. The common white potato is actually a storage stem; each eye is a lateral bud, ready to send up a branch next year, using the energy stored as starch in the potato to power the growth of the branch.

6 What Is the Structure of Leaves and What Do They Do?

Leaves are the major photosynthetic structures of most plants. To carry out photosynthesis and make valuable sugar, leaves require water, carbon dioxide, and sunlight. Sunlight is best gathered by leaves with large surface area, and a porous leaf would best permit carbon dioxide (CO_2) to diffuse into the leaf from the air. But a large, porous leaf would allow too much water, which is transported from the soil to the leaf through the xylem, to evaporate, so a leaf must be reasonably waterproof as well. The leaves of flowering plants represent an elegant compromise among these conflicting demands.

Leaves Have Two Major Parts

A typical leaf of a flowering plant consists of a broad, flat portion, the blade, connected to the stem by a stalk. The stalk positions the blade in space, usually orienting the leaf for maximum exposure to the sun. Inside the stalk are xylem and phloem that are continuous with the vascular tissues in the stem, root, and blade. Within the blade, the vascular tissues branch, forming the veins of the leaf.

The leaf epidermis consists of a layer of nonphotosynthetic, transparent cells that secrete a cuticle on their outer surfaces that reduces evaporation (Fig. 17-15). The epidermis and its cuticle are pierced by adjustable pores, the stomata, which regulate the diffusion of CO_2 and water into and out of the leaf. Each stoma is surrounded by two sausage-shaped **guard cells**, which regulate the size of the opening into the interior of the leaf.

Beneath the epidermis lies the loosely packed cells of the **mesophyll**. These cells contain chloroplasts and perform most of the photosynthesis of the leaf. The

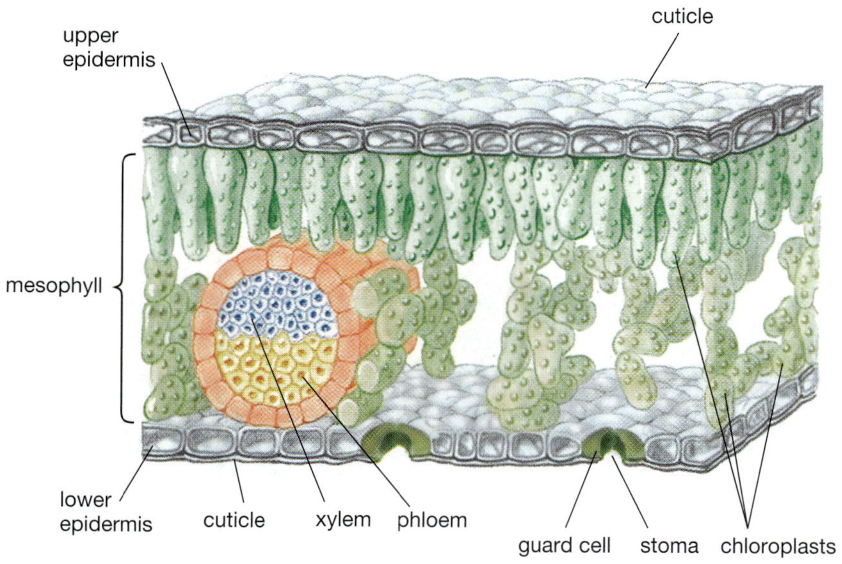

Figure 17-15 The structure of a typical dicot leaf The cells of the epidermis lack chloroplasts and are transparent, allowing sunlight to penetrate to the chloroplast-containing mesophyll cells beneath. The stomata that pierce the epidermis and the loose, open arrangement of the mesophyll cells ensure that CO_2 can diffuse into the leaf from the air and reach all the photosynthetic cells.

openness of the leaf interior allows CO_2 to diffuse easily to all the mesophyll cells. The branching vascular tissues are embedded within the mesophyll tissue, with fine veins reaching very close to each photosynthetic cell. Thus, each mesophyll cell receives energy from sunlight transmitted through the clear epidermis; carbon dioxide from the air, diffusing through the stomata; and water from the xylem. The sugars it produces are carried away to the rest of the plant by the phloem.

Specialized Leaves May Provide Support, Store Food, or Even Capture Insects

In some plants, modified leaves serve functions unrelated to photosynthesis or water conservation. The common edible pea, for example, grasps fences, mailbox posts, or other plants with clinging tendrils. Unlike the tendrils of grapes, which are derived from branches, pea tendrils are slender, supple leaflets. Some plants, such as onions, daffodils, and tulips, use thick, fleshy leaves as food-storage organs. A few plants have turned the table on the animals and have become predators. For example, sundews (Fig. 17-16) have leaves that are modified into snares for trapping unwary insects.

Figure 17-16 A specialized leaf
A sundew leaf grasps a lacewing fly with sticky, enzyme-laden hairs. Nutrients from the insect's body help sundews to thrive in nitrogen-poor soils.

7 How Do Plants Acquire Nutrients?

Nutrients are elements essential to normal life; they differ for different organisms. Plants require relatively large quantities of the following nutrients: carbon (obtained from CO_2), hydrogen (from water), oxygen (from air and water), phosphorus (from phosphate ions in soil), nitrogen (from nitrate and ammonium ions in soil), and magnesium, calcium, and potassium (as ions in soil). Plants also require very small quantities of such nutrients as iron, chlorine, copper, manganese, zinc, boron, and molybdenum. Carbon dioxide and oxygen usually enter a plant by diffusion from the air into leaves, stem, and roots. Roots extract water and all other nutrients, collectively called **minerals**, from the soil.

Roots Take Up Minerals Dissolved in Water

Soil consists of bits of pulverized rock, air, water, and organic matter. Although both the rock particles and the organic matter contain essential minerals, only minerals dissolved in the soil water are accessible to roots. The concentration of minerals in the soil water is very low, usually much lower than the concentration within plant cells and fluids. For example, the concentration of potassium (K^+) in root cells is at least 10 times greater than that in soil water, so diffusion cannot move potassium into the root. In general, most minerals are moved into a root against their concentration gradients by active transport, using energy from sugar transported from the leaves to the roots. (Recall from Chapter 3 that the movement of molecules from areas of low concentration to areas of high concentration requires energy.)

Fungi and Bacteria Help Plants Acquire Nutrients

Many minerals are not abundant enough in soil water to support plant growth, even though plenty of minerals may be bound up in the surrounding rock particles. One nutrient—nitrogen—is almost always in short supply both in rock particles and in soil water. Most plants have evolved beneficial relationships with other organisms that help the plants acquire scarce nutrients such as nitrates and phosphates and minerals. Examples include root–fungus relationships, called mycorrhizae, and root–bacteria relationships formed in nodules of legumes.

Fungal Mycorrhizae Help Plants Acquire Minerals

Most land plants form symbiotic relationships with fungi to form root–fungus complexes called **mycorrhizae** (singular, **mycorrhiza**), which help the plant extract and absorb minerals. Fungal strands intertwine between the root cells and extend out into the soil (Fig. 17-17). In some way that is not yet understood, the fungus renders nutrients, such as phosphorus and certain minerals, accessible for uptake by the roots, perhaps by converting rock-bound minerals into simple soluble compounds that root plasma membranes can transport. The fungus, in return, receives sugars, amino acids, and vitamins from the plant. Both the fungus and the plant can thereby grow in places where neither could survive alone, including deserts and in high-altitude, rocky soils that are low in nutrients.

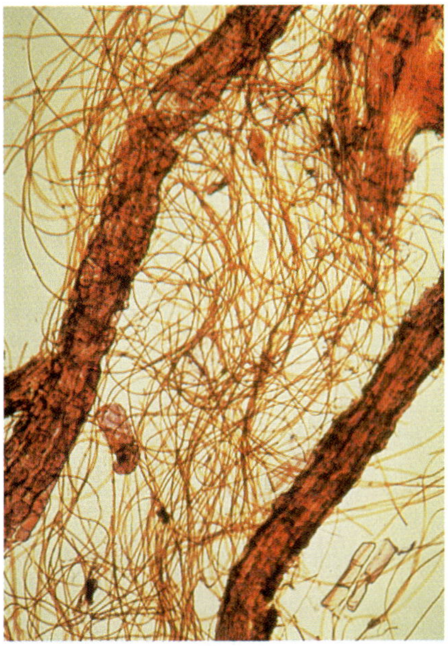

Figure 17-17 Mycorrhizae, a root–fungus symbiosis
A tangled meshwork of fungal strands surrounds and penetrates into the roots.

Bacteria-Filled Nodules on Their Roots Help Legumes Acquire Nitrogen

Amino acids, nucleic acids, and chlorophyll all contain nitrogen, so plants need large amounts of this element. Although nitrogen gas (N_2) makes up about 79% of the atmosphere and diffuses from the atmosphere into the air spaces in soil, it cannot be used by plants. Plants can take up nitrogen only in the form of ammonium ion (NH_4^+) or nitrate ion (NO_3^-), but they don't have the enzymes needed to carry out **nitrogen fixation**, the conversion of N_2 into ammonium or nitrate. Various **nitrogen-fixing bacteria,** however, do have these enzymes. Some of these bacteria are free-living in the soil, but they don't routinely manufacture a lot of extra ammonium and liberate it into the soil.

Some plants, particularly the legumes (peas, clover, and soybeans), enter into a mutually beneficial relationship with certain species of nitrogen-fixing bacteria. By secreting chemicals into the soil, legumes attract nitrogen-fixing bacteria to their roots. Once there, the bacteria enter the root hairs. The bacteria then digest channels through the cytoplasm of the epidermal cells and into underlying cortex cells. As both bacteria and their host cortex cells multiply, a nodule, or swelling, that houses the root–bacteria complex forms (Fig. 17-18). A cooperative relationship develops. The plant transports sugars from its leaves down to the cortex in its roots, just as it normally would for storage. The bacteria within the cortex cells take up some of the sugar and use its energy for all of their metabolic processes, including nitrogen fixation. The bacteria obtain so much energy that they produce more ammonium than they need. The surplus ammonium diffuses into the cytoplasm of their host cells, providing the plant with a steady supply of usable nitrogen. Surplus ammonium also diffuses into the surrounding soil, making it better able to support other types of plants. Farmers plant legumes not only for their commercial value but also to enrich the soil with ammonium for future crops.

Earth Watch

Plants Help Regulate the Distribution of Water

Probably the most important environmental factor influencing the distribution of plants is water: Cacti inhabit deserts because they can withstand drought; orchids and mahogany trees need the frequent drenching rains of the rain forest. However, people often overlook the flip side of the relationship between plants and water: Plants, through transpiration, help control the amount and distribution of rainfall, soil water, and even river flow.

Consider the Amazon rain forest (Fig. E17-1). An acre of soil supports hundreds of towering trees, each bearing millions of leaves. The surface area of the leaves dwarfs the surface area of the soil, so up to 75% of all the water evaporating from the acre of forest is due to leaf transpiration. This transpiration raises the humidity of the air and causes rain to fall. In fact, about half of the water transpired from the leaves falls again as rain; as a result about one-third of the total rainfall consists of water recycled by transpiration. Thus, the high humidity and frequent showers that the rain forest needs to survive are partly created by the forest itself. If large areas of rain forest are cut down, less water evaporates in that area, so less rain falls, and new rainforest tree seedlings cannot grow. An entirely different plant community would probably become established on the disturbed land, possibly becoming a permanent new community.

Figure E17-1 The Amazon rain forest
The rainforest community helps mold its own environment.

Plant transpiration might even have a moderating influence on the global warming caused by rising levels of atmospheric carbon dioxide (CO_2), due mostly to burning fossil fuels (such as coal and oil) during industrial activity and to cutting down forests (as we shall see in Chapter 29). Higher CO_2 levels will raise temperatures on Earth, and a warmer planet will increase water evaporation. Increased evaporation, in turn, could lead to drier soils and the expansion of deserts. Some researchers, however, suggest that soil moisture and river flow might actually *increase* in response to increased atmospheric CO_2. Why? Because the stomata of leaves tend to close in response to high CO_2 levels within their guard cells. Elevated atmospheric CO_2 levels would also raise CO_2 within the guard cells and cause partial closing of the stomata. Further, a recent study found that plants grown in an atmosphere with high CO_2 levels have fewer stomata per unit area of leaf than do plants grown at low, preindustrial CO_2 levels. If there are fewer stomata, and they are chronically partially closed, then we might expect less transpiration, and hence less loss of soil moisture.

The responses of plants to human activities can have major impacts on ecosystems, because plants wield an enormous influence on nonliving aspects of the biosphere, such as humidity, rainfall, soil water, and river flow. However, the exact consequences of plant responses to human alterations of the environment are not easily predictable.

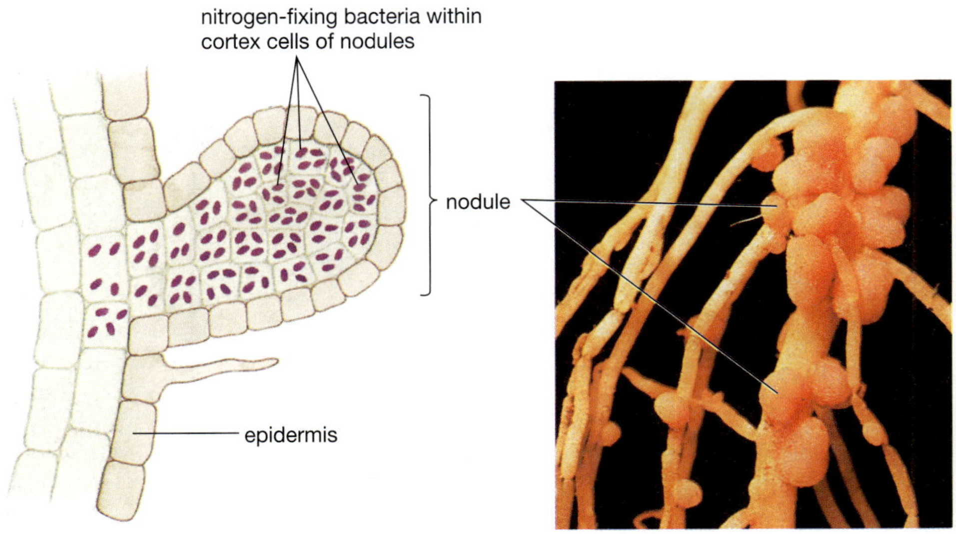

Figure 17-18 Nitrogen fixation in legumes
A diagram and photograph of root nodules, which contain nitrogen-fixing bacteria.

The Big Picture

Partnerships between plants and microorganisms play a huge role in maintaining Earth's ability to sustain life. Plant photosynthesis provides much of the planet's oxygen gas, and plants could not thrive without the beneficial bacteria and fungi that inhabit the soil around their roots.

MEDIATUTOR 17.3 Plant Transport Mechanisms

8 How Do Plants Acquire Water and Transport Water and Minerals?

Nearly 99% of the water absorbed by the roots of plants is evaporated through the stomata of leaves and, to a lesser extent, through the stomata of stems in a process called **transpiration**. Transpiration drives the movement of water through the plant body by pulling water up through the xylem of the roots and stem and into the leaves.

Transpiration Provides the Force for Water Movement in Xylem

After entering the root xylem, water and dissolved minerals still must be moved to the uppermost reaches of the plant. (In redwood trees, the distance may be more than 300 feet.) How do plants overcome the force of gravity and make water flow upward? Transpiration provides the necessary force. As a leaf transpires, the concentration of water in its mesophyll tissue drops. This drop causes water to move by osmosis from the leaf's xylem into the mesophyll cells. The water molecules leaving the xylem are attached to other water mol-

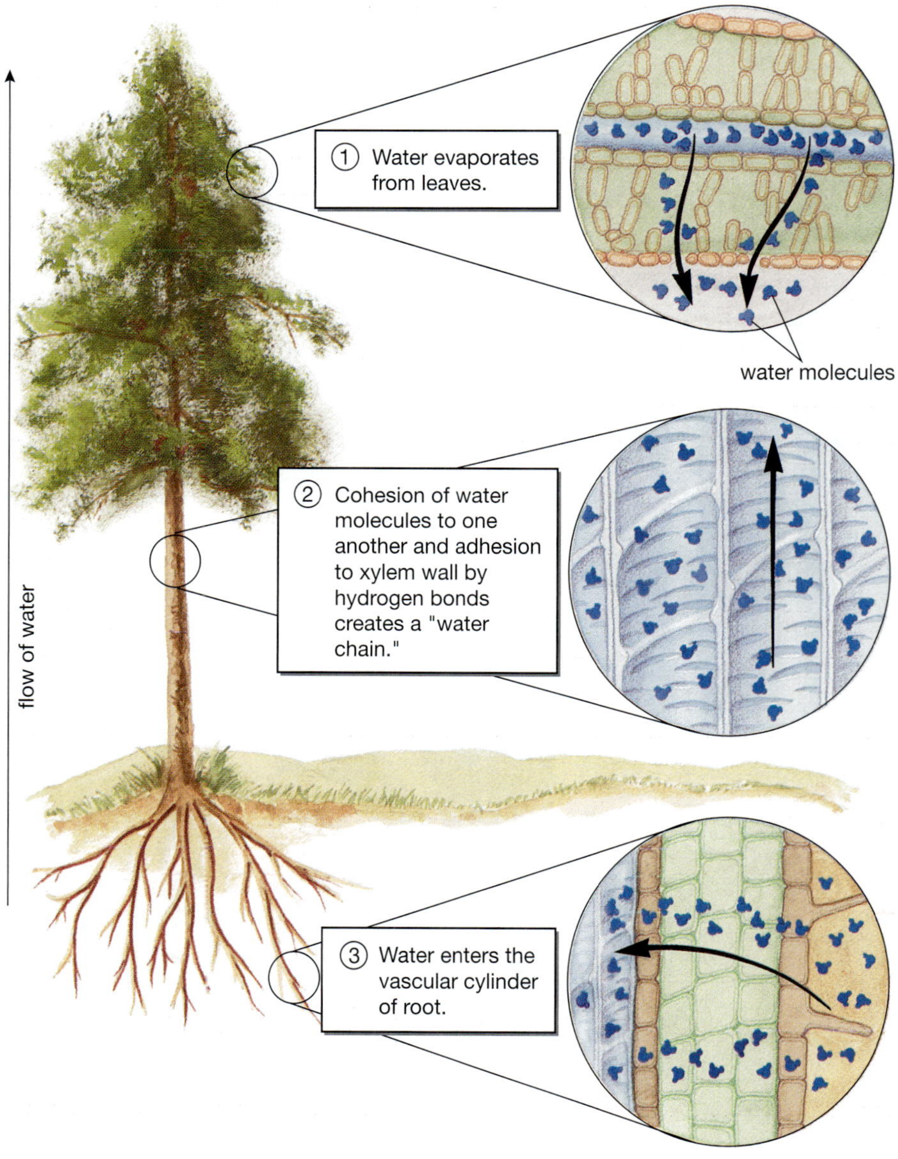

Figure 17-19 Transpiration powers the movement of water

ecules in the same xylem tube by hydrogen bonds. Therefore, when one water molecule leaves, it pulls adjacent water molecules up the xylem. As these water molecules move upward, other water molecules farther down the tube move up to replace them. This process continues all the way to the roots, where water is pulled into the xylem (Fig. 17-19).

This transpiration-driven upward movement of water is made possible by the strength with which the water in the xylem holds together. Just as individually weak cotton threads together make the strong fabric of your jeans, the network of individually weak hydrogen bonds in water collectively produces a very high cohesion, or tendency to resist being separated. Experiments have demonstrated that the column of water within the xylem is at least as strong—and as unbreakable—as a steel wire of the same diameter. Supplementing the cohesion between water molecules is the attraction of water molecules to the cell walls of the thin xylem tubes. This force helps the water creep upward, just as water is pulled upward into a very narrow glass tube.

Water Enters Roots Mainly by Pressure Differences Created by Transpiration

The upward movement of water ultimately causes soil water to move into the roots. Water moves first into the epidermis and outer cortex layers of the root between the cells along a pathway created by their porous cell walls. From there, the water crosses the endodermal cells by osmosis and moves into the vascular cylinder. The main force powering this flow of water is reduced water pressure in the vascular cylinder created by water losses due to transpiration from the leaves. The force generated by the evaporation of water from the leaves, transmitted down the xylem to the roots, is so strong that water can be absorbed even from quite dry soils.

Adjustable Stomata Control the Rate of Transpiration

Although transpiration provides the force that transports water and minerals to the leaves at the top of the plant, it is also the plant's largest source of water loss—a loss that may threaten the very survival of the plant, especially in hot, dry weather. Most water is transpired through the stomata of the leaves and stem, so you might think that a plant could prevent water loss simply by closing its stomata. Don't forget, however, that photosynthesis requires carbon dioxide from the air, which diffuses into the leaf primarily through open stomata. A plant that simply closed its stomata would be unable to photosynthesize. Instead, a plant must balance carbon dioxide uptake and water loss by opening and closing its stomata. In general, stomata open during the day, when sunlight allows photosynthesis, and close at night, conserving water. They will also close in the sunlight if the plant is losing too much water.

CASE STUDY REVISITED

Hunting for Medical Treasures

The progress of bioprospecting has been hampered by disputes over the ownership of biological resources. Although settling ownership disputes between drug companies and governments is a political, rather than a technical, problem, two scientific agencies have nonetheless stepped in with a plan to help break the impasse. The National Institutes of Health and the National Science Foundation, government agencies that fund scientific research in the United States, have established a program to promote bioprospecting. The agencies have provided funds to establish a number of International Cooperative Biodiversity Groups (ICBG), which are intended to foster bioprospecting in poorer countries.

ICBG projects have been established in African, Asian, and Latin American countries. Their goal is to negotiate access to a country's forests, usually with the involvement of local healers' associations. In return, the healers are paid a fee, and they are offered help with setting up apprenticeship programs in traditional healing and with testing the ingredients of their remedies. If possible, local laboratories are used to analyze plants for potentially useful compounds. Further analysis and clinical trials on promising compounds are carried out in the United States. If a compound leads to a successful drug, most of the royalties are invested in the host country. For example, one of the African programs specifies that 20% of royalties goes to the drug company, 50% into a community development trust fund run by local people, and 30% toward research into tropical diseases.

A number of tropical countries have agreed to host an ICBG, and the program's sponsors hope that the resulting examples of cooperation will demonstrate to private-sector companies that investment in bioprospecting can be rewarding to all parties. Because bioprospecting at its best promotes both human health and forest preservation, this would indeed be a promising outcome.

Summary of Key Concepts

1) How Are Plant Bodies Organized?

The body of a plant consists of root and shoot. The underground roots anchor the plant in the soil, absorb and transport water and minerals from the soil, and store surplus photosynthetic products. The aboveground shoot consists of stem, leaves, buds, flowers, and fruit. Shoot functions include photosynthesis, transport of materials, and reproduction. Flowering plants comprise the monocots and dicots.

2) How Do Plants Grow?

Plant bodies are composed of meristem cells and differentiated cells. Meristem cells retain the capacity for mitotic cell division. Differentiated cells arise from divisions of meristem cells, become specialized for particular functions, and usually do not divide. Meristem cells are located in apical meristems at the tips of roots and shoots and in lateral meristems in stems and roots. Primary growth (growth in length and differentiation of parts) results from the division and differentiation of apical meristem cells; secondary growth (growth in diameter) results from the division and differentiation of lateral meristem cells.

3) What Are the Tissues and Cell Types of Plants?

Plant bodies consist of dermal, ground, and vascular tissues. Dermal tissue forms the outer covering of the plant body. The dermal tissue of leaves and of primary roots and stems is a single cell layer of epidermis. Dermal tissue after secondary growth is a multilayered covering of cork.

Ground tissue consists of cell types that are involved in photosynthesis, support, or storage. Ground tissue makes up most of a young plant during primary growth. During secondary growth of stems and roots, ground tissue becomes an increasingly small part of the plant body.

Vascular tissue consists of xylem, which transports water and nutrients from the roots to the shoots, and phloem, which transports water, sugars, amino acids, and hormones throughout the plant body.

4) How Do Roots Grow and What Do They Do?

Primary growth in roots produces an outer epidermis, an inner vascular cylinder of conducting tissues, and cortex between the two. The apical meristem near the tip of the root is protected by the root cap. Cells of the root epidermis absorb water and nutrients from the soil. Root hairs are extensions of epidermal cells that increase the surface area for absorption. Most cortex cells store surplus sugars (usually in the form of starch) produced by photosynthesis. The innermost layer of cortex cells is the endodermis, which controls the movement of water and nutrients from the soil into the vascular cylinder. The vascular cylinder contains the conducting tissues—xylem and phloem.

5) How Do Stems Grow and What Do They Do?

Primary growth in dicot stems results in a structure consisting of an outer, waterproof epidermis; supporting and photosynthetic cells of cortex beneath the epidermis; vascular tissues of xylem and phloem; and supporting and storage cells of pith at the center. Leaves and lateral buds grow at intervals along the surface of the stem. Under appropriate conditions, lateral buds may sprout into a branch.

Secondary growth in stems results from cell divisions in the vascular cambium and cork cambium. Vascular cambium produces secondary xylem and secondary phloem, increasing the stem's diameter. Cork cambium produces waterproof cork cells that cover the outside of the stem.

6) What Is the Structure of Leaves and What Do They Do?

Leaves are the main photosynthetic organs of plants. The blade of a leaf consists of a waterproof outer epidermis surrounding mesophyll cells, which have chloroplasts and which carry out photosynthesis, and the xylem and phloem, which carry water, minerals, and photosynthetic products to and from the leaf. The epidermis is perforated by adjustable stomata that regulate the exchange of gases and water.

7) How Do Plants Acquire Nutrients?

Most minerals are taken up from the soil water by active transport into the root hairs. These minerals diffuse into the root. There they are actively transported into the extracellular space of the vascular cylinder. The minerals diffuse from the extracellular space into the xylem.

Many plants have fungal mycorrhizae associated with their roots that help absorb soil nutrients. Nitrogen can be absorbed only as ammonium or nitrate, both of which are scarce in most soils. Legumes have evolved a cooperative relationship with nitrogen-fixing bacteria that invade legume roots. The plant provides the bacteria with sugars, and the bacteria use some of the energy in those sugars to convert atmospheric nitrogen to ammonium, some of which is then absorbed by the plant.

8) How Do Plants Acquire Water and Transport Water and Minerals?

The water within xylem tubes holds together almost as if it were a solid chain, because water molecules are attracted to one another by hydrogen bonds. As water molecules evaporate from the leaves during transpiration, the hydrogen bonds pull other water molecules up the xylem to replace them. This movement is transmitted down the xylem to the root, where water loss from the vascular cylinder promotes water movement across the endodermis from the soil water by osmosis. Water moves by osmosis across the plasma membranes of the endodermal cells into the extracellular space of the vascular cylinder. The water pressure gradient caused by loss of water through transpiration is the primary force drawing water into the root.

Key Terms

apical meristem *p. 316*
cambium *p. 316*
cork cambium *p. 323*
cork cell *p. 317*
cortex *p. 319*
cuticle *p. 316*
dermal tissue *p. 316*
dicot *p. 314*
differentiated cell *p. 316*
endodermis *p. 320*
epidermis *p. 316*
fibrous roots *p. 318*
ground tissue *p. 316*
guard cell *p. 324*
lateral bud *p. 322*
lateral meristem *p. 316*
leaf *p. 324*
meristem cell *p. 314*
mesophyll *p. 324*
mineral *p. 325*
monocot *p. 314*
mycorrhiza *p. 326*
nitrogen fixation *p. 326*
nitrogen-fixing bacterium *p. 326*
phloem *p. 317*
primary growth *p. 316*
root *p. 314*
root cap *p. 318*
root hair *p. 319*
secondary growth *p. 316*
stem *p. 314*
stoma *p. 320*
taproot *p. 318*
terminal bud *p. 320*
transpiration *p. 328*
vascular cambium *p. 322*
vascular cylinder *p. 320*
vascular tissue *p. 316*
xylem *p. 317*

Thinking Through the Concepts

Multiple Choice

1. *In a mycorrhizal relationship, a plant root has a mutually beneficial relationship with*
   a. a fungus that helps obtain minerals from the soil
   b. a fungus that helps in the fixation of nitrogen from the air into a form usable by the plant
   c. bacteria that help obtain minerals from the soil
   d. bacteria that help in the fixation of nitrogen from the air into a form usable by the plant
   e. an alga that helps the plant photosynthesize
2. *Which of the following is involved in water movement in plants?*
   a. the presence of hydrogen bonds that hold water molecules together
   b. the attraction of water molecules to the walls of the xylem
   c. the diffusion of water from cells in the root to cells in the shoot
   d. the evaporation of water through the stomata
   e. All of the above are involved in this theory.
3. *You and a friend carve your initials 5 feet above the ground on a tree on campus. The tree is now 40 feet tall. When you come back for your twenty-fifth reunion, the tree will be 100 feet tall. How high above the ground should you look for your initials?*
   a. 5 feet
   b. 40 feet
   c. 60 feet
   d. 95 feet
   e. 100 feet
4. *If you wanted to show a friend how much you had learned in biology class, which of the following would you NOT use to identify a plant as a dicot?*
   a. leaves with veins arranged like a net
   b. hand-shaped leaves
   c. six petals
   d. a taproot
   e. a seed with two cotyledons
5. *When water enters a plant root, it is forced to travel through the endodermal cells by the waxy composition of part of the*
   a. cuticle
   b. epidermis
   c. root cap
   d. xylem
   e. endodermis

? Review Questions

1. Describe the locations and functions of the three main types of tissues systems in plants.
2. Distinguish between primary growth and secondary growth, and describe the cell types involved in each.
3. Distinguish between meristem cells and differentiated cells. Which meristems cause primary growth? Which ones cause secondary growth? Where is each type located?
4. Diagram the internal structure of a root after primary growth, labeling and describing the function of epidermis, cortex, endodermis, xylem, and phloem. What tissues are located in the vascular cylinder?
5. How do xylem and phloem differ?
6. What are the main functions of roots, stems, and leaves?
7. What types of cells form root hairs? What is the function of root hairs?
8. Diagram the internal structure of leaves. What structures regulate water loss and CO_2 absorption by a leaf?

Applying the Concepts

1. A goal of molecular botanists is to insert the genes for nitrogen fixation into crop plants such as corn or wheat (see Chapter 12). Why would the insertion of such genes be useful? What changes in farming practices would this technique allow?
2. We learned in Chapter 2 about the peculiar characteristics of water. Discuss several ways in which the evolution of vascular plants has been greatly influenced by water's special characteristics.
3. A major environmental problem is desertification, in which overgrazing by cattle or other animals leaves too few plants in an area. Show how what you know about the movement of water through plants enables you to understand this process, in which there is less water in the atmosphere, less rain, and thus dry, desert-like conditions.
4. The tropical rain forest contains a large number of as-yet unidentified plants, many of which may have uses as medicines or food. If you were given the job of searching a particular portion of the rain forest for useful products, how would you use the information you gained from this chapter to help you narrow your search? What kinds of plant tissues or organs would be most likely to contain such products?
5. The desert tends to have two categories of plants—small grasses or herbs, and shrubs or small trees. The grasses and herbs typically have fibrous roots. The shrubs and trees have taproots. What advantages can you think of for each type of root? How does each type of root allow for survival in a desert environment?
6. Grasses (monocots) form their primary meristem near the ground surface rather than at the tips of branches the way dicots do. How does this feature allow you to grow a lawn and mow it every week in the summer? What would happen if you had a dicot lawn and tried to mow it?

Group Activity

Like other large perennials, palms can make new leaves and roots if needed because of damage—for example, by freezing. Nevertheless, palms mostly grow only in the Tropics because they cannot recover from freezing-cold winters. Take three minutes to write your own explanation of why the structure of palms prevents them from living in areas with cold winters. Pass your explanation to the person on your right. Take another three minutes to arrive at an explanation different from the one suggested by the person on your left. Write your alternative beneath their suggested explanation. Discuss all the possible explanations with the rest of your class.

For More Information

Baskin, Y. "Forests in the Gas." *Discover*, October 1994. As carbon dioxide increases in the atmosphere, the relationships of plants in the natural environment will change.

Joyce, C. *Earthly Goods: Medicine-Hunting in the Rainforest.* Boston: Little, Brown, 1994. Science and adventure combine in this account of prospecting for new medicines and the people who do it.

Zimmer, C. "The Processing Plant." *Discover*, September 1995. The purple pitcher plant supplements its diet by digesting flies, with the surprising help of insect larvae that thrive in the pitcher plant's "digestive chamber."

Zimmer, C. "The Web Below." *Discover*, November 1997. An underground web of mycorrhizae transfers nutrients between trees and helps maintain forest health.

Answers to Multiple-Choice Questions

1. a 2. e 3. a 4. c 5. e

MediaTutor

CD Activities

Activity 17.1: Plant Anatomy

Estimated time: 5 minutes

This tutorial will allow you to move about the plant body, zooming in and out to view and become more familiar with the plant body. Click on the plant structures to view an enlarged illustration of that structure's anatomy.

Activity 17.2: Primary and Secondary Growth

Estimated time: 5 minutes

In this tutorial we will explore the relationship between primary and secondary growth in plants. Plant stems and roots grow in length by primary growth and in width by secondary growth. Secondary growth produces wood and bark. The focus here is on stems; roots show similar development patterns.

Activity 17.3: Plant Transport Mechanisms

Estimated time: 5 minutes

In this tutorial you will explore two mechanisms of transport in plants. Section one illustrates the mechanism of water and mineral transport in the xylem, and section two illustrates the mechanism of sugar transport in the phloem.

Start the MediaTutor Student CD-ROM and enter the activity number in the Quick Search box to be taken directly to that activity.

Web Investigation

Hunting for Medical Treasures

Estimated time: 15 minutes

As discussed in the chapter and in the Case Study: Hunting for Medical Treasures, plants are believed to hold a potential "treasure trove" of treatments and cures for various human diseases. Plants have been used as "herbal remedies" for centuries. Many of these cures remain locked away within the cells of the plant structure simply awaiting their discovery. Despite the promise of finding medically useful compounds in plants, thousands of species are being driven to extinction each year due to the loss of habitat. This activity will further investigate the medical potential of the Plant Kingdom.

Go to http://www.prenhall.com/audesirk3, the Audesirk Companion Web site. Select Chapter 17 and Web Investigation to begin.

The indescribably delicious taste of a fresh, ripe tomato is a sensation rarely experienced by those who buy their produce in a supermarket. New research on plant hormones may help change that sad situation.

18 The Plant Life Cycle

AT A GLANCE

CASE STUDY

Tasty Tomatoes

A great new product is coming soon to a supermarket near you: a tomato that never ripens. You are perhaps under the impression that your supermarket already has plenty of unripe tomatoes. Those, however, are nothing more than normal tomatoes that were picked green and hard to avoid damage during shipping. A tomato's condition on arrival at your supermarket is rather unpredictable; plant physiologists and produce companies have not had much success in controlling post-harvest ripening of tomatoes. So not only might your supermarket tomato turn out to be either too green and hard or too soft and squishy, it almost certainly won't taste as good as it would have if it had ripened normally on the vine. Tomato growers and sellers would love to improve the quality of supermarket tomatoes, and that's where the never-ripe tomato comes in.

Tomato plants that carry the mutation known as *never-ripe* bear fruit in which ripening—the process that leads to softening of the fruit, conversion of starch to sugar, loss of chlorophyll, and synthesis of red pigments and aromatic compounds—proceeds extremely slowly. Never-ripe tomatoes can be left on the vine much longer and can be shipped and stored without refrigeration. These qualities by themselves, however, will not solve the problems of the tomato industry. Consumers obviously have no use for a permanently unripe tomato. Researchers are nonetheless optimistic that, once they fully understand the mechanism by which mutations such as *never-ripe* have their effect on ripening, it will be possible to exert fine control over the ripening process. For example, it may become possible to turn genes that control ripening on or off at will (expression of particular genes can often be controlled with regulatory chemicals). In this scenario, tomatoes could remain on the vine until they were red and tasty, at which point ripening could be halted so that no further softening occurred during harvest and shipping. At home, you could leave your tomatoes on the windowsill for weeks without fear of spoilage.

Sophisticated control of ripening is no longer an idle dream, because scientists have already made significant progress in understanding how *never-ripe* and similar mutations work. We will describe some of their findings later in this chapter, after some exploration of the plant life cycle. ■

1 What Is the Life Cycle of Plants?

Many plants can reproduce either sexually or asexually. In asexual reproduction, part of a single plant (say, a stem) gives rise to a new plant. The offspring plant is produced by mitotic division of the parent plant's cells, so the offspring is genetically identical to the parent. Asexual reproduction is efficient, allowing plants to quickly colonize an area. Nonetheless, most plants reproduce sexually at least some of the time.

Two Types of Plant Bodies Alternate in the Sexual Life Cycle

The life cycle of a sexually reproducing plant includes two distinct types of bodies, the diploid sporophyte and the haploid gametophyte. The sporophyte and gametophyte forms give rise to each other, so the plant life cycle is called **alternation of generations.** As the cycle proceeds, diploid plants (the sporophyte generation) alternate with haploid plants (the gametophyte generation).

Alternation of generations is most easily observed in plants, such as mosses and ferns, that do not produce flowers. For example, consider the life cycle of a fern (Fig. 18-1). The adult **sporophyte** bears reproductive cells. These cells undergo meiosis to produce haploid cells that are **spores**, not gametes. The difference between spores and gametes is that spores do not fuse together to form a diploid cell. Instead, wind blows a fern spore off the parent leaf, and the spore lands on the soil. There the spore begins to grow and develop. It divides repeatedly by mitosis to form a multicellular, haploid organism. Because this organism produces gametes, it is called the **gametophyte**. The gametophyte's cells are haploid and it produces sperm and eggs without further meiosis. Sperm and egg fuse to form a zygote that develops into a new diploid sporophyte plant.

Alternation of Generations Is Less Obvious in Flowering Plants

In primitive land plants, including mosses and ferns, the gametophytes are independent, although generally small, plants that produce both sperm and eggs. They liberate sperm that, to reach an egg, either swim through thin films of water or are splashed by raindrops from one plant to the next. This need for water limits fern and moss reproduction to moist habitats.

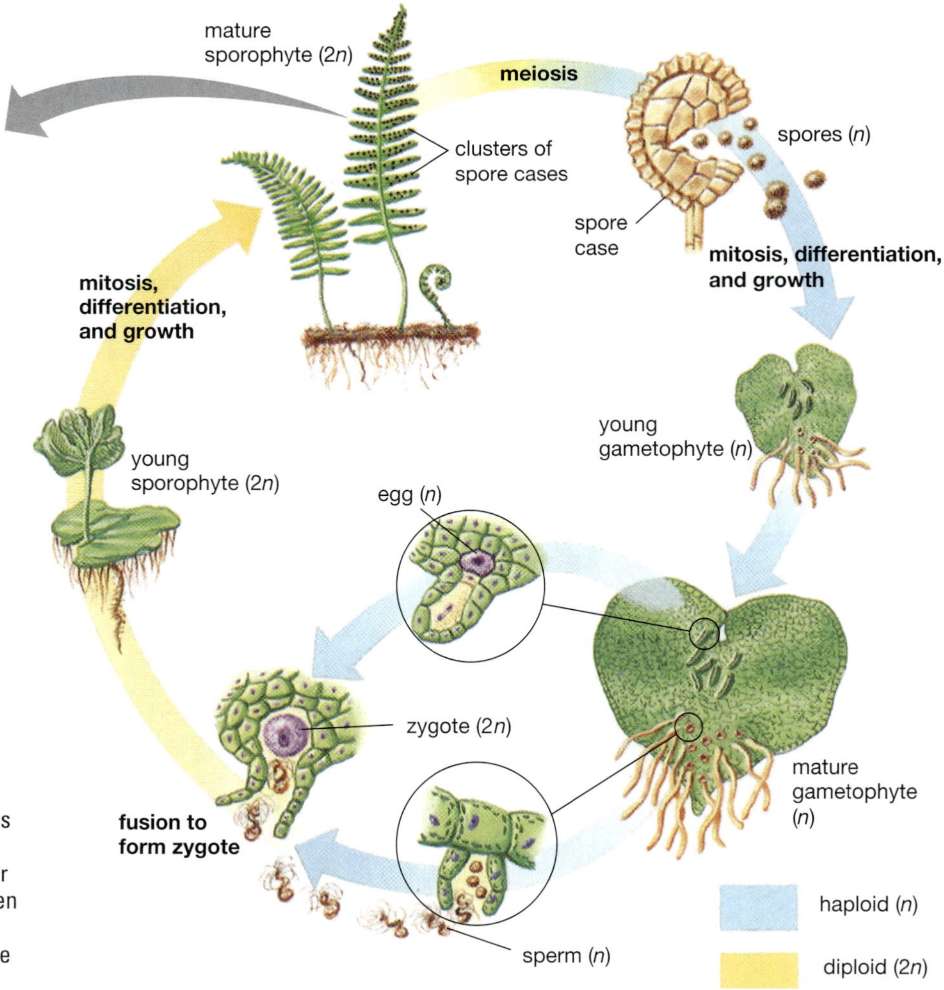

Figure 18-1 The life cycle of a fern—a nonflowering plant
Ferns typify the alternation-of-generations life cycle found in all plants. One generation is the haploid gametophyte. The other generation is the diploid sporophyte. When you see ferns, look on the undersides of their leaves for clusters of brownish spore cases (see photo).

MediaTutor 18.1 Fern Life Cycle

In the flowering plants, gametophytes are microscopic and do not live independently of the sporophyte. Instead, the gametophytes develop within **flowers**, the reproductive structures of flowering plants. The gametophytes grow from two types of spores that are formed by meiosis within the flowers borne by the sporophyte generation (Fig. 18-2). One type of spore, the megaspore, undergoes a few mitotic divisions and develops into the female gametophyte, a small cluster of cells called an **embryo sac**, that is permanently retained within the flower and includes an egg cell. The other type of spore, the microspore, develops into the male gametophyte: a tough, watertight **pollen grain** that enables flowering plants to reproduce in dry environments. The pollen grain, which contains the sperm, drifts on the wind or is carried by an animal from one flower to another. On the recipient flower, the pollen grain elongates, burrowing through the flower's tissues to the female gametophyte within. This miniature male gametophyte liberates its sperm inside the female gametophyte, where fertilization takes place. The zygote develops into an embryonic plant that becomes enclosed in a drought-resistant **seed**, which also includes a food reserve and a protective outer coating.

The Big Picture

Alternation of generations is a mode of reproduction that is unique to plants and distinguishes them from the members of all other kingdoms.

2 What Is the Structure of Flowers?

The flower is actually a sexual display that enhances a plant's reproductive success. By enticing animals to transfer pollen from one plant to another, flowers enable stationary plants to "court" distant members of their own species. This critical advantage has allowed the flowering plants to become the dominant plants on land.

Flowers Evolved from Leaves

Evolution produces new structures by modifying old ones, and flower parts are actually highly modified leaves, shaped by natural selection into structures that foster the transfer of pollen from one plant to another. **Complete flowers**, such as those of petunias, roses, and lilies, consist of a central axis on which four successive

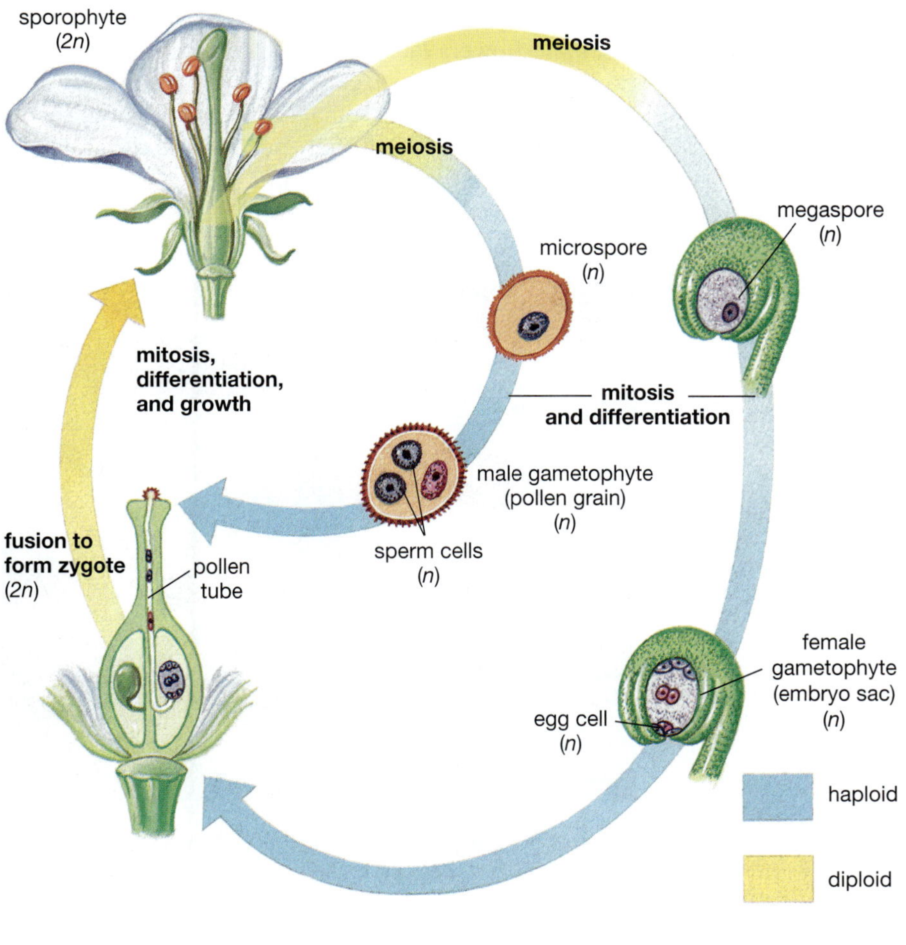

Figure 18-2 The life cycle of a flowering plant
Although this cycle shows the same basic stages as the life cycle of a fern (see Fig. 18-1), the haploid stages are much smaller and cannot live independently of the diploid plant.

MEDIATUTOR 18.2 Flower Structure and Function

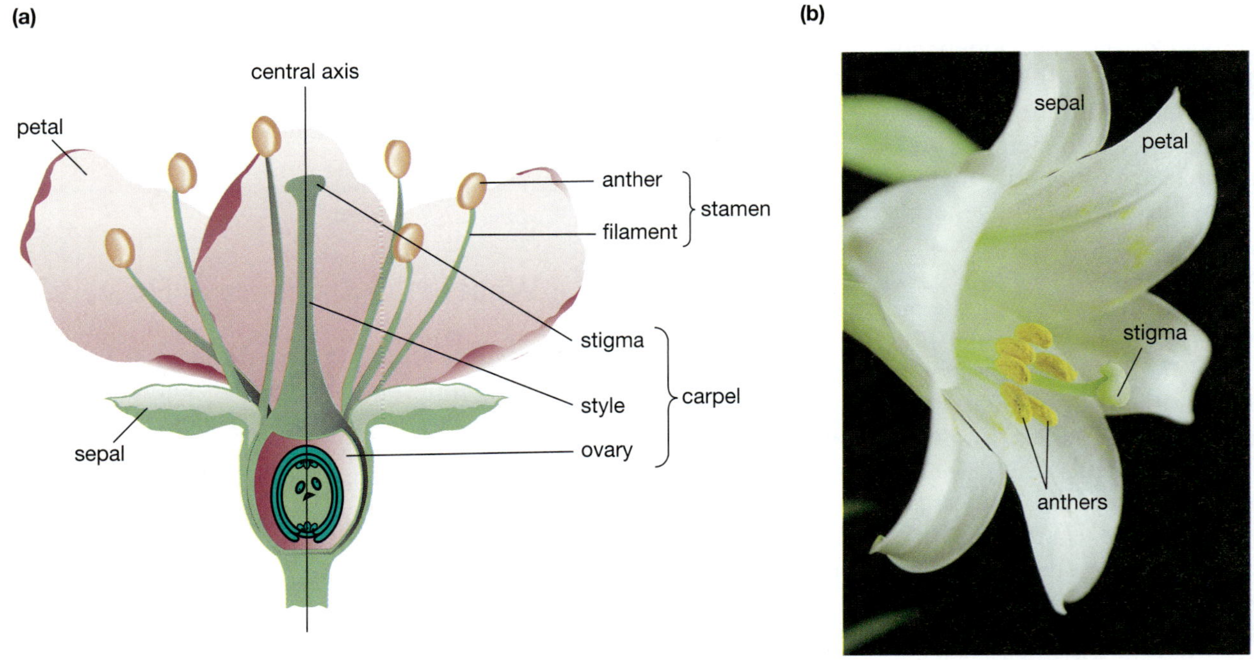

Figure 18-3 A complete flower
(a) A complete flower has four parts: sepals, petals, stamens (the male reproductive structures), and at least one carpel (the female reproductive structure). This drawing shows a complete dicot flower. *(b)* The lily is a complete monocot flower, with three sepals (virtually identical to the petals), three petals, six stamens, and three fused carpels.

sets of modified leaves are attached (Fig. 18-3). These modified leaves are the sepals, petals, stamens, and carpels. The **sepals** are located at the base of the flower. In dicots, the sepals are typically green and leaflike; in monocots, most sepals resemble the petals (see Fig. 18-3b). In either case, sepals enclose and protect the flower bud as the other three structures develop. Just above the sepals are the **petals**, which are usually brightly colored and fragrant, advertising the location of the flower.

Flowers Incorporate Male and Female Reproductive Structures

The male reproductive structures, the **stamens**, are attached just above the petals. Most stamens consist of a long and slender **filament** that supports an **anther**, the structure that produces pollen. The female reproductive structures, the **carpels**, occupy the uppermost position in the flower. An idealized carpel is somewhat vase-shaped, with a sticky **stigma** for catching pollen mounted atop an elongated **style**. The style connects the stigma with the bulbous **ovary**. Inside the ovary are one or more **ovules**, in which the female gametophytes develop. When mature, each ovule will become a seed, and the ovary will develop into a protective, adhesive, or edible enclosure, the **fruit**.

As you may know from your own gardening experience, not all flowers are complete. **Incomplete flowers** lack one or more of the four floral parts. For example, incomplete flowers may lack sepals or petals. In addition, many plants have separate male and female flowers; male flowers lack carpels, and female flowers lack stamens. In some species, such as cucumbers and squashes, the male and female flowers are borne on the same plant (Fig. 18-4). Other species have the male and female flowers on different plants.

Figure 18-4 Some plants have separate male and female flowers
Plants of the squash family, such as these zucchinis, bear separate female (left) and male (right) flowers.

3 What Are the Gametophytes of Flowering Plants?

The familiar flowering plant seen in meadows, gardens, and farms is the diploid sporophyte, as Figure 18-2 illustrates. The pollen grain (male) and the embryo sac (female) are the haploid gametophytes; they develop within the flowers of the sporophyte.

The Pollen Grain Is the Male Gametophyte

Each anther produces thousands of haploid **microspores**, each of which divides once, by mitosis, to produce a pollen grain. As the pollen grain matures, it produces two haploid sperm cells. A tough surface coat develops around the pollen grain and protects it during its journey to the carpel (Fig. 18-5). When the pollen grains have matured, they are released from the anther. In wind-pollinated flowers, such as those of grasses and oaks, the pollen grains spill out and are carried widely by wind currents; a few of those grains reach and pollinate other flowers of the same species. In animal-pollinated flowers, the pollen grains adhere weakly to the anther case until a pollinator comes along and brushes or picks them off.

The Embryo Sac Is the Female Gametophyte

Within the ovules in the ovary, large haploid **megaspores** undergo an unusual set of mitotic divisions. Three nuclear divisions produce a total of eight haploid nuclei. Plasma membranes then divide the cytoplasm into seven, not eight, cells: three small cells at each end, each containing one nucleus, and one remaining large cell in the middle containing two nuclei, called **polar nuclei**. This seven-celled organism, called the embryo sac (because it is where the embryo will develop), is the haploid female gametophyte. The **egg** is the central small cell at the bottom of the embryo sac, located near an opening in the ovule.

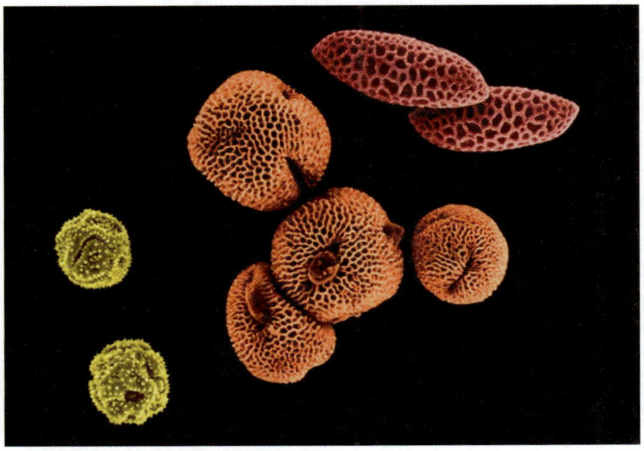

Figure 18-5 Pollen grains
The tough outer coverings of many pollen grains are elaborately sculptured in species-specific shapes and patterns. The pollen grains in this color-enhanced SEM photo are from a geranium (orange), a tiger lily (fuchsia), and a dandelion (yellow).

4 How Does Pollination Lead to Fertilization?

When a pollen grain lands on the stigma of a flower of the same species, a remarkable chain of events begins (Fig. 18-6). A cell of the pollen grain elongates to form a tube, growing down the style toward an ovule in the ovary. If all goes well, the pollen tube reaches the pore in the ovule and breaks into the embryo sac. The tube's tip ruptures, releasing the two sperm. One sperm fertilizes the egg, forming the diploid zygote that will develop into the embryo and eventually into a new sporophyte. The second sperm enters the large central cell and its nucleus fuses with both polar nuclei, forming a triploid nucleus (having three sets of chromosomes). Through repeated mitotic divisions, this cell will develop into the triploid ($3n$) **endosperm**, a food-storage tissue within the seed. The fusion of the egg with one sperm and the fusion of the polar nuclei with the second sperm are called **double fertilization**, a process unique to flowering plants. The other five cells of the embryo sac degenerate soon after fertilization.

The distinction between pollination and fertilization is important. **Pollination** occurs when a pollen grain lands on a stigma; **fertilization** is the fusion of sperm and egg. Although pollination is necessary for fertilization, these are two separate events. For example, pollination will not lead to fertilization if the pollen tube fails to grow properly, if the embryo sac has no egg, or if a sperm from another pollen grain has already reached the egg.

5 How Do Seeds and Fruits Develop?

After fertilization, certain parts of flowers develop into seeds, which protect and nourish the developing embryos, and fruits, which help the seeds disperse to new locations distant from the parent plant.

The Seed Develops from the Ovule and Embryo Sac

Drawing on the resources of the parent plant, the embryo sac and the surrounding cells of the ovule develop into a seed. The outer layers of the ovule develop into the **seed coat**, the outer covering of the seed. Meanwhile, within the seed coat, development proceeds: First, the triploid endosperm cell divides rapidly. Its daughter cells absorb nutrients from the parent plant, forming a large,

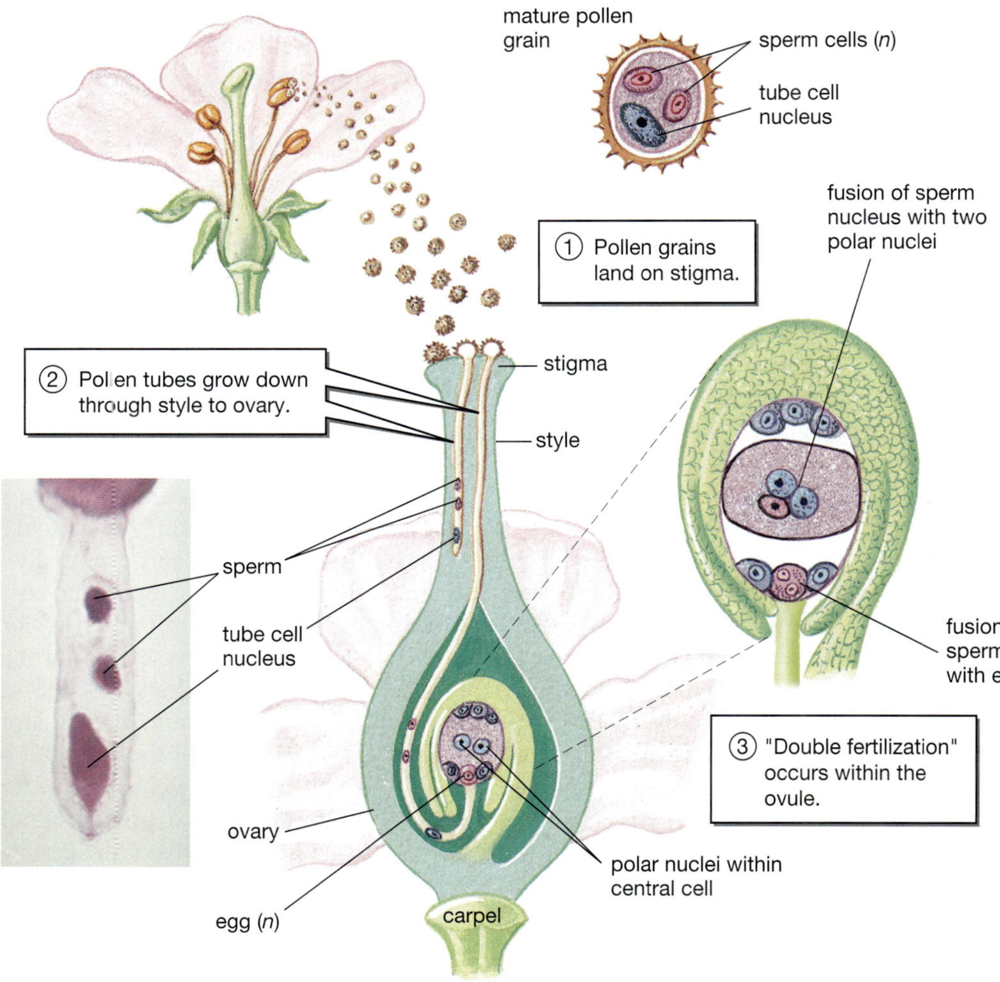

Figure 18-6 Pollination and fertilization of a flower

food-filled endosperm. Second, the zygote develops into the embryo (Fig. 18-7a). Both dicot and monocot embryos consist of three parts: the shoot, the root, and the **cotyledons**, or seed leaves (Fig. 18-7b). The cotyledons absorb food molecules from the endosperm and transfer them to other parts of the embryo.

The Fruit Develops from the Ovary Wall

The seed is surrounded by the ovary, which develops into a fruit (Fig. 18-8). As the fruit enlarges, stamens and petals shrivel and fall away, having already served their functions of producing pollen and attracting pollinators. Fruits assume a huge variety of shapes and sizes, with outer layers that are variously fleshy, hard, winged, or even spiked like a medieval mace. The ways in which they assist seed dispersal are also varied (see "Evolutionary Connections: Adaptations for Pollination and Seed Dispersal").

The Big Picture

The human attraction to flowers and fruits is shared by many nonhuman members of the animal kingdom. In fact, animal attraction to these plant structures is essential to the continued survival of flowering plants. Reproduction and dispersal of flowering plants depends heavily on their intimate partnership with animals.

6 How Do Seeds Germinate and Grow?

The stage at which a seed or spore begins to grow and develop—when it sprouts—is known as **germination**. Many seeds do not germinate immediately after they mature, but instead enter a period of **dormancy**, during which they will not germinate. Dormancy may last for months or years, as the seed awaits conditions favorable for growth.

Seed Dormancy Helps Ensure Germination at an Appropriate Time

Seed dormancy avoids two problems. First, if a seed germinated while still enclosed in a fruit and hanging from a tree or vine, it might exhaust its food reserves before it ever touched the ground. Second, environmental conditions that are suitable for seedling growth (such as adequate moisture and temperatures) may not coincide

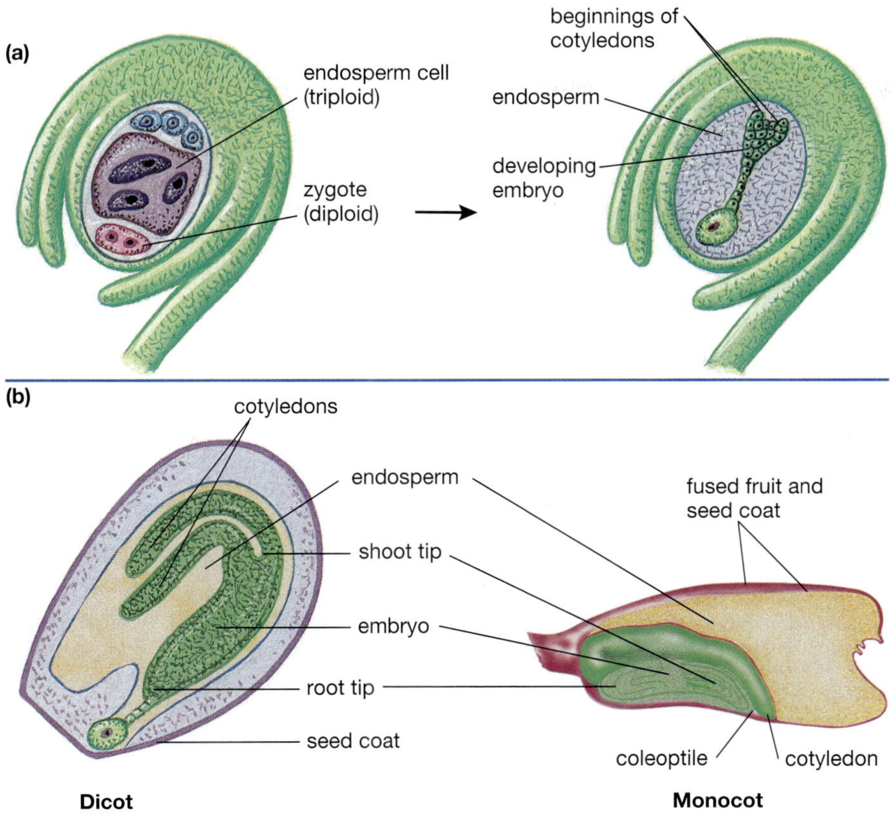

Figure 18-7 Seed development
(a) The endosperm develops first, absorbing nutrients from the parent plant. The embryo develops later, absorbing nutrients from the endosperm to fuel its growth. ***(b)*** Monocot and dicot seeds differ in the number of cotyledons and the fate of the endosperm. *(left)* A dicot seed, such as the shepherd's purse, has two cotyledons, which usually absorb most of the endosperm as the seed develops; hence, the mature seed is mostly cotyledons. *(right)* A monocot seed, such as the corn kernel, retains a large endosperm. (Cornmeal is the ground-up endosperm of corn seeds.) The embryo produces a single cotyledon. As the seed germinates, the cotyledon absorbs the food reserves of the endosperm and transfers them to the growing embryo.

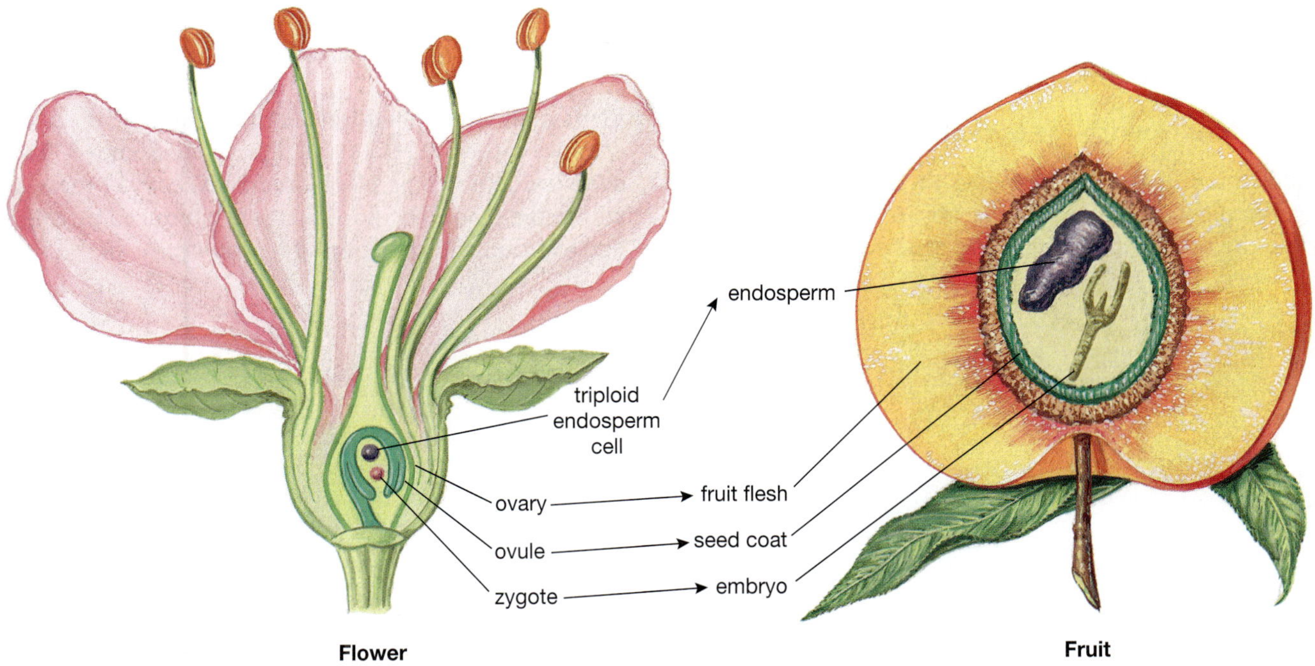

MEDIATUTOR
18.3 Fruit and Seed Structure and Development

Figure 18-8 Development of fruit and seed from flower parts
The fruit and seed coat are derived from the parent sporophyte plant. The ovary wall ripens into the fruit flesh, and the wall of the ovule forms the seed coat. Inside the developing seed, the triploid endosperm cell becomes the endosperm.

with seed maturation. Seeds that mature in the late summer in temperate climates, for example, face the harsh winter to come. Spending the winter as a dormant seed is clearly preferable to death by freezing as a tender young sprout. In the warm, moist Tropics, seed dormancy is much less common than in temperate regions because environmental conditions are suitable for germination throughout the year.

The Root Emerges First

During germination, the embryo breaks dormancy and emerges from the seed. The embryo absorbs water, which makes it swell and burst its seed coat. The root is usually the first structure to emerge from the seed coat, growing rapidly and absorbing water and minerals from the soil. Much of the water is transported to cells in the shoot. As its cells elongate, the stem lengthens, pushing up through the soil.

The Shoot Tip Must Be Protected

The growing shoot faces a serious difficulty: It must push through the soil without scraping away the apical meristem and tender leaflets at its tip. Unlike roots, which must always contend with tip abrasion and whose apical meristem is protected by a root cap (see Fig. 17-8), shoots spend most of their time in the air and do not develop permanent protective caps. Instead, germinating shoots have other mechanisms that cope with the abrasion of sprouting. In monocots, the **coleoptile**, a tough sheath, encloses the shoot tip like a glove (Fig. 18-9). The coleoptile "glove" pushes aside the soil particles as it grows. Once out in the air, the coleoptile tip degenerates, allowing the tender shoot "finger" to emerge. Dicots do not have coleoptiles. Instead, the dicot shoot forms a hook. The bend of the hook, encased in epidermal cells with tough cell walls, leads the way through the soil, clearing the path for the downward-pointing apical meristem with its delicate new leaves.

Cotyledons Nourish the Sprouting Seed

Food stored in the seed provides the energy for sprouting. In dicots, the cotyledons absorb the endosperm while the seed is developing and are swollen and full of food by the time the seed germinates. In many dicots, the elongating shoot carries the cotyledons out of the soil into the air, where they become green and photosynthetic and transfer both previously stored food and newly synthesized sugars to the shoot (Fig. 18-10). In other dicots, the cotyledons stay below the ground, shriveling up as the embryo absorbs their stored food. Monocots retain most of their food reserve in the endosperm until germination, when it is digested and absorbed by the cotyledon as the embryo grows. The cotyledon remains below ground.

Throughout the Life Cycle, Development Is Regulated

Once out in the air, the shoot rapidly spreads its leaves to the sun. Simultaneously, the root system delves into the soil. The apical meristem cells of shoot and root divide, giving rise to the mature structures discussed in Chapter 17. Eventually this plant, too, will mature, flower, and set seed, renewing the cycle of life. Regulation of this cycle—ensuring that shoots grow upward while roots grow downward, that plants produce flowers at the proper time of year, and that fruits ripen—is accomplished by plant hormones.

7 What Are Plant Hormones, and How Do They Act?

Plant cells are miniature factories filled with diverse chemicals that regulate the life cycle. Some of these chemicals, known as **plant hormones**, convey messages within the plant. The hormones are produced by cells in one location and transported to other parts of the body,

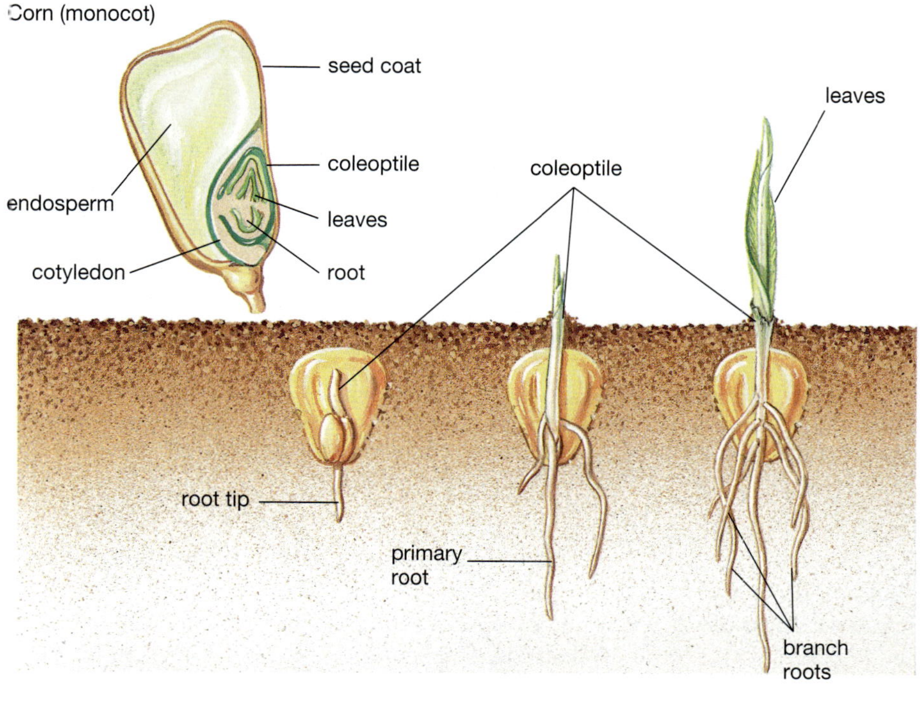

Figure 18-9 Seed germination
The delicate shoot tip of a monocot is protected within a tough coleoptile, which pushes up through the soil when the seed sprouts.

Figure 18-10 Cotyledons nourish the developing plant
In the squash family, the cotyledons expand into photosynthetic leaves (the pair of smooth oval leaves). The first true leaf (crinkled single leaf) develops a little later. Eventually, the cotyledons shrivel up and die.

where they exert specific effects. Each hormone can elicit a variety of responses from plant cells, depending on factors such as a plant's species, the type of target cell, the developmental stage of the plant, the concentration of the hormone, and the presence of other hormones.

Plant physiologists have identified five major classes of plant hormones: (1) auxins, (2) gibberellins, (3) cytokinins, (4) ethylene, and (5) abscisic acid (Table 18-1).

Auxins promote the elongation of cells in coleoptiles and other parts of the shoot; high concentrations cause cells to elongate (see "Scientific Inquiry: How Were Plant Hormones Discovered?"). In roots, which differ from stems in their response to auxin, low concentrations stimulate elongation, while slightly higher concentrations inhibit elongation. Auxin also affects many other aspects of plant development, such as the differentiation of conducting tissues (xylem and phloem) and the development of fruits. Auxin may also prevent sprouting by lateral buds. It stimulates root branching and can be used to cause stems of plants to grow roots, which is useful when growing a new plant from a cutting.

Gibberellins, like auxin, promote the elongation of cells in stems. In some plants, gibberellins stimulate flowering, fruit development, seed germination, and bud sprouting.

Cytokinins promote cell division in many plant tissues; consequently, they stimulate the sprouting of buds and the development of fruit, endosperm, and the embryo. Cytokinins also stimulate plant metabolism, preventing or at least delaying the aging of plant parts, especially leaves.

Ethylene is an unusual plant hormone in that it is a gas at normal environmental temperatures. Ethylene is best known, and most commercially valuable, for its ability to cause fruit to ripen. It also regulates the timing of leaf, flower, and fruit drop.

Abscisic acid is a hormone that helps plants withstand unfavorable environmental conditions. For example, it inhibits the activity of gibberellins, thus helping maintain dormancy in buds and seeds at times when germination would be dangerous.

In the rest of this chapter, we describe a year in the life of a plant, illustrating how hormones regulate its growth and development.

Table 18-1 Hormone Actions in Plants

Hormone	Functions
Abscisic acid	Maintenance of seed dormancy and bud dormancy
Auxins	Elongation of cells in coleoptiles and shoots; phototropism; gravitropism in shoots and roots; root growth and branching; development of vascular tissue; fruit development; retarding senescence in leaves and fruit; ethylene production in fruit
Cytokinins	Promotion of sprouting of lateral buds; prevention of leaf senescence; promotion of cell division; stimulation of fruit, endosperm, and embryo development
Ethylene	Ripening of fruit; abscission of fruits, flowers, and leaves
Gibberellins	Germination of seeds and sprouting of buds; elongation of stems; stimulation of flowering; development of fruit

The Big Picture

The dramatic changes that mark the life of a plant are controlled by chemical messages that move from one part of the plant body to another, often in response to changes in the surrounding environment.

Scientific Inquiry
How Were Plant Hormones Discovered?

Everyone who keeps houseplants on a windowsill knows that the plants bend toward the window as they grow, in response to the sunlight streaming in. More than a hundred years ago, Charles Darwin and his son Francis studied this phenomenon of growth toward light, or phototropism.

First, the Darwins Determined the Direction of Information Transfer

The Darwins illuminated grass coleoptiles (protective sheaths surrounding monocot seedlings) from various angles. They noted that a region of the coleoptile a few millimeters below the tip bent toward the light, causing the tip to point toward the light source.

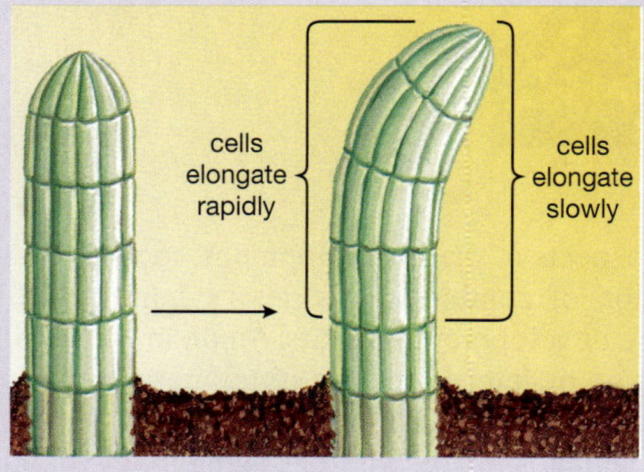

When they covered the tip with an opaque cap, the coleoptile didn't bend.

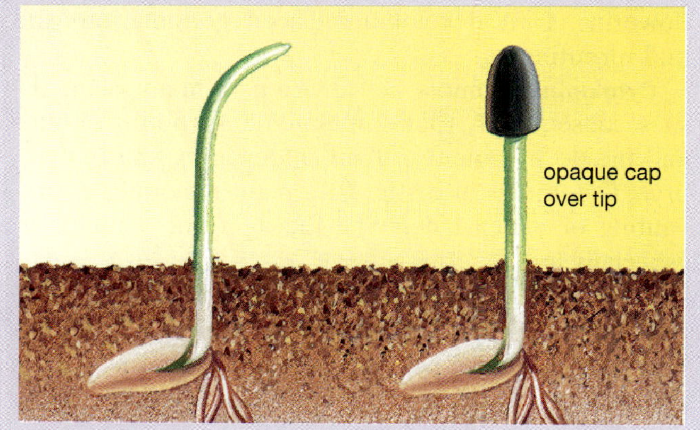

A clear cap allowed the stem to bend, as did covering the bending region below the tip with an opaque sleeve.

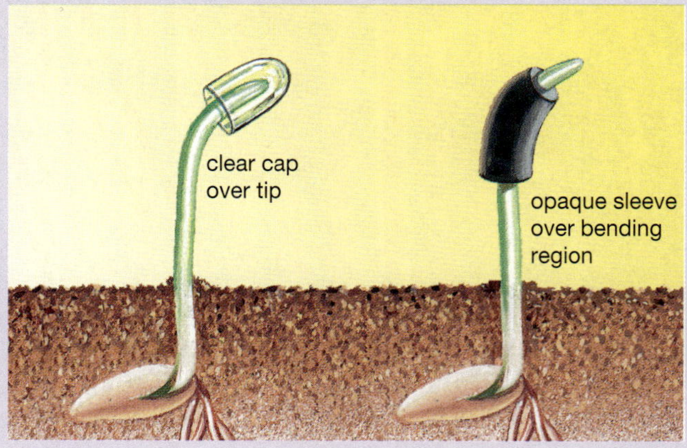

The Darwins concluded that (1) the tip of the coleoptile detects the direction of light and (2) the bending occurs farther down the coleoptile; therefore, (3) the tip transmits information about the light direction down to the bending region. How does the coleoptile bend? Although the Darwins didn't know this, the coleoptile bends as a result of unequal elongation of cells on opposite sides of the shaft. The cells on the darker side elongate faster than those facing the light, bending the shaft toward the light. Information transmitted from the tip to the bending region causes unequal cell elongation.

Next, Peter Boysen-Jensen Demonstrated That the Information Is Transferred by Chemicals

About 30 years after the Darwins' experiments, Peter Boysen-Jensen cut the tips off coleoptiles and found that the remaining stump neither elongated nor bent toward the light. If he replaced the tip and placed the patched-together coleoptile in the dark, it elongated straight up. In the light, it showed normal phototropism. When he inserted a thin layer of porous gelatin that prevented direct contact but permitted diffusion of substances between the severed tip and the stump, he still observed elongation and bending. In contrast, an impenetrable barrier eliminated these responses.

8 How Do Hormones Regulate the Plant Life Cycle?

At each stage in its life cycle, a plant produces a distinctive set of hormones that interact with one another in directing the growth of the plant body.

Abscisic Acid Maintains Seed Dormancy

A seed maturing within a juicy fruit on a warm autumn day has ideal conditions for germination, yet it remains dormant until the following spring. In many seeds, this dormancy is enforced by abscisic acid. Abscisic acid slows down the metabolism of the embryo within the seed, preventing its growth. The seeds of some desert plants contain high concentrations of abscisic acid; only a very hard rain can wash out the abscisic acid, freeing the embryo from the hormone's inhibitory effects and allowing the seed to germinate. Seeds of most high-latitude plants require a prolonged period of cold, like that of winter, to break dormancy; in these seeds, chilling induces the destruction of abscisic acid.

Gibberellin Stimulates Germination

Germination is stimulated by other hormones, especially gibberellin. The same environmental conditions that cause the breakdown of abscisic acid also promote the

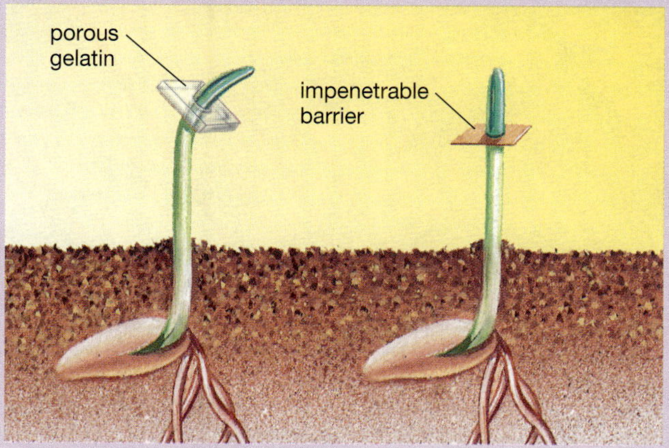

Boysen-Jensen concluded that a chemical is produced in the tip and moves down the shaft, causing cell elongation. In the dark, the chemical that causes the cells to elongate diffuses straight down from the tip and causes the coleoptile to elongate straight up. Presumably, light causes the chemical to become more concentrated on the "shady" far side of the shaft, so cells on the shady side elongate faster than do cells on the "sunny" near side, causing the shaft to bend toward the light.

Finally, the Chemical Auxin Was Identified

The next step was to isolate and identify the chemical. In the 1920s Frits Went devised a way to collect the elongation-promoting chemical. He cut off the tips of oat coleoptiles and placed them on a block of agar (a porous, gelatinous material) for a few hours. Went hypothesized that the chemical would migrate out of the coleoptiles into the agar.

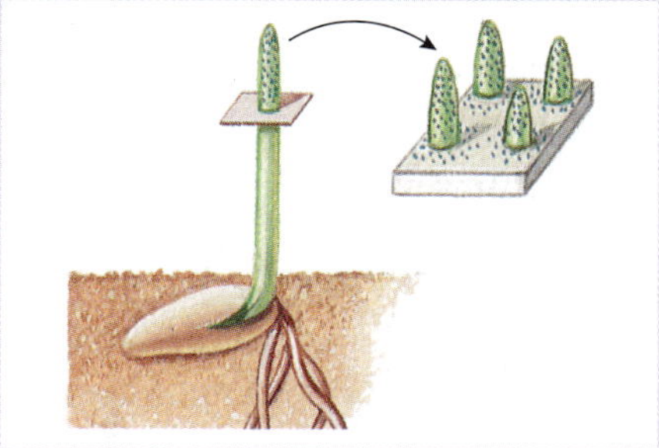

He then cut up the agar, now presumably loaded with the chemical, and placed small pieces on the tops of coleoptile stumps growing in darkness. When he put a piece of agar squarely atop a stump, the stump elongated straight up. All the stump cells received equal amounts of the chemical and elongated at the same rate. If he placed a piece on one side of a cut stump, the stump would invariably bend away from the side with the agar.

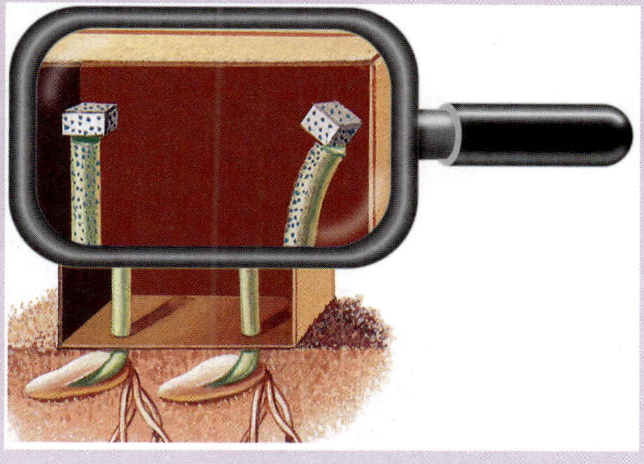

It was apparent that cells on the side under the agar received more of the chemical and were more stimulated to elongate. Went called the chemical "auxin," from a Greek word meaning "to increase." Kenneth Thimann later purified auxin and determined its molecular structure.

synthesis of gibberellin. In germinating seeds, gibberellin initiates the synthesis of enzymes that digest the food reserves of the endosperm and cotyledons, making sugars, lipids, and amino acids available to the growing embryo.

Auxin Controls the Orientation of the Sprouting Seedling

When the growing embryo breaks out of the seed coat, it immediately faces a crucial problem: Which way is up? The roots must burrow downward, while the shoot must grow upward to find the light. Auxin controls the responses of both roots and shoots to light and gravity.

Auxin Stimulates Shoot Elongation Away from Gravity and Toward Light

Let's look at the growth of a shoot as it first emerges from the seed, buried underground. Auxin is synthesized in shoot tips, moves down the shaft of the stem, and stimulates cell elongation. If the stem is not exactly vertical, organelles in the cells of the stem detect the direction of gravity and cause auxin to accumulate on the stem's lower side (Fig. 18-11a). Therefore, the lower cells elongate rapidly, forcing the stem to bend upward. When the shoot tip is vertical, the auxin distribution becomes symmetrical. The stem then grows straight up, emerging from the soil into the light (Fig. 18-11c). This is an example of **gravitropism**—directional growth with respect to gravity.

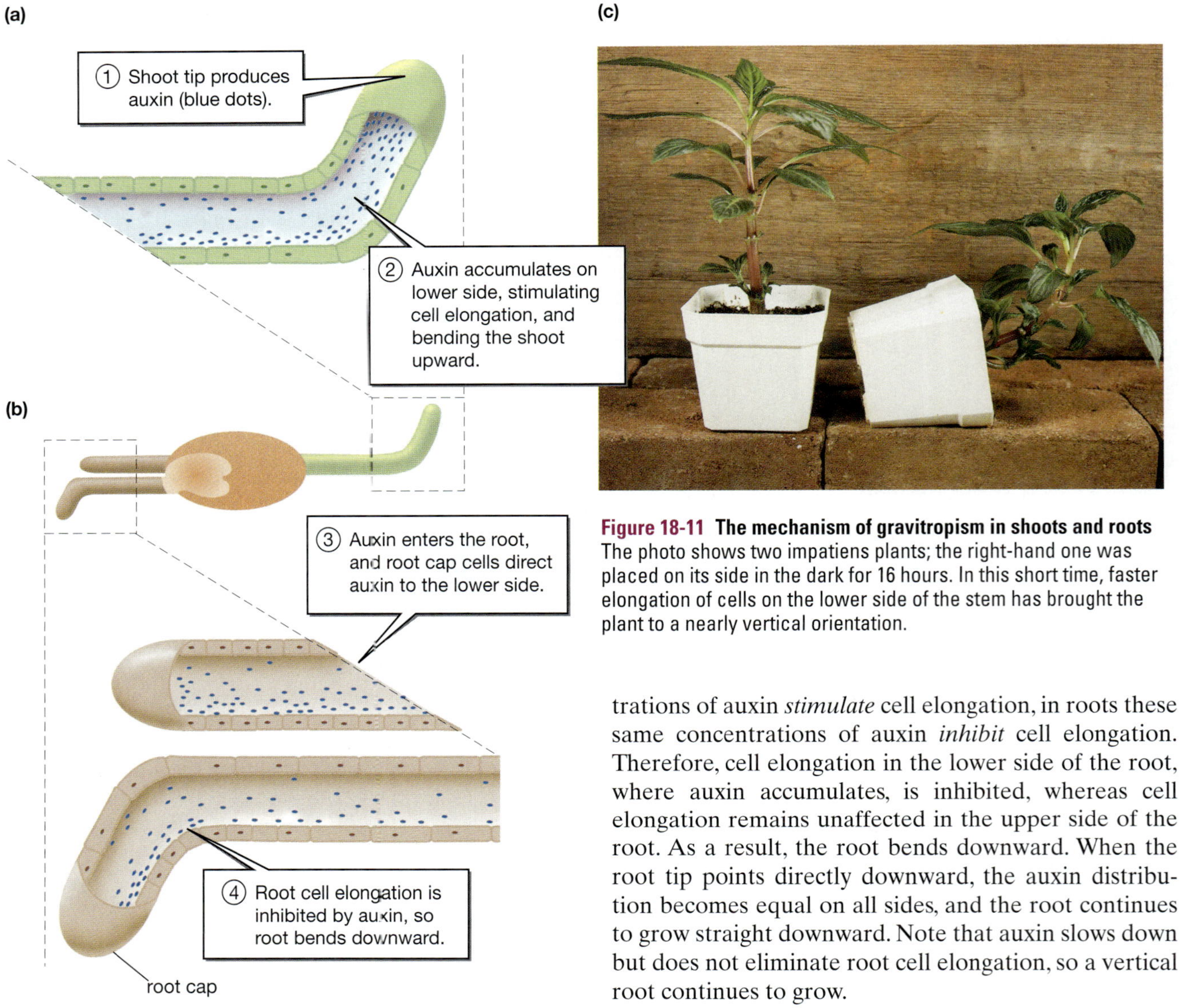

Figure 18-11 The mechanism of gravitropism in shoots and roots The photo shows two impatiens plants; the right-hand one was placed on its side in the dark for 16 hours. In this short time, faster elongation of cells on the lower side of the stem has brought the plant to a nearly vertical orientation.

Auxin also influences a growing shoot's **phototropism** (tendency to grow toward light). Ordinarily, the distribution of auxin caused by light is the same as the distribution caused by gravity, because the direction of brightest light (the sun) is roughly opposite that of gravity. For example, if a young shoot still buried underground is close enough to the surface that some light penetrates down to it, both light and gravity cause auxin to be transported to the lower side of the shoot and promote upward bending. Thus, under normal conditions gravitropism and phototropism add to each other's effects.

Auxin May Control the Direction of Root Growth

Gravitropism in roots is less well understood than it is in stems. According to one hypothesis, auxin controls the direction of root growth (Fig. 18-11b). Auxin is transported from the shoot down to the root. If the root is not vertical, the root cap senses the direction of gravity and causes the auxin to accumulate on the lower side of the root. Unlike shoots, in which moderate concentrations of auxin *stimulate* cell elongation, in roots these same concentrations of auxin *inhibit* cell elongation. Therefore, cell elongation in the lower side of the root, where auxin accumulates, is inhibited, whereas cell elongation remains unaffected in the upper side of the root. As a result, the root bends downward. When the root tip points directly downward, the auxin distribution becomes equal on all sides, and the root continues to grow straight downward. Note that auxin slows down but does not eliminate root cell elongation, so a vertical root continues to grow.

Stem Branching Is Influenced by Auxin and Cytokinin

Auxin by itself appears to inhibit the sprouting of lateral buds, whereas auxin and cytokinin together stimulate bud sprouting. Auxin is produced by the shoot tip and transported down the stem, gradually decreasing in concentration. Cytokinin is produced by the roots and is transported up the stem. Therefore, the relative concentrations of these two hormones will vary along the length of the stem, and buds at different positions will experience differing hormonal influences. The lateral buds closest to the shoot tip receive a great deal of auxin, probably enough to inhibit their growth, but receive very little cytokinin because they are so far from the roots. Therefore, they remain dormant. Lower buds receive less auxin while receiving much more cytokinin. They are stimulated by optimal concentrations of both hormones, so they sprout (Fig. 18-12). In many types of plants, this interaction between auxin and cytokinin produces an orderly progression of bud sprouting from the bottom to the top of the shoot.

Gradients of Auxin and Cytokinin Create a Balance Between the Root and Shoot Systems

Root and shoot systems control each other's growth through the interaction of auxin and cytokinin. This mutual control is important, because the two systems supply complementary nutrients. As a root system grows, it synthesizes a large amount of cytokinin, which stimulates lateral buds on the stem to sprout. But if the root system is growing too slowly to keep up with the growth of the shoot system, the roots produce less cytokinin, and the growth of the shoot system is slowed. At the same time, the shoot system produces lots of auxin as it grows, which stimulates root branching and growth. Thus, neither root nor shoot can outgrow the other, and the plant is adequately supplied with all its needs.

Daylength Controls Flowering

Eventually, the plant matures enough to reproduce. The timing of flowering and seed production is finely tuned to the physiology of the plant and the rigors of its environment. In temperate climates, plants must flower early enough so that their seeds can mature before the killing frosts of autumn. Depending on how quickly the seed and fruit develop, a plant may flower in spring, as do oaks; in summer, as does lettuce; or even in autumn, as do asters.

What environmental cues do plants use to determine the season? Most cues, such as temperature or water availability, are quite variable: October can be warm, a late snow may fall in May, or the summer might be unusually cool and wet. The only reliable cue is daylength. Increasing daylength means that spring and summer are coming; decreasing daylength foretells the onset of autumn and winter.

Different Species Respond Differently to Changing Daylength

With respect to flowering, plants are classified as day-neutral, long-day, or short-day (Fig. 18-13). A day-neutral plant flowers as soon as it has sufficiently grown and developed, regardless of the length of the day. Day-neutral plants include tomatoes, corn, and snapdragons. A long-day plant flowers when the daylength is longer than some species-specific critical value. A short-day plant flowers when the daylength is shorter than some species-specific critical value. Thus, spinach is classified

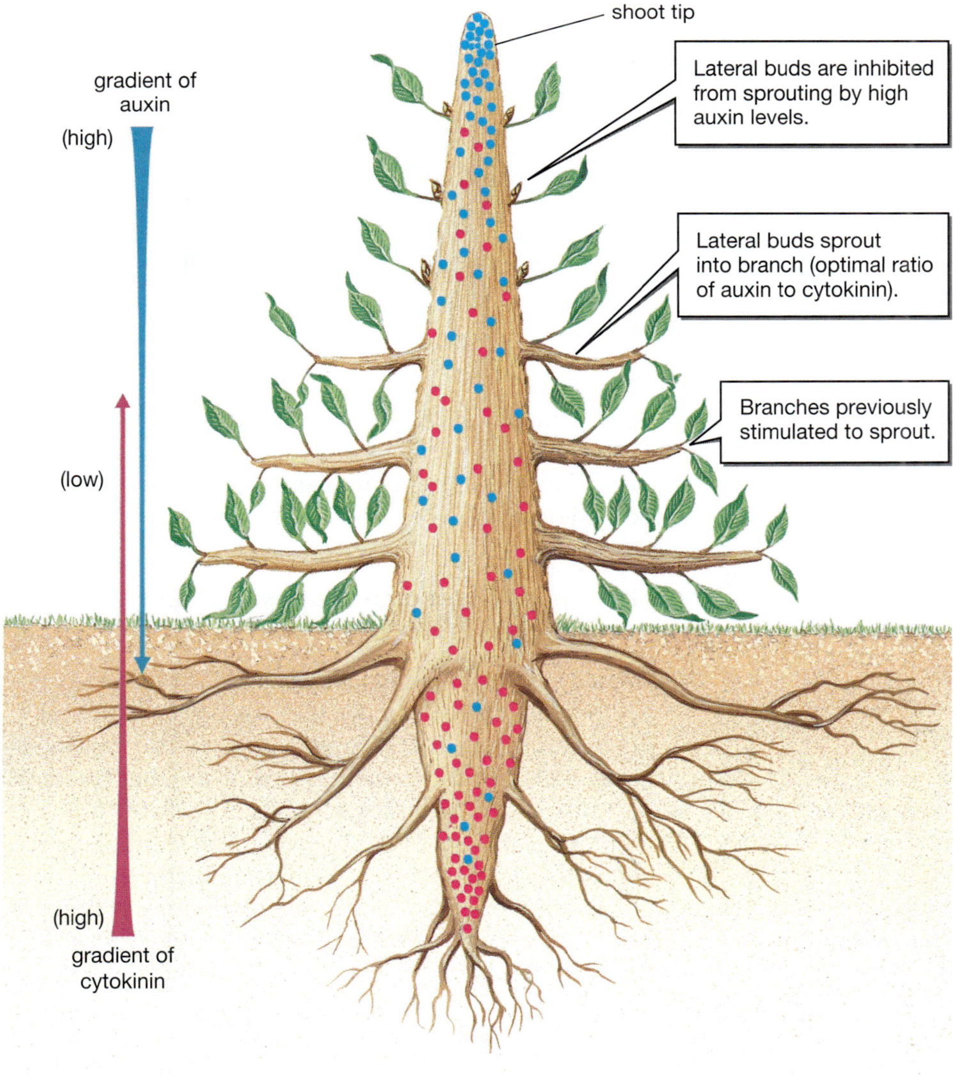

Figure 18-12 Role of auxin and cytokinin in lateral bud sprouting
A simplified diagram of the interplay of auxin (blue dots) and cytokinin (red dots) in the control of sprouting of lateral buds. Auxin is produced by shoot tips and moves downward; cytokinin is produced by root tips and moves upward.

MEDIATUTOR 18.4 Auxin Action

Figure 18-13 The effects of daylength on flowering

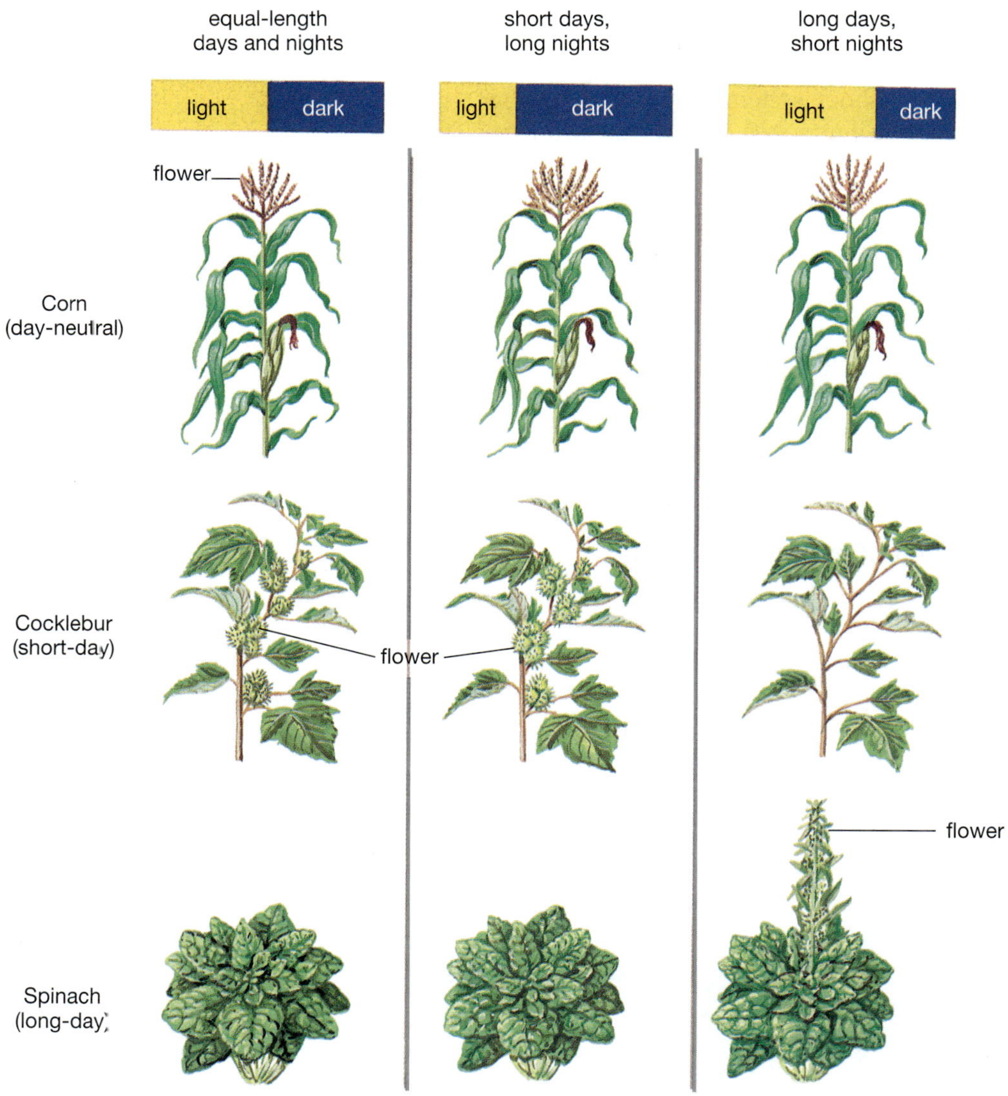

MediaTutor 18.5 Is It Time to Grow?

as a long-day plant because it flowers only if the day is *longer* than 13 hours, and cockleburs are short-day plants because they flower only if the day is *shorter* than 15.5 hours. Note that both will flower with 14 hours of light. Spinach, however, will also flower in much longer daylengths (for example, 16 hours of light), but cocklebur will not. Conversely, cocklebur will flower in much shorter daylengths (for example, 12 hours), but spinach will not.

The Hormones That Control Flowering Remain Undescribed

Grafting a branch from a (short-day) cocklebur plant maintained on a short-day schedule onto a cocklebur plant on a long-day schedule will induce flowering. In fact, plant physiologists have been able to induce flowering in the cocklebur by exposing a single leaf to short days in a special chamber while the rest of the plant was experiencing long days. Clearly, a signal must be traveling from leaf to bud. Studies on a variety of plants suggest that still-unidentified substances can both trigger and inhibit flowering. These hypothetical substances, which may differ among plant species, are collectively called florigens.

Hormones Coordinate the Development of Seeds and Fruit

When a flower is pollinated, the pollen releases auxin or gibberellin that stimulates the ovary to begin developing into a fruit. If fertilization follows, the developing seeds release still more auxin or gibberellin (or both) into the surrounding ovary tissues. Cells of the ovary multiply and grow larger, commonly storing starches and other food materials, forming a mature fruit.

When the seeds mature, the fruit ripens: It becomes softer, sweeter (as starches are converted to sugar), and more brightly colored, making it more noticeable and attractive to animals (Fig. 18-14). If you walk in a field in the autumn, look around at the wild fruits on the dying plants. A few may be blue, and many are red. Almost none are green when ripe. Or look around the produce section of your supermarket at all the brightly colored fruits—adapted to attract animal seed dispersers.

Ripening is stimulated by ethylene, which also causes the breakdown of green chlorophyll, revealing the attractive pigments that signal a ripe fruit. Ethylene is synthesized by fruit cells in response to a surge of auxin that is released by the seeds (a mechanism by which seed and fruit development are coordinated). Because ethylene is a gas, a ripe fruit continually leaks ethylene into the air. In nature this probably doesn't make much difference. When you store fruit in a closed container, however, ethylene released from one fruit will hasten ripening in the rest. The discovery of the role of ethylene in ripening revolutionized modern fruit and vegetable marketing. For example, bananas grown in Central America can be picked green and tough and shipped to North American markets. By exposing them to ethylene at their destination, grocers can market perfectly ripe fruit. Unfortunately, not all fruits seem to ripen properly when separated from the plant. Grapes and strawberries, for example, are not sensitive to ethylene and must be allowed to ripen on the plant. Shipping them to markets without damage is a difficult challenge.

Senescence and Dormancy Prepare the Plant for Winter

By autumn, the time has come to let any uneaten fruits drop to the ground. For perennial broadleaf plants, the leaves must be shed as well, because they would be a liability in winter, unable to photosynthesize but still allowing water to evaporate. Leaves, fruits, and flowers undergo a rapid aging called **senescence**, and drop from the plant.

As winter approaches, cytokinin production in the roots slows down, and fruits and leaves produce less auxin. Perhaps driven by these hormonal changes, much of the organic material in leaves is broken down to simple molecules that are transported to the roots for winter storage. A layer of thin-walled cells called an **abscission layer** forms at the base of the leaf stalk (Fig. 18-15). Ethylene, released by both aging leaves and ripening fruit, stimulates the abscission layer to produce an enzyme that digests the cell wall holding a leaf, fruit, or flower to the stem. When the abscission layers weaken, leaves and fruits fall from the branches.

Other changes also prepare the plant for winter. Buds, which developed into new leaves and branches during spring and summer, now become tightly wrapped up and dormant, waiting out the winter. Dormancy in buds, as in seeds, is enforced by abscisic acid. Metabolism

Figure 18-14 Ripe fruit is attractive to animal seed dispersers The prickly pear cactus fruit is green, hard, and bitter before it ripens, so animals are discouraged from eating it. After the seeds mature, the fruit becomes soft, red, and tasty, attracting animals such as this desert tortoise.

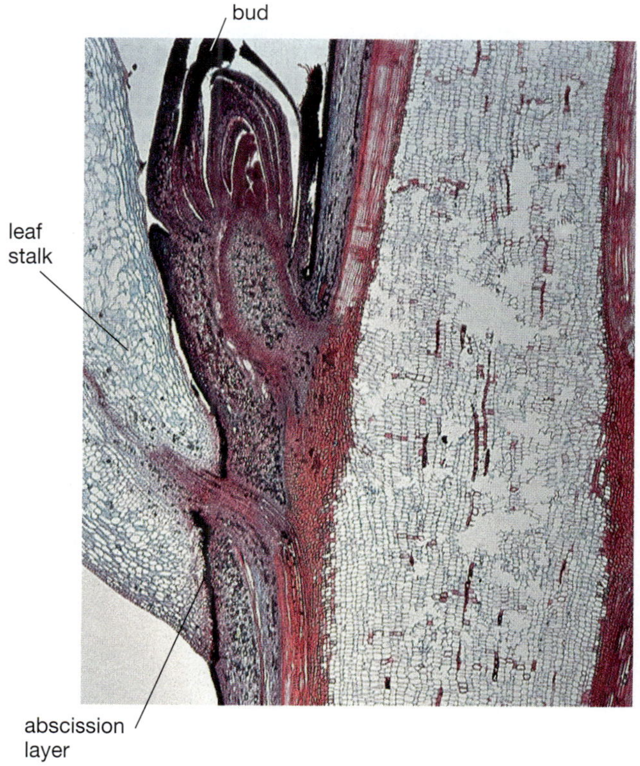

Figure 18-15 The abscission layer This cross section shows the abscission layer forming at the base of a maple leaf. A new leaf bud is visible above the dying leaf.

slows to a crawl, and the plant enters its long winter sleep, waiting for the signals of warmth and longer days in spring before awakening once again.

Evolutionary Connections

Adaptations for Pollination and Seed Dispersal

Evolutionary success depends on reproductive success, and natural selection has led to a fascinating array of adaptations that increase the odds that a plant's pollen will reach its target and that its seeds will be spread far and wide.

Pollinators May Be Attracted by Feeding Opportunities

Most wind-pollinated flowers, such as those of grasses and oaks, are inconspicuous and unscented, many of them scarcely more than naked stamens that liberate pollen to the wind (Fig. 18-16). The beautiful flowers so admired by humans, however, are pollinated by animals. Many animal-pollinated flowers attract pollinators by offering food to foraging animals such as beetles, bees, moths, butterflies, and hummingbirds. In return, the animals unwittingly distribute pollen from flower to flower.

Figure 18-16 A wind-pollinated flower
The flowers of grasses and many deciduous trees are wind-pollinated, with anthers (yellow structures) exposed to the wind.

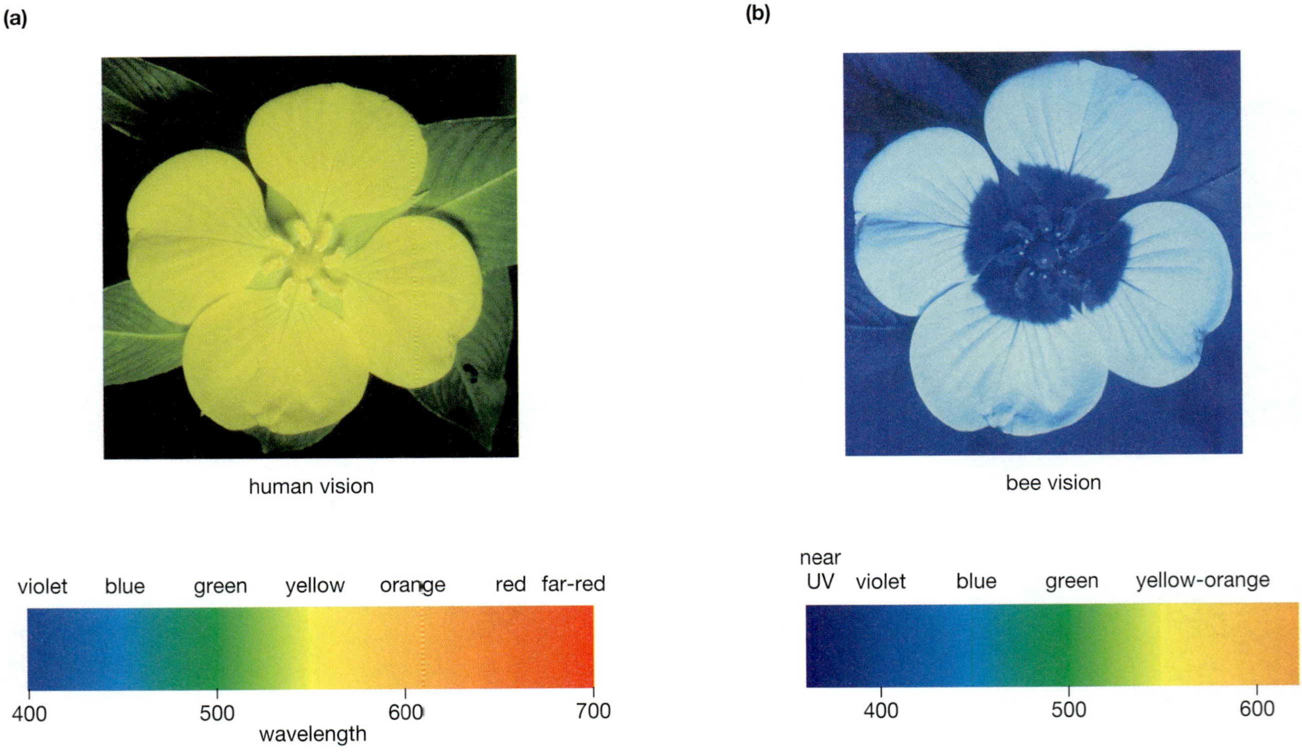

Figure 18-17 Ultraviolet patterns guide bees to nectar
The spectra of color vision for ***(a)*** humans and ***(b)*** bees overlap considerably in the blue, green, and yellow ranges but differ on the edges. Humans are sensitive to red, which bees do not perceive; bees can see UV light, which is invisible to the human eye. Many flowers photographed under ordinary daylight (in part *a*) and under UV light (in part *b*) show striking differences in color patterns. Bees can see the UV patterns that presumably lead them to the nectar- and pollen-containing centers of the flowers.

Some Flowers Have Adaptations That Help Ensure Pollination by Insects

We can thank the bees for most of the sweet-smelling flowers, because sweet "flowery" odors attract these pollinators. Bees also have good color vision and can recognize nectar-producing flowers by color. Bees, however, do not see exactly the same range of colors that humans do (Fig. 18-17). Typically, the flowers that attract bees are ones we see as white, blue, yellow, or orange, and many have other markings, such as central spots or lines pointing toward the center, that we do not see because they reflect UV light, which is invisible to our eyes.

Many flowers that are pollinated by moths or butterflies have nectar-containing tubes that accommodate the long tongues of moths and butterflies. Flowers pollinated by night-flying moths are open at night; most are white and some give off strong, musky odors that help the moths locate the flowers in the dark.

Many beetles prefer to feed on animal material, and beetle-pollinated flowers typically smell like dung or rotting flesh, which attracts these scavenging insects. Their names, including "carrion flower," "skunk cabbage," and "corpse flower" (Fig. 18-18), describe their odor. The corpse flower and its relative, the skunk cabbage common in U.S. swamps, are related. Both share another adaptation: their flowers heat up! By metabolizing stored food, mostly fat, eastern skunk cabbage flowers can reach temperatures up to 25 °C higher than the air around them. The heat probably attracts pollinators and certainly helps broadcast foul-smelling scents. These flowers deceive their pollinators—they smell like nutrient-rich rotting meat, but offer no food at all.

Figure 18-18 The corpse flower is overwhelming in size and scent
Scavenging insects such as flies and beetles are attracted to the odor of rotting meat produced by some of the largest flowers in the world. To date, only 11 corpse flowers (*Amorphophallus titanum*, a species native to Sumatra, also called "devil's tongue") have ever been coaxed into bloom in the United States, the first of which grew in 1937. Thousands of visitors flocked to see and smell this amazing specimen (nearly 6 feet tall) that bloomed in California's Huntington Botanical Gardens in August 1999. The "bloom" actually contains hundreds of separate male and female flowers, which are pollinated (in Sumatra) by large carrion beetles.

Birds Can Be Pollinators, Too

Hummingbirds are among the few important vertebrate pollinators. Many birds have poor senses of smell, and most hummingbird-pollinated flowers do not synthesize fragrant chemicals. However, these flowers often produce more nectar than other flowers do, because hummingbirds need more energy than insects and will favor flowers that provide it. These flowers protect their large nectar supplies from insects, which would become sated on the abundant sugar and fail to transfer pollen to another flower. Adaptations to hummingbird pollination include flower shapes that provide no place for insects to land and rest while dining, and a deep, tubular structure that matches the long bills and tongues of hummers. In addition, most hummingbird-pollinated flowers are red or orange. Red is particularly attractive to hummingbirds but appears gray or black to a bee.

Fruits Help Disperse Seeds

A plant species will be most successful if its seeds are dispersed far enough that the young plants don't compete with the parents for light and nutrients. Also, a plant species will be widespread if its members send seeds to distant habitats. In flowering plants, seed dispersal is accomplished by fruits. A wide variety of fruits has evolved, each type dispersing seeds in a different way.

Lightweight Fruits Can Be Dispersed by Wind

Dandelions and maples (Fig. 18-19) produce lightweight fruits with surfaces that catch the wind. (Yes, each individual hairy tuft on a dandelion ball is a separate fruit!) Each fruit typically contains a single small seed; having only one seed reduces weight and lets the fruit remain aloft longer. These featherweight fruits aid the seed in traveling away from the parent plant, from a few yards for maples to miles for a milkweed or dandelion on a windy day.

Floating Fruits Can Be Dispersed by Water

Many fruits can float on water for a time and may be dispersed by streams or rivers. The coconut fruit, however, is the ultimate floater. Round, buoyant, and watertight, the coconut drops off its parent palm, rolls to the sea, and floats for weeks or months until it washes

(a) (b)

Figure 18-19 Wind-dispersed fruits
(a) Dandelion fruits have filamentous tufts that catch the breezes. *(b)* Maple fruits are miniature glider-helicopters, silently whirling away from the tree as they fall.

Figure 18-20 Water-dispersed fruit
After a long journey at sea, this coconut was washed high onto a beach by a storm. The massive food reserves of coconuts support seedling growth on barren, sandy beaches.

ashore on some distant isle (Fig. 18-20). There it germinates, perhaps establishing a new coconut colony on a formerly barren island.

Clingy or Tasty Fruits Can Be Dispersed by Animals

Perhaps the majority of fruits use animals as agents of seed dispersal. Two quite distinct strategies have evolved for dispersal by animals: grabbing an animal as it passes by, or enticing it to eat the fruit but not digest the seeds.

Anyone who takes a long-haired dog on a walk through an abandoned field knows about fruits that hitchhike on animal fur. Burdocks, burr clover, foxtails, and sticktights all develop fruits with prongs, hooks, spines, or adhesive hairs (Fig. 18-21). The parent plants hold these fruits very loosely, so even slight contact with fur pulls the fruit free of the plant and leaves it stuck on the animal. Some of these fruits may fall off the next time the animal brushes against a tree or rock or come off when the animal grooms its fur.

Unlike hitchhiker fruits, edible fruits benefit both animal and plant. The plant stores sugars and tasty flavors in a fleshy fruit that surrounds the seeds, enticing animals to eat the fruit. Some fruits, such as peaches and plums, contain large, hard seeds that animals usually do not eat. After consuming the flesh of the fruit, the animals discard the seeds. Other fruits, including blackberries, raspberries, strawberries, and tomatoes, have small seeds that are swallowed along with the fruit flesh. The seeds then pass through the animal's digestive tract without harm. In addition to transport away from its parent, a seed that is swallowed and excreted benefits in another way: It ends up with its own supply of fertilizer!

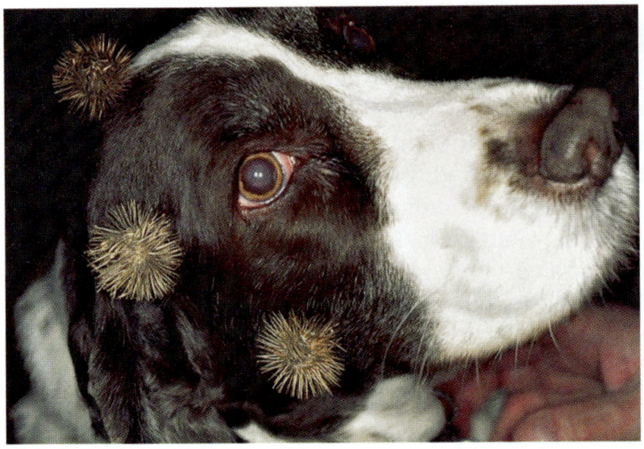

Figure 18-21 The cocklebur fruit uses hooked spines to hitch a ride on furry animals

CASESTUDYREVISITED

Tasty Tomatoes

As you may have guessed, the peculiar properties of *never-ripe* tomatoes are related to the mutant plants' response to ethylene. Although the plants produce normal amounts of ethylene, their fruits fail to respond to the hormonal signal. This failure to respond is due to a lack of *receptors* for ethylene. (Receptors are special protein molecules that are situated on the surface of target cells and initiate the appropriate response when a hormone molecule binds to them.) Tomato plants with the *never-ripe* mutation do not manufacture the ethylene-receptor protein.

The gene that codes for the ethylene-receptor protein was first discovered in the mustard plant *Arabidopsis*. Researchers working with a strain of *Arabidopsis* that did not respond to ethylene traced the defect to a mutation in a gene known as *ETR1*. They then determined that the normal function of the gene was to produce receptors for ethylene. This discovery in 1995 was the first time that a plant hormone receptor had been characterized. Later, researchers working with *never-ripe* tomatoes found that these mutants contained a gene very similar to *ETR1* that probably acts by a similar mechanism. In fact, it seems increasingly likely that many plant species possess a similar mechanism for response to ethylene and that, using the techniques of biotechnology (see Chapter 12), ethylene-resistance genes could be introduced into crop plants.

As you learned in this chapter, ethylene plays a role in a number of different physiological processes, and plant breeders plan to make use of ethylene-resistance genes for purposes beyond improved control of fruit ripening. For example, ethylene's role in promoting the onset of flower senescence means that the flowers of ethylene-resistant plants retain their shape and fresh appearance for much longer before wilting. Researchers have successfully transferred a tomato's *never-ripe* gene into petunia plants; the resulting transgenic flowers lasted two to three times longer than those of nonmanipulated petunia plants. The tomato that never ripens may lead the way to a whole range of flowers that last longer in a vase and fruits that last longer in the refrigerator.

Summary of Key Concepts

1) What Is the Life Cycle of Plants?

The sexual life cycle of plants, called alternation of generations, includes both a multicellular diploid form (the sporophyte generation) and a multicellular haploid form (the gametophyte generation).

In seed plants, the gametophyte is greatly reduced. The male gametophyte is the pollen grain, a drought-resistant structure that can be carried from plant to plant by wind or animals. The female gametophyte is also reduced and is retained within the body of the sporophyte. In this way, seed plants can reproduce independently of liquid water.

2) What Is the Structure of Flowers?

Complete flowers consist of four parts: sepals, petals, stamens (male reproductive structures), and carpels (female reproductive structures). The sepals form the outer covering of the flower bud. Most petals (and in some cases, the sepals) are brightly colored and attract pollinators to the flower. The stamen consists of a filament that bears at its tip an anther, in which pollen grains (the male gametophytes) develop. The carpel consists of the ovary, in which one or more embryo sacs (the female gametophytes) develop, and a style. The style bears at its end a sticky stigma, to which pollen adheres during pollination.

Incomplete flowers lack one or more of the four floral parts.

3) What Are the Gametophytes of Flowering Plants?

Pollen grains develop in the anthers. Each haploid microspore divides mitotically to form pollen grains. A pollen grain contains two sperm cells. The embryo sac develops within the ovules of the ovary. A haploid megaspore undergoes three sets of mitotic divisions to produce the eight nuclei of the embryo sac. These eight nuclei come to reside in only seven cells. One of these cells, with a single nucleus, is the egg cell.

4) How Does Pollination Lead to Fertilization?

Pollination is the transfer of pollen from anther to stigma. When a pollen grain lands on a stigma, it grows a tube that extends through the style to the embryo sac. The two sperm cells travel down the style within the tube, eventually entering the embryo sac. One sperm fuses with the egg to form a diploid zygote, which will give rise to the embryo. The other sperm fuses with the large cell containing the polar nuclei to produce a triploid cell. This cell will give rise to the endosperm, a food-storage tissue within the seed.

5) How Do Seeds and Fruits Develop?

The embryo develops a root, shoot, and cotyledons. Cotyledons digest and absorb food from the endosperm and transfer it to the growing embryo. The seed is enclosed within a fruit, which develops from the ovary wall. The fruit helps disperse the seeds away from the parent plant.

6) How Do Seeds Germinate and Grow?

Seeds typically remain dormant for some time after fruit ripening. Seed germination requires warmth and moisture. Energy for germination comes from food stored in the endosperm; that food is transferred to the embryo by the cotyledons.

7) What Are Plant Hormones, and How Do They Act?

Plant hormones are chemicals that are produced by cells in one part of a plant body and transported to other parts of the plant, where they exert specific effects. The five major classes of plant hormones are auxins, gibberellins, cytokinins, ethylene, and abscisic acid. The major functions of these hormones are summarized in Table 18-1.

8 How Do Hormones Regulate the Plant Life Cycle?

Dormancy in seeds is enforced by abscisic acid. Falling levels of abscisic acid and rising levels of gibberellin trigger germination. As the seedling develops, it grows directionally with respect to light (phototropism) and gravity (gravitropism). Auxin is responsible for phototropism and gravitropism in shoots and gravitropism in roots. In shoots, auxin stimulates the elongation of cells. In roots, similar concentrations of auxin inhibit elongation.

Branching in stems results from the interplay of two hormones, auxin (produced in shoot tips and transported downward) and cytokinin (synthesized in roots and transported upward). High concentrations of auxin inhibit the growth of lateral buds. An optimum concentration of both auxin and cytokinin stimulates the growth of lateral buds.

The timing of flowering is normally controlled by daylength. Flowering is both stimulated and inhibited by hormones called florigens.

Developing seeds produce auxin, which diffuses into the surrounding ovary tissues and causes the production of a fruit. A surge of auxin as the seed matures stimulates fruit cells to release another hormone, ethylene, which causes the fruit to ripen. Ripening includes the conversion of starches to sugars, softening of the fruit, development of bright colors, and, commonly, the formation of an abscission layer at the base of the fruit stalk.

Several changes prepare perennial plants of temperate zones for winter. Leaves and fruits undergo a rapid aging process called *senescence*, including the formation of an abscission layer. Senescence occurs as a result of a fall in levels of auxin and cytokinin and, perhaps, a rise in ethylene concentrations. Other parts of the plant, including buds, become dormant. Dormancy in buds is enforced by high concentrations of abscisic acid.

Key Terms

abscisic acid *p. 343*
abscission layer *p. 349*
alternation of generations *p. 336*
anther *p. 338*
auxin *p. 343*
carpel *p. 338*
coleoptile *p. 342*
complete flower *p. 337*
cotyledon *p. 340*
cytokinin *p. 343*
dormancy *p. 340*
double fertilization *p. 339*
egg *p. 339*
embryo sac *p. 337*
endosperm *p. 339*
ethylene *p. 343*
fertilization *p. 339*
filament *p. 338*
flower *p. 337*
fruit *p. 338*
gametophyte *p. 336*
germination *p. 340*
gibberellin *p. 343*
gravitropism *p. 345*
incomplete flower *p. 338*
megaspore *p. 339*
microspore *p. 339*
ovary *p. 338*
ovule *p. 338*
petal *p. 338*
phototropism *p. 346*
plant hormone *p. 342*
polar nuclei *p. 339*
pollen grain *p. 337*
pollination *p. 339*
seed *p. 337*
seed coat *p. 339*
senescence *p. 349*
sepal *p. 338*
spore *p. 336*
sporophyte *p. 336*
stamen *p. 338*
stigma *p. 338*
style *p. 338*

Thinking Through the Concepts

Multiple Choice

1. *When a bee gets nectar from a flower, the bee picks up pollen from the ____________ and carries it to another flower.*
   a. stigma
   b. ovary
   c. sepals
   d. anthers
   e. filament

2. *In plants, the gametophyte produces eggs and sperm by*
   a. mitosis
   b. meiosis
   c. spore formation
   d. fertilization
   e. germination

3. *Reproduction in flowering plants is known as double fertilization because*
   a. one tube nucleus fuses with one egg, and one sperm nucleus fuses with another egg
   b. two sperm nuclei fuse with one egg
   c. one sperm fuses with two eggs
   d. one sperm nucleus fuses with one egg, and another sperm nucleus fuses with the two polar nuclei
   e. two polar nuclei fuse with one egg nucleus

4. *If you put an underripe banana in a bag with an apple, the banana will quickly ripen because of the hormone ____________ produced by the apple.*
   a. auxin
   b. cytokinin
   c. gibberellin
   d. abscisic acid
   e. ethylene

5. *In your warm house, poinsettias act as short-day plants that turn red when the daylength is less than 12 hours. If you want to get a poinsettia ready for Christmas, which of the following would work best?*
   a. Give it continuous light.
   b. Give it continuous darkness.
   c. Keep it in the dark, but shine a light on it for 1 hour every 13 hours.
   d. Keep it in the light except for turning off the light once a day for 1 hour.
   e. None of these would get it to turn red.

6. *When a seed is first formed in the fall, it typically will not germinate because _________ must first be washed from the seed by a hard rain or broken down by cycles of freezing and thawing.*
   a. auxin
   b. cytokinin
   c. gibberellin
   d. abscisic acid
   e. ethylene

? Review Questions

1. Diagram the plant life cycle, comparing ferns with flowering plants. Which stages are haploid, and which are diploid? At which stage are gametes formed?
2. What are the advantages of the reduced gametophytes of flowering plants, compared with the larger gametophytes of ferns?
3. Diagram a complete flower. Where are the male and female gametophytes formed? What are these gametophytes called?
4. How does an egg develop within an embryo sac? How does this structure allow double fertilization to occur?
5. What does it mean when we say that pollen is the male gametophyte? How is pollen formed?
6. What is the endosperm? From which cell of the embryo sac is it derived?
7. What did the Darwins, Boysen-Jensen, and Went each contribute to our understanding of phototropism? Do their experiments truly prove that auxin is the hormone controlling phototropism? What other experiments would you like to see?
8. How can one hormone, an auxin, cause shoots to grow up and roots to grow down?
9. Which hormones cause fruit development? From where do these hormones come? Which hormone causes fruit ripening?
10. Which hormone(s) is (are) involved in leaf and fruit drop? In bud dormancy?

Applying the Concepts

1. In areas where farms have been left uncultivated for many years, it is often possible to see certain kinds of trees growing in straight lines, which mark old fences where birds sat and deposited seeds they had eaten. Why are such seeds more likely to germinate than are those of the same species that have not passed through a bird's digestive tract? How might an anthropologist use such lines of trees to study past inhabitants of an area?
2. Charles Darwin once described a flower that produced nectar at the bottom of a tube 25 centimeters (10.5 inches) deep. He predicted that there must be a moth or other animal with a 25-centimeter-long tongue to match; he was right. Such specialization almost certainly means that this particular flower could be pollinated only by that specific moth. What are the advantages and disadvantages of such specialization?
3. Many plants that we call weeds were brought from another continent either accidentally or purposefully. In a new environment they have few competitors or animal predators, so they tend to grow in such large amounts that they come to be considered weeds. Think of all the ways you can in which humans disperse plants. To what degree do you think humans have changed the distributions of plants? In what ways is this change helpful to humans? In what ways is it a disadvantage?
4. A student reporting on a project said that one of her seeds did not grow properly because it was planted upside down so that it got confused and tried to grow down. Do you think the teacher accepted this explanation? Why or why not? Which plant hormone or hormones would be involved?
5. Agent Orange, a combination of two synthetic auxins, was used in Vietnam to defoliate the rain forest during the Vietnam War. Being similar to natural growth hormones, how can synthetic auxins harm or kill plants? What do you think would happen if natural auxins were used in excess quantities on plants?
6. Bean sprouts such as those you might eat in a salad have to be grown in the dark for them to form the long, yellowish stems that you see. We call such stems "etiolated." If they are grown in the light, they will be short and green. Why do seedlings grow etiolated in the dark? Under what conditions does etiolation occur in nature? How do plant hormones enable these seedlings to form this shape?
7. Suppose that on July 4, you discover that both a short-day plant and a long-day plant have bloomed in your garden. Discuss how it is possible for both to bloom.

Group Activity

Form groups of three. Take three scraps of paper, and mark one "SDP," another "LDP," and the third "DNP" (for short-day, long-day, and day-neutral plant, respectively). Shuffle the papers; each of you should randomly choose one. Then take three minutes to write down the advantages and disadvantages of being that type of plant. Be creative but not frivolous. Take another two minutes to answer: If you could choose the flowering category you'd most like to become, which would you choose, and why? Then discuss your decisions with your group.

For More Information

Eiseley, L. "How Flowers Changed the World." *National Wildlife*, April–May 1996. The late philosopher/naturalist Loren Eiseley eloquently explains how the evolution of flowers has changed the history of life on Earth. Beautifully written and lavishly illustrated.

Handel, S. N., and Beattie, A. J. "Seed Dispersal by Ants." *Scientific American*, August 1990. Many plants depend on ants for seed dispersal, even producing fat deposits on the outside of the seed as a lure for the ants. Dispersal is not the only benefit to the seed: Being "planted" in an anthill seems to provide an ideal environment for germination and growth as well.

Kearns, C. A., and Inouye, D. W. "Pollinators, Flowering Plants, and Conservation Biology." *Bioscience*, May 1997. To preserve plant species, we must preserve their pollinators and the web of life that sustains these pollinators.

Milius, S. "The Science of Big, Weird, Flowers." *Science News*, September 11, 1999. Many giant flowers deceive flies and beetles into pollinating them by smelling like carrion.

Simons, P. "An Explosive Start for Plants." *New Scientist*, January 2, 1993. Explosive movements in some plants spread seeds and spores and snap pollen onto bees.

Sunquist, F. "Blessed Are the Fruit Eaters." *International Wildlife*, May–June 1992. Especially in the Tropics, fruit-eating mammals and birds are crucial to the dispersal and germination of seeds. Overhunting threatens not only the animals but also the plants that depend on them.

Answers to Multiple-Choice Questions

1.d 2.a 3.d 4.e 5.c 6.d

MediaTutor

CD Activities

Activity 18.1: Fern Life Cycle

Estimated time: 5 minutes

In this tutorial you will explore the different stages and structures in the life cycle of the fern.

Activity 18.2: Flower Structure and Function

Estimated time: 5 minutes

The following tutorial will help you to explore basic flower structure and function.

Activity 18.3: Fruit and Seed Structure and Development

Estimated time: 5 minutes

In this activity you will first examine the basic structure of fruits and seeds and then explore the process of fruit and seed development.

Activity 18.4: Auxin Action

Estimated time: 10 minutes

In this simulation you can explore some of the effects of auxin on plant growth and development by altering the source and concentration of auxin in the plant.

Activity 18.5: Is It Time to Grow?

Estimated time: 10 minutes

In this simulation, you will review the conversion of phytochrome by red and far-red light to initiate biological actions. You will be able to explore the effects of different types of light and the length of dark and light periods on different types of plants, and then you will see the response of the plant to the stimulus.

Start the MediaTutor Student CD-ROM and enter the activity number in the Quick Search box to be taken directly to that activity.

Web Investigation

Tasty Tomatoes

Estimated time: 15 minutes

If you have paid attention to the news over the past few years, you are no doubt familiar with the term "biotechnology revolution" and some of the new "bio-technologies" that have resulted such as cloning, gene therapy, and genetic engineering. As discussed in this and previous chapters, the process of genetic engineering applied to agriculture has tremendous potential for helping to alleviate some of the burden the human population has placed on the world's shrinking supply of agricultural land. This investigation will explore some of the dynamics of plant physiology and how genetic engineering can be used to modify the growth and development of many crops.

Go to http://www.prenhall.com/audesirk3, the Audesirk Companion Web site. Select Chapter 18 and Web Investigation to begin.

UNIT FIVE

Animal Anatomy and Physiology

The animal body is an exquisite expression of the elegance with which evolution has linked form to function.

"As a physiologist I am especially impressed by the common ignorance of bodily organs and their functions. It seems to me now, as it seemed to Robert Boyle nearly 300 years ago, that it is 'highly dishonorable for a Reasonable Soul to live in so Divinely built a Mansion as the Body [he or] she resides in, altogether unacquainted with the exquisite structure of it.'"

W. B. Cannon in* The Way of an Investigator *(1945)

Despite drying winds, subzero temperatures, and an absence of shelter, the environment inside the bodies of these chinstrap penguins remains moist, warm, and protected.

19 Homeostasis and the Organization of the Animal Body

AT A GLANCE

CASESTUDY

Surviving a Dive

An emperor penguin can dive to an underwater depth of more than 1,500 feet and remain submerged for up to 20 minutes. How is this aquatic bird able to accomplish such feats, which are far beyond the abilities of even the best-conditioned human athlete? To appreciate the difficulties faced by a diving emperor penguin, consider what would happen to you if you decided to do some deep diving. A human diver faces two main problems: lack of oxygen and the effects of pressure.

Every cell in an animal body needs oxygen, and humans, like most land-dwelling animals, have no means of extracting oxygen from water. Instead, a diver must rely on whatever oxygen was inhaled at the surface before diving. That oxygen lasts the average person only a few moments; even the world's best breath-holders must resurface after less than 7 minutes of lying motionless under water. The need to breathe quickly forces a human diver to the surface.

A diver's body also faces increasing pressure during a descent. Eventually, the pressure becomes great enough to collapse a diver's lungs. High pressure at depth also forces air to move out of the lungs and into the blood, where it dissolves. When the diver ascends and pressure decreases, the dissolved gases emerge from the blood—rapidly. The rapid exit of these gases from the blood can be quite dangerous, especially if bubbles form in blood vessels. The consequences of ascending from a deep dive can injure or even kill a human diver.

The greatest depth a human diver has ever managed to reach without using an air tank was 436 feet. Divers taking extraordinary precautions and breathing from tanks of special mixtures of gases can attain depths of 600 feet or more. Although humans penetrate these relatively modest depths only with heroic effort and high risk, the emperor penguin routinely dives to far greater depths each day in search of a meal. How does this deep-diving penguin supply its cells with oxygen and resist the ill effects of high pressure and subsequent reduction of pressure? We will examine some of the special adaptations of deep-diving penguins following an introductory overview of the animal body and how it functions. ■

1 How Do Animals Maintain Internal Constancy?

Animals inhabit almost every environment on Earth, from scorching deserts to icy polar seas, from windy mountaintops to underground rivers. Regardless of an animal's surroundings, however, the cells inside its body are bathed in warm, slightly salty liquid. This wet, salty internal environment is required for the cells' survival and must be maintained even if the animal's external environment consists of dry air or fresh, salt-free water. What's more, it's not enough that the internal environment simply be wet and salty. Instead, cells require a very specific, complex mixture of dissolved substances and a narrow temperature range. Because the body's cells cannot survive if the conditions of the internal environment deviate from a small range of acceptable states, the cells devote a large portion of their energy to actions that serve to keep the environment within that range.

This constancy of the internal environment was first recognized by French physiologist Claude Bernard in the mid-nineteenth century; the term **homeostasis** was later coined to describe it. Although the word "homeostasis" (derived from Greek words meaning "to stay the same") may suggest a static, unchanging state, the internal environment in fact seethes with activity as the body continuously adjusts to internal and external changes. The end result of all this activity, however, is that conditions are kept within the narrow range that cells require to function. Conditions are maintained within the appropriate range by a host of mechanisms that are collectively known as feedback systems.

Negative Feedback Reverses the Effects of Changes

The most important mechanism governing homeostasis is **negative feedback**, in which the response to a change is to counteract the change. In other words, an input causes an output response that "feeds back" to the initial input and decreases its effects. Because the initial change triggers a response that reverses its effects, the overall result is to return the system to its original condition. This kind of feedback is called "negative" because it reverses or negates the initial change.

A familiar example of negative feedback is your home thermostat. In a thermostat, an input—the temperature's dropping below a set point, the thermostat setting—is detected by a thermometer. The thermometer responds with an output, signalling a control device that switches on a heater. The heater restores the temperature to the set point, and the heater is switched off. The response to the initial change counteracted that change. Continuously repeated on-off cycles keep your home's temperature near the set point. Note that the thermostat's negative feedback mechanism requires a *control center* with a set point, a *sensor* (the thermometer), and an *effector* (the furnace), which accomplishes the change (Fig. 19-1, top).

How do people maintain their internal temperature within a narrow range despite extreme fluctuations in the temperature around them? Negative feedback keeps a person's body temperature close to 98.6 °F (37 °C). The control center of the temperature control system is in the hypothalamus, a region of the brain (see Fig. 19-1). Nerve endings in the abdomen, spinal cord, skin, and large veins act as temperature sensors and transmit this information to the hypothalamus. When your body temperature drops, the hypothalamus activates various effector mechanisms that raise your body temperature, including shivering (which generates heat through muscular activity), constriction of blood supply to the skin (which reduces heat loss), and elevation of metabolic rate (which generates heat). When normal body temperature is restored, sensors signal the hypothalamus to switch off these temperature control mechanisms.

Negative feedback mechanisms abound in physiological systems. In the chapters that follow, you will find many examples in which homeostasis is maintained by negative feedback, including blood oxygen content, water balance, blood sugar levels, and many other components of the internal environment.

Positive Feedback Drives Events to a Conclusion

In contrast to a negative feedback system, a change in a **positive feedback** system produces a response that intensifies the original change. The end result is that change tends to proceed in the same direction rather than reversing to return to a set point (see Fig. 19-1). Positive feedback, as you can imagine, tends to create chain reactions. For example, in nuclear fission, each particle that is split from an atom triggers the splitting of another atom, the pieces of which trigger the splitting of other atoms, and so on. When controlled, the chain reaction supplies nuclear power. When deliberately set out of control, it produces an atomic explosion. A familiar biological example of positive feedback is population growth; each offspring gives rise to still more offspring. Ecologist Paul Ehrlich coined the apt expression "population bomb" to describe unchecked population growth.

Positive feedback mechanisms are rare in animal physiology. One example of positive feedback, however, controls events during childbirth (see Fig. 19-1, bottom). The early contractions of labor begin to force the baby's head against the cervix, located at the base of the uterus; this pressure causes the cervix to dilate (open). Stretch receptors in the cervix respond to this expansion by signaling the hypothalamus, which responds by triggering the release of a hormone (oxytocin) that

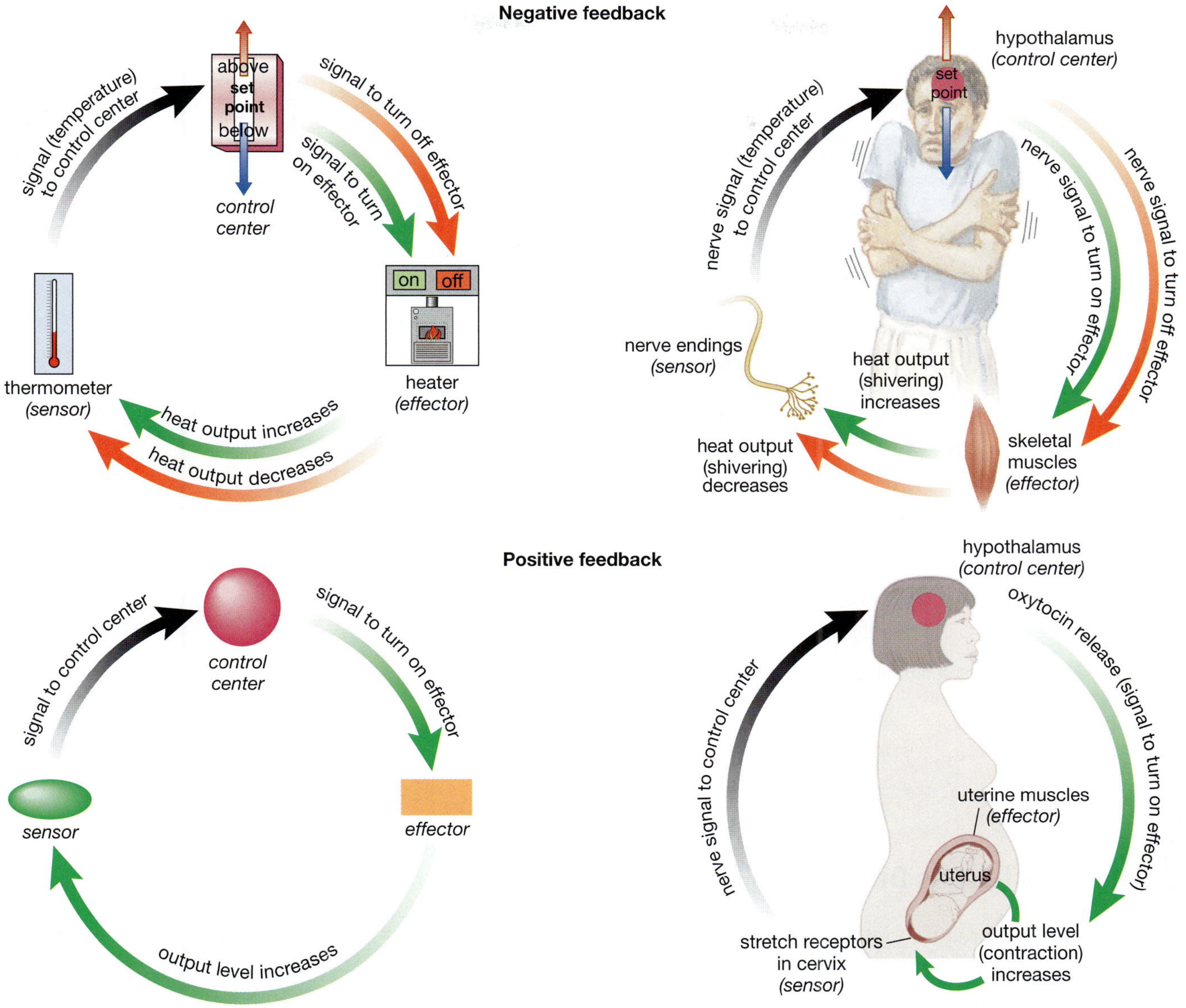

Figure 19-1 Negative and positive feedbacks
(top) Both our homes (*left*) and our bodies (*right*) use negative feedback to maintain appropriate temperatures. ***(bottom)*** Positive feedback is seen in this hypothetical model (*left*) and in childbirth (*right*).

MediaTutor 19.1 Feedback Loops and Homeostasis

stimulates more and stronger uterine contractions. Stronger contractions create further pressure on the cervix, which in turn prompts the release of more hormone. The feedback cycle is finally terminated by the expulsion of the baby and its placenta.

2 How Is the Animal Body Organized?

The animal body is an engineering marvel. From a beginning of simple, free-living cells, evolutionary change has produced astonishingly complex, self-regulating systems consisting of trillions of specialized cells that accomplish hundreds of functions simultaneously. The parts fit together with a degree of precision and integration of which human engineers can only dream. This exquisite machine is made up of **tissues**, each tissue composed of dozens to billions of structurally similar cells that act in concert to perform a particular function (Fig. 19-2). Tissues are the building blocks of **organs**, such as the stomach, small intestine, kidneys, and urinary bladder. Organs in turn are organized into **organ systems**; for example, the digestive system includes the stomach, small intestine, large intestine, and other organs.

The Big Picture

A wide variety of structures has evolved in the protected environments inside animal bodies. The diversity of a body's structures reflects the diverse activities that the body performs. All body structures, however, share a common property: Their form is intimately related to their function.

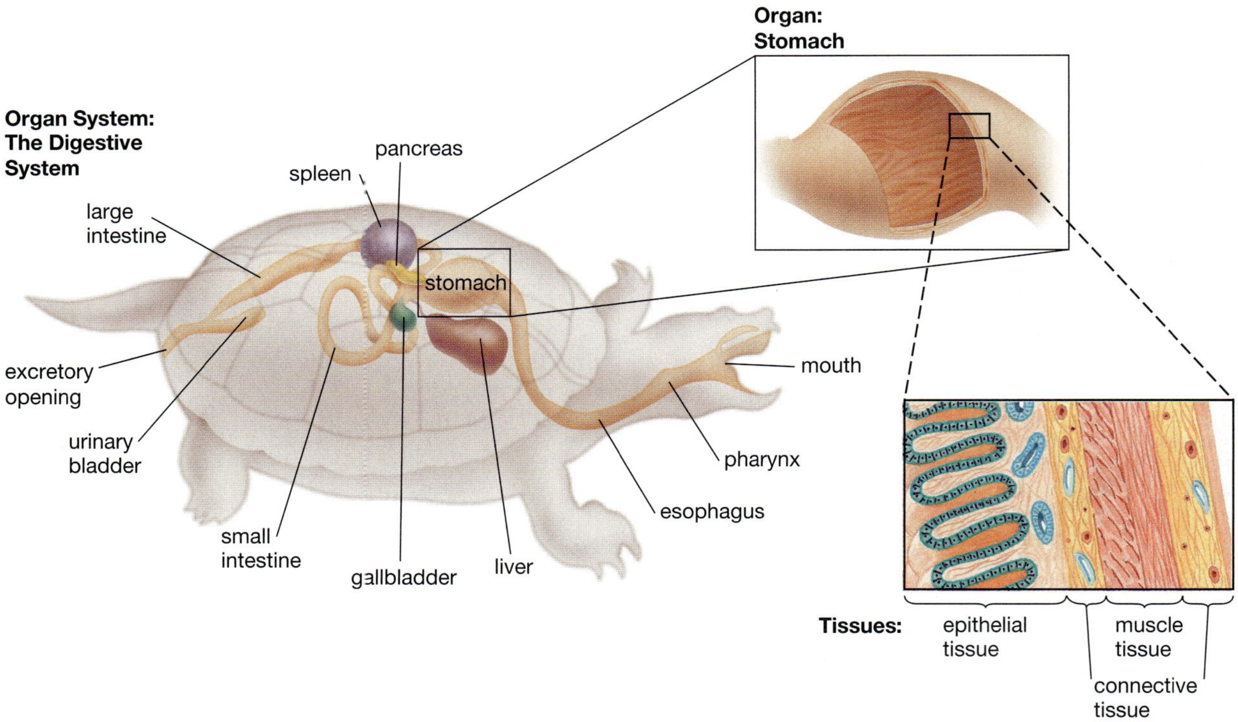

Figure 19-2 Organ systems, organs, and tissues
The animal body contains organ systems, each composed of several organs, which are in turn composed of several types of tissue.

3 How Do Tissues Differ?

Although you will read about many different animal tissues as you make your way through Unit 5, all of them can be grouped into four tissue types. Here we present a brief overview of the four types of animal tissue: epithelial tissue, connective tissue, muscle tissue, and nerve tissue.

Epithelial Tissue Covers the Body and Lines Its Cavities

Epithelial tissue (also called epithelium) forms sheets that cover the body and line body cavities such as the mouth, the stomach, and the bladder. The cells of epithelial tissues are packed closely together, creating barriers (such as the skin) that resist the movement of substances across them or barriers (such as the lining of the small intestine) that allow the movement of specific substances across them. There are many types of epithelial tissues; the structure of each type is related to its function. For example, the epithelium that lines the lungs, where gas exchange takes place, consists of thin, flattened cells arranged in a single layer, across which gas molecules can pass easily (Fig. 19-3a). Another form of epithelium consists of elongated cells, often with cilia, that can secrete mucus (Fig. 19-3b). This type of epithelium lines the passage leading to the lungs; the mucus traps dust particles and the cilia transport them away from the lungs. It also lines tubes of the reproductive organs, where the cilia transport gametes to their destinations.

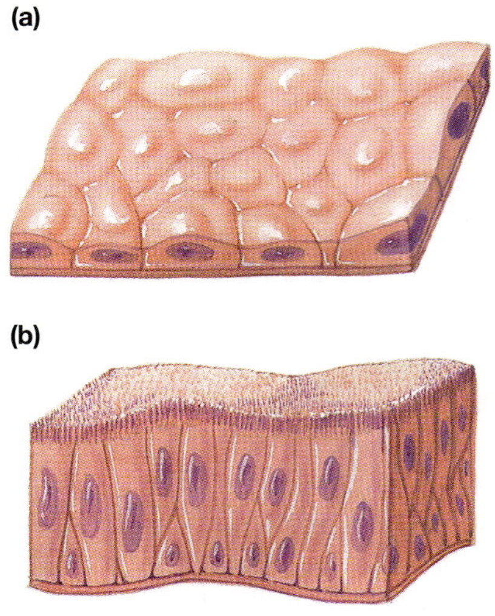

Figure 19-3 Epithelial tissue
(a) Thin, flattened cells in a single layer form the epithelial tissue that lines lungs and blood vessels, where exchange of materials is important. *(b)* Elongated epithelial cells bearing cilia and capable of secreting mucus line the passage to the lungs and tubes of the reproductive organs.

An important property of epithelial tissues is that they are continuously lost and replaced by mitotic cell division. For example, consider the abuse suffered by the epithelium that lines your mouth. Scalded by coffee and scraped by corn chips, it would be destroyed within a few days if it were not continuously replacing itself. The stomach lining, abraded by food and attacked by acids and protein-digesting enzymes, is completely replaced every two to three days. Your skin's outer surface is renewed about twice a month.

Some Epithelial Tissues Form Glands

During development, the cells of some epithelial tissues change shape and function to form **glands**, clusters of cells that are specialized to secrete (release) substances. Glands are classified into two broad categories: exocrine glands and endocrine glands. **Exocrine glands** remain connected to the epithelium by a passageway, or duct. Examples of exocrine glands are sweat glands and sebaceous (oil-secreting) glands; both types are found in the skin. Exocrine glands called salivary glands release saliva into the mouth, and still other exocrine glands line the stomach, where they secrete a protective layer of mucus. **Endocrine glands** become separated from the epithelium that produced them. Most products of endocrine glands are hormones, which are secreted into the extracellular fluid that surrounds the glands and then diffuse into nearby capillaries. Endocrine glands and their hormones are covered in detail in Chapter 23.

Connective Tissues Have Diverse Structures and Functions

Connective tissues serve mainly to support and bind other tissues. Most connective tissues include large quantities of extracellular substances, typically secreted by the connective tissue cells themselves. Some connective tissues that support and nourish epithelial tissues consist mainly of loosely woven fibers. For example, a connective tissue called the dermis lies beneath the epithelial tissue of the skin and contains capillaries and fluid-filled spaces that nourish the epithelium (see Fig. 19-10). Other fibrous connective tissues known as tendons and ligaments attach muscles to bones and bones to bones, respectively. The strength necessary for these functions comes from densely packed, parallel strands of an extracellular protein called collagen.

Cartilage is a flexible and resilient connective tissue that consists of widely spaced cells surrounded by a thick, nonliving matrix. The matrix is made of collagen secreted by the cartilage cells (Fig. 19-4). Cartilage covers the ends of bones at joints, provides the supporting framework for the passages through which we breathe, supports the ear and nose, and forms shock-absorbing pads between the vertebrae. **Bone** resembles cartilage but is hardened by deposits of calcium phosphate. Bone forms in concentric circles around a central canal, which contains a blood vessel (Fig. 19-5).

Long-term energy storage in the animal body is provided by fat cells, collectively called adipose tissue (Fig. 19-6a). Adipose tissue is especially important in the physiology of animals adapted to cold environments, because not only does it store energy, it also serves as insulation (Fig. 19-6b).

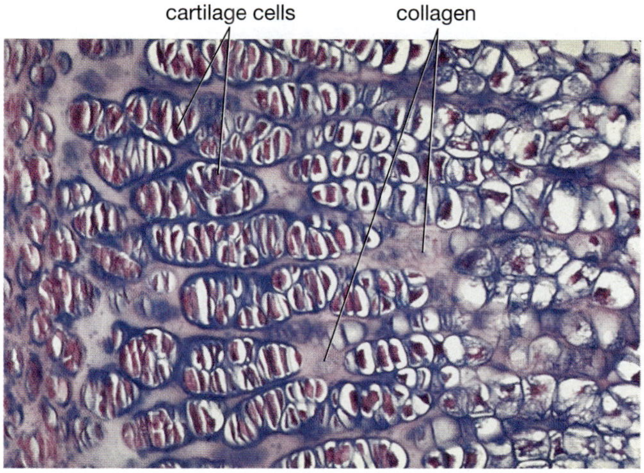

Figure 19-4 Cartilage
This specialized connective tissue supports the body and protects soft tissues. It is more flexible and less rigid than bone. The plasma membrane of the cartilage cells is stained dark purple. The material stained pale purple is the matrix of collagen secreted by the cartilage cells.

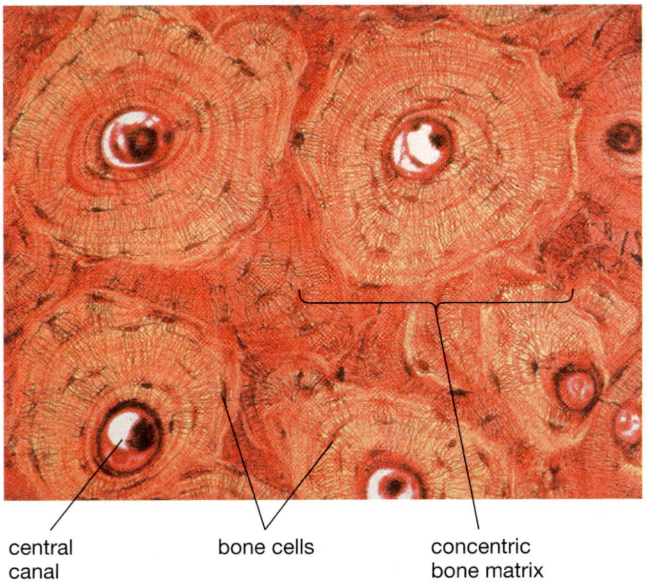

Figure 19-5 Bone consists of cells embedded in a hard matrix
Concentric circles of bone, deposited around a central canal that contains a blood vessel, are clearly visible in this micrograph. Bone cells appear as dark spots trapped in small chambers within the hard matrix that the cells themselves deposit.

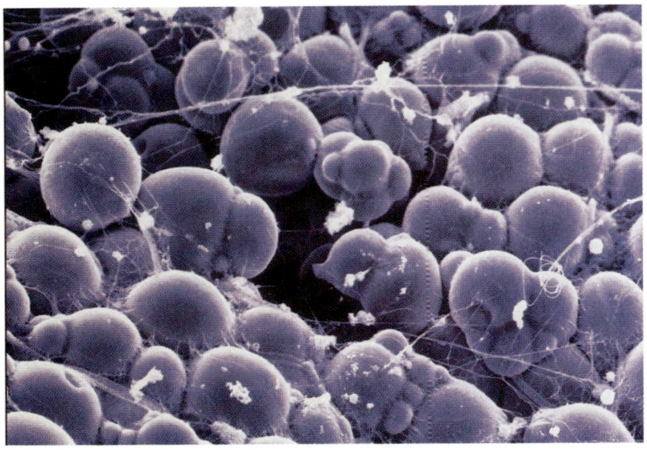

(a)

(b)

Figure 19-6 Adipose tissue
(a) Adipose tissue, such as that shown here from a mouse, is made up almost exclusively of fat cells; very little extracellular material is present. Each fat cell contains a droplet of oil that takes up most of the volume of the cell. ***(b)*** A hooded seal at four days of age has doubled her birth weight by drinking mother's milk that consists of 61% fat. At 100 pounds, she is almost too fat to move. The fat will feed and insulate the pup as the ice floes break up and she dives into icy water to learn to hunt and feed on her own.

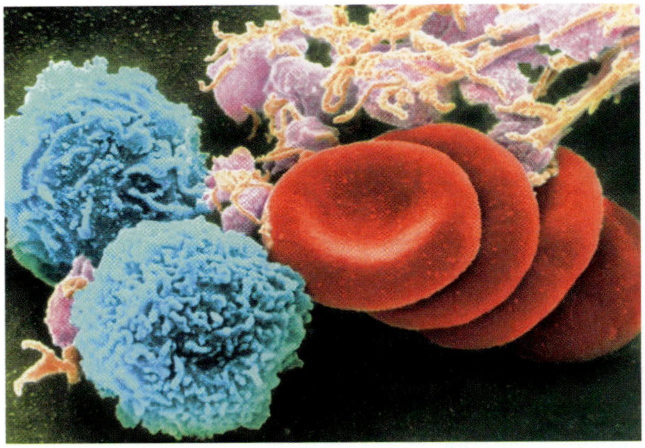

Figure 19-7 Blood
Blood contains three types of cellular components. In this color-enhanced scanning electron micrograph, red blood cells, which carry oxygen to tissues throughout the body, appear as red concave disks. White blood cells, important in the immune response, appear here as fuzzy blue balls, and platelets, cell fragments that aid in blood clotting, are lavender.

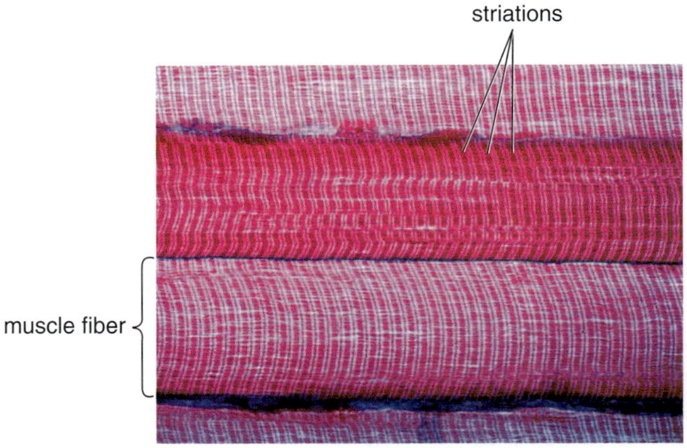

Figure 19-8 Muscle tissue consists of contractile cells called muscle fibers
A regular arrangement of fibrous proteins inside the muscle fibers of skeletal muscle gives this muscle tissue stripes or "striations" when viewed under a microscope.

Although they are liquids, blood and lymph are considered connective tissues because they are composed largely of extracellular fluids. The cellular portion of blood consists of red and white blood cells and cell fragments called platelets in extracellular fluid called plasma (Fig. 19-7). Blood carries dissolved oxygen and other nutrients to cells and carries away carbon dioxide and other cellular waste products. Blood also carries hormones throughout the body. Lymph consists largely of fluid that has leaked out of blood vessels; it is returned to the blood after traveling within lymphatic vessels.

Muscle Tissue Has the Ability to Contract

The long, thin cells of **muscle tissue** contract (shorten) when stimulated, then relax passively (Fig. 19-8). There are three types of muscle tissue: skeletal, cardiac, and smooth. Skeletal muscle is generally under voluntary control. As its name implies, its main function is to move the skeleton, as when you walk or turn the pages of this text. Cardiac muscle is located only in the heart. Unlike skeletal muscle, it is spontaneously active and not under voluntary control. Cardiac muscle cells are interconnected such that electrical signals spread rapidly through the heart, stimulating the cardiac muscle cells to contract in a coordinated fashion. Smooth muscle, so named because it lacks the orderly arrangement of fibrous proteins seen in cardiac and skeletal muscles, is embedded in the walls of the digestive tract, the uterus, the bladder, and large blood vessels. It produces slow, sustained contractions that are normally involuntary.

Nerve Tissue Transmits Electrical Signals

You owe your ability to sense and respond to the world to **nerve tissue**, which makes up the brain, the spinal cord, and the nerves that travel from them to all parts of the body. Nerve tissue is composed of two types of cells: neurons and glial cells. **Neurons** are specialized to generate electrical signals and to conduct these signals to other neurons, muscles, or glands. A neuron has four major parts, each with a specialized function (Fig. 19-9). The dendrites of a neuron receive signals from other neurons or from the external environment. The cell body directs the maintenance and repair of the cell. The axon conducts the electrical signal to its target cell, and the synaptic terminals transmit the signal to the target cell. Glial cells surround, support, and protect neurons and regulate the composition of the extracellular fluid, allowing neurons to function optimally. We shall discuss nerve tissue more fully in Chapter 24.

4 How Are Tissues Combined into Organs?

An organs is formed from at least two tissue types that function together. Some organs include all four tissue types. In this section, we examine the components and functions of a representative animal organ, the skin.

Skin Is an Organ That Contains All Four Tissue Types

In animal skin, an outer epithelium sits atop connective tissue that contains a blood supply, nerves, in some cases muscle, and glandular structures derived from the epithelium.

The epidermis, or outer layer of the skin, is a specialized epithelial tissue (Fig. 19-10). It is covered by a protective layer of dead cells produced by underlying living epidermal cells. These dead cells are packed with the protein keratin, which helps keep the skin both airtight and relatively waterproof. Immediately beneath the epidermis lies a layer of connective tissue, the dermis. The loosely packed cells of the dermis are permeated by blood vessels, which carry the blood that nourishes both the dermal and epidermal tissue. Lymph vessels within the dermis collect and carry off extracellular fluid. Various nerve endings responsive to temperature, touch, pressure, vibration, and pain are scattered throughout the dermis and epidermis.

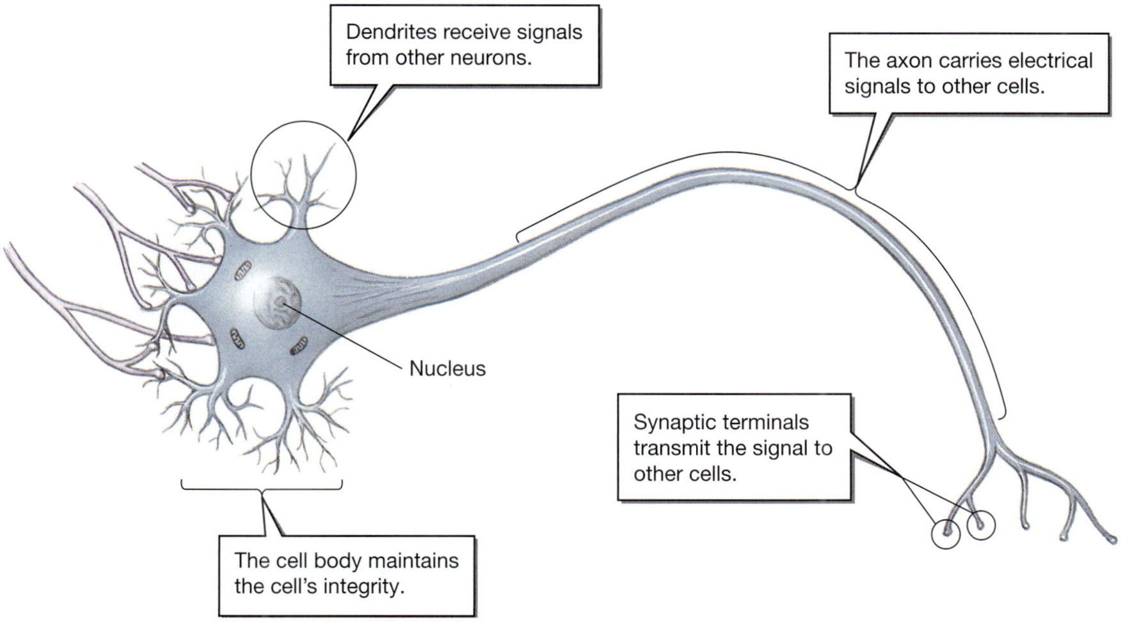

Figure 19-9 A nerve cell
A nerve cell has four major parts, each specialized for a specific function. Nerve cells vary tremendously in shape, depending on their functions in the nervous system.

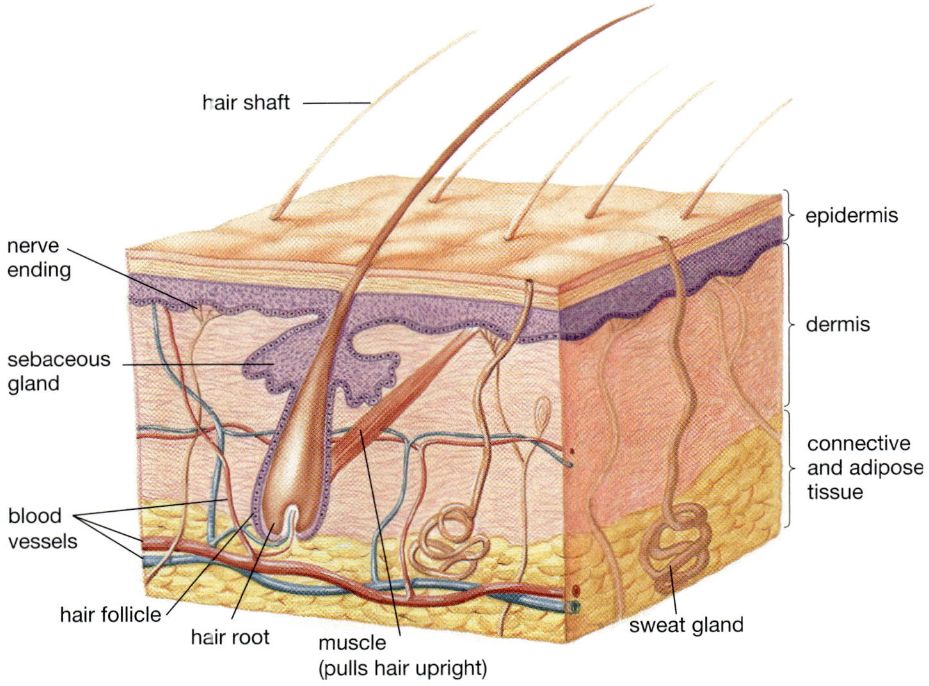

Figure 19-10 Skin
Mammalian skin, a representative organ, in cross section.

The dermis is also packed with glands derived from epithelial tissue. Glands called hair follicles produce hair from protein-containing secretions. Sweat glands produce watery secretions that cool the skin and excrete substances such as salts and urea. Sebaceous glands secrete an oily substance that lubricates the epithelium.

In addition to the epithelial, connective, and nerve tissues, some areas of the skin also contain muscle tissue. Tiny muscles attached to the hair follicles can cause the hairs of the skin to "stand on end" in response to signals from neurons. Although this reaction is useless for retaining heat in humans, most mammals are able to increase the thickness of their insulating fur in cold weather by erecting individual hairs

Organ Systems Consist of Two or More Interacting Organs

Organ systems consist of two or more individual organs (in some cases located in different regions of the body) that work together, performing a common function. For example, the skin is part of the integumentary system, which also includes the hair and the nails and which serves as a barrier between the environment and the inside of the body. Another example is the digestive system, in which the mouth, esophagus, stomach, intestines, and other organs that supply digestive enzymes, such as the liver and pancreas, all function together to convert food into nutrient molecules (see Fig. 19-2). The major organ systems of the vertebrate body and their representative organs and functions are listed in Table 19-1. The anatomy and physiology of these organ systems are the subjects of the next five chapters.

The Body's Organ Systems Act in Concert

The organ systems of the body work together in a coordinated manner to maintain the internal environment in the narrow range required for life. The job of regulating a multitude of factors throughout the body cannot be accomplished by a few feedback mechanisms operating independently within different organ systems. Numerous feedback mechanisms are constantly at work, responding to inputs that continuously change as the animal's activities and external environment change. If these control mechanisms worked independently, each without regard to the activities of the other organ systems, it would not be possible to maintain homeostasis in the body as a whole.

Fortunately, evolution has ensured that the different organ systems work together. For example, the systems that take substances into the body (for example, the digestive system) act in concert with the systems responsible for transporting substances within the body (for example, the circulatory system) and systems that remove substances from the body (such as the excretory

Table 19-1 Major Vertebrate Organ Systems

Organ System	Major Structures	Physiological Role	Organ System	Major Structures	Physiological Role
Circulatory system 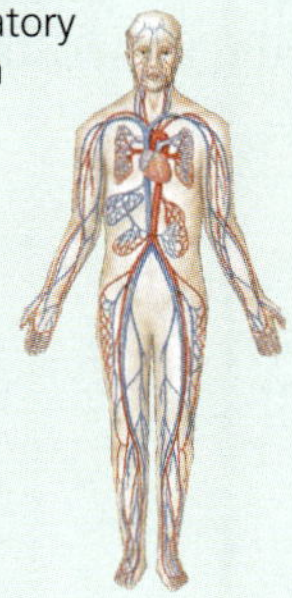	Heart, blood vessels, blood	Transports nutrients, gases, hormones, metabolic wastes; also assists in temperature control	Endocrine system 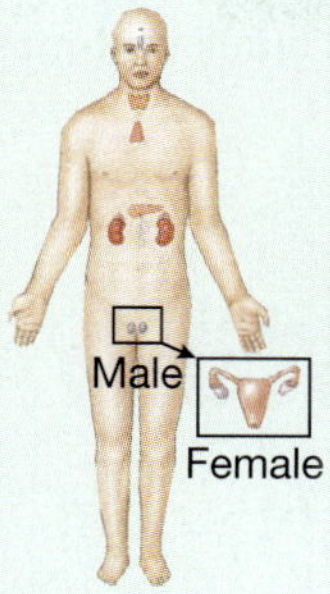	A variety of hormone-secreting glands, including the hypothalamus, pituitary, thyroid, pancreas, and adrenals	Controls physiological processes, typically in conjunction with the nervous system
Lymphatic/ immune system 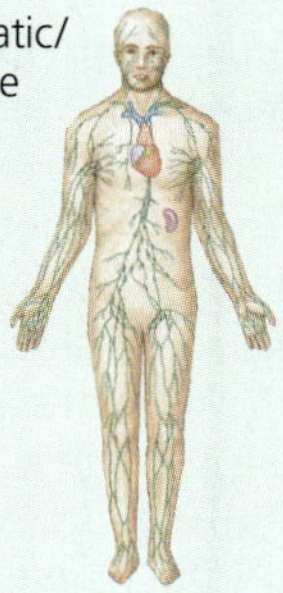	Lymph, lymph nodes and vessels, white blood cells	Carries fat and excess fluids to blood; destroys invading microbes	Nervous system 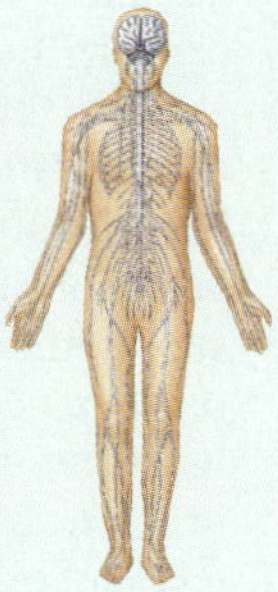	Brain, spinal cord, peripheral nerves	Controls physiological processes in conjunction with the endocrine system; senses the environment, directs behavior
Digestive system 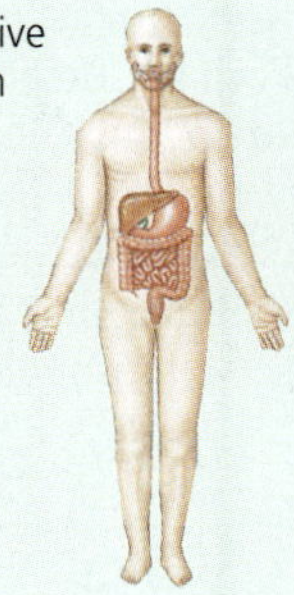	Mouth, esophagus, stomach, small and large intestines, glands producing digestive secretions	Supplies the body with nutrients that provide energy and materials for growth and maintenance	Muscular system 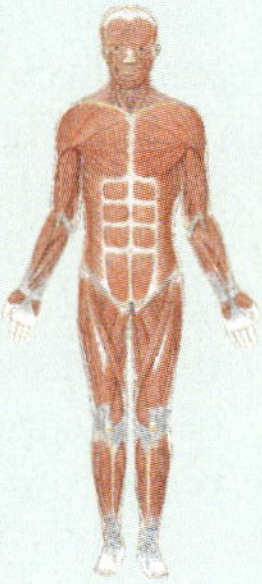	Skeletal muscle	Moves the skeleton
				Smooth muscle	Controls movement of substances through hollow organs (digestive tract, large blood vessels)
				Cardiac muscle	Initiates and implements heart contractions
Excretory system 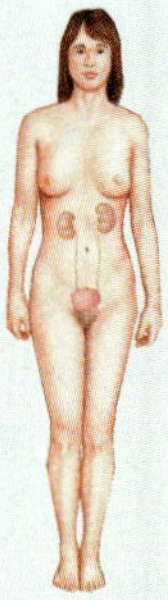	Kidneys, ureters, bladder, urethra	Maintains homeostatic conditions within bloodstream; filters out cellular wastes, certain toxins, and excess water and nutrients	Skeletal system 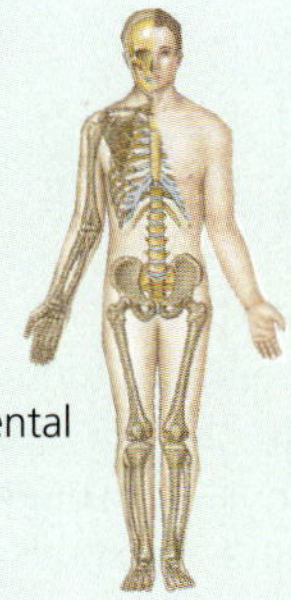	Bones, cartilage, tendons, ligaments	Provides support for the body, attachment sites for muscles, and protection for internal organs
			Integumental system	Skin, hair, nails	Protects internal body parts from injury, infection, and drying out
Respiratory system 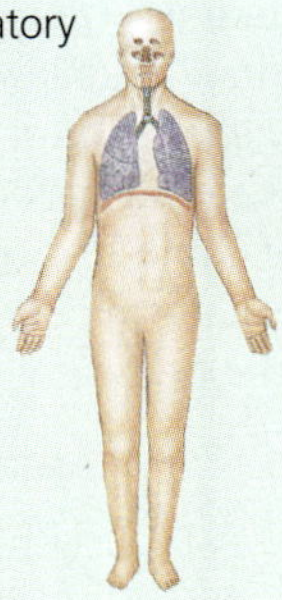	Nose, trachea, lungs (mammals, birds, reptiles, amphibians), gills (fish and some amphibians)	Provides an area for gas exchange between the blood and the environment; allows oxygen acquisition and carbon dioxide elimination	Reproductive system 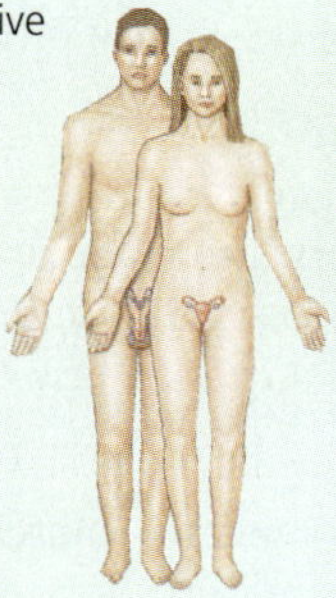	Males: testes, seminal vesicles, penis Female (mammal): ovaries, oviducts, uterus, vagina, mammary glands	Male: produces sperm, inseminates female Female (mammal): produces egg cells, nurtures developing offspring

system). This kind of coordinated action is possible because the body has mechanisms for sending signals from one part of the body to another. Each cell in the body is connected to all the others by an elaborate network of vessels and nerves that carries molecules and messages to the appropriate locations. Molecules are transported to the sites where they are needed and away from locations where their presence would be harmful. Messages travel from sensors to effectors and back again, allowing feedback mechanisms to maintain homeostasis.

CASE STUDY REVISITED

Surviving a Dive

What adaptations make it possible for an emperor penguin to dive to a depth of more than 1,500 feet? One important feature is the bird's ultra-streamlined body. The penguin's streamlined shape, along with its distinctively modified feathers that move through the water with little friction, allows the bird to experience only a small amount of drag as it swims through the water. The penguin thus uses comparatively little energy to power its dive, reducing demand for oxygen. Conservation of oxygen makes it possible to dive deeper.

In addition to conserving oxygen by swimming more efficiently than landlubbers such as humans, emperor penguins also carry more oxygen with them when they dive. Researchers have calculated that, per pound of body weight, an emperor penguin's body contains almost three times as much oxygen as a human's body. Perhaps the emperor penguin's key adaptation for diving, this extraordinary capacity to store high levels of oxygen, is attained by way of extremely high concentrations of the protein myoglobin. Like hemoglobin, myoglobin binds oxygen molecules that are later released to supply the metabolic needs of cells. This protein occurs in emperor penguins at levels ten times higher than those found in terrestrial birds. An abundance of myoglobin acts as a tremendous reservoir of oxygen, allowing the emperor penguin to remain submerged through long dives in deep water.

Streamlining and concentrated myoglobin help account for an emperor penguin's ability to remain submerged for long dives, but they do not explain how the bird's body is able to resist the effects of pressure and the subsequent reduction of pressure. In fact, the penguin's mechanisms for coping with the effects of pressure remain poorly understood. Some clues, however, come from studies of a deep-diving mammal, the Weddel seal. This seal can dive to 2,200 feet and stay submerged for more than an hour. During even its deepest dives, the level of nitrogen dissolved in the seal's blood remains low. Apparently, the animal's lungs collapse completely during a dive, rendering impossible any exchange of gases between the lungs and the blood. The level of dissolved gases in the blood therefore stays low, and the worst effects of ascending are avoided. Perhaps emperor penguins have an analogous method of preventing the movement of gases from the lungs to the blood (although researchers do not believe that the penguin's lungs collapse so thoroughly).

Summary of Key Concepts

1) How Do Animals Maintain Internal Constancy?

Conditions in an animal's internal environment tend to remain within a narrow range that permits the continuation of life. This homeostasis is maintained mostly through negative feedback, in which a change triggers a response that counteracts the change and restores conditions to a set point. Positive feedback, in which a change initiates events that intensify the change, is rare in physiological processes.

2) How Is the Animal Body Organized?

The animal body is composed of organ systems consisting of two or more organs. Organs, in turn, are made up of tissues. A tissue is a group of cells that forms a structural and functional unit that is specialized for a specific task.

3) How Do Tissues Differ?

Animal tissues include epithelial, connective, muscle, and nerve tissue. Epithelial tissue covers internal and external body surfaces and also gives rise to glands. Connective tissue contains considerable extracellular material. Muscle tissue is specialized for movement. Nerve tissue is specialized for the generation and conduction of electrical signals.

4) How Are Tissues Combined into Organs?

Skin is an organ that includes epidermis (epithelial tissue), dermis (connective tissue), nerve endings (nerve tissue), and tiny muscles (muscle tissue). Animal organ systems include the digestive, excretory, immune, respiratory, circulatory/lymphatic, nervous, muscular, skeletal, endocrine, and reproductive systems, summarized in Table 19-1. The organ systems contain many feedback mechanisms that work in concert to maintain homeostasis in the body as a whole.

Key Terms

bone *p. 363*
cartilage *p. 363*
connective tissue *p. 363*
endocrine glands *p. 363*
epithelial tissue *p. 362*
exocrine gland *p. 363*
gland *p. 363*
homeostasis *p. 360*
muscle tissue *p. 364*
negative feedback *p. 360*
nerve tissue *p. 365*
neuron *p. 365*
organ *p. 361*
organ system *p. 361*
positive feedback *p. 360*
tissue *p. 361*

Thinking Through the Concepts

Multiple Choice

1. *The skin contains*
   a. epithelial tissue
   b. connective tissue
   c. nerve tissue
   d. muscle tissue
   e. all of the above
2. *Glands that become separated from the epithelium that produced them are called ________ glands.*
   a. sebaceous
   b. sweat
   c. exocrine
   d. endocrine
   e. saliva
3. *Epithelial tissues*
   a. cover the body
   b. line body cavities
   c. may create barriers that alter the movement of certain substances
   d. are continuously replaced by cell division
   e. all of the above
4. *All of the following are examples of connective tissue EXCEPT*
   a. tendons
   b. ligaments
   c. blood
   d. muscle
   e. adipose tissue
5. *Which of the following statements about muscle is true?*
   a. Smooth muscle is important in locomotion.
   b. Skeletal muscle is not under voluntary control.
   c. Cardiac muscle conducts electrical signals.
   d. Smooth muscle is called voluntary muscle.
   e. Smooth muscle moves the skeleton.
6. *All of the following are found in the dermis EXCEPT*
   a. blood vessels
   b. sensory nerve endings
   c. hair follicles
   d. sebaceous glands
   e. cells packed with keratin

? Review Questions

1. Define homeostasis, and explain how negative feedback helps maintain it. Explain one example of homeostasis in the human body.
2. Explain positive feedback, and provide one physiological example. Explain why this type of feedback is rare in physiological processes.
3. Explain why body temperature in humans cannot be maintained at *exactly* 98.6 °F (37 °C) at all times.
4. Describe the structure and functions of epithelial tissue.
5. What property distinguishes connective tissue from all other tissue types? List five types of connective tissue, and briefly describe the function of each type.
6. Describe the skin as an organ. Include the various tissues that compose it and the role of each of these.

Applying the Concepts

1. Why does life on land present more difficulties in maintaining homeostasis than does life in water? What made it evolutionarily advantageous for organisms to colonize dry land?
2. The majority of homeostatic regulatory mechanisms in animals are "autonomic"—that is, not requiring conscious control. Discuss several reasons why this type of regulation is more advantageous to the animal than is conscious regulation of homeostatic controls.
3. Third-degree burns are usually painless. Skin regenerates only from the edges of these wounds. Second-degree burns regenerate from cells located at the burn edges, in

hair follicles and in sweat glands. First-degree burns are painful but heal rapidly from undamaged epidermal cells. From this information, draw the depth of first-, second-, and third-degree burns on Figure 19-10.

4. A coroner dictates the following description during an autopsy: "The tissue I am looking at forms part of the fetal skeleton. The extracellular matrix appears transparent. Fibers of collagen are present but are small and evenly dispersed in the extracellular matrix. Blood vessels have not yet penetrated the matrix." What tissue is the coroner describing?

5. Imagine you are a health-care professional teaching a prenatal class for fathers. Make a design for a machine with sensors, electrical currents, motors, and so on that would illustrate feedback relationships involved in the initiation of labor that a layperson could understand.

Group Activity

The physiological systems that arise as a result of natural selection often seem as if they have been designed to achieve their functions. This seemingly intentional design in natural systems is an illusion, but the human process of intentional design can nonetheless help us understand how physiological systems work. To try this for yourself, form a group of four students. Your group is part of a larger team that is trying to design a self-regulating robot vehicle. You've been given two tasks. First, design a system that will keep the temperature of the lubricant in the vehicle's wheel bearings between 100 °F and 110 °F, no matter how fast the vehicle is going. Second, design a system that will ensure that the vehicle swerves to avoid obstacles. Your designs may include any imaginable sensors, effectors, and control centers but must be totally self-contained. Submit a diagram and a brief written description of your designs.

For More Information

Bruemmer. F. "Five Days with Fat Hoods." *International Wildlife*, January–February 1999. The rapid growth and prodigious fat-storing ability of the hooded seal adapts it to maintaining homeostasis under the extreme conditions of the far north.

Kooyman, G. L., and Ponganis, P. J. "The Challenges of Diving to Depth." *American Scientist*, November–December 1997. Summary of recent research on the physiological adaptations of deep-diving birds and mammals.

Nuland, S. *The Wisdom of the Body*. New York: Alfred A. Knopf, 1997. Human physiology as seen through a surgeon's eyes. A firsthand account of the beauty and power of the body's mechanisms for maintaining homeostasis.

Storey, K. B., and Storey, J. M. "Frozen and Alive." *Scientific American*, December 1990. Some animals have special adaptations that allow them to withstand freezing.

Answers to Multiple-Choice Questions

1. e 2. d 3. e 4. d 5. c 6. e

MEDIATUTOR

CD Activities

Activity 19.1: Feedback Loops and Homeostasis

Estimated time: 5 minutes

In animals, homeostasis is maintained through feedback systems. In this tutorial, you will discover the nature of the feedback system and explore some examples.

Start the MediaTutor Student CD-ROM and enter the activity number in the Quick Search box to be taken directly to that activity.

Web Investigation

Surviving a Dive

Estimated time: 15 minutes

The physiology of organisms is often a function of their anatomy. As discussed in the chapter, form is very closely tied to function when it comes to the tissues, organs, and overall structure of an animal. This investigation will explore some intriguing aspects of this relationship, including an in-depth look at the physiology of deep-sea diving.

Go to http://www.prenhall.com/audesirk3, the Audesirk Companion Web site. Select Chapter 19 and Web Investigation to begin.

In the future, pigs may be raised not just for bacon and ham but also for hearts and lungs that will be transplanted to human recipients.

20 Circulation and Respiration

AT A GLANCE

CASESTUDY

Organ Farmers

In a brightly lit operating room, a pig lies on an operating table. Working quickly, surgeons remove the animal's heart and lungs, which are packed in ice and rushed to a hospital a short distance away. There, a different surgical team has opened the chest cavity of a human patient. The patient has heart disease, and his lungs are failing as well. A heart-lung transplant is the only option. When the pig's organs arrive, the surgeons use them to replace the diseased ones in the human patient.

The scenario outlined above is impossible now, but it could become reality in the not-too-distant future. Pressure is building to develop *xenotransplants*, transplants of organs and tissues from one species to another. The pressure arises because the need for transplants is large and growing. Heart and lung transplants, for example, save many lives, because diseases of the heart and lungs are among the leading causes of death in industrialized countries. But the demand for transplants far exceeds the supply of available human organs. In the United States alone, more than 8000 people are currently on waiting lists for hearts or lungs. Many of these people will die before a usable organ becomes available. Overall, about 6000 Americans die each year while on waiting lists for organs.

To many people, the logical way out of this crisis is xenotransplantation of animal organs. Most governments, however, currently prohibit xenotransplants, mainly because of the complex ethical and practical questions raised by cross-species transplants. For example, some researchers worry that xenotransplants of primate organs might provide a pathway for primate diseases to move into human populations. (The virus that causes AIDS in humans probably originated in wild chimpanzees.) Furthermore, the idea of raising captive apes and monkeys to be sacrificed for organ transplants is, for many, a disturbing prospect.

In light of low enthusiasm for primate donors, xenotransplant advocates have turned to pigs as the donor animal of choice. But even for those who are willing to sacrifice pigs to save human lives, a number of thorny technical difficulties remain. We will consider some of them at the end of this chapter, after reviewing some of the organ systems whose failure can lead to a need for transplants. ■

1 What Are the Major Features and Functions of Circulatory Systems?

Billions of years ago, the first cells were nurtured by the sea in which they had evolved. The seawater carried dissolved nutrients, which diffused into the cells and washed away the wastes that had diffused out. Today, microorganisms and some simple multicellular animals still rely almost exclusively on diffusion for the exchange of nutrients and wastes with the environment. Sponges, for example, rely on seawater that circulates through pores in their bodies, bringing the environment close to each cell. In larger, more complex animals, however, many cells are more distant from the outside world. This distance suggests a question: How can nutrients reach each cell and how can the cell avoid being poisoned by its own wastes? The answer lies in the circulatory system, a sort of internal sea that serves the same purpose that the actual sea did for the first cells. The circulatory system transports to each cell a fluid rich in food and oxygen and carries away wastes produced by the cells.

Circulatory systems have three main components. A fluid, **blood**, serves as a medium of transport. A system of channels, **blood vessels**, conducts blood throughout the body. A pump, the **heart**, keeps the blood circulating. In the next few sections, we look at the different parts of the vertebrate circulatory system, focusing on humans as a representative vertebrate.

The Vertebrate Circulatory System Transports Many Substances

Your circulatory system helps sustain all the other organ systems in your body by performing a variety of transport functions:

- Transporting oxygen from the lungs to the tissues
- Transporting carbon dioxide from the tissues to the lungs
- Transporting nutrients from the digestive system to the tissues
- Transporting waste products and toxic substances to the liver (where many of them are detoxified) and to the kidney for excretion
- Transporting hormones from the glands and organs that produce them to the tissues on which they act

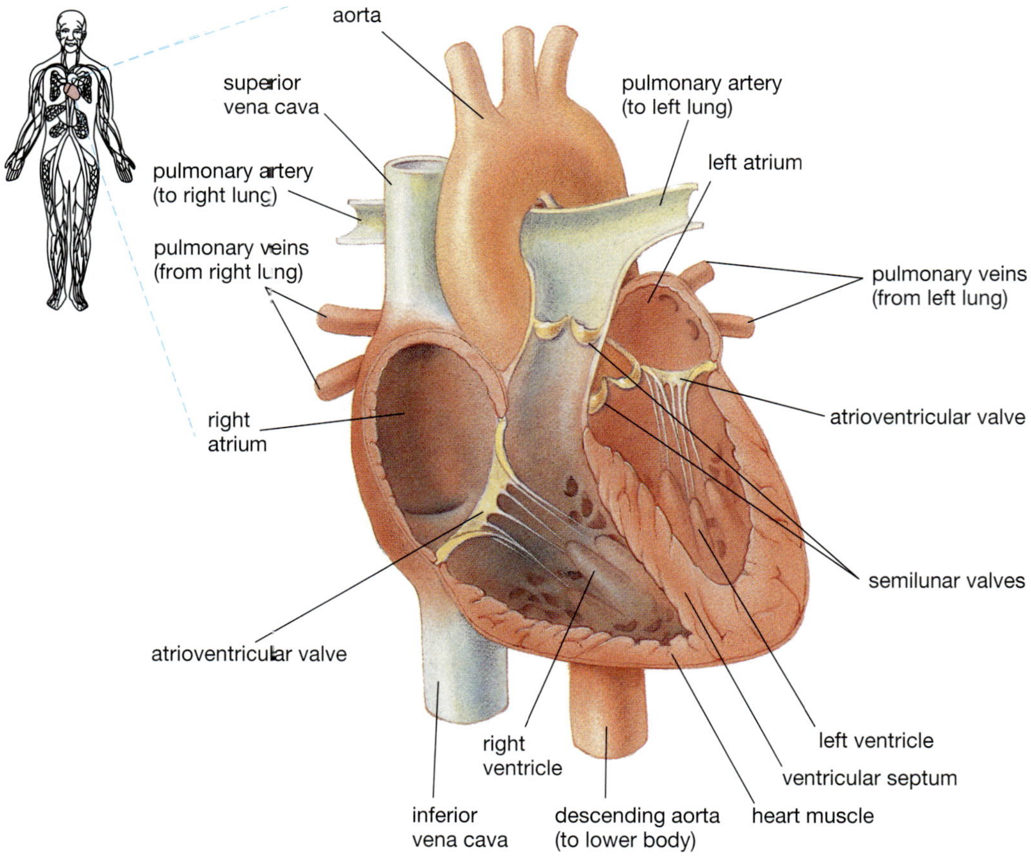

Figure 20-1 The human heart and its valves and vessels
The heart appears here as if it were in a body facing you, so that right and left appear reversed. Note the thickened walls of the left ventricle, which must pump blood much farther through the body than does the right ventricle, which propels blood to the lungs. One-way valves, called semilunar valves, are located between the aorta and the left ventricle, and between the pulmonary artery and the right ventricle. Atrioventricular valves separate the atria and ventricles.

MediaTutor 20.3 Heart Structure and Function

In addition, the circulatory system helps regulate body temperature by adjusting blood flow and helps protect the body from bacteria and viruses by circulating components of the immune system.

2 How Does the Vertebrate Heart Work?

The circulatory system requires a dependable pump. Blood must be moved through the body continuously throughout an animal's life. In vertebrates, blood movement is driven by muscle contractions in the heart.

The Vertebrate Heart Consists of Muscular Chambers

In vertebrate hearts, muscular chambers called **atria** (singular, **atrium**) collect blood, which moves from the atria into the **ventricles**, chambers whose contractions circulate blood through the body. Human hearts (and those of other mammals and birds) can be thought of as two separate pumps. In each pump, an atrium receives and briefly stores the blood and then passes it to a ventricle that propels it through the body (Fig. 20-1).

One pump, consisting of the right atrium and right ventricle, deals with oxygen-depleted blood, pumping it to the lungs for replenishment. Oxygen-depleted blood moves from the body to the right atrium through a large **vein** (a vessel that carries blood *toward* the heart) called the superior vena cava. The right atrium contracts, forcing the blood into the right ventricle. Then, the right ventricle contracts, sending the oxygen-depleted blood to the lungs via pulmonary **arteries** (vessels that carry blood *away from* the heart).

The other pump, consisting of the left atrium and ventricle, deals with oxygen-rich blood. Oxygen-rich blood from the lungs enters the left atrium through pulmonary veins and is then squeezed into the left ventricle. Strong contractions of the left ventricle, the heart's most muscular chamber, send the oxygenated blood coursing out through a major artery, the *aorta*, to the rest of the body.

The Atria and Ventricles Contract in a Coordinated Cycle

Your heart beats about 100,000 times each day, with the chambers alternately contracting and relaxing. The two atria contract at the same time, emptying their contents into the ventricles. A fraction of a second later, the two ventricles contract simultaneously, forcing blood into arteries that exit the heart. Both atria and ventricles then relax briefly before the cycle repeats (Fig. 20-2). At a normal resting heart rate, this cycle lasts just under 1 second. In determining blood pressure (Fig. 20-3), systolic pressure (the higher of the two readings) is measured *during* ventricular contraction and diastolic pressure is measured *between* contractions.

Valves Prevent Blood from Moving in the Wrong Direction

Coordinating the activity of the heart's four chambers to maintain blood flow presents challenges. First, when the ventricles contract, the blood must be directed out through the arteries and not back up into the atria. Then, once blood has entered the arteries, it must be prevented from flowing back as the heart relaxes. These needs are met by four one-way *valves* (see Figs. 20-1 and 20-2). Pressure in one direction opens the valves, but reverse pressure forces them tightly closed. **Atrioventricular valves**

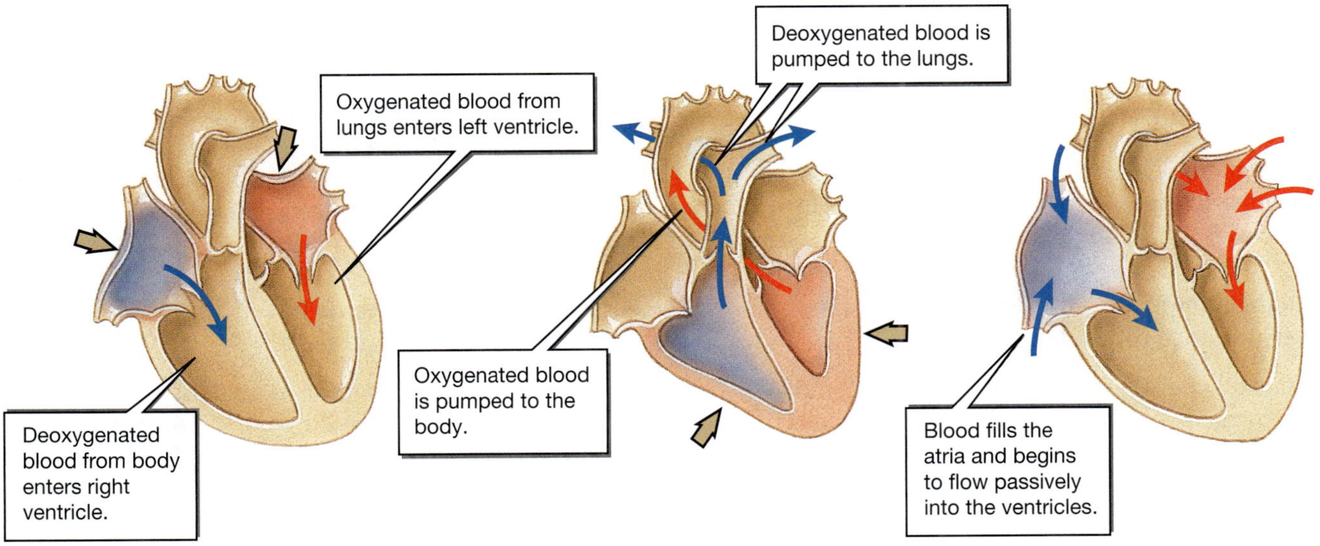

(a) Atria contract, forcing blood into the ventricles.

(b) Then the ventricles contract, forcing blood through arteries to the lungs and the rest of the body.

(c) The cycle ends as the heart relaxes.

Figure 20-2 The cardiac cycle

MEDIATUTOR 20.4 Cardiac Cycle

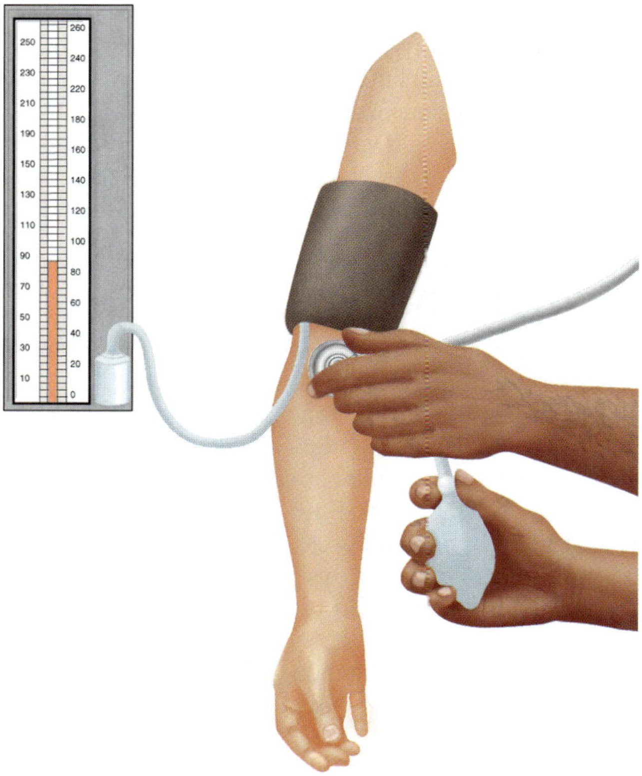

Figure 20-3 Measuring blood pressure
Blood pressure is measured with an inflatable blood pressure cuff and a stethoscope. The cuff is inflated until its pressure closes off the arm's main artery, blood ceases to flow, and no pulse can be detected below the cuff. Then the pressure is gradually reduced. When the pulse is first audible in the artery, the pressure pulses created by the contracting left ventricle are just overcoming the pressure in the cuff and blood is flowing. This is the upper reading: the systolic pressure. Cuff pressure is then further reduced until no pulse is audible, indicating that blood is flowing continuously through the artery and that the pressure between ventricular contractions is just overcoming the cuff pressure. This is the lower reading: the diastolic pressure. The numbers are in millimeters of mercury, a standard measure of pressure also used in barometers.

separate the atria from the ventricles. **Semilunar valves** allow blood to enter the pulmonary artery and the aorta when the ventricles contract but prevent it from returning as the ventricles relax.

Electrical Impulses Coordinate the Sequence of Contractions

A second challenge is to coordinate the contractions of all four chambers. The atria must contract first and empty their contents into the ventricles so that the atria can refill while the ventricles contract. Thus, there must be a delay between the contractions of the atria and those of the ventricles. The contraction of the heart is initiated and coordinated by a **pacemaker**, a cluster of specialized heart muscle cells that produce spontaneous electrical signals at a regular rate. The electrical signals pass rapidly between heart cells and cause them to contract.

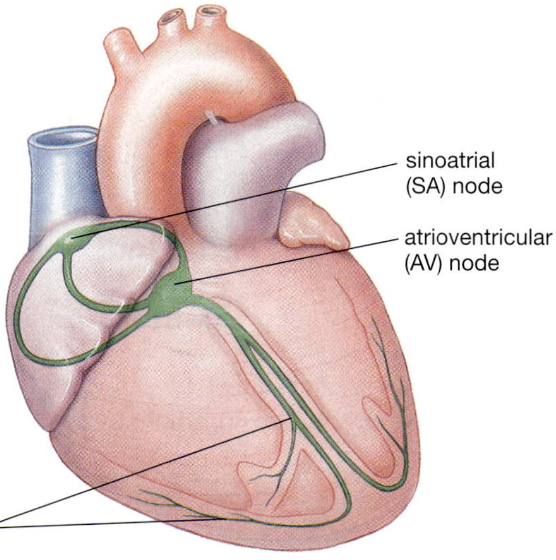

Figure 20-4 The heart's pacemaker and its connections
The sinoatrial (SA) node serves as the heart's pacemaker. Its signal to contract spreads through the muscle fibers of both atria to the atrioventricular (AV) node. The AV node then transmits the signal through bundles of excitable fibers that stimulate the ventricular muscles to contract.

The heart's primary pacemaker is the **sinoatrial (SA) node**, located in the wall of the right atrium (Fig. 20-4). From the SA node, an electrical impulse creates a wave of contraction that sweeps through the muscles of the right and left atria, which contract together. Then the signal reaches a barrier of unexcitable tissue between the atria and the ventricles. There the excitation is channeled through the **atrioventricular (AV) node**, a small mass of specialized muscle cells located in the floor of the right atrium (see Fig. 20-4). The impulse is delayed at the AV node, postponing the ventricular contraction for about 0.1 second after the atria contract. This delay gives the atria time to complete the transfer of blood into the ventricles before ventricular contraction begins. From the AV node, the signal to contract spreads to the base of the two ventricles along tracts of excitable fibers. The impulse then travels rapidly from these fibers through the muscle cells, causing the ventricles to contract in unison.

The Heart's Contractions Result from Movement of Filaments in Muscle Cells

The cells that make up heart muscle tissue are known as *muscle fibers*. Each muscle fiber contains many *myofibrils*, cylindrical structures that extend from one end of the fiber to the other (Fig. 20-5a). The myofibrils are composed of subunits called *sarcomeres*, which are aligned end to end along the length of the myofibril (Fig. 20-5b). Within each sarcomere is a beautifully precise arrangement of filaments of the proteins actin and myosin. Actin molecules form the *thin filaments*; myosin forms the *thick filaments*. The thick filaments are sus-

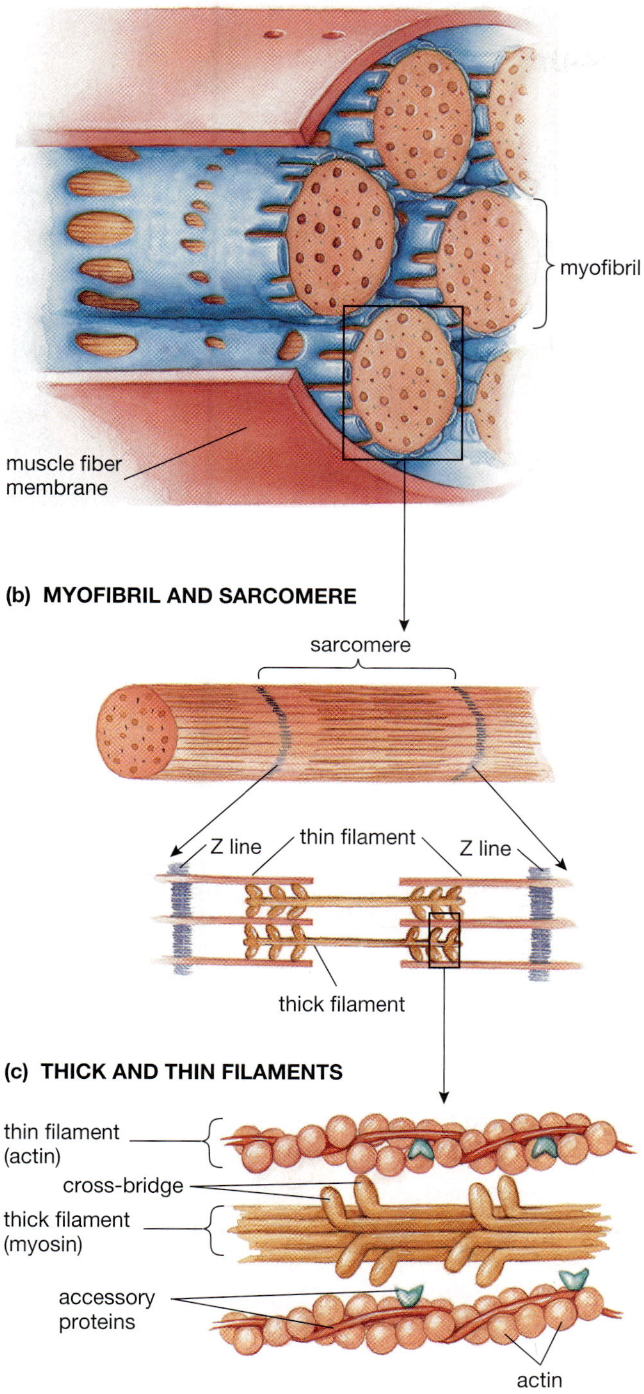

Figure 20-5 A muscle fiber
(a) Each muscle fiber is surrounded by plasma membrane and contains numerous myofibrils. *(b)* Each myofibril consists of sarcomeres, attached end to end by proteins called Z lines. *(c)* Within each sarcomere thick and thin filaments alternate and can be connected by cross-bridges.

pended between the thin filaments and are capable of linking temporarily to the thin filaments using small myosin projections called *cross-bridges* (Fig. 20-5c). The thin filaments are attached to fibrous protein bands called Z lines, which separate adjacent sarcomeres. The regular arrangement of thick and thin filaments in all the myofibrils gives heart (and skeletal) muscle fibers their striped appearance.

Muscle contraction is controlled by a process that depends on the molecular structure of the thin filament. The actin protein that makes up most of the thin filament is formed from a double chain of subunits, and resembles a twisted double strand of pearls. Each of the subunits has a binding site for a myosin cross-bridge. In a relaxed muscle cell, however, these binding sites are blocked by molecules of accessory proteins (see Fig. 20-5c). These accessory proteins prevent the myosin cross-bridges from attaching to the thin filament.

When a muscle contracts, the accessory proteins of the thin filament move aside, exposing the binding sites on the actin. As soon as the sites are exposed, myosin binds the actin, forming cross-bridges. Using energy from the splitting of adenosine triphosphate (ATP), the cross-bridges repeatedly bend, release, and reattach farther along, much like a sailor pulling in an anchor line hand over hand (Fig. 20-6a). The thin filaments are pulled past the thick filaments, shortening the sarcomere and contracting the muscle (Fig. 20-6b).

Injury and Disease Can Impair Heart Function

Certain disorders interfere with the events that produce smooth, regular heartbeats. If the pacemaker fails, uncoordinated, irregular contractions called fibrillation occur. Fibrillation of the ventricles can be fatal, because blood is not pumped out of the heart to the brain and other organs but is merely sloshed around. A defibrillating machine applies a jolt of electricity to the heart, synchronizing the contractions of the ventricular muscles and sometimes allowing the pacemaker to resume its normal coordinating function.

MEDIATUTOR 20.1 Muscle Structure

Heart valves may be attacked and damaged by bacterial infections, allowing blood to flow inappropriately between the chambers. In such cases valve transplants, some using pig heart valves, can be lifesaving. Some diseases attack the heart muscle itself. Damaged heart muscle cells die, and the heart walls thin and weaken. Under these conditions, the patient becomes a candidate for a heart transplant.

3 What Is Blood?

Blood, which has been called the "river of life," transports dissolved nutrients, gases, hormones, and wastes through the body. It has two major components: a fluid called **plasma**, and cellular components (red blood cells, white blood cells, and platelets) that are suspended in the plasma. On average, the cellular components of blood account for 40% to 45% of its volume; the other 55% to 60% is plasma. The average human has 5 to 6 liters of blood, constituting about 8% of the total body weight.

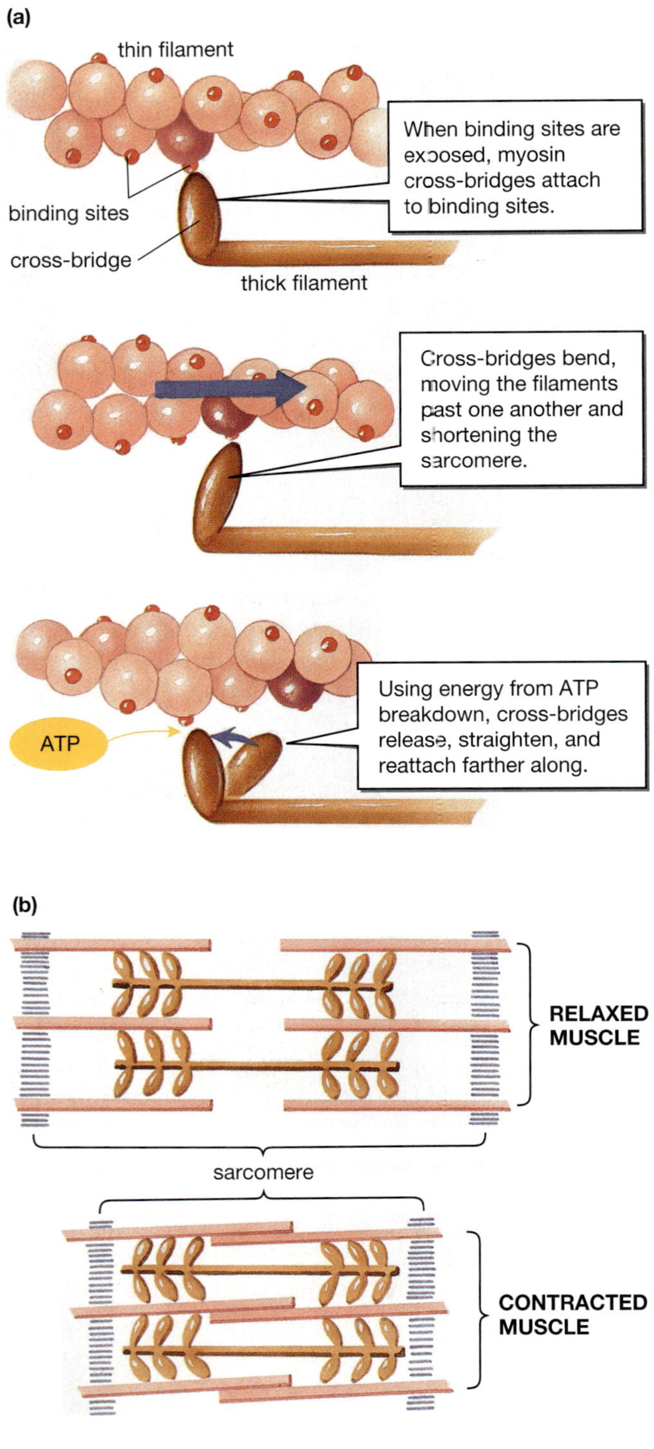

Figure 20-6 Muscle contraction
(a) The cycle of cross-bridge attachment, bending, release, and reattachment repeats over and over again. ***(b)*** The thick and thin filaments slide past one another, shortening the muscle cell.

MEDIATUTOR 20.2 Muscle Movement

Plasma Is Primarily Water in Which Substances Are Dissolved

Water makes up about 90% of the straw-colored plasma. Dissolved in the plasma are proteins, hormones, nutrients (glucose, vitamins, amino acids, lipids), gases (carbon dioxide, oxygen), salts (sodium, calcium, potassium, magnesium), and wastes, such as urea. Plasma proteins are the most abundant of the dissolved substances.

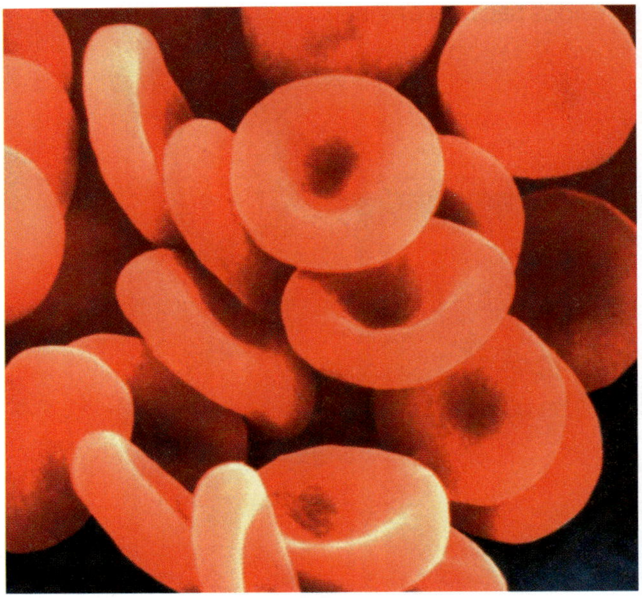

Figure 20-7 Red blood cells
This false-color scanning electron micrograph shows the shape of red blood cells.

Red Blood Cells Carry Oxygen from the Lungs to the Tissues

The most abundant cells in the blood are the oxygen-carrying **red blood cells**, or erythrocytes. In fact, red blood cells make up about 99% of all blood cells. The shape of a red blood cell resembles a ball of clay that has been squeezed between your thumb and forefinger (Fig. 20-7). This shape provides a larger surface area than would a spherical cell of the same volume, and increases the cell's ability to absorb and release oxygen through its plasma membrane.

Red blood cells get their red color from **hemoglobin**, an iron-containing protein that accounts for about one-third the weight of each red blood cell and carries about 97% of the blood's oxygen. One hemoglobin molecule can bind and carry up to four molecules of oxygen, permitting blood to hold far more oxygen than would be possible if all the gas were simply dissolved in plasma. Hemoglobin picks up oxygen in the blood vessels of the lungs, where oxygen concentration is high, and releases it in other tissues of the body, where the oxygen concentration is lower. After releasing its oxygen, a hemoglobin molecule may pick up carbon dioxide from the tissues for transport back to the lungs.

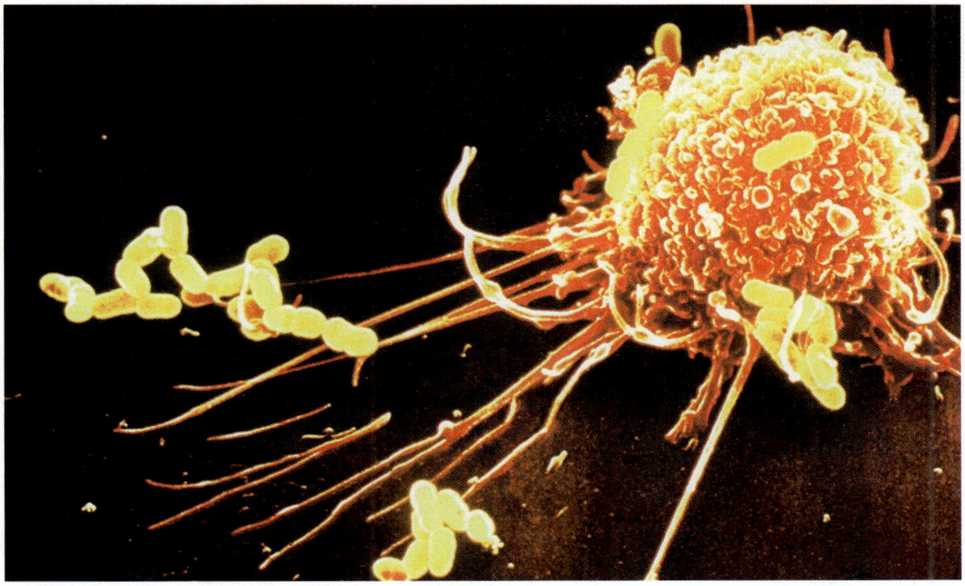

Figure 20-8 A white blood cell attacks bacteria
An amoeba-like white blood cell captures bacteria (the small yellow rods). These bacteria are *Escherichia coli*, intestinal bacteria that can cause disease if they enter the bloodstream.

White Blood Cells Help Defend the Body Against Disease

White blood cells are your body's first line of defense when it is invaded by disease-causing microbes. There are five common types of white blood cells, or leukocytes, which together make up less than 1% of all blood cells. Most white blood cells function in some way to protect the body against foreign invaders and use the circulatory system to travel to the site of invasion. For example, some travel through blood vessels to wounds where bacteria have gained entry, and then ooze out through narrow openings in the capillary walls. After leaving the capillaries, some white blood cells change into amoeba-like cells that engulf foreign particles, including bacteria and cancer cells (Fig. 20-8). They typically die in the process, and their dead bodies accumulate and contribute to the white substance, called *pus*, seen at infection sites.

Platelets Are Cell Fragments That Aid in Blood Clotting

Platelets, which are crucial to blood clotting, are pieces of large cells called megakaryocytes. Megakaryocytes remain in the bone marrow, where they pinch off membrane-enclosed chunks of their cytoplasm that we call platelets (Fig. 20-9). The platelets then enter the blood and play a central role in blood clotting.

Blood clotting is a complex process that keeps us from bleeding to death from normal wear and tear on the body. Clotting starts when platelets contact an irregular surface, such as a damaged blood vessel. The ruptured surface of an injured blood vessel causes platelets to adhere and partially block the opening. The adhering platelets and the injured tissues initiate a complex sequence of events among circulating plasma proteins. These events result in the production of a fibrous network of molecules of the protein fibrin. This protein web immobilizes the fluid portion of the blood, causing it to solidify in much the same way that gelatin does as it cools. The web traps red blood cells, further increasing the density of the clot (Fig. 20-10). Platelets adhere to the fibrous mass and send out sticky projections that attach to one another. Within half an hour, the platelets contract, pulling the mesh tighter and forcing liquid out. This action creates a denser, stronger clot (on the skin it is called a *scab*) and also constricts the wound, pulling the damaged surfaces closer together in a way that promotes healing.

The Big Picture

The circulatory system is the body's highway network. It connects to all parts of the body and provides pathways for the transport of any substance that needs to be moved from one body location to another.

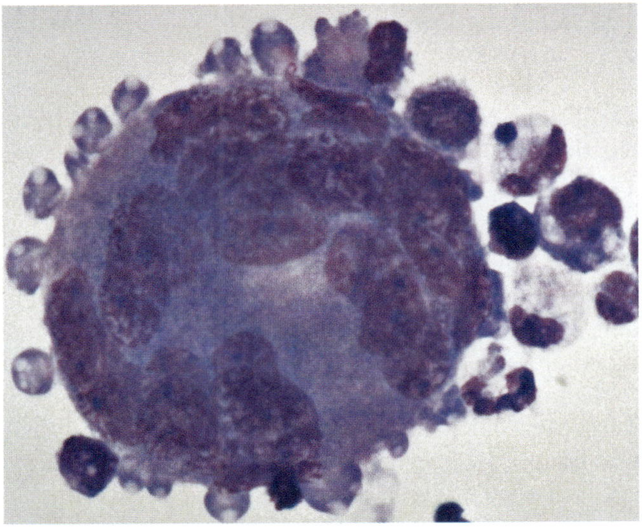

Figure 20-9 The production of platelets
Here a single megakaryocyte is budding off dozens of platelets.

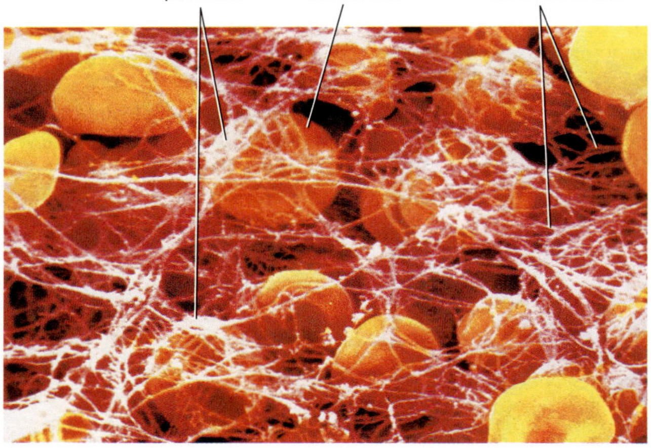

Figure 20-10 Blood clotting
Threadlike fibrin proteins produce a tangled sticky mass that traps red blood cells and eventually forms a clot.

4 What Are the Types and Functions of Blood Vessels?

The river of life flows in well-defined channels called blood vessels. Some of the major blood vessels of the human circulatory system are diagrammed in Figure 20-11. As it leaves the heart, blood travels from arteries to arterioles to capillaries to venules to veins, which return it finally to the heart. These vessels are shown in Figure 20-12. The consequences of impaired vessel function are discussed in "Health Watch: Matters of the Heart."

Arteries and Arterioles Carry Blood Away from the Heart

Arteries carry blood away from the heart. These vessels have thick walls embedded with smooth muscle and elastic tissue (see Fig. 20-12). With each surge of blood from the ventricles, the arteries expand slightly, like thick-walled balloons. As their elastic walls recoil between heartbeats, the arteries actually help pump the blood and maintain a steady flow through the smaller vessels. Arteries branch into vessels of smaller diameter called **arterioles**.

Capillaries Are Microscopic Vessels Through Which Nutrients and Wastes Are Exchanged

The entire circulatory system is an elaborate device for allowing each cell of the body to exchange nutrients and wastes by diffusion. The diffusion occurs in the **capillaries**, the smallest of all blood vessels. Here wastes, nutrients, gases, and hormones are exchanged between the blood and the body cells. Capillaries are finely adapted to their role of exchange, with walls that are only one cell thick. Most nutrients, oxygen, and carbon dioxide diffuse readily through capillary plasma membranes. Salts and small charged molecules (including some small proteins) move through the capillary plasma membrane or be-

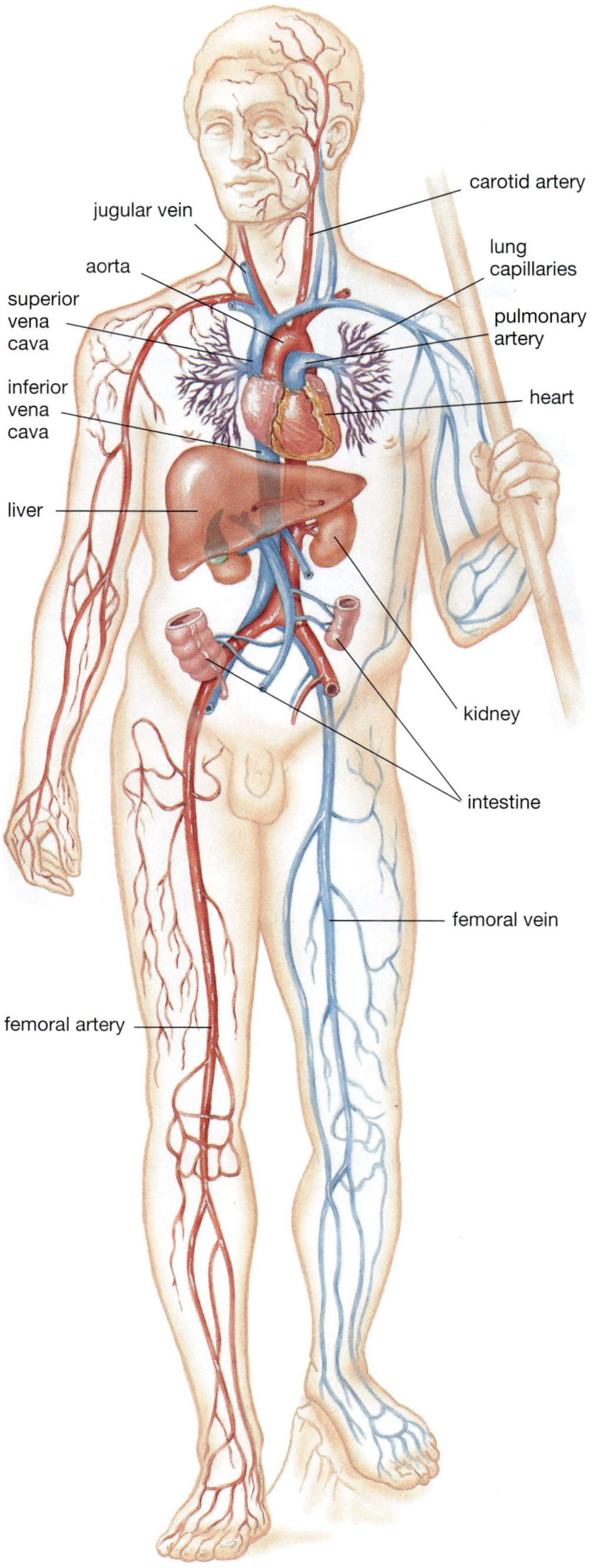

Figure 20-11 The human circulatory system
Most veins (shown on the figure's right side) carry deoxygenated blood to the heart, and most arteries (shown on left) conduct oxygenated blood away from the heart. The pulmonary veins (carrying oxygenated blood) and arteries (carrying deoxygenated blood) are exceptions.

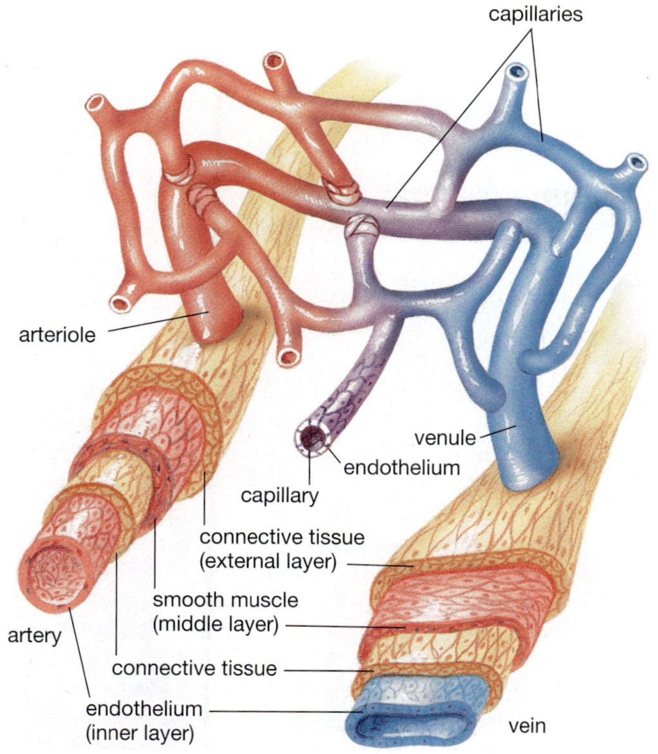

Figure 20-12 Structures and interconnections of blood vessels Oxygenated blood moves from arteries to arterioles to capillaries. Capillaries empty deoxygenated blood into venules, which empty into veins.

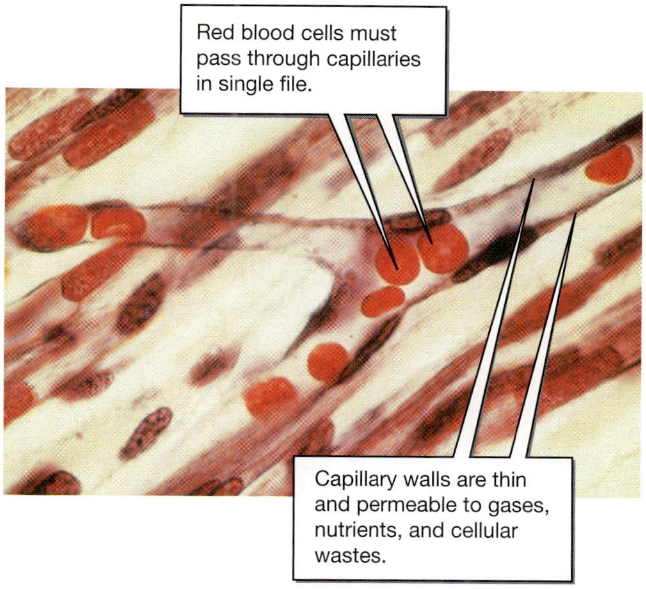

Figure 20-13 Red blood cells flow through a capillary

tween adjacent capillary cells. White blood cells can ooze through the openings between capillary cells. However, large plasma proteins, red blood cells, and platelets are unable to leave the capillaries because they are too large to fit through the channels in the plasma membrane.

The pressure within capillaries causes fluid to leak continuously from the blood plasma into the spaces that surround the capillaries and tissues. This fluid, known as the **interstitial fluid**, consists primarily of water in which are dissolved nutrients, hormones, gases, wastes, and small proteins from the blood. Materials are exchanged between capillary blood and nearby cells through this interstitial fluid, which bathes nearly all the body's cells.

Capillaries are so narrow that red blood cells must pass through them in single file (Fig. 20-13). Consequently, all the blood is sure to pass very close to the capillary walls, where exchange occurs. In addition, capillaries are so numerous that no body cell is more than 100 micrometers (about as thick as four pages of this book) from a capillary. A person has about 50,000 miles (80,600 kilometers) of capillaries, enough to encircle the globe twice! The speed of blood flow drops very quickly as it moves through this narrow, almost endless capillary network, allowing more time for diffusion.

Venules and Veins Carry Blood Back to the Heart

After picking up carbon dioxide and other cellular wastes from cells, capillary blood drains into larger vessels called **venules**, which empty into still larger veins (see Fig. 20-12). Veins provide a low-resistance pathway by which blood can return to the heart. The walls of veins are thinner, less muscular, and more expandable than those of arteries, although both contain a layer of smooth muscle. To prevent blood from flowing away from the heart, veins are equipped with one-way valves that allow blood to flow in only one direction (Fig. 20-14).

Because blood pressure in the veins is low, contractions of skeletal muscle during exercise and breathing help return blood to the heart by squeezing the veins and forcing blood through them. When you sit or stand for long periods, the lack of muscular activity allows blood to accumulate in the veins of the lower legs. This is why you may find your feet swollen after a long airplane flight. Long periods of inactivity can also contribute to varicose veins, in which the valves become stretched and weakened, and the veins become permanently swollen and often visible through the skin.

5 How Does the Lymphatic System Work with the Circulatory System?

The **lymphatic system** consists of a network of lymph capillaries and larger vessels that empty into the circulatory system, numerous small lymph nodes, and two additional organs: the thymus and the spleen (Fig. 20-15). Although not strictly part of the circulatory system, the lymphatic system is closely associated with it. The lymphatic system removes excess fluid and dissolved substances that leak from the capillaries, transports fats from the small intestine to the bloodstream, and defends the body by exposing bacteria and viruses to white blood cells.

Health Watch
Matters of the Heart

Cardiovascular disorders, as disorders of the heart and blood vessels are known, are the leading cause of death in the United States, killing nearly 1 million Americans annually. And no wonder. Your heart must contract vigorously more than 2.5 billion times during your lifetime without once stopping to rest, forcing blood through vessels whose total length would encircle the globe twice. Because these vessels may become constricted, weakened, or clogged, the cardiovascular system is a prime candidate for malfunction.

Hypertension Strains the Heart

High blood pressure, also called *hypertension*, is caused by the constriction of arterioles, causing resistance to blood flow and strain on the heart. For most of the 50 million Americans with this condition, the cause is unknown. Hypertension may cause the heart to increase in size, but its blood supply may not increase proportionately. The heart muscle is then inadequately supplied with blood, especially during exercise. Lack of sufficient oxygen to the heart can cause chest pain called *angina*. Hypertension also contributes to "hardening of the arteries," or atherosclerosis, described below. Hypertension, in conjunction with hardened arteries, can lead to the rupture of an artery and internal bleeding. The rupture of vessels supplying the brain causes *stroke*, in which brain function is lost in the area deprived of blood and of the vital oxygen and nutrients the blood carries.

Mild hypertension is sometimes alleviated by weight reduction, exercise, stress management, and (for some individuals) reduction of dietary salt. For severe cases, doctors may prescribe drugs to reduce fluid in the body, reduce the heart rate, and expand the arteries and arterioles.

Atherosclerosis Obstructs Blood Vessels

Atherosclerosis causes the walls of the large arteries to thicken and lose their elasticity. This is caused by deposits called *plaques*, which are composed of cholesterol and other fatty substances as well as calcium and fibrin. Plaques are deposited within the wall of the artery between the smooth muscle and the endothelium (Fig. E20-1a). A plaque may rupture through the endothelium into the interior of the vessel, stimulating platelets to adhere to the vessel wall and initiate blood clots. These clots further obstruct the artery and may completely block it, or a clot may break loose and clog a narrower artery "downstream." Arterial clots are responsible for the most serious consequences of atherosclerosis: heart attacks and strokes.

A *heart attack* occurs when one of the coronary arteries is blocked (coronary arteries supply the heart muscle itself; one that is nearly blocked is shown in Fig. E20-1b). Deprived of nutrients and oxygen, the heart muscle supplied by the blocked artery dies rapidly and painfully. Heart attacks are the major cause of death from atherosclerosis; about 1.5 million Americans suffer heart attacks each year, and about half a million people die from them. But atherosclerosis also causes plaques and clots to form in arteries throughout the body. If a clot or plaque obstructs an artery that supplies the brain, it can cause a stroke.

As with hypertension, the exact cause of atherosclerosis is unclear, but it is promoted by hypertension, cigarette smoking, genetic predisposition, obesity, diabetes, lack of exercise, and high blood levels of a certain type of cholesterol bound to a carrier molecule called low-density lipoprotein (LDL). If LDL-bound cholesterol levels are too high, cholesterol can be deposited in arterial walls. In contrast, cholesterol bound to high-density lipoprotein (HDL) is metabolized or excreted and hence is often called "good" cholesterol. (Cholesterol is discussed in "Health Watch: Cholesterol—Friend and Foe" in Chapter 2.)

Traditional treatment of atherosclerosis includes the use of drugs or changes in diet and lifestyle to lower blood pressure and blood cholesterol levels. Plaques are sometimes treated by inserting and inflating a tiny balloon, which squashes the

Lymphatic Vessels Resemble the Capillaries and Veins of the Circulatory System

Like blood capillaries, lymph capillaries form a complex network of microscopically narrow, thin-walled vessels into which substances can move readily. Unlike the walls of blood capillaries, lymph capillary walls have openings between the cells that act as one-way valves. These openings allow relatively large particles to be carried into lymph capillaries along with fluid. Also unlike blood capillaries, which form a continuous connected network, lymph capillaries dead-end in the body's tissues (Fig. 20-16). Materials collected by the lymph capillaries flow into larger lymph vessels. Large lymph vessels have somewhat muscular walls, but, as in blood veins, most of the force for lymph flow comes from the contraction of nearby muscles, such as those used in breathing and walking. As in blood veins, the direction of flow is regulated by one-way valves.

The Lymphatic System Returns Fluids to the Blood

As described earlier, dissolved substances are exchanged between the capillaries and body cells through the interstitial fluid that bathes nearly all the body's cells. In an average person, the amount of fluid that leaves the blood capillaries each day exceeds the amount that is reabsorbed by them by about 3 quarts. The lymphatic system returns this excess fluid to the blood. As interstitial fluid accumulates, its pressure forces the fluid through the openings in the lymph capillaries (see Fig. 20-16). The lymphatic system transports this fluid, now called **lymph**, back to the circulatory system.

plaques flat; by shaving off the artery wall with a miniature drill; or by vaporizing the plaque with a laser. In a new treatment of coronary artery blockage called laser revascularization, a laser is used to shoot 15 to 30 small (1-mm diameter) channels through the heart's ventricular walls. Blood clots form on the outside of the ventricle, preventing blood from leaking out. As the heart beats, blood from inside the ventricles flows in and out of the channels, supplying oxygen to the ventricular muscle and partially replacing the function of coronary arteries clogged by plaques. Coronary bypass surgery, performed on more than 350,000 individuals each year in the United States, consists of bypassing one or more obstructed coronary arteries with a piece of vein (usually obtained from the patient's leg) or with an artery (often from the patient's forearm). Researchers are working to perfect artificial vessels so the patient's own vessels can be spared.

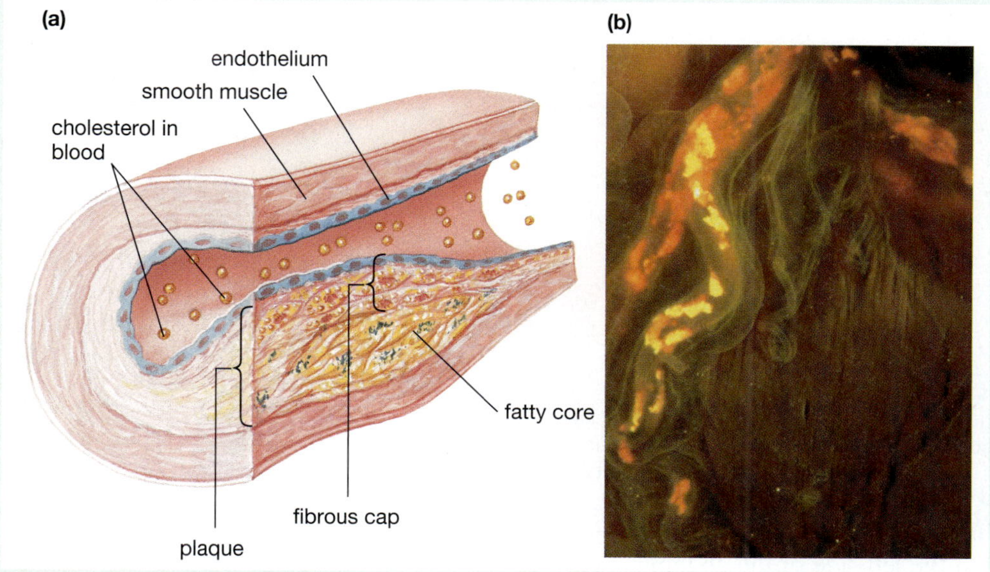

Figure E20-1 Plaques clog arteries
(a) Diagram of an artery with a plaque. ***(b)*** In this remarkable photo of coronary arteries, plaques are seen in glowing yellow. If they block a coronary artery, a heart attack will occur.

The Lymphatic System Transports Fats from the Small Intestine to the Blood

After a fatty meal, the cells of the small intestine absorb globules of digested fat. These globules are released from the intestinal cells into the interstitial fluid, but they are too large to diffuse into blood capillaries. They can, however, easily move through the openings between lymph capillary cells. Once in the lymph, they are dumped into the vena cava, a large vein that enters the heart.

The Lymphatic System Helps Defend the Body Against Disease

In addition to its other roles, the lymphatic system helps defend the body against foreign invaders, such as bacteria and viruses. Inserted at intervals within the large lymph vessels are kidney-bean-shaped structures about 1 inch (2.5 centimeters) long called **lymph nodes** (see Fig. 20-15b). In the nodes, lymph is forced through channels lined with masses of white blood cells that recognize and destroy foreign particles, such as bacteria and viruses. The white blood cells are killed in the process of destroying particles, and the painful swelling of lymph nodes that accompanies certain diseases is largely a result of the accumulation of dead white blood cells and dead virus-infested cells they have engulfed.

The thymus and the spleen are often considered part of the lymphatic system (see Fig. 20-15a). The **thymus**, which is an organ that produces white blood cells, is located beneath the breastbone slightly above the heart. The thymus is particularly active in infants and young children but decreases in size and importance in early adulthood. The **spleen**, which is another white-blood-cell-producing

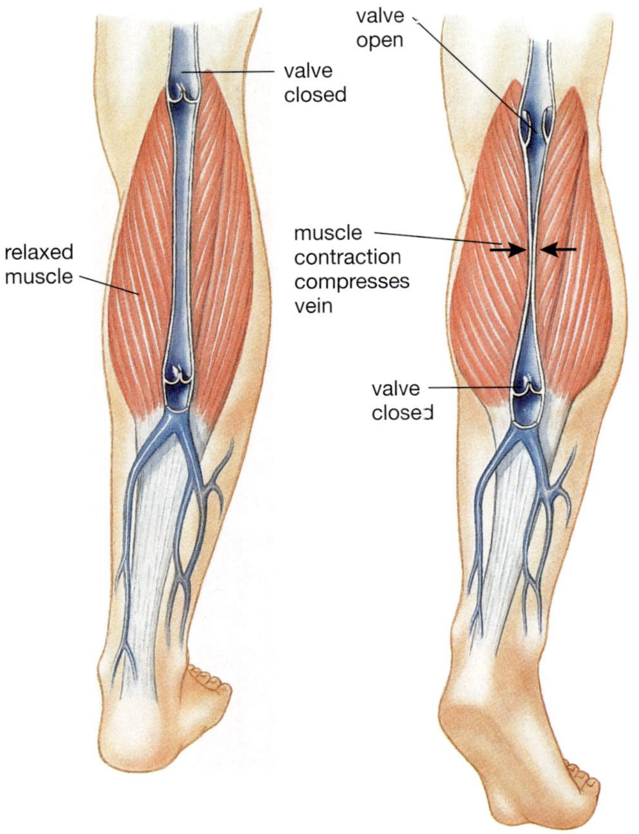

Figure 20-14 Valves direct the flow of blood in veins
Veins and venules have one-way valves that maintain blood flow in the proper direction. When the vein is compressed by nearby muscles, the valves allow blood to flow toward the heart but clamp shut to prevent backflow.

organ, is located in the left side of the abdominal cavity, between the stomach and diaphragm. Just as the lymph nodes filter lymph, the spleen filters blood, exposing it to white blood cells that destroy foreign particles and aged red blood cells.

6 How Are Oxygen and Carbon Dioxide Exchanged in Animal Bodies?

One of the key functions of circulatory systems is to transport oxygen to the tissues and carbon dioxide away from the tissues. To accomplish this function, circulatory systems must be able to acquire oxygen gas from the surrounding environment and deliver carbon dioxide gas there. This process of gas exchange is called **respiration**, and the organ systems that exchange gases are called *respiratory systems*. As you might expect, respiratory systems are closely connected to circulatory systems.

Animal respiratory systems are diverse, and include both systems in which gases are exchanged with the atmosphere and systems in which the exchanged gases are dissolved in water. In all cases, however, respiratory systems rely on diffusion of gases across a respiratory surface. Respiratory surfaces are large and moist. They must have a large surface area in contact with the environment to allow adequate gas exchange, and they must remain moist, as gases must be dissolved in water to diffuse into or out of cells.

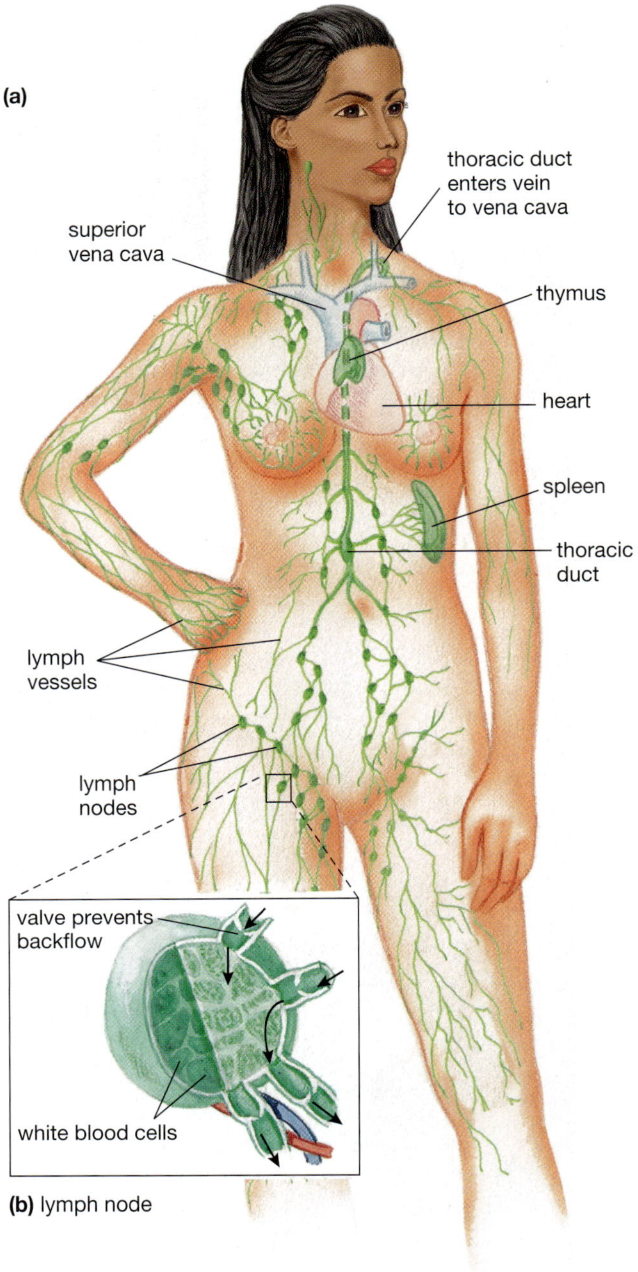

Figure 20-15 The human lymphatic system
(a) Lymph vessels, lymph nodes, and two auxiliary lymph organs, the thymus and spleen. Lymph is returned to the circulatory system by way of the thoracic duct, which empties into the vena cava, a large vein. ***(b)*** A cross section of a lymph node. The node is filled with channels lined with white blood cells that attack foreign matter in the lymph.

Aquatic Animals May Have Gills

Some very small animals that live in moist environments exchange gases without specialized respiratory systems. The outside of their bodies provides an adequate surface area for the diffusion of gases. Most larger

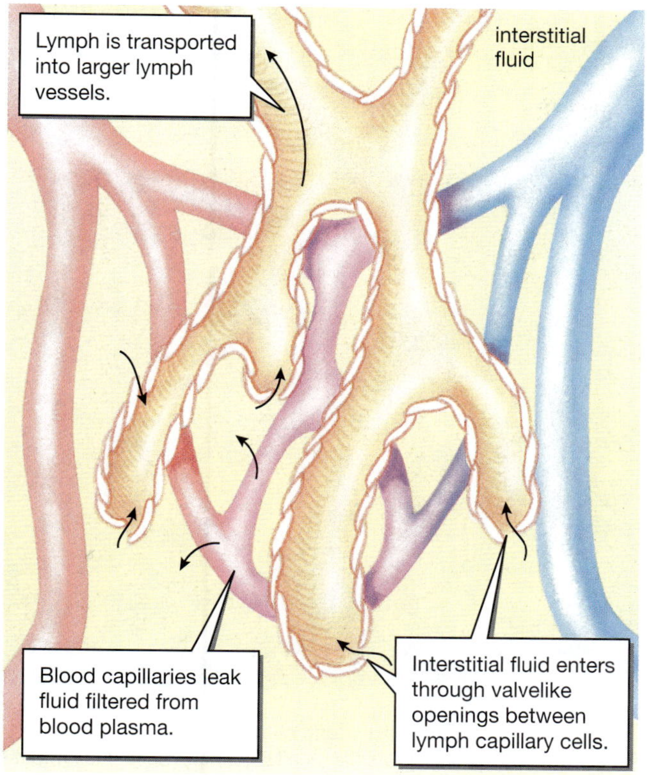

Figure 20-16 Lymph capillary structure
Lymph capillaries end blindly in the body tissues, where pressure from the accumulation of interstitial fluid forces the fluid into the lymph capillaries.

animals, however, have specialized respiratory structures. For example, many animals that live in water have respiratory organs called **gills**. Gills usually have elaborately branched or folded shapes; the branches and folds maximize surface area. A dense profusion of capillaries lies just beneath a gill's delicate outer membrane, so that blood comes to the gill surface, where gases are exchanged. Though gills must remain in constant contact with the surrounding water, the delicate gill tissue is housed within protective structures in many species (Fig. 20-17).

Terrestrial Animals Have Internal Respiratory Structures

Land-dwelling animals live surrounded by air, which has a far higher oxygen concentration than does water. Extracting oxygen from dry air, however, presents special challenges. Respiratory surfaces must remain moist, but moist surfaces on the outside of the body lose water continuously by evaporation and tend to dry out. Land animals have evolved internal structures in which respiratory surfaces are protected from drying. In land-dwelling vertebrates, the respiratory structures are lungs. **Lungs** are chambers containing moist, delicate respiratory surfaces that are protected within the body, where water loss is minimized and the body wall provides support.

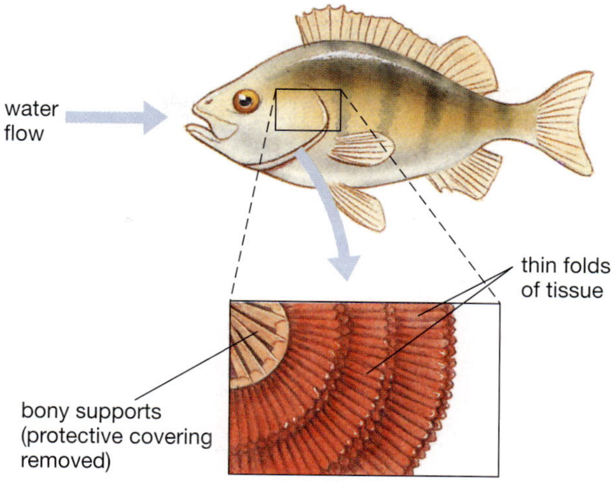

Figure 20-17 Gills exchange gases with water
The gill reaches its greatest complexity in the fish, where it is made of thin folds of tissue and is protected under a bony flap. A one-way flow of water is maintained over the gill by pumping water in through the mouth and out through an opening behind the gills.

The Big Picture

The respiratory and circulatory systems work in intimate association to ensure that all of the body's cells are supplied with oxygen, and that carbon dioxide is safely returned to the environment.

7 How Does the Human Respiratory System Work?

The respiratory system in humans and other lung-breathing vertebrates has two parts: the conducting portion and the gas-exchange portion. The conducting portion consists of a series of passageways that carry air into and out of the gas-exchange portion, where gases move to and from the blood.

The Conducting Portion Carries Air to the Lungs

The conducting portion of the respiratory system brings air to the lungs. Air enters through the nose or the mouth, passes through the nasal cavity or oral cavity into the **pharynx**, and then travels through the **larynx** (Fig. 20-18). The opening to the larynx is guarded by the epiglottis, a flap of tissue supported by cartilage. During normal breathing, the epiglottis is tilted upward, as shown in Figure 20-18, allowing air to flow freely into the larynx. During swallowing, the epiglottis tilts downward and covers the larynx, directing substances into the esophagus instead. If a person attempts to inhale and swallow at the same time, the epiglottis may not tilt downward and food may become lodged in the larynx, blocking air from entering the lungs. What should you

Health Watch
Smoking—A Life and Breath Decision

An estimated 430,000 people in the United States die of smoking-related diseases each year. The American Cancer Society estimates that nearly 160,000 individuals die of lung cancer annually and that 87% of those deaths are due to smoking. The rest of the smoking-related deaths are from emphysema, chronic bronchitis, heart disease, stroke, and other forms of cancer. The cost of health care and lost productivity due to smoking-related ailments in the United States is more than $100 billion per year.

Tobacco smoke has a dramatic impact on the human respiratory tract. As smoke is inhaled, toxic substances such as nicotine and sulfur dioxide paralyze the cilia that line the respiratory tract. These ciliary sweepers remove inhaled particles, and smoking inhibits them just when they are most needed. With the cilia out of action, the billions of microscopic carbon particles that make up the visible portion of cigarette smoke stick to the walls of the respiratory tract and enter the lungs. In response to the resulting irritation, the respiratory tract produces more mucus in an effort to trap the foreign particles. But without the cilia to sweep it along, the mucus builds up and can obstruct the airways; the familiar "smoker's cough" is an attempt to expel it.

Microscopic smoke particles accumulate in the alveoli over the years until the lungs of a heavy smoker are literally blackened. Adhering to the particles are about 200 different toxic substances, of which more than a dozen are known or probable carcinogens (cancer-causing substances). The longer the delicate tissues of the lungs are exposed to the carcinogens on the trapped particles, the greater the chance that cancer will develop (Fig. E20-2).

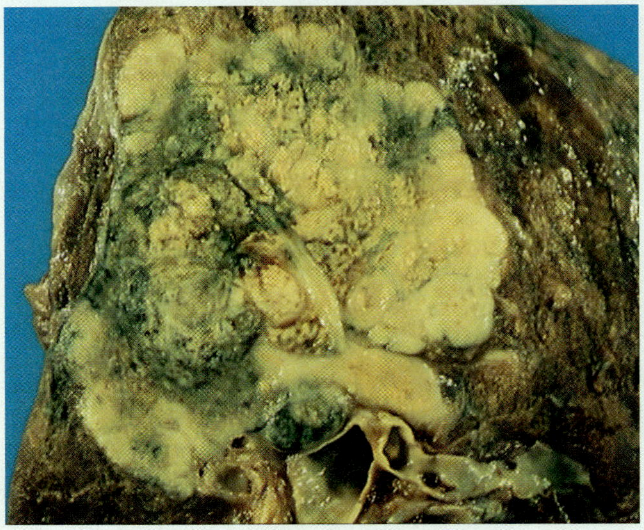

Figure E20-2 Smoking causes lung cancer
A tumor of lung cancer is visible as a large, pale mass; the lung tissue surrounding it is blackened by trapped smoke particles.

Chronic bronchitis and emphysema together kill about 96,000 people each year in the United States, and smoking is the leading cause of both of these diseases. Chronic bronchitis is a persistent lung infection characterized by coughing, swelling of the lining of the respiratory tract, an increase in mucus production, and a decrease in the number and activity of cilia. The result: a decrease in air flow to the alveoli. Emphysema develops when toxic substances in cigarette smoke cause the body to produce substances that lead to brittle and ruptured alveoli. The lung gradually loses its normal sponge-cake appearance (Fig. E20-3a) and more closely resembles blackened Swiss cheese (Fig. E20-3b). The loss of the alveoli, where gas exchange occurs, leads to oxygen deprivation of all body tissues. In an individual with emphysema, breathing becomes labored and grows increasingly worse until death.

Carbon monoxide, present in high levels in cigarette smoke, binds tenaciously to red blood cells in place of oxygen. This binding reduces the blood's oxygen-carrying capacity and thereby increases the work the heart must do. Chronic bronchitis and emphysema compound this problem. Smoking also causes atherosclerosis (see "Health Watch: Matters of the Heart"). As a result, smokers are 70% more likely than nonsmokers to die of heart disease. The carbon monoxide in cigarette smoke may also contribute to the reproductive problems of women who smoke, including infertility, miscarriage, lower birth weight of their babies, and, for their children, more learning and behavioral problems.

do if you see this happen? The Heimlich maneuver (Fig. 20-19) is easy to perform and has saved countless lives.

Inhaled air continues past the larynx into the **trachea**, a flexible tube whose walls are reinforced with semicircular bands of stiff cartilage. Within the chest, the trachea splits into two large branches called **bronchi** (singular, **bronchus**), one leading to each lung. Inside the lung, each bronchus branches repeatedly into ever-smaller tubes called **bronchioles**. Bronchioles lead finally to the microscopic **alveoli** (singular, **alveolus**), tiny air sacs where gas exchange occurs (see Fig. 20-15). The microscopic (0.2 millimeter in diameter) alveoli give magnified lung tissue the appearance of sponge cake (see Fig. E20-3a).

During its passage through the conducting portion, air is warmed and moistened. Much of the dust and bacteria it carries is trapped in mucus secreted by cells that line the respiratory passages. The mucus, with its trapped debris, is continuously swept upward toward the pharynx by cilia that line the bronchioles, bronchi, and trachea. Upon reaching the pharynx, the mucus is coughed up or swallowed. Smoking interferes with this cleansing process by paralyzing the cilia (see "Health Watch: Smoking—A Life and Breath Decision").

(a)

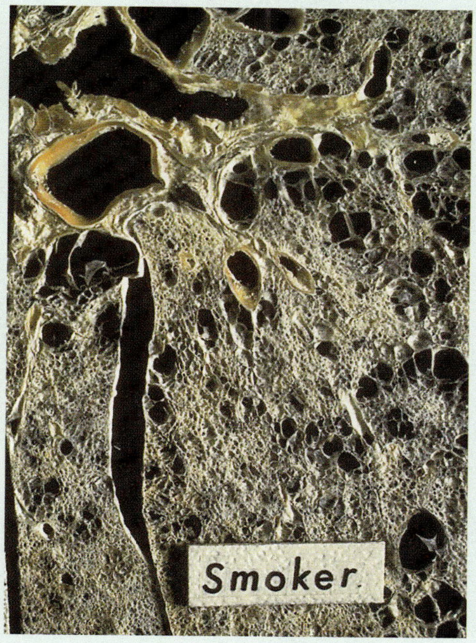

(b)

Figure E20-3 Smoking causes emphysema
(a) Normal lung tissue from a nonsmoker has nearly invisible small openings, the alveoli, surrounded by healthy tissue. *(b)* The lung of a smoker suffering from emphysema is full of large holes, each caused by the rupture of hundreds of alveoli.

"Passive smoking," or breathing secondhand smoke, poses health hazards for both children and adults. Children whose parents smoke are more likely to contract bronchitis, pneumonia, ear infections, coughs, and colds and to have decreased lung capacity. Children who grow up with smokers are more likely to develop asthma and allergies; for children with asthma, the number and severity of asthma attacks are increased by secondhand smoke. Among adults, nonsmoking spouses of smokers face a 30% higher risk of both heart attack and lung cancer than do spouses of nonsmokers. A recent study links even relatively infrequent exposure to secondhand smoke with atherosclerosis. Government agencies report that secondhand smoke is responsible for an estimated 3000 lung-cancer deaths and 37,000 deaths from heart disease in nonsmokers in the United States each year.

For smokers who quit, healing begins immediately and the chances of heart attack, lung cancer, and numerous other smoking-related illnesses gradually diminish. Atherosclerosis may, however, continue to progress.

Gas Exchange Occurs in the Alveoli

The lung has an enormous moist surface for gas exchange. Each lung is packed with 1.5 million to 2.5 million alveoli, which together provide a huge surface area for diffusion—about 800 square feet (75 square meters) in an adult human, or roughly 80 times the total skin surface area. The alveoli, which cluster about the end of each bronchiole like a bunch of grapes, are entirely enmeshed in capillaries (see Fig. 20-18b). Because both the alveolar wall and the adjacent capillary walls are only one cell thick, the air in the lungs is extremely close to the blood in the capillaries. A thin layer of watery fluid lines each alveolus. Gases dissolve in this fluid and diffuse through the alveolar and capillary membranes (Fig. 20-20).

Blood flows to the lungs after returning from its passage though the body's tissues, so the blood surrounding the alveoli is low in oxygen (which has been used up by the body cells) and high in carbon dioxide (which is released by the cells). Thus, oxygen diffuses from the air in the alveoli, where its concentration is high, into the blood, where its concentration is low. Conversely, carbon dioxide diffuses out of the blood, where its concentration

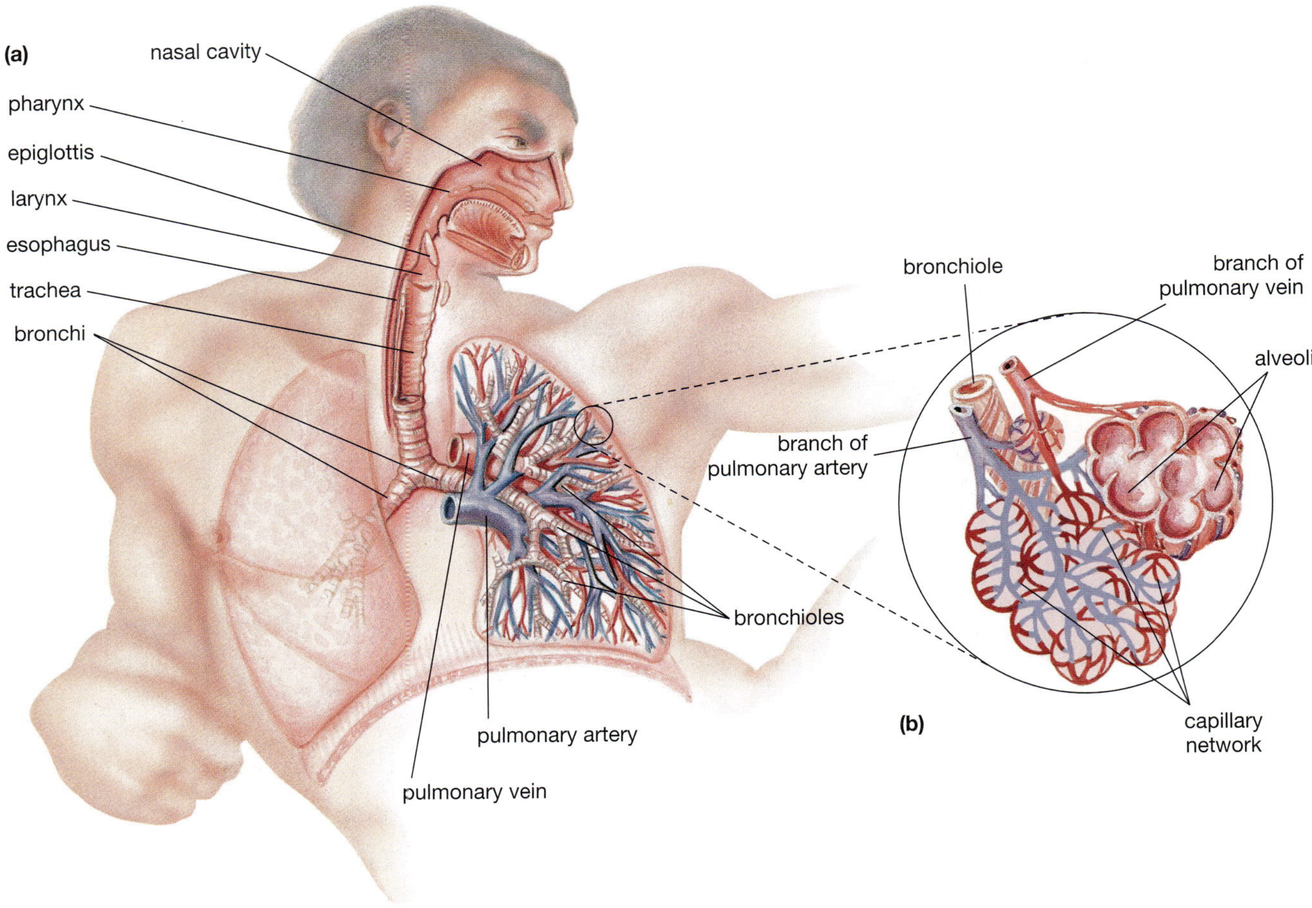

MEDIATUTOR 20.5 Human Respiratory Anatomy

Figure 20-18 The human respiratory system
(a) The passages and vessels of the respiratory system and its interface with the circulatory system. *(b)* Close-up of alveoli (their interiors are shown in this cut-away section) and their surrounding capillaries.

is high, into the air in the alveoli, where its concentration is low. The blood, now oxygenated and purged of carbon dioxide, returns to the heart, which pumps it to the body tissues (Fig. 20-21).

The Lungs Are Protected by an Airtight Cavity

The chest cavity that surrounds the lungs is airtight—bounded by neck muscles and connective tissue on top and by the dome-shaped, muscular **diaphragm** on the bottom. Within the wall of the chest, the rib cage surrounds and protects the lungs. Lining the rib cage and surrounding the lungs is a double layer of membranes. These membranes contribute to the airtight seal between the lungs and the chest wall.

Air Is Inhaled Actively and Exhaled Passively

We breathe in two stages: **inhalation**, when air is actively drawn into the lungs, and **exhalation**, when it is passively expelled from the lungs. Inhalation is accomplished by enlarging the chest cavity. To do so, the diaphragm muscles contract, drawing the diaphragm downward. The rib muscles also contract, lifting the ribs up and outward (Fig. 20-22). When the chest cavity is expanded, the lungs expand with it as surface tension holds them tightly against the moist inner wall of the chest. As the lungs

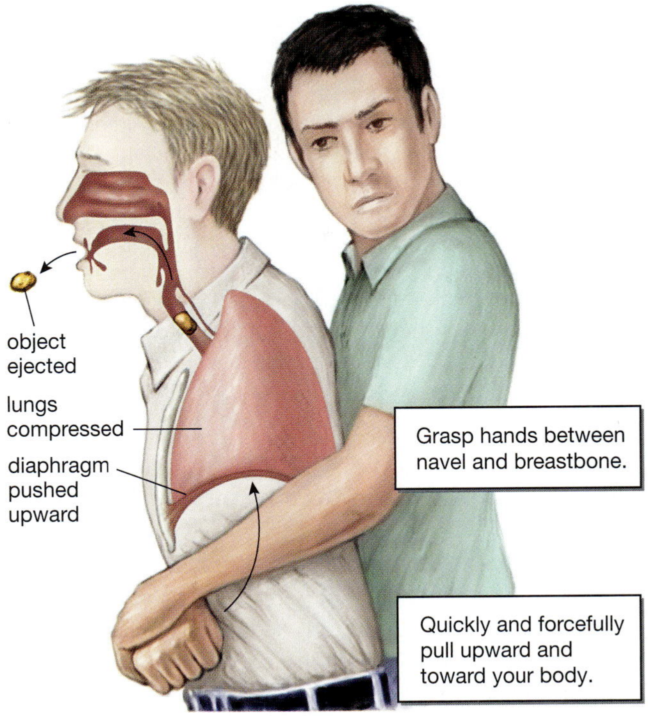

Figure 20-19 The Heimlich maneuver can save lives
If a person has choked on food or some other object and is unable to breathe, the Heimlich maneuver will push upward on the victim's diaphragm and force air out of his or her lungs, possibly dislodging the object.

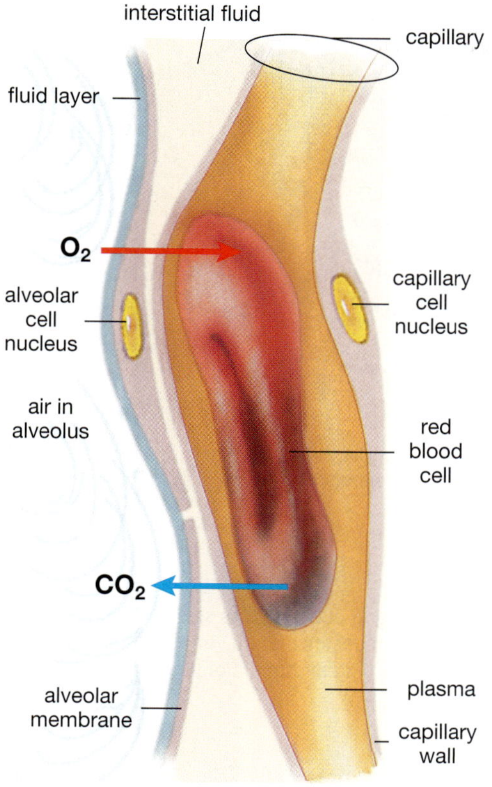

Figure 20-20 Gas exchange between alveoli and capillaries
The alveoli and capillary walls are only one cell thick, very close to one another, and the cells are coated in a thin layer of fluid. This allows gases to dissolve and diffuse easily between the lungs and circulatory system.

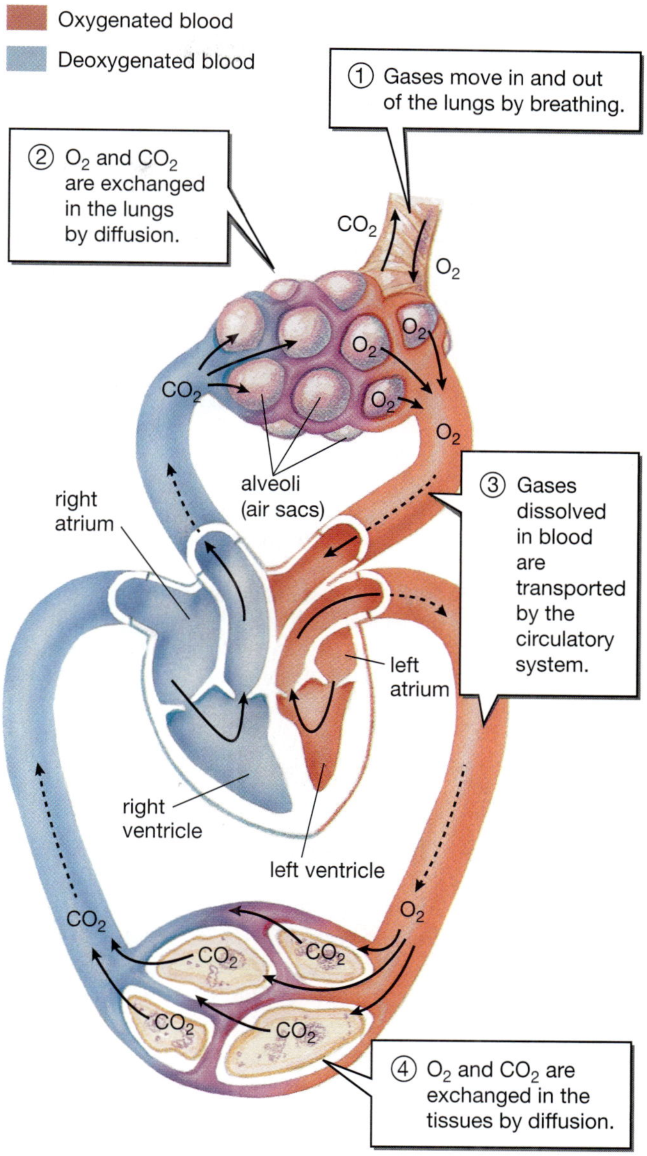

Figure 20-21 An overview of gas exchange

MEDIATUTOR 20.6 Respiration: Gas Exchange

expand, their increased volume decreases the air pressure inside them to below atmospheric pressure. Because gases always move from an area of higher pressure to an area of lower pressure, the outside air, which is at atmospheric pressure, moves into the lungs.

Exhalation occurs automatically when the muscles that cause inhalation are relaxed. The relaxed diaphragm domes upward, and the ribs move down and inward, decreasing the size of the chest cavity and forcing air out of the lungs. More air can be forced out by contracting the abdominal muscles. After exhalation, the lungs still contain some air. This air prevents the thin alveoli from collapsing and fills the space within the conducting portion of the respiratory system.

Breathing Rate Is Controlled by the Respiratory Center of the Brain

We breathe automatically, without conscious thought. The muscles used in breathing, however, are not self-activating like the heart muscle. Instead, each contraction is stimulated by impulses from nerve cells. These impulses originate in the **respiratory center**, which is located in the medulla of the brain, just above the spinal cord. Nerve cells in the respiratory center generate cyclic bursts of impulses that cause the alternating contraction and relaxation of the respiratory muscles.

The respiratory center receives input from several sources and adjusts breathing rate and volume to meet the body's changing needs. The respiratory rate is regulated to maintain a constant level of carbon dioxide in the blood, as monitored by carbon dioxide receptors in the medulla. For example, an elevated level of carbon dioxide caused by an increase in cellular activity, such as that experienced by your muscle cells when you run up stairs, signals a need for more oxygen, causing the receptors to stimulate an increase in the rate and depth of breathing. These receptors are extremely sensitive; an increase in carbon dioxide of only 0.3% can double the breathing rate.

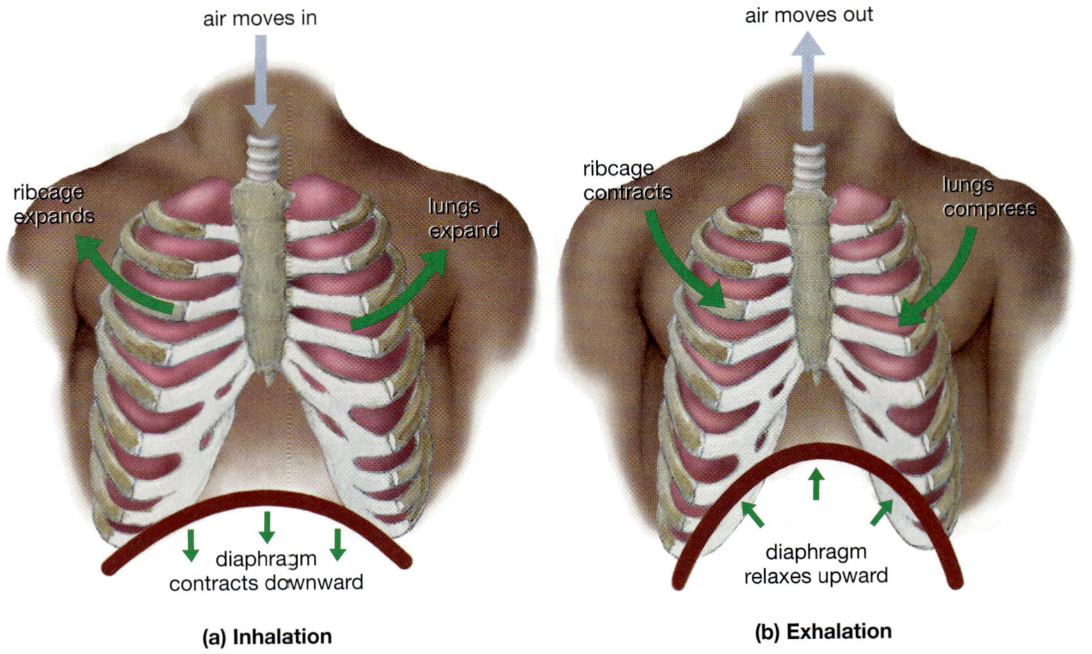

Figure 20-22 The mechanics of breathing
(a) During inhalation, the diaphragm is pulled downward and the ribs move up and outward. The size of the chest cavity increases, causing air to rush in. *(b)* During exhalation, relaxation of the muscles that moved the diaphragm and ribs allows the diaphragm to dome upward and the rib cage to collapse, forcing air out of the lungs.

CASE STUDY REVISITED

Organ Farmers

A person cannot live without functional systems for circulation and respiration, so organs to transplant will continue to be needed for patients whose hearts or lungs are damaged by disease. The promise of organs from genetically engineered pigs has raised hope that cross-species transplants will one day be used to treat some of the most devastating human diseases. First, though, some difficult problems will have to be solved. One is the phenomenon known as hyperacute rejection. Our blood contains antibody molecules that ordinarily serve to attack infectious microbes, but these antibodies will also attack and destroy foreign tissues such as those introduced in a xenotransplant. A pig's heart transplanted to a human would be destroyed in minutes by the body's natural defense mechanisms.

One way of coping with the rejection problem is to administer drugs that suppress the body's normal defenses. Such drugs may prevent rejection of the transplanted organ, but they leave the body vulnerable to infections. Genetic engineering, however, may offer another solution. In recent years, researchers have managed to transfer certain human genes to pigs. In these "humanized" pigs, organ surfaces include molecules that inhibit the human immune response, essentially fooling the immune system into treating the pig tissue as human tissue. Will organs from these genetically modified pigs avoid hyperacute rejection after a xenotransplant? We'll probably find out soon, because the developers of this process have received permission to begin more research that may lead to a clinical test on humans.

As you can imagine, not everyone is ready to jump on the xenotransplantation bandwagon. Animal-rights activists believe that xenotransplantation is unethical, because it exploits and harms animals. Some xenotransplantation researchers urge caution as well. Of particular concern is the possibility that xenotransplantation could provide a means for disease-causing organisms to jump from other species to humans. Should research on xenotransplantation go forward? The issues involved in determining how best to proceed are complex and difficult. Where do you stand?

Summary of Key Concepts

1) What Are the Major Features and Functions of Circulatory Systems?

Circulatory systems transport blood rich in dissolved nutrients and oxygen close to each cell, where nutrients can be released and wastes absorbed by diffusion. All circulatory systems have three major parts: (1) blood, a fluid; (2) vessels, a system of channels to conduct the blood; and (3) a heart, a pump to circulate the blood. Vertebrate circulatory systems transport gases, hormones, and wastes; distribute nutrients; help regulate body temperature; and defend the body against disease.

2) How Does the Vertebrate Heart Work?

In the mammalian four-chambered heart, blood is pumped separately to the lungs and through the body, maintaining complete separation of oxygenated and deoxygenated blood. Deoxygenated blood is collected from the body in the right atrium and passed to the right ventricle, which pumps it to the lungs. Oxygenated blood from the lungs enters the left atrium, is passed to the left ventricle, and is pumped to the rest of the body.

The heartbeat consists of two stages: (1) atrial contraction, followed by (2) ventricular contraction. The direction of blood flow is maintained by valves within the heart. The contractions of the heart are initiated and coordinated by the sinoatrial node, the heart's pacemaker.

3) What Is Blood?

Blood is composed of both fluid and cellular materials. The fluid plasma consists of water that contains proteins, hormones, nutrients, gases, and wastes. Red blood cells, or erythrocytes, are packed with a large iron-containing protein called hemoglobin, which carries oxygen. White blood cells, or leukocytes, fight infection. Platelets are important for blood clotting.

4) What Are the Types and Functions of Blood Vessels?

Blood leaving the heart travels (in sequence) through arteries, arterioles, capillaries, venules, veins, and then back to the heart. Each vessel is specialized for its role. Elastic, muscular arteries help pump the blood. The thin-walled capillaries are the sites of exchange of materials between the body cells and the blood. Veins provide a path of low resistance back to the heart, with one-way valves that maintain the direction of blood flow.

5) How Does the Lymphatic System Work with the Circulatory System?

The human lymphatic system consists of lymphatic vessels, lymph nodes, and the thymus and spleen. The lymphatic system removes excess interstitial fluid that leaks through blood capillary walls. It transports fats to the bloodstream from the small intestine and fights infection by filtering the lymph through lymph nodes, where white blood cells ingest foreign invaders, such as viruses and bacteria. The thymus, which is most active in young children, produces white blood cells that function in immunity. The spleen filters blood past white blood cells, which remove bacteria and damaged blood cells.

6) How Are Oxygen and Carbon Dioxide Exchanged in Animal Bodies?

Respiration makes possible the exchange of oxygen and carbon dioxide between the body and the environment by diffusion of these gases across a moist surface. In moist environments, animals with small bodies may rely exclusively on diffusion through the body surface. Larger, more active animals have evolved specialized respiratory systems. Animals in aquatic environments have evolved gills. Terrestrial vertebrates have evolved lungs that afford internal protection to moist respiratory surfaces.

7) How Does the Human Respiratory System Work?

The human respiratory system consists of a conducting portion and a gas-exchange portion. Air passes first through the conducting portion, consisting of the nose and mouth, pharynx, larynx, trachea, bronchi, and bronchioles, and then into the gas-exchange portion, composed of alveoli (microscopic sacs). Blood within a dense capillary network surrounding the alveoli releases carbon dioxide and absorbs oxygen from the air.

Most of the oxygen in the blood is bound to hemoglobin within red blood cells. Hemoglobin binds oxygen at concentrations typical of those in alveolar capillaries and releases it at the lower oxygen concentrations of the body tissues. Carbon dioxide diffuses into the blood from the body tissues and is transported bound to hemoglobin.

To inhale, we actively draw air into the lungs by contracting the diaphragm and the rib muscles, which expand the chest cavity. The subsequent automatic relaxation of these muscles causes the chest cavity to collapse and the air to be exhaled. Respiration is controlled by nerve impulses that originate in the medulla's respiratory center. The respiration rate is modified by receptors in the medulla that monitor carbon dioxide levels in the blood.

Key Terms

Thinking Through the Concepts

Multiple Choice

1. *The sites of exchange of wastes, nutrients, gases, and hormones between the blood and body cells are the*
   a. arteries
   b. arterioles
   c. capillaries
   d. veins
   e. all blood vessels
2. *What produces systolic blood pressure?*
   a. contraction of the right atrium
   b. contraction of the right ventricle
   c. contraction of the left atrium
   d. contraction of the left ventricle
   e. the pause between heartbeats
3. *Lymph most closely resembles which of the following?*
   a. blood
   b. urine
   c. plasma
   d. interstitial fluid
   e. water
4. *The gas-exchange portion of the human respiratory system is the*
   a. larynx
   b. trachea
   c. bronchi
   d. pharynx
   e. alveoli
5. *How is most of the oxygen transported in the blood?*
   a. dissolved in plasma
   b. bound to hemoglobin
   c. in the form of CO_2
   d. as bicarbonate
   e. dissolved in water
6. *Which of the following statements regarding cigarette smoking is true?*
   a. Cigarette smoke damages the alveoli.
   b. Cigarette smoke decreases the amount of oxygen in the blood.
   c. Cigarette smoke causes many types of cancer.
   d. Cigarette smoke can lead to heart damage.
   e. all of the above

? Review Questions

1. Trace the flow of blood through the circulatory system, starting and ending with the right atrium.
2. What are five functions of the vertebrate circulatory system?
3. Describe three important functions of the lymphatic system.
4. Distinguish among plasma, interstitial fluid, and lymph.
5. Describe veins, capillaries, and arteries, noting their similarities and differences.
6. Explain in detail what causes the vertebrate heart to beat.
7. Describe the formation of an atherosclerotic plaque. What are the risks associated with atherosclerosis?
8. Trace the route taken by air in the vertebrate respiratory system, listing the structures through which it flows and the point at which gas exchange occurs.
9. What events occur during human inhalation? Exhalation? Which of these is always an active process?
10. Trace the pathway of an oxygen molecule in the human body, starting with the nose and ending with a body cell.
11. Describe the effects of smoking on the human respiratory system.
12. Explain how the structure and arrangement of alveoli make them well suited for their role in gas exchange.

Applying the Concepts

1. Discuss the steps you can take now and in the future to reduce your risks of developing heart disease.
2. Heart surgeons have attempted to transplant baboon hearts into humans whose hearts were failing. Discuss the implications of this operation from as many angles as you can think of.
3. Joe, a 45-year-old executive of a major corporation, has been diagnosed with mild hypertension. What treatments or lifestyle changes might Joe's physician recommend? If Joe's hypertension becomes more severe, what treatments might Joe's physician use? Should Joe be concerned about mild hypertension? Explain your answer.
4. Heart–lung transplants are performed in some cases, but donors are scarce. On the basis of your knowledge of the respiratory and circulatory systems and of lifestyle factors that might damage them, what criteria would you use in selecting a recipient for such a transplant?
5. Nicotine is a drug in tobacco that supplies the effects that smokers crave. Discuss the advantages and disadvantages of low-nicotine cigarettes.
6. Mary, a strong-willed 3-year-old, threatens to hold her breath until she dies if she doesn't get her way. Can she carry out her threat? Explain.

Group Activity

Form groups of three or four to address the following issue: The level of oxygen in the atmosphere is currently 21%, but evidence from gases trapped in amber 25 million to 40 million years old indicates that the level was once as high as 30%. If the level were to drop slowly over the next few centuries due to pollution of the oceans, where much of Earth's oxygen is produced, what might happen? What short-term adaptations would humans and other mammals have to make to adjust to this change? What sorts of evolutionary changes in mammalian respiratory systems would be favored in this new environment? As the level of oxygen in the atmosphere dropped, which types of present-day organisms would you expect to be affected most, which least, and why?

For More Information

Gibbons, R., and associates. "Waiting for Organ Transplantation." *Science,* January 14, 2000. The statistics of the need, availability, and waiting time for organ transplantation suggest that new sources of organs are needed.

Hajjar, D. P., and Nicholson, A. C. "Atherosclerosis." *American Scientist*, September–October 1995. Scientists are learning more about the biochemistry of plaque formation, an understanding that will lead to more-effective treatments in the future.

Houston, C. "Mountain Sickness." *Scientific American*, October 1992. The mechanisms of potentially fatal altitude sickness are explained.

Nucci, M. L., and Abuchowski, A. "The Search for Blood Substitutes." *Scientific American*, February 1998. As the authors summarize: "The threat of blood shortages and fears about contamination have hastened attempts to find life-sustaining alternatives."

Radetsky, P. "The Mother of All Blood Cells." *Discover*, March 1995. Discusses the discovery by Irving Weissman of a "stem cell" that gives rise to all other types of blood cells.

Seppa, N. "Secondary Smoke Carries High Price." *Science News*, January 17, 1998. New research suggests that smoking causes atherosclerosis, that the damage continues after a smoker quits, and that exposure to secondhand smoke significantly increases the buildup of plaque in the carotid artery of nonsmokers.

Answers to Multiple-Choice Questions

1. c 2. d 3. d 4. e 5. b 6. e

MediaTutor

CD Activities

Activity 20.1: Muscle Structure

Estimated time: 10 minutes

The functional contractile unit of a heart or skeletal muscle cell is a structure called a sarcomere. In this tutorial, you will explore the various levels of structure found within muscle and begin to learn how the sarcomere functions to produce a contraction.

Activity 20.2: Muscle Movement

Estimated time: 5 minutes

The contraction of heart and skeletal muscle involves the interaction of several proteins found within a sarcomere. This process is described as a sliding-filament mechanism due to the fact that the thin filament slides by the thick filament as the sarcomere shortens. In this tutorial, you will explore the events that occur at the molecular level of muscle contraction.

Activity 20.3: Heart Structure and Function

Estimated time: 5 minutes

The vertebrate heart is a complex organ. In this tutorial, you will explore the various structures of the vertebrate heart and the functions of each of those structures.

Activity 20.4: The Cardiac Cycle

Estimated time: 10 minutes

The alternating contraction and relaxation of the heart chambers is called the cardiac cycle. In this tutorial, you will view an animation of this process and explore the various steps involved in the cycle.

Activity 20.5: Human Respiratory Anatomy

Estimated time: 5 minutes

Human respiratory anatomy represents a highly efficient gas-exchange system. In this tutorial, you will explore the various structures of the human respiratory system and the functions of each of those structures.

Activity 20.6: Gas Exchange

Estimated time: 5 minutes

All animals have evolved specialized respiratory surfaces that work together with the circulatory system so that gases can be exchanged between the body's cells and the environment. In this tutorial, you will explore how the human respiratory system accomplishes this important physiological task.

Start the MediaTutor Student CD-ROM and enter the activity number in the Quick Search box to be taken directly to that activity.

Web Investigation

Organ Farmers

Estimated time: 15 minutes

Sophisticated surgical procedures have become quite common. Unfortunately, technical expertise is often ahead of careful consideration of ethical issues. One hotly debated issue is xenotransplantation. This exercise examines the question of whether xenotransplantation should become standard medical practice.

Go to http://www.prenhall.com/audesirk3, the Audesirk Companion Web site. Select Chapter 20 and Web Investigation to begin.

Despite the proliferation of plans, pills, and programs for losing weight, the best way to avoid obesity remains a sensible diet and plenty of exercise.

21 Nutrition, Digestion, and Excretion

AT A GLANCE

CASESTUDY

The Magic Bullet for Obesity?

Do you worry about your weight? If so, you're not alone. The *Wall Street Journal* reports that Americans spend more than $30 billion each year on efforts to control body weight. Although many people wish to lose weight just to look better, true obesity is a major health problem. Obesity, defined as a body weight that is 20% or more above the ideal weight for a given size and build, is associated with diabetes, heart disease, cancer, and other ailments. More than a third of all Americans are obese, and the Centers for Disease Control estimates that obesity contributes to 300,000 deaths per year in the United States alone.

Whether for reasons of health or fashion, many people would like to lose some weight. The best prescription for success is simple: Eat a diet with fewer calories and get plenty of exercise. This straightforward solution to obesity is, as many of us know all too well, easy to understand but difficult to accomplish. It is therefore perhaps inevitable that many people will search for easier solutions.

The idea of a diet pill has an especially firm grip on the imaginations of the overweight. The dream of a simple pill that causes the pounds to melt away as if by magic has been around since the dawn of pharmacology. None of the resulting pills, however, has proved to be a satisfactory remedy. Amphetamines, for example, were routinely prescribed for weight loss in the 1950s and 1960s, but these appetite-suppressing stimulants turned out to be addictive and harmful to health. The 1990s saw the rise of "fen-phen," a wildly popular combination of the drugs fenfluramine and phentermine. Fen-phen was a very effective appetite suppressant, but its golden age came to an abrupt end in 1997, with the publication of a study that linked the drug combination to potentially lethal damage to heart valves.

In the end, every attempt to find the magic bullet for weight loss has ended in failure and, often, danger to hopeful users. This legacy of failure, however, has not discouraged obesity researchers or the pharmaceutical industry. The quest for the perfect diet pill continues. ■

1 What Nutrients Do Animals Need?

Nutrition is the process of acquiring nutrients and processing them into usable form; **nutrients** are substances that must be supplied in the diet. Animal nutrients fall into five major categories: lipids, carbohydrates, proteins, minerals, and vitamins. These substances provide the body with its basic needs:

- energy to fuel cellular metabolism and activity;
- chemical building blocks needed to construct complex molecules;
- minerals and vitamins needed for metabolic reactions.

The Primary Sources of Energy Are Carbohydrates and Lipids

Cells require a continuous supply of energy to stay alive and perform their functions. In animals, the energy is provided by three kinds of nutrients: lipids, carbohydrates and, to a lesser extent, proteins. These molecules are broken down during cellular respiration, and the energy derived from them is used to produce adenosine triphosphate (ATP; see Chapter 7).

The energy in food is measured in Calories. Each **Calorie** (with a capital *C*) contains 1000 **calories** (with a lowercase *c*). A calorie is the amount of energy required to raise the temperature of 1 gram of water by 1 degree Celsius. The average human body at rest burns about 1550 Calories per day (somewhat more for males and less for females). Exercise significantly boosts caloric requirements: Well-trained athletes temporarily raise their calorie consumption during vigorous exercise from a resting rate of about 1 Calorie per minute to nearly 20 Calories per minute, and all of us burn more Calories when exercising than when resting (Table 21-1).

Lipids Include Fats, Phospholipids, and Cholesterol

Lipids are a diverse group of molecules that generally contain long chains of carbon atoms and are insoluble in water. The principal types of lipids are fats, phospholipids, and cholesterol (see Chapter 2). Although fats are sometimes viewed as an enemy in our overweight society, they are in fact essential nutrients that are used primarily as a source of energy. Phospholipids are important components of cellular membranes and also make up the outer layer of nerve cells. Cholesterol is used in the manufacture of cellular membranes, sex hormones, and bile (which aids in fat breakdown). Some animal species can synthesize all the particular lipids they need. Others must acquire specific types of lipid building blocks, called **essential fatty acids**, from their food. For example, humans are unable to synthesize linoleic acid, which is required for the synthesis of certain phospholipids, so we need to obtain this essential fatty acid from our diet.

Fats Store Energy in Concentrated Form

Humans and most other animals store energy primarily in fat. When an animal's diet provides more energy than is expended through metabolic activities, most of the excess is converted to fat for storage. Fats have two major advantages as energy-storage molecules. First, they are the most concentrated energy source, containing more than twice the energy per unit weight of either carbohydrates or protein (about 9 Calories per gram for fats compared with about 4 per gram for proteins and carbohydrates). Second, lipids are hydrophobic—that is, they do not mix with water. Fat deposits, therefore, do not cause any extra accumulation of water in the body. For both these reasons, fats store more calories with less weight than do other molecules.

Table 21-1 Approximate Energy Consumed by a 150-Pound Person for Different Activities

		Time to "Work Off"			

Activity	Calories/hr	500 Calories Cheeseburger	300 Calories Ice Cream Cone	70 Calories Apple	40 Calories 1 Cup Broccoli
Running (6 mph)	700	43 min	26 min	6 min	3 min
Cross-country skiing (moderate)	560	54 min	32 min	7.5 min	4 min
Roller skating	490	1 hr 1 min	37 min	8.6 min	5 min
Bicycling (11 mph)	420	1 hr 11 min	43 min	10 min	6 min
Walking (3 mph)	250	2 hr	1 hr 12 min	17 min	10 min
Frisbee® playing	210	2 hr 23 min	1 hr 26 min	20 min	11 min
Studying	100	5 hr	3 hr	42 min	24 min

Figure 21-1 Fat provides insulation
These walruses can withstand the icy waters of the polar seas because they are insulated with a thick layer of fat beneath the skin.

Fats Provide Insulation
In mammals, which maintain an elevated body temperature, fat deposits do double duty by providing insulation as well as storing energy. Fat is typically stored in a layer beneath the skin, where it insulates the body because it conducts heat at only one-third the rate of other body tissues. Mammals who live near the North or South Pole or in cold ocean waters are particularly dependent on this insulating layer, which reduces the amount of energy they must expend to keep warm (Fig. 21-1).

Carbohydrates Are a Source of Quick Energy

Carbohydrates include simple sugars, as well as longer chains of sugars called polysaccharides (see Chapter 2). The simple sugar glucose is the primary source of energy for body cells, but the typical diet contains little glucose. During digestion, glucose is derived from the breakdown of more complex carbohydrates (as well as proteins and fats).

Animals, including humans, store the carbohydrate **glycogen** (a large, highly branched chain of glucose molecules) in the liver and muscles. This store of glycogen serves as a source of quick energy during exercise. We store less than half a pound of glycogen, so when activity is prolonged, as in the case of a marathon runner, the stored glycogen can become totally depleted. The expression "hitting the wall" describes the extreme fatigue that long-distance runners may experience after exhausting their supply of glycogen.

Proteins Provide Amino Acid Building Blocks

Protein in the diet serves mainly as a source of amino acids to make new molecules. Dietary protein is broken down in the digestive tract to yield amino acid subunits. The amino acids then move to body cells, where they are linked in specific sequences to form the many different proteins required by the body. Any excess protein in the diet is broken down to extract energy for immediate use or storage in fat.

Table 21-2 The Essential Amino Acids and Their Plant Sources

Beans and Other Legumes	Both Legumes and Grains	Corn and Other Grains
Isoleucine	Valine	Tryptophan
Lysine	Histidine	Methionine
	Threonine	
	Phenylalanine	
	Leucine	

In humans, dietary protein from foods such as meat, milk, eggs, corn, and beans is the only source of **essential amino acids**. Our bodies can synthesize the other amino acids used to build proteins, but the nine essential amino acids must be supplied by the diet. Because many plant proteins are deficient in some of the essential amino acids, vegetarians risk protein deficiency unless their diets include plants whose proteins together provide all nine of the essential amino acids (Table 21-2).

Minerals Are Elements and Small Inorganic Molecules

You may have seen a display of minerals in a museum—beautiful, brightly colored crystals. Many **minerals**, which are elements and inorganic molecules, play crucial roles in animal nutrition (Table 21-3). Minerals must be obtained through the diet, either from food or dissolved in drinking water, because the body cannot manufacture them. Required minerals include calcium, magnesium, and phosphorus, which are major constituents of bones and teeth. Others, such as sodium and potassium, are essential for muscle contraction and the conduction of nerve impulses. Iron is used in the production of hemoglobin, and iodine is found in hormones produced by the thyroid gland. In addition, trace amounts of several other minerals, including zinc, copper, and selenium, are required, typically as parts of enzymes.

Vitamins Play Many Roles in Metabolism

Vitamins are a diverse group of organic compounds that animals require in small amounts. In general, the body cannot synthesize vitamins (or cannot synthesize them in adequate amounts), so they must be obtained from food. Our modern diet is so different from the natural diet on which we evolved that many people find vitamin supplements a prudent way to compensate for possible vitamin deficiencies. The vitamins considered essential in human nutrition are listed in Table 21-4.

Table 21-3 Major Mineral Sources and Functions for Humans

Mineral	Dietary Sources	Major Functions in Body	Deficiency Symptoms
Calcium	Milk, cheese, green vegetables, legumes	Bone and tooth formation Blood clotting Nerve impulse transmission	Stunted growth Rickets, osteoporosis Convulsions
Phosphorus	Milk, cheese, meat, poultry, grains	Bone and tooth formation Acid–base balance	Weakness Demineralization of bone Loss of calcium
Potassium	Meats, milk, fruits	Acid–base balance Body water balance Nerve function	Muscular weakness Paralysis
Chlorine	Table salt	Formation of gastric juice Acid–base balance	Muscle cramps Apathy Reduced appetite
Sodium	Table salt	Acid–base balance Body water balance Nerve function	Muscle cramps Apathy Reduced appetite
Magnesium	Whole grains, green leafy vegetables	Activation of enzymes in protein synthesis	Growth failure Behavioral disturbances Weakness, spasms
Iron	Eggs, meats, legumes, whole grains, green vegetables	Constituent of hemoglobin and enzymes involved in energy metabolism	Iron-deficiency anemia (weakness, reduced resistance to infection)
Fluorine	Fluoridated water, tea, seafood	Maintenance of teeth and probably bone structure	High frequency of tooth decay
Zinc	Widely distributed in foods	Constituent of enzymes involved in digestion	Growth failure Small sex glands
Iodine	Seafish and shellfish, dairy products, many vegetables, iodized salt	Constituent of thyroid hormones	Goiter

Some Vitamins Dissolve in Water

Vitamins required by humans are often grouped into two categories, water-soluble and fat-soluble. Water-soluble vitamins include vitamin C and 11 compounds that make up the B-vitamin complex. These substances dissolve in the water of the blood plasma and are excreted by the kidneys. They are not stored in the body in any appreciable amounts. Water-soluble vitamins generally work in conjunction with enzymes to promote chemical reactions that supply energy or synthesize materials. Because each vitamin participates in several metabolic processes, a deficiency of a single vitamin can have wide-ranging effects (see Table 21-4).

Other Vitamins Dissolve in Fat

The fat-soluble vitamins are A, D, E, and K. Their roles are different from those of the water-soluble vitamins. Vitamin K, for example, helps regulate blood clotting, and vitamin A is used to produce visual pigment (the light-capturing molecule in the retina of the eye). Fat-soluble vitamins can be stored in body fat and may accumulate in the body over time. For this reason, some fat-soluble vitamins may be toxic if consumed in excessively high doses.

Nutritional Guidelines Help People Obtain a Balanced Diet

Most people in the United States are fortunate to live amidst an abundance of food. However, the diversity of foods in a typical U.S. supermarket and the easy availability of "fast food" can lead to poor nutritional choices. To help people make informed choices, the U.S. government has developed recommendations and goals that are summarized in Table 21-5.

Further help is provided by the Food Guide Pyramid, designed by the U.S. Department of Agriculture, which illustrates the relative abundances of different food groups in an optimal diet (Fig. 21-2). Another source of information is the nutritional labeling now required on commercially packaged foods. These labels provide complete information about Calorie, fiber, fat, sugar, and vitamin content (Fig. 21-3). Also, some fast-food chains have fliers available that list nutritional information about their products.

Are You Too Heavy?

A simple way to determine whether your weight is likely to pose a health risk is by calculating your body mass index (BMI). The BMI takes into account your weight and height to arrive at an estimate of body fat. The original formula is: weight (in kg)/ height^2 (in meters), but you can calculate the same number by multiplying your weight in pounds by 703, then dividing by your height in inches, squared. A BMI between 20 and 24 is considered healthy. Between 25 and 30, your BMI tells you that you are overweight, and you are obese if your BMI exceeds 30.

Table 21-4 Vitamin Sources and Functions for Humans			
Vitamin	**Dietary Sources**	**Functions in Body**	**Deficiency Symptoms**
Water soluble			
Vitamin B_1 (thiamin)	Milk, meat, bread	Coenzyme in metabolic reactions	Beriberi (muscle weakness, peripheral nerve changes, edema, heart failure)
Vitamin B_2 (riboflavin)	Widely distributed in foods	Constituent of coenzymes in energy metabolism	Reddened lips, cracks at corner of mouth, lesions of eye
Niacin	Liver, lean meats, grains, legumes	Constituent of two coenzymes in energy metabolism	Pellagra (skin and gastrointestinal lesions; nervous, mental disorders)
Vitamin B_6 (pyridoxine)	Meats, vegetables, whole-grain cereals	Coenzyme in amino acid metabolism	Irritability, convulsions, muscular twitching, dermatitis, kidney stones
Pantothenic acid	Milk, meat	Constituent of coenzyme A, with a role in energy metabolism	Fatigue, sleep disturbances, impaired coordination
Folic acid	Legumes, green vegetables, whole wheat	Coenzyme involved in nucleic and amino acid metabolism	Anemia, gastrointestinal disturbances, diarrhea, retarded growth, birth defects
Vitamin B_{12}	Meats, eggs, dairy products	Coenzyme in nucleic acid metabolism	Pernicious anemia, neurological disorders
Biotin	Legumes, vegetables, meats	Coenzymes required for fat synthesis, amino acid metabolism, and glycogen formation	Fatigue, depression, nausea, dermatitis, muscular pains
Choline	Egg yolk, liver, grains, legumes	Constituent of phospholipids, precursor of the neuro-transmitter acetylcholine	None reported in humans
Vitamin C (ascorbic acid)	Citrus fruits, tomatoes, green peppers	Maintenance of cartilage, bone, and dentin (hard tissue of teeth); collagen synthesis	Scurvy (degeneration of skin, teeth, blood vessels; epithelial hemorrhages)
Fat soluble			
Vitamin A (retinol)	Beta-carotene in green, yellow, and red vegetables Retinol added to dairy products	Constituent of visual pigment Maintenance of epithelial tissues	Night blindness, permanent blindness
Vitamin D	Cod-liver oil, eggs, dairy products	Promotes bone growth and mineralization Increases calcium absorption	Rickets (bone deformities) in children; skeletal deterioration
Vitamin E (tocopherol)	Seeds, green leafy vegetables, margarines, shortenings	Antioxidant, prevents cellular damage	Possibly anemia
Vitamin K	Green leafy vegetables Product of intestinal bacteria	Important in blood clotting	Bleeding, internal hemorrhages

2 How Is Digestion Accomplished?

Digestion refers to physically grinding up, then chemically breaking down food. The digestive systems of animals take in complex molecules of food and then digest them into simpler molecules that can be absorbed. Material that cannot be used or broken down is then expelled from the body.

An Overview of Digestion

Animals eat the bodies of other organisms, bodies that may resist becoming food. The plant body, for example, armors each cell with a wall of indigestible cellulose. Animal bodies may be covered with equally indigestible fur, scales, or feathers. In response to these defenses, however, animal digestive systems have evolved the means to consume and obtain nutrients from even the

Table 21-5 Dietary Changes Advocated by the U.S. Government

Dietary Component	Percentage of Total Daily Energy Intake: Average U.S. Diet	Percentage of Total Daily Energy Intake: Dietary Goals
Carbohydrates	46	58
Lipids	38	30
Proteins	16	12

Summary of Recommendations:

1. Increase consumption of fruits, vegetables, and whole grains.
2. Decrease consumption of refined sugars.
3. Decrease consumption of fats; replace saturated with unsaturated fats.
4. Decrease consumption of animal fats by selecting lean meats, poultry, and fish.
5. Decrease consumption of high-cholesterol foods, such as butter and eggs.
6. Decrease consumption of salt and foods high in salt content.
7. Decrease caloric intake to maintain desirable weight.

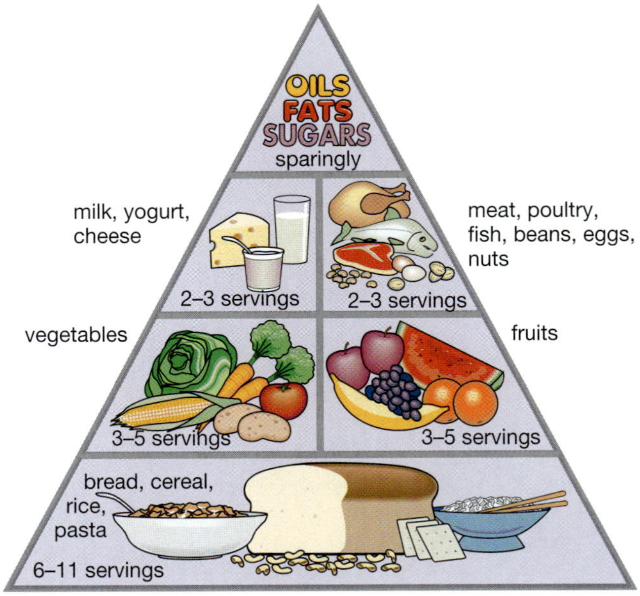

Figure 21-2 The Food Guide Pyramid
This chart shows the numbers of daily servings recommended by the U.S. Department of Agriculture.

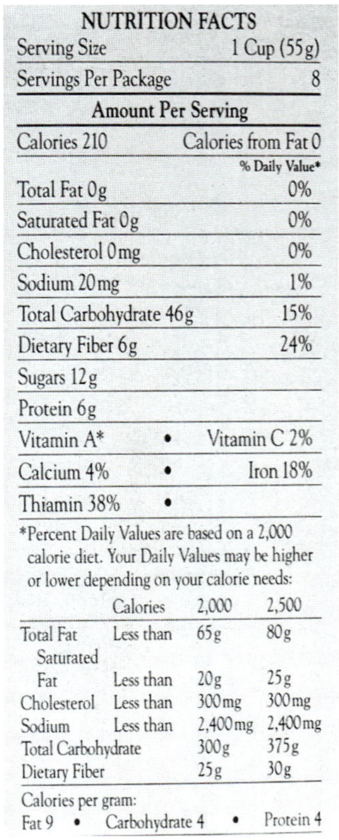

NUTRITION FACTS

Serving Size	1 Cup (55g)
Servings Per Package	8
Amount Per Serving	
Calories 210	Calories from Fat 0
	% Daily Value*
Total Fat 0g	0%
Saturated Fat 0g	0%
Cholesterol 0mg	0%
Sodium 20mg	1%
Total Carbohydrate 46g	15%
Dietary Fiber 6g	24%
Sugars 12g	
Protein 6g	
Vitamin A* •	Vitamin C 2%
Calcium 4% •	Iron 18%
Thiamin 38% •	

*Percent Daily Values are based on a 2,000 calorie diet. Your Daily Values may be higher or lower depending on your calorie needs:

	Calories	2,000	2,500
Total Fat	Less than	65g	80g
Saturated Fat	Less than	20g	25g
Cholesterol	Less than	300mg	300mg
Sodium	Less than	2,400mg	2,400mg
Total Carbohydrate		300g	375g
Dietary Fiber		25g	30g

Calories per gram:
Fat 9 • Carbohydrate 4 • Protein 4

Figure 21-3 Complete food labels
The U.S. government now requires more complete nutritional labeling of foods, such as in this sample. At the bottom of the label, the total recommended number of grams of key nutrients is listed for a 2000- and a 2500-Calorie diet.

most well-defended bodies. The different types of animals, considered together, are able to exploit every conceivable food source because they have diverse types of digestive tracts in which to process the nutrients. Amid this diversity, however, are certain tasks that all digestive systems must accomplish:

1. *Ingestion.* The food must be brought into the digestive tract through an opening, usually called a mouth.
2. *Mechanical breakdown.* The food must be physically broken down into smaller pieces. The particles produced by mechanical breakdown provide a large surface area for attack by digestive enzymes.
3. *Chemical breakdown.* The particles of food must be exposed to digestive enzymes and other digestive fluids that break down large molecules into smaller subunits.
4. *Absorption.* The small molecules must be transported out of the digestive cavity and into cells.
5. *Elimination.* Indigestible materials must be expelled from the body.

A Sac with One Opening Forms the Simplest Digestive System

In almost all animals, digestion takes place in a chamber within the body where chunks of food are broken down by enzymes that act outside the cells. One of the simplest of these chambers is found in cnidarians, such as sea anemones, hydra, and jellyfish. These animals possess a digestive sac with a single opening through which food is ingested and wastes are ejected (Fig. 21-4). Although it is generally referred to as the mouth, this opening also serves as an anus. Food captured by stinging tentacles is moved into the digestive sac, where enzymes break it down. Cells lining the cavity absorb the nutrients and engulf small food particles. The undigested remains are eventually expelled through the same opening that they entered.

Health Watch
Eating Disorders—Betrayal of the Body

No animal can survive without eating, and natural selection has provided animals with mechanisms to ensure that the impulse to eat arises when nutrients are needed. In humans, however, these natural impulses, so crucial to our health and well-being, can go terribly awry. In recent decades we have seen an increase in the number of people with eating disorders, ailments characterized by the disruption of normal eating behavior.

Eating disorders include two particularly debilitating illnesses, anorexia nervosa and bulimia nervosa. People with *anorexia nervosa* experience an intense fear of gaining weight, and they achieve extreme weight loss by eating very little food. This drastically reduced eating is commonly accompanied by self-induced vomiting, laxative intake, and excessive exercise. The consequences are disastrous. Anorexics become emaciated, losing both fat and muscle. This emaciation can in turn disrupt cardiac, digestive, endocrine, and reproductive functions. Up to 18% of anorexics die as a result of their disorder.

The main symptom of *bulimia nervosa* is repeated bouts of binge eating, in which extraordinarily large amounts of food are consumed rapidly. Binge eating by bulimics is typically followed by the same kinds of purging behaviors common in anorexics, such as vomiting and laxative consumption. In most cases, the health consequences of bulimia nervosa are not as dire as are those of anorexia nervosa, because there is no drastic weight loss.

Girls and young women are most at risk for eating disorders (Fig. E21-1). More than 90% of diagnosed eating disorders are in females; in most of these cases, onset of the illness takes place between 13 and 20 years of age. The causes of eating disorders are not well understood. In a body wracked by the near-starvation of anorexia nervosa, the effects of the illness are hard to distinguish from a defect that might be a cause. No clear genetic link to the disorders has been found thus far, so suspicion has fallen on environmental factors, especially on societal pressures to be thin.

Unfortunately, it is difficult to treat eating disorders. Victims are usually hospitalized to restore nutritional health and counseled, but they often fail to recover completely. Antidepressant drugs are helpful in some cases but are more likely to be effective for bulimia nervosa than for the more dangerous anorexia nervosa. Hope may be on the horizon, however, owing to a recent explosion of research findings about the hormones that control appetite. For example, researchers have discovered a class of hormones known as orexins, which, when injected into mice, bind to receptors in the brain and cause food consumption to increase dramatically. Such new discoveries raise hopes that as our understanding of the physiological control of appetite grows, we may yet be able to devise chemical treatments for eating disorders.

Figure E21-1 Eating disorders
The late Princess Diana made headlines when she announced that she had been suffering from bulimia.

Digestion in a Tube Allows More Frequent Meals

Most animal species, including humans and other vertebrates, worms, mollusks, and arthropods, have a digestive system that is basically a tube through the body, beginning with a mouth and ending with an anus. Such digestive systems generally consist of a series of specialized regions that process food in an orderly sequence, first grinding it up, then breaking it down with enzymes, and then absorbing the nutrient molecules into body cells.

In the earthworm, the tube consists of a series of compartments, each with a specific role in the breakdown of food, which consists of organic material found in soil (Fig. 21-5). A tubular digestive system is essential to the earthworm, which continuously ingests soil as it burrows through the earth, taking the soil in at one end of its body and passing it out at the other. The worm's muscular pharynx is a cavity that connects the mouth with the esophagus, a muscular passageway. Ingested soil and bits of vegetation are passed through the pharynx and the esophagus to the crop, a thin-walled storage organ. The crop collects the food and gradually passes it to the gizzard. There, bits of sand and the contraction of muscles grind the food into smaller particles. The food then travels to the intestine, where enzymes break it down into simple molecules that can be absorbed by the cells lining the intestine. The soil itself passes out through the anus.

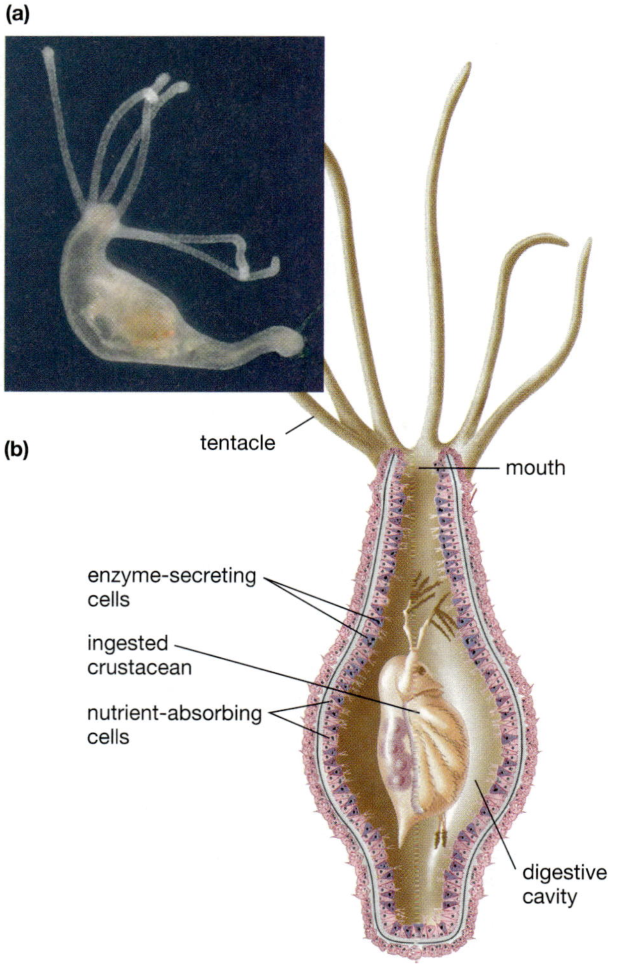

Figure 21-4 Digestion in a sac
(a) A *Hydra* (a cnidarian) has just captured and ingested a waterflea (*Daphnia*, a microscopic crustacean). ***(b)*** Within the digestive cavity of *Hydra*, enzymes digest the prey into smaller particles and nutrients.

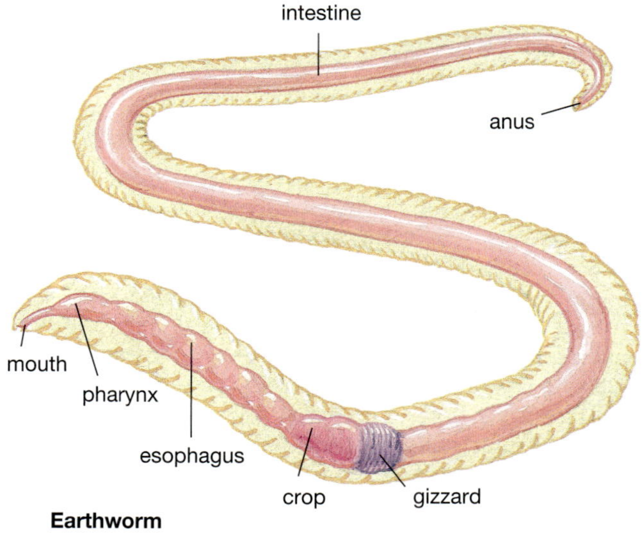

Figure 21-5 Tubular digestive tracts
The earthworm has a one-way digestive system that passes food through a series of compartments, each specialized to play a specific role in breaking down food and absorbing it.

The Big Picture

One way to think of a body is as a community of millions of cells that have come together "under one roof" for mutual benefit. An important job that an animal body can do for its cells is to convert comparatively large pieces of tissue into molecules that can be used by cells. That is the task of the digestive system.

3 How Do Humans Digest Food?

Like earthworms, humans have tubular digestive tracts with several compartments in which food is broken down—first physically, then chemically—before being absorbed by individual cells. The human digestive system (Fig. 21-6) is adapted for processing a wide variety of foods.

Food travels in a continuous tube from mouth to anus. In the course of traveling this route, the food is subjected to a carefully orchestrated succession of digestive operations. By the time its passage is complete, the food has been chopped, mashed, mixed, churned, and bathed in powerful chemicals. Everything of value has been extracted, and the residue is ejected. This stepwise taking apart of food requires coordinated action from the structures that make up the digestive system.

The Mechanical and Chemical Breakdown Begins in the Mouth

You take a bite, your mouth waters, and you begin chewing. This begins both the mechanical and the chemical breakdown of food. In humans and other mammals, the mechanical work is done mostly by teeth. In adult humans, 32 teeth of varying shapes and sizes cut and grind food into small pieces. *Incisors* at the front of the mouth snip off pieces of food; the pointed *canine* teeth beside them are useful for tearing the pieces apart; and the *premolars* and *molars* at the back of the mouth have flat surfaces that can grind food to a paste (Fig. 21-7).

While the food is being pulverized by the teeth, the first phase of chemical digestion begins as three pairs of salivary glands pour saliva into the mouth. Saliva contains the digestive enzyme amylase, which begins the breakdown of starches into sugar. Saliva has other functions as well. It contains a bacteria-killing enzyme and antibodies that help guard against infection. Saliva also lubricates the food to facilitate swallowing and dissolves some food molecules such as acids and sugars, carrying them to taste buds on the tongue. The taste buds are sensory receptors that help identify the type and quality of the food.

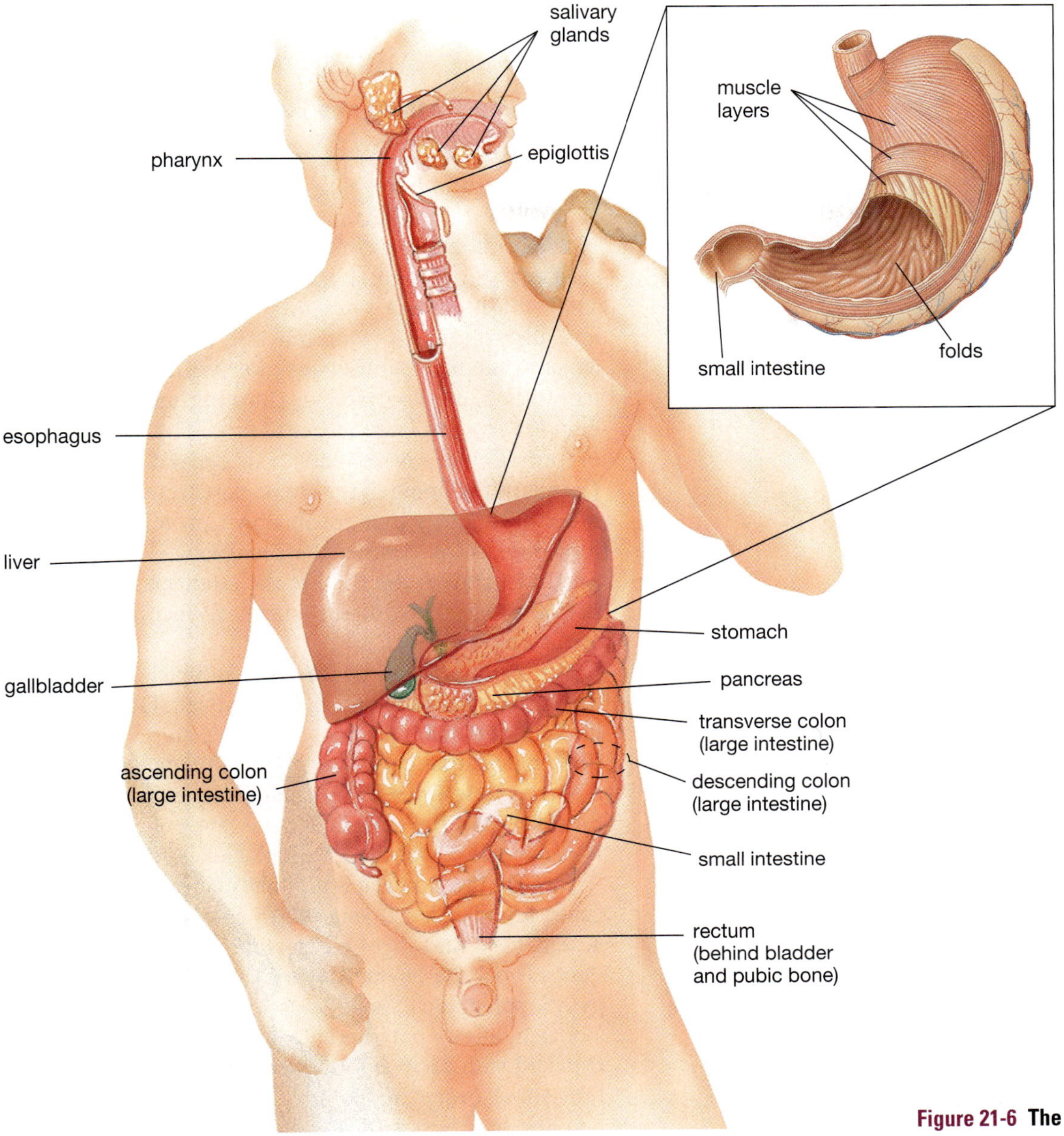

Figure 21-6 The human digestive tract

The Pharynx Connects the Mouth to the Rest of the Digestive System

With the help of the muscular tongue, the food is manipulated into a mass and pressed backward into the **pharynx**, a muscular cavity connecting the mouth with the esophagus (Fig. 21-8a). Via the larynx, the pharynx also connects the nose and mouth with the trachea, which conducts air to the lungs. This arrangement occasionally causes problems, as anyone who has ever choked on a piece of food can attest. Normally, however, the swallowing reflex (triggered by food entering the pharynx) elevates the larynx, which moves the epiglottis (a flap of tissue) so that it blocks off the respiratory passage. Food is thus directed into the esophagus rather than into the larynx (Fig. 21-8b).

The Esophagus Conducts Food to the Stomach

Swallowing forces food into the **esophagus**, a muscular tube that propels food from the mouth to the stomach. Muscles surrounding the esophagus contract in sequence

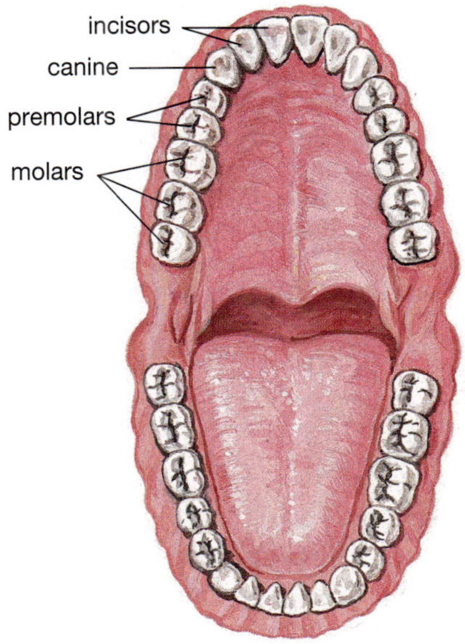

Figure 21-7 Teeth begin the mechanical breakdown of food Humans have an unspecialized set of teeth that includes flat incisors for biting, pointed canines for tearing, premolars for grinding, and molars for crushing and chewing.

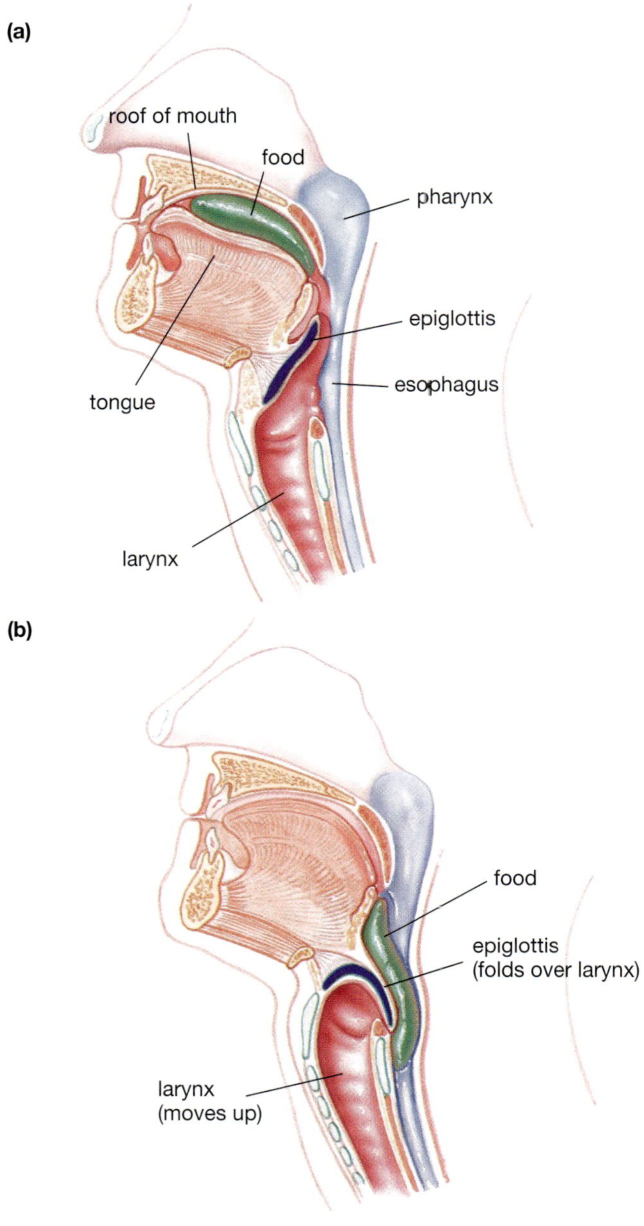

Figure 21-8 The challenge of swallowing
(a) Swallowing is complicated by the fact that both the esophagus (part of the digestive system) and the larynx (part of the respiratory system) open into the pharynx. *(b)* During swallowing, the larynx moves upward beneath the epiglottis. The epiglottis folds down over the larynx, sealing off the opening to the respiratory system and directing food down the esophagus instead.

above the swallowed food mass, squeezing it down toward the stomach. This muscular action, called **peristalsis**, also occurs in the stomach and intestines, where it helps move food along the digestive tract. Peristalsis is so effective that a person can actually swallow when upside down. Mucus secreted by cells that line the esophagus helps protect it from abrasion; the mucus also lubricates food during its passage.

The Stomach Stores and Further Breaks Down Food

The **stomach** in humans is an expandable muscular sac capable of holding as much as a gallon of food and liquids. The stomach has three major functions. First, the stomach stores food and releases it gradually into the small intestine at a rate suitable for proper digestion and absorption. A second function of the stomach is to assist in the mechanical breakdown of food. Its muscular walls make contracting, churning movements that help break apart large pieces of food. The third function of the stomach is the chemical breakdown of food. Glands in the lining of the stomach secrete enzymes and other substances that break down proteins and other food molecules. Among the secretions is hydrochloric acid, which gives the fluid in the stomach a very acidic pH of 1 to 3; the stomach enzymes function best in an acidic environment.

As you may have noticed, the stomach produces all the ingredients necessary to digest itself. However, cells lining the stomach normally produce a large quantity of thick mucus that coats the stomach lining and serves as a barrier to self-digestion. The protection is not perfect, however, and the cells lining the stomach are digested to some extent and must be replaced every few days.

Food in the stomach is gradually converted to a thick, acidic liquid called **chyme**, which consists of partially digested food and digestive secretions. Peristaltic waves then propel the chyme toward the small intestine. The pyloric sphincter, a ring of muscle at the lower end of the stomach, allows only about a teaspoon of chyme to be expelled with each contraction, at a rate of one contraction about every 20 seconds. It takes 2 to 6 hours, depending on the size of the meal, to empty the stomach completely. The continued churning movements of an empty stomach are felt as "hunger pangs."

Most Digestion Occurs in the Small Intestine

The **small intestine** is narrow (about 1 to 2 inches in diameter in an adult human) but, with a length of 10 feet, it is the longest part of the digestive tract. The small intestine digests food into small molecules and absorbs these molecules into the bloodstream. The first role of the small intestine—digestion—is accomplished with the aid of digestive secretions from three sources: the liver, the pancreas, and the cells of the small intestine itself.

The Liver and Gallbladder Provide Bile

The **liver** is perhaps the most versatile organ in the body. It has many diverse functions, but its role in digestion is to produce **bile**, a liquid stored and concentrated in the **gallbladder** and released into the small intestine through a tube called the bile duct (see Fig. 21-6). Bile is a complex mixture composed of **bile salts**, water, other salts, and cholesterol. Bile salts are synthesized in the liver from cholesterol and amino acids. Although they assist in the breakdown of lipids, bile salts are not enzymes. Rather, they act as detergents or emulsifying agents, dispersing globs of fat in the chyme into microscopic particles. These particles expose a large surface area for attack by lipid-digesting enzymes produced by the pancreas.

The Pancreas Secretes Digestive Substances

The **pancreas** lies in the loop between the stomach and small intestine (see Fig. 21-6). It consists of two major types of cells. One type produces hormones that help regulate blood sugar (as we shall see in Chapter 23), and the other produces a digestive secretion called **pancreatic juice**, which is released into the small intestine. About a quart of pancreatic juice is released into the small intestine each day. This secretion contains water, sodium bicarbonate, and several digestive enzymes. Sodium bicarbonate (the active ingredient in baking soda) neutralizes the acidic chyme in the small intestine, producing a slightly basic pH. The digestive enzymes in pancreatic juice require this basic pH for proper functioning, in contrast to the stomach's digestive enzymes, which require an acidic pH. The pancreatic digestive enzymes break down carbohydrates, lipids, and proteins.

The Intestinal Wall Completes the Digestive Process

The wall of the small intestine is studded with cells that complete the digestive process and absorb the small molecules that result. These cells have various enzymes on their external membranes, which form the lining of the small intestine. Because these enzymes are actually embedded in the membranes of the cells that line the small intestine, this final phase of digestion occurs as the nutrient is being absorbed into the cell. Like the stomach, the small intestine is protected from digesting itself by large amounts of mucous secretions from specialized cells in its lining.

Most Absorption Occurs in the Small Intestine

The small intestine is not only the principal site of chemical digestion, it is also the major site of nutrient **absorption** into the blood. The small intestine has numerous foldings and projections that give it an internal surface area that is 600 times that of a smooth tube of the same length. Minute, fingerlike projections called **villi** (singular, **villus**) cover the entire folded surface of the intestinal wall. Villi, which range from 0.5 to 1.5 millimeters in length, give the intestinal lining a velvety appearance to the naked eye. They move gently back and forth in the chyme that passes through the intestine. This movement increases the exposure of the villi to the molecules to be digested and absorbed. Further, each individual cell of the villi bears a fringe of microscopic projections called **microvilli**. Taken together, these projections of the small intestine wall give it a surface area of more than 2200 square feet, almost the size of a tennis court (Fig. 21-9).

Contractions of the circular muscles of the intestine slosh the chyme back and forth, bringing nutrients into contact with the absorptive surface of the small intestine. Water, simple sugars, amino acids, fatty acids, vitamins, and minerals are absorbed, mostly by processes in which energy is expended to transport the nutrients into intestinal cells. The nutrients then diffuse out of the intestinal cells into the interstitial fluid. From the interstitial fluid, the nutrients diffuse into the blood capillaries within the villi. When absorption is complete, coordinated peristaltic waves conduct the leftovers into the large intestine.

Water Is Absorbed and Feces Are Formed in the Large Intestine

The **large intestine** in an adult human is about 5 feet long and about 3 inches in diameter, wider and shorter than the small intestine. The large intestine has two parts: For most of its length it is called the **colon**, but its final 6-inch compartment is called the **rectum**. Into the large intestine flow the leftovers of digestion: a mixture of water, undigested fats and proteins, and indigestible fibers, such as the cell walls of vegetables and fruits. The large intestine contains a flourishing bacterial population that lives on unabsorbed nutrients. These bacteria earn their keep by synthesizing vitamin B_{12}, thiamin, riboflavin, and, most important, vitamin K, which would otherwise be deficient in a normal diet. Cells lining the large intestine absorb these vitamins as well as leftover water and salts.

After absorption is complete, the result is the semisolid **feces**. Feces consist of indigestible wastes and the dead bodies of bacteria, which account for about one-third of the dry weight of feces. The feces are transported by peristaltic movements until they reach the rectum. Expansion of this chamber stimulates the desire to defecate.

Digestion Is Controlled by the Nervous System and Hormones

When you begin to eat, your body coordinates events that convert food into nutrients circulating in your blood. As your mouth responds to the first bite, your stomach must be warned that food is on the way. The enzymes of the stomach and small intestine require different environments to function properly (highly acidic in the stomach, slightly basic in the small intestine), and secretions into various parts of the digestive tract correspond with the arrival of food. Not surprisingly, both nerves and hormones coordinate the secretions and activity of the digestive tract.

Sensory Signals Initiate Digestion

The initial phase of digestion is controlled by the nervous system, which transmits signals that originate in the head. These signals include the sight, smell, taste,

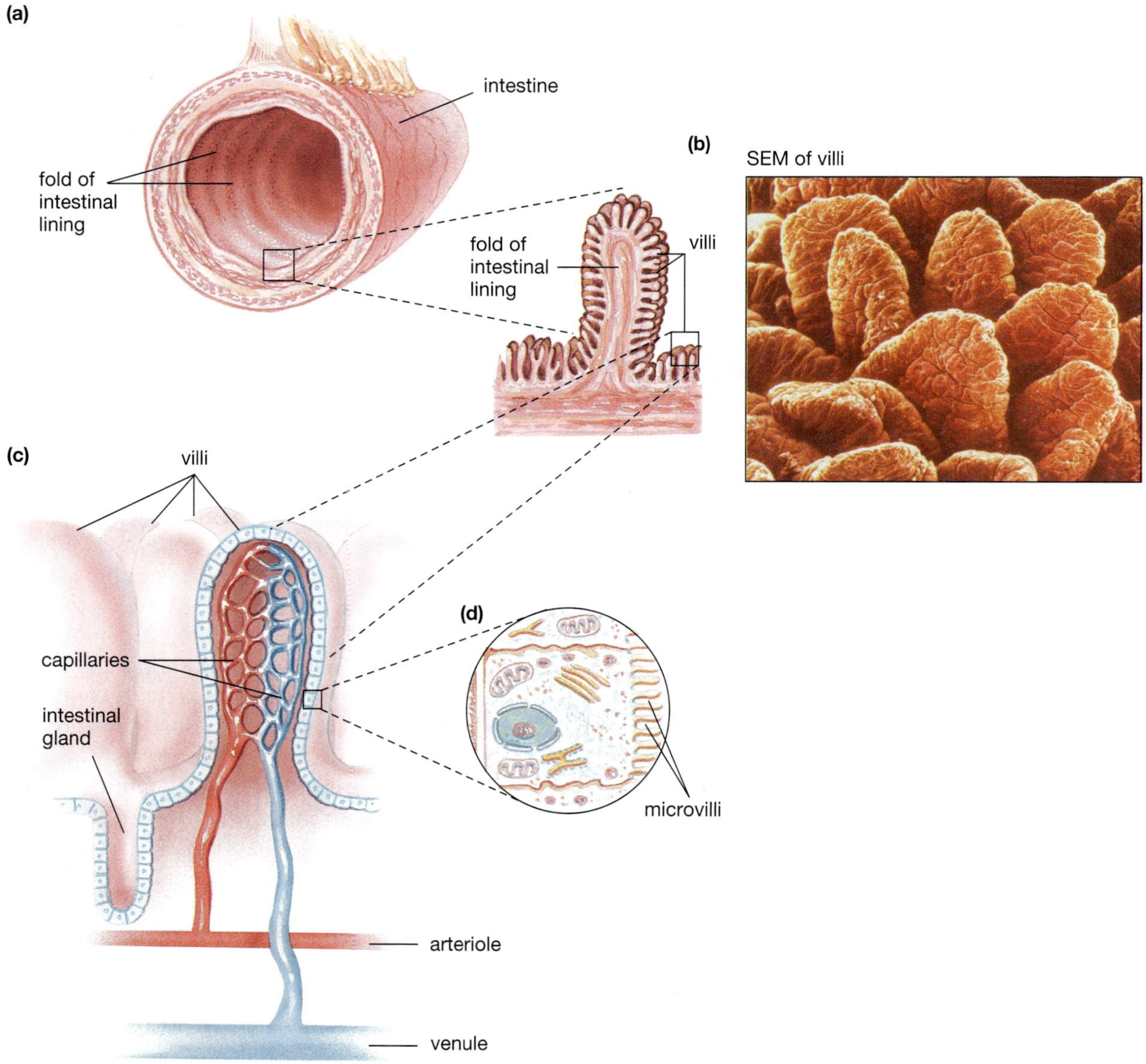

Figure 21-9 The small intestine
The folds of the small intestine maximize the surface area available to absorb nutrients. ***(a)*** Large folds in the intestinal lining are themselves carpeted in ***(b)*** tiny, fingerlike projections called *villi*, which enclose ***(c)*** a network of capillaries. ***(d)*** If we use a microscope to zoom in on one villus, we see that the epithelial cells on its surface are sheathed in plasma membranes that have yet another level of microscopic projections, microvilli.

and sometimes the thought of food, as well as the muscular activity of chewing. In response to these signals, the salivary glands secrete saliva into the mouth while nervous-system signals to the stomach walls cause secretions to begin there.

The Stomach Responds to the Presence of Food

Food arriving in the stomach triggers the second phase of digestion. As the stomach wall is stretched, its cells produce a large quantity of mucus, which protects the stomach from digesting itself. Other cells release the hormone gastrin, which stimulates the secretion of hydrochloric acid. Still other cells secrete pepsinogen, an inactive form of the protein-digesting enzyme pepsin. The acidity of the stomach converts pepsinogen to its active form, pepsin, which begins protein digestion.

Chyme Stimulates Digestion in the Small Intestine

As the liquid chyme is gradually released into the small intestine, its acidity stimulates the release of another hormone, secretin, into the bloodstream by cells of the upper small intestine. Secretin causes the pancreas to pour sodium bicarbonate into the small intestine. Sodium bicarbonate neutralizes the acidity of the incoming chyme, creating an environment in which the pancreatic enzymes can function. Other hormones stimulate the pancreas to release digestive enzymes into the small intestine, and stimulate the gallbladder to contract, squeezing bile through the bile duct to the small intestine.

4 What Are the Functions of Vertebrate Urinary Systems?

When an animal ingests and digests food, the digestive system retains the useful components of the food and eliminates useless or harmful components from the body. The vertebrate urinary system performs a similar function with regard to the intake and output of water. In particular, the urinary system ensures that the volume of water inside an animal's body remains relatively constant. This function is important because if the volume of water inside body cells fluctuates too much, the chemicals dissolved in internal fluids will become too concentrated or too dilute, with disastrous consequences for the animal.

Different Urinary Systems Are Effective in Different Environments

Each animal species has evolved a urinary system that is effective in the species' own environment. For example, a trout, which spends its life bathed in fresh water, must get rid of the water that is driven into its body by osmosis. The trout urinates more or less constantly, excreting an amount of urine equivalent to its total blood volume every 2 or 3 hours. At the other extreme, a kangaroo rat spends its life bathed in dry desert air, where the main regulatory task is to prevent water from escaping its body. A kangaroo rat excretes only a few milliliters of urine over the course of a day.

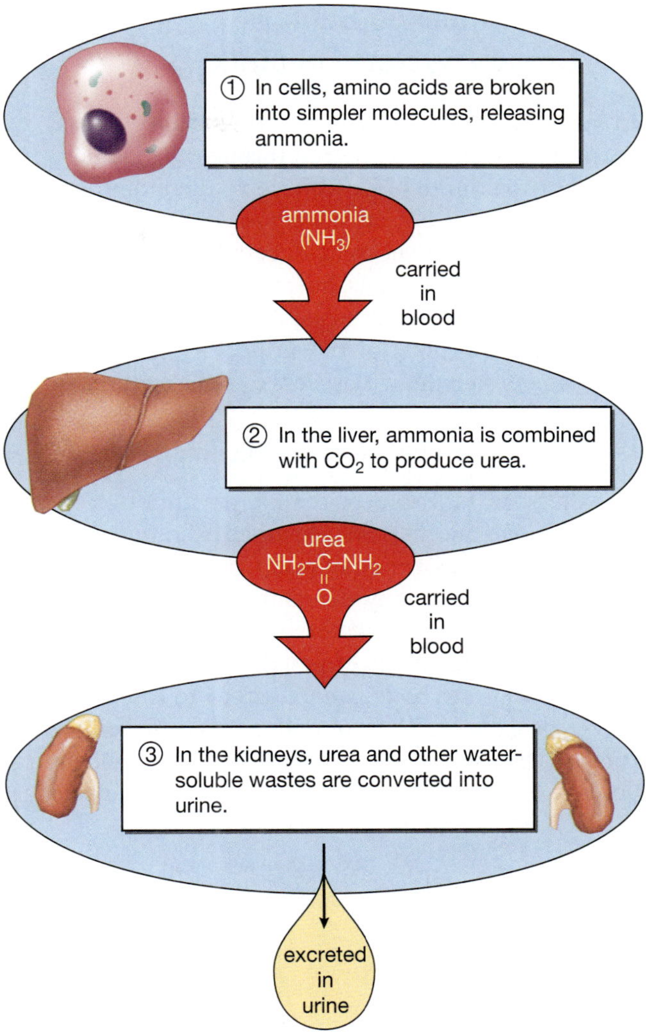

Figure 21-10 A flow diagram showing the formation and excretion of urea

The Urinary System Regulates the Concentrations of Dissolved Substances

The organs of the urinary system not only regulate the volume of water in the body but also ensure that the chemical composition of the blood and extracellular fluid remains within the bounds required for cellular metabolism. The urinary system regulates blood levels of ions such as sodium, potassium, chloride, and calcium; regulates the water content of the blood; and maintains the proper pH of the blood. In addition, it ensures that important dissolved nutrients such as glucose and amino acids are retained and not eliminated.

The Urinary System Eliminates Dissolved Wastes

Another vital function of the urinary system is to eliminate cellular waste products, especially the products of amino acid metabolism. When amino acids are used in cells as a source of energy or for the synthesis of new molecules, ammonia (NH_3) is produced as a by-product. Ammonia is very toxic and, in mammals, it is carried in the blood to the liver, where it is converted to **urea**, a far less toxic substance (Fig. 21-10). Urea is filtered from the blood and excreted in **urine**, a fluid consisting of water and dissolved wastes and some excess nutrients. By excreting waste in the form of urea, mammals avoid damage from ammonia's toxicity.

The Big Picture

Animal urinary systems perform a delicate balancing act, ridding the body of harmful wastes while restricting the escape of substances useful to the body. When conditions inside or outside the body change, urinary systems quickly adjust, so that homeostasis is maintained.

5 How Does the Human Urinary System Function?

In vertebrates, the functions of the urinary system are performed as blood is filtered through the **kidneys**. The fluid portion of the blood collects in the kidneys, where water and important nutrients are reabsorbed into the blood, while toxic substances, cellular waste products, and excess vitamins, salts, hormones, and water are left behind to be excreted as urine. The kidneys produce the urine; other portions of the urinary system transport, store, and eliminate it, a process called **excretion**.

Kidney Function Balances Conflicting Demands

The different functions of the kidney are often in conflict with one another. For example, excreting dissolved wastes requires excretion of water, but the need to maintain water volume may demand that the water be retained. Furthermore, water that contains dissolved wastes is very likely also to contain dissolved nutrients and salts that the body cannot afford to lose. How can wastes be excreted without undue loss of water and nutrients? The urinary system's method of producing urine, described in the following sections, meets this challenge.

The Urinary System Produces, Transports, and Excretes Urine

Human kidneys are paired, kidney-bean-shaped organs located on either side of the spinal column, slightly above the waist (Fig. 21-11). Each is approximately 5 inches long, 3 inches wide, and 1 inch thick. Blood carrying dissolved cellular wastes enters each kidney through a renal artery. After the blood has been filtered, it exits through the renal vein. Urine leaves each kidney through a narrow, muscular tube called the **ureter**. Using peristaltic contractions, the ureters transport urine to the **bladder**. This hollow, muscular chamber collects and stores the urine. Urine completes its journey to the outside through the **urethra**, a single narrow tube about 1.5 inches long in the female and about 8 inches long in the male.

Urine Forms in the Kidneys

Each kidney contains a solid outer layer where urine forms, and a subdivided inner chamber (the renal pelvis) that collects urine and funnels it into the ureter. Microscopic examination of the outer layer of the kidney reveals an array of tiny individual filters, or **nephrons**, richly supplied with blood vessels (Fig. 21-12). More

(a)

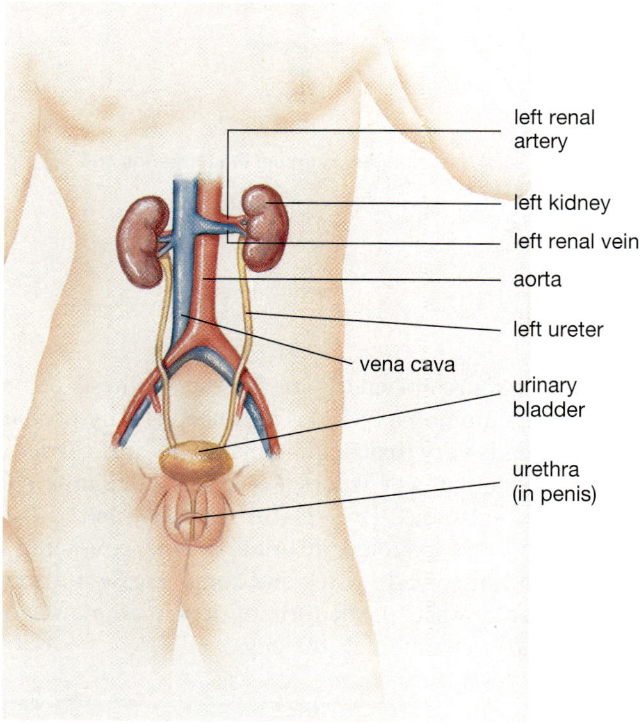

(b)

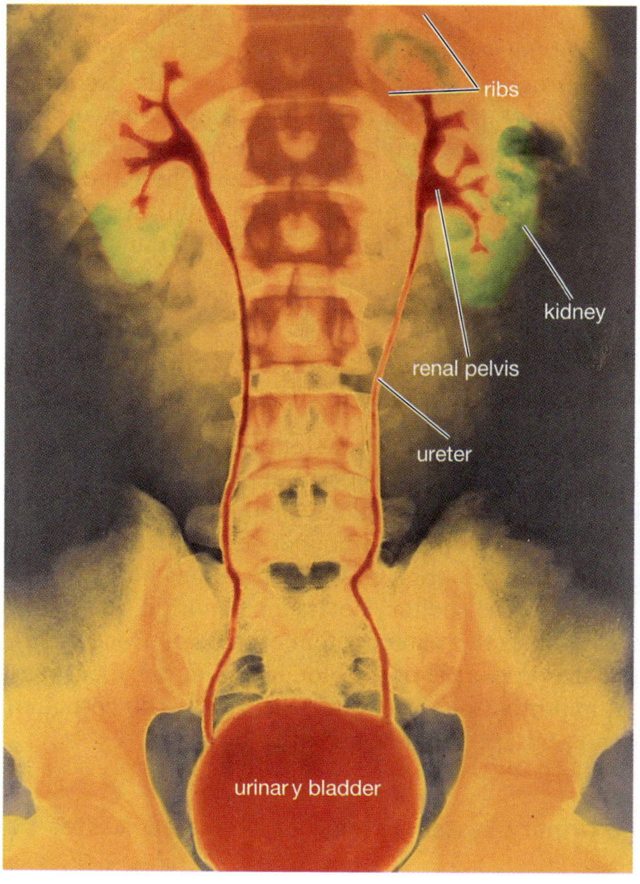

Figure 21-11 The human urinary system
(a) Diagrammatic view of the human urinary system and its blood supply. *(b)* Color-enhanced X-ray view (called a pyelogram).

MediaTutor 21.2 Urinary System Anatomy

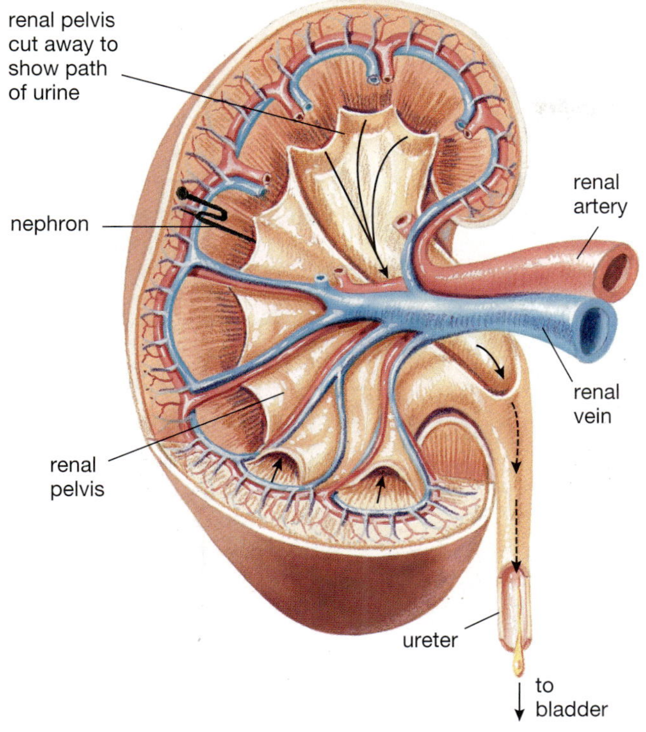

Figure 21-12 Cross section of a kidney
The cross section shows the blood supply and internal structure of a kidney. A nephron, considerably enlarged, is drawn to show its orientation in the kidney.

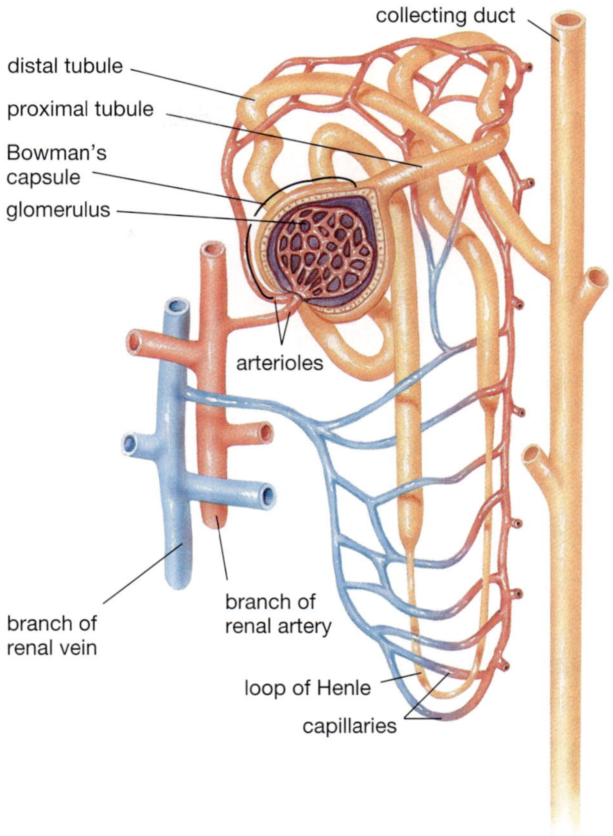

Figure 21-13 An individual nephron and its blood supply

than 1 million nephrons are packed into the outer layer of each kidney.

The nephron has three major parts: (1) the **glomerulus**, a dense knot of capillaries from which fluid from the blood is collected into (2) a surrounding cuplike structure, called **Bowman's capsule**, and (3), a long, twisted **tubule**. The tubule is further subdivided into first the **proximal tubule**, then the **loop of Henle**, and finally the **distal tubule**, which leads to the **collecting duct**, which conducts urine into the renal pelvis (Fig. 21-13). In the tubule, nutrients are selectively reabsorbed from the filtered fluid back into the blood, while wastes and some of the water are left behind to form urine. Additional wastes are also secreted into the tubule from the blood. Different portions of the tubule selectively modify the fluid as it travels through.

Blood Is Filtered by the Glomerulus

Blood is conducted to each nephron by an arteriole that branches from the renal artery. Within a cup-shaped portion of the nephron—Bowman's capsule—the arteriole branches into numerous microscopic capillaries that form the intertwined mass of the glomerulus (see Fig. 21-13). The walls of the glomerular capillaries are extremely permeable to water and small dissolved molecules, but they prevent the movement of most of the large proteins found in blood. Beyond the glomerulus, the capillaries reunite to form an arteriole whose diameter is smaller than that of the incoming arteriole. Because the diameters of the outgoing arterioles are smaller, pressure builds within the glomerulus, driving water and many of the dissolved substances from the blood through the capillary walls. This process is called **filtration** (Fig. 21-14, step ①), and the resulting fluid is called the **filtrate**. The watery filtrate, resembling blood plasma without its plasma proteins, is collected in Bowman's capsule for transport through the nephron.

The Filtrate Is Converted to Urine in the Nephron

Now for the major challenge—the filtrate collected in Bowman's capsule contains a mixture of wastes and essential nutrients, including most of the blood's water. The nephron must restore the nutrients and most of the water to the blood, while retaining wastes for elimination. This is accomplished by two processes: tubular reabsorption and tubular secretion.

Water and Nutrients Are Returned to the Blood by the Tubule

From Bowman's capsule the filtrate passes through the proximal tubule, which is surrounded by capillaries that make intimate contact with it. At these points of contact, water and nutrients in the filtrate are returned to the blood, a process called **tubular reabsorption** (Fig. 21-14,

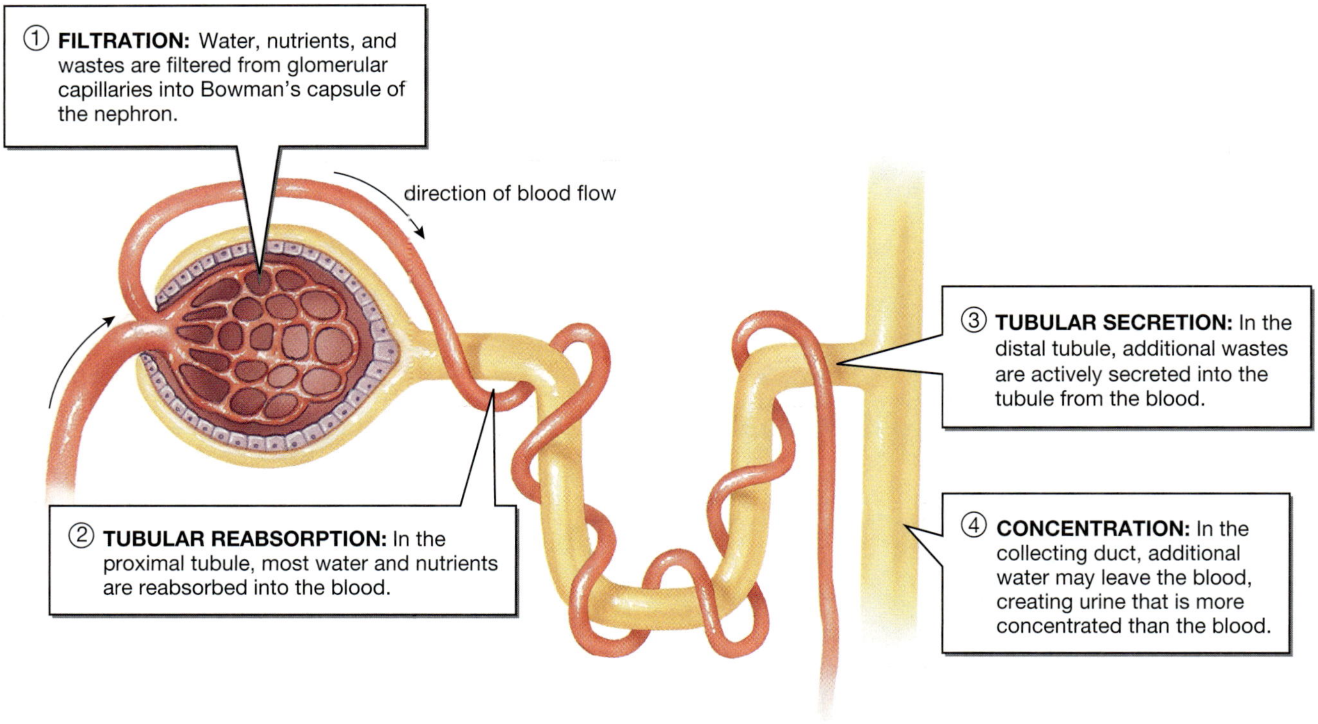

Figure 21-14 Urine formation in the nephron

Health Watch
When the Kidneys Collapse

If the kidneys fail, a person dies rapidly, usually within 2 weeks. Each year in the United States, about 90,000 people die as a result of urinary system disease. The kidneys are vulnerable to attack from several sources. Diabetes is the most common cause of kidney failure, followed by high blood pressure. Both of these disorders damage the glomerular capillaries. Infections, particularly by intestinal bacteria that reach the kidney via the urethra, can also harm the kidneys, as can overdoses of painkilling medicines. Kidney failure is typically treated by hemodialysis.

First used in 1945, hemodialysis operates on the simple principle that substances diffuse from areas of higher concentration to areas of lower concentration across a permeable membrane. During hemodialysis (Fig. E21-2), the patient's blood is diverted from the body and pumped through narrow tubing made of a cellophane membrane suspended in fluid. This membrane has pores too small to permit the passage of blood cells or large proteins but large enough to allow passage of small molecules, such as water, sugar, salts, amino acids, and urea. The composition of the surrounding fluid is adjusted to have normal blood levels of salts and nutrients, so only molecules whose concentration is higher than normal will diffuse into the dialyzing fluid. Urea, for example, is present in a relatively high concentration in the blood of a dialysis patient and is absent in the dialyzing fluid, so it diffuses out.

The patient must remain attached to the dialysis machine for 4 to 6 hours, three times a week. The average period for which people in their 40s can be kept alive on dialysis is 10 years or less, and their lives are far from normal. Their diet and fluid intake must be carefully regulated. Even so, their blood composition fluctuates and toxic substances reach higher than normal levels between sessions. Nonetheless, hemodialysis keeps alive many who would not otherwise survive and, given the continuing shortage of donor kidneys for transplant, is likely to remain a key treatment for kidney disease.

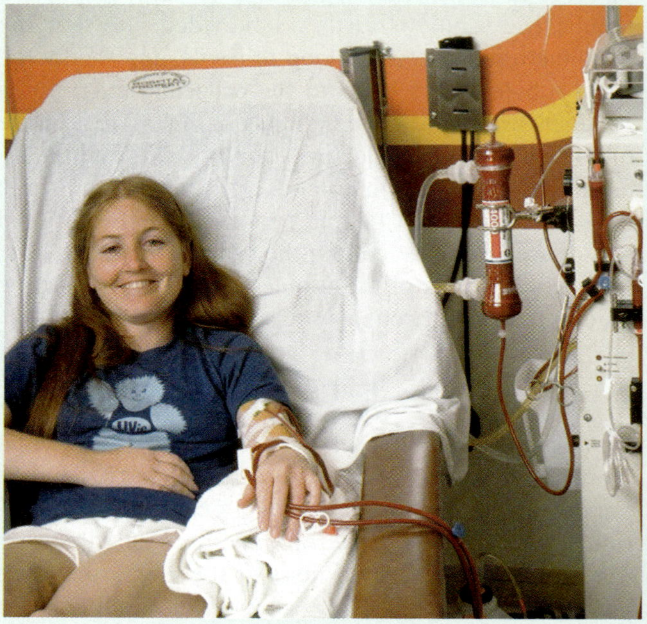

Figure E21-2 A patient undergoes hemodialysis

step ②). In tubular reabsorption, the cells of the proximal tubule transport salts and other nutrients, such as amino acids and glucose, out of the tubule by active transport (that is, the cells expend energy to transport the nutrients). The nutrient molecules then enter adjacent capillaries by diffusion. Water follows the nutrients out of the tubule by osmosis. Wastes such as urea remain in the tubule and become more concentrated as water leaves.

The Tubule Also Removes Some Substances

In **tubular secretion**, wastes and excess substances that were not filtered out into Bowman's capsule are removed from the blood (Fig. 21-14, step ③). These wastes are secreted into the distal tubule by tubule cells. Secreted substances include hydrogen and potassium ions, ammonia, and many drugs.

Urine Becomes Concentrated in the Collecting Duct

Mammals and birds can conserve water when necessary by producing urine that has a higher concentration of dissolved materials than their blood does. Urine can become concentrated because there is a concentration gradient of salts and urea in the interstitial fluid that surrounds the nephron and the collecting duct (Fig. 21-14, step ④). The concentration gradient is created and maintained by the loops of Henle, which are oriented such that the fluid becomes more concentrated with increasing distance from the outer edge of the kidney. Thus, as filtrate travels through the collecting duct to the renal pelvis, it passes through increasingly concentrated interstitial fluid, eventually reaching fluid that is much more concentrated than blood. As the difference in concentration between the filtrate and the surrounding fluid increases, water may leave the filtrate by osmosis and be carried off by the surrounding capillaries. The filtrate in the collecting duct, now called urine, can become as concentrated as the surrounding fluid. Because the rest of the excretory system does not allow water to enter or urea to escape, the urine remains concentrated.

Negative Feedback Regulates the Water Content of the Blood

Maintaining the proper volume of water in the body is a key function of the urinary system, but the proper amount of water to retain changes continually as conditions change. Our kidneys must make adjustments; they reabsorb more water when a person is perspiring heavily, and less water when the person has just consumed several beverages.

The amount of water reabsorbed into the blood is controlled by a negative feedback mechanism (see Chapter 19) based on the amount of antidiuretic hormone (ADH) circulating in the blood. This hormone increases the permeability to water of the distal tubule and the collecting duct, thereby allowing more water to be reabsorbed from the urine.

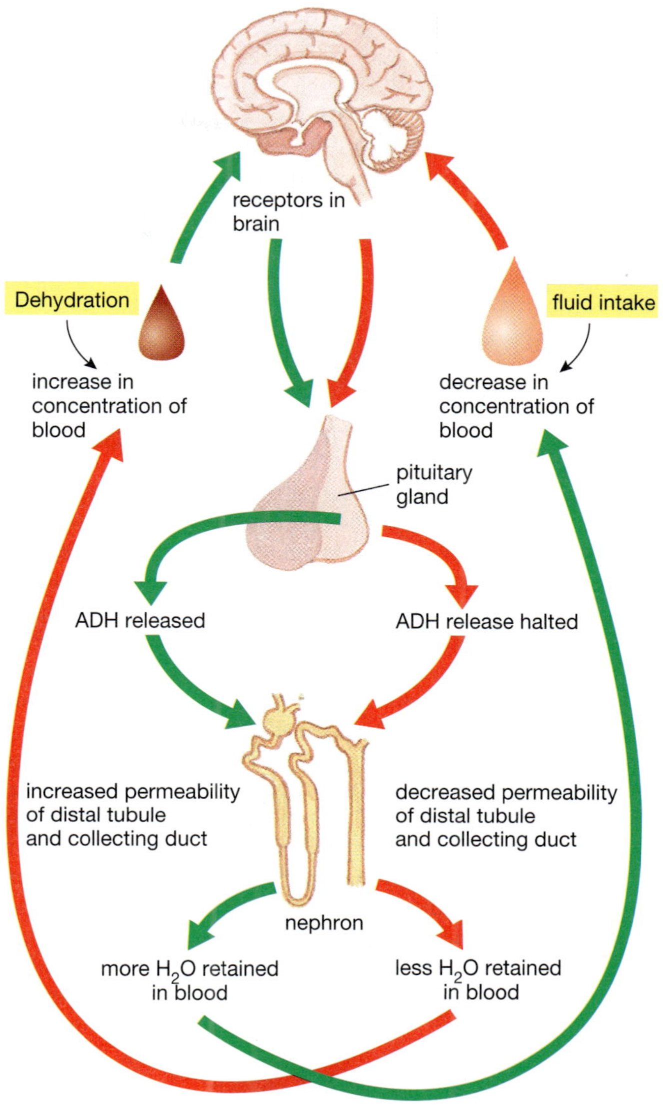

Figure 21-15 Regulation of the water content of the blood The water content of the blood is regulated in part by a negative feedback mechanism. Sensors in the brain gauge the concentration of solutes in the blood; deviations from a set point either turn on or turn off a release of the hormone ADH. Any perturbation of the system (for example, by dehydration) triggers a change that will restore blood solute concentration to the appropriate level.

The release of ADH is regulated by receptor cells in the brain that monitor the concentration of the blood and by receptors in the heart that monitor blood volume. For example, as the lost traveler staggers under the searing desert sun, the tissues of his body become dehydrated. With the loss of water, the concentration of his blood rises and his blood volume falls, triggering the release of more ADH from the pituitary gland (Fig. 21-15). The release of ADH increases water reabsorption and produces urine more concentrated than the blood.

In contrast, a partygoer overindulging in liquid refreshment experiences a decrease in blood concentration and an increase in blood volume, and her receptors cause ADH output to decrease. Reduced ADH concentration makes the distal tubule and collecting duct less permeable to water. When the ADH level is very low, little water is reabsorbed after the urine leaves the loop of Henle, and the urine produced is more dilute than the blood. In extreme cases, urine flow may exceed 1 quart per hour. As the proper water level is restored, the increased concentration of the blood and decreased blood volume stimulates increased ADH production, which in turn stimulates increased water reabsorption.

The Kidneys Are Important Organs of Homeostasis

Each drop of blood in your body passes through a kidney about 350 times a day, giving the kidney ample opportunity to fine-tune the composition of the blood and help maintain homeostasis. The importance of this task is illustrated by the fact that kidney failure results in death within a short time (see "Health Watch: When the Kidneys Collapse").

CASE STUDY REVISITED

The Magic Bullet for Obesity?

The demise of fen-phen in the late 1990s left a void in the diet pill market—one that marketers have rushed to fill. Pharmaceutical companies hope that newly approved appetite-suppressing drugs such as sibutramine (sold under the brand name Meridia) and orlistat (sold as Xenical) will prove to be free of the harmful side effects that doomed fen-phen. Amazingly, phentermine (one of compounds in fen-phen) still accounts for more than half of all diet-pill prescriptions, based on the presumption that it was the less-dangerous component of fen-phen. Many consumers have turned to over-the-counter herbal remedies for obesity. Herbal diet pills commonly include leaves of the herbaceous plants *Hypericum* (St. John's wort) and *Ephedra* (Ma Huang). Neither St. John's wort nor Ma Haung has been tested in a clinical setting for either safety or effectiveness, though the FDA did issue a caution to consumers of Ma Huang after some users of the herb died suddenly of heart malfunctions.

How do diet pills work? Fen-phen works by causing the body to increase production of the neurotransmitter serotonin. This increase mimics the body's natural response to consumption of large quantities of carbohydrates; escalation of serotonin seems to turn off the craving for more carbohydrates. Meridia takes advantage of the same principle, but instead of increasing serotonin production, it blocks a cell's ability to remove previously secreted serotonin from the bloodstream. Xenical takes a different approach, disabling the digestive enzymes that break down fat, so that fat in the diet is not absorbed by the digestive system but instead passes through undigested. The mechanisms by which St. John's wort and Ma Huang act remain unknown.

Some drug companies are hoping that the protein leptin will prove to be an effective remedy for obesity. Researchers have found that mice with a mutation that prevents the production of leptin invariably become obese, but if the obese mutants receive injections of leptin, they eat less and lose weight. This response suggests that leptin functions as a signal that tells the brain to turn off the appetite, and it raises hopes that the appetites of obese humans could be controlled by injections of leptin. Unfortunately, it turns out that obese humans have a large amount of leptin in their blood, so a shortage of leptin is apparently not the source of human obesity. Still, researchers have not given up hope. Some obese mice also have high levels of circulating leptin; the problem for these mice is not that they lack leptin, but that they are insensitive or resistant to its effects. Accordingly, researchers are now focusing on new drugs and ways of administering leptin that could suppress appetite by overcoming this resistance.

Summary of Key Concepts

1) What Nutrients Do Animals Need?

Each animal species has specific nutritional requirements. These requirements include molecules that can be broken down to liberate energy, such as lipids, carbohydrates, and protein; chemical building blocks used to construct complex molecules; and minerals and vitamins needed for the chemical reactions of metabolism.

2) How Is Digestion Accomplished?

Digestive systems must accomplish five tasks: ingestion, mechanical and chemical breakdown of food, absorption, and elimination of wastes. Digestive systems convert the complex molecules of the bodies of ingested organisms into simpler molecules that can be used. The simplest digestive system is a saclike cavity such as that of cnidarians. More complex animals have a tubular digestive system with specialized parts where food is processed in a well-defined sequence.

3) How Do Humans Digest Food?

In humans, digestion begins in the mouth, where food is mechanically broken down by chewing, and chemical digestion is initiated by saliva. Food is then conducted to the stomach by peristaltic waves of the esophagus. In the acidic environment of the stomach, food is churned into smaller particles, and protein digestion begins. Gradually, the liquefied food, now called chyme, is released to the small intestine. There it is neutralized by sodium bicarbonate from the pancreas. Secretions from the pancreas, liver, and the cells of the intestine complete the breakdown of proteins, fats, and carbohydrates. In the small intestine, the simple molecular products of digestion are absorbed into the bloodstream for distribution to the body cells. The large intestine absorbs the remaining water and converts indigestible material to feces.

Digestion is regulated by the nervous system and hormones. The smell and taste of food and the action of chewing trigger the secretion of saliva in the mouth and the production of gastrin by the stomach. Gastrin stimulates stomach acid production. As chyme enters the small intestine, additional hormones are produced by intestinal cells, including secretin, which causes sodium bicarbonate production to neutralize the acidic chyme. Other hormones stimulate bile release and cause the pancreas to secrete digestive enzymes into the small intestine.

4) What Are the Functions of Vertebrate Urinary Systems?

The urinary system plays a crucial role in homeostasis, regulating the water and ion contents of the blood and blood pH. The organs of the urinary system help retain nutrients and eliminate cellular wastes and toxic substances.

5) How Does the Human Urinary System Function?

The human urinary system consists of kidneys, ureters, bladder, and urethra. Kidneys produce urine, which is conducted by the ureters to the bladder, a storage organ. Urine passes out of the body through the urethra.

Each kidney contains more than a million individual nephrons in its outer layer. Urine formed in the nephrons enters collecting ducts that empty into the renal pelvis, from which it is funneled into the ureter.

Each nephron is served by an arteriole that branches from the renal artery. The arteriole further branches into a mass of porous-walled capillaries called the glomerulus. There, water and dissolved substances are filtered from the blood by pressure. The filtrate is collected in the cup-shaped Bowman's capsule and conducted along the tubular portion of the nephron. During tubular reabsorption, nutrients are actively pumped out of the filtrate through the walls of the tubule. Nutrients then enter capillaries that surround the tubule, and water follows by osmosis. Some wastes remain in the filtrate; others are pumped into the tubule by tubular secretion. The tubule forms the loop of Henle, which creates a salt concentration gradient surrounding it. After completing its passage through the tubule, the filtrate enters the collecting duct, which passes through the concentration gradient. Final passage of the filtrate through this gradient via the collecting duct concentrates the urine.

The kidneys are important organs of homeostasis. The water content of the blood is regulated by antidiuretic hormone (ADH), produced in the brain. Low blood volume and high osmotic concentration of the blood signal dehydration and stimulate the release of ADH into the bloodstream. The ADH increases the permeability to water of the distal tubule and the collecting duct, allowing more water to be reabsorbed into the blood.

Key Terms

absorption *p. 405*
bile *p. 404*
bile salt *p. 404*
bladder *p. 408*
Bowman's capsule *p. 409*
calorie *p. 396*
Calorie *p. 396*
chyme *p. 404*
collecting duct *p. 409*
colon *p. 405*
digestion *p. 399*
distal tubule *p. 409*
esophagus *p. 403*
essential amino acid *p. 397*
essential fatty acid *p. 396*
excretion *p. 408*
feces *p. 405*
filtrate *p. 409*
filtration *p. 409*
gallbladder *p. 404*
glomerulus *p. 409*
glycogen *p. 397*
kidney *p. 408*
large intestine *p. 405*
liver *p. 404*
loop of Henle *p. 409*
microvillus *p. 405*
mineral *p. 397*
nephron *p. 408*
nutrient *p. 396*
nutrition *p. 396*
pancreas *p. 405*
pancreatic juice *p. 405*
peristalsis *p. 404*
pharynx (in vertebrates) *p. 403*
proximal tubule *p. 409*
rectum *p. 405*
small intestine *p. 404*
stomach *p. 404*
tubular reabsorption *p. 409*
tubular secretion *p. 411*
tubule *p. 409*
urea *p. 407*
ureter *p. 408*
urethra *p. 408*
urine *p. 407*
villus *p. 405*
vitamin *p. 397*

Thinking Through the Concepts

Multiple Choice

1. *An acidic mixture of partially digested food that moves from the stomach into the small intestine is called*
   a. pancreatic juice
   b. bile
   c. lymph
   d. secretin
   e. chyme

2. *A sudden increase in the amount of secretin circulating in the blood is an indication that food has recently been introduced to the*
   a. mouth
   b. pharynx
   c. stomach
   d. small intestine
   e. large intestine

3. *Which of the following is NOT characteristic of bile?*
   a. It is produced in the gallbladder.
   b. It is a mixture of special salts, water, and cholesterol.
   c. It acts as a detergent or emulsifying agent.
   d. It helps expose a large surface area of lipid for attack by lipases.
   e. It works in the small intestine.

4. *The walls of the ___________ are made more or less permeable to water, depending on the need to conserve water.*
   a. ureter
   b. urethra
   c. proximal tubule
   d. collecting duct
   e. glomerulus

5. *The function of the glomerulus and Bowman's capsule of the nephron is to*
   a. reabsorb water into the blood
   b. eliminate ammonia from the body
   c. reabsorb salts and amino acids
   d. filter the blood and capture the filtrate
   e. concentrate the urine

6. *Which of the following processes does NOT occur in the nephron and collecting duct?*
   a. filtration
   b. elimination of urea from the body
   c. reabsorption of nutrients
   d. tubular secretion
   e. concentration of urine

? Review Questions

1. List four general types of nutrients, and describe the role of each in nutrition.
2. List and describe the function of the three principal secretions of the stomach.
3. List the substances secreted into the small intestine, and describe the origin and function of each.
4. Name four structural or functional adaptations of the human small intestine that ensure good digestion and absorption.
5. Explain the two major functions of excretory systems.
6. Trace a urea molecule from the bloodstream to the external environment.
7. Describe and compare the processes of filtration, tubular reabsorption, and tubular secretion.

Applying the Concepts

1. The food label on a soup can shows that the product contains 10 grams of protein, 4 grams of carbohydrate, and 3 grams of fat. How many Calories are in this soup?
2. Control of the human digestive tract involves several feedback loops and messages that coordinate activity in one chamber with those taking place in subsequent chambers. List the coordinating events you discovered in this chapter, in order, beginning with tasting, chewing, and swallowing a piece of meat and ending with residue that enters the large intestine. What turns on and what shuts off each process?
3. Trace a ham and cheese with lettuce sandwich through the human digestive system, discussing what happens to each part of the sandwich as it passes through each region of the digestive tract.
4. One of the common remedies for constipation (difficulty eliminating feces) is a laxative solution that contains magnesium salts. In the large intestine, magnesium salts are absorbed very slowly by the intestinal wall, remaining in the intestinal tract for long periods of time. Thus, the salts affect water movement in the large intestine. On the basis of this information, explain the laxative action of magnesium salts.
5. Discuss the differences in function of the two major capillary beds in the kidneys: the glomerular capillaries and those surrounding the tubules.
6. Some "quick weight loss" diets require the ingestion of much protein-rich food and the elimination of carbohydrates. Two side effects of such diets are increased thirst and increased urination. Explain the connections between the diets and the side effects.

Group Activity

While rummaging in the basement of a natural history museum, a curator comes across a case of preserved animals that were collected in the early 1900s by a British expedition to Brazil. None of the museum's experts can determine the species-level identity of any of the specimens; the labels have been lost, and many of the animals are now extinct or very rare. As part of the effort to study the specimens, the museum director asks you to form teams of three to examine the digestive systems of the mysterious animals. Your first group of three specimens includes:

- An invertebrate, 3 inches long, with a very flat body; no appendages (limbs); no eyes; and a long, narrow digestive tube with a single opening to the outside world.
- A legged vertebrate, 2 feet long, with 30 teeth (all of which are either dagger-shaped or long and narrow with sharp edges); a single stomach that seems large for the animal's size; a short, straight small intestine; a prominent gallbladder; and very thick, muscular walls of the digestive tube, especially the stomach.
- A legless vertebrate, 3 feet long, with sharp incisor teeth at the very front of its mouth; teeth at the back of its mouth that have flat, wide, bumpy surfaces; a stomach divided into three separate chambers; a long, highly coiled small intestine; and no gallbladder.

Formulate a specific, well-reasoned speculation about the food and feeding habits of each of these long-dead animals. Be sure to account for all of the information provided above for each specimen.

For More Information

Greenberg, A. (ed.). *Primer of Kidney Disease*. San Diego: Academic Press, 1998. An up-to-date, comprehensive review of kidney diseases, written by a variety of experts.

Martin, R. J., White, B. D., and Hulsey, M. G. "The Regulation of Body Weight." *American Scientist*, November–December 1991. Describes some of the complex mechanisms that control eating, obesity, and the body's energy balance.

Mayfield, E. "A Consumer's Guide to Fats." *FDA Consumer*, May 1994. Commonsense information about fat, cholesterol, diet, and health.

Moog, F. "The Lining of the Small Intestine." *Scientific American*, November 1981. A description of this intricate tissue, which is responsible for absorbing nutrients into the body.

O'Brien, C. "Lucky Break for Kidney Disease Gene." *Science*, June 24, 1994. Discusses research on a dominant gene that causes kidney disease.

Taylor, C. R. "The Eland and the Oryx." *Scientific American*, January 1969. An account of the physiological adaptations by which two species of African antelopes cope with the hot, dry environments in which they live.

Vogel, S. "Why We Get Fat." *Discover,* April 1999. Americans are getting fatter for many reasons—environmental, genetic, and behavioral.

Willett, W. C. "Diet and Health: What Should We Eat?" *Science,* April 22, 1994. Summarizes studies that suggest that diet can play a major role in the prevention of disease.

Answers to Multiple-Choice Questions

1. e 2. d 3. a 4. d 5. d 6. b

MediaTutor

CD Activities

Activity 21.1: The Digestive System

Estimated time: 5 minutes

The digestive system is where food molecules are broken down into their basic building blocks so that they can be absorbed. In this tutorial you will be introduced to the anatomy of the digestive system, and explore the role of the various organs in the digestion of the carbohydrates, lipids, and proteins found in our food.

Activity 21.2: Urinary System Anatomy

Estimated time: 5 minutes

The urinary system plays an important role in maintaining homeostasis. In addition to ridding the body of wastes in the form of urine, this system plays a crucial role in water balance. In this tutorial, you will explore the anatomy of the human urinary system.

Activity 21.3: Urine Formation

Estimated time: 10 minutes

The nephron is the functional unit of the kidney and is responsible for modifying the composition of blood and producing urine. The process of urine formation involves filtration, reabsorption, secretion, and concentration. In this tutorial you will explore these processes in more detail.

Start the MediaTutor Student CD-ROM and enter the activity number in the Quick Search box to be taken directly to that activity.

Web Investigation

Magic Bullet for Obesity?

Estimated time: 20 minutes

More than half of Americans are overweight or obese. Many of these people try to lose weight with drugs or fad diets, but such measures are often ineffective. At the same time, other people seem able to eat just about anything without gaining an ounce. Perhaps a person's weight is somehow related to his or her genetic makeup. This activity explores how genes and nutrition affect weight, and also explores disorders of the digestive system.

Go to http://www.prenhall.com/audesirk3, the Audesirk Companion Web site. Select Chapter 21 and Web Investigation to begin.

"If you wish to be well and keep well, take Braggs Vegetable Charcoal and Charcoal Biscuits. Absorbs all impurities in the stomach and bowels, effectively warding off cholera, smallpox, typhoid, and all malignant fevers. Eradicate worms in children. Sweeten the breath."

From an early 1900s newspaper ad

Ticks are well adapted to their parasitic mode of life, possessing an array of weapons for penetrating the defenses of their hosts. These weapons also help pave the way for transmission of tick-borne diseases.

22 Defenses Against Disease: The Immune Response

AT A GLANCE

CASE STUDY

Tick Saliva and Lyme Disease

When a tick finds a mammalian host and begins to chew through its skin to reach some nourishing blood, the host's body responds rapidly. The blood in the wound begins to coagulate to seal the rupture, blood components known as platelets (see Chapter 20) collect around the wound to plug the hole, and nearby blood vessels constrict to reduce blood loss. Special cells and chemical agents move to the wound to inflame it and destroy any invading bacteria.

The tick's saliva counters all of these defensive efforts. As the tick chews toward a blood vessel, its saliva dribbles into the wound, carrying a chemical mixture that includes painkillers to prevent the host from noticing the attack, anti-coagulants that prevent blood clots, platelet inhibitors that prevent the platelets from clumping, vasodilators that prevent blood vessels from constricting, anti-inflammatories that dampen inflammation of the wound, and even substances that interfere with chemical signaling among the cells of the host's defense system. This salivary arsenal is all part of the tick's effort to remain aboard its host and to keep the blood flowing. Unfortunately for the host, the tick's chemical weaponry also creates ideal conditions for the entry of microscopic parasites that can cause disease.

For many of these disease-causing microorganisms, the tick itself is the vehicle for transportation from victim to victim. For example, the bacterium *Borrelia burgdorferi* lives in the bodies of mice, deer, and humans. If a deer tick takes a meal from an infected mouse or deer, it may ingest some *Borrelia* bacteria. The bacteria can survive in the tick's gut and, if the tick later latches onto a human, the bacteria can move from the tick's gut into the human host. An infected human may exhibit a bull's-eye-shaped rash, fever, and flu-like symptoms. This malady, known as Lyme disease, can usually be treated effectively with antibiotics but, if undiagnosed or untreated, it can cause nerve damage and chronic arthritis.

The incidence of Lyme disease has been increasing in recent years, inspiring development of a vaccine against it. Like all vaccines, the Lyme disease vaccine takes advantage of the body's own mechanisms for combating disease. Before describing the new vaccine, let's get acquainted with the body's defenses against invasion. ■

1 How Does the Body Defend Against Invasion?

Our bodies are under constant attack from viruses, bacteria, fungi, and protists. Should these microbes have the opportunity to take up residence and multiply, they can harm or even kill us. Fortunately for us, the body has several lines of defense against microbial attack: First, external barriers to invasion keep microbes out of the body. Second, nonspecific internal defenses combat invading microbes. Third, the immune system directs its assault—an immune response—against specific microbes (Fig. 22-1).

The Skin and Mucous Membranes Are Barriers to Invasion

The best defensive strategy is to prevent microbes from entering the body in the first place. In animal bodies, this first line of defense is performed by the two surfaces that are exposed to the environment: the skin and the mucous membranes of the digestive, respiratory, and urogenital tracts.

Skin Is an Inhospitable Environment for Microbial Growth

The outer surface of the **skin** consists of dry, dead cells filled with tough proteins similar to those in hair and nails. Most microbes that land on the skin cannot obtain there the water and nutrients they need to survive. Those few that gain a foothold on skin are likely to be shed before they can do harm, because skin cells are constantly sloughed off and replaced from below. The skin is also protected by secretions from sweat glands and oil glands. These secretions contain natural antibiotics, such as lactic acid, that inhibit the growth of bacteria and fungi. These multiple defenses make the unbroken skin an extremely effective barrier against microbial invasion.

Secretions, Mucus, and Ciliary Action Defend Against Microbes

The warm, moist **mucous membranes** that line the digestive, respiratory, and urogenital tracts are much more hospitable to microbes than is the dry, oily skin, but these membranes nonetheless possess effective defense mechanisms. In particular, the membranes secrete mucus that contains antibacterial enzymes such as lysozyme, which destroys bacterial cell walls. The mucus also physically traps microbes that enter the body through the nose or mouth (Fig. 22-2). Cilia on the membranes sweep up the

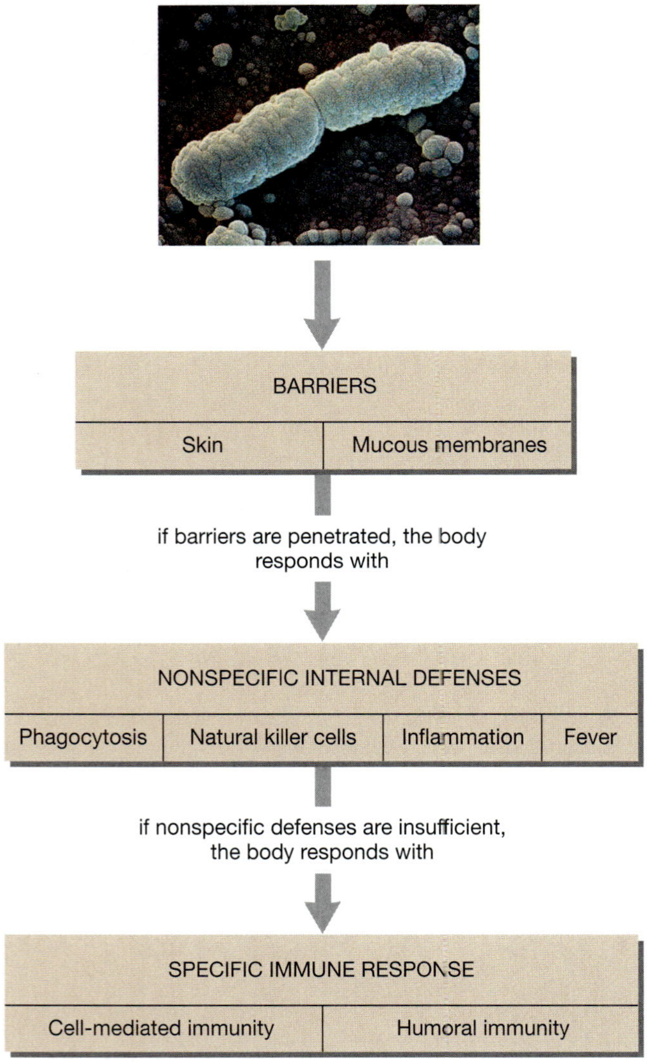

Figure 22-1 Levels of defense against infection

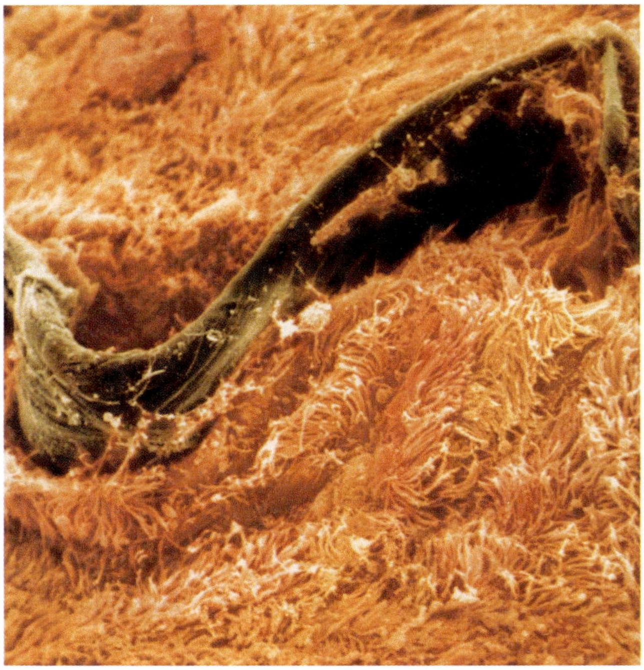

Figure 22-2 The protective function of mucus
Mucus traps microbes and debris in the respiratory tract (a brownish green strand of dirt is shown caught in mucus atop the orange cilia). The cilia lining the walls of the respiratory tract then sweep both mucus and foreign matter out of the body.

mucus, microbes and all, until it is either coughed or sneezed out of the body, or swallowed. If microbes are swallowed, they enter the stomach, where they encounter both extreme acidity (pH 1 to 3) and protein-digesting enzymes, both of which can kill microbes. Farther along in the digestive tract, the intestine contains bacteria that are harmless in their intestinal habitat but secrete substances that destroy invading bacteria or fungi. Despite these defenses, many disease organisms manage to enter the body through the mucous membranes.

Nonspecific Internal Defenses Combat Invaders

Invading microbes that penetrate the first line of defense at the skin and mucous membranes encounter an array of internal defenses. Some of these defenses are nonspecific—that is, they attack a wide variety of microbes rather than targeting specific invaders. Nonspecific defenses fall into three main categories:

- The body has a standing army of **phagocytic cells**, which directly destroy microbes, and **natural killer cells**, which destroy body cells that have been infected by viruses. The steady trickle of microbes that make it past the body's external barriers is mostly mopped up by these cells.
- An injury, with its combination of tissue damage and relatively massive invasion of microbes, provokes an **inflammatory response**. The inflammatory response simultaneously recruits new members of the army of phagocytic cells and natural killer cells and walls off the injured area, isolating the infected tissue from the rest of the body (Fig. 22-3a,b).
- If a population of microbes succeeds in establishing a major infection, the body may produce a **fever**, which both slows down microbial reproduction and enhances the body's own fighting abilities.

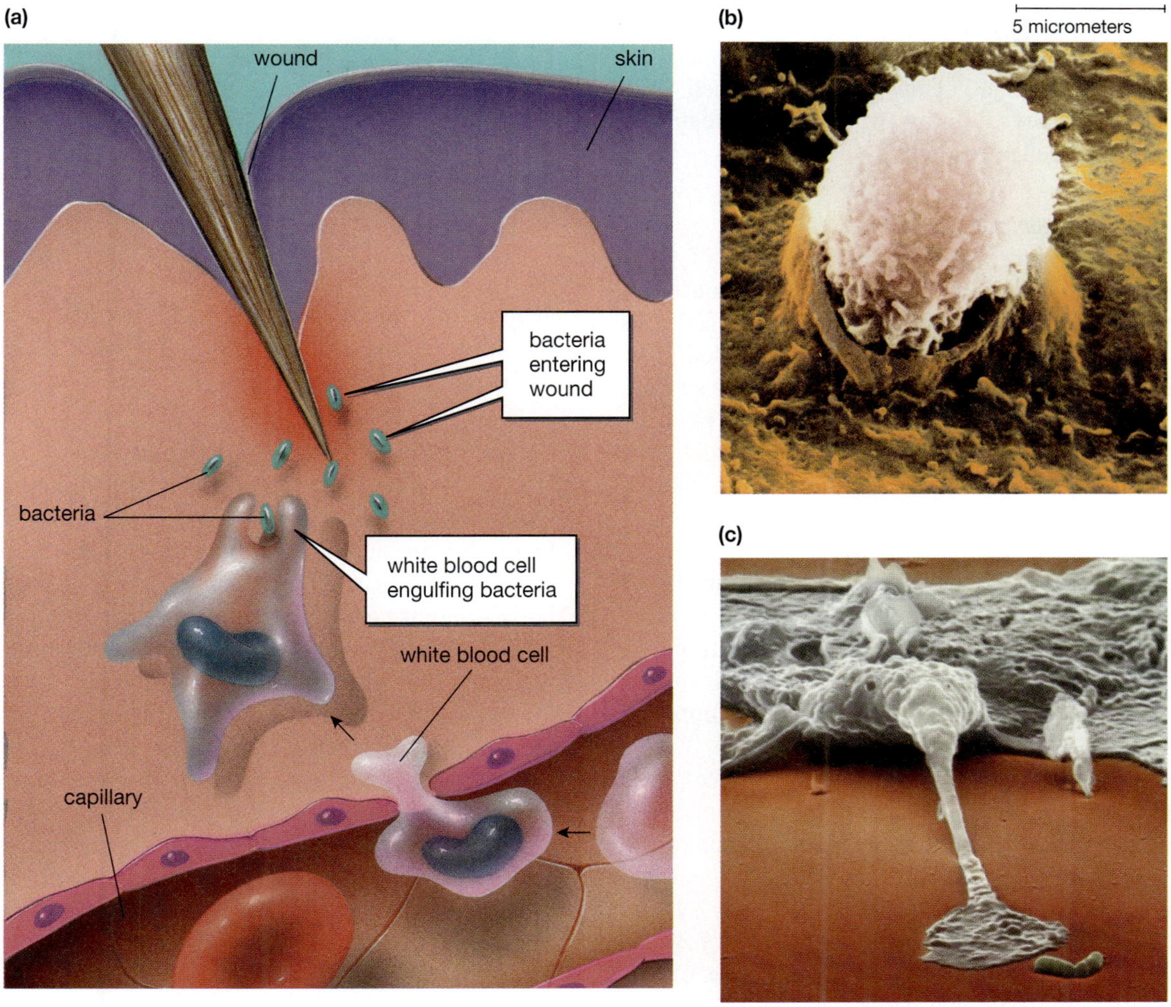

Figure 22-3 The inflammatory response
(a) During an inflammatory response triggered by a cut in the skin, white blood cells squeeze through gaps in the capillary wall and enter the wounded tissue, where they engulf microbes, dirt, and debris from damaged cells. ***(b)*** A white blood cell leaving a capillary to join the fray against foreign matter that has entered a cut. ***(c)*** A macrophage advances toward a rod-shaped bacterium; this photo was taken on an artificial surface for clarity.

MEDIATUTOR
22.1 The Immune System: Inflammation

Phagocytic Cells and Natural Killer Cells Destroy Invading Microbes

The body contains several types of phagocytic white blood cells that can engulf and digest microbes. One important type are the **macrophages**, white blood cells that crawl around in the blood and extracellular fluid (Fig. 22-3c). Macrophages ingest microbes by phagocytosis, consuming some of the bacteria and foreign substances that penetrate the mucous membranes. As we shall discuss later, macrophages also play a crucial role in the immune response.

Natural killer cells are another class of white blood cells. In general, natural killer cells strike at the body's own cells that have been invaded by viruses. Cells that have been infected by a virus have some viral proteins on their surfaces. Natural killer cells seek out and recognize these proteins and kill the infected cell. Natural killer cells also recognize and kill cancerous cells. Rather than eating their victims, natural killer cells secrete proteins into the plasma membrane of infected or cancerous cells. These proteins form a ring (much like making a barrel out of individual staves of wood), thereby opening up a large pore in the membrane of the target cell (Fig. 22-4). Killer cells also secrete enzymes that break up some of the molecules of the target cell. Shot full of holes and chewed on by enzymes, the infected or cancerous target cell soon dies.

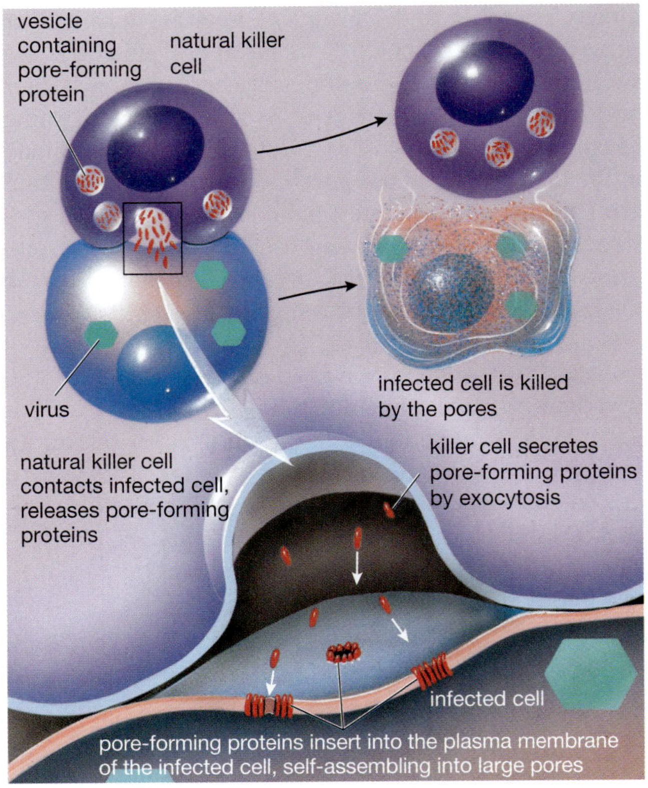

Figure 22-4 How natural killer cells kill their targets
A natural killer cell releases pore-forming proteins, which assemble to create large holes in the target cell's plasma membrane. The target cell's contents leak out, and it dies.

The Inflammatory Response Defends Against Localized Infection

An injury to the skin or mucous membranes causes an inflammatory response that begins when damaged cells release the chemical histamine into the wounded area. Histamine increases blood flow by relaxing the smooth muscle that surrounds arterioles, and also makes capillary walls leaky. With extra blood flowing through leaky capillaries, fluid seeps from the capillaries into the tissues around the wound. The wound becomes red, swollen, and warm.

Meanwhile, other chemicals released by injured cells initiate blood clotting (see Chapter 20). The clotting both blocks damaged blood vessels, preventing microbes from entering the bloodstream, and seals off the wound from the outside world, limiting the entry of more microbes.

Still other chemicals, some released by wounded cells and others produced by the microbes themselves, attract macrophages and other phagocytic cells to the wound. Some of the phagocytic cells emerge from local tissues; others arrive via the circulatory system and squeeze out through the leaky capillary walls. After entering the wound, the phagocytic cells engulf microbes, dirt, and damaged cells (see Fig. 22-3). Unfortunately, each phagocyte can eat just so many microbes, and then it dies. If tissue damage is too severe or the wound is too dirty, the phagocytes may be unable to complete the cleanup job. In that case, the fluid around the wound turns into pus, a thick mixture of microbes, tissue debris, and living and dead white blood cells.

Fever Combats Large-Scale Infections

If invaders breach other defenses and mount a large-scale infection, they may trigger a fever. Although high fevers can be dangerous, and even moderate ones can be very unpleasant, fever is an effective part of the body's defense against infection.

The onset of fever is controlled by the hypothalamus, the part of the brain that contains the temperature-sensing nerve cells that are the body's thermostat. Normally, the thermostat is set at about 98.6 °F (37 °C). Certain macrophages, as they respond to the infection, release hormones called endogenous pyrogens. Pyrogens travel in the bloodstream to the hypothalamus and raise the thermostat's set point. The body responds with shivering, increased fat metabolism, constriction of surface blood vessels, and behaviors such as burrowing under blankets—activities that increase body temperature.

Fever both enhances the body's normal defenses and harms invading microbes. The higher temperature increases the activity of the phagocytic white blood cells, including their secretion of pyrogens, which reduce the concentration of iron in the blood. Many bacteria need more iron at higher temperatures, and a deficiency of it curtails their growth. Fever also helps fight viral infections by increasing infected cells' production of the protein interferon, which travels to uninfected cells and increases their resistance to viral attack.

The Big Picture

To survive, animal bodies must acquire gases, nutrients, and water from the environment. The pathways provided for these substances, however, also provide entryways for hostile invaders. The body erects barriers against entry, but invaders frequently overcome them and enter the body.

2 What Are the "Soldiers" of the Immune Response?

Phagocytic cells, natural killer cells, the inflammatory response, and fever are all *nonspecific* defenses; their role is to prevent or overcome *any* microbial invasion of the body. Unfortunately, however, these nonspecific defenses are not impregnable. When they fail to do the job, the body's **immune system** mounts a response directed specifically against the particular organism that has successfully invaded the body. The immune system is an unusual system. Unlike the digestive system, for example, it is not composed of physically attached structures. Instead, as befits its mission of patrolling the entire body for microbial invaders, the immune system consists of an army of separate cells. This army is highly coordinated, and coordinating it requires complex communications involving hormones, receptors, cells, antigens, and antibodies.

The essential features of the immune response to infection were recognized more than 2000 years ago by Greek historian Thucydides. He observed that occasionally someone would contract a disease, recover, and never catch that particular disease again—the person had become immune. With rare exceptions, however, immunity to one disease confers no protection against other diseases. Thus, the immune system attacks one type of microbe, overcomes it, and provides future protection against that microbe but no others. This is why we refer to the immune response as a defense against the invasion of *specific* microbes or toxins.

White Blood Cells Are Key Components of the Immune System

Your immune system includes about 2 trillion white blood cells called lymphocytes. Many lymphocytes are clustered in the thymus, lymph nodes, and spleen; others are distributed throughout the body in the blood and lymph. The **immune response** arises from interactions among the various types of lymphocytes and the molecules that they produce, with the result that foreign molecules or microbes are eliminated from the body. The theater of the immune response has a large cast of characters and is difficult to follow without a program. Table 22-1 provides a brief overview of the major cellular actors and their roles.

Table 22-1 The Cells of the Immune Response

Macrophages	White blood cells that both engulf invading microbes and help alert other immune cells to the invasion
Natural killer cells	White blood cells that destroy infected or cancerous cells
B cells	Lymphocytes that produce antibodies
Plasma cells	Offspring of B cells that secrete antibodies into the bloodstream
Memory B cells	Offspring of B cells that provide future immunity against invasion by the same antigen
T cells	Lymphocytes that regulate the immune response or kill certain types of cells
Cytotoxic T cells	Offspring of T cells that destroy specific targeted cells, normally either foreign eukaryotic cells, infected body cells, or cancerous body cells
Helper T cells	Offspring of T cells that stimulate immune responses by both B cells and cytotoxic T cells
Suppressor T cells	Offspring of T cells that inhibit immune responses by other lymphocytes after an infection is conquered
Memory T cells	Offspring of T cells that provide future immunity against invasion by the same antigen

Immune Cells Originate in the Bone Marrow

The main actors in the immune response are two types of lymphocytes, called **B cells** and **T cells**. Like all white blood cells, B cells and T cells arise from lymphocyte precursor cells in the bone marrow (marrow is found in the interior of certain bones). Some of these lymphocyte precursors are released into the bloodstream and travel to the thymus, where they complete their differentiation into T (for thymus) cells. In contrast, B cells differentiate in the bone marrow. The two cell types play quite different roles in the immune response, but immune responses produced by both B cells and T cells consist of the same three fundamental steps. The immune system *first* recognizes the invader; *second*, launches an attack; and *third*, retains a memory of the invader that can be used to ward off future infections.

3 How Does the Immune System Recognize Invaders?

To understand how the immune system recognizes invading microbes and initiates a response, we must answer three related questions: How do immune cells recognize foreign molecules? How can immune cells produce specific responses to so many different types of molecules? How do immune cells avoid mistaking the body's own cells and molecules for invaders?

Invaders' Cell Surface Molecules Are Recognized

Immune cells can identify microbes or foreign molecules because these invaders either consist of, or bear on their surfaces, molecules that act as **antigens.** In general, antigens are large, complex molecules such as proteins, polysaccharides, and proteins. Antigens can be located on the surfaces of cells (for example, on invading microbes) or dissolved in blood or extracellular fluid (snake venom and bacterial toxins are dissolved antigens).

The immune system produces two kinds of large proteins that interact with antigens: **antibodies**, produced by B cells, and **T-cell receptors**, produced by T cells. Antibodies either remain attached to the surfaces of the B cells that produced them or are released by the B cells and become dissolved in the blood plasma. T-cell receptors always remain attached to the surfaces of the T cells that produced them; they are never secreted into the plasma.

Antibodies Recognize and Bind to Antigens

Antibodies are Y-shaped molecules composed of two pairs of peptide chains: one pair of identical large (heavy) chains and one pair of identical small (light) chains (Fig. 22-5). Both heavy and light chains consist of a **constant region**, which is similar in all antibodies of the same type, and a **variable region**, which differs among individual antibodies. The combination of light and heavy chains makes an antibody with two functional parts: the "arms" and the "stem" of the Y. The variable regions at the tips of an antibody's arms form binding sites for specific antigens. These binding sites are a lot like the active sites of enzymes (Chapter 5): Each binding site has a particular shape and electrical charge, so only certain molecules can fit in and bind. The binding sites are so specific that each antibody can bind, at most, a few types of antigen molecules—perhaps only one.

In an antibody whose stem remains attached to the B cell that produced it, the two arms of the antibody protrude outward from the B cell, sampling the blood and lymph for antigen molecules (Fig. 22-6a). When the arms of the antibody encounter an antigen with a compatible chemical structure, they bind to it. This binding triggers a response in the B cell.

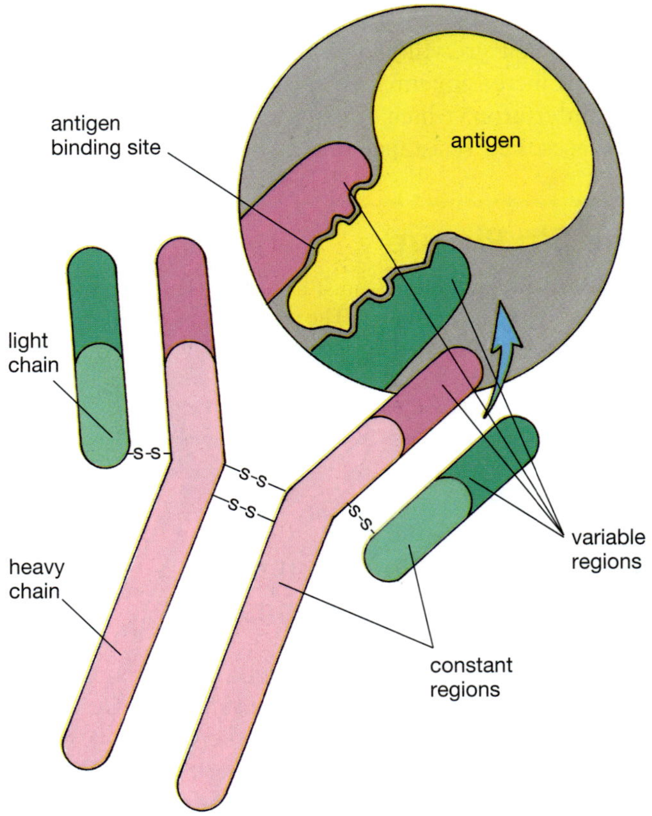

Figure 22-5 Antibody structure
Antibodies are proteins composed of two pairs of peptide chains (light chains and heavy chains) arranged like the letter Y. Different antibodies have different variable regions that form unique binding sites.

Other antibodies circulate in the bloodstream, where they neutralize poisonous antigens, or destroy microbes that bear antigens, or attract macrophages that eat the antigen-bearing microbes (Fig. 22-6b). These functions of antibodies are described in more detail in the following sections.

T-Cell Receptors Also Recognize and Bind Antigens

T-cell receptors, found only on the surfaces of T cells, have both similarities to and differences from antibodies. Like antibodies, they consist of two peptide chains and the ends of the two chains protrude out from the T cell, forming highly specific binding sites for an antigen. Unlike antibodies, T-cell receptors are not released into the bloodstream. A T-cell receptor triggers a response in the T cell only when it encounters an antigen on the surface of a cancerous or infected cell, or on the surface of a macrophage that has ingested an invading microbe. Also unlike antibodies, the T-cell receptors do not directly contribute to the destruction of invading microbes or toxic molecules.

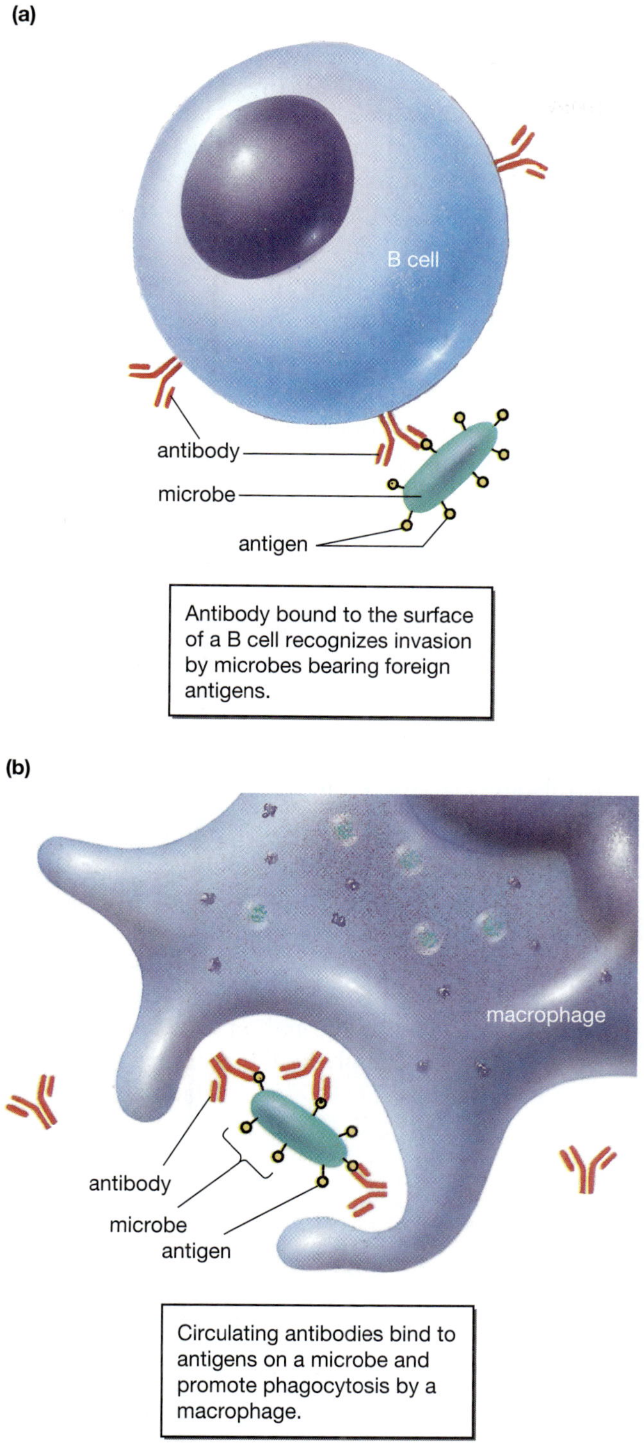

Figure 22-6 Antibodies at work

Your Immune System Can Recognize Millions of Different Molecules

During your lifetime, your body will be challenged by a multitude of different invaders. Your classmates will sneeze cold and flu viruses into the air you breathe. Roadside weeds and trees will release pollen that will find its way to your lungs. Your food may play host to a bacterial population or to toxins released by bacteria. You may drink water that contains the *Giardia* parasite or the microbes that cause dysentery. As you relax outdoors, a tick that carries the bacteria responsible for Lyme disease may bite you. There is no escape from the assaults of the invaders. Fortunately, your immune system recognizes and responds to virtually all of the millions of antigens that you might encounter, because your immune cells produce millions of different antibodies and T-cell receptors, each capable of binding a different antigen. These amazing recognition abilities are one of the main reasons you don't get sick more often.

The mechanisms by which the immune system recognizes so many antigens are complex. Antibodies and T-cell receptors are, after all, proteins, and proteins are encoded by genes. It would seem that, to recognize millions of different antigens, each person would have to possess millions of different genes for antibodies and T-cell receptors. But there are probably fewer than 75,000 genes in the entire human genome. How, then, can a subset of these genes code for millions of antibodies and T-cell receptors? The answer lies in two distinct but complementary mechanisms that join forces to produce an enormous diversity of antibodies and T-cell receptors from a comparatively small number of genes. Our description explains how antibody diversity develops in B cells; similar phenomena produce a large diversity of T-cell receptors.

An Antibody Is Pieced Together from Fragments

There are no genes for entire antibody molecules. Instead, the genome includes genes that encode many different antibody fragments; these fragments can be pieced together in many millions of combinations. As each B cell develops, it retains only a few randomly selected fragment genes (including genes for both light and heavy chains and both constant and variable regions); the rest of the fragment genes are cut out from the DNA as the cell matures. Therefore, each B cell produces a single type of antibody, determined by random combination of antibody-fragment genes. The random selection of antibody-fragment genes from a large pool of choices yields a vast number of possible unique combinations, so each B cell produces an antibody that is different from the one produced by any other B cell (except its own daughter cells). It may be helpful to think of antibody genes in terms of card playing. Each B cell is dealt a "hand" of two variable-region genes: one for the light chain and one for the heavy chain, randomly chosen from two large "decks" of genes. With each deck containing hundreds of "cards" (genes), virtually every cell will synthesize its own unique antibody.

Antibodies Are Not Tailor-made for Antigens

The end result of mutation and gene recombination in B cells is that, at any given time, a human body contains an army of perhaps 100 million different antibodies. With that much diversity, an antigen will almost always be confronted by an antibody that can bind it. But it is important to recognize that the immune system does not design antibodies (or T-cell receptors) to fit invading antigens, as a tailor might design custom clothes for a customer. Instead, the immune system randomly synthesizes millions of different antibodies. Like clothes in a department store, the array of antibodies is simply there, waiting. The fact that virtually every antigen can be bound by at least a few antibodies is due simply to the immense numbers of different antibodies present in the body. Given enough racks of clothing from which to choose, each of us will find something suitable that fits.

The Immune System Distinguishes "Self" from "Non-Self"

The surfaces of the body's own cells bear large proteins and polysaccharides, just as those of microbes do. So, why doesn't your immune system destroy your own cells? The key seems to be the continuous presence of the body's own antigens during the time when the immune cells mature. As an embryo develops, some differentiating immune cells do indeed produce antibodies or T-cell receptors that can bind the body's own proteins and polysaccharides, treating them as antigens. However, if these *immature* immune cells contact molecules that bind to their antibodies or T-cell receptors, the immune cells are destroyed. In this way, potentially dangerous immune cells, which would attack the body's own cells, are eliminated during immune system development. Thus, the immune system distinguishes "self" from "non-self" by retaining only those immune cells that do not take the body's own molecules to be invaders.

Some cell-surface proteins, collectively called the major histocompatibility complex (MHC), are unique to each person (except identical twins, who have the same genes and hence the same MHC proteins). Because they are different from those of everyone else, your MHC proteins act as antigens in other people's bodies. That is why transplants are rejected; the recipient's immune system recognizes MHC proteins on the donor's cells as foreign and destroys the transplanted tissue.

4 How Does the Immune System Launch an Attack?

If the body is invaded, the immune system mounts two types of attack: **Humoral immunity** is provided by B cells and the circulating antibodies they secrete into the bloodstream; invaders are attacked before they can enter body cells. **Cell-mediated immunity** is produced by T cells, which attack invaders that have made their way into body cells. Although humoral and cell-mediated immunity are not completely independent, we consider them separately for ease of understanding.

Humoral Immunity Is Produced by Antibodies Dissolved in Blood

Each B cell bears a specific type of antibody on its surface. During an infection, the antibodies borne by a few B cells are able to bind to antigens on the invader. This antigen–antibody binding causes these B cells to divide rapidly. This process is called **clonal selection** because the resulting population of cells is composed of "clones" (genetically identical to the parent B cells) that have been "selected" to multiply by the presence of particular invading antigens (Fig. 22-7). The daughter cells differentiate into two cell types: memory B cells and

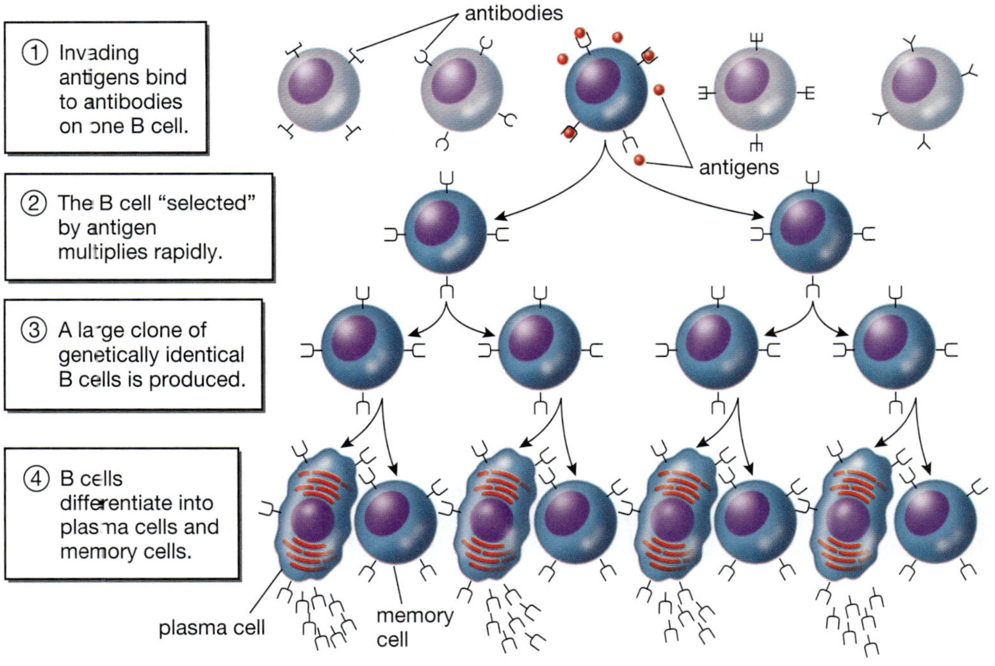

Figure 22-7 Clonal selection among B cells by invading antigens

MEDIATUTOR 22.2 The Immune System: Clonal Selection

plasma cells. **Memory B cells** do not release antibodies but do play an important role in future immunity to the particular invader that stimulated their production (as we shall soon see). **Plasma cells** become enlarged and packed with endoplasmic reticulum (Fig. 22-8); these cells churn out huge quantities of their own specific antibodies. These antibodies are released into the bloodstream (hence the name "humoral" immunity; to the ancient Greeks, blood was one of the four "humors," or body fluids).

Humoral Antibodies Have Multiple Modes of Action

Humoral immunity produced by antibodies in the blood may destroy foreign chemicals or microbes in several ways. First, the circulating antibodies can bind to a foreign molecule and render it inactive. Alternately, the antibodies may coat the surface of an invading cell, or each antibody may bind two microbes and cause them to clump together. Both of these tactics encourage macrophages to engulf the microbes. Finally, an antibody–antigen complex on the surface of an invading cell may trigger a series of reactions, the end result of which is the attraction of phagocytic white blood cells to the site.

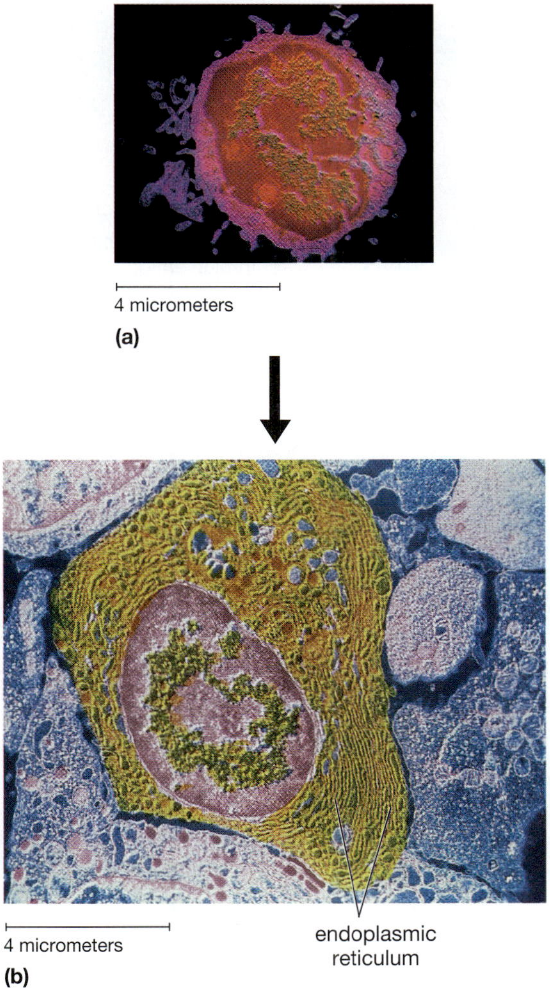

Figure 22-8 A B cell becomes a plasma cell
Colorized micrographs of B cells *(a)* before and *(b)* after conversion to plasma cells. The plasma cell is much larger than the B cell and has a great amount of rough endoplasmic reticulum that synthesizes antibodies.

Humoral Immunity Fights Invaders That Are Outside Cells

Because B-cell antibodies circulate in the bloodstream, humoral immunity can only defend against invaders that are in the blood or extracellular fluid. Bacteria (most of which never enter the body's cells), bacterial toxins, and some fungi and protists are therefore vulnerable to the humoral immune response. Invaders that penetrate into the body's cells, as viruses do, are safe from antibody attack as long as they are within the cytoplasm of a body cell. Fighting viral infections requires the help of cell-mediated immune responses, described in the following section.

Cell-Mediated Immunity Is Produced by T Cells

Cell-mediated immunity is the body's primary defense against cells that are cancerous or have been infected by viruses. Four types of T cells contribute to cell-mediated immunity: cytotoxic T cells, helper T cells, suppressor T cells, and memory T cells.

Cytotoxic T cells release proteins that disrupt the infected cell's plasma membrane (Fig. 22-9). This attack is activated when T-cell receptors on the cytotoxic T cell's plasma membrane bind to antigens on the surface of an infected cell. In some instances, the proteins released by cytotoxic T cells are similar or identical to those that natural killer cells use to create giant holes in the target cell's membrane. The receptors on **helper T cells** bind to microbial antigens presented on the plasma membranes of macrophages that have engulfed the invader or on

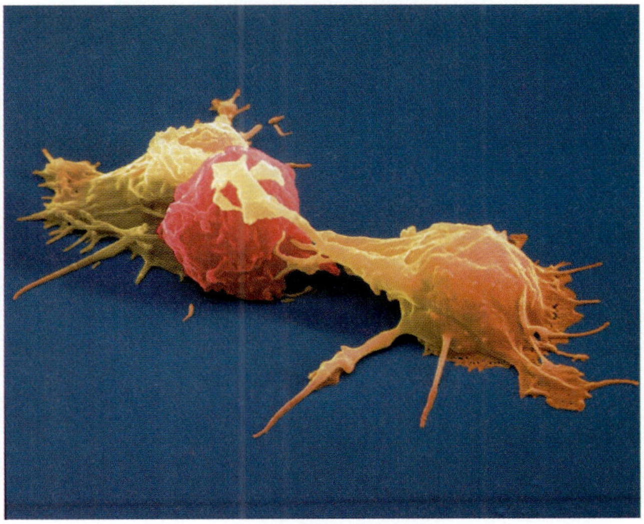

Figure 22-9 Cell-mediated immunity defends against cancer
Cytotoxic T cells (yellow) attack a cancer cell (red) in this colorized scanning electron micrograph.

the surfaces of infected cells. Upon binding an antigen, the helper T cells release chemicals that stimulate cell division and differentiation in both B cells and cytotoxic T cells that respond to the same microbial invasion. In fact, very little immune response, either cell-mediated or humoral, can occur without the chemical boost provided by helper T cells. That is why AIDS, which destroys helper T cells, is such a deadly disease. **Suppressor T cells** act after an infection has been conquered. They help shut off the immune response in both B and cytotoxic T cells. After an infection is over, some suppressor T cells and helper T cells persist and function as **memory T cells.** Like memory B cells, these help protect the body against future infection. Figure 22-10 compares the humoral immune response with the cell-mediated immune response.

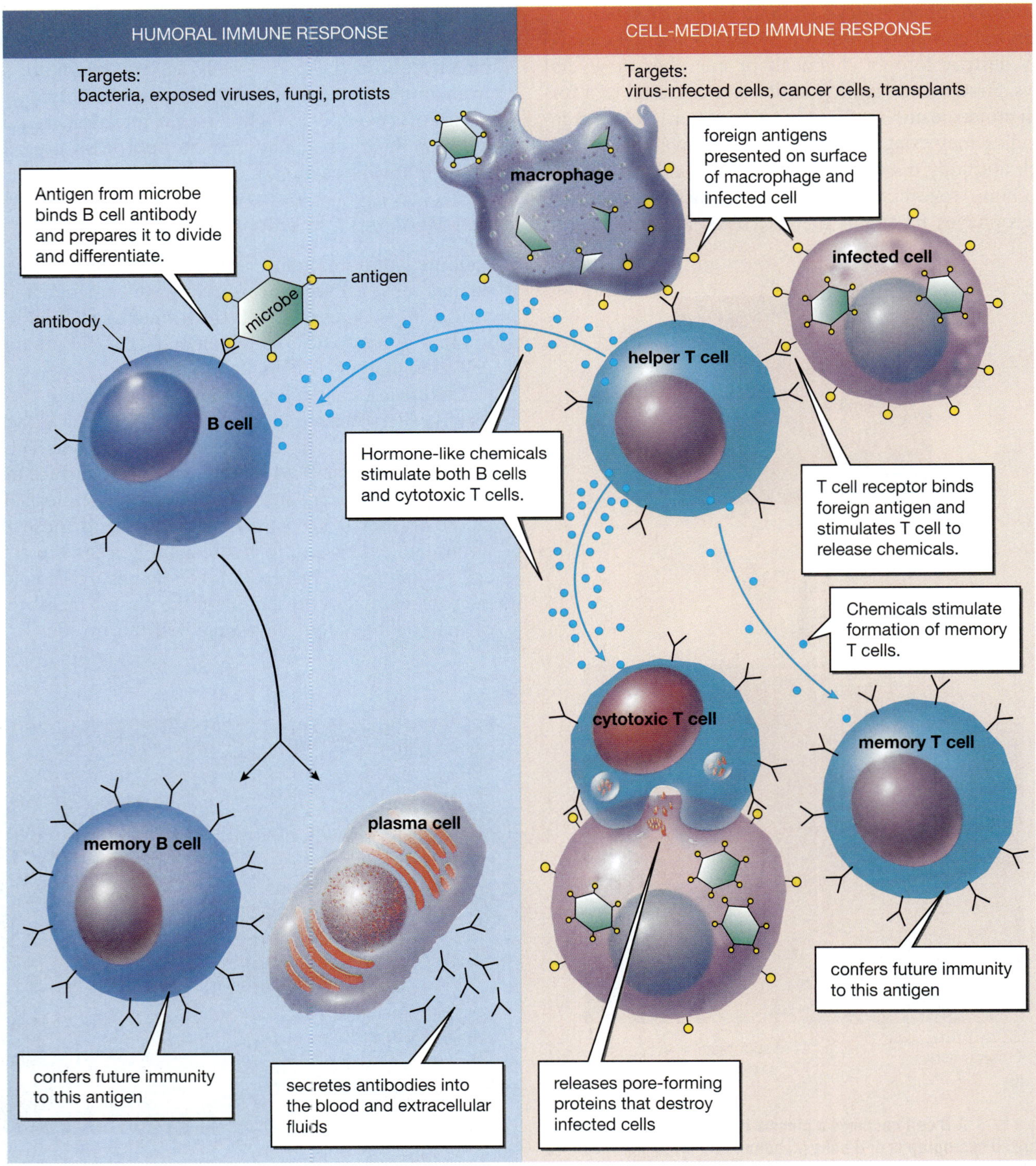

Figure 22-10 A summary of humoral and cell-mediated immune responses

The Big Picture

The immune system is complex, multi-layered, and intricate. Its complexity stems from a long evolutionary battle between the animal body and its invaders. As a huge number of different attackers have evolved ever more diverse and sophisticated means of overcoming the body's defenses, those defenses have in turn evolved complex responses to counter the attacks.

5 How Does the Immune System Remember Its Past Victories?

After you have recovered from a disease, you will remain immune to the particular strain of microbe that made you ill for many years, perhaps a lifetime. Retaining immunity is the function of memory cells. Plasma cells and cytotoxic T cells directly fight disease organisms, but they normally live only a few days. Each of these, however, leaves behind hundreds or thousands of memory cells that may survive for many years. If the body is re-invaded by microbes to which the immune system has already mounted a response, the appropriate memory cells will recognize the invaders. These memory cells will then multiply rapidly and produce a second immune response by generating huge populations of plasma cells and cytotoxic T cells. Because there are so many memory cells primed to respond, the body is able to fend off a second attack before it gains a foothold.

Despite the presence of memory cells, some diseases, such as colds and flu, do recur in the same person. Colds and flu are able to overcome the immune system's memory because this year's colds and flu aren't caused by the same viruses that you fought off last year (see "Health Watch: Can We Beat the Flu Bug?").

6 How Does Medical Care Augment the Immune Response?

We may have made the immune system sound invincible, but as you know, this is not the case. Unfortunately, the body provides ideal conditions for the growth and reproduction of disease microbes. An infection thus becomes a race between the invading microbes and the immune response. If the initial infection is massive or if the microbes produce particularly toxic products, full activation of the immune response may come too late. Further, many microbes have evolved defenses that help them evade the immune system. In the end, many infections kill their victims.

For most of human evolutionary history, the battle against disease was fought by the immune system alone. Now, however, the immune system has a powerful assistant: medical treatment. Scientific and medical advances have greatly reduced the toll taken by infectious diseases. We describe here two of the most important medical tools, antibiotics and vaccinations.

Antibiotics Slow Down Microbial Reproduction

Antibiotics are drugs that help combat infection by slowing down the growth and multiplication of many invaders, including bacteria, fungi, and protists (but not viruses). Although antibiotics usually don't destroy every single disease-causing microbe in a body, they can kill a sufficiently large number to give the immune system time to finish the job. One problem with antibiotics, however, is that they are potent agents of natural selection. The occasional mutant microbe that is resistant to an antibiotic will pass on the genes for resistance to its offspring. The result: Resistant mutants proliferate while susceptible microbes die off. Eventually, many antibiotics become ineffective in treating diseases.

Until very recently, little could be done about viral infections except to treat the symptoms and hope that the immune system would be successful. Now, a new class of drugs called neuraminidase inhibitors has entered the physician's arsenal. These anti-viral drugs work by blocking the ability of viruses to enter cells, which is necessary for the viruses to replicate. In coming years we may see the immune system receiving powerful help from these drugs.

Vaccinations Stimulate the Development of Memory Cells

As early as A.D. 1000, people in India, China, and Africa deliberately exposed themselves to mild cases of smallpox to acquire immunity to the disease. Today, immunization is achieved by injecting weakened or dead microbes into healthy people, a process called **vaccination.** The weakened microbes do not cause disease (or at least not a severe case). However, they bear antigens that elicit a vigorous immune response, including production of an army of memory cells that confer immunity against subsequent exposure to the living, dangerous microbes.

Today, many diseases, including polio, diphtheria, typhoid fever, and measles, are controlled through vaccination. Smallpox, one of the deadliest diseases of all, has been completely eradicated since 1980 as a result of a massive vaccination program sponsored by the World Health Organization. Despite eradication of the disease, however, the smallpox virus is not extinct. Two facilities, one in Siberia and one in Atlanta, legally store what health officials hope are the last smallpox viruses on Earth. One compelling reason to maintain these stocks is as a source of genetic material that will allow scientists to make additional smallpox vaccines. A generation of people has never been vaccinated against smallpox, and some authorities believe that potential

Health Watch
Can We Beat the Flu Bug?

Every winter, a wave of influenza, or flu, sweeps across the world. Thousands of the elderly, the newborn, and those already suffering from illness perish, while hundreds of millions more suffer the respiratory distress, fever, and muscle aches of milder cases. Occasionally, devastating flu varieties appear. In the great flu pandemic (worldwide epidemic) of 1918, the worldwide toll was 20 million dead in one winter. In 1968 the Hong Kong flu infected 50 million people in the United States, causing 70,000 deaths in 6 weeks.

Flu Viruses Flu is caused by viruses that invade the cells of the respiratory tract, turning each cell into a factory for manufacturing new viruses. The outer surface of a flu virus is studded with proteins, some of which are recognized by the immune system as antigens. Most people survive the flu because their immune systems eventually inactivate the viruses or kill off virus-infected body cells. Why then, don't our memory T and B cells make us immune to future outbreaks of flu?

The answer lies in a flu virus's amazing ability to change. Flu virus genes are made up of RNA and lack the proofreading mechanisms that reduce mutations in genes made of DNA. Therefore, flu virus genes mutate rapidly. Four or five mutations in a type of flu virus may change its surface antigens enough that the immune system doesn't fully recognize the virus as the same old flu that was beaten off last year. So your memory cells provide only partial protection, and you get the flu again this year.

Deadly New Strains On rare occasions, dramatically new and deadly flu viruses appear, as in the pandemic of 1918, the Asian flu of 1957, the Hong Kong flu of 1968, and the Russian flu of 1977. In these viruses, entirely new antigens appear suddenly with distinctive structures that the human immune system has never before encountered. In response, the immune system must "start from scratch," selecting out entirely new lines of B cells and T cells to attack the intruder. But in the meantime, the virus multiplies so rapidly that many individuals die or become so weakened that they contract some other fatal disease. Where do the genes that encode these new flu antigens come from? Believe it or not, they come from viruses that infect birds and pigs. The intestinal tracts of birds, especially migratory waterfowl such as ducks, may carry viruses strikingly similar to human flu viruses, although the infected waterfowl show no symptoms. While human flu viruses aren't known to infect birds, *both* human and bird viruses can infect pigs, so occasionally, both human and bird viruses will meet within the same pig cell. Once in a great while, new viruses that spring from a doubly infected pig cell end up with a mixture of genes from human and bird viruses (Fig. E22-1). Some of these hybrid viruses combine the worst genes of each (at least from our perspective): From the human virus, the deadly new viruses pick up the genes needed to subvert human cellular metabolism to produce new viruses; from the bird virus, they pick up genes for new surface antigens. The hybrid viruses can move easily from pigs to humans, because pigs live near humans and, like people, pigs cough when they have the flu. Most flu pandemics are thought to originate in China, where farming practices tend to bring pigs, ducks, and humans in close proximity.

Swatting the Flu Bug Flu viruses mutate so rapidly that they may be impossible to eradicate. Fortunately, we now have two lines of defense. One is vaccinations. Each year, World Health Organization officials collect new samples of mutated flu virus from more than 100 sites throughout the world, and identify three strains they believe are most likely to spread widely. These viruses are allowed to multiply in fertilized chicken eggs, and the viral proteins are then isolated and injected into people to stimulate an immune response that protects against subsequent infections by these strains. Flu shots are often quite effective, especially if the selected strains do become widespread. For people who contract flu anyway, cleverly designed new drugs, first marketed in the United States in 2000, promise to disarm the disease. These new drugs (neuraminidase inhibitors) attack certain regions of the spikelike proteins that stud the flu virus. These crucial regions bind to our cell membranes and allow the virus to invade our cells; mutations that change these regions would

"bioterrorists" may possess their own stocks of smallpox virus.

Through genetic engineering (see Chapter 12), we now enjoy the prospect of manufacturing tailor-made vaccines. One method is to synthesize the antigenic proteins of disease-causing microbes. These antigens can then be used as vaccines without having to raise, isolate, and weaken the disease microbes themselves, or even to inject people with microbes. A vaccine against anthrax, an often fatal disease of livestock and humans that is also a potential biological weapon, has been manufactured by this procedure. A second technique is to insert genes that encode the antigens of, for example, herpes into the genome of harmless microbes, such as the cowpox virus. These "designer" microbes produce herpes antigens without being able to cause the disease and have been shown to be effective in experiments with animals.

7 What Happens When the Immune System Malfunctions?

The immune system is always on guard, protecting the body against threats to health. In some cases, however, the immune system attacks inappropriately, undermining health instead of promoting it. In addition, the

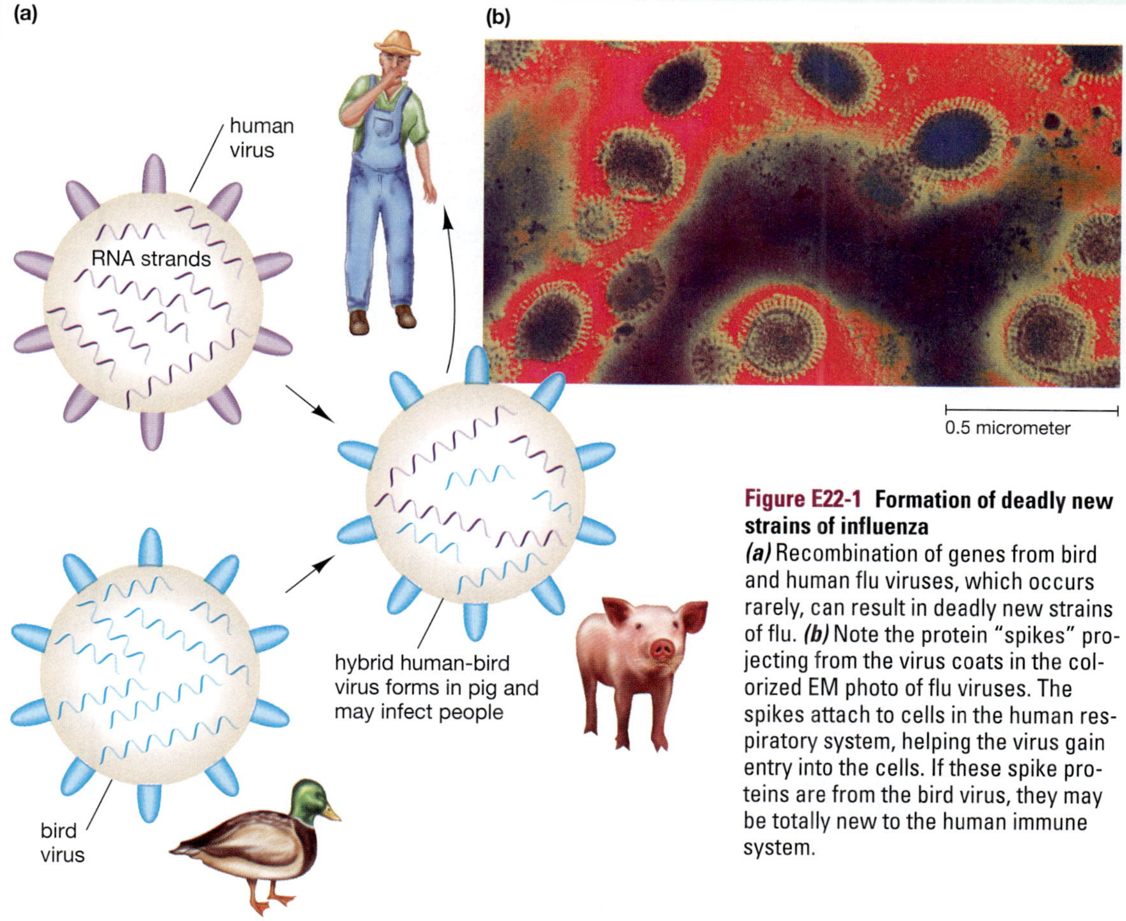

Figure E22-1 Formation of deadly new strains of influenza
(a) Recombination of genes from bird and human flu viruses, which occurs rarely, can result in deadly new strains of flu. ***(b)*** Note the protein "spikes" projecting from the virus coats in the colorized EM photo of flu viruses. The spikes attach to cells in the human respiratory system, helping the virus gain entry into the cells. If these spike proteins are from the bird virus, they may be totally new to the human immune system.

render a virus incapable of reproduction and so are never present in dangerous flu strains. Neuraminidase inhibitors bind the key regions of the spikes tightly and prevent the viruses from entering human cells, the only place where they can multiply. Administered after flu symptoms begin, the drugs hasten recovery. Researchers hope that soon people will be tested for flu virus before symptoms appear; treatment at this early stage might prevent the symptoms entirely. We may never eradicate the flu virus, but it may soon lose its ability to disrupt our lives.

immune system, like other body systems, is itself subject to diseases and malfunctions that decrease its effectiveness.

Allergies Are Misdirected Immune Responses

More than 35 million Americans suffer from allergies, adverse reactions to substances that are not harmful in themselves and to which many other people do not respond. Common allergies include those to pollen, dust, mold spores, animal dander, and bee stings. Allergies are actually a form of immune response. A foreign substance, such as a pollen grain, enters the bloodstream and is recognized as an antigen by a B cell. This B cell proliferates, producing plasma cells that pour out antibodies against the pollen antigen.

The stems of the antibodies attach to the plasma membranes of mast cells, histamine-containing cells located in the respiratory and digestive tracts. If pollen antigens later encounter and bind to the attached antibodies in the respiratory tract, they trigger the release of histamine, which (as you learned earlier) causes leaky capillaries and other symptoms of inflammation (Fig. 22-11). In the respiratory tract, histamine also increases mucus secretion. Thus, airborne substances such as pollen grains, which typically enter the nose and throat, often trigger allergic reactions that include the runny nose, sneezing, and

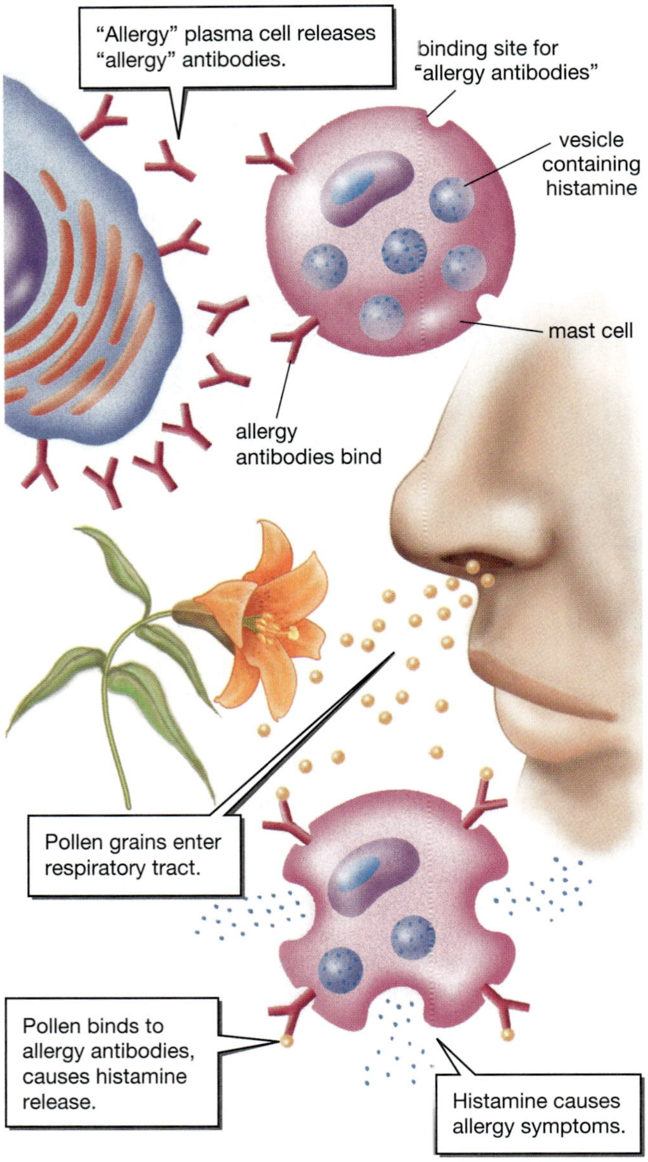

Figure 22-11 Allergic reaction

congestion of "hay fever." Food allergies cause analogous symptoms, including cramps and diarrhea, in the digestive tract.

An Autoimmune Disease Is an Immune Response Against the Body's Own Molecules

Fortunately, our immune systems rarely mistake our own cells for invaders. Occasionally, however, something goes awry, and "anti-self" antibodies are produced. The result is an **autoimmune disease**, in which the immune system attacks a component of one's own body. Some types of anemia, for example, are caused by antibodies that destroy a person's red blood cells. Many cases of juvenile-onset diabetes begin when the immune system attacks the insulin-secreting cells of the pancreas. Multiple sclerosis results when immune cells launch a misdirected attack against the insulating fatty sheath that coats parts of neurons in the brain and spinal cord. Rheumatoid arthritis is also caused by inappropriate immune activation. Unfortunately, at present there are no known cures for autoimmune diseases. For some diseases, replacement therapy can alleviate the symptoms—for instance, by administering insulin to diabetics or blood transfusions to anemics. Alternatively, the autoimmune response can be suppressed with drugs. Immune suppression, however, also reduces immune responses to the everyday assaults of disease microbes, so this therapy has major drawbacks.

An Immune Deficiency Disease Eliminates the Immune Response

On rare occasions, a child is born with severe combined immune deficiency (SCID), a defect in which few or no immune cells are formed. A child with SCID may survive the first few months of postnatal life, protected by antibodies acquired from the mother during pregnancy or in her milk. Once these antibodies are lost, however, common bacterial infections can prove fatal, because there is no immune response. One form of therapy is to transplant bone marrow (from which immune cells arise) from a healthy donor into the child. In some children, marrow transplants have resulted in some antibody production, occasionally enough to confer normal immune responses. In 1990 researchers at the National Institutes of Health began clinical trials of injecting genetically engineered bone marrow cells into children with SCID. The therapy has been somewhat successful but is not yet in widespread use.

AIDS Is a Devastating Immune Deficiency Disease

The most common and widespread immune deficiency disease is **acquired immune deficiency syndrome**, or **AIDS**. Two viruses, named **human immunodeficiency viruses** 1 and 2 (HIV-1 and HIV-2), cause AIDS. These viruses undermine the immune system by infecting and destroying the helper T cells, which stimulate both the cell-mediated and humoral immune responses. AIDS does not kill people directly, but AIDS victims become increasingly susceptible to other diseases as their helper T-cell populations decline.

HIV Is a Retrovirus That Destroys Helper T Cells

Both HIV-1 and HIV-2 are **retroviruses**—viruses that have RNA as their genetic material. Retroviruses reproduce by transcribing their RNA into DNA and then inserting the DNA into the chromosomes of a host cell. HIV consists of an outer envelope, taken from an infected cell's plasma membrane as the virus leaves the cell, and two protein layers. The innermost protein

(a)

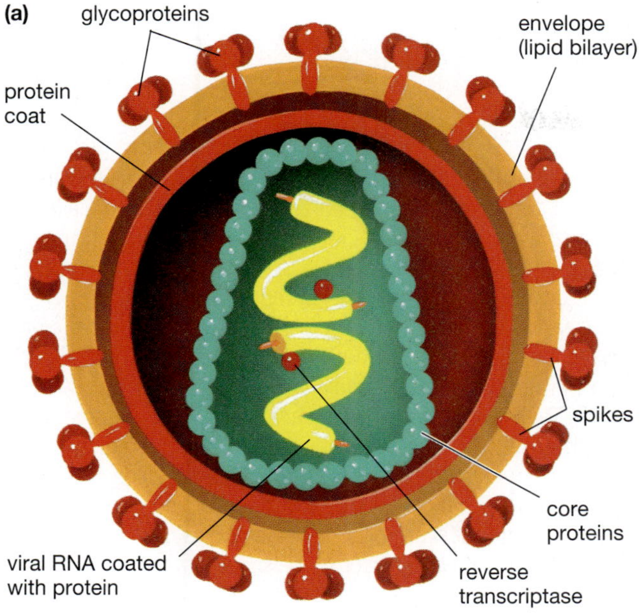

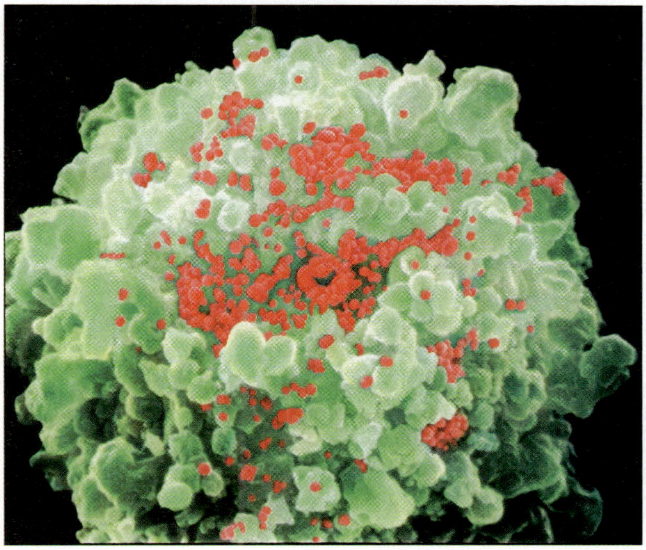

(b)

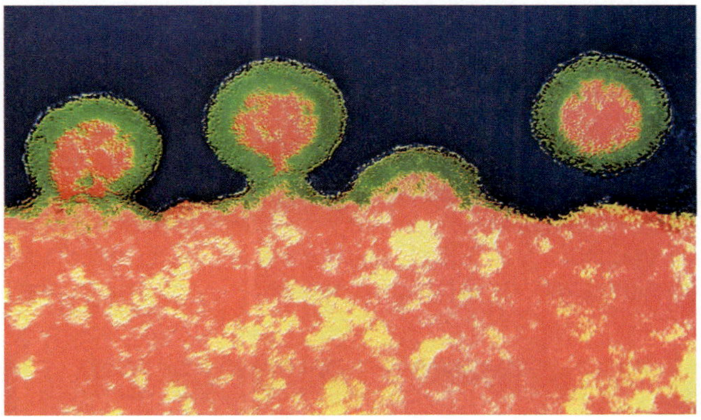

(c)

Figure 22-12 The virus that causes AIDS
(a) HIV consists of an outer envelope, a protein coat, and an inner protein capsule that contains RNA and reverse transcriptase. Proteins protruding through the envelope attach to the plasma membranes of helper T cells. ***(b)*** The red specks in this colorized scanning electron micrograph are HIV that have just emerged from the large, green helper T cell. ***(c)*** In this colorized transmission electron micrograph, HIV are emerging from the helper T cell and acquiring an outer envelope of plasma membrane (green) from the infected T cell.

MEDIATUTOR 22.3 Retrovirus Replication

layer encloses the viral RNA and an enzyme called **reverse transcriptase** (Fig. 22-12). When HIV attacks a helper T cell, the virus's outer envelope binds to the plasma membrane of the cell and allows the virus to enter the cell. Once the virus is inside, its reverse transcriptase catalyzes the transcription of the virus's RNA genome into DNA (a process known as reverse transcription, because it reverses the normal DNA-to-RNA transcription of cells). The resulting DNA then travels to the nucleus and is inserted into the T cell's genome, directing it to make more HIV particles.

In the early stages of an HIV infection, the patient may have a fever, rash, muscle aches, headaches, and enlarged lymph nodes as the immune system fights the infection. After about 6 months, the rate of viral replication slows and stabilizes at a relatively low level. Enough T cells remain that the patient is able to resist disease and may feel quite well. This stage can persist for several years. Eventually, however, the number of helper T cells begins to drop. When only 200 helper T cells per milliliter of blood remain (one-fourth of the normal number), the patient is described as having AIDS. From this point, in an untreated patient, HIV levels skyrocket, helper T cell numbers continue to fall, and the patient falls victim to other infections. The life expectancy for untreated AIDS victims is then about 1 to 2 years.

HIV Is Transmitted by Body Fluids

HIV cannot survive for very long outside the body. The virus can be transmitted only by the direct contact of broken skin or mucous membranes with virus-laden body fluids—including blood, semen, vaginal secretions, and breast milk. HIV infection can be spread by sexual activity, by sharing of needles among intravenous drug users, by blood transfusions (this is rare since it became standard practice to screen all donated blood for anti-HIV antibodies), or from mother to child during pregnancy, childbirth, or breastfeeding. Nearly all people infected with HIV eventually develop AIDS.

In the early days of the AIDS epidemic in the United States, most AIDS patients were homosexual men or intravenous drug users. It is now obvious that AIDS can also be transmitted through heterosexual contact.

Earth Watch
Dams, Deforestation, and Diseases

As researchers work in laboratories to devise ways to assist the immune system's fight against diseases, human activities are simultaneously altering the environment in ways that promote the spread of diseases, some of them deadly. A now-classic example is the construction of the Aswan High Dam on the Nile River in Egypt (Fig. E22-2a). Completed in 1968, the dam was built to provide irrigation water and hydroelectric power for Egypt's rapidly expanding population. Officials did not foresee that the stagnant waters of the newly created Lake Nassar and the irrigation ditches fed by the dam would enormously increase the habitat for a snail that carries the parasitic worm responsible for schistosomiasis—a chronic, debilitating, and potentially fatal disease. Human infection rates in the region around the dam increased from 5% in 1968 to 77% in 1993. The stagnant waters also created a mosquito-breeding paradise, leading to several epidemics of a mosquito-transmitted viral infection called Rift Valley fever.

Meanwhile, China is now building the largest dam in the world on the Yangtze River. Twice as tall as the Aswan Dam, the 660-foot-high Three Gorges Dam will create a 400-mile-long reservoir. Experts are predicting an explosion in the population of local schistosomiasis-carrying snails, as occurred in Egypt, with major adverse effects on human health.

Throughout the Tropics, rain forests are being felled for agriculture, with devastating consequences—not only for the rare species that inhabit them, but for local human populations as well. As forests are cleared and original habitats destroyed, new types of habitats are created for malaria-carrying mosquitoes (Fig. E22-2b). For example, a malaria-carrying *Anopheles* mosquito in Africa prefers to breed in the open areas created by clear-cuts, while the warmer surface temperatures created by removing trees speeds up the life cycle of the mosquito and the development of the malaria parasite. Malaria, which kills more than 1 million people each year, is also spreading in South America as the Brazilian rain forest is cleared.

Global warming caused by human activities (see Chapter 29) is also hastening the spread of mosquito-borne diseases such as malaria, dengue fever, yellow fever, and west Nile virus. Warmer temperatures not only increase the speed of the reproduction of mosquitoes and the disease-causing organisms they carry, but mosquitoes bite more frequently at warmer temperatures, increasing the rate at which they spread diseases. In addition, warmer global temperatures permit the spread of the disease-causing organisms into areas from which they were previously excluded by cold winter weather. Already, disease-carrying mosquitoes and the diseases they spread have been found at many high elevation sites from which they were historically absent. Many researchers fear that as global temperatures continue their expected rise, these diseases will make their way to temperate regions that have been so far free of them.

Figure E22-2 Environmental damage spreads disease
(a) The Aswan High Dam produced Lake Nassar and filled irrigation ditches with stagnant water. The snails that carry the schistosomiasis worm thrive in these quiet waters. ***(b)*** Clear-cutting rain forests has increased the habitat for the *Anopheles* mosquito that carries and spreads the malaria protist.

Heterosexual intercourse is the most common means of infection in Africa and Asia, where the largest number of AIDS victims live.

There Are Treatments, But No Cures, for AIDS

For persons infected with AIDS, there are two categories of therapy. First, infections that result from the impaired immune system can be treated as they would be in any patient. Second, new drugs that target retroviruses can slow the progress of AIDS. These drugs fall into two general classes: reverse transcriptase inhibitors and protease inhibitors.

Reverse transcriptase inhibiting drugs such as AZT and ddI inhibit viral reproduction by blocking the production of DNA from RNA. Protease inhibitors neutralize an enzyme that helps assemble viruses from the components produced by the host cell's pirated genetic machinery. Unfortunately, these drugs are not completely successful, partly because the HIV virus often mutates and some mutants are resistant to the drugs. In some patients the drugs have severe side effects. Often two or more reverse transcriptase inhibitors are given in combination with two or more protease inhibitors, attacking simultaneously at two points in the HIV life cycle. This multi-drug treatment may require patients to take up to 24 pills daily and cost $10,000 to $12,000 each year.

Clearly, the best solution would be to develop an AIDS vaccine. This is a major challenge, because the HIV antibodies produced naturally by people with AIDS do not halt the infection. Therefore, a vaccine would have to evoke a more effective immune response than does the HIV infection itself. Further, HIV has an incredibly high mutation rate, perhaps a thousand times faster than that of flu viruses, so infected individuals may harbor multiple strains of HIV. Nevertheless, trials of AIDS vaccines are under way in several countries, though none yet appear to be particularly promising.

AIDS Is a Widespread, Lethal Disease

We are currently in the midst of a worldwide AIDS epidemic that has killed about 22 million people. UNAIDS, the United Nations organization responsible for tracking AIDS cases, estimates that 40 million people are infected with HIV and that each day 7000 people die from AIDS and 14,500 become infected with HIV. Most of those infected (95%) live in the developing world, especially in Sub-Saharan Africa, which currently has the highest infection rate and where AIDS is the leading cause of death. Southern Africa has been especially hard hit; for example, more than 30% of pregnant women in Botswana are infected with HIV. In the United States, about 40,000 people are infected with HIV each year, about one-third of whom are women. There are about 750,000 AIDS cases in the United States, and perhaps as many more people harbor the virus but have not developed AIDS symptoms.

AIDS is one the world's deadliest infectious diseases, rivaling other notorious killers such as malaria and tuberculosis in terms of annual deaths. How concerned should you be about AIDS, personally? Because the disease is deadly and the incidences of other sexually transmitted diseases are on the rise, "safer-sex" practices are important for everyone. Further, health-care workers should (and now routinely do) take extra precautions when handling blood products and used needles.

Cancer Can Evade or Overwhelm the Immune Response

Cancer is one of the most dreaded words in the English language, and with good reason. The disease kills more than 500,000 U.S. citizens each year and about 6 million people worldwide. The proportion of U.S. citizens who will eventually contract cancer is about 40%. Advances in detection and treatment have reduced the death rates for some forms of cancer, but the overall cancer death rate has increased since the mid-1970s. Despite decades of intensive research, the quest for a cancer cure remains unfulfilled. Why can't we cure or prevent cancer? We can, after all, prevent smallpox and polio and can cure dozens of other diseases. What makes cancer different?

Unlike most other diseases, cancer is not a straightforward invasion of the body by a foreign organism. Instead, cancer is essentially a failure of the mechanisms that control the growth of the body's own cells—a disease in which the body destroys itself. In particular, **cancer** is characterized by the unchecked growth of malignant tumor cells. A tumor is a population of cells that has escaped from normal regulatory processes and grows at an abnormal rate. The cells in a *benign* tumor, or polyp, normally remain in one area, but the cells in a *malignant* tumor grow uncontrollably and spread to other areas of the body. As a malignant tumor grows, it uses increasing amounts of the body's energy and nutrient supplies and literally squeezes out vital organs nearby.

The Immune System Defends Against Cancerous Cells

Cancer cells form in our bodies every day. Even the best of preventive measures cannot eliminate cancer completely. We cannot avoid some carcinogens, such as gamma rays from the sun, radioactivity from the rocks beneath our feet, and naturally produced carcinogens in our food. Fortunately, natural killer cells and cytotoxic T cells screen the body for cancer cells and destroy nearly all of them before they have a chance to proliferate and spread. Cancer cells are, of course, "self" cells (the body's own cells), and the immune system does not respond to "self." How are cancer cells weeded out? It seems likely that the processes that cause cancer also cause new and slightly different proteins to appear on the surfaces of cancer cells. Cytotoxic T cells encounter these new proteins, recognize them as "non-self" antigens, and destroy the cancer cells (see Fig. 22-9). However, some cancer cells may not be recognized as threats by the immune system. Even tumors that are attacked by the immune system may develop variant cell types that are resistant to immune attack. In other cases, tumors can actively suppress the immune system or can simply grow so fast that the immune response can't keep up. If the immune system is thwarted by any of these means, the tumor grows and spreads. At this point, the individual's health depends on medical treatment. Unfortunately, the available treatments are only partially effective and have serious drawbacks. The rate of cure is increasing, but it is still scarcely a third of all cancers.

Treatments for Cancer Selectively Kill Cancerous Cells

The three main forms of medical treatment for cancer are surgery, radiation, and chemotherapy. The surgical removal of the tumor is the first step in the treatment of many cancers. Unfortunately, the surgeon may not be able to see and remove very small patches of cancer cells that may extend from the main tumor. And surgery cannot be used to treat cancer that has begun to spread throughout the body. Alternatively, cancerous cells can be bombarded with radiation in an effort to kill them. Unlike surgery, radiation may destroy even microscopic clusters of cancer cells. Like surgery, however, radiation cannot be used to treat widespread cancers, because to irradiate the whole body would damage a great deal of healthy tissue. Both surgery and radiation therapy can be traumatic and dangerous.

Chemotherapy, or drug treatment, is commonly used to supplement surgery and radiation or to treat cancers that cannot be treated with surgery or radiation. Chemotherapy drugs attack the machinery of cell division. Because cancer cells divide much more frequently than do normal cells, the hope is that drug attacks on dividing cells will selectively kill cancer cells. Unfortunately, other cells of the body divide too, and chemotherapy inevitably kills some healthy cells. It is damage to dividing cells in the hair follicles and intestinal lining that produces the well-known side effects of hair loss, nausea, and vomiting.

A tremendous amount of research has been devoted to the search for cancer treatments that are more effective and have fewer unpleasant side effects. Developing a "cancer vaccine" is a high priority. Other approaches include developing therapies that would stimulate the immune system to attack tumors, and developing antibody molecules that recognize tumor cells and could be used to deliver drugs and other treatments directly to tumor cells without affecting healthy cells. Research and clinical trials continue, however, and cancer patients may one day soon reap the benefits of innovative new treatments.

Each of us can reduce his or her own chances of developing cancer by avoiding cigarette smoke and by eating a diet high in fruits and vegetables and low in red meat and saturated fats. Other behaviors that lower the risk of cancer include avoiding excessive alcohol consumption, using sunscreen, and getting regular exercise.

CASE STUDY REVISITED

Tick Saliva and Lyme Disease

The search for a vaccination against Lyme disease focused on several proteins that lie on the cell surface of *Borrelia burgdorferi*, the tick-borne bacterium that causes the disease. Researchers thought that if the host's body could be induced to mount an immune response to these proteins, then antibodies would attack any *B. burgdorferi* that later managed to arrive via a deer tick. A protein known as OspA emerged as a vaccine candidate; mice injected with OspA and then with *B. burgdorferi* did not develop symptoms of Lyme disease and were free of the bacteria.

The researchers realized, however, that the successful immune response of vaccinated mice to *B. burgdorferi* injected by a needle was not necessarily a good predictor of how the mice would respond to bacteria transferred by ticks, with their array of salivary immune suppressors. To make a more realistic test of the effectiveness of OspA injections against Lyme disease, researchers allowed infected deer ticks to bite healthy mice, some of which had received OspA injections and some of which had not. At the close of the experiment, the vaccinated mice were free of infection and disease symptoms, but most of the unvaccinated mice were sick and infected. The OspA vaccination worked in mice, and researchers began trials with human subjects. In late 1998 the FDA approved the new vaccine for public use.

The experiment with vaccinated mice had an odd and interesting postscript. The researchers concluded the experiment by testing all the "carrier" ticks that had been used to infect the mice with *B. burgdorferi* bacteria. All the ticks that had been feeding on unvaccinated mice remained infected, but the ticks that had been feeding on vaccinated mice were free of *B. burgdorferi*. The researchers were quite surprised to find bacteria-free ticks. Apparently, when the deer ticks ingested blood that contained antibodies to OspA, the bacteria in the tick's gut were killed. The vaccination works not by killing bacteria in the human host, but by killing them before they can leave the tick.

Summary of Key Concepts

1) How Does the Body Defend Against Invasion?

The human body has three lines of defense against invasion by microbes: (1) the barriers of skin and mucous membranes; (2) nonspecific internal defenses, including phagocytosis, killing by natural killer cells, inflammation, and fever; and (3) the immune response.

If microbes do enter the body, white blood cells travel to the site of entry and engulf the invading cells. Natural killer cells secrete proteins that kill infected or cancerous cells. Injuries stimulate the inflammatory response, which attracts phagocytic white blood cells, increases blood flow, and makes capillaries leaky. Later, blood clots wall off the injury site. Fever is caused by raising the set point of the body's thermostat. High temperature from a fever inhibits bacterial growth and accelerates the immune response.

2) What Are the "Soldiers" of the Immune Response?

B cells, T cells, and other lymphocytes interact to carry out the immune response. Immune responses have three steps: (1) recognition, (2) attack, and (3) memory.

3) How Does the Immune System Recognize Invaders?

Antibodies (on B cells) and T-cell receptors (on T cells) recognize antigens and trigger the immune response. Antibodies are Y-shaped proteins composed of a constant region and a variable region. Antibodies both detect and actively destroy antigens. Each B cell synthesizes only one type of antibody, unique to that particular cell and its progeny. The diversity of antibodies arises from gene shuffling and the mutation of antibody genes during B-cell development. Each antibody has specific sites that bind only one or a few types of antigen.

4) How Does the Immune System Launch an Attack?

Antibodies attack the invaders. Antigens from an invader bind to and activate only those B and T cells with the complementary antibodies or T-cell receptors. In humoral immunity, the presence of antigens stimulates B cells with complementary antibodies to divide rapidly to produce plasma cells that synthesize massive quantities of the antibody. Antibodies circulating in the blood destroy antigens and antigen-bearing microbes by coating them, causing them to clump together, or promoting phagocytosis by white blood cells. In cell-mediated immunity, T cells with the proper receptors bind antigens and divide rapidly. Cytotoxic T cells bind to antigens on microbes, infected cells, or cancer cells and then kill the cells. Helper T cells stimulate, and suppressor T cells turn off, both the B-cell and cytotoxic-T-cell responses.

5) How Does the Immune System Remember Its Past Victories?

Some progeny cells of both B and T cells are long-lived memory cells. If the same antigen reappears in the bloodstream, these memory cells are immediately activated, divide rapidly, and cause an immune response that is much faster and more effective than the original response.

6) How Does Medical Care Augment the Immune Response?

Antibiotics kill cellular microbes or slow down their reproduction, thus allowing the immune system more time to respond and exterminate the invaders. Vaccinations are injections of antigens from disease organisms, typically the weakened or dead microbes themselves. An immune response is evoked by the antigens, providing memory and a rapid response should an infection occur.

7) What Happens When the Immune System Malfunctions?

Allergies are immune responses to normally harmless foreign substances, such as pollen or dust. Mast cells respond to the presence of these substances by releasing histamine, which causes a local inflammatory response. Autoimmune diseases arise when the immune system destroys some of the body's own cells. Immune deficiency diseases occur when the immune system cannot respond strongly enough to ward off normally minor diseases.

Infection with either HIV 1 or 2 can lead to AIDS. These viruses invade and destroy helper T cells. Without helper T cells to stimulate the immune responses of B cells and cytotoxic T cells, a person with AIDS is extremely susceptible to infections, which are eventually fatal.

Cancer is a population of the body's cells that divides without control. Cancerous cells may be recognized as "different" by the immune system and destroyed by natural killer cells and cytotoxic T cells. A few cancerous cells evolve the capacity to evade the immune system, some attack immune cells, and others multiply too fast for the immune system to keep up. In these cases, cancer develops and requires aggressive medical treatment.

Key Terms

acquired immune deficiency syndrome (AIDS) *p. 430*
antibody *p. 422*
antigen *p. 422*
autoimmune disease *p. 430*
B cell *p. 421*
cancer *p. 433*
cell-mediated immunity *p. 424*
clonal selection *p. 424*
constant region *p. 422*
cytotoxic T cell *p. 425*
fever *p. 419*
helper T cell *p. 425*
human immunodeficiency virus (HIV) *p. 430*
humoral immunity *p. 424*
immune response *p. 421*
immune system *p. 421*
inflammatory response *p. 419*
macrophage *p. 420*
memory B cell *p. 425*
memory T cell *p. 426*
mucous membrane *p. 418*
natural killer cell *p. 419*
phagocytic cell *p. 419*
plasma cell *p. 425*
retrovirus *p. 430*
reverse transcriptase *p. 430*
skin *p. 418*
suppressor T cell *p. 426*
T cell *p. 421*
T-cell receptor *p. 422*
vaccination *p. 427*
variable region *p. 422*

Thinking Through the Concepts

Multiple Choice

1. *In addition to the immune system, we are protected from disease by*
   a. the skin
   b. mucous membranes
   c. natural secretions such as acids, protein-digesting enzymes, and antibiotics
   d. cilia
   e. all of the above

2. *The first line of defense against body cells infected by viruses is produced by*
   a. antibodies
   b. phagocytes
   c. natural killer cells
   d. histamines
   e. natural antibiotics

3. *Fevers*
   a. decrease histamine production
   b. decrease the concentration of iron in the blood
   c. decrease the activity of phagocytes
   d. increase the reproduction rate of invading bacteria
   e. do all of the above
4. *T cells and B cells are*
   a. lymphocytes
   b. macrophages
   c. natural killer cells
   d. red blood cells
   e. phagocytes
5. *During allergic responses*
   a. the foreign substance binds to antibodies on B cells
   b. B cells produce antibodies to the foreign substance
   c. antibodies to the foreign substance bind to that substance and to mast cells
   d. stimulation of mast cells causes them to release histamine
   e. all of the above
6. *What shuts off the immune response in T cells and B cells after an infection has been conquered?*
   a. cytotoxic T cells
   b. histamine
   c. pyrogens
   d. natural killer cells
   e. suppressor T cells

? Review Questions

1. List the human body's three lines of defense against invading microbes. Which are nonspecific (that is, act against all types of invaders), and which are specific (act only against a particular type of invader)? Explain your answer.
2. How do natural killer cells and cytotoxic T cells destroy their targets?
3. Describe humoral immunity and cell-mediated immunity. Include in your answer the types of immune cells involved in each, the location of antibodies and receptors that bind foreign antigens, and the mechanisms by which invading cells are destroyed.
4. How does the immune system construct so many different antibodies?
5. How does the body distinguish "self" from "non-self"?
6. Diagram the structure of an antibody. What parts bind to antigens? Why does each antibody bind only to a specific antigen?
7. What are memory cells? How do they contribute to long-lasting immunity to specific diseases?
8. What is a vaccine? How does it confer immunity to a disease?
9. Compare and contrast an inflammatory response with an allergic reaction from the standpoint of cells involved, substances produced, and symptoms experienced.
10. Distinguish between autoimmune diseases and immune deficiency diseases, and give one example of each.
11. Describe the causes and eventual outcome of AIDS. How do AIDS treatments work? How is the HIV virus spread?
12. Why is cancer sometimes fatal?

Applying the Concepts

1. Why is it essential that antibodies and T-cell receptors bind only relatively large molecules (such as proteins) and not relatively small molecules (such as amino acids)?
2. There are smallpox virus stocks in two laboratories—one in the United States and one in Russia. A debate is raging about whether these stocks should be destroyed. In brief, one side argues that having smallpox around is too dangerous. The other side argues that we may be able to learn things from smallpox, answers to questions we don't know enough to ask yet, that may help us conquer future diseases. Do you believe the smallpox stocks should be destroyed? Why or why not?
3. The essay "Health Watch: Can We Beat the Flu Bug?" states that the flu virus is different each year. If that is true, what good is it to get a "flu shot" each winter?
4. Organ transplant patients typically receive the drug cyclosporine. This drug inhibits the production of interleukin-2, a regulatory molecule that stimulates helper T cells to proliferate. How does cyclosporine prevent the rejection of transplanted organs? Some patients who received successful transplants many years ago are now developing various kinds of cancers. Propose a hypothesis to explain this phenomenon.

Group Activity

This chapter has introduced you to some of the mechanisms that have evolved to defend the vertebrate body against attack. The attackers, however, have not stood idly by in the face of evolutionary improvements in our defenses. Many viruses, bacteria, fungi, and invertebrate animals have themselves evolved strategies for evading and resisting the immune response. Can you guess what some of these strategies might be? Form a group of four students, and try to design the structures and behaviors that might have emerged in these invaders through natural selection. Assign two members of your group to design a structural solution and a behavioral solution that an invader might use to avoid being recognized by the immune system. The other two members of the group should design one structure and one behavior that might resist immune attack even after recognition. Next, bring your entire group together and discuss the strengths and drawbacks of your proposed solutions. Your final product should be a "designer organism" that has multiple ways of evading immune recognition and attack. Be sure to specify your organism's size, life cycle, structure, and food source.

For More Information

"Defeating AIDS: What Will It Take?" *Scientific American*, July 1998. A series of nine articles covering all aspects of AIDS.

Laver, W. G., Bischofberger, N., and Webster, R. G. "Disarming Flu Viruses." *Scientific American*, January 1999. Presents a clear description of how flu viruses work, how new strains arise, and the development of the new neuraminidase inhibitors that may stop them.

Lichtenstein, L. M. "Allergy and the Immune System." *Scientific American*, September 1993. Allergic responses, evolved as protection against parasites, turn against us when we respond violently to harmless pollen and foods.

Marrack, P., and Kappler, J. W. "How the Immune System Recognizes the Body." *Scientific American*, September 1993. How the immune system's potential weapons against the body are disarmed.

Paul, W. E. "Infectious Diseases and the Immune System." *Scientific American*, September 1993. Microbes and the immune system engage in constant evolutionary warfare.

Steinman, L. "Autoimmune Disease." *Scientific American*, September 1993. An estimated 5% of Americans suffer from some type of autoimmune disorder.

Answers to Multiple-Choice Questions

1. e 2. c 3. b 4. a 5. e 6. e

MediaTutor

CD Activities

Activity 22.1: Inflammation

Estimated time: 5 minutes

The immune system includes both nonspecific and specific defenses to invading organisms. One of the nonspecific defenses is the inflammatory response. The symptoms of inflammation (heat, redness, swelling, and pain) indicate that the body is fighting an invasion. In this tutorial you will explore this process in more detail.

Activity 22.2: Clonal Selection

Estimated time: 10 minutes

The production of antibodies by B cells is an important part of the specific immune response. The antibodies that are produced are highly specific for one particular foreign antigen. Before a B cell can begin its production of these important immune proteins, it must be activated by the foreign antigen in a process called *clonal selection*. In this tutorial you will explore this process in more detail.

Activities 22.3: Retrovirus Replication

Estimated time: 5 minutes

Viruses take over our cellular machinery and use it to produce more viruses. Understanding how viruses replicate may provide a way to prevent or cure many diseases. In this activity, you will look at how viruses such as HIV multiply in human cells.

Start the MediaTutor Student CD-ROM and enter the activity number in the Quick Search box to be taken directly to that activity.

Web Investigation

Tick Saliva and Lyme Disease

Estimated time: 20 minutes

"Case Study: Tick Saliva and Lyme Disease" describes some of the consequences associated with contracting Lyme disease. Recently, scientists have been able to develop a new vaccine against this growing disease. This activity will explore how the immune system works and how it can be manipulated through the process of vaccination/immunization.

Go to http://www.prenhall.com/audesirk3, the Audesirk Companion Web site. Select Chapter 22 and Web Investigation to begin.

Most physiological functions are controlled by hormones, the chemical messengers of the animal body. Biologists working with animals that inhabit polluted environments have discovered that environmental contaminants can disrupt these chemical messengers.

23 Chemical Control of the Animal Body: The Endocrine System

AT A GLANCE

CASESTUDY

Gender Bending

In 1980 an industrial accident near Florida's Lake Apopka spilled a large quantity of the pesticide kelthane into the lake. Shortly thereafter, biologists began noticing ill effects on the lake's wildlife. Alligators were especially hard hit. By 1988 only a small fraction of the alligator eggs laid were hatching successfully. This very low reproductive success was disturbing, but not necessarily unexpected in such a badly polluted environment. Researchers were surprised, however, by the condition of the surviving juvenile alligators. In particular, the researchers noticed that many of the young males were feminized: their penises were abnormally small, and their testes were deformed. It seemed that something in Lake Apopka was bending the gender of the local alligators.

Strange distortions of the sexual characteristics of animals in polluted waters have also been documented elsewhere. In England, male fish living near municipal sewage outlets were found to be producing vitellogenin, an egg-yolk protein normally produced only by females. Conversely, female fish living in waters near paper mills have developed male sex organs and exhibited male courtship behaviors. Female mollusks living on or near ships or boats that have been treated with an anti-fouling chemical also develop male sexual characteristics. These and other observations have led some scientists to hypothesize that chemicals in these polluted environments are somehow disrupting the animals' normal sexual development. More specifically, these researchers propose that the pollutants interfere with the body's internal system for sending chemical signals.

The hypothesis that substances in the environment can distort sexual development has led to an intensive search for the chemicals that might be responsible for the changes as well as for the physiological mechanisms underlying disruption of normal development. By extension, the connection between environmental contaminants and abnormal reproductive systems also raises questions about human health. If environmental chemicals are indeed disrupting chemical signaling in animals, could they be doing the same thing to humans? And if humans are susceptible to such disruption, what kinds of health effects might result? ■

1 What Are the Characteristics of Animal Hormones?

Hormones are chemical messages, secreted by specialized cells, that initiate changes in other cells. The hormonal messages are normally sent in response to an environmental or physiological stimulus, and travel by way of the circulatory system to **target cells**, cells that are specialized to respond to the messages. The changes induced by hormonal messages may be prolonged and irreversible, as in the onset of puberty or the transformation of a caterpillar into a butterfly. More typically, the induced changes are temporary and reversible and help control and regulate the physiological systems of the animal body. The **endocrine system** (consisting of hormones and the various cells that secrete and receive them) can be viewed as the postal system of the body, moving information and instructions between cells that may be some distance apart.

Many Different Molecules Serve As Hormones

A chemical is classified as a hormone based on the type of function it performs; the term does not refer to a particular chemical structure. A wide variety of different molecules can function as hormones. Nonetheless, vertebrate hormones fall into three general classes: **peptide hormones**, which consist of chains of amino acids; *amino-acid-based hormones*, which are synthesized from single amino acids; and **steroid hormones**, most of which are synthesized from cholesterol. Most animal hormones are peptide hormones.

Hormones Bind to Specific Receptors on Target Cells

Because nearly all cells have a blood supply, a hormone that is released into the bloodstream will reach nearly every cell of the body. Nonetheless, only certain target cells respond to a circulating hormone. The target cells that respond have receptors for molecules of the circulating hormone; cells that lack the appropriate receptor cannot respond to the hormonal message (Fig. 23-1). Receptors may be located on the plasma membranes of target cells, or inside the cells, normally within the nucleus.

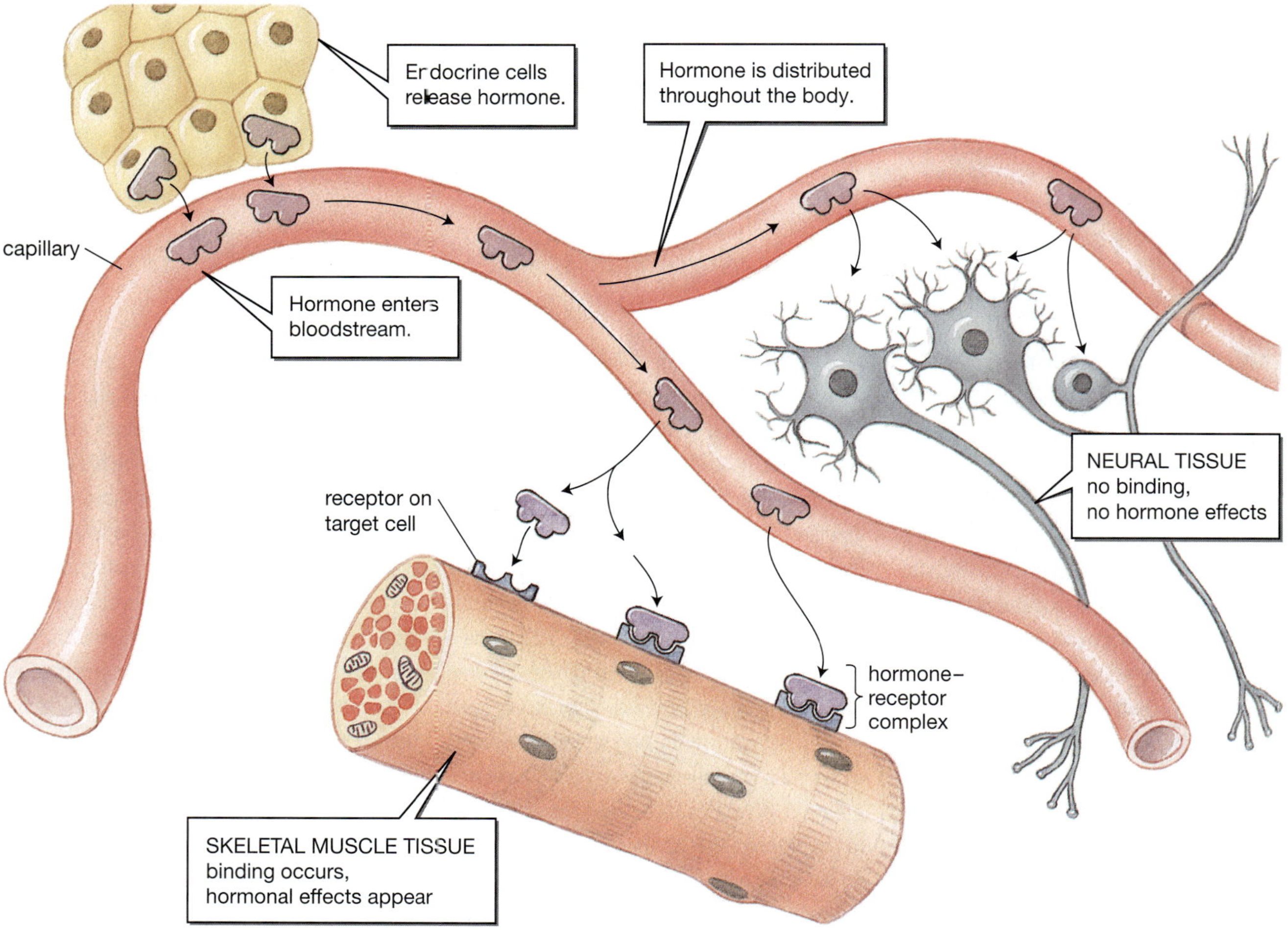

Figure 23-1 Endocrine glands, hormones, and target cells
The cells of endocrine glands secrete hormones into the extracellular fluid, from which they diffuse into the capillaries. A hormone binds to (and influences) only those cells with receptors for it. Muscle cells but not nerve cells have receptors for the particular hormone shown here.

Some Hormones Bind to Receptors on the Cell Surface

Peptide and amino-acid-based hormones tend to be soluble in water but not in lipid, and so cannot penetrate plasma membranes (which are composed largely of phospholipids). Instead, these hormones bind to receptor molecules on the target cell's plasma membrane. This reaction stimulates production inside the cell of a **second messenger**, a molecule that transfers a signal from the first messenger—the hormone—to other molecules within the cell. A common second messenger is cyclic AMP (cAMP), a nucleotide that regulates many cellular activities (see Chapter 2). The second messenger in turn initiates a chain of biochemical reactions (Fig. 23-2a). The end result of the reactions varies depending on the hormone and the target cell; in many cases, the target cells synthesize or secrete substances.

Other Hormones Bind to Receptors Within the Cell

In contrast to peptide hormones, some hormones, such as the steroid hormones, are lipid soluble and are therefore able to pass through the plasma membrane. Once inside the cell, these hormones bind to receptors, typically to receptors in the nucleus. The hormone–receptor complex then binds to DNA and stimulates particular genes to transcribe messenger RNA, which moves to the cytoplasm and directs the synthesis of proteins that alter the cell's activity (Fig. 23-2b). Hormones that bind to within-cell receptors may take minutes or even days to exert their full effects, in contrast to the relatively rapid but short-lived effects of hormones that bind to surface receptors.

Hormone Release Is Regulated by Feedback Mechanisms

A signal that is always turned on cannot carry an instruction for change. If a hormone is to be useful for physiological control, there must be a way to turn its message on and off. In animals, the switching mechanism usually involves negative feedback: The secretion of a hormone stimulates a response in target cells that inhibits further secretion of the hormone.

Most hormones exert such powerful effects on the body that it would be harmful to have too much hormone working for too long; therefore, control of hormone release by negative feedback is especially important. Suppose you have jogged a few miles on a hot, sunny day and have lost a pint of water through perspiration. In response to the loss of water from your bloodstream, your pituitary gland releases antidiuretic hormone (ADH), which causes your kidneys to reabsorb water and to produce very concentrated urine (see Chapter 21). If, however, you arrive home and drink a quart of Gatorade, you will more than replace the water you lost in sweat. Continued retention of this excess water could raise blood pressure and possibly damage your heart. Negative feedback ensures that ADH secretion is turned off when the water content of your blood returns to normal, and your kidneys begin eliminating the excess water (see Fig. 21-15). Look for more examples of negative feedback as you read through this chapter.

In a few cases, positive feedback controls hormone release. For example, as described in Chapter 19, contractions of the uterus early in childbirth cause the release of the hormone oxytocin. Oxytocin stimulates stronger contractions of the uterus, which cause more oxytocin to be released, creating a positive feedback cycle.

2 What Are the Structures of the Mammalian Endocrine System?

Some Glands Are Not Part of the Endocrine System

Although this chapter focuses on endocrine glands, you should be aware that glands come in two basic types: exocrine glands and endocrine glands. **Exocrine glands** produce secretions that are released outside the body or into the digestive tract (a hollow tube whose lining is continuous with the outside surface of the body). The secretions of exocrine glands are released through tubes or openings called **ducts**. Among the exocrine glands are the sweat glands and oil-producing (sebaceous) glands of the skin, the tear-producing (lacrimal) glands of the eye, the milk-producing (mammary) glands, as well as glands, such as the salivary glands, that produce digestive secretions.

Hormone-Producing Glands Are Enmeshed in Blood Vessels

Endocrine glands, sometimes called *ductless* glands, release their hormones within the body. An endocrine gland generally consists of clusters of hormone-producing cells embedded within a network of capillaries. The cells secrete their hormones into the extracellular fluid surrounding the capillaries (see Fig. 23-1). The hormones then enter the capillaries by diffusion and are carried throughout the body by the bloodstream.

The sections that follow discuss the functions of the major endocrine glands and organs: the hypothalamus–pituitary complex, the thyroid gland, the pancreas, the sex organs, and the adrenal glands (Fig. 23-3). Table 23-1 shows the hormones that these structures produce.

The Big Picture

The body's ability to function as a single, coordinated system depends on continuous communication among its many different tissues and organs. The chemical messengers of the endocrine system provide much of that communication.

MEDIATUTOR
23.2 Modes of Action of Hormones

Figure 23-2 Modes of action of hormones
***(a)* Peptide and amino acid hormones**, which are not soluble in lipids,
① bind to receptors on the outside of the target cell's plasma membrane.
② Hormone-receptor binding activates an enzyme that triggers the synthesis of cyclic AMP (cAMP) from ATP.
③ Cyclic AMP activates enzymes that
④ promote specific cellular reactions, producing new products.

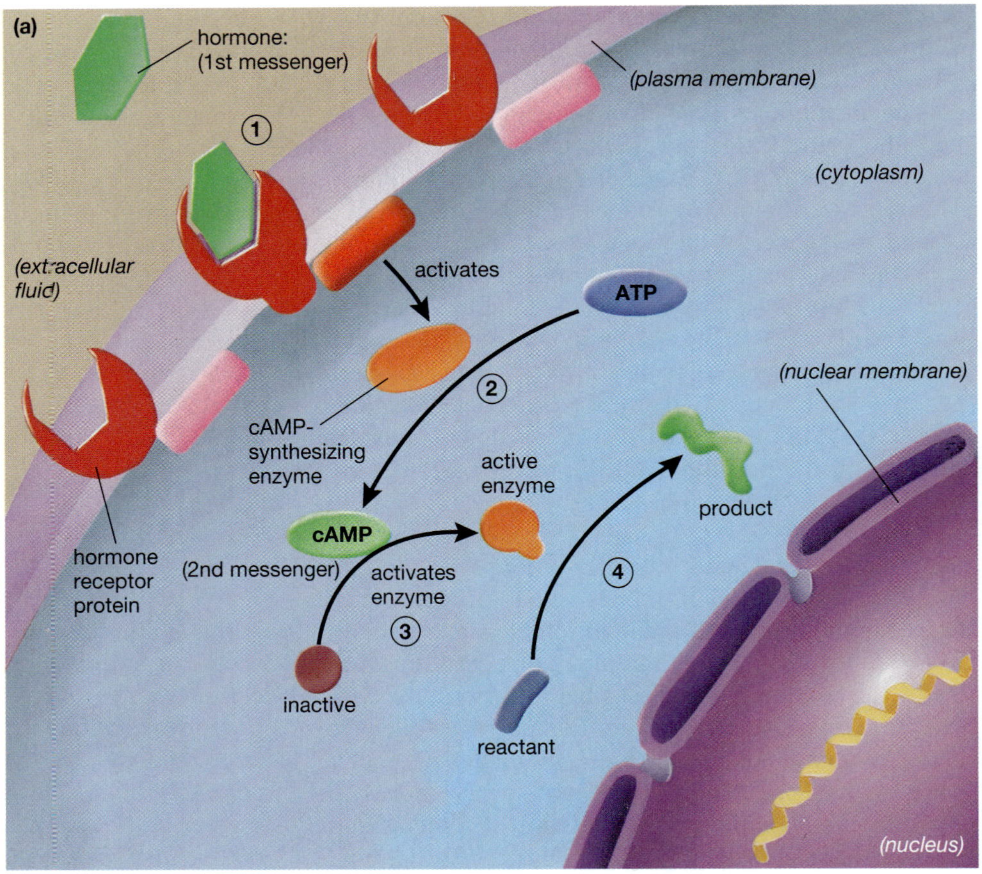

***(b)* Lipid-soluble steroid hormones**
① diffuse readily through the plasma membrane into the target cell and into the nucleus,
② where they combine with a receptor molecule.
③ The hormone-receptor complex binds to DNA and facilitates the binding of RNA polymerase to promoter sites on specific genes, accelerating
④ the transcription of DNA into messenger RNA (mRNA), which moves out into the cytoplasm.
⑤ The mRNA then directs protein synthesis.

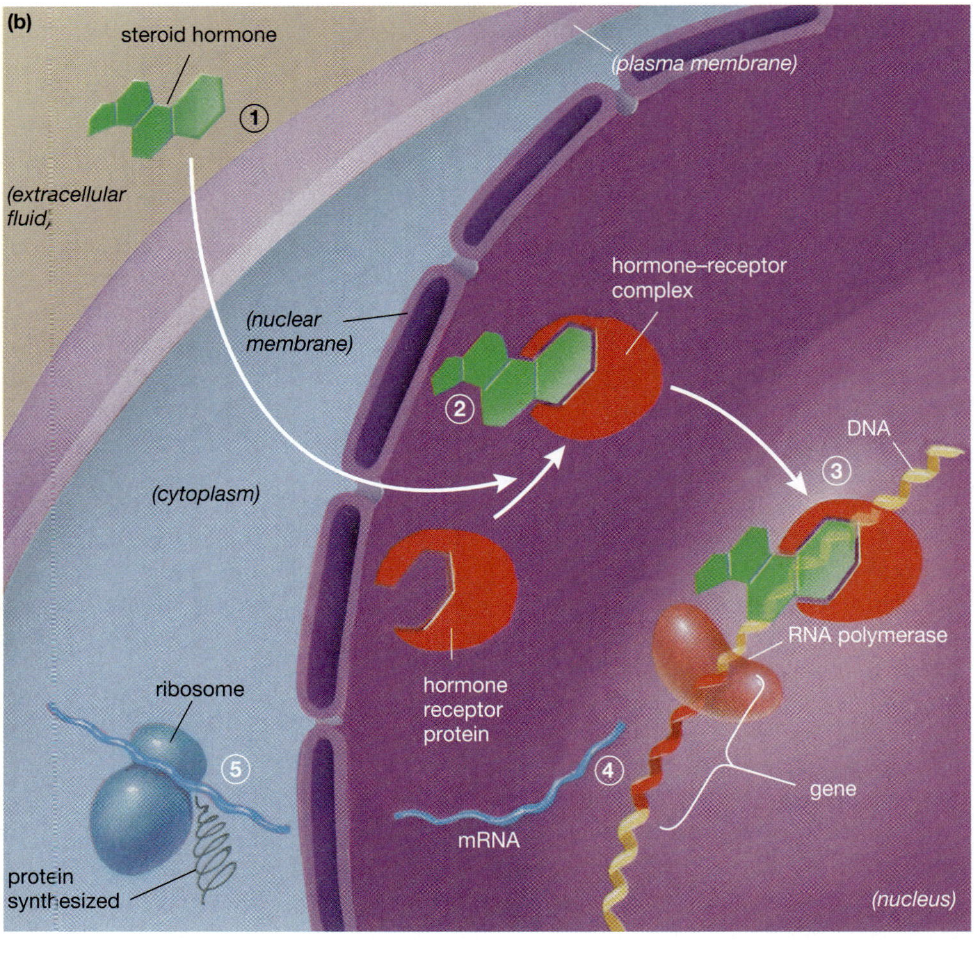

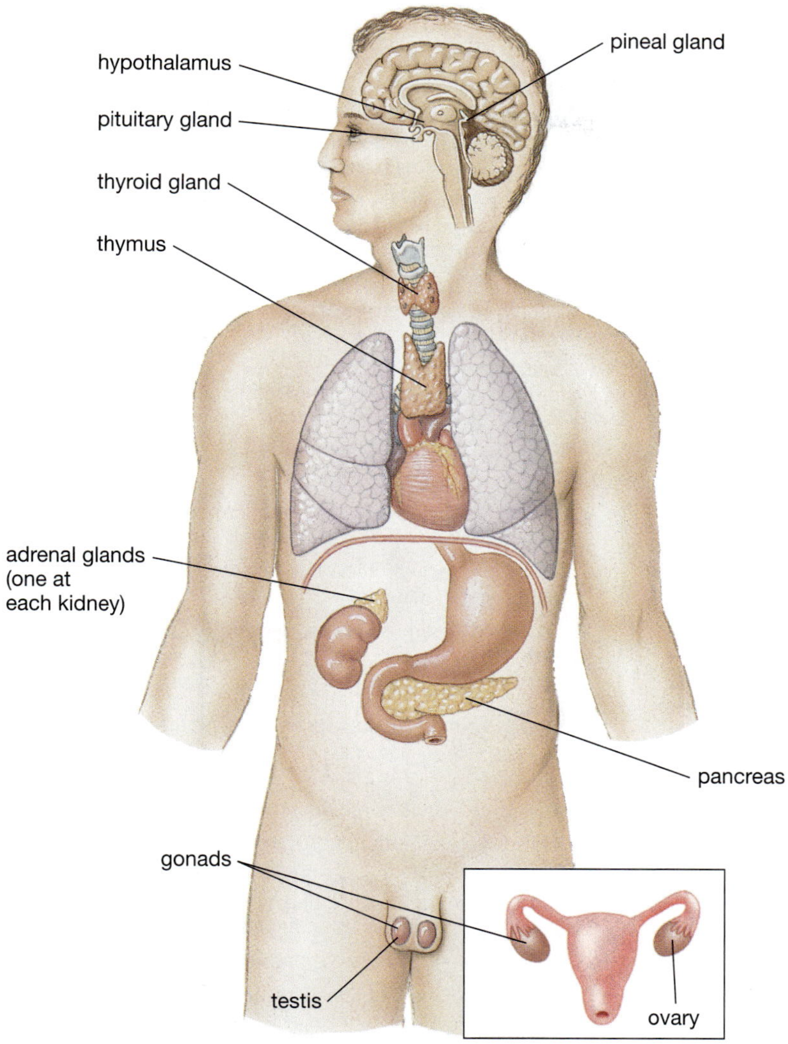

Figure 23-3 Major mammalian endocrine glands

MEDIATUTOR 23.1 The Endocrine System: Anatomy

3 What Are the Functions of the Hypothalamus and Pituitary Gland?

If the endocrine system is the body's postal service, then the hypothalamus is the main post office and the pituitary gland is the administrative headquarters. Together, these structures coordinate the action of many key hormonal messaging systems. The **hypothalamus** is a part of the brain that contains clusters of specialized nerve cells called **neurosecretory cells**. Neurosecretory cells synthesize peptide hormones, store them, and release them when stimulated. The **pituitary gland** is a pea-sized gland that dangles from the hypothalamus by a stalk (see Fig. 23-3). Anatomically, the pituitary consists of two distinct parts: the anterior pituitary and the posterior pituitary. The anterior pituitary is a true endocrine gland, composed of several types of hormone-secreting cells enmeshed in a network of capillaries. The posterior pituitary, however, is derived from an outgrowth of the hypothalamus. The release of hormones from both parts of the pituitary is controlled by the hypothalamus.

Hormones from the Hypothalamus Control the Anterior Pituitary

Neurosecretory cells of the hypothalamus produce at least nine peptide hormones that regulate the release of hormones from the anterior pituitary. These hormones are synthesized in nerve cells in the hypothalamus, secreted into a capillary bed in the lower portion of the hypothalamus, and travel a short distance through blood vessels to a second capillary bed that surrounds the endocrine cells of the anterior pituitary. There, they diffuse out of the capillaries and influence pituitary hormone secretion (Fig. 23-4).

The Anterior Pituitary Produces and Releases Several Hormones

The anterior pituitary produces several peptide hormones. Four of these regulate hormone production in other endocrine glands. *Follicle-stimulating hormone* (FSH) and *luteinizing hormone* (LH) stimulate the production of sperm and testosterone in males, and of eggs,

Table 23-1 Mammalian Endocrine Glands and Hormones

Endocrine Gland	Hormone	Type of Chemical	Principal Function
Hypothalamus (via posterior pituitary)	Antidiuretic hormone (ADH)	Peptide	Promotes reabsorption of water from kidneys; constricts arterioles
	Oxytocin	Peptide	In females, stimulates contraction of uterine muscles during childbirth, milk ejection, and maternal behaviors; in males, causes sperm ejection
Hypothalamus (to anterior pituitary)	Releasing and inhibiting hormones	Peptides	At least nine hormones; releasing hormones stimulate release of hormones from anterior pituitary; inhibiting hormones inhibit release of hormones from anterior pituitary
Anterior pituitary	Follicle-stimulating hormone (FSH)	Peptide	In females, stimulates growth of follicle, secretion of estrogen, and perhaps ovulation; in males, stimulates development of sperm
	Luteinizing hormone (LH)	Peptide	In females, stimulates ovulation, growth of corpus luteum, and secretion of estrogen and progesterone; in males, stimulates secretion of testosterone
	Thyroid-stimulating hormone (TSH)	Peptide	Stimulates thyroid to release thyroxine
	Adrenocorticotropic hormone (ACTH)	Peptide	Stimulates adrenal cortex to release hormones, especially glucocorticoids
	Growth hormone	Peptide	Stimulates growth, protein synthesis, and fat metabolism; inhibits sugar metabolism
	Prolactin	Peptide	Stimulates milk synthesis in and secretion from mammary glands
	Endorphins	Peptides	Reduce perception of pain
	Melanocyte-stimulating hormone (MSH)	Peptide	Promotes synthesis of brown skin pigment, melanin
Thyroid	Thyroxine	Amino acid derivative	Increases metabolic rate of most body cells; increases body temperature; regulates growth and development
	Calcitonin	Peptide	Inhibits release of calcium from bones
Pancreas	Insulin	Peptide	Decreases blood glucose levels by increasing uptake of glucose into cells and converting glucose to glycogen, especially in liver; regulates fat metabolism
	Glucagon	Peptide	Converts glycogen to glucose, raising blood glucose levels
Ovaries[a]	Estrogen	Steroid	Causes development of female secondary sexual characteristics and maturation of eggs; promotes growth of uterine lining
	Progesterone	Steroid	Stimulates development of uterine lining and formation of placenta
Testes[a]	Testosterone	Steroid	Stimulates development of genitalia and male secondary sexual characteristics; stimulates development of sperm
Adrenal medulla	Epinephrine (adrenaline) and norepinephrine (noradrenaline)	Amino acid derivative	Increase levels of sugar and fatty acids in blood; increase metabolic rate; increase rate and force of contractions of the heart; constrict some blood vessels
Adrenal cortex	Glucocorticoids	Steroid	Increase blood sugar; regulate sugar, lipid, and fat metabolism; anti-inflammatory effects
	Aldosterone	Steroid	Increases reabsorption of salt in kidney
	Testosterone	Steroid	Causes masculinization of body features, growth
		Other Sources of Hormones	
Pineal gland	Melatonin	Amino acid derivative	Regulates seasonal reproductive cycles and sleep–wake cycles; may regulate onset of puberty
Thymus	Thymosin	Peptide	Stimulates maturation of cells of immune system
Kidney	Renin	Peptide	Acts on blood proteins to produce hormone (angiotensin) that regulates blood pressure
	Erythropoietin	Peptide	Stimulates red blood cell synthesis in bone marrow
Heart	Atrial natriuretic peptide (ANP)	Peptide	Increases salt and water excretion by kidneys; lowers blood pressure

(continued at top of next page)

Table 23-1 (*Continued*)			
Endocrine Gland	**Hormone**	**Type of Chemical**	**Principal Function**
Digestive tract[b]	Secretin, gastrin, cholecystokinin, and others	Peptides	Control secretion of mucus, enzymes, and salts in digestive tract; regulate peristalsis
Fat cells	Leptin	Peptide	Regulates appetite; stimulates immune function; promotes blood vessel growth; required for onset of puberty

[a]See Chapter 25.

[b]See Chapter 21.

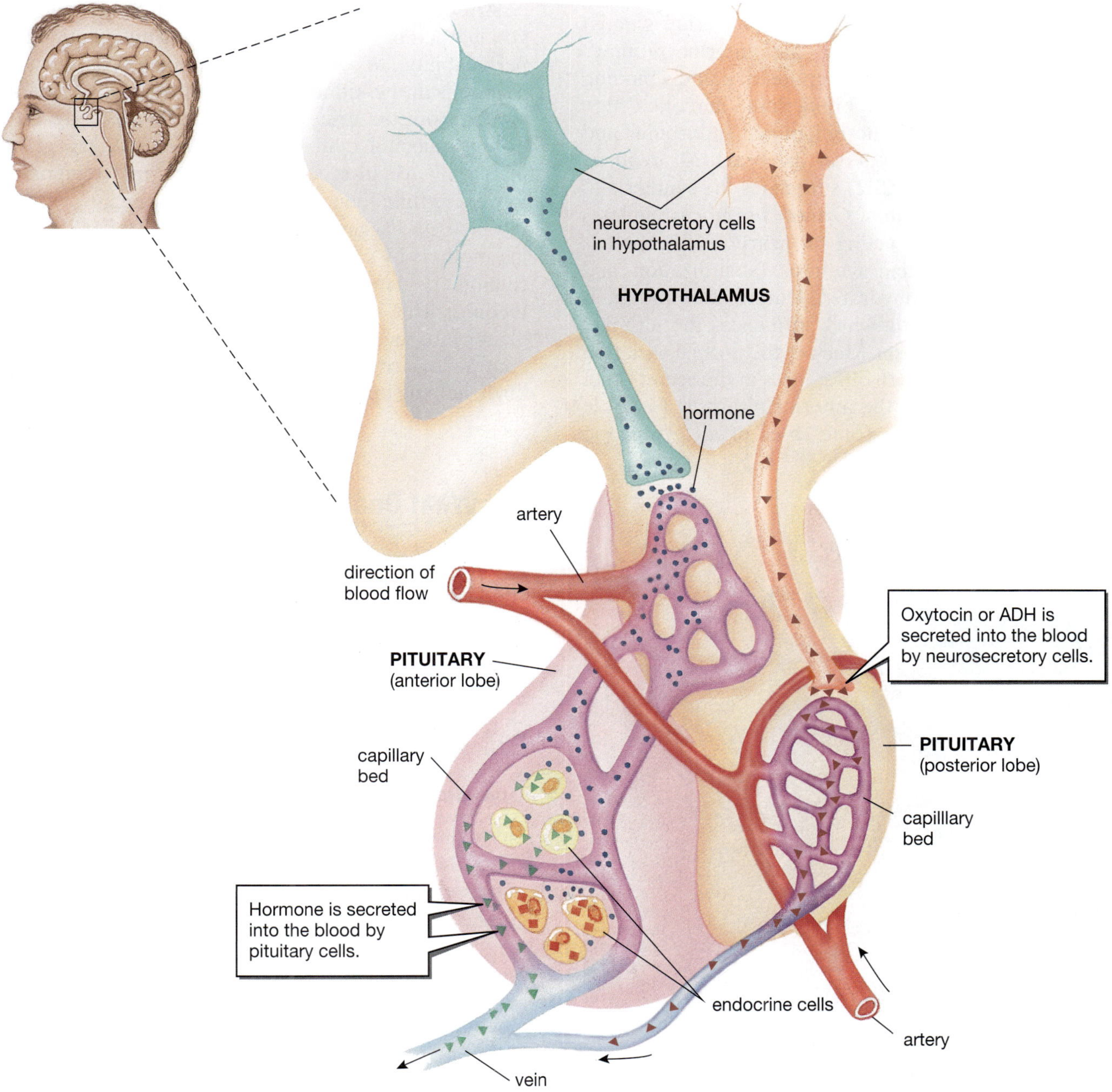

Figure 23-4 The hypothalamus controls the pituitary
Neurosecretory cells of the hypothalamus control hormone release in the anterior lobe of the pituitary by producing hormones (left). These hormones are secreted into a capillary network that carries them to the anterior pituitary. There, each hormone from the hypothalamus stimulates endocrine cells with appropriate receptors to secrete that hormone while leaving other types unaffected. The posterior lobe of the pituitary (right) is an extension of the hypothalamus. Neurosecretory cells in the hypothalamus have cell endings on a capillary bed in the posterior lobe, where the cells release oxytocin or antidiuretic hormone (ADH).

estrogen, and progesterone in females. We shall discuss the roles of FSH and LH in more detail in Chapter 25. *Thyroid-stimulating hormone* (TSH) stimulates the thyroid gland to release its hormones, and ACTH, or *adrenocorticotropic hormone*, causes the release of hormones from the adrenal cortex. We discuss the effects of thyroid and adrenal cortical hormones later in this chapter.

The remaining hormones of the anterior pituitary do not act on other endocrine glands. *Prolactin*, in conjunction with other hormones, stimulates the development of the mammary glands (which are exocrine glands) during pregnancy. *Endorphins* are anterior pituitary hormones that inhibit the perception of pain by binding to certain receptors in the brain. Some narcotic drugs, including morphine and heroin, mimic endorphins and bind to the same receptors in the brain. *Melanocyte-stimulating hormone* (MSH) stimulates the synthesis of the skin pigment melanin. *Growth hormone* regulates the body's growth by acting on nearly all the body's cells—increasing protein synthesis, fat utilization, and the storage of carbohydrates. During maturation, it stimulates bone growth, which influences the ultimate size of the adult organism. Much of the normal variation in human height is due to differences in the secretion of growth hormone from the anterior pituitary. Too little growth hormone causes some cases of *dwarfism*; too much can cause *gigantism* (Fig. 23-5).

Figure 23-5 When the anterior pituitary malfunctions
An improperly functioning anterior pituitary produces either too much or too little growth hormone. Too little causes dwarfism; too much causes gigantism.

The Posterior Pituitary Releases Hormones from the Hypothalamus

The posterior pituitary contains the endings of two types of neurosecretory cells whose cell bodies are located in the hypothalamus. These neurosecretory cell endings are enmeshed in a capillary bed into which they release hormones to be carried into the bloodstream (see Fig. 23-4). Two peptide hormones are synthesized in the hypothalamus and released from the posterior pituitary: *antidiuretic hormone* (ADH) and *oxytocin*.

Antidiuretic hormone helps prevent dehydration. As you learned in Chapter 21, ADH acts by increasing the permeability to water of the collecting ducts of nephrons in the kidney, which causes water to be reabsorbed from the urine and retained in the body. Interestingly, alcohol inhibits the release of ADH, greatly increasing urination, so a beer drinker may temporarily lose more fluid than he or she has taken in.

Oxytocin triggers the "milk letdown reflex" in nursing mothers by causing muscle tissue within the breasts (mammary glands) to contract during lactation (breastfeeding). This reflex ejects milk from the saclike milk glands into the nipples (Fig. 23-6). Oxytocin also causes contractions of the muscles of the uterus during childbirth, helping expel the fetus from the womb.

4 What Does the Thyroid Gland Do?

Lying at the front of the neck, nestled just below the larynx (Fig. 23-7a), the **thyroid gland** produces two major hormones: *thyroxine* and *calcitonin*. Thyroxine, often referred to as "thyroid hormone," is an iodine-containing modified amino acid that raises the metabolic rate of most body cells. Calcitonin is a peptide important in calcium metabolism.

Thyroxine Influences Metabolism

Thyroxine influences most of the cells in the body, elevating their metabolic rate and stimulating the synthesis of enzymes that break down glucose and provide energy. These cellular effects cause the body to increase oxygen consumption and heart rate. Thyroxine's influence on metabolic rate helps regulate body temperature and response to stress. Exposure to cold, for example, greatly increases thyroid hormone production, which increases metabolic rate, warming the body.

In juvenile animals, including humans, thyroxine helps regulate growth by stimulating both metabolic rate and the development of the nervous system. Undersecretion of thyroxine early in life causes *cretinism*, a condition characterized by mental retardation and dwarfism. Fortunately, early diagnosis and thyroxine supplementation can prevent this condition. Conversely, over-secretion of thyroxine in developing vertebrates

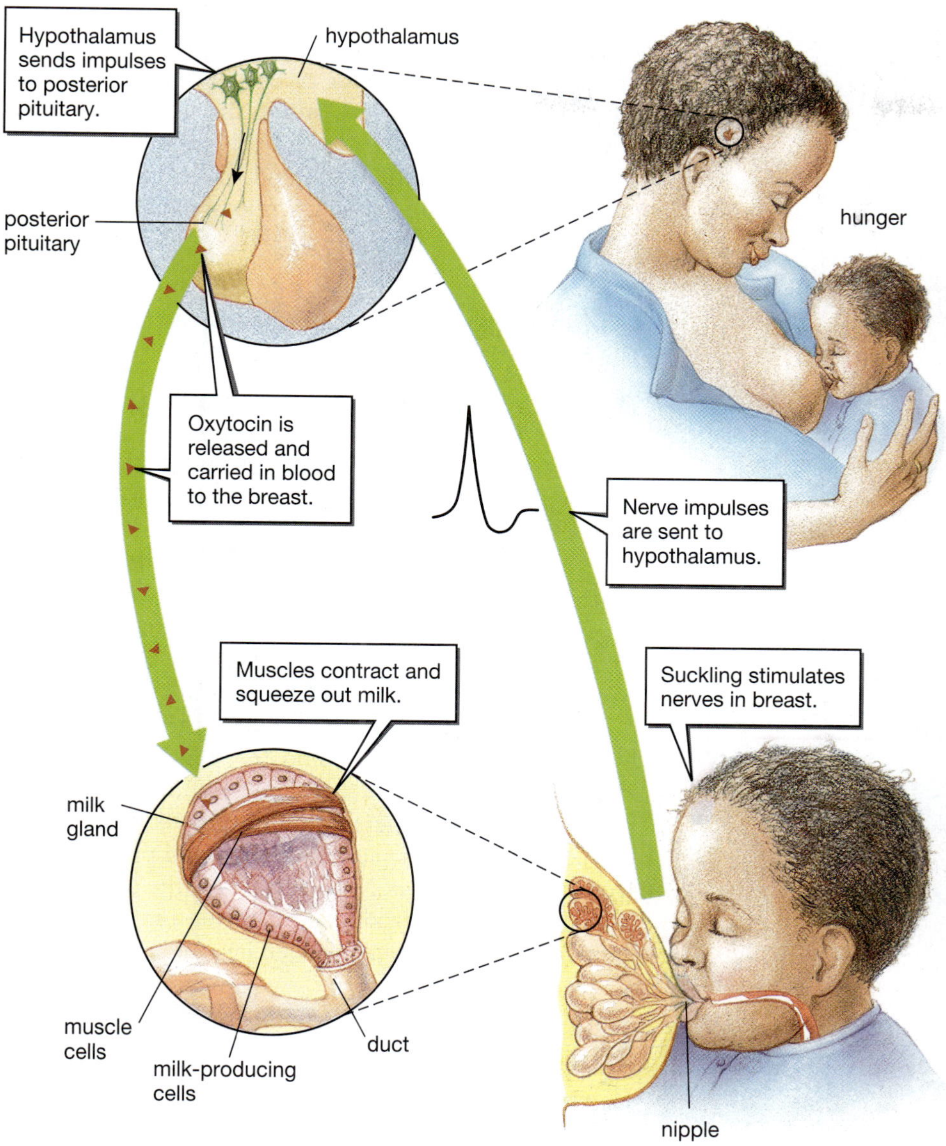

Figure 23-6 Hormones and breastfeeding
The control of milk letdown by oxytocin during breastfeeding is regulated by feedback between an infant and its mother. The cycle begins with the infant's suckling and continues until the infant is full and stops suckling. With the nipple no longer being stimulated, oxytocin release stops, the muscles relax, and milk flow ceases.

can trigger precocious development. In 1912, in one of the first demonstrations of hormone action, a physiologist discovered that thyroxine can induce early metamorphosis in tadpoles (see "Evolutionary Connections: The Evolution of Hormones").

Thyroxine Release Is Controlled by the Hypothalamus

Levels of thyroxine in the bloodstream are finely tuned by negative feedback. Thyroxine release is stimulated by thyroid-stimulating hormone (TSH) from the anterior pituitary, which in turn is stimulated by a hormone from the hypothalamus. The amount of TSH released from the pituitary is regulated by thyroxine levels in the blood: High levels of thyroxine inhibit the secretion of TSH, thus inhibiting further release of thyroxine from the thyroid.

A diet deficient in iodine can reduce the production of thyroxine and trigger a feedback mechanism that acts to restore normal hormone levels by dramatically increasing the number of thyroxine-producing cells. This compensating mechanism leads to excessive growth of the thyroid gland; the enlarged gland may bulge from the neck, producing a condition called goiter (Fig. 23-7b). Goiter was once common in some regions of the United States, but widespread use of iodized salt has now all but eliminated this condition in developed countries.

5 What Does the Pancreas Do?

The **pancreas** is a gland that produces both exocrine and endocrine secretions. The exocrine portion synthesizes digestive secretions that are released into the *pancreatic duct* and flow into the small intestine (see Chapter 21). The endocrine portion consists of clusters of cells called **islet cells**, which produce peptide hormones. One type of islet cell produces the hormone *insulin*; another type produces the hormone *glucagon*.

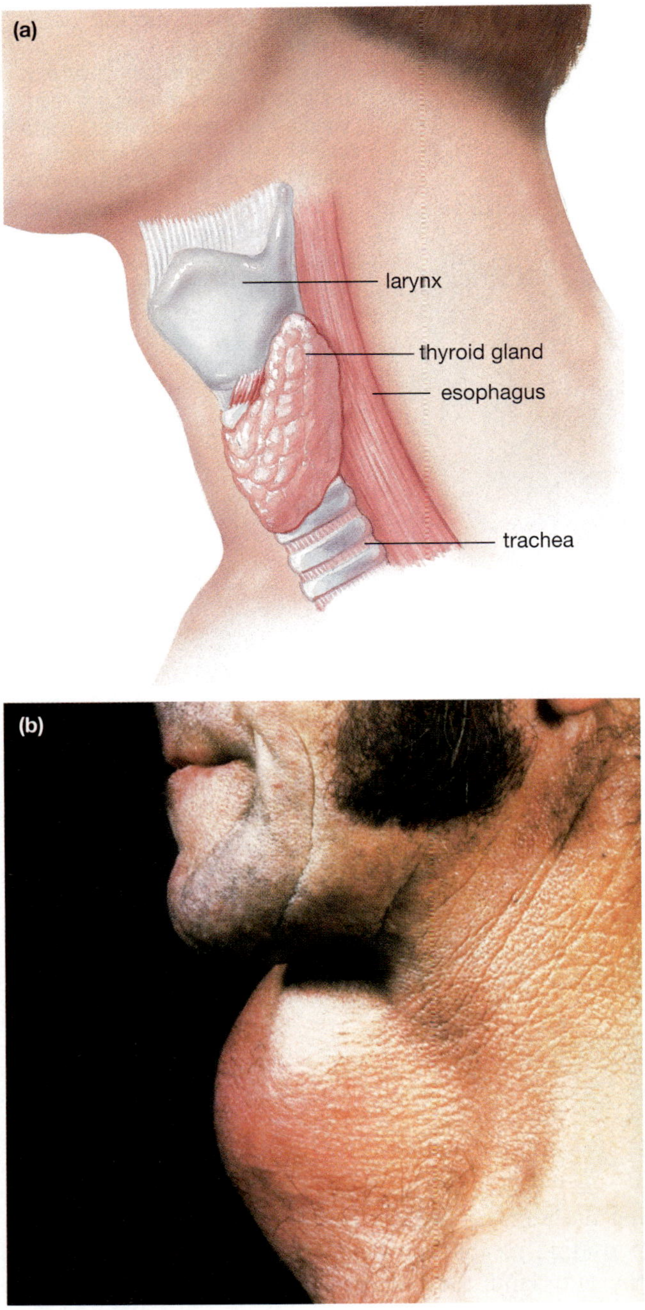

Figure 23-7 The thyroid gland
(a) The thyroid gland wraps around the front of the larynx in the neck. ***(b)*** Goiter, a condition in which the thyroid gland becomes greatly enlarged, is caused by an iodine-deficient diet.

Pancreatic Hormones Control Glucose Levels in the Blood

Insulin and glucagon work in opposition to regulate carbohydrate and fat metabolism: Insulin reduces the blood glucose level; glucagon increases it (Fig. 23-8). Together the two hormones help keep the blood glucose level nearly constant. When blood glucose rises (for example, after you've eaten), insulin is released. Insulin causes body cells to take up glucose and either metabolize it for energy or convert it to fat or glycogen (a starchlike molecule) for storage. When blood glucose levels drop (for example, after you've skipped breakfast or have run a 10-kilometer race), glucagon is released. Glucagon activates an enzyme in the liver that breaks down glycogen (which is stored primarily in the liver), releasing glucose into the blood. Glucagon also promotes lipid breakdown, which releases fatty acids that can be metabolized for energy.

Malfunctions of the Insulin Control System Harm Health

Defects in insulin production, release, or reception by target cells result in *diabetes mellitus*, a condition in which blood glucose levels are high and fluctuate wildly with sugar intake. The lack of functional insulin in diabetics causes the body to rely much more heavily on fats as an energy source, leading to high circulating levels of lipids, including cholesterol. Severe diabetes causes fat deposits in the blood vessels, resulting in high blood pressure and heart disease; diabetes is an important cause of heart attacks in the United States. The fatty deposits in small vessels can also damage the retina of the eye, leading to blindness, and the kidneys, leading to kidney failure. Until recently, the insulin supplements required to treat diabetes were formulated from insulin extracted from the pancreas of cows and pigs, obtained from slaughterhouses. Now, however, large quantities of human insulin are produced by bacteria into which the gene for human insulin has been inserted.

The Big Picture

Hormones play a key day-to-day role in the ongoing control mechanisms that maintain stability and homeostasis, but they are also called into action to initiate the crucial physiological changes that accompany reproduction and development.

6 What Are the Endocrine Functions of Sex Organs?

The sex organs do far more than produce sperm or eggs. The **testes**, in males, and **ovaries**, in females, are important endocrine organs. The testes secrete several steroid hormones. The most important of these is *testosterone*. The ovaries secrete two types of steroid hormones: *estrogen* and *progesterone*. The role of the sex hormones in development, the menstrual cycle, and pregnancy is discussed in Chapter 25.

Sex Hormone Levels Increase During Puberty

The sex hormones also play a key role in *puberty*, the phase of life that links immaturity and maturity. Puberty begins when, for reasons not fully understood, the hy-

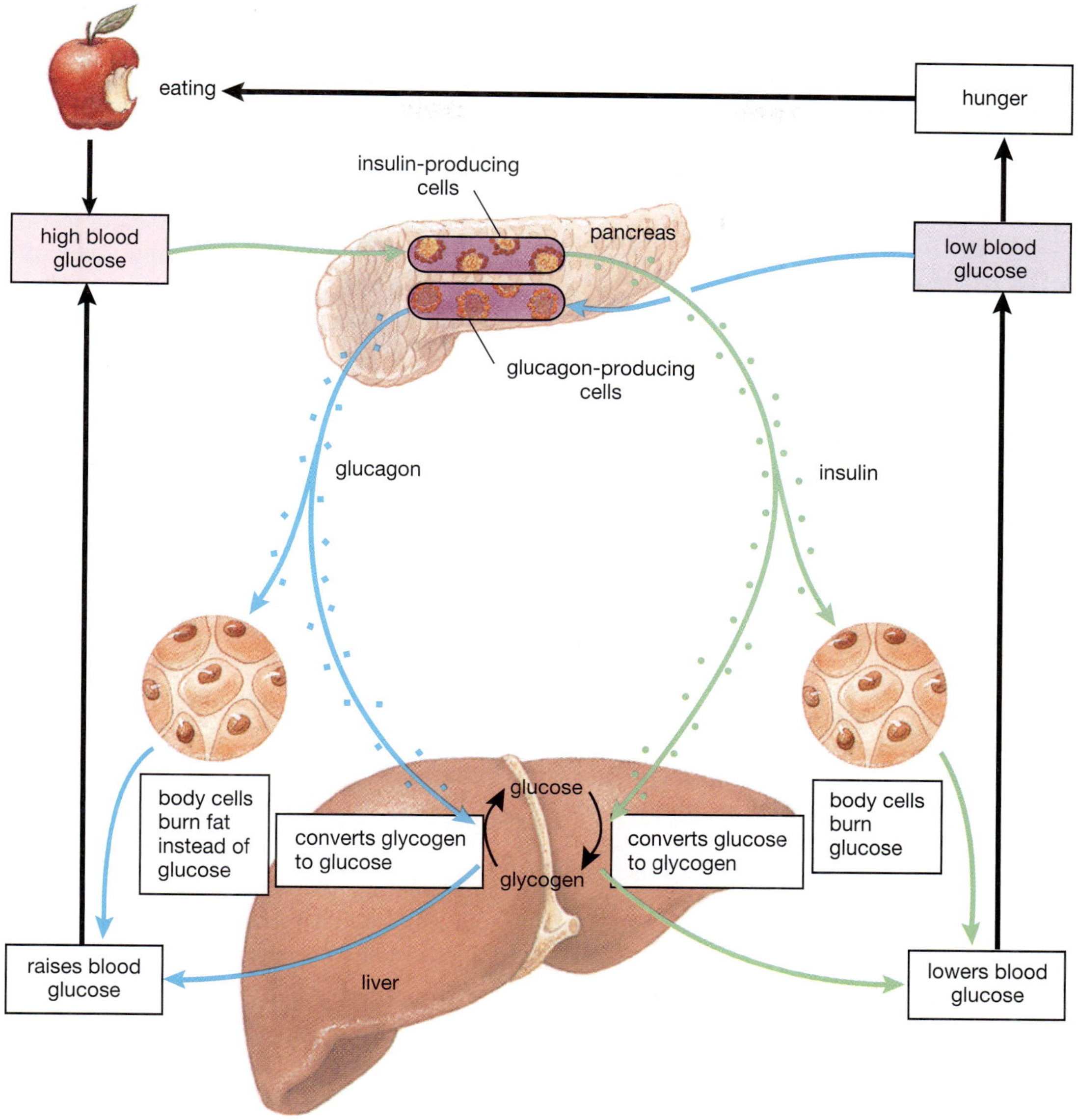

Figure 23-8 The pancreas controls blood glucose levels
The pancreatic islet cells contain two populations of hormone-producing cells: one producing insulin; the other, producing glucagon. These two hormones cooperate in a two-part negative-feedback loop to control blood glucose concentrations.

pothalamus starts to secrete increasing amounts of hormones that stimulate the anterior pituitary to secrete more lutenizing hormone (LH) and follicle-stimulating hormone (FSH) into the bloodstream. Both LH and FSH stimulate target cells in the testes or ovaries to produce higher levels of sex hormones. The resulting increase in circulating sex hormones ultimately affects tissues throughout the body that bear the appropriate receptors. Puberty also leads to the activation of the adult reproductive system, to the new growth underlying secondary sexual characteristics, and to the behavioral changes that make puberty such an interesting time for parents and teenagers.

Humans Develop External Indicators of Sexual Maturity

Mature mammals are capable of reproduction, and puberty encompasses the physiological changes that endow immature individuals with this reproductive capacity. In humans, the development of full reproductive competence is accompanied by a growth spurt and by the development of secondary sexual characteristics. Both sexes develop pubic and underarm hair; females develop breasts, while males acquire facial hair, larger muscles, and a prominent larynx ("Adam's apple"), which in turn lowers the voice. Although there is a surge of sex hormone production at puberty, sex hormones are present from the fetal stage onward. They influence brain development in both sexes, and they continue to have an effect on both behavior and cognitive processes throughout life.

7 What Do the Adrenal Glands Do?

Imagine how your body feels when you are startled, afraid, or angry. These physical reactions are the result of hormones produced by the adrenal glands, which act in conjunction with your nervous system. Like the pituitary gland and pancreas, the **adrenal glands** are two glands in one: the adrenal medulla and the adrenal cortex (Fig. 23-9).

The Hormones of the Adrenal Medulla Prepare the Body for Flight or Fight

The *adrenal medulla* is located in the center of each gland, and produces two hormones—*epinephrine* and *norepinephrine* (also called *adrenaline* and *noradrenaline*, respectively)—in response to stress. These hormones, which are amino acid derivatives, prepare the body for emergency action. They increase the heart and respiratory rates, cause blood glucose levels to rise, and direct blood flow away from the digestive tract and toward the brain and muscles. They also cause the air passages of the lungs to expand, allowing more efficient exchange of gases. For this reason, epinephrine is administered to asthmatics, whose airways become constricted during an attack. The adrenal medulla is activated by the nervous system, which prepares the body to respond to emergencies, as we shall describe in Chapter 24.

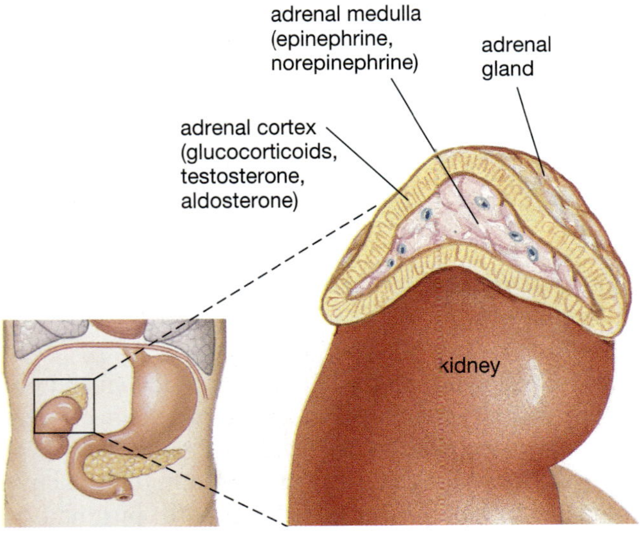

Figure 23-9 The adrenal glands
Atop each kidney sits an adrenal gland, which is a two-part gland composed of very dissimilar cells. The outer cortex secretes steroid homones; the inner medulla secretes epinephrine and norepinephrine.

The Hormones of the Adrenal Cortex Raise Blood Glucose in Response to Stress

The outer layer of the adrenal gland forms the *adrenal cortex*. The cortex secretes three types of steroid hormones, called *glucocorticoids*. These hormones help control glucose metabolism. Their release is stimulated by ACTH from the anterior pituitary. ACTH release is stimulated by hormones produced by the hypothalamus in response to stressful stimuli, including trauma, infection, or exposure to temperature extremes. In some respects, the glucocorticoids act similarly to glucagon, raising blood glucose levels by stimulating glucose production and promoting the use of fats instead of glucose for energy production.

The adrenal cortex also secretes the hormone *aldosterone*, which regulates the sodium content of the blood. Sodium ions, derived from salt in the diet, are the most abundant positive ions in blood and extracellular fluid. The sodium ion gradient across plasma membranes (high outside, low inside) is a factor in many cellular events, including the production of electrical signals by nerve cells. If blood sodium falls, the adrenal cortex releases aldosterone, which causes the kidneys and sweat glands to retain sodium. Then salt and other sources of dietary sodium, combined with aldosterone-induced sodium conservation, raise blood sodium levels, and shut off further aldosterone secretion.

8 What Are Some Other Sources of Hormones?

The **pineal gland** is located between the two hemispheres of the brain, just above and behind the hypothalamus (see Fig. 23-3). Named for its resemblance to a pine cone, the pineal gland is smaller than a pea. In 1646 philosopher René Descartes described it as "the seat of the rational soul." Since then, scientists have learned more about this organ, but many of its functions are still poorly understood. The pineal produces the hormone *melatonin*, an amino-acid derivative. Melatonin is secreted in a daily rhythm, which in mammals is regulated by light falling on the eyes. In some vertebrates, such as the frog, the pineal itself contains photoreceptive cells, and the skull above it is thin, so the pineal can detect sunlight and thus day length. By responding to day lengths characteristic of different seasons, the pineal appears to regulate the seasonal reproductive cycles of many mammals. Despite years of research, the function in humans of the pineal gland and melatonin is still unclear. One hypothesis is that the pineal gland and melatonin secretion influence sleep–wake cycles.

The **thymus** is located in the chest cavity behind the breastbone. In addition to producing white blood cells, the thymus produces the hormone *thymosin*, which stimulates the development of specialized white blood cells (T cells) that play an important role in the immune system (see Chapter 22). The thymus is extremely large in infants but, under the influence of sex hormones, begins decreasing in size after puberty.

Some organs whose major function is not endocrine nonetheless also produce horomones. For example, the kidney secretes many of the hormones that regulate blood volume and blood pressure. Similarly, the stomach and small intestine produce peptide hormones that help regulate digestion. Even the heart releases a peptide hormone that causes a reduction of blood volume by decreasing the release of both ADH and aldosterone.

Can fat be an endocrine organ? In 1995 researchers described the peptide hormone *leptin*, which is released by adipose (fat) cells. In an experiment, mice missing the gene for leptin became obese (Fig. 23-10), and leptin injections caused them to lose weight. The researchers hypothesized that adipose tissue, by releasing leptin, tells the body how much fat it has stored and therefore how much to eat. However, researchers have discovered additional functions for leptin and are finding leptin receptors in unexpected places, such as blood vessels and white blood cells. Leptin appears to stimulate the growth of new capillaries and to speed wound healing. It also stimulates the immune system and is required for the onset of puberty and the development of secondary sexual characteristics.

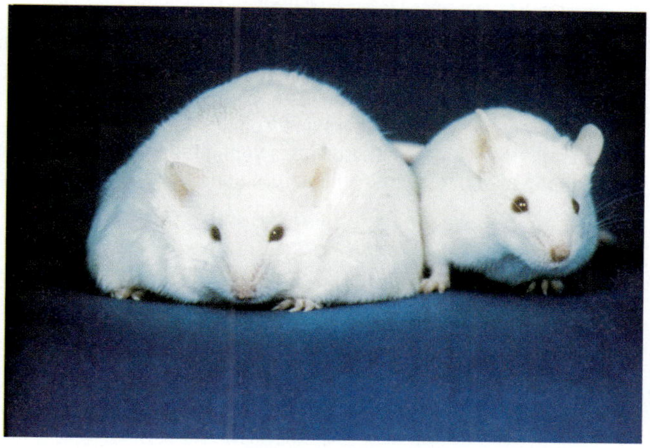

Figure 23-10 Leptin helps regulate body fat
The mouse on the left has been genetically engineered to lack the gene for the hormone leptin.

Evolutionary Connections

The Evolution of Hormones

Not long ago, vertebrate endocrine systems were considered unique to our phylum, and the endocrine chemicals were thought to have evolved only after the origin of the vertebrates. In recent years, however, physiologists have discovered that hormones are evolutionarily ancient. Insulin, for example, is found not only in vertebrates but also in protists, fungi, and bacteria, although research has not yet determined the function of insulin in most of those organisms. Protists also manufacture ACTH, even though they have no adrenal glands to stimulate. Yeasts have receptors for estrogen but no ovaries. Thyroid hormones have been found in certain invertebrates, such as worms, insects, and mollusks. Even among vertebrates, the effects of chemically identical hormones, secreted by the same glands, can vary dramatically from species to species. Let's look briefly at the diverse effects that the thyroid hormone thyroxine has on several different vertebrate species.

Some fish undergo radical physiological changes during their lifetimes. A salmon, for example, begins life in fresh water, migrates to the ocean, and returns to fresh water to spawn. In the stream where the salmon hatched, fresh water tends to enter the fish's tissues by osmosis; in salt water, osmosis causes the fish to lose water and become dehydrated. The salmon's migrations, therefore, require complete revamping of salt and water control. In salmon, one of the functions of thyroxine is to produce the metabolic changes necessary to go from life in streams to life in the ocean and back.

In amphibians, thyroxine has the dramatic effect of triggering metamorphosis. This effect was first demonstrated in 1912, in one of the first demonstrations of the action of any hormone. Tadpoles that were fed minced horse thyroid metamorphosed prematurely into miniature adult frogs, showing that some component of the thyroid gland (later determined to be thyroxine) stimulates metamorphosis (Fig. 23-11). This function of thyroxine has had interesting effects on the evolution of some species. For example, in high mountain lakes in Mexico, where the water is deficient in the iodine needed to synthesize thyroxine, natural selection has produced a species of salamander that reproduces while still in its juvenile form.

Thyroxine also regulates seasonal molting in many vertebrates. From snakes to birds to the family dog, surges of thyroxine stimulate the shedding of skin, feathers, or hair. In humans (who don't migrate regularly, metamorphose, or molt), thyroxine regulates growth and metabolism.

The presence of hormones across a wide range of species demonstrates that the use of chemicals to regulate cellular activity evolved early in the history of life. Living things use a relatively small number of different chemicals to coordinate activities within single cells and among groups of cells. Many of these chemicals were presumably present in an ancient common ancestor of today's species, but different branches of the tree of life have evolved different systems for delivering the chemicals and different responses to them.

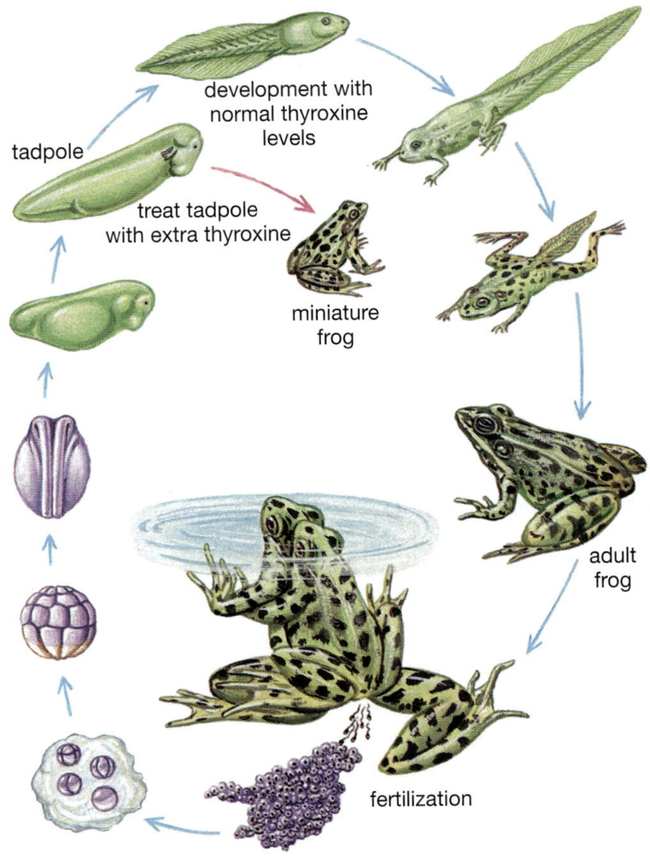

Figure 23-11 Thyroxine controls metamorphosis in amphibians The life cycle of the frog begins with fertilization of the eggs (bottom). The fertilized eggs develop into an aquatic, fishlike tadpole, which grows and ultimately metamorphoses into an adult frog. Metamorphosis is triggered by a surge of thyroxine from the tadpole's thyroid gland. If injected with extra thyroxine, a young tadpole will metamorphose ahead of schedule into a miniature adult frog.

CASE STUDY REVISITED

Gender Bending

The chemicals responsible for the feminization of male alligators in Lake Apopka and for other gender-bending effects are collectively known as endocrine disruptors. These chemicals interfere with the normal action of sex hormones in at least two ways. They can block access to receptors, preventing cells from responding to a hormone signal, or they can mimic hormones by binding to receptors, causing the cells to respond as if to hormone molecules.

Numerous chemicals have been shown in the laboratory to be endocrine disruptors. Pesticide residues and chemicals used in plastics are especially likely to block or mimic sex hormone activity. Unfortunately, these chemicals are widespread in the environment and tend to persist for quite some time, both in the environment and in animal bodies.

Can endocrine disruptors in the environment harm us? At this point, no one can say for sure. Scientists have begun to test some suspected endocrine disruptors on laboratory animals, and have shown, for example, that turtle or alligator eggs exposed to estrogen-mimicking chemicals almost always develop into females, rarely into males. Still, even if particular chemicals are ultimately proven to have hormone-like effects in living animals, it remains unclear if such effects would necessarily be harmful to health.

Much of the current concern about the health effects of endocrine disruptors stems from earlier experience with diethylstilbestrol (DES), an estrogen-mimic that was used as a drug. In the 1950s and '60s, DES was widely administered to pregnant women to prevent miscarriages. Unfortunately, children of women who took the drug have increased rates of reproductive cancers and other disorders of the reproductive system. Subsequent experimental studies of mice have shown DES to be a potent carcinogen and disrupter of normal development.

The harmful effects of DES have led some observers to suggest that the widespread endocrine disruptors in the environment might be responsible for such phenomena as an increase in the rate of testicular cancer and a decrease in sperm counts in some countries. It is difficult to know, however, whether findings about DES can be extended to the case of environmental estrogen mimics, because DES is a far more potent estrogen-mimic than any known environmental contaminant. And we're exposed to environmental contaminants in doses far smaller than those experienced in utero by the "DES babies." Still, the experience with DES was frightening enough that many scientists believe that we should take steps to control further proliferation of the endocrine-disrupting chemicals that now pervade our environment. It's true that we don't yet understand exactly how environmental endocrine disrupters affect our bodies, and we certainly don't yet know the exact nature of the health risk. The alligators of Lake Apopka, however, suggest that caution may be the best strategy.

Summary of Key Concepts

1) What Are the Characteristics of Animal Hormones?

A hormone is secreted in one part of the body and transported in the bloodstream to another part of the body, where it affects specific target cells. Hormones are synthesized either from amino acids (amino-acid-based hormones and peptides) or from lipids (steroids).

Most hormones act on their target cells in one of two ways: (1) Peptide hormones and amino acid derivatives bind to receptors on the surfaces of target cells and activate intracellular second messengers, such as cyclic AMP. The second messengers then alter the metabolism of the cell. (2) Steroid hormones can diffuse through the plasma membranes of their target cells and bind with receptors in the nucleus or cytoplasm. In the nucleus, the hormone–receptor complex promotes the transcription of specific genes.

Hormone action is commonly regulated through negative feedback, a process in which a hormone causes changes that inhibit further secretion of that hormone.

2) What Are the Structures of the Mammalian Endocrine System?

Hormones are produced by endocrine glands, which are clusters of cells embedded within a network of capillaries. Hormones are secreted into the extracellular fluid and diffuse into the capillaries. The major endocrine glands of the human body are the hypothalamus–pituitary complex, the thyroid gland, the pancreas, the sex organs, and the adrenal glands.

3-7) The Hypothalamus and Pituitary Gland, the Thyroid Gland, the Pancreas, the Sex Organs, and the Adrenal Glands

The functions of the hormones released by the major endocrine glands of the human body are summarized in Table 23-1.

8) What Are Some Other Sources of Hormones?

Other structures that produce hormones include the pineal gland, thymus, kidneys, heart, stomach, small intestine, and fat cells.

Key Terms

adrenal gland *p. 450*
duct *p. 441*
endocrine gland *p. 440*
endocrine system *p. 440*
exocrine gland *p. 441*
hormone *p. 440*
hypothalamus *p. 443*
islet cell *p. 447*
neurosecretory cell *p. 443*
ovary *p. 448*
pancreas *p. 447*
peptide hormone *p. 440*
pineal gland *p. 450*
pituitary gland *p. 443*
second messenger *p. 441*
steroid hormone *p. 440*
target cell *p. 440*
testis *p. 448*
thymus *p. 451*
thyroid gland *p. 446*

Thinking Through the Concepts

Multiple Choice

1. *Steroid hormones*
   a. alter the activity of genes
   b. trigger rapid, short-term responses in cells
   c. work via second messengers
   d. cannot cross the plasma membrane
   e. bind to cell-surface receptors

2. *Examples of posterior pituitary hormones are*
   a. FSH and LH
   b. prolactin and thyroxine
   c. secretin and cholecystokinin
   d. melatonin and prostaglandin
   e. ADH and oxytocin

3. *Negative feedback to the hypothalamus controls the level of _________ in the blood.*
   a. thyroxine
   b. estrogen
   c. glucocorticoids
   d. insulin
   e. all of the above

4. *The primary targets for FSH are cells in the*
   a. hypothalamus
   b. ovary
   c. thyroid gland
   d. adrenal medulla
   e. pituitary gland

5. *The ovary is a source of*
   a. thyroxine and antidiuretic hormone
   b. calcitonin and oxytocin
   c. estrogen and progesterone
   d. endorphins and epinephrine
   e. glucagon and glucocorticoids

? Review Questions

1. What are the three types of molecules used as hormones in vertebrates? Give an example of each.
2. What is the difference between an endocrine gland and an exocrine gland? Which type releases hormones?

3. When peptide hormones attach to target cell receptors, what cellular events follow? How do steroid hormones behave?
4. Diagram the process of negative feedback, and give an example of it in the control of hormone action.
5. What are the major endocrine glands in the human body, and where are they located?
6. Describe the structure of the hypothalamus–pituitary complex. Which pituitary hormones are neurosecretory? What are their functions?
7. Name the hormones of the anterior pituitary, and give one function of each.
8. Describe how the hormones of the pancreas act together to regulate the concentration of glucose in the blood.
9. Compare the adrenal cortex and adrenal medulla by answering the following questions: Where are they located within the adrenal gland? Which hormones do they produce? Which organs do their hormones target? What homeostatic processes regulate blood levels of the respective hormones?

Applying the Concepts

1. A student decides to do a science project on the effect of the thyroid gland on frog metamorphosis. She sets up three aquaria with tadpoles. She adds thyroxine to the water of one, the drug thiouracil to a second, and nothing to the third. Thiouracil reacts with thyroxine in tadpoles to produce an ineffective compound. Assuming that the student uses physiologically appropriate concentrations, predict what will happen.
2. If you were obese, would you consider injections of leptin? What pros and cons can you think of? Defend your decision.
3. Suggest a hypothesis about the endocrine system to explain why many birds lay their eggs in the spring and why poultry farmers keep lights on at night in their egg-laying operations.
4. Anabolic steroids, used by risk-taking athletes and bodybuilders, are chemically related to testosterone. They increase bone mass and muscle mass and seem to improve athletic performance. But anabolic steroids can cause liver problems, heart attacks, strokes, testicular atrophy, and personality changes in males. Females on anabolic steroids also have liver and circulatory problems. In addition, their voices deepen, their bodies develop more hair, and their menstrual cycles are disturbed. Explain how the same compound can produce different effects in males and females.
5. Some parents who are interested in college sports scholarships for their children are asking physicians to prescribe growth hormone treatments. Farmers also have an economic incentive to treat cows with growth hormone, which can now be produced in large quantities by genetic-engineering techniques. What biological and ethical problems do you foresee for parents, children, physicians, coaches, college scholarship boards, food consumers, farmers, the U.S. Food and Drug Administration, and biotechnology companies?
6. Argue for and against banning or restricting the use of common endocrine disruptors, such as plasticizers and certain pesticides. What compromises can you suggest?

Group Activity

In 1995 researchers discovered that if young mice are injected with a protein called leptin, they begin puberty earlier than do mice that received no leptin. On the basis of this finding, the researchers believe that the activation of genes that encode leptin may be the first step in the onset of puberty. Receptors for leptin have been found on cells in the hypothalamus, but they've also been found on cells in the reproductive organs of both males and females. Which receptors are responsible for leptin's role in initiating puberty? Form a group of four students to begin answering this question. Two members of your group should develop a hypothetical mechanism by which leptin could initiate puberty by binding with receptors on the hypothalamus. Your proposed mechanism should be as detailed as possible (specify all glands, organs, and hormones that are involved, and make a drawing that shows all pathways and regulatory feedback mechanisms). The other two members of your group should develop an equally detailed hypothesis of how leptin could initiate puberty by binding to cells in the reproductive system. After your two competing hypotheses are complete, reconvene the group. Discuss the strengths and weaknesses of each hypothesis, and agree on one key experiment that, if done, would enable you to determine which hypothesis is most likely to be correct.

For More Information

Atkinson, M., and MacLaren, N. "What Causes Diabetes?" *Scientific American*, July 1991. An in-depth look at what's behind a common but serious disease.

Beardsley, T. "Melatonin Mania." *Scientific American*, April 1996. The hormone melatonin has been touted as a treatment for ailments ranging from insomnia to cancer. What do scientists think of the evidence?

Berridge, M. J. "The Molecular Basis of Communication Within the Cell." *Scientific American*, October 1985. Both the "classical" cyclic AMP and more recently discovered second messengers convey information from cell-surface receptors to DNA and cellular metabolism.

Gold, C. "Hormone Hell." *Discover*, September 1996. Environmental pollutants that mimic the chemical activity of hormones may be playing havoc with animal (including human) physiology.

McLachlan, J., and Arnold, S. "Environmental Estrogens." *American Scientist*, September–October 1996. A sober evaluation of the effects of estrogen-like substances in the environment, written by two of the leading researchers in the field.

Raloff, J. "Macho Waters: Some River Pollution Spawns Body-Altering Steroids." *Science News*, January 6, 2001. A summary of research on gender-bending pollutants in streams near paper mills.

Sapolsky, R. "Testosterone Rules." *Discover*, March 1997. An engaging inquiry into the question of whether testosterone causes aggression.

Snyder, S. H. "The Molecular Basis of Communication Between Cells." *Scientific American*, October 1985. The similarities and differences between neural and hormonal control systems in the body.

Vogel, S. "The Mouse on the Left Needs Leptin." *Discover*, January 1996. The hormone leptin regulates appetite and fat deposition in mice. Can it become a weight-loss treatment for humans?

Answers to Multiple-Choice Questions

1. a 2. e 3. e 4. b 5. c

MediaTutor

CD Activities

Activity 23.1: Endocrine System Anatomy

Estimated time: 10 minutes

The organs of the endocrine system secrete hormones that act as chemical messengers to cells that may be some distance away. The effect of hormones on distant cells may be prolonged and irreversible, or transient and reversible. In this tutorial, you will explore the anatomy of the human endocrine system and the hormones that are produced by the glands that make up this physiological system.

Activity 23.2: Modes of Action of Hormones

Estimated time: 5 minutes

Hormones exert their characteristic effects on target cells by altering the activity of those cells. The way in which this task is accomplished depends on the type of hormone and the type of cell that is responsive to that hormone. In this tutorial, you will explore two different modes of action for hormones through an interactive animation.

Start the MediaTutor Student CD-ROM and enter the activity number in the Quick Search box to be taken directly to that activity.

Web Investigation

Gender Bending

Estimated time: 15 minutes

"Case Study: Gender Bending" describes how the presence of certain chemicals in the environment can disrupt the normal development of organisms. These chemicals can interfere with the normal operation of the endocrine system by either blocking or mimicking the normal action of certain sex hormones. This activity will explore the function of the endocrine system and its role in regulating development.

Go to http://www.prenhall.com/audesirk3, the Audesirk Companion Web site. Select Chapter 23 and Web Investigation to begin.

"Know thyself."

—Inscribed above the entrance to the Temple of Apollo, home of the Oracle of Apollo at Delphi

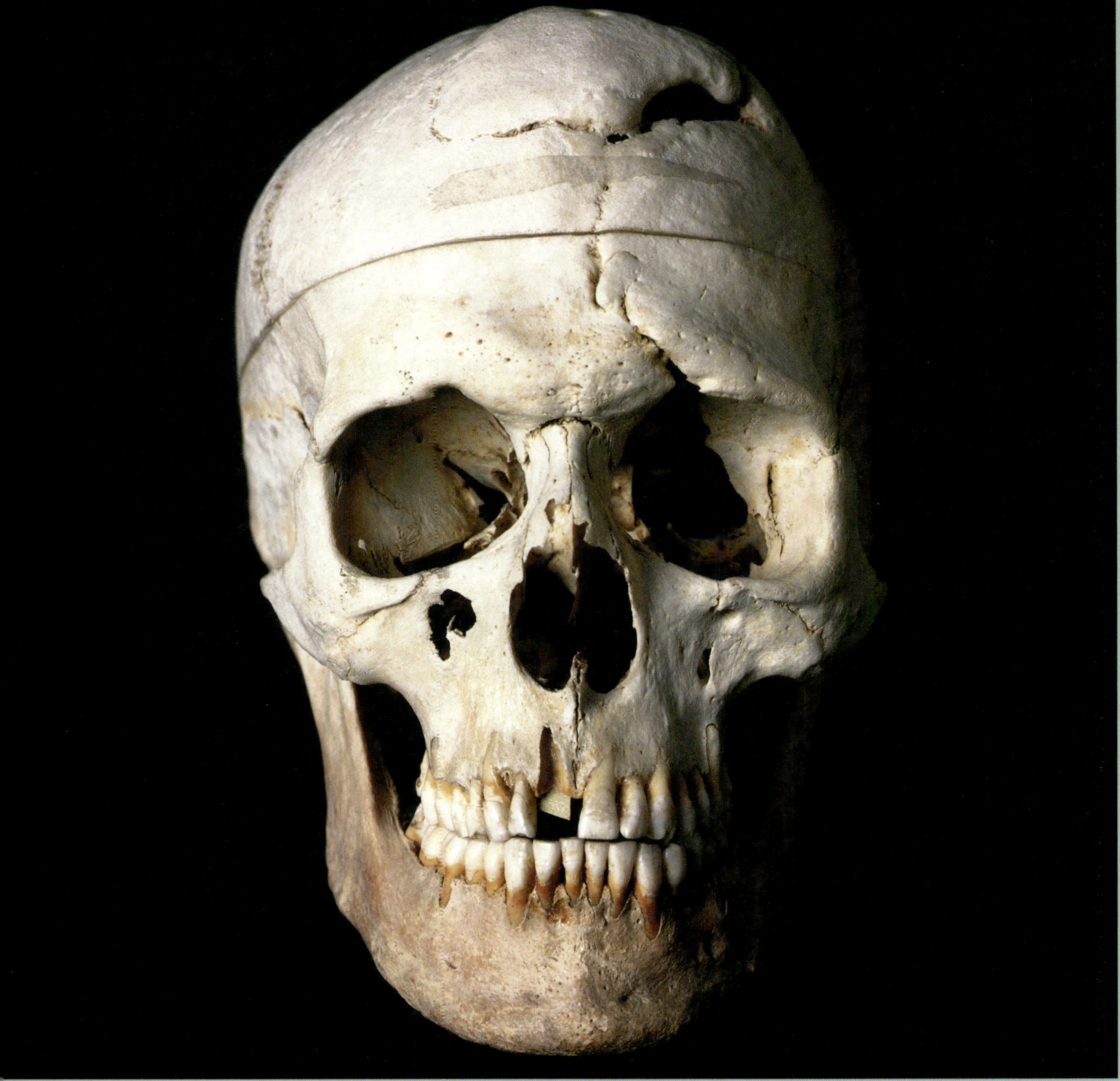

The damaged skull of Phineas Gage, a nineteenth-century construction foreman who survived severe brain trauma, is a reminder of the role that accidental injury has played in advancing our understanding of how the brain and nervous system function.

24 The Nervous System and the Senses

AT A GLANCE

CASESTUDY

A Bad Day for Phineas Gage

On a warm September day in 1848, a railway construction foreman named Phineas Gage was busy planting explosive charges in the ground. Each charge had to be tamped down with a metal rod. As Gage tamped one of the charges, it exploded prematurely and blew the heavy tamping iron upward into his face. The force of the explosion was so great that the rod, 3 feet long and more than an inch in diameter at its widest point, traveled right through Gage's head and landed 10 yards behind him. Miraculously, the injury did not kill Gage. He lived, and was even able to move and speak immediately after the accident.

A local physician cleaned and dressed Gage's gaping wound. The patient was reported to be calm and lucid throughout the treatment, asking when he would be able to return to work. Despite the severity of the wound, it healed within weeks and Gage quickly recovered his health.

That Gage could survive and function despite the destruction of so much brain tissue was seen at the time as evidence that the brain is an undifferentiated organ like the liver or pancreas. If a portion of an undifferentiated organ is destroyed, the remainder can still accomplish the organ's functions. This simple interpretation of the brain, however, became impossible to defend as additional facts emerged about the effects of the wound on Phineas Gage.

Although Gage had survived, he was far from unchanged. In fact, his personality had undergone a radical transformation after the accident. Before the accident Gage had been a polite, hard-working, and respectful man, diligent and dependable at work. After his unfortunate encounter with the tamping rod, however, Gage became rude, hostile, and sullen—a social misfit incapable of inhibiting his aggressive impulses.

The strange case of Phineas Gage provided strong evidence that different mental functions are localized in different areas of the brain. A severe brain injury changed Gage's emotions and social behavior, even though other mental activities such as speech, memory, and motor control remained intact. What part of Gage's brain was responsible for the dramatic changes in his character and personality? This question can best be answered after we describe some basic elements of nervous system structure and function. ■

1 How Do Nerve Cells Work?

Our study of the nervous system begins with the individual nerve cell, or **neuron**. As the fundamental unit of the nervous system, each neuron must perform four functions:

- Receive information from the internal or external environment or from other neurons
- Integrate the information it receives and produce an appropriate output signal
- Conduct the signal to its terminal, which may be some distance away
- Transmit the signal to other neurons or to glands or muscles

Although neurons vary enormously in structure, a typical vertebrate neuron has four distinct regions, each of which performs one the functions listed above. These regions are the dendrites, the cell body, the axon, and the synaptic terminals.

Dendrites Respond to Stimuli

Dendrites, branched tendrils that extend outward from the cell body, perform the "receive information" function (Fig. 24-1). Their branched shape provides a large surface area for receiving signals, which may come from other neurons or from the external environment. The signals to which dendrites respond may be chemicals released by other neurons or stimuli from the environment, such as pressure, odor, light, or heat.

The Cell Body Receives Signals from Dendrites

When dendrites are stimulated, electrical signals travel down them and converge on the neuron's **cell body**, which performs the "integrate information" function (see Fig. 24-1). To carry out this function, the cell body adds up the various signals from the dendrites. If the sum of these signals is sufficiently large and positive, the cell body produces an **action potential**, the electrical output signal of the neuron. The cell body also carries on the routine activities common to most body cells, such as synthesizing complex molecules and coordinating the metabolic activities of the cell.

Axons Are the Neuron's Distribution Lines

In a typical neuron, a long, thin fiber called an **axon** extends outward from the cell body. The axon performs the "conduct the signal" function, conducting action potentials from the cell body to the synaptic terminal located at its far end; an axon may branch near its end and have several synaptic terminals (see Fig. 24-1). In vertebrates, axons emerge from the brain and spinal cord and extend out to all regions of the body. Single axons may stretch from your spinal cord to your toes, a distance of about 3 feet, making neurons the longest cells in the body. Axons are normally bundled together into **nerves**, much like bundles of wires in an electrical cable. However, unlike wires, in which energy is lost along the way as electricity moves from source to destination, axons conduct action potentials undiminished from the cell bodies to the synaptic terminals. Some axons are wrapped with insulation called *myelin*, which allows more rapid conduction of the electrical signal. Myelin is formed from non-neuronal cells that wrap around the axon (see Fig. 24-1).

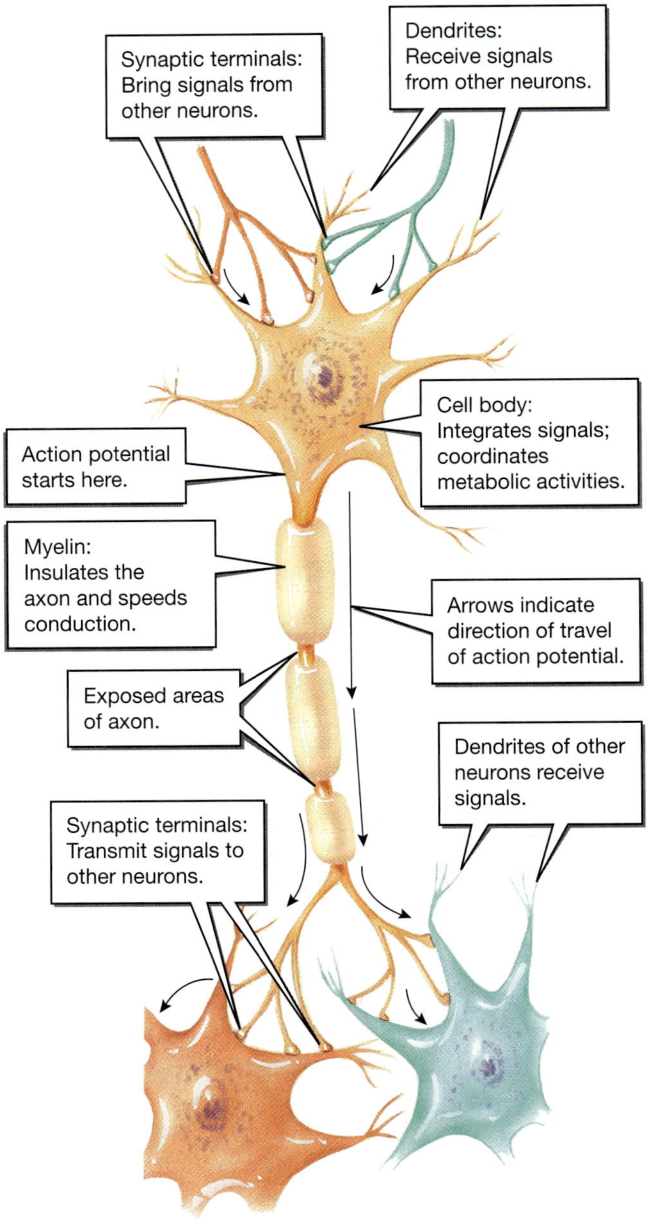

Figure 24-1 A neuron, showing its specialized parts and their functions

Signals Are Transmitted at Synaptic Terminals

The transmission of signals to other cells occurs at **synaptic terminals**, swellings at the ends of axons. Most synaptic terminals contain a **neurotransmitter**, a chemical that is released when an action potential reaches the terminal. The synaptic terminals of a neuron may transmit signals to a gland, a muscle cell, or the dendrites or cell body of another neuron. The output of the first cell becomes the input to the second cell. The site at which synaptic terminals communicate with other cells is called the **synapse**.

2 How Do Nerve Cells Produce and Transmit Signals?

The electrical charge inside a living cell differs from the charge outside the cell, and this difference gives rise to a voltage difference, or *potential*, across the plasma membrane. In an unstimulated, inactive neuron, this potential is called the **resting potential**. The inside of the neuron is negatively charged compared to the outside, so the resting potential in the cell is negative; its magnitude can be measured with special instruments and ranges from –40 to –90 millivolts (mV; thousandths of a volt).

The Potential Inside a Neuron Changes During an Action Potential

When a neuron is stimulated, the negative potential inside the neuron changes; it can become more negative or less negative. If the potential becomes sufficiently less negative that it reaches a level called **threshold**, an action potential is triggered (Fig. 24-2). During an action potential, the potential inside the neuron rises rapidly to about +50 mV. The positive charge of the action potential flows rapidly down the axon to the synaptic terminal. Action potentials last a few milliseconds (thousandths of a second), after which the cell's negative resting potential is restored.

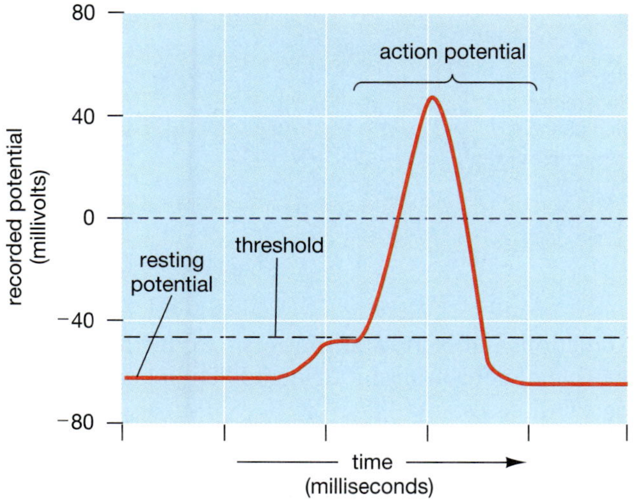

Figure 24-2 An action potential makes the inside of a neuron positive
A recording (by oscilloscope) of the electrical events in a neuron. The resting potential is about 60 millivolts negative with respect to the outside. When the neuron is stimulated to threshold, the inside rapidly becomes positively charged. Shortly thereafter, the resting potential is reestablished.

Neurons Communicate at Synapses

Once an action potential reaches the synaptic terminal of a neuron, the signal is transmitted across a synapse to another cell, such as another neuron, a muscle cell, or a gland cell. For simplicity, we will describe only neuron-to-neuron synapses. There are many synapses on the dendrites and the cell bodies of neurons.

A synapse includes the synaptic terminal of one neuron, an area of specialized plasma membrane on the surface of a second neuron, and a tiny gap between the two. The first neuron is the **presynaptic neuron**; the second is the **postsynaptic neuron**. The synaptic part of the plasma membrane of the postsynaptic neuron contains receptors for neurotransmitters.

When an action potential reaches a synaptic terminal, the inside of the terminal becomes positively charged. This charge causes storage vesicles in the terminal to release a neurotransmitter into the gap between the neurons. The neurotransmitter molecules rapidly diffuse across the gap and bind briefly to receptors on the postsynaptic neuron (Fig. 24-3).

Excitatory and Inhibitory Potentials Are Produced at Synapses

The binding of neurotransmitter molecules to receptors in the postsynaptic plasma membrane causes ion channels in the membrane to open, allowing ions to flow across the membrane along their concentration gradients. The flow of ions into the postsynaptic neuron causes a small, brief change in electrical charge: the **postsynaptic potential** (**PSP**). Depending on the type of channels that open and the type of ions that flow, PSPs can be either *excitatory* (EPSPs) or *inhibitory* (IPSPs). EPSPs make the neuron less negative inside and thus more likely to produce an action potential; IPSPs make the neuron more negative and less likely to produce an action potential.

MediaTutor 24.1 Ions and Electrical Signals

Excitatory and Inhibitory Potentials Are Integrated in the Cell Body

Postsynaptic potentials are small, rapidly fading signals, but they travel far enough to reach the cell body, where they determine whether an action potential will be produced. How? The dendrites and cell body of a single neuron may receive EPSPs and IPSPs from the synaptic terminals of thousands of presynaptic neurons. The PSPs that reach the postsynaptic cell body at the same time are "added up," or integrated. If the excitatory and

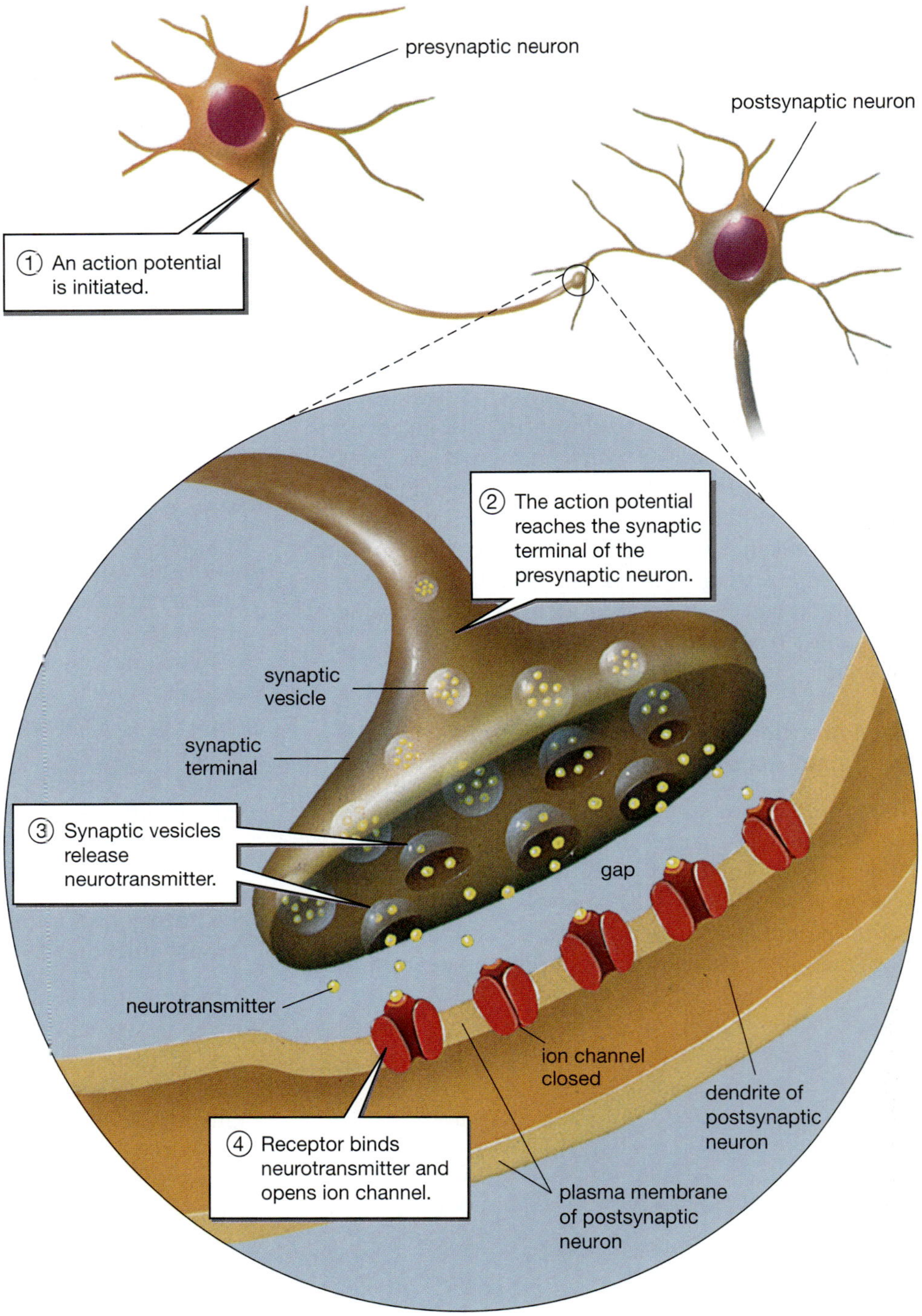

Figure 24-3 The structure and operation of the synapse
The synaptic terminal contains numerous vesicles that enclose a neurotransmitter for which the postsynaptic neuron has membrane receptors. When the neurotransmitter binds to postsynaptic receptors, ion channels open, causing a postsynaptic potential in the postsynaptic cell.

MEDIATUTOR 24.2 Synapses

inhibitory potentials, when added together, raise the electrical potential inside the neuron above threshold, the postsynaptic cell produces an action potential.

The Nervous System Uses Many Neurotransmitters

Over the past few decades, researchers have become increasingly aware that the brain is a teeming cauldron; its neurons synthesize and respond to a vast array of chemicals, including many hormones. At least 50 neurotransmitters have been identified, and the list is growing. In Table 24-1 we list a few well-known neurotransmitters and some of their functions. In "Health Watch: Drugs, Addiction, and Neurotransmitters," we further explore the role of neurotransmitters in addiction and in neurological diseases.

3 How Do Nervous Systems Process Information?

A neuron can respond to a stimulus only by generating and transmitting action potentials. Yet the simple responses of neurons somehow endow animal nervous systems with the ability to initiate and coordinate diverse, complex, and intricate behaviors. One key to the

Health Watch
Drugs, Addiction, and Neurotransmitters

Chances are, you know someone who is addicted—perhaps to nicotine, to alcohol, or maybe to cocaine. How do these substances exert their profound influence over people's lives? A big part of the answer lies in the effects these drugs have on neurotransmitters, and in the way the nervous system responds to those effects. For example, cocaine affects synapses in the brain that contribute to a person's energy level and overall sense of well-being. These synapses use the neurotransmitters *dopamine*, *serotonin*, or *norepinephrine*. Normally, the presynaptic neuron, after releasing one of these neurotransmitters, immediately starts pumping it back in, thus limiting its effects. Cocaine blocks the pumping mechanism, so when a person takes cocaine, the neurotransmitters remain in the synapses much longer and reach higher levels than normal. The effects of the neurotransmitters are thus enhanced, and the cocaine user feels not just energetic but euphoric.

If cocaine is frequently present in the brain, the nervous system undergoes changes to reduce the impact of the drug. In particular, postsynaptic neurons decrease the number of receptors for dopamine, serotonin, and norepinephrine. With fewer receptors present, the high levels of neurotransmitter induced by cocaine are now *necessary* for the user to feel normal. If cocaine is withdrawn, the postsynaptic neurons are inadequately stimulated, and the user experiences an emotional "crash" that can be relieved only by more cocaine. Increasing amounts of the drug are required to produce the euphoric effects. The user has become an addict (Fig. E24-1).

Alcohol stimulates receptors for the neurotransmitter GABA (gamma amino butyric acid), increasing inhibitory neuronal signals, and blocks receptors for glutamate, reducing excitatory signals. When a person drinks frequently, the brain compensates for these changes by decreasing the number of GABA receptors and increasing the number of glutamate receptors. So, an alcoholic deprived of alcohol experiences a smaller number of GABA-induced inhibitory signals and a larger number of excitatory signals from glutamate, and feels overstimulated—jittery, nervous, and trembling. In extreme cases, withdrawal from alcohol can cause convulsions.

To overcome addictions, drug users must undergo the misery caused by a nervous system that is deprived of a drug to which it has adapted. Although receptors eventually return to normal levels, drug cravings often recur periodically for unknown reasons.

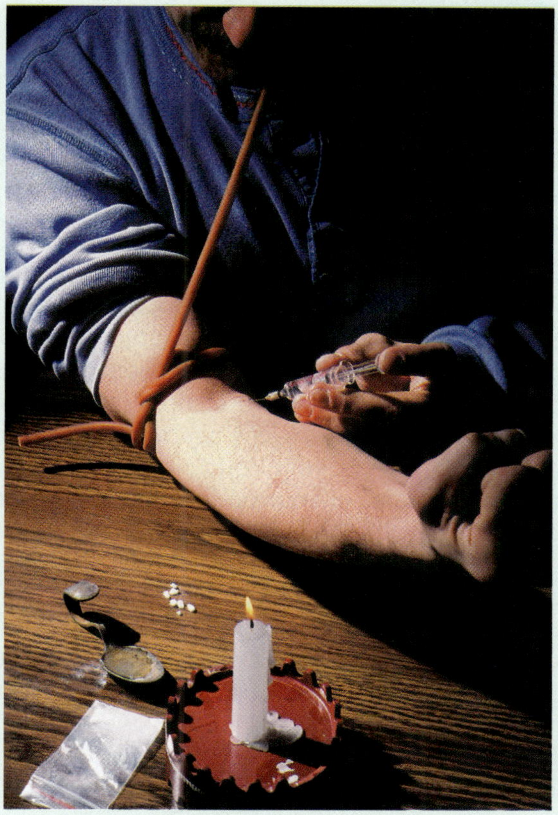

Figure E24-1 Drug addiction
An addict experiences extreme physical and emotional distress without his or her drug because the nervous system has adapted to it.

Table 24-1 Some Important Neurotransmitters

Neurotransmitter	Location in Nervous System	Some Functions
Acetylcholine	Motor neuron-to-muscle synapse; autonomic nervous system, brain	Activates skeletal muscles; activates target organs of parasympathetic nervous system
Dopamine	Midbrain	Important in control of movement
Epinephrine (adrenaline)	Sympathetic nervous system	Activates target organs of sympathetic nervous system
Serotonin	Midbrain, pons, and medulla	Influences mood, sleep
Glutamate	Brain and spinal cord	Major excitatory neurotransmitter in CNS
Glycine	Spinal cord	Major inhibitory neurotransmitter in spinal cord
GABA (gamma amino butyric acid)	Throughout brain	Major inhibitory neurotransmitter in brain
Endorphins	Brain and spinal cord	Influence mood, reduce pain sensations
Nitric oxide	Brain	Important in forming memories

versatility of nervous systems is the presence of complex networks containing thousands to billions of neurons. As in computers, simple elements can perform amazing feats when connected properly.

Information processing in animals requires four basic operations. To respond appropriately to a stimulus, a nervous system must:

- Determine the type of stimulus.
- Determine the intensity of a stimulus.
- Integrate information from many sources.
- Initiate and direct the response.

Let's examine how nervous systems perform each of these operations.

The Type of Stimulus Is Distinguished by Wiring Patterns in the Brain

The nervous system can distinguish among different types of stimuli—for example, light, touch, and sound—despite the fact that all action potentials are similar electrical signals and reveal nothing about the kind of stimulus that elicited them. To identify a stimulus, the nervous system monitors *which* neurons are producing action potentials. Thus, your brain interprets action potentials in the axons of your optic nerves (which originate in the eye and travel to the brain) as the sensation of light, action potentials in olfactory nerves (which originate in receptors in the nose and travel to a different region of the brain) as odors, and so on. In this way, you have no trouble distinguishing the action potentials caused by the music from your stereo from those caused by the bitter taste of coffee on your tongue.

This wiring system, however, may occasionally yield false information. Being poked in the eye may trigger action potentials in the optic nerve. The stimulus is mechanical, but your brain interprets all action potentials in the optic nerve as light, so you "see stars." For the same reason, a blow to the head can make your ears "ring."

The Intensity of a Stimulus Is Revealed by the Frequency of Signals

Because all action potentials are of roughly the same magnitude and duration, no information about the strength, or *intensity*, of a stimulus (for example, the loudness of a sound) can be encoded in a single action potential. Instead, intensity is coded in two other ways. First, intensity can be signaled by the rate at which action potentials are produced in a single neuron. The more intense the stimulus, the faster the neuron fires (produces action potentials). Second, intensity can be signaled by the number of neurons that fire at the same time. Multiple neurons may respond to the same stimulus, and stronger stimuli excite more neurons, whereas weaker stimuli excite fewer. The loud wail of a fire alarm activates many of your auditory neurons and causes them to fire action potentials very rapidly. In most cases, the two types of intensity coding work together: The intensity of a stimulus is signaled by both the rate at which individual neurons produce action potentials and by the number of neurons that do so (Fig. 24-4).

Information from Many Sources Converges in Certain Neurons

Your nervous system is continuously bombarded by stimuli and must filter all the inputs, determine which ones are important, and decide how to respond. Nervous systems integrate information much as individual neu-

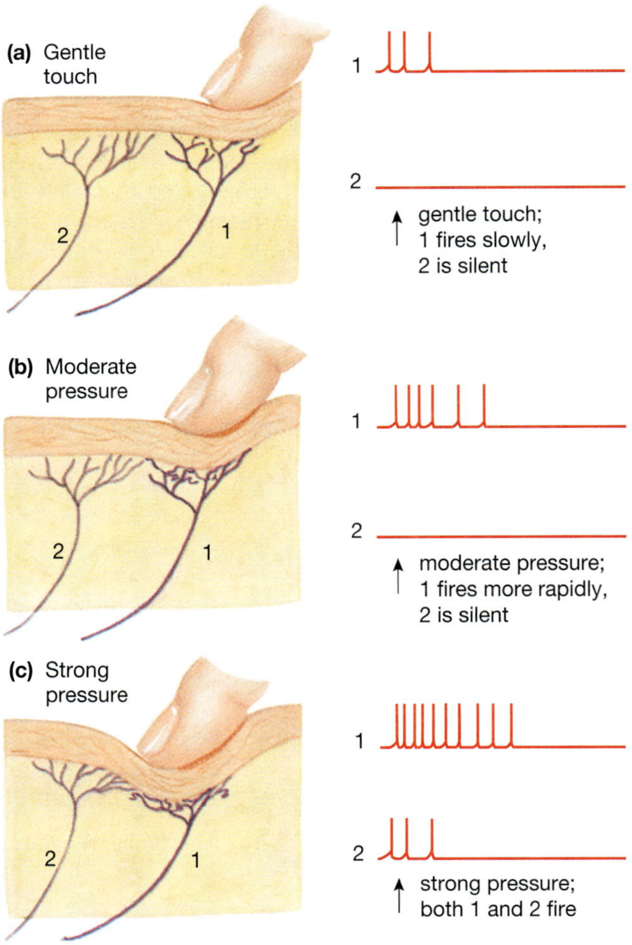

Figure 24-4 Determining the intensity of a stimulus
Consider two touch receptors that have endings in adjacent patches of skin. ***(a)*** A gentle touch elicits only a few action potentials from one sensory neuron but none from another. ***(b)*** Moderate pressure stimulates only the first receptor, which now fires faster, informing the brain that the touch is more intense than before. ***(c)*** Strong pressure activates both receptors, firing the first one very fast and the other more slowly, thus signaling that the pressure is very intense and centered over the faster-firing receptor.

rons do, through *convergence*. In this process, many neurons funnel their signals to fewer neurons. For example, many sensory neurons may send signals to a smaller number of brain neurons. The brain neurons, which can be viewed as *decision-making cells*, add up the postsynaptic potentials that result from the activity of the sensory neurons and produce appropriate outputs.

Complex Responses Require Divergence of Signals

The output of the decision-making cells initiates activity. The resulting actions may involve many parts of the body. Such actions require *divergence*, the flow of electrical signals from a relatively small number of decision-making cells to many different neurons that control the activity of muscles or glands.

4 How Do Nervous Systems Direct Behavior?

Most behaviors are controlled by neuron-to-muscle pathways composed of four elements:

- **Sensory neurons** respond to stimuli that originate inside or outside the body.
- **Association neurons** receive signals from many sources, including sensory neurons, hormones, neurons that store memories, and many others. Based on this input, they activate motor neurons.
- **Motor neurons** receive instructions from association neurons and activate muscles or glands.
- **Effectors** are the muscles and glands that respond to signals from motor neurons.

The simplest type of behavior is the *reflex*, a largely involuntary movement of a body part in response to a stimulus. Examples of human reflexes include the familiar knee-jerk and pain-withdrawal reflexes. Nearly all animals, however, are capable of behavior much more subtle and varied than a simple reflex. These more complex behaviors are organized by interconnected neural pathways, in which several types of sensory input (along with memories, hormones, and other factors) converge on association neurons. By integrating the postsynaptic potentials from several sources, the association neurons can "decide" what to do and can stimulate the motor neurons to direct the appropriate activity in muscles and glands.

The Big Picture

An important attribute of the nervous system is its ability to transmit signals at very high speed. High-speed transmission is the key to such quintessential animal characteristics as locomotion, sensory perception, and communication.

5 How Is the Human Nervous System Organized?

The human nervous system can be divided into two parts, central and peripheral, each of which has further subdivisions (Fig. 24-5). The **central nervous system** (**CNS**) consists of the **brain** and the **spinal cord**, which extends down the back of the torso. The **peripheral nervous system** (**PNS**) consists of nerves that connect the central nervous system to the rest of the body.

The Peripheral Nervous System Links the Central Nervous System to the Body

The nerves of the peripheral nervous system link the brain and spinal cord to muscles, sensory organs, and the organs of the digestive, respiratory, excretory, and circulatory systems. The peripheral nerves contain axons of sensory neurons that bring sensory information *to* the central nervous system from all parts of the body, and they also contain the axons of motor neurons that carry signals *from* the central nervous system to the organs and muscles. The motor portion of the peripheral nervous system can be subdivided into two parts: the **somatic nervous system** and the **autonomic nervous system**.

Somatic System Neurons Control Voluntary Movement

Motor neurons of the somatic nervous system form synapses on skeletal muscles and control voluntary movement. As you take notes, lift a coffee cup, or adjust your stereo, your somatic nervous system is in charge. The cell bodies of somatic motor neurons are located in the spinal cord, and their axons go directly to the muscles they control.

Autonomic System Neurons Control Involuntary Actions

Motor neurons of the autonomic nervous system form synapses on the heart, smooth muscle, and glands, and control involuntary responses. The autonomic nervous system is controlled by the medulla and the hypothalamus, parts of the brain that are described later in this chapter. It consists of two divisions: the *sympathetic division* and the *parasympathetic division*. The two divisions of the autonomic nervous system generally make synaptic contacts with the same organs but normally produce opposite effects.

Sympathetic Neurons Prepare the Body for Action

The sympathetic division acts on organs in ways that prepare the body for stressful or highly energetic activity, such as fighting or escaping. During such "fight-or-flight" activities, the sympathetic nervous system reduces activity of the digestive tract, redirecting some of its blood supply to the muscles of the arms and legs.

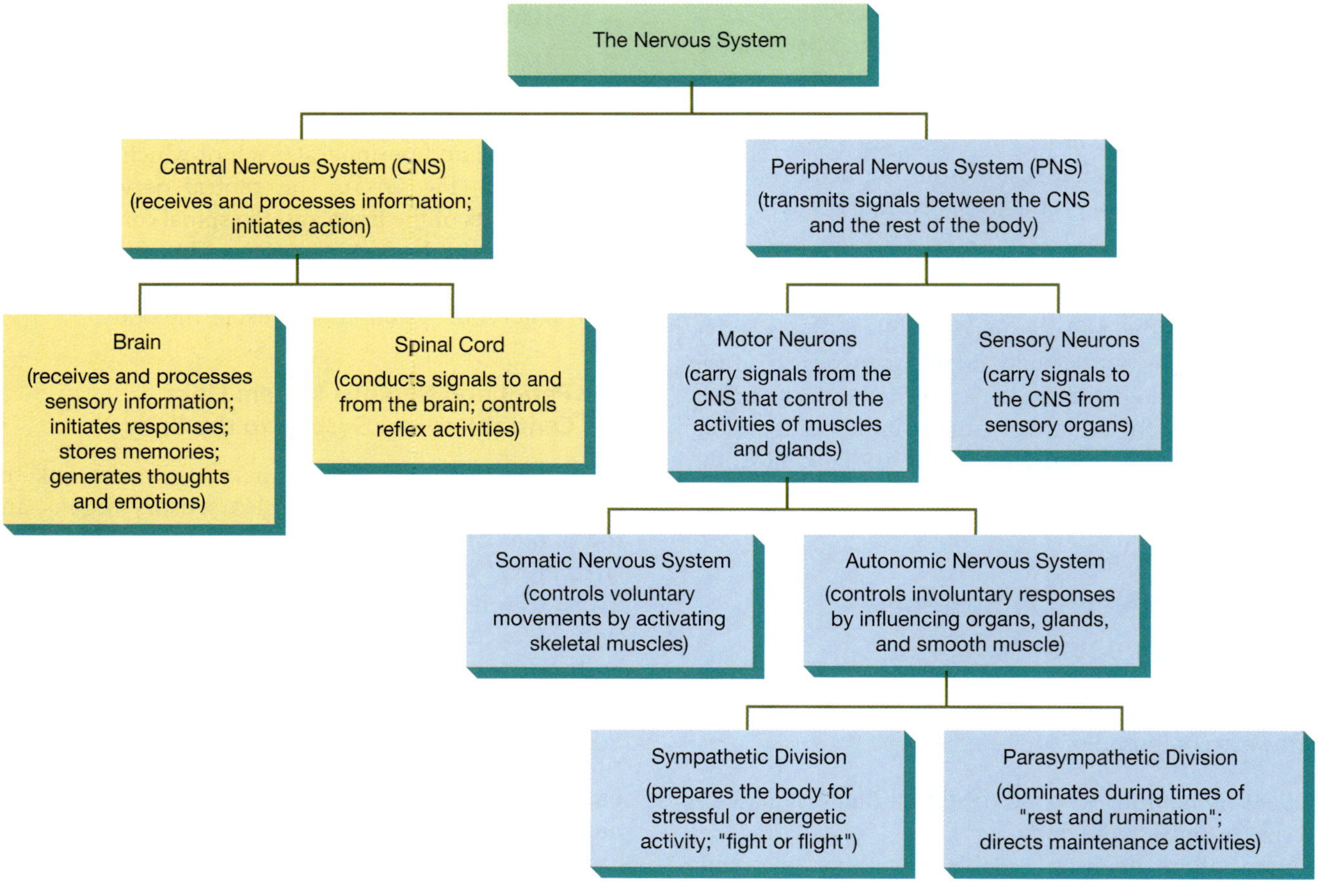

Figure 24-5 The organization and functions of the human nervous system

The heart rate accelerates. The pupils of the eyes open wider, admitting more light, and the air passages in the lungs expand, accommodating more air.

Parasympathetic Neurons Prepare the Body for Rest

The parasympathetic division, in contrast, dominates during maintenance activities that can be carried on at leisure. Under parasympathetic control, the digestive tract becomes active, the heart rate slows, and air passages in the lungs constrict. As you read this text, your parasympathetic nervous system is probably dominating your unconscious functions.

The Central Nervous System Consists of the Spinal Cord and Brain

The spinal cord and brain make up the central nervous system. The CNS receives and processes sensory information, generates thoughts, and directs responses. It consists primarily of association neurons—somewhere between 10 billion and 100 billion of them! The brain and spinal cord are protected by a bony armor, consisting of the *skull*, which surrounds the brain, and the backbone, or *vertebral column*, a chain of bones (the vertebrae) that protects the spinal cord. Beneath the bones of the skull and backbone lies a triple layer of connective tissue called the *meninges*. Between the layers of the meninges, the cerebrospinal fluid, a clear, lymphlike liquid, cushions the brain and spinal cord as it nourishes the cells of the CNS.

The Spinal Cord Is a Cable of Axons Protected by the Backbone

The spinal cord is a neural cable, about as thick as your little finger, that extends from the base of the brain to the lower back. Although the spinal cord is both strong and flexible, it can be damaged by violent trauma. (To learn what happens when the spinal cord is injured and what treatments are under study, see "Health Watch: Healing the Spinal Cord.")

Between the vertebrae, nerves carrying axons of sensory neurons emerge from the dorsal (back side of body) part of the spinal cord, and nerves carrying axons of motor neurons emerge from the ventral (front) part. The nerves from the two sides of the spinal cord merge to form the peripheral nerves of the spinal cord (part of

Health Watch
Healing the Spinal Cord

Actor Christopher Reeve was a real-life athlete whose athleticism and muscular physique made him well suited for such movie roles as Superman. But on Memorial Day in 1995, his life changed abruptly at an equestrian jumping competition. Racing up to a jump, his horse suddenly balked, throwing Reeve headfirst onto the jumping rail. The impact shattered the first two vertebrae in his neck and crushed his spinal cord, interrupting the flow of signals from his brain to the rest of his body and from his body to his brain (Fig. E24-2). Unable to move or breathe, Reeve's life was saved by the fast action of paramedics, who began pumping oxygen into his lungs within 3 minutes of the fall. If one more minute had elapsed, the lack of oxygen could have caused irreversible brain damage.

Whether from auto accidents, bullets, disease, falls, or sporting mishaps, nearly 8000 people in the United States sustain spinal cord injuries each year, and about 400,000 people are living with the disabling effects of these injuries. Why are spinal cord injuries so devastating? When the spinal cord is crushed, axons are severed. Swelling and inflammation in the cord squashes more axons and cell bodies, adding to the destruction. Damaged cells release substances that are toxic to neighboring cells, further enlarging the neuron death zone. The cut ends of axons that are still attached to living cell bodies begin to regrow, but their progress is thwarted because they cannot grow into the surrounding region of dead tissue and scar tissue. The scar tissue, in addition to creating a physical barrier, secretes proteins that inhibit axon growth. Even the insulating myelin in the CNS produces proteins that deter axon regeneration.

Research on methods to overcome these obstacles to spinal cord regeneration is progressing on several fronts. Medical researchers are investigating drugs that limit swelling and inflammation. Researchers are also working to identify genes that code for the proteins that inhibit axon growth in the CNS, hoping to devise ways to block the action of those proteins or to prevent their synthesis. To help increase the success of regenerating axons, scientists are experimenting with grafting tissue, such as myelin from the PNS, that will promote and channel the regrowth of axons. Neural transplants to replace dead neurons are also under investigation. All of these approaches have yielded promising results in laboratory animals.

The recent discovery that unspecialized embryonic cells, and even some adult cells, can be induced to develop into neurons has caused great excitement among researchers. Such cells may one day be used as transplants to replace damaged neurons in spinal cords or even in the brain. You'll learn more about these cells, called *stem cells*, in Chapter 25. Finally, researchers are programming computers to stimulate muscles directly with electrical currents, causing even muscles whose neural connection to the brain has been destroyed to contract in specific sequences to perform useful functions. A device that stimulates muscles that control the hand will soon be tested on humans. There will be no quick or complete cures, but Christopher Reeve and other individuals with spinal cord injuries have reason to be cautiously optimistic.

Figure E24-2 A famous spinal-cord-injury survivor
Christopher Reeve discusses spinal cord injuries at the National Press Club in Washington, D.C., in December 1999.

the peripheral nervous system; Fig. 24-6). In the center of the spinal cord is a butterfly-shaped area of *gray matter*. Gray matter consists of the cell bodies of several types of neurons that control voluntary muscles and the autonomic nervous system, plus neurons that communicate with the brain and other parts of the spinal cord. The gray matter is surrounded by *white matter*, formed of myelin-coated axons of neurons that extend up or down the spinal cord. These axons carry sensory signals from internal organs, muscles, and the skin up to the brain and carry signals from the brain that direct the motor portions of the peripheral nervous system.

Reflex Pathways Include Neurons in the Spinal Cord

In addition to relaying neural signals between the brain and the rest of the body, the spinal cord contains the neural pathways for certain simple behaviors, such as reflexes. For example, in the pain-withdrawal reflex a painful stimulus at the skin causes a signal to be transmitted along sensory neurons whose cell bodies are just

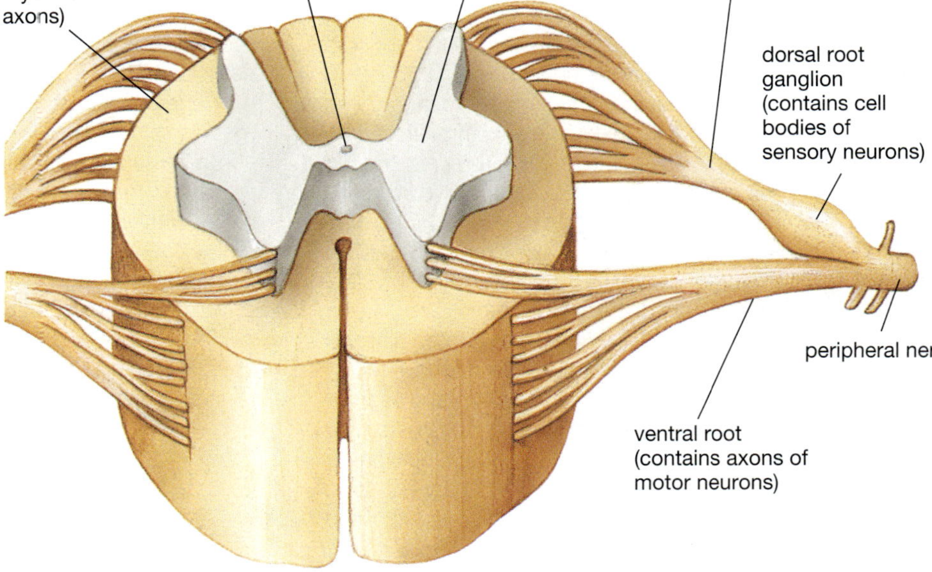

Figure 24-6 The spinal cord in cross section

outside the spinal cord, in clusters of neurons called *dorsal root ganglia* (see Fig. 24-6). From the dorsal root ganglia, the signal is transmitted a short distance to association neurons in the gray matter at the center of the spinal cord, and then to nearby cell bodies of motor neurons, which direct muscles to withdraw from the painful stimulus (Fig. 24-7). The association neurons also have axons that extend up to the brain, along which signals are sent that alert the brain to the painful event. The brain, in turn, sends impulses down axons in the white matter to cells in the gray matter. These signals can modify spinal reflexes. With sufficient motivation, you can suppress the pain-withdrawal reflex; to rescue a child from a burning crib, for example, you could reach into the flames.

Some Complex Actions Are Coordinated by the Spinal Cord

The entire "software" for operating some fairly complex activities resides within the spinal cord. All the neurons and interconnections needed for the basic movements of walking and running, for example, are contained in the spinal cord. The advantage of this semi-independent arrangement between brain and spinal cord is probably an increase in speed and coordination, because messages do not have to travel all the way up to the brain and back down again merely to swing forward one leg. The brain's role in these "semi-automatic" behaviors is to initiate, guide, and modify the activity of spinal motor neurons, based on conscious decisions (Where are you going? How fast should you walk?). To maintain balance, the brain also uses sensory input from the muscles to command motor neurons to adjust the way the muscles move.

The Brain Consists of Many Parts

All vertebrate brains have the same general structure, consisting of three regions, the hindbrain, midbrain, and the forebrain, each of which is in turn made up of specialized parts with specific functions. In non-mammalian vertebrates, each region is prominent. In adult mammals, however, and particularly in humans, the hindbrain and the midbrain are relatively small, and the forebrain is greatly enlarged. Some of the major structures of the human brain are shown in Figure 24-8.

The Hindbrain Controls Breathing, Sleep, and Movement

In humans, the **hindbrain** is represented by the medulla, the pons, and the cerebellum. In both structure and function, the *medulla* is much like an enlarged extension of the spinal cord. Like the spinal cord, the medulla has neuron cell bodies at its center, surrounded by a layer of myelin-covered axons. The medulla controls several automatic functions, such as breathing, heart rate, blood pressure, and swallowing. Certain neurons in the *pons*, located above the medulla, appear to influence transitions between sleep and wakefulness and between stages of sleep. Other neurons in the pons

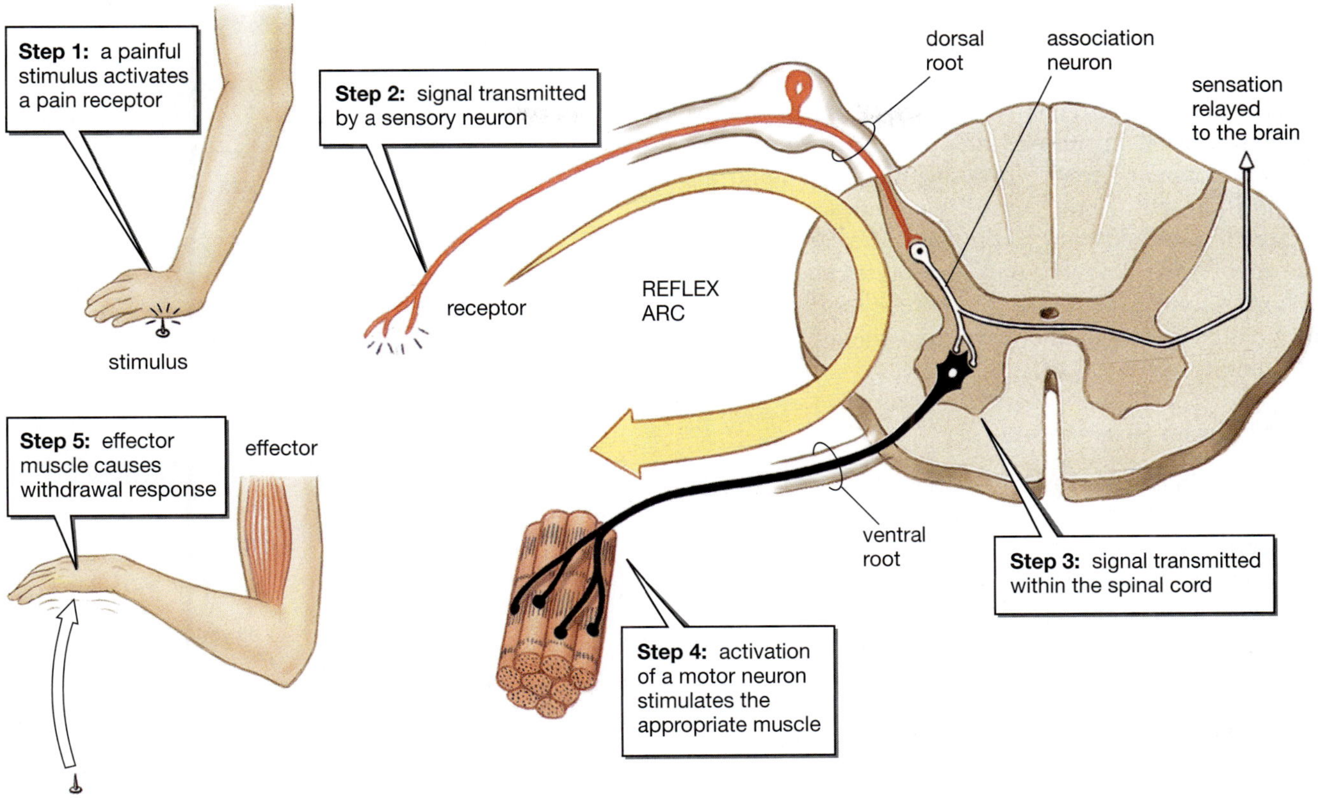

Figure 24-7 The pain-withdrawal reflex
This simple reflex circuit includes a sensory neuron that has pain-sensitive endings in the skin and a long fiber leading to the spinal cord. The sensory neuron stimulates an association neuron in the spinal cord, which in turn stimulates a motor neuron in the cord. An association neuron also sends the signal to the brain, informing it of the danger.

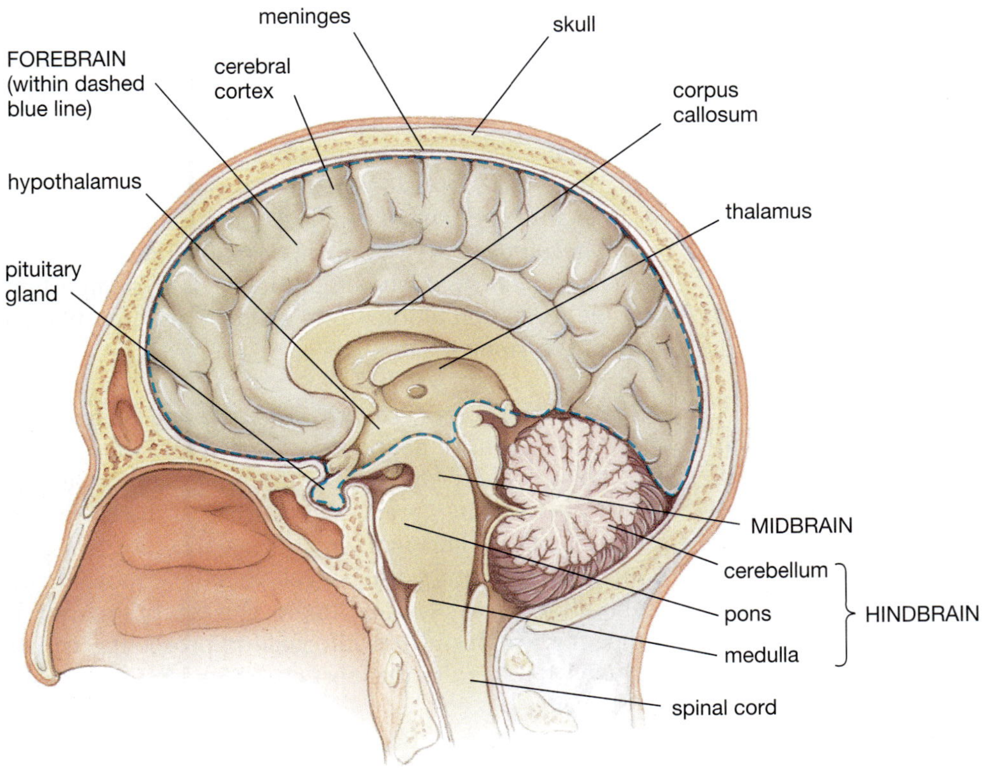

Figure 24-8 The human brain
A section through the midline of the human brain reveals some of its major structures.

influence the rate and pattern of breathing. The *cerebellum* is crucial in coordinating movements of the body. It receives information from command centers in the areas of the brain that control movement and also from position sensors in muscles and joints. By comparing what the command centers order with information from the position sensors, the cerebellum guides smooth, accurate motions and body positioning. The cerebellum is also involved in learning and memory storage for behaviors. As you take notes, your cerebellum is instructing your brain about the order and timing of muscle movements in your hand.

The Midbrain Filters Input

The **midbrain** is quite small in humans (see Fig. 24-8). It contains an auditory relay center and a center that controls reflex movements of the eyes. Another important relay center, the *reticular formation*, passes through it. The neurons of the reticular formation extend all the way from the central core of the medulla up into lower regions of the forebrain. It receives input from virtually every sense, from every part of the body, and from many areas of the brain as well. Your reticular formation filters sensory inputs before they reach the conscious regions of the brain, so that you can read and concentrate in the presence of such distracting stimuli as the music from your stereo and the smell of coffee. The fact that a mother wakens upon hearing the faint cry of her infant but sleeps through loud traffic noise outside her window testifies to the effectiveness of the reticular formation in screening inputs to the brain.

The Forebrain Gives Us Our Emotions, Thoughts, and Perceptions

The **forebrain**, also called the *cerebrum*, includes the thalamus, the limbic system, and the cerebral cortex.

The Thalamus Relays Signals

The *thalamus* is a complex relay station that channels sensory information from all parts of the body to the limbic system and cerebral cortex. Signals from the cerebellum and limbic system to the cerebral cortex are also channeled through this busy thoroughfare.

The Limbic System Produces Emotions

The *limbic system* is a diverse group of structures located in an arc between the thalamus and cerebral cortex. These structures work together to produce our most basic emotions and drives, including fear, rage, tranquility, hunger, thirst, pleasure, and sexual responses. Portions of the limbic system are also important in the formation of memories. The limbic system includes the hypothalamus, the amygdala, and the hippocampus as well as nearby regions of the cerebral cortex (Fig. 24-9).

The *hypothalamus* is a major coordinating center, controlling body temperature, hunger, the menstrual cycle, water balance, and the sleep–wake cycle. Some of its neurons are neurosecretory cells that release hormones into the blood; others control the release of hormones from the pituitary gland (see Chapter 23). Other regions of the hypothalamus direct the activities of the autonomic nervous system.

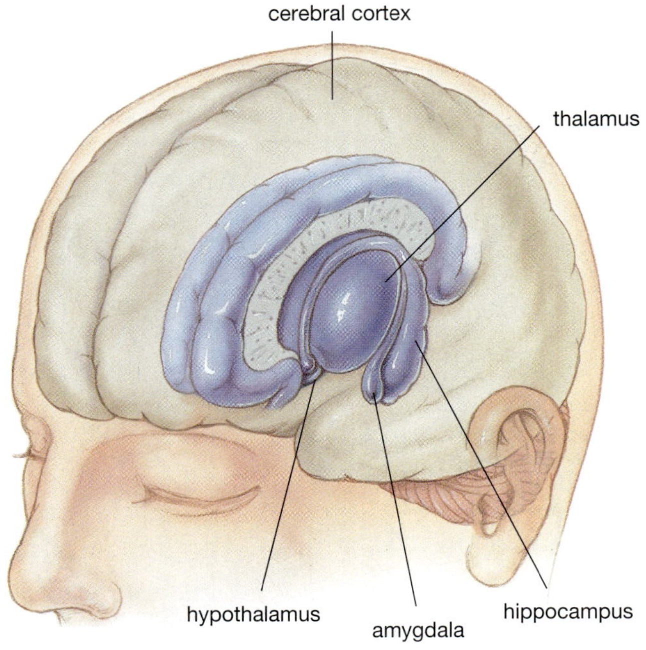

Figure 24-9 The limbic system and thalamus
In this view of the brain, the left cerebral cortex is transparent.

The *amygdala* produces sensations of pleasure, punishment, or sexual arousal when stimulated. Conscious humans whose amygdalae are electrically stimulated have reported feelings of rage or fear. Recent studies have revealed that damage to the amygdala early in life eliminates the ability both to feel fear and to recognize fearful facial expressions in other people.

As with the amygdala, electrical stimulation of the *hippocampus* elicits emotional responses, including rage and sexual arousal. The hippocampus is also important in the formation of long-term memory and thus plays a role in learning.

The Cerebral Cortex Produces Thoughts and Perceptions

Tens of billions of neurons are packed into the *cerebral cortex*, the outer layer of the forebrain and the largest part of the human brain. The neurons of the cortex receive sensory information, process it, store some in memory for future use, direct voluntary movements, and allow us to plan and think. The cerebral cortex is the most sophisticated information-processing center known, but it is also the most poorly understood area of the brain.

The cerebral cortex is divided into four regions: the *frontal*, *parietal*, *occipital*, and *temporal lobes*. In addi-

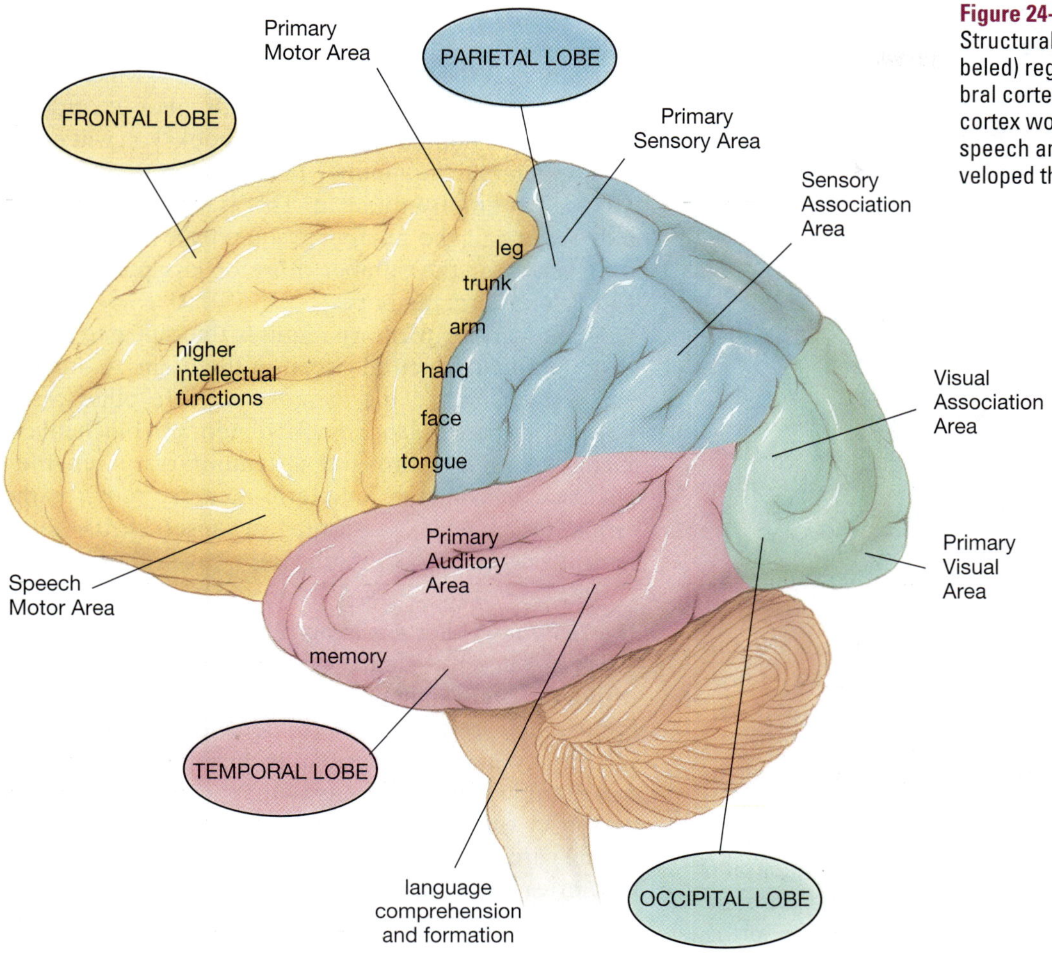

Figure 24-10 The cerebral cortex Structural (colored) and functional (labeled) regions of the human left cerebral cortex. A map of the right cerebral cortex would be similar, except that speech and language are less well developed there.

tion, the cortex and underlying parts of the forebrain are divided into two halves, called *cerebral hemispheres* (Fig. 24-10). These halves communicate with each other by means of a large band of axons, the corpus callosum.

Different areas of the cortex perform different functions (see Fig. 24-10). Some portions serve as *primary sensory areas*, regions where signals from sensory organs such as the eyes and ears are received and converted into subjective perceptions such as light and sound. Nearby *association areas* interpret the perceptions as, for example, speech or music, or as recognizable objects or words on a page. Association areas also link the stimuli with memories previously stored in the cortex, and generate commands to produce speech. The association area of the frontal lobe is involved in complex reasoning functions such as decision making, predicting the consequences of actions, controlling aggression, and planning for the future. *Primary motor areas* generate commands for movements of the body. These commands control the motor neurons that form synapses with muscles, allowing you to walk to class, throw a Frisbee®, or type a term paper.

Damage to the cortex from trauma, stroke, or a tumor results in specific deficits, such as problems with speech, difficulty reading, or the inability to sense or move specific parts of the body. Most brain cells of adults cannot be replaced, so if a brain region is destroyed, these deficits may be permanent. Fortunately, however, training can sometimes allow undamaged regions of the cortex to take control and restore some of the lost functions.

6 How Does the Brain Produce the Mind?

Many people have found it difficult to accept the idea that a few pounds of soft tissue in the skull produce the human mind, with its thoughts, emotions, memories, and self-awareness. This "mind–brain problem" has intrigued generations of philosophers and, more recently, neurobiologists. Scientific exploration of the problem began with observations of people with head injuries and progressed to systematic experimentation and observation with sophisticated instruments. Today, a picture of how the brain creates the mind is beginning to emerge. Here, we explore a few of the insights provided by scientific examination of the mind.

The "Left Brain" and "Right Brain" Have Different Specialties

Although the two hemispheres of the forebrain are similar in appearance, it has been known since the early 1900s that this symmetry does not extend to brain function. Much of what is known of the differences in hemisphere function comes from studies of accident or stroke victims with localized damage to one hemisphere. A great deal has also been learned from studies of patients who have had their corpus callosum (which connects the two hemispheres) severed, a surgical procedure that is performed in certain cases of uncontrollable epilepsy to prevent the spread of seizures through the brain.

Roger Sperry, a neurobiologist working at the California Institute of Technology, studied people whose hemispheres had been surgically separated by cutting the corpus callosum. In his studies, Sperry took advantage of the pathway followed by the axons that conduct signals from the eye to the brain; this pathway causes the left half of each visual field to be projected onto the right cerebral hemisphere and the right half to be projected onto the left hemisphere. Using an ingenious device that projected different images onto the left and right visual fields, Sperry sent different visual signals to each hemisphere.

If Sperry projected an image of a nude figure onto just the left visual field, the subjects would blush and smile but would claim to have seen nothing, because the image had reached only the nonverbal right side of the brain. The same figure projected onto the right visual field was readily described verbally. These experiments, begun in the 1950s, and other studies since then, have revealed that in right-handed people, the left hemisphere is dominant in speech, reading, writing, language comprehension, mathematical ability, and logical problem solving. The right side of the brain is superior to the left in musical skills, artistic ability, recognition of faces, spatial visualization, and the ability to recognize and express emotions.

Recent experiments, however, indicate that the left–right dichotomy is not as rigid as was once believed. An individual who has suffered a stroke that disrupted the blood supply to his or her left hemisphere typically shows symptoms such as loss of speaking ability. Commonly, however, training can partially overcome these speech or reading deficits, even though the left hemisphere itself has not recovered. This observation suggests that the right hemisphere has some latent language capabilities.

The Mechanisms of Learning and Memory Are Poorly Understood

Although the cellular mechanisms of learning and memory remain poorly understood, biologists have nonetheless made significant progress in understanding some aspects of these phenomena.

Memory May Be Brief or Long Lasting

Experiments have shown that learning has two phases: it begins with the use of *working memory*, which is followed by the use of *long-term memory*. For example, if you look up a number in the phone book, you will probably remember the number long enough to dial it but will then promptly forget it. This is working memory. But if you call the number frequently, eventually you will remember the number more or less permanently. This is long-term memory.

Some working memory requires the repeated activity of a particular neural circuit in the brain. As long as the circuit is active, the memory stays. In other cases, working memory depends on temporary biochemical changes within neurons that temporarily strengthen synapses. In contrast, long-term memory requires the formation of new, long-lasting synapses or the long-term strengthening of existing but weak synapses.

The Temporal Lobes Are Important for Memory

There is strong evidence that the hippocampus, deep within the brain's temporal lobe, is involved in learning. For example, researchers have detected intense electrical activity in the hippocampus during learning. Even more striking are the results of hippocampal damage. A person in whom both hippocampi (one in each hemisphere) are destroyed retains much of his or her former memories but is unable to learn new information after the loss. For example, one patient whose hippocampi were surgically removed in an attempt to control seizures remains unable to recall his address or find his way home after many years at the same residence. He can entertain himself indefinitely by reading the same magazine over and over, and people who he sees daily require reintroduction at each encounter. This and other examples suggest that the hippocampus is responsible for transferring information from working memory to long-term memory.

The temporal lobes of the cerebral hemispheres seem to be important in recall of long-term memories. In a famous series of experiments in the 1940s, neurosurgeons electrically stimulated the temporal lobes of conscious patients undergoing brain surgery. The patients did not merely recall memories but felt that they were experiencing the past events right there in the operating room!

Insights on How the Brain Creates the Mind Come from Diverse Sources

Until about 100 years ago, the mind was considered to be a subject more appropriate for philosophers than for scientists, in part because tools for studying the brain did not yet exist. New tools and new discoveries, however, are rapidly changing our view of how the mind works.

During recent decades, we have begun to understand the neural bases of at least some psychological phenomena. Many forms of mental illness, such as schizophrenia, manic depression, and autism, once thought to be caused by childhood trauma or inept parenting, are now recognized as the results of biochemical imbalances or structural abnormalities in the brain. Studies have also revealed that many traits that were once considered to be entirely learned, such as shyness and alcoholism, can be inherited (and hence have a biological basis).

Study of patients with brain injuries has revealed that some parts of the brain are highly specialized. One patient with very localized damage to the left frontal lobe of the cerebral cortex was unable to name fruits and vegetables (although he could name everything else). Describing this patient, one writer quipped, "Does the brain have a produce section?" Other victims of brain damage are unable to recognize faces, or to recognize any object that is *not* a face, suggesting that the brain has a region specialized to recognize faces that is separate from the region that allows it to recognize objects in general.

In the past, the insights that could be gained from brain-damaged patients were often limited, because the exact extent of the damage remained unknown until revealed by autopsy. Today, however, new techniques for visualizing brain structure and activity, such as the PET (positron emission tomography) and fMRI (functional magnetic resonance imaging) scans, permit insight into the functioning of normal, as well as diseased, brains. These techniques, and increasingly sophisticated ones that will be developed in future, will create ever-larger windows into the "black box" that is the human brain and a clearer understanding of how the brain generates the human mind.

The Big Picture

The nervous system does much more than just transmit signals. Its largest component, the brain, is the body's command center, computer, data storage unit, and, at least in humans, the seat of consciousness.

7 How Do Sensory Receptors Work?

The word "receptor" is used in several contexts in biology. In this chapter, we discuss **sensory receptors**, which consist of an entire specialized cell (typically, a neuron) that produces an electrical response to a particular stimulus.

Sensory receptors translate sensory stimuli into the language of the nervous system, producing electrical signals called **receptor potentials** when stimulated. Each receptor is specialized to produce a receptor potential only in response to a particular type of environmental stimulus; receptor cells are named after the stimulus to which they respond (Table 24-2). In many cases, sensory receptors are clustered into sensory organs, such as the eye, ear, skin, or tongue. Their electrical activity, after being processed by the brain, gives rise to the subjective perceptions of light, sound, touch, and taste that we describe as our "senses."

Different types of sensory receptors have different ways of influencing postsynaptic neurons. In some sensory receptor neurons, a receptor potential brings the cell above threshold, producing an action potential. In other sensory receptors, receptor potentials cause neurotransmitters to be released onto postsynaptic neurons, which in turn produce action potentials that travel to the central nervous system.

Sensory receptors are our links with the world around us. As Aristotle observed in the fourth century B.C., "Nothing is understood by the intellect which is not first perceived by the senses." In the following sections, we shall focus on some senses that determine the way we as humans perceive the world.

8 How Is Sound Sensed?

Sound is produced by any vibrating object—a drum, a motor, vocal cords, or the speaker of your stereo. These vibrations, or waves, are transmitted through air

Table 24-2 Some Vertebrate Receptor Types

Type of Receptor	Sensory Cell Type	Stimulus	Location
Thermoreceptor	Free nerve ending	Heat, cold	Skin
Mechanoreceptor	Hair cell	Vibration, motion, gravity	Inner ear
	Nerve endings in skin	Vibration, pressure, touch	Skin
	Nerve endings in muscles or joints	Stretch	Muscles, tendons
Photoreceptor	Rod, cone	Light	Retina of eye
Chemoreceptor	Olfactory receptor	Odor (airborne molecules)	Nasal cavity
	Taste receptor	Taste (waterborne molecules)	Tongue
Pain receptor	Free nerve ending	Chemicals released by tissue injury	Widespread in body

and intercepted by our ears, which convert them to signals that our brain interprets as the location, pitch, and loudness of sound.

The Ear Captures and Transmits Sound

The ear of humans and most other vertebrates consists of three parts: the outer, middle, and inner ear (Fig. 24-11a). The *outer ear* consists of the *external ear* and the *auditory canal*. The external ear, with its fleshy folds, modifies sound waves in ways that the brain uses to determine the location of the sound source. The air-filled auditory canal conducts the sound waves to the *middle ear*, consisting of the *tympanic membrane*, or *eardrum*; three tiny bones called the *hammer*, *anvil*, and *stirrup*; and the *Eustachian tube*. The Eustachian tube connects the middle ear to the pharynx and equalizes the air pressure between the middle ear and the atmosphere.

Within the middle ear, sound vibrates the tympanic membrane, which in turn vibrates the hammer, the anvil, and the stirrup. These bones transmit vibrations to the fluid-filled hollow bones of the *inner ear*, which contains the spiral-shaped *cochlea*. The stirrup bone transmits vibrations to the fluid within the cochlea by vibrating a membrane in the cochlea called the *oval window*. The *round window* is a second membrane that allows fluid within the cochlea to shift back and forth as the stirrup bone vibrates the oval window.

Sound Is Converted into Electrical Signals in the Cochlea

If we were to unroll the cochlea, it would consist of two fluid-filled tubes: an outer U-shaped canal and a straight central canal (Fig. 24-11b). The floor of the central canal consists of the *basilar membrane*, on top of which are the receptors, or *hair cells*, which are a type of mechanoreceptor. Protruding into the central canal is the *tectorial membrane*, a gelatinous structure in which the "hairs" of the hair cells are embedded.

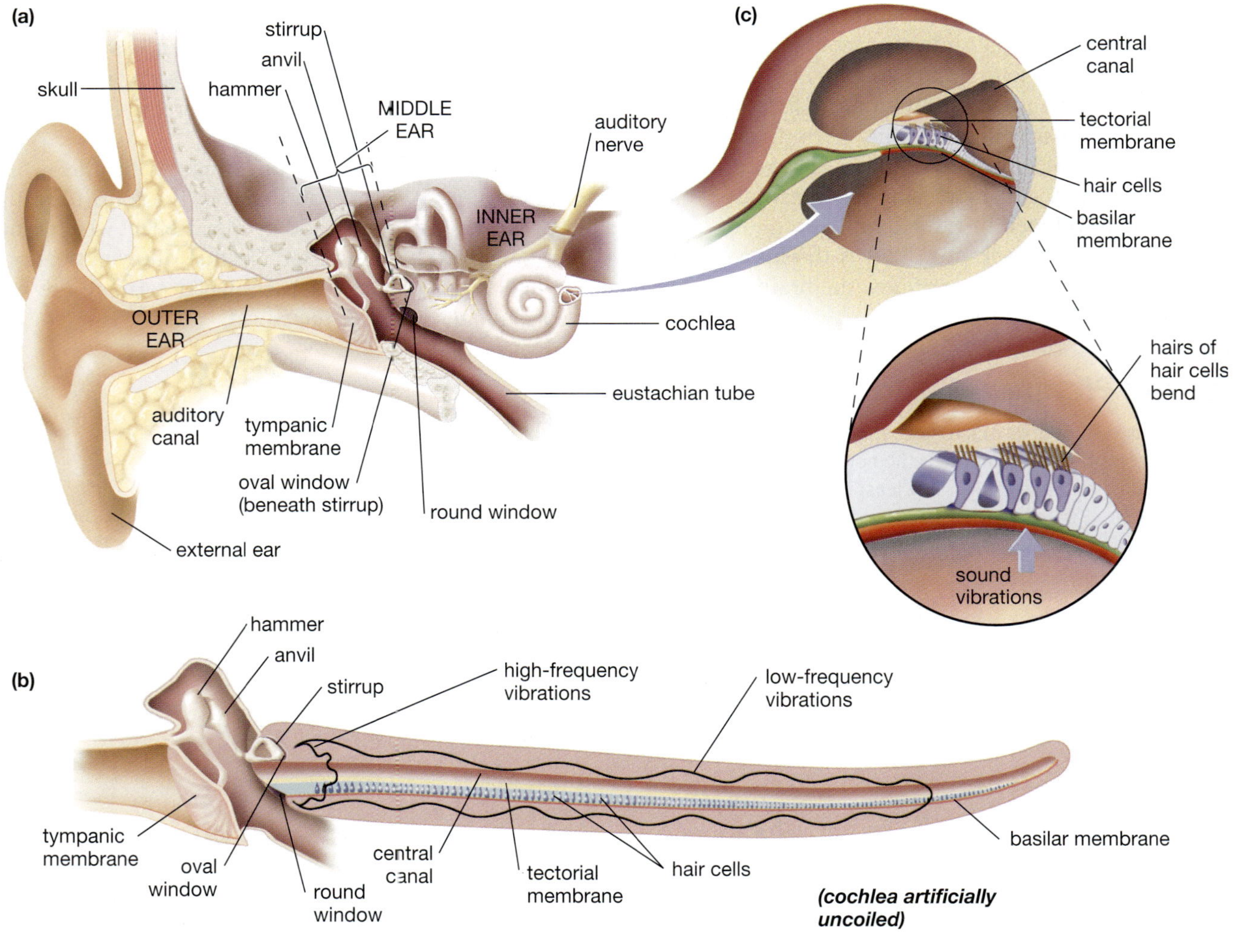

Figure 24-11 The human ear
(a) Overall anatomy of the ear. ***(b)*** Uncoiled, the cochlea consists of an outer tube surrounding the central canal. Different frequencies of sound waves vibrate the basilar membrane at different locations. ***(c)*** The hairs of hair cells span the gap between the membranes in the central canal. Sound vibrations move the membranes relative to one another, bending the hairs and producing a receptor potential in the hair cells.

How do these structures allow the perception of sound? The oval window passes vibrations to the fluid in the cochlea, which in turn vibrates the basilar membrane, causing it to move up and down. This movement bends the hairs of the hair cells, producing receptor potentials (Fig. 24-11c). The hair cells then release neurotransmitters onto neurons of the auditory nerve to the brain. Action potentials triggered in axons of the auditory nerve travel to the brain.

The structures of the inner ear allow us to perceive *loudness* (the magnitude of sound vibrations) and *pitch* (the frequency of sound vibrations). Weak sounds cause small vibrations, which bend the hairs only slightly and result in a low rate of action potentials in axons of the auditory nerve. Loud sounds cause large vibrations, which cause greater bending of the hairs and a larger receptor potential, producing a high rate of action potentials in the auditory nerve. Loud sounds sustained for a long time can damage the hair cells (Fig. 24-12a), resulting in hearing loss, a fate suffered by many rock musicians and their fans. In fact, many sounds in our everyday environment have the potential to damage hearing, especially if exposure to them is prolonged (Fig. 24-12b).

The perception of pitch is a little more complex. The basilar membrane is stiff and narrow at the end near the oval window but more flexible and wider near the other end of the cochlea. This progressive change in structure causes each portion of the membrane to vibrate in synchrony with a particular frequency of sound (see Fig. 24-11b). The brain interprets signals from receptors near the oval window as high-pitched sound, whereas signals from receptors farther along the cochlea are interpreted as lower in pitch. Humans can detect vibration frequencies from about 30 vibrations per second (very low pitched) to about 20,000 vibrations per second (very high pitched).

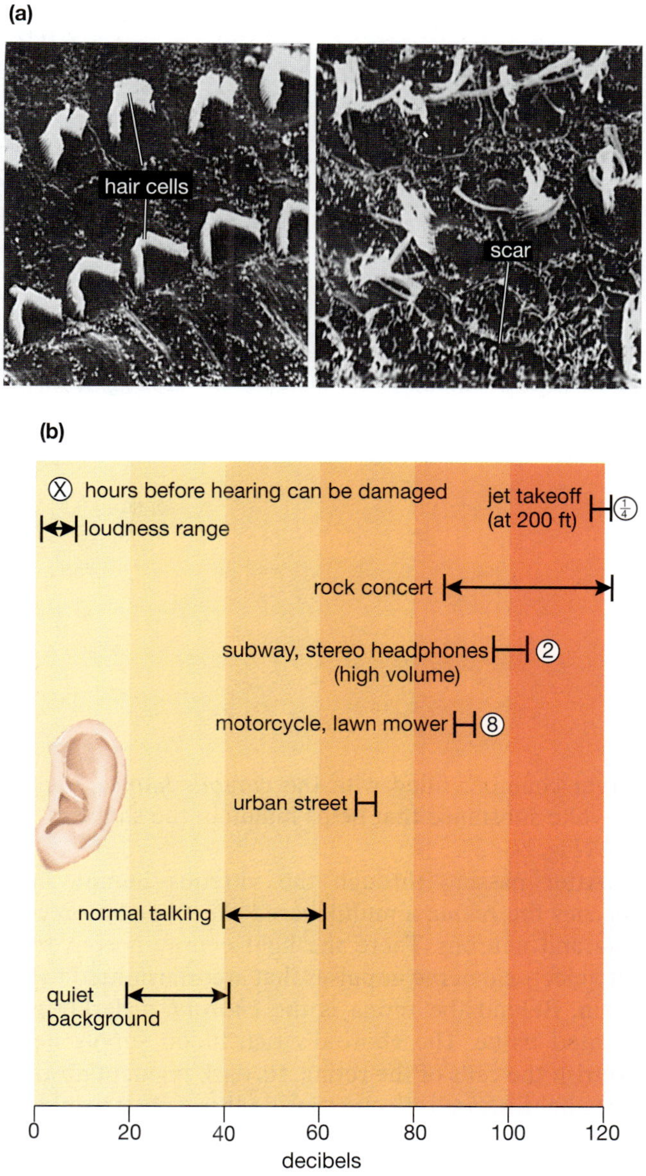

Figure 24-12 Loud sounds can damage hair cells
(a) Scanning electron micrographs show the effect of intense sound on the hair cells of the inner ear. (left) The hairs of hair cells in a normal guinea pig; hairs emerge from each receptor in a V-shaped pattern. (right) After 24-hour exposure to a sound level approached by loud rock music (2000 vibrations per second at 120 decibels), many of the hairs are damaged or missing, leaving "scars." Hair cells in humans do not regenerate, so such hearing loss is permanent. [SEMs by Robert S. Preston, courtesy of Professor J. E. Hawkins, Kresge Hearing Research Institute, University of Michigan Medical School.] ***(b)*** Sound levels of everyday noises and their potential to damage hearing. Sound intensity is measured in *decibels* on a logarithmic scale; a 10-decibel sound is 10 times as loud as a 1-decibel sound, and a 20-decibel sound is 100 times as loud. You feel pain at sound intensities above 120 decibels. [Source: Deafness Research Foundation, National Institute on Deafness and Other Communication Disorders]

9 How Is Light Sensed?

Vision depends on *photoreceptors*. These sensory cells contain molecules called *photopigments*, which change chemically when they absorb light. This chemical change alters ion channels in the receptor cell membrane, producing a receptor potential.

The Eye Collects and Focuses Light

As you read this, light reflected from the page first encounters the *cornea*, a transparent covering over the front of the eyeball. Behind the cornea, light passes through a chamber filled with a watery fluid called *aqueous humor*, which provides nourishment for both the lens and cornea. The amount of light entering the eye is adjusted by the *iris*, pigmented muscular tissue. The iris regulates the size of the *pupil*, a circular opening in the center of the iris. Light passing through the pupil encounters the *lens*, a structure that is composed of transparent protein fibers and shaped like a flattened sphere. The lens is suspended behind the pupil by muscles that regulate its shape. Behind the lens is a much

Figure 24-13 The human eye

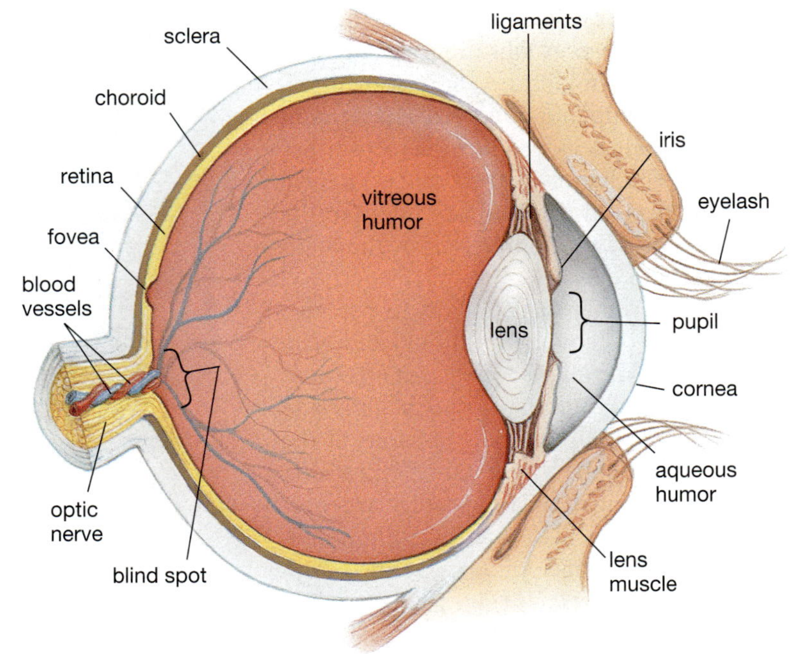

larger chamber filled with the *vitreous humor*, a clear jellylike substance that helps maintain the shape of the eye (Fig. 24-13).

After passing through the vitreous humor, light reaches the *retina*, a multilayered sheet of photoreceptors and neurons. There the light energy is converted into electrical nerve impulses that are transmitted to the brain. Behind the retina is the *choroid*, a darkly pigmented tissue. The choroid's rich blood supply helps nourish the cells of the retina. Its dark pigment absorbs stray light whose reflection inside the eyeball would interfere with clear vision. Surrounding the outer portion of the eyeball is the *sclera*, a tough connective tissue layer that is visible as the white of the eye and is continuous with the cornea.

The Adjustable Lens Can Focus on Distant and Nearby Objects

Although focusing begins at the cornea, whose rounded contour bends light rays, the lens is responsible for final, sharp focusing. The shape of the lens is adjusted by its encircling muscle. When viewed from the side, the lens is either rounded, to focus on nearby objects, or flattened, to focus on distant objects (Fig. 24-14a).

If your eyeball is even a millimeter too long, you will be nearsighted—unable to focus on distant objects. Farsighted people, whose eyeballs are slightly too short, cannot focus on nearby objects. These conditions can be corrected by external lenses of the appropriate shape (Fig. 24-14b,c). Nearsightedness can also be corrected by surgery that slightly flattens the cornea. The lens stiffens as humans age, causing farsightedness, because the lens can no longer round up enough to focus on nearby objects. By their mid-forties, most people require glasses for close work such as reading.

Light Striking the Retina Is Captured by Photoreceptors

The photoreceptors of the eye, called *rods* and *cones* after their shapes, have their light-gathering parts farthest from the light, at the rear of the retina (Fig. 24-15). Between the receptors and incoming light are several layers of neurons that process the signals from the photoreceptors. The retinal layer nearest the vitreous humor consists of *ganglion cells*, whose axons make up the *optic nerve*. Ganglion cell axons pass through the retina at a location called the *blind spot* (see Fig. 24-13). This area lacks receptors; objects focused there seem to disappear. You can demonstrate your blind spot to yourself by using the star and circle below: Close your left eye and focus steadily on the star with your right eye. Start with the book about a foot away and gradually move it closer. The circle will disappear when its image falls on your blind spot.

The receptor potential from the photoreceptors is processed by other retinal neurons in ways that enhance our ability to detect edges, movement, dim light, and changes in light intensity. The much-modified signals from the photoreceptors are finally converted to action

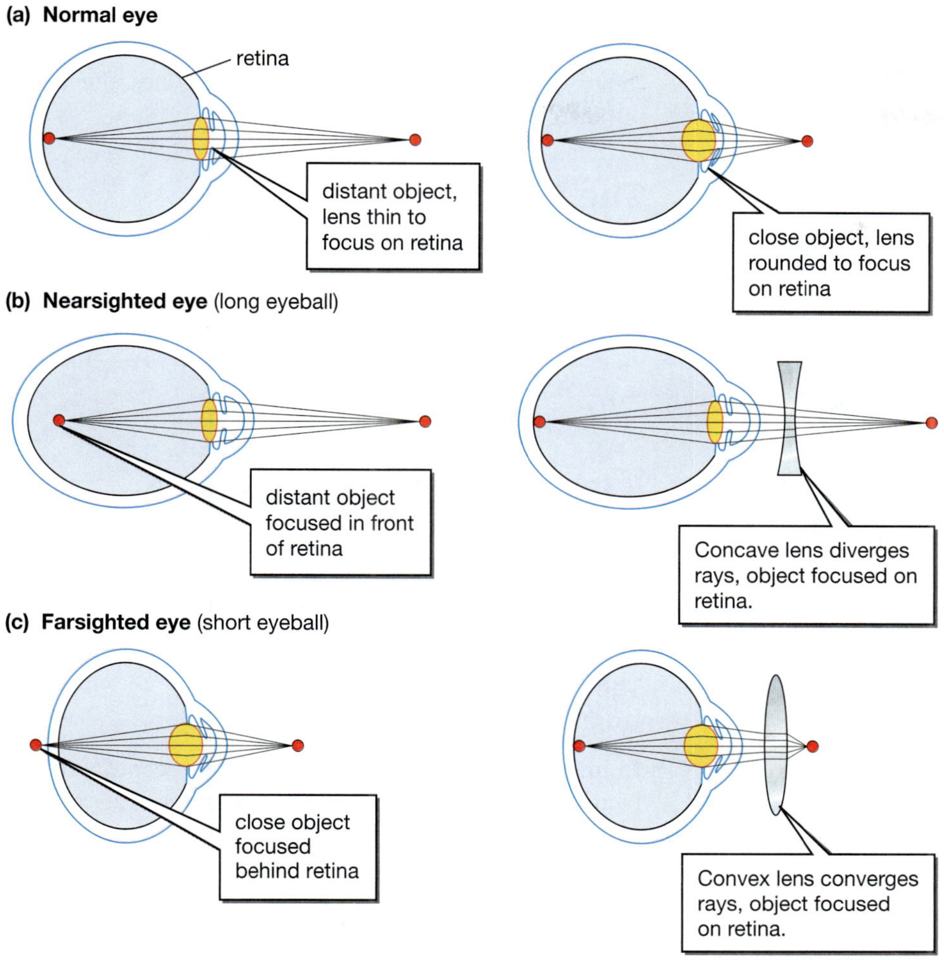

Figure 24-14 Focusing in the human eye
(a) The lens changes shape to focus on objects at different distances.
(b) Nearsightedness is corrected by eyeglasses with concave lenses.
(c) Farsightedness is corrected by eyeglasses with convex lenses.

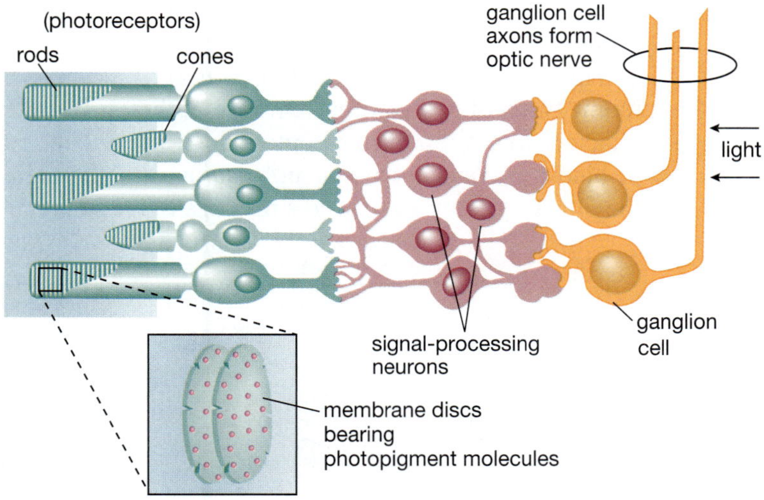

Figure 24-15 The layers of the retina
The human retina has rods and cones (photoreceptors), signal-processing neurons, and ganglion cells. Each rod and cone bears a long extension; the photopigments are embedded in the plasma membranes of the extensions.

potentials, which are carried by the optic nerve to the brain. In the brain, further processing ultimately results in the sensation of vision.

Rods and Cones Differ in Distribution and Light Sensitivity

Photoreception in both rods and cones begins with the absorption of light by photopigment molecules that are embedded in the plasma membranes of the photoreceptors (see Fig. 24-15). Light hitting a photopigment molecule causes a change in the receptor membrane's permeability to ions, producing a receptor potential in the photoreceptor.

Although cones are located throughout the retina, they are concentrated in the *fovea*, the area of the retina where the lens focuses images most sharply (Fig. 24-16; see also Fig. 24-13). The fovea appears as a depression

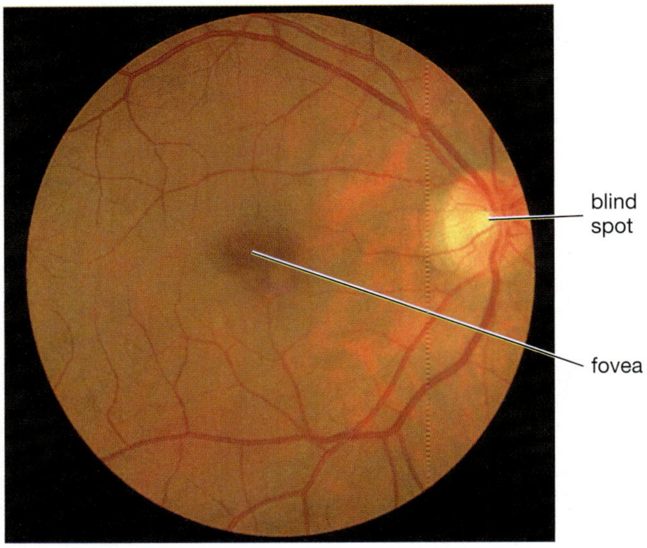

Figure 24-16 The human retina
A portion of the human retina, photographed through the cornea and lens of a living person. The blind spot and fovea are visible. Blood vessels supply oxygen and nutrients; notice that they are dense over the blind spot (where they don't interfere with vision) and scarcer near the fovea.

near the center of the retina, because the layers of signal-processing neurons are pushed aside there. This arrangement allows light to reach the cones of the fovea with relatively little interference.

Human eyes have three varieties of cones, each containing a slightly different photopigment. Each type of photopigment is most strongly stimulated by a particular wavelength of light, corresponding roughly to red, green, or blue. The brain distinguishes color according to the relative intensity of stimulation of different cones. For example, the sensation of yellow is caused by fairly equal stimulation of red and green cones.

Rods dominate in the peripheral portions of the retina. Rods are far more sensitive to light than are cones and so are largely responsible for our vision in dim light. Unlike cones, rods do not distinguish colors; in moonlight, which is too dim to activate the cones, the world appears in shades of gray.

The Big Picture

We tend to think of sights, sounds, smells, and tastes as essential features of the world around us, but these sensations are in fact creations of our brains, which interpret interactions between our sensory receptors and light, vibrations, and chemicals.

10 How Are Chemicals Sensed?

Humans have two separate chemical senses: one for airborne molecules, called smell, or *olfaction*, and one for chemicals dissolved in water or saliva, the sense of *taste*.

The Ability to Smell Arises from Olfactory Receptors

In humans and most other vertebrates, receptors for olfaction are neurons located in a patch of mucus-covered epithelial tissue in the upper portion of each nasal cavity (Fig. 24-17). The human olfactory epithelium is small compared with that of many other mammals (dogs, for example) whose sense of smell is hundreds of times more acute than ours. Olfactory receptors have hairlike dendrites that protrude into the nasal cavity and are embedded in a layer of mucus. Odorous molecules in the air diffuse into the mucus layer and bind with receptor protein molecules on the dendrites. A thousand or more different types of receptor proteins are embedded in the olfactory dendrites. Each receptor protein is specialized to bind a particular type of molecule and to stimulate the olfactory receptor to send a message to the brain.

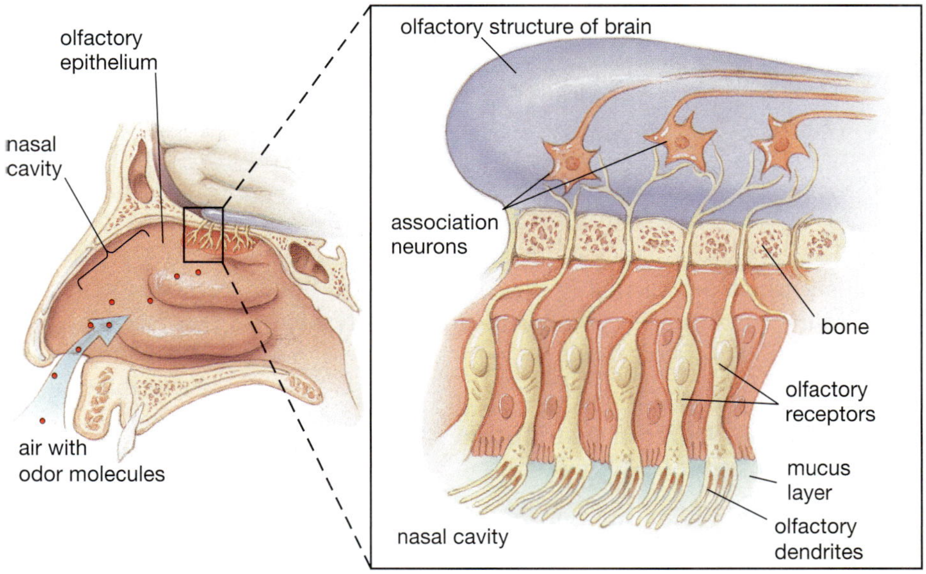

Figure 24-17 Human olfactory receptors
The receptors for olfaction in humans are neurons bearing hairlike projections that protrude into the nasal cavity. The projections are embedded in a mucus layer in which odorous molecules dissolve before contacting the receptors.

Taste Receptors Are Clustered on the Tongue

The human tongue bears about 10,000 *taste buds*, structures embedded in small bumps that cover the surface of the tongue (Fig. 24-18a). Each taste bud is a cluster of 60 to 80 taste receptors surrounded by supporting cells in a small pit. The cells in the pit communicate with the mouth through a taste pore (Fig. 24-18b). Microvilli (thin membrane projections) of taste receptors protrude through the pore. Dissolved chemicals enter the pore and bind to receptor molecules on the microvilli, producing a receptor potential.

Four major types of taste receptors have long been known: sweet, sour, salty, and bitter; a fifth type, *umami* (a Japanese word loosely translated as "delicious"), has recently been identified. The umami receptor responds to glutamate, an ingredient in MSG, which is sometimes used as a seasoning and enhances the flavor of foods. Although it was once believed that each type of taste bud is concentrated on specific areas of the tongue, recent research has shown the different types are more or less evenly distributed.

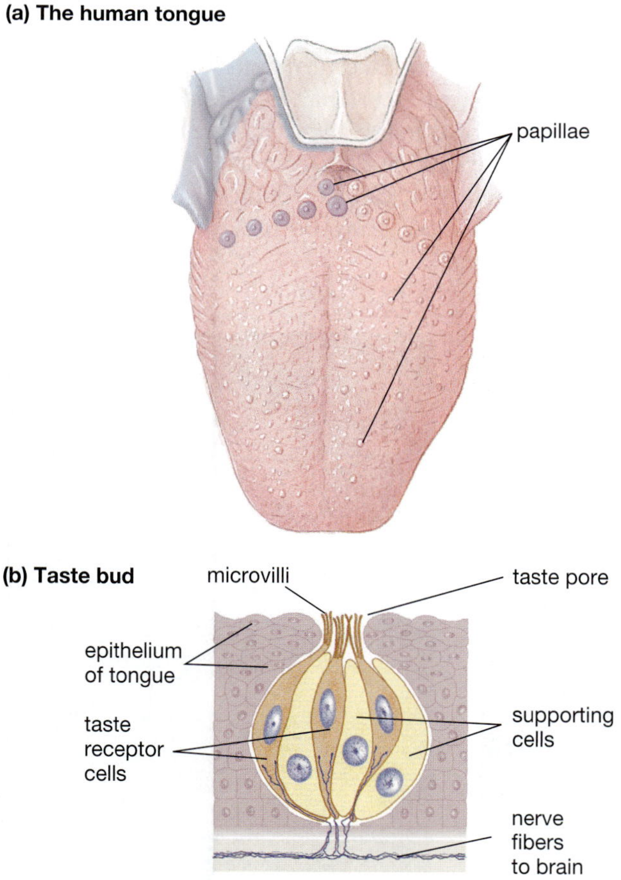

Figure 24-18 Human taste receptors
(a) The human tongue is covered with papillae, bumps in which taste buds are embedded. Small papillae are located on the front two-thirds of the tongue; larger ones with more taste buds are far in the back. ***(b)*** Each taste bud consists of supporting cells surrounding taste receptors, whose microvilli have protein receptors that bind tasty molecules, producing a receptor potential.

With only five types of taste receptor, how do we perceive such a wide variety of different flavors? First, a particular substance may stimulate two or more receptor types to different degrees, making it taste "salty–sweet," for example. More importantly, a substance on the tongue typically releases molecules into the air inside the mouth. These odorous molecules diffuse to the olfactory receptors, which contribute an odor component to the basic flavor. (Recall that the mouth and nasal passages are connected.) The importance of olfaction to taste perception is illustrated when you have a bad cold that blocks your sense of smell, and normally tasty foods seem bland and unappealing.

Pain Is a Specialized Chemical Sense

Researchers have found that pain perception is actually a special kind of chemical sense (Fig. 24-19). For example, when cells and capillaries are damaged by a cut or a burn, their contents flow into the extracellular fluid. The cell contents include potassium ions, which produce receptor potentials in *pain receptors*. Damaged cells also release enzymes that convert certain blood proteins into a chemical called *bradykinin*, another stimulus that produces receptor potentials in pain receptors.

Each part of the body has a separate set of pain receptor neurons that provide input to particular brain cells. Hence, the brain can identify the location of the

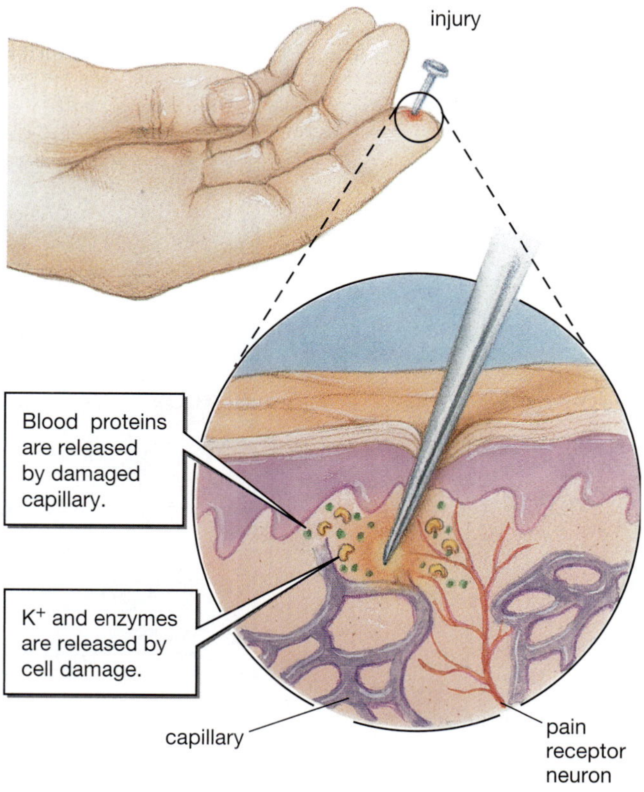

Figure 24-19 Pain perception
An injury, such as a tack stab, damages both cells and blood vessels. The damaged cells and vessels release substances that activate pain receptor neurons.

pain. Drugs that provide pain relief, such as morphine or Demerol, block synapses in the pain pathways of the brain or spinal cord. As we mention in "Health Watch: Drugs, Addiction, and Neurotransmitters," the brain can modulate its perception of pain through its own narcotic-like endorphins.

Evolutionary Connections

Uncommon Senses

We have reviewed the "common" senses of sound, sight, smell, taste, and pain. But had this text been written for bats, it would undoubtedly have included a large section on *echolocation* and would not have covered vision at all! Here we review a few of the "uncommon" senses that animals have evolved.

Echolocation

Some animals are able to hunt in darkness or murky water because they have evolved a type of sonar called *echolocation*, similar to the navigational system used by ships. Using sound waves and echolocation, bats can navigate and hunt insect prey in total darkness. An echolocating bat emits pulses of noise at ultrasonic frequencies (higher than can be detected by the human ear). The pulses bounce off nearby objects, and the patterns of returning sound convey accurate information about the size, shape, surface texture, and location of objects in the environment. Little brown bats can detect wires only 1 millimeter thick from a distance of more than 6 feet. Several adaptations contribute to this remarkable sensitivity. The bat's large external ears collect the returning echoes and help the bat locate the source of the echoes (Fig. 24-20a). As the bat emits its cry, muscles attached to the bones of the middle ear contract briefly, reducing the bones' vibrations and preventing the bat from being deafened by its own calls. The tympanic membrane and bones of the middle ear are exceptionally light and easily vibrated by the faint returning echoes.

Porpoises produce ultrasonic clicks within their nasal passages and emit the clicks through the front of their heads (Fig. 24-20b). A large, flexible, oil-filled sac in the porpoise's head directs the sound forward in a broad beam (for navigation) or a narrow beam (for prey location). A porpoise detects reflected sound waves mainly through its lower jaw bone, which contains a fat-filled cavity that transmits vibrations to the inner ear. Porpoises also have small external ear openings just behind each eye, but little, if any, sound is transmitted through these openings. An echolocating porpoise can locate a pea-sized object on the floor of its tank and can distinguish among species of fish.

Detecting Electrical Fields

Some fish species use electrical fields for *electrolocation* in much the same way that bats and porpoises use sound waves for echolocation. These fish produce high-frequency electrical signals from an electric organ just in front of their tails; they detect these signals with *electroreceptors* located along both sides of their bodies

(a)

(b)

Figure 24-20 Echolocation
(a) A long-eared bat, showing the enormous size and elaborate folds of the external ears that help it localize returning echoes. ***(b)*** Bottlenose porpoises focus ultrasonic clicks by using oil-filled sacs in the front of their heads.

(Fig. 24-21). Objects near the fish distort the electrical field that surrounds the fish. This distortion is detected by the electroreceptors, which send an altered pattern of action potentials to the brain. The fish uses this information to detect and localize nearby objects.

Detecting Magnetic Fields

Homing pigeons are famous for their ability to fly home after being released some distance away. They can accurately locate their home roost even when few landmarks are present and clouds obscure the sun. If a researcher straps a small magnet to a pigeon's back, however, the bird finds its way home only in sunny weather; it loses its way when there are clouds. Apparently, pigeons can navigate either by the sun or, if the sun is hidden, by magnetic fields. In cloudy weather, the experimental magnets confuse the pigeon's internal magnetic compass. How do pigeons detect magnetic fields? Pigeons have deposits of magnetite (a magnetic iron compound) just beneath their skull—a built-in magnet that is used to tell direction.

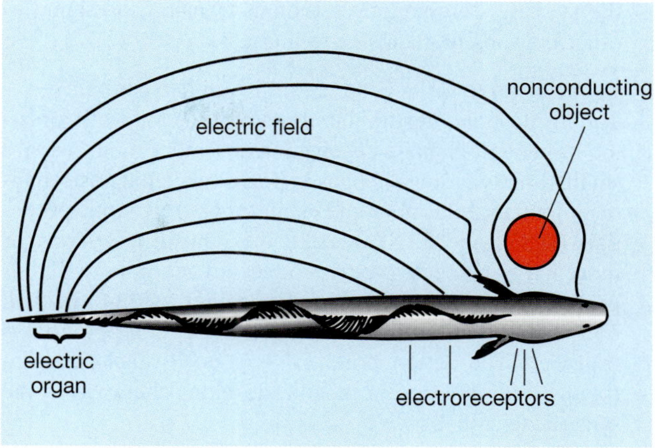

Figure 24-21 Electrolocation
Fish of some species locate nearby objects by producing an electric field and then sensing distortions in the field caused by the objects.

CASESTUDYREVISITED

A Bad Day for Phineas Gage

As you've probably guessed by now, the blow that disrupted the personality of Phineas Gage did so by damaging that most human of brain regions, the cerebral cortex. More specifically, the tamping bar that struck Gage entered below his left eye and passed upward, destroying a large piece of the frontal lobe of his cortex before exiting through the top of his head. Present-day researchers have managed to further pinpoint the zone of destruction, using Gage's skull (exhumed after his death and donated to the Warren Medical Museum at Harvard). Working from the original skull, neurobiologists Hanna and Antonio Damasio used computer modeling techniques to create a three-dimensional image of a brain that nestled nicely inside an image of Gage's skull. The Damasios were then able to re-create the path through the brain that the tamping iron projectile must have followed to leave the particular holes that it did.

The Damasios concluded that the most likely trajectory of the rod would have resulted in severe damage to the ventromedial region of the frontal lobes (the area on the underside of the lobes and near the center of the head), with especially heavy damage to the left lobe. Damage in this same area is observed today in accident victims and patients with brain tumors who exhibit personality changes similar to those of Phineas Gage. Furthermore, damage that was confined mostly to the ventromedial area would be expected to leave intact the frontal lobe areas now known to be associated with language and motor function. So the Damasios' retroactive diagnosis of Gage's injury is consistent with modern understanding of localization in the frontal lobe.

Gage's accident and its aftermath marked the beginning of scientific study of the manner in which the brain divides its work. But the mishap also pointed the way for a new, and still evolving, view of human consciousness. In the words of neurologist Frank Vertosick, Jr., "the lesson of Phineas Gage, a man traumatically converted from saint to sinner, still haunts us. In this enlightened age, we readily accept that neurons govern our mundane actions—walking, talking, urinating. With a little prodding, we can even accept that neurons are responsible for the Sistine Chapel and general relativity. But what happened to Phineas Gage showed us that our sense of right and wrong, of moral restraint and societal convention, also derives from the brain, or, more specifically, the human frontal lobes."

Summary of Key Concepts

1) How Do Nerve Cells Work?

Nervous systems are composed of billions of neurons. A neuron has four major specialized functions, which are reflected in its structure: (1) Dendrites receive information from the environment or from other neurons. (2) The cell body adds together electrical signals from the dendrites and from synapses on the cell body itself and "decides" whether to produce an action potential. (The cell body also coordinates the neuron's metabolic activities.) (3) The axon conducts the action potential to its output terminal:

the synapse. (4) Synaptic terminals transmit the signal to other neurons, to glands, or to muscles.

2 How Do Nerve Cells Produce and Transmit Signals?

The inside of an unstimulated neuron has a negative resting potential. Signals received from other neurons are small, rapidly fading changes in potential called postsynaptic potentials. Inhibitory and excitatory postsynaptic potentials (IPSPs and EPSPs) make the neuron less likely or more likely, respectively, to produce an action potential. If postsynaptic potentials, added together within the cell body, bring the neuron to threshold, an action potential is triggered. The action potential is a positive charge that travels, undiminished in magnitude, along the axon to the synaptic terminal.

At a synapse, neurotransmitters from the presynaptic neuron, released in response to an action potential, bind to receptors in the postsynaptic neuron's plasma membrane, opening ion channels there. Ions flow, producing either an EPSP or an IPSP, depending on the type of ion channels opened.

3 How Do Nervous Systems Process Information?

Information processing in the nervous system requires four operations. The nervous system must (1) determine the type of stimulus, (2) signal the intensity of the stimulus, (3) integrate information from many sources, and (4) initiate and direct the response. The nervous system collects and processes sensory information from many sources. These sensory stimuli may converge on fewer neurons whose activity in response to the stimuli determines the action taken. The "decision" to act may then be transmitted to many more neurons (divergence), which direct the activity.

4 How Do Nervous Systems Direct Behavior?

Behavior is controlled by neural pathways that have (1) sensory neurons, (2) association neurons, (3) motor neurons, and (4) effectors. The convergence of many pathways on association neurons organizes complex behaviors.

5 How Is the Human Nervous System Organized?

The nervous system consists of the central nervous system and the peripheral nervous system. The peripheral nervous system is further subdivided into sensory and motor portions. The motor portions consist of the somatic nervous system (which controls voluntary movement) and the autonomic nervous system (which directs involuntary responses).

Within the central nervous system, the spinal cord contains (1) neurons controlling voluntary muscles and the autonomic nervous system and neurons communicating with the brain and other parts of the spinal cord; (2) axons leading to and from the brain; and (3) neural pathways for reflexes and certain simple behaviors. The brain consists of three regions: the hindbrain, midbrain, and forebrain, each further subdivided into distinct parts.

The hindbrain in humans consists of the medulla and pons, which control involuntary functions (such as breathing), and the cerebellum, which coordinates complex motor activities (such as typing). In humans, the small midbrain contains the reticular formation, a filter and relay for sensory stimuli. The forebrain includes the thalamus, a sensory relay station that shuttles information to and from conscious centers in the forebrain. The diverse structures of the limbic system of the forebrain are involved in emotion, learning, and the control of instinctive behaviors such as sex, feeding, and aggression. The cerebral cortex of the forebrain is the center for information processing, memory, and initiation of voluntary actions. It includes primary sensory and motor areas and association areas that analyze sensory information and plan movements.

6 How Does the Brain Produce the Mind?

Each cerebral hemisphere is specialized. The left hemisphere dominates speech, reading, writing, language comprehension, mathematical ability, and logical problem solving. The right hemisphere specializes in recognizing faces and spatial relationships, artistic and musical abilities, and recognition and expression of emotions.

Working memory is electrical or chemical and lasts for only a short time. Long-term memory is acquired through structural changes that increase the effectiveness of synapses. The hippocampus is an important site for learning and for the transfer of information into long-term memory. The temporal lobes are important for memory.

7 How Do Sensory Receptors Work?

Receptors are specialized cells that produce receptor potentials or neurotransmitters in response to sensory stimuli. The receptor potentials or neurotransmitters in turn produce action potentials in postsynaptic neurons.

8 How Is Sound Sensed?

In the vertebrate ear, air vibrates the tympanic membrane, which transmits vibrations to the bones of the middle ear and then to the oval window of the fluid-filled cochlea. Within the cochlea, vibrations bend the hairs of hair cells, which are receptors located between the basilar and tectorial membranes. This bending produces receptor potentials in the hair cells that cause action potentials in the axons of the auditory nerve, which leads to the brain.

9 How Is Light Sensed?

In the vertebrate eye, light enters the cornea and passes through the pupil to the lens, which focuses an image on the fovea of the retina. Two types of photoreceptor, rods and cones, are located deep in the retina. They produce receptor potentials in response to light. These signals are processed through several layers of neurons in the retina and are translated into action potentials in the optic nerve, which leads to the brain. Rods are more abundant and more light-sensitive than are cones and provide vision in dim light. Cones, which are concentrated in the fovea, provide color vision.

10 How Are Chemicals Sensed?

Humans detect chemicals in the environment by olfaction and taste. Each type of olfactory or taste receptor responds to only one or a few specific types of molecules, allowing discrimination among tastes and odors. Olfactory neurons of vertebrates are located in epithelial tissue that lines the nasal cavity. Taste receptors are located in clusters called taste buds on the tongue. Pain is a special chemical sense in which pain receptors respond to chemicals released by damaged cells.

Key Terms

action potential *p. 458*
association neuron *p. 463*
autonomic nervous system *p. 463*
axon *p. 458*
brain *p. 463*
cell body *p. 458*
central nervous system *p. 463*
dendrite *p. 458*
effector *p. 463*
forebrain *p. 468*
hindbrain *p. 466*
midbrain *p. 468*
motor neuron *p. 463*
nerve *p. 458*
neuron *p. 458*
neurotransmitter *p. 458*
peripheral nervous system *p. 463*
postsynaptic neuron *p. 459*
postsynaptic potential (PSP) *p. 459*
presynaptic neuron *p. 459*
receptor potential *p. 471*
resting potential *p. 459*
sensory neuron *p. 463*
sensory receptor *p. 471*
somatic nervous system *p. 463*
spinal cord *p. 463*
synapse *p. 459*
synaptic terminal *p. 458*
threshold *p. 459*

Thinking Through the Concepts

Multiple Choice

1. __________ *are integration centers in neurons.*
   a. dendrites
   b. axons
   c. cell bodies
   d. ion channels
   e. synapses

2. *The role of the axon is to*
   a. integrate signals from the dendrites
   b. release neurotransmitter
   c. conduct the action potential to the synaptic terminal
   d. synthesize cellular components
   e. stimulate a muscle, gland, or another neuron

3. *Which of the following statements is INCORRECT?*
   a. The hindbrain contains the medulla, pons, and cerebellum.
   b. The thalamus is an important sensory relay structure.
   c. The cerebral cortex in humans is the largest part of the brain.
   d. The hippocampus has an important role in recognizing and experiencing fear.
   e. The hypothalamus coordinates the activities of the autonomic nervous system.

4. *The fovea is*
   a. the blind spot
   b. a clear area in front of the pupil and iris
   c. the tough outer covering of the eyeball
   d. the substance that gives the eyeball its shape
   e. the central focal region of the vertebrate retina

5. *Light entering the eye and striking the choroid would travel first, in order, through the*
   a. lens, vitreous humor, cornea, aqueous humor, and retina
   b. retina, aqueous humor, lens, vitreous humor, and cornea
   c. cornea, aqueous humor, retina, vitreous humor, and lens
   d. cornea, aqueous humor, lens, vitreous humor, and retina
   e. lens, aqueous humor, cornea, vitreous humor, and retina

6. *Sound reception is carried out by hair cells, which are a special type of*
   a. chemoreceptor
   b. photoreceptor
   c. mechanoreceptor
   d. magnetoreceptor
   e. thermoreceptor

? Review Questions

1. List four major parts of a neuron, and explain the specialized function of each part.
2. Diagram a synapse. How are signals transmitted from one neuron to another at a synapse?
3. How does the brain perceive the intensity of a stimulus? The type of stimulus?
4. What are the four elements of a simple nervous pathway? Describe how these elements function in the human pain-withdrawal reflex.
5. Draw a cross section of the spinal cord. What types of neurons are located in the spinal cord? Explain why severing the cord paralyzes the body below the level where it is severed.
6. Describe the functions of the following parts of the human brain: medulla, cerebellum, reticular formation, thalamus, limbic system, and cerebrum.
7. What structure connects the two cerebral hemispheres? Describe the evidence that the two hemispheres are specialized for distinct intellectual functions.
8. Distinguish between long-term memory and working memory.
9. What are the names of the specific receptors used for taste, vision, hearing, smell, and touch?
10. Why are we apparently able to distinguish hundreds of different flavors if we have only five types of taste receptors? How are we able to distinguish so many different odors?
11. Describe the structure and function of the various parts of the human ear by tracing the route of a sound wave from the air outside the ear to the cells that cause action potentials in the auditory nerve.

12. How does the structure of the inner ear allow for the perception of pitch? Of sound intensity?
13. Diagram the overall structure of the human eye. Label the cornea, iris, lens, retina, and choroid. Describe the function of each structure.
14. How does the lens change shape to allow focusing of distant objects? What defect makes focusing on distant objects impossible, and what is this condition called? What type of lens can be used to correct it, and how does it do so?
15. List the similarities and differences between rods and cones.
16. Distinguish between taste and olfaction.
17. Describe how pain is signaled by tissue damage.

Applying the Concepts

1. Argue for or against the statement "Consciousness by its nature is incomprehensible; the brain will never understand the mind."
2. In Parkinson's disease, which afflicts several million Americans, the cells that produce the neurotransmitter dopamine degenerate in a small part of the brain that is important in the control of movement. Some physicians have reported improvement after injecting cells taken from the same general brain region of an aborted fetus into appropriate parts of the brain of a Parkinson's patient. Discuss this type of surgery from as many viewpoints as possible: ethical, financial, practical, and so on. On the basis of your responses, is fetal transplant surgery the answer to curing Parkinson's disease?
3. If the axons of human spinal cord neurons were unmyelinated, would the spinal cord be larger or smaller? Would you move faster or slower? Explain your answer.
4. What is the adaptive value of reflexes? If Christopher Reeve was accidentally pricked on the toe by a thumbtack, would he withdraw his leg? Explain your answer.
5. Explain the statement "Your sensory perceptions are purely a creation of your brain." Discuss the implications for communicating with other humans, with other animals, and with intelligent life from another universe.
6. Corneal transplants can help restore vision and greatly improve the recipient's quality of life. What properties of the cornea make it an excellent candidate for transplantation? Suggest some ways in which society could improve the availability of corneal and other tissues for transplantation.

Group Activity

Form groups of three or four for the following exercise: Devise and carry out a simple experiment to demonstrate that there are fewer cones and less acute vision in the peripheral portions of the human retina than in and near the fovea. (Recall the different properties and abilities of rods and cones.) Try this experiment on all members of the group. Estimate the percentage of your field of vision that has relatively few cones in it.

For More Information

Axel, R. "The Molecular Logic of Smell." *Scientific American*, October 1995. Describes research that uncovers some of the mechanisms by which the nose and brain decipher scents.

Beardsley, T. "The Machinery of Thought." *Scientific American*, August 1997. Using PET and MRI on monkeys and humans, researchers are learning more about where working memory resides.

Bower, B. "Creatures in the Brain." *Science News*, April 13, 1996. Using imaging techniques, scientists have discovered clues as to how different regions of the brain are specialized for different concepts.

Dolnick, E. "Obsessed." *Health*, September 1994. Researchers have linked obsessive-compulsive disorders to imbalances in neurotransmitters.

Freedman, D. H. "In the Realm of the Chemical." *Discover*, June 1993. Describes how smell and taste help us experience our world.

Gutin, J. C. "Good Vibrations." *Discover*, June 1993. Research uncovers mechanisms of hearing.

Holloway, M. "Rx for Addiction." *Scientific American*, March 1991. By studying drug addiction, neurobiologists are learning more about the brain and new ways to counteract addictions.

Kimura, D. "Sex Differences in the Brain." *Scientific American*, September 1992. Hormonal differences between the sexes influence brain development.

Koretz, J. F., and Handelman, G. H. "How the Human Eye Focuses." *Scientific American*, July 1988. Describes focusing in the human eye with an emphasis on the loss of focusing ability that occurs with age.

McKean, K. "Pain." *Discover*, October 1986. Describes the discovery of bradykinin and its role in pain perception.

Nathans, J. "The Genes for Color Vision." *Scientific American*, February 1989. Good discussion of the physiology and genetics of color blindness.

Raichle, M. E. "Visualizing the Mind." *Scientific American*, April 1994. Brain imaging techniques partially open the "black box" of the mind.

Scientific American, special issue on Mind and Brain, September 1992. Articles describe the development of the brain, the biochemical basis of learning, major neurological disorders, and aging effects on the brain.

Answers to Multiple-Choice Questions

1. c 2. c 3. d 4. e 5. d 6. c

MEDIATUTOR

CD Activities

Activity 24.1: Ions and Electrical Signals

Estimated time: 5 minutes

The electrical signal that is generated by a neuron involves a flow of ions across the membrane of the cell. As these ions flow, the electrical potential (or charge) across the membrane of the cell changes. The result is the production of an action potential, which is the electrical output signal of the neuron. In this tutorial, you will explore the events that generate an action potential.

Activity 24.2: Synapses

Estimated time: 5 minutes

Neurons communicate at junctions called synapses. This communication involves the action of a group of chemicals called neurotransmitters. In this tutorial, you will view an animation that simulates the action that occurs at a synapse and the role of the neurotransmitter in this action.

Start the MediaTutor Student CD-ROM and enter the activity number in the Quick Search box to be taken directly to that activity.

Web Investigation

A Bad Day for Phineas Gage

Estimated time: 20 minutes

The human nervous system is a highly complex, intricate system that controls our movements, emotions, and cognition. When part of this intricate system fails, the result can be any of a number of diseases. This exercise will investigate the various mechanisms of some of the better-known diseases that involve the nervous system.

Go to http://www.prenhall.com/audesirk3, the Audesirk Companion Web site. Select Chapter 24 and Web Investigation to begin.

As the frequency of multiple births grows ever higher, more parents are facing challenges like those experienced by Bobbi and Kenny McCaughey. The McCaugheys had septuplets in 1997.

25 Animal Reproduction and Development

AT A GLANCE

CASE STUDY

Multiplicity

Media coverage was intense in November 1997 when Bobbi McCaughey of Carlisle, Iowa, gave birth to seven babies. Hundreds of newspaper and television reporters converged on the Des Moines hospital where the septuplets were born. As news of the septuplets' arrival spread, gifts, donations, and offers of help for the busy parents began to pour in. The public was captivated by the saga of the seven tiny siblings (four boys and three girls), the first septuplets born in the United States since 1985 and the only set in which all the babies survived.

The birth of septuplets is obviously exceedingly rare, but the phenomenon of multiple births has become increasingly common. Over the past decade, the proportion of births that gave rise to twins has increased by more than 30%, and the number of births involving triplets or higher-order multiples has nearly quadrupled since the 1970s. Most experts attribute these dramatic increases to two main factors: a trend for women to delay childbearing (the chances of a multiple birth increase with the age of the mother) and the increasingly common use of drugs for enhancing fertility. Sometimes, the fertility drugs work too well, and multiple births result.

Most parents are happy to have more than one baby added to their families. Multiple births are not, however, without problems. Babies born in multiple births tend to be smaller and premature. The McCaughey septuplets were born nine weeks prematurely, and none weighed more that 3¼ pounds at birth. The last septuplet did not leave the hospital until four months after its birth. Even today, several of the McCaughey children have serious, ongoing health problems. In general, multiple-birth babies (and their mothers) are at greater risk for poor health.

The ultimate biological causes of multiple births remain mysterious, but we can trace part of the sequence of events that leads to multiples. The events will be easier to understand following an exploration of the processes of animal reproduction and development. ■

1) How Do Animals Reproduce?

Animals reproduce either sexually or asexually. In **sexual reproduction**, adults produce haploid gametes by meiosis, and two gametes (usually from separate parents) fuse to form a diploid offspring. Offspring receive some genes from each parent, so their genetic make-up differs from that of either parent. In contrast, offspring generated by **asexual reproduction** are genetically identical to the parent, because the offspring arise through mitosis of adult body cells.

In most animal species, an individual is either male or female. The sexes are defined by the type of haploid gamete that each produces. Females produce **eggs**, which are large, nonmotile cells containing food reserves.

Males produce small, motile **sperm**, which have almost no cytoplasm and hence no food reserves. The union of sperm and egg to form a diploid cell is known as **fertilization**.

In Asexual Reproduction, There Is No Fusion of Sperm and Egg

Because humans reproduce sexually, we tend to regard sexual reproduction as the normal, best method. In reality, sexual reproduction is a rather inefficient method of producing offspring. Asexual reproduction is far more efficient, because there is no need to find a mate, court, and fend off rivals, and there is no waste of gametes that fail to unite to form an offspring. Many animals do reproduce asexually, at least some of the time.

Budding Produces a Miniature Version of the Adult

One common method of asexual reproduction in animals is **budding**. Many sponges and cnidarians, for example, reproduce in this manner (Fig. 25-1). A miniature version of the animal—a bud—grows directly on the body of the adult, drawing nourishment from its parent. When it has grown large enough, the bud breaks off and becomes independent.

Eggs May Develop Without Fertilization

The females of some animal species can reproduce by a process in which haploid egg cells develop into adults without being fertilized. Some species of fish and lizards, in fact, have completely done away with males. Their populations consist entirely of asexually reproducing females. Still other animals, such as the aphid, can reproduce either sexually or asexually, depending on environmental factors such as the season or the availability of food (Fig. 25-2).

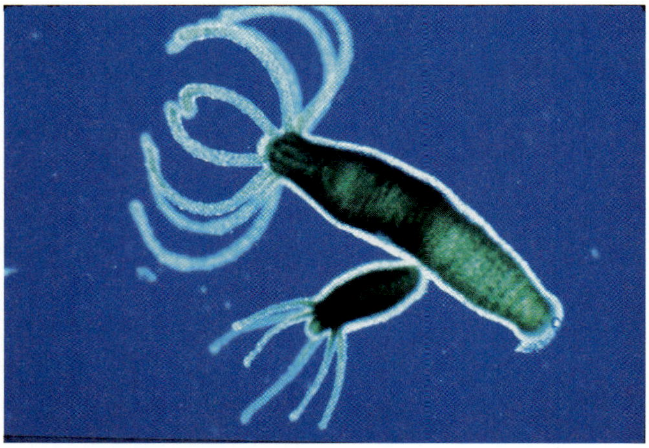

Figure 25-1 Budding
The offspring of some cnidarians, such as the *Hydra* shown here, grow as buds that appear as miniature adults sprouting from the body of the parent. When sufficiently developed, the buds break off and assume independent existence.

Sexual Reproduction Requires the Union of Sperm and Egg

Given the obvious efficiency of asexual reproduction, no one is sure why sex arose. But it did, and sexual reproduction is now a prominent feature in the lives of most animals. It also has an important outcome. The genetic recombination that results from the union of sperm and egg tends to create novel genotypes—and therefore novel phenotypes—that are an important source of variation on which natural selection may act.

Although in most animal species an individual is either male or female, in some, such as earthworms and many snails, single individuals produce both sperm and eggs. Such individuals are commonly called *hermaphrodites* (after Hermaphroditos, a male Greek god whose body was merged with that of a female water nymph, producing a half-male and half-female being). In most hermaphroditic species, reproduction requires a mutual exchange of sperm between individuals. In some hermaphroditic species, however, individuals can fertilize their own eggs if necessary. These animals, including tapeworms and some types of snails, are relatively immobile and may find themselves isolated from other members of their species. Obviously, the ability to fertilize oneself is advantageous under these circumstances.

Gametes Are Brought Together for Fertilization in Various Ways

For species with separate sexes and for hermaphrodites that cannot self-fertilize, successful reproduction requires that sperm and eggs from different animals be brought together for fertilization. How the union of

Figure 25-2 A female aphid gives live birth
In spring and early summer, when food is abundant, aphid females reproduce asexually. In fall, they reproduce sexually, as the females mate with males. Aphids have thus evolved the ability to exploit the advantages of both asexual reproduction (rapid population growth during times of abundant food, no energy spent in seeking a mate, no wasted gametes) and sexual reproduction (genetic recombination).

sperm and egg is accomplished depends on the mobility of the animals and on whether they breed in water or on land.

Sperm and Egg May Meet Outside the Parents' Bodies

In *external fertilization*, sperm and eggs unite outside the bodies of the parents. In species that use external fertilization to breed in water, gametes are released into the water, and the sperm swim to reach the eggs. This type of breeding is called *spawning*. In species that spawn, the reproductive behavior of males and females must be synchronized, so that sperm and eggs are in the same place at the same time.

Most spawning animals rely at least in part on environmental cues to synchronize breeding. Cues such as seasonal changes in daylength typically stimulate the physiological changes that lead to readiness for breeding. As a result, breeding is restricted to certain times of year. More precise synchrony, however, is required to coordinate the actual release of sperm and egg. For example, grunion, fish that inhabit coastal waters off southern California, perform their reproductive rituals only at a particular season, time of day, and phase of the moon (Fig. 25-3). On autumn nights with a full moon (which coincides with the highest tides), grunion swim up onto sandy beaches. Writhing masses of males and females release their gametes into the wet sand and then swim back out to sea on the next wave.

Reproduction may still be inefficient even if the timing of spawning is synchronized. Corals, sea stars, and mussels all waste enormous quantities of sperm and eggs because the gametes are released relatively far apart. In more mobile species, mating behaviors can help ensure synchrony in both time *and* place. Most fish, for example, have a courtship ritual in which the male and female come very close together and release their gametes in the same place and at the same time (Fig. 25-4). Frogs and toads carry this ritual one step further by assuming a characteristic mating pose (Fig. 25-5). At the edges of ponds and lakes, the male mounts the female and stimulates her to release her eggs, which the male fertilizes by releasing a cloud of sperm above them.

Figure 25-3 Environmental cues may synchronize spawning
At the highest tides of fall, grunion swarm ashore on the few undeveloped beaches left in southern California. The fish burrow slightly into the sand and release sperm and eggs.

Eggs May Be Fertilized Within the Female Body

In *internal fertilization*, sperm and eggs unite within the body of the female. Because sperm must be bathed in fluid until they reach the eggs, internal fertilization is a

Figure 25-4 Courtship rituals synchronize release of sperm and eggs
During spawning, male and female seahorses orient their bodies so that the female can deposit her eggs in the male's pouch. Sperm fertilize the eggs as they enter the pouch.

Figure 25-5 Golden toads mating
By clutching the female, the smaller male stimulates her to release eggs.

key adaptation to life on land. Even in aquatic environments, internal fertilization increases the likelihood of successful fertilization, because the sperm and eggs are confined together in a small space rather than relying on encounters within a large volume of water.

Internal fertilization usually occurs by **copulation**, a behavior in which the male deposits sperm directly into the reproductive tract of the female (Fig. 25-6). Copulation, however, is not the only method by which internal fertilization can be accomplished. In some species, sperm are packaged in containers called *spermatophores*. Males of many spermatophore-producing species, including some scorpions, grasshoppers, and salamanders, simply drop their spermatophores on the ground, and if a female finds one, she inserts it into her reproductive cavity, where the enclosed sperm are liberated. Other species have evolved more reliable ways to transfer a spermatophore. For example, in some scorpion species the male attaches his spermatophore to the ground, then engages his mate in a "push-pull dance," during which he maneuvers her back and forth until she is positioned over the spermatophore.

Simply depositing sperm into the body of the female does not guarantee fertilization. To be fertilized, a mature egg must be released into the female reproductive tract during the limited time when sperm are present. So even when fertilization is internal, the mating behavior of males and females must be synchronized, just as with spawning animals. Most mammals, for example, copulate only at certain seasons of the year or when the female signals readiness to mate. The season or signal typically coincides with the release of the egg cell from the ovary. An alternative method, employed by many female snails and insects, is to store sperm for days, weeks, or even months, thus ensuring a supply of sperm whenever eggs are ready.

(a) (b) (c)

Figure 25-6 Internal fertilization is essential for reproduction on land
(a) Beetles mate on a dandelion flower. ***(b)*** South American tortoises must cope with confining shells. ***(c)*** King penguins mate comfortably in the snow.

2 How Does the Human Reproductive System Work?

Like other mammals, humans have separate sexes and reproduce sexually, with internal fertilization. Unlike most other mammal species, however, human reproduction is not restricted by season. Men produce sperm more or less continuously, and women ovulate (release a mature egg cell) about once a month. The central structures of the reproductive system are the **gonads**, paired organs that produce sperm or eggs.

The Male Reproductive System Includes the Testes and Other Structures

The male gonads are the **testes** (singular, **testis**). The male reproductive system also includes glands and ducts that secrete substances that activate and nourish

Table 25-1 Structures and Functions of the Human Male Reproductive Tract

Structure	Type of Organ	Function
Testis	Gonad	Produces sperm and testosterone
Epididymis and vas deferens	Ducts	Store sperm; conduct sperm from testes to penis
Urethra	Duct	Conducts semen from vas deferens and urine from urinary bladder to the tip of the penis
Penis	External "appendage"	Deposits sperm in female reproductive tract
Seminal vesicles	Glands	Secrete fluids that contain fructose (energy source); fluids may wash sperm out of ducts of male reproductive tract into vagina
Prostate	Gland	Secretes fluids that are basic (neutralize acidity of vagina) and contain factors that enhance sperm motility
Bulbourethral glands	Glands	Secrete mucus (may lubricate penis in vagina)

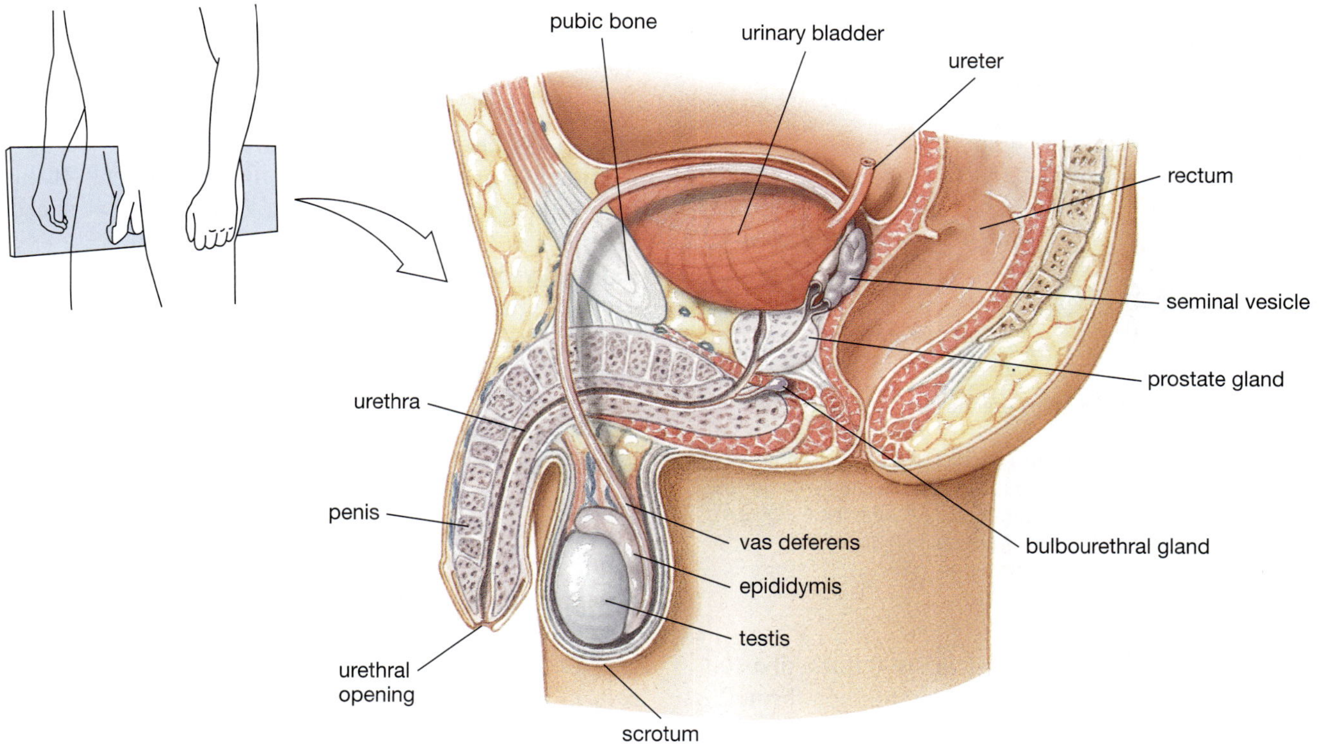

Figure 25-7 The human male reproductive tract

MEDIATUTOR 25.1 Human Reproductive Anatomy

the sperm, store it, and conduct it to the female reproductive tract (Table 25-1 and Fig. 25-7).

Sperm Are Produced in the Testes

The testes, which produce both sperm and male sex hormones, are located in the **scrotum**, a pouch that hangs outside the main body cavity. This location keeps the testes about 7 °F cooler than the core of the body and provides the optimal temperature for sperm development. (Tight jeans push the scrotum up against the body, raising the temperature of the testes. Some researchers believe that wearing tight pants may decrease fertility by reducing the number of sperm. This is not, however, a reliable means of birth control!) Coiled, hollow *seminiferous tubules*, in which sperm are produced, nearly fill each testis (Fig. 25-8a). In the spaces between the tubules are cells that produce the male hormone testosterone.

Just inside the wall of each seminiferous tubule lie **spermatogonia** (singular, **spermatogonium**), diploid cells from which sperm arise, and much larger *Sertoli cells*, which nourish developing sperm and regulate their growth (Fig. 25-8b). Each spermatogonium that divides takes one of two developmental paths. In one

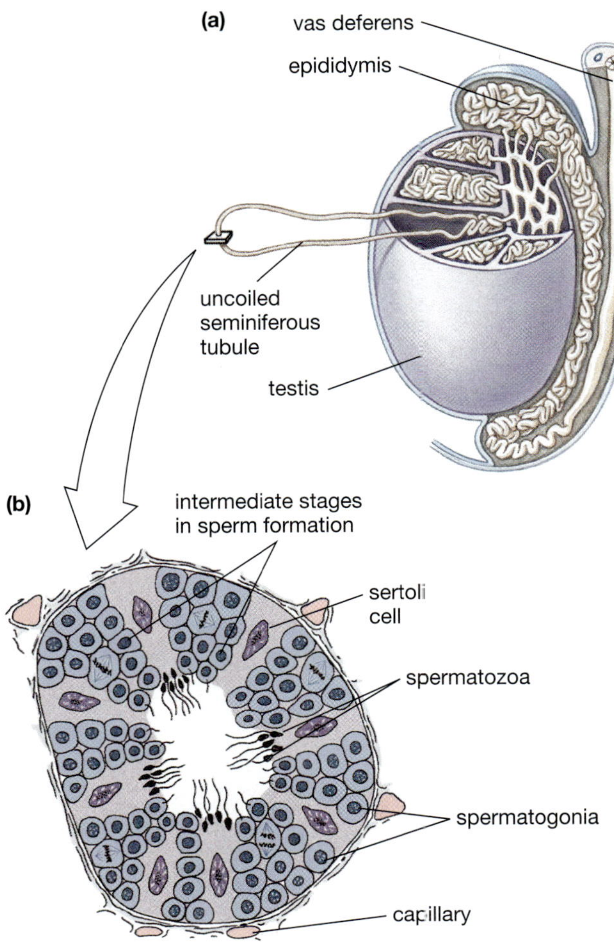

Figure 25-8 Seminiferous tubules are the sites of sperm production
(a) A section of the testis. ***(b)*** Cross section of a seminiferous tubule. The walls of the seminiferous tubules are lined with Sertoli cells and spermatogonia. As spermatogonia divide, the daughter cells move inward. Mature sperm are freed into the central cavity of the tubules for transport to the penis.

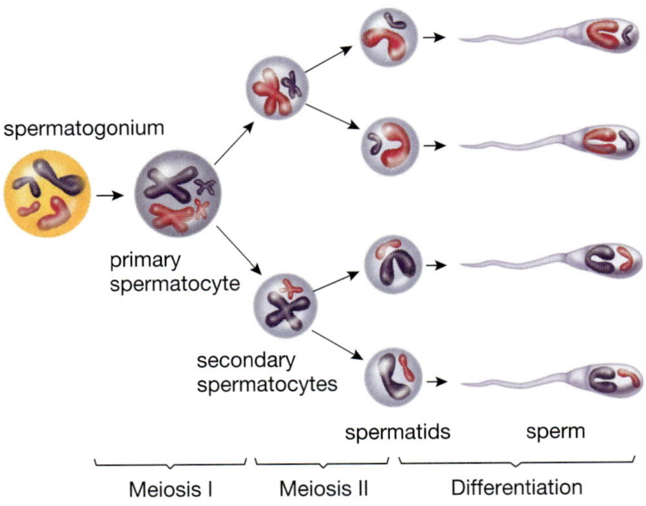

Figure 25-9 Sperm are produced by meiosis
Spermatogenesis is accomplished by meiotic divisions followed by differentiation, producing haploid sperm. Although 4 chromosomes are shown for clarity, in humans the diploid number is 46 and the haploid number is 23.

path, the cell simply divides by mitosis; mitotic division ensures a steady supply of new spermatogonia throughout a male's life. Alternatively, the spermatogonium may undergo **spermatogenesis**—a series of developmental events that produces haploid sperm.

Sperm Develop in Several Steps

Spermatogenesis begins with the growth and differentiation of spermatogonia into *primary spermatocytes*, which are large diploid cells. Each primary spermatocyte then undergoes meiosis (Fig. 25-9; see also Chapter 10). The two diploid cells produced by meiosis I are called *secondary spermatocytes*. Each secondary spermatocyte divides again, during meiosis II, to produce two haploid *spermatids*, for a total of four spermatids per primary spermatocyte. The spermatids undergo further development and differentiation to become sperm. The mature sperm are liberated into the central cavity of the seminiferous tubule.

Sperm Cells Are Specialized for Fertilization

A human sperm is unlike any other human cell. It has very little cytoplasm and a haploid nucleus that nearly fills the *head* (Fig. 25-10). Atop the nucleus lies a specialized lysosome called the *acrosome*. The acrosome contains enzymes that dissolve protective layers around the egg, enabling the sperm to enter and fertilize it. Behind the head is the *midpiece*, which is packed with mitochondria. The mitochodria provide the energy needed to move the *tail*, which protrudes out the back. Whiplike movements of the tail, which is really a long flagellum, propel the sperm through the female reproductive tract.

Sperm Production Begins at Puberty

Immature males do not produce sperm. Spermatogenesis does not begin until *puberty*, a period of rapid growth and transition to sexual maturity. At puberty, cells in the testes begin to produce testosterone, a hormone that acts to initiate spermatogenesis. Testosterone also stimulates the development of *secondary sexual characteristics* (such as the growth of facial hair in males), maintains sexual drive, and is required for successful *intercourse* (a term we will use for human copulation).

Tubules and Ducts Form a Pathway to the Outside

The seminiferous tubules merge to form the *epididymis*, a single continuous, folded duct (see Fig. 25-8a). The epididymis leads into a duct called the **vas deferens**, which leaves the scrotum and enters the abdominal cavity. Most of the hundreds of millions of sperm produced each day are stored in the epididymis and vas deferens. The vas deferens joins the **urethra**, which connects the bladder to the tip of the **penis**. This final common path is shared, at different times, by both

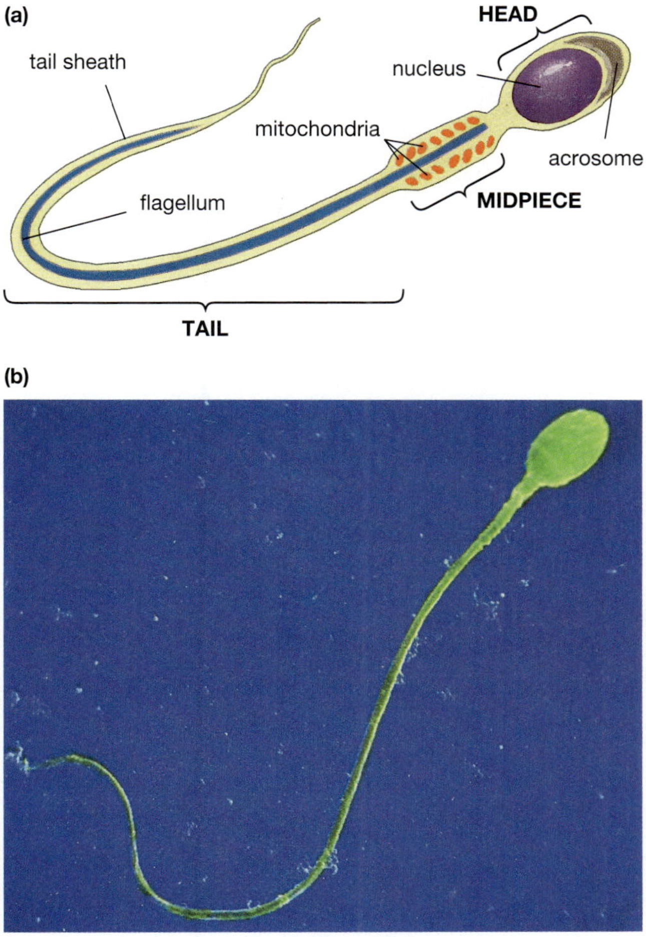

Figure 25-10 A human sperm cell
(a) A mature sperm is a cell equipped only with the elements needed for locomotion and fertilization. ***(b)*** False-color electron micrograph of a human sperm.

urine (during urination) and sperm (during ejaculation—a reflex, caused by sexual stimulation, that forces sperm out through the penis).

Glands Produce Semen, Which Activates Sperm

The fluid ejaculated from the penis, called *semen*, consists of sperm mixed with secretions from three glands that empty into the vas deferens or urethra: the *seminal vesicles*, the *prostate gland*, and the *bulbourethral glands*. The secretions activate swimming by the sperm, provide energy for swimming, and neutralize the acidic fluids of the vagina that normally inhibit bacterial growth (see Fig. 25-7 and Table 25-1).

The Female Reproductive System Includes the Ovaries and Other Structures

The female reproductive tract is almost entirely contained within the abdominal cavity (Table 25-2 and Fig. 25-11). It consists of paired gonads—the **ovaries** (Fig. 25-12a)—and structures that accept sperm, conduct the sperm to the egg, and nourish the developing **embryo** (as an organism is called in its early stages of development).

Egg Production in the Ovaries Begins Before Birth

Oogenesis, the formation of egg cells, begins in the developing ovaries of a female **fetus** (an embryo that is sufficiently developed to be recognizably human). The process begins with the formation of precursor egg cells called **oogonia** (singular, **oogonium**). By the end of the third month of fetal development, no oogonia remain, as they have all divided by mitosis and grown into *primary oocytes*. Each oocyte is surrounded a layer of much smaller cells that, together with the oocyte, make up a **follicle** (Fig. 25-12b). As fetal development continues, meiosis begins in all primary oocytes but is halted at prophase of meiosis I. By birth, a lifetime supply of primary oocytes is already in place, and no new ones are generated later in life. The ovaries start out with about 2 million primary oocytes. Many primary oocytes die each day, until at puberty (usually 11 to 14 years of age) only about 400,000 remain. That number is more than enough, because only a few oocytes resume meiosis during each month of a woman's reproductive span (from puberty to *menopause* at about age 50).

Egg Development Resumes After Puberty

After puberty, oocytes mature into eggs, but only at times when the uterus is prepared to receive and nourish a fertilized egg. Egg maturation and release is coordinated with preparation of the uterus by a **menstrual cycle**, which is regulated by hormonal interactions

Table 25-2 Structures and Functions of the Human Female Reproductive Tract

Structure	Type of Organ	Function
Ovary	Gonad	Produces oocytes, estrogen, and progesterone
Fimbria	Mouth of duct	Cilia sweep egg into uterine tube
Uterine tube	Duct	Conducts egg to uterus; site of fertilization
Uterus	Muscular chamber	Site of development of fetus
Cervix	Connective tissue ring	Closes off lower end of uterus, supports fetus, and prevents foreign material from entering uterus
Vagina	Large duct	Receptacle for semen; birth canal

MediaTutor
25.1 Human Reproductive Anatomy

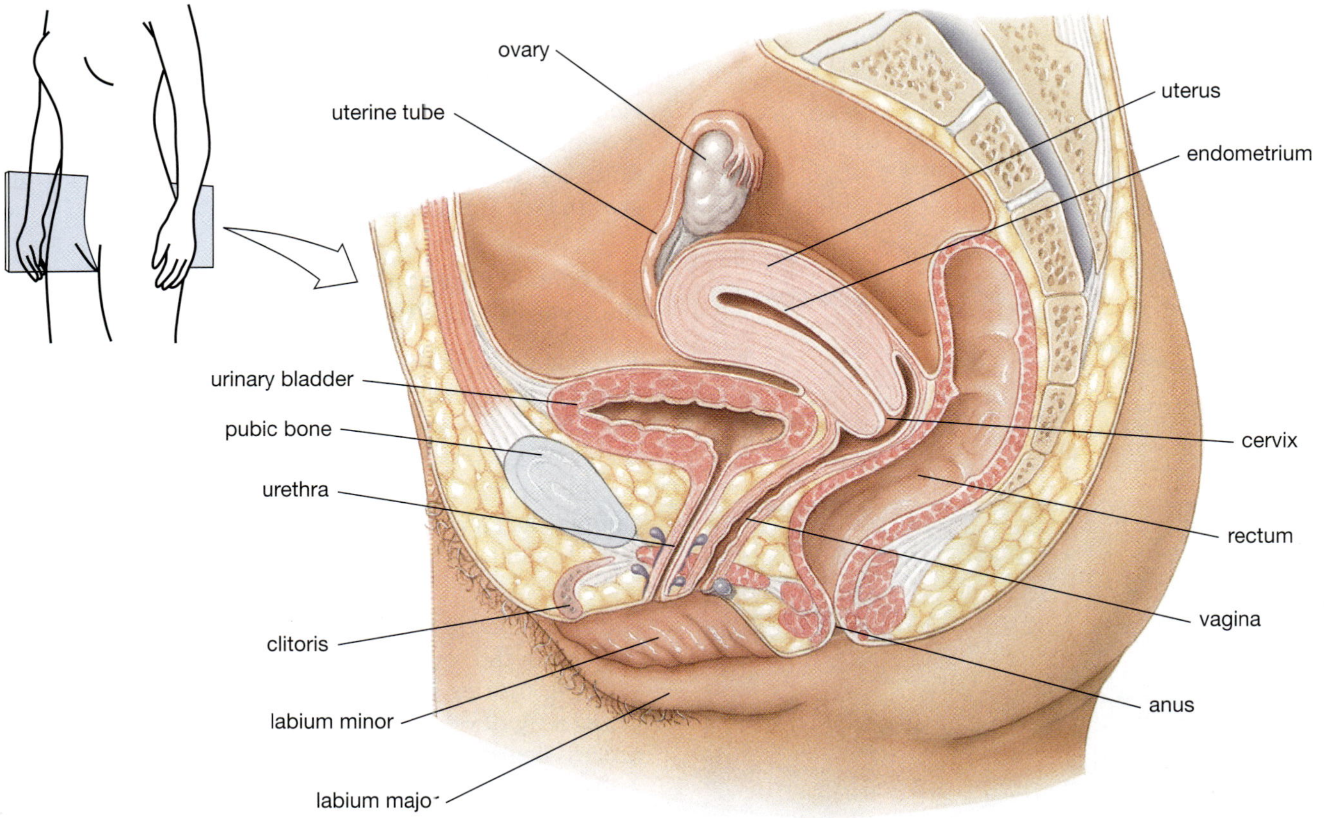

Figure 25-11 The human female reproductive tract

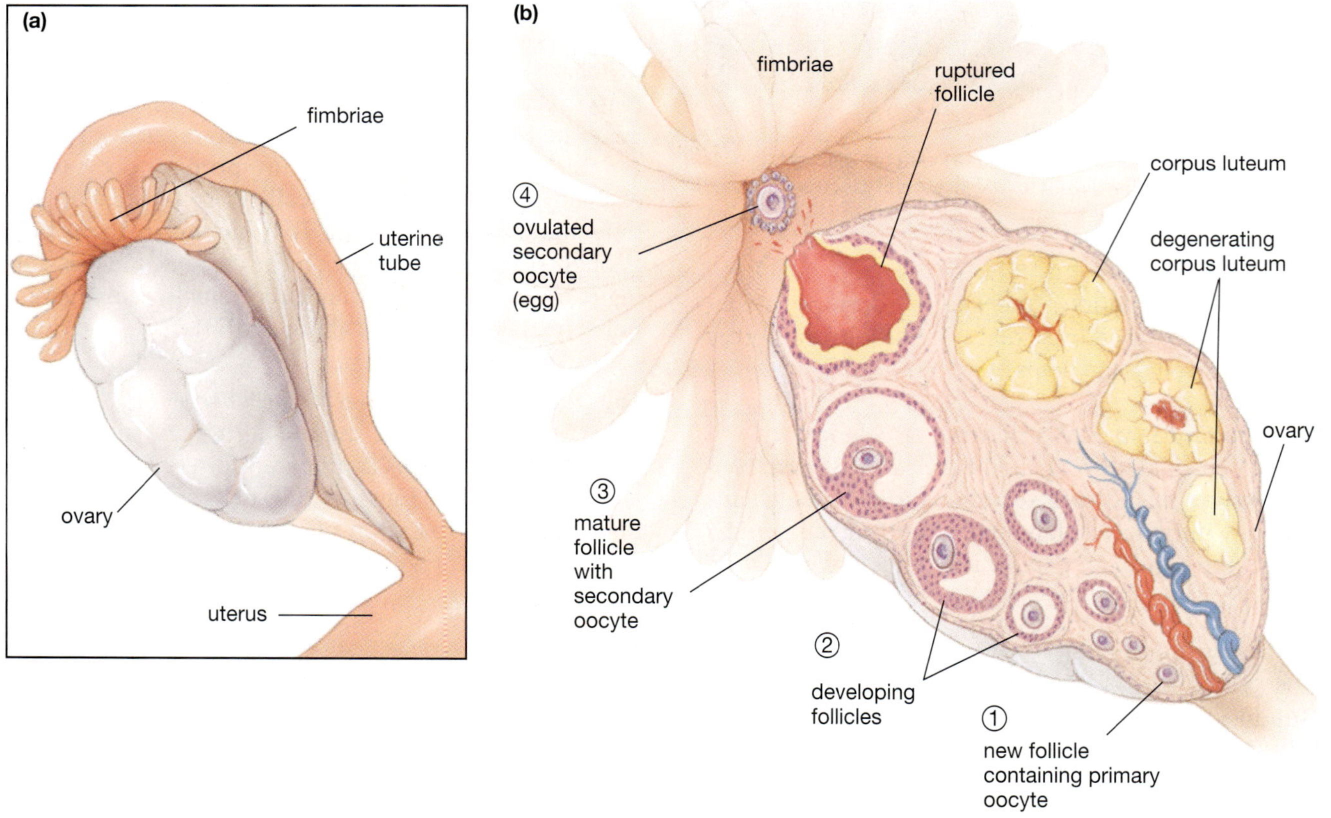

Figure 25-12 Ovaries are the sites of egg production
(a) External view of the ovary and uterine tube. ***(b)*** The development of follicles in an ovary, portrayed in a time sequence (clockwise from the lower right). ① A primary oocyte begins development within a follicle. ②, ③ The follicle grows, providing both hormones and nourishment for the enlarging oocyte. ④ At ovulation, the secondary oocyte, or egg, bursts through the ovary wall, surrounded by some follicle cells. The remaining follicle cells develop into the corpus luteum, which secretes hormones. If the egg is not fertilized, the corpus luteum breaks down after a few days.

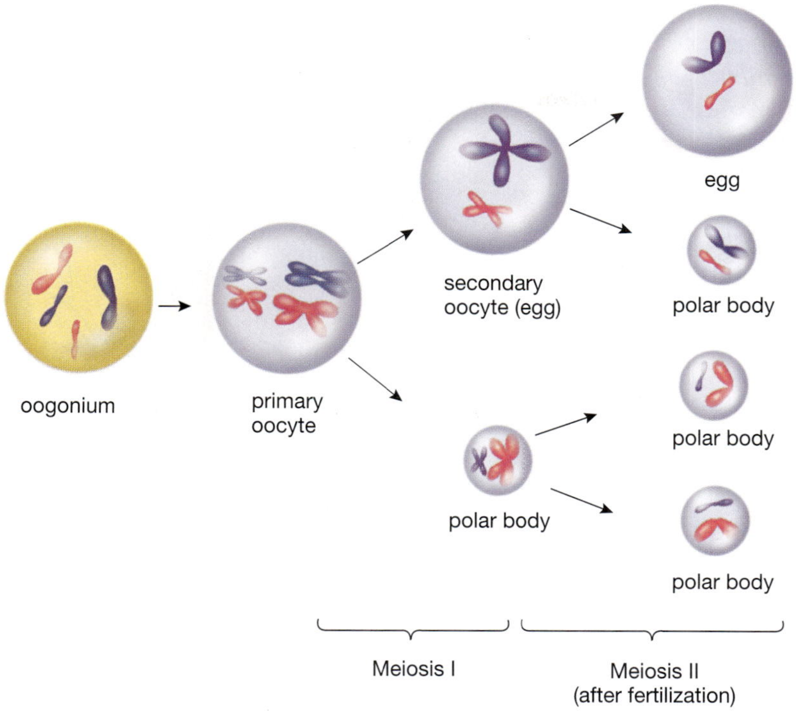

Figure 25-13 Egg cells are formed by meiosis The stages of oogenesis. The oogonium enlarges to form the primary oocyte. At meiosis I, almost all the cytoplasm is included in one daughter cell, the secondary oocyte. The other daughter cell is a small polar body that contains chromosomes but little cytoplasm. At meiosis II, almost all the cytoplasm of the secondary oocyte is included in the egg, and a second small polar body discards the remaining "extra" chromosomes. The first polar body may also undergo the second meiotic division. In humans, meiosis II does not occur until a sperm penetrates into the egg.

among the hypothalamus, pituitary gland, and ovary. Approximately once a month during a woman's reproductive years, hormones stimulate the development of a dozen or more oocyte-containing follicles, although normally only one follicle matures completely.

A Mature Oocyte Leaves the Ovary

As follicle development resumes, the primary oocyte completes meiosis I (which was suspended during fetal development), dividing into a single *secondary oocyte* and a *polar body*, which is little more than a discarded set of chromosomes (Fig. 25-13). Meanwhile, the small cells of the follicle multiply and secrete the hormone estrogen. As the follicle matures, it grows, eventually erupting through the surface of the ovary and releasing the secondary oocyte, a process called **ovulation** (Fig. 25-14). The secondary oocyte then travels through the tube positioned above the ovary, called the **uterine tube** or *oviduct.* For convenience, we will refer to the ovulated secondary oocyte as the *egg*.

Some of the follicle cells accompany the egg, but most remain in the ovary, where they enlarge, forming the **corpus luteum** (see Fig. 25-12b). The corpus luteum secretes the two hormones estrogen and progesterone. If the egg is not fertilized, the corpus luteum breaks down a few days later.

An Egg May Be Fertilized in the Uterine Tube

Each ovary is adjacent to, but not attached to, a uterine tube (see Fig. 25-12a). The open end of the uterine tube is fringed with ciliated "fingers" called *fimbriae* (singular, *fimbria*), which nearly surround the ovary. The cilia create a current that sweeps the egg into the mouth of the uterine tube; once inside the tube, it may encounter a sperm and be fertilized. The **zygote**, as the fertilized egg is now called, is swept down the uterine tube by beating cilia and released into the pear-shaped **uterus**. There it will develop for 9 months.

The Uterus Has Two Functions

The wall of the uterus has two layers that correspond to its dual functions of nourishment and childbirth. The inner lining, or *endometrium*, is richly supplied with blood vessels. (This lining will form the mother's contribution

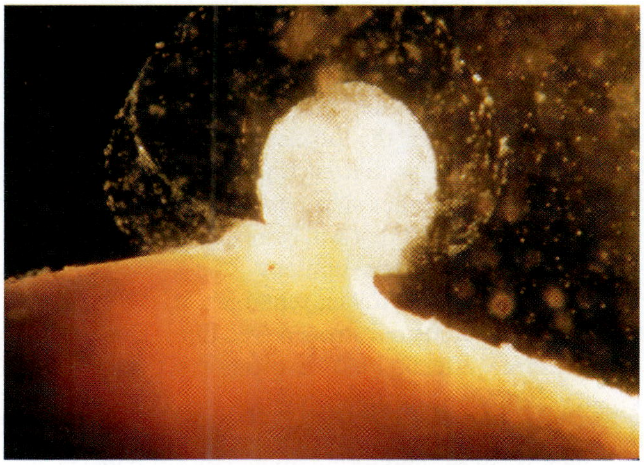

Figure 25-14 A follicle erupts from the ovary The mature follicle grows so large and is filled with so much fluid that it moves to the surface of the ovary and literally bursts through the ovary wall like a miniature volcano. It then releases the secondary oocyte into the uterine tube.

to the **placenta**, the structure that transfers oxygen, carbon dioxide, nutrients, and wastes between fetus and mother.) The outer muscular wall of the uterus, the *myometrium*, contracts strongly during delivery, expelling the infant out into the world. The outer end of the uterus is nearly closed off by the **cervix**, a ring of connective tissue. The cervix holds a developing baby in the uterus, expanding only at the onset of labor to permit passage of the child. Beyond the cervix is the **vagina**, which opens to the outside. The vagina serves both as the receptacle for the penis during intercourse and as the birth canal (see Fig. 25-11).

Uterine Tissues Are Shed If the Egg Is Unfertilized

If an egg is fertilized, it encounters a rich environment for growth, because hormones secreted by developing follicles and the corpus luteum stimulate the lining of the uterus to grow an extensive network of blood vessels and nutrient-producing glands. If the egg is not fertilized, however, the corpus luteum disintegrates, hormone levels fall, and the overgrown endometrium disintegrates as well. The uterus contracts, squeezing out the excess endometrial tissue (perhaps causing menstrual cramps in the process). The resulting flow of tissue and blood is called **menstruation**.

The Menstrual Cycle Is Controlled by Hormones

Changes in the levels of hormones circulating in the blood regulate the events of the menstrual cycle. These hormonal changes are illustrated in Figure 25-15; the numbers in the figure correspond to those in the descriptions below. Note that gonadotropin-releasing hormone (GnRH) is released spontaneously and continuously by cells in the hypothalamus, unless the release is suppressed by other hormones.

MediaTutor 25.2 Hormonal Control of the Menstrual Cycle

① GnRH stimulates the anterior pituitary to release follicle-stimulating hormone (FSH; purple line) and lutenizing hormone (LH; green line).

② Both FSH and LH circulate in the bloodstream and initiate the development of several follicles within the ovaries. The cells of these developing follicles secrete estrogen. Under the combined influences of FSH, LH, and estrogen, the follicles grow. As they grow, the primary oocyte within each follicle enlarges, storing both food and other substances (mostly proteins and messenger RNA) that will be needed by the fertilized egg during early development.

③ As the maturing follicle enlarges, it secretes ever-greater amounts of estrogen (red line). This estrogen has three effects. First, it promotes the continued development of the follicle and of the primary oocyte within it (second panel). Second, it stimulates the growth of the uterine lining (fourth panel).

④ The third effect of high levels of estrogen is to stimulate both the hypothalamus and pituitary, resulting in a surge of LH and FSH at about the 12th day of the cycle. The function of the peak in FSH concentration is not well understood, but the surge of LH has three important consequences: First, it triggers the resumption of meiosis I in the oocyte, resulting in the formation of the secondary oocyte and the first polar body.

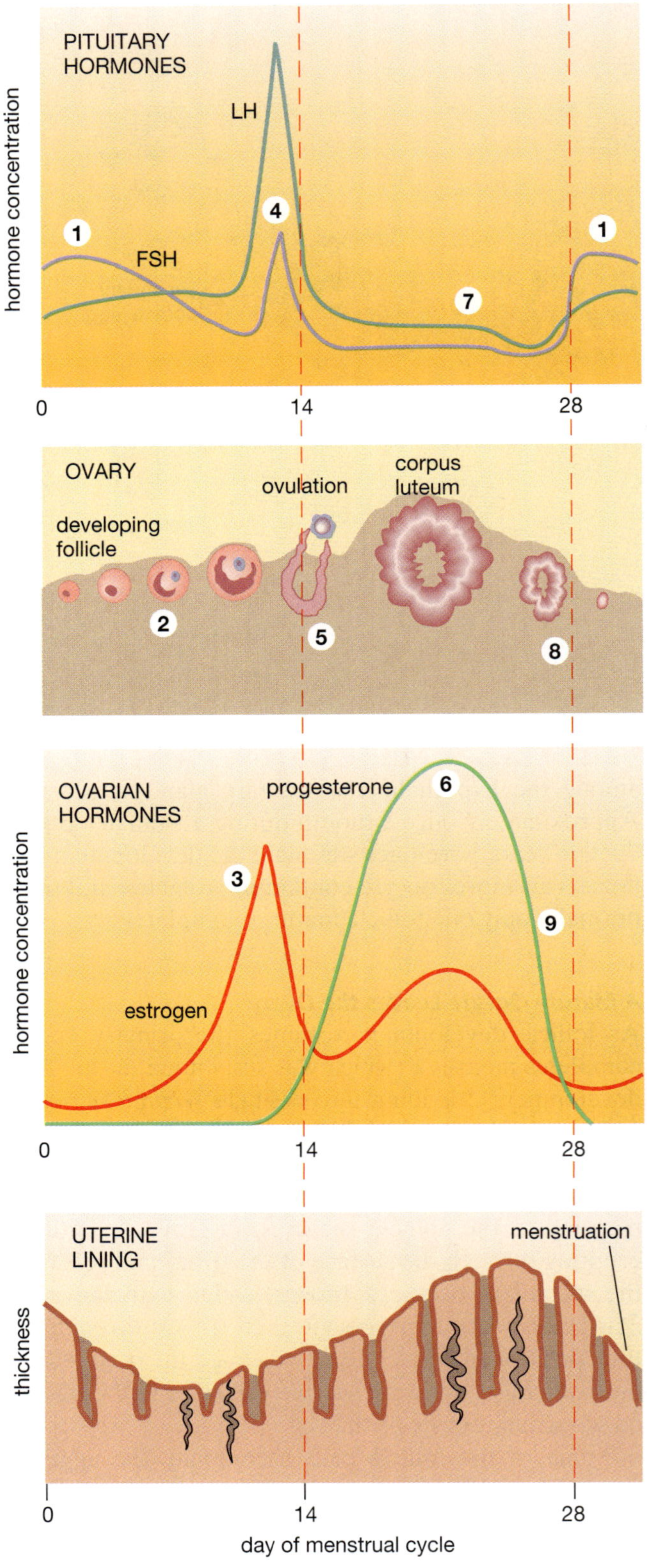

Figure 25-15 Hormonal control of the menstrual cycle
The menstrual cycle is generated by interactions among the hormones of the hypothalamus, the anterior pituitary, and the ovaries. The circled numbers refer to the interactions discussed in the text.

⑤ Second, the LH surge causes the final explosive growth of the follicle, culminating in ovulation, and third, it transforms the remnants of the follicle into the corpus luteum.
⑥ The corpus luteum secretes both estrogen (red line) and progesterone (green).
⑦ The combination of estrogen and progesterone inhibits the hypothalamus and pituitary, shutting down the release of FSH and LH, thereby preventing the development of more follicles. Simultaneously, estrogen and progesterone stimulate further growth of the uterine lining (fourth panel), which eventually becomes about 5 millimeters thick.
⑧ If the egg is not fertilized, the corpus luteum starts to disintegrate about 1 week after ovulation. This disintegration is caused by the corpus luteum itself, which secretes the progesterone that shuts down LH secretion. Because the corpus luteum can persist only while it is stimulated by LH (or by a similar hormone released by the developing embryo, described below), it essentially causes its own destruction.
⑨ With the corpus luteum gone, estrogen and progesterone levels plummet. Deprived of stimulation by estrogen and progesterone, the endometrium of the uterus also dies, and its blood and tissue are shed. This shedding forms the menstrual flow that begins about the 28th day of the cycle. The reduced level of circulating progesterone no longer inhibits the hypothalamus and pituitary, so the spontaneous release of GnRH from the hypothalamus resumes. The release of GnRH in turn stimulates the release of FSH and LH (back to step 1), initiating the development of a new set of follicles, restarting the cycle.

During pregnancy, the embryo itself prevents these changes. Shortly after the ball of cells formed by the dividing fertilized egg embeds itself in the endometrium, it starts secreting an LH-like hormone called chorionic gonadotropin (CG). This hormone travels in the bloodstream to the ovary, where it prevents the breakdown of the corpus luteum. The corpus luteum continues to secrete estrogen and progesterone, and the uterine lining continues to grow, nourishing the embryo. The embryo releases so much CG that the hormone is excreted in the mother's urine. In fact, most pregnancy tests use the presence of CG in urine to determine pregnancy.

Copulation Allows Internal Fertilization

As terrestrial mammals, humans use internal fertilization to deposit sperm into the moist environment of the female's reproductive tract. The penis is inserted into the vagina, where sperm are released during ejaculation. The sperm swim upward in the female reproductive tract, from the vagina through the opening of the cervix into the uterus, and up into the uterine tubes. If the female has ovulated within the past day or so, the sperm will meet an egg in one of the uterine tubes. Only one sperm can succeed in fertilizing the egg and thus begin the development of a new human being.

During Copulation, Sperm Are Deposited in the Female's Vagina

The male role in copulation begins with erection of the penis. Before erection, the penis is relaxed (flaccid), because the arterioles that supply it are constricted, allowing little blood flow (Fig. 25-16a). Under psychological and physical stimulation, the arterioles dilate and blood flows into spaces in the tissue within the penis. As these tissues swell, they squeeze off the veins that drain the penis (Fig. 25-16b). Blood pressure increases, causing an erection. After the penis is inserted into the vagina, movements further stimulate touch receptors on the

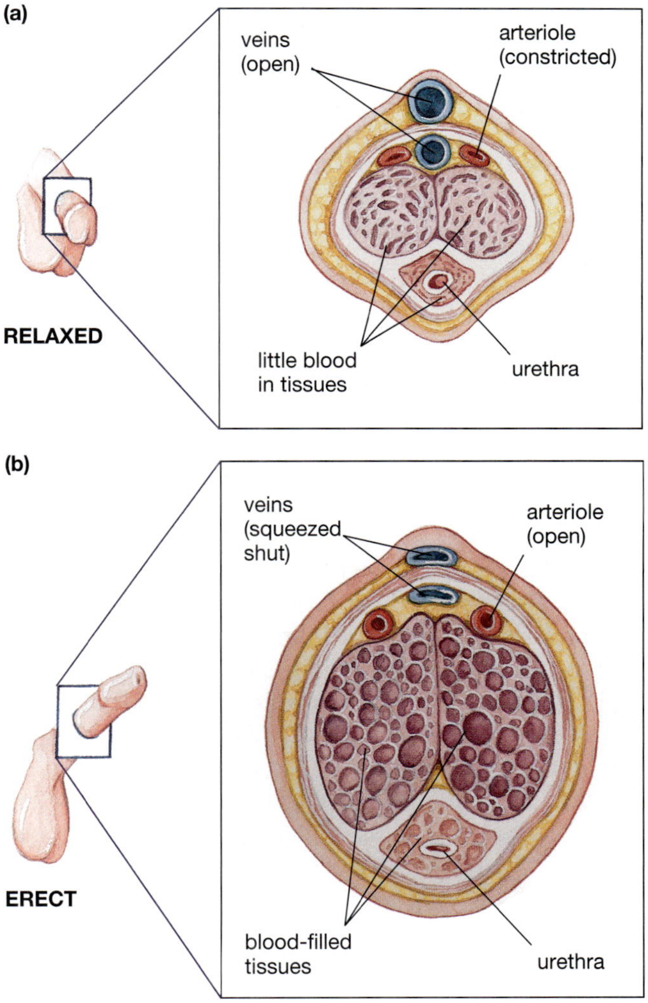

Figure 25-16 Changes in blood flow within the penis cause erection
(a) Most of the time, the smooth muscles that encircle the arterioles leading into the penis are contracted, limiting blood flow. *(b)* During sexual excitement, these muscles relax, and blood flows into spaces within the penis. The swelling penis closes the veins through which blood leaves the penis, thereby increasing the pressure exerted by fluids within the penis and causing it to become elongated and firm.

penis, triggering male orgasm, a feeling of intense pleasure and release that is accompanied by ejaculation. Ejaculation forces semen out of the penis and into the vagina when muscles encircling the epididymis, vas deferens, and urethra contract. On average, 3 or 4 milliliters of semen, containing 300 million to 400 million sperm, are ejaculated.

Copulation induces similar changes in females. Sexual excitement causes increased blood flow to the vagina and to the external parts of the reproductive tract. These external parts consist of the **labia** (singular, **labium**)—paired folds of skin known as the *labia minor* and *labia major*—and the **clitoris**, a rounded projection (see Fig. 25-11). The clitoris becomes erect. and stimulation by the penis may result in female orgasm, a series of rhythmic contractions of the vagina and uterus accompanied by sensations of pleasure and release. Female orgasm is not necessary for fertilization.

Intimate contact during copulation creates a situation in which disease organisms can readily be transmitted. Ever since the "sexual revolution," which began in the 1960s, many people have had multiple sexual partners, and the incidence of *sexually transmitted diseases* (STDs) has greatly increased (see "Health Watch: Sexually Transmitted Diseases").

During Fertilization, the Sperm and Egg Nuclei Unite

Neither sperm nor egg lives very long on its own. Sperm, under ideal conditions, may live for two days, and an unfertilized egg remains viable for only a day. Therefore, fertilization can succeed only if copulation occurs within a couple of days of ovulation.

When it leaves the ovary, the egg is surrounded by follicle cells. These cells, now called the *corona radiata*, form a barrier between the sperm and the egg (Fig. 25-17a). A second barrier, the jellylike *zona pellucida*, lies between the corona radiata and the egg. Recent research suggests that the human egg releases a chemical attractant that lures the sperm toward it.

In the uterine tube, hundreds of sperm (out of the hundreds of millions that were ejaculated) reach the egg and encircle the corona radiata, each sperm releasing enzymes from its acrosome (Fig. 25-17b). These enzymes weaken both the corona radiata and the zona pellucida, allowing the sperm to wriggle through to the egg. When the first sperm contacts the surface of the egg, the plasma membranes of egg and sperm fuse, and the sperm's head is drawn into the egg's cytoplasm. As the sperm enters, it triggers two vital changes in the egg. First, vesicles near the surface of the egg release chemicals into the zona pellucida, reinforcing it and preventing additional sperm from entering the egg. Second, the egg undergoes its second meiotic division, producing a haploid gamete at last. The haploid nuclei of sperm and egg fuse, forming a diploid nucleus that contains all the genes of a new human being.

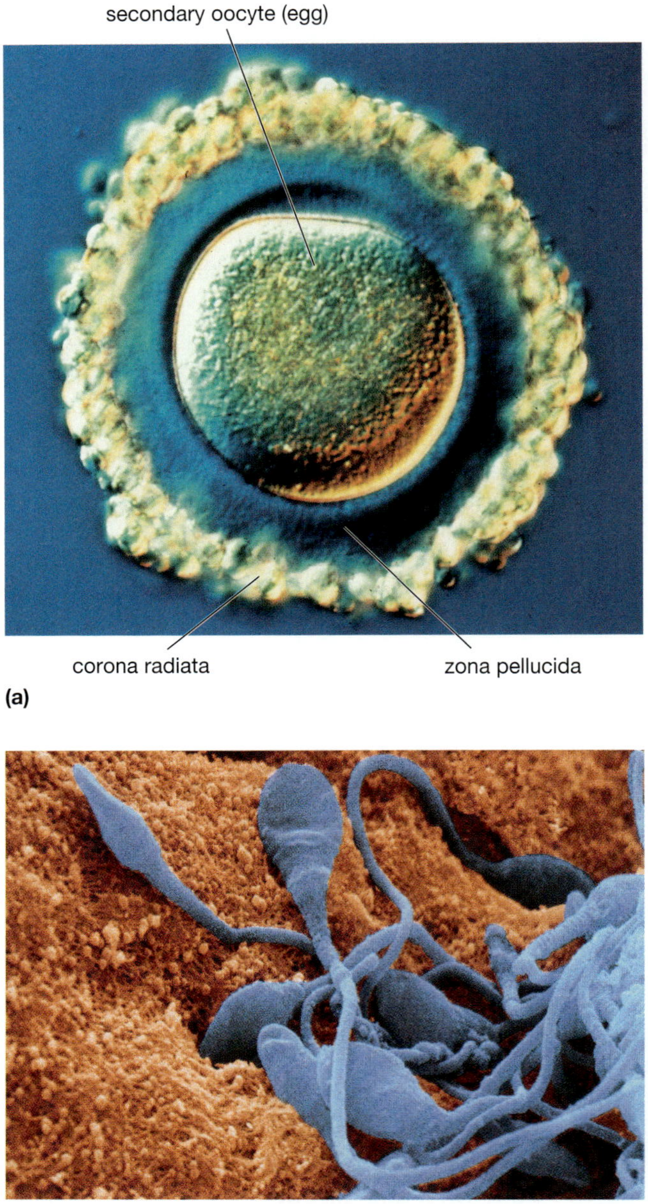

Figure 25-17 The secondary oocyte and fertilization
(a) A human secondary oocyte shortly after ovulation. Sperm must digest their way through the small follicular cells of the corona radiata and the clear zona pellucida to reach the oocyte itself. ***(b)*** Sperm on the surface of an oocyte, attacking its defensive barriers.

The Big Picture

In evolutionary terms, reproduction is the primary function of an animal, and animals have evolved many structures and behaviors that are directly related to reproduction. Animals have adaptations that enable them to produce gametes, ensure that the gametes fuse with those from individuals of the opposite sex, protect and nourish the developing embryo, and provide resources that help offspring survive to adulthood.

Health Watch
Sexually Transmitted Diseases

Sexually transmitted diseases (STDs), which are caused by viruses, bacteria, protists, or arthropods that infect the sexual organs and reproductive tract, are a serious and growing health problem worldwide. The World Health Organization estimates that there are 250 million new STD cases each year. As the name implies, these diseases are transmitted either exclusively or primarily through sexual contact. Here we discuss some of the more common STDs.

Gonorrhea, an infection of the genital and urinary tract, is a very common STD in the United States, with nearly 2 million cases reported annually. The bacterium that causes gonorrhea, *Neisseria gonorrhoeae*, is transmitted almost exclusively by intimate contact. It penetrates the membranes that line the urethra, anus, cervix, uterus, uterine tubes, and throat. In males, inflammation of the urethra results in painful urination and a discharge of pus from the penis. About 10% of infected males and 50% of infected females do not seek treatment, because they have no symptoms or only mild ones, and they become carriers who can readily spread the disease. Gonorrhea can lead to infertility by blocking the uterine tubes with scar tissue. Treatment is by antibiotics. Infants born to infected mothers can acquire the bacterium during delivery. The bacterium attacks the eyes of newborns and was once a major cause of blindness. Today, most newborns are immediately given antibiotic eyedrops to kill any bacteria that might be present.

Syphilis is caused by a spiral-shaped bacterium, *Treponema pallidum*, which enters the mucous membranes of the genitals, lips, anus, or breasts. Like gonorrhea, it is readily killed by exposure to air and is spread only by intimate contact. Syphilis begins with a sore at the site of infection and can be cured with antibiotics. If untreated, syphilis bacteria spread through the body, multiplying and damaging many organs, including the skin, kidneys, heart, and brain, in some cases with fatal results. Syphilis can be transmitted to the fetus during pregnancy; and the skin, teeth, bones, liver, and central nervous system of such infants may be damaged.

Chlamydia causes inflammation of the urethra in males and the urethra and cervix in females. In many cases there are no obvious symptoms, so the infection goes untreated and is spread. Like the gonorrhea bacterium, the chlamydia bacterium, *Chlamydia trachomatis*, can infect and block the uterine tubes, resulting in sterility. Chlamydial infection can cause eye inflammation in infants born to infected mothers.

Acquired immune deficiency syndrome, or *AIDS*, is caused by the HIV virus, discussed in Chapter 22. Because the virus does not survive exposure to air, it is spread primarily by sexual activity and by contaminated blood and needles. The HIV virus attacks the immune system, leaving the victim vulnerable to infections, which eventually prove fatal. Children born to mothers with AIDS can become infected before or during birth. There is no cure, but certain drugs, such as AZT and protease inhibitors, can prolong life.

Genital herpes, another viral infection, causes painful blisters on the genitals and surrounding skin and is transmitted primarily when blisters are present. The herpes virus never leaves the body but resides in certain nerve cells, emerging unpredictably, possibly in response to stress. The first outbreak is the most severe; subsequent outbreaks produce fewer blisters and can be quite infrequent. The drug acyclovir, which inhibits viral DNA replication, can reduce the severity of outbreaks. Pregnant women with an active case of genital herpes can transmit the virus to the developing fetus, causing severe mental or physical disability or stillbirth. Herpes can also be transmitted from mother to infant if the infant contacts blisters during childbirth.

Genital warts are growths or bumps that appear on the genitals, in or around the vagina or anus, or on the cervix in females, and on the penis, scrotum, groin, or thigh in males. The warts, which are normally painless, are caused by the *human papillomavirus* (*HPV*), which is transmitted by skin-to-skin contact during sex. The most common treatment is removal of the warts, typically by freezing them with liquid nitrogen. Antiviral drugs can also be effective. Like the herpes virus, the human papillomavirus resides inside cells and is therefore difficult to cure. Genital warts are considered to be a risk factor for certain types of cervical cancer.

Trichomoniasis is caused by *Trichomonas*, a flagellated protist that colonizes the mucous membranes that line the urinary tract and genitals of both males and females. The symptoms are a discharge caused by inflammation in response to the parasite. The protist is spread by intercourse but can also be acquired through contaminated clothing and toilet articles. Lengthy untreated infections can result in sterility.

Crab lice, also called *pubic lice*, are microscopic arachnids (relatives of spiders) that live and lay their eggs in pubic hair. Their mouthparts are adapted for penetrating the skin and sucking blood and body fluids, a process that causes severe itching. "Crabs" are not only irritating; they can also spread infectious diseases. They can be controlled through careful hygiene and chemical treatments.

3 How Does Animal Development Proceed?

Development, the transformation from fertilized egg—a single cell—to a multicellular organism, is a beautiful, nearly magical process that has been carefully described for a number of animal species. In textbooks, the changes in the developing embryo are generally depicted in stages, but the stages are just convenient "snapshots" for the purpose of illustration. The actual development is a smoothly continuous process.

The Fertilized Egg Develops into a Hollow Ball of Cells

Development begins with an egg that contains reserves of **yolk**, a substance rich in lipids and protein that provides

MEDIATUTOR 25.3 Stages of Animal Development

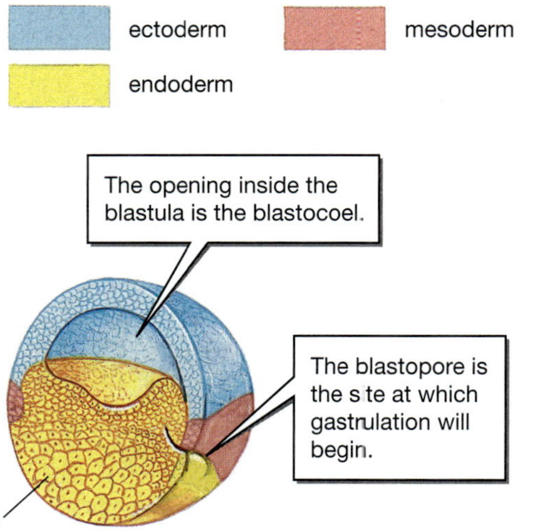

(a) The blastula just before gastrulation.

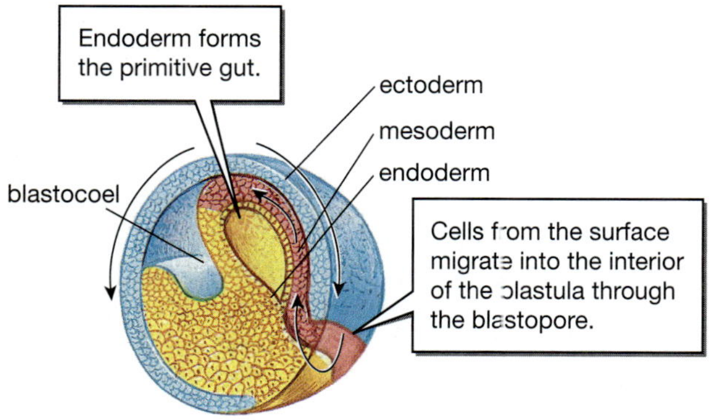

(b) Cells migrate at the start of gastrulation. The migrating cells will form the endoderm and mesoderm layers of the gastrula; the cells remaining on the surface will form ectoderm.

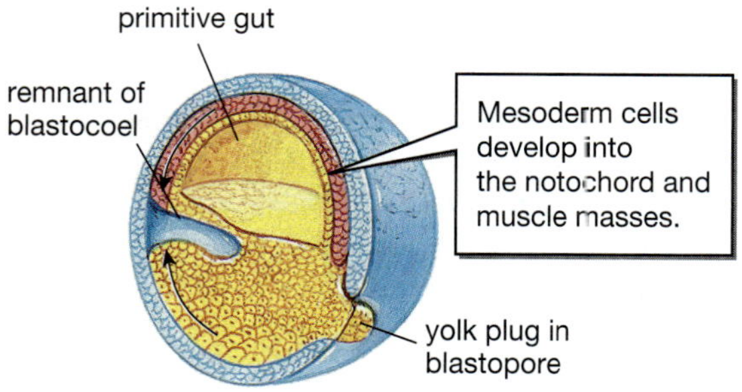

(c) Mesoderm differentiates.

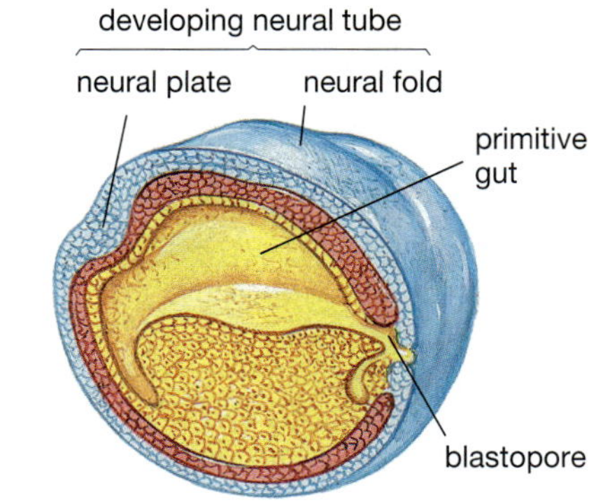

(d) Ectoderm cells lying directly above the notochord form the neural tube.

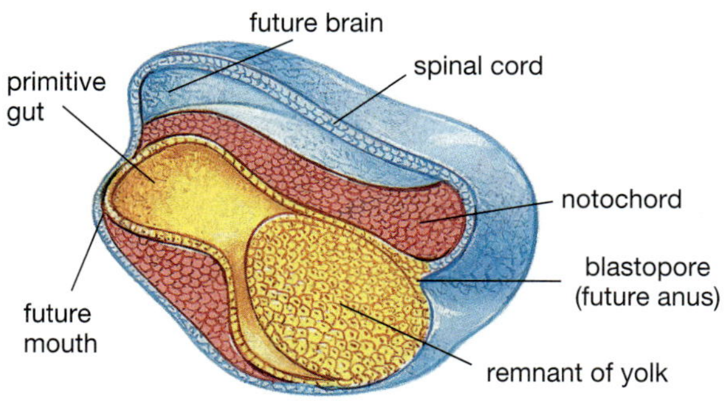

(e) Further development. The neural tube differentiates into brain and spinal cord. A future mouth is produced when the opening formed by the primitive gut breaks through at the end of the embryo opposite the blastopore. The blastopore is the future anus.

Figure 25-18 Gastrulation in the frog

nourishment for the early embryo. The formation of an embryo begins with **cleavage**, a series of mitotic divisions of the fertilized egg. An egg is a very large cell, and cleavage reduces cell size even as it increases cell number. Unlike most mitotic cell divisions—in which cells divide, grow, duplicate genetic material, then divide again—cell divisions during cleavage skip the growth phase. Consequently, as cleavage progresses, the cytoplasm is split up into ever-smaller cells whose sizes approach those of cells in the adult organism. Finally, a solid ball of small cells, the **morula**, is formed. The morula as a whole is about the same size as the fertilized egg. As cleavage continues, a cavity opens within the morula. The cells of the morula then become the outer covering of a hollow (typically spherical) structure, the **blastula**. The space inside the blastula is called the *blastocoel* (Fig. 25-18a).

Next, Three Tissue Layers Form

In the next stage of development, many cells of the blastula migrate to new locations, a process called **gastrulation**. Gastrulation begins when an indentation called the **blastopore** forms on one side of the blastula. Blastula cells then migrate through the blastopore in a continuous sheet, much as if you punched in an underinflated basketball. The resulting indentation enlarges to form a cavity that will eventually become the digestive tract (Fig. 25-18b).

During gastrulation, three embryonic tissue layers are formed. The cells that move through the blastopore to line the future digestive cavity become **endoderm**. The cells remaining on the outside will form the epidermis of the skin and the nervous system and are called **ectoderm**. Some cells migrate between the endoderm and ectoderm, forming a third layer, the **mesoderm**. Mesoderm gives rise to muscles, the circulatory system, and the skeleton, including the **notochord**, a supporting rod found at some stage in all chordates. At this point, the three-layered embryo is called a **gastrula** (Fig. 25-18c).

Next, ectoderm cells just above the developing notochord form a plate that soon folds inward, rolling up to form the *neural tube* (Fig 27-18d), which will develop into the brain and spinal cord. Other parts of the gastrula will form the openings of the digestive system; the blastopore becomes the anus and one end of the endoderm-lined gut cavity will break through the outer ectoderm layer to form the mouth (Fig. 27-18e).

Eventually, Adult Structures Form

Gradually, the ectoderm, mesoderm, and endoderm rearrange themselves into organs and other body structures that will be present in the adult (Table 25-3). This process is called **organogenesis.**

In some cases, adult structures are "sculpted" by the death of excess cells produced during embryonic development. For example, humans pass through developmental stages in which the embryo has a tail and webbed hands and feet; these features can be seen clearly in a 5-week-old human embryo. Two weeks later, the webbing cells have died, leaving separate fingers, and the tail cells are dying, shortening the tail.

At least two mechanisms seem to cause cell death during development. Some cells die unless they receive a signal that promotes survival. Motor neurons, for example, are programmed to die unless they form a synapse with a skeletal muscle, which releases a chemical that prevents the death of the neuron. For other cells, the situation is just the reverse; they live unless they receive a signal that induces death.

The Big Picture

The question of how form arises from formlessness is one of the central mysteries of biology. The transformation of a single cell, the zygote that forms from the fusion of sperm and egg, into a complex organism is a process that has been extensively described but poorly understood. In recent years, however, a flood of new discoveries about development has opened a new era of increased understanding and potentially revolutionary new medical applications.

4 How Do Humans Develop?

Figure 25-19 summarizes the stages of human embryonic and fetal development. You may want to refer to this figure as we describe human development in the next few sections.

In the First 2 Months, Differentiation and Growth Are Rapid

Usually, a human egg is fertilized in the uterine tube and undergoes a few cleavage divisions on its way to the uterus, a journey that takes about 4 days (Fig. 25-20). By about 1 week after fertilization, the zygote has developed into a hollow ball of cells, known as the **blastocyst** (the mammalian version of a blastula). One section of the blastocyst wall is thickened to form a group of cells called the *inner cell mass*. The sticky outer wall of the blastocyst adheres to the uterus and burrows into the endometrium, a process called **implantation**.

Following Implantation, Two Cavities Form

After implantation, the blastocyst grows and the inner cell mass splits to form two fluid-filled sacs. One sac, bounded by a membrane called the **amnion**, forms the amniotic cavity (Fig. 25-21). The amnion eventually grows around the embryo, providing a barrier to contain the watery environment needed by all animal embryos. The second sac is called the *yolk sac*, although in humans it contains no yolk. The two sacs are separated by a double layer of cells called the *embryonic disc*, which will eventually develop into the embryo. The side of the embryonic disc facing the amniotic cavity consists of cells that will become ectoderm, and the side facing the yolk sac consists of future endoderm cells.

After Two Weeks, the Stage Is Set for Gastrulation

On about the 15th day after fertilization, gastrulation begins. The layers of the embryonic disc split apart slightly. A slit, the **primitive streak** (corresponding to the

Table 25-3 Derivation of Adult Tissues and Organs from Embryonic Cell Layers

Embryonic Layer	Adult Tissue
Ectoderm	Epidermis of skin; lining of mouth and nose; hair; glands of skin (sweat, sebaceous, and mammary glands); nervous system; lens of eye; inner ear
Mesoderm	Dermis of skin; muscle, skeleton; circulatory system; gonads; kidneys; outer layers of digestive and respiratory tracts
Endoderm	Lining of digestive and respiratory tracts; liver; pancreas

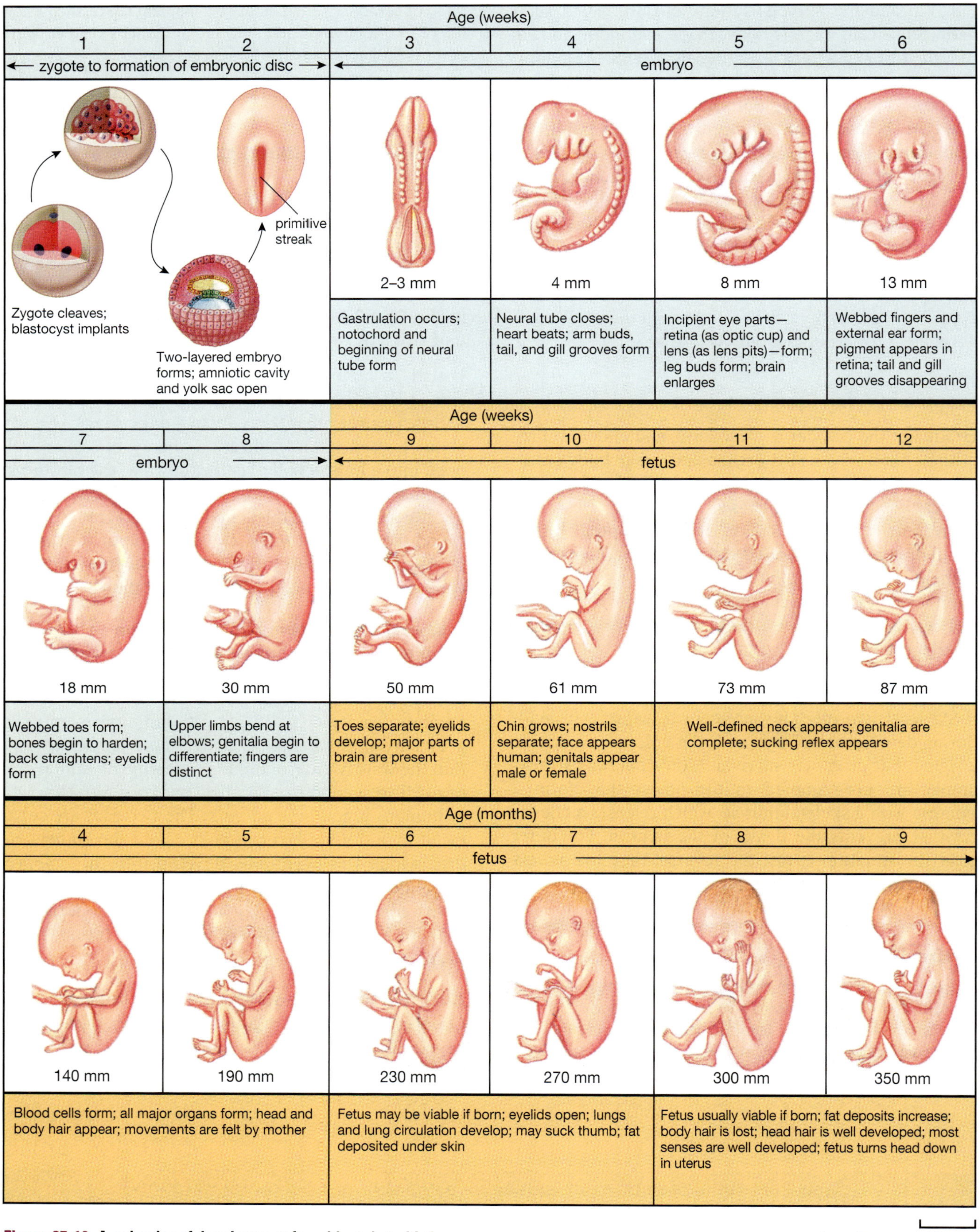

Figure 25-19 **A calendar of development from blastula to birth**

MEDIATUTOR 25.4 Human Development

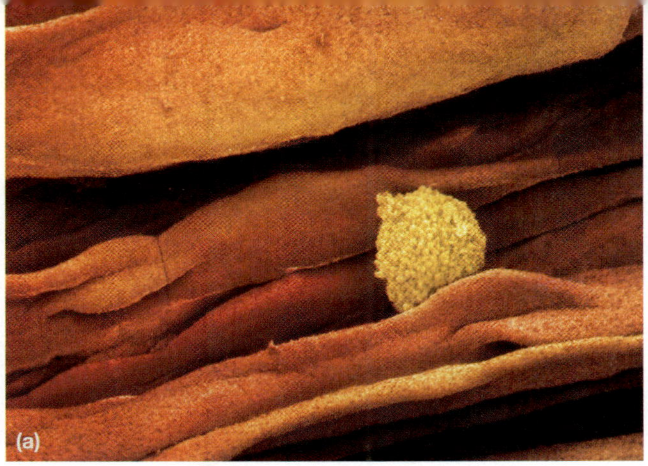

Figure 25-20 The journey of the egg
(a) The egg, surrounded by cells from the follicle (the corona radiata), travels down the uterine tube toward the uterus. ***(b)*** The egg is fertilized in the uterine tube and slowly travels down to the uterus. Along the way, the zygote divides a few times, until a hollow blastocyst is formed. When the blastocyst reaches the uterine endometrium, it burrows in.

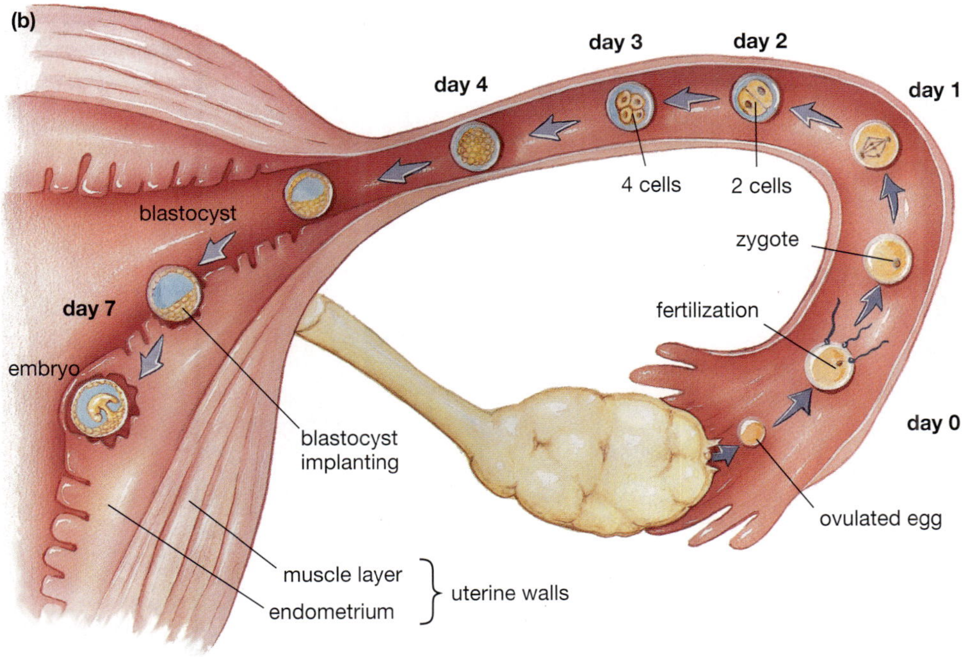

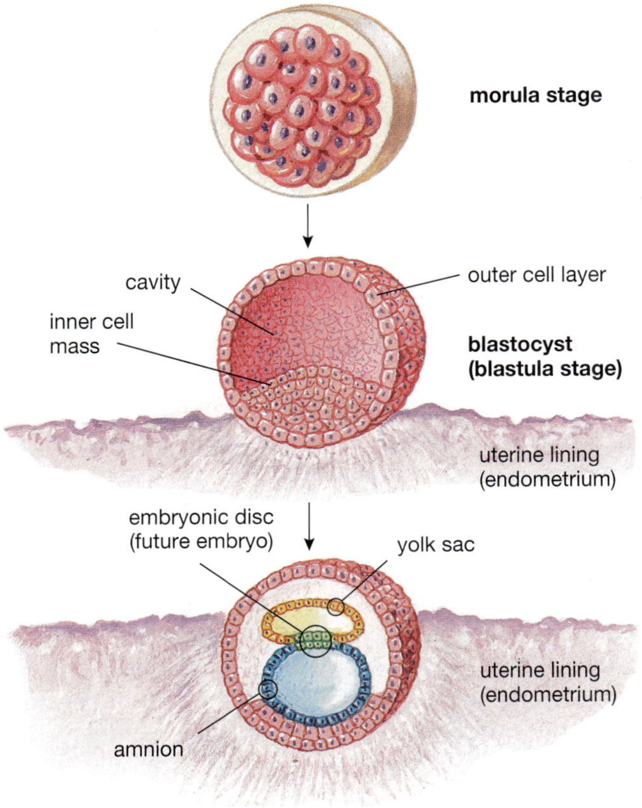

Figure 25-21 Human development during the first and second weeks
As it travels through the uterine tube, the fertilized egg undergoes cleavage, forming a morula. In the uterus, the morula becomes a blastocyst. As the blastocyst implants in the uterine lining, the inner cell mass develops into the yolk sac, the amnion, and the embryonic disc.

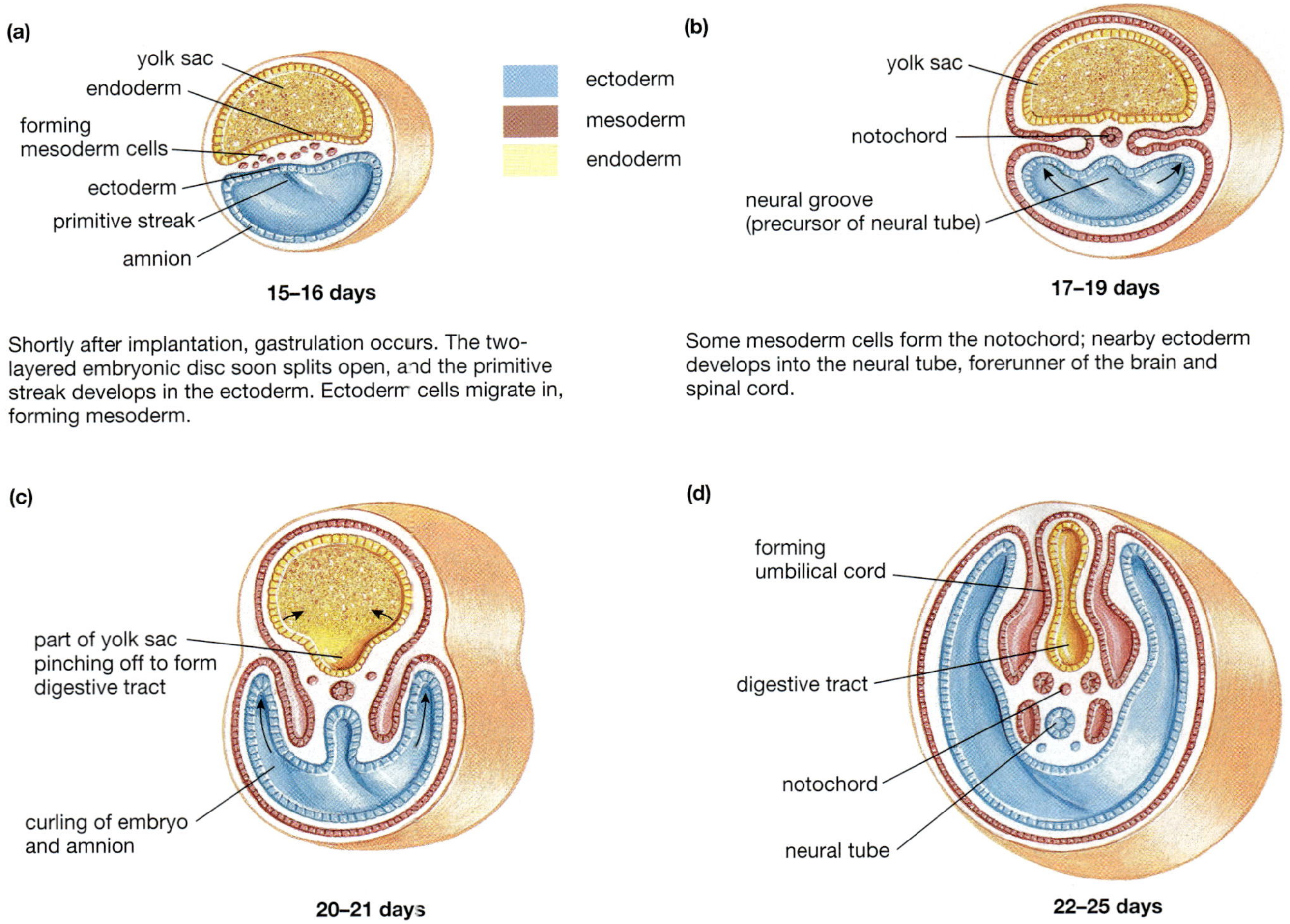

(a) Shortly after implantation, gastrulation occurs. The two-layered embryonic disc soon splits open, and the primitive streak develops in the ectoderm. Ectoderm cells migrate in, forming mesoderm.

(b) Some mesoderm cells form the notochord; nearby ectoderm develops into the neural tube, forerunner of the brain and spinal cord.

(c) During the third and fourth weeks of development, the embryo curls toward the yolk sac, forming a tubelike embryo typical of vertebrates.

(d) As the embryo curls, part of the (empty) yolk sac pinches together to form the digestive tract. The amnion curls with the embryo, eventually completely enclosing it, except where the umbilical cord extends through.

Figure 25-22 Human development during the third and fourth weeks

blastopore), appears in the ectoderm layer of the disc (Fig. 25-22a). Some cells of this layer then migrate through the primitive streak into the interior of the embryo, forming mesoderm. Some mesoderm-derived structures begin to form, including the notochord (Fig. 25-22b).

In Weeks Three to Six, Body Structures Begin to Form

During the third week of development, the embryo and its amniotic sac begin to curl toward the yolk sac (Fig. 25-22c). As the embryo grows, the endoderm forms a tube that will become the digestive tract (Fig. 25-22d). Simultaneously, a groove forms in ectoderm near the notochord, and then folds inward and closes over to become the **neural tube**, the forerunner of the brain and spinal cord (see Fig. 25-18d).

By the end of the fourth week, the amnion completely surrounds the embryo. The amnion is punctured only by the **umbilical cord**, which connects the embryo to the placenta (Fig. 25-23a).

By the end of the sixth week, the embryo already has the rudimentary beginnings of eyes, a beating heart, and separating fingers and toes on its tiny hands and feet (Fig. 25-23b). Especially notable at this stage is the rapid growth of the brain, which is nearly as large as the rest of the body. Many of the structures of the adult brain are already recognizable. The six-week-old embryo also has a notochord, a prominent tail, and *gill grooves*. These structures are reminders that we share ancestry with other vertebrates that retain them in adulthood, but in humans they disappear as development continues.

After Two Months, the Embryo Is Recognizably Human

As the second month draws to an end, nearly all the major organs have formed. The gonads appear and develop into testes or ovaries, depending on the presence or absence of the Y chromosome. Sex hormones—either testosterone from the testes or estrogen from the ovaries—are secreted. These hormones will affect the future development of the embryonic organs, including not only the reproductive organs but also certain regions of the brain. After the second month of development, the embryo is called a **fetus**, a label signifying that the developing embryo has taken on a generally human appearance (Fig. 25-23c).

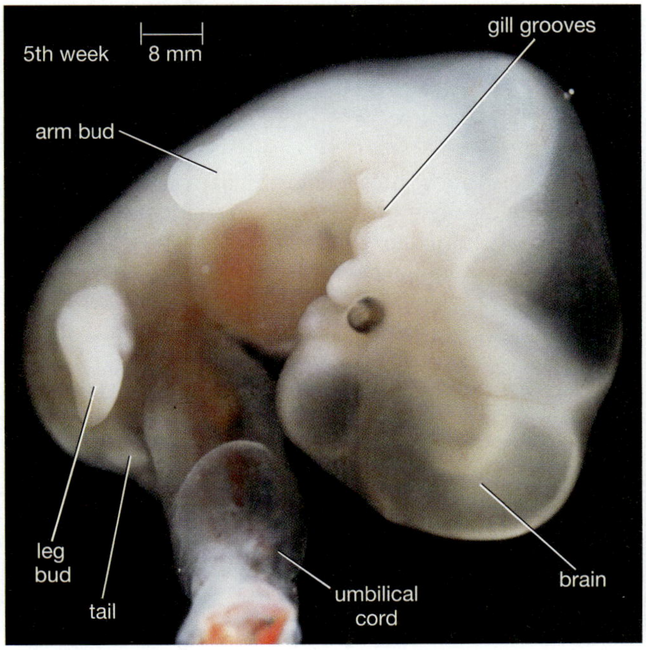

(a)

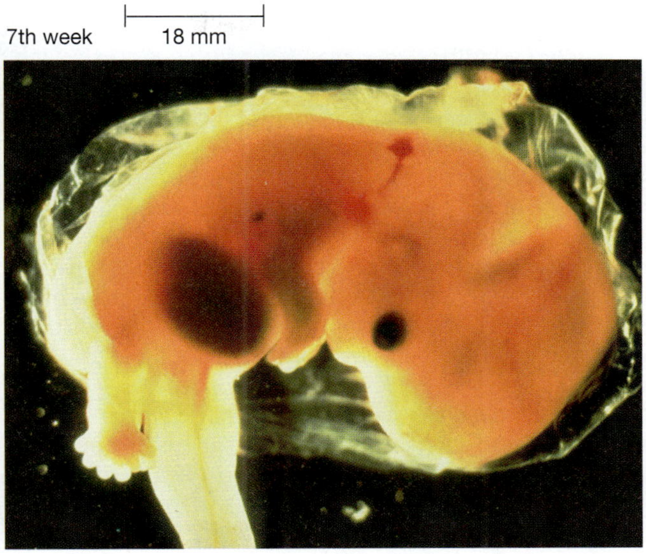

(b)

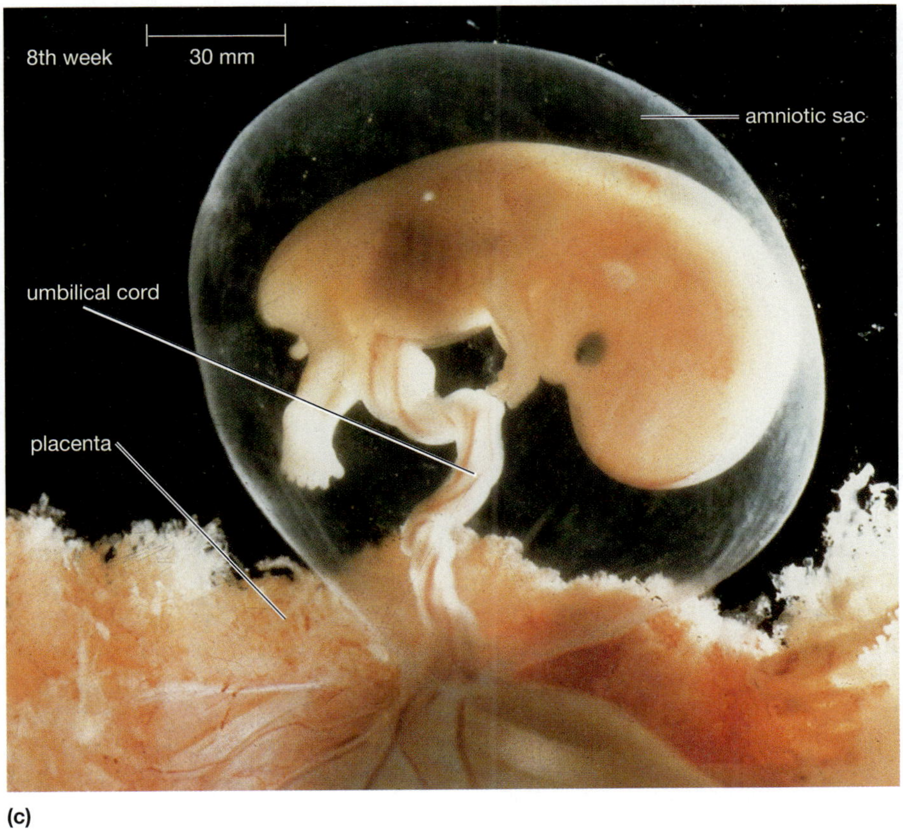

(c)

Figure 25-23 Human development during the fifth through eighth weeks ***(a)*** At the end of the fifth week, the human embryo is about half head. The feet and hands have begun to develop digits. A tail and gill grooves are clearly visible, evidence of our evolutionary relationship to other vertebrates. ***(b)*** By the seventh week, the human form has been more clearly defined by the selective death of cells that form the tail and connect the fingers and toes. The tail has nearly disappeared, and fingers and toes are separated from one another. ***(c)*** At the end of the eighth week, the embryo is clearly human in appearance and is now called a fetus. Most of the major organs of the adult body have begun to develop.

These first 2 months of pregnancy are times of extremely rapid growth of the embryo and also times of considerable danger to it. Although the fetus is vulnerable throughout development, rapidly developing organs are especially sensitive to environmental insults, such as drugs (including alcohol) or certain medications taken by the mother.

Growth and Development Continue During the Last 7 Months

The fetus continues to grow and develop for another 7 months. The brain continues to develop rapidly and the head remains disproportionately large (though the size of the rest of the body does begin to catch up). As

Scientific Inquiry
The Promise of Stem Cells

The ability of a single cell to give rise to the 200 or so types of cells in an adult is one of the wonders of the living world. Scientists have long known that each cell nucleus contains the entire genetic blueprint for an organism. They have also known that the fate of any cell—whether it becomes part of a muscle or a bone or a brain—is determined by complex factors in the cell's environment. If scientists could harness the potential of cells to give rise to any type of tissue, they might create new tissues from a person's own donor cells—tissues that wouldn't be rejected by the immune system—to replace those that are damaged or missing. For example, specific types of neurons might be transplanted to replace those that die from Parkinson's disease or to bridge the gap in a crushed spinal cord. Likewise, bone cells might fill in a shattered fracture, or heart muscle cells might replace those killed by a heart attack.

The cells most likely to prove useful for producing customized tissues for transplant are **embryonic stem cells**. These cells form in the hours immediately after an egg is fertilized and give rise to all the specialized cell types that eventually compose the adult body. In November 1998, James Thompson and coworkers at the University of Wisconsin reported the first success in isolating human embryonic stem cells and inducing them to grow in culture. Transplanted under the skin of mice, the cells differentiated into a wide variety of human tissues.

Despite the potential medical benefits of embryonic stem cells, their use is controversial. As their name implies, they are derived from human embryonic tissue. The tissue may come from aborted embryos but is more often derived from leftover blastocysts obtained from *in vitro* fertilization clinics. These clinics often destroy leftover embryos, each consisting of about 100 cells, after a client has become pregnant.

The ability to culture human stem cells has generated a flurry of research and tremendous excitement in the scientific community. While researchers of human aging wonder if someday worn-out organs might be replaced, biotech firms are scrambling to find new ways to produce stem cells quickly and efficiently for use as tissue transplants. At one company, researchers are implanting nuclei from human skin cells into cow's eggs, and then jolting the eggs with electricity to trick them into developing as if they had been fertilized. If all goes well, the eggs will begin dividing and give rise to blastocysts that could be used as a source of stem cells (Fig. E25-1). The crucial next step is to develop techniques for causing stem cells to differentiate into the desired tissues. To accomplish this goal, researchers must first identify the crucial signals that send the messages to produce specific tissues in a developing embryo. Once the signals are identified, the next step will be to figure out how to deliver similar signals to cells growing in culture dishes. Alternatively, it might be possible to introduce undifferentiated stem cells into a specific part of the body—say, the brain or the bone marrow—and let the normal chemical environment there guide differentiation. Other hurdles must also be overcome, including finding ways to ensure that transplanted stem cells do not grow out of control and form tumors.

The technical and ethical challenges of medical use of embryonic stems cells are formidable. The potential rewards, however, are so far-reaching that both research and passionate debate are likely to continue.

the brain and spinal cord grow, they begin to generate behavior. As early as the third month of pregnancy, the fetus begins to move and respond to stimuli. Some instinctive behaviors appear, including sucking, which will have obvious importance soon after birth. Structures that the fetus will need when it emerges from the uterus, such as the lungs, stomach, intestine, and kidneys, enlarge and become functional, though they will not be used until after birth.

The Placenta Exchanges Materials Between Mother and Embryo

During the first few weeks of pregnancy, embryonic cells obtain nutrients directly from the nearby cells of the endometrium. Eventually, however, the embryo is nourished via the placenta, which develops from interlocking tissues of the embryo and the mother. The embryo's contribution to the placenta comes from an embryonic membrane called the *chorion*, which penetrates the mother's endometrium with fingerlike projections called *chorionic villi* (singular, *villus*). The chorionic villi contain a dense network of capillaries and are bathed in pools of maternal blood (Fig. 25-24). This arrangement permits many small molecules to diffuse between fetal blood and maternal blood. Oxygen diffuses from maternal blood to fetal blood, and carbon dioxide from fetal blood to maternal blood. Nutrients, some aided by active transport, travel from mother to fetus. Fetal urea diffuses into the mother's blood, to be filtered out by the mother's kidneys.

Note that the placenta does not allow fetal and maternal blood to mix. The membranes of the fetal capillaries and chorionic villi allow exchange by diffusion but also act as barriers to the passage of some large proteins and most cells. Despite this barrier, some disease-causing organisms and many harmful chemicals, such as alcohol, can penetrate the placental barrier.

Development Culminates in Labor and Delivery

During the last months of pregnancy, the fetus usually becomes positioned head downward in the uterus with the crown of the skull resting against (and being held up

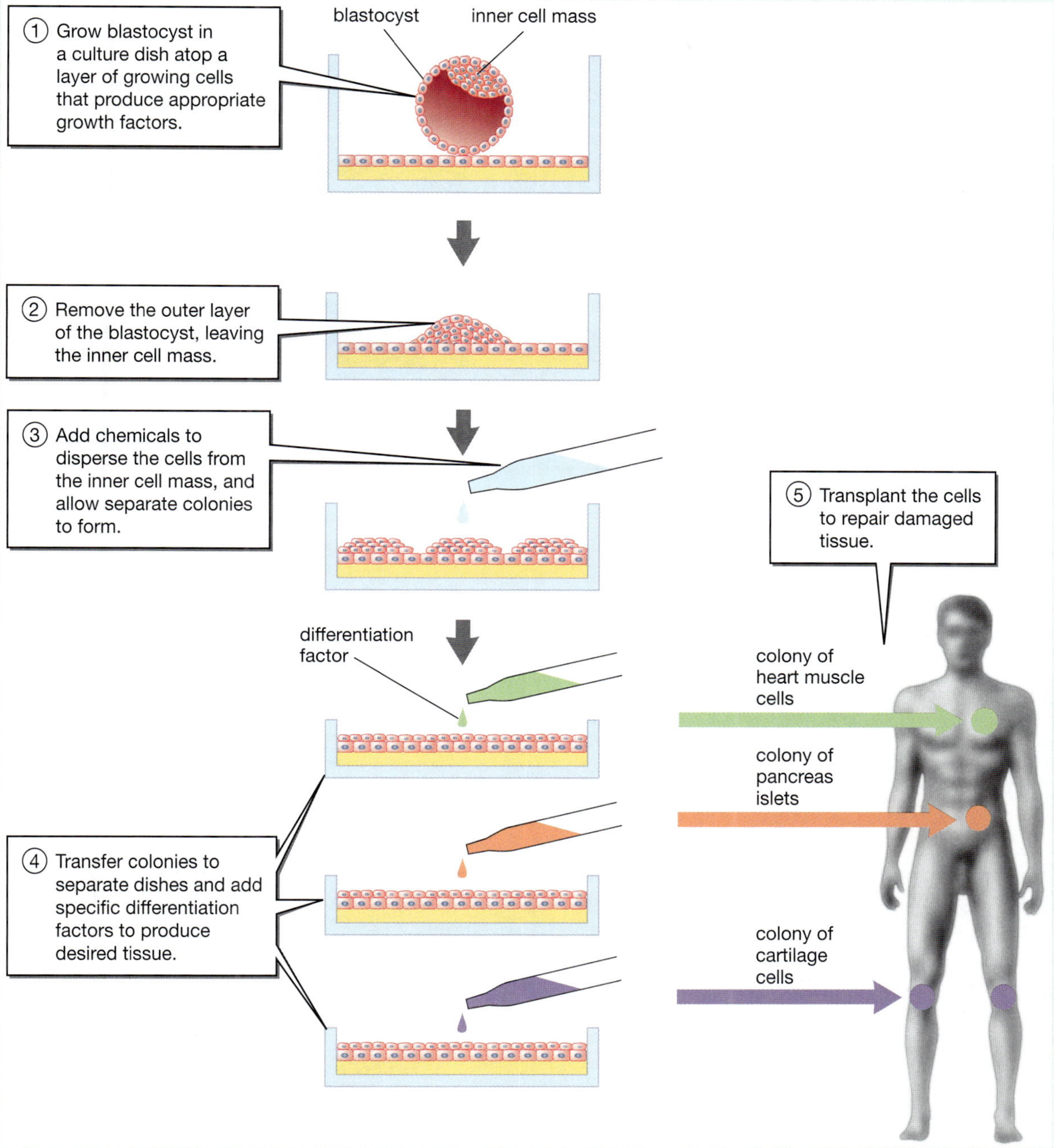

Figure E25-1 Culturing stem cells from a blastocyst

by) the cervix. Near the end of the ninth month, give or take a few weeks, the process of birth normally begins (Fig. 25-25). Birth is the result of a complex interplay between uterine stretching caused by the growing fetus and fetal and maternal hormones that finally trigger **labor** (contractions of the uterus that result in birth).

Unlike skeletal muscles, uterine muscles can contract spontaneously, and stretching enhances these contractions. As the baby grows, it stretches the uterine muscles, which contract occasionally even weeks before delivery. The final trigger for labor is probably provided by hormones secreted by the fetus; these hormones

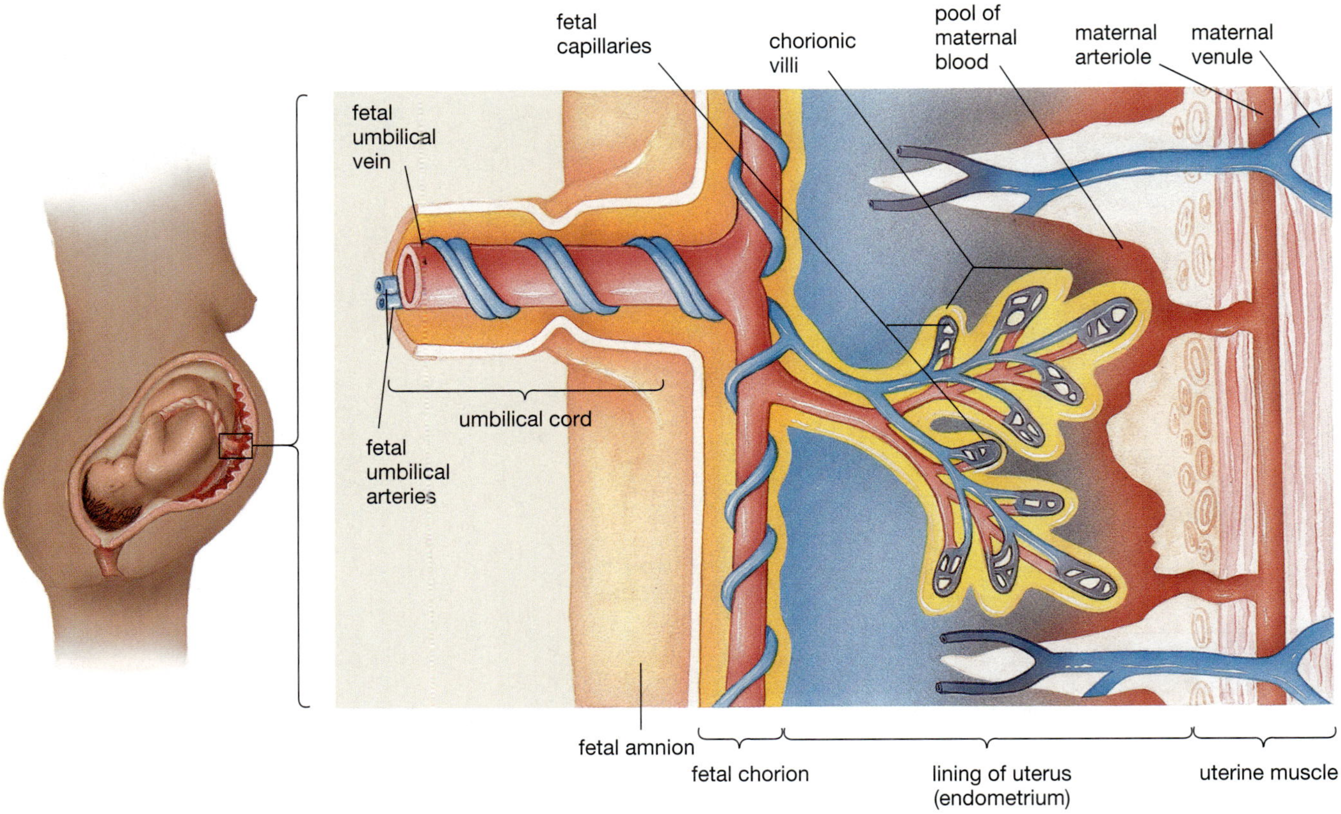

Figure 25-24 The placenta
The placenta forms from both the chorion of the embryo and the endometrium of the mother. Capillaries of the endometrium break down, releasing blood to form pools within the placenta. Meanwhile, the chorion develops projections (the chorionic villi) that extend into these pools of maternal blood. Blood vessels from the umbilical cord branch extensively within the villi. The maternal and fetal blood supplies are separate. Umbilical arteries carry deoxygenated blood from the fetus to the placenta, and umbilical veins carry oxygenated blood to the fetus.

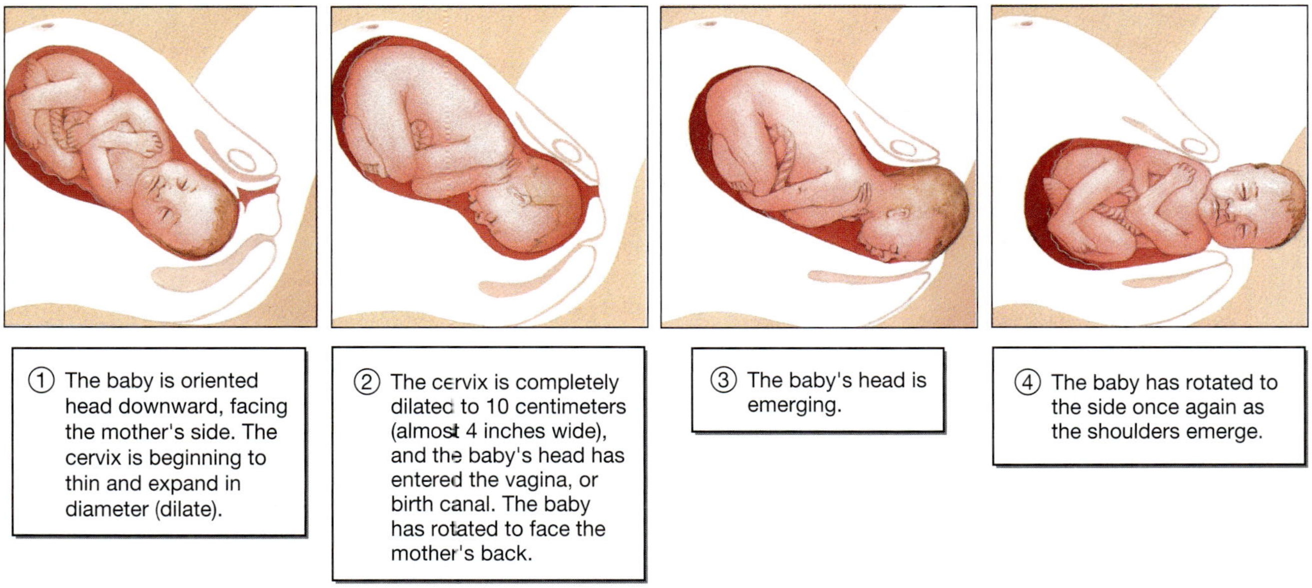

Figure 25-25 Birth

make the uterus even more likely to contract. When the combination of hormones and stretching activate the uterus beyond some critical point, strong contractions begin, signaling the onset of labor. As the contractions proceed, the baby's head pushes against the mother's cervix, which responds by dilating (expanding in diameter). Stretch receptors in the walls of the cervix send signals to the hypothalamus, triggering oxytocin release. Under the additional stimulation of oxytocin, the uterus contracts even more strongly. This positive feedback cycle continues until the baby emerges from the vagina.

After a brief rest, uterine contractions resume, causing the uterus to shrink remarkably. During these contractions, the placenta is sheared from the uterus and expelled through the vagina as the *afterbirth.* The muscles that surround fetal blood vessels in the umbilical cord contract, shutting off blood flow to the placenta. (Tying off the cord is standard practice but is not usually necessary.) A new human being has been born.

Milk Secretion Is Stimulated by Hormones of Pregnancy

As the fetus grows, nourished by nutrients diffusing through the placenta, changes in the mother's breasts prepare her to continue nourishing her child after it is born. During pregnancy, large quantities of estrogen and progesterone (acting together with several other hormones) stimulate **mammary glands**, milk-producing glands in the breasts, to grow, branch, and develop the capacity to secrete milk (Fig. 25-26). Each mammary gland has a milk duct that leads to the *nipple*, a projection of epithelial tissue. The actual secretion of milk, called *lactation*, is promoted by the pituitary hormone prolactin (see Chapter 23).

5 How Can People Limit Fertility?

Natural selection favors individuals that reproduce successfully and whose offspring survive long enough to reproduce themselves. During most of human evolution, child mortality was high, and natural selection favored people who produced enough children to offset this high mortality rate. Today, however, most humans no longer need to have many children to ensure that a few will survive to adulthood, but we nonetheless retain the reproductive drives with which evolution has endowed us. As a result, each passing week sees nearly 1.6 million new people added to our increasingly overcrowded planet, and the control of birthrates has become an environmental necessity. On the individual level, birth control allows people to plan their families and provide the best opportunities for themselves and their children.

Historically, limiting fertility has not been easy. In the past, some cultures have tried such inventive, if bizarre, techniques as swallowing froth from the mouth of a camel or placing crocodile dung in the vagina. Since the 1970s, however, several effective techniques have been developed for **contraception**, or the prevention of pregnancy. These techniques are described next, and their reliability and some possible side effects are summarized in Table 25-4. The choice of a contraceptive should be made in consultation with a health professional, who can provide more complete information.

Sterilization Provides Permanent Contraception

In the long run, the most effortless method of contraception is *sterilization*, in which the pathways through which sperm or egg must travel are interrupted. In men, the vas deferens leading from each testis may be severed in an operation called a *vasectomy*. Sperm are still produced, but they cannot reach the penis during ejaculation. The surgery is performed under a local anesthetic, and vasectomy has no known effects on health or sexual performance. The slightly more complex operation of *tubal ligation* renders a woman infertile by cutting her uterine tubes. Eggs are still produced, but sperm cannot travel to the egg; nor can the egg reach the uterus. Sterilization is generally permanent. Sometimes, however, in a delicate and expensive operation, a surgeon can reconnect the vas deferens or uterine tubes.

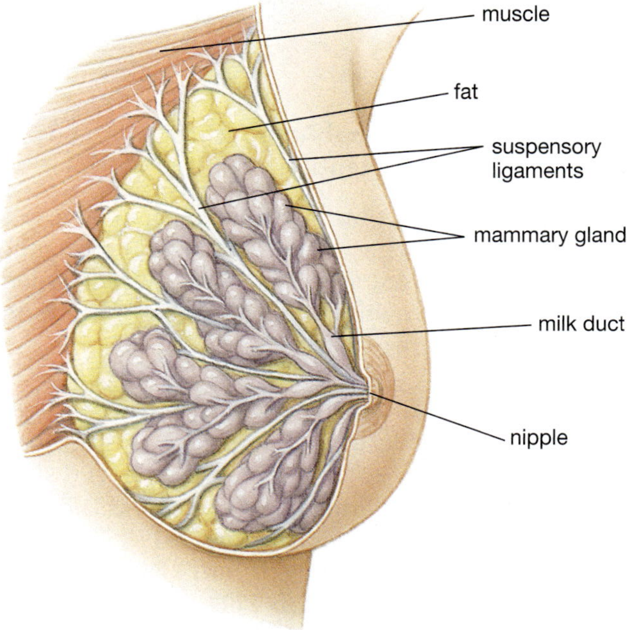

Figure 25-26 The structure of the mammary glands
During pregnancy, both fatty tissue and the milk-secreting glands and ducts increase in size.

Table 25-4 Nonpermanent Contraceptive Techniques

Method	Technique and Mechanism	Failure Rate*	Protection from STD
Abstinence	Deciding not to be sexually active.	0%	Protects against STD
Rhythm	Using cervical mucus changes, and body temperature measurements to estimate the time of ovulation and avoiding intercourse during the fertile period.	2% to 20% (almost never done correctly)	No protection
Spermicide	Spermicidal foam (such as non-oxynol-9) is placed in vagina prior to intercourse, forming a chemical barrier to sperm.	6% to 21%	Partial protection; may not protect against HIV
Contraceptive Sponge	Soft, disposable sponge impregnated with spermicide inserted in vagina. Acts as sperm-killing barrier.	10% to 15%	Partial protection
Diaphragm/ Cervical Cap	Flexible, reusable domed rubber or rubber-like barriers; spermicide is placed within the dome, and the diaphragm (larger) or cap (smaller) is fitted over the cervix prior to intercourse.	6% to 18%	Partial protection
Condom (male)	Thin, disposable latex sheath placed over penis prior to intercourse. May be lubricated with spermicide. Catches sperm and prevents it from entering vagina.	2% to 12%	Best protection against STD for sexually active people. Should be combined with other methods of birth control for STD protection.
Condom (female)	Lubricated polyurethane pouch inserted into vagina, catches sperm before they enter cervix.	5% to 25%	Partial protection
Birth Control Pill	Pill containing either estrogen and synthetic progesterone (combination pill) or progesterone only (minipill). Must be taken daily. Prevents ovulation.	0.1% to 3%	No protection
Norplant®	Six flexible match-stick-sized tubes of synthetic progesterone are inserted under the skin. Gradual release of hormone prevents ovulation for 5 years.	0.04%	No protection
Depo-Provera®	Injection of synthetic progesterone that blocks ovulation for 3 months. Repeated at 3-month intervals.	0.3%	No protection
IUD (intrauterine device)	Small plastic device treated with hormones or copper and placed in the uterus by physician. Prevents sperm from reaching egg or, if sperm do reach egg, prevents implantation of embryo. The copper IUD may be left in for 8 to 10 years.	0.6% to 2.6%	No protection

*Percentage of women becoming pregnant per year. The low and high numbers, respectively, indicate the differences between consistent and correct use and use in a more typical way that is not always consistent or correct.

There Are Three General Approaches to Temporary Contraception

Temporary contraception techniques fall into three general categories: (1) preventing ovulation, (2) preventing sperm and egg from meeting when ovulation does occur, and (3) preventing the implantation of a fertilized egg in the uterus.

Synthetic Hormones Can Prevent Ovulation

During a normal menstrual cycle, ovulation is triggered by a mid-cycle surge of lutenizing hormone. This surge can be suppressed by providing a continuous supply of estrogen and progesterone (see Fig. 25-15). These two hormones (generally in synthetic form) are the components of *birth control pills*. "The Pill" is an extremely effective form of birth control, but it must be taken daily, usually for 21 days each menstrual period.

Two long-term contraceptives, Norplant and Depo-Provera, have been approved for use in the United States since 1990. Both contain synthetic hormones resembling progesterone that prevent ovulation. Norplant consists of six slim, 1.3-inch-long silicone rubber rods inserted under the skin of the upper arm. The rods provide a gradual, steady diffusion of hormone into the bloodstream for 5 years. In extensive tests, Norplant has proved slightly more effective than the birth control pill. Most women using Norplant become fertile within months after the capsule is removed. Depo-Provera is injected once every 3 months.

Barrier Methods Prevent Sperm from Reaching the Egg

There are several effective *barrier methods*, which prevent the encounter of sperm and egg. The *diaphragm* and *cervical cap* are rubber caps that fit snugly over the cervix, preventing sperm from entering the uterus. In conjunction with a *spermicide*, diaphragms and cervical caps are very effective and have no serious known side effects. Alternatively, the male can wear a *condom* over the penis, preventing sperm from being deposited in the vagina. A female condom is now available that completely lines the vagina. Diaphragms and condoms must be applied shortly before intercourse, typically at a time when the participants are preoccupied with other matters. Furthermore, if a diaphragm or condom has even a small hole, sperm can still enter the uterus and fertilize the egg.

Other, less effective birth control procedures include the use of spermicides alone, *withdrawal* (the removal of the penis from the vagina just before ejaculation), and *douching* (washing sperm out of the vagina before, it is hoped, they have entered the uterus). Spermicides have some contraceptive effect, but withdrawal and douching are essentially useless. A final method of preventing fertilization is the *rhythm method*: abstinence from intercourse during the ovulatory period of the menstrual cycle. In practice, the rhythm method has a high failure rate, because of lack of discipline on the part of the users or inaccuracies in determining the menstrual cycle, which varies somewhat from month to month. Because a slight rise in body temperature and changes in the discharge of mucus from the cervix can be used to predict ovulation, the rhythm method can be made more effective if these indicators are monitored and sexual activity is regulated accordingly.

Intrauterine Devices and "Morning After" Pills May Prevent Implantation

Even if an egg is fertilized, pregnancy will result only if the blastocyst implants in the uterus, so contraception can be achieved by preventing implantation. One approach to this kind of contraception is an *intrauterine device* (*IUD*), a small copper or plastic loop, squiggle, or shield that is inserted into the uterus by a physician. Highly effective (if it stays in place), the IUD seems to work by preventing implantation of the embryo in the uterine lining, although it may also prevent sperm from reaching the egg in the first place.

A second method of preventing implantation is the "morning after" pill, which contains hormones similar to birth control pills, only in larger dosages. If a woman has intercourse without protection, these pills may be taken within 3 days ("morning after" is a misnomer) to help avoid pregnancy. The pills may stop or delay ovulation or prevent the embryo from implanting. Pregnancy may also be prevented after unprotected intercourse by inserting a copper IUD within 5 days.

Abortion Removes the Embryo from the Uterus

When contraception fails, pregnancy can be terminated by *abortion*. Abortion commonly involves dilating the cervix and removing the embryo and placenta by suction. Most abortions are performed during the first 3 months of pregnancy. Alternatively, abortion can be induced during the first 7 weeks of pregnancy by the drug RU-486 (mifepristone), which has been used extensively abroad and which received U.S. FDA approval in 2000. RU-486 blocks the action of progesterone, which is required for maintenance of the uterine lining during pregnancy. Under the supervision of a physician, RU-486 is taken as a pill. Two days later, the physician administers a synthetic prostaglandin that induces uterine contractions, causing the embryo to be expelled.

Abortion is, of course, a controversial procedure. Although science can describe fetal development during pregnancy, it cannot provide judgments about when a fetus becomes a "person" or about the relative merits of fetal rights versus maternal rights.

CASE STUDY REVISITED

Multiplicity

About 2.5% of all births in the United States are multiple births. All but 5% or so of multiple births are twins, and most of these are *fraternal twins*. Fraternal twins develop when, for unknown reasons, two eggs (instead of the usual single egg) are released simultaneously by the ovary into the oviduct. If both eggs are fertilized at about the same time and both successfully implant into the endometrium, then two embryos develop. These two embryos are called fraternal twins to emphasize that, because they result from two independent fertilizations, they are no more closely related than any other pair of siblings.

Most twins are fraternal, but about one-third of twin births are of *identical twins*, also known as *monozygotic* twins. As the latter name implies, monozygotic twins arise from a single fertilized egg. After its first division, a two-cell zygote can, on rare occasions, split in half such that each half implants separately into the endometrium. At this very early stage of development, each half is still capable of giving rise to a complete, viable embryo. Because the two embryos are derived from the same zygote, monozygotic twins are genetically identical. They are thus invariably of the same sex and very similar in appearance (any differences are due to environmental factors during development and growth).

Higher-order multiple births (triplets or greater) can be fraternal, monozygotic, or a combination of the two types. As with twins, however, most multiples result from simultaneous fertilizations of multiple eggs (in other words, they're fraternal). At least one set of living monozygotic quintuplets is known, however.

Although science has documented some of the mechanics of twinning and other multiple births, we do not really understand the underlying causes. Many puzzles remain. Why, for example, does a tendency to produce fraternal multiples run in families, while monozygotic multiples have no such inheritable component and arise at random in all human populations? Perhaps the inexplicable nature of multiple births is for the best, as it enables us to more fully enjoy our fascination with a mysterious phenomenon.

Summary of Key Concepts

1) How Do Animals Reproduce?

Animals reproduce either sexually or asexually. In sexual reproduction, haploid gametes, usually from two separate parents, unite and produce an offspring that is genetically different from either parent. In asexual reproduction, offspring tend to be genetically identical to the parent.

Among animals that engage in sexual reproduction, the female produces large, nonmotile eggs and the male produces small, motile sperm. Fertilization occurs, depending on the species, outside the bodies of the animals (external fertilization) or inside the body of the female (internal fertilization). For external fertilization, the animals must be in water so that the sperm can swim to meet the egg. In many species, internal fertilization requires copulation, in which the male deposits sperm directly into the female's reproductive tract.

2) How Does the Human Reproductive System Work?

The male reproductive system consists of paired testes, which produce sperm and testosterone, and structures that conduct the sperm to the female's reproductive tract and secrete fluids that activate swimming by the sperm and provide energy. Spermatogenesis and testosterone production are nearly continuous, beginning at puberty and lasting until death.

The female reproductive system consists of paired ovaries, which produce eggs as well as the hormones estrogen and progesterone, and structures that conduct sperm to the egg and receive and nourish the embryo during prenatal development. In human females, oogenesis, hormone production, and development of the lining of the uterus vary in a monthly menstrual cycle. The cycle is controlled by hormones from the hypothalamus (GnRH), anterior pituitary (FSH and LH), and ovaries (estrogen and progesterone).

During copulation, the male ejaculates semen into the female's vagina. The sperm move through the vagina and uterus into the uterine tube, where fertilization usually takes place. The unfertilized egg is surrounded by two barriers, the corona radiata and the zona pellucida. Enzymes released from the acrosomes in the head of sperm digest these layers, permitting sperm to reach the egg. Only one sperm enters the egg and fertilizes it.

3) How Does Animal Development Proceed?

Animals develop through several stages. *Cleavage:* The fertilized egg undergoes cell divisions with little intervening growth, so the egg cytoplasm is partitioned into smaller cells. Cleavage divisions form the morula, a solid ball of cells. A cavity then opens up within the morula, forming the blastula, a hollow ball of cells. *Gastrulation:* A dimple forms in the blastula, and cells migrate from the surface into the interior of the ball, eventually forming a three-layered gastrula. The three cell layers of ectoderm, mesoderm, and endoderm give rise to all the adult tissues (see Table 25-3). *Organogenesis:* The cell layers of the gastrula form organs characteristic of the animal species.

4) How Do Humans Develop?

A fertilized egg (zygote) develops into a hollow blastocyst and implants in the endometrium. The outer wall will become the chorion and will form the embryonic contribution to the placenta; the inner cell layers of the embryonic disc develop into the embryo. During gastrulation, cells migrate and differentiate into ectoderm, mesoderm, and endoderm. During the third week, the endoderm forms a tube that will become the digestive tract and the notochord induces the formation of the neural tube, the forerunner of the brain and spinal cord. By the end of the second month, the major organs have formed, and the embryo appears human and is called a fetus. In the next 7 months, until birth, the fetus continues to grow, and the lungs, stomach, intestine, and kidneys enlarge and become functional. Human embryonic development follows the same principles as the development of other mammals. The stages of human development are summarized in Figure 25-19. During pregnancy, mammary glands in the mother's breasts enlarge under the influence of estrogen, progesterone, and other hormones. After about 9 months, uterine contractions are triggered by a complex interplay of uterine stretch and prostaglandin and oxytocin release. As a result, the uterus expels the baby and then the placenta.

5) How Can People Limit Fertility?

Pregnancy can be prevented by abstinence or by sterilization: severing the vas deferens in males (vasectomy) or the uterine tubes in females (tubal ligation). Temporary contraception techniques include those that prevent ovulation: birth control pills, Norplant, and Depo-Provera. Barrier methods, which prevent sperm and egg from meeting, include the diaphragm, the cervical cap, and the condom, accompanied by spermicide. Spermicide alone is less effective. Withdrawal and douching are ineffective techniques. The rhythm method requires abstinence around the time of ovulation. Intrauterine devices may block sperm and also prevent implantation of the early embryo. Abortion causes the expulsion of the developing embryo.

Key Terms

follicle *p. 491*
gastrula *p. 499*
gastrulation *p. 498*
gonad *p. 488*
implantation *p. 499*
labium *p. 496*
labor *p. 505*
mammary glands *p. 507*
menstrual cycle *p. 491*
menstruation *p. 494*
mesoderm *p. 499*
morula *p. 498*
neural tube *p. 502*
notochord *p. 499*
oogenesis *p. 491*
oogonium *p. 491*
organogenesis *p. 499*
ovary *p. 491*
ovulation *p. 493*
penis *p. 490*
placenta *p. 494*
primitive streak *p. 499*
scrotum *p. 489*
sexual reproduction *p. 485*
sexually transmitted disease (STD) *p. 497*
sperm *p. 486*
spermatogenesis *p. 490*
spermatogonium *p. 489*
testis *p. 488*
umbilical cord *p. 502*
urethra *p. 490*
uterine tube *p. 493*
uterus *p. 493*
vagina *p. 494*
vas deferens *p. 490*
yolk *p. 497*
zygote *p. 493*

Thinking Through the Concepts

Multiple Choice

1. *All of the following are barrier contraceptive devices EXCEPT the*
   a. IUD
   b. female condom
   c. cervical cap
   d. diaphragm
   e. male condom
2. *In humans, spermatogenesis yields ______ sperm for each diploid sex cell, and oogenesis yields ______ secondary oocyte(s) for each sex cell.*
   a. one, four
   b. two, one
   c. one, two
   d. four, one
   e. four, two
3. *Which structure adds the final secretions to semen as it moves out of the male reproductive tract?*
   a. epididymis
   b. bulbourethral gland
   c. seminal vesicle
   d. prostate gland
   e. Sertoli cells
4. *The corpus luteum*
   a. accompanies the egg as it enters the oviduct
   b. is the fertilized egg
   c. secretes prostaglandin
   d. forms in the uterus
   e. secretes both estrogen and progesterone
5. *The endoderm gives rise to the*
   a. lining of the digestive tract
   b. epidermis of skin
   c. skeletal system
   d. muscles
   e. nervous system
6. *In human development, ectoderm cells migrate through the primitive streak to form*
   a. endoderm
   b. mesoderm
   c. the chorion
   d. the yolk sac
   e. the amnion

? Review Questions

1. List the advantages and disadvantages of asexual reproduction, sexual reproduction, external fertilization, and internal fertilization, including an example of an animal that uses each type.
2. Compare the structures of the egg and sperm. What structures do sperm have that facilitate movement, energy use, and digestion?
3. What is the role of the corpus luteum in a menstrual cycle? In early pregnancy? What determines its survival after ovulation?
4. List the structures, in order, through which a sperm passes on its way from the seminiferous tubules of the testis to the uterine tube of the female.
5. Diagram the menstrual cycle, and describe the interactions among hormones secreted by the pituitary gland and ovaries that produce the cycle.
6. What is gastrulation? Describe gastrulation in frogs and in humans.
7. Name two structures derived from each of the three embryonic tissue layers—endoderm, ectoderm, and mesoderm.
8. How does cell death contribute to development?
9. Where is the human egg fertilized, and what stages of development does the fertilized egg complete before reaching the uterus?
10. Explain how the structure of the placenta prevents mixing of fetal and maternal blood while allowing the exchange of substances between the mother and the fetus.
11. How do changes in the breast prepare a mother to nurse her newborn? How do hormones influence these changes and stimulate milk production?
12. Describe the events that lead to the expulsion of the baby and the placenta from the uterus. Explain why this is an example of positive feedback.

Applying the Concepts

1. *Pelvic endometriosis* is a relatively common disease of women in which bits of the endometrial lining find their way onto abdominal organs and respond in typical ways to hormones during a menstrual cycle. When the uterine lining bleeds during menstruation, so do these implants. Common treatments are oral contraceptives, Danazol™ (a compound that inhibits gonadotropins), and synthetic GnRH analogues that are, paradoxically, powerful inhibitors of FSH and LH. How does each of these compounds provide relief?
2. Would contraceptive drugs that block cell receptors for FSH be useful in males and/or females? Explain. What side effects would such drugs have?
3. Fertility drugs have greatly increased the incidence of multiple births. When more than two embryos share the uterus, the incidence of premature birth and developmental problems increases substantially. The costs of caring for multiple premature infants is staggering. When fertility drugs produce multiple embryos, the physician can selectively eliminate some of these embryos early in development, so the remaining few have a better chance to develop fully and normally. Given these facts, discuss the ethical implications of taking fertility drugs.
4. On the basis of your knowledge of genetics (Unit Two) and evolution (Unit Three), explain why the human embryo passes through a developmental stage in which it has gill grooves and a tail.
5. If the nuclei of adult cells can be transplanted into eggs from which the nucleus has been removed to produce clones of the parent, is it theoretically possible to produce human clones? Would such clones yield offspring that are *exactly* identical to the parents who supplied the nuclei? Explain.
6. Estimates of the number of women in the United States who self-medicate with over-the-counter (OTC) drugs during pregnancy range from 65% to 95%. Evidently, the public doesn't perceive nonprescription substances as drugs. Many subtle disorders, especially of the nervous system, will probably be connected to over-the-counter use in coming years. If you were engaged in pharmaceutical research, which OTC drugs would you focus on as possible culprits? Which months of pregnancy would you study closely? Who would you choose as your study population?

Group Activity

The president of the National Academy of Sciences has chosen you to serve on a panel that will try to answer the question, "Why is there sex?" The panel will meet next week, so form a group of four students to develop a working hypothesis. The problem you face is this: Sexual reproduction is inefficient. To illustrate this, have your group imagine a population of 20 insects (10 male, 10 female) in which each female can produce two offspring in her lifetime. The females in this population, however, are of two types. Five females reproduce asexually, giving birth only to more females. The other five reproduce sexually and give birth to males and females in equal proportions. After five generations, how many females of each type will be in the population? Clearly, sexual reproduction puts individuals at a severe disadvantage in the contest to reproduce. Although it's true that sexual recombination of genes can have advantages over the very long term, the theory of natural selection tells us that a trait can persist only if it helps individuals survive and reproduce in the very short term. So we're faced with a mystery. What direct advantage was gained by the first sexually reproducing organisms? Was a series of intermediate steps required? Were particular environmental conditions required? Did sex originally have a function other than reproduction? The Academy is expecting you to produce a detailed, plausible scenario for the origin of sex, so gather your group for critical discussion, and submit your hypothesis.

For More Information

Alexander, N. "Future Contraceptives." *Scientific American*, September 1995. A review of the new techniques that may be used for contraception in the future.

Caldwell, M. "How Does a Single Cell Become a Whole Body?" *Discover*, November 1992. Explores the miracles of development in layperson's terms.

Christiansen, D. "Male Choice." *Science News*, September 30, 2000. Describes the problems and progress in developing contraceptives for men.

Crews, D. "Animal Sexuality." *Scientific American*, January 1994. Explores the wide range of mechanisms that control male and female sexual development among different types of animals.

Morell, V. "A Clone of One's Own." *Discover*, May 1998. Someday, someone will probably clone a person. What will be the outcome?

Nüsslein-Volhard, C. "Gradients That Organize Embryo Development." *Scientific American*, August 1996. A Nobel laureate describes developmental experiments that have revealed a great deal about how chemical gradients in embryos help guide cell differentiation.

Pedersen, R. A. "Embryonic Stem Cells for Medicine." *Scientific American*, April 1999. Embryonic stem cells hold promise for regenerating damaged tissues.

Riddle, J. M., and Estes, J. W. "Oral Contraceptives in Ancient and Medieval Times." *American Scientist*, May–June 1992. How did women control their fertility before modern medicine stepped in?

Smith, B. R. "Visualizing Human Embryos." *Scientific American*, March 1999. Magnetic resonance microscopy allows computerized reconstuction of three-dimensional structures within human embryos.

Wassarman, P. W. "Fertilization in Mammals." *Scientific American*, December 1988. Describes the events leading to the penetration of the egg by the sperm, with emphasis on the role and composition of the zona pellucida.

Wright, K. "Human in the Age of Mechanical Reproduction." *Discover*, May 1998. The child's question "Where do babies come from?" has become far more difficult to answer as technology has provided more than a dozen ways to have a child.

Answers to Multiple-Choice Questions

1. a 2. d 3. b 4. e 5. a 6. b

MediaTutor

CD Activities

Activity 25.1: Human Reproductive Anatomy

Estimated time: 10 minutes

The human reproductive system is complex and contains a majority of the differences between male and female anatomies. In this tutorial, you will explore the structure and function of the human reproductive system. Your exploration will include a comparison of the male and female anatomies and an introduction to the process of gametogenesis, which is the formation of sperm in the male and an egg in the female.

Activity 25.2: Hormonal Control of the Menstrual Cycle

Estimated time: 10 minutes

The female menstrual cycle is controlled by a variety of complex hormonal interactions that affect both the ovaries and the uterus. In this tutorial, you will explore the events of the menstrual cycle and the function of the various reproductive hormones that are linked to these events.

Activity 25.3: Stages of Animal Development

Estimated time: 10 minutes

The stages of animal embryonic development are remarkably similar in a wide range of animals, including amphibians, birds and reptiles, and mammals. One developmental process shared by all animals is the formation of three tissue layers (endoderm, mesoderm, and ectoderm) during gastrulation. In this tutorial, you will explore the various stages in the development of an animal embryo, with special emphasis on the process of gastrulation.

Activity 25.4: Human Development

Estimated time: 10 minutes

Human development is controlled by the same mechanisms that control development of other animals, and many of the early events of human development resemble those seen in other animals. This similarity is a product of our evolutionary heritage. In this tutorial, you will explore the key events of human embryonic and fetal development; then you can play an interactive game that will test your knowledge and skill.

Start the MediaTutor Student CD-ROM and enter the activity number in the Quick Search box to be taken directly to that activity.

Web Investigation

Multiplicity

Estimated time: 15 minutes

"Case Study: Multiplicity" describes an emerging trend seen in the human population: multiple births. As members of our society increasingly put parenthood off until later in life, many hopeful parents are forced to seek the assistance of medical professionals to assist with fertility problems. This activity will explore human fertility and the steps we take to control our reproductive capacity.

Go to http://www.prenhall.com/audesirk3, the Audesirk Companion Web site. Select Chapter 25 and Web Investigation to begin.

"I find studying the behavior of animals in their natural surroundings a fascinating hobby. It allows one to live out of doors and in beautiful scenery; it gives free scope to one's urges to observe and to reflect; and it leads to new discoveries. Even the most trivial discovery gives intense delight."

Niko Tinbergen in* Curious Naturalists *(1958)

Both this male scorpionfly and this male human are exceptionally attractive to females of their species. The secret to their sex appeal may be that both have highly symmetrical bodies.

26 Animal Behavior

AT A GLANCE

CASE STUDY

Sex and Symmetry

For a male Japanese scorpionfly, finding a mate can be a real struggle. Female scorpionflies will mate only with males that can offer a tasty meal (usually a dead insect). Competition for dead insects is fierce and, typically, once a male finds an insect, he must defend it from other males. The competition for insects often erupts in bitter combat, characterized by repeated head-butting and grappling with sharp-pointed genital claspers.

In the competition to gain access to mates, not all male Japanese scorpionflies can be equally successful. The most successful males have qualities that help them defeat other males in combat and that are especially attractive to females. Biologist Randy Thornhill has found that one quality in particular accurately predicts the mating success of male scorpionflies: symmetry. In Thornhill's experiments and observations, the most successful males were those whose left and right wings were equal or nearly equal in length. Males with one wing longer than the other were less likely to win fights or to copulate; the greater the difference between the two wings, the lower the likelihood of success.

Thornhill's work with scorpionflies led him to wonder if the advantages of symmetry also extend to humans. Working with psychologist Steven Gangestad, he devised some fascinating studies that suggest that male symmetry does indeed play an important role in human sexual relationships. Women find symmetrical men more attractive, and highly symmetrical human males do indeed appear to gain some of the advantages that accrue to symmetrical scorpionflies. The preferences of human females resemble, at least in this one respect, those of female scorpionflies. One could almost say that insects and humans share a standard of beauty. ■

MediaTutor 26.1 Innate Animal Behavior

1 How Do Innate and Learned Behaviors Differ?

Behavior is any observable activity of a living animal. A moth flies toward a bright light, a honeybee flies toward a cup of sugar-water, and a housefly flies toward a piece of rotting meat. Bluebirds sing, wolves howl, and frogs croak. Mountain goats butt heads in ritual combat, chimpanzees groom one another, ants attack a termite that approaches an anthill. Humans smoke cigarettes, play tennis, and plant gardens. Even the most casual observer sees many examples of animal behavior each day, and a careful observer encounters a virtually limitless number of fascinating behaviors.

Innate Behaviors Can Be Performed Without Prior Experience

An **innate behavior** is one that is performed in reasonably complete form even the first time an animal of the right age and motivational state encounters a particular stimulus. (The proper motivational state for feeding, for example, would be hunger.) Scientists can demonstrate that a behavior is innate by depriving an animal of the opportunity to learn it. For example, red squirrels, which in the wild bury nuts in the fall for retrieval during the winter, can be raised from birth in a bare cage on a liquid diet, providing them with no experience of nuts, digging, or burying. Nonetheless, such a squirrel will, when presented with nuts for the first time, carry one to the corner of its cage, and then make covering and patting motions with its forefeet. Nut burying is therefore an innate behavior.

Innate behaviors can also be recognized by their occurrence immediately after birth, before there is any opportunity for learning. The cuckoo, for example, lays its eggs in the nest of another bird species, to be raised by the unwitting adoptive parent. Immediately after hatching, the cuckoo chick shoves the nest owner's eggs (or baby birds) out of the nest, eliminating its competitors for food (Fig. 26-1).

Learned Behaviors Are Modified by Experience

Natural selection may favor innate behaviors in many circumstances. For instance, it is clearly to the advantage of a herring gull chick to be able to peck at its parent's bill as soon as possible after hatching, because pecking stimulates the parent to feed the chick. But in other circumstances, rigidly fixed behavior patterns may be less useful. For example, a male red-winged blackbird that is presented with a stuffed female blackbird will often attempt to copulate with the stuffed bird, a behavior that obviously will produce no offspring. In many situations, a greater degree of behavioral flexibility is advantageous.

The capacity to make changes in behavior on the basis of experience is called **learning**. This deceptively simple definition encompasses a vast array of different phenomena. A toad learns to avoid distasteful insects (Fig. 26-2), a baby shrew learns which adult is its mother, a human learns to speak a language, a blackbird learns to use the stars for navigation. Each of the many

(a) (b)

Figure 26-1 Innate behavior
(a) The cuckoo chick, just hours after it hatches and before its eyes have opened, evicts the eggs of its foster parents from the nest. *(b)* The parents, responding to the stimulus of the cuckoo chick's wide-gaping mouth, feed the chick, unaware that it is not related to them.

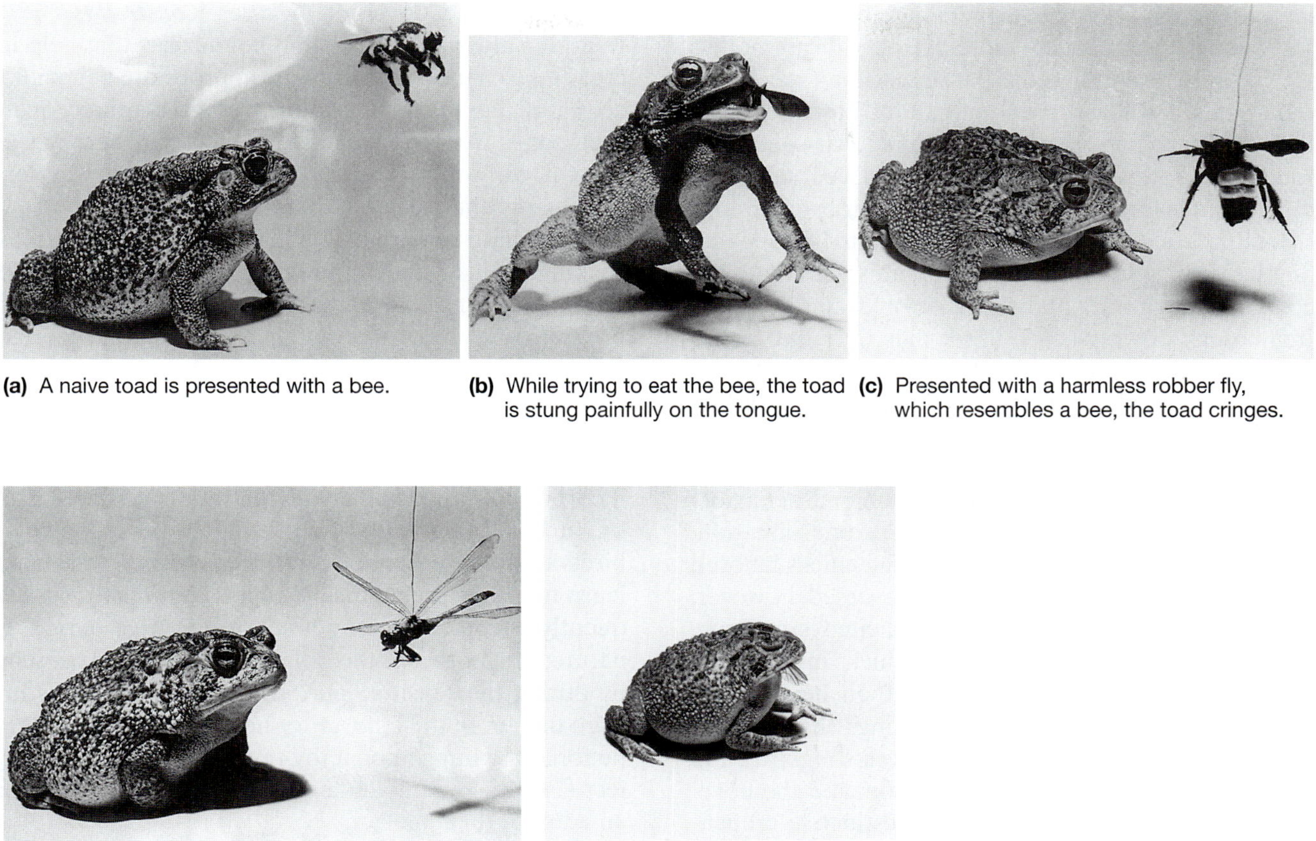

(a) A naive toad is presented with a bee.

(b) While trying to eat the bee, the toad is stung painfully on the tongue.

(c) Presented with a harmless robber fly, which resembles a bee, the toad cringes.

(d) The toad is presented with a dragonfly.

(e) The toad immediately eats the dragonfly, demonstrating that the learned aversion is specific to bees and insects resembling them.

Figure 26-2 Learning in a toad

examples of animal learning represents the outcome of a unique evolutionary history, so learning is as diverse as animals themselves.

There Is No Sharp Distinction Between Innate and Learned Behaviors

Although the terms *innate* and *learned* can be useful tools in describing and understanding behaviors, these words also have the potential to lull us into an oversimplified view of animal behavior. In practice, no behavior is unambiguously innate or unequivocally learned; all behaviors are an intimate mixture of the two.

Seemingly Innate Behavior Can Be Modified by Experience

Behaviors that seem to be performed correctly on the first attempt without prior experience can later be modified by experience. For example, a newly hatched herring gull chick is able to peck at a red spot on its parent's beak (Fig. 26-3), a stereotyped behavior that causes the parent to regurgitate food for the chick to eat. Biologist Niko Tinbergen studied this pecking behavior and found that the pecking response of very young chicks was triggered by the long, thin shape and red color of the parent's bill. In fact, when Tinbergen offered newly hatched chicks a thin, red rod with white stripes painted on it, they pecked at it more often than at a real beak. Within a few days, however, the chicks

Figure 26-3 Innate behaviors can be modified by experience A herring gull chick pecks at the red spot on its mother's bill, causing her to regurgitate food.

learned enough about the appearance of their parents that they began pecking more frequently at models more closely resembling the parents. After 1 week, the young gulls had learned enough about how their parents looked to prefer models of their own species to models of a closely related species. Eventually, the young birds learned to beg only from their own parents.

Habituation (a decline in response to a repeated stimulus) can also fine-tune an organism's innate responses to environmental stimuli. For example, young birds crouch down when a hawk flies over but ignore harmless birds such as geese. Early observers hypothesized that only the very specific shape of predatory birds provoked crouching. Using an ingenious model, Niko Tinbergen and Konrad Lorenz (two of the founding fathers of **ethology**, the study of animal behavior) tested this hypothesis. When moved in one direction, the model resembled a goose, which the chicks ignored (Fig. 26-4). When its movement was reversed, however, the model resembled a hawk and elicited crouching. Further research revealed that naive chicks instinctively crouch when *any* object moves over their heads. Over time, their response habituates to things that soar by harmlessly and frequently, such as leaves, songbirds, and geese. Predators are much less common, and the novel shape of a hawk continues to elicit instinctive crouching. Thus, learning modifies the innate response, making it more advantageous.

Learning May Be Governed by Innate Constraints

Learning always occurs within boundaries that help increase the chances that only the appropriate behavior is acquired. For example, young robins routinely learn their songs from the adult robins that they hear but almost never copy the songs of the sparrows, warblers, finches, or other species that they also hear.

The innate constraints on learning are perhaps most strikingly illustrated by **imprinting**, a special form of learning in which an animal's nervous system is rigidly programmed to learn a certain thing only at a certain period of development. This causes a strong association to be formed during a particular stage, called a *sensitive period*, in the animal's life. During this stage, the animal is primed to learn specific information, which is then incorporated into a behavior that is not easily altered by further experience. Imprinting is best known in birds such as geese, ducks, and chickens. These birds learn to follow the animal or object that they most frequently encounter during an early sensitive period. In nature, the mother is the object most likely to be nearby during the sensitive period, so the young birds imprint on her. In the laboratory, however, these birds can be forced to imprint on a toy train or other moving object (Fig. 26-5). If given a choice, however, they select an adult of their species.

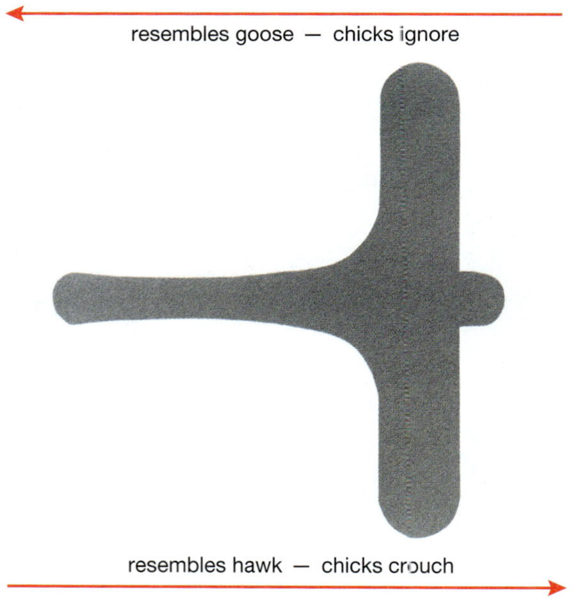

Figure 26-4 Habituation modifies innate responses
The model used by Konrad Lorenz and his student Niko Tinbergen to investigate the response of chicks to the shape of objects flying overhead. The chicks' response depended on which direction the model moved. Moving toward the right, the model resembles a predatory hawk, but when it moves left, it resembles a harmless goose.

Figure 26-5 Konrad Lorenz and imprinting
Konrad Lorenz, known as the "father of ethology," is followed by goslings that imprinted on him shortly after they hatched. They follow him as they would their mother.

All Behavior Arises Out of Interaction Between Genes and Environment

Many early ethologists saw innate behaviors as rigidly controlled by genetic factors and viewed learned behaviors as determined exclusively by an animal's environment. Today, however, this false dichotomy has given way to the realization that, just as no behavior is wholly innate or wholly learned, no behavior can be caused strictly by genes or strictly by the environment. Instead, all behavior develops out of an interaction between genes and their environment. The relative contributions of heredity and learning vary among animal species and among behaviors within an individual.

The precise nature of the link among genes, environments, and behaviors is not well understood in most cases. The chain of events between the transcription of genes and the performance of a behavior may be so complex that we will never be able to decipher it fully. Nonetheless, a great deal of evidence demonstrates the existence of both genetic and environmental components in the development of behaviors. For example, consider the case of bird migration. Even though it is well known that migratory birds must learn by experience how to navigate by using celestial cues, this learning is not the only factor involved.

Bird Migration Behavior Has an Inherited Component

At the close of the summer, many birds disappear from the habitats where they've spent the breeding season and head for winter quarters hundreds or even thousands of miles to the south. Many of the migrating birds are traveling for the first time, because they were hatched only a few months earlier. Amazingly, these naïve birds depart at the proper time and in the proper direction and locate the proper wintering location, even though they do not simply follow more-experienced birds (which typically depart a few weeks in advance of the first-year birds). Somehow, these young birds are able to execute a very difficult task the first time they try it. It seems that birds must be born with the ability to migrate; it must be "in their genes." And indeed, birds that are experimentally hatched and raised in isolation indoors still orient in the proper migratory direction when autumn comes, apparently without the need for any learning or experience.

The conclusion that birds must have a genetically controlled ability to migrate in the right direction was further supported by hybridization experiments with blackcap warblers. This species breeds in Europe and migrates to Africa, but populations from different areas travel by different routes. Blackcaps from western Europe travel in a southwesterly direction to reach Africa, whereas birds from eastern Europe must travel to the southeast (Fig. 26-6). If birds from the two populations are crossbred in captivity, however, the hybrid offspring exhibit migratory orientation toward due south, intermediate between that of the two parents. This result suggests that the parents had genes that influenced migratory direction and that the hybrids inherited a mixture of these genes.

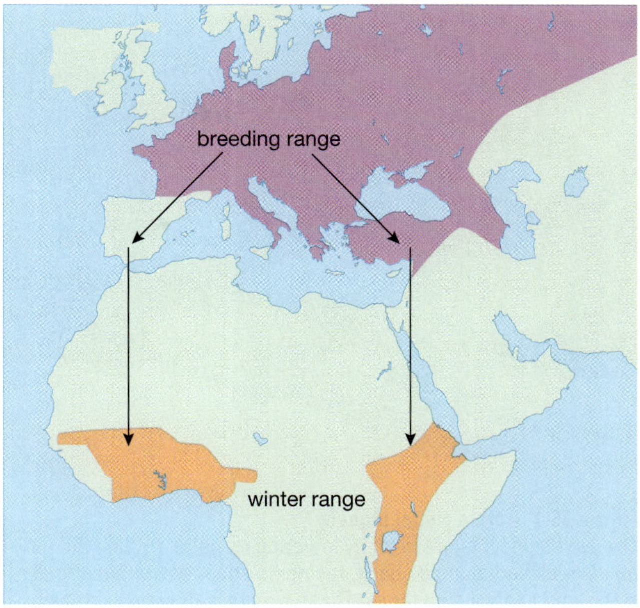

Figure 26-6 Genes influence migratory behavior
Blackcap warblers from western Europe begin their fall migration by flying in a southwesterly direction, but those from eastern Europe fly to the southeast when they begin migrating.

2 How Do Animals Communicate?

Animals frequently make information available for sharing. The sounds uttered, movements made, and chemicals emitted by animals can reveal the animals' physical location, level of aggression, readiness to mate, and so on. If revealing such information evokes a response from other individuals, and if that response tends to benefit the sender and the receiver, then a communication channel can form. **Communication** is defined as the production of a signal by one organism that causes another organism to change its behavior in a way beneficial to one or both.

Although animals of different species may communicate (picture a cat, its tail erect and bushy, hissing at a dog), most communication is between members of the same species. Potential mates must communicate, as must parents and offspring. At the same time, members of the same species compete most directly with one another for food, space, and mates. Communication is often used to resolve conflicts caused by competition.

The mechanisms by which animals communicate are astonishingly diverse and use all of the senses. In the following sections, we look at communication by visual displays, sound, chemicals, and touch.

Figure 26-7 Active visual signals
The wolf signals aggression by lowering its head, ruffling the fur on its neck and along its back, facing its opponent with a direct stare, and exposing its fangs. These signals can vary in intensity, communicating different levels of aggression.

Figure 26-8 A passive visual signal
The female mandrill's colorfully swollen buttocks serve as a passive visual signal that she is fertile and ready to mate.

Visual Communication Is Most Effective over Short Distances

Animals with well-developed eyes, from insects to mammals, use visual signals to communicate. Visual signals can be *active*, in which a specific movement (such as baring fangs) or posture (such as lowering the head) conveys a message (Fig. 26-7). Alternatively, visual signals may be *passive*, in which case the size, shape, or color of the animal conveys important information, commonly about its sex and reproductive state. For example, when female mandrills become sexually receptive, they develop a large, brightly colored swelling on their buttocks (Fig. 26-8). Active and passive signals can be combined, as illustrated by the lizard in Figure 26-9.

Like all forms of communication, visual signals have both advantages and disadvantages. On the plus side, they are instantaneous, and active signals can be rapidly changed to convey a variety of messages in a short period. Visual communication is quiet and unlikely to alert distant predators, although the signaler does make itself conspicuous to those nearby. On the negative side, visual signals are generally ineffective in darkness and in dense vegetation, although female fireflies signal potential mates by using species-specific patterns of flashes. Finally, visual signals are limited to close-range communication.

Figure 26-9 Active and passive visual signals combined
The South American *Anolis* lizard raises his head high in the air (an active visual signal), revealing a brilliantly colored throat pouch (a passive visual signal) that warns others to keep their distance.

Communication by Sound Is Effective over Longer Distances

The use of sound overcomes many of the shortcomings of visual displays. Like visual displays, sound signals reach receivers almost instantaneously. But unlike visual signals, sound can be transmitted through darkness, dense forests, and murky water. Sound signals can also be effective over longer distances than visual signals. For example, the low, rumbling calls of African elephants can be heard by elephants several miles distant, and the songs of humpback whales are audible hundreds of miles away from the singer. Likewise, the howls of a wolf pack carry for miles on a still night. Even the small kangaroo rat produces a sound (by striking the desert floor with its hind feet) that is audible 150 feet (45 meters) away.

Sound signals are similar to visual displays in that they can be varied to convey rapidly changing messages. (Think of words and of the emotional nuances conveyed by the human voice during a conversation.) Changes in motivation can be signaled by a change in the loudness or pitch of the sound. An individual can convey different messages by variations in the pattern, volume, and pitch of the sound produced. Ethologist Thomas Struhsaker studied vervet monkeys in Kenya in the 1960s and found that they produced different calls in response to threats from each of their major predators: snakes, leopards, and eagles. In 1980 other researchers reported that the response of other vervet monkeys to each of these calls is appropriate to the particular predator. The "bark" that warns of a leopard or other four-legged carnivore causes monkeys on the ground to take to trees and those in trees to climb higher. The "rraup" call signaling an eagle or other hunting bird causes monkeys on the ground to look upward and take cover, whereas monkeys already in trees drop to the shelter of lower, denser branches. The "chutter" call that indicates presence of a snake causes the monkeys to stand up and search the ground for this predator.

The use of sound is by no means limited to birds and mammals. Male crickets produce species-specific songs that attract female crickets of the same species. The annoying whine of the female mosquito as she prepares to bite alerts nearby males that she may soon have the blood meal necessary for laying eggs. Male water striders vibrate their legs, sending species-specific patterns of vibrations through the water, attracting mates and repelling other males (Fig. 26-10). From these rather simple signals to the virtuoso performance of human language, sound is one of the most important forms of communication.

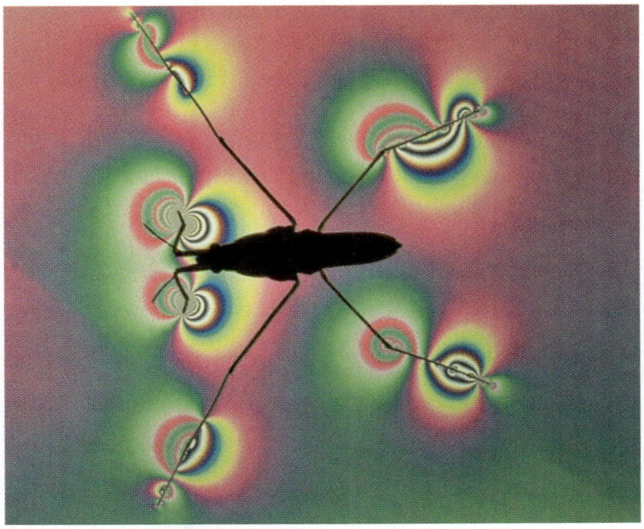

Figure 26-10 Communication by vibration
The light-footed water strider relies on the surface tension of water to support its weight. By vibrating its legs, the water strider sends signals that radiate out over the surface of the water. These vibrations advertise the strider's species and sex to others nearby.

Chemical Messages Persist Longer But Are Hard to Vary

Chemical substances that are produced by individuals and that influence the behavior of other members of the species are called **pheromones**. Pheromones can carry messages over long distances, and, unlike sound, take very little energy to produce. Pheromones may not even be detected by other species, including predators who might be attracted to visual or sound signals. Like a signpost, a pheromone persists over time and can convey a message after the animal has departed. Wolf packs, hunting over areas as large as 385 square miles (1000 square kilometers), warn other packs of their presence by marking the boundaries of their travels with urine that contains pheromones. As anyone who has walked a dog can attest, the domesticated dog reveals its wolf ancestry by staking out its neighborhood with urine that carries a chemical message: "I live in this area."

Such communication requires animals to synthesize a different chemical for each message. As a result, chemical signaling systems communicate fewer different messages than do sight- or sound-based systems. In addition, pheromone signals cannot easily convey rapidly changing messages. Nonetheless, chemicals effectively convey a few simple but critical messages.

Many pheromones cause an immediate change in the behavior of the animal that detects them. For example, foraging termites that discover food lay a trail of pheromones from the food to the nest, and other termites follow the trail (Fig. 26-11). Other pheromones convey messages such as "This area is mine" or "I am ready to mate."

Figure 26-11 Communication by chemical messages
A trail of pheromones, secreted by termites from their own colony, orients foraging termites toward a source of food.

(a)

(b)

Figure 26-12 Communication by touch *(a)* An adult olive baboon grooms a juvenile. Grooming both reinforces social relationships and removes debris and parasites from the fur. *(b)* Touch is also important in sexual communication. These land snails (*Helix*) engage in courtship behavior that will culminate in mating.

Communication by Touch Helps Establish Social Bonds

Communication by physical contact often serves to establish and maintain social bonds among group members. This function is especially apparent in humans and other primates, which have many gestures—including kissing, nuzzling, patting, petting, and grooming—with important social functions (Fig. 26-12a). Touch may even be essential to human well-being. Recent research showed that when the limbs of premature human infants were stroked and moved for 45 minutes daily, the infants were more active, responsive, and emotionally stable and gained weight more rapidly than did premature infants who received standard hospital treatment.

Communication by touch is not limited to primates, but is widespread. In many other mammal species, close physical contact helps cement the bond between parent and offspring. Species in which sexual activity is preceded or accompanied by physical contact can be found across the animal kingdom (Fig. 26-12b).

The Big Picture

An animal's behavior is part of its phenotype. Like other aspects of phenotype, such as body structure, behavior has evolved. The behaviors we observe today arose because they enhanced the survival and reproduction of the individuals that possessed them.

3 How Do Animals Compete for Resources?

The Darwinian contest to survive and reproduce stems from the scarcity of resources relative to the reproductive potential of populations. The resulting competition underlies many of the most frequent types of interactions between animals.

Aggressive Behavior Helps Secure Resources

One of the most obvious manifestations of competition for resources such as food, space, or mates is **aggression**, or antagonistic behavior, between members of the same species. Although the expression "survival of the fittest" evokes images of the strongest animal emerging triumphantly from among the dead bodies of its competitors, in reality most aggressive encounters between members of the same species end without physical damage to the participants. Natural selection has favored the evolution of symbolic displays or rituals for resolving conflicts. During fighting, even the victorious animal can be injured, so fighters might not survive to pass on their genes. Aggressive *displays*, in contrast, allow the competitors to assess each other and acknowledge a winner on the basis of its size, strength, and motivation rather than on the wounds it can inflict.

During aggressive displays, animals may exhibit weapons, such as claws and fangs (Fig. 26-13a), and often behave in ways that make them appear larger (Fig. 26-13b). Competitors often stand upright and erect their fur, feathers, ears, or fins (see Fig. 26-7). The displays are typically accompanied by intimidating sounds (growls, croaks, roars, chirps) whose loudness can help decide the winner. Fighting tends to be a last resort when displays fail to resolve a dispute.

In addition to aggressive visual and vocal displays, many animal species engage in ritualized combat. Deadly weapons may clash harmlessly (Fig. 26-14) or may not be used at all. In many cases these encounters involve shoving rather than slashing. The ritual thus allows contestants to assess the strength and the motivation of their rivals, and the loser slinks away in a submissive posture that minimizes the size of its body.

(a)

(b)

Figure 26-13 Aggressive displays
(a) Threat display of the male baboon. Despite the potentially lethal fangs so prominently displayed, aggressive encounters between baboons rarely cause injury. ***(b)*** The aggressive display of many male fish (here, striped grunts) includes elevating the fins and flaring the gill covers, thus making the body appear larger.

Figure 26-14 Displays of strength
Ritualized combat of fiddler crabs. Oversized claws, which could severely injure another animal, grasp harmlessly. Eventually one crab, sensing greater vigor in his opponent, retreats unharmed.

Figure 26-15 A dominance hierarchy
The dominance hierarchy of the male bighorn sheep is signaled by the size of the horns; these rams increase in status from right to left. The backward-curving horns, clearly not designed to inflict injury, are used in ritualized combat.

Dominance Hierarchies Help Manage Aggressive Interactions

Aggressive interactions use a lot of energy, can cause injury, and can disrupt other important tasks, such as finding food, watching for predators, or raising young. Thus, there are advantages to resolving conflicts with minimal aggression. In a **dominance hierarchy**, each animal establishes a rank that determines its access to resources. Although aggressive encounters occur frequently while the dominance hierarchy is being established, once each animal learns its place in the hierarchy, disputes are minimized and the dominant individuals obtain most access to the resources needed for reproduction, including food, space, and mates. For example, domestic chickens, after squabbling, sort themselves into a reasonably stable "pecking order." Thereafter, all birds in the group defer to the dominant bird, all but the dominant bird give way to the second, and so on. In wolf packs, one member of each sex is the dominant, or "alpha," individual to whom all others of that sex are subordinate. Among male bighorn sheep, dominance is reflected in horn size (Fig. 26-15).

Figure 26-16 Jane Goodall observing chimps at play

Figure 26-17 A feeding territory
Acorn woodpeckers live in communal groups that excavate acorn-sized holes in dead trees, stuffing the holes with green acorns for dining during the lean winter months. The group defends the trees vigorously against other groups of acorn woodpeckers and against acorn-eating birds of other species, such as jays.

Perhaps the most thoroughly studied dominance hierarchy is that of chimpanzees. Ethologist Jane Goodall (Fig. 26-16) has devoted more than 30 years to making meticulous observations of chimpanzee behavior in the field at Gombe National Park in Tanzania, and she has described and documented the animals' complex social organization. Chimps live in groups, and dominance hierarchies among males are a key aspect of their social life. Many males devote a significant amount of time to maintaining their position in the hierarchy, largely by way of an aggressive *charging display* in which a male rushes forward, throws rocks, leaps up to shake vegetation, and otherwise seeks to intimidate rival males. The advantages that accrue to dominant males, however, are not clear. According to Goodall, low-ranking males can still secure access to food and successful copulations (albeit not quite as easily as higher-ranking males). In Goodall's view, little evolutionary advantage accrues to a dominant male, and the function of chimpanzee dominance hierarchies remains unexplained.

Animals May Defend Territories That Contain Resources

In many animal species, competition for resources takes the form of **territoriality**, the defense of an area where important resources are located. The defended area may include places to mate, raise young, feed, or store food. Territorial animals generally restrict most or all of their activities to the defended area and advertise their presence there. Territories may be defended by males, females, a mated pair, or entire social groups (as in the defense of the nest by social insects). However, territorial behavior is most commonly seen in adult males, and territories are normally defended against members of the same species, who compete most directly for the resources being protected. Territories are as diverse as the animals defending them. For example, a territory can be a tree where a woodpecker stores acorns (Fig. 26-17), small depressions in the sand used as nesting sites by cichlid fish, a hole in the sand used as a home by a crab, or an area of forest providing food for a squirrel.

Territories Reduce Aggression

Acquiring and defending a territory require considerable time and energy, yet territoriality is seen in animals as diverse as worms, arthropods, fish, birds, and mammals. The fact that organisms as distantly related as worms and humans independently evolved similar behavior suggests that territoriality provides some important advantages. Although the particular benefits depend on the species and the type of territory it defends, some broad generalizations are possible. First (as with dominance hierarchies), once a territory is established through aggressive interactions, relative peace prevails as boundaries are recognized and respected. The saying "good fences make good neighbors" also applies to nonhuman territories.

Competition for Mates May Be Based on Territories

For males of many species, successful territorial defense has a direct impact on reproductive success. In these species, males defend territories, and females are attracted to high-quality territories, which might have features such as large size, abundant food, and secure nesting areas. Males who successfully defend the best territories have the greatest chance of mating and passing on their genes. For example, experiments have shown that male stickleback fish that defend large territories are more successful in attracting mates than are males that defend small territories. Females that select males with the best territories increase their own reproductive success and pass their genetic traits (typically including their mate-selection preferences) to their offspring.

Animals Advertise Their Occupancy

Territories are advertised through sight, sound, and smell. If the territory is small enough, the owner's mere presence, reinforced by aggressive displays at intruders, can be a sufficient defense. A mammal that has a territory but cannot always be present may use pheromones to scent-mark the boundaries of its territory. Male rabbits use pheromones secreted by glands in their chin and by anal glands to mark their territories. Hamsters rub the areas around their dens with secretions from special glands in their flanks.

Vocal displays are a common form of territorial advertisement. Male sea lions defend a strip of beach by swimming up and down in front of it, calling continuously. Male crickets produce a specific pattern of chirps to warn other males away from their burrows. Birdsong is a striking example of territorial defense. The husky trill of the male seaside sparrow is part of an aggressive display, warning other males to steer clear of his territory (Fig. 26-18). In fact, male sparrows that are unable to sing are unable to defend territories. The importance of singing to seaside sparrows' territorial defense was elegantly demonstrated by ornithologist M. Victoria McDonald, who captured territorial males and performed an operation that left them temporarily unable to sing but still able to utter the other, shorter and quieter signals in their vocal repertoires. The songless males were unable to defend territories or attract mates but regained their lost territories when they recovered their singing ability.

4 How Do Animals Find Mates?

In many sexually reproducing animal species, mating involves copulation or other close contact between male and female. Before animals can successfully mate, however, they must identify one another as members of the same species, as members of the opposite sex, and as being sexually receptive. In many species, finding an appropriate potential partner is only the first step, because the male must demonstrate his quality before the female will accept him as a mate. The need to fulfill all of these requirements has resulted in the evolution of a diverse and fascinating array of courtship behaviors.

Signals Encode Sex, Species, and Individual Quality

Individuals that waste energy and gametes by mating with members of the wrong sex or wrong species are at a disadvantage in the contest to reproduce. Thus, natural selection favors behaviors by which animals communicate their sex and species to potential mates.

Many Mating Signals Are Acoustic

One result of selection for signals that advertise sex and species is that many campers have been kept awake by a raucous nighttime chorus of male tree frogs, each singing a species-specific song. Male grasshoppers and crickets also advertise their sex and species by their calls, as does the female mosquito with her high-pitched whine.

The signals that advertise sex and species may also be used by potential mates in comparisons among rival suitors. For example, the male bellbird uses its deafening song to defend large territories and attract females from great distances. The females fly from one territory to another, alighting near each male in his tree. The male, beak gaping, leans directly over the flinching female and utters an earsplitting note. The female apparently endures this noise to compare the songs of the various males, perhaps choosing the loudest as a mate.

Visual Mating Signals Are Also Common

Many species use visual displays for courting. The firefly, for example, flashes a message that identifies its sex and species. Male fence lizards bob their heads in a

Figure 26-18 Defense of a territory by song
A male seaside sparrow announces ownership of his territory.

Figure 26-19 Sexual displays
(a) During courtship, the male satin bowerbird builds a bower out of twigs and decorates it with colorful items that he gathers. ***(b)*** A male frigate bird inflates his scarlet throat pouch to attract passing females.

species-specific rhythm, and females distinguish and prefer the rhythm of their own species. The elaborate construction projects of the male satin bowerbird and the scarlet throat of the male frigate bird serve as flashy advertisements of sex, species, and male quality (Fig. 26-19). Sending these extravagant signals must be risky, as they make it much easier for predators to locate the sender. For males, the added risk is an evolutionary necessity, as females won't mate with males that lack the appropriate signal. Females, in contrast, typically do not need to attract males or assume the risk associated with a conspicuous signal, so in many species females are drab in comparison to males (Fig. 26-20).

The intertwined functions of sex recognition and species recognition, advertisement of individual quality, and synchronization of reproductive behavior commonly require a complex series of signals, both active and passive, by both sexes. Such signals are beautifully illustrated by the complex underwater "ballet" executed by the male and female three-spined stickleback fish (Fig. 26-21).

Figure 26-20 Sexual dimorphism in guppies
As in many animal species, the male guppy (left) is brighter and more colorful than the female.

Chemical Signals Can Also Bring Mates Together

Pheromones can play an important role in reproductive behavior. A sexually receptive female silk moth, for example, sits quietly and releases a chemical message so powerful that it can be detected by males up to 3 miles (5 kilometers) away. The exquisitely sensitive and selective receptors on the antennae of the male silk moth respond to just a few molecules of the substance, allowing him to travel upwind along a concentration gradient to find the female (Fig. 26-22a).

Water is an excellent medium for dispersing chemical signals, and fish commonly use a combination of pheromones and elaborate courtship movements to ensure the synchronous release of gametes. Mammals, with their highly developed sense of smell, often rely on pheromones released by the female during her fertile periods to attract males (Fig. 26-22b).

5 What Kinds of Societies Do Animals Form?

Sociality is a widespread feature of animal life. Most animals interact at least a little with other members of their species; many spend the bulk of their lives in the company of others; and a few species have developed

(a) A male, inconspicuously colored, leaves the school of males and females to establish a breeding territory.

(b) As his belly takes on the red color of the breeding male, he displays aggressively at other red-bellied males, exposing his red underside.

(c) Having established a territory, the male begins nest construction by digging a shallow pit that he will fill with bits of algae cemented together by a sticky secretion from his kidneys.

(d) After he tunnels through the nest to make a hole, his back begins to take on the blue courting color that makes him attractive to females.

(e) An egg-carrying female displays her enlarged belly to him by assuming a head-up posture. Her swollen belly and his courting colors are passive visual displays.

(f) Using a zigzag dance, he leads her to the nest.

(g) After she enters, he stimulates her to release eggs by prodding at the base of her tail.

(h) He enters the nest as she leaves and deposits sperm, which fertilize the eggs.

Figure 26-21 Courtship of the three-spined stickleback

complex, highly structured societies. Social interaction can be cooperative or competitive and is typically a mixture of the two.

Group Living Has Advantages and Disadvantages

Living in a group has both costs and benefits, and a species will not evolve social behavior unless the benefits of doing so outweigh the costs.

On the negative side, social animals may encounter:

- Increased competition within the group for limited resources
- Increased risk of infection from contagious diseases
- Increased risk that offspring will be killed by other members of the group
- Increased risk of being spotted by predators

(a)

(b)

Figure 26-22 Pheromone detectors
(a) Male moths find females not by sight but by following airborne pheromones released by females. These odors are sensed by receptors on the male's huge antennae, whose enormous surface area maximizes the chances of detecting the female scent. ***(b)*** When dogs meet, they typically sniff each other near the base of the tail. Scent glands there broadcast information about the bearer's sex and interest in mating.

Benefits to social animals include:

- Increased abilities to detect, repel, and confuse predators
- Increased hunting efficiency or increased ability to spot localized food resources
- Advantages resulting from the potential for division of labor within the group
- Increased likelihood of finding mates

The Extent of Sociality Varies Among Species

The degree to which animals of the same species cooperate varies from one species to the next. Some types of animals, such as the mountain lion, are basically solitary; interactions between adults consist of brief aggressive encounters and mating. Other types of animals cooperate on the basis of changing needs. For example, the coyote is solitary when food is abundant but hunts in packs in the winter when food becomes scarce.

Loose social groups, such as pods of dolphins, schools of fish, flocks of birds, and herds of musk oxen (Fig. 26-23), can provide benefits. For example, the characteristic spacing of fish in schools or the V-pattern of geese in flight provides a hydrodynamic or aerodynamic advantage for each individual in the group, reducing the energy required for swimming or flying. Some biologists hypothesize that herds of antelope or schools of fish confuse predators—their myriad bodies make it difficult for the predator to focus on and pursue a single individual.

A Few Species Form Complex Societies

At the other end of the social spectrum are a few highly integrated cooperative societies, primarily among the insects and mammals. As you read the following section,

Figure 26-23 Cooperation in loosely organized social groups
A herd of musk oxen functions as a unit when threatened by predators such as wolves. Males form a circle, horns pointed outward, around the females and young.

you may notice that some cooperative societies are based on behavior that seems to sacrifice the individual for the good of the group. There are many examples: Young, mature Florida scrub jays may remain at their parents' nest and help them raise subsequent broods instead of breeding; worker ants often die in defense of their nest; Belding ground squirrels may sacrifice their own lives to warn the rest of their group of an approaching predator. These behaviors are examples of **altruism**—behavior that decreases the reproductive success of one individual to the benefit of another.

Forming Groups with Relatives Fosters the Evolution of Altruism

How could altruistic behavior evolve? When individuals perform self-sacrificing deeds, why aren't the alleles that contribute to this behavior eliminated from the gene pool? One possibility is that other members of the group are close relatives of the altruistic individual. Because close relatives share alleles, the altruistic individual may promote the survival of its own alleles through behaviors that maximize the survival of its close relatives. When an altruistic behavior benefits close relatives, the penalty suffered by the altruist may be offset if the behavior helps the relatives pass to the next generation more copies of the shared alleles than the altruist sacrifices by virtue of its selfless behavior. This concept is called **kin selection**. Kin selection helps explain the self-sacrificing behaviors that contribute to the success of cooperative societies. Cooperative behavior is illustrated in the following sections, which describe two examples of complex societies, one in an insect species and one in a mammal species.

Honeybees Live Together in Rigidly Structured Societies

Perhaps the most perplexing of all animal societies are those of the bees, ants, and termites. Scientists have long struggled to explain the evolution of a social structure in which most individuals never breed but instead labor slavishly to feed and protect the offspring of a different individual. Whatever its evolutionary explanation, the intricate organization of a social insect colony makes a compelling story. In these communities, the individual is a mere cog in an intricate, smoothly running machine and could not survive by itself.

Individual social insects are born into one of several castes within the society. These castes are groups of similar individuals that perform a specific function. Honeybees emerge from their larval stage into one of three major preordained roles. One role is that of *queen*. Only one queen is tolerated in a hive at any time. Her functions are to produce eggs (up to 1000 per day for a lifetime of 5 to 10 years) and to regulate the lives of the workers. Male bees, called *drones*, serve merely as mates for the queen. Lured by her sex pheromones, drones mate with the queen, perhaps as many as 15 times, during her first week of life. This relatively brief "orgy" supplies her with sperm that will last a lifetime, enough to fertilize more than 3 million eggs. Their sexual chore accomplished, the drones become superfluous and are eventually driven out of the hive or killed.

The hive is run by the third class of bees, sterile female *workers*. A worker's tasks are determined by her age and conditions in the colony. A newly emerged worker starts life as a waitress, carrying food such as honey and pollen to the queen, to other workers, and to developing larvae. As she matures, special glands begin wax production, and she becomes a builder, constructing perfectly hexagonal cells of wax in which the queen deposits her eggs and the larvae develop. She also takes shifts as a maid, cleaning the hive and removing the dead, and as a guard, protecting the hive against intruders. Her final role in life is that of a forager, gathering pollen and nectar, food for the hive. She spends nearly half of her 2-month life in this role. Acting as a forager scout, she seeks new and rich sources of nectar and, having found one, returns to the hive and communicates its location to other foragers. She communicates by means of the **waggle dance**, an elegant form of symbolic expression (Fig. 26-24).

Pheromones play a major role in regulating the lives of social insects. Honeybee drones are drawn irresistibly to the queen's sex pheromone (*queen substance*), which she releases during her mating flights. Back at the hive, she uses the same substance to maintain her position as the only fertile female. The queen substance is licked off her body and passed among all the workers, rendering them sterile. The queen's presence and health are signaled by her continuing production of queen substance; a decrease in production (which occurs normally in the spring) alters the behavior of the workers. Almost immediately they begin building extra-large "royal cells." The workers also feed the larvae that develop in these cells a special glandular secretion known as "royal jelly." This unique food alters the development of the growing larvae so that, instead of a worker, a new queen emerges from the royal cell. The old queen then leaves the hive, taking a swarm of workers with her to establish residence elsewhere. If more than one new queen emerges, a battle to the death ensues, with the victorious queen taking over the hive.

Naked Mole Rats Form a Complex Vertebrate Society

The nervous systems of vertebrates are far more complex than those of insects, and we might therefore expect vertebrate societies to be proportionately more complex. With the exception of human society, however, they are not. Perhaps the most bizarre society among nonhuman mammals is that of the naked mole rat. These nearly blind, nearly hairless relatives of guinea pigs live in large underground colonies in southern Africa and have an antlike form of social organization not known to exist in any other mammalian society. The

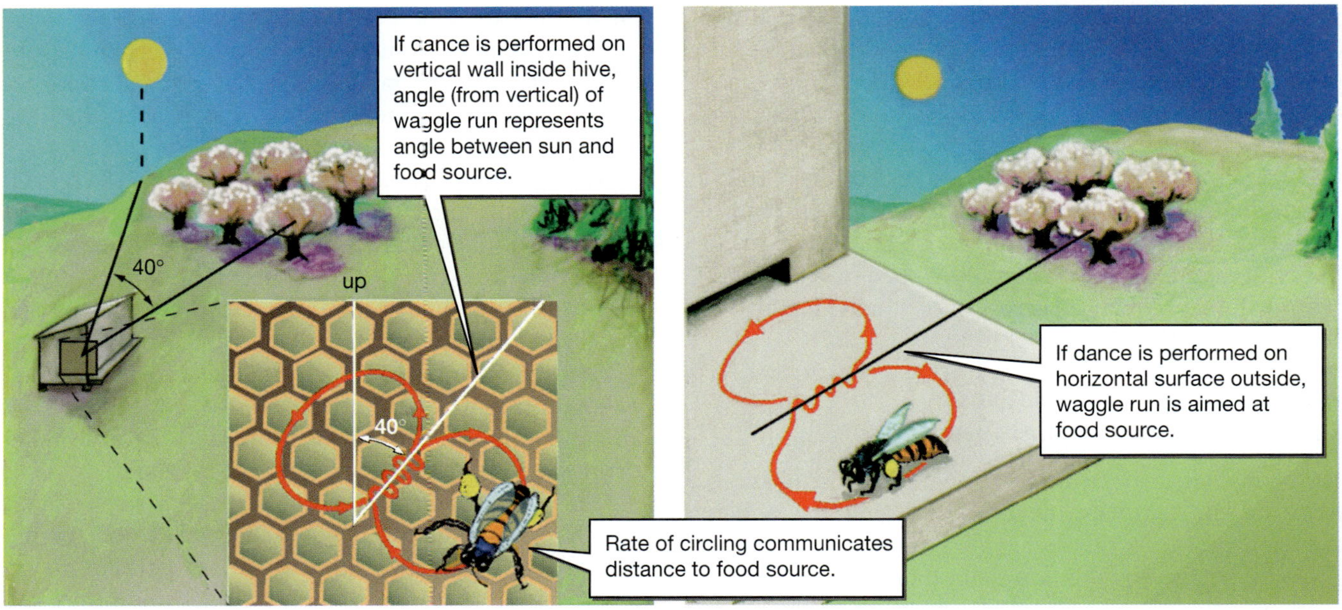

Figure 26-24 Bee language: the waggle dance
A forager, returning from a rich source of nectar, performs a waggle dance that communicates the distance and direction of the food source, as other foragers crowd around her, touching her with their antennae. The bee moves in a straight line while shaking her abdomen back and forth ("waggling") and buzzing her wings. She repeats this over and over in the same location, circling back in alternating directions.

colony is dominated by the queen, a single reproducing female to whom all other members are subordinate.

The queen is the largest individual in the colony and maintains her status by aggressive behavior, particularly shoving. She prods and shoves lazy workers, stimulating them to become more active (Fig. 26-25). As in honeybee hives, there is a division of labor among the workers, in this case based on size. Small, young rats clean the tunnels, gather food, and tunnel. Tunnelers line up head to tail and pass excavated dirt along the completed tunnel to an opening. Just below the opening, a larger mole rat flings the dirt into the air, adding it to a cone-shaped mound. Biologists observing this behavior from the surface dubbed it "volcanoing." In addition to volcanoing, large mole rats defend the colony against predators and members of other colonies.

Figure 26-25 The naked mole rat queen
Encountering a lazy worker in one of the colony's underground tunnels, the queen shoves it and knocks it over to stimulate the worker to greater efforts.

If another female begins to become fertile, the queen apparently senses changes in the estrogen levels of the subordinate female's urine. The queen then selectively shoves the would-be breeder, causing stress that prevents her rival from ovulating. Large males are more likely to mate with the queen than are small ones, although all adult males are fertile. When the queen dies, a few of the females gain weight and begin shoving one another. The aggression may escalate until a rival is killed. Ultimately, a single female becomes dominant. Her body lengthens; she assumes the queenship and begins to breed. Litters averaging 14 pups are produced about four times a year. During the first month, the queen nurses her pups, and the workers feed the queen. Then the workers begin feeding the pups solid food.

6 Can Biology Explain Human Behavior?

Because humans are animals whose behaviors, like those of all other animals, have an evolutionary history, the techniques and concepts of ethology can help us understand and explain human behavior. Human ethology, however, is, and will remain, a less rigorous science than animal ethology. We cannot treat people as laboratory animals, devising experiments that control and manipulate the factors that influence their attitudes and actions. In addition, some observers argue that human culture has been freed from the constraints of its evolutionary past for so long that we cannot explain our behavior in terms of biological evolution. Nevertheless, many scientists have taken an ethological, evolutionary approach to human behavior, and their work has had a major impact on our view of ourselves.

The Behavior of Newborn Infants Has a Large Innate Component

Because newborn infants have not had time to learn, we can assume that much of their behavior is innate. The rhythmic movement of an infant's head in search of its mother's breast is an innate behavior that is expressed in the first days after birth. Sucking, which can be observed even in the human fetus, is also innate (Fig. 26-26). Other behaviors seen in newborns and even premature infants include grasping with the hands and feet and making walking movements when the body is supported. Another example is smiling, which can occur soon after birth. Initially, smiling can be induced by almost any object looming over the newborn. This initial indiscriminate response, however, is soon modified by experience. Infants up to 2 months old will smile in response to a stimulus consisting of two dark, eye-sized spots on a light background, which at that stage of development is a more potent stimulus for smiling than is an accurate representation of a human face. But as the child's development continues, learning and further development of the nervous system interact to limit the response to more-correct representations of a face.

Newborns in their first 3 days of life can be conditioned to produce certain rhythms of sucking when their mother's voice is used as reinforcement. In an experiment performed by William Fifer, infants preferred their own mothers' voices to other female voices, as indicated by their responses (Fig. 26-27). The infant's ability to learn his or her mother's voice and respond positively to it within days of birth has strong parallels to imprinting and may help initiate bonding with the mother.

Young Humans Acquire Language Easily

One of the most important insights to have emerged from studies of animal learning is that animals of a given species tend to have an inborn predilection for specific types of learning that are important to that species' mode of life. In humans, one such inborn predilection seems to be for the acquisition of language.

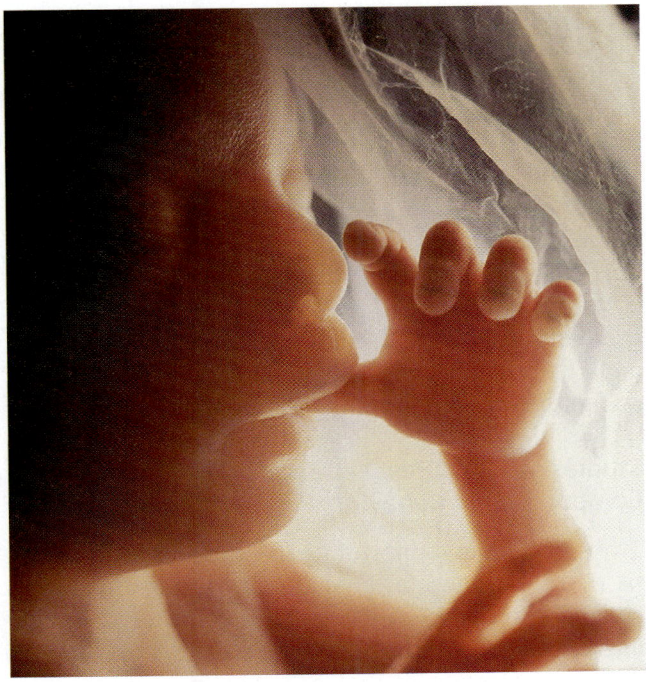

Figure 26-26 A human instinct
Thumb sucking is a difficult habit to discourage in young children, because sucking on appropriately sized objects is an instinctive, food-seeking behavior. This fetus sucks its thumb at about 4 months of development.

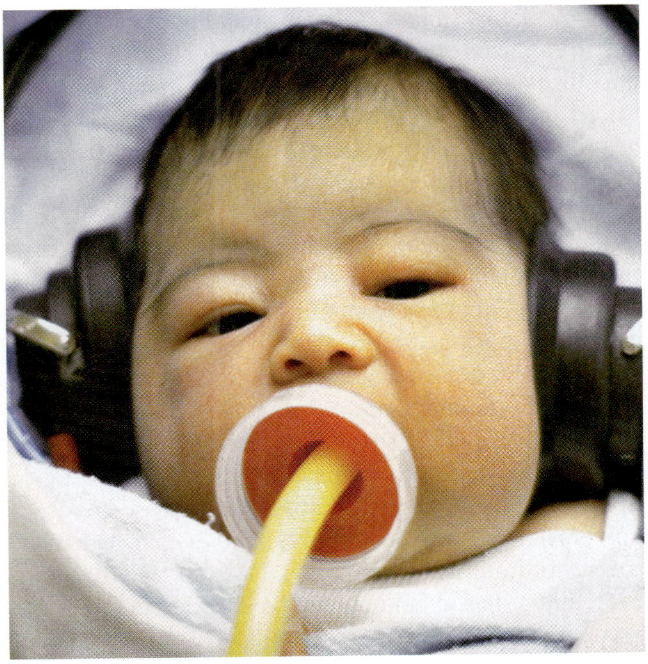

Figure 26-27 Newborns prefer their mother's voices
Using a nipple connected to a computer that plays audio tapes, researcher William Fifer demonstrated that newborns can be conditioned to suck at specific rates for the privilege of listening to their own mothers' voices through headphones. For example, if the infant sucks faster than normal, her mother's voice is played; if she sucks more slowly, another woman's voice is played. Researchers found that infants easily learned and were willing to work hard at this task just to listen to their own mothers' voices, presumably because they had become used to her voice in the womb.

Young children are able to acquire language rapidly and nearly effortlessly; they typically acquire a vocabulary of 28,000 words before the age of 8. A large body of research suggests that we are born with a brain that is already primed for this early facility with language. For example, the human fetus begins responding to sounds during the third trimester of pregnancy, and researchers have demonstrated that infants are able to distinguish among consonant sounds by 6 weeks after birth. The demonstration used the sucking-measurement technique to gauge an infant's ability to discriminate sounds. In the demonstration, an infant responded to various consonant sounds by sucking on a pacifier that contained a force transducer to record the sucking rate. When one sound (such as "ba") was presented repeatedly, the infant became habituated and decreased her sucking rate. But when a new sound (such as "pa") was presented, sucking rate increased, revealing that the infant perceived the new sound as different.

Behaviors Shared by Diverse Cultures May Be Innate

Another way to study the innate bases of human behavior is to compare simple acts performed by people from diverse cultures. This comparative approach, pioneered by ethologist Irenaus Eibl-Eibesfeldt, has revealed several gestures that seem to form a universal, and therefore probably innate, human signaling system. Such gestures include facial expressions for pleasure, rage, and disdain, and greeting movements such as an upraised hand or the "eye flash" (in which the eyes are widely opened and the eyebrows rapidly elevated). The evolution of "hard-wiring" for these gestures presumably depended on the advantages that accrued to both senders and receivers from sharing information about the emotional state and intentions of the sender. A species-wide method of communication was perhaps especially important before the advent of language, and later remained useful during encounters between people who shared no common language.

Certain complex social behaviors are widespread among diverse cultures. For example, the incest taboo (avoidance of mating with close relatives) seems to be universal across human cultures (and even across many species of nonhuman primates). It seems unlikely, however, that a shared belief could be encoded in our genes. Some biologists have suggested that the taboo is instead a cultural expression of an evolved, adaptive behavior. According to this hypothesis, close contact among family members early in life suppresses sexual desire, and this effect arose because of the negative consequences of inbreeding (for example, higher incidence of genetic diseases). The hypothesis does not require us to assume an innate social belief, but rather proposes that we inherit a learning program that causes us to undergo a kind of imprinting early in life.

Humans May Respond to Pheromones

Although the main channels of human communication are through the eyes and ears, humans also seem to respond to chemical messages. The possible existence of human pheromones was hinted at in the early 1970s, when biologist Martha McClintock found that the menstrual cycles of roommates and close friends tended to become synchronized. McClintock suggested that the synchrony resulted from some chemical signal between the women, but almost 30 years passed before she and her colleagues uncovered more conclusive evidence that a pheromone was at work.

In 1998 McClintock's research group asked nine female volunteers to wear cotton pads in their armpits for eight hours per day over the course of a menstrual cycle. The pads were then disinfected with alcohol, and swabbed above the upper lips of another set of 20 female subjects (who reported that they could detect no odors other than alcohol on the pads). The subjects were exposed to the pads in this way each day for two months, with half the group sniffing secretions from women in the early (pre-ovulation) part of the menstrual cycle, while the other half was exposed to secretions from later in the cycle (post-ovulation). Women exposed to early-cycle secretions had shorter-than-usual menstrual cycles, and women exposed to late-cycle secretions had delayed menstruation. It appears that women release different pheromones, with different effects on receivers, at different points in the menstrual cycle.

Although the McClintock experiment offers strong evidence for the existence of human pheromones, little else is known about chemical communication in humans. The actual molecules that caused the effects documented by McClintock remain unknown, as does their function (what benefit would a woman gain by influencing the menstrual cycles of other women?). Receptors for chemical messages have not yet been found in humans, and we don't know if the "menstrual pheromones" are the first known example of an important communication system or merely an isolated case of a vestigial ability. Despite the hopeful advertisements for "sex attraction pheromones" on late-night television, chemical communication in humans is a scientific mystery awaiting a solution.

Studies of Twins Reveal Genetic Components of Behavior

Twins present an opportunity to examine the hypothesis that differences in human behavior are related to genetic differences. If a particular behavior is heavily influenced by genetic factors, we would expect to find similar expression of that behavior in *identical twins* (which arise from a single fertilized egg and have identical genes) but not in *fraternal twins* (which arise from two individual eggs and are no more similar genetically than are other siblings). Data from twin studies, and from other within-family investigations, have tended to confirm the heritability of many human behavioral traits. These studies have documented a significant genetic component for traits such as activity level, alcoholism, sociability, anxiety, intelligence, dominance, and even political attitudes. On the basis of tests designed to measure many aspects of personality, identical twins are about twice as similar in personality as are fraternal twins.

The most fascinating, although less rigorous, twin findings come from observations of identical twins separated soon after birth, reared in different environments, and reunited for the first time as adults. Identical twins reared apart have been found to be as similar in personality as those reared together, indicating that the differences in their environments had little influence on their personality development. They have been found to share nearly identical taste in jewelry, clothing, humor, food, and names for children and pets. Personal idiosyncrasies such as giggling, nail biting, drinking patterns, hypochondria, and mild phobias are in some cases shared by these separated twins.

Biological Investigation of Human Behavior Is Controversial

The field of human behavioral genetics is controversial, especially among non-scientists, because it challenges the long-held belief that environment is the most important determinant of human behavior. As discussed earlier in this chapter, we now recognize that all behavior has some genetic basis and that complex behavior in nonhuman animals typically combines elements of both innate and learned behaviors. Thus, it seems likely that our own behavior is influenced by both our evolutionary history and our cultural heritage. The debate over the relative importance of heredity and environment in determining human behavior continues and is unlikely ever to be fully resolved. Human ethology is not yet recognized as a rigorous science, and it will always be hampered because we can neither view ourselves with detached objectivity nor treat other people as if they were laboratory rats. Despite these limitations, there is much to be learned about the interaction of learning and innate tendencies in humans.

The Big Picture

Like all animal behavior, human behavior is the product of evolution. We can therefore begin to understand the origin of our current behavior by asking how it might have affected the reproductive success of our ancestors. Understanding the biological basis of human behavior could be a first step toward eliminating (or at least controlling) ancient behaviors such murder, rape, and war.

Evolutionary Connections

Why Do Animals Play?

Pigface, a giant 50-year-old African softshell turtle, spends hours each day batting a ball around his aquatic home in the National Zoo in Washington, D.C., to the delight of thousands of visitors and the puzzlement of behavioral biologists. Play has always been somewhat of a mystery. It has been observed in many birds and in most mammals, but, until zookeepers tossed Pigface a ball a few years ago, it had never been seen in animals as evolutionarily ancient as turtles.

Animals at play are fascinating. Pygmy hippopotamuses push one another, shake and toss their heads, splash in the water, and pirouette on their hind legs. Otters delight in elaborate acrobatics. Bottlenose dolphins balance fish on their snouts, throw objects, and carry them in their mouths while swimming. Baby vampire bats chase, wrestle, and slap each other with their wings. Even octopuses have been seen playing a game: pushing objects away from themselves and into a current, then waiting for the objects to drift back, only to push them back into the current to start the cycle over again.

Play can be solitary, and this typically consists of a single animal manipulating an object, such as a cat with a ball of yarn, or the dolphin with its fish, or a macaque monkey making and playing with a snowball. Or, play can be social. Often young of the same species play together, but parents may join them (Fig. 26-28a). Social play typically includes chasing, fleeing, wrestling, kicking, and gentle biting (Fig. 26-28b,c).

What are the features of play? Play seems to lack any clear immediate function and is abandoned in favor of feeding, courtship, and escaping from danger. Young animals play more frequently than do adults. Play typically borrows movements from other behaviors (attacking, fleeing, stalking, and so on) and uses considerable energy. Also, play is potentially dangerous. Many young humans and other animals are injured, and some are killed, during play. In addition, play can distract an animal from the presence of danger while making it conspicuous to predators. So, why do animals play?

The most logical conclusion is that play must have survival value and that natural selection has favored those individuals who engage in playful activities. One of the best explanations for the survival value of play is the "practice theory," first proposed by K. Groos in 1898. He suggested that play allows young animals to gain experience in behaviors that they will use as adults. By performing these acts repeatedly in play, the animal practices skills that will later be important in hunting, fleeing, or social interactions.

More-recent research supports and extends Groos's proposal. Play is most intense early in life when the

(a)

(b)

(c)

Figure 26-28 **Young animals at play**

brain is developing and crucial neural connections are forming. John Byers, a zoologist at the University of Idaho, has observed that animals with larger brains tend to be more playful than are animals with smaller brains. Because larger brains are generally linked to increased learning ability, this relationship supports the idea that adult skills are learned during juvenile play. Watch children roughhousing or playing tag, and you will see how strength and coordination are fostered by play and how skills are developed that might have helped our hunting ancestors survive. Quiet play with other children, with dolls, blocks, and other toys helps children prepare to interact socially, nurture their own children, and deal with the physical world.

Although Shakespeare tells us "play needs no excuse," there is good evidence that the tendency to play has evolved as an adaptive behavior in animals capable of learning. Play is quite literally "serious fun"!

CASE STUDY REVISITED

Sex and Symmetry

To assess whether male body symmetry is correlated with male "mating success" in humans, Randy Thornhill, Steven Gangestad, and their colleagues began by measuring symmetry in some young adult males. Each man's degree of symmetry was assessed by measurements of his ear length and the width of his foot, ankle, hand, wrist, elbow, and ear. From these measurements, the researchers derived an index that summarized the degree to which the size of these features differed between the right and left sides of the body.

The researchers found that, among the males in their sample, the most symmetrical men were judged (by a panel of female observers who were unaware of the nature of the study) to be more attractive than other men. A survey of the study subjects revealed that the more symmetrical men also tended to begin having sex earlier in life and to have had a larger number of different sexual partners. Apparently, a man's sexual activity and attractiveness to women are correlated with his symmetry.

Why would male body symmetry affect mating success? The most likely explanation is that symmetry is an indicator of good physical condition. As developmental biologists have long known, disruptions of normal embryological development can cause bodies to be asymmetrical, so a highly symmetrical body indicates healthy, normal development. It also may indicate a high-quality genotype that was able to overcome any disturbances (such as diseases or exposure to toxic substances) during development. Females that mate with individuals whose health and vitality are announced by their symmetrical bodies might have offspring that are similarly healthy and vital. Natural selection would thus favor females who chose to mate with symmetrical males, and that kind of mating behavior would come to predominate.

Summary of Key Concepts

1) How Do Innate and Learned Behaviors Differ?

Although all animal behavior is influenced by both genetic and environmental factors, biologists can distinguish between behaviors whose development is not highly dependent on external factors and behaviors that require more extensive environmental stimuli in order to develop. Behaviors in the first category are sometimes designated as innate and can be performed properly the first time an animal encounters the appropriate stimulus. Behavior that changes in response to input from an animal's social and physical environment is said to be learned. Learning can modify innate behavior to make it more appropriate.

Although the distinction between innate and learned behavior is conceptually useful, the distinction is not sharp in naturally occurring behaviors. Learning allows animals to modify innate responses so that they occur only with appropriate stimuli. Imprinting, a form of learning with innate constraints, is only possible at a certain time in an animal's development.

2) How Do Animals Communicate?

Communication allows animals of the same species to interact effectively in their quest for mates, food, shelter, and other resources. Animals communicate through visual signals, sound, chemicals (pheromones), and touch. Visual communication is quiet and can convey rapidly changing information. Visual signals are active (body movements) or passive (body shape and color). Sound communication can also convey rapidly changing information, and it is effective where vision is impossible. Pheromones can be detected after the sender has departed, conveying a message over time. Physical contact reinforces social bonds and is a part of mating.

3) How Do Animals Compete for Resources?

Although many competitive interactions are resolved through aggression, serious injuries are rare. Most aggressive encounters are settled by means of displays that communicate the motivation, size, and strength of the combatants.

Some species establish dominance hierarchies that minimize aggression. On the basis of initial aggressive encounters, each animal acquires a status in which it defers to more dominant individuals and dominates subordinates. When resources are limited, dominant animals obtain the largest share and are most likely to reproduce.

Territoriality, a behavior in which animals defend areas where important resources are located, also minimizes aggressive encounters. In general, territorial boundaries are respected, and the best-adapted individuals defend the richest territories and produce the most offspring.

4) How Do Animals Find Mates?

Successful reproduction requires that animals recognize the species, sex, and sexual receptivity of potential mates. In many species, animals also assess the quality of potential mates. These requirements have contributed to the evolution of sexual displays that use all forms of communication.

5) What Kinds of Societies Do Animals Form?

Social living has both advantages and disadvantages, and species vary in the degree to which their members cooperate. Some species form cooperative societies. The most rigid and highly organized are those of the social insects such as the honeybee, in which the members follow rigidly defined roles throughout life. These roles are maintained through both genetic programming and the influence of certain pheromones. Naked mole rats exhibit the most complex and rigid vertebrate social interactions, resembling those of social insects.

6) Can Biology Explain Human Behavior?

The degree to which human behavior is genetically influenced is highly controversial. Because we cannot freely experiment on humans, and because learning plays a major role in nearly all human behavior, investigators must rely on studies of newborn infants and comparative cultural studies, correlations between certain behaviors and physiology (which suggest a role for pheromones), and studies of identical and fraternal twins. Evidence is mounting that our genetic heritage plays a role in personality, intelligence, simple universal gestures, our responses to certain stimuli, and our tendency to learn specific things such as language at particular stages of development.

Key Terms

aggression *p. 522*
altruism *p. 529*
behavior *p. 516*
communication *p. 519*
dominance hierarchy *p. 523*
ethology *p. 518*
habituation *p. 518*
imprinting *p. 518*
innate behavior *p. 516*
kin selection *p. 529*
learning *p. 516*
pheromone *p. 521*
territoriality *p. 524*
waggle dance *p. 529*

Thinking Through the Concepts

Multiple Choice

1. *In general, animal behaviors*
a. are determined mainly by genetic factors
b. are determined mainly by environmental factors
c. arise from an interaction between genes and the environment
d. do not affect an individual's chances of survival

2. *Which of the following is NOT a feature of play?*
a. Young animals play more often than adults.
b. It may borrow movements from other behaviors.
c. It uses considerable energy.
d. It always has a clear, immediate function.
e. It is potentially dangerous.

3. *Naked mole rats are unique among mammals whose behavior has been studied, in that*
a. they live in colonies
b. most of their life is spent underground
c. even adults have little body hair
d. most individuals never reproduce
e. they use pheromones to communicate

4. *In which of the following ways does an individual benefit from being in a social group with other animals?*
a. increased ability to detect, repel, or confuse predators
b. increased hunting efficiency
c. increased likelihood of finding mates
d. increased efficiency due to division of labor
e. all of the above

5. *Which of the following pairs of forms of communication and advantages is NOT accurate?*
a. pheromones, long-lasting
b. visual displays, instantaneous
c. sound communication, effective at night
d. pheromones, convey rapidly changing information
e. touch, maintains social bonds

6. *In an insect society, such as the honeybee society,*
a. the division of labor is based on biologically determined castes
b. all adult members share labor equally
c. all adult members have the opportunity to reproduce
d. reproduction is altered seasonally among the adults
e. the organization of the society is flexible and adaptable

? Review Questions

1. Explain why neither "innate" nor "learned" adequately describes the behavior of any given organism.

2. Explain why animals play. Include the features of play in your answer.

3. List four senses through which animals communicate, and give one example of each form of communication. After each, present both advantages and disadvantages of that form of communication.

4. A bird will ignore a squirrel in its territory but will act aggressively toward a member of its own species. Explain why.

5. Why are most aggressive encounters among members of the same species relatively harmless?

6. Discuss advantages and disadvantages of group living.

7. In what ways do naked mole rat societies resemble those of the honeybee?

Applying the Concepts

1. Male mosquitoes orient toward the high-pitched whine of the female, and female mosquitoes (only females suck blood) are attracted to the warmth, humidity, and carbon dioxide exuded by their prey. Using this information, design a mosquito trap or killer that exploits a mosquito's innate behaviors. Now design one for moths.
2. You raise honeybees but are new at the job. Trying to increase honey production, you introduce several queens into the hive. What is the likely outcome? What different things could you do to increase production?
3. Describe and give an example of a dominance hierarchy. What role does it play in social behavior? Give a human parallel, and describe its role in human society. Are the two roles similar? Why or why not? Repeat this exercise for territorial behavior in humans and in another animal.
4. You are manager of an airport. Planes are being endangered by large numbers of flying birds, which can be sucked into engines, disabling the engines. Without harming the birds, what might you do to discourage them from nesting and flying near the airport and its planes?

Group Activity

To see how playing games can help us understand the evolution of behavior, try playing the following game. Imagine a population of animals in which two behavior patterns are present. The first pattern is called cooperate, and cooperators always behave helpfully during encounters with another individual of the same species. The second behavior pattern is called defect, and defectors always harm other individuals to gain an advantage. Whenever two individuals meet and interact, each individual gets a payoff, which can be positive, negative, or neutral. These payoffs are summarized in the accompanying "payoff matrix." If two cooperators meet, each gets two points, and so on according to the matrix.

To play the game, form a group of four students. Choose one member of the group to be a cooperator (always cooperates) and a second member to be a defector (always defects). In each round, the remaining two members will either cooperate or defect, depending on the results of a coin flip (heads cooperate, tails defect; each person flips separately). Next, on each of four slips of paper, write the name of a group member, and place the slips of paper in a hat. To play a round of the game, each member of the group chooses a slip of paper and "interacts" with the member whose name is chosen. (Try again if you choose your own name, and be sure that both coin-flippers have selected a new role before each round.) Each group will thus have four interactions per round, with points generated according to the payoff matrix. Play 20 rounds, with each player keeping track of his or her own score. Which strategy scored the highest number of points in your group? Tabulate the results of all the groups in your class to determine the best overall strategy.

Next, play another 20 rounds of the game, but this time each member of the group should use the same strategy: "If my game partner cooperated last round, I'll cooperate; if my partner defected last round, I'll defect." (For the first round, players should choose a role by coin flip.) What happens to the scores now? Gather your group and, on the basis of the results of your two games, develop a hypothesis about the conditions under which cooperative behavior could evolve.

	Cooperate	**Defect**
Cooperate	Mutual help; each earns 2 points	Defector earns 1 point, Cooperator loses 1 point
Defect	Defector earns 1 point, Cooperator loses 1 point	Stand-off; no points earned or lost

For More Information

Brown, S. L. "Animals at Play." *National Geographic*, December 1994. Clearly written description of why animals play, accompanied by beautiful photographs that capture wild animals at play.

Dawkins, R. *The Selfish Gene*, 2nd ed. New York: Oxford University Press, 1989. Exceptionally clear explanation of the evolutionary basis of altruistic behavior, written for both laypeople and scientists.

de Waal, F. "The End of Nature versus Nurture." *Scientific American*, December 1999. An eminent ethologist makes the case for taking a nuanced biological approach to understanding human behavior.

Dugatkin, L., and Godin, J. "How Females Choose Their Mates." *Scientific American*, April 1998. An entertaining look at how biologists study the processes by which females choose among available males.

Gould, J. L., and Gould, C. G. "The Instinct to Learn." *Science 81*, May 1981. Birds, bees, and perhaps even humans are genetically programmed to learn specific things at particular stages in life.

Kirchner, W. H., and Towne, W. F. "The Sensory Basis of the Honeybee Dance Language." *Scientific American*, June 1994. Combines a historical look at the work of Karl von Frisch with a modern update on recent research that has almost fully elucidated the mechanisms of honeybee communication during the waggle dance.

Lorenz, K. *King Solomon's Ring: New Light on Animal Ways.* New York: Thomas Y. Crowell, 1952. Beautifully written and filled with interesting anecdotes; provides important insights into early ethology.

Macdonald, D., and Brown, R. "The Smell of Success." *New Scientist*, May 1985. Describes the amazing diversity of mammalian pheromones.

Phillips, K. "School Riddles." *International Wildlife*, March–April 1995. Explores research on fish schooling behavior and how it helps avoid predation.

Pinker, S. *The Language Instinct.* New York: William Morrow, 1994. An entertaining account of linguists' current understanding of how we develop the ability to use language and how that ability may have evolved.

Seeley, T. D. "The Honeybee Colony as a Superorganism." *American Scientist*, November–December 1989. The honeybee society exemplifies the concept of kin selection in which the society, rather than the individual, serves as a "vehicle for the survival of genes."

Sherman, P., and Alcock, J. *Exploring Animal Behavior.* Sunderland, MA: Sinauer, 1998. A collection of articles in which behavioral biologists describe their research and their conclusions about the mechanisms, function, and evolution of behavior.

Sherman, P. W., Jarvis, J. U. M., and Braude, S. H. "Naked Mole Rats." *Scientific American*, August 1992. Describes the newly investigated society of the naked mole rat, a vertebrate whose social behavior resembles that of some social insects.

Weiner, J. *Time, Love, Memory: A Great Biologist and His Quest for the Origins of Behavior.* New York: Knopf, 2000. A wonderful science writer chronicles the study of the genetics of behavior, especially the work of a great pioneer of the field, Seymour Benzer.

Wilson, E. O. "Empire of the Ants." *Discover*, March 1990. This entertaining article examines the diversity of ant behavior that has contributed to their enormous success.

Answers to Multiple-Choice Questions

1. c 2. d 3. d 4. e 5. d 6. a

MediaTutor

CD Activities

Activity 26.1: Innate Animal Behavior

Estimated time: 15 minutes

A behavior is any observable activity of a living animal. Each behavior arises through an interaction between an animal's genetic make-up and its environment, but some behaviors appear to be more heavily influenced by genetic factors. These innate behaviors can typically be performed very early in life without the benefit of prior experience. In this tutorial, you will explore the evolutionary significance of an innate behavior that was discovered by Nobel-Prize-winning scientist Niko Tinbergen. You will have an opportunity to repeat some of Tinbergen's experiments with digger wasps and to draw conclusions based on your observations.

Start the MediaTutor Student CD-ROM and enter the activity number in the Quick Search box to be taken directly to that activity.

Web Investigation

Sex and Symmetry

Estimated time: 15 minutes

The pioneering studies conducted by Niko Tinbergen helped establish the field of ethology (the study of animal behavior) as a true science (see chapter for additional details). These investigations produced one of the most interesting of all behavioral questions: Are behaviors the product of an individual's genes or of the environment in which they develop? This activity will investigate the role that behavior plays in a variety of animal populations.

Go to http://www.prenhall.com/audesirk3, the Audesirk Companion Web site. Select Chapter 26 and Web Investigation to begin.

UNIT SIX

Ecology

The beauty and interdependence of Earth's biosphere is illustrated in "Paradise," Part One of "The Trilogy of the Earth" by Suzanne Duranceau/Illustratrice, Inc.

"There is no exception to the rule that every organic being naturally increases at so high a rate that, if not destroyed, the Earth would soon be covered by the progeny of a single pair."

Charles Darwin in* On the Origin of Species *(1859)

If they could speak, the mysterious statues of Easter Island would tell the tale of a population that outgrew the capacity of its environment to sustain it.

27 Population Growth

AT A GLANCE

CASESTUDY

The Mystery of Easter Island

More than 200 beautiful stone statues dominate the landscape of Easter Island. These enormous representations of human heads stand as high as 33 feet and are arranged on stone platforms that are themselves up to 500 feet long and 10 feet high. The first Europeans to reach Easter Island (in the early 1700s) were impressed by the beauty and craftsmanship of the statues, but were perplexed by the question of how the islanders could have succeeded in transporting and erecting such huge and heavy structures.

The quarry from which the statues' stone was cut was 6 miles from their display site. Surely the task of moving these heavy objects such a long distance, and then maneuvering them into an upright position, would have required long ropes and strong timbers. Easter Island, however, seemed incapable of supplying these materials; its landscape was virtually barren, devoid of anything that could have furnished wood or fibers for rope. The predominant vegetation was grass, and the island held not a single tree. None of the very few shrubs on the island grew more than 10 feet high. Nor could wood and fiber have been imported from elsewhere, given the island's extreme isolation (2000 miles from the South American mainland and 1400 miles from the nearest habitable island).

In the centuries since the arrival of Europeans, the mystery of Easter Island's statues has continued to intrigue and perplex scholars. It is difficult to imagine that a culture capable of organizing the construction of magnificent artifacts could have thrived in such a desolate environment. In recent years, however, the mystery has begun to yield to the efforts of archeologists and paleontologists, who have begun to piece together the early history of the island. The picture that has emerged is both fascinating and frightening.

It is now known that the earliest human settlers of Easter Island found a forested isle that held abundant natural resources and a diverse array of plant and animals species useful to humans. Over time, however, the islanders' population grew until it exceeded the environment's capacity to support it. As a result, the island's human population declined dramatically, and its society came apart. Some observers have argued that the fate of Easter Island will be the fate of Earth as a whole, that the link between depleted resources, a degraded environment, and social disintegration is now being played out on a larger canvas. Is Easter Island indeed a microcosm of our planet? ■

This chapter begins our unit on **ecology**, the study of organisms' interactions with each other and with the nonliving part of their environment. These interactions can be examined at different levels of organization; different perspectives can yield different insights. Ecologists work at three main levels of organization:

- A **population** consists of all the members of a particular species who live in a particular area.
- A **community** includes all the interacting populations of different species that coexist in a particular area.
- An **ecosystem** consists of all the communities within a defined area, along with their nonliving environment.

In Chapters 27 through 30, we will proceed through these levels of organization, starting with populations, moving to the communities that contain the populations, and then on to the ecosystems that contain the communities. We will close with an exploration of the diverse ecosystems that make up the *biosphere*, which encompasses all life on Earth.

1 How Do Populations Grow?

Studies of undisturbed ecosystems show that the size of some populations tends to remain relatively constant over time, others fluctuate in a cyclical pattern, and still others vary sporadically in response to environmental changes. Earth's human population, in contrast to most nonhuman populations, has grown steadily for centuries.

Birth, Deaths, and Migration Determine Population Growth

A population remains stable if, on average, as many individuals join as leave. Individuals join a population through birth or immigration (migration in), and leave it through death or emigration (migration out). Thus, a population grows when the number of births plus immigrants exceeds the number of deaths plus emigrants. Populations shrink when the reverse occurs. A simple equation for the change in population size within a given time period is as follows:

$$(\text{births} - \text{deaths}) + (\text{immigrants} - \text{emigrants}) = \text{change in population size}$$

In many natural populations, organisms moving in and out contribute relatively little to population change, leaving birth and death rates as the primary factors that influence population growth.

Population Growth Can Be Expressed as a Rate

The **per capita growth rate** (r) of a population is a measure of how fast a population grows, expressed as a change in population size per individual per unit of time. This value is determined by subtracting the per capita death rate (d) from the per capita birth rate (b):

$$\underset{\text{(birth rate)}}{b} - \underset{\text{(death rate)}}{d} = \underset{\text{(growth rate)}}{r}$$

For example, to calculate the annual growth rate of a human population of 10,000 in which there are 1500 births and 500 deaths each year, we first calculate the annual per capita birth rate:

$$b = 1500 \text{ births} / 10{,}000 \text{ people} = 0.15 \text{ births per person per year}$$

Next, the annual per capita death rate:

$$d = 500 \text{ deaths} / 10{,}000 \text{ people} = 0.05 \text{ deaths per person per year}$$

Finally, we calculate the per capita growth rate by subtracting the death rate from the birth rate:

$$r = 0.15 \text{ births per person per year} - 0.05 \text{ deaths per person per year} = 0.10 \text{ (that is, 10\% increase per year)}$$

Note that if the death rate exceeds the birth rate, the growth rate will be negative, and the population will decline; here we will focus on growing populations.

A Constant Growth Rate Increases Population Size Rapidly

To calculate the number of individuals added to a population in a year, the per capita growth rate (r) is multiplied by the original population size (N):

$$\text{population growth} = rN$$

In our example, population growth in the first year (rN) is $0.10 \times 10{,}000 = 1000$ people. Thus, at the end of year 1, the population contains 11,000 people. If the per capita growth rate remains the same during year 2, then population growth in the second year is $0.10 \times 11{,}000 = 1100$, and the population size at the end of year 2 is 12,100. In the third year, the populations grows by 1210 people, and so on in subsequent years. As you can see, if the per capita growth rate r remains constant, the number of people added to the population increases each year. This pattern of continuously accelerating increase in population size is **exponential growth**.

A Population's Growth Rate Depends on Patterns of Reproduction

A population experiences growth during periods when births consistently exceed deaths. Growth will persist if, on average, each individual produces more than one surviving offspring during its lifetime. Each individual, of course, has the potential to replace itself many times over during its lifetime; an oyster may have millions of offspring and even the most slowly reproducing species, such as elephants, may have several offspring over their lifetimes. So every species has a built-in capacity for population growth, but the speed of this potential growth varies among species, depending on several factors, including the following:

- The age at which each organism first reproduces.
- The frequency with which each organism reproduces.
- The average number of offspring produced each time.
- The length of each organism's reproductive life span.
- The death rate of the organisms under ideal conditions.

Let's look at some examples that illustrate how these factors affect population growth.

Some Species Produce Large Numbers of Offspring Quickly

The bacterium *Staphylococcus* is normally a harmless resident in and on the human body. If transferred to an ideal environment, such as warm custard, each bacterial cell can divide every 20 minutes, doubling the population three times each hour. The larger the population grows, the more cells there are to divide. The potential population growth of bacteria is so great that, hypothetically, the offspring of a single bacterium could swamp Earth in a layer 7 feet deep within 48 hours.

Other Species Produce Fewer But Longer-lived Offspring

In contrast to *Staphylococcus*, the bald eagle is a relatively long-lived, rather slowly reproducing species. Figure 27-1 compares the potential population growth of eagles to that of bacteria, assuming that there are no deaths in either population during the time graphed. Although the time scale and the number-of-individuals scale differ tremendously for the two species, note that all three curves have the same shape—the J-shape that represents exponential growth. That is, the bacteria and eagle populations grow at vastly different rates and reach vastly different sizes, but both populations have constant growth rates and are growing exponentially.

Delayed Onset of Reproduction Slows Population Growth

Figure 27-1b also shows what happens if there is a difference in the age at which individuals in a population begin reproduction. The red line in the graph represents population growth if each bald eagle lives for 30 years, reaches sexual maturity at 4 years of age, and produces one offspring annually for its remaining 26 years (that is, each mated pair of eagles produces two offspring per year). Each bald eagle in the population whose growth is shown by the green line also lives for 30 years and produces one offspring each year, but it does not begin reproducing until it is 6 years old. Note that growth in both populations is exponential, but the population that begins reproduction later takes longer to reach a particular population size. This result has important implications for the human population: Delayed childbearing significantly slows population growth. For example, if each woman bears three children in her early teens, the population will grow much faster than if each woman bears five children but begins at age 30.

Death Rate Also Influences Growth Rate

So far we have looked mainly at the effects of birth rates. Even under ideal conditions, however, deaths are inevitable. To illustrate the effect of differing death rates, in Figure 27-2 we compare three bacterial populations. Again, the shapes of the curves are the same. In each case the population eventually approaches infinite size, but the time required to reach any given population size is increased by increased mortality.

The Big Picture

Populations have an infinite capacity for growth. Though the particular rate at which a given population can grow depends on the reproductive behavior and average life span of its members, all populations will expand in the absence of constraints on growth.

2 How Is Population Growth Regulated?

Our discussion up to this point has focused on **biotic potential**, the maximum rate at which a population can increase, assuming ideal conditions that allow the highest possible birth rate and the lowest possible death rate. In fact, however, the ultimate size of a population is also affected by limits that oppose this potential for growth. These limits are set by the living and nonliving environments, and are collectively known as **environmental resistance**. Environmental resistance is imposed by the availability of food and space, and by interactions with competitors, predators, and disease-causing organisms. Natural catastrophes such as storms, fires, freezing weather, floods, and droughts also work to limit population size. Environmental resistance can both increase death rates and decrease birth rates. For example, a drought might restrict growth of an animal population both by increasing the number of deaths from starvation and by causing malnutrition that reduces the number of births.

Rapid Growth Cannot Continue Indefinitely

In nature, populations grow rapidly only under certain circumstances and only for a limited time. For example, short-term explosive growth is seen in populations that undergo regular cycles in which rapid population growth is followed by a sudden, massive die-off. These **boom-and-bust cycles** occur in a variety of organisms.

(a)

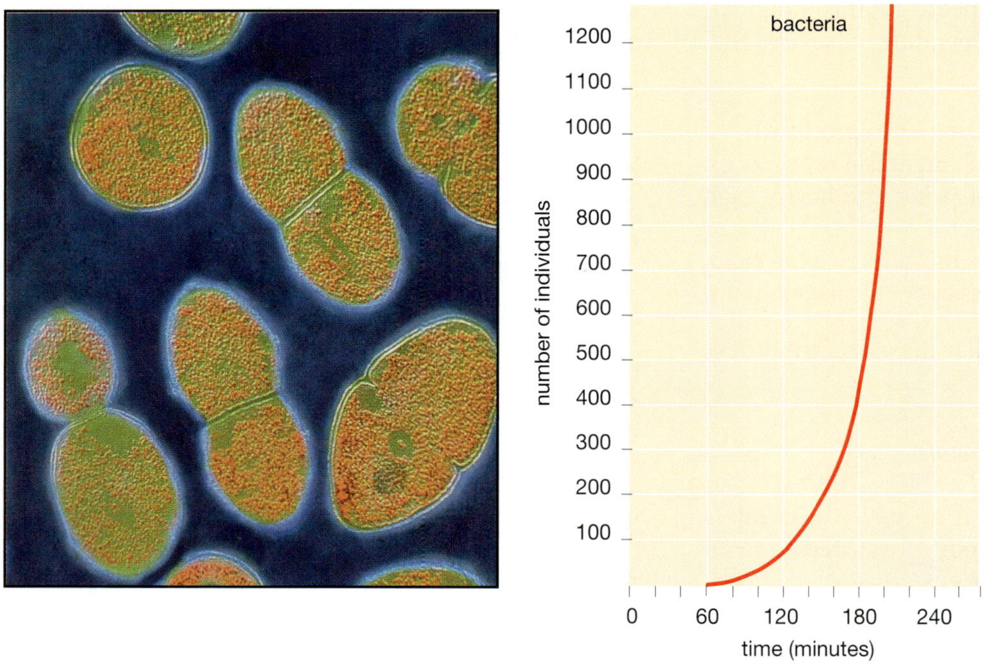

time (minutes)	number of bacteria
0	1
20	2
40	4
60	8
80	16
100	32
120	64
140	128
160	256
180	512
200	1024
220	2048

(b)

time (years)	number of eagles (i)	number of eagles (ii)
0	2	2
2	2	2
4	4	2
6	8	4
8	14	8
10	28	12
12	52	18
14	100	32
16	190	54
18	362	86
20	630	142
22	1314	238
24	2504	392
26	4770	644
28	9088	1066
30	17314	1764

MEDIATUTOR 27.1 Population Growth and Regulation

Figure 27-1 Exponential growth curves
All such curves share a similar J shape; they differ in the scales of the axes. ***(a)*** Growth of a population of bacteria, starting with a single individual and with a doubling time of 20 minutes. ***(b)*** Growth of two populations of eagles (*i* and *ii*), each starting with a single pair. The age at first reproduction of population *i* is 4 years (red line) and that of population *ii* is 6 years (green line). Notice that after 26 years, the population that began reproducing at age 4 is nearly seven times the size of the population that began reproducing at age 6.

Many short-lived, rapidly reproducing species, from cyanobacteria to insects, have seasonal population cycles that are linked to predictable changes in rainfall, temperature, or nutrient availability (Fig. 27-3). Many insect populations in temperate climates grow rapidly during the spring and summer and then crash with the killing hard frosts of winter.

Some Predator and Prey Populations Fluctuate Cyclically

More complex factors produce roughly 4-year population cycles for small rodents such as voles and lemmings, and much longer cycles in hares, muskrats, and grouse. Lemming populations, for example, may grow until the lemmings overgraze their fragile arctic tundra ecosystem. Lack of food, increasing populations of

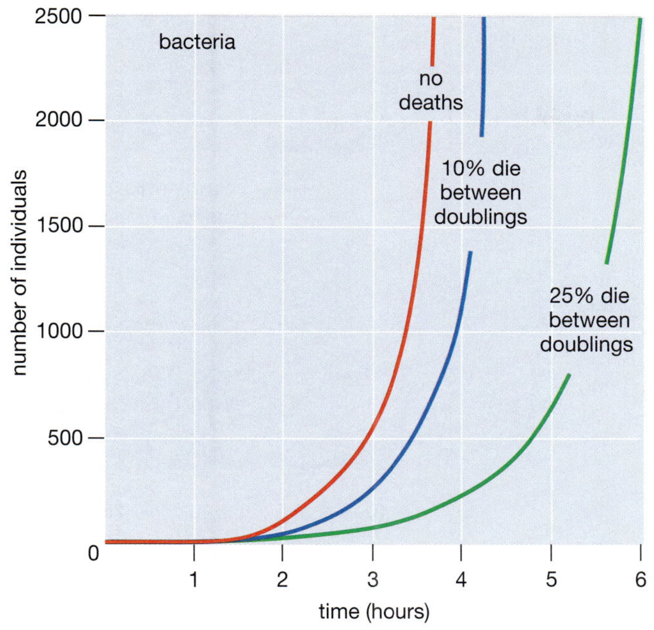

Figure 27-2 The effect of death rates on population growth
The graphs assume that bacterial populations double every 20 minutes. Notice that the population in which a quarter of the bacteria die every 20 minutes reaches a size of 2500 2 hours and 20 minutes later than the one in which there are no deaths.

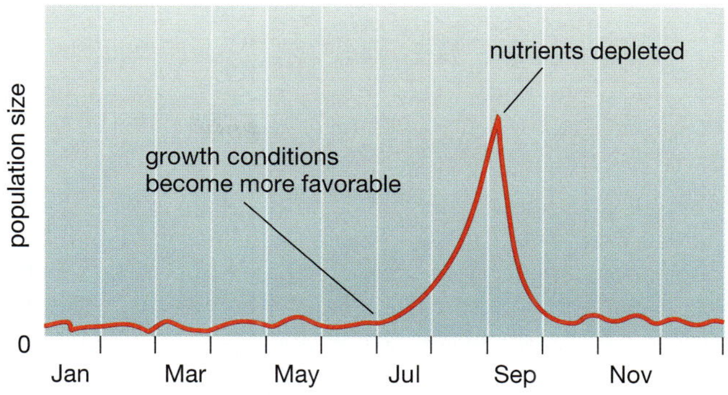

Figure 27-3 A boom-and-bust population cycle
A small number of cyanobacteria in a lake survive through the fall, winter, and spring. Early in July, conditions become favorable for growth, and the population grows rapidly through August. Nutrients soon become depleted, and the population "goes bust."

predators, and social stress caused by overpopulation all contribute to a sudden high death rate. Many lemmings die as they emigrate from regions of high population density. During these dramatic mass movements, lemmings are easy targets for predators. Many drown trying to cross a body of water. The reduced lemming population eventually contributes to a decline in predator numbers (see "Scientific Inquiry: Cycles in Predator and Prey Populations") and a recovery of the plant community on which the lemmings feed. These responses, in turn, set the stage for another round of rapid growth in the lemming population (Fig. 27-4).

Ecosystem Changes May Allow Temporary Rapid Growth

In populations that do not experience boom-and-bust cycles, growth may temporarily be rapid under special circumstances—for example, if population-controlling factors, such as predators, are eliminated or if the food supply is increased. Growth can also be explosive when individuals invade a new habitat in which conditions are favorable and competition is scarce. Invasion of new habitats and subsequent explosive growth of the invader's population may occur when people introduce foreign, or **exotic**, species into ecosystems, in many cases with very damaging results (see "Earth Watch: Exotic Invaders" in Chapter 28).

Environmental Resistance Limits Population Growth

All populations that undergo rapid growth must eventually stabilize or crash (decrease rapidly). As new individuals join a population, competition for resources

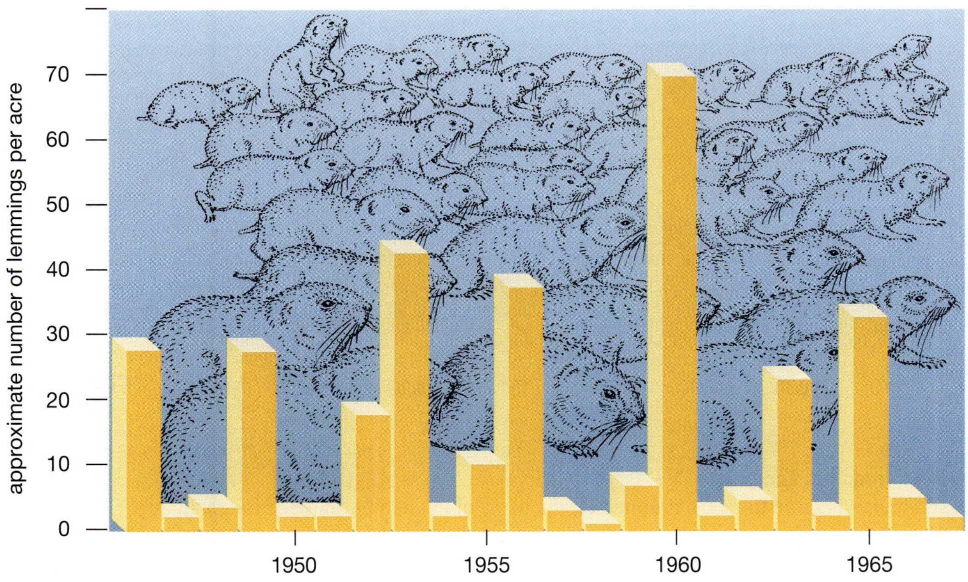

Figure 27-4 Lemming population cycles
The size of lemming populations cycles every 2–4 years (data from Point Barrow, Alaska).

Scientific Inquiry
Cycles in Predator and Prey Populations

If we assume that prey, such as hares, are eaten exclusively by a particular predator, such as the lynx, it seems logical that both populations might show cyclic changes, with changes in the size of the predator population lagging behind changes in the size of the prey population. For example, a large hare population would provide abundant food for lynx and their offspring, which would then survive in large numbers. The increased lynx population would eat more hares, reducing the hare population. With fewer prey, fewer lynx would survive and reproduce, so the lynx population would decline a short time later.

Does this out-of-phase population cycling of predators and prey actually occur in nature? One example of such a cycle was demonstrated by using the numbers of pelts of northern Canada lynx and snowshoe hares purchased from trappers by the Hudson Bay Company between 1845 and 1935. The availability of pelts (which presumably reflects population size) showed dramatic, closely linked population cycles of these predators and their prey (Fig. E27-1). Unfortunately, many uncontrolled variables could have influenced the relationship between the numbers of hares and lynx. One problem is that hare populations sometimes fluctuate even without lynx present, possibly because in the absence of predators hares overshoot their carrying capacity and reduce their food supply. Further, lynx do not feed exclusively on hares but also eat other small mammals. Environmental variables such as exceptionally severe winters could have adversely affected both populations and produced similar cycles. Recently, researchers tested the hare/predator relationship more rigorously by fencing off 1-kilometer-square areas in northern Canada to exclude predators. The crash of the hare population was lessened both by providing extra food and by excluding predators, but by far the greatest success in preventing the crash of the hare population was achieved when the researchers excluded predators *and* provided extra food.

To test the predator–prey cycle hypothesis in an even more controlled manner, investigators turned to laboratory studies on populations of small predators and their prey. One study used wasps as predators and bean weevils as prey. As predicted, the sizes of the two populations showed regular cycles, with the predator population increasing and decreasing slightly later than the prey population. The wasps lay their eggs on weevil larvae, which provide food for the newly hatched wasps. A large weevil population ensures a high survival rate for wasp offspring, increasing the predator population. Then, as the predators feed, the weevil population plummets, reducing the population size of the next generation of wasps. Reduced predation then allows the weevil population to increase rapidly, and so on (Fig. E27-2).

It is highly unlikely that such a straightforward example will ever be found in nature, but this type of predator–prey interaction clearly can contribute to the fluctuations observed in many natural populations.

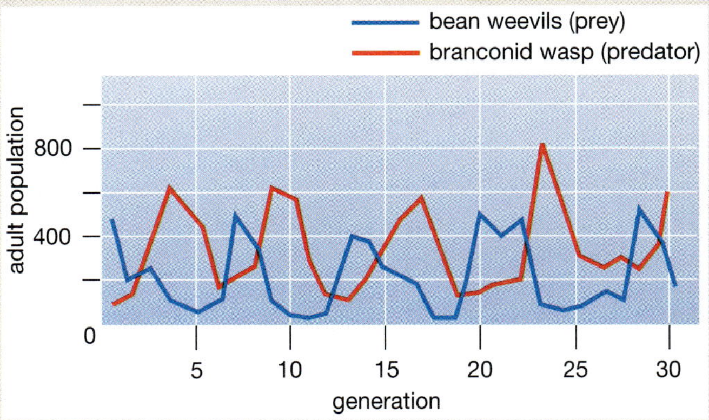

Figure E27-2 Experimental predator–prey cycles
Out-of-phase fluctuations in laboratory populations of bean weevils and their wasp predators.

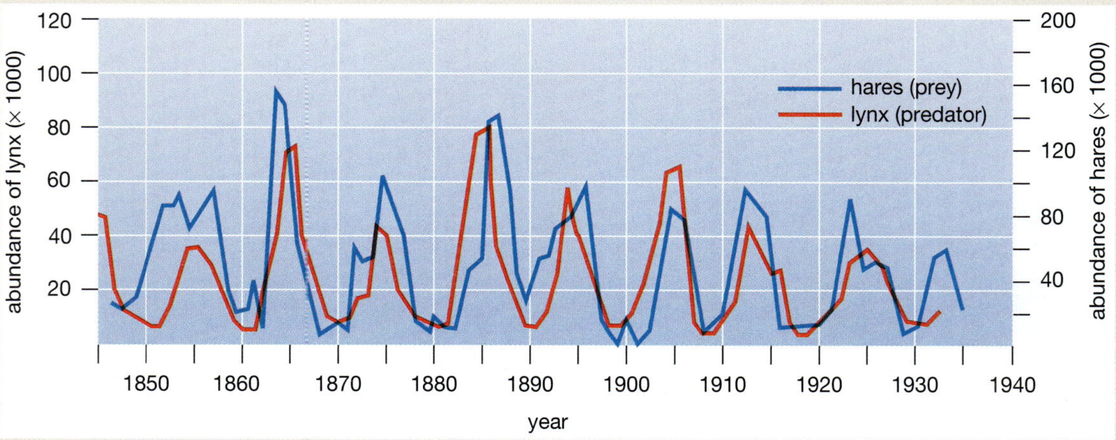

Figure E27-1 Population cycles in predators and prey
Numbers of snowshoe hares and their lynx predators are graphed on the basis of the number of pelts received by the Hudson Bay Company.

intensifies. For example, as dandelion plants invade an abandoned farm field and their numbers grow, competition for space, water, sunlight, and soil nutrients increases until further expansion is impossible. In addition to increased competition, rapidly expanding populations may also experience increased predation. Population growth in animals such as prairie dogs can cause predators such as hawks to increase in number or make this newly abundant prey a larger part of their diet. Similarly, diseases spread more readily through growing populations, in which individuals are more crowded and may be weaker from lack of food or from stress caused by crowding.

Most Populations Reach and Remain at a Stable Size

Under pressure from increased competition, predation, and disease, the death rate increases (as does emigration in species that can move). Consequently, after a period of rapid growth, populations tend to stabilize at or below their ecosystem's **carrying capacity**, the maximum population of particular species that an ecosystem can support indefinitely. As population size approaches the carrying capacity, the rate of growth gradually declines and reaches a long-term state of equilibrium, fluctuating around a growth rate of zero. In this equilibrium, the birth rate is balanced by the death rate and population size is stabilized. This type of population growth, which is typical of long-lived organisms that colonize a new area, is represented graphically by an S-shaped growth curve (Fig. 27-5).

The factors that maintain populations at or below the carrying capacity of their environment can be classified into two broad categories. **Density-independent** factors limit population size regardless of the population density (number of individuals per given area). **Density-dependent** factors increase in effectiveness as the population density increases. In the following sections, we will look more closely at these factors and how they control population growth.

Density-Independent Factors Limit Population Size

Natural events, including hurricanes, droughts, floods, and fire can have profound effects on local populations; these effects are independent of the population density. Perhaps the most important natural density-independent factor is weather. For example, many insects and annual plant populations are limited in size by the number of individuals that can be produced before the first hard freeze. Such populations typically do not reach their carrying capacity, because density-independent factors intervene first.

Human activities can also limit the growth of natural populations in ways that are independent of population density. Pesticides and pollutants can cause drastic declines in natural populations. Before it was banned in the 1970s, the pesticide DDT had drastically reduced populations of predatory birds, such as eagles, ospreys, and pelicans. Habitat destruction caused by the construction of farms, roads, and housing developments has eliminated most of the large prairie dog towns and pushed such diverse species as the California condor and the giant panda to the brink of extinction.

Density-Dependent Factors Have Greater Effect as Population Density Increases

For long-lived species, the most important elements of environmental resistance are density-dependent factors. Because density-dependent factors become increasingly effective as population density increases, they tend to limit population size. The larger a population grows, the more changes are triggered that counteract its growth. The most important density-dependent factors are predation and competition. These factors are discussed below and in Chapter 28.

Predators Exert Density-Dependent Controls on Populations

Predation is one organism feeding on another and harming it in the process. Often, one organism (the **predator**) kills another (its **prey**) in order to eat it (Fig. 27-6), but

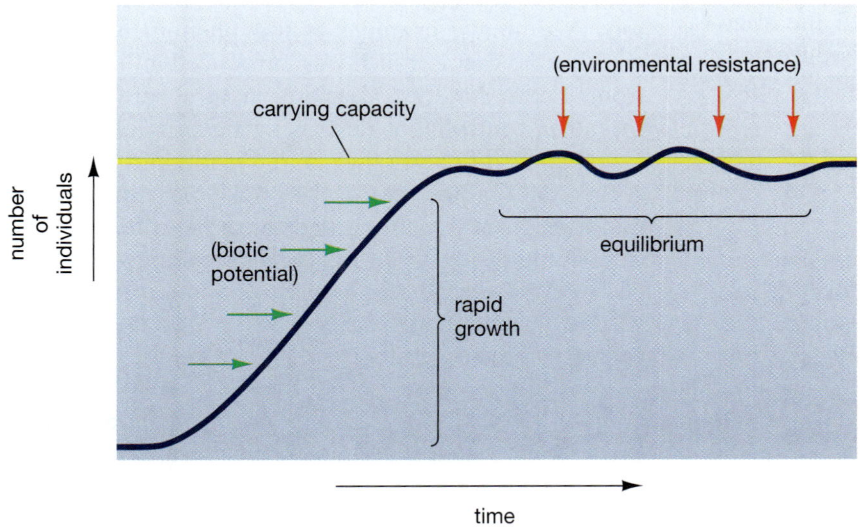

Figure 27-5 The S-curve of population growth The population grows rapidly at first, then fluctuates around the carrying capacity.

Figure 27-6 Predators help control prey populations
Grey wolves, hunting in a pack, have brought down an elk, who may have been weakened by old age or parasites.

the prey is not always killed. For example, when deer browse the buds of a maple tree, the tree is harmed but not killed. In addition, the prey is not necessarily killed in the special form of predation known as *parasitism*. In parasitism, the predator (called the **parasite** in this case) lives on or inside another organism (its **host**; usually a much larger organism) and feeds on the host's body without killing it—at least not immediately.

Predation becomes an increasingly important factor in population control as prey populations increase, simply because the more abundant the prey, the more often predators encounter them. Many predators eat a variety of prey, depending on what is most abundant and easiest to find. Coyotes might eat more mice when the mouse population is high but switch to eating more ground squirrels as the mouse population declines, incidentally allowing the mouse population to recover.

Predators can also exert density-dependent effects by increasing in number as their prey population grows. For instance, predators that feed heavily on lemmings, such as the arctic fox and snowy owl, regulate the number of offspring they produce according to the abundance of lemmings. The snowy owl might produce up to 13 chicks when lemmings are abundant but not reproduce at all in years when lemmings are scarce.

Parasites Spread Faster When Host Population Density Is High

Many parasites cause disease—including some bacteria, fungi, worms, ticks, and protists. Like other forms of predation, parasitism is density-dependent. Most parasites have limited motility and spread more readily from host to host at high host-population densities. For example, plant diseases spread readily through acres of densely planted crops, and childhood diseases spread rapidly through schools and day-care centers. Although parasites do not benefit from the death of their hosts, their activities may nonetheless kill them. Parasites also influence population size by weakening their hosts and making them more susceptible to death from other causes. For example, parasites can contribute to deaths by making the host organisms weaker and more vulnerable to other predators.

Populations Can Soar or Crash When Predator-Prey Relationships Are Disrupted

The population balance in ecosystems can be destroyed when predators are introduced into regions where they did not evolve and where local prey species have had no opportunity to evolve defenses against them. For example, the smallpox virus, inadvertently carried by traveling Europeans, ravaged the native population of Hawaii, the Amerindians of Argentina, and the Aborigines of Australia. Similarly, introduced rats, snakes, and mongooses have exterminated many of Hawaii's native bird populations.

Conversely, prey populations can grow out of control when introduced to areas where they have no predators. A dramatic example of this phenomenon is the prickly pear cactus, which was introduced into Australia from Latin America. Lacking natural predators, it spread uncontrollably, destroying millions of acres of valuable pasture and range land. Finally, in the 1920s, a cactus moth (a predator of the prickly pear cactus) was imported from Argentina and released to feed on the cacti. Within a few years, the cacti were almost eliminated. Today the moth continues to maintain its prey at very low population densities, well below the carrying capacity of the ecosystem.

Competition for Resources Helps Curb Population Size

The resources that determine carrying capacity (space, nutrients, water, light) are not unlimited. A given resource may not be present in sufficient amounts to sup-

port all the organisms present. As a result, organisms must compete for access to limited resources. This **competition** among individuals limits population size in a density-dependent manner.

There are two major forms of competition: **interspecific competition** among individuals of different species (described further in Chapter 28), and **intraspecific competition** among individuals of the same species. Because the resource requirements of members of the same species are almost identical, intraspecific competition is more intense than interspecific competition.

Competition May Be Indirect or Direct

Some organisms, including most plants and many insects, engage in **scramble competition**, a kind of free-for-all in which individuals independently seek resources without interacting directly. For example, when a plant disperses its seeds in a small area, hundreds may germinate. As they grow, however, some plants will grow taller and gain access to more sunlight, and some will have more extensive root systems and will absorb most of the water. Eventually, the less successful individuals wither and die.

Many animals (and even a few plants) have evolved **contest competition**, in which individuals interact to contest access to important resources. For example, as you may recall from Chapter 26, territorial species—such as wolves, many fish, rabbits, and songbirds—defend an area that contains important resources such as food or nesting sites. Only the best competitors are able to defend adequate territories. Poor competitors fail to secure territories and are unlikely to reproduce or perhaps even to survive.

The Big Picture

Despite the enormous growth potential of populations, no population can grow indefinitely. The resources available to support populations are limited, and populations stop growing when the limits are reached.

3 How Are Populations Distributed in Space and Time?

Organisms arrange themselves in space in many different ways. They might live in large flocks or herds, in small groups, as pairs, or as solitary individuals. They might spread themselves evenly through their habitat or cluster around a resource such as a water hole. Their spatial arrangment might vary with time, changing with the breeding season, for example. Ecologists recognize three major types of spatial distribution: clumped, uniform, and random.

Individuals in Many Populations Clump Together in Groups

The members of many populations live clumped together in groups (Fig. 27-7a). Examples of such **clumped distributions** include social groups such as elephant herds, wolf packs, lion prides, flocks of birds, and schools of fish. What are the advantages of clumping? In groups there are many eyes that can search for localized food, such as a tree full of fruit or a lake with fish. Membership in a large school of fish or flock of birds may reduce the odds that any one individual will be killed by an attacking predator. A temporary group that forms for mating may increase individuals' chances of finding a mate. Alternatively, plant or animal populations may cluster not for social reasons, but because resources are localized. Cottonwood trees, for example, cluster along streams and rivers in grasslands.

Some Individuals Disperse Themselves Evenly

Populations with **uniform distribution** maintain relatively constant distances between individuals. This type of distribution is most common among animals that defend territories to protect scarce resources. Male Galapagos iguanas, for example, establish regularly spaced breeding territories. Shorebirds are often found in evenly spaced nests during the breeding season. Other territorial species, such as the tawny owl, continuously occupy well-defined, relatively uniformly spaced territories. (For animals that remain together to raise their young, spacing often refers to pairs rather than to individuals.) Some plants, such as sage, release chemicals into the soil around them that inhibit germination of other plants and thus space themselves relatively evenly (Fig. 27-7b).

In a Few Populations, Individuals Are Distributed at Random

Populations with **random distributions** are relatively rare. Individuals in such populations do not form social groups. The resources they need are more or less equally available throughout the area they inhabit, and those resources are not scarce enough to require territorial spacing. Trees and other plants in rain forests come close to being randomly distributed (Fig. 27-7c).

4 How Is the Human Population Changing?

Compare the graph of human population growth in Figure 27-8 with the exponential growth curves in Figure 27-1. The time spans are different, but each has the J-shape characteristic of explosive growth. It took more than 100,000 years for the human population to

(a)

(b)

(c)

Figure 27-7 Population distributions
(a) Clumped: a gathering of caterpillars. *(b)* Uniform: creosote bushes in the desert. *(c)* Random: trees and other plants in a rain forest.

reach 1 billion, but the second billion was added in just 100 years, the third billion in 30 years, the fourth billion in 15 years, and the fifth billion in 12 years, and the sixth billion in another 12 years. In the year 1750, the human population was 0.75 billion. It doubled in 150 years, reaching 1.6 billion in 1900. In contrast, the 1965 population of 3.5 billion will double to 7 billion in less than 50 years. World population currently grows by about 1.4% or 82 million yearly—more than 224,000 people are added every day, nearly 1.6 million every week! Why hasn't environmental resistance put an end to our tremendous population growth?

The Big Picture

The human population has never before grown as rapidly as it has grown since the middle of the 20th century. Never before has the human population been as large as it is now.

Technological Advances Have Increased Earth's Carrying Capacity for Humans

Like all populations, humans have encountered environmental resistance, but unlike other populations, we have responded to environmental resistance by devising ways to overcome it. As a result, the human population has grown rapidly for an unprecedented time span. To accommodate our growing numbers, we have altered the face of the globe. Human population growth has been spurred by a series of "revolutions" that conquered various aspects of environmental resistance and increased Earth's carrying capacity for people.

Early Tools Allowed Human Populations to Spread and Prosper

Prehistoric people produced a *cultural revolution* when they discovered fire, invented tools and weapons, built shelters, and designed protective clothing. Tools and weapons allowed more effective hunting and thereby increased the supply of food. Shelter, fire, and clothing allowed humans to move into previously uninhabitable areas of the globe. Together, these developments set the stage for future population growth.

Agriculture Fostered Population Growth

By about 8000 B.C., cultivation of domesticated crops and animals had replaced hunting and gathering as humans' primary means of acquiring food. This *agricultural revolution* provided people with a larger, more dependable food supply and further increased Earth's carrying capacity for humans. Increased food resulted in a longer life span and more childbearing years, and the human population began to climb steadily.

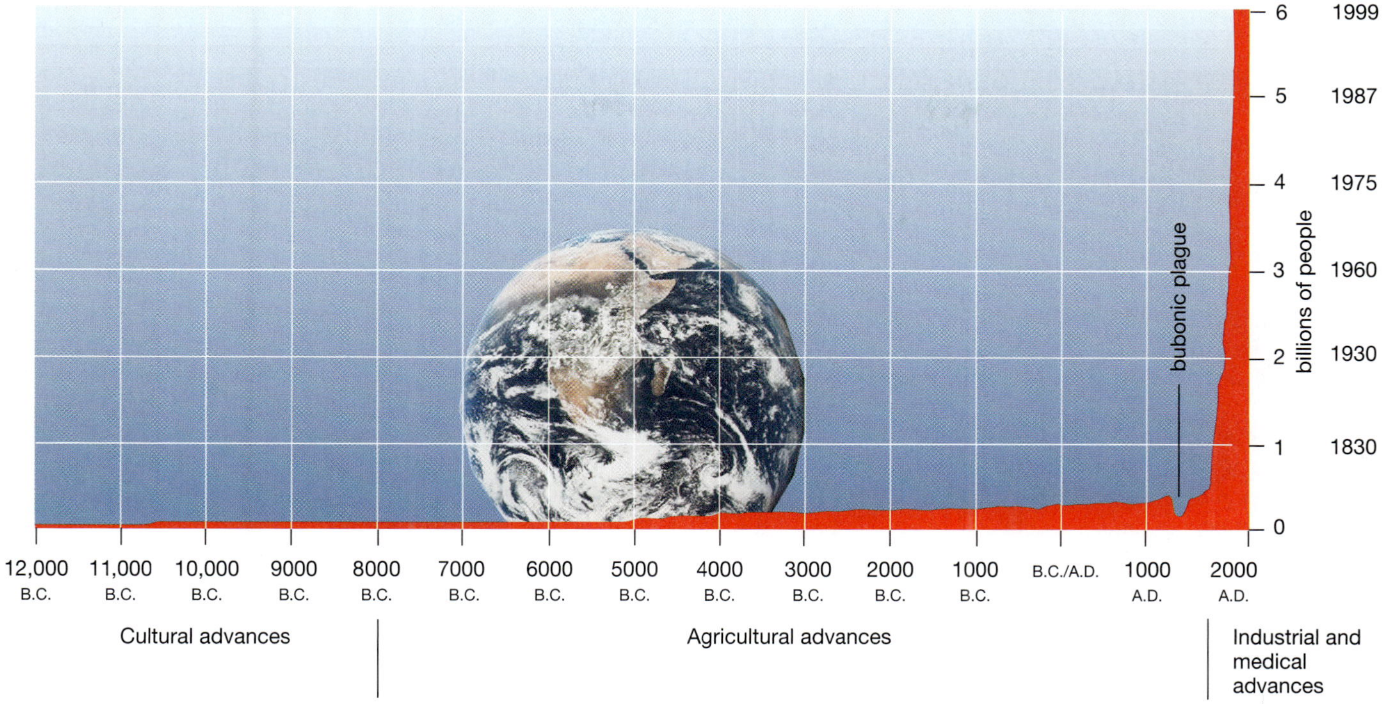

Figure 27-8 Human population growth
The human population from the Stone Age to the present has shown continual rapid growth. Note the dip in the fourteenth century caused by the bubonic plague.

MediaTutor 27.2 Population Growth and Regulation

Explosive Growth Began with the Industrial Revolution
Human population growth continued slowly for thousands of years until the *industrial–medical revolution* began in England in the mid-eighteenth century and spread through Europe and North America in the nineteenth century. This revolution marked the beginning of a period of extremely rapid population growth, as medical advances dramatically decreased the death rate. These advances included the discovery of bacteria and their role in infection, leading to the control of bacterial diseases through improved sanitation. Another advance was the discovery of viruses, leading to the development of vaccines for diseases such as smallpox. The medical revolution (and its accompanying explosive population growth) continues today as research proceeds on vaccines against major killers such as malaria and AIDS, and sophisticated medical procedures, such as coronary bypass operations and organ transplants, are improved and new procedures are developed to extend the human life span.

Population Growth Continues Today But Is Unevenly Distributed

In many developed countries, such as those of Western Europe, the population explosion fostered by the declining death rates of the industrial–medical revolution was eventually followed by a decline in birth rates. The tendency for birth rates to decline in affluent societies is not fully understood but has been attributed to many factors, including better education, increased availability of contraceptives, a shift to a primarily urban lifestyle, and more career options for women. In developed countries, population sizes have more or less stabilized, except for the effects of immigration.

In developing countries, such as those of Central and South America, Asia, and Africa, medical advances have decreased death rates and increased the life span (though the AIDS epidemic threatens this progress in some areas). Birth rates, however, remain high, as these countries have not experienced the increased wealth associated with the birth rate decline in developed countries. In agricultural societies, children are an important source of labor and may be the only support for elderly parents. Social traditions may offer prestige to men who father many children and to women who bear them. Although many developing nations have concluded that their population growth rates are too high, social traditions, lack of education, and lack of access to contraceptives impede progress in curbing population growth.

Of the more than 6 billion people on Earth today, close to 5 billion reside in developing countries. For the

Earth Watch
Have We Exceeded Earth's Carrying Capacity?

In Côte d'Ivoire, a small country on the west African coast, the government is waging a battle to protect some of its rapidly dwindling tropical rain forest from thousands of illegal hunters, farmers, and loggers. Officials burn the homes of the squatters, who immediately return and rebuild. One illegal resident is Sep Djekoule. "I have 10 children and we must eat," he explains. "The forest is where I can provide for my family, and everybody has that right." His words illustrate the conflict between population growth and environmental protection. A glance at the age structure of developing countries, where most of the world's population resides, shows a tremendous momentum for growth. World population in the year 2010 is predicted to be about 7 billion and growing. A conservative United Nations projection is that the human population will reach 8.9 billion by the year 2050, and will still be increasing.

How many people can Earth support? No one knows. Estimates of Earth's carrying capacity for humans have ranged from 3 billion to 44 billion people. Each of these estimates was based on different assumptions regarding technological developments and lifestyles.

The upper limit of the planet's carrying capacity is determined by the ability of its plant life to harvest the energy from sunlight and produce high-energy molecules that other organisms can use as food. As humans appropriate an ever larger share of this captured energy, less is available for the rest of the planet's species. Worse, the total amount of captured energy may be declining. Stanford University biologist Peter Vitousek estimates that human activities have already reduced the productivity of Earth's forests and grasslands by 12%. Each year, overgrazing and deforestation further decreases the productivity of land, especially in developing countries (Fig. E27-3).

In a world where more than 800 million people are chronically undernourished (Fig. E27-3, inset), the quest for more agricultural land leads to deforestation. More than 30 million acres of rain forest are destroyed each year. Yet even as people in developing nations desperately clear land for farming, the United States loses nearly half a million acres of prime farmland each year to development of homes, shopping malls, and roads. As our own growing population spreads onto our farmland, our ability to export grain to help sustain other nations will diminish, while their needs increase. Worldwide, the amount of cropland per person has diminished by half in the past 50 years.

Although we take fresh water for granted, in many developing countries, water supplies are badly polluted and underground water supplies, called *aquifers*, are being depleted and not replaced. Both India and China are rapidly depleting aquifers to supply the needs of their growing cities and to irrigate their cropland, a policy that may result in reduced grain harvests in the near future.

The demand for wood in developing countries is far outstripping production, providing yet another source of deforestation. Deforestation in turn causes erosion of precious topsoil (two-thirds of the world's agricultural land is suffering moderate to severe erosion), as well as runoff of much-needed fresh water and the spread of deserts.

The total world fish harvest peaked in 1989 and has been declining since, despite increased investments in fishing equipment and fish farming. Almost 70% of commercial ocean fish populations have been fully exploited or overfished, and many fisheries, such as the New England cod harvest, have collapsed because of overfishing. These are clear indications that our present population, at its present level of technology, is "overgrazing" the world's ocean ecosystems.

Taken together, the observations described above strongly suggest that we are severely stressing Earth's ability to support us. Furthermore, when we estimate how many people Earth can support, we must keep in mind that humans desire more out of life than a minimum caloric intake each day. Do we want to be able to eat meat, drive a car, live in a single-family dwelling, walk in a wilderness, and know that somewhere eagles, pandas, and elephants are living free? For people living at Earth's maximum carrying capacity, these will probably be unattainable luxuries. William Rees of the University of British Columbia has estimated that for all of Earth's 6 billion people to live as people in the United States do would require the resources of three additional Earth-sized planets.

Hope for the future lies in using the intelligence that has allowed us to overcome environmental resistance to act before we have irrevocably damaged our biosphere. The human population *will* stop its explosive growth. Either we will voluntarily reduce our birth rate, or various forces of environmental resistance such as disease and starvation will dramatically increase human death rates. The choice is ours. Facing the problem of how to limit births is politically and emotionally difficult, but continued failure to do so will be disastrous. Our dignity, our intelligence, and our role as self-appointed stewards of life on Earth demand that we make the decision to halt population growth before we have permanently reduced Earth's ability to support all life, including our own.

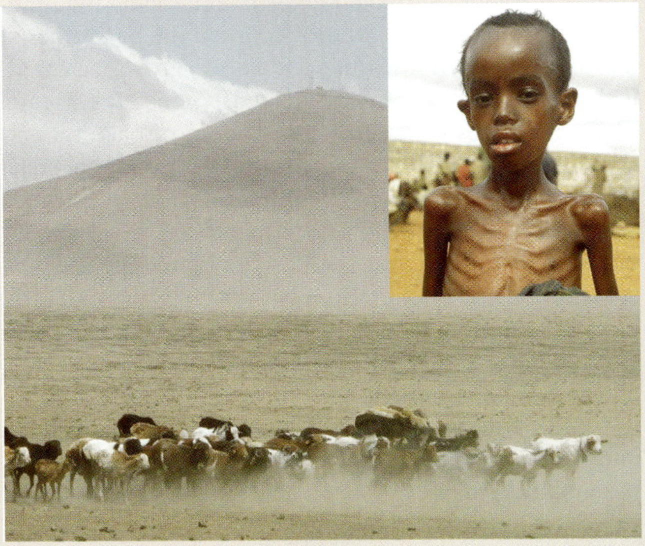

Figure E27-3 Deforestation can lead to the spread of deserts Human activities, including overgrazing livestock, deforestation, and poor agricultural practices, may sometimes convert once-productive land into barren desert. (Inset) An expanding human population, coupled with a loss of productive land, can lead to tragedy.

year 2050, the United Nations predicts that there will be nearly 9 billion people, with 8 billion living in the developing nations.

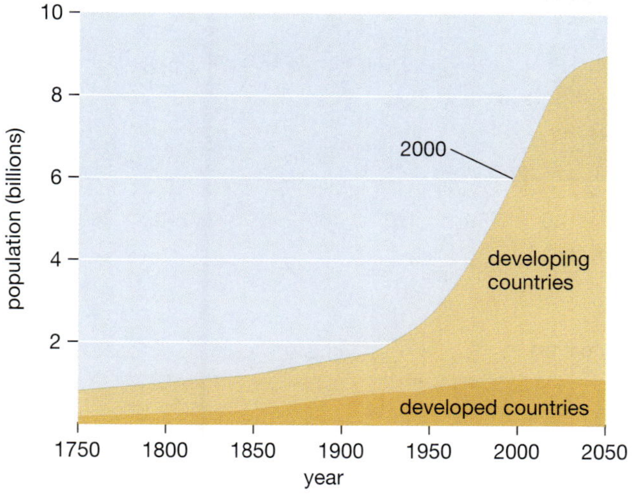

Modified from the Population Reference Bureau: 1999 World Population Data Sheet.

The prospects for population stabilization—*zero population growth*—in the near future are nil. In Africa, AIDS infects 25 million people and has taken an enormous toll in human lives, but the United Nations predicts that Africa will contribute 34% of the population growth in developing countries until the year 2050. You can understand why by examining the age structures of the developing countries and comparing them with those of countries with stable populations, as we describe next.

The Age Structure of a Population Predicts Its Future Growth

The **age structure** of a population is the numbers of males and females of each age group, which can be shown graphically in an age-structure diagram. The vertical axis of an age-structure diagram shows the age of people, while the horizontal axis shows numbers of individuals in each age group, with males and females graphed in opposite directions from the central axis. All age-structure diagrams rise to a peak at the top, because few people live into their nineties. The shape of the rest of the diagram, however, shows at a glance whether the population is expanding, stable, or shrinking. The shape is determined by the relative numbers of reproductive age adults (ages 15 to 45) and their children (ages 0 to 14). If the adults of reproductive age are producing just enough children to replace themselves, the population is said to have **replacement level fertility (RLF)**. The age structure diagram of such a population has relatively straight sides. If children in the various age classes exceed the number of reproductive-age adults, the population is exceeding RLF and is expanding as a result. Such expanding populations have a pyramidal shape. In shrinking populations, there are fewer children than reproducing adults and the age-structure diagram narrows at the base.

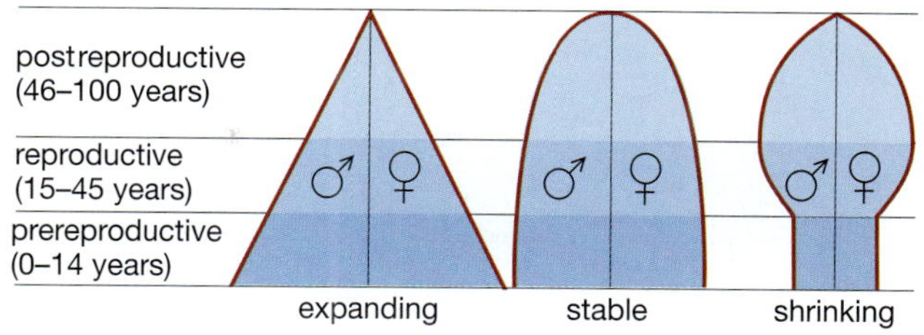

Almost All Future Population Growth Will Be in Developing Countries

Figure 27-9 shows the average age structures of the populations of developed and developing countries. The red boundaries represent the projected population structure for the year 2025; the green areas give actual values for 1995. Each graph has been divided into three parts to show individuals who are prereproductive (ages 0 to 14), reproductive (15 to 45), and postreproductive (46 and older). In 2000 the developing countries (in Asia, Africa, India, and South and Central America) had an average annual *natural increase* (population growth based on the births minus deaths; this does not include migration) of 1.7% and a doubling time (the time required for the population to double) of 42 years. In contrast, developed countries (in North America, Europe, Australia, Japan, and New Zealand) showed an average annual natural increase of 0.1% and a doubling time of 809 years. In the developed countries, projections for the year 2025 are only slightly higher than present levels (see Fig. 27-9a). Because developing countries are home to the vast majority of people, doubling time for the world population as a whole is currently 51 years.

More than 95% of the growth of the human population occurs in the developing countries, where increasing numbers of people enter their reproductive years and give birth to an ever-increasing base of infants. Even if these countries were immediately to reach replacement-level fertility, population growth would continue for decades. The growth would come as the children of the large families of the recent past enter their reproductive years and begin having children—even if each couple has only 2 children. This entry into the reproductive age classes fuels China's population growth of 1% annually, even though China's fertility rate is below replacement level. The percentage growth per year caused by natural increase for various parts of the world is shown in Figure 27-10.

Fertility in Europe Is Below Replacement Level

In Europe, the fertility rate is well below RLF, and much of eastern Europe is experiencing a slight decline in population, as poor economic conditions prompt women to delay or forego having children. This raises

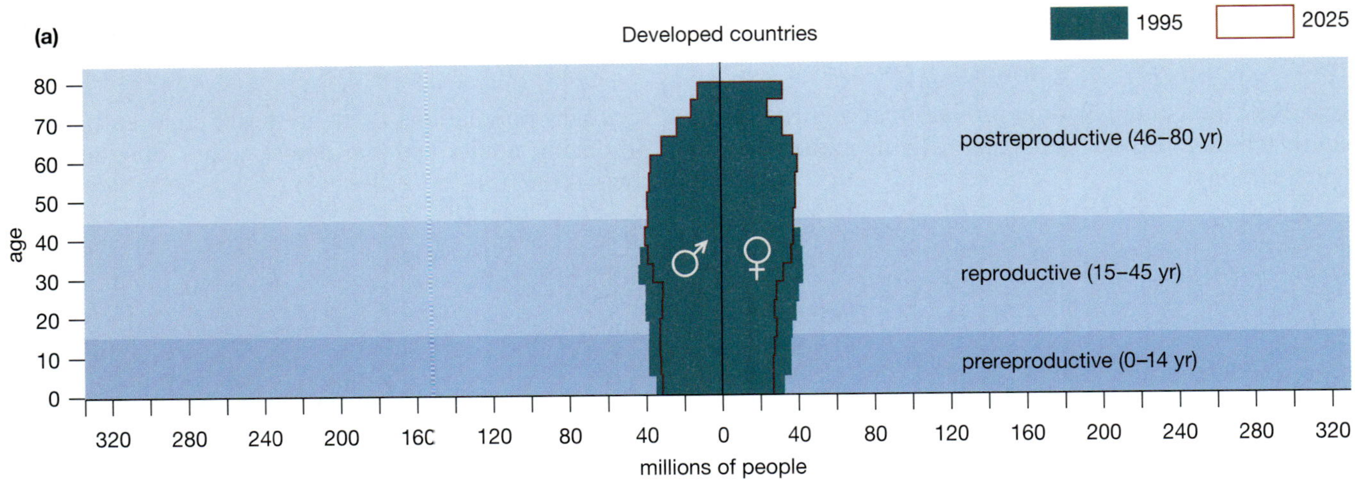

Figure 27-9 Age structures compared

Figure 27-10 Population growth by world regions Growth rates shown are due to natural increase (births – deaths) expressed as the percentage increase per year. The bar that represents Europe includes the former Soviet Union. These figures do not include immigration or emigration.

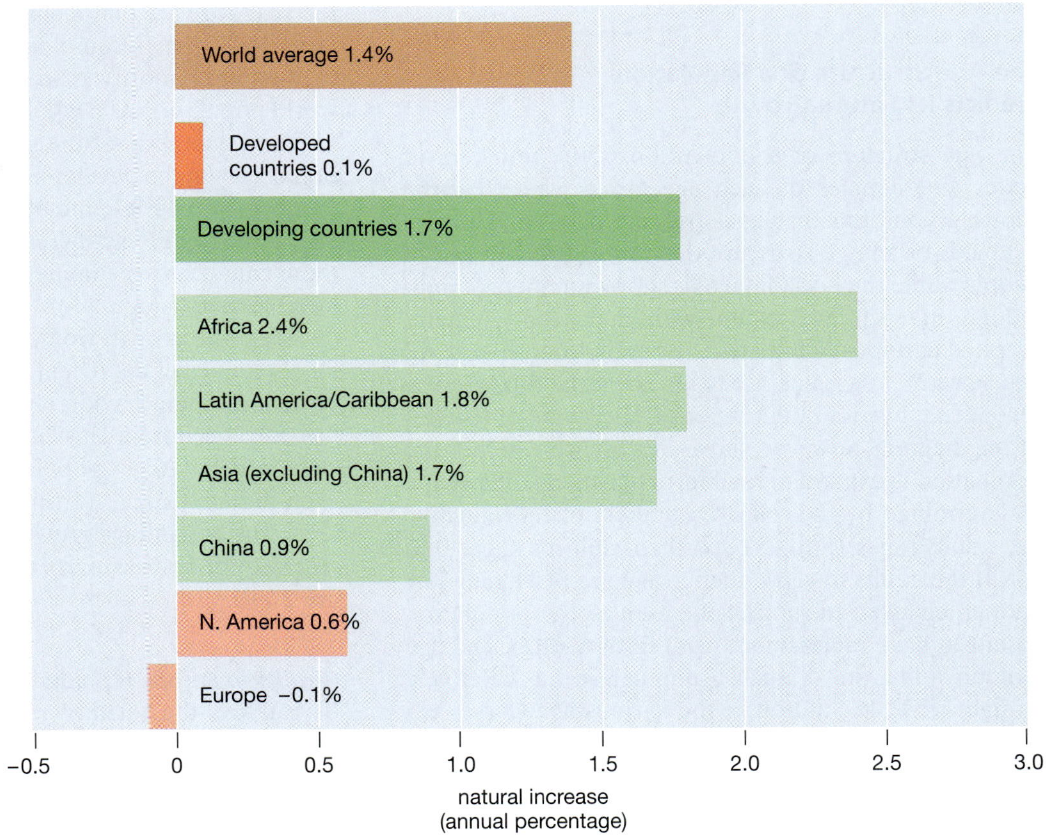

economic concerns about the availability of future workers and taxpayers to support the increasing percentage of elderly people. Japan, a country the size of Montana, has 127 million people (slightly less than half of the entire U.S. population). Nonetheless, the low Japanese fertility rate is causing concern about slowed economic growth and an aging population. The Japanese government provides subsidies that encourage people to have larger families, and recently a Japanese toy company began offering $10,000 bonuses to their employees for having more than 2 children. It is clear that current economic structures are based on growing populations. Although a reduced and eventually stable population may offer tremendous benefits in the future, the transition is difficult.

The U.S. Population Is Growing Rapidly

As shown in Figure 27-11, the United States is the fastest growing industrialized country in the world. With a natural increase of 0.6%, the U.S. population grows at six times the average rate of developed countries. Adding immigration to this picture, U.S. annual growth rate approaches 1%. During 1999, for example, the U.S. population increased by almost 2.6 million; the United States adds nearly 300 people to its population every hour. The population in 2000 was more than 275 million. Recall the following equation:

$$\text{population change} = (\text{births} - \text{deaths}) + (\text{immigrants} - \text{emigrants})$$

Let's examine each component of the equation to see why the population of the United States is growing so rapidly. If, on average, each woman in the United States had 2.1 children, we would have replacement-level fertility. RLF is slightly higher than 2 because parents must replace both themselves and children who die before reaching maturity. The U.S. fertility rate in 2000 was 2.1, just at RLF, so why does the U.S. population continue to grow? Two factors are contributing to the rapid growth of the U.S. population: the "baby boom" and immigration.

Part of the current U.S. growth rate is a legacy of the recent past. Parents of the late 1940s through the 1960s had families that were larger than replacement level, resulting in a "baby boom" and a spurt in population growth that has not yet subsided. Even though women are averaging 2.1 children each, because more women are having children, the U.S. population swells.

A second crucial component of the U.S. population equation is immigration. Legal immigration is well over 800,000 per year, causing about 30% of the total U.S. population growth. Illegal immigration adds an estimated 275,000 people each year. The U.S. Census Bureau projects that by the year 2050, there will be 404 million people in the United States (about 130 million more than today) and the population will still be growing. Even with RLF, immigration will guarantee continued U.S. population growth.

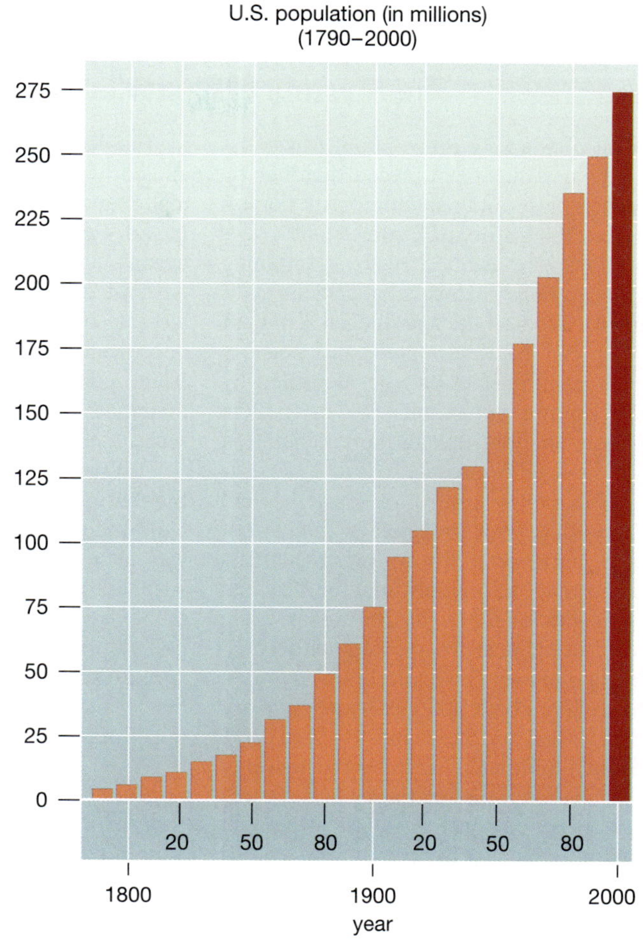

Figure 27-11 U.S. population growth
Since 1790, U.S. population growth has shown the J-shaped curve characteristic of rapid growth.

The rapid growth of the U.S. population has major environmental implications both for us and for the world. The average person in the United States uses five times as much energy as the global average, so with less than 5% of the world's population, we account for 25% of world energy use. Because people in developing countries use considerably less energy than the global average, the 2.6 million people added to the United States each year contribute more to climate change through the release of CO_2 and other greenhouse gases than do the 32 million people added annually by China and India combined.

When and how will human numbers ever stabilize? How many people can Earth support? There are no certain answers to these questions, but in "Earth Watch: Have We Exceeded Earth's Carrying Capacity?", we explore them in more detail.

CASE STUDY REVISITED

The Mystery of Easter Island

Most of what we know about Easter Island's past has been pieced together by archeologists and paleontologists. For example, the history of Easter Island's vegetation was established by examining pollen grains in layers of ancient sediments. Because the age of each sediment layer can be determined, and because each plant species can be identified by the unique appearance of its pollen, pollen analysis reveals how vegetation has changed over time.

The pollen record of Easter Island shows that, prior to human arrival, the island supported a diverse forest, including plants that could supply fiber for rope, and trees with long, straight trunks that would have been ideal for moving statues and making canoes. The forest, however, began to decline about 400 years after humans first settled on the island. By A.D. 1400, about a thousand years after the arrival of humans, almost all of Easter Island's tree species were extinct. A thousand years of clearing land for agriculture and cutting trees for firewood and construction materials had destroyed the forest.

The destruction of the forest was paralleled by a corresponding wave of animal extinctions. Fossils show that Easter Island was once home to parrots, owls, herons, and numerous other bird species. At least 25 species of seabirds, including albatrosses, boobies, and frigate birds, once nested on the island. Today, Easter Island contains no native land birds or mammals. No animal larger than an insect is a permanent resident there.

Forest destruction no doubt hastened the disappearance of Easter Island's animals, but direct consumption by humans was also responsible. Archeologists investigating early sites of human settlement have found large quantities of bones from most of the extinct bird species in the sites' garbage heaps, evidence that the islanders made many meals of the local birds.

When the forest and its animals disappeared, the island's human population probably lacked adequate food. With no wood for canoes or boats, there was no way to leave the island. Ominously, garbage heaps from the post-forest period contain gnawed human bones, testimony to the cannibalism that emerged in this period. By the time Europeans arrived, the small number of people left on Easter Island lived impoverished lives and had no record or memory of the culture that built the magnificent statues.

What can we learn from Easter Island? According to biologist and author Jared Diamond, "the meaning of Easter Island for us should be chillingly obvious. Easter Island is Earth writ small. Today, again, a rising population confronts shrinking resources. We too have no emigration valve, because all human societies are linked by international transport, and we can no more escape into space than the Easter Islanders could flee into the ocean. If we continue to follow our present course, we shall have exhausted the world's major fisheries, tropical rain forests, fossil fuels, and much of our soil by the time my sons reach my current age."

Summary of Key Concepts

1) How Do Populations Grow?

Individuals join populations through birth or immigration and leave through death or emigration. All organisms have the potential to more than replace themselves over their lifetimes, so all populations have potential for growth. A population's growth rate depends on the age at which individuals begin to reproduce, the number of offspring produced by each individual, and the life spans of individuals. Populations in which the growth rate remains constant over time grow exponentially, with increasing numbers of individuals added during each successive time period.

2) How Is Population Growth Regulated?

The ultimate size of a stable population is the result of interactions between biotic potential (the maximum possible growth rate) and environmental resistance (which limits population growth). Populations cannot grow indefinitely; they either stabilize or undergo periodic boom-and-bust cycles as a result of environmental resistance. Environmental resistance restrains population growth by increasing the death rate or decreasing the birth rate. The maximum size at which a population may be sustained indefinitely by an ecosystem is termed the carrying capacity, determined by limited resources such as space, nutrients, and light. Environmental resistance generally maintains populations at or below the carrying capacity.

Population growth is restrained by density-independent forms of environmental resistance (such as weather) and density-dependent forms of resistance (including predation, parasitism, and competition).

3) How Are Populations Distributed in Space and Time?

The distribution of the members of a population may be clumped, uniform, or random. Distributions may be clumped for social reasons or around limited resources. Uniform distributions are typically the result of territorial spacing based on resources. Distributions are random only when individuals do not interact socially and when resources are abundant and evenly distributed. Random distributions are rare.

4) How Is the Human Population Changing?

The human population has exhibited rapid growth for an unprecedented time by overcoming certain aspects of environmental resistance and increasing Earth's carrying capacity for humans. This has been accomplished by the use of tools, agriculture, industry, and medical advances. Age-structure diagrams depict numbers of males and females in various age groups that a population comprises. Expanding populations have pyramidal age structures, stable populations have straight-sided age structures, and shrinking populations have age structures that are constricted at the base.

Today, most of the world's people live in developing countries with rapidly expanding populations, where social and cultural conditions encourage people to have large families. The United States is the fastest growing of the developed countries, owing to high immigration rates and the baby boom of the 1940s through the 1960s. Earth's carrying capacity for humans is unknown, but with a population of more than 6 billion, resources are already too limited for all to be supported at a high standard of living. A steady decline in productive land, fresh water, and in wood and fish harvests suggests that we are already damaging our world ecosystem and decreasing its ability to sustain us.

Key Terms

age structure *p. 553*
biotic potential *p. 543*
boom-and-bust cycle *p. 543*
carrying capacity *p. 547*
clumped distribution *p. 549*
community *p. 542*
competition *p. 548*
contest competition *p. 549*
density-dependent *p. 547*
density-independent *p. 547*
ecology *p. 542*
ecosystem *p. 542*
environmental resistance *p. 543*
exotic species *p. 545*
exponential growth *p. 542*
host *p. 548*
interspecific competition *p. 549*
intraspecific competition *p. 549*
parasite *p. 548*
per capita growth rate *p. 542*
population *p. 542*
predator *p. 547*
prey *p. 547*
random distribution *p. 549*
replacement-level fertility (RLF) *p. 553*
scramble competition *p. 549*
uniform distribution *p. 549*

Thinking Through the Concepts

Multiple Choice

1. *Which of the following factors is NOT an example of density-dependent environmental resistance?*
   a. weather
   b. competition
   c. predation
   d. parasitism
   e. lack of food
2. *Population growth may be exponential when*
   a. there is no mortality
   b. there are no density-independent limits
   c. the birth rate exceeds the death rate and both remain constant
   d. a species is reproducing rapidly
   e. the species is an exotic invader in an ecosystem
3. *Which of the following currently contributes most to human population growth within the United States?*
   a. the consequences of the "baby boom"
   b. immigration
   c. a birth rate above RLF
   d. both a and b
   e. all of the above
4. *Which is the most common type of spatial distribution?*
   a. territorial
   b. uniform
   c. random
   d. exponential
   e. clumped
5. *Which continent has the highest rate of natural increase in the human population?*
   a. North America
   b. Africa
   c. Asia
   d. Australia
   e. South America
6. *If a population exceeds its carrying capacity,*
   a. it must immediately crash
   b. it can remain stable at this level indefinitely
   c. it will continue to increase for the indefinite future
   d. it must decline sooner or later
   e. the food supply will increase to support it

? Review Questions

1. Explain *biotic potential* and *environmental resistance*.
2. Draw the growth curve of a population before it encounters significant environmental resistance. What is the name of this type of growth, and what is its distinguishing characteristic?
3. Distinguish between density-independent and density-dependent environmental resistance.
4. Describe (or draw a graph illustrating) what is likely to happen to a population that far exceeds the carrying capacity of its ecosystem. Explain your answer.
5. List three density-dependent forms of environmental resistance, and explain why each is density-dependent.
6. Given that the U.S. birth rate is currently at replacement-level fertility, why is our population growing?
7. Discuss some reasons why making the transition from a growing to a stable population can be economically difficult.

Applying the Concepts

1. Explain natural selection in terms of biotic potential and environmental resistance.
2. The United States has a long history of accepting large numbers of immigrants. Discuss the implications of immigration for population stabilization.
3. What factors encourage rapid population growth in developing countries? What will it take to change this growth?
4. Contrast age structure in rapidly growing versus stable human populations. Why is there a momentum in population growth built into a population that is above RLF?
5. Why is the concept of carrying capacity difficult to apply to human populations? In reference to human population, should the concept be modified to include quality of life?

Group Activity

Working in groups of three or four, develop a proposed policy on U.S. population. In the policy, include current information on population statistics and patterns (use the textbook, the Internet, and other sources of information). Include also a statement of your views on immigration (how much, how little, how come, . . .) and on government-sponsored or voluntary family planning and education. Address the following questions:

- Should the United States limit its population size? Why or why not?
- Should there be mandatory or voluntary birth control? Why or why not?
- How would you convince people to limit births?
- What is the importance to population growth of education, particularly of girls and women?
- Should immigration be limited? Why or why not? (If it should be limited, how you would do this?)
- Will your plan address issues such as limited resources, quality of life, environmental disasters, and pollution? Explain your plan's implications, on the basis of the projections noted in the textbook, in terms of age structure and numbers of people expected by 2025.

Although there will undoubtedly be disagreement in your group, come to a consensus to prepare your statement. (You can agree to disagree for now.) The accompanying graph gives projections for the U.S. population for both a two-child and a three-child family growth rate.

When you are finished, meet with another group. For your group presentation, each group first reads its statement while the other group listens and takes notes. After each group has read its statement, questions can be asked of the other group and clarification sought. Compare areas where there is agreement or disagreement. After discussion, prepare a brief report of the similarities and differences between your group and the other, and submit that report with your own report to your instructor.

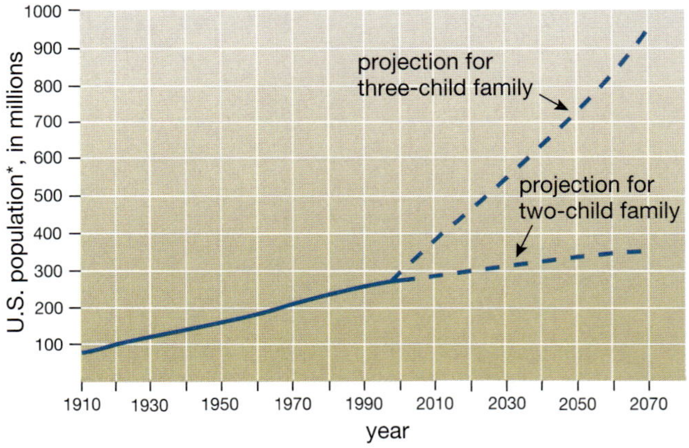

*Data based on "Population: the U.S. Problem, the World Crisis," Supplement to *The New York Times*, August 14, 1974.

For More Information

Bender, W., and Smith, M. "Population, Food, and Nutrition." *Population Bulletin*, February 1997. A comprehensive summary of issues regarding food supply and the growing human population.

Bongaarts, J. "Can the Growing Human Population Feed Itself?" *Scientific American*, March 1994. A balanced account of the arguments between optimists and pessimists concerning the future world food supply.

Diamond, J. "Easter's End." *Discover*, August 1995. An excellent scientific storyteller describes the historical, archeological, and paleontological evidence that provides clues about rise and fall of the Easter Island's civilization.

Kates, R. W. "Sustaining Life on the Earth." *Scientific American*, October 1994. Provides a prescription for an environmentally sustainable future.

Korpimaki, E., and Krebs, C. J. "Predation and Population Cycles of Small Mammals." *BioScience*, November 1996. A review of recent studies designed to evaluate cycles of predators and their prey.

Myers, N. "Biotic Holocaust." *International Wildlife*, March–April 1999. Human activities are causing extinction of species unprecedented since the disappearance of the dinosaurs. How can we reverse this trend?

Potts, M. "The Unmet Need for Family Planning." *Scientific American*, January 2000. Reducing population growth and increasing the quality of life requires increased access to contraceptives in developing countries.

Raloff, J. "The Human Numbers Crunch." *Science News*, June 22, 1996. Reviews the challenges for the next 50 years as humans grow in number.

Safina, C. "The World's Imperiled Fish." *Scientific American*, November 1995. The collapse of many fisheries clearly illustrates that wild fish populations cannot sustain themselves against modern fishing techniques.

Answers to Multiple-Choice Questions

1. a 2. c 3. d 4. e 5. b 6. d

MediaTutor

CD Activities

Activity 27.1: Population Growth and Regulation

Estimated time: 10 minutes

In this activity you will investigate wolf populations in Yellowstone National Forest to explore the concept of biotic potential, the relationship of predator–prey interactions as they apply to population biology, and the graphical portrayal of population growth over time.

Activity 27.2: Population Growth and Regulation in Human Populations

Estimated time: 5 minutes

In this activity you will run a series of tests to determine how the delayed onset of childbearing by humans affects human population growth.

Start the MediaTutor Student CD-ROM and enter the activity number in the Quick Search box to be taken directly to that activity.

Web Investigation

The Mystery of Easter Island

Estimated time: 15 minutes

Overpopulation on our planet is a major problem that faces all countries around the globe. Environmentalists warn that we have already met or exceeded the "carrying capacity" of our planet. Diminishing food supplies and natural resources pose a very real threat to society as we know it. This activity will explore the dynamics of population growth and what can go wrong when populations grow out of control.

Go to http://www.prenhall.com/audesirk3, the Audesirk Companion Web site. Select Chapter 27 and Web Investigation to begin.

A past brush with extinction revealed the sea otter's key role in coastal ecological communities. Today, a new threat to the sea otter is revealing still more about the threads that bind ecological communities together.

28 Community Interactions

AT A GLANCE

CASESTUDY

The Search for the Missing Sea Otters

In the nineteenth century, large numbers of sea otters lived in nearshore waters of the Pacific Ocean, over a range that extended from Baja California northward to Alaska and across the Bering Strait to the eastern coast of Russia. By the early 1900s, however, that range had been drastically reduced. Market demand for thick, luxurious otter fur gave rise to commercial hunting that drove sea otters nearly to extinction. Before sea otters disappeared completely, however, an international treaty was signed, protecting them from hunting, and their population began to rebound.

By the 1970s, sea otters were again abundant. The recovery, however, was uneven. Otter populations in some areas were large, but other areas held only small populations, and some locations had no sea otters at all. This uneven distribution presented an opportunity to ecologists, who were able to compare similar coastal habitats that held varying numbers of sea otters. What they found was rather unexpected. Areas with many sea otters also had extensive, lush kelp forests. (Long blades of kelp, anchored to the seafloor and floating upward toward the surface, play an ecological role similar to that of trees in a terrestrial forest, providing food, shelter, and protection to a diverse array of other organisms.) Areas without otters tended to lack kelp forests, and in areas with small otter populations, kelp forest coverage grew as the otter population grew. Clearly, otters and kelp forests were connected in some way, but what was the connection?

The connection between sea otters and kelp forests turned out to be sea urchins. These spiny, herbivorous animals are a favorite food of sea otters, so sea urchin populations were small where otters were common. But in areas with few or no otters, sea urchins were abundant and consumed kelp faster than it could grow. A healthy kelp forest was possible only where otters were present in sufficient numbers to limit the density of the sea urchin population.

The crucial role of sea otters in maintaining kelp forest ecosystems would likely have remained undiscovered had sea otters not been nearly extinguished. If sea otters had indeed become extinct, kelp forests (and the species diversity that they contain) might also have disappeared. The almost accidental discovery of the crucial ecological role of the sea otter serves as a warning that the effects of removing even a single species from a community are difficult to predict and can be profoundly disruptive. ■

1 Why Are Interactions in Ecological Communities Important?

An ecological **community** consists of all the interacting populations within an ecosystem. Populations in an ecological community interact in various ways, the most important of which are competition, predation, parasitism, and mutualism. The distinctions between these types of interactions are based on whether the interactions are harmful or beneficial to the participating organisms (Table 28-1).

Community Interactions Help Limit Population Size

The interacting web of life that forms a community tends to maintain a balance between resources and the numbers of individuals consuming them. For example,

Table 28-1 Interactions Within Communities

Type of Interaction	Effect on Organism A	Effect on Organism B
Competition between A and B	Harms	Harms
Predation by A on B	Benefits	Harms
Symbiosis		
Parasitism by A on B	Benefits	Harms
Commensalism of A with B	Benefits	No effect
Mutualism between A and B	Benefits	Benefits

as you learned in Chapter 27, community interactions such as predation and competition help limit the size of populations. The finely tuned interactions that maintain balance among populations developed as a result of prolonged coexistence among populations, but the balance can be fragile and susceptible to disruption. Such disruption is especially likely when organisms are introduced into an ecosystem where they did not evolve, as described in "Earth Watch: Exotic Invaders."

Community Interactions Influence Evolutionary Change

When members of different populations interact with one another, they may influence each other's ability to survive and reproduce. Community interactions thus serve as agents of natural selection. Predators, for example, tend to kill the members of prey populations that are easiest to catch, leaving behind the individuals with better defenses against predation. These individuals therefore tend to survive longer and leave the most offspring, so that over time the inheritable characteristics that help them avoid predation become more common in the prey population. At the same time, the increasingly effective defenses of prey favor the survival and reproduction of those members of the predator population that are best able to overcome the defenses. Thus, even as predator–prey interactions limit population size, they also shape the bodies and behaviors of the interacting populations. This process, in which interacting species influence one another's evolution by acting as mutual agents of natural selection, is called **coevolution**.

2 What Are the Effects of Competition Among Species?

During competition among members of different species, called **interspecific competition**, two or more species attempt to use the same limited resources. In interspecific competition, each species is harmed, because access to resources is reduced for both. The intensity of interspecific competition depends on how similar the requirements of the competing species are. Ecologists express this idea by saying that the degree of competition depends on the degree to which the *ecological niches* of the competing species overlap.

Each Species Has a Unique Place in Its Ecosystem

Each species occupies a unique **ecological niche** that encompasses all aspects of its way of life. A species' niche includes the particular type of habitat in which it lives, the environmental factors necessary for its survival, and the methods by which it acquires its supply of nutrients. Among the particular factors that may define a species' niche are: the range of temperatures under which it can survive, the amount of moisture it requires, and the types of nutrients needed to sustain it. For an animal species, we can add to that list items such as the types of food it eats, the time of day at which it forages for food, and the type of shelter it needs. A plant species' niche might include the range of soil pH it can tolerate, the amount of sunlight it needs, and the amount of soil nutrients it requires. A species' predators, prey, and competitors are also considered to be elements of its niche. In sum, a species' ecological niche is the combination of all factors that define how the species "makes a living" within its ecosystem.

The Ecological Niches of Coexisting Species Never Overlap Completely

Although different species may share some aspects of their niches, no two species ever occupy exactly the same ecological niche. Just as no two organisms can occupy exactly the same physical space at the same time, no two species can simultaneously inhabit the same niche. This important concept, often called the **competitive exclusion principle**, was formulated in 1934 by the Russian microbiologist G. F. Gause.

Gause demonstrated competitive exclusion in an experiment with two species of the protist *Paramecium*, *P. aurelia*, and *P. caudatum*. Grown separately in laboratory flasks and provided with bacteria to eat, each species flourished (Fig. 28-1a). But because both species used the same food source and sought food in the same part of the flask, Gause hypothesized that their niches were too similar for them to coexist. Sure enough, in trials in which Gause placed the two species together in a single flask, one species always died out (Fig. 28-1b). Gause then repeated the experiment, replacing *P. caudatum* with a different species, *P. bursaria*, which tended to feed in a different part of the flask. In this case, the two species of *Paramecium* were able to coexist indefinitely, because their niches did not overlap completely.

Competitive Exclusion Helps Determine How Populations Are Distributed

A classic study of competitive exclusion under natural conditions was performed by the ecologist J. Connell. Connell experimented with barnacles of the genera *Chthamalus* and *Balanus* (barnacles are crustaceans

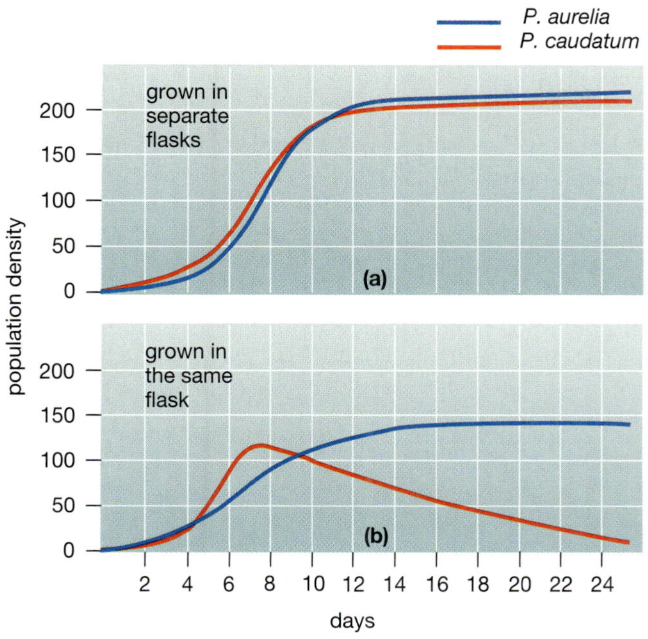

Figure 28-1 Competitive exclusion
(a) Raised separately with a constant food supply, both *Paramecium aurelia* and *P. caudatum* show the S-curve typical of a population that initially grows rapidly, then stabilizes. ***(b)*** When the two species are raised together and forced to occupy the same niche, *P. aurelia* consistently outcompetes *P. caudatum* and causes that population to die off. (Modified from G. F. Gause, *The Struggle for Existence*. Baltimore: Williams & Wilkins, 1934.)

that attach permanently to rocks and other surfaces). Both genera live in the *intertidal zone*, an area of the shore that is alternately covered and exposed by the tides. Connell found that *Chthamalus* dominates the upper shore and *Balanus* dominates the lower. When he scraped off *Balanus*, however, the *Chthamalus* population increased, spreading downward into the area its competitor had once inhabited. So *Chthamalus* is clearly able to thrive in the areas that are inhabited by *Balanus*. Connell concluded that *Balanus*, by virtue of growing faster, competitively excluded *Chthamalus* from the lower zone. *Chthamalus* survives in to the drier upper zone, where only the highest tides submerge the barnacles, because it is better than *Balanus* at tolerating the conditions there. As this example illustrates, interspecific competition can limit the distribution (and therefore the size) of populations.

Species Evolve to Reduce Niche Overlap

The ecologist Robert MacArthur documented the effects of competitive exclusion by observing five warbler species in the wild. All five species live in the same habitat, and all forage for insects and build their nests in the same spruce trees, so it appears at first glance that their niches overlap considerably. MacArthur, however, found that the different species search for food in different portions of the tree, employ different hunting tactics, and nest at slightly different times. The five species in effect divide the resource, thereby reducing niche overlap and reducing interspecific competition (Fig. 28-2).

Figure 28-2 Reducing niche overlap
Each of these five species of North American warblers searches for insects in a different region of spruce trees.

Earth Watch
Exotic Invaders

In 1989 residents of Monroe, Michigan, a town on the shores of Lake Erie, suddenly found themselves without water. Their schools, industries, and businesses were closed for two days while workers labored to fix the problem: Zebra mussels had clogged their water-treatment plant. Their problem was not unique; at one treatment plant on Lake Erie, zebra mussel populations reached 600,000 per square yard (Fig. E28-1a). Where did the mussels come from? Sometime in 1985 or 1986, a trading vessel bringing cargo from Europe discharged fresh water into Lake St. Clair, located between Lake Huron and Lake Erie at the border between Ontario and Michigan. The water, used for ballast during the ship's transatlantic voy-

Figure E28-1 Exotic species
(a) Workers blast jets of hot water at zebra mussels coating the interior of a water treatment plant. *(b)* The Japanese vine kudzu rapidly covers entire trees and even houses. *(c)* The beauty that became a beast—water hyacinths, originally from South America, today clog waterways in our southern states. *(d)* Importing the rosy wolf snail proved disastrous for Hawaii's native snails, one of which is being attacked in this photo.

(a) (b) (c) (d)

age, contained stowaways—millions of larvae of the zebra mussel. Although these mollusks are native to the Caspian and Black seas (large lakes in southeastern Europe), they found ideal conditions here. Now they inhabit all the Great Lakes and the entire Mississippi and Ohio River drainage systems and are spreading throughout eastern waterways. Their microscopic mussel larvae can be carried in currents for hundreds of miles. The inch-sized adults use sticky threads to attach themselves to nearly any underwater surface, including piers, pipes, machinery, underwater debris, boat hulls, and even sand and silt. Because they can survive out of water for days, mussels clinging to small boats may be transported to other lakes and rivers, where they quickly move in. An adult female can produce 50,000 eggs each year, and the mussel menace has proven unstoppable.

The zebra mussel is an **exotic species**, introduced into an ecosystem where it did not evolve. Such species sometimes encounter no predators or parasites in their new environment, and their population grows unchecked. As their population explodes, the invaders may seriously damage the ecosystem as they displace, outcompete, and prey on native species. For example, starlings and house sparrows, deliberately introduced to the eastern United States in the 1890s, have displaced native songbirds, with which they compete for nesting sites. Similarly, red fire ants from South America, accidentally introduced into Alabama on shiploads of lumber in the 1930s, have spread throughout the southern United States, with its hospitable warm, moist climate. Fire ants displace and may kill other insects, birds, and even small mammals. Their mounds can ruin farm fields, and their fiery stings and aggressive temperament can make backyards uninhabitable. Another insect invader, the gypsy moth, was introduced from Europe in 1866 and still poses a serious threat to fruit and forest trees in North America. Now a new invader, the Asian long-horned beetle, is devouring hardwood trees in New York City and Chicago. The beetle, which officials believe may pose the biggest threat to forests since the gypsy moth, arrived around 1996 in wooden pallets and boxes shipped from China.

Exotic plants also threaten natural communities. In the 1940s a Japanese vine called *kudzu* was planted extensively in the southern United States to control erosion. Today kudzu is a major pest, overgrowing and killing trees and underbrush and occasionally engulfing small houses (Fig. E28-1b). The water hyacinth, introduced from South America as an ornamental plant, now clogs about 2 million acres of southern lakes and waterways, slowing boat traffic and displacing natural vegetation (Fig. E28-1c). In the northwestern Mediterranean Sea, an alien algae (*Caulerpa*) recently escaped from a saltwater aquarium and now blankets the seafloor like a carpet of living Astroturf, smothering natural communities. In the summer of 2000, a large patch of the algae was spotted in a lagoon off the California coast near San Diego. The lagoon was quarantined and poisoned with herbicide, and divers and boaters were alerted to keep a lookout for more of the noxious seaweed. Will diligent efforts prevent the spread of *Caulerpa*, or will the coastal marine ecosystems of the United States also be smothered? Meanwhile, imported Old World climbing ferns, probably introduced by nurseries, are blanketing treetops throughout southern Florida and rapidly invading the Everglades, killing trees by depriving them of light.

Ecologists estimate that more than 6500 exotic species have established themselves in the United States. The cost of attempting to control destructive invaders, when combined with the cost of the damage they do, is estimated at more than $120 billion annually in the United States. By evading the checks and balances imposed by millennia of coevolution, exotic species are wreaking havoc on natural ecosystems throughout the world. Recently, wildlife officials have made cautious attempts to reestablish these checks and balances by importing predators or parasites to attack selected exotic species.

In Australia a dozen European rabbits introduced in the 1840s had exploded in number to 300 million by the mid-1990s, wiping out native vegetation and competing for food with native species such as kangaroos. In 1996, officials released a virus deadly to rabbits at hundreds of sites across the continent. This controversial experiment has worked well so far. The virus has not spread to other species, rabbit populations have been reduced by 95% in some areas, and native plants and animals are rebounding. This type of control is fraught with danger, however, because introducing more exotic predators or parasites into an ecosystem can have unpredicted and possibly disastrous consequences for native species. In 1958 a large predatory Florida snail, the rosy wolf snail (Fig. E28-1d), was imported into Hawaii to control another exotic, the giant African snail, which was a menace to the native vegetation. The rosy wolf snail did not do much damage to the giant African snails, but its population nonetheless grew quickly, as it preyed upon many native snail species, threatening them with extinction. Unable to kill the Florida invader without also killing native species, Hawaiian officials have resorted to enclosing tracts of land that are home to endangered native snails with corrugated aluminum barriers surrounded by a "moat" of salt crystals to deter the predators.

Despite the risks of importing predators and parasites to control invasive species, there often seems to be little alternative, since poisons intended for exotics will kill natives as well. Thus, biologists are now considering importing sea slugs to feed on *Caulerpa* in the Mediterranean. U.S. Department of Agriculture scientists plan to release a fly whose larvae feed on fire ants and are researching possible insects to import that will feed on the Old World climbing fern—without, it is hoped, attacking native plants.

(a) (b) (c)

Figure 28-3 Predation
(a) A cow grazes on prairie grass. *(b)* A long-eared bat uses echolocation to hunt a moth. *(c)* A red-tailed hawk prepares to eat a smaller bird.

As MacArthur discovered, when two species with similar requirements coexist, they typically occupy a smaller niche than either would if it were by itself. This phenomenon, called **resource partitioning**, is an outcome of the coevolution of species with extensive niche overlap. Because natural selection favors individuals with fewer competitors, competing species evolve physical and behavioral characteristics that reduce their competitive interactions. A dramatic example of resource partitioning is found among the Darwin's finches of the Galapagos Islands. The different finch species have evolved different bill sizes and shapes and different feeding behaviors that reduce competition among them (see Fig. 13-4).

3 What Are the Effects of Predator–Prey Interactions?

An organism that consumes other living organisms is a predator (Fig. 28-3). Many predators kill the organisms they eat, but other predators are parasites (organisms that feed on a host's body) or herbivores (animals that eat plants), and do not necessarily kill the prey on which they feed.

Predator and Prey Coevolve

To survive, predators must feed and prey must avoid becoming food. Therefore, predator and prey populations exert intense environmental pressure on one another. As prey become more difficult to catch, predators must become more adept at hunting, and vice versa. Coevolution has thus endowed the cheetah with speed and camouflage spots and its zebra prey with speed and camouflage stripes. It has produced the keen eyesight of the hawk, and the poisons and bright colors of the poison arrow frog and the coral snake.

Bats and moths provide an excellent example of how both physical structures and behaviors are molded by coevolution. Most bats are nighttime hunters that navigate and locate insect prey by echolocation. They emit pulses of extremely high-frequency sound and, by analyzing the returning echoes, create an "image" of their surroundings. In response to this highly effective prey-detection system, certain moth species (a favored prey of bats) have evolved ears that are particularly sensitive to the frequencies used by echolocating bats. When these moths hear a bat, they take evasive action, flying erratically or dropping to the ground. The bats may counter this defense by switching the frequency of their sound pulses away from the moth's sensitivity range. Some moth species have even evolved a way to interfere with the bats' echolocation by producing their own high-frequency clicks. In response, when hunting a clicking moth, a bat may turn off its own sound pulses temporarily and zero in on the moth by following the moth's clicks.

In the following sections, we examine some of the other evolutionary results of predator–prey interactions.

Camouflage Conceals Both Predators and Their Prey

Both predators and prey have evolved colors, patterns, and shapes that resemble their surroundings. Such disguises, called **camouflage**, render animals inconspicuous even when they are in plain sight (Fig. 28-4). Camouflaged prey animals may resemble specific objects such as leaves, twigs, seaweed, thorns, or even bird droppings (Fig. 28-5a–c). Such animals tend to remain motionless when they need to be concealed; a moving "bird dropping" would be quite conspicuous.

Predators may also be camouflaged (Fig. 28-6). For example, a spotted cheetah becomes inconspicuous in the grass as it watches for grazing mammals. The frog-

(a)

(b)

Figure 28-4 Camouflage by blending in
(a) The sand dab is a flattened, bottom-dwelling ocean fish with a mottled color that closely resembles the sand on which it rests. ***(b)*** This nightjar on its nest in Belize is barely visible among the surrounding leaf litter.

fish closely resembles the algae-covered rocks and algae on which it sits motionless, dangling a small lure from a threadlike appendage on its upper lip. Small fish notice only the lure and are engulfed as they approach it.

Bright Colors Often Warn of Danger

In contrast to camouflaged animals, some animal species have evolved bright, conspicuous **warning coloration**. Animals with warning coloration are usually distasteful, and many are poisonous (Fig. 28-7). Clearly, the defensive value of being poisonous is increased if potential predators can recognize dangerous prey *before* trying to eat them. Predators learn from their mistakes; a single unpleasant experience is often sufficient to teach a predator to avoid similarly colored prey in the future.

(a)

(b)

(c)

Figure 28-5 Camouflage by resembling specific objects
(a) A moth whose color and shape resemble a bird dropping sits motionless on a leaf. ***(b)*** The leafy sea dragon (an Australian fish) has evolved extensions of its body that mimic the algae in which it often hides. ***(c)*** Florida treehopper insects avoid detection by resembling thorns on a branch.

(a)

(b)

Figure 28-6 Camouflage assists predators
(a) As it waits for prey, a cheetah blends into grass in which it stands. ***(b)*** A frogfish waits in ambush, its camouflaged body matching the sponge-encrusted rock on which it rests. A lure that resembles a small fish dangles above the frogfish's mouth; the lure attracts small predators that the frogfish eats.

Figure 28-7 Warning coloration
The South American poison arrow frog advertises its unpleasant taste and poisonous skin with bright, contrasting color patterns.

Some Organisms Gain Protection Through Mimicry

In **mimicry**, one species evolves to resemble another. The evolution of mimicry stems from added protection from predators gained by mimics; this benefit can be gained in several different ways:

- Harmless animals may evolve to resemble poisonous ones, thereby gaining some protection from predators that avoid the poisonous species. For example, the harmless mountain king snake avoids predation because it mimics the brilliant warning coloration of the deadly coral snake (Fig. 28-8a,b).
- A distasteful species may evolve to resemble another distasteful species, such that each species benefits from predators' painful experience with the other. The shared color pattern leads to faster learning by predators—and less predation on all similarly colored species. For example, predators rapidly learn to avoid the conspicuous stripes on bees, hornets, and yellow jackets. Similarly, the poisonous and distasteful monarch butterfly has wing patterns strikingly similar to those of the equally distasteful viceroy butterfly (Fig. 28-8c,d).
- Prey species may evolve to mimic predators. A fascinating example of this type if mimicry is seen in snowberry flies, which are hunted by jumping spiders. When a fly detects an approaching spider, it spreads its wings, moving them back and forth in a jerky dance. A spider that sees this display is likely to flee from the harmless fly. Why? The markings on the fly's wings closely resemble the legs of another jumping spider. The jerky movements of the fly mimic those of a jumping spider when it drives another spider from its territory (Fig. 28-9). Natural selection has finely tuned both the behavior and the appearance of the fly to avoid predation by jumping spiders.
- Prey species may evolve **startle coloration**, patterns that disrupt predation by distracting the predator. For example, some animal species, including several insects and a few vertebrates, have evolved patterns of color that closely resemble the eyes of a much larger animal (Fig. 28-10). If a predator gets close, the prey suddenly flashes its eye-spots, startling the predator and allowing the prey to escape.

Some Animal Predators and Prey Engage in Chemical Warfare

Some predators and some prey have evolved toxic chemicals for attack and defense. Perhaps the best-known examples of such chemicals are the venoms that spiders and poisonous snakes use to paralyze prey and deter predators. In addition, certain mollusks, including squid, octopus, and some sea slugs, emit clouds of ink when attacked. These colorful chemical "smoke screens" confuse predators and mask the prey's escape. An especially dramatic example of chemical defense is seen in the bombardier beetle. In response to the bite of an ant,

(a) (b) (c) (d)

Figure 28-8 Warning mimicry
The warning coloration of *(a)* the poisonous coral snake is mimicked by *(b)* the harmless mountain king snake. Mutual mimicry of warning coloration by the distasteful monarch butterfly *(c)* and the equally distasteful viceroy *(d)* protects both species.

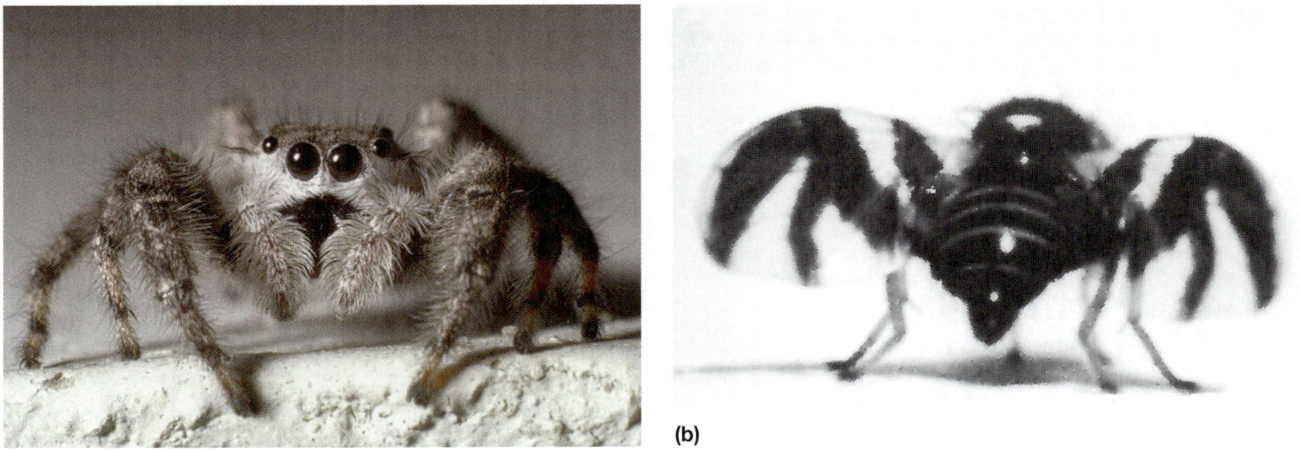

(a) (b)

Figure 28-9 Visual and behavioral mimicry
(a) In response to the approach of a jumping spider, *(b)* the snowberry fly spreads its wings, revealing a pattern that resembles spider legs. The fly enhances the effect by performing a jerky, side-to-side dance that resembles the leg-waving display of a jumping spider defending its territory.

(a) (b) (c)

Figure 28-10 Startle coloration
(a) When threatened, the false-eyed frog raises its rump, which resembles the eyes of a larger predator. ***(b)*** The peacock moth from Trinidad is well camouflaged, but should a predator approach, it opens its wings to reveal spots resembling large eyes. ***(c)*** Predators of this swallowtail butterfly caterpillar are deterred by its resemblance to a snake. Note that the caterpillar's head is the "snake's nose."

(a)

(b)

Figure 28-11 Chemical warfare
(a) The bombardier beetle sprays a hot, toxic brew in response to a leg pinch. ***(b)*** A monarch caterpillar feeds on milkweed that contains a powerful toxin.

the beetle releases secretions from special glands into an abdominal chamber. There, enzymes catalyze an explosive chemical reaction that shoots a toxic, boiling-hot spray onto the attacker (Fig. 28-11a).

Plants Have Defenses Against Herbivores

To an even greater extent than animals, plants have evolved chemical adaptations that deter their herbivorous predators. In particular, many plant species synthesize toxic and distasteful chemicals. Lupine plants, for example, produce chemicals called *alkaloids*, which deter attack by the blue butterfly, whose larvae feed on the lupine's buds. Plants may also defend themselves by producing compounds that toughen their edible parts. For example, many grasses have tough, glassy substances in their leaves, so grass blades are difficult for large herbivores to tear off and chew. Other plant species deter predation with physical structures such as thorns or thick bark.

Herbivores Have Adaptations for Overcoming Plant Defenses

Herbivorous predators have coevolved with the chemical defenses of plants. For example, the tough leaves of grasses have fostered the evolution of strong, grinding teeth and powerful jaws in grazing animals. Similarly, certain insect species have evolved increasingly efficient ways to detoxify or even use the toxic chemicals present in may plant tissues. In fact, nearly every toxic plant is eaten by at least one type of insect. For example, monarch butterflies lay their eggs on milkweed; when their larvae hatch, they consume the toxic plant (Fig. 28-11b). The caterpillars not only tolerate the milkweed poison but even store it in their tissues, where it serves as a defense against their own predators. The stored toxin is retained in the metamorphosed monarch butterfly.

4 What Is Symbiosis?

Symbiosis is an intimate, prolonged interaction between organisms of different species. In a symbiotic relationship, one species always benefits, but the second species may be unaffected, harmed, or helped by the relationship (see Table 28-1). A symbiotic relationship that benefits one species while neither helping nor harming the other is said to be *commensal*. For example, barnacles that attach themselves to the skin of a whale benefit by getting a free ride through nutrient-rich waters, and the whale is not affected. The sections below discuss symbiotic relationships in which one species suffers harm (parasitic interactions), and those in which both species benefit (mutualistic interactions).

Parasites Harm, But Do Not Immediately Kill, the Host

In parasitic symbiosis, one organism benefits by feeding on another. Parasites live in or on their prey, which are called *hosts*, normally harming or weakening them but not immediately killing them. Parasites are generally much smaller and more numerous than their hosts. Familiar parasites include tapeworms, fleas, and numerous disease-causing protists, bacteria, and viruses. There are a few parasitic vertebrates, such as the lamprey, which attaches to a host fish and sucks its blood.

Like other predator–prey interactions, parasite–host interactions are a powerful source of coevolutionary change. The effects of this mutual selection are evident in the precision and complexity of the human immune system (see Chapter 22) and in the adaptations that help disease-causing parasites overcome the immune response. Consider, for example, the protist that causes malaria. This parasite takes up residence inside human red blood cells. Ordinarily, red blood cells live for a few days before being swept into the spleen, where they are destroyed. Such destruction would also kill the malaria parasite, but the parasite resists this fate by producing a protein knob that extends from the surface of the infected red blood cell and firmly attaches the cell to the wall of a blood vessel. The immune system in turn recognizes the foreign protein on the blood cell and sends its forces to attack. But the malaria parasite has evolved yet another defense: it can produce thousands of different versions of the sticky protein that anchors it to the blood vessel. By the time the immune system recognizes one version, the parasite has already switched to another.

In Mutualistic Interactions, Both Species Benefit

Many symbiotic relationships are mutualistic, beneficial to both participants. For example, certain species of algae and fungi form mutualistic associations so intimate that the resulting structure appears to be a single organism: a lichen (Fig. 28-12). The fungus in a lichen provides support and protection for the photosynthetic alga, which in turn provides sugar molecules for the fungus. Many animals benefit from mutualistic associations with microorganisms that inhabit their digestive systems. Cows and termites, for example, are able to digest cellulose-containing plant tissues only with the help of protists and bacteria that live in their guts. In these associations, the animal provides a stable home and regular supply of food for the microorganisms, which return the favor by breaking cellulose down into smaller nutrient molecules that the animal can absorb. Another important example of mutualistic symbiosis is the interaction between legume plants and the nitrogen-fixing bacteria that inhabit special chambers on their roots. These bacteria obtain food and shelter from the plant and in return convert nitrogen to a form that the plant can use.

Figure 28-12 Mutualism
Each of these brightly colored lichens is a mutualistic association between an alga and a fungus.

The Big Picture

Each species in a community interacts with many other species, creating an intricate web of interlocking interdependencies among species. The complexity of communities makes them difficult to understand, so we are often surprised by unexpected consequences when communities are disrupted by human activities. We depend on healthy, functioning communities for our own survival and disrupt them at our peril.

5 What Are Keystone Species?

In some communities, a **keystone species** plays a major role in determining community structure—a role that is out of proportion to its abundance in the community. Removal of the keystone species dramatically alters the community. For example, the ecologist Robert Paine performed an experiment in which he removed the predatory starfish *Pisaster* from sections of the rocky intertidal coast of Washington (Fig. 28-13a). In areas lacking *Pisaster*, mussels (mollusks that are a favored prey of *Pisaster*) became so abundant that they outcompeted algae and other invertebrates that normally coexist in intertidal communities.

It is not necessarily easy to predict which of a community's species is a keystone species. For example, the keystone role played by sea otters in Pacific coastal waters of North America (see this chapter's Case Study) is played on the Atlantic coast by a completely different kind of animal: the lobster. Overfishing of lobsters off the east coast of Canada allowed sea urchins, a prey of lobsters, to increase in numbers. The population explosion of sea urchins nearly eliminated certain types of algae on which the urchins prey, leaving large expanses of bare rock where a diverse community once existed.

In the African savanna, the African elephant is a keystone predator. By grazing on small trees and bushes, elephants prevent the encroachment of forests and help maintain the grassland community (Fig. 28-13b).

MEDIATUTOR 28.1 Effects of Keystone Species Extinction

(a)

(b)

Figure 28-13 Keystone species
(a) The starfish *Pisaster* is a keystone species along the rocky coast of the Pacific Northwest. ***(b)*** The elephant is a keystone species on the African savanna.

6 How Does a Community Change over Time?

Mature communities do not spring fully formed from bare rock or naked soil but rather emerge in stages over a long period. This gradual process is called **succession** because different communities replace one another in a sequence that is somewhat predictable. For example, freshwater ponds and lakes tend to undergo a series of changes that, through time, transform them first into marshes and eventually to dry land. Coastal sand dunes tend to be stabilized by creeping plants and undergo changes that eventually lead to a forest. Different environments have different, characteristic pattern of succession.

We are able to observe succession because events frequently disrupt existing communities. Volcanic eruptions may create new islands ripe for colonization or leave behind a nutrient-rich environment that is rapidly invaded by new life (Fig. 28-14a,b). Forest fires, while destroying an existing community, also release nutrients and create conditions that favor rapid succession (Fig. 28-14c).

The precise changes during succession are as diverse as the environments in which succession occurs, but most succession follows the same overall pattern. It begins with a few hardy invaders called **pioneers** and, if

Figure 28-14 New life emerges after disruption
(a) Left: The Hawaiian volcano Mount Kilauea has erupted repeatedly since 1983, sending rivers of lava over the surrounding countryside. Right: A pioneer fern takes root on hardened lava. ***(b)*** Left: On May 18, 1980, the explosion of Mount St. Helens in Washington State devastated the pine forest ecosystem on its sides. Right: Twenty years later, life abounds on the once-barren landscape. ***(c)*** Left: In the summer of 1988, extensive fires swept through the forests of Yellowstone National Park in Wyoming. Right: Trees and flowering plants sprung up in the sunlight, and wildlife populations are rebounding as secondary succession occurs. (Photo taken in 1998.)

undisturbed, progresses to a diverse and relatively stable **climax community**, the end point of succession.

Our discussion of succession in the following sections focuses largely on plant species. Keep in mind, however, that each of the communities we describe also includes animals, fungi, and microorganisms in addition to the plants we mention.

There Are Two Main Types of Succession

Succession takes two major forms: primary and secondary. During **primary succession**, a community gradually colonizes bare rock, sand, or a clear glacial pool where there is no trace of a previous community. The generation of a community "from scratch" is a process that typically requires thousands or even tens of thousands of years. During **secondary succession**, a new community develops after an existing ecosystem is disturbed; the disturbance might be a forest fire or the abandonment of a farm field. Secondary succession happens more rapidly than does primary succession, because the previous community has left its mark in the form of soil and seeds.

Primary Succession Can Begin on Bare Rock

Figure 28-15 illustrates primary succession on Isle Royale, Michigan, an island in Lake Superior. Succession begins with bare rock, such as that exposed by a retreating glacier or cooled from molten lava. Over time, the rock *weathers* and releases nutrients such as minerals. In weathering, cracks form as the rock alternately freezes and thaws and as chemical reactions further break down the surface.

The weathered rock provides a place for lichens to settle; they have no competitors and plenty of sunlight. The lichens manufacture food by photosynthesis and obtain minerals by dissolving some of the rock with an acid they secrete. As the pioneering lichens spread over the rock, drought-resistant, sun-loving mosses begin growing in the cracks. Fortified by nutrients liberated by the lichens, the moss forms a dense mat that traps dust, tiny rock particles, and bits of organic debris. Any moss body that dies decomposes, and thus adds to a growing nutrient base. The moss mat acts like a sponge, trapping moisture. Within the mat, seeds of larger plants, such as bluebell and yarrow, germinate. Later, these plants die and their bodies contribute to a growing layer of soil. Eventually, woody shrubs such as blueberry and juniper take advantage of the newly formed soil; the moss and lichens are shaded out and buried by decaying leaves and vegetation. Ultimately, trees such as jack pine, blue spruce, and aspen take root in the deeper crevices, and the sun-loving shrubs are themselves shaded out. On the forest floor, shade-tolerant seedlings of taller or faster-growing trees, such as balsam fir, paper birch, and white spruce, thrive. In time they tower over and replace the original trees, which are intolerant of shade. After a thousand years or more, a tall climax forest thrives on what was once bare rock.

Abandoned Farmland May Undergo Secondary Succession

Figure 28-16 illustrates secondary succession on an abandoned southeastern farm. The pioneer species are fast-growing annual plants (species that live for a single growing season) such as crabgrass, ragweed, and sorrel, which sprout in the rich soil already present and thrive in direct sunlight. A few years later, longer-lived perennial plants such as asters, goldenrod, and broom sedge grass invade, along with woody shrubs such as blackberry. These plants multiply rapidly and dominate for the next few decades. Eventually, they are replaced by pines and fast-growing deciduous trees, such as tulip poplar and sweet gum, which sprout from windblown seeds. These trees become prominent after about 25 years, and a pine forest dominates the field for the rest of the first century. Meanwhile, shade-resistant, slow-growing hardwoods such as oak and hickory take root beneath the pines. After the first century, these begin to tower over and shade the pines, which eventually die from lack of sun. A relatively stable climax forest dominated by oak and hickory is present by the end of the second century.

Ponds and Lakes Eventually Fill In

Freshwater ponds or lakes undergo gradual succession to dry land (Fig. 28-17). This transition is influenced by both changes within the pond or lake and by sediments and nutrients carried in by runoff from the surrounding land. In forest habitats, lakes may undergo succession to produce meadows. The lake fills in from the edges, and grasses colonize the newly exposed soil. As the

Figure 28-15 Primary succession
Changes over a thousand years, beginning with bare rock, in upper Michigan.

MEDIATUTOR 28.2 Primary Succession

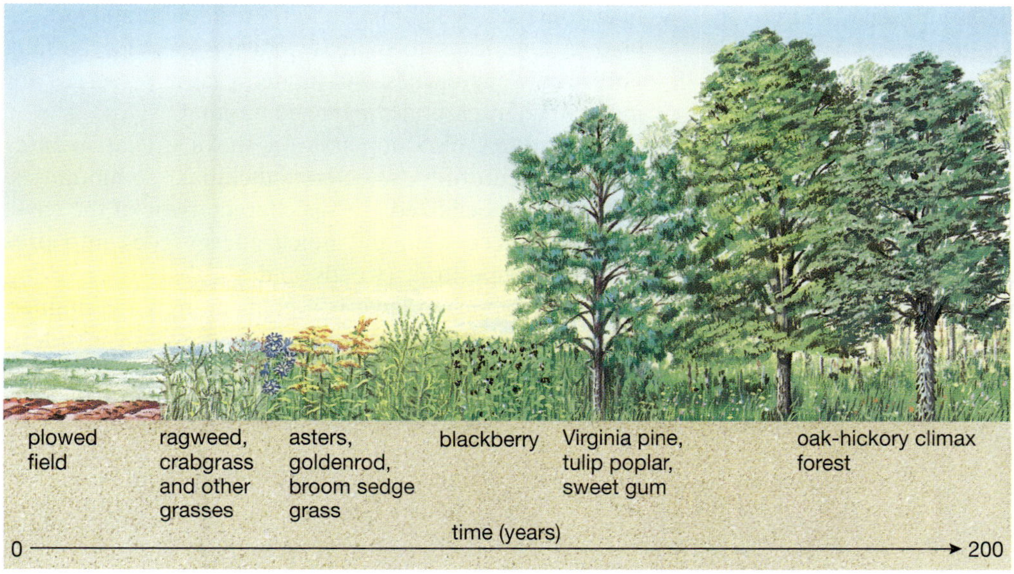

Figure 28-16 Secondary succession
Changes over two hundred years, beginning with the abandonment of a southeastern farm field.

lake shrinks and the meadow grows, trees encroach around the meadow's edges. If you return to a forest lake 20 years after your first visit, it probably will be a bit smaller.

Succession Culminates in the Climax Community

Succession ends with a relatively stable climax community, which perpetuates itself as long as it is not disturbed by external forces (such as fire, invasion of an introduced species, or human activities). The species in a climax community have ecological niches that allow them to coexist without replacing one another. In general, climax communities have more species and more community interactions than do early stages of succession. The plant species that dominate climax communities are generally longer-living and larger than pioneer species; this trend is particularly evident in ecosystems where forest is the climax community.

In your travels, you have undoubtedly noticed that the type of climax community varies dramatically from one area to the next. For example, if you drive through Colorado, you will see a shortgrass prairie climax community on the eastern plains (in those areas where it has not been replaced by farms), pine-spruce forests in the mountains, tundra on their uppermost reaches, and a sagebrush-dominated climax community in the western valleys. The exact nature of a climax community is determined by numerous geological and climatic variables, including temperature, rainfall, elevation, latitude, type of rock (which determines the type of nutrients available), exposure to sun and wind, and many more.

Human activities may dramatically alter the climax vegetation. Large stretches of grasslands in the west, for

Figure 28-17 Succession in a freshwater pond
(a) In this small pond, dissolved minerals carried by runoff from the surrounding land support aquatic plants, whose seeds or spores were carried in by winds or by birds and other animals. ***(b)*** Over time, the decaying bodies of aquatic plants build up soil that provides anchorage for more terrestrial plants. ***(c)*** Finally, the pond is entirely converted to dry land.

example, are now dominated by sagebrush due to overgrazing. The grass that usually outcompetes sagebrush is selectively eaten by cattle, allowing the sagebrush to prosper.

Some Ecosystems Are Maintained in a Subclimax State

Some ecosystems are not allowed to reach the climax stage but are maintained in a **subclimax** stage. The tallgrass prairie that once covered northern Missouri and Illinois is a subclimax of an ecosystem whose climax community is deciduous forest. The prairie was maintained by periodic fires, some set by lightning and others deliberately set by Native Americans to increase grazing land for buffalo. Forest now encroaches, and limited prairie preserves are maintained by carefully managed burning.

Agriculture also depends on the artificial maintenance of carefully selected subclimax communities. Grains are specialized grasses characteristic of the early stages of succession, and much energy goes into preventing competitors (weeds and shrubs) from taking over. The suburban lawn is a painstakingly maintained subclimax ecosystem. Mowing destroys woody invaders, and the selective herbicides that some homeowners use kill pioneers such as crabgrass and dandelions.

CASE STUDY REVISITED

The Search for the Missing Sea Otters

Around 1990, sea otter populations in Alaska unexpectedly began a rapid decline. By 1997, the number of sea otters in western Alaskan waters had dropped to about 10 percent of the number present in 1990. As you might expect, sea urchin abundance increased correspondingly, and the area covered by kelp forests shrunk dramatically.

What could have changed to decimate the previously thriving Alaskan otter population? Dr. James Estes and his colleagues believe that otters are disappearing because killer whales are eating them. Although killer whales and sea otters have shared these waters for hundreds, if not thousands, of years, the researchers believe that whales did not begin to eat otters until the early 1990s.

The researchers first saw a whale kill and eat an otter in 1991, and more attacks were observed in subsequent years. Whales had begun to prey on otters. But was this new behavior responsible for declining otter populations? To answer this question, the researchers performed a clever natural experiment. They placed radio-tracking devices on sea otters in two different locations: Clam Lagoon, a cove accessible only through a shallow passage that whales could not pass, and Kuluk Bay, a nearby open coastal area. After one year, the otter population in Kuluk Bay had declined by 65 percent. The population in Clam Lagoon, out of the reach of killer whales, remained stable. Whales, it seemed, were quite capable of reducing otter populations.

Why have killer whales, which have long coexisted with sea otters, suddenly begun to eat them? The researchers hypothesize that the whales have shifted their predation because of reductions in the populations of their more typical prey species, especially Steller sea lions and harbor seals. These marine mammals are much larger than sea otters and, unlike sea otters, possess a thick layer of blubber, which makes them more nutritious. Deprived of access to these preferred prey, the killer whales have turned to the less desirable, but perhaps more available, sea otters.

And why have the sea lions and seals declined? The prevailing view is that their decline is due to vastly reduced numbers of the fish on which they depend for food. These open-ocean fish stocks, in turn, have declined for reasons that probably include overfishing by commercial fleets.

The otter researchers have uncovered a chain of events in which human disruption of an oceanic community initiated a cascade of changes that ultimately resulted in deterioration of nearshore kelp forests. Different ecological communities may be linked by species that move between them, and disturbance of one community can have unpredictable consequences in unexpected places.

Summary of Key Concepts

1) Why Are Interactions in Ecological Communities Important?

Community interactions influence population size, and the interacting populations within communities act on one another as agents of natural selection. Thus, community interactions also shape the evolution of the interacting populations.

2) What Are the Effects of Competition Among Species?

A species' ecological niche includes all aspects of the species' habitat and interactions with its living and nonliving environments. Each species occupies a unique ecological niche. When the niches of two populations within a community overlap, there is interspecific competition between the populations. When two species are forced under laboratory conditions to occupy the same ecological niche, one species always outcompetes the other. Species within natural communities have evolved in ways that reduce niche overlap, with adaptations that foster resource partitioning.

3) What Are the Effects of Predator–Prey Interactions?

Predators eat other organisms. Predators and prey act as strong agents of selection on one another. Some prey animals have evolved protective colorations that render them inconspicuous (camouflage) or startling (startle coloration) to their predators. Some prey are poisonous and exhibit warning coloration by which they are readily recognized and avoided. Some prey species have color patterns that mimic those of other species. Some predators and some prey have evolved toxic chemicals for attack and defense. Many plants have evolved chemical and physical defenses against predation. These defenses, in turn, have selected for predators that can overcome them.

4) What Is Symbiosis?

Two species in a symbiotic relationship interact closely over an extended time. Symbiosis includes parasitism, in which the parasite feeds on a larger, less abundant host, usually harming it but not killing it immediately. In commensal symbiotic relationships, one species benefits, typically by finding food more easily in the presence of the other species, which is not affected by the association. Mutualism benefits both symbiotic species.

5) What Are Keystone Species?

Keystone species have a greater influence on community structure than can be predicted by their numbers. Removal of a keystone species radically alters community structure to an extent that would not be predicted on the basis of the species' abundance.

6) How Does a Community Change over Time?

Over time, succession changes the species that compose the community in a particular place. Primary succession, which may take thousands of years, begins where there is no remnant of a previous community (such as on rock scraped bare by a glacier or cooled from molten lava, a sand dune, or in a newly formed glacial lake). Secondary succession proceeds more rapidly, because it builds on the remains of a disrupted community in, for example, an abandoned field or burned forest. Secondary succession is initiated by fast-growing, readily dispersing pioneer plants, which are eventually replaced by longer-lived, generally larger species. Uninterrupted succession ends with a climax community, which tends to be self-perpetuating unless acted on by outside forces, such as fire or human activities. Some ecosystems, including tallgrass prairie and farm fields, are maintained in relatively early stages of succession by periodic disruptions.

Key Terms

camouflage *p. 566*
climax community *p. 573*
coevolution *p. 562*
community *p. 561*
competitive exclusion principle *p. 562*
ecological niche *p. 562*
exotic species *p. 565*
interspecific competition *p. 562*
keystone species *p. 572*
mimicry *p. 568*
pioneer *p. 572*
primary succession *p. 573*
resource partitioning *p. 566*
secondary succession *p. 574*
startle coloration *p. 568*
subclimax *p. 576*
succession *p. 572*
symbiosis *p. 571*
warning coloration *p. 567*

Thinking Through the Concepts

Multiple Choice

1. *Which of the following is usually NOT true of climax communities?*
a. They have more species than do pioneer communities.
b. They have longer-lived species than do early successional stages.
c. Climax communities are relatively stable.
d. Some climax communities are maintained by fire.
e. Climax communities vary with geographic location.

2. *What were the differences in the niches of the* Paramecium *in Gause's second experiment that allowed them to coexist?*
a. food eaten
b. body size
c. feeding area
d. preferred water temperature
e. preferred pH

3. *Which of the following is a mutualistic relationship?*
a. flowering plants and their pollinators
b. a caterpillar eating a tomato plant
c. bats and moths
d. monarch and viceroy butterflies
e. lupines and blue butterflies

4. *What of the following is a function of mimicry?*
a. to hide a prey from a predator
b. to warn a predator that a prey is likely to escape
c. to warn a predator that a prey is distasteful
d. to hide a predator from its prey
e. to startle a prey when it sees a predator

5. *What is coevolution?*
a. two species acting as agents of natural selection on each other
b. individuals of two species living together
c. the presence of two species in the same community
d. two species evolving separately
e. individuals of two species learning how to coexist

6. *By the competitive exclusion principle, coexisting species*
a. can use the same resources
b. cannot eat exactly the same things
c. cannot have identical ecological interactions
d. cannot be exactly the same size
e. cannot be closely related to one another

? Review Questions

1. Define ecological community, and list three important types of community interactions.
2. Describe four different ways in which specific plants or animals protect themselves against being eaten. In each, describe an adaptation that might evolve in predators of these species that would overcome their defenses.
3. List two important types of symbiosis; define and provide an example of each.
4. Which type of succession would occur on a clear-cut (a region in which all the trees have been removed by logging) in a national forest, and why?
5. List two subclimax and two climax communities. How do they differ?
6. Define succession, and explain why it occurs.

Applying the Concepts

1. Consider the interaction between the tarantula spider and the tarantula hawk wasp: This wasp attacks a tarantula, paralyzing it with venom from its stinger. The wasp then lays eggs on the paralyzed spider. The eggs hatch, and the young eat the living, immobilized tissues of the spider. Discuss whether this relationship between spider and wasp exemplifies parasitism or predation.
2. An ecologist visiting an island finds two very closely related species of birds, one of which has a slightly larger bill than the other. Interpret this finding with respect to the competitive exclusion principle and the ecological niche, and explain both concepts.
3. Think about the case of the camouflaged frogfish and its prey. As the frogfish sits camouflaged on the ocean floor, wiggling its lure, a small fish approaches the lure and is eaten, while a very large predatory fish fails to notice the frogfish. Describe all the possible types of community interactions and adaptations these organisms have been selected for. Remember that predators can also be prey and that community interactions are complex!
4. Design an experiment to determine whether the kangaroo is a keystone species in the Australian outback.
5. Why is it difficult to study succession? Suggest some ways you would approach this challenge for a few different ecosystems.

Group Activity

In groups of three, divide up the accompanying case studies. Each student should focus on one case study and present to the group an overview of the case and the answers to the questions. Then, as a group, discuss each case and agree on a response to it.

Case 1

A plateau in Arizona is covered mostly by evergreen forests and grasslands. The plateau is surrounded by rugged canyons, including the Grand Canyon, which form natural barriers to the movement of most animals. The animals on the plateau include deer, mountain lions, coyotes, bobcats, rabbits, ground squirrels, chipmunks, and wood rats. Mountain lions mainly eat deer. Coyotes and bobcats eat some deer but depend on rabbits and small rodents for most of their food. The rodents, rabbits, and deer eat grasses and low shrubs. The deer also eat the buds and tender branches of trees they can reach.

Before the arrival of Caucasians in the late 1800s, Native Americans lived on the plateau. They killed about 800 deer each year for food and hides. They seldom killed the deers' predators. During this time the deer population did not change greatly from year to year.

A study of the plateau animals was made in 1906. The deer population was estimated to be about 4000. About this time, the plateau was declared a game reserve, and the killing of deer was prohibited. A campaign was begun to kill the predators. As a result, about 10,000 predators were killed between 1906 and 1939. By 1924 there were about 100,000 deer. However, much of the plant life had been eaten or destroyed. Thousands of deer died of starvation or disease during the next few years. By 1930, the number of deer was down to about 30,000.

Suppose that you have been called in by government officials who need your advice on the management of the plateau's wildlife. How many deer would you suggest the plateau could reasonably support? What actions would you suggest to keep the deer population at the size you suggested? Give reasons for your answers to both questions.

Case 2

High mountain lakes seldom support many large fish. An ecologist decided to add more fish food so that there would be more and larger fish. One summer, she dumped fertilizer into the lake. The plants grew and supported more insects, which caused fish populations to increase. By summer's end, there were so many plants in the lake that from above it looked green rather than the blue it used to be.

The ecologist seemed to think that she had done the right thing, so she decided to fertilize more mountain lakes the next summer. However, when she flew over the lake country early the following summer, she could not find the green lake, because it had turned blue again. In fact, the plants, insects, and fish were scarcer than they had been before the lake was fertilized. Explain why the plants died and why there were fewer insects and fish than before.

Case 3

To save a forest that was being destroyed by insects, a governmental agency used sprays to kill them. Because the area was a nature preserve, the agency thought it should explain to the public why it used the sprays. A special video was made telling about the harmful effects of insects, the need for the use of sprays, and how carefully the sprays had been applied. Furthermore, the video narrator said that the sprays had killed only the insects and that no other wildlife had been affected.

Explain why spraying would affect animals other than insects even if the sprays did directly harm only the insects. Then explain how the sprays might also affect natural selection of the insects being sprayed.

For More Information

Amos, W. H. "Hawaii's Volcanic Cradle of Life." *National Geographic*, July 1990. A naturalist explores succession on lava flows.

Enserink, M. "Biological Invaders Sweep In;" Kaiser, J. "Stemming the Tide of Invading Species;" and Malkoff, D. "Fighting Fire with Fire." *Science*, September 17, 1999. A series of articles covering the problems posed by exotic species.

Power, M., et al. "Challenges in the Quest for Keystones." *Bioscience*, September 1996. A comprehensive review of the importance of keystone species and the challenges of studying them.

Answers to Multiple-Choice Questions

1. d 2. c 3. a 4. c 5. a 6. c

MediaTutor

CD Activities

Activity 28.1: Effects of Keystone Species Extinction

Estimated time: 10 minutes

In this simple simulation you will explore how removing a keystone species, in this case a fruit bat of the family Pteropidae, affects the community of which it is a member.

Activity 28.2: Primary Succession

Estimated time: 10 minutes

Rapid glacial retreat in the last 200 years at Glacier Bay, Alaska, has created a natural laboratory for the study of primary succession. In this tutorial you will explore how succession varies in different sections of the bay and will have an opportunity to view data that indicate how the pattern of invasion by tree species varies among sites.

Start the MediaTutor Student CD-ROM and enter the activity number in the Quick Search box to be taken directly to that activity.

Web Investigation

The Search for the Missing Sea Otters

Estimated time: 15 minutes

An ecological community consists of all the populations that interact and coexist within an ecosystem. Often times the relationships shared between the species within an ecological community are extremely delicate. This activity will explore some of the ways in which these delicate relationships can be disrupted and the effects that this can have on the ecosystem as a whole.

Go to http://www.prenhall.com/audesirk3, the Audesirk Companion Web site. Select Chapter 28 and Web Investigation to begin.

"A thing is right, when it tends to preserve the integrity, stability, and beauty of the biotic community. It is wrong when it tends otherwise."

Aldo Leopold in* A Sand County Almanac *(1949)

This tropical rain forest is not in Brazil, Indonesia, or the Congo. It is in the Arizona desert, inside the confines of Biosphere 2, an enclosed, artificial ecosystem.

29 How Do Ecosystems Work?

AT A GLANCE

CASESTUDY

A World in a Bubble

Like all organisms, humans depend on functioning ecosystems to provide food, clean air, drinkable water, and all the other necessities of life. Unlike other organisms, however, humans may one day wish to sustain life on the surface of Mars or the moon or on a permanent space station. Or, if some pessimistic predictions are correct, humans may someday need to build enclosures to protect against a hopelessly polluted atmosphere or the effects of prolonged global warming. Some futurists argue that, whether in space colonies or on our own planet's surface, humans eventually will need to build self-contained systems that can supply everything that is currently provided by natural ecosystems.

But aren't ecosystems so complex and varied that we can never hope to fully understand how they work, much less acquire the skill to reproduce one? Many observers hold this view, and some interpret efforts to create self-contained ecosystems as acts of technological arrogance, a waste of effort that would be better spent on finding ways to ensure that we don't undermine the ability of natural ecosystems to sustain life.

Regardless of where you stand in this debate, it's hard to avoid being intrigued by the Biosphere 2 project. Biosphere 2 is a 3-acre glass and steel enclosure in the Arizona desert, built in the late 1980s at a cost of more than $150 million by Texas multimillionaire Edward Bass. Inside are 3800 species in several artificial ecosystems, including a rain forest, a tropical grassland, a desert, and a 900,000-gallon ocean. Electronic sensors, pumps, and fans control temperature and humidity within the giant greenhouse. The whole system can be isolated from the surrounding environment (although large inputs of energy are required to keep it running).

Biosphere 2 was designed to be a self-contained, self-sustaining environment, and ecologists of all stripes watched with interest when, in 1991, eight intrepid people entered the enclosure. They intended to remain sealed inside for 2 years, with nothing but energy coming from the outside. Biosphere 2's internal environment would, they hoped, supply their needs for food, air, and water. By demonstrating that an artificial ecosystem that includes humans could sustain itself, the creators of Biosphere 2 hoped to show that our knowledge of ecological science is sufficient to prepare us for whatever crises and opportunities the future might hold. ■

1 How Do Ecosystems Obtain Energy and Nutrients?

The activities of life, from the flight of the eagle to active transport of molecules through a cell membrane, are powered by the energy of sunlight. Solar energy continuously bombards Earth, is captured in ecosystems, transformed in the chemical reactions that power life, and ultimately converted to heat energy that radiates back into space. Energy thus moves through ecosystems in a continuous one-way flow and is constantly replenished from an outside source, the sun.

In contrast to energy, nutrients are not replenished. **Nutrients** are elements and small molecules that form

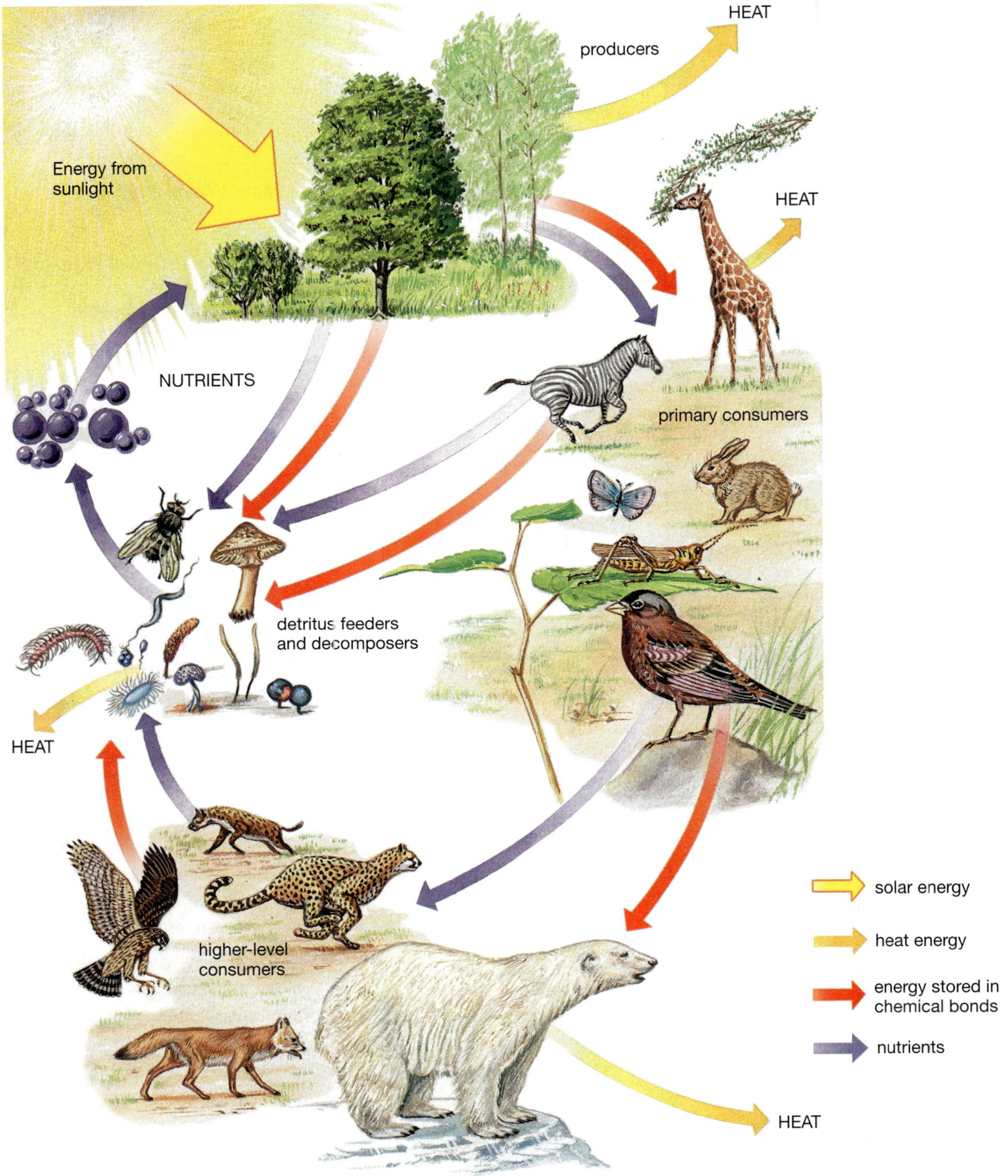

Figure 29-1 Energy flow, nutrient cycling, and feeding relationships in ecosystems
Note that nutrients (purple) neither enter nor leave the cycle. Energy (yellow), continuously supplied to producers as sunlight, is captured in chemical bonds and transferred through various levels of organisms (red). At each level, some energy is lost as heat (orange).

MEDIATUTOR 29.1 Building a Food Web

the chemical building blocks of life. Although these chemicals may be transported, redistributed, or converted to different molecular forms, they do not leave Earth, nor is their supply increased (except for small amounts in meteorites that reach Earth). Instead, nutrients are constantly recycled within and among ecosystems.

Ecological study at the level of the ecosystem focuses in large measure on the flow of energy and nutrients. Ecosystem ecologists follow the pathways of energy and nutrients in order to understand the factors that shape the complex interactions within communities and between communities and their nonliving environment. Figure 29-1 provides a preview of our study of ecosystems.

2 How Does Energy Flow Through Ecosystems?

Millions of miles from Earth, hydrogen molecules in the sun fuse to form helium molecules, releasing tremendous amounts of energy. A tiny fraction of this energy reaches Earth, arriving in the form of electromagnetic waves, including heat, light, and ultraviolet energy. Of the energy that reaches Earth, much is absorbed or reflected back into space by Earth's atmosphere, clouds, and surface. Overall, only about 1% of the solar energy that reaches Earth makes it to the surface as light that is available to power life. Of this 1%, green plants, algae, and other photosynthetic organisms capture 3% or less. Thus, the teeming life on this planet is supported by less than 0.03% of the energy that reaches Earth from the sun.

Energy Enters Ecosystems Through Photosynthesis

During photosynthesis (described in detail in Chapter 6), solar energy powers reactions that store energy in the chemical bonds of sugar and other high-energy molecules (Fig. 29-2). Photosynthetic organisms, from mighty oaks to the zucchini plants in your garden to single-celled diatoms in the ocean, are called **autotrophs** or **producers**; they produce food for themselves using nonliving nutrients and sunlight. In doing so, they directly or indirectly produce food for nearly all other organisms as well. Organisms that cannot photosynthesize, called **heterotrophs** or **consumers**, must acquire energy and many of their nutrients from the bodies of other organisms.

Energy Captured by Producers Is Available to the Ecosystem

The amount of life that a particular ecosystem can support is determined by the amount of energy captured by the producers in that ecosystem. The energy that photosynthetic organisms store and make available to other organisms is called **net primary productivity**. Net primary productivity is measured as the energy stored or biomass (dry weight) added by producers per unit area over a given time span.

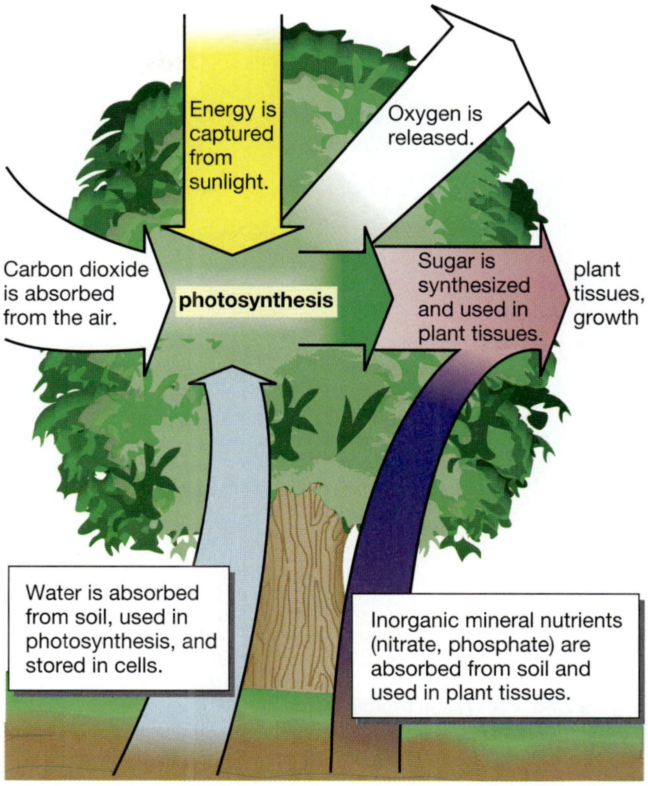

Figure 29-2 Photosynthesis
Photosynthetic organisms capture solar energy. They also remove inorganic nutrients from their nonliving reservoirs and incorporate these nutrients into living tissue. They ultimately provide all the energy and most of the nutrients for other organisms.

The net primary productivity of an ecosystem is influenced by many environmental variables, including the amount of nutrients available to the producers, the amount of sunlight reaching them, the availability of water, and the temperature. In the desert, for example, lack of water limits productivity; in the open ocean, light is limited in deep water and nutrients are limited in surface water. In ecosystems in which resources are abundant, productivity is high—for example, in tropical rain forests and estuaries (where rivers meet the ocean, carrying nutrients washed from the land). Some average productivities of ecosystems are shown in Figure 29-3.

Energy Passes from One Trophic Level to Another

Energy flows through communities from photosynthetic producers through several levels of consumers. Each category of organism is called a **trophic level**. The producers are the first trophic level, obtaining their energy directly from sunlight (see Fig. 29-1). Consumers occupy several

Figure 29-3 Ecosystem productivities compared
The numbers give the average net primary productivity, in grams of organic material per square meter per year. Notice the enormous differences in productivity among the ecosystems.

trophic levels. Some consumers feed directly on producers, the most abundant living energy source in any ecosystem. These **primary consumers**, or **herbivores**, ranging from grasshoppers to giraffes, form the second trophic level. The third trophic level consists of **carnivores** (meat eaters), predators such as the spider, eagle, and wolf that feed primarily on primary consumers. Carnivores are also called **secondary consumers**. Some carnivores occasionally eat other carnivores. When doing so, they occupy the fourth trophic level, **tertiary consumers**.

Feeding Relationships Within Ecosystems Form Chains and Webs

To illustrate who feeds on whom in an ecosystem, it is common to identify a representative of each trophic level such that each representative species eats another on the level below it. This linear feeding relationship is called a **food chain**. As illustrated in Figure 29-4, different ecosystems have radically different food chains.

Natural communities rarely contain well-defined groups of primary, secondary, and tertiary consumers, however. Instead, the actual feeding relationships form a **food web** of many interconnecting food chains (Fig. 29-5). Some animals, such as raccoons, bears, rats, and humans, are **omnivores**—that is, at different times they act as primary, secondary, and tertiary consumers. Many carnivores eat both herbivores and other carnivores, thus acting as both secondary and tertiary consumers. An owl, for instance, is a secondary consumer when it eats a mouse, which feeds on plants, but a tertiary consumer when it eats a shrew, which feeds on insects. When the shrew eats carnivorous insects, it is a tertiary consumer, and the owl that feeds on it is a quaternary (fourth-level) consumer. A carnivorous plant, such as a Venus flytrap, that digests a spider is simultaneously a producer and a secondary consumer!

Detritus Feeders and Decomposers Release Nutrients for Reuse

Among the most important strands in a food web are the detritus feeders and decomposers. The **detritus feeders** are an army of mostly small and often unnoticed animals and protists that live on the refuse of life: molted exoskeletons, fallen leaves, wastes, and dead bodies. (Detritus means "debris.") The network of detritus feeders is complex; in terrestrial ecosystems it includes earthworms, mites, protists, centipedes, some insects, land-dwelling crustaceans called a pillbugs, nematode worms, and even a few large vertebrates such as vultures. These organisms consume dead organic matter, extract some of the energy stored in it, and excrete it in a further decomposed state. Their excretory products serve as food for other detritus feeders and for decomposers.

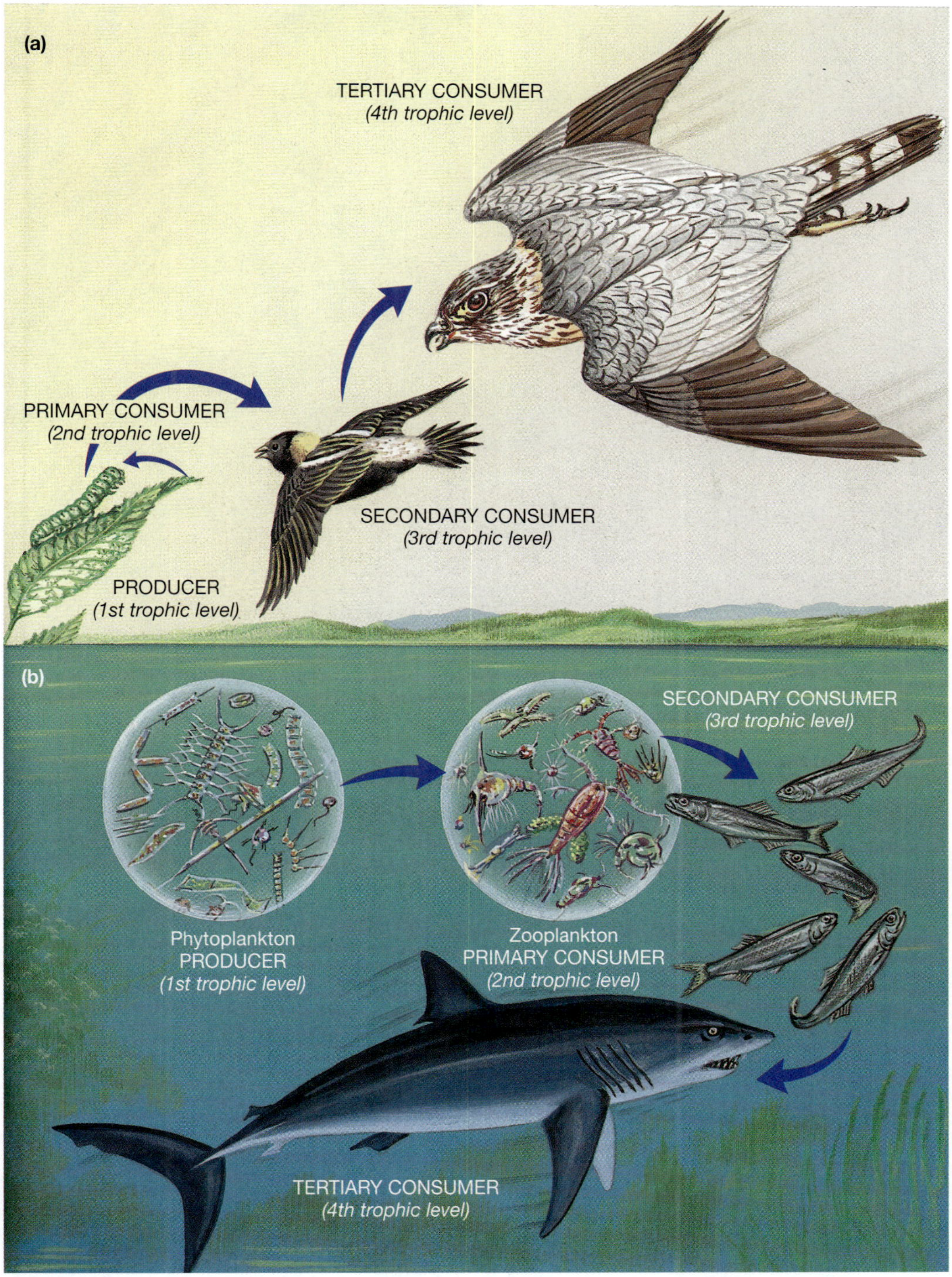

Figure 29-4 Food chains
(a) A simple terrestrial food chain. ***(b)*** A simple marine food chain.

The **decomposers** are primarily fungi and bacteria (the black coating or gray fuzz you may notice on decaying tomatoes and bread crusts are fungal decomposers). Decomposers digest food outside their bodies by secreting digestive enzymes into the environment. They absorb the nutrients they need, and the remaining nutrients are released to the environment.

Detritus feeders and decomposers are absolutely essential to life on Earth. Through their activities, the bodies and wastes of other organisms are reduced to simple molecules, such as carbon dioxide, water, minerals, and organic molecules, that return to the atmosphere, soil, and water. By liberating nutrients for reuse, detritus feeders and decomposers perform a

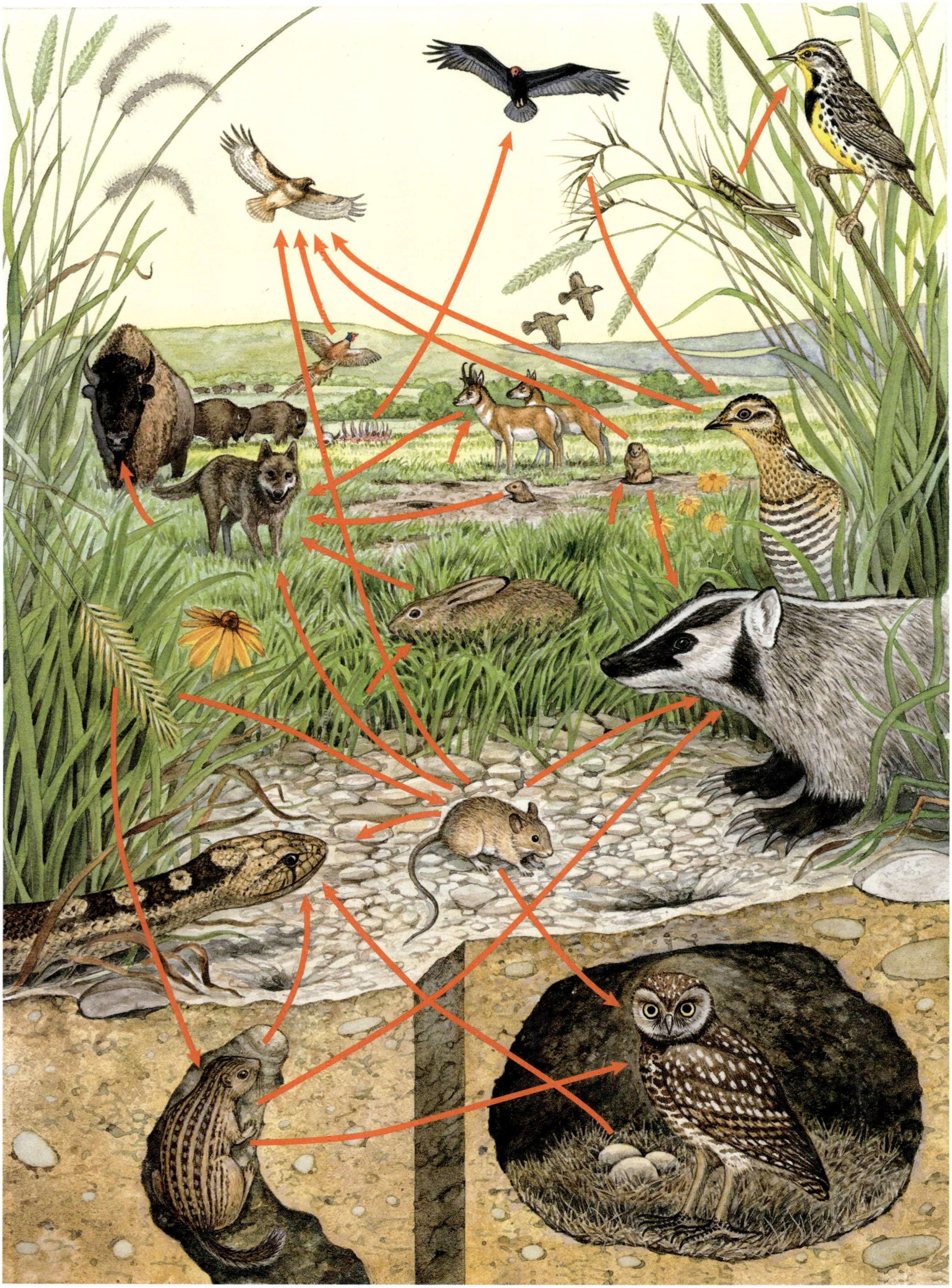

Figure 29-5 A food web in a short-grass prairie

MEDIATUTOR
29.1 Building a Food Web

vital role in recycling nutrients. If detritus feeders and decomposers were to disappear, ecosystems would gradually be smothered by accumulated wastes and dead bodies, and the nutrients stored in these bodies would remain "locked-up" and unavailable for use by other organisms.

Energy Transfer Through Trophic Levels Is Inefficient

As we discussed in Chapter 5, a basic law of thermodynamics is that energy use is never completely efficient. For example, as your car burns gasoline, about 75% of the energy released is immediately lost as heat and cannot be used to do the work of moving the car. The same inefficiency also applies in living systems. For example, splitting the chemical bonds of ATP to cause muscular contraction produces heat as a by-product; this is why walking briskly on a cold day warms you. Waste heat is produced by all the biochemical reactions that keep cells alive.

Energy transfer from one trophic level to the next is also inefficient. When a caterpillar (a primary consumer) eats the leaves of a tomato plant (a producer), only a portion of the solar energy originally trapped by the plant is available to the insect. Some of the original energy was used by the plant for growth and maintenance, and more was lost as heat during these processes. Some energy was converted into the chemical bonds of molecules such as cellulose, which the caterpillar cannot digest. Therefore, only a fraction of the energy captured by the first trophic level is available to organisms in the second trophic level. Some of the energy consumed by the caterpillar is in turn used to power crawling and chewing. Some is used to construct the indigestible exoskeleton, and much is given off as heat. All this energy is unavailable to the songbird in the third trophic level when it eats the caterpillar. The bird loses energy as body heat, uses some to fly, and converts a considerable amount into indigestible feathers, beak, and bone. All this energy is unavailable to the hawk that catches the bird. A simplified model of energy flow through the trophic levels in a deciduous forest ecosystem is illustrated in Figure 29-6.

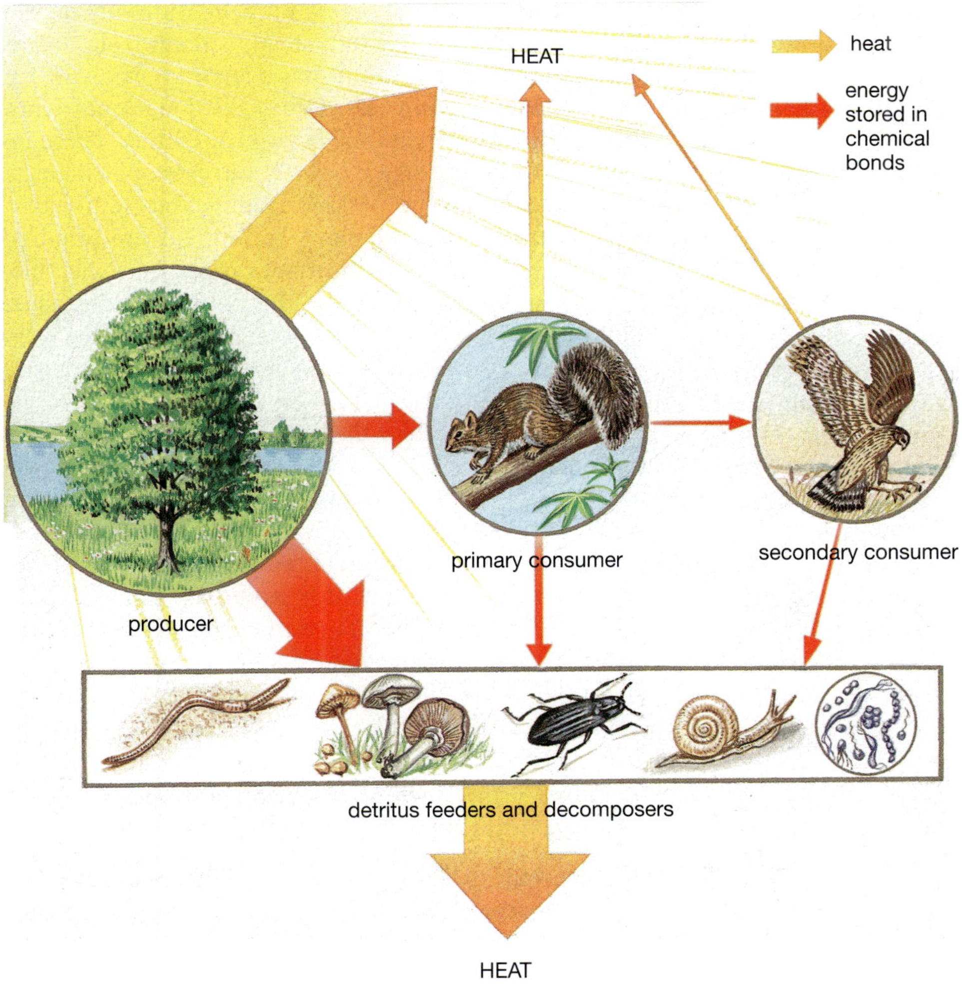

Figure 29-6 Energy transfer and loss in a forest
The width of the arrows is roughly proportional to the quantity of energy transferred (red) or lost (orange).

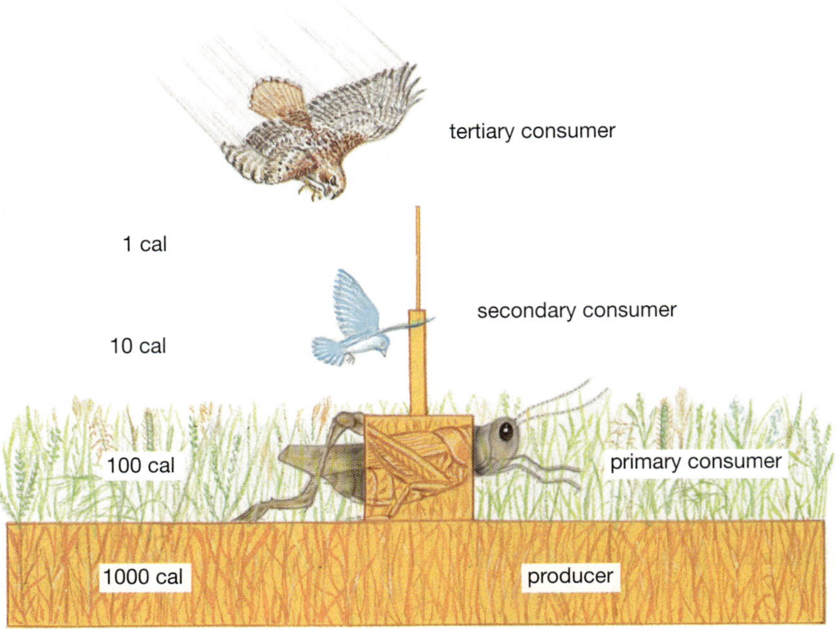

Figure 29-7 An energy pyramid for a prairie The width of each rectangle is proportional to the energy stored at that trophic level.

Energy Pyramids Illustrate Energy Transfer Between Trophic Levels

Studies of ecosystems have shown that, in general, about 90% of the energy stored at a trophic level is lost in the transfer to the next level. Thus, the energy stored by an ecosystem's primary consumers (herbivores) is only about 10% of the energy stored in the bodies of producers. In turn, the bodies of secondary consumers possess roughly 10% of the energy stored in primary consumers. In other words, for every 100 calories of solar energy captured by grass, only about 10 calories are converted into herbivores, and only 1 calorie into carnivores. This inefficient energy transfer between trophic levels is called the "10% law." An **energy pyramid**, which shows maximum energy at the base and steadily diminishing amounts at higher levels, illustrates the energy relationships between trophic levels (Fig. 29-7). Energy pyramids also show why primary producers are the most abundant organisms in an ecosystem, and carnivores the rarest.

An unfortunate side effect of the inefficiency of energy transfer is that certain persistent toxic chemicals produced by humans become concentrated in the bodies of carnivores, as described in "Earth Watch: Food Chains Magnify Toxic Substances."

The Big Picture

Ecosystems capture only a tiny fraction of available solar energy. One consequence of this inefficiency is that humans rely on energy stored long ago in fossil fuels to support an exploding population. This supply of stored energy will eventually be exhausted, and methods to capture a greater share of the sun's energy will become increasingly important.

3 How Do Nutrients Move Within and Among Ecosystems?

In contrast to the energy of sunlight, nutrients do not flow down onto Earth in a steady stream from above. The same pool of nutrients has been supporting life for more than 3 billion years. Some, called **macronutrients**, are required by organisms in large quantities. These include carbon, hydrogen, oxygen, nitrogen, phosphorus, sulfur, and calcium. **Micronutrients**, including zinc, molybdenum, iron, selenium, and iodine, are required only in trace quantities. **Nutrient cycles**, also called **biogeochemical cycles**, describe the pathways substances follow as they move through communities to the nonliving portions of ecosystems and then back again to communities.

As nutrients move through biogeochemical cycles, they may accumulate in one portion of the cycle and remain there for long periods; these storage sites are called **reservoirs**. Most major reservoirs are in the nonliving part of an ecosystem. For example, carbon has several major reservoirs: It is stored as carbon dioxide gas in the atmosphere, in dissolved form in oceans, and as fossil fuels underground.

In the following sections, we briefly describe the cycles of carbon, nitrogen, phosphorus, and water.

Carbon Cycles Through the Atmosphere, Oceans, and Communities

Carbon enters the living part of the ecosystem when producers capture carbon dioxide (CO_2) during photosynthesis, and incorporate its carbon atoms in organic molecules. On land, producers acquire CO_2 from the atmosphere, where it represents 0.036% of the total

Earth Watch
Food Chains Magnify Toxic Substances

In the 1940s, the new insecticide DDT seemed close to miraculous. In the Tropics, DDT saved millions of lives by killing the mosquitoes that spread malaria. DDT also destroyed agricultural insect pests, increasing crop yields and saving millions from starvation. What's more, DDT is long-lasting, so a single application keeps on killing. DDT's discoverer, the Swiss chemist Paul Müller, was awarded the 1948 Nobel Prize for Medicine and Physiology, and people looked forward to a new age of freedom from insect pests. Little did they realize that indiscriminate use of this pesticide was unraveling the complex web of life.

For example, in the mid-1950s the World Health Organization sprayed DDT on the island of Borneo to control malaria. One unintended casualty was a predatory wasp species that fed on caterpillars. The wasp population was wiped out, but its main prey, a caterpillar that fed on the thatched roofs of houses, was relatively unaffected. Within a short period, thatched roofs were collapsing, eaten by the uncontrolled population of caterpillars. At the same time, gecko lizards that ate poisoned insects accumulated high concentrations of DDT in their bodies. Both the geckos and the village cats that ate them died of DDT poisoning. With the cats eliminated, the rat population exploded. The human population was then threatened with an outbreak of plague, carried by the uncontrolled rats. The outbreak was avoided only by airlifting new cats to the villages.

During the 1950s and 1960s in the United States, wildlife biologists witnessed an alarming decline in populations of several predatory birds, especially fish-eaters, such as bald eagles, cormorants, ospreys, and brown pelicans. These top predators are never abundant, and the decline pushed some, including the brown pelican and the bald eagle, close to extinction. The cause of the declining bird populations was eventually traced to DDT, which had been sprayed on the aquatic ecosystems supporting these birds. Scientists were amazed to find concentrations of DDT in the bodies of predatory birds that were up to *one million* times the concentration present in the water. This led to the discovery of **biological magnification**, the process by which toxic substances accumulate at increasingly higher concentrations in progressively higher trophic levels. Because the transfer of energy from lower to higher trophic levels is extremely inefficient, a pesticide that has been sprayed on plants is consumed in large amounts by herbivores, which must eat large quantities of plant material to survive. Carnivores, in turn, must eat many herbivores, along with the pesticide contained in the bodies.

Man-made substances that undergo biological magnification have two properties that make them dangerous: they are not **biodegradable**—that is, they cannot readily be broken down into harmless substances by decomposer organisms; and they are fat soluble, but not water soluble. Thus, they persist for a long time in the environment, and they accumulate in the bodies of animals, particularly in fat, rather than being broken down and excreted in the watery urine. Because these substances remain in the body, a predator accumulates the poison from its prey over many years.

Prior to the advent of pollution control measures, the Great Lakes built up relatively high levels of bioaccumulating compounds, including DDT and PCBs (compounds used in electrical transformers and cables), which continue to affect the health of local wildlife. In the Great Lakes region, populations of fish-eating river otters have declined sharply, and fish-eating birds produce some deformed offspring and eggs that never hatch (Fig. E29-1).

Figure E29-1 The price of pollution
Deformities such as the twisted beak of this double-crested cormorant from Lake Michigan have been linked to bioaccumulating chemicals.

The movement of pollutants through food webs presents a hazard to human health as well as to wildlife. Many of the children born to women who consumed PCB-contaminated fish from Lake Michigan exhibited slowed intellectual development. Exposure to high levels of pesticides and other persistent pollutants has been linked to certain forms of cancer, infertility, heart disease, and suppressed immune function. Our vulnerability is increased because we often "feed high on the food chain." When we eat tuna or swordfish, for example, we act as tertiary or even quaternary consumers, and are especially vulnerable to bioaccumulating substances. In addition, our long life span provides ample time for substances stored in our bodies to accumulate to toxic levels.

Today, DDT, PCBs, and many other bioaccumulating pollutants are banned in the United States, but they are still used in developing countries. Even in the United States, the banned chemicals persist in ecosystems from past usage. In any case, many other bioaccumulating chemicals remain in widespread use, and are released to the environment by plastics, detergents, pesticides, and as by-products of manufacturing processes. Once released into the environment, man-made chemicals are transported worldwide by wind, water, and commerce. As a result, even Inuit people living north of the Arctic Circle have high levels of PCBs and chlordane (a banned pesticide similar to DDT) in their bodies from consuming the blubber (fat) of whales and other marine predators.

gases. For producers in lakes, rivers, and oceans, there is abundant CO_2 dissolved in the water. Some of the carbon taken up by producers is returned as CO_2 to the atmosphere and ocean during cellular respiration; the rest is incorporated into the producers' bodies. Primary consumers, such as cows, shrimp, and grasshoppers, eat the producers and acquire the carbon stored in their tissues. These herbivores also release some carbon through respiration and store the rest, which may then be consumed by organisms in higher trophic levels. Both producers and consumers eventually die, and their bodies are broken down by detritus feeders and decomposers. Cellular respiration by these organisms returns CO_2 to the reservoirs in the atmosphere and oceans (Fig. 29-8).

Some carbon cycles much more slowly. For example, mollusks and marine protists extract CO_2 dissolved in water and combine it with calcium to form calcium carbonate ($CaCO_3$), from which they construct their shells. After death, the shells of these organisms collect in undersea deposits, are buried, and may eventually be converted to limestone. Geological events may expose the limestone, which dissolves gradually as water runs over it, making the carbon available to living organisms once again.

Another slow-moving part of the carbon cycle incorporates carbon into fossil fuels, a major reservoir for carbon. **Fossil fuels** form from the buried remains of plants and animals. Over millions of years, the carbon in the organic molecules of these organisms is transformed by high temperatures and pressures into coal, oil, or natural gas. When people burn fossil fuels to use the energy they contain, CO_2 is released into the atmosphere. By freeing carbon from the fossil fuel

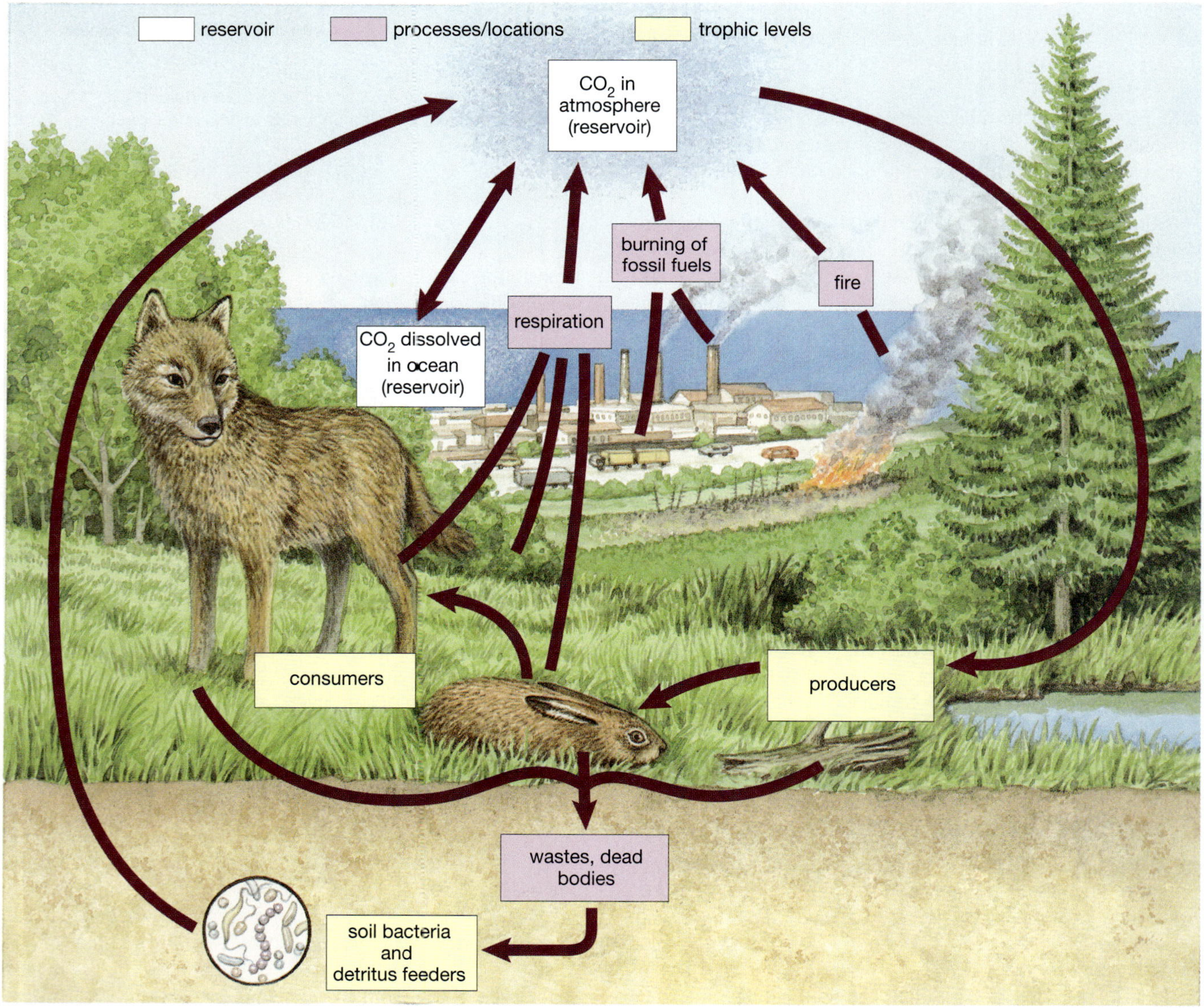

Figure 29-8 The carbon cycle

reservoir, humans are increasing the amount of CO_2 in the atmosphere, as we shall describe later in this chapter.

The Major Reservoir for Nitrogen Is the Atmosphere

The atmosphere contains about 79% nitrogen gas (N_2) and is thus the major reservoir for this important nutrient. Neither plants nor animals, however, can extract nitrogen from the atmosphere. Instead, nitrogen enters the food web mainly through certain bacteria in soil and water that engage in **nitrogen fixation**, a process in which nitrogen and hydrogen are combined to form ammonia (NH_3). The nitrogen-fixing bacteria do not use all the ammonia they produce, and the excess is excreted into the surrounding soil or water. Much of this ammonia is in turn converted by other bacteria to nitrate (NO_3^-); nitrate also forms when atmospheric nitrogen and oxygen react during electrical storms and when fossil fuels are burned.

Plants (as well as some bacteria and fungi) can absorb ammonia and nitrate, and incorporate the nitrogen from these molecules into amino acids, proteins, nucleic acids, and vitamins. These nitrogen-containing molecules in plants are eventually consumed, either by primary consumers, detritus feeders, or decomposers. As nitrogen passes through the food web, some is returned to soil or water when decomposer bacteria convert wastes and dead bodies back to nitrate and ammonia. Atmospheric nitrogen is replenished by **denitrifying bacteria**, which break down nitrate, releasing nitrogen gas back to the atmosphere (Fig. 29-9).

In human-dominated ecosystems, such as farm fields, gardens, and suburban lawns, ammonia and nitrate are supplied by chemical fertilizers. These fertilizers are produced by using the energy in fossil fuels to artificially "fix" atmospheric nitrogen. In fact, human activities now dominate the nitrogen cycle, a cause for concern. By burning fossil fuels, by burning forests and draining wetlands that store nitrogen, by cultivating legumes

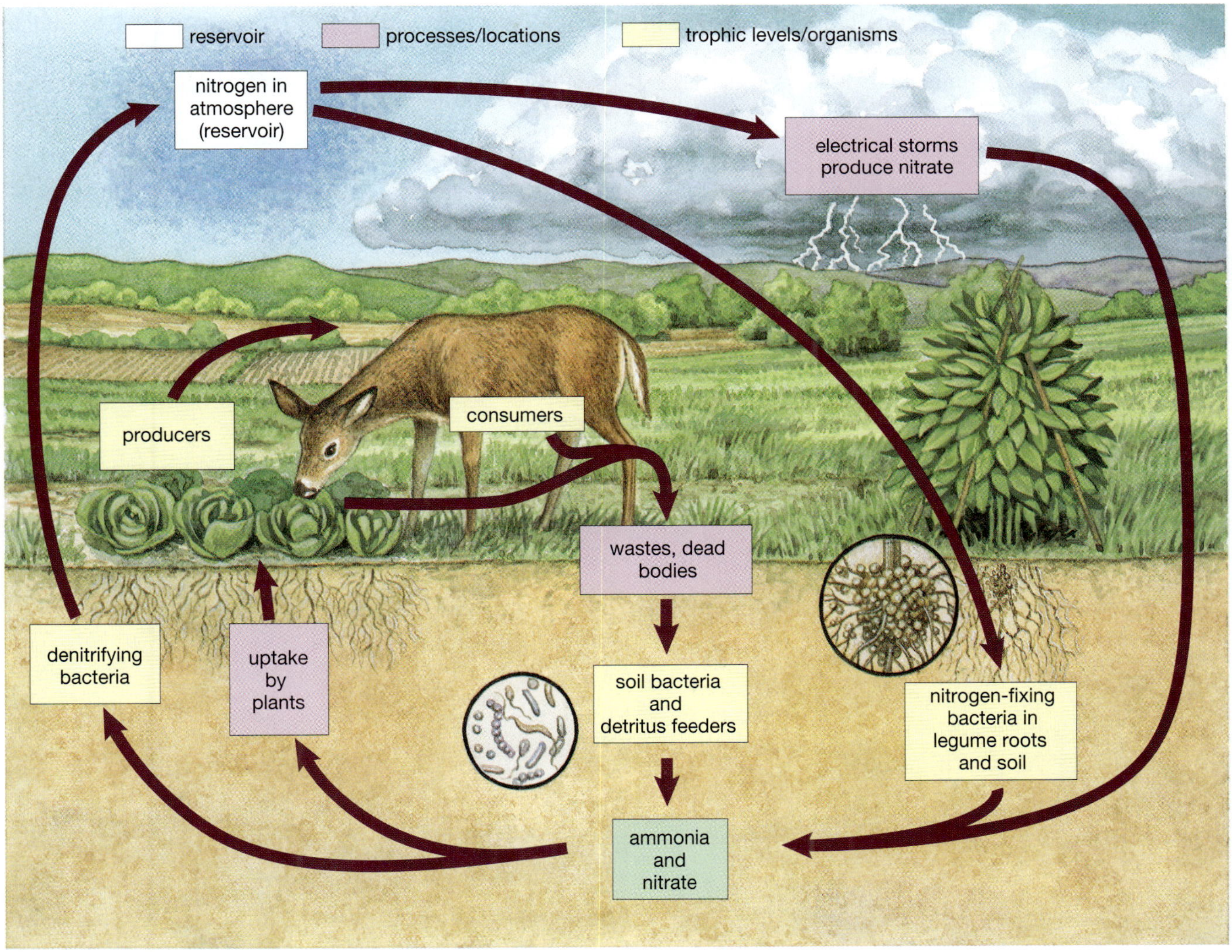

Figure 29-9 The nitrogen cycle

(plants that have nitrogen-fixing bacteria associated with their roots), and by producing fertilizers, humans each year convert an estimated 140 million metric tons of nitrogen from its gaseous form to ammonia and nitrogen oxides (nitrogen combined with oxygen). These nitrogen oxides contribute to global warming, destroy the protective ozone layer, and acidify precipitation. Runoff from fertilized farm fields pours tremendous quantities of ammonia and nitrates into waterways, disrupting aquatic ecosystems. For example, runoff into the Mississippi River is carried to the Gulf of Mexico, where scientists believe it is the cause of a 7000-square-mile "dead zone" that appears there each summer.

The Major Reservoir for Phosphorus Is Rock

Phosphorus is a crucial component of biological molecules, including the energy transfer molecules (ATP and NADP), nucleic acids, and the phospholipids of cell membranes. It is also a major component of vertebrate teeth and bones. In contrast to the cycles of carbon and nitrogen, the phosphorus cycle has no atmospheric component. The reservoir of phosphorus in ecosystems is rock, where it is bound to oxygen in the form of phosphate. As phosphate-rich rocks are exposed and eroded, rainwater dissolves the phosphate. Dissolved phosphate is readily absorbed through the roots of plants and by other autotrophs, such as photosynthetic protists and cyanobacteria. From these producers, phosphorus passes through food webs (Fig. 29-10). At each level, excess phosphate is excreted. Ultimately, decomposers return the phosphorus from dead bodies back to the soil and water in the form of phosphate. It may then be again absorbed by autotrophs, or it may become bound to sediment and eventually reincorporated into rock.

Some of the phosphate that dissolves in fresh water is carried to the oceans. Although much of this phosphate ends up in marine sediments, some is absorbed by marine producers and eventually incorporated into the bodies of invertebrates and fish. Some of these, in turn, are consumed by seabirds, which excrete large quantities of phosphorus back onto the land. At one time, the guano (droppings) deposited by seabirds along the western coast of South America was mined as a source of phosphorus for fertilizer and industrial use.

Soil that erodes from fertilized fields carries large quantities of phosphates into lakes, streams, and the ocean, where it stimulates the growth of producers. In lakes, phosphorus-rich runoff from land can stimulate such a rich growth of algae and bacteria that the natural interactions among populations in the lake are disrupted.

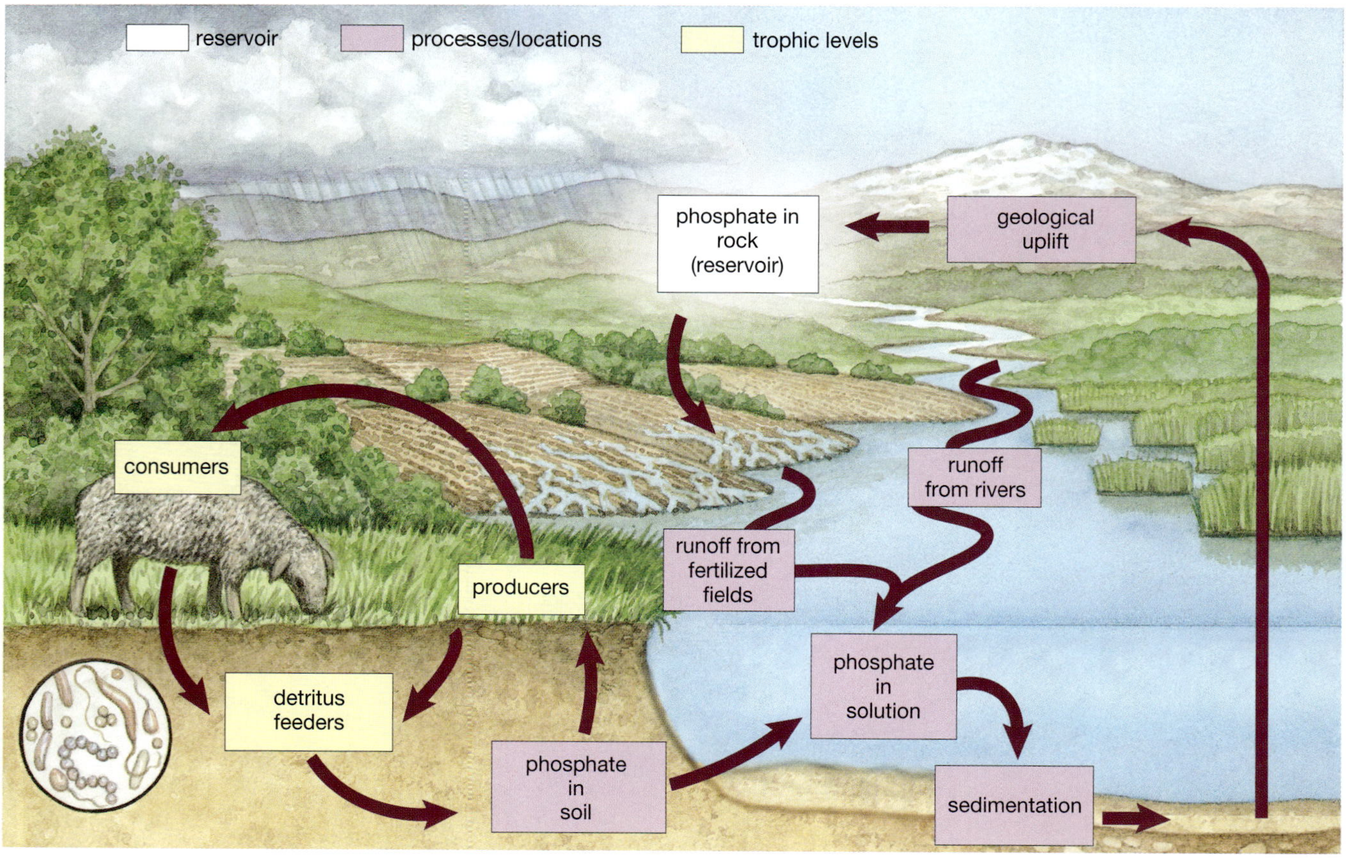

Figure 29-10 The phosphorus cycle

Water Remains Unchanged During the Water Cycle

The water cycle, or **hydrologic cycle** (Fig. 29-11), differs from most other biogeochemical cycles in that most water is not used in the synthesis of new molecules during the cycle; rather, it remains in the form of water throughout the cycle. The major reservoir of water is the ocean, which covers about three-quarters of Earth's surface and contains more than 97% of the available water. The hydrologic cycle is driven by solar energy, which evaporates water, and by gravity, which draws the water back to Earth in the form of precipitation (rain, snow, sleet, and dew). Most evaporated water comes from the oceans, and much water returns directly to them by precipitation. Water falling on land takes various paths. Some water evaporates from the soil, lakes, and streams. A portion runs off the land back to the oceans, and a small amount enters underground reservoirs.

Some water enters the living communities of ecosystems. It is absorbed by the roots of plants; most of the absorbed water evaporates back to the atmosphere from leaves, but a small amount is combined with carbon dioxide during photosynthesis to produce high-energy molecules. Eventually these molecules are broken down during cellular respiration, releasing water back to the environment. Consumers get water from their food or by drinking.

As the human population has grown, fresh water has become scarce in many regions of the world where the human demand for it is high. This scarcity limits food production, as about 40% of the world's food is grown on irrigated cropland. Over the past quarter century farmers have relied increasingly on aquifers—natural underground reservoirs—from which they pump water to irrigate their crops. Unfortunately, in many areas of the world, including China, India, Northern Africa, and the western United States, this groundwater is being "mined"; that is, it is removed faster than it is replenished. The drop in groundwater level is particularly alarming in India, where loss of groundwater places up to 25% of the grain harvest in jeopardy.

The Big Picture

Earth's supply of nutrients is finite, so ecosystem health depends on finely tuned systems for recycling nutrients. Ecosystems can thus be disrupted by human activities that relocate massive quantities of carbon, nitrogen, and other key nutrients.

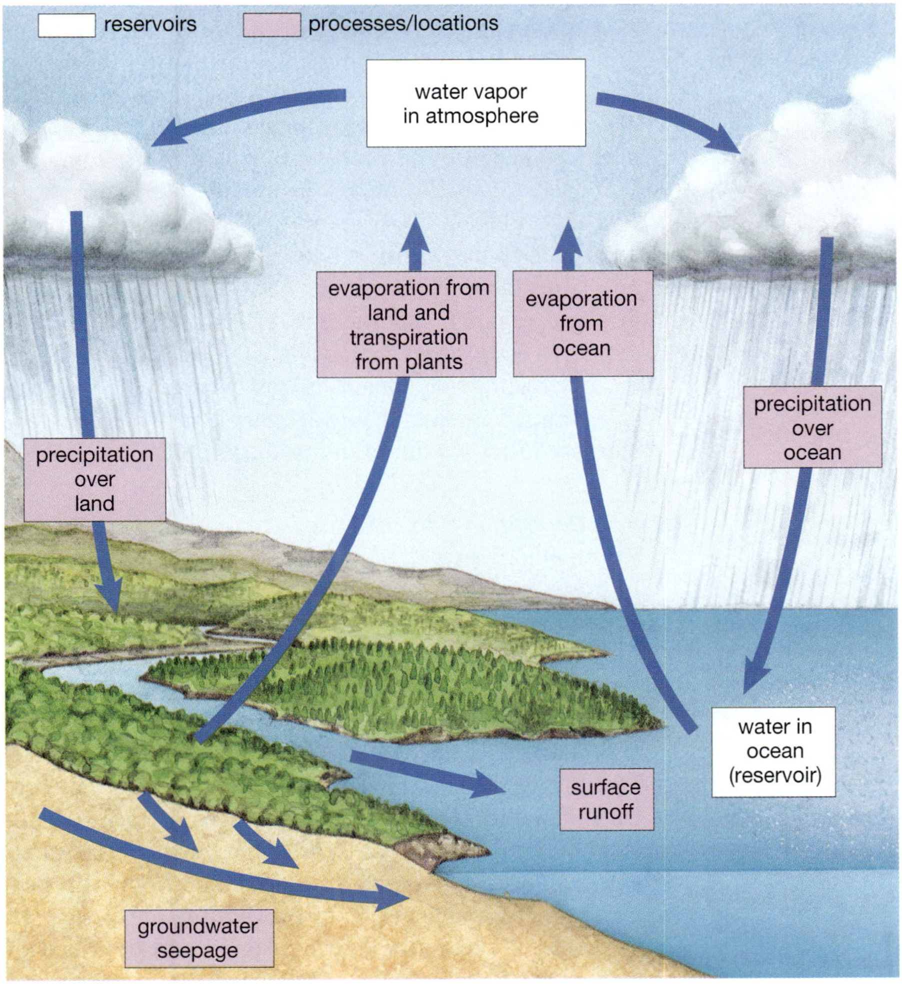

Figure 29-11 The hydrologic cycle

4 What Is Causing Acid Rain and Global Warming?

Many of the environmental problems that beset modern society are caused by human disruption of ecosystems. Pre-industrial peoples were sustained solely by the energy flowing from the sun, and their wastes were readily absorbed by nutrient cycles. But as the human population grew and technology progressed, we began to act more and more independently of these natural processes. Instead of sunlight, we now rely on energy from fossil fuels for heat, light, transportation, industry, and even agriculture. Our industrial and agricultural activities often produce more nutrients than nutrient cycles can efficiently process. In addition, we have mined substances, such as lead, arsenic, cadmium, mercury, oil, and uranium, that are rare or absent from natural ecosystems and that are toxic to many organisms (Fig. 29-12). In our factories, we synthesize substances never before found on Earth: pesticides, solvents, and other chemicals harmful to many forms of life.

Figure 29-12 A natural substance out of place
This bald eagle was killed by an oil spill off the coast of Alaska.

In the following sections, we describe two global environmental problems that are a direct result of human reliance on fossil fuels: acid rain and global warming.

Overloading the Nitrogen and Sulfur Cycles Causes Acid Rain

Each year, the United States discharges about 20 million tons of sulfur dioxide into the atmosphere, two-thirds of it from power plants burning coal or oil. The rest is largely a by-product of industrial boilers, smelters, and refineries. Although volcanoes and hot springs also release sulfur dioxide, human industrial activities account for 90% of the sulfur dioxide in the atmosphere. In addition, 24 million tons of nitrogen oxides are released from vehicles, power plants, and industry in the United States annually. Excess production of these substances was identified in the late 1960s as the cause of a growing environmental threat: acid rain.

Acid Rain Is Acid Deposition

Acid rain is more accurately called **acid deposition**. When combined with water vapor in the atmosphere, nitrogen oxides are converted to nitric acid, and sulfur dioxide is converted to sulfuric acid. Days later, and often hundreds or thousands of miles from the source, the acids fall, eating away at statues and buildings (Fig. 29-13), damaging trees and crops, and rendering lakes lifeless. In the United States, the Northeast, Florida, the Mid-Atlantic region, the upper Midwest, and the high-elevation West are the areas most vulnerable to acid deposition, because the rocks and soils that predominate there are not able to buffer the acidity.

The Acid Rain Program of the U.S. EPA, implemented in 1995, has substantially reduced sulfur dioxide emissions from power plants. Air quality has improved in some regions, and there is cautious optimism that damaged ecosystems may begin the slow road to recovery. Nitrogen oxide emissions, however, remain high despite improvements at power plants, partly because of increased fuel consumption by automobiles.

Acid Deposition Damages Life in Lakes, Farms, and Forests

In New York's Adirondack Mountains, acid rain has made about 25% of the lakes and ponds too acidic to support fish. But even before the fish die, much of the food web that sustains them is destroyed. Clams, snails, crayfish, and insect larvae die first, then amphibians, and finally fish. The result is a crystal-clear lake—beautiful but dead. The impact is not limited to aquatic organisms. Acid rain also interferes with the growth and yield of many farm crops by leaching out essential nutrients such as calcium and potassium and killing decomposer microorganisms, thus preventing the return of nutrients to the soil. Plants, poisoned and deprived of nutrients, become weak and vulnerable to infection and

Figure 29-13 Acid deposition is corrosive
This limestone statue at Rheims Cathedral in France is being dissolved by acid deposition.

Figure 29-14 Acid deposition can destroy forests
Acid rain and fog have destroyed this forest atop Mount Mitchell in North Carolina.

insect attack. High in the Green Mountains of Vermont, scientists have witnessed the death of about half of the red spruce and beech trees and one-third of the sugar maples since 1965. The snow, rain, and heavy fog that commonly cloak these eastern mountaintops are highly acidic. At a monitoring station atop Mount Mitchell in North Carolina, the pH of fog has been recorded at 2.9, more acidic than vinegar (Fig. 29-14).

Acid deposition also increases the exposure of organisms to toxic metals, including aluminum, lead, mercury, and cadmium, which are far more soluble in acidified water than in water of neutral pH. Aluminum dissolved from rock may inhibit plant growth and kill fish. Drinking water in some households has been found to be dangerously contaminated with lead dissolved by acidic water from lead solder in old pipes. Fish in acidified water can have levels of mercury in their bodies that are dangerous to both the fish and the animals (including humans) that consume them.

Disruption of the Carbon Cycle Contributes to Global Warming

Between 345 million and 280 million years ago, huge quantities of carbon were diverted from the carbon cycle when millions of dead plant and animal bodies were buried in sediments, escaping decomposition. Over time, heat and pressure converted the bodies into fossil fuels, such as coal, oil, and natural gas. Without human intervention, this carbon would have remained untouched. Beginning with the Industrial Revolution, however, we have increasingly relied on the energy stored in these fuels. When we burn them in our power plants, factories, and cars, CO_2 is released into the atmosphere. Since 1850, the CO_2 content of the atmosphere has increased more than 30%, from 280 parts per million (ppm) to 370 ppm, bringing concentrations to their highest levels in more than 400,000 years. Atmospheric CO_2 continues to increase at a rate of 1.5 ppm per year, mainly as a result of fossil fuel combustion.

A second source of added atmospheric CO_2 is global **deforestation**, which eliminates tens of millions of acres of forests annually. Deforestation is greatest in the Tropics, where rain forests are rapidly being converted to marginal agricultural land. The carbon stored in the massive trees of these forests returns to the atmosphere after they are cut and the wood decomposes or is burned.

Greenhouse Gases Trap Heat in the Atmosphere

Even at its current elevated levels, CO_2 is still only a tiny fraction of Earth's atmosphere. Atmospheric CO_2, however, acts something like the glass in a greenhouse: It allows solar energy to pass through and reach Earth's surface, but it absorbs and is heated by the longer-wavelength energy that then radiates from the surface back into the atmosphere (Fig. 29-15). This **greenhouse effect** in effect traps the sun's energy as heat and keeps Earth's atmosphere warm enough to support life.

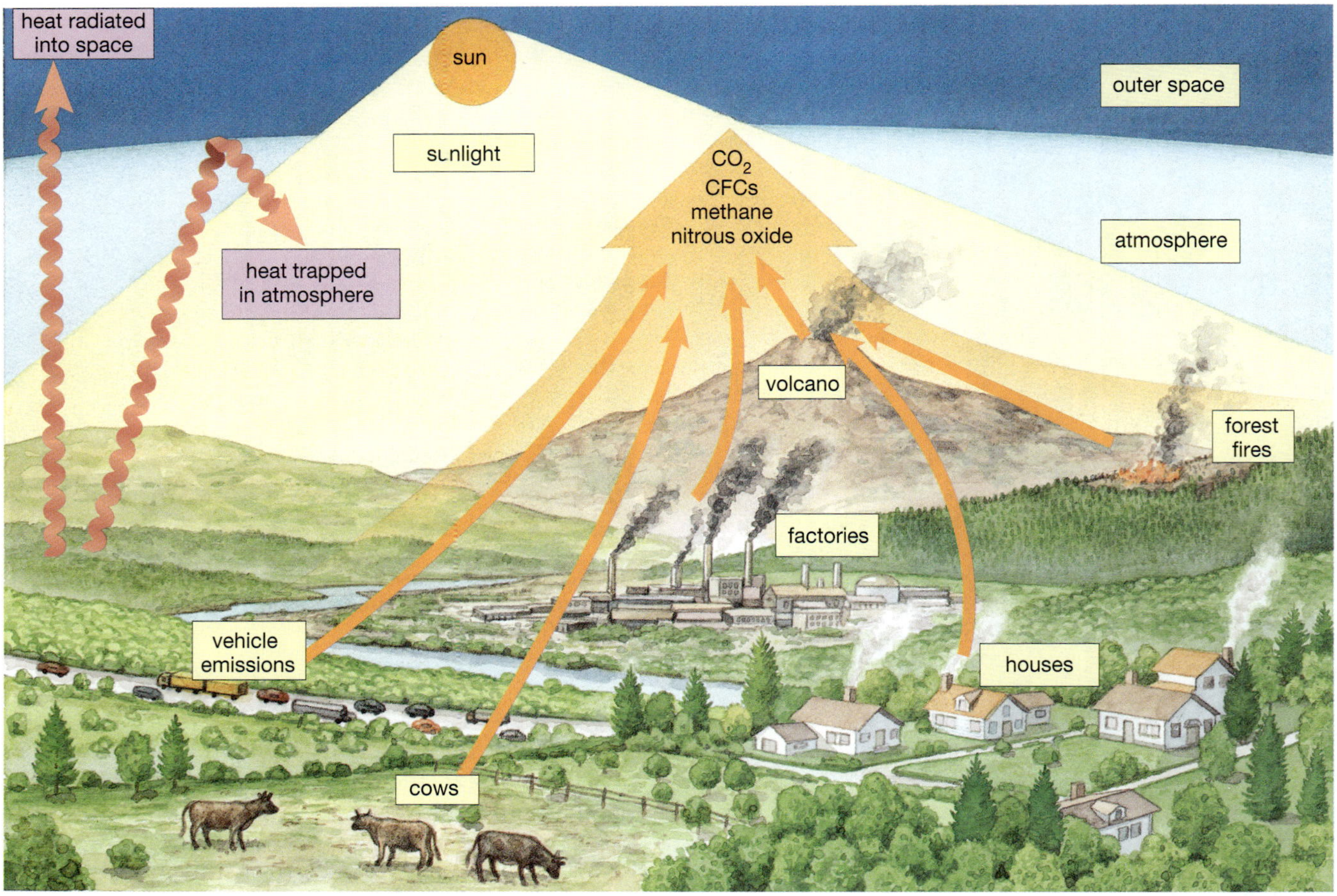

Figure 29-15 Increases in greenhouse gas emissions contribute to global warming
Incoming sunlight warms Earth's surface and is radiated back to the atmosphere. Greenhouse gases absorb some of this radiated energy, trapping it in the atmosphere.

Although the greenhouse effect is a natural phenomenon, most climate scientists believe that it has been amplified by human activities that produce CO_2 and other **greenhouse gases** such as methane, chlorofluorocarbons (CFCs), water vapor, and nitrous oxide. Historical temperature records have revealed a pattern of **global warming**; the average global temperature has increased 0.5 to 1.0 °C since 1860, paralleling the increase in atmospheric CO_2 (Fig. 29-16). The 1990s were the warmest decade on record; the 1980s were the second warmest.

The factors that influence climate are extremely numerous and complex, so predicting future climate change is difficult. Nonetheless, a large, international group of climate scientists known as the Intergovernmental Panel on Climate Change (IPPC) predicts that, if greenhouse gas emissions are not curtailed, the average surface air temperature of Earth will increase by 2.5 to 10.4 °F (1.4 to 5.8 °C) by 2100. To put this change in perspective, average air temperatures during the peak of the last Ice Age (20,000 years ago) were only about 5 °C lower than at present.

Global Warming May Have Severe Consequences

Though the precise consequences of global warming are difficult to predict, disruptions that affect living things are likely. For example, polar ice caps and glaciers will probably melt in response to atmospheric warming, causing sea levels to rise. Expanding oceans will threaten coastal cities and flood coastal wetlands. Another likely outcome of warming is the spread of tropical diseases, as the organisms that cause and transmit these diseases, such as malaria parasites and the mosquitoes that carry them, move into areas where winters were previously too cold for their survival.

Weather patterns will be altered by a warmer world. Even small temperature changes can dramatically alter the paths of major air and ocean currents, changing precipitation patterns in unpredictable ways. Some land might become too hot and dry for agriculture, while other areas might become warmer, wetter, and more productive. Overall, U.S. crop productivity might increase. As the world warms, however, experts predict longer, more severe droughts and more extremes in rainfall, leading to more frequent crop failure and flooding. Agricultural disruption could be disastrous for some nations that are already barely able to feed themselves.

The impact of global warming on ecosystems could be profound. In many forest ecosystems, overall growth may increase, but species distributions will change. For example, sugar maples may disappear from northeastern U.S. forests, while southeastern forests could be partially

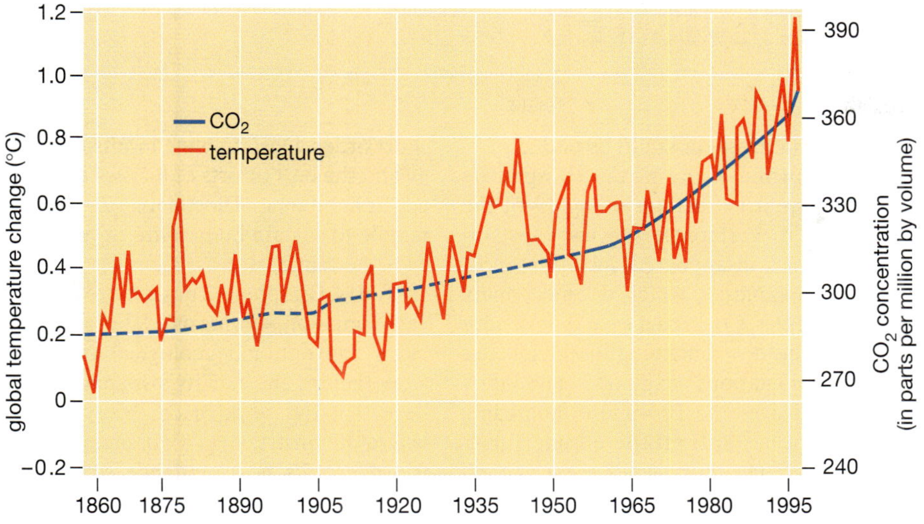

Figure 29-16 Global warming parallels CO_2 increases
The CO_2 concentration of the atmosphere (blue line) has increased steadily since 1860. The dashed portion of that curve represents measurements made from air trapped in ice cores; the solid portion reflects direct measurements made at Mauna Loa, Hawaii. Average global temperatures (red line) show a parallel increase. (With thanks to Drs. Kevin Trenberth and Jim Hurrell of the National Center for Atmospheric Research.)

replaced by grasslands. Coral reefs, already stressed by human activities, are likely to suffer further damage from warmer waters, as reef-building corals are harmed by even small changes in water temperature.

Biologists worldwide are documenting changes related to warming. Scientists in Germany, for example, analyzed data from sites throughout Europe and concluded that the length of the growing season had increased by more than 10 days over the past 40 years. Mexican Jays in southern Arizona are nesting 10 days earlier than they did in 1971. Many species of butterflies have shifted their ranges northward. In the Monteverde Cloud Forest Preserve in Costa Rica, rising ocean temperatures have caused more mist-free days, with adverse effects on local amphibians, birds, and reptiles. Some species have shifted their ranges to higher elevations; other populations have disappeared. The cumulative weight of data from diverse sources worldwide is strong evidence that biological changes caused by warming have begun. These data also provide researchers with a glimpse of what is to come as global warming continues.

Our Decisions Make a Difference

Americans make a disproportionately large contribution to acid rain and global warming. Despite accounting for less than 5% of the world's population, the U.S. generates more than 20% of world's greenhouse gases, 6.6 tons per person each year. What can we as individuals do to help reduce this imact? A car with a fuel efficiency of 20 miles per gallon releases 1 pound of CO_2 per mile, so we can substantially reduce emissions of CO_2 by using fuel-efficient vehicles, car pools, and public transportation. As electricity is generated in fossil fuel-fired power plants, so are tremendous quantities of CO_2, sulfur dioxide, and nitrogen oxides. We can reduce these emissions by supporting the efforts of utility companies to use renewable energy sources such as wind and solar power. To conserve electricity, we can purchase more efficient appliances, turn off unused computers and lights, and replace incandescent with fluorescent lightbulbs. Insulating and weatherproofing our homes, incorporating solar energy features into new homes, and planting deciduous trees near our houses to provide for summer shade and winter sun will significantly reduce fuel consumption while cutting back on heating and air conditioning costs. Recycling is also a tremendous energy saver—for example, 95% of the energy used to produce an aluminum can from raw materials is conserved when the can is recycled. We can also support reforestation efforts to replace trees both in tropical rain forests and in our communities.

Although continued global warming is inevitable, its speed can be reduced by human decisions. Global Earth Summit conferences, such as the one held in Kyoto, Japan, in 1997 offer a ray of hope. There, representatives from 160 nations proposed a treaty whose goal is to curb CO_2 emissions. If ratified, the treaty will commit industrialized nations to reducing CO_2 emissions to about 5% below 1990 levels by the year 2012. Developing countries, such as China and India, are not bound by the Kyoto agreement. Many observers are concerned that the benefits of the treaty could be overwhelmed by emissions from such nations, which are home to more than one-third of Earth's people, as they strive to industrialize and increase their standard of living. Although the obstacles are considerable, the Kyoto conference is evidence of a worldwide recognition of global warming and a concern for its consequences, the first steps toward change.

CASE STUDY REVISITED

A World in a Bubble

The mission of the eight "Biospherians" who attempted to live for 2 years in the closed ecosystem of Biosphere 2 turned out to be a failure. It proved impossible to maintain livable conditions inside the enclosure. A year or so after the enclosure was sealed, the oxygen concentration in the atmosphere had dropped from 21% to 14% (the level ordinarily found atop a 17,500-foot mountain), and the level of carbon dioxide soared. Flourishing populations of microbes in Biosphere 2's topsoil had consumed the atmospheric oxygen and released huge quantities of carbon dioxide. The internal atmosphere also became polluted by high concentrations of nitrous oxide, which reached levels that could have led to brain damage in the inhabitants. To protect the safety of the researchers inside, the seal was broken so that fresh air could be added.

Meanwhile, many of the Biosphere's species went extinct, including 19 of the original 25 vertebrate species. Most insect species disappeared, including most of the pollinators, which doomed many plant species. Cockroaches and ants were among the few remaining insect species; their populations exploded and they swarmed over the Biosphere landscape. Other species also prospered uncontrollably. Morning glory vines grew aggressively, benefiting from the high carbon dioxide concentration. Continuous weeding was required to protect food crops from being smothered by the vines. Food was in short supply and the human inhabitants lost weight. Water systems became polluted with excess nutrients. Overall, the closed ecosystem unraveled in fairly short order. Despite huge inputs of energy and the best efforts of its very dedicated inhabitants, Biosphere 2 proved unable to sustain human life or maintain ecosystem integrity.

Today, Biosphere 2 has a more conservative mission, serving as a laboratory for more conventional ecological research. Nonetheless, the facility's managers believe that the work of the Biospherians was quite instructive. Two members of Biosphere 2's new guidance committee published a report in which they concluded that "the major retrospective conclusion that can be drawn is simple. At present there is no demonstrated alternative to maintaining the viability of the Earth. . . . Despite its mysteries and hazards, Earth remains the only known home that can sustain life."

Summary of Key Concepts

1) How Do Ecosystems Obtain Energy and Nutrients?

Ecosystems are sustained by a continuous flow of energy from sunlight and a constant recycling of nutrients.

2) How Does Energy Flow Through Communities?

Energy enters ecosystems when it is harnessed by autotrophs during photosynthesis. The amount of energy that autotrophs store in a given unit of area during a given period is the ecosystem's net primary productivity.

Trophic levels describe feeding relationships in ecosystems. Autotrophs are the producers, the lowest trophic level. Herbivores occupy the second level as primary consumers. Carnivores act as secondary consumers when they prey on herbivores and as tertiary or higher-level consumers when they eat other carnivores.

Feeding relationships in which each trophic level is represented by one organism are called food chains. In natural ecosystems, feeding relationships are more complex and are described as food webs. Detritus feeders and decomposers, which digest dead bodies and wastes, use and release the stored energy and free up nutrients for recycling. In general, only about 10% of the energy captured by organisms at one trophic level is converted to the bodies of organisms at the next level. The higher the trophic level, the less energy is available to sustain it. As a result, plants are more abundant than herbivores, and herbivores are more abundant than carnivores. The storage of energy at successive trophic levels is illustrated graphically as an energy pyramid.

3) How Do Nutrients Move Within and Among Ecosystems?

A nutrient cycle depicts the movement of a particular nutrient from its reservoir (usually in the nonliving portion of the ecosystem) through the living portion of the ecosystem and back to its reservoir, where it is again available to producers. Carbon reservoirs include the oceans, atmosphere, and fossil fuels. Carbon enters producers through photosynthesis. From autotrophs it is passed through the food web and released to the atmosphere as CO_2 during cellular respiration.

The major reservoir of nitrogen is the atmosphere. Nitrogen gas is converted by bacteria and human industrial processes into ammonia and nitrate, which can be used by plants. Nitrogen passes from producers to consumers and is returned to the environment through excretion and the activities of detritus feeders and decomposers.

The reservoir of phosphorus is in rocks as phosphate, which dissolves in rainwater. Phosphate is absorbed by photosynthetic organisms, then passed through food webs. Some is excreted, and the rest is returned to the soil and water by decomposers. Some is carried to the oceans, where it is deposited in marine sediments. Humans mine phosphate-rich rock to produce fertilizer.

The major reservoir of water is the oceans. Water is evaporated by solar energy and returned to Earth as precipitation. The water flows into lakes and underground reservoirs and in rivers, which flow to the oceans. Water is absorbed directly by plants and animals and is also passed through food webs. A small amount is combined with CO_2 during photosynthesis to form high-energy molecules.

4) What Is Causing Acid Rain and Global Warming?

Environmental problems arise when human activities interfere with the natural functioning of ecosystems. Human industrial processes release toxic substances and produce more nutrients than nutrient cycles can efficiently process. Through massive consumption of fossil fuels, we have increased the flow of energy and disrupted the natural cycles of carbon, sulfur, and nitrogen, causing acid rain and global warming (an amplification of the greenhouse effect).

Key Terms

acid deposition *p. 594*
autotroph *p. 583*
biodegradable *p. 589*
biogeochemical cycle *p. 588*
biological magnification *p. 589*
carnivore *p. 584*
consumer *p. 583*
decomposer *p. 585*
deforestation *p. 595*
denitrifying bacterium *p. 591*
detritus feeder *p. 584*
energy pyramid *p. 588*
food chain *p. 584*
food web *p. 584*
fossil fuel *p. 590*
global warming *p. 596*
greenhouse effect *p. 595*
greenhouse gas *p. 596*
herbivore *p. 584*
heterotroph *p. 583*
hydrologic cycle *p. 593*
macronutrient *p. 588*
micronutrient *p. 588*
net primary productivity *p. 583*
nitrogen fixation *p. 591*
nutrient *p. 582*
nutrient cycle *p. 588*
omnivore *p. 584*
primary consumer *p. 584*
producer *p. 583*
reservoir *p. 588*
secondary consumer *p. 584*
tertiary consumer *p. 584*
trophic level *p. 583*

Thinking Through the Concepts

Multiple Choice

1. *The major source(s) of carbon for living things is (are)*
   a. coal, oil, and natural gas
   b. plants
   c. CO_2 in the atmosphere and oceans
   d. methane in the atmosphere
   e. carbon in animal bodies
2. *Why would you expect there to be a smaller biomass of big predators (lions, leopards, hunting dogs, etc.) than of grazing mammals (gazelles, zebra, elephants, etc.) in the African savanna?*
   a. too little cover for the predators to hide in
   b. the inefficiency of energy transfer between trophic levels
   c. like domestic cats, large predators are susceptible to hair balls
   d. many predators occur in social groups, thus limiting their numbers
   e. because grazers are better adapted to moving long distances, they can better follow the rains
3. *What is the one group of organisms that is able to fix atmospheric nitrogen into forms usable by living organisms?*
   a. plants b. fungi
   c. insects d. bacteria
   e. viruses
4. *The biological process by which carbon is returned to its reservoir is*
   a. photosynthesis b. denitrification
   c. carbon fixation d. glycolysis
   e. cellular respiration
5. *As a black widow spider consumes her mate, what is the lowest trophic level she could be occupying?*
   a. third b. first
   c. second d. fourth
   e. fifth

? Review Questions

1. What makes the flow of energy through ecosystems fundamentally different from the flow of nutrients?
2. What is an autotroph? What trophic level does it occupy, and what is its importance in ecosystems?
3. Define *primary productivity*. Would you predict higher net primary productivity in a farm pond or an alpine lake? Defend your answer.
4. List the first three trophic levels. Among the consumers, which are most abundant? Why would you predict that there will be a greater biomass of plants than herbivores in any ecosystem? Relate your answer to the "10% law."
5. How do food chains and food webs differ? Which is the more accurate representation of actual feeding relationships in ecosystems?
6. Define *detritus feeders* and *decomposers*, and explain their importance in ecosystems.
7. Trace the movement of carbon from its reservoir through the biotic community and back to the reservoir. How have human activities altered the carbon cycle, and what are the implications for future climate?
8. Explain how nitrogen gets from the air to a plant.
9. Trace the movement of a phosphorus molecule from a phosphate-rich rock into the DNA of a carnivore. How does the phosphorus cycle differ from the carbon and nitrogen cycles?
10. Trace the movement of a water molecule from the moment it leaves the ocean until it eventually reaches a plant root, then a plant stoma, and then makes its way back to the ocean.

Applying the Concepts

1. What could your college or university do to reduce its contribution to acid rain and global warming? Be specific and, if possible, offer practical alternatives to current practices.
2. Relate fossil fuel consumption to (a) the loss of aquatic life in lakes in the Northeast and Canada and (b) the lengthening of the growing season in Europe. Trace each step from the burning of gasoline in a car or power plant to the change in question.
3. Define and give an example of biological magnification. What qualities are present in materials that undergo biological magnification? In which trophic level are the problems worst, and why?
4. Discuss the contribution of population growth to (a) acid rain and to (b) the greenhouse effect.
5. Describe what would happen to a population of deer if all predators were removed and hunting was banned. Include effects on vegetation as well as on the deer population itself. Relate your answer to carrying capacity as discussed in Chapter 27.

Group Activity

How do humans affect food webs? In the 1950s the island of Borneo had a severe problem with malaria, which is carried by mosquitoes. The World Health Organization (WHO) decided to spray a pesticide called DDT to rid the island of mosquitoes. The accompanying diagram (adapted from The Center for Applied Research in Education, 1991) shows a simplified food chain of the ecosystem affected by the DDT. Additional facts:

- DDT causes nerve damage, slow reflexes, and eventual death in animals.
- An animal that consumes another animal that has consumed DDT will be affected by the DDT.
- DDT kills mosquitoes; it also affects cockroaches but does not kill them.

In groups of three, predict the effects DDT will have on each member of the food web. Your group should be prepared to share with the class and to compare your findings to the actual events.

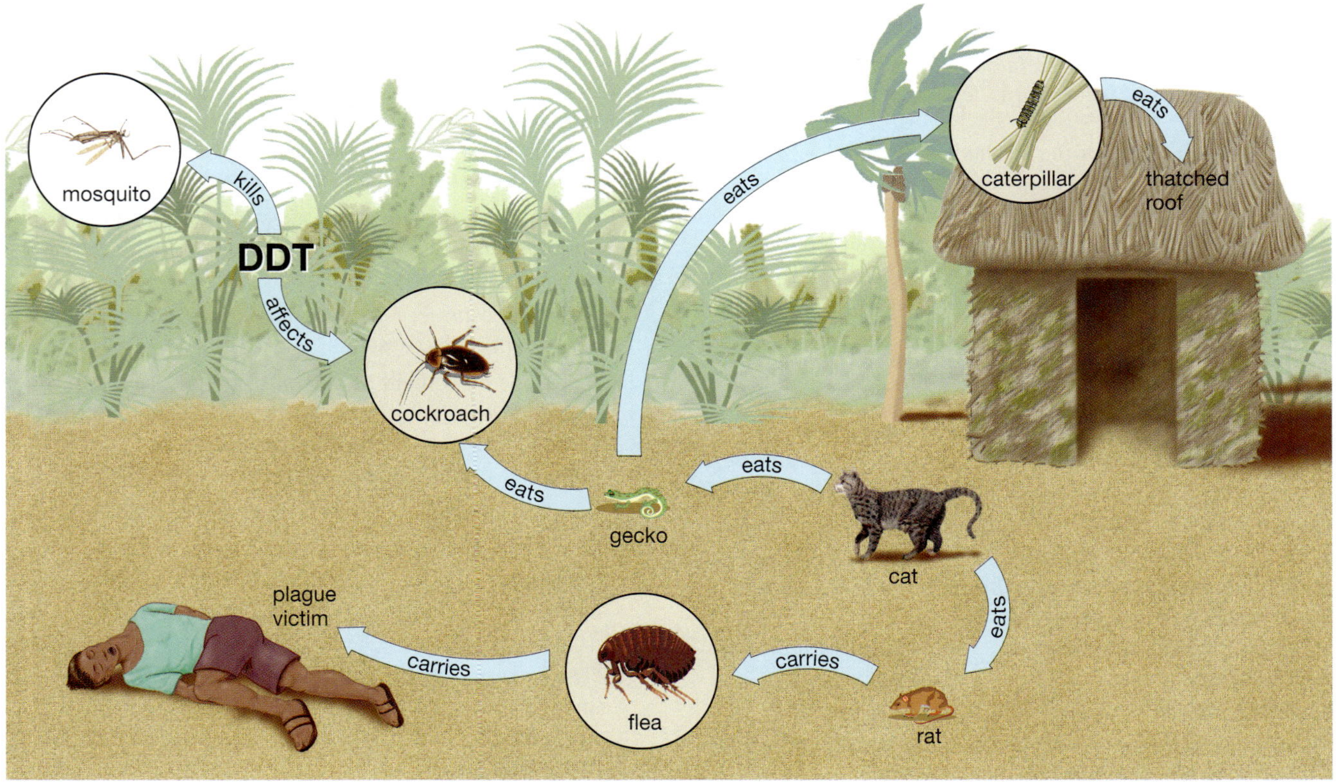

For More Information

Epstein, P. R. "Is Global Warming Harmful to Health?" *Scientific American*, August 2000. Global warming may increase the incidence of disease as the climate becomes more hospitable for disease-causing organisms and their vectors.

Holloway, M. "The Glass House in the Desert." *Scientific American*, January 2002. A nice description of Biosphere 2, complete with information on how to arrange a visit.

Holmes, B. "Case of the Dwindling Cloud Forest." *International Wildlife*, July–August 2000. The unique and rare animals and plants inhabiting the cloud forest of Monteverde, Costa Rica, are in trouble as their habitat becomes increasingly dry. Global warming may be to blame.

Karl, T. R., Nicholls, N., and Gregory, J. "The Coming Climate." *Scientific American*, May 1997. What changes in weather patterns can we expect in a warmer world?

Mlot, C. "Tallying Nitrogen's Increasing Toll." *Science News*, February 15, 1997. Human activities now dominate the nitrogen cycle, with increasingly disruptive effects.

Post, W. M., and others. "The Global Carbon Cycle." *American Scientist*, July–August 1990. A readable and comprehensive look at the natural carbon cycle and how humans have influenced it.

Wuethrich, B. "How Climate Change Alters Rhythms in the Wild." *Science*, February 4, 2000. The distribution, behavior, and ecology of many animal populations are already being influenced by global warming.

Answers to Multiple-Choice Questions

1. c 2. b 3. d 4. e 5. d

MediaTutor

CD Activities

Activity 29.1: Building a Food Web

Estimated time: 10 minutes

In this exercise you will have the opportunity to learn about food webs for a variety of ecosystems. You can also reinforce your understanding of trophic interactions by identifying the roles that organisms play within the ecosystem.

Start the MediaTutor Student CD-ROM and enter the activity number in the Quick Search box to be taken directly to that activity.

Web Investigation

A World in A Bubble

Estimated time: 20 minutes

Our modern environment is a delicate balance between nature's forces and mankind's influence. "Case Study: A World in a Bubble" describes a somewhat futile attempt at creating a self-contained, man-made mini-ecosystem. This activity will investigate this project in more depth and explore some of the dangers currently facing our environment.

Go to http://www.prenhall.com/audesirk3, the Audesirk Companion Web site. Select Chapter 29 and Web Investigation to begin.

"We must consider our planet to be on loan from our children, rather than being a gift from our ancestors."

Gro Harlem Brundtland, former Prime Minister of Norway

At the Cajari River reserve in Brazil, people are allowed to settle at low density and to extract renewable resources. This kind of extractive reserve may represent the most viable option for preserving what remains of the world's tropical rain forests.

30 Earth's Diverse Ecosystems

AT A GLANCE

CASESTUDY

A Rubber Tapper's Legacy

Tropical rain forests are among the most productive, most diverse ecosystems on Earth. Unfortunately, these ecological jewels are being destroyed at an alarming rate. The forces behind the destruction are largely economic. For example, in Brazil, home to much of the world's remaining rain forest, forested land is cleared for its valuable timber, to create pasture for cattle ranches, and to open new fields for subsistence farmers. Brazil's growing population and expanding economy foster an ever-increasing pressure to destroy forests.

Fortunately for our planet's ecological health, not everyone believes that the economic well-being of Brazil's population necessarily requires the sacrifice of the country's magnificent forests. One person who refused to accept the necessity of forest destruction was Chico Mendes. Mendes, poor, with little formal education, began working in Brazil's rain forests at age 9 as a rubber tapper. Rubber tappers make their living by collecting and selling the sap of the tree *Hevea brasiliensis*, which is rich in latex, the raw material from which rubber is manufactured. These rubber trees are scattered throughout the forest; the tappers seek them out, make a shallow cut in the bark, and hang a bucket to collect the sap that drips out.

Rubber tappers are dependent on the forest for their livelihood, and Chico Mendes was among those who grew increasingly alarmed as loggers and cattle ranchers cleared more and more of the forest around his hometown of Xapuri. Mendes believed that the local economy would be harmed rather than helped by forest destruction; he saw that an intact, healthy forest could provide jobs and income. He began to organize rubber tappers to resist the wholesale clearing of forests. Under his leadership, the newly formed National Council of Rubber Tappers mounted a series of dramatic, peaceful protests that garnered worldwide attention.

Mendes understood that protests and publicity alone could not protect the forest. He therefore argued that the government should establish *extractive reserves*. Mendes envisioned these reserves as areas of publicly owned forest that would be protected from destruction. Residents would be permitted to pursue low-impact economic exploitation of the resources by gathering latex, nuts, and fruits, and by very limited hunting, fishing, and timber-cutting. The genius of the extractive reserve concept is that it protects forests not by asking local residents to forgo economic development but by tying their economic well-being to the health of the forest. The idea was nonetheless bitterly opposed by timber and cattle-ranching interests. Could Chico Mendes and his union of rubber tappers persuade the Brazilian government to try this new approach to forest conservation? ■

1 What Factors Influence Earth's Climate?

The distribution of life, particularly on land, is strongly influenced by weather and climate. **Weather** refers to short-term fluctuations in temperature, humidity, cloud cover, wind, and precipitation over periods of hours or days. **Climate**, in contrast, refers to patterns of weather that prevail from year to year and even century to century in a particular region. Climate is determined by a region's range of temperatures and amount of sunlight and water, and it influences how species are distributed.

Both Climate and Weather Are Driven by the Sun

Both climate and weather are powered by a great thermonuclear engine: the sun. Solar energy reaches Earth in a range of wavelengths, from short, high-energy ultraviolet (UV) rays, through visible light, to longer infrared wavelengths (see Fig. 6-2). The solar energy that reaches Earth drives the wind, ocean currents, and the global water cycle. Before it reaches Earth's surface, however, sunlight is modified by the atmosphere. Much of the sun's high-energy UV radiation, which can damage biological molecules, is absorbed by the **ozone layer**, a region relatively rich in ozone (O_3) that is located in the middle atmosphere (see "Earth Watch: The Ozone Hole—A Puncture in Our Protective Shield"). Similarly, energy at infrared wavelengths is selectively absorbed by carbon dioxide, water vapor, methane, and other *greenhouse gases*, trapping heat in the atmosphere. In addition, dust, water vapor, and clouds in the atmosphere scatter light, reflecting some of the energy back into space.

Only about half the solar energy that reaches the atmosphere actually strikes Earth's surface. Of this, a small fraction is immediately reflected directly back into space, another small fraction is captured by photosynthetic plants and microorganisms and used to power photosynthesis, and the rest is absorbed and converted to heat. Eventually, nearly all the incoming solar energy is returned to space, either as light or as heat. The solar energy temporarily absorbed and stored as heat by the atmosphere and by Earth's surface maintains Earth's relative warmth.

The forces driven by sun's energy interact with the physical structures of Earth to produce different climates in different places. Among the factors that influence climate are latitude (a measure of a location's distance north or south of the equator, expressed in degrees), air currents, ocean currents, and the presence of mountains and irregularly shaped continents.

Sunlight Strikes Earth at Various Angles

The amount of sunlight that strikes a given area of Earth's surface has a major effect on the average yearly temperatures of that area. At the equator, sunlight hits Earth's surface nearly at a right angle, making the weather there consistently warm. At higher latitudes farther north or south, the sun's rays strike Earth's surface at a greater slant. This angle spreads the same amount of sunlight over a larger area, producing lower overall temperatures.

Because Earth is tilted on its axis, the angle of the sunlight at higher latitudes varies as the planet makes its yearly trip around the sun. This variation causes pronounced seasons. For example, when the Northern Hemisphere is tilted toward the sun, it receives sunlight more directly and has summer. Six months later, when the Northern Hemisphere is tilted away from the sun and having winter, the Southern Hemisphere is at its closest to the sun (Fig. 30-1). The tilting hardly affects the angle

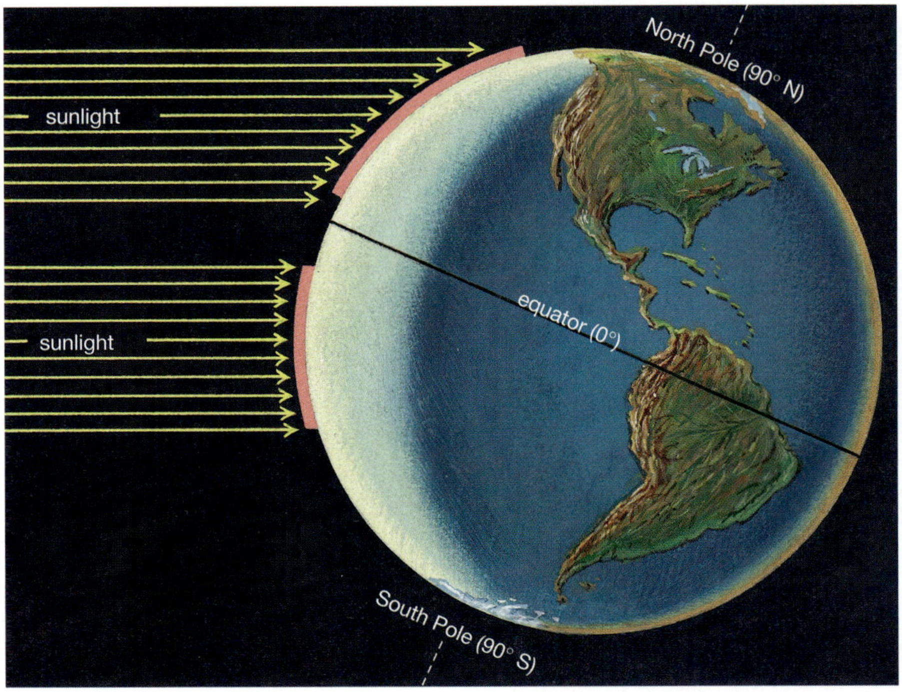

Figure 30-1 Earth's curvature and tilt produce seasons and climate
Here, we see the angles of sunlight during winter in the Northern Hemisphere and summer in the Southern Hemisphere. Near the equator, sunlight falls nearly perpendicular to Earth's surface, so its warmth is concentrated on a relatively small area. Closer to the poles, the same amount of sunlight falls over a much larger surface area. Temperatures are thus highest at the equator and lowest at the poles. The temperatures in each hemisphere are at their lowest when that hemisphere tilts away from the sun.

Earth Watch

The Ozone Hole—A Puncture in Our Protective Shield

A small fraction of the radiant energy produced by the sun, called *ultraviolet*, or UV, radiation, is so highly energetic that it can damage biological molecules. In small quantities, UV radiation helps human skin produce vitamin D and causes tanning in fair-skinned people. But in larger doses, UV causes sunburn, premature aging of skin, skin cancer, and cataracts, a condition in which the lens of the eye becomes cloudy.

Fortunately, most UV radiation is filtered out by ozone in the stratosphere, a layer of atmosphere extending from 6 to 30 miles above Earth. In pure form, ozone (O_3) is a bluish, explosive, and highly poisonous gas. In the stratosphere, the normal concentration of ozone is about 0.1 part per million (ppm), compared with 0.02 ppm in the lower atmosphere. The ozone-enriched layer in the stratosphere is called the *ozone layer.* UV radiation is converted to heat when it strikes the ozone layer, but the overall level of ozone remains reasonably constant—or did until humans intervened.

In 1985 British atmospheric scientists published a startling discovery: The springtime levels of stratospheric ozone over Antarctica had declined by more than 40% since 1977. A hole had been pierced in Earth's protective shield. In the *ozone hole* over Antarctica, ozone now dips during part of the year to about one-third of its pre-depletion levels (Fig. E30-1). Although ozone layer depletion is most severe over Antarctica, the ozone layer is somewhat reduced over most of the world, including the continental United States. Recently, springtime ozone levels over Greenland, Scandinavia, and western Siberia were reduced up to 40%. Evidence of ozone thinning over the Arctic has led researchers to predict that we will soon have an Arctic, as well as an Antarctic, ozone hole.

Satellite data reveal that, since the early 1970s, UV radiation reaching Earth's surface has increased by nearly 7% per decade in the Northern Hemisphere and by 10% or more per decade in the Southern Hemisphere. In New Zealand, researchers have charted a 12% rise during the past decade. Skin cancer rates are expected to increase as a result. Studies of mice show that exposure to UV light in doses similar to a day at the beach without sunscreen can depress their immune systems, making them more vulnerable to disease. Human health effects are only one cause for concern. Photosynthesis by phytoplankton, the producers for marine ecosystems, is reduced under the ozone hole above Antarctica. Some types of trees and crops are also harmed by increased UV radiation.

The thinning of the ozone layer is caused by increasing levels of chlorofluorocarbons (CFCs), gaseous compounds that contain chlorine, fluorine, and carbon. Developed in 1928, these gases were widely used as coolants in refrigerators and air conditioners, aerosol spray propellants, in the production of foam plastic, and as cleansers for electronic parts. CFCs are very stable and were considered safe. Their stability, however, proved to be a major problem, because they remain chemically unchanged as they gradually rise into the stratosphere. There, under intense bombardment by UV light, the CFCs break down, releasing chlorine atoms. Chlorine catalyzes the breakdown of ozone to oxygen gas (O_2) while remaining unchanged itself.

Figure E30-1 Satellite image of the Antarctic ozone hole A record-sized ozone hole is seen in purple on this image based on satellite data from September 10, 2000. Notice that the hole extends over the tip of South America. (Image courtesy of NASA.)

Fortunately, we have taken the first steps toward "plugging" the ozone hole. In an almost unprecedented example of global cooperation and concern, talks beginning in 1985 led to treaties in 1990 and 1992 in which industrialized nations throughout the world agreed to rapidly phase out ozone-depleting chemicals, eliminating CFCs by 1996. These compounds are no longer used in the production of plastic foam, and CFC substitutes have been found for use in spray cans, refrigerators, and car air conditioners as well. Ground-level atmospheric chlorine levels (an indicator of CFC use) peaked in 1994. In 1996 scientists measured a slight reduction in chlorine in the lower regions of the atmosphere. By 1999 they detected chlorine reductions in the stratosphere as well, the result of gradual mixing that transports ground-level air upward. But use of some types of CFCs has increased dramatically in developing countries, and a large reservoir of CFCs remains stored in old refrigeration units. Because these highly stable compounds can persist 50 to 100 years and take a decade or more to ascend into the stratosphere, our current release of CFCs, along with the millions of tons already released, will continue to erode the protective ozone shield, with possibly severe consequences. For example, in September 2000, the Antarctic ozone hole was larger than ever before, extending over the tip of Chile and exposing residents of the city of Punta Arenas to high levels of UV radiation (see Fig. E30-1). Some scientists are optimistic, however, that the ozone hole may begin to get smaller by the year 2010 and might be eliminated by 2050 if global cooperation continues.

of the equator to the sun's rays, so there is little seasonal variation there.

Air Currents Produce Regional Climates

Air currents are generated by Earth's rotation and by differences in temperature between different air masses. As the sun's direct rays fall on the equator, the air there heats and rises, because warm air is less dense than cold air. The warm air near the equator is also laden with water evaporated by solar heat (Fig. 30-2a). As the water-saturated air rises, it cools somewhat. Cool air cannot hold as much moisture as can warm air, so water condenses from the rising air and falls as rain. This rainfall, along with the direct rays of the sun, creates a warm, wet zone around the equator, called the *Tropics*.

The cooler, dry air that remains after the moisture has fallen from the equatorial air then flows north and south from the equator. Near 30° N and 30° S latitudes, the cooled air becomes dense enough to sink. As it sinks, it is warmed by heat radiated from Earth. By the time it reaches the surface, it is both warm and very dry. Not surprisingly, the major deserts of the world are found at these latitudes (Fig. 30-2a,b). This air then flows back toward the equator, completing a circular flow. Farther north and south, this general circulation pattern is repeated, dropping moisture at around 60° N and 60° S and creating extremely dry conditions at the North and South Poles.

Ocean Currents Moderate Near-Shore Climates

Ocean currents are driven by Earth's rotation, winds, and the direct heating of the water by the sun. Continents interrupt the currents, breaking them into roughly circular patterns that rotate clockwise in the Northern Hemisphere and counterclockwise in the Southern Hemisphere (Fig. 30-3). Ocean currents influence the climates of coastal areas by transporting water over large distances. For example, a current in the Atlantic Ocean brings warm water (the Gulf Stream) from equatorial regions north along the eastern coast of North America, creating a warmer, moister climate than is found farther inland. It then carries the still-warm water farther north and east, warming the western coast of Europe before returning south. More generally, coastal areas tend to have less-variable climates than do areas near the center of continents. Because water both heats and cools more slowly than land or air, ocean currents moderate temperature extremes.

Continents and Mountains Complicate Weather and Climate

If Earth's surface were uniform, climate would be determined entirely by the movements of air masses, and climate zones would occupy neat bands corresponding to

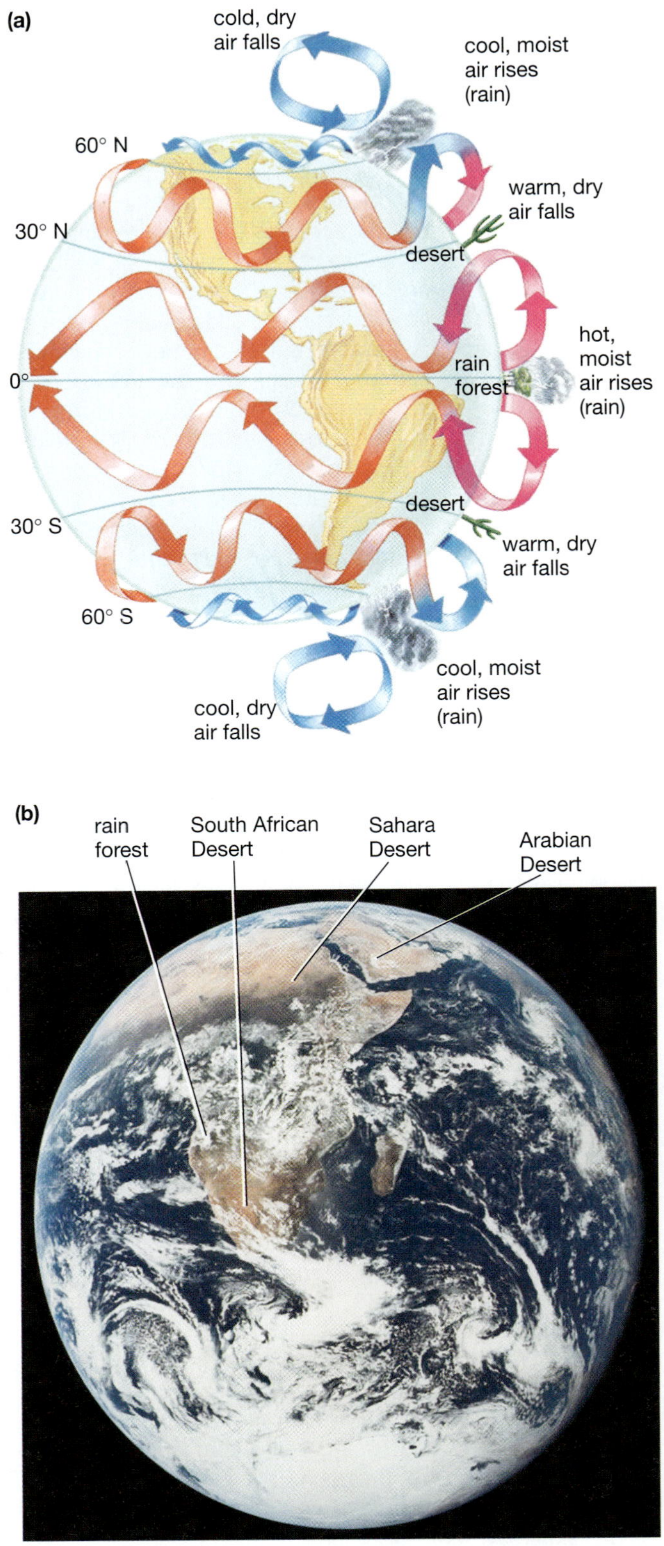

Figure 30-2 Distribution of air currents and climatic regions ***(a)*** Earth's rotation and the distribution of temperatures interact to create air currents that rise and fall predictably with latitude, producing broad climatic regions. ***(b)*** A photograph of the African continent taken from *Apollo 11*. Along the equator are heavy clouds that drop moisture on the central African rain forests. Note the lack of clouds over the Sahara and Arabian Deserts near 30° N and the South African Desert near 30° S.

MediaTutor 30.1 Climate Builds Biomes

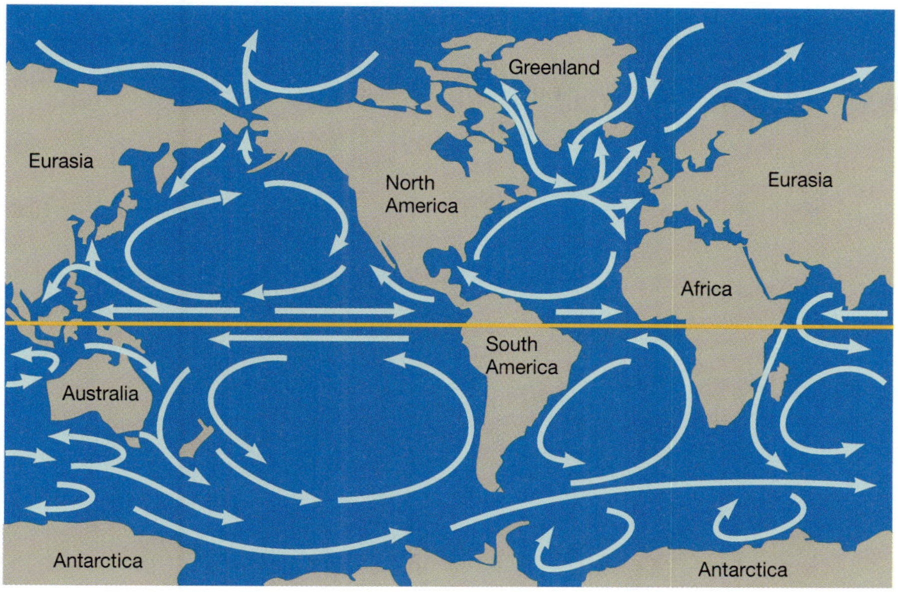

Figure 30-3 Ocean circulation patterns
Currents run in circles that travel clockwise in the Northern Hemisphere and counterclockwise in the Southern Hemisphere, distributing warmth from the equator to northern and southern coastal areas.

latitude. However, the presence of irregularly shaped continents (which heat and cool relatively quickly) surrounded by oceans (which heat and cool more slowly) alters the flow of wind and water and contributes to the irregular distribution of climate and ecosystems.

Variations in elevation within continents further complicate the situation. As elevation increases, the atmosphere becomes less dense and retains less heat. The temperature drops approximately 3.5 °F (2 °C) for every 1000 feet (305 meters) of increased elevation. This phenomenon explains why snow-capped mountains are found even in the Tropics (Fig. 30-4).

Mountains also modify rainfall patterns. When water-laden air is forced to rise as it meets a mountain, it cools.

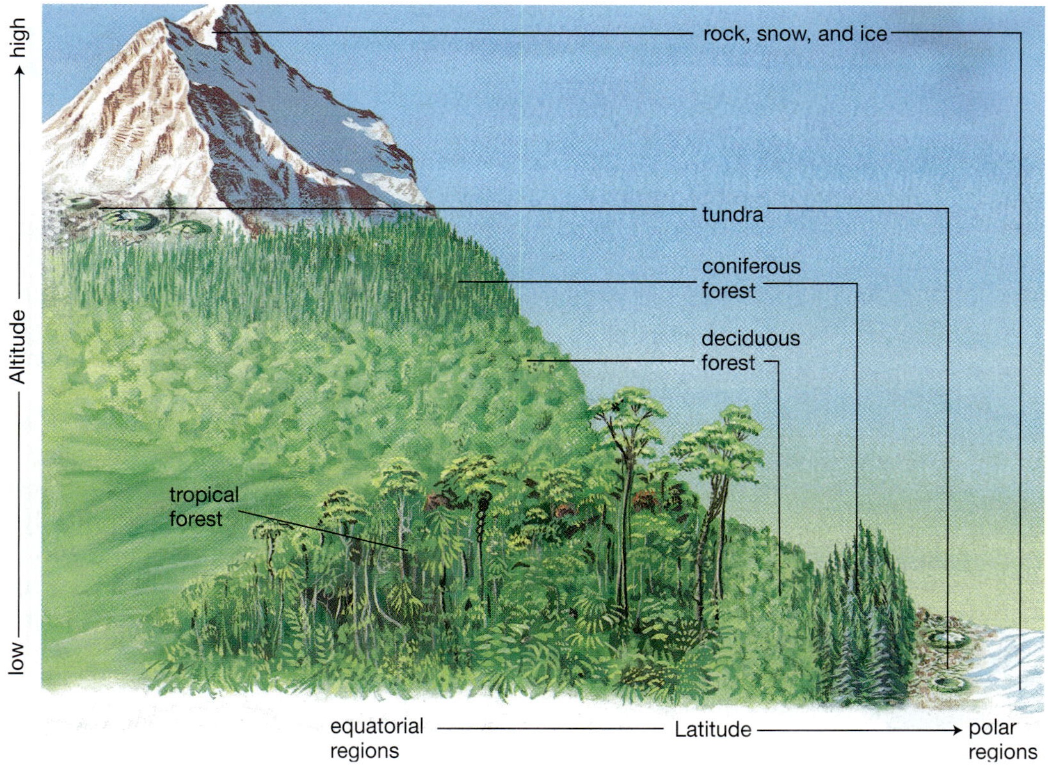

Figure 30-4 Effects of elevation on temperature
Climbing a mountain in some ways is like going toward one of Earth's poles; increasingly cool temperatures produce a similar series of biomes in both cases.

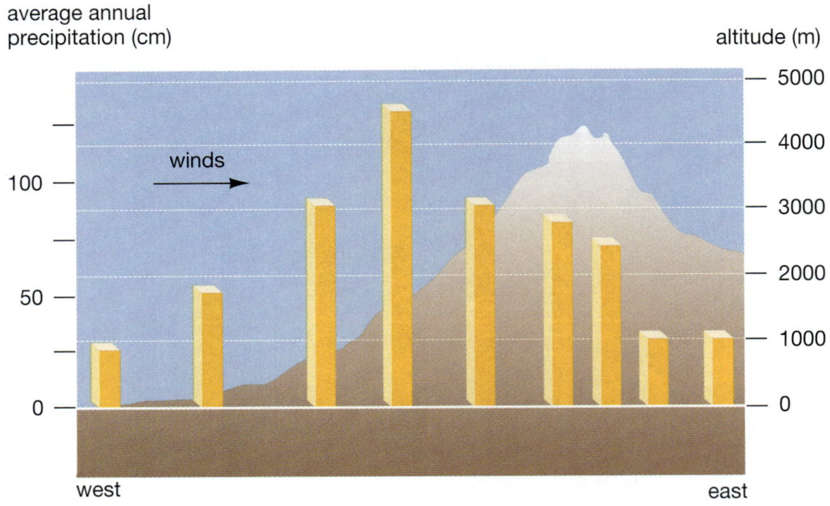

Figure 30-5 The rain shadow of the Sierra Nevada This graph shows how rainfall changes at different locations as westerly (eastward-moving) winds deposit their moisture on the western slope of a mountain, leaving a desert to the east.

Cooling reduces the air's ability to retain water, and the water condenses as rain or snow on the windward (near) side of the mountain. The cool, dry air is warmed again as it travels down the far side of the mountain and absorbs water from the land, creating a local dry area called a **rain shadow**. For example, mountain ranges such as the Sierra Nevada of the western United States wring the moisture from the westerly winds that come off the Pacific Ocean, leaving deserts in the rain shadow on their eastern sides (Fig. 30-5).

2 What Conditions Does Life Require?

From bare Arctic rocks to steamy tropical rivers to the the pressure-cooker conditions of a deep-sea vent, Earth's habitats teem with life. Underlying the diversity of habitats is a common ability to provide four fundamental resources required for life:

- Nutrients from which to construct living tissue
- Energy to power that construction
- Liquid water in which metabolic reactions can take place
- Appropriate temperatures

As we shall see in the following sections, these resources are very unevenly distributed over Earth. Their relative availability determines the types of organisms that live in Earth's various terrestrial and aquatic ecosystems.

The community that is characteristic of each ecosystem contains species that are adapted to particular environmental conditions. The desert community, for example, is dominated by plants with adaptations to hot, dry conditions. These adaptations tend to take similar forms wherever hot, dry conditions prevail. Thus, the cacti of the Mojave Desert of the American Southwest are strikingly similar to the euphorbia of the deserts of Africa, although these plants are in separate families and are only distantly related. Their spinelike leaves and thick, green, water-storing stems are adaptations for water conservation (Fig. 30-6). Likewise, the plants of the arctic tundra and those of the alpine tundra of the Rocky Mountains show growth patterns clearly recognizable as adaptations to a cold, dry, windy climate. Thus, in regions where the environmental conditions are similar, similar types of organisms are organized into similar types of communities.

3 How Is Life on Land Distributed?

The distribution of terrestrial organisms is determined largely by temperature and the availability of water. Other resources tend be comparatively abundant; terrestrial ecosystems receive plenty of light, even on an overcast day, and the soil provides ample nutrients. Water, however, is limited and very unevenly distributed, both in place and in time. Terrestrial organisms must be able to obtain water when it is available and to conserve it when it is scarce.

Like water, temperatures favorable to life are very unevenly distributed in place and time. At the South Pole, even in summer, the average temperature is usually well below freezing; not surprisingly, life is scarce there. Conversely, the Tropics have a uniformly warm, moist climate, and life abounds there. Between these extremes are places that have favorable temperatures during only part of the year.

Terrestrial Biomes Have Characteristic Plant Communities

Terrestrial communities are dominated and defined by their plant life. Large land areas with similar environmental conditions and characteristic plant communities are called **biomes** (Fig. 30-7). Biomes are generally named after the major type of vegetation found there.

Figure 30-6 Environmental demands mold physical characteristics Evolution in response to similar environments has molded the bodies of *(a)* American cacti and *(b)* Canary Islands euphorbia into nearly identical shapes, although they are in different families.

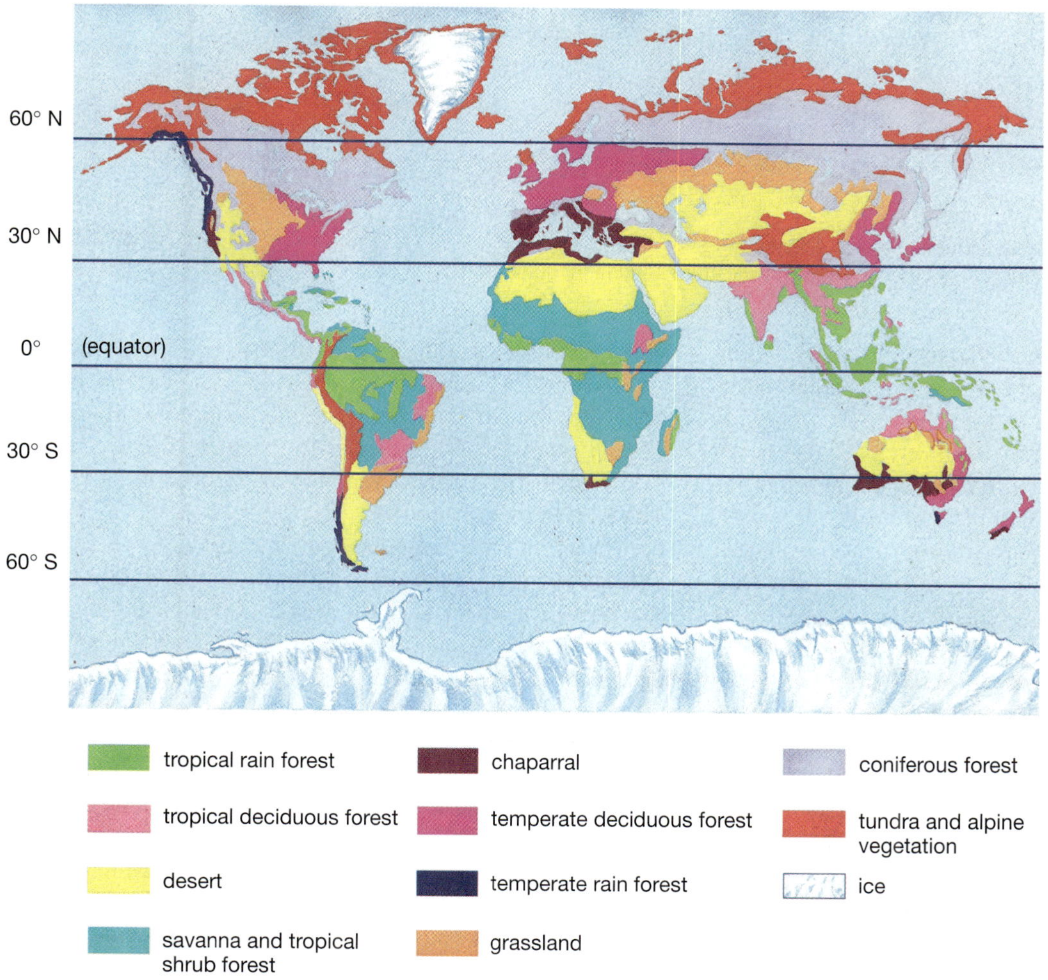

Figure 30-7 The distribution of biomes Although the distribution pattern of biomes is complex, note the overall consistencies. Tundras and coniferous forests are in the northernmost parts of the Northern Hemisphere, whereas the deserts of Mexico, the Sahara, Saudi Arabia, South Africa, and Australia are located about 20° to 30° N and S.

MEDIATUTOR
30.2 Biomes of the World

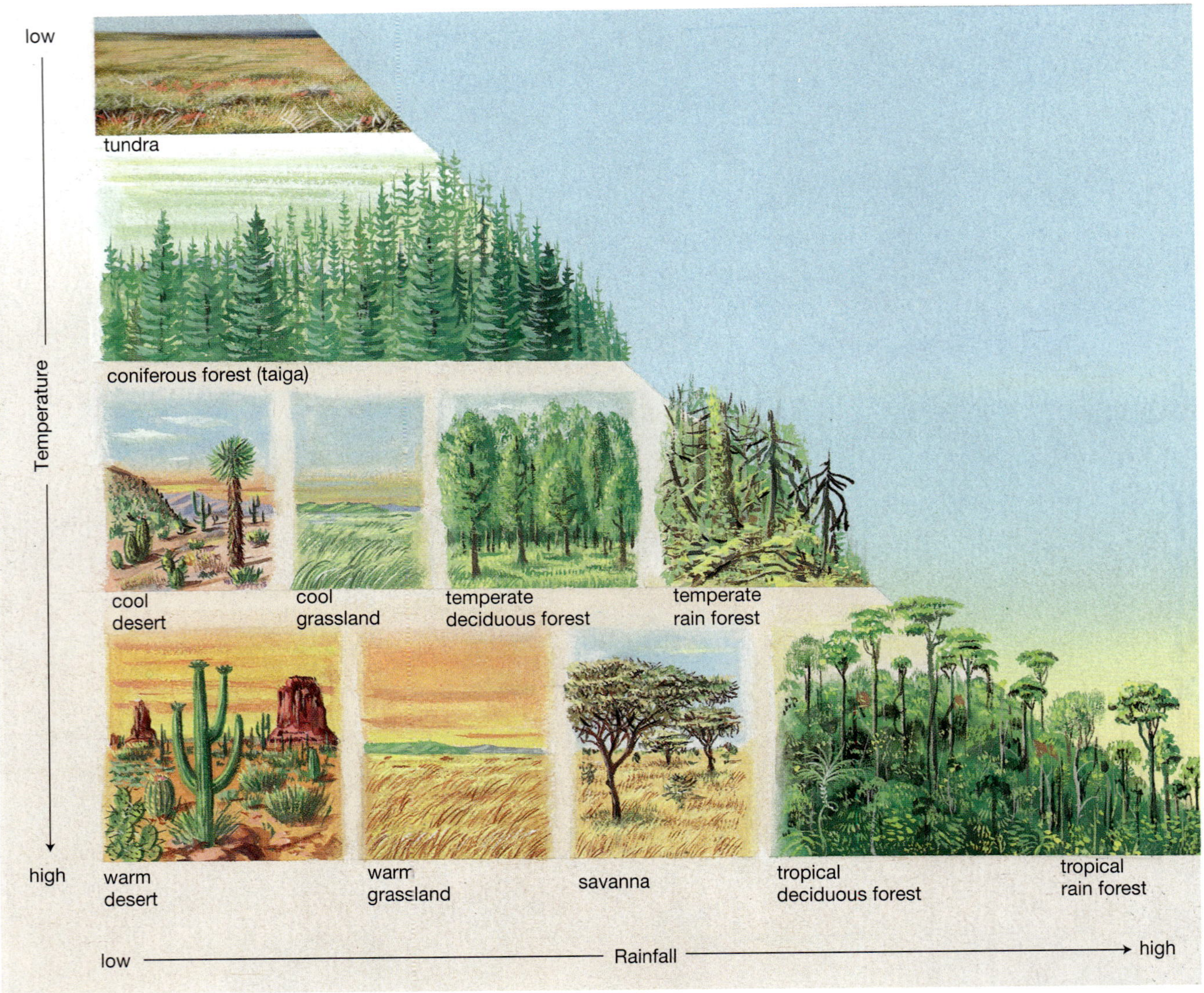

Figure 30-8 Rainfall and temperature influence biome distribution
Together, rainfall and temperature determine the available soil moisture needed for plant growth.

The predominant vegetation of each biome is determined by the complex interplay of rainfall and temperature (Fig. 30-8). Among the factors that affect the types of plants that can grow in a particular area are the area's total amount of rainfall, overall yearly average temperature, and distribution of rain and temperature throughout the year. Plants in the arctic tundra, for example, must survive in marshy conditions in the early summer but cold and desertlike conditions for much of the rest of the year, when water is solidly frozen and unavailable.

In the following sections we discuss the major biomes, beginning at the equator and working our way poleward. We also discuss some of the effects human activities have on these biomes.

Tropical Rain Forests

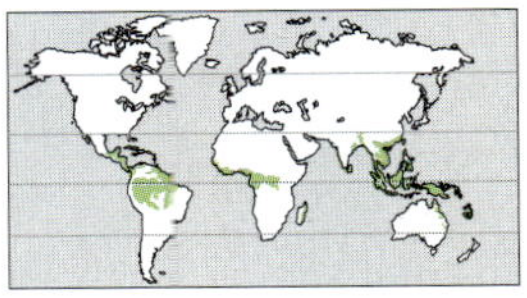

Large areas of South America, Africa, and southeast Asia lie along the equator. There the temperature averages between 77 °F and 86 °F (25 °C and 30 °C) with little seasonal variation, and rainfall ranges from 100 to 160 inches (250 to 400 centimeters) annually. These evenly warm, evenly moist conditions create the most diverse biome on Earth, the **tropical rain forest**, dominated by huge broadleaf evergreen trees (Fig. 30-9).

Although rain forests cover only 6% of Earth's total land area, ecologists estimate that they are home to half to two-thirds of the world's species. **Biodiversity** refers to the total number of species within an ecosystem and the resulting complexity of interactions among them; in short, it defines the biological "richness" of an ecosystem. Rain forests have the highest biodiversity of any ecosystem on Earth. For example, a recent survey of a 2.5-acre site in the upper Amazon basin revealed 283 species of trees, most of which were represented by a single individual. In a 5-square-kilometer (about 3-square-mile) tract of rain forest in Peru, scientists counted more than 1300 species of butterfly and 600 species of bird. In comparison, the entire

Figure 30-9 The tropical rain forest biome
(a) Towering trees draped with vines reach for the light in the dense tropical rain forest. Amid their branches dwells the most diverse assortment of organisms on Earth, including ***(b)*** a fruit-eating toucan, ***(c)*** a golden-eyed leaf frog, ***(d)*** a tree-climbing orchid, and ***(e)*** a howler monkey.

United States is home to only 400 butterfly species and 700 bird species.

Tropical rain forests typically have several layers of vegetation. The tallest trees reach 150 feet (50 meters) and tower above the rest of the forest. Below is a fairly continuous canopy of treetops at about 90 to 120 feet (30 to 40 meters). Another layer of shorter trees typically stands below the canopy. Huge woody vines, commonly 325 feet (100 meters) or more in length, grow up the trees, reaching the sunlight far above. These layers of vegetation block out most of the sunlight, and many of the plants that live in the dim light that filters through to the forest floor have enormous leaves to gather the small amount of available energy.

Because edible plant material close to the ground in tropical rain forests is relatively scarce, much of the animal life, including numerous birds, monkeys, and insects, is *arboreal* (living in the trees). Competition for the nutrients that do reach the ground is intense among both animals and plants. Even such unlikely sources of food as the droppings of monkeys are in great demand. For example, dung beetles feed and lay their eggs on monkey droppings. When ecologists attempted to collect droppings of the South American howler monkeys to find out what the monkeys had been eating, they found themselves in a race with the beetles, hundreds of which would arrive within minutes after a dropping hit the ground!

Almost as soon as bacteria or fungi release nutrients from dead plants or animals into the soil, rainforest trees and vines absorb the nutrients. This is one of the reasons why, despite the teeming vegetation, agriculture is very risky and destructive in rainforest soils. Because the nutrients in a rain forest are tied up in the vegetation, the soil is very infertile. If the trees are carried away for lumber, few nutrients remain to support crops. Further, even if the nutrients are released by burning the vegetation, the heavy year-round rainfall quickly washes them away, leaving the soil infertile after a few seasons of cultivation. The exposed soil, which is rich in iron and aluminum, then takes on an impenetrable, bricklike quality as it bakes in the tropical sun. As a result, secondary succession on cleared rainforest land is slow. Even small forest cuttings take about 70 years to regenerate.

Human Impact Despite their poor suitability for agriculture, rain forests are being felled for lumber or burned down for ranching or agriculture at a rate of at least 38 million acres each year, or more than 70 acres each minute, accounting for 90% of the world's total loss of forests (Fig. 30-10). Rain forest left standing often consists only of fragments too small to allow reproduction of trees and provide adequate habitat for larger animals. Fires in clearcut areas also spread along the ground into adjacent forest, killing young trees and disrupting community interactions. In West Africa, burning the forests is causing acid rain, which is damaging the remaining trees. Further, much of the rainfall in a rain forest comes from water that evaporates from the forest's leaves. As large tracts of rain forests disappear, the region becomes drier, more stressed, and more susceptible to fire.

At least 40% of the world's rain forests are now gone. Harvard University biologist Edward O. Wilson estimates that the destruction of tropical rain forests may be driving 27,000 species to extinction annually, most of them extinguished before they have even been discovered and named. With this loss of biodiversity, humanity loses access to a wealth of potential drugs and raw materials. For example, in Malaysia, the sap of a rare species of tree yielded a compound called *calanolide A*, which offers promise as an anti-AIDS drug. A synthetic version is now being tested in human clinical trials. A compound newly isolated from an African rain forest leaf fungus may eventually help control diabetes, a health problem that is becoming more common in the United States.

Although disastrous losses continue, some areas have been set aside as protected preserves. Local residents in some places are becoming more involved in conservation efforts, as illustrated in this chapter's Case Study. These efforts are steps toward the ultimate solution, which is tragically slow in coming: sustainable use of all forests. Sustainable use means deriving benefits from an ecosystem, whether from tourism or harvesting products, in a way that can be sustained indefinitely without continuing damage to the ecosystem.

Tropical Deciduous Forests

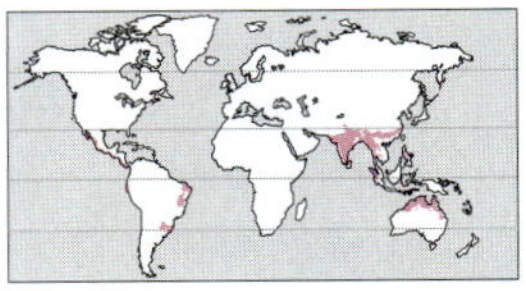

Slightly farther from the equator, the rainfall is not nearly as constant, and there are pronounced wet and dry seasons. In these areas, which include India, much of Southeast Asia, and parts of South and Central America, **tropical deciduous forests** grow. During the dry season, the trees

Figure 30-10 Fire engulfs a tropical rain forest in Brazil
The cleared area will be converted to ranching or agriculture, even though soil quality is poor. Spreading fires and the smoke they produce threaten the uncut forests.

cannot get enough water from the soil to compensate for evaporation from their leaves. As a result, the plants cope with the dry season by shedding their leaves, thereby minimizing water loss. If the rains fail to return on schedule, the trees delay growth of new leaves until the drought passes.

Savannas

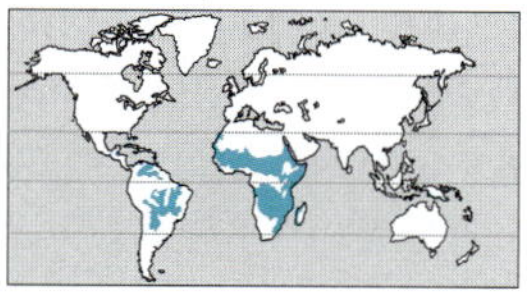

Along the edges of the tropical deciduous forest, the trees gradually become more widely spaced, with grasses growing between them. Eventually, grasses become the dominant vegetation, with only scattered trees and thorny scrub forests here and there; this biome is the **savanna** (Fig. 30-11). Savannas typically have a rainy season during which virtually all of the year's precipitation falls—12 inches (30 centimeters) or less. When the dry season arrives, it comes with a vengeance. Rain might not fall for months, and the soil becomes hard, dry, and dusty. Grasses are well adapted to this type of climate, growing very rapidly during the rainy season and dying back to drought-resistant roots during the arid times. Only a few specialized trees, such as the thorny acacia and the water-storing baobab, can survive the devastating savanna dry seasons. In areas in which the dry season becomes even more pronounced, virtually no trees can grow, and the savanna imperceptibly grades into tropical grassland.

The African savanna has probably the most diverse and impressive array of large mammals on Earth. These mammals include numerous herbivores such as antelope, wildebeest, water buffalo, elephants, and giraffes and such carnivores as the lion, leopard, hyena, and wild dog.

Human Impact Africa's rapidly expanding human population threatens the wildlife of the savanna. Poaching has driven the black rhinoceros to the brink of extinction (Fig. 30-12) and endangers the African elephant. The abundant grasses that make the savanna a suitable habitat for so much wildlife also make it suitable for grazing domestic cattle. As the human population of East Africa increases, so does cattle grazing on the savanna. Fences increasingly disrupt the migration of the great herds of herbivores, which move in search of food and water. Ecologists have discovered that the native herbivores are much more efficient at converting grass

Figure 30-11 The African savanna
(a) Elephants roam while ***(b)*** a cheetah feasts on its prey. ***(c)*** Large herds of grazing animals, such as zebras, can still be seen on African preserves. The herds of herbivores provide food for the greatest assortment of large carnivores on Earth.

Figure 30-12 Poaching threatens African wildlife
Rhinoceros horns, believed by some to have aphrodisiac properties, fetch staggering prices and encourage poaching. The black rhino is now nearly extinct.

into meat than are cattle. Perhaps the future African savanna will support herds of domesticated antelope and other large native grazers in place of cattle.

Deserts

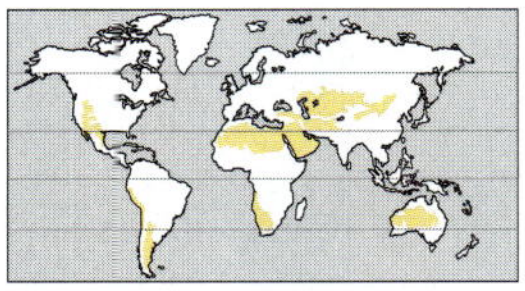

Even drought-resistant grasses need at least 10 to 20 inches (25 to 50 centimeters) of rain a year. Where less than 10 inches of rain fall, **deserts** are found. Although we tend to think of deserts as hot, they are defined by lack of rain rather than by temperature. In the Gobi desert of Asia, for example, temperatures average below freezing for half the year, while summer temperatures average 105 °F to 110 °F (41 °C to 43 °C). The desert biome is found on every continent, mostly around 20° to 30° N and S latitude and also in the rain shadows of major mountain ranges. The desert category encompasses a range of environments. At one extreme are certain parts of the Sahara Desert and deserts in Chile, where it virtually never rains and no vegetation grows (Fig. 30-13a). More commonly, deserts are characterized by widely spaced vegetation and large areas of bare ground. The plants tend to be spaced evenly, as if planted by hand (Fig. 30-13b). In many cases, the perennial plants are bushes or cacti with large, shallow root systems. The shallow roots quickly soak up the soil moisture after the infrequent desert storms. The rest of the plant is typically covered with a waterproof, waxy coating to prevent evaporation of precious water. Water is stored in the thick stems of cacti and other succulents. The spines of cacti are leaves modified for protection and water conservation, presenting almost no surface area for evaporation. In many deserts, all the rain falls in just a few storms, and specialized annual wildflowers take advantage of the brief period of moisture to race through germination, growth, flowering, and seed production in a month or less (Fig. 30-14).

The animals of the deserts, like the plants, are adapted to survive on little water. Most deserts appear to be devoid of animal life during summer days because the animals seek relief from the sun and heat in cool, underground burrows. After dark, when deserts cool down considerably, lizards, snakes, and other reptiles emerge to feed, as do mammals such as the kangaroo rat (Fig. 30-13c) and birds such as the burrowing owl. Most of the smaller animals survive without ever drinking, getting all the water they need from their food and from that produced during cellular respiration in their tissues. Larger animals, such as desert bighorn sheep, are dependent on permanent water holes during the driest times of the year.

Human Impact Desert ecosystems are fragile. Ecologists studying the soil of the Mojave desert in southern California recently found treadmarks left by

(a) (b) (c)

Figure 30-13 The desert biome
(a) Under the most extreme conditions of heat and drought, as in these sand dunes of the Sahara Desert in Africa, deserts can be almost devoid of life. ***(b)*** Throughout much of Utah and Nevada, the Great Basin Desert presents a landscape of widely spaced shrubs, such as sagebrush and greasewood. These shrubs often secrete a growth inhibitor from their roots, preventing germination of nearby plants and thus reducing competition for water. ***(c)*** The kangaroo rat is an elusive inhabitant of the deserts of North America.

Figure 30-14 The Sonoran Desert
In spring, this Arizona desert is carpeted with annual wildflowers. Through much of the year, the wildflower seeds lie dormant, waiting for the spring rains to fall.

tanks in 1940 when General Patton trained tank crews in preparation for entry into World War II. The desert soil here is stabilized and enriched by microscopic cyanobacteria whose filaments intertwine among sand grains. Vehicles, such as the numerous off-road vehicles that careen about the desert for recreation, destroy this crucial network, allowing the soil to erode and reducing nutrients available to the desert's slow-growing plants. Ecologists estimate that the desert soil may require hundreds of years to fully recover from heavy vehicle use.

Scientists once believed that human activities were causing a progressive southward spread of the Sahara desert in Africa, a process called desertification. Recently published data from satellite photographs taken at intervals since 1980 indicate that the Sahara Desert has spread southward and then has retreated back repeatedly over this 20-year period in accordance with changes in the amount of rainfall. Although the dry savanna south of the Sahara Desert has been significantly overgrazed and degraded by a human population

Figure 30-15 The chaparral biome
This biome is limited primarily to coastal mountains in dry regions, such as the San Gabriel Mountains in southern California. Chaparral is maintained by frequent fires set by summer lightning. Although the tops of the plants may be burned off, the roots send up new sprouts the following spring.

that is above its sustainable carrying capacity, the desert biome is not spreading as a result, as was formerly believed.

Chaparral

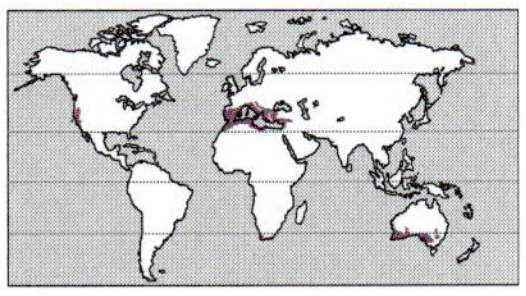

In many coastal regions that border on deserts, such as southern California and much of the Mediterranean, we find a distinctive type of vegetation called **chaparral**. The annual rainfall in these regions is up to 30 inches, nearly all of which falls during cool, wet winters that alternate with hot, dry summers. The proximity of the sea lengthens the winter rainy season and causes frequent fogs during the dry season, reducing evaporation. Chaparral consists of small trees or large bushes such as sages and evergreen oak, with thick waxy or fuzzy leaves that conserve water. These hardy shrubs are also able to withstand the frequent summer fires started by lightning (Fig. 30-15).

Grasslands

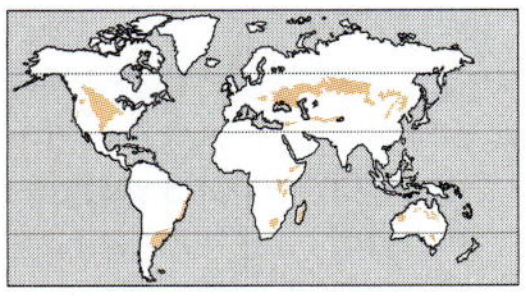

Most **grassland**, or **prairie**, biomes are located in the centers of continents, where they receive 10 to 30 inches (25 to 75 centimeters) of rain annually (note the large grasslands

Figure 30-16 Tallgrass prairie
In the central United States, moisture-bearing winds out of the Gulf of Mexico produce summer rains, allowing a lush growth of tall grasses and wildflowers such as these coneflowers. Periodic fires, now carefully managed, prevent encroachment of forest.

in the centers of the North American and Eurasian continents). In general, they have a continuous cover of grass and virtually no trees, except along rivers. From the tallgrass prairies of Iowa, Missouri, and Illinois (Fig. 30-16) to the shortgrass prairies of eastern Colorado, Wyoming, and Montana (Fig. 30-17), the North American grassland once stretched across almost half the continent. These grasslands, on which grasses have grown and decomposed for thousands of years, contain what may be the most fertile soil in the world.

Water and fire are the crucial factors in the competition between grasses and trees. The hot, dry summers and frequent droughts of the shortgrass prairies can be tolerated by grass but are fatal to trees. Forests are the climax ecosystems in the more eastern tallgrass prairies, but historically the trees were destroyed by frequent fires, often set by Native Americans to maintain grazing land for the bison. Trees are killed outright by fire, but the root systems of grasses usually survive, even if the tops of the grasses are destroyed.

The grasslands of North America once supported huge herds of bison—as many as 60 million in the early nineteenth century. Pronghorn antelope can still be

Figure 30-17 Shortgrass prairie
The lands east of the Rocky Mountains receive relatively little rainfall and support ***(a)*** shortgrass prairie, which is characterized by low-growing bunch grasses such as buffalo grass and grama grass. ***(b)*** Pronghorn antelope, ***(c)*** prairie dogs, and ***(d)*** protected bison herds occupy this biome, in which ***(e)*** wildflowers such as this coneflower abound.

seen in some prairies of the western United States; bobcats and coyotes are the major large predators there (a prairie food web is illustrated in Fig. 29-5).

Human Impact When people developed plows that could break through the dense grass turf, the stage was set for converting the midwestern U.S. prairies into the "breadbasket" of North America, so named because enormous quantities of grain are cultivated in its fertile soil. The tallgrass prairie has been converted to agricultural land, except for tiny protected remnants, which are maintained by periodic controlled burning.

On the dry western shortgrass prairie, cattle have replaced the bison and pronghorn antelope. As a result of their overgrazing the grasses, the boundary between the cool deserts and the grassland has commonly been altered in favor of desert plants. Much of the sagebrush desert of the American West is actually overgrazed shortgrass prairie (Fig. 30-18). Cattle prefer grass to sagebrush, so heavy grazing destroys the grass. Consequently, moisture that the grass would have absorbed is left in the soil, encouraging the growth of the woody sagebrush. Thus, the prairie grasses are replaced by plants characteristic of the cool desert.

Figure 30-18 Sagebrush desert or shortgrass prairie?
The shortgrass prairie field on the right has been overgrazed by cattle, causing the grasses to be replaced by sagebrush.

Temperate Deciduous Forests

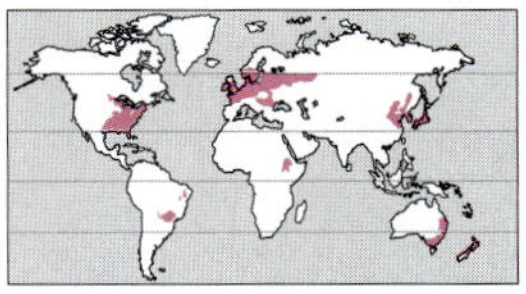

At their eastern edge, the North American grasslands merge into the **temperate deciduous forest** biome, also found in Western Europe and East Asia (Fig. 30-19).

Figure 30-19 The temperate deciduous forest biome
(a) In temperate deciduous forests of the eastern United States, ***(b)*** the white-tailed deer is the largest herbivore, and ***(c)*** birds such as the blue jay are abundant. ***(d)*** In spring, a profusion of woodland wildflowers (such as these hepaticas) blooms briefly before the trees produce leaves.

Figure 30-20 The temperate rain forest biome
(a) Temperate rain forest in Olympic National Park. Ferns, mosses, and wildflowers grow in the pale green light of the forest floor. *(b)* The dead feed the living, as new trees grow from the decay of this fallen giant, called a "nurse log," and *(c)* flowering foxglove and *(d)* fungi find ideal conditions amid the moist, decaying vegetation.

Precipitation is higher than in grasslands (30 to 60 inches, or 75 to 150 centimeters) and, in particular, more rain falls during the summer. The soil retains enough moisture for trees to grow, and the resulting forest shades out grasses. In contrast to the tropical forests, the temperate deciduous forest biome has cold winters, usually with at least several hard frosts and often with long periods of below-freezing weather. Winter in this biome has an effect on the trees similar to that of the dry season in the tropical deciduous forests: During periods of subfreezing temperatures, liquid water is not available to the trees. To reduce evaporation when water is in short supply, the trees drop their leaves in the fall. They produce leaves again in the spring, when liquid water becomes available. During the brief time in spring when the ground has thawed but the leaves of trees have not yet blocked off all the sunlight, abundant wildflowers grace the forest floor.

Insects and other arthropods are numerous and conspicuous in deciduous forests. The decaying leaf litter on the forest floor also provides food and habitat for bacteria, earthworms, fungi, and small plants. Many arthropods feed on these or on each other. Vertebrate animals, including mice, shrews, squirrels, raccoons, deer, bear, and many species of birds, dwell in the deciduous forests.

Human Impact Large predatory mammals such as black bear, wolves, bobcats, and mountain lions were formerly common in deciduous forests, but hunting and habitat loss have eliminated the wolves and severely reduced the populations of the others. In many deciduous forests, deer are abundant because few of their predators remain. Clearing for lumber, agriculture, and housing has dramatically reduced deciduous forests in the United States from their original extent, and virgin deciduous forests are now almost nonexistent. Over the past 50 years, however, Forest Service data show that forest cover in the United States (both evergreen and deciduous) has increased due to regrowth of forests on abandoned farms, paper recycling that decreases demand for wood pulp, more efficient lumber milling and tree farming techniques, and the use of alternative building materials.

Figure 30-21 The taiga (or northern coniferous forest) biome
(a) The small needles and pyramidal shape of conifers allow them to shed heavy snows. *(b)* Winter is a challenge not only for the trees but also for animals such as this snowshoe hare and the bobcat that preys on it. The hare is also prey for *(c)* the great horned owl.

Temperate Rain Forests

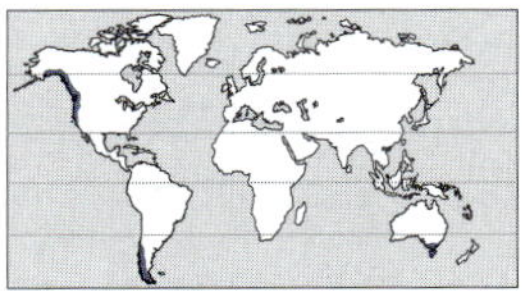

On the U.S. Pacific coast, from the lowlands of the Olympic Peninsula in Washington State to southeast Alaska, lies the **temperate rain forest** biome (Fig. 30-20). Temperate rain forests, which are relatively rare, are also located along the southeastern coast of Australia and the southwestern coast of New Zealand. As in the tropical rain forest, there is ample liquid water year-round. Water is abundant for two reasons. First, there is a tremendous amount of rain. The Hoh River rain forest in Olympic National Park receives more than 160 inches (400 centimeters) of rain annually, more than 24 inches (60 centimeters) in the month of December alone. Second, the moderating influence of the Pacific Ocean keeps temperatures mild along the coast, so the ground seldom freezes and liquid water remains available.

The abundance of water means that the trees have no need to shed their leaves in the fall, and almost all the trees are evergreens. In contrast to the broadleaf evergreen trees of the Tropics, temperate rain forests are dominated by conifers. The ground and the trunks of the trees are covered with mosses and ferns. As in tropical rain forests, so little light reaches the forest floor that tree seedlings usually cannot become established. Whenever one of the forest giants falls, however, it opens up a patch of light, and new tree seedlings quickly sprout, commonly right atop the fallen log. This event produces a "nurse log" (Fig. 30-20b).

Taiga

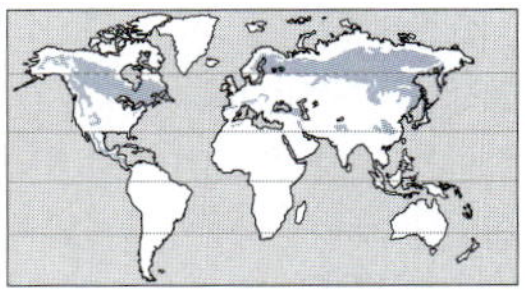

North of the grasslands and temperate forests, the **taiga**, also called the northern coniferous forest (Fig. 30-21), stretches across all of North America and Eurasia, including parts of the northern United States and much of southern Canada. Conditions in the taiga are harsher than those in the temperate deciduous forest. In the taiga, the winters are longer and colder, and the growing season is shorter. The few months of warm weather are too short to allow trees the luxury of regrowing leaves in the spring. As a result, the trees of the taiga are almost all evergreen conifers with narrow, waxy needles. The waxy coating and small surface area of the needles reduce water loss by

Figure 30-22 Clear-cutting
The bare mountains of this Oregon forest have been clear-cut. Clear-cutting is relatively simple and cheap—but its environmental costs are high. Erosion will diminish the fertility of the soil, slowing new growth. Further, the dense stands of same-age trees that typically regrow are more vulnerable to attack by parasites than a natural stand of trees of various ages would be.

evaporation during the cold months, and the leaves remain on the trees year-round. Thus, the trees are instantly ready to take advantage of good growing conditions when spring arrives, and they can continue slow growth late into the fall.

Because of the harsh climate in the taiga, the diversity of life there is much lower than in many other biomes. Vast stretches of central Alaska, for example, are covered by a somber forest that consists almost exclusively of black spruce and an occasional birch. Large mammals such as the wood bison, grizzly bear, moose, and wolf, which have mostly been eradicated in the southern regions of their original range, still roam the taiga, as do smaller animals such as the wolverine, fox, snowshoe hare, and deer.

Human Impact The taiga is a major source of lumber for construction. *Clear-cutting*, the removal of all the trees in a given area, has destroyed huge expanses of forest, in both Canada and the Pacific Northwest in the United States (Fig. 30-22). Owing to the remoteness of the northernmost taiga and the severity of its climate, a greater percentage of the taiga remains in undisturbed condition than any other North American biome except the tundra.

Tundra

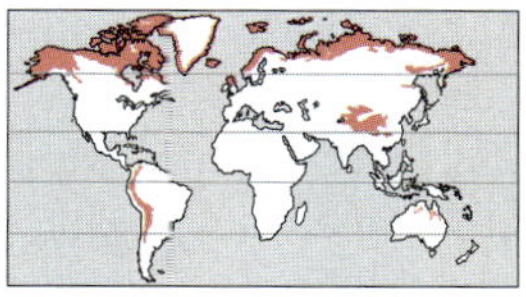

The last biome we encounter before reaching the polar ice cap is the arctic **tundra**, a vast treeless region bordering the Arctic Ocean (Fig. 30-23). Conditions in the tundra are severe. Winter temperatures often reach –40 °F (–55 °C) or below, winds howl at 30 to 60 miles (50 to 100 kilometers) an hour, and precipitation averages 10 inches (25 centimeters) or less each year, making this a "freezing desert." Even during the summer, the temperature can drop to freezing, and the growing season may last only a few weeks before a hard frost occurs. Somewhat less cold but similar conditions produce alpine tundra on mountaintops above the elevation where trees can grow.

The cold climate of the arctic tundra results in **permafrost**, a permanently frozen layer of soil typically no more than about 1.5 feet (45 centimeters) below the surface. As a result, when summer thaws come, the water from melted snow and ice cannot soak into the ground, and the tundra becomes a huge marsh. Trees cannot survive in the tundra; the permafrost limits root growth to the topmost layer of soil.

Nevertheless, the tundra supports a surprising abundance and variety of life. The ground is carpeted with small perennial flowers and dwarf willows no more than a few centimeters tall and is often covered with a large lichen called "reindeer moss," a favorite food of caribou. The standing water provides a superb habitat for mosquitoes. The mosquitoes and other insects provide food for numerous birds, most of which migrate long distances to nest and raise their young during the brief, insect-filled summer. The tundra vegetation supports lemmings, which are eaten by wolves, snowy owls, arctic foxes, and even grizzly bears.

Human Impact The tundra is among the most fragile of all the biomes because of its short growing season. A willow 4 inches (10 centimeters) high may have a trunk 3 inches (7 centimeters) in diameter and be 50 years old. Human activities in the tundra leave scars that persist for centuries. Fortunately for the tundra inhabitants, the impact of civilization is localized around oil drilling sites, pipelines, mines, and military bases.

Rainfall and Temperature Determine the Vegetation a Biome Can Support

Terrestrial biomes are greatly influenced by both temperature and rainfall, whose effects interact. Temperature strongly influences the effectiveness of rainfall in providing soil moisture for plants and standing water for animals to drink. The hotter it is, the more rapidly water evaporates, both from the ground and from plants. As a result of the interaction of temperature with rainfall (and to a lesser extent, the distribution of rain throughout the year), areas that receive almost exactly the same rainfall can have startlingly different vegetation, all the way from desert to taiga. Now that you are familiar with biomes, take a trip with us from southern Arizona to central Alaska as we visit ecosystems that each receive about 12 inches (28 centimeters) of rain annually.

The Sonoran Desert near Tucson, Arizona (see Fig. 30-14), has an average annual temperature of 68 °F (20 °C) and receives about 12 inches (28 centimeters) of rain each year. The landscape is dominated by giant

Figure 30-23 The tundra biome
(a) Life on the tundra is adapted to cold. ***(b)*** Plants such as dwarf willows and perennial wildflowers (such as this dwarf clover) grow low to the ground, escaping the chilling tundra wind. Tundra animals, such as ***(c)*** caribou and ***(d)*** arctic foxes, can regulate blood flow in their legs, keeping them just warm enough to prevent frostbite, while preserving precious body heat for the brain and vital organs.

saguaro cactus and low-growing, drought-resistant bushes. Going north for 900 miles (1500 kilometers) brings you into eastern Montana, where rainfall is about the same, but the vegetation is shortgrass prairie (see Fig. 30-17), largely because the average temperature is much lower, about 45 °F (7 °C). Much farther north, central Alaska receives about the same annual rainfall (12 inches), yet it is covered with taiga forest (see Fig. 30-21). As a result of the low average annual temperature (about 25 °F, or –4 °C), permafrost underlies much of the ground here. During the summer thaw, the taiga earns its Russian name "swamp forest," although its rainfall is about the same as that of the Sonoran Desert.

4 How Is Life in Water Distributed?

Saltwater oceans and seas are the largest ecosystems on Earth, covering about 71% of its surface. Freshwater ecosystems, in contrast, cover less than 1%.

The unique properties of water lend some common features to aquatic ecosystems. First, because water is slower to heat and cool than air, temperatures in aquatic ecosystems are more moderate than are those in terrestrial ecosystems. Second, although water may appear quite transparent, it absorbs a considerable amount of the light energy that sustains life. Even in the clearest water, the intensity of light decreases rapidly with depth. At depths of 650 feet (200 meters) or more, little light is left to power photosynthesis. If the water is at all cloudy—for example, because of suspended sediment or microorganisms—the depth to which light can penetrate is greatly reduced.

Of the four requirements for life, aquatic ecosystems provide abundant water and appropriate temperatures. Thus, the major factors that determine the quantity and type of life in aquatic ecosystems are the remaining two factors: energy and nutrients. Nutrients in aquatic ecosystems tend to be concentrated near the bottom sediments where light levels are often too low to support photosynthesis. This separation of energy and nutrients limits aquatic life.

There are many different kinds of aquatic ecosystems. Freshwater ecosystems encompass rivers, streams, ponds, lakes, and marshes; marine (saltwater) ecosystems include estuaries, tide pools, coral reefs, the open ocean, and hydrothermal vent communities. In the following sections, we look more closely at some of these important aquatic ecosystems.

Freshwater Lakes Have Distinct Regions of Life

Freshwater lakes vary tremendously in size, depth, and nutrient content. Although each lake is unique, moderate to large lakes in temperate climates share some common features, including distinct zones of life.

Earth Watch
Humans and Ecosystems

The expanding human population has left relatively few ecosystems undisturbed (Fig. E30-2). Human impacts on natural ecosystems are so diverse and wide ranging that they far exceed the scope of this book. We can, however, identify some general characteristics of ecosystems dominated by humans and contrast them with the characteristics of undisturbed ecosystems. Listed below are six characteristic differences between these two types of ecosystems.

Ecosystems dominated by humans tend to be simpler—to have fewer species and fewer community interactions—than do undisturbed ecosystems. As humans enter an ecosystem, animals in the highest trophic levels are the first to go. Even in undisturbed ecosystems, carnivores are relatively rare, and their specialized needs are easily disrupted. Many big carnivores, such as the wolf and mountain lion, require large, undisturbed hunting territories. Humans often selectively destroy large predators to reduce the threat they pose to people or livestock. As a result, even in relatively undisturbed areas many large carnivores have been eliminated. Grizzly bears and wolves no longer roam the southern Rocky Mountains. Smaller animals can also fall victim to human encroachment. On the western shortgrass prairie, prairie dog towns disappear as human towns and ranches rise. The black-footed ferret, a predator of prairie dogs, faces possible extinction. Agriculture is another human activity that simplifies ecosystems. Farm fields are deliberately simplified relative to the original biome to eliminate competition and predation and allow the maximum possible productivity of a single crop species.

There is probably no practical way to restore to human-dominated ecosystems the great diversity found in undisturbed areas, nor is doing so always desirable from a human standpoint. The simplified ecosystems of agriculture are necessary to support human populations at their current levels. Cities concentrate human activities and culture and may lessen the human impact on the surrounding countryside. Even as we recognize the benefits of artificially simplified ecosystems, however, we must be aware of the need to preserve intact as many natural communities as possible. The undisturbed forest traps and purifies water and can reduce air pollution. Swamps and estuaries contain a wealth of detritus feeders and decomposers that purify water. Coastal wetlands are breeding places for millions of birds and spawning sites for the majority of our commercially important fish and crustacean species. In undisturbed, diverse ecosystems we find aesthetic pleasure as well as a storehouse of species whose commercial or medicinal values are not yet recognized. For example, nearly half the medicines in use today were originally discovered in plants, and we have examined only a small fraction of existing plants for possible medical uses.

Figure E30-2 Habitat destruction
Loss of habitat due to human activities is a major threat to most of Earth's wildlife.

Natural ecosystems run on sunlight, but human ecosystems are dependent on nonrenewable energy from fossil fuels. From the suburbanite pouring gas into a lawn mower to the farmer driving a tractor, managing a simplified ecosystem is an energy-intensive proposition. Energy must be expended to oppose the tendency of the natural system to restore complexity.

Agriculture is among the human activities that require large inputs of fossil fuel. To counteract this trend, some farmers are returning to organic farming. Instead of using fertilizers that require large amounts of energy to produce, organic farmers use plant and animal waste to maintain soil fertility. In addition, the practice of alternating legume crops, such as soybeans and alfalfa, with other crops helps maintain nitrogen in the soil. Mulching can help retain fertility, reduce water loss, and control weeds. The use of natural insect predators and limited pesticide spraying at critical times during a pest's life cycle can dramatically reduce the need for poisons. Organic farms can be as productive as more conventional farms while using 15% to 50% less energy from fossil fuels.

Life Zones Are Determined Light and Nutrients

The distribution of life in lakes depends largely on access to light, nutrients, and, in some cases, a place for attachment (the bottom). The life zones of lakes correspond to specific locations within the lake. We recognize three such zones: the littoral zone, the limnetic zone, and the profundal zone (Fig. 30-24).

Near the shore is the **littoral zone**. In this zone, the water is shallow, and plants find abundant light, anchorage, and adequate nutrients from the bottom sediments. Not surprisingly, littoral-zone communities are the most diverse. Cattails and bulrushes abound nearest the shore, and water lilies and entirely submerged vascular plants and algae may flourish at the deepest

In our homes and commercial buildings, better insulation and increased use of solar heat can result in dramatic energy savings. By increasing our use of renewable energy sources, such as wind and especially sunlight, we can conserve fossil fuels, dramatically reduce pollution, and move human ecosystems a step closer to those that occur naturally.

Natural ecosystems recycle nutrients; human ecosystems tend to lose nutrients. Walk around a suburban neighborhood on trash-pickup day. Grass clippings and leaves are packed in plastic bags to be hauled away. To compensate for this loss of natural nutrients, many suburbanites heavily fertilize their lawns and gardens. Modern farming exhibits a similar trend: Exposed soil is eroded away by wind and water, removing crucial nutrients and requiring large inputs of fertilizer to replace them. Runoff from fields, carrying fertile topsoil and artificial fertilizers, may pollute nearby rivers, streams, and lakes. Pesticides may kill detritus feeders and decomposers, further disrupting natural nutrient cycles. Thus, whereas fertile soil accumulates in many natural ecosystems, it tends to be lost in those dominated by humans. Some 3 billion tons of topsoil are eroded annually from farms in the United States. The Mississippi River alone carries off 40 tons each hour.

Changes in farm techniques can help reverse this trend too. Farmers can counteract erosion by using contour planting, in which row crops are oriented so that they slow the flow of water instead of funneling it downslope. Row crops can also be alternated in strips with dense, soil-catching crops such as wheat. Planting rows of trees as windbreaks helps prevent soil loss from blowing wind, and it creates a more diverse ecosystem and a nesting place for insect-eating birds. Farm fields need not be plowed in the fall and left unplanted to erode during the winter; in fact, an increasing number of farmers are planting crops in the stubble from the previous season, reducing erosion.

Natural ecosystems tend to store water and purify it through biological processes, while human ecosystems tend to pollute water and shed it rapidly. A thundershower strikes a forest, an adjacent city, and a farm. The rich soil of the forest sponges up the water, which gradually drains into the ground, filtered by the soil and purified by decomposers that break down organic contaminants. In the nearby city, water pours from sidewalks, rooftops, and streets, picking up soot, silt, oil, heavy metals, and garbage. It races down gutters into storm sewers, and a weakly toxic soup gushes into the nearest stream or river. Farm runoff carries priceless topsoil, expensive fertilizer, and animal manure into rivers and lakes, where these potential resources become pollutants. If we take steps to reduce erosion, we will simultaneously conserve water and reduce water pollution from farm runoff.

Natural ecosystems tend to remain stable over time, but human ecosystems tend to be unstable. In natural ecosystems, populations of herbivorous insects are controlled by predators, including other insects, birds, and mammals. Insect populations are also limited because their preferred food species are scattered among many plants of many other species, rather than growing in a pure stand, as on a farm. On farms, both pest insects and their natural predators are exposed to pesticides, but if the pest species evolves resistance to the poison and the predator does not, the pest population may explode. Simplified communities are also more vulnerable to disruption by unfavorable weather or the introduction of exotic species.

Farmers can counteract such instability by planting smaller fields with a wider variety of crops. Alternating crops helps maintain soil fertility and also helps prevent the proliferation of diseases and insect pests that are specialized for a particular crop. Populations of corn borers, for example, will die off during years when the field is planted in alfalfa. The use of biological controls such as natural insect predators and insect diseases can reduce reliance on pesticides.

Sizes of nonhuman populations in natural ecosystems are relatively constant, but the human population grows continuously. As our population expands, the spread of human-dominated ecosystems presents a growing threat to the diversity of species and to the delicate balance that has evolved over the 3.5-billion-year history of life on Earth.

In summary, natural ecosystems tend to be complex, stable, and self-sustaining, powered by solar energy and nourished by recycled nutrients. They provide diverse habitats for wildlife, purify contaminants through the action of decomposers, and build up nutrient-rich soil. Modern human ecosystems are relatively simple and are sustained by large inputs of energy from fossil fuels. They tend to reduce wildlife habitat, contaminate soil and water, and lose nutrients and soil. These problems are compounded by continued population growth, which is causing the expansion of human-dominated ecosystems at the expense of undisturbed ones.

Human ecosystems need not be as disruptive and alien to the operation of natural ecosystems as we have allowed them to become. Through understanding, education, commitment, appropriate use of technology, and stabilization of our population, we can reverse many of these destructive trends.

reaches of the littoral zone. The plants of the littoral zone trap sediments carried in by streams and by runoff from the surrounding land, increasing the nutrient content in this region. Living among the anchored plants are microscopic organisms called plankton. There are two categories of plankton: autotrophic (photosynthesizing) **phytoplankton**, including protists, bacteria, and algae, and heterotrophic, nonphotosynthetic **zooplankton**, such as protists and tiny crustaceans. The greatest diversity of animals in the lake is also found in this zone. Littoral invertebrate animals include small crustaceans, insect larvae, snails, flatworms, and hydra; littoral vertebrates include frogs, minnows, and aquatic snakes and turtles.

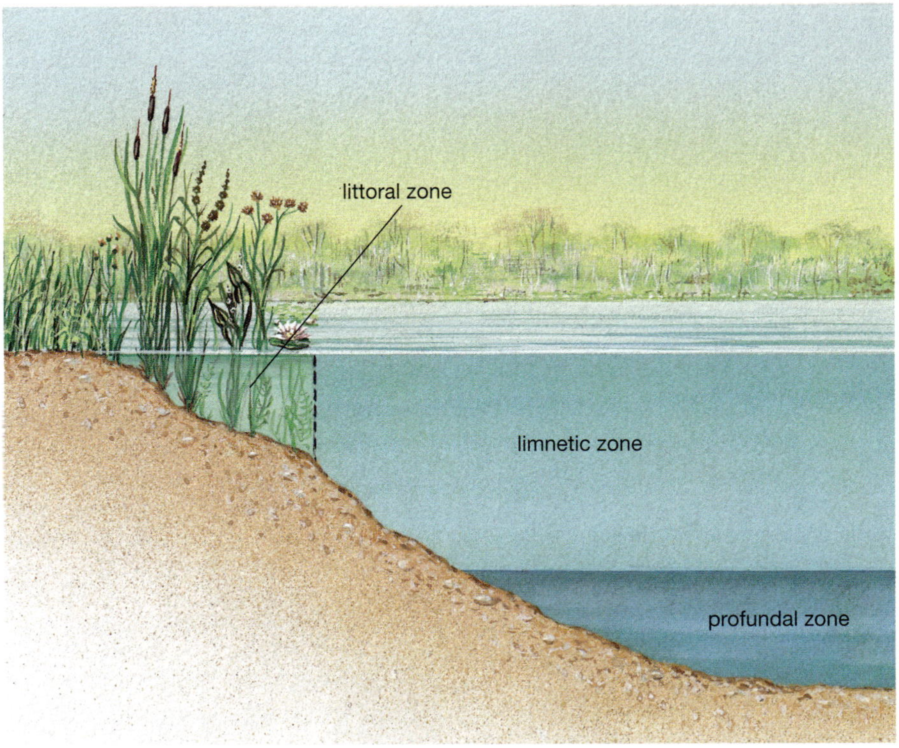

Figure 30-24 Life zones of a lake
There are three life zones in a typical lake: a near-shore littoral zone with rooted plants, an open-water limnetic zone, and a deep, dark profundal zone.

As the water increases in depth farther from shore, plants are unable to anchor to the bottom and still collect enough light for photosynthesis. This open-water area is divided into two regions: the upper limnetic zone and the lower profundal zone (see Fig. 30-24). In the **limnetic zone**, enough light penetrates to support photosynthesis. Here phytoplankton including cyanobacteria serve as producers. These are eaten by zooplankton and small crustaceans, which in turn are consumed by fish. Below the limnetic zone lies the **profundal zone**, where light is insufficient to support photosynthesis. This area is nourished mainly by detritus that falls from the littoral and limnetic zones and by incoming sediment. It is inhabited primarily by decomposers and detritus feeders, such as bacteria, snails and insect larvae, and by fish that swim freely among the different zones.

Freshwater Lakes Are Classified by Nutrient Content

Although each lake is unique, freshwater lakes can be classified on the basis of their nutrient content as either oligotrophic or eutrophic.

Oligotrophic lakes are very low in nutrients. Many are formed by glaciers that scrape depressions in bare rock, and they are fed by mountain streams carrying little sediment. Because there is little sediment or microscopic life to cloud the water, oligotrophic lakes are clear, and light penetrates deeply. Therefore, photosynthesis (and by extension oxygenation, as oxygen is produced as a by-product of photosynthesis) in deeper water is possible, and the limnetic zone may extend all the way to the center of the lake. Oxygen-loving fish, such as trout, thrive in oligotrophic lakes.

Eutrophic lakes receive larger inputs of sediments, organic material, and inorganic nutrients (such as phosphorus) from their surroundings, allowing them to support dense communities. They are murkier, from both suspended sediment and dense phytoplankton populations, so the lighted limnetic zone is shallower. Dense "blooms" of algae occur seasonally in the limnetic zone. Their dead bodies fall into the profundal zone, where they are consumed by decomposers. The metabolic activities of the decomposers use oxygen, depleting the oxygen content of the profundal zone in eutrophic lakes.

Lakes are transient ecosystems, although very large lakes may persist for millions of years. Over time, they gradually fill with sediment, undergoing succession to dry land (see Chapter 28). As nutrient-rich sediment accumulates, oligotrophic lakes tend to become eutrophic, a process called *eutrophication*.

Human Impact Human activities can greatly accelerate the process of eutrophication, because nutrients are carried into lakes from farms, feedlots, sewage, and even fertilized suburban lawns. Overenriched lakes become clogged with microorganisms whose dead bodies are attacked by bacteria that deplete the water of oxygen. Normal community interactions are disrupted as organisms in higher trophic levels are smothered. In Seattle, Washington, in the early 1960s local sewage plants released 20 million gallons of effluent into Lake Washington daily. Nutrients from human waste and high-phosphate detergents eutrophied the lake, causing foul odors, murky water, and dead fish. Citizen concern led Seattle to divert sewage effluent out of the lake starting in 1963. By 1975 the lake had fully recovered.

The Great Lakes, whose shores hold paper mills and other polluting industries and whose waters carry vessels from distant countries, have been heavily affected by human activities. Exotic species such as the zebra mussel have invaded and altered the structure of the lake community. Imported parasitic sea lampreys and overfishing devastated lake trout populations in the 1940s. Although fishing is now regulated and the lampreys are controlled, the trout, repeatedly reintroduced, are still unable to reproduce because of persistent endocrine-disrupting chemicals (discussed in Chapter 23) that have accumulated in the water and sediments. Recently, officials were puzzled to find that the beleaguered trout carry high levels of toxiphene, which was used heavily as a substitute for the pesticide DDT (discussed in Chapter 29). The EPA banned toxiphene in 1982 after discovering that it could cause birth defects and cancer. Some scientists hypothesize that the toxiphene is still being carried in the atmosphere from fields in the south, where it was heavily sprayed on cotton 20 years ago, and deposited in the Great Lakes.

Acid rain, caused primarily by burning fossil fuels, poses a very different threat, particularly to small freshwater lakes and ponds. In the Adirondack Mountains of New York State, 500 lakes and ponds (almost 20% of the total) have been rendered nearly lifeless by acid rain, and the problem is growing. (Acid rain is discussed in Chapter 29.)

Marine Ecosystems Cover Much of Earth

In the oceans, the upper layer of water to a depth of about 650 feet (200 meters), where the light is strong enough to support photosynthesis, is called the **photic zone**. Below the photic zone lies the **aphotic zone**, where the only energy comes from the excrement and bodies of organisms that sink or swim down there (Fig. 30-25).

As in lakes, most of the nutrients in the oceans are at or near the bottom, where there is not enough light for photosynthesis. Nutrients dissolved in the water of the photic zone are quickly incorporated into the bodies of living organisms. When these organisms die, some sink into the aphotic zone, providing the organisms there with energy and nutrients. This constant drain of nutrients from the photic zone would eventually end life there, if no additional nutrients entered the zone.

Fortunately, there are two sources of nutrients to the photic zone: the land, from which rivers constantly remove nutrients and carry them to the oceans, and **upwelling**, an upward flow that brings cold, nutrient-laden

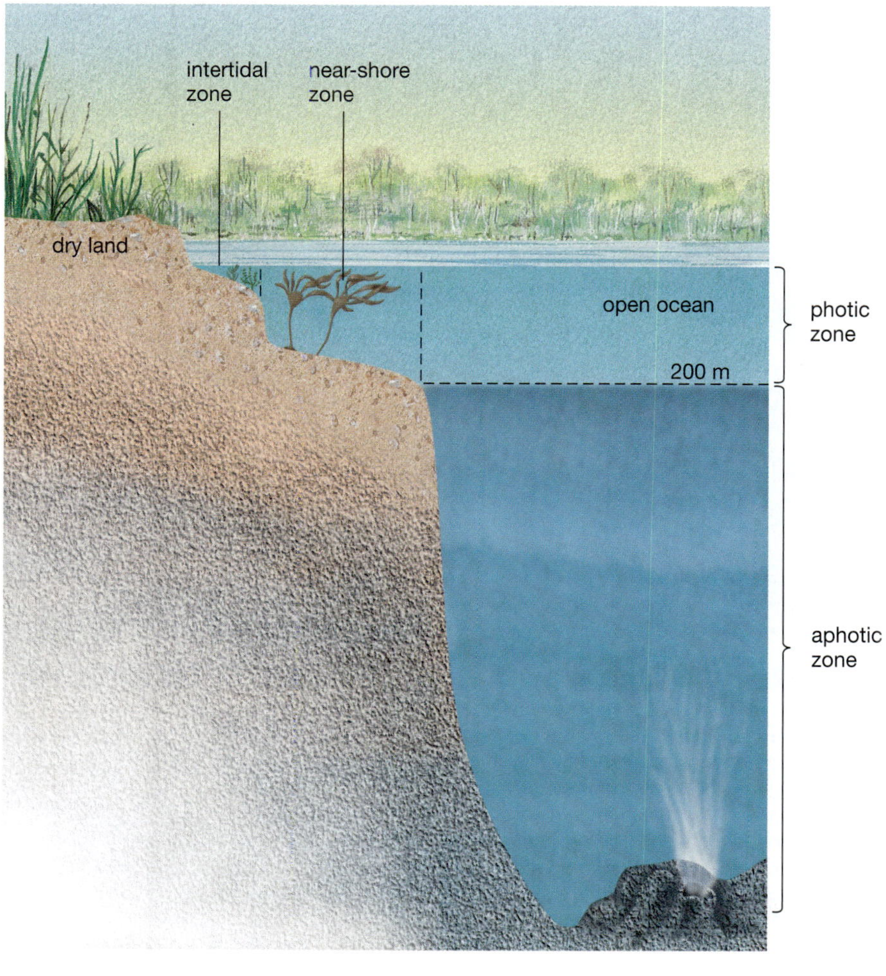

Figure 30-25 Ocean life zones
There is sufficient light for photosynthesis only in the photic zone, which includes the intertidal and near-shore zones and the upper waters of the open ocean. Life in the aphotic zone relies on energy-rich material that drifts down from the photic zone or (in hydrothermal vent communities) on energy that is stored in hydrogen sulfide.

water from the ocean depths to the surface. Upwelling occurs along western coastlines, as in California, Peru, and West Africa, where prevailing winds displace surface water, causing it to be replaced by water from below. Upwelling also occurs around Antarctica. Not surprisingly, the major concentrations of life in the oceans are found where abundant light is combined with a source of nutrients, a combination most commonly found in regions of upwelling and in shallow coastal waters.

Coastal Waters

The greatest abundance of life in the oceans is found in a narrow zone surrounding Earth's landmasses, where the water is shallow and a steady flow of nutrients washes off the land. Coastal waters include two main types of habitat. The **intertidal zone** is the area that is alternately covered and uncovered by water with the rising and falling of the tides. To the seaward side of the intertidal zone is the **near-shore zone**, which includes relatively shallow but constantly submerged areas, including bays and coastal wetlands such as salt marshes and **estuaries**, which are wetlands formed where rivers meet the oceans (Fig. 30-26). The near-shore zone is the only part of the ocean where large plants or seaweeds can grow anchored to the bottom. In addition, the abundance of nutrients and sunlight in this zone promotes the growth of a veritable soup of photosynthetic phytoplankton. Associated with the plants and protists are animals from nearly every phylum: annelid worms, sea anemones, jellyfish, sea urchins, sea stars, mussels, snails, fish, and sea otters, to name just a few. A large number and variety of organisms live permanently in coastal waters, but many that spend most of their lives in the

Figure 30-26 Near-shore ecosystems
(a) A salt marsh in the eastern United States. Expanses of shallow water fringed by marsh grass (*Spartina*) provide excellent habitat and breeding grounds for many marine organisms and shorebirds. ***(b)*** Although shifting sands in the sandy intertidal zone present a challenge to life, grasses stabilize them, and animals such as (inset) this *Emerita* crab burrow in them. ***(c)*** A rocky intertidal shore in Oregon, where animals and algae grip the rock against the pounding waves and resist drying during low tide. (Inset) Colorful sea stars cling to the rocks, surrounded by a seaweed, *Fucus*. ***(d)*** Towering kelp sway through the clear water off southern California, providing the basis for a diverse community of invertebrates, fishes, and (inset) an occasional sea otter.

open ocean come into the coastal waters to reproduce. Bays, salt marshes, and estuaries in particular are the breeding grounds for crabs, shrimp, and fish, including most commercially important species. Off the coast of southern California, great undersea forests of kelp provide food and shelter for a rich assemblage of fish and invertebrates that in turn provide food for otters and seals (Fig. 30-26d).

Human Impact Coastal regions are of great importance not only to the organisms that live or breed there but also to humans who use them for food sources, recreation, mineral and oil extraction, or living places. Wetlands of all types once covered 6% of Earth's surface, but nearly half of them have been dredged or filled in. In the United States, only about 100 million acres remain out of an original 215 million acres of wetlands. As populations increase in coastal states, and resources such as oil become increasingly scarce, the conflict between preservation of coastal wetlands and development of these areas for housing, harbors and marinas, and energy extraction will become increasingly intense. Although conservation efforts have slowed the loss of wetlands, the United States still loses well over 100,000 acres yearly. Estuaries are also threatened, especially by runoff from farming operations. Pig farms often collect concentrated animal wastes in large holding ponds. When leaks or floods transport this material into local rivers and estuaries, it can be fatal to the estuarian community of phytoplankton, invertebrates, and fish. Because so much of the life of the ocean depends on the well-being of the coastal waters, it is essential to protect these fragile, vital areas.

Coral Reefs

In warm, tropical waters with just the right combination of bottom depth, wave action, and nutrients, specialized algae and corals (types of cnidarians) build reefs from their own calcium carbonate skeletons. **Coral reefs** are most abundant in tropical waters of the Pacific and Indian Oceans, the Caribbean, and the Gulf of Mexico as far north as southern Florida, where the average water temperatures range between 72 °F and 82 °F (22 °C and 28 °C).

Reef-building corals thrive within the photic zone at depths of less than 130 feet (40 meters), where light penetrates the clear water and provides energy for photosynthesis. The corals require light because they harbor photosynthetic unicellular algae called *dinoflagellates* within their tissues in a mutualistic relationship. The algae benefit from the high nitrogen, phosphorus, and carbon dioxide levels in the coral tissues and, in return, provide food for the coral and help produce calcium carbonate, which forms the coral skeleton. The algae represent up to half the weight of the coral polyp and give the corals their diverse, brilliant colors (Fig. 30-27).

The skeletons of corals accumulate over thousands of years, providing an anchoring place for diverse forms of algae and supplying shelter and food for the most diverse collection of invertebrates and fish in the oceans (see Fig. 30-27). In some ways, coral reefs might be considered the ocean equivalent of rain forests, since they are home to more than 93,000 known species with probably ten times that number yet to be identified. The Great Barrier Reef in Australia supports more than 200 species of coral alone, and a single reef may harbor 3000 identified species of fish, invertebrates, and algae.

Human Impact Coral reefs are extremely sensitive to disturbance. Anything that diminishes the clarity of the water harms the coral's photosynthetic partners and hinders coral growth. When sewage and agricultural runoff pollute coastal waters, eutrophication reduces both sunlight and oxygen. When people farm, log, and develop coastal land, erosion carries silt into the water. Silt has ruined several reefs near Honolulu, Hawaii. In the Philippines, rainforest logging has dramatically increased erosion; destroying the rain forests is also destroying the local coral reefs.

Overfishing threatens reef communities. In many tropical countries, mollusks, turtles, fish, crustaceans, and the corals themselves are being harvested from reefs faster than they can reproduce. In some areas, fish are harvested with dynamite, which kills indiscriminately and damages the reef. Most of the harvest is for food, but much is for sale to shell enthusiasts and aquarium owners in developed countries. Some collectors for the aquarium trade use poison to stun the fish before collecting them, leaving most dead. Removing predatory fish and invertebrates from reefs may disrupt the ecological balance of the community, allowing an explosion in populations of coral-eating sea urchins or sea stars.

Throughout the Tropics, scientists and concerned citizens have observed a disturbing trend—coral reefs and their communities are becoming "sick." Infections, tumors, and lesions with unknown causes are causing massive mortality of corals and their communities throughout the world. Teams of researchers at Harvard and at the University of New Hampshire who are studying these trends suggest that the declining health of coral reefs stems from complex causes, including both human disturbances and global warming. Higher water temperature combined with runoff from land encourage explosions of microscopic algal populations in nearshore waters. The algae release toxins that damage the health of the entire reef community. Bleaching is another symptom of poor health observed in corals worldwide. When waters become too warm, the corals expel their colorful symbiotic algae, leaving the corals a deathly white. Without their algal partners, the corals will eventually starve. Continued global warming poses a severe threat to the world's coral reefs.

Florida's famous coral reefs are in trouble. Warm water from a nuclear power plant, nutrients from sewage effluent, and silt carried in from rivers cloud the

Figure 30-27 Coral reefs
(a) Coral reefs, composed of the bodies of corals and algae, provide habitat for an extremely diverse community of extravagantly colored animals. *(b)* Many fish, including this blue tang, feed on coral (note the bright yellow corals in background). A vast array of invertebrates such as *(c)* this sponge and *(d)* blue-ringed octopus live among the corals of Australia's Great Barrier Reef. This tiny octopus (6 inches, fully extended) is one of the world's most venomous creatures.

water and promote the growth of harmful algae. More than a million divers and snorkelers visit the reefs annually and accidentally (or sometimes deliberately) damage the coral. Boat anchors claw the reefs, and boat engines release oil into formerly pristine waters. Infections are increasingly common, and the two warmest decades in recorded history have caused extensive coral bleaching. But there is also some good news—the world's first marine park was established off Florida in 1960, and several other marine sanctuaries have been established in Florida in more recent years. A ban on fishing over small coral reef reserves in the Florida Keys, imposed in 1997, has already increased populations of several important marine species such as lobsters and grouper fish.

As with any ecosystem, both protection and sustainable use are crucial to the survival of these fragile and diverse underwater treasures. Carefully regulated harvesting and tourism produce far greater and longer-lasting economic benefits than do activities that destroy reefs. Establishing preserves is crucial, but preserves will only be effective if we can prevent human activities on land from damaging the quality of the water on which reef communities rely.

The Open Ocean

Beyond the coastal regions lie vast areas of ocean. Most life in the open ocean (Fig. 30-28a) is limited to the photic zone, where organisms are **pelagic**—that is, free-swimming or floating—for their entire lives (Fig. 30-28b–d). The food web of the open ocean is dependent on phytoplankton consisting of microscopic photosynthetic protists, mainly diatoms and dinoflagellates (Fig. 30-28e). These organisms are consumed by zooplankton, such as tiny crustaceans that are relatives of crabs and lobsters (Fig. 30-28f). Zooplankton in turn serve as food for larger invertebrates, small fish, and even marine mammals such as the humpback whale (Fig. 30-28d).

One challenge faced by inhabitants of the open ocean is to remain afloat in the photic zone, where sunlight and food are abundant. Many members of the planktonic community have elaborate flotation de-

Figure 30-28 The open ocean
(a) The open ocean supports abundant life in the photic zone, where light is available. ***(b)*** Porpoises skim the surface, ***(c)*** fish such as the blue jack swim, and ***(d)*** rare humpback whales leap clear of the water. ***(e)*** The photosynthetic phytoplankton are the producers on which most other marine life ultimately depends. Phytoplankton are eaten by ***(f)*** zooplankton, represented by this microscopic copepod, a crustacean. The spiny projections on these planktonic creatures help keep them from sinking below the photic zone.

vices, such as oil droplets in their cells or long projections, to slow their rate of sinking (see Fig. 30-28e). Most fish have swim bladders that can be filled with gas to regulate their buoyancy. Some animals, and even some of the phytoplankton, actively swim to stay in the photic zone. Many small crustaceans migrate to the surface at night to feed, then sink into the dark depths during daylight, thus avoiding visual predators such as fish. The amount of pelagic life varies tremendously from place to place. The blue clarity of tropical waters is a result of a lack of nutrients, which limits the concentration of plankton in the water. Nutrient-rich waters that support a large plankton community are greenish and relatively murky.

Below the photic zone, the only available energy in most regions comes from the excrement and dead bodies that drift down from above. Nevertheless, a surprising quantity and variety of life exist in the aphotic zone, including fishes of bizarre shapes, worms, sea cucumbers, sea stars, and mollusks.

Human Impact Two major threats to the open ocean are pollution and overfishing. Open-ocean pollution takes several forms. Oceangoing vessels such as cruise ships dump millions of plastic containers overboard daily, and plastic six-pack holders, foam cups, and packing material wash and blow off the land, collecting on parts of the ocean surface. The plastic looks like food to unsuspecting sea turtles, gulls, porpoises, seals, and whales, many of which die after trying to consume it. Until 1992, New York City placed its refuse and sewage sludge on barges and towed them out to sea, creating a heavily contaminated area covering 40 square miles of open-ocean floor. The open ocean has also served as a dumping ground for radioactive wastes. Oil contaminates the open ocean from many sources, including oil tanker spills, runoff from improper disposal on land, leakage from offshore oil wells, and natural seepage. In the Gulf of Mexico, the Mississippi River deposits nutrient-laden sediment from agricultural runoff onto the seafloor, creating an expanding "dead zone" where oxygen is depleted and the normal marine community is eliminated. The dead zone now covers 7000 square miles during the warmer months of each year.

The increasing demand for fish to feed a growing human population, coupled with increasingly efficient fishing technologies including satellite technology, radar, and sonar, has depleted many important fisheries. Most of the world's fisheries are harvesting fewer fish despite the advanced technology—evidence that fish are being taken in unsustainable numbers. Many governments heavily subsidize their fishing industries, a practice that promotes continued harvest of a diminishing resource. The cod fishery of the northeastern United States has collapsed from overfishing. Natural populations of lobsters, salmon, haddock, king crab, swordfish, and many other types of seafood have also declined dramatically because of overfishing. While dredging for scallops, fishermen often scrape entire marine communities from the seafloor. "Dolphin safe" tuna netting techniques have dramatically reduced accidental dolphin catches but have trapped more sea turtles and sharks.

In 1994 a marine reserve was established off the coast of Massachusetts. Within 5 years, scallop and fish populations were rebounding. Fisheries researchers have now proposed protection of about three dozen key regions off the New England coast to allow fish populations to recover. It has been politically difficult to establish no-fishing zones due to intense opposition from the fishing industry, but this approach holds the most promise for saving endangered fish populations throughout the world. Fish farming, or *aquaculture*, can also help meet the demand for some types of seafood, including shrimp, but fish and shrimp farms must be carefully designed and managed to avoid damaging local ecosystems.

Hydrothermal Vents

In 1977 a new and unusual source of nutrients, forming the basis of a spectacular undersea community, was discovered in the deep ocean. Geologists exploring the Galapagos Rift (an area of the Pacific floor where plates that form Earth's crust are separating) found vents spewing superheated water, black with sulfur and minerals. Surrounding these vents was a rich community of pink fish, blind white crabs, enormous mussels, giant white clams, sea anemones, and giant tube worms (Fig. 30-29). Twenty-two previously undescribed families and 284 new species of organisms have been found in these **hydrothermal vent communities**. Scientists have now identified vent communities in many deep-sea areas where plates are separating and material from Earth's interior is spewing forth to form new crust.

In this ecosystem, sulfur bacteria serve as the primary producers. They harvest energy from an unlikely source that is deadly to most other forms of life—hydrogen sulfide spewed from cracks in Earth's crust. This process, called *chemosynthesis*, replaces photosynthesis in these vent communities, which flourish more than a mile below the ocean surface. Bacteria and archaea proliferate in the hot water surrounding the vents, covering nearby rocks with thick, matlike colonies. These colonies provide the food on which the animals of the vent community thrive. Many vent animals consume the microorganisms directly. Others, such as the giant tube worm (which lacks a digestive system), harbor the bacteria within special organs in their bodies. In this mutualistic association, the bacteria provide high-energy carbon compounds and the tube worm provides hydrogen sulfide. The worm, which can reach a length of 12 feet, derives its red color from a unique form of hemoglobin that transports hydrogen sulfide, rather than oxygen, to the symbiotic bacteria.

Figure 30-29 Hydrothermal vent communities Located in the ocean depths, vent communities include giant tube worms nearly 12 feet long. Parts of these worms are red with oxygen-trapping hemoglobin. They have no digestive tract but are host to sulfur bacteria, which provide the worms with energy by oxidizing hydrogen sulfide.

The bacteria and archaea that inhabit the vent communities hold the record for survival at high temperatures. One species stops reproducing at water temperatures below 194 °F (90 °C) because it gets too cold; another can survive water temperatures of 248 °F (106 °C; water at this depth can reach temperatures much higher than boiling because of the tremendous pressure). Scientists are investigating how the enzymes and other proteins of these heat-loving microbes can continue to function at temperatures that would destroy the proteins in our bodies.

The world still holds wonders and mysteries for those who seek them. We have only begun to explore the versatility and diversity of life on Earth.

CASESTUDYREVISITED

A Rubber Tapper's Legacy

On December 22, 1988, Chico Mendes, rubber tapper and advocate for extractive reserves in the Brazilian rain forest, was shot and killed outside his home. His assassins are believed to have been hired by cattle ranchers hostile to his goal of slowing the destruction of his country's forests. But Mendes did live long enough to see his dream realized. Shortly before his death, the Brazilian government established its first extractive reserve. Today, there are nine such reserves in Brazil, protecting a total of 5.5 million acres of forest and inhabited by 28,000 people. The largest reserve, which protects almost 2.5 million acres in the state of Acre, is named in honor of Chico Mendes. And the National Council of Rubber Tappers remains a powerful voice of advocacy for the forests' human inhabitants.

Mendes's legacy, the idea that environmental protection and the well-being of local populations are inextricably linked, represents one of the more promising strategies in the struggle to protect Earth's endangered ecosystems. Protected environments are much more likely to remain protected if the people who live there have an economic stake in preservation. The extractive reserve concept, however, has been applied to only 1% of the remaining Brazilian rain forest, and it has not been implemented at all in most other countries with rain forests within their boundaries. Furthermore, recent years have seen dramatic declines in the prices paid for natural resources such as latex, brazil nuts, and many of the other products on which the residents of extractive reserves depend. As a result, residents have been forced to increase their dependence on higher-impact activities like hunting, logging, and agriculture. Even with these problems, however, extractive reserves remain a source of hope for the future of Earth's tropical rain forests.

Summary of Key Concepts

1) What Factors Influence Earth's Climate?

The availability and distribution through the year of sunlight and water determine the climate of a given region. Each unit of solar energy is spread over a smaller area at the equator than farther north and south, making the equator relatively warm, whereas higher latitudes have lower overall temperatures. Earth's tilt on its axis causes dramatic seasonal variation at northern and southern latitudes.

Circular air currents consistently carry moist or dry air to particular latitudes, producing areas of low and high precipitation. These broad rainfall patterns are modified locally by the topography of continents and by ocean currents.

2) What Conditions Does Life Require?

The requirements for life include nutrients, energy, and liquid water. Differences between locations in the types and abundance of living things are largely attributable to differences in distribution (in time and space) and abundance of these resources, and on temperature.

3) How Is Life on Land Distributed?

On land, the crucial limiting factors are temperature and liquid water. A biome comprises regions that have similar climates and vegetation owing to the interaction of temperature and rainfall or the availability of water.

Tropical forest biomes, located near the equator, are warm and wet and dominated by huge broadleaf evergreen trees. Most nutrients are tied up in vegetation, and most animal life is arboreal. Rain forests, home to at least 50% of all species, are rapidly being cut for agriculture, although the soil is extremely poor.

The African savanna, a grassland studded with scattered trees, has pronounced wet and dry seasons. It is home to the world's most diverse and extensive herds of large mammals.

Most deserts are located between 20° and 30° north and south latitude and in the rain shadows of mountain ranges. In deserts, plants, if there are any, are widely spaced and have adaptations to conserve water. Most animals are small and nocturnal.

Chaparral exists in desertlike conditions that are moderated by proximity to a coastline, allowing small trees and bushes to thrive. Grasslands, concentrated in the centers of continents, have a continuous grass cover without trees. They have the world's richest soils and have largely been converted to agriculture.

Temperate deciduous forests, whose broadleaf trees drop their leaves in winter to conserve moisture, dominate the eastern half of the United States and are also found in Western Europe and East Asia. Precipitation is higher than in grasslands. The wet temperate rain forests, dominated by evergreens, are on the northern Pacific coast of the United States. The taiga, or northern coniferous forest, covers much of the northern United States, southern Canada, and northern Eurasia. It is dominated by conifers whose small, waxy needles are adapted for water conservation and year-round photosynthesis.

The tundra is a frozen desert where permafrost prevents the growth of trees, and bushes remain stunted. Nonetheless, diverse arrays of animal life and perennial plants flourish in this fragile biome, which is found on mountain peaks and in the Arctic.

4) How Is Life in Water Distributed?

Energy and nutrients are the major limiting factors in the distribution and abundance of life in aquatic ecosystems. Nutrients are found in bottom sediments and are washed in from surrounding land, concentrating them near shore and in deep water.

Freshwater lakes have three life zones. The littoral zone, near shore, is rich in energy and nutrients and supports the most diverse community. The limnetic zone is the lighted region of open water where photosynthesis can occur. The profundal zone is the deep water, where light is inadequate for photosynthesis and the community is dominated by heterotrophic organisms. Oligotrophic lakes are clear and low in nutrients, and they support sparse communities. Eutrophic lakes are rich in nutrients and support dense communities. During succession to dry land, lakes tend to go from an oligotrophic to a eutrophic condition.

Most life in the oceans is found in shallow water, where sunlight can penetrate, and is concentrated near the continents and in areas of upwelling, where nutrients are most plentiful. Coastal waters, consisting of the intertidal zone and the near-shore zone, contain the most abundant life. Producers consist of aquatic plants anchored to the bottom and photosynthetic protists called phytoplankton. Coral reefs are confined to warm, shallow seas. The calcium carbonate reefs form a complex habitat supporting the most diverse undersea ecosystem, threatened by silt, overfishing, and global warming.

In the open ocean, most life is found in the photic zone, where light supports phytoplankton. In the aphotic zone, life is supported by nutrients that drift down from the photic zone. Many ocean fisheries have been overexploited. Hydrothermal vent communities, supported by chemosynthetic bacteria, thrive at great depths in the superheated waters where Earth's crustal plates are separating.

Key Terms

aphotic zone *p. 625*
biodiversity *p. 610*
biome *p. 608*
chaparral *p. 615*
climate *p. 604*
coral reef *p. 627*
desert *p. 614*
estuary *p. 626*
eutrophic lake *p. 624*
grassland *p. 615*
hydrothermal vent community *p. 630*
intertidal zone *p. 626*
limnetic zone *p. 624*
littoral zone *p. 622*
near-shore zone *p. 626*
oligotrophic lake *p. 624*
ozone layer *p. 604*
pelagic *p 628*
permafrost *p. 620*
photic zone *p. 625*
phytoplankton *p. 623*
prairie *p. 615*
profundal zone *p. 624*
rain shadow *p. 608*
savanna *p. 613*
taiga *p. 619*
temperate deciduous forest *p. 617*
temperate rain forest *p. 619*
tropical deciduous forest *p. 612*
tropical rain forest *p. 610*
tundra *p. 620*
upwelling *p. 625*
weather *p. 604*
zooplankton *p. 623*

Thinking Through the Concepts

Multiple Choice

1. *Most plant species fit into only a few morphological types. Which type would you think is most restricted by temperature and rainfall?*
a. trees
b. shrubs
c. grasses
d. perennial herbs
e. annual weeds

2. *In which biome is the smallest fraction of carbon and nutrients present in the soil?*
a. tropical rain forest
b. savanna
c. tundra
d. grassland
e. coniferous forest

3. *How do mountain ranges create deserts?*
a. by lifting land up into colder, drier air
b. by completely blocking the flow of air into desert areas, thus preventing clouds from getting there
c. by forcing air to first rise and then fall, thus causing rain on one side of the mountains and desert on the other
d. by causing the global wind patterns that make certain latitudes very dry
e. by causing very steep slopes that are subject to erosion

4. *What is the primary reason that plants from distant, but climatically similar, places commonly look the same?*
a. common ancestry
b. adaptation to the same physical conditions
c. adaptation to similar herbivores
d. continental drift
e. effects of past climate change

5. *Which of these biomes has been increased in area by human activities?*
   a. savanna
   b. temperate rain forest
   c. grassland
   d. coniferous forest
   e. desert
6. *What biome has the richest soil and has largely been converted to agriculture?*
   a. tundra
   b. coniferous forest
   c. grassland
   d. tropical rain forest
   e. deciduous forest

? Review Questions

1. Explain how air currents contribute to the formation of the Tropics and the large deserts.
2. What effect do the roughly circular ocean currents have on climate, and where is that effect strongest?
3. What are the four major requirements for life? Which two are most often limiting in terrestrial ecosystems? In ocean ecosystems?
4. Explain why traveling up a mountain takes you through biomes similar to those you would encounter traveling for a long distance toward one of Earth's poles.
5. Where are the nutrients of the tropical forest biome concentrated? Why is animal life in the tropical rain forest concentrated high above the ground?
6. Explain two undesirable effects of agriculture in the tropical rain forest biome.
7. List some adaptations to heat and drought of (a) desert plants and (b) desert animals.
8. What human activities damage deserts?
9. How are trees of the taiga adapted to a lack of water and a short growing season?
10. How do deciduous and coniferous biomes differ?
11. What single environmental factor best explains why there is shortgrass prairie in Colorado, tallgrass prairie in Illinois, and deciduous forest in Ohio?
12. Where are the world's largest populations of large herbivores and carnivores located?
13. Where is life in the oceans most abundant, and why?
14. Why is the diversity of life so high in coral reefs? What human impacts threaten them?
15. Distinguish among the limnetic, littoral, and profundal zones of lakes in terms of their location and the communities they support.
16. Distinguish between oligotrophic and eutrophic lakes. Describe (a) a natural scenario and (b) a human-created scenario under which an oligotrophic lake might be converted to a eutrophic lake.
17. Distinguish between the photic and aphotic zones. How do organisms in the photic zone obtain nutrients? How are nutrients obtained in the aphotic zone?
18. What unusual primary producer forms the basis for hydrothermal vent communities?
19. On the basis of the location of the worst atmospheric ozone depletion, which biomes are likely to be most affected by increased UV penetration?

Applying the Concepts

1. List at least six differences between human-dominated and undisturbed ecosystems, and discuss how these differences can be minimized.
2. In which terrestrial biome is your college or university located? Discuss similarities and differences between your location and the general description of that biome in the text. If you are living in a city, how has the urban environment modified the human interaction with the biome?
3. During the 1960s and 1970s, many parts of the United States and Canada banned the use of detergents containing phosphates. Until that time, almost all laundry detergents and many soaps and shampoos had high concentrations of phosphates. What environmental concern do you think prompted these bans, and what ecosystem has benefited most from the bans?
4. Because ozone depletion is expected to get worse, not better, for decades to come, biologists have tried to assess which types of species will be most susceptible to increased UV radiation. Two of the groups that may be most vulnerable—ocean plankton and long-lived birds and mammals—are quite different from each other. What makes each of these groups susceptible to increased UV radiation?
5. Understanding the ways in which the four basic requirements for life determine where different biomes occur can help us predict the consequences of global warming. Global warming is expected to make most areas warmer, but it is also expected to change rainfall in ways that are hard to predict—some areas will get wetter, and others drier. Our ignorance of how rainfall will change is not very important in understanding shifts in more northerly biomes, but it is very important for our understanding of changes in tropical areas. Look at Figure 30-8 and explain why this is true.
6. More-northerly forests are better able to regenerate after logging than are tropical rain forests. Explain why this is true. HINT: The cold soils of northern climates greatly slow down decomposition rates.

Group Activity

Work in groups of three or four to answer the following question: How can humans become more fully integrated into the biosphere? Use "Earth Watch: Humans and Ecosystems" on page 622 to present a realistic answer to this question. Include the following components:

- Awareness and Assessment—What problems in ecosystems have been caused by humans?
- Analysis—What effects (both positive and negative) might your plan have?
- Education—How will you inform others of problems and possible solutions?
- Politics—How can you get others involved?
- Action—How will the actions taken be monitored and evaluated to ensure success?

For More Information

Chadwick, D. H. "Blue Refuges." *National Geographic*, March 1998. Beautiful photographs highlight 12 national marine sanctuaries set aside to preserve a variety of marine ecosystems along the coastal U.S.

Helmuth, L. "Can This Swamp Be Saved?" *Science News*, April 17, 1999. Describes the ambitious protection plan for the Florida Everglades.

Hinrichsen, D. "Requiem for Reefs?" *International Wildlife*, March–April 1997. Stunning photographs and a compelling text describe the beauty of Earth's imperiled coral reefs.

Holloway, M. "Sustaining the Amazon." *Scientific American*, July 1993. Describes the threats to the Amazon rain forest and innovative ways to preserve tropical forests and their biodiversity.

Holmes, B. "Case of the Dwindling Cloud Forests." *International Wildlife*, July–August 2000. Global warming may be threatening the unique and diverse cloud forests of Costa Rica.

Kusler, J. A., Mitsch, W. J., and Larson, J. S. "Wetlands." *Scientific American*, January 1994. Discusses the characteristics and values of these threatened but highly important ecosystems.

Mallin, M. A., "Impacts of Industrial Animal Production on Rivers and Estuaries." *American Scientist*, January–February 2000. Lagoons designed to store waste from large numbers of farm animals pose a threat to nearby freshwater ecosystems.

Mares, M. A. "Desert Rodents, Seed Consumption, and Convergence: Evolutionary Shuffling of Adaptations." *BioScience*, Vol. 43, 1993. An example of how animal as well as plant species and communities living in similar climates can show convergence.

Mitchell, J. G. "Our National Forests." *National Geographic*, March 1997. Stunning photographs and engaging text describe the threats to forests in the United States.

Monks, V. "How Did the Poison Get into the Trout?" *National Wildlife,* August–September 1998. Toxic substances accumulating in the Great Lakes pose a hazard to wildlife and a puzzle to scientists.

Prather, M., Midgley, P., Rowland, F. S., and Storlarski, R. "The Ozone Layer: The Road Not Taken." *Nature*, June 13, 1996. Reviews the progress toward slowing ozone depletion and explores what would have happened had the ozone-protection treaties not been passed.

Toon, B., and Turco, R. P. "Polar Stratospheric Clouds and Ozone Depletion." *Scientific American*, June 1991. Describes the unique chemistry of the Antarctic clouds that causes ozone breakdown.

Tunnicliffe, V. "Hydrothermal-Vent Communities of the Deep Sea." *American Scientist*, July–August 1992. A comprehensive look at the structure and function of undersea vent communities, with excellent illustrations.

Williams, T. "What Good Is a Wetland?" *Audubon*, November–December 1996. In addition to supporting a diversity of wildlife, purifying drinking water, and protecting people from floods, wetlands are beautiful.

Answers to Multiple-Choice Questions

1. a 2. a 3. c 4. b 5. e 6. c

MEDIATUTOR

CD Activities

Activity 30.1: Climate Builds Biomes

Estimated time: 5 minutes

Hadley cells are a major feature of atmospheric circulation that strongly influence climate and community type in the Tropics and subtropics. Warm tropical air rises over the equator and produces the wet, warm conditions that support communities like tropical rain forests. The same air circulates poleward and eventually cools enough that it sinks. As it sinks, it warms again. The sinking air is quite dry (since warm air can hold more moisture than cool air can), so it produces the warm, dry conditions responsible for the existence of the major subtropical deserts of the world (such as the Kalahari).

Activity 30.2: Biomes of the World

Estimated time: 10 minutes

In this tutorial you will explore one of the critical factors involved in determining biomes. You will then be able to predict the different characteristics of various biomes, including climatic (abiotic) factors as well as the typical representative plants and animals (biotic factors) of those respective biomes.

Start the MediaTutor Student CD-ROM and enter the activity number in the Quick Search box to be taken directly to that activity.

Web Investigation

A Rubber Tapper's Legacy

Estimated time: 20 minutes

The diversity of life present on our Earth is a direct reflection of the diverse array of ecosystems found on our planet. Conservation efforts designed to protect these biomes have intensified in recent years as a response to the centuries of damage caused by humans. "Case Study: A Rubber Tapper's Legacy" highlights one such effort that started in the Brazilian rain forest. This activity will explore this and other conservation efforts that have been part of this movement over the past few decades.

Go to http://www.prenhall.com/audesirk3, the Audesirk Companion Web site. Select Chapter 30 and Web Investigation to begin.

Glossary

abdomen: the body segment at the posterior end of an animal with segmentation; contains most of the digestive structures.

abiotic (ā-bī-ah´-tik): nonliving; the abiotic portion of an ecosystem includes soil, rock, water, and the atmosphere.

abortion: the procedure for terminating pregnancy; the cervix is dilated, and the embryo and placenta are removed.

abscisic acid (ab-sis´-ik): a plant hormone that generally inhibits the action of other hormones, enforcing dormancy in seeds and buds and causing the closing of stomata.

abscission layer: a layer of thin-walled cells, located at the base of the petiole of a leaf, that produces an enzyme that digests the cell wall holding leaf to stem, allowing the leaf to fall off.

absorption: the process by which nutrients are taken into cells.

accessory pigments: colored molecules other than chlorophyll that absorb light energy and pass it to chlorophyll.

acellular slime mold: a type of funguslike protist that forms a multinucleate structure that crawls in amoeboid fashion and ingests decaying organic matter; also called *plasmodial slime mold.*

acetylcholine (ah-sēt´-il-kō´-lēn): a neurotransmitter in the brain and in synapses of motor neurons that innervate skeletal muscles.

acid: a substance that releases hydrogen ions (H^+) into solution; a solution with a pH of less than 7.

acid deposition: the deposition of nitric or sulfuric acid, either dissolved in rain (acid rain) or in the form of dry particles, as a result of the production of nitrogen oxides or sulfur dioxide through burning, primarily of fossil fuels.

acidic: with an H^+ concentration exceeding that of OH^-; releasing H^+.

acquired immune deficiency syndrome (AIDS): an infectious disease caused by the human immunodeficiency virus (HIV); attacks and destroys T cells, thus weakening the immune system.

acrosome (ak´-rō-sōm): a vesicle, located at the tip of an animal sperm, that contains enzymes needed to dissolve protective layers around the egg.

actin (ak´-tin): a major muscle protein whose interactions with myosin produce contraction; found in the thin filaments of the muscle fiber; see also *myosin.*

action potential: a rapid change from a negative to a positive electrical potential in a nerve cell. This signal travels along an axon without a change in intensity.

activation energy: in a chemical reaction, the energy needed to force the electron shells of reactants together, prior to the formation of products.

active site: the region of an enzyme molecule that binds substrates and performs the catalytic function of the enzyme.

active transport: the movement of materials across a membrane through the use of cellular energy, normally against a concentration gradient.

adaptation: a trait that increases the ability of an individual to survive and reproduce compared to individuals without the trait.

adaptive radiation: the rise of many new species in a relatively short time as a result of a single species that invades different habitats and evolves under different environmental pressures in those habitats.

adenine: a nitrogenous base found in both DNA and RNA; abbreviated as *A*.

adenosine diphosphate (a-den´-ō-sēn dī-fos´-fāt; ADP): a molecule composed of the sugar ribose, the base adenine, and two phosphate groups; a component of ATP.

adenosine triphosphate (a-den´-ō-sēn trī-fos´-fāt; ATP): a molecule composed of the sugar ribose, the base adenine, and three phosphate groups; the major energy carrier in cells. The last two phosphate groups are attached by "high-energy" bonds.

adipose tissue (a´-di-pōs): tissue composed of fat cells.

adrenal cortex: the outer part of the adrenal gland, which secretes steroid hormones that regulate metabolism and salt balance.

adrenal gland: a mammalian endocrine gland, adjacent to the kidney; secretes hormones that function in water regulation and in the stress response.

adrenal medulla: the inner part of the adrenal gland, which secretes epinephrine (adrenaline) and norepinephrine (noradrenaline).

adrenocorticotropic hormone (a-drēn-ō-kor-tik-ō-trō´-pik; ACTH): a hormone, secreted by the anterior pituitary, that stimulates the release of hormones by the adrenal glands, especially in response to stress.

aerobic: using oxygen.

age structure: the distribution of males and females in a population according to age groups.

agglutination (a-gloo-tin-ā´-shun): the clumping of foreign substances or microbes, caused by binding with antibodies.

aggression: antagonistic behavior, normally among members of the same species, often resulting from competition for resources.

aggressive mimicry (mim´ik-rē): the evolution of a predatory organism to resemble a harmless animal or part of the environment, thus gaining access to prey.

aldosterone: a hormone, secreted by the adrenal cortex, that helps regulate ion concentration in the blood by stimulating the reabsorption of sodium by the kidneys and sweat glands.

alga (al´-ga; pl., **algae, al´-jē):** any photosynthetic member of the eukaryotic Kingdom Protista.

allantois (al-an-tō´-is): one of the embryonic membranes of reptiles, birds, and mammals; in reptiles and birds, serves as a waste-storage organ; in mammals, forms most of the umbilical cord.

allele (al-ēl´): one of several alternative forms of a particular gene.

allele frequency: for any given gene, the relative proportion of each allele of that gene in a population.

allergy: an inflammatory response produced by the body in response to invasion by foreign materials, such as pollen, that are themselves harmless.

allopatric speciation (al-ō-pat´-rik): speciation that occurs when two populations are separated by a physical barrier that prevents gene flow between them (geographical isolation).

allosteric regulation: the process by which enzyme action is enhanced or inhibited by small organic molecules that act as regulators by binding to the enzyme and altering its active site.

alternation of generations: a life cycle, typical of plants, in which a diploid sporophyte (spore-producing) generation alternates with a haploid gametophyte (gamete-producing) generation.

altruism: a type of behavior that may decrease the reproductive success of the individual performing it but benefits that of other individuals.

alveolus (al-vē´-ō-lus; pl., **alveoli):** a tiny air sac within the lungs, surrounded by capillaries, where gas exchange with the blood occurs.

amino acid: the individual subunit of which proteins are made, composed of a central carbon atom bonded to an amino group ($-NH_2$), a carboxyl group ($-COOH$), a hydrogen atom, and a variable group of atoms denoted by the letter *R*.

ammonia: NH_3; a highly toxic nitrogen-containing waste product of amino acid breakdown. In the mammalian liver, it is converted to urea.

amniocentesis (am-nē-ō-sen-tē´-sis): a procedure for sampling the amniotic fluid surrounding a fetus: A sterile needle is inserted through the abdominal wall, uterus, and amniotic sac of a pregnant woman; 10 to 20 milliliters of amniotic fluid is withdrawn. Various tests may be performed on the fluid and the fetal cells suspended in it to provide information on the developmental and genetic state of the fetus.

amnion (am´-nē-on): one of the embryonic membranes of reptiles, birds, and mammals; encloses a fluid-filled cavity that envelops the embryo.

amniote egg (am-nē-ōt´): the egg of reptiles and birds; contains an amnion that encloses the embryo in a watery environment, allowing the egg to be laid on dry land.
amoeba: a type of animal-like protist that uses a characteristic streaming mode of locomotion by extending a cellular projection called a *pseudopod.*
amoeboid cell: a protist or animal cell that moves by extending a cellular projection called a *pseudopod.*
amphibian: member of a class of vertebrate tetrapods whose members generally lack scales, lay eggs in watery environments, respire with gills as larvae and with lungs as adults. Frogs, toads, and salamanders are amphibians.
amplexus (am-plek´-sus): in amphibians, a form of external fertilization in which the male holds the female during spawning and releases his sperm directly onto her eggs.
ampulla: a muscular bulb that is part of the water-vascular system of echinoderms; controls the movement of tube feet, which are used for locomotion.
amygdala (am-ig´-da-la): part of the forebrain of vertebrates that is involved in the production of appropriate behavioral responses to environmental stimuli.
amylase (am´-i-lās): an enzyme, found in saliva and pancreatic secretions, that catalyzes the breakdown of starch.
anaerobe: an organism whose respiration does not require oxygen.
anaerobic: not using oxygen.
analogous structures: structures that have similar functions and superficially similar appearance but very different anatomies, such as the wings of insects and birds. The similarities are due to similar environmental pressures rather than to common ancestry.
anaphase (an´-a-fāz): in mitosis, the stage in which the sister chromatids of each chromosome separate from one another and are moved to opposite poles of the cell; in meiosis I, the stage in which homologous chromosomes, consisting of two sister chromatids, are separated; in meiosis II, the stage in which the sister chromatids of each chromosome separate from one another and are moved to opposite poles of the cell.
androgen: a male sex hormone.
androgen insensitivity: a rare condition in which an individual with XY chromosomes is female in appearance because the body's cells don't respond to the male hormones that are present.
angina (an-jī´-nuh): chest pain associated with reduced blood flow to the heart muscle, caused by the obstruction of coronary arteries.
angiosperm (an´-jē-ō-sperm): a flowering vascular plant.
angiotensin (an-jē-ō-ten´-sun): a hormone that functions in water regulation in mammals by stimulating physiological changes that increase blood volume and blood pressure.
annual ring: a pattern of alternating light (early) and dark (late) xylem of woody stems and roots, formed as a result of the unequal availability of water in different seasons of the year, normally spring and summer.
antagonistic muscles: a pair of muscles, one of which contracts and in so doing extends the other; an arrangement that makes possible movement of the skeleton at joints.
anterior: the front, forward, or head end of an animal.
anterior pituitary: a lobe of the pituitary gland that produces prolactin and growth hormone as well as hormones that regulate hormone production in other glands.
anther (an´-ther): the uppermost part of the stamen, in which pollen develops.
antheridium (an-ther-id´-ē-um): a structure in which male sex cells are produced, found in the bryophytes and certain seedless vascular plants.
antibiotic resistance: the ability of a mutated pathogen to resist the effects of an antibiotic that normally kills it.
antibody: a protein, produced by cells of the immune system, that combines with a specific antigen and normally facilitates the destruction of the antigen.
anticodon: a sequence of three bases in transfer RNA that is complementary to the three bases of a codon of messenger RNA.
antidiuretic hormone (an-tē-dī-ūr-et´-ik; ADH): a hormone produced by the hypothalamus and released into the bloodstream by the posterior pituitary when blood volume is low; increases the permeability of the distal tubule and the collecting duct to water, allowing more water to be reabsorbed into the bloodstream.
antigen: a complex molecule, normally a protein or polysaccharide, that stimulates the production of a specific antibody.
aphotic zone: the region of the ocean below 200 m, where sunlight does not penetrate.
apical dominance: the phenomenon whereby a growing shoot tip inhibits the sprouting of lateral buds.
apical meristem (āp´-i-kul mer´-i-stem): the cluster of meristematic cells at the tip of a shoot or root (or one of their branches).
appendicular skeleton (ap-pen-dik´-ū-lur): the portion of the skeleton consisting of the bones of the extremities and their attachments to the axial skeleton; the pectoral and pelvic girdles, the arms, legs, hands, and feet.
aqueous humor (ā´-kwē-us): the clear, watery fluid between the cornea and lens of the eye.
Archaea: one of life's three domains; consists of prokaryotes that are only distantly related to members of the domain Bacteria.
archegonium (ar-ke-gō´-nē-um): a structure in which female sex cells are produced; found in the bryophytes and certain seedless vascular plants.
arteriole (ar-tēr´-ē-ōl): a small artery that empties into capillaries. Contraction of the arteriole regulates blood flow to various parts of the body.
artery (ar´-tuh-rē): a vessel with muscular, elastic walls that conducts blood away from the heart.
arthropod: member of an animal phylum whose members generally have segmented bodies, stiff exoskeletons, and paired, jointed appendages. Insects, arachnids, and crustaceans are arthropods.
artificial selection: a selective breeding procedure in which only those individuals with particular traits are chosen as breeders; used mainly to enhance desirable traits in domestic plants and animals; may also be used in evolutionary biology experiments.
ascus (as´-kus): a saclike case in which sexual spores are formed by members of the fungal division Ascomycota.
asexual reproduction: reproduction that does not involve the fusion of haploid sex cells. The parent body may divide and new parts regenerate, or a new, smaller individual may form as an attachment to the parent, to drop off when complete.
association neuron: in a neural network, a nerve cell that is postsynaptic to a sensory neuron and presynaptic to a motor neuron. In actual circuits, there may be many association neurons between individual sensory and motor neurons.
atherosclerosis (ath´-er-ō-skler-ō´-sis): a disease characterized by the obstruction of arteries by cholesterol deposits and thickening of the arterial walls.
atom: the smallest particle of an element that retains the properties of the element.
atomic nucleus: the membrane-bound organelle of eukaryotic cells that contains the cell's genetic material.
atomic number: the number of protons in the nuclei of all atoms of a particular element.
atrial natriuretic peptide (ā´-trē-ul nā-trē-ū-ret´-ik; ANP): a hormone, secreted by cells in the mammalian heart, that reduces blood volume by inhibiting the release of ADH and aldosterone.
atrioventricular (AV) node (ā´-trē-ō-ven-trik´-ū-lar nōd): a specialized mass of muscle at the base of the right atrium through which the electrical activity initiated in the sinoatrial node is transmitted to the ventricles.
atrioventricular valve: a heart valve that separates each atrium from each ventricle, preventing the backflow of blood into the atria during ventricular contraction.
atrium (ā´-trē-um): a chamber of the heart that receives venous blood and passes it to a ventricle.
auditory canal (aw´-di-tor-ē): a canal within the outer ear that conducts sound from the external ear to the tympanic membrane.
auditory nerve: the nerve leading from the mammalian cochlea to the brain, carrying information about sound.
autoimmune disease: a disorder in which the immune system produces antibodies against the body's own cells.
autonomic nervous system: the part of the peripheral nervous system of vertebrates that synapses on glands, internal organs, and smooth muscle and produces largely involuntary responses.
autosome (aw´-tō-sōm): a chromosome that occurs in homologous pairs in both males and females and that does not bear the genes determining sex.

autotroph (aw´-tō-trōf): "self-feeder"; normally, a photosynthetic organism; a producer.
auxin (awk´-sin): a plant hormone that influences many plant functions, including phototropism, apical dominance, and root branching; generally stimulates cell elongation and, in some cases, cell division and differentiation.
axial skeleton: the skeleton forming the body axis, including the skull, vertebral column, and rib cage.
axon: a long extension of a nerve cell, extending from the cell body to synaptic endings on other nerve cells or on muscles.
bacillus (buh-sil´-us; pl., **bacilli):** a rod-shaped bacterium.
Bacteria: one of life's three domains; consists of prokaryotes that are only distantly related to members of the domain Archaea.
bacterial conjugation: the exchange of genetic material between two bacteria.
bacteriophage (bak-tir´-ē-ō-fāj): a virus specialized to attack bacteria.
bacterium (bak-tir´-ē-um; pl., **bacteria):** an organism consisting of a single prokaryotic cell surrounded by a complex polysaccharide coat.
balanced polymorphism: the prolonged maintenance of two or more alleles in a population, normally because each allele is favored by a separate environmental pressure.
ball-and-socket joint: a joint in which the rounded end of one bone fits into a hollow depression in another, as in the hip; allows movement in several directions.
bark: the outer layer of a woody stem, consisting of phloem, cork cambium, and cork cells.
Barr body: an inactivated X chromosome in cells of female mammals, which have two X chromosomes; normally appears as a dark spot in the nucleus.
basal body: a structure resembling a centriole that produces a cilium or flagellum and anchors this structure within the plasma membrane.
base: (1) a substance capable of combining with and neutralizing H^+ ions in a solution; a solution with a pH of more than 7; (2) in molecular genetics, one of the nitrogen-containing, single- or double-ringed structures that distinguish one nucleotide from another. In DNA, the bases are adenine, guanine, cytosine, and thymine.
basic: with an H^+ concentration less than that of OH^-; combining with H^+.
basidiospore (ba-sid´-ē-ō-spor): a sexual spore formed by members of the fungal division Basidiomycota.
basidium (bas-id´-ē-um): a diploid cell, typically club-shaped, formed by members of the fungal division Basidiomycota; produces basidiospores by meiosis.
basilar membrane (bas´-eh-lar): a membrane in the cochlea that bears hair cells that respond to the vibrations produced by sound.
basophil (bas´-ō-fil): a type of white blood cell that releases both substances that inhibit blood clotting and chemicals that participate in allergic reactions and in responses to tissue damage and microbial invasion.
B cell: a type of lymphocyte that participates in humoral immunity; gives rise to plasma cells, which secrete antibodies into the circulatory system, and to memory cells.
behavior: any observable activity of a living animal.
behavioral isolation: the lack of mating between species of animals that differ substantially in courtship and mating rituals.
bilateral symmetry: a body plan in which only a single plane through the central axis will divide the body into mirror-image halves.
bile (bīl): a liquid secretion, produced by the liver, that is stored in the gallbladder and released into the small intestine during digestion; a complex mixture of bile salts, water, other salts, and cholesterol.
bile salt: a substance that is synthesized in the liver from cholesterol and amino acids and that assists in the breakdown of lipids by dispersing them into small particles on which enzymes can act.
binary fission: the process by which a single bacterium divides in half, producing two identical offspring.
binocular vision: the ability to see objects simultaneously through both eyes, providing greater depth perception and more-accurate judgment of the size and distance of an object from the eyes.
biodegradable: able to be broken down into harmless substances by decomposers.
biodiversity: the total number of species within an ecosystem and the resulting complexity of interactions among them.
biogeochemical cycle: also called a *nutrient cycle*, the process by which a specific nutrient in an ecosystem is transferred between living organisms and the nutrient's reservoir in the nonliving environment.
biological clock: a metabolic timekeeping mechanism found in most organisms, whereby the organism measures the approximate length of a day (24 hours) even without external environmental cues such as light and darkness.
biological magnification: the increasing accumulation of a toxic substance in progressively higher trophic levels.
biomass: the dry weight of organic material in an ecosystem.
biome (bī´-ōm): a terrestrial ecosystem that occupies an extensive geographical area and is characterized by a specific type of plant community: for example, deserts.
biosphere (bī´-ō-sfēr): that part of Earth inhabited by living organisms; includes both living and nonliving components.
biotechnology: any industrial or commercial use or alteration of organisms, cells, or biological molecules to achieve specific practical goals.
biotic (bī-ah´-tik): living.
biotic potential: the maximum rate at which a population could increase, assuming ideal conditions that allow a maximum birth rate and minimum death rate.
birth control pill: a temporary contraceptive method that prevents ovulation by providing a continuing supply of estrogen and progesterone, which in turn suppresses LH release; must be taken daily, normally for 21 days of each menstrual cycle.
bladder: a hollow muscular storage organ for storing urine.
blade: the flat part of a leaf.
blastocyst (blas´-tō-sist): an early stage of human embryonic development, consisting of a hollow ball of cells, enclosing a mass of cells attached to its inner surface, which becomes the embryo.
blastopore: the site at which a blastula indents to form a gastrula.
blastula (blas´-tū-luh): in animals, the embryonic stage attained at the end of cleavage, in which the embryo normally consists of a hollow ball with a wall one or several cell layers thick.
blind spot: the area of the retina at which the axons of the ganglion cell merge to form the optic nerve; the blind spot of the retina.
blood: a fluid consisting of plasma in which blood cells are suspended; carried within the circulatory system.
blood–brain barrier: relatively impermeable capillaries of the brain that protect the cells of the brain from potentially damaging chemicals that reach the bloodstream.
blood clotting: a complex process by which platelets, the protein fibrin, and red blood cells block an irregular surface in or on the body, such as a damaged blood vessel, sealing the wound.
blood vessel: a channel that conducts blood throughout the body.
body mass index (BMI): a number derived from an individual's weight and height used to estimate body fat. The formula is: weight (in kg)/height^2 (in meters^2).
bone: a hard, mineralized connective tissue that is a major component of the vertebrate endoskeleton; provides support and sites for muscle attachment.
book lung: a structure composed of thin layers of tissue, resembling pages in a book, that are enclosed in a chamber and used as a respiratory organ by certain types of arachnids.
boom-and-bust cycle: a population cycle characterized by rapid exponential growth followed by a sudden massive die-off, seen in seasonal species and in some populations of small rodents, such as lemmings.
Bowman's capsule: the cup-shaped portion of the nephron in which blood filtrate is collected from the glomerulus.
bradykinin (brā´-dē-kī´-nin): a chemical, formed during tissue damage, that binds to receptor molecules on pain nerve endings, giving rise to the sensation of pain.
brain: the part of the central nervous system of vertebrates that is enclosed within the skull.
branch root: a root that arises as a branch of a preexisting root, through divisions of pericycle cells and subsequent differentiation of the daughter cells.
bronchiole (bron´-kē-ōl): a narrow tube, formed by repeated branching of the bronchi, that conducts air into the alveoli.
bronchus (bron´-kus): a tube that conducts air from the trachea to each lung.

bryophyte (brī´-ō-fīt): a simple nonvascular plant of the division Bryophyta, including mosses and liverworts.
bud: in animals, a small copy of an adult that develops on the body of the parent and eventually breaks off and becomes independent; in plants, an embryonic shoot, normally very short and consisting of an apical meristem with several leaf primordia.
budding: asexual reproduction by the growth of a miniature copy, or bud, of the adult animal on the body of the parent. The bud breaks off to begin independent existence.
buffer: a compound that minimizes changes in pH by reversibly taking up or releasing H^+ ions.
bulbourethral gland (bul-bō-ū-rē´-thrul): in male mammals, a gland that secretes a basic, mucus-containing fluid that forms part of the semen.
bulk flow: the movement of many molecules of a gas or fluid in unison from an area of higher pressure to an area of lower pressure.
bundle-sheath cell: one of a group of cells that surround the veins of plants; in C_4 (but not in C_3) plants, bundle-sheath cells contain chloroplasts.
C_3 cycle: the cyclic series of reactions whereby carbon dioxide is fixed into carbohydrates during the light-independent reactions of photosynthesis; also called *Calvin-Benson cycle*.
C_4 pathway: the series of reactions in certain plants that fixes carbon dioxide into oxaloacetic acid, which is later broken down for use in the C_3 cycle of photosynthesis.
calcitonin (kal-si-tōn´-in): a hormone, secreted by the thyroid gland, that inhibits the release of calcium from bone.
calorie (kal´-ō-rē): the amount of energy required to raise the temperature of 1 gram of water by 1 degree Celsius.
Calorie: a unit of energy, in which the energy content of foods is measured; the amount of energy required to raise the temperature of 1 liter of water 1 degree Celsius; also called a *kilocalorie,* equal to 1000 calories.
Calvin-Benson cycle: see *C_3 cycle.*
cambium (kam´-bē-um; pl., cambia): a lateral meristem, parallel to the long axis of roots and stems, that causes secondary growth of woody plant stems and roots. See *cork cambium*; *vascular cambium.*
camouflage (cam´-a-flaj): coloration and/or shape that renders an organism inconspicuous in its environment.
cancer: a disease in which some of the body's cells escape from normal regulatory processes and divide without control.
capillary: the smallest type of blood vessel, connecting arterioles with venules. Capillary walls, through which the exchange of nutrients and wastes occurs, are only one cell thick.
capsule: a polysaccharide or protein coating that some disease-causing bacteria secrete outside their cell wall.
carbohydrate: a compound composed of carbon, hydrogen, and oxygen, with the approximate chemical formula $(CH_2O)_n$; includes sugars and starches.
carbon fixation: the initial steps in the C_3 cycle, in which carbon dioxide reacts with ribulose bisphosphate to form a stable organic molecule.
cardiac cycle (kar´-dē-ak): the alternation of contraction and relaxation of the heart chambers.
cardiac muscle (kar´-dē-ak): the specialized muscle of the heart, able to initiate its own contraction, independent of the nervous system.
carnivore (kar´-neh-vor): literally, "meat eater"; a predatory organism that feeds on herbivores or on other carnivores; a secondary (or higher) consumer.
carotenoid (ka-rot´-en-oid): a red, orange, or yellow pigment, found in chloroplasts, that serves as an accessory light-gathering molecule in thylakoid photosystems.
carpel (kar´pel): the female reproductive structure of a flower, composed of stigma, style, and ovary.
carrier: an individual who is heterozygous for a recessive condition; displays the dominant phenotype but can pass on the recessive allele to offspring.
carrier protein: a membrane protein that facilitates the diffusion of specific substances across the membrane. The molecule to be transported binds to the outer surface of the carrier protein; the protein then changes shape, allowing the molecule to move across the membrane through the protein.
carrying capacity: the maximum population size that an ecosystem can support indefinitely; determined primarily by the availability of space, nutrients, water, and light.
cartilage (kar´-teh-lij): a form of connective tissue that forms portions of the skeleton; consists of chondrocytes and their extracellular secretion of collagen; resembles flexible bone.
Casparian strip (kas-par´-ē-un): a waxy, waterproof band, located in the cell walls between endodermal cells in a root, that prevents the movement of water and minerals into and out of the vascular cylinder through the extracellular space.
catalyst (kat´-uh-list): a substance that speeds up a chemical reaction without itself being permanently changed in the process; lowers the activation energy of a reaction.
catastrophism: the hypothesis that Earth has experienced a series of geological catastrophes, probably imposed by a supernatural being that accounts for the multitude of species, both extinct and modern, and preserves creationism.
cell: the smallest unit of life, consisting, at a minimum, of an outer membrane that encloses a watery medium containing organic molecules, including genetic material composed of DNA.
cell body: the part of a nerve cell in which most of the common cellular organelles are located; typically a site of integration of inputs to the nerve cell.
cell cycle: the sequence of events in the life of a cell, from one division to the next.
cell division: splitting of one cell into two; the process of cellular reproduction.
cell-mediated immunity: an immune response in which foreign cells or substances are destroyed by contact with T cells.
cell plate: in plant cell division, a series of vesicles that fuse to form the new plasma membranes and cell wall separating the daughter cells.
cellular respiration: the oxygen-requiring reactions, occurring in mitochondria, that break down the end products of glycolysis into carbon dioxide and water while capturing large amounts of energy as ATP.
cellular slime mold: a funguslike protist consisting of individual amoeboid cells that can aggregate to form a sluglike mass, which in turn forms a fruiting body.
cellulase: an enzyme that catalyzes the breakdown of the carbohydrate cellulose into its component glucose molecules; almost entirely restricted to microorganisms.
cellulose: an insoluble carbohydrate composed of glucose subunits; forms the cell wall of plants.
cell wall: a layer of material, normally made up of cellulose or cellulose-like materials, that is outside the plasma membrane of plants, fungi, bacteria, and some protists.
central nervous system: in vertebrates, the brain and spinal cord.
central vacuole: a large, fluid-filled vacuole occupying most of the volume of many plant cells; performs several functions, including maintaining turgor pressure.
centriole (sen´-trē-ōl): in animal cells, a short, barrel-shaped ring consisting of nine microtubule triplets; a microtubule-containing structure at the base of each cilium and flagellum; gives rise to the microtubules of cilia and flagella and is involved in spindle formation during cell division.
centromere (sen´-trō-mēr): the region of a replicated chromosome at which the sister chromatids are held together until they separate during cell division.
cephalization (sef-ul-ī-zā´-shun): the tendency of sensory organs and nervous tissue to become concentrated in the head region over evolutionary time.
cerebellum (ser-uh-bel´-um): the part of the hindbrain of vertebrates that is concerned with coordinating movements of the body.
cerebral cortex (ser-ē´-brul kor´-tex): a thin layer of neurons on the surface of the vertebrate cerebrum, in which most neural processing and coordination of activity occurs.
cerebral hemisphere: one of two nearly symmetrical halves of the cerebrum, connected by a broad band of axons, the corpus callosum.
cerebrospinal fluid: a clear fluid, produced within the ventricles of the brain, that fills the ventricles and cushions the brain and spinal cord.
cerebrum (ser-ē´-brum): the part of the forebrain of vertebrates that is concerned with sensory processing, the direction of motor output, and the coordination of most bodily activities; consists of two nearly symmetrical halves (the

hemispheres) connected by a broad band of axons, the corpus callosum.

cervical cap: a birth control device consisting of a rubber cap that fits over the cervix, preventing sperm form entering the uterus.

cervix (ser´-viks): a ring of connective tissue at the outer end of the uterus, leading into the vagina.

channel protein: a membrane protein that forms a channel or pore completely through the membrane and that is usually permeable to one or to a few water-soluble molecules, especially ions.

chaparral: a biome that is located in coastal regions but has very low annual rainfall.

chemical bond: the force of attraction between neighboring atoms that holds them together in a molecule.

chemical equilibrium: the condition in which the "forward" reaction of reactants to products proceeds at the same rate as the "backward" reaction from products to reactants, so that no net change in chemical composition occurs.

chemical reaction: the process that forms and breaks chemical bonds that hold atoms together.

chemiosmosis (ke-mē-oz-mō´-sis): a process of ATP generation in chloroplasts and mitochondria. The movement of electrons down an electron transport system is used to pump hydrogen ions across a membrane, thereby building up a concentration gradient of hydrogen ions across the membrane; the hydrogen ions diffuse back across the membrane through the pores of ATP-synthesizing enzymes; the energy of their movement down their concentration gradient drives ATP synthesis.

chemoreceptor: a sensory receptor that responds to chemicals from the environment; used in the chemical senses of taste and smell.

chemosynthetic (kēm´-ō-sin-the-tik): capable of oxidizing inorganic molecules to obtain energy.

chemotactic (kēm-ō-tak´-tik): moving toward chemicals given off by food or away from toxic chemicals.

chiasma (kī-as´-muh; pl., **chiasmata):** a point at which a chromatid of one chromosome crosses with a chromatid of the homologous chromosome during prophase I of meiosis; the site of exchange of chromosomal material between chromosomes.

chitin (kī´-tin): a compound found in the cell walls of fungi and the exoskeletons of insects and some other arthropods; composed of chains of nitrogen-containing, modified glucose molecules.

chlamydia (kla-mid´-ē-uh): a sexually transmitted disease, caused by a bacterium, that causes inflammation of the urethra in males and of the urethra and cervix in females.

chlorophyll (klor´-ō-fil): a pigment found in chloroplasts that captures light energy during photosynthesis; absorbs violet, blue, and red light but reflects green light.

chloroplast (klor´-ō-plast): the organelle in plants and plantlike protists that is the site of photosynthesis; surrounded by a double membrane and containing an extensive internal membrane system that bears chlorophyll.

cholecystokinin (kō´-lē-sis-tō-ki´-nin): a digestive hormone, produced by the small intestine, that stimulates the release of pancreatic enzymes.

chondrocyte (kon´-drō-sīt): a living cell of cartilage. With their extracellular secretions of collagen, chondrocytes form cartilage.

chorion (kor´-ē-on): the outermost embryonic membrane in reptiles, birds, and mammals; in birds and reptiles, functions mostly in gas exchange; in mammals, forms most of the embryonic part of the placenta.

chorionic gonadotropin (CG): a hormone, secreted by the chorion (one of the fetal membranes), that maintains the integrity of the corpus luteum during early pregnancy.

chorionic villus (kor-ē-on-ik; pl., **chorionic villi):** in mammalian embryos, a fingerlike projection of the chorion that penetrates the uterine lining and forms the embryonic portion of the placenta.

chorionic villus sampling (CVS): a procedure for sampling cells from the chorionic villi produced by a fetus: A tube is inserted into the uterus of a pregnant woman, and a small sample of villi are suctioned off for genetic and biochemical analyses.

choroid (kor´-oid): a darkly pigmented layer of tissue, behind the retina, that contains blood vessels and pigment that absorbs stray light.

chromatid (krō´-ma-tid): one of the two identical strands of DNA and protein that forms a replicated chromosome. The two sister chromatids are joined at the centromere.

chromatin (krō´-ma-tin): the complex of DNA and proteins that makes up eukaryotic chromosomes.

chromosome (krō´-mō-sōm): a single DNA double helix together with proteins that help to organize the DNA.

chronic bronchitis: a persistent lung infection characterized by coughing, swelling of the lining of the respiratory tract, an increase in mucus production, and a decrease in the number and activity of cilia.

chyme (kīm): an acidic, souplike mixture of partially digested food, water, and digestive secretions that is released from the stomach into the small intestine.

ciliate (sil´-ē-et): a protozoan characterized by cilia and by a complex unicellular structure, including harpoonlike organelles called *trichocysts.* Members of the genus *Paramecium* are well-known ciliates.

cilium (sil´-ē-um; pl., **cilia):** a short, hairlike projection from the surface of certain eukaryotic cells that contains microtubules in a 9 + 2 arrangement. The movement of cilia may propel cells through a fluid medium or move fluids over a stationary surface layer of cells.

circadian rhythm (sir-kā´-dē-un): an event that recurs with a period of about 24 hours, even in the absence of environmental cues.

citric acid cycle: see *Krebs cycle.*

class: the taxonomic category composed of related genera. Closely related classes form a division or phylum.

cleavage: the early cell divisions of embryos, in which little or no growth occurs between divisions; reduces the cell size and distributes gene-regulating substances to the newly formed cell.

climate: patterns of weather that prevail from year to year and even from century to century in a given region.

climax community: a diverse and relatively stable community that forms the endpoint of succession.

clitoris: an external structure of the female reproductive system; composed of erectile tissue; a sensitive point of stimulation during sexual response.

clonal selection: the mechanism by which the immune response gains specificity; an invading antigen elicits a response from only a few lymphocytes, which proliferate to form a clone of cells that attack only the specific antigen that stimulated their production.

clone: offspring that are produced by mitosis and are therefore genetically identical to each other.

cloning: the process of producing many identical copies of a gene; also the production of many genetically identical copies of an organism.

closed circulatory system: the type of circulatory system, found in certain worms and vertebrates, in which the blood is always confined within the heart and vessels.

club fungus: a fungus of the division Basidiomycota, whose members (which include mushrooms, puffballs, and shelf fungi) reproduce by means of basidiospores.

clumped distribution: the distribution characteristic of populations in which individuals are clustered into groups; may be social or based on the need for a localized resource.

cnidocyte (nīd´-ō-sīt): in members of the phylum Cnidaria, a specialized cell that houses a stinging apparatus.

cochlea (kahk´-lē-uh): a coiled, bony, fluid-filled tube found in the mammalian inner ear; contains receptors (hair cells) that respond to the vibration of sound.

codominance: the relation between two alleles of a gene, such that both alleles are phenotypically expressed in heterozygous individuals.

codon: a sequence of three bases of messenger RNA that specifies a particular amino acid to be incorporated into a protein; certain codons also signal the beginning or end of protein synthesis.

coelom (sē´-lōm): a space or cavity that separates the body wall from the inner organs.

coenzyme: an organic molecule that is bound to certain enzymes and is required for the enzymes' proper functioning; typically, a nucleotide bound to a water-soluble vitamin.

coevolution: the evolution of adaptations in two species due to their extensive interactions with one another, such that each species acts as a major force of natural selection on the other.

cohesion: the tendency of the molecules of a substance to stick together.

cohesion–tension theory: a model for the transport of water in xylem, by which water is pulled up the xylem tubes, powered by the force of evaporation of water from the leaves (producing tension) and held together by hydrogen bonds between nearby water molecules (cohesion).

coleoptile (kō-lē-op´-tīl): a protective sheath surrounding the shoot in monocot seeds, allowing the shoot to push aside soil particles as it grows.

collagen (kol´-uh-jen): a fibrous protein in connective tissue such as bone and cartilage.

collar cell: a specialized cell lining the inside channels of sponges. Flagella extend from a sievelike collar, creating a water current that draws microscopic organisms through the collar and into the body, where they become trapped.

collecting duct: a conducting tube, within the kidney, that collects urine from many nephrons and conducts it through the renal medulla into the renal pelvis. Urine may become concentrated in the collecting ducts if ADH is present.

collenchyma (kōl-en´-ki-muh): an elongated, polygonal plant cell type with irregularly thickened primary cell walls that is alive at maturity and that supports the plant body.

colon: the longest part of the large intestine, exclusive of the rectum.

colostrum (kō-los´-trum): a yellowish fluid, high in protein and containing antibodies, that is produced by the mammary glands before milk secretion begins.

commensalism (kum-en´-sal-iz-um): a symbiotic relationship in which one species benefits while another species is neither harmed nor benefited.

communication: the act of producing a signal that causes another animal, normally of the same species, to change its behavior in a way that is beneficial to one or both participants.

community: all the interacting populations within an ecosystem.

compact bone: the hard and strong outer bone; composed of osteons.

companion cell: a cell adjacent to a sieve-tube element in phloem, involved in the control and nutrition of the sieve-tube element.

competition: interaction among individuals who attempt to utilize a resource (for example, food or space) that is limited relative to the demand for it.

competitive exclusion principle: the concept that no two species can simultaneously and continuously occupy the same ecological niche.

competitive inhibition: the process by which two or more molecules that are somewhat similar in structure compete for the active site of an enzyme.

complement: a group of blood-borne proteins that participate in the destruction of foreign cells to which antibodies have bound.

complementary base pair: in nucleic acids, bases that pair by hydrogen bonding. In DNA, adenine is complementary to thymine and guanine is complementary to cytosine; in RNA, adenine is complementary to uracil, and guanine to cytosine.

complement reaction: an interaction among foreign cells, antibodies, and complement proteins that results in the destruction of the foreign cells.

complement system: a series of reactions in which complement proteins bind to antibody stems, attracting to the site phagocytic white blood cells that destroy the invading cell that triggers the reactions.

complete flower: a flower that has all four floral parts (sepals, petals, stamens, and carpels).

compound: a substance whose molecules are formed by different types of atoms; can be broken into its constituent elements by chemical means.

compound eye: a type of eye, found in arthropods, that is composed of numerous independent subunits called *ommatidia.* Each ommatidium apparently contributes a piece of a mosaiclike image perceived by the animal.

concentration: the number of particles of a dissolved substance in a given unit of volume.

concentration gradient: the difference in concentration of a substance between two parts of a fluid or across a barrier such as a membrane.

conclusion: the final operation in the scientific method; a decision made about the validity of a hypothesis on the basis of experimental evidence.

condensation: compaction of eukaryotic chromosomes into discrete units in preparation for mitosis or meiosis.

condom: a contraceptive sheath worn over the penis during intercourse to prevent sperm from being deposited in the vagina.

conducting portion: the portion of the respiratory system in lung-breathing vertebrates that carries air to the lungs.

cone: a cone-shaped photoreceptor cell in the vertebrate retina; not as sensitive to light as are the rods. The three types of cones are most sensitive to different colors of light and provide color vision; see also *rod*.

conifer (kon´-eh-fer): a member of a class of tracheophytes (Coniferophyta) that reproduces by means of seeds formed inside cones and that retains its leaves throughout the year.

connective tissue: a tissue type consisting of diverse tissues, including bone, fat, and blood, that generally contain large amounts of extracellular material.

constant region: the part of an antibody molecule that is similar in all antibodies.

consumer: an organism that eats other organisms; a heterotroph.

contest competition: a type of intraspecific competition in which competing individuals interact directly with one another—for example, by fighting.

contraception: the prevention of pregnancy.

contractile vacuole: a fluid-filled vacuole in certain protists that takes up water from the cytoplasm, contracts, and expels the water outside the cell through a pore in the plasma membrane.

control: that portion of an experiment in which all possible variables are held constant; in contrast to the "experimental" portion, in which a particular variable is altered.

convergence: a condition in which a large number of nerve cells provide input to a smaller number of cells.

convergent evolution: the independent evolution of similar structures among distantly related organisms as a result of similar environmental pressures; see *analogous structures.*

convolution: a folding of the cerebral cortex of the vertebrate brain.

copulation: reproductive behavior in which the penis of the male is inserted into the body of the female, where it releases sperm.

coral reef: a biome created by animals (reef-building corals) and plants in warm tropical waters.

cork cambium: a lateral meristem in woody roots and stems that gives rise to cork cells.

cork cell: a protective cell of the bark of woody stems and roots; at maturity, cork cells are dead, with thick, waterproofed cell walls.

cornea (kor´-nē-uh): the clear outer covering of the eye, in front of the pupil and iris.

corona radiata (kuh-rō´-nuh rā-dē-a´-tuh): the layer of cells surrounding an egg after ovulation.

corpus callosum (kor´pus kal-ō´-sum): the band of axons that connect the two cerebral hemispheres of vertebrates.

corpus luteum (kor´-pus loo´-tē-um): in the mammalian ovary, a structure that is derived from the follicle after ovulation and that secretes the hormones estrogen and progesterone.

cortex: the part of a primary root or stem located between the epidermis and the vascular cylinder.

cotyledon (kot-ul-ē´don): a leaflike structure within a seed that absorbs food molecules from the endosperm and transfers them to the growing embryo; also called *seed leaf.*

coupled reaction: a pair of reactions, one exergonic and one endergonic, that are linked together such that the energy produced by the exergonic reaction provides the energy needed to drive the endergonic reaction.

covalent bond (kō-vā´-lent): a chemical bond between atoms in which electrons are shared.

crab lice: an arthropod parasite that can infest humans; can be transmitted by sexual contact.

creationism: the hypothesis that all species on Earth were created in essentially their present form by a supernatural being and that significant modification of those species—specifically, their transformation into new species—cannot occur by natural processes.

crista (kris´-tuh; pl., **cristae):** a fold in the inner membrane of a mitochondrion.

crop: an organ, found in both earthworms and birds, in which ingested food is temporarily stored before being passed to the gizzard, where it is pulverized.

cross-bridge: in muscles, an extension of myosin that binds to and pulls on actin to produce muscle contraction.

cross-fertilization: the union of sperm and egg from two individuals of the same species.

crossing over: the exchange of corresponding segments of the chromatids of two homologous chromosomes during meiosis.

cultural evolution: changes in the behavior of a population of animals, especially humans, by learning behaviors acquired by members of previous generations.
cuticle (kū´-ti-kul): a waxy or fatty coating on the exposed surfaces of epidermal cells of many land plants, which aids in the retention of water.
cyanobacterium: a photosynthetic prokaryotic cell that utilizes chlorophyll and releases oxygen as a photosynthetic by-product; sometimes called *blue-green algae*.
cyclic AMP: a cyclic nucleotide, formed within many target cells as a result of the reception of amino acid derivatives or peptide hormones, that causes metabolic changes in the cell; often called a *second messenger*.
cyclic nucleotide (sik´-lik noo´-klē-ō-tīd): a nucleotide in which the phosphate group is bonded to the sugar at two points, forming a ring; serves as an intracellular messenger.
cyst (sist): an encapsulated resting stage in the life cycle of certain invertebrates, such as parasitic flatworms and roundworms.
cystic fibrosis: an inherited disorder characterized by the buildup of salt in the lungs and the production of thick, sticky mucus that clogs the airways, restricts air exchange, and promotes infection.
cytokinesis (sī-tō-ki-nē´-sis): the division of the cytoplasm and organelles into two daughter cells during cell division; normally occurs during telophase of mitosis.
cytokinin (sī-tō-kī´-nin): a plant hormone that promotes cell division, fruit growth, and the sprouting of lateral buds and prevents the aging of plant parts, especially leaves.
cytoplasm (sī´-tō-plaz-um): the material contained within the plasma membrane of a cell, exclusive of the nucleus.
cytosine: a nitrogenous base found in both DNA and RNA; abbreviated as C.
cytoskeleton: a network of protein fibers in the cytoplasm that gives shape to a cell, holds and moves organelles, and is typically involved in cell movement.
cytotoxic T cell: a type of T cell that, upon contacting foreign cells, directly destroys them.
day-neutral plant: a plant in which flowering occurs as soon as the plant has grown and developed, regardless of daylength.
decomposer: an organism, normally a fungus or bacterium, that digests organic material by secreting digestive enzymes into the environment, in the process liberating nutrients into the environment.
deductive reasoning: the process of generating hypotheses about how a specific experiment or observation will turn out.
deforestation: the excessive cutting of forests, primarily rain forests in the Tropics, to clear space for agriculture.
dehydration synthesis: a chemical reaction in which two molecules are joined by a covalent bond with the simultaneous removal of a hydrogen from one molecule and a hydroxyl group from the other, forming water; the reverse of hydrolysis.
deletion mutation: a mutation in which one or more pairs of nucleotides are removed from a gene.
denature: to disrupt the secondary and/or tertiary structure of a protein while leaving its amino acid sequence intact. Denatured proteins can no longer perform their biological functions.
dendrite (den´-drīt): a branched tendril that extends outward from the cell body of a neuron; specialized to respond to signals from the external environment or from other neurons.
denitrifying bacterium (dē-nī´-treh-fī-ing): a bacterium that breaks down nitrates, releasing nitrogen gas to the atmosphere.
density-dependent: referring to any factor, such as predation, that limits population size more effectively as the population density increases.
density-independent: referring to any factor that limits a population's size and growth regardless of its density.
deoxyribonucleic acid (dē-ox-ē-rī-bō-noo-klā´-ik; DNA): a molecule composed of deoxyribose nucleotides; contains the genetic information of all living cells.
dermal tissue: plant tissue that makes up the outer covering of the plant body.
dermis (dur´-mis): the layer of skin beneath the epidermis; composed of connective tissue and containing blood vessels, muscles, nerve endings, and glands.
desert: a biome in which less than 25 to 50 centimeters (10 to 20 inches) of rain falls each year.
desertification: the spread of deserts by human activities.
desmosome (dez´-mō-sōm): a strong cell-to-cell junction that attaches adjacent cells to one another.
detritus feeder (de-trī´-tus): one of a diverse group of organisms, ranging from worms to vultures, that live off the wastes and dead remains of other organisms.
deuterostome (doo´-ter-ō-stōm): an animal with a mode of embryonic development in which the coelom is derived from outpocketings of the gut; characteristic of echinoderms and chordates.
development: the process by which an organism proceeds from fertilized egg through adulthood to eventual death.
diabetes mellitus (dī-uh-bē´-tēs mel-ī´-tus): a disease characterized by defects in the production, release, or reception of insulin; characterized by high blood glucose levels that fluctuate with sugar intake.
dialysis (dī-al´-i-sis): the passive diffusion of substances across an artificial semipermeable membrane.
diaphragm (dī´-uh-fram): in the respiratory system, a dome-shaped muscle forming the floor of the chest cavity that, when it contracts, pulls itself downward, enlarging the chest cavity and causing air to be drawn into the lungs; in a reproductive sense, a contraceptive rubber cap that fits snugly over the cervix, preventing the sperm from entering the uterus and thereby preventing pregnancy.
diatom (dī´-uh-tom): a protist that includes photosynthetic forms with two-part glassy outer coverings; important photosynthetic organisms in fresh water and salt water.
dicot (dī´-kaht): short for dicotyledon; a type of flowering plant characterized by embryos with two cotyledons, or seed leaves, modified for food storage.
differentially permeable: referring to the ability of some substances to pass through a membrane more readily than can other substances.
differential reproduction: differences in reproductive output among individuals of a population, normally as a result of genetic differences.
differentiated cell: a mature cell specialized for a specific function; in plants, differentiated cells normally do not divide.
differentiation: the process whereby relatively unspecialized cells, especially of embryos, become specialized into particular tissue types.
diffusion: the net movement of particles from a region of high concentration of that particle to a region of low concentration, driven by the concentration gradient; may occur entirely within a fluid or across a barrier such as a membrane.
digestion: the process by which food is physically and chemically broken down into molecules that can be absorbed by cells.
digestive system: a group of organs responsible for ingesting and then digesting food substances into simple molecules that can be absorbed and then expelling undigested wastes from the body.
dinoflagellate (dī-nō-fla´-jel-et): a protist that includes photosynthetic forms in which two flagella project through armorlike plates; abundant in oceans; can reproduce rapidly, causing "red tides."
dioecious (dī-ē´-shus): pertaining to organisms in which male and female gametes are produced by separate individuals rather than in the same individual.
diploid (dip´-loid): referring to a cell with pairs of homologous chromosomes.
direct development: a developmental pathway in which the offspring is born as a miniature version of the adult and does not radically change in body form as it grows and matures.
directional selection: a type of natural selection in which one extreme phenotype is favored over all others.
disaccharide (dī-sak´-uh-rīd): a carbohydrate formed by the covalent bonding of two monosaccharides.
disruptive selection: a type of natural selection in which both extreme phenotypes are favored over the average phenotype.
distal tubule: in the nephrons of the mammalian kidney, the last segment of the renal tubule through which the filtrate passes just before it empties into the collecting duct; a site of selective secretion and reabsorption as water and ions pass between the blood and the filtrate across the tubule membrane.
disulfide bridge: the covalent bond formed between the sulfur atoms of two cysteines in a protein; typically causes the protein to fold by bringing otherwise distant parts of the protein close together.

divergence: a condition in which a small number of nerve cells provide input to a larger number of cells.
divergent evolution: evolutionary change in which the differences between two lineages become more pronounced with the passage of time.
division: the taxonomic category contained within a kingdom and consisting of related classes of plants, fungi, bacteria, or plantlike protists.
DNA: see *deoxyribonucleic acid.*
DNA–DNA hybridization: a technique by which DNA from two species is separated into single strands and then allowed to re-form; hybrid double-stranded DNA from the two species can occur where the sequence of nucleotides is complementary. The greater the degree of hybridization, the closer the evolutionary relatedness of the two species.
DNA fingerprinting: the use of restriction enzymes to cut DNA segments into a unique set of restriction fragments from one individual that can be distinguished from the restriction fragments of other individuals by gel electrophoresis.
DNA helicase: an enzyme that helps unwind the DNA double helix during DNA replication.
DNA library: a readily accessible, easily duplicable complete set of all the DNA of a particular organism, normally cloned into bacterial plasmids.
DNA ligase: an enzyme that joins the sugars and phosphates in a DNA strand to create a continuous sugar-phosphate backbone.
DNA polymerase: an enzyme that bonds DNA nucleotides together into a continuous strand, using a preexisting DNA strand as a template.
DNA probe: a sequence of nucleotides that is complementary to the nucleotide sequence in a gene under study; used to locate a given gene within a DNA library.
DNA replication: the copying of the double-stranded DNA molecule, producing two identical DNA double helices.
DNA sequencing: the process of determining the chemical composition of a DNA molecule (in particular, the order in which the molecule's constituent nucleic acids are arranged).
domain: the broadest category for classifying organisms; organisms are classified into three domains: Bacteria, Archaea, and Eukarya.
dominance hierarchy: a social arrangement in which a group of animals, usually through aggressive interactions, establishes a rank for some or all of the group members that determines access to resources.
dominant: an allele that can determine the phenotype of heterozygotes completely, such that they are indistinguishable from individuals homozygous for the allele; in the heterozygotes, the expression of the other (recessive) allele is completely masked.
dopamine (dōp´-uh-mēn): a transmitter in the brain whose actions are largely inhibitory. The loss of dopamine-containing neurons causes Parkinson's disease.
dormancy: a state in which an organism does not grow or develop; usually marked by lowered metabolic activity and resistance to adverse environmental conditions.
dorsal (dor´-sul): the top, back, or uppermost surface of an animal oriented with its head forward.
dorsal root ganglion: a ganglion, located on the dorsal (sensory) branch of each spinal nerve, that contains the cell bodies of sensory neurons.
double covalent bond: a covalent bond in which two atoms share two pairs of electrons.
double fertilization: in flowering plants, the fusion of two sperm nuclei with the nuclei of two cells of the female gametophyte. One sperm nucleus fuses with the egg to form the zygote; the second sperm nucleus fuses with the two haploid nuclei of the primary endosperm cell, forming a triploid endosperm cell.
double helix (hē´-liks): the shape of the two-stranded DNA molecule; like a ladder twisted lengthwise into a corkscrew shape.
doubling time: the time it would take a population to double in size at its current growth rate.
douching: washing the vagina; after intercourse, an attempt to wash sperm out of the vagina before they enter the uterus; an ineffective contraceptive method.
Down syndrome: a genetic disorder caused by the presence of three copies of chromosome 21; common characteristics include mental retardation, distinctively shaped eyelids, a small mouth with protruding tongue, heart defects, and low resistance to infectious diseases; also called *trisomy 21.*
duct: a tube or opening through which exocrine secretions are released.
duplicated chromosome: a eukaryotic chromosome following DNA replication; consists of two sister chromatids joined at the centromeres.
echolocation: the use of ultrasonic sounds, which bounce back from nearby objects, to produce an auditory "image" of nearby surroundings; used by bats and porpoises.
ecological isolation: the lack of mating between organisms belonging to different populations that occupy distinct habitats within the same general area.
ecological niche (nitch): the role of a particular species within an ecosystem, including all aspects of its interaction with the living and nonliving environments.
ecology (ē-kol´-uh-jē): the study of the interrelationships of organisms with each other and with their nonliving environment.
ecosystem (ē´kō-sis-tem): all the organisms and their nonliving environment within a defined area.
ectoderm (ek´-tō-derm): the outermost embryonic tissue layer, which gives rise to structures such as hair, the epidermis of the skin, and the nervous system.
effector (ē-fek´-tor): a part of the body (normally a muscle or gland) that carries out responses as directed by the nervous system.
egg: the haploid female gamete, normally large and nonmotile, containing food reserves for the developing embryo.
electrolocation: the production of high-frequency electrical signals from an electric organ in front of the tail of weak electrical fish; used to detect and locate nearly objects.
electron: a subatomic particle, found in an electron shell outside the nucleus of an atom, that bears a unit of negative charge and very little mass.
electron carrier: a molecule that can reversibly gain or lose electrons. Electron carriers generally accept high-energy electrons produced during an exergonic reaction and donate the electrons to acceptor molecules that use the energy to drive endergonic reactions.
electron micrograph: a photographic image of an object viewed through an electron microscope.
electron shell: a region within which electrons orbit that corresponds to a fixed energy level at a given distance from the atomic nucleus of an atom.
electron transport system: a series of electron carrier molecules, found in the thylakoid membranes of chloroplasts and the inner membrane of mitochondria, that extract energy from electrons and generate ATP or other energetic molecules.
element: a substance that cannot be broken down, or converted, to a simpler substance by ordinary chemical means.
embryo: in animals, the stages of development that begin with the fertilization of the egg cell and end with hatching or birth; in mammals in particular, the early stages in which the developing animal does not yet resemble the adult of the species.
embryonic disc: in human embryonic development, the flat, two-layered group of cells that separates the amniotic cavity from the yolk sac.
embryonic stem cell: a cell derived from an early embryo that is capable of differentiating into any of the adult cell types.
embryo sac: the haploid female gametophyte of flowering plants.
emergent property: an intangible attribute that arises as the result of complex ordered interactions among individual parts.
emigration (em-uh-grā´shun): migration of individuals out of an area.
emphysema (em-fuh-sē´-muh): a condition in which the alveoli of the lungs become brittle and rupture, causing decreased area for gas exchange.
endergonic (en-der-gon´-ik) reaction: pertaining to a chemical reaction that requires an input of energy to proceed; an "uphill" reaction.
endocrine disruptors: environmental pollutants that interfere with endocrine function, often by disrupting the action of sex hormones.
endocrine gland: a ductless, hormone-producing gland consisting of cells that release their secretions into the extracellular fluid from which the secretions diffuse into nearby capillaries.
endocrine system: an animal's organ system for cell-to-cell communication, composed of hormones and the cells that secrete them and receive them.

endocytosis (en-dō-sī-tō´-sis): the process in which the plasma membrane engulfs extracellular material, forming membrane-bound sacs that enter the cytoplasm and thereby move material into the cell.
endoderm (en´-dō-derm): the innermost embryonic tissue layer, which gives rise to structures such as the lining of the digestive and respiratory tracts.
endodermis (en-dō-der´-mis): the innermost layer of small, close-fitting cells of the cortex of a root that form a ring around the vascular cylinder.
endogenous pyrogen: a chemical, produced by the body, that stimulates the production of a fever.
endometrium (en-dō-mē´-trē-um): the nutritive inner lining of the uterus.
endoplasmic reticulum (ER) (en-dō-plaz´-mik re-tik´-ū-lum): a system of membranous tubes and channels within eukaryotic cells; the site of most protein and lipid syntheses.
endorphin (en-dor´-fin): one of a group of peptide neuromodulators in the vertebrate brain that, by reducing the sensation of pain, mimics some of the actions of opiates.
endoskeleton (en´-dō-skel´-uh-tun): a rigid internal skeleton with flexible joints to allow for movement.
endosperm: a triploid food storage tissue in the seeds of flowering plants that nourishes the developing plant embryo.
endospore: a protective resting structure of some rod-shaped bacteria that withstands unfavorable environmental conditions.
endosymbiont hypothesis: the hypothesis that certain organelles, especially chloroplasts and mitochondria, arose as mutually beneficial associations between the ancestors of eukaryotic cells and captured bacteria that lived within the cytoplasm of the pre-eukaryotic cell.
energy: the capacity to do work.
energy-carrier molecule: a molecule that stores energy in "high-energy" chemical bonds and releases the energy to drive coupled endothermic reactions. In cells, ATP is the most common energy-carrier molecule.
energy level: the specific amount of energy characteristic of a given electron shell in an atom.
energy pyramid: a graphical representation of the energy contained in succeeding trophic levels, with maximum energy at the base (primary producers) and steadily diminishing amounts at higher levels.
entropy (en´-trō-pē): a measure of the amount of randomness and disorder in a system.
environmental estrogens: chemicals in the environment that mimic some of the effects of estrogen in animals.
environmental resistance: any factor that tends to counteract biotic potential, limiting population size.
enzyme (en´zīm): a protein catalyst that speeds up the rate of specific biological reactions.
eosinophil (ē-ō-sin´-ō-fil): a type of white blood cell that converges on parasitic invaders and releases substances to kill them.
epicotyl (ep´-ē-kot-ul): the part of the embryonic shoot located above the cotyledons but below the tip of the shoot.
epidermal tissue: dermal tissue in plants that forms the epidermis, the outermost cell layer that covers young plants.
epidermis (ep-uh-der´-mis): in animals, specialized epithelial tissue that forms the outer layer of skin; in plants, the outermost layer of cells of a leaf, young root, or young stem.
epididymis (e-pi-di´-di-mus): a series of tubes that connect with and receive sperm from the seminiferous tubules of the testis.
epiglottis (ep-eh-glah´-tis): a flap of cartilage in the lower pharynx that covers the opening to the larynx during swallowing; directs food down the esophagus.
epinephrine (ep-i-nef´-rin): a hormone, secreted by the adrenal medulla, that is released in response to stress and that stimulates a variety of responses, including the release of glucose from skeletal muscle and an increase in heart rate.
epithelial cell (eh-puh-thē´-lē-ul): a flattened cell that covers the outer body surfaces of a sponge.
epithelial tissue (eh-puh-thē´-lē-ul): a tissue type that forms membranes that cover the body surface and line body cavities, and that also gives rise to glands.
equilibrium population: a population in which allele frequencies and the distribution of genotypes do not change from generation to generation.
erythroblastosis fetalis (eh-rith´-rō-blas-tō´-sis fē-tal´-is): a condition in which the red blood cells of a newborn Rh-positive baby are attacked by antibodies produced by its Rh-negative mother, causing jaundice and anemia. Retardation and death are possible consequences if treatment is inadequate.
erythrocyte (eh-rith´-rō-sīt): a red blood cell, active in oxygen transport, that contains the red pigment hemoglobin.
erythropoietin (eh-rith´-rō-pō-ē´-tin): a hormone produced by the kidneys in response to oxygen deficiency that stimulates the production of red blood cells by the bone marrow.
esophagus (eh-sof´-eh-gus): a muscular passageway that conducts food from the pharynx to the stomach in humans and other mammals.
essential amino acid: an amino acid that is a required nutrient; the body is unable to manufacture essential amino acids, so they must be supplied in the diet.
essential fatty acid: a fatty acid that is a required nutrient; the body is unable to manufacture essential fatty acids, so they must be supplied in the diet.
estrogen: in vertebrates, a female sex hormone, produced by follicle cells of the ovary, that stimulates follicle development, oogenesis, the development of secondary sex characteristics, and growth of the uterine lining.
estuary: a wetland formed where a river meets the ocean; the salinity there is quite variable but lower than in sea water and higher than in fresh water.
ethology (ē-thol´-ō-jē): the study of animal behavior in natural or near-natural conditions.
ethylene: a plant hormone that promotes the ripening of fruits and the dropping of leaves and fruit.
euglenoid (ū´-gle-noid): a protist characterized by one or more whiplike flagella that are used for locomotion and by a photoreceptor that detects light. Euglenoids are photosynthetic, but if deprived of chlorophyll, some are capable of heterotrophic nutrition.
Eukarya: one of life's three domains; includes all organisms whose cells contain a cell nucleus.
eukaryotic (ū-kar-ē-ot´-ik) cell: referring to cells of organisms of the domain Eukarya (kingdoms Protista, Fungi, Plantae, and Animalia). Eukaryotic cells have genetic material enclosed within a membrane-bound nucleus and contain other membrane-bound organelles.
Eustachian tube (ū-stā´-shin): a tube connecting the middle ear with the pharynx; allows pressure between the middle ear and the atmosphere to equilibrate.
eutrophic lake: a lake that receives sufficiently large inputs of sediments, organic material, and inorganic nutrients from its surroundings to support dense communities; murky with poor light penetration.
evergreen: a plant that retains green leaves throughout the year.
evolution: the descent of modern organisms with modification from preexisting life-forms; strictly speaking, any change in the proportions of different genotypes in a population from one generation to the next.
excretion: the elimination of waste substances from the body; can occur from the digestive system, skin glands, urinary system, or lungs.
excretory pore: an opening in the body wall of certain invertebrates, such as the earthworm, through which urine is excreted.
exergonic (ex-er-gon´-ik) reaction: pertaining to a chemical reaction that liberates energy (either as heat or in the form of increased entropy); a "downhill" reaction.
exhalation: the act of releasing air from the lungs, which results from a relaxation of the respiratory muscles.
exocrine gland: a gland that releases its secretions into ducts that lead to the outside of the body or into the digestive tract.
exocytosis (ex-ō-sī-tō´-sis): the process in which intracellular material is enclosed within a membrane-bound sac that moves to the plasma membrane and fuses with it, releasing the material outside the cell.
exon: a segment of DNA in a eukaryotic gene that codes for amino acids in a protein (see also *intron*).
exoskeleton (ex´-ō-skel´-uh-tun): a rigid external skeleton that supports the body, protects the internal organs, and has flexible joints that allow for movement.
exotic/exotic species: a foreign species introduced into an ecosystem where it did not evolve; such species may flourish and outcompete native species.

experiment: the third operation in the scientific method; the testing of a hypothesis by further observations, leading to a conclusion.
exponential growth: a continuously accelerating increase in population size.
extensor: a muscle that straightens a joint.
external ear: the fleshy portion of the ear that extends outside the skull.
external fertilization: the union of sperm and egg outside the body of either parent.
extinction: the death of all members of a species.
extracellular digestion: the physical and chemical breakdown of food that occurs outside a cell, normally in a digestive cavity.
extraembryonic membrane: in the embryonic development of reptiles, birds, and mammals, either the chorion, amnion, allantois, or yolk sac; functions in gas exchange, provision of the watery environment needed for development, waste storage, and storage of the yolk, respectively.
eyespot: a simple, lensless eye found in various invertebrates, including flatworms and jellyfish. Eyespots can distinguish light from dark and sometimes the direction of light, but they cannot form an image.
facilitated diffusion: the diffusion of molecules across a membrane, assisted by protein pores or carriers embedded in the membrane.
fairy ring: a circular pattern of mushrooms formed when reproductive structures erupt from the underground hyphae of a club fungus that has been growing outward in all directions from its original location.
family: the taxonomic category contained within an order and consisting of related genera.
farsighted: the inability to focus on nearby objects, caused by the eyeball being slightly too short.
fat (molecular): a lipid composed of three saturated fatty acids covalently bonded to glycerol; solid at room temperature.
fat (tissue): adipose tissue; connective tissue that stores the lipid fat; composed of cells packed with triglycerides.
fatty acid: an organic molecule composed of a long chain of carbon atoms, with a carboxylic acid (COOH) group at one end; may be saturated (all single bonds between the carbon atoms) or unsaturated (one or more double bonds between the carbon atoms).
feces: semisolid waste material that remains in the intestine after absorption is complete and is voided through the anus. Feces consist of indigestible wastes and the dead bodies of bacteria.
feedback inhibition: in enzyme-mediated chemical reactions, the condition in which the product of a reaction inhibits one or more of the enzymes involved in synthesizing the product.
fermentation: anaerobic reactions that convert the pyruvic acid produced by glycolysis into lactic acid or alcohol and CO_2.
fertilization: the fusion of male and female haploid gametes, forming a zygote.
fetal alcohol syndrome (FAS): a cluster of symptoms, including retardation and physical abnormalities, that occur in infants born to a mother who consumed large amounts of alcoholic beverages during pregnancy.
fetus: the later stages of mammalian embryonic development (after the second month for humans), when the developing animal has come to resemble the adult of the species.
fever: an elevation in body temperature caused by chemicals (pyrogens) that are released by white blood cells in response to infection.
fibrillation: rapid, uncoordinated, and ineffective contractions of heart muscle cells.
fibrin (fī´-brin): a clotting protein formed in the blood in response to a wound; binds with other fibrin molecules and provides a matrix around which a blood clot forms.
fibrous roots: a root system, commonly found in monocots, characterized by many roots of approximately the same size arising from the base of the stem.
filament: in flowers, the stalk of a stamen, which bears an anther at its tip.
filtrate: the fluid produced by filtration; in the kidneys, the fluid produced by the filtration of blood through the glomerular capillaries.
filtration: within Bowman's capsule in each nephron of a kidney, the process by which blood is pumped under pressure through permeable capillaries of the glomerulus, forcing out water, dissolved wastes, and nutrients.
fimbria (fim´-brē-uh; pl., **fimbriae):** in female mammals, the ciliated, fingerlike projections of the oviduct that sweep the ovulated egg from the ovary into the oviduct.
first law of thermodynamics: the principle of physics that states that within any isolated system, energy can be neither created nor destroyed but can be converted from one form to another.
fission: asexual reproduction by dividing the body into two smaller, complete organisms.
fitness: the reproductive success of an organism, usually expressed in relation to the average reproductive success of all individuals in the same population.
flagellum (fla-jel´-um; pl., **flagella):** a long, hairlike extension of the plasma membrane; in eukaryotic cells, it contains microtubules arranged in a 9 + 2 pattern. The movement of flagella propel some cells through fluids.
flame cell: in flatworms, a specialized cell, containing beating cilia, that conducts water and wastes through the branching tubes that serve as an excretory system.
flexor: a muscle that flexes (decreases the angle of) a joint.
florigen: one of a group of plant hormones that can both trigger and inhibit flowering; daylength is a stimulus.
flower: the reproductive structure of an angiosperm plant.
flowering plant: member of a group of vascular plants whose seeds are formed inside protective chambers called ovaries; flowering plants are also known as angiosperms.
fluid: a liquid or gas.
fluid mosaic model: a model of membrane structure; according to this model, membranes are composed of a double layer of phospholipids in which various proteins are embedded. The phospholipid bilayer is a somewhat fluid matrix that allows the movement of proteins within it.
follicle: in the ovary of female mammals, the oocyte and its surrounding accessory cells.
follicle-stimulating hormone (FSH): a hormone, produced by the anterior pituitary, that stimulates spermatogenesis in males and the development of the follicle in females.
food chain: a linear feeding relationship in a community, using a single representative from each of the trophic levels.
food vacuole: a membranous sac, within a single cell, in which food is enclosed. Digestive enzymes are released into the vacuole, where intracellular digestion occurs.
food web: a representation of the complex feeding relationships (in terms of interacting food chains) within a community, including many organisms at various trophic levels, with many of the consumers occupying more than one level simultaneously.
foraminiferan (for-am-i-nif´-er-un): an aquatic (largely marine) protist characterized by a typically elaborate calcium carbonate shell.
forebrain: during development, the anterior portion of the brain. In mammals, the forebrain differentiates into the thalamus, the limbic system, and the cerebrum. In humans, the cerebrum contains about half of all the neurons in the brain.
fossil: the remains of a dead organism, normally preserved in rock—may be petrified bones or wood; shells; impressions of body forms, such as feathers, skin, or leaves; or markings made by organisms, such as footprints.
fossil fuel: a fuel such as coal, oil, and natural gas, derived from the remains of ancient organisms.
founder effect: a type of genetic drift in which an isolated population founded by a small number of individuals may develop allele frequencies that are very different from those of the parent population as a result of chance inclusion of disproportionate numbers of certain alleles in the founders.
fovea (fō´-vē-uh): in the vertebrate retina, the central region on which images are focused; contains closely packed cones.
free-living: not parasitic.
free nerve ending: on some receptor neurons, a finely branched ending that responds to touch and pressure, to heat and cold, or to pain; produces the sensations of itching and tickling.
free nucleotides: nucleotides that have not been joined together to form a DNA or RNA strand.
fruit: in flowering plants, the ripened ovary (plus, in some cases, other parts of the flower), which contains the seeds.
fruiting body: a spore-forming reproductive structure of certain protists, bacteria, and fungi.
functional group: one of several groups of atoms commonly found in an organic mole-

cule, including hydrogen, hydroxyl, amino, carboxyl, and phosphate groups, that determine the characteristics and chemical reactivity of the molecule.

gallbladder: a small sac, next to the liver, in which the bile secreted by the liver is stored and concentrated. Bile is released from the gallbladder to the small intestine through the bile duct.

gamete (gam´-ēt): a haploid sex cell formed in sexually reproducing organisms.

gametic incompatibility: the inability of sperm from one species to fertilize eggs of another species.

gametophyte (ga-mēt´-ō-fīt): the multicellular haploid stage in the life cycle of plants.

ganglion (gang´-lē-un): a cluster of neurons.

ganglion cell: a type of cell, of which the innermost layer of the vertebrate retina is composed, whose axons form the optic nerve.

gap junction: a type of cell-to-cell junction in animals in which channels connect the cytoplasm of adjacent cells.

gas-exchange portion: the portion of the respiratory system in lung-breathing vertebrates where gas is exchanged in the alveoli of the lungs.

gastric inhibitory peptide: a hormone, produced by the small intestine, that inhibits the activity of the stomach.

gastrin: a hormone, produced by the stomach, that stimulates acid secretion in response to the presence of food.

gastrovascular cavity: a saclike chamber with digestive functions, found in simple invertebrates; a single opening serves as both mouth and anus, and the chamber provides direct access of nutrients to the cells.

gastrula (gas´-troo-luh): in animal development, a three-layered embryo with ectoderm, mesoderm, and endoderm cell layers. The endoderm layer normally encloses the primitive gut.

gastrulation (gas-troo-la´-shun): the process whereby a blastula develops into a gastrula, including the formation of endoderm, ectoderm, and mesoderm.

gel electrophoresis: a technique in which molecules (such as DNA fragments) are placed on restricted tracks in a thin sheet of gelatinous material and exposed to an electric field; the molecules then migrate at a rate determined by certain characteristics, such as length.

gene: a unit of heredity that encodes the information needed to specify the amino acid sequence of proteins and hence particular traits; a functional segment of DNA located at a particular place on a chromosome.

gene flow: the movement of alleles from one population to another owing to the migration of individual organisms.

gene pool: the total of all alleles of all genes in a population; for a single gene, the total of all the alleles of that gene that occur in a population.

generative cell: in flowering plants, one of the haploid cells of a pollen grain; undergoes mitosis to form two sperm cells.

genetic code: the collection of codons of mRNA, each of which directs the incorporation of a particular amino acid into a protein during protein synthesis.

genetic drift: a change in the allele frequencies of a small population purely by chance.

genetic engineering: the modification of genetic material to achieve specific goals.

genetic equilibrium: a state in which the allele frequencies and the distribution of genotypes of a population do not change from generation to generation.

genetic recombination: the generation of new combinations of alleles on homologous chromosomes due to the exchange of DNA during crossing over.

genital herpes: a sexually transmitted disease, caused by a virus, that can cause painful blisters on the genitals and surrounding skin.

genital warts: a sexually transmitted disease, caused by a virus, that forms growths or bumps on the external genitalia, in or around the vagina or anus, or on the cervix in females or penis, scrotum, groin, or thigh in males.

genome (jē´-nōm): the entire set of genes carried by a member of any given species.

genotype (jēn´-ō-tīp): the genetic composition of an organism; the actual alleles of each gene carried by the organism.

genus (jē-nus): the taxonomic category contained within a family and consisting of very closely related species.

geographical isolation: the separation of two populations by a physical barrier.

germ layer: a tissue layer formed during early embryonic development.

germination: the growth and development of a seed, spore, or pollen grain.

gibberellin (jib-er-el´-in): a plant hormone that stimulates seed germination, fruit development, and cell division and elongation.

gill: in aquatic animals, a branched tissue richly supplied with capillaries around which water is circulated for gas exchange.

gizzard: a muscular organ, found in earthworms and birds, in which food is mechanically broken down prior to chemical digestion.

gland: a cluster of cells that are specialized to secrete substances such as sweat or hormones.

glial cell: a cell of the nervous system that provides support and insulation for neurons.

global warming: a gradual rise in global atmospheric temperature as a result of an amplification of the natural greenhouse effect due to human activities.

glomerulus (glō-mer´-ū-lus): a dense network of thin-walled capillaries, located within the Bowman's capsule of each nephron of the kidney, where blood pressure forces water and dissolved nutrients through capillary walls for filtration by the nephron.

glucagon (gloo´-ka-gon): a hormone, secreted by the pancreas, that increases blood sugar by stimulating the breakdown of glycogen (to glucose) in the liver.

glucocorticoid (gloo-kō-kor´-tik-oid): a class of hormones, released by the adrenal cortex in response to the presence of ACTH, that make additional energy available to the body by stimulating the synthesis of glucose.

glucose: the most common monosaccharide, with the molecular formula $C_6H_{12}O_6$; most polysaccharides, including cellulose, starch, and glycogen, are made of glucose subunits covalently bonded together.

glyceraldehyde-3-phosphate (G3P): a three-carbon sugar that is an important component of the Calvin-Benson cycle in the light-independent reactions of photosynthesis.

glycerol (glis´-er-ol): a three-carbon alcohol to which fatty acids are covalently bonded to make fats and oils.

glycogen (glī´-kō-jen): a long, branched polymer of glucose that is stored by animals in the muscles and liver and metabolized as a source of energy.

glycolysis (glī-kol´-i-sis): reactions, carried out in the cytoplasm, that break down glucose into two molecules of pyruvic acid, producing two ATP molecules; does not require oxygen but can proceed when oxygen is present.

glycoprotein: a protein to which a carbohydrate is attached.

goiter: a swelling of the neck caused by iodine deficiency, which affects the functioning of the thyroid gland and its hormones.

Golgi complex (gōl´-jē): a stack of membranous sacs, found in most eukaryotic cells, that is the site of processing and separation of membrane components and secretory materials.

gonad: an organ where reproductive cells are formed; in males, the testes, and in females, the ovaries.

gonadotropin-releasing hormone (GnRH): a hormone produced by the neurosecretory cells of the hypothalamus, which stimulates cells in the anterior pituitary to release FSH and LH. GnRH is involved in the menstrual cycle and in spermatogenesis.

gonorrhea (gon-uh-rē´-uh): a sexually transmitted bacterial infection of the reproductive organs; if untreated, can result in sterility.

gradient: a difference in concentration, pressure, or electrical charge between two regions.

Gram stain: a stain that is selectively taken up by the cell walls of certain types of bacteria (gram-positive bacteria) and rejected by the cell walls of others (gram-negative bacteria); used to distinguish bacteria on the basis of their cell wall construction.

granum (gra´-num; pl., **grana):** a stack of thylakoids in chloroplasts.

grassland: a biome, located in the centers of continents, that supports grasses; also called *prairie*.

gravitropism: growth with respect to the direction of gravity.

gray crescent: in frog embryonic development, an area of intermediate pigmentation in the fertilized egg; contains gene-regulating substances required for the normal development of the tadpole.

gray matter: the outer portion of the brain and inner region of the spinal cord; composed largely of neuron cell bodies, which give this area a gray color.

greenhouse effect: the process in which certain gases such as carbon dioxide and methane trap sunlight energy in a planet's atmosphere as heat; the glass in a greenhouse does the same. The result, global warming, is being enhanced by the production of these gases by humans.
greenhouse gas: a gas, such as carbon dioxide or methane, that traps sunlight energy in a planet's atmosphere as heat; a gas that participates in the greenhouse effect.
ground tissue: plant tissue consisting of parenchyma, collenchyma, and sclerenchyma cells that makes up the bulk of a leaf or young stem, excluding vascular or dermal tissues. Most ground tissue cells function in photosynthesis, support, or carbohydrate storage.
growth hormone: a hormone, released by the anterior pituitary, that stimulates growth, especially of the skeleton.
growth rate: a measure of the change in population size per individual per unit of time.
guanine: a nitrogenous base found in both DNA and RNA; abbreviated as *G*.
guard cell: one of a pair of specialized epidermal cells surrounding the central opening of a stoma of a leaf, which regulates the size of the opening.
gymnosperm (jim´-nō-sperm): a nonflowering seed plant, such as a conifer, cycad, or gingko.
gyre (jīr): a roughly circular pattern of ocean currents, formed because continents interrupt the currents' flow; rotates clockwise in the Northern Hemisphere and counterclockwise in the Southern Hemisphere.
habituation (heh-bich-oo-ā´-shun): simple learning characterized by a decline in response to a harmless, repeated stimulus.
hair cell: the type of receptor cell in the inner ear; bears hairlike projections, the bending of which causes the receptor potential between two membranes.
hair follicle: a gland in the dermis of mammalian skin, formed from epithelial tissue, that produces a hair.
halophile (hā´-lō-fīl): literally, "salt-loving"; a type of archaen that thrives in concentrated salt solutions.
haploid (hap´-loid): referring to a cell that has only one member of each pair of homologous chromosomes.
Hardy-Weinberg principle: a mathematical model proposing that, under certain conditions, the allele frequencies and genotype frequencies in a sexually reproducing population will remain constant over generations.
Haversian system (ha-ver´-sē-un): see *osteon*.
head: the anteriormost segment of an animal with segmentation.
heart: a muscular organ responsible for pumping blood within the circulatory system throughout the body.
heart attack: a severe reduction or blockage of blood flow through a coronary artery, depriving some of the heart muscle of its blood supply.
heartwood: older xylem that contributes to the strength of a tree trunk.
heat of fusion: the energy that must be removed from a compound to transform it from a liquid into a solid at its freezing temperature.
heat of vaporization: the energy that must be supplied to a compound to transform it from a liquid into a gas at its boiling temperature.
heliozoan (hē-lē-ō-zō´-un): an aquatic (largely freshwater) animal-like protist; some have elaborate silica-based shells.
helix (hē´-liks): a coiled, springlike secondary structure of a protein.
helper T cell: a type of T cell that helps other immune cells recognize and act against antigens.
hemocoel (hē´-mō-sēl): a blood cavity within the bodies of certain invertebrates in which blood bathes tissues directly; part of an open circulatory system.
hemodialysis (hē-mō-dī-al´-luh-sis): a procedure that simulates kidney function in individuals with damaged or ineffective kidneys; blood is diverted from the body, artificially filtered, and returned to the body.
hemoglobin (hē´mō-glō-bin): the iron-containing protein that gives red blood cells their color; binds to oxygen in the lungs and releases it to the tissues.
hemophilia: a recessive, sex-linked disease in which the blood fails to clot normally.
herbivore (erb´-i-vor): literally, "plant-eater"; an organism that feeds directly and exclusively on producers; a primary consumer.
hermaphrodite (her-maf´-ruh-dīt´): an organism that possesses both male and female sexual organs.
hermaphroditic (her-maf´-ruh-dit´-ik): possessing both male and female sexual organs. Some hermaphroditic animals can fertilize themselves; others must exchange sex cells with a mate.
heterotroph (het´-er-ō-trōf´): literally, "other-feeder"; an organism that eats other organisms; a consumer.
heterozygous (het-er-ō-zī´-gus): carrying two different alleles of a given gene; also called *hybrid*.
hindbrain: the posterior portion of the brain, containing the medulla, pons, and cerebellum.
hinge joint: a joint at which one bone is moved by muscle and the other bone remains fixed, such as in the knee, elbow, or fingers; allows movement in only two dimensions.
hippocampus (hip-ō-kam´-pus): the part of the forebrain of vertebrates that is important in emotion and especially learning.
histamine: a substance released by certain cells in response to tissue damage and invasion of the body by foreign substances; promotes the dilation of arterioles and the leakiness of capillaries and triggers some of the events of the inflammatory response.
homeobox (hō´-mē-ō-boks): a sequence of DNA coding for special, 60-amino-acid proteins, which activate or inactivate genes that control development; these sequences specify embryonic cell differentiation.
homeostasis (hōm-ē-ō-stā´sis): the maintenance of a relatively constant environment required for the optimal functioning of cells, maintained by the coordinated activity of numerous regulatory mechanisms, including the respiratory, endocrine, circulatory, and excretory systems.
hominid: a human or a prehistoric relative of humans, beginning with the Australopithecines, whose fossils date back at least 4.4 million years.
homologous structures: structures that may differ in function but that have similar anatomy, presumably because the organisms that possess them have descended from common ancestors.
homologue (hō-´mō-log): a chromosome that is similar in appearance and genetic information to another chromosome with which it pairs during meiosis; also called *homologous chromosome*.
homozygous (hō-mō-zī´-gus): carrying two copies of the same allele of a given gene; also called *true-breeding*.
hormone: a chemical that is synthesized by one group of cells, secreted, and then carried in the bloodstream to other cells, whose activity is influenced by reception of the hormone.
host: the prey organism on or in which a parasite lives; is harmed by the relationship.
host cell: a cell that has been invaded by a virus and is usually harmed by the invasion.
human immunodeficiency virus (HIV): a pathogenic retrovirus that causes acquired immune deficiency syndrome (AIDS) by attacking and destroying the immune system's T cells.
humoral immunity: an immune response in which foreign substances are inactivated or destroyed by antibodies that circulate in the blood.
Huntington disease: an incurable genetic disorder, caused by a dominant allele, that produces progressive brain deterioration, resulting in the loss of motor coordination, flailing movements, personality disturbances, and eventual death.
hybrid: an organism that is the offspring of parents differing in at least one genetically determined characteristic; also used to refer to the offspring of parents of different species.
hybrid infertility: reduced fertility (typically, complete sterility) in the hybrid offspring of two species.
hybrid inviability: the failure of a hybrid offspring of two species to survive to maturity.
hybridoma: a cell produced by fusing an antibody-producing cell with a myeloma cell; used to produce monoclonal antibodies.
hydrogen bond: the weak attraction between a hydrogen atom that bears a partial positive charge (due to polar covalent bonding with another atom) and another atom, normally oxygen or nitrogen, that bears a partial negative charge; hydrogen bonds may form between atoms of a single molecule or of different molecules.
hydrologic cycle: the water cycle, driven by solar energy; a nutrient cycle in which the main reservoir of water is the ocean and most of the water remains in the form of water throughout the cycle (rather than being used in the synthesis of new molecules).

hydrolysis (hī-drol´-i-sis): the chemical reaction that breaks a covalent bond by means of the addition of hydrogen to the atom on one side of the original bond and a hydroxyl group to the atom on the other side; the reverse of dehydration synthesis.
hydrophilic (hī-drō-fil´-ik): pertaining to a substance that dissolves readily in water or to parts of a large molecule that form hydrogen bonds with water.
hydrophobic (hī-drō-fō´-bik): pertaining to a substance that does not dissolve in water.
hydrophobic interaction: the tendency for hydrophobic molecules to cluster together when immersed in water.
hydrostatic skeleton (hī-drō-stat´-ik): a body type that uses fluid contained in body compartments to provide support and mass against which muscles can contract.
hydrothermal vent community: a community of unusual organisms, living in the deep ocean near hydrothermal vents, that depends on the chemosynthetic activities of sulfur bacteria.
hypertension: arterial blood pressure that is chronically elevated above the normal level.
hypertonic (hī-per-ton´-ik): referring to a solution that has a higher concentration of dissolved particles (and therefore a lower concentration of free water) than has the cytoplasm of a cell.
hypha (hī´-fuh; pl., **hyphae):** a threadlike structure that consists of elongated cells, typically with many haploid nuclei; many hyphae make up the fungal body.
hypocotyl (hī´-pō-kot-ul): the part of the embryonic shoot located below the cotyledons but above the root.
hypothalamus (hī-pō-thal´-a-mus): a region of the brain that controls the secretory activity of the pituitary gland; synthesizes, stores, and releases certain peptide hormones; directs autonomic nervous system responses.
hypothesis (hī-poth´-eh-sis): the second operation in the scientific method; a supposition based on previous observations that is offered as an explanation for the observed phenomenon and is used as the basis for further observations, or experiments.
hypotonic (hī-pō-ton´-ik): referring to a solution that has a lower concentration of dissolved particles (and therefore a higher concentration of free water) than has the cytoplasm of a cell.
immigration (im-uh-grā´-shun): migration of individuals into an area.
immune response: a specific response by the immune system to the invasion of the body by a particular foreign substance or microorganism, characterized by the recognition of the foreign substance by immune cells and its subsequent destruction by antibodies or by cellular attack.
immune system: cells such as macrophages, B cells, and T cells and molecules such as antibodies that work together to combat microbial invasion of the body.
imperfect fungus: a fungus of the division Deuteromycota; no species in this division has been observed to form sexual reproductive structures.
implantation: the process whereby the early embryo embeds itself within the lining of the uterus.
imprinting: the process by which an animal forms an association with another animal or object in the environment during a sensitive period of development.
inclusive fitness: the reproductive success of all organisms that bear a given allele, normally expressed in relation to the average reproductive success of all individuals in the same population; compare with *fitness*.
incomplete dominance: a pattern of inheritance in which the heterozygous phenotype is intermediate between the two homozygous phenotypes.
incomplete flower: a flower that is missing one of the four floral parts (sepals, petals, stamens, or carpels).
independent assortment: see *law of independent assortment*.
indirect development: a developmental pathway in which an offspring goes through radical changes in body form as it matures.
induction: the process by which a group of cells causes other cells to differentiate into a specific tissue type.
inductive reasoning: the process of creating a generalization based on many specific observations that support the generalization, coupled with an absence of observations that contradict it.
inflammatory response: a nonspecific, local response to injury to the body, characterized by the phagocytosis of foreign substances and tissue debris by white blood cells and by the walling off of the injury site by the clotting of fluids that escape from nearby blood vessels.
inhalation: the act of drawing air into the lungs by enlarging the chest cavity.
inheritance: the genetic transmission of characteristics from parent to offspring.
inheritance of acquired characteristics: the hypothesis that organisms' bodies change during their lifetimes by use and disuse and that these changes are inherited by their offspring.
inhibiting hormone: a hormone, secreted by the neurosecretory cells of the hypothalamus, that inhibits the release of specific hormones from the anterior pituitary.
initiation complex: a structure that forms as the first step in assembly of the ribosome; includes a small ribosomal subunit, a methionine tRNA, a methionine amino acid, and several proteins.
innate (in-āt´) behavior: behavior that can be performed on the first attempt, without the need for any experience or learning.
inner cell mass: in human embryonic development, the cluster of cells, on one side of the blastocyst, that will develop into the embryo.
inner ear: the innermost part of the mammalian ear; composed of the bony, fluid-filled tubes of the cochlea and the vestibular apparatus.
inorganic: describing any molecule that does not contain both carbon and hydrogen.
insertion: the site of attachment of a muscle to the relatively movable bone on one side of a joint.
insertion mutation: a mutation in which one or more pairs of nucleotides are inserted into a gene.
insight learning: a complex form of learning that requires the manipulation of mental concepts to arrive at adaptive behavior.
instinctive: innate; inborn; determined by the genetic makeup of the individual.
insulin: a hormone, secreted by the pancreas, that lowers blood sugar by stimulating the conversion of glucose to glycogen in the liver.
integration: in nerve cells, the process of adding up electrical signals from sensory inputs or other nerve cells to determine the appropriate outputs.
integument (in-teg´-ū-ment): in plants, the outer layers of cells of the ovule that surrounds the embryo sac; develops into the seed coat.
intensity: the strength of stimulation or response.
interferon: a protein released by certain virus-infected cells that increases the resistance of other, uninfected, cells to viral attack.
intermediate filament: part of the cytoskeleton of eukaryotic cells that probably functions mainly for support and is composed of several types of proteins.
intermembrane compartment: the fluid-filled space between the inner and outer membranes of a mitochondrion.
internal fertilization: the union of sperm and egg inside the body of the female.
internode: the part of a stem between two nodes.
interphase: the stage of the cell cycle between cell divisions; the stage in which chromosomes are replicated and other cell functions occur, such as growth, movement, and acquisition of nutrients.
interspecific competition: competition among individuals of different species.
interstitial cell (in-ter-sti´-shul): in the vertebrate testis, a testosterone-producing cell located between the seminiferous tubules.
interstitial fluid (in-ter-sti´-shul): fluid, similar in composition to plasma (except lacking large proteins), that leaks from capillaries and acts as a medium of exchange between the body cells and the capillaries.
intertidal zone: an area of the ocean shore that is alternately covered and exposed by the tides.
intervertebral disc (in-ter-ver-tē´-brul): a pad of cartilage between two vertebrae that acts as a shock absorber.
intracellular digestion: the chemical breakdown of food within single cells.
intraspecific competition: competition among individuals of the same species.
intrauterine device (IUD): a small copper or plastic loop, squiggle, or shield that is inserted in the uterus; a contraceptive method that works by irritating the uterine lining so that it cannot receive the embryo.
intron: a segment of DNA in a eukaryotic gene that does not code for amino acids in a protein.

invertebrate (in-vert´-uh-bret): an animal that never possesses a vertebral column.
ion (ī´-on): a charged atom or molecule; an atom or molecule that has either an excess of electrons (and hence is negatively charged) or has lost electrons (and is positively charged).
ionic bond: a chemical bond formed by the electrical attraction between positively and negatively charged ions.
iris: the pigmented muscular tissue of the vertebrate eye that surrounds and controls the size of the pupil, through which light enters.
islet cell: a cluster of cells in the endocrine portion of the pancreas that produce insulin and glucagon.
isolating mechanism: a morphological, physiological, behavioral, or ecological difference that prevents members of two species from interbreeding.
isotonic (ī-sō-ton´-ik): referring to a solution that has the same concentration of dissolved particles (and therefore the same concentration of free water) as has the cytoplasm of a cell.
isotope: one of several forms of a single element, the nuclei of which contain the same number of protons but different numbers of neutrons.
J-curve: the J-shaped growth curve of an exponentially growing population in which increasing numbers of individuals join the population during each succeeding time period.
joint: a flexible region between two rigid units of an exoskeleton or endoskeleton, allowing for movement between the units.
karyotype: a preparation showing the number, sizes, and shapes of all chromosomes within a cell and, therefore, within the individual or species from which the cell was obtained.
keratin (ker´-uh-tin): a fibrous protein in hair, nails, and the epidermis of skin.
keystone species: a species whose influence on community structure is greater than its abundance would suggest.
kidney: one of a pair of organs of the excretory system that is located on either side of the spinal column and filters blood, removing wastes and regulating the composition and water content of the blood.
kinetic energy: the energy of movement; includes light, heat, mechanical movement, and electricity.
kinetochore (ki-net´-ō-kor): a protein structure that forms at the centromere regions of chromosomes; attaches the chromosomes to the spindle.
kingdom: the second broadest taxonomic category, contained within a domain and consisting of related phyla or divisions. This textbook recognizes four kingdoms within the domain Eukarya: Protista, Fungi, Plantae, and Animalia.
kin selection: a type of natural selection that favors a certain allele because it increases the survival or reproductive success of relatives that bear the same allele.
Klinefelter syndrome: a set of characteristics typically found in individuals who have two X chromosomes and one Y chromosome; these individuals are phenotypically males but are sterile and have several femalelike traits, including broad hips and partial breast development.
Krebs cycle: a cyclic series of reactions, occurring in the matrix of mitochondria, in which the acetyl groups from the pyruvic acids produced by glycolysis are broken down to CO_2, accompanied by the formation of ATP and electron carriers; also called *citric acid cycle.*
kuru: a degenerative brain disease, first discovered in the cannibalistic Fore tribe of New Guinea, that is caused by a prion.
labium (pl., **labia**)**:** one of a pair of folds of skin of the external structures of the mammalian female reproductive system.
labor: a series of contractions of the uterus that result in birth.
lactation: the secretion of milk from the mammary glands.
lacteal (lak-tēl´): a single lymph capillary that penetrates each villus of the small intestine.
lactose (lak´-tōs): a disaccharide composed of glucose and galactose; found in mammalian milk.
large intestine: the final section of the digestive tract; consists of the colon and the rectum, where feces are formed and stored.
larva (lar´-vuh): an immature form of an organism with indirect development prior to metamorphosis into its adult form; includes the caterpillars of moths and butterflies and the maggots of flies.
larynx (lar´-inks): that portion of the air passage between the pharynx and the trachea; contains the vocal cords.
lateral bud: a cluster of meristematic cells at the node of a stem; under appropriate conditions, it grows into a branch.
lateral meristem: a meristematic tissue that forms cylinders parallel to the long axis of roots and stems; normally located between the primary xylem and primary phloem (vascular cambium) and just outside the phloem (cork cambium); also called *cambium.*
law of independent assortment: the independent inheritance of two or more distinct traits; states that the alleles for one trait may be distributed to the gametes independently of the alleles for other traits.
law of segregation: Gregor Mendel's conclusion that each gamete receives only one of each parent's pair of genes for each trait.
laws of thermodynamics: the physical laws that define the basic properties and behavior of energy.
leaf: an outgrowth of a stem, normally flattened and photosynthetic.
leaf primordium (pri-mor´-dē-um; pl., **primordia):** a cluster of meristem cells, located at the node of a stem, that develops into a leaf.
learning: an adaptive change in behavior as a result of experience.
legume (leg´-ūm): a member of a family of plants characterized by root swellings in which nitrogen-fixing bacteria are housed; includes soybeans, lupines, alfalfa, and clover.
lens: a clear object that bends light rays; in eyes, a flexible or movable structure used to focus light on a layer of photoreceptor cells.
leptin: a peptide hormone. One of the functions of leptin, which is released by fat cells, is to help the body monitor its fat stores and regulate weight.
leukocyte (loo´-kō-sīt): any of the white blood cells circulating in the blood.
lichen (lī´-ken): a symbiotic association between an alga or cyanobacterium and a fungus, resulting in a composite organism.
life cycle: the events in the life of an organism from one generation to the next.
ligament: a tough connective tissue band connecting two bones.
light-dependent reactions: the first stage of photosynthesis, in which the energy of light is captured as ATP and NADPH; occurs in thylakoids of chloroplasts.
light-harvesting complex: in photosystems, the assembly of pigment molecules (chlorophyll and accessory pigments) that absorb light energy and transfer that energy to electrons.
light-independent reactions: the second stage of photosynthesis, in which the energy obtained by the light-dependent reactions is used to fix carbon dioxide into carbohydrates; occurs in the stroma of chloroplasts.
lignin: a hard material that is embedded in the cell walls of vascular plants and provides support in terrestrial species; an early and important adaptation to terrestrial life.
limbic system: a diverse group of brain structures, mostly in the lower forebrain, that includes the thalamus, hypothalamus, amygdala, hippocampus, and parts of the cerebrum and is involved in basic emotions, drives, behaviors, and learning.
limnetic zone: a lake zone in which enough light penetrates to support photosynthesis.
linkage: the inheritance of certain genes as a group because they are parts of the same chromosome. Linked genes do not show independent assortment.
lipase (lī´-pās): an enzyme that catalyzes the breakdown of lipids such as fats.
lipid (li´-pid): one of a number of organic molecules containing large nonpolar regions composed solely of carbon and hydrogen, which make lipids hydrophobic and insoluble in water; includes oils, fats, waxes, phospholipids, and steroids.
littoral zone: a lake zone, near the shore, in which water is shallow and plants find abundant light, anchorage, and adequate nutrients.
liver: an organ with varied functions, including bile production, glycogen storage, and the detoxification of poisons.
lobefin: a member of a group of bony fishes, characterized by paired rounded fins, that gave rise to amphibians; extinct except for the coelacanths.
locus: the physical location of a gene on a chromosome.
long-day plant: a plant that will flower only if the length of daylight is greater than some species-specific duration.

long-term memory: the second phase of learning; a more-or-less permanent memory formed by a structural change in the brain, brought on by repetition.
loop of Henle (hen´-lē): a specialized portion of the tubule of the nephron in birds and mammals that creates an osmotic concentration gradient in the fluid immediately surrounding it. This gradient in turn makes possible the production of urine more osmotically concentrated than blood plasma.
lung: a paired respiratory organ consisting of inflatable chambers within the chest cavity in which gas exchange occurs.
luteinizing hormone (LH): a hormone, produced by the anterior pituitary, that stimulates testosterone production in males and the development of the follicle, ovulation, and the production of the corpus luteum in females.
lymph (limf): a pale fluid, within the lymphatic system, that is composed primarily of interstitial fluid and lymphocytes.
lymphatic system: a system consisting of lymph vessels, lymph capillaries, lymph nodes, and the thymus and spleen; helps protect the body against infection, absorbs fats, and returns excess fluid and small proteins to the blood circulatory system.
lymph node: a small structure that filters lymph; contains lymphocytes and macrophages, which inactivate foreign particles such as bacteria.
lymphocyte (lim´-fō-sīt): a type of white blood cell important in the immune response.
lysosome (lī´-sō-sōm): a membrane-bound organelle containing intracellular digestive enzymes.
macronutrient: a nutrient needed in relatively large quantities (often defined as making up more than 0.1% of an organism's body).
macrophage (mak´-rō-fāj): a type of white blood cell that engulfs microbes and destroys them by phagocytosis; also presents microbial antigens to T cells, helping stimulate the immune response.
magnetotactic: able to detect and respond to Earth's magnetic field.
major histocompatibility complex (MHC): proteins, normally located on the surfaces of body cells, that identify the cell as "self"; also important in stimulating and regulating the immune response.
maltose (mal´-tōs): a disaccharide composed of two glucose molecules.
mammary gland (mam´-uh-rē): a milk-producing gland used by female mammals to nourish their young.
mantle (man´-tul): an extension of the body wall in certain invertebrates, such as mollusks; may secrete a shell, protect the gills, and, as in cephalopods, aid in locomotion.
marsupial (mar-soo´-pē-ul): a mammal whose young are born at an extremely immature stage and undergo further development in a pouch while they remain attached to a mammary gland; includes kangaroos, opossums, and koalas.
mass extinction: the extinction of an extraordinarily large number of species in a short period of geologic time. Mass extinctions have recurred periodically throughout the history of life.
mast cell: a cell of the immune system that synthesizes histamine and other molecules used in the body's response to trauma and that are a factor in allergic reactions.
matrix: the fluid contained within the inner membrane of a mitochondrion.
mechanical incompatibility: the inability of male and female organisms to exchange gametes, normally because their reproductive structures are incompatible.
mechanoreceptor: a receptor that responds to mechanical deformation, such as that caused by pressure, touch, or vibration.
medulla (med-ū´-luh): the part of the hindbrain of vertebrates that controls automatic activities such as breathing, swallowing, heart rate, and blood pressure.
medusa (meh-doo´-suh): a bell-shaped, typically free-swimming stage in the life cycle of many cnidarians; includes jellyfish.
megakaryocyte (meg-a-kar´-ē-ō-sīt): a large cell type that remains in the bone marrow, pinching off pieces of itself that then enter the circulation as platelets.
megaspore: a haploid cell formed by meiosis from a diploid megaspore mother cell; through mitosis and differentiation, develops into the female gametophyte.
megaspore mother cell: a diploid cell, within the ovule of a flowering plant, that undergoes meiosis to produce four haploid megaspores.
meiosis (mī-ō´-sis): a type of cell division, used by eukaryotic organisms, in which a diploid cell divides twice to produce four haploid cells.
meiotic cell division: meiosis followed by cytokinesis.
melanocyte-stimulating hormone (me-lan´-ō-sīt): a hormone, released by the anterior pituitary, that regulates the activity of skin pigments in some vertebrates.
melatonin (mel-uh-tōn´-in): a hormone, secreted by the pineal gland, that is involved in the regulation of circadian cycles.
membrane: in multicellular organism, a continuous sheet of epithelial cells that covers the body and lines body cavities; in a cell, a thin sheet of lipids and proteins that surrounds the cell or its organelles, separating them from their surroundings.
memory B cell: a type of white blood cell that is produced as a result of the binding of an antibody on a B cell to an antigen on an invading microorganism. Memory B cells persist in the bloodstream and provide future immunity to invaders bearing that antigen.
memory T cell: a type of white blood cell that is produced as a result of the binding of a receptor on a T cell to an antigen on an invading microorganism. Memory T cells persist in the bloodstream and provide future immunity to invaders bearing that antigen.
meninges (men-in´-jēz): three layers of connective tissue that surround the brain and spinal cord.
menstrual cycle: in human females, a complex 28-day cycle during which hormonal interactions among the hypothalamus, pituitary gland, and ovary coordinate ovulation and the preparation of the uterus to receive and nourish the fertilized egg. If pregnancy does not occur, the uterine lining is shed during menstruation.
menstruation: in human females, the monthly discharge of uterine tissue and blood from the uterus.
meristem cell (mer´-i-stem): an undifferentiated cell that remains capable of cell division throughout the life of a plant.
mesoderm (mēz´-ō-derm): the middle embryonic tissue layer, lying between the endoderm and ectoderm, and normally the last to develop; gives rise to structures such as muscle and skeleton.
mesoglea (mez-ō-glē´-uh): a middle, jellylike layer within the body wall of cnidarians.
mesophyll (mez´-ō-fil): loosely packed parenchyma cells beneath the epidermis of a leaf.
messenger RNA (mRNA): a strand of RNA, complementary to the DNA of a gene, that conveys the genetic information in DNA to the ribosomes to be used during protein synthesis; sequences of three bases (codons) in mRNA specify particular amino acids to be incorporated into a protein.
metabolic pathway: a sequence of chemical reactions within a cell, in which the products of one reaction are the reactants for the next reaction.
metabolism: the sum of all chemical reactions that occur within a single cell or within all the cells of a multicellular organism.
metamorphosis (met-a-mor´-fō-sis): in animals with indirect development, a radical change in body form from larva to sexually mature adult, as seen in amphibians (tadpole to frog) and insects (caterpillar to butterfly).
metaphase (met´-a-fāz): the stage of mitosis in which the chromosomes, attached to spindle fibers at kinetochores, are lined up along the equator of the cell.
methanogen (me-than´-ō-jen): a type of anaerobic archaean capable of converting carbon dioxide to methane.
microbe: a microorganism.
microevolution: change over successive generations in the composition of a population's gene pool.
microfilament: part of the cytoskeleton of eukaryotic cells that is composed of the proteins actin and (in some cases) myosin; functions in the movement of cell organelles and in locomotion by extension of the plasma membrane.
micronutrient: a nutrient needed only in small quantities (often defined as making up less than 0.01% of an organism's body).
microsphere: a small, hollow sphere formed from proteins or proteins complexed with other compounds.
microspore: a haploid cell formed by meiosis from a microspore mother cell; through mitosis and differentiation, develops into the male gametophyte.

microspore mother cell: a diploid cell contained within an anther of a flowering plant, which undergoes meiosis to produce four haploid microspores.
microtubule: a hollow, cylindrical strand, found in eukaryotic cells, that is composed of the protein tubulin; part of the cytoskeleton used in the movement of organelles, cell growth, and the construction of cilia and flagella.
microvillus (mī-krō-vi´-lus; pl., **microvilli):** a microscopic projection of the plasma membrane of each villus; increases the surface area of the villus.
midbrain: during development, the central portion of the brain; contains an important relay center, the reticular formation.
middle ear: the part of the mammalian ear composed of the tympanic membrane, the Eustachian tube, and three bones (hammer, anvil, and stirrup) that transmit vibrations from the auditory canal to the oval window.
middle lamella: a thin layer of sticky polysaccharides, such as pectin, and other carbohydrates that separates and holds together the primary cell walls of adjacent plant cells.
mimicry (mim´-ik-rē): the situation in which a species has evolved to resemble something else—typically another type of organism.
mineral: an inorganic substance, especially one in rocks or soil.
mitochondrion (mī-tō-kon´-drē-un): an organelle, bounded by two membranes, that is the site of the reactions of aerobic metabolism.
mitosis (mī-tō´-sis): a type of nuclear division, used by eukaryotic cells, in which one copy of each chromosome (already duplicated during interphase before mitosis) moves into each of two daughter nuclei; the daughter nuclei are therefore genetically identical to each other.
mitotic cell division: mitosis followed by cytokinesis.
molecule (mol´-e-kūl): a particle composed of one or more atoms held together by chemical bonds; the smallest particle of a compound that displays all the properties of that compound.
molt: to shed an external body covering, such as an exoskeleton, skin, feathers, or fur.
monoclonal antibody: an antibody produced in the laboratory by the cloning of hybridoma cells; each clone of cells produces a single antibody.
monocot: short for monocotyledon; a type of flowering plant characterized by embryos with one seed leaf, or cotyledon.
monoecious (mon-ē´-shus): pertaining to organisms in which male and female gametes are produced in the same individual.
monomer (mo´-nō-mer): a small organic molecule, several of which may be bonded together to form a chain called a *polymer*.
monophyletic: referring to a group of species that contains all the known descendents of an ancestral species.
monosaccharide (mo-nō-sak´-uh-rīd): the basic molecular unit of all carbohydrates, normally composed of a chain of carbon atoms bonded to hydrogen and hydroxyl groups.
monotreme: a mammal that lays eggs; for example, the platypus.
morula (mor´-ū-luh): in animals, an embryonic stage during cleavage, when the embryo consists of a solid ball of cells.
motor neuron: a neuron that receives instructions from the association neurons and activates effector organs, such as muscles or glands.
motor unit: a single motor neuron and all the muscle fibers on which it forms synapses.
mouth: the opening of a tubular digestive system into which food is first introduced.
mucous membrane: the lining of the inside of the respiratory and digestive tracts.
multicellular: many-celled; most members of the kingdoms Fungi, Plantae, and Animalia are multicellular, with intimate cooperation among cells.
multiple alleles: as many as dozens of alleles produced for every gene as a result of different mutations.
muscle fiber: an individual muscle cell.
muscle tissue: tissue that consists mainly of muscle fibers, and whose function is accomplished when the tissue contracts.
mutation: a change in the base sequence of DNA in a gene; normally refers to a genetic change significant enough to alter the appearance or function of the organism.
mutualism (mū´-choo-ul-iz-um): a symbiotic relationship in which both participating species benefit.
mycelium (mī-sēl´-ē-um): the body of a fungus, consisting of a mass of hyphae.
mycorrhiza (mī-kō-rī´zuh; pl., **mycorrhizae):** a symbiotic relationship between a fungus and the roots of a land plant that facilitates mineral extraction and absorption.
myelin (mī´-uh-lin): a wrapping of insulating membranes of specialized nonneural cells around the axon of a vertebrate nerve cell; increases the speed of conduction of action potentials.
myofibril (mī-ō-fī´-bril): a cylindrical subunit of a muscle cell, consisting of a series of sarcomeres; surrounded by sarcoplasmic reticulum.
myometrium (mī-ō-mē´-trē-um): the muscular outer layer of the uterus.
myosin (mī´-ō-sin): one of the major proteins of muscle, the interaction of which with the protein actin produces muscle contraction; found in the thick filaments of the muscle fiber; see also *actin*.
natural causality: the scientific principle that natural events occur as a result of preceding natural causes.
natural killer cell: a type of white blood cell that destroys some virus-infected cells and cancerous cells on contact; part of the immune system's nonspecific internal defense against disease.
natural selection: the unequal survival and reproduction of organisms due to environmental forces, resulting in the preservation of favorable adaptations. Usually, natural selection refers specifically to differential survival and reproduction on the basis of genetic differences among individuals.
near-shore zone: the region of coastal water that is relatively shallow but constantly submerged; includes bays and coastal wetlands and can support large plants or seaweeds.
nearsighted: the inability to focus on distant objects caused by an eyeball that is slightly too long.
negative feedback: a situation in which a change initiates a series of events that tend to counteract the change and restore the original state. Negative feedback in physiological systems maintains homeostasis.
nephridium (nef-rid´-ē-um): an excretory organ found in earthworms, mollusks, and certain other invertebrates; somewhat resembles a single vertebrate nephron.
nephron (nef´-ron): the functional unit of the kidney; where blood is filtered and urine formed.
nephrostome (nef´-rō-stōm): the funnel-shaped opening of the nephridium of some invertebrates such as earthworms; coelomic fluid is drawn into the nephrostome for filtration.
nerve: a bundle of axons of nerve cells, bound together in a sheath.
nerve cord: a paired neural structure in most animals that conducts nervous signals to and from the ganglia; in chordates, a nervous structure lying along the dorsal side of the body; also called *spinal cord*.
nerve net: a simple form of nervous system, consisting of a network of neurons that extend throughout the tissues of an organism such as a cnidarian.
nerve tissue: the tissue that make up the brain, spinal cord, and nerves; consists of neurons and glial cells.
net primary productivity: the energy stored in the autotrophs of an ecosystem over a given time period.
neural tube: a structure, derived from ectoderm during early embryonic development, that later becomes the brain and spinal cord.
neuromuscular junction: the synapse formed between a motor neuron and a muscle fiber.
neuron (noor´-on): a single nerve cell.
neuropeptide: a small protein molecule with neurotransmitter-like actions.
neurosecretory cell: a specialized nerve cell that synthesizes and releases hormones.
neurotransmitter: a chemical that is released by a nerve cell close to a second nerve cell, a muscle, or a gland cell and that influences the activity of the second cell.
neutral mutation: a mutation that has little or no effect on the function of the encoded protein.
neutralization: the process of covering up or inactivating a toxic substance with antibody.
neutron: a subatomic particle that is found in the nuclei of atoms, bears no charge, and has a mass approximately equal to that of a proton.
nicotinamide adenine dinocleotide phosphate (NADPH): an energy-carrier molecule produced by the light-dependent reactions of photosynthesis; transfers energy to the carbon-fixing reactions (light-independent) reactions.

nitrogen fixation: the process that combines atmospheric nitrogen with hydrogen to form ammonium (NH_4^+).
nitrogen-fixing bacterium: a bacterium that possesses the ability to remove nitrogen (N_2) from the atmosphere and combine it with hydrogen to produce ammonium (NH_4^+).
node: in plants, a region of a stem at which leaves and lateral buds are located; in vertebrates, an interruption of the myelin on a myelinated axon, exposing naked membrane at which action potentials are generated.
nodule: a swelling on the root of a legume or other plant that consists of cortex cells inhabited by nitrogen-fixing bacteria.
nondisjunction: an error in meiosis in which chromosomes fail to segregate properly into the daughter cells.
nonpolar covalent bond: a covalent bond with equal sharing of electrons.
nonpolar molecule: a molecule bound by covalent bonds in which electrical charge is symmetrically distributed, so that no portion of the molecule is electrically charged relative to other portions.
norepinephrine (nor-ep-i-nef-rin´): a neurotransmitter, released by neurons of the parasympathetic nervous system, that prepares the body to respond to stressful situations; also called *noradrenaline.*
northern coniferous forest: a biome with long, cold winters and only a few months of warm weather; populated almost entirely by evergreen coniferous trees; also called *taiga.*
notochord (nōt´-ō-kord): a stiff but somewhat flexible, supportive rod found in all members of the phylum Chordata at some stage of development.
nuclear envelope: the double-membrane system surrounding the nucleus of eukaryotic cells; the outer membrane is typically continuous with the endoplasmic reticulum.
nucleic acid (noo-klā´-ik): an organic molecule composed of nucleotide subunits; the two common types of nucleic acids are ribonucleic acid (RNA) and deoxyribonucleic acid (DNA).
nucleoid (noo-klē-oid): the location of the genetic material in prokaryotic cells; not membrane-enclosed.
nucleolus (noo-klē´-ō-lus): the region of the eukaryotic nucleus that is engaged in ribosome synthesis; consists of the genes encoding ribosomal RNA, newly synthesized ribosomal RNA, and ribosomal proteins.
nucleotide: a subunit of which nucleic acids are composed; a phosphate group bonded to a sugar (deoxyribose in DNA), which is in turn bonded to a nitrogen-containing base (adenine, guanine, cytosine, or thymine in DNA). Nucleotides are linked together, forming a strand of nucleic acid, as follows: Bonds between the phosphate of one nucleotide link to the sugar of the next nucleotide.
nucleotide substitution: a mutation that replaces one nucleotide in a DNA molecule with another; for example, a change from an adenine to a guanine.
nucleus (atomic): the central region of an atom, consisting of protons and neutrons.
nucleus (cellular): the membrane-bound organelle of eukaryotic cells that contains the cell's genetic material.
nutrient: a substance acquired from the environment and needed for the survival, growth, and development of an organism.
nutrient cycle: a description of the pathways of a specific nutrient (such as carbon, nitrogen, phosphorus, or water) through the living and nonliving portions of an ecosystem.
nutrition: the process of acquiring nutrients from the environment and, if necessary, processing them into a form that can be used by the body.
observation: the first operation in the scientific method; the noting of a specific phenomenon, leading to the formulation of a hypothesis.
oil: a lipid composed of three fatty acids, some of which are unsaturated, covalently bonded to a molecule of glycerol; liquid at room temperature.
olfaction (ōl-fak´-shun): a chemical sense, the sense of smell; in terrestrial vertebrates, the result of the detection of airborne molecules.
oligotrophic lake: a lake that is very low in nutrients and hence clear with extensive light penetration.
ommatidium (ōm-ma-tid´-ē-um): an individual light-sensitive subunit of a compound eye; consists of a lens and several receptor cells.
omnivore: an organism that consumes both plants and other animals.
one gene, one protein rule: the premise that each gene encodes the information for the synthesis of a single protein.
oogenesis: the process by which egg cells are formed.
oogonium (ō-ō-gō´-nē-um; pl., **oogonia):** in female animals, a diploid cell that gives rise to a primary oocyte.
open circulatory system: a type of circulatory system found in some invertebrates, such as arthropods and mollusks, that includes an open space (the hemocoel) in which blood directly bathes body tissues.
operant conditioning: a laboratory training procedure in which an animal learns to perform a response (such as pressing a lever) through reward or punishment.
operculum: an external flap, supported by bone, that covers and protects the gills of most fish.
opioid (ōp´-ē-oid): one of a group of peptide neuromodulators in the vertebrate brain that mimic some of the actions of opiates (such as opium) and also seem to influence many other processes, including emotion and appetite.
optic nerve: the nerve leading from the eye to the brain, carrying visual information.
order: the taxonomic category contained within a class and consisting of related families.
organ: a structure (such as the liver, kidney, or skin) composed of two or more distinct tissue types that function together.
organelle (or-guh-nel´): a structure, found in the cytoplasm of eukaryotic cells, that performs a specific function; sometimes refers specifically to membrane-bound structures, such as the nucleus or endoplasmic reticulum.
organic molecule: a molecule that contains both carbon and hydrogen.
organism (or´-guh-niz-um): an individual living thing.
organogenesis (or-gan-ō-jen´-uh-sis): the process by which the layers of the gastrula (endoderm, ectoderm, mesoderm) rearrange into organs.
organ system: two or more organs that work together to perform a specific function; for example, the digestive system.
origin: the site of attachment of a muscle to the relatively stationary bone on one side of a joint.
osmosis (oz-mō´-sis): the diffusion of water across a differentially permeable membrane, normally down a concentration gradient of free water molecules. Water moves into the solution that has a lower concentration of free water from a solution with the higher concentration of free water.
osmotic pressure: the pressure required to counterbalance the tendency of water to move from a solution with a higher concentration of free water molecules into a solution with a lower concentration of free water molecules.
osteoblast (os´-tē-ō-blast): a cell type that produces bone.
osteoclast (os´-tē-ō-klast): a cell type that dissolves bone.
osteocyte (os´-tē-ō-sīt): a mature bone cell.
osteon: a unit of hard bone consisting of concentric layers of bone matrix, with embedded osteocytes, surrounding a small central canal that contains a capillary.
osteoporosis (os´-tē-ō-por-ō´-sis): a condition in which bones become porous, weak, and easily fractured; most common in elderly women.
outer ear: the outermost part of the mammalian ear, including the external ear and auditory canal leading to the tympanic membrane.
oval window: the membrane-covered entrance to the inner ear.
ovary: in animals, the gonad of females; in flowering plants, a structure at the base of the carpel that contains one or more ovules and develops into the fruit.
oviduct: in mammals, the tube leading from the ovary to the uterus.
ovulation: the release of a secondary oocyte, ready to be fertilized, from the ovary.
ovule: a structure within the ovary of a flower, inside which the female gametophyte develops; after fertilization, develops into the seed.
oxaloacetate: a four-carbon molecule that participates in the Krebs cycle of glucose metabolism and also in the carbon-fixing pathway of C_4 photosynthesis.
oxytocin (oks-ē-tō´-sin): a hormone, released by the posterior pituitary, that stimulates the contraction of uterine and mammary gland muscles.
ozone layer: the ozone-enriched layer of the upper atmosphere that filters out some of the sun's ultraviolet radiation.
pacemaker: a cluster of specialized muscle cells in the upper right atrium of the heart that produce spontaneous electrical signals at a regular rate; the sinoatrial node.

pain receptor: a receptor that has extensive areas of membranes studded with special receptor proteins that respond to light or to a chemical.

palisade cell: a columnar mesophyll cell, containing chloroplasts, just beneath the upper epidermis of a leaf.

pancreas (pān´-krē-us): a combined exocrine and endocrine gland located in the abdominal cavity next to the stomach. The endocrine portion secretes the hormones insulin and glucagon, which regulate glucose concentrations in the blood. The exocrine portion secretes enzymes for fat, carbohydrate, and protein digestion into the small intestine and neutralizes the acidic chyme.

pancreatic juice: a mixture of water, sodium bicarbonate, and enzymes released by the pancreas into the small intestine.

parasite (par´-uh-sīt): an organism that lives in or on a larger prey organism, called a *host*, weakening it.

parasitism: a symbiotic relationship in which one organism (commonly smaller and more numerous than its host) benefits by feeding on the other, which is normally harmed but not immediately killed.

parasympathetic division: the division of the autonomic nervous system that produces largely involuntary responses related to the maintenance of normal body functions, such as digestion.

parathormone: a hormone, secreted by the parathyroid gland, that stimulates the release of calcium from bones.

parathyroid gland: one of a set of four small endocrine glands, embedded in the surface of the thyroid gland, that produces parathormone, which (with calcitonin from the thyroid gland) regulates calcium ion concentration in the blood.

parenchyma (par-en´-ki-muh): a plant cell type that is alive at maturity, normally with thin primary cell walls, that carries out most of the metabolism of a plant. Most dividing meristem cells in a plant are parenchyma.

parthenogenesis (par-the-nō-jen´uh-sis): a specialization of sexual reproduction, in which a haploid egg undergoes development without fertilization.

passive transport: the movement of materials across a membrane down a gradient of concentration, pressure, or electrical charge without using cellular energy.

pathogenic (path´-ō-jen-ik): capable of producing disease; refers to an organism with such a capability (a pathogen).

pedigree: a diagram showing genetic relationships among a set of individuals, normally with respect to a specific genetic trait.

pelagic (puh-la´-jik): free-swimming or floating.

penis: an external structure of the male reproductive and urinary systems; serves to deposit sperm into the female reproductive system and delivers urine to the exterior.

peptide (pep´-tīd): a chain composed of two or more amino acids linked together by peptide bonds.

peptide bond: the covalent bond between the amino group's nitrogen of one amino acid and the carboxyl group's carbon of a second amino acid, joining the two amino acids together in a peptide or protein.

peptide hormone: a hormone consisting of a chain of amino acids; includes small proteins that function as hormones.

peptidoglycan (pep-tid-ō-glī´-kan): a component of prokaryotic cell walls that consists of chains of sugars cross-linked by short chains of amino acids called *peptides*.

pericycle (per´-i-sī-kul): the outermost layer of cells of the vascular cylinder of a root.

periderm: the outer cell layers of roots and a stem that have undergone secondary growth, consisting primarily of cork cambium and cork cells.

peripheral nerve: a nerve that links the brain and spinal cord to the rest of the body.

peripheral nervous system: in vertebrates, the part of the nervous system that connects the central nervous system to the rest of the body.

peristalsis: rhythmic coordinated contractions of the smooth muscles of the digestive tract that move substances through the digestive tract.

permafrost: a permanently frozen layer of soil in the arctic tundra that cannot support the growth of trees.

petal: part of a flower, typically brightly colored and fragrant, that attracts potential animal pollinators.

petiole (pet´-ē-ōl): the stalk that connects the blade of a leaf to the stem.

phagocytic cell (fa-gō-sit´-ik): a type of immune system cell that destroys invading microbes by using phagocytosis to engulf and digest the microbes.

phagocytosis (fa-gō-sī-tō´-sis): a type of endocytosis in which extensions of a plasma membrane engulf extracellular particles and transport them into the interior of the cell.

pharyngeal gill slit (far-in´-jē-ul): an opening, located just posterior to the mouth, that connects the digestive tube to the outside environment; present (at some stage of life) in all chordates.

pharynx (far´-inks): in vertebrates, a chamber that is located at the back of the mouth and is shared by the digestive and respiratory systems; in some invertebrates, the portion of the digestive tube just posterior to the mouth.

phenotype (fēn´-ō-tīp): the physical characteristics of an organism; can be defined as outward appearance (such as flower color), as behavior, or in molecular terms (such as glycoproteins on red blood cells).

pheromone (fer´-uh-mōn): a chemical produced by an organism that alters the behavior or physiological state of another member of the same species.

phloem (flō´-um): a conducting tissue of vascular plants that transports a concentrated sugar solution up and down the plant.

phosphoenolpyruvate (PEP): a three-carbon molecule that reacts with carbon dioxide in the carbon-fixing reaction of C_4 photosynthesis.

phosphoglyceric acid (PGA): a three-carbon sugar that is an important component of the Calvin-Benson cycle in the light-independent reactions of photosynthesis.

phospholipid (fos-fō-li´-pid): a lipid consisting of glycerol bonded to two fatty acids and one phosphate group, which bears another group of atoms, typically charged and containing nitrogen. A double layer of phospholipids is a component of all cellular membranes.

phospholipid bilayer: a double layer of phospholipids that forms the basis of all cellular membranes. The phospholipid heads, which are hydrophilic, face the water of extracellular fluid or the cytoplasm; the tails, which are hydrophobic, are buried in the middle of the bilayer.

photic zone: the region of the ocean where light is strong enough to support photosynthesis.

photon (fō´-ton): the smallest unit of light energy.

photopigment (fō´-tō-pig-ment): a chemical substance in photoreceptor cells that, when struck by light, changes in molecular conformation.

photoreceptor: a receptor cell that responds to light; in vertebrates, rods and cones.

photorespiration: a series of reactions in plants in which O_2 replaces CO_2 during the C_3 cycle, preventing carbon fixation; this wasteful process dominates when C_3 plants are forced to close their stomata to prevent water loss.

photosynthesis: the complete series of chemical reactions in which the energy of light is used to synthesize high-energy organic molecules, normally carbohydrates, from low-energy inorganic molecules, normally carbon dioxide and water.

photosystem: in thylakoid membranes, a light-harvesting complex and its associated electron transport system.

phototactic: capable of detecting and responding to light.

phototropism: growth with respect to the direction of light.

pH scale: a scale, with values from 0 to 14, used for measuring the relative acidity of a solution; at pH 7 a solution is neutral, pH 0 to 7 is acidic, and pH 7 to 14 is basic; each unit on the scale represents a tenfold change in H^+ concentration.

phycocyanin (fī-kō-sī´-uh-nin): a blue or purple pigment that is located in the membranes of chloroplasts and is used as an accessory light-gathering molecule in thylakoid photosystems.

phylogeny (fī-lah´-jen-ē): the evolutionary history of a group of species.

phylum (fī-lum): the taxonomic category of animals and animal-like protists that is contained within a kingdom and consists of related classes.

phytochrome (fī´-tō-krōm): a light-sensitive plant pigment that mediates many plant responses to light, including flowering, stem elongation, and seed germination.

phytoplankton (fī´-tō-plank-ten): photosynthetic protists that are abundant in marine and freshwater environments.

pilus (pil´-us; pl., **pili):** a hairlike projection that is made of protein, located on the surface of certain bacteria, and is typically used to attach a bacterium to another cell.

pineal gland (pī-nē´-al): a small gland within the brain that secretes melatonin; controls the seasonal reproductive cycles of some mammals.

pinocytosis (pi-nō-sī-tō´-sis): the nonselective movement of extracellular fluid, enclosed within a vesicle formed from the plasma membrane, into a cell.

pioneer: an organism that is among the first to colonize an unoccupied habitat in the first stages of succession.

pit: an area in the cell walls between two plant cells in which secondary walls did not form, such that the two cells are separated only by a relatively thin and porous primary cell wall.

pith: cells forming the center of a root or stem.

pituitary gland: an endocrine gland, located at the base of the brain, that produces several hormones, many of which influence the activity of other glands.

placenta (pluh-sen´-tuh): in mammals, a structure formed by a complex interweaving of the uterine lining and the embryonic membranes, especially the chorion; functions in gas, nutrient, and waste exchange between embryonic and maternal circulatory systems and secretes hormones.

placental (pluh-sen´-tul): referring to a mammal, possessing a placenta (that is, species that are not marsupials or monotremes).

plankton: microscopic organisms that live in marine or freshwater environments; includes phytoplankton and zooplankton.

plant hormone: the plant-regulating chemicals auxin, gibberellin, cytokinin, ethylene, and abscisic acid; somewhat resemble animal hormones in that they are chemicals produced by cells in one location that influence the growth or metabolic activity of other cells, typically some distance away in the plant body.

plaque (plak): a deposit of cholesterol and other fatty substances within the wall of an artery.

plasma: the fluid, noncellular portion of the blood.

plasma cell: an antibody-secreting descendant of a B cell.

plasma membrane: the outer membrane of a cell, composed of a bilayer of phospholipids in which proteins are embedded.

plasmid (plaz´-mid): a small, circular piece of DNA located in the cytoplasm of many bacteria; normally does not carry genes required for the normal functioning of the bacterium but may carry genes that assist bacterial survival in certain environments, such as a gene for antibiotic resistance.

plasmodesma (plaz-mō-dez´-muh; pl., **plasmodesmata):** a cell-to-cell junction in plants that connects the cytoplasm of adjacent cells.

plasmodial slime mold: see *acellular slime mold.*

plasmodium (plaz-mō´-dē-um): a sluglike mass of cytoplasm containing thousands of nuclei that are not confined within individual cells.

plastid (plas´-tid): in plant cells, an organelle bounded by two membranes that may be involved in photosynthesis (chloroplasts), pigment storage, or food storage.

platelet (plāt´-let): a cell fragment that is formed from megakaryocytes in bone marrow and lacks a nucleus; circulates in the blood and plays a role in blood clotting.

plate tectonics: the process by which huge plates of Earth's crust move about the planet's surface; has resulted in "continental drift," the movement of continental landmasses over the course of Earth's history.

pleated sheet: a form of secondary structure exhibited by certain proteins, such as silk, in which many protein chains lie side-by-side, with hydrogen bonds holding adjacent chains together.

pleiotropy (plē´-ō-trō-pē): a situation in which a single gene influences more than one phenotypic characteristic.

pleural membrane: a membrane that lines the chest cavity and surrounds the lungs.

point mutation: a mutation in which a single base pair in DNA has been changed.

polar body: in oogenesis, a small cell, containing a nucleus but virtually no cytoplasm, produced by the first meiotic division of the primary oocyte.

polar covalent bond: a covalent bond with unequal sharing of electrons, such that one atom is relatively negative and the other is relatively positive.

polar molecule: a molecule bound by covalent bonds in which electrical charge is asymmetrically distributed, so that electrical charge differs in different portions of the molecule.

polar nucleus: in flowering plants, one of two nuclei in the primary endosperm cell of the female gametophyte; formed by the mitotic division of a megaspore.

pollen/pollen grain: the male gametophyte of a seed plant.

pollination: in flowering plants, when pollen grains land on the stigma of a flower of the same species; in conifers, when pollen grains land within the pollen chamber of a female cone of the same species.

polygenic inheritance: a pattern of inheritance in which the interactions of two or more functionally similar genes determine phenotype.

polymer (pah´-li-mer): a molecule composed of three or more (perhaps thousands) smaller subunits called *monomers*, which may be identical (for example, the glucose monomers of starch) or different (for example, the amino acids of a protein).

polymerase chain reaction (PCR): a method of producing virtually unlimited numbers of copies of a specific piece of DNA, starting with as little as one copy of the desired DNA.

polyp (pah´-lip): the sedentary, vase-shaped stage in the life cycle of many cnidarians; includes hydra and sea anemones.

polypeptide: a short polymer of amino acids; often used as a synonym for protein.

polyploidy (pahl´-ē-ploid-ē): having more than two homologous chromosomes of each type.

polysaccharide (pahl-ē-sak´-uh-rīd): a large carbohydrate molecule composed of branched or unbranched chains of repeating monosaccharide subunits, normally glucose or modified glucose molecules; includes starches, cellulose, and glycogen.

pons: a portion of the hindbrain, just above the medulla, that contains neurons that influence sleep and the rate and pattern of breathing.

population: all the members of a particular species within an ecosystem, found in the same time and place and actually or potentially interbreeding.

population bottleneck: a form of genetic drift in which a population becomes extremely small; may lead to differences in allele frequencies as compared with other populations of the species and to a loss in genetic variability.

population cycle: out-of-phase cyclical patterns of predator and prey populations.

population genetics: the study of the frequency, distribution, and inheritance of alleles in a population.

positive feedback: a situation in which a change initiates events that tend to amplify the original change.

post-anal tail: a tail that extends beyond the anus; exhibited by all chordates at some stage of development.

posterior: the tail, hindmost, or rear end of an animal.

posterior pituitary: a lobe of the pituitary gland that is an outgrowth of the hypothalamus and that releases antidiuretic hormone and oxytocin.

post-mating isolating mechanism: any structure, physiological function, or developmental abnormality that prevents organisms of two different populations, once mating has occurred, from producing vigorous, fertile offspring.

postsynaptic neuron: at a synapse, the nerve cell that changes its electrical potential in response to a chemical (the neurotransmitter) released by another (presynaptic) cell.

postsynaptic potential (PSP): an electrical signal produced in a postsynaptic cell by transmission across the synapse; it may be excitatory (EPSP), making the cell more likely to produce an action potential, or inhibitory (IPSP), tending to inhibit an action potential.

potential energy: "stored" energy, normally chemical energy or energy of position within a gravitational field.

prairie: a biome, located in the centers of continents, that supports grasses; also called *grassland.*

preadaptation: a feature evolved under one set of environmental conditions that, purely by chance, helps an organism adapt to new environmental conditions.

prebiotic evolution: evolution before life existed; especially, the abiotic synthesis of organic molecules.

precapillary sphincter (sfink´-ter): a ring of smooth muscle between an arteriole and a capillary that regulates the flow of blood into the capillary bed.

predation (pre-dā´-shun): the act of killing and eating another living organism.
predator: an organism that kills and eats other organisms.
pre-mating isolating mechanism: any structure, physiological function, or behavior that prevents organisms of two different populations from exchanging gametes.
pressure-flow theory: a model for the transport of sugars in phloem, by which the movement of sugars into a phloem sieve tube causes water to enter the tube by osmosis, while the movement of sugars out of another part of the same sieve tube causes water to leave by osmosis; the resulting pressure gradient causes the bulk movement of water and dissolved sugars from the end of the tube into which sugar is transported toward the end of the tube from which sugar is removed.
presynaptic neuron: a nerve cell that releases a chemical (the neurotransmitter) at a synapse, causing changes in the electrical activity of another (postsynaptic) cell.
prey: organisms that are killed and eaten by another organism.
primary cell wall: cellulose and other carbohydrates secreted by a young plant cell between the middle lamella and the plasma membrane.
primary consumer: an organism that feeds on producers; an herbivore.
primary endosperm cell: the central cell of the female gametophyte of a flowering plant, containing the polar nuclei (normally two); after fertilization, undergoes repeated mitotic divisions to produce the endosperm of the seed.
primary growth: growth in length and development of the initial structures of plant roots and shoots, due to the cell division of apical meristems and differentiation of the daughter cells.
primary oocyte (ō´-ō-sīt): a diploid cell, derived from the oogonium by growth and differentiation that undergoes meiosis, producing the egg.
primary phloem: phloem in young stems produced from an apical meristem.
primary root: the first root that develops from a seed.
primary spermatocyte (sper-ma´-tō-sīt): a diploid cell, derived from the spermatogonium by growth and differentiation, that undergoes meiosis, producing four sperm.
primary structure: the amino acid sequence of a protein.
primary succession: succession that occurs in an environment, such as bare rock, in which no trace of a previous community was present.
primary xylem: xylem in young stems produced from an apical meristem.
primate: a mammal characterized by the presence of an opposable thumb, forward-facing eyes, and a well-developed cerebral cortex; includes lemurs, monkeys, apes, and humans.
primitive streak: in reptiles, birds, and mammals, the region of the ectoderm of the two-layered embryonic disc through which cells migrate, forming mesoderm.
prion (prē´-on): a protein that, in mutated form, acts as an infectious agent that causes certain neurodegenerative diseases, including kuru and scrapie.
producer: a photosynthetic organism; an autotroph.
product: an atom or molecule that is formed from reactants in a chemical reaction.
profundal zone: a lake zone in which light is insufficient to support photosynthesis.
progesterone (prō-ge´-ster-ōn): a hormone, produced by the corpus luteum, that promotes the development of the uterine lining in females.
prokaryote: an organism whose cells lack a cell nucleus and other organelles; all prokaryotes are single-celled.
prokaryotic (prō-kar-ē-ot´-ik) cell: cells of the domains Bacteria or Archaea. Prokaryotic cells have genetic material that is not enclosed in a membrane-bound nucleus; they lack other membrane-bound organelles.
prolactin: a hormone, released by the anterior pituitary, that stimulates milk production in human females.
promoter: a specific sequence of DNA to which RNA polymerase binds, initiating gene transcription.
prophase (prō´-fāz): the first stage of mitosis, in which the chromosomes first become visible in the light microscope as thickened, condensed threads and the spindle begins to form; as the spindle is completed, the nuclear envelope breaks apart, and the spindle fibers invade the nuclear region and attach to the kinetochores of the chromosomes. Also, the first stage of meiosis: In meiosis I, the homologous chromosomes pair up and exchange parts at chiasmata; in meiosis II, the spindle re-forms and chromosomes attach to the microtubules.
prostaglandin (pro-stuh-glan´-din): a family of modified fatty acid hormones manufactured by many cells of the body.
prostate gland (pros´-tāt): a gland that produces part of the fluid component of semen; prostatic fluid is basic and contains a chemical that activates sperm movement.
protease (prō´-tē-ās): an enzyme that digests proteins.
protein: polymer of amino acids joined by peptide bonds.
protist: member of the kingdom Protista, which consists of eukaryotic, mostly single-celled organisms. Algae, amoebas, slime molds, and ciliates are examples of protists.
Protista (prō-tis´-tuh): a taxonomic kingdom including unicellular, eukaryotic organisms.
protocell: the hypothetical evolutionary precursor of living cells, consisting of a mixture of organic molecules within a membrane.
proton: a subatomic particle that is found in the nuclei of atoms, bears a unit of positive charge, and has a relatively large mass, roughly equal to the mass of the neutron.
protonephridium (prō-tō-nef-rid´-ē-um; pl., **protonephridia):** an excretory system consisting of tubules that have an external opening but lack internal openings; for example, the flame-cell system of flatworms.
protostome (prō´-tō-stōm): an animal with a mode of embryonic development in which the coelom is derived from splits in the mesoderm; characteristic of arthropods, annelids, and mollusks.
protozoan (prō-tuh-zō´-an; pl., **protozoa):** a nonphotosynthetic or animal-like protist.
proximal tubule: in nephrons of the mammalian kidney, the portion of the renal tubule just after the Bowman's capsule; receives filtrate from the capsule and is the site where selective secretion and reabsorption between the filtrate and the blood begins.
pseudocoelom (soo´-dō-sēl´-ōm): "false coelom"; a body cavity that has a different embryological origin than a coelom but serves a similar function; found in roundworms.
pseudoplasmodium (soo´-dō-plaz-mō´-dē-um): an aggregation of individual amoeboid cells that form a sluglike mass.
pseudopod (sood´-ō-pod): an extension of the plasma membrane by which certain cells, such as amoebae, locomote and engulf prey.
Punnett square method: an intuitive way to predict the genotypes and phenotypes of offspring in specific crosses.
pupa: a developmental stage in some insect species in which the organism stops moving and feeding and may be encased in a cocoon; occurs between the larval and the adult phases.
pupil: the adjustable opening in the center of the iris, through which light enters the eye.
pyloric sphincter (pī-lor´-ik sfink´-ter): a circular muscle, located at the base of the stomach, that regulates the passage of chyme into the small intestine.
pyruvate: a three-carbon molecule that is formed by glycolysis and then used in fermentation or cellular respiration.
quaternary structure (kwat´-er-nuh-rē): the complex three-dimensional structure of a protein composed of more than one peptide chain.
queen substance: a chemical, produced by a queen bee, that can act as both a primer and a pheromone.
radial symmetry: a body plan in which any plane along a central axis will divide the body into approximately mirror-image halves. Cnidarians and many adult echinoderms have radial symmetry.
radioactive: pertaining to an atom with an unstable nucleus that spontaneously disintegrates, with the emission of radiation.
radiolarian (rā-dē-ō-lar´-ē-un): an aquatic protist (largely marine) characterized by typically elaborate silica shells.
radula (ra´-dū-luh): a ribbon of tissue in the mouth of gastropod mollusks; bears numerous teeth on its outer surface and is used to scrape and drag food into the mouth.
rain shadow: a local dry area created by the modification of rainfall patterns by a mountain range.
random distribution: distribution characteristic of populations in which the probability of finding an individual is equal in all parts of an area.

reactant: an atom or molecule that is used up in a chemical reaction to form a product.
reaction center: in the light-harvesting complex of a photosystem, the chlorophyll molecule to which light energy is transferred by the antenna molecules (light-absorbing pigments); the captured energy ejects an electron from the reaction center chlorophyll, and the electron is transferred to the electron transport system.
receptor: a cell that responds to an environmental stimulus (chemicals, sound, light, pH, and so on) by changing its electrical potential; also, a protein molecule in a plasma membrane that binds to another molecule (hormone, neurotransmitter), triggering metabolic or electrical changes in a cell.
receptor-mediated endocytosis: the selective uptake of molecules from the extracellular fluid by binding to a receptor located at a coated pit on the plasma membrane and pinching off the coated pit into a vesicle that moves into the cytoplasm.
receptor potential: an electrical potential change in a receptor cell, produced in response to the reception of an environmental stimulus (chemicals, sound, light, heat, and so on). The size of the receptor potential is proportional to the intensity of the stimulus.
receptor protein: a protein, located on a membrane (or in the cytoplasm), that recognizes and binds to specific molecules. Binding by receptor proteins typically triggers a response by a cell, such as endocytosis, increased metabolic rate, or cell division.
recessive: an allele that is expressed only in homozygotes and is completely masked in heterozygotes.
recognition protein: a protein or glycoprotein protruding from the outside surface of a plasma membrane that identifies a cell as belonging to a particular species, to a specific individual of that species, and in many cases to one specific organ within the individual.
recombinant DNA: DNA that has been altered by the recombination of genes from a different organism, typically from a different species.
recombination: the formation of new combinations of the different alleles of each gene on a chromosome; the result of crossing over.
rectum: the terminal portion of the vertebrate digestive tube, where feces are stored until they can be eliminated.
red blood cell: the most common type of cell in vertebrate blood; active in oxygen transport; contains the red pigment hemoglobin.
reflex: a simple, stereotyped movement of part of the body that occurs automatically in response to a stimulus.
regeneration: the regrowth of a body part after loss or damage; also, asexual reproduction by means of the regrowth of an entire body from a fragment.
releasing hormone: a hormone, secreted by the hypothalamus, that causes the release of specific hormones by the anterior pituitary.
renal artery: the artery carrying blood to each kidney.
renal cortex: the outer layer of the kidney; where nephrons are located.
renal medulla: the layer of the kidney just inside the renal cortex; where loops of Henle produce a highly concentrated interstitial fluid, important in the production of concentrated urine.
renal pelvis: the inner chamber of the kidney; where urine from the collecting ducts accumulates before it enters the ureter.
renal vein: the vein carrying cleansed blood away from each kidney.
renin: an enzyme that is released (in mammals) when blood pressure and/or sodium concentration in the blood drops below a set point; initiates a cascade of events that restores blood pressure and sodium concentration.
replacement-level fertility (RLF): the average birthrate at which a reproducing population exactly replaces itself during its lifetime.
replication bubble: the unwound portion of the two parental DNA strands, separated by DNA helicase, in DNA replication.
reproductive isolation: the failure of organisms of one population to breed successfully with members of another; may be due to premating or post-mating isolating mechanisms.
reptile: member of a class of vertebrates whose members typically have scales, shelled amniote eggs, respiration by lungs, and ectothermy. Snakes, lizards, turtles, and crocodiles are reptiles.
reservoir: the major source and storage site of a nutrient in an ecosystem, normally in the abiotic portion.
resource partitioning: the coexistence of two species with similar requirements, each occupying a smaller niche than either would if it were by itself; a means of minimizing their competitive interactions.
respiration: the process by which an organism exchanges gases with the environment.
respiratory center: a cluster of neurons, located in the medulla of the brain, that sends rhythmic bursts of nerve impulses to the respiratory muscles, resulting in breathing.
resting potential: a negative electrical potential in unstimulated nerve cells.
restriction enzyme: an enzyme, normally isolated from bacteria, that cuts double-stranded DNA at a specific nucleotide sequence; the nucleotide sequence that is cut differs for different restriction enzymes.
restriction fragment: a piece of DNA that has been isolated by cleaving a larger piece of DNA with restriction enzymes.
restriction fragment length polymorphism (RFLP): a difference in the length of restriction fragments, produced by cutting samples of DNA from different individuals of the same species with the same set of restriction enzymes; the result of differences in nucleotide sequences among individuals of the same species.
reticular formation (reh-tik´-ū-lar): a diffuse network of neurons extending from the hindbrain, through the midbrain, and into the lower reaches of the forebrain; involved in filtering sensory input and regulating what information is relayed to conscious brain centers for further attention.
retina (ret´-in-uh): a multilayered sheet of nerve tissue at the rear of camera-type eyes, composed of photoreceptor cells plus associated nerve cells that refine the photoreceptor information and transmit it to the optic nerve.
retrovirus: a virus that uses RNA as its genetic material. When it invades a eukaryotic cell, a retrovirus "reverse transcribes" its RNA into DNA, which then directs the synthesis of more viruses, using the transcription and translation machinery of the cell.
reverse transcriptase: an enzyme found in retroviruses that catalyzes the synthesis of DNA from an RNA template.
Rh factor: a protein on the red blood cells of some people (Rh-positive) but not others (Rh-negative); the exposure of Rh-negative individuals to Rh-positive blood triggers the production of antibodies to Rh-positive blood cells.
rhizoid (rī´-zoid): a rootlike structure found in bryophytes that anchors the plant and absorbs water and nutrients from the soil.
rhizome (rī´-zōm): an undergound stem, usually horizontal, that stores food.
rhythm method: a contraceptive method involving abstinence from intercourse during ovulation.
ribonucleic acid (rī-bō-noo-klā´-ik; RNA): a molecule composed of ribose nucleotides, each of which consists of a phosphate group, the sugar ribose, and one of the bases adenine, cytosine, guanine, or uracil; transfers hereditary instructions from the nucleus to the cytoplasm; also the genetic material of some viruses.
ribosomal RNA (rRNA): a type of RNA that combines with proteins to form ribosomes.
ribosome: an organelle consisting of two subunits, each composed of ribosomal RNA and protein; the site of protein synthesis, during which the sequence of bases of messenger RNA is translated into the sequence of amino acids in a protein.
ribozyme: an RNA molecule that can catalyze certain chemical reactions, especially those involved in the synthesis and processing of RNA itself.
ribulose biphosphate (RUBP): a six-carbon molecule that reacts with carbon dioxide in the carbon-fixing reaction of C_3 photosynthesis; an important participant in the Calvin-Benson cycle.
RNA polymerase: in RNA synthesis, an enzyme that catalyzes the bonding of free RNA nucleotides into a continuous strand, using RNA nucleotides that are complementary to those of a strand of DNA.
rod: a rod-shaped photoreceptor cell in the vertebrate retina, sensitive to dim light but not involved in color vision; see also *cone*.
root: the part of the plant body, normally underground, that provides anchorage, absorbs water and dissolved nutrients and transports them to the stem, produces some hormones, and in some plants serves as a storage site for carbohydrates.

root cap: a cluster of cells at the tip of a growing root, derived from the apical meristem; protects the growing tip from damage as it burrows through the soil.

root hair: a fine projection from an epidermal cell of a young root that increases the absorptive surface area of the root.

root system: the part of a plant, normally below ground, that anchors the plant in the soil, absorbs water and minerals, stores food, transports water, minerals, sugars, and hormones, and produces certain hormones.

rough endoplasmic reticulum: endoplasmic reticulum lined on the outside with ribosomes.

runner: a horizontally growing stem that may develop new plants at nodes that touch the soil.

sac fungus: a fungus of the division Ascomycota, whose members form spores in a saclike case called an *ascus*.

sapwood: young xylem that transports water and minerals in a tree trunk.

saprobe (sap´-rōb): an organism that derives its nutrients from the bodies of dead organisms.

sarcodine (sar-kō´-dīn): a nonphotosynthetic protist (protozoan) characterized by the ability to form pseudopodia; some sarcodines, such as amoebae, are naked, whereas others have elaborate shells.

sarcomere (sark´-ō-mēr): the unit of contraction of a muscle fiber; a subunit of the myofibril, consisting of actin and myosin filaments and bounded by Z lines.

sarcoplasmic reticulum (sark´-ō-plas´-mik re-tik´-ū-lum): the specialized endoplasmic reticulum in muscle cells; forms interconnected hollow tubes. The sarcoplasmic reticulum stores calcium ions and releases them into the interior of the muscle cell, initiating contraction.

saturated: referring to a fatty acid with as many hydrogen atoms as possible bonded to the carbon backbone; a fatty acid with no double bonds in its carbon backbone.

savanna: a biome that is dominated by grasses and supports scattered trees and thorny scrub forests; typically has a rainy season in which all the year's precipitation falls.

scientific method: a rigorous procedure for making observations of specific phenomena and searching for the order underlying those phenomena; consists of four operations: observation, hypothesis, experiment, and conclusion.

scientific name: the name of an organism formed from the two smallest major taxonomic categories—the genus and the species.

scientific theory: a general explanation of natural phenomena developed through extensive and reproducible observations; more general and reliable than a hypothesis.

sclera: a tough, white connective tissue layer that covers the outside of the eyeball and forms the white of the eye.

sclerenchyma (skler-en´-ki-muh): a plant cell type with thick, hardened secondary cell walls that normally dies as the last stage of differentiation and both supports and protects the plant body.

scramble competition: a free-for-all scramble for limited resources among individuals of the same species.

scrotum (skrō´-tum): the pouch of skin containing the testes of male mammals.

S-curve: the S-shaped growth curve that describes a population of long-lived organisms introduced into a new area; consists of an initial period of exponential growth, followed by decreasing growth rate, and, finally, relative stability around a growth rate of zero.

sebaceous gland (se-bā´-shus): a gland in the dermis of skin, formed from epithelial tissue, that produces the oily substance sebum, which lubricates the epidermis.

secondary cell wall: a thick layer of cellulose and other polysaccharides secreted by certain plant cells between the primary cell wall and the plasma membrane.

secondary consumer: an organism that feeds on primary consumers; a carnivore.

secondary growth: growth in the diameter of a stem or root due to cell division in lateral meristems and differentiation of their daughter cells.

secondary oocyte (ō´-ō-sīt): a large haploid cell derived from the first meiotic division of the diploid primary oocyte.

secondary phloem: phloem produced from the cells that arise toward the outside of the vascular cambium.

secondary spermatocyte (sper-ma´-tō-sīt): a large haploid cell derived by meiosis I from the diploid primary spermatocyte.

secondary structure: a repeated, regular structure assumed by protein chains held together by hydrogen bonds; for example, a helix.

secondary succession: succession that occurs after an existing community is disturbed—for example, after a forest fire; much more rapid than primary succession.

secondary xylem: xylem produced from cells that arise at the inside of the vascular cambium.

second law of thermodynamics: the principle of physics that states that any change in an isolated system causes the quantity of concentrated, useful energy to decrease and the amount of randomness and disorder (entropy) to increase.

second messenger: an intracellular chemical, such as cyclic AMP, that is synthesized or released within a cell in response to the binding of a hormone or neurotransmitter (the first messenger) to receptors on the cell surface; brings about specific changes in the metabolism of the cell.

secretin: a hormone, produced by the small intestine, that stimulates the production and release of digestive secretions by the pancreas and liver.

seed: the reproductive structure of a seed plant; protected by a seed coat; contains an embryonic plant and a supply of food for it.

seed coat: the thin, tough, and waterproof outermost covering of a seed, formed from the integuments of the ovule.

segmentation (seg-men-tā´-shun): an animal body plan in which the body is divided into repeated, typically similar units.

segmentation movement: a contraction of the small intestine that results in the mixing of partially digested food and digestive enzymes. Segmentation movements also bring nutrients into contact with the absorptive intestinal wall.

segregation: see *law of segregation.*

selectively permeable: refers to membranes across which some substances may pass freely while other substances cannot pass.

self-fertilization: the union of sperm and egg from the same individual.

selfish gene: the concept that genes promote their own survival in individuals through innate self-sacrificing behavior that enhances the survival of others that carry the same genes; helps explain the evolution of altruism.

semen: the sperm-containing fluid produced by the male reproductive tract.

semiconservative replication: the process of replication of the DNA double helix; the two DNA strands separate, and each is used as a template for the synthesis of a complementary DNA strand. Consequently, each daughter double helix consists of one parental strand and one new strand.

semilunar valve: a paired valve between the ventricles of the heart and the pulmonary artery and aorta; prevents the backflow of blood into the ventricles when they relax.

seminal vesicle: in male mammals, a gland that produces a basic, fructose-containing fluid that forms part of the semen.

seminiferous tubule (sem-i-ni´-fer-us): in the vertebrate testis, a series of tubes in which sperm are produced.

senescence: in plants, a specific aging process, typically including deterioration and the dropping of leaves and flowers.

sensitive period: the particular stage in an animal's life during which it imprints.

sensory neuron: a nerve cell that responds to a stimulus from the internal or external environment.

sensory receptor: a cell (typically, a neuron) specialized to respond to particular internal or external environmental stimuli by producing an electrical potential.

sepal (sē´-pul): the set of modified leaves that surround and protect a flower bud, typically opening into green, leaflike structures when the flower blooms.

septum (pl., septa): a partition that separates the fungal hypha into individual cells; pores in septa allow the transfer of materials between cells.

serotonin (ser-uh-tō´-nin): in the central nervous system, a neurotransmitter that is involved in mood, sleep, and the inhibition of pain.

Sertoli cell: in the seminiferous tubule, a large cell that regulates spermatogenesis and nourishes the developing sperm.

sessile (ses´-ul): not free to move about, usually permanently attached to a surface.

severe combined immune deficiency (SCID): a disorder in which no immune cells, or very few, are formed; the immune system is incapable of responding properly to invading disease organisms, and the individual is very vulnerable to common infections.

sex chromosomes: the pair of chromosomes that usually determines the sex of an organism; for example, the X and Y chromosomes in mammals.

sex-linked: referring to a pattern of inheritance characteristic of genes located on one type of sex chromosome (for example, X) and not found on the other type (for example, Y); also called X-linked. In sex-linked inheritance, traits are controlled by genes carried on the X chromosome; females show the dominant trait unless they are homozygous recessive, whereas males express whichever allele is on their single X chromosome.

sexually transmitted disease (STD): a disease that is passed from person to person by sexual contact.

sexual recombination: during sexual reproduction, the formation of new combinations of alleles in offspring as a result of the inheritance of one homologous chromosome from each of two genetically distinct parents.

sexual reproduction: a form of reproduction in which genetic material from two parent organisms is combined in the offspring; normally, two haploid gametes fuse to form a diploid zygote.

sexual selection: a type of natural selection in which the choice of mates by one sex is the selective agent.

shoot system: all the parts of a vascular plant exclusive of the root; normally aboveground, consisting of stem, leaves, buds, and (in season) flowers and fruits; functions include photosynthesis, transport of materials, reproduction, and hormone synthesis.

short-day plant: a plant that will flower only if the length of daylight is shorter than some species-specific duration.

sickle-cell anemia: a recessive disease caused by a single amino acid substitution in the hemoglobin molecule. Sickle-cell hemoglobin molecules tend to cluster together, distorting the red blood cell's shape and causing them to break and clog capillaries.

sieve plate: in plants, a structure between two adjacent sieve-tube elements in phloem, where holes formed in the primary cell walls interconnect the cytoplasm of the elements; in echinoderms, the opening through which water enters the water-vascular system.

sieve tube: in phloem, a single strand of sieve-tube elements that transports sugar solutions.

sieve-tube element: one of the cells of a sieve tube, which form the phloem.

simple diffusion: the diffusion of water, dissolved gases, or lipid-soluble molecules through the phospholipid bilayer of a cellular membrane.

single covalent bond: a covalent bond in which two atoms share one pair of electrons.

sink: in plants, any structure that uses up sugars or converts sugars to starch and toward which phloem fluids will flow.

sinoatrial (SA) node (sī´-nō-āt´-rē-ul): a small mass of specialized muscle in the wall of the right atrium; generates electrical signals rhythmically and spontaneously and serves as the heart's pacemaker.

skeletal muscle: the type of muscle that is attached to and moves the skeleton and is under the direct, normally voluntary, control of the nervous system; also called *striated muscle*.

skeleton: a supporting structure for the body, on which muscles act to change the body configuration; may be external or internal.

skin: the tissue that makes up the outer surface of an animal body.

slime layer: a sticky polysaccharide or protein coating that some disease-causing bacteria secrete outside their cell wall; helps the cells aggregate and stick to smooth surfaces.

small intestine: the portion of the digestive tract, located between the stomach and large intestine, in which most digestion and absorption of nutrients occur.

smooth endoplasmic reticulum: endoplasmic reticulum without ribosomes.

smooth muscle: the type of muscle that surrounds hollow organs, such as the digestive tract, bladder, and blood vessels; normally not under voluntary control.

sodium–potassium pump: in nerve cell plasma membranes, a set of active-transport molecules that use the energy of ATP to pump sodium ions out of the cell and potassium ions in, maintaining the concentration gradients of these ions across the membrane.

solvent: a liquid capable of dissolving (uniformly dispersing) other substances in itself.

somatic nervous system: that portion of the peripheral nervous system that controls voluntary movement by activating skeletal muscles.

source: in plants, any structure that actively synthesizes sugar and away from which phloem fluid will be transported.

spawning: a method of external fertilization in which male and female parents shed gametes into the water, and sperm must swim through the water to reach the eggs.

speciation: the process of species formation, in which a single species splits into two or more species.

species (spē´-sēs): the basic unit of taxonomic classification, consisting of a population or series of populations of closely related and similar organisms. In sexually reproducing organisms, a species can be defined as a population or series of populations of organisms that interbreed freely with one another under natural conditions but that do not interbreed with members of other species.

specific heat: the amount of energy required to raise the temperature of 1 gram of a substance by 1 °C.

sperm: the haploid male gamete, normally small, motile, and containing little cytoplasm.

spermatid: a haploid cell derived from the secondary spermatocyte by meiosis II; differentiates into the mature sperm.

spermatogenesis: the process by which sperm cells form.

spermatogonium (pl., **spermatogonia**)**:** a diploid cell, lining the walls of the seminiferous tubules, that gives rise to a primary spermatocyte.

spermatophore: in a variation on internal fertilization in some animals, the males package their sperm in a container that can be inserted into the female reproductive tract.

spermicide: a sperm-killing chemical; used for contraceptive purposes.

spicule (spik´-ūl): a subunit of the endoskeleton of sponges that is made of protein, silica, or calcium carbonate.

spinal cord: the part of the central nervous system of vertebrates that extends from the base of the brain to the hips and is protected by the bones of the vertebral column; contains the cell bodies of motor neurons that form synapses with skeletal muscles, the circuitry for some simple reflex behaviors, and axons that communicate with the brain.

spindle microtubules: microtubules organized in a spindle shape that separate chromosomes during mitosis or meiosis.

spiracle (spi´-ruh-kul): an opening in the abdominal segment of insects through which air enters the tracheae.

spirillum (spi´-ril-um; pl., **spirilla):** a spiral-shaped bacterium.

spleen: an organ of the lymphatic system in which lymphocytes are produced and blood is filtered past lymphocytes and macrophages, which remove foreign particles and aged red blood cells.

spongy bone: porous, lightweight bone tissue in the interior of bones; the location of bone marrow.

spongy cell: an irregularly shaped mesophyll cell, containing chloroplasts, located just above the lower epidermis of a leaf.

spontaneous generation: the proposal that living organisms can arise from nonliving matter.

sporangium (spor-an´-jē-um; pl., **sporangia):** a structure in which spores are produced.

spore: a haploid reproductive cell capable of developing into an adult without fusing with another cell; in the alternation-of-generation life cycle of plants, a haploid cell that is produced by meiosis and then undergoes repeated mitotic divisions and differentiation of daughter cells to produce the gametophyte, a multicellular, haploid organism.

sporophyte (spor´-ō-fīt): the diploid form of a plant that produces haploid, asexual spores through meiosis.

sporozoan (spor-ō-zō´-un): a parasitic protist with a complex life cycle, typically involving more than one host; named for their ability to form infectious spores. A well-known sporozoan (genus *Plasmodium*) causes malaria.

stabilizing selection: a type of natural selection in which those organisms that display extreme phenotypes are selected against.

stamen (stā´-men): the male reproductive structure of a flower, consisting of a filament and an anther, in which pollen grains develop.

starch: a polysaccharide that is composed of branched or unbranched chains or glucose molecules; used by plants as a carbohydrate-storage molecule.

start codon: the first AUG codon in a messenger RNA molecule.

startle coloration: a form of mimicry in which a color pattern (in many cases resembling large eyes) can be displayed suddenly by a prey organism when approached by a predator.

stem: the portion of the plant body, normally located above ground, that bears leaves and reproductive structures such as flowers and fruit.

sterilization: a generally permanent method of contraception in which the pathways through which the sperm (vas deferens) or egg (oviducts) must travel are interrupted; the most common form of contraception.

steroid: see *steroid hormone*.

steroid hormone: a class of hormone whose chemical structure (four fused carbon rings with various functional groups) resembles cholesterol; steroids, which are lipids, are secreted by the ovaries and placenta, the testes, and the adrenal cortex.

stigma (stig´-muh): the pollen-capturing tip of a carpel.

stoma (stō´-muh; pl., **stomata):** an adjustable opening in the epidermis of a leaf, surrounded by a pair of guard cells, that regulates the diffusion of carbon dioxide and water into and out of the leaf.

stomach: the muscular sac between the esophagus and small intestine where food is stored and mechanically broken down and in which protein digestion begins.

stop codon: a codon in messenger RNA that stops protein synthesis and causes the completed protein chain to be released from the ribosome.

strand: a single polymer of nucleotides; DNA is composed of two strands.

striated muscle: see *skeletal muscle*.

stroke: an interruption of blood flow to part of the brain caused by the rupture of an artery or the blocking of an artery by a blood clot. Loss of blood supply leads to rapid death of the area of the brain affected.

stroma (strō´-muh): the semi-fluid material inside chloroplasts in which the grana are embedded.

style: a stalk connecting the stigma of a carpel with the ovary at its base.

subatomic particle: the particles of which atoms are made: electrons, protons, and neutrons.

subclimax: a community in which succession is stopped before the climax community is reached and is maintained by regular disturbances—for example, tallgrass prairie maintained by periodic fires.

substrate: the atoms or molecules that are the reactants for an enzyme-catalyzed chemical reaction.

subunit: a small organic molecule, several of which may be bonded together to form a larger molecule. See also *monomer*.

succession (suk-seh´-shun): a structural change in a community and its nonliving environment over time. Community changes alter the ecosystem in ways that favor competitors, and species replace one another in a somewhat predictable manner until a stable, self-sustaining climax community is reached.

sucrose: a disaccharide composed of glucose and fructose.

sugar: a simple carbohydrate molecule, either a monosaccharide or a disaccharide.

sugar-phosphate backbone: a major feature of DNA structure, formed by attaching the sugar of one nucleotide to the phosphate from the adjacent nucleotide in a DNA strand.

suppressor T cell: a type of T cell that depresses the response of other immune cells to foreign antigens.

surface tension: the property of a liquid to resist penetration by objects at its interface with the air, due to cohesion between molecules of the liquid.

survivorship curve: a curve resulting when the number of individuals of each age in a population is graphed against their age, usually expressed as a percentage of their maximum life span.

symbiosis (sim´-bī-ō´sis): a close interaction between organisms of different species over an extended period. Either or both species may benefit from the association, or (in the case of parasitism) one of the participants is harmed. Symbiosis includes parasitism, mutualism, and commensalism.

symbiotic: referring to an ecological relationship based on symbiosis.

sympathetic division: the division of the autonomic nervous system that produces largely involuntary responses that prepare the body for stressful or highly energetic situations.

sympatric speciation (sim-pat´-rik): speciation that occurs in populations that are not physically divided; normally due to ecological isolation or chromosomal aberrations (such as polyploidy).

synapse (sin´-aps): the site of communication between nerve cells. At a synapse, one cell (presynaptic) normally releases a chemical (the neurotransmitter) that changes the electrical potential of the second (postsynaptic) cell.

synaptic terminal: a swelling at the branched ending of an axon; where the axon forms a synapse.

syphilis (si´-ful-is): a sexually transmitted bacterial infection of the reproductive organs; if untreated, can damage the nervous and circulatory systems.

systematics: the branch of biology concerned with reconstructing phylogenies and with naming and classifying species.

taiga (tī´-guh): a biome with long, cold winters and only a few months of warm weather; dominated by evergreen coniferous trees; also called *northern coniferous forest*.

taproot system: a root system, commonly found in dicots, that consists of a long, thick main root and many smaller lateral roots, all of which grow from the primary root.

target cell: a cell on which a particular hormone exerts its effect.

taste: a chemical sense for substances dissolved in water or saliva; in mammals, perceptions of sweet, sour, bitter, or salt produced by the stimulation of receptors on the tongue.

taste bud: a cluster of taste receptor cells and supporting cells that is located in a small pit beneath the surface of the tongue and that communicates with the mouth through a small pore. The human tongue has about 10,000 taste buds.

taxis (taks´-is; pl., **taxes):** an innate behavior that is a directed movement of an organism toward or away from a stimulus such as heat, light, or gravity.

taxonomy (tax-on´-uh-mē): the science by which organisms are classified into hierarchically arranged categories that reflect their evolutionary relationships.

Tay-Sachs disease: a recessive disease caused by a deficiency in enzymes that regulate lipid breakdown in the brain.

T cell: a type of lymphocyte that recognizes and destroys specific foreign cells or substances or that regulates other cells of the immune system.

T-cell receptor: a protein receptor, located on the surface of a T cell, that binds a specific antigen and triggers the immune response of the T cell.

tectorial membrane (tek-tor´-ē-ul): one of the membranes of the cochlea in which the hairs of the hair cells are embedded. In sound reception, movement of the basilar membrane relative to the tectorial membrane bends the cilia.

telophase (tēl´-ō-fāz): in mitosis, the final stage, in which a nuclear envelope re-forms around each new daughter nucleus, the spindle fibers disappear, and the chromosomes relax from their condensed form; in meiosis I, the stage during which the spindle fibers disappear and the chromosomes normally relax from their condensed form; in meiosis II, the stage during which chromosomes relax into their extended state, nuclear envelopes re-form, and cytokinesis occurs.

temperate deciduous forest: a biome in which winters are cold and summer rainfall is sufficient to allow enough moisture for trees to grow and shade out grasses.

temperate rain forest: a biome in which there is no shortage of liquid water year-round and that is dominated by conifers.

template strand: the strand of the DNA double helix from which RNA is transcribed.

temporal isolation: the inability of organisms to mate if they have significantly different breeding seasons.

tendon: a tough connective tissue band connecting a muscle to a bone.

tendril: a slender outgrowth of a stem that coils about external objects and supports the stem; normally a modified leaf or branch.

tentacle (ten´-te-kul): an elongate, extensible projection of the body of cnidarians and cephalopod mollusks that may be used for grasping, stinging, and immobilizing prey, and for locomotion.

terminal bud: meristem tissue and surrounding leaf primordia that are located at the tip of the plant shoot.

territoriality: the defense of an area in which important resources are located.

tertiary consumer (ter´-shē-er-ē): a carnivore that feeds on other carnivores (secondary consumers).

tertiary structure (ter´-shē-er-ē): the complex three-dimensional structure of a single peptide chain; held in place by disulfide bonds between cysteines.
test cross: a breeding experiment in which an individual showing the dominant phenotype is mated with an individual that is homozygous recessive for the same gene. The ratio of offspring with dominant versus recessive phenotypes can be used to determine the genotype of the phenotypically dominant individual.
testis (pl., testes): the gonad of male mammals.
testosterone: in vertebrates, a hormone produced by the interstitial cells of the testis; stimulates spermatogenesis and the development of male secondary sex characteristics.
thalamus: the part of the forebrain that relays sensory information to many parts of the brain.
theory: in science, an explanation for natural events that is based on a large number of observations and is in accord with scientific principles, especially causality.
thermoacidophile (ther-mō-a-sid´-eh-fīl): an archaean that thrives in hot, acidic environments.
thermoreceptor: a sensory receptor that responds to changes in temperature.
thick filament: in the sarcomere, a bundle of myosin that interacts with thin filaments, producing muscle contraction.
thin filament: in the sarcomere, a protein strand that interacts with thick filaments, producing muscle contraction; composed primarily of actin, with accessory proteins.
thorax: the segment between the head and abdomen in animals with segmentation; the segment to which structures used in locomotion are attached.
thorn: a hard, pointed outgrowth of a stem; normally a modified branch.
threshold: the electrical potential (less negative than the resting potential) at which an action potential is triggered.
thrombin: an enzyme produced in the blood as a result of injury to a blood vessel; catalyzes the production of fibrin, a protein that assists in blood clot formation.
thylakoid (thī´-luh-koid): a disk-shaped, membranous sac found in chloroplasts, the membranes of which contain the photosystems and ATP-synthesizing enzymes used in the light-dependent reactions of photosynthesis.
thymine: a nitrogenous base found only in DNA; abbreviated as *T.*
thymosin: a hormone, secreted by the thymus, that stimulates the maturation of cells of the immune system.
thymus (thī´-mus): an organ of the lymphatic system that is located in the upper chest in front of the heart and that secretes thymosin, which stimulates lymphocyte maturation; begins to degenerate at puberty and has little function in the adult.
thyroid gland: an endocrine gland, located in front of the larynx in the neck, that secretes the hormones thyroxine (affecting metabolic rate) and calcitonin (regulating calcium ion concentration in the blood).
thyroid-stimulating hormone (TSH): a hormone, released by the anterior pituitary, that stimulates the thyroid gland to release hormones.
thyroxine (thī-rox´-in): a hormone, secreted by the thyroid gland, that stimulates and regulates metabolism.
tight junction: a type of cell-to-cell junction in animals that prevents the movement of materials through the spaces between cells.
tissue: a group of (normally similar) cells that together carry out a specific function; for example, muscle; may include extracellular material produced by its cells.
tonsil: a patch of lymphatic tissue consisting of connective tissue that contains many lymphocytes; located in the pharynx and throat.
trachea (trā´-kē-uh): in birds and mammals, a rigid but flexible tube, supported by rings of cartilage, that conducts air between the larynx and the bronchi; in insects, an elaborately branching tube that carries air from openings called *spiracles* near each body cell.
tracheid (trā´-kē-id): an elongated xylem cell with tapering ends that contains pits in the cell wall; forms tubes that transport water.
tracheophyte (trā´-kē-ō-fīt): a plant that has conducting vessels; a vascular plant.
transcription: the synthesis of an RNA molecule from a DNA template.
transducer: a device that converts signals from one form to another. Sensory receptors are transducers that convert environmental stimuli, such as heat, light, or vibration, into electrical signals (such as action potentials) recognized by the nervous system.
transfer RNA (tRNA): a type of RNA that binds to a specific amino acid by means of a set of three bases (the anticodon) on the tRNA that are complementary to the mRNA codon for that amino acid; carries its amino acid to a ribosome during protein synthesis, recognizes a codon of mRNA, and positions its amino acid for incorporation into the growing protein chain.
transformation: a method of acquiring new genes, whereby DNA from one bacterium (normally released after the death of the bacterium) becomes incorporated into the DNA of another, living, bacterium.
transgenic: referring to an animal or a plant that expresses DNA derived from another species.
translation: the process whereby the sequence of bases of messenger RNA is converted into the sequence of amino acids of a protein.
transpiration (trans´-per-ā-shun): the evaporation of water through the stomata of a leaf.
transport protein: a protein that regulates the movement of water-soluble molecules through the plasma membrane.
trial-and-error learning: a process by which adaptive responses are learned through rewards or punishments provided by the environment.
trichomoniasis (trik-ō-mō-nī´-uh-sis): a sexually transmitted disease, caused by the protist *Trichomonas*, that causes inflammation of the mucous membranes than line the urinary tract and genitals.
tricuspid valve: the valve between the right ventricle and the right atrium of the heart.
triglyceride (trī-glis´-er-īd): a lipid composed of three fatty-acid molecules bonded to a single glycerol molecule.
triple covalent bond: a covalent bond that occurs when two atoms share three pairs of electrons.
trisomy 21: see *Down syndrome*.
trisomy X: a condition of females who have three X chromosomes instead of the normal two; most such women are phenotypically normal and are fertile.
trophic level: literally, "feeding level"; the categories of organisms in a community, and the position of an organism in a food chain, defined by the organism's source of energy; includes producers, primary consumers, secondary consumers, and so on.
tropical deciduous forest: a biome with pronounced wet and dry seasons and plants that must shed their leaves during the dry season to minimize water loss.
tropical rain forest: a biome with evenly warm, evenly moist conditions; dominated by broadleaf evergreen trees; the most diverse biome.
true-breeding: pertaining to an individual all of whose offspring produced through self-fertilization are identical to the parental type. True-breeding individuals are homozygous for a given trait.
T tubule: a deep infolding of the muscle plasma membrane; conducts the action potential inside a cell.
tubal ligation: a surgical procedure in which a woman's oviducts are cut so that the egg cannot reach the uterus, making her infertile.
tube cell: the outermost cell of a pollen grain; digests a tube through the tissues of the carpel, ultimately penetrating into the female gametophyte.
tube foot: a cylindrical extension of the water-vascular system of echinoderms; used for locomotion, grasping food, and respiration.
tubular reabsorption: the process by which cells of the tubule of the nephron remove water and nutrients from the filtrate within the tubule and return those substances to the blood.
tubular secretion: the process by which cells of the tubule of the nephron remove additional wastes from the blood, actively secreting those wastes into the tubule.
tubule (toob´-ūl): the tubular portion of the nephron; includes a proximal portion, the loop of Henle, and a distal portion. Urine is formed from the blood filtrate as it passes through the tubule.
tumor: a mass that forms in otherwise normal tissue; caused by the uncontrolled growth of cells.
tundra: a biome with severe weather conditions (extreme cold and wind and little rainfall) that cannot support trees.
turgor pressure: pressure developed within a cell (especially the central vacuole of plant cells) as a result of osmotic water entry.

Turner syndrome: a set of characteristics typical of a woman with only one X chromosome: sterile, with a tendency to be very short and to lack normal female secondary sexual characteristics.
tympanic membrane (tim-pan´-ik): the eardrum; a membrane, stretched across the opening of the ear, that transmits vibration of sound waves to bones of the middle ear.
unicellular: single-celled; most members of the domains Bacteria and Archaea and the kingdom Protista are unicellular.
umbilical cord: structure that connects the circulatory systems of a mammalian mother and fetus.
uniform distribution: the distribution characteristic of a population with a relatively regular spacing of individuals, commonly as a result of territorial behavior.
uniformitarianism: the hypothesis that Earth developed gradually through natural processes, similar to those at work today, that occur over long periods of time.
unsaturated: referring to a fatty acid with fewer than the maximum number of hydrogen atoms bonded to its carbon backbone; a fatty acid with one or more double bonds in its carbon backbone.
upwelling: an upward flow that brings cold, nutrient-laden water from the ocean depths to the surface; occurs along western coastlines.
uracil: a nitrogenous base found in RNA; abbreviated as *U.*
urea (ū-rē´-uh): a water-soluble, nitrogen-containing waste product of amino acid breakdown; one of the principal components of mammalian urine.
ureter (ū´-re-ter): a tube that conducts urine from each kidney to the bladder.
urethra (ū-rē´-thruh): the tube leading from the urinary bladder to the outside of the body; in males, the urethra also receives sperm from the vas deferens and conducts both sperm and urine (at different times) to the tip of the penis.
uric acid (ūr´-ik): a nitrogen-containing waste product of amino acid breakdown; a relatively insoluble white crystal excreted by birds, reptiles, and insects.
urine: the fluid produced and excreted by the urinary system of vertebrates; contains water and dissolved wastes, such as urea.
uterine tube: also called the *oviduct,* the tube leading out of the ovary to the uterus, into which the secondary oocyte (egg cell) is released.
uterus: in female mammals, the part of the reproductive tract that houses the embryo during pregnancy.
vaccination: an injection into the body that contains antigens characteristic of a particular disease organism and that stimulates an immune response.
vacuole (vak´-ū-ōl): a vesicle that is typically large and consists of a single membrane enclosing a fluid-filled space.
vagina: the passageway leading from the outside of a female mammal's body to the cervix of the uterus.
variable: a condition, particularly in a scientific experiment, that is subject to change.
variable region: the part of an antibody molecule that differs among antibodies; the ends of the variable regions of the light and heavy chains form the specific binding site for antigens.
vascular: describing tissues that contain vessels for transporting liquids.
vascular bundle (vas´-kū-lar): a strand of xylem and phloem in leaves and stems; in leaves, commonly called a *vein.*
vascular cambium: a lateral meristem that is located between the xylem and phloem of a woody root or stem and that gives rise to secondary xylem and phloem.
vascular cylinder: the centrally located conducting tissue of a young root, consisting of primary xylem and phloem.
vascular plant: a plant that has specialized structures (vessels) for transporting water and nutrients though its body. Horsetails, ferns, gymnosperms and flowering plants are examples of vascular plants.
vascular tissue: plant tissue consisting of xylem (which transports water and minerals from root to shoot) and phloem (which transports water and sugars throughout the plant).
vas deferens (vaz de´-fer-enz): the tube connecting the epididymis of the testis with the urethra.
vasectomy: a surgical procedure in which a man's vas deferens are cut, preventing sperm from reaching the penis during ejaculation, thereby making him infertile.
vector: a carrier that introduces foreign genes into cells.
vein: in vertebrates, a large-diameter, thin-walled vessel that carries blood from venules back to the heart; in vascular plants, a vascular bundle, or a strand of xylem and phloem in leaves.
ventral (ven´-trul): the lower side or underside of an animal whose head is oriented forward.
ventricle (ven´-tre-kul): the lower muscular chamber on each side of the heart, which pumps blood out through the arteries. The right ventricle sends blood to the lungs; the left ventricle pumps blood to the rest of the body.
venule (ven´-ūl): a narrow vessel with thin walls that carries blood from capillaries to veins.
vertebral column (ver-tē´-brul): a column of serially arranged skeletal units (the vertebrae) that enclose the nerve cord in vertebrates; the backbone.
vertebrate: an animal that possesses a vertebral column.
vesicle (ves´-i-kul): a small, membrane-bound sac within the cytoplasm.
vessel: a tube of xylem composed of vertically stacked vessel elements with heavily perforated or missing end walls, leaving a continuous, uninterrupted hollow cylinder.
vessel element: one of the cells of a xylem vessel; elongated, dead at maturity, with thick, lignified lateral cell walls for support but with end walls that are either heavily perforated or missing.
vestigial structure (ves-tij´-ē-ul): a structure that serves no apparent purpose but is homologous to functional structures in related organisms and provides evidence of evolution.
villus (vi´-lus; pl., **villi):** a fingerlike projection of the wall of the small intestine that increases the absorptive surface area.
viroid (vī´-roid): a particle of RNA that is capable of infecting a cell and of directing the production of more viroids; responsible for certain plant diseases.
virus (vī´-rus): a noncellular parasitic particle that consists of a protein coat surrounding a strand of genetic material; multiplies only within a cell of a living organism (the host).
vitamin: one of a group of diverse chemicals that must be present in trace amounts in the diet to maintain health; used by the body in conjunction with enzymes in a variety of metabolic reactions.
vitreous humor (vit´-rē-us): a clear, jellylike substance that fills the large chamber of the eye between the lens and the retina.
vocal cord: one of a pair of bands of elastic tissue that extend across the opening of the larynx and produce sound when air is forced between them. Muscles alter the tension on the vocal cords and control the size and shape of the opening, which in turn determines whether sound is produced and what its pitch will be.
waggle dance: a symbolic form of communication used by honeybee foragers to communicate the location of a food source to their hivemates.
warning coloration: bright coloration that warns predators that the potential prey is distasteful or even poisonous.
water mold: a funguslike protist that includes some pathogens, such as the downy mildew, which attacks grapes.
water-vascular system: a system in echinoderms that consists of a series of canals through which seawater is conducted and is used to inflate tube feet for locomotion, grasping food, and respiration.
wax: a lipid composed of fatty acids covalently bonded to long-chain alcohols.
weather: short-term fluctuations in temperature, humidity, cloud cover, wind, and precipitation in a region over periods of hours to days.
Werner syndrome: a rare condition in which a defective gene causes premature aging; caused by a mutation in the gene that codes for DNA replication/repair enzymes.
white blood cell: cellular components of blood; not as numerous as red blood cells; most function as part of the immune system and help defend the body against invaders.
white matter: the portion of the brain and spinal cord that consists largely of myelin-covered axons and that give these areas a white appearance.
withdrawal: the removal of the penis from the vagina just before ejaculation in an attempt to

avoid pregnancy; an ineffective contraceptive method.

working memory: the first phase of learning; short-term memory that is electrical or biochemical in nature.

xylem (zī-lum): a conducting tissue of vascular plants that transports water and minerals from root to shoot.

yolk: protein-rich or lipid-rich substances contained in eggs that provide food for the developing embryo.

yolk sac: one of the embryonic membranes of reptilian, bird, and mammalian embryos; in birds and reptiles, a membrane surrounding the yolk in the egg; in mammals, forms part of the umbilical cord and the digestive tract but is empty.

Z line: a fibrous protein structure to which the thin filaments of skeletal muscle are attached; forms the boundary of a sarcomere.

zona pellucida (pel-oo´-si-duh): a clear, noncellular layer between the corona radiata and the egg.

zooflagellate (zō-ō-fla´-jel-et): a nonphotosynthetic protist that moves by using flagella.

zooplankton: nonphotosynthetic protists that are abundant in marine and freshwater environments.

zoospore (zō´-ō-spor): a nonsexual reproductive cell that swims by using flagella; formed by members of the protistan division Oomycota.

zygospore (zī´-gō-spor): a fungal spore, produced by the division Zygomycota, that is surrounded by a thick, resistant wall and forms from a diploid zygote.

zygote (zī´-gōt): in sexual reproduction, a diploid cell (the fertilized egg) formed by the fusion of two haploid gametes.

zygote fungus: a fungus of the division Zygomycota, which includes the species that cause fruit rot and bread mold.

Photo Credits

Chapter 1
Opener NASA Headquarters
1-1a Andrew Syred/Science Photo Library/Photo Researchers, Inc.
1-1b Craig Tuttle/Corbis/Stock Market
1-1c Kim Taylor/Bruce Coleman Inc.
1-2 Johnny Johnson/DRK Photo
1-3 Kim Taylor/Bruce Coleman Inc.
1-4 Lawrence Livermore National Laboratory/Photo Researchers, Inc.
1-9 Teresa and Gerald Audesirk
E1-1 Luiz C. Marigo/Peter Arnold, Inc.
E1-2 inset Gunter Ziesler/Peter Arnold, Inc.

Unit One
Opener Manfred Kage/Peter Arnold, Inc.

Chapter 2
Opener Tom Hubbard/Stockphoto.com
2-7a Stephen Dalton/Photo Researchers, Inc.
2-7b Teresa and Gerald Audesirk
2-12a Jean-Michel Labat/Auscape International Pty. Ltd.
2-12b Donald Specker/Animals Animals/Earth Scenes
2-15a Robert Pearcy/Animals Animals/Earth Scenes
2-15b Jeff Foott/DRK Photo
2-15c Nuridsany et Perennou/Photo Researchers, Inc.
E2-1 Calvin Larsen/Photo Researchers, Inc.

Chapter 3
Opener Oliver Meckes & Nicole Ottawa/Photo Researchers, Inc.
3-5aR Joseph Kurantsin-Mills, The George Washington University Medical Center
3-5bR Joseph Kurantsin-Mills, The George Washington University Medical Center
3-5cR Joseph Kurantsin-Mills, The George Washington University Medical Center
3-8b L.A. Hufnagel, Ultrastructural Aspects of Chemoreception in Ciliated Protists (Ciliophora), Journal of Electron Microscopy Technique, 1991. Photomicrograph by Jurgen Bohmer and Linda Hufnagel, University of Rhode Island.
3-9 Stephen J. Krasemann/DRK Photo

Chapter 4
Opener Charles A. Vacanti, M.D., UMass Memorial Health Care
4-4b E. Guth, T. Hashimoto, and S.F. Conti
4-5 Carolina Biological Supply Company/Phototake NYC
4-7UR Barry F. King/Biological Photo Service
4-7WR Barry F. King/Biological Photo Service
4-8M Don W. Fawcett/Photo Researchers, Inc.
E4-1aUL Biophoto Associates/Photo Researchers, Inc.
E4-1aWR Cecil Fox/Science Source/Photo Researchers, Inc.
E4-1b Jean-Claude Revy/Phototake NYC
E4-2a M.I. Walker/Photo Researchers, Inc.
E4-2b David M. Phillips/Visuals Unlimited
E4-2c Manfred Kage/Peter Arnold, Inc.
E4-2d National Library of Medicine

Chapter 5
Opener Sylvain Grandadam/Getty Images Inc.
5-1 Photograph by Dr. Harold E. Edgerton. © Harold & Esther Edgerton Foundation, 1999, courtesy of Palm Press, Inc.

Chapter 6
Opener Joe Tucciarone
6-1a Ken W. Davis/Tom Stack & Associates, Inc.
E6-1 Fred Bruemmer/Peter Arnold, Inc.

Chapter 7
Opener Frank Pedrick/The Image Works
7-6a1 Teresa Audesirk
7-6a2 Teresa Audesirk
7-6b Paul J. Sutton/Duomo Photography Incorporated

Unit Two
Opener USDA-Agricultural Research Service Information Staff

Chapter 8
Opener Christopher Brown/Stock Boston
E8-1 A. Barrington Brown/Photo Researchers, Inc.
E8-2 Providence Creative Group

Chapter 9
Opener Aaron Haupt/Photo Researchers, Inc.
9-4a Oscar L. Miller, Jr., University of Virginia
E9-2 Howard W. Jones, Jr., M.D., Eastern Virginia Medical School
E9-3 Reproduced from C.J. Epstein et al., Werner's syndrome: A review of its symptomatology, natural history, pathologic features, genetics, and relationship to the natural aging process. Medicine 45(3):177-221 (1966). Copyright © 1966 by Lippincott Willia

Chapter 10
Opener University of Massachusetts News Office, Amherst
10-1a Michael Abbey/Photo Researchers, Inc.
10-1b John Durham/Science Photo Library/Photo Researchers, Inc.
10-1c Carolina Biological Supply Company/Phototake NYC
10-2a CC Studio/Science Photo Library/Photo Researchers, Inc.
10-2b Pascal Goetgheluck/Science Photo Library/Photo Researchers, Inc.
10-2c Neil Harding/Getty Images Inc
10-2d G.W. Willis/Biological Photo Service/Biological Photo Service
10-5 Biophoto Associates/Photo Researchers, Inc.
10-6 CNRI/Science Photo Library/Photo Researchers, Inc.
10-7a Dr. Andrew S. Bajer, University of Oregon
10-7b Dr. Andrew S. Bajer, University of Oregon
10-7cv1 Dr. Andrew S. Bajer, University of Oregon
10-7cv2 Dr. Andrew S. Bajer, University of Oregon
10-7d Dr. Andrew S. Bajer, University of Oregon
10-7ev1 Dr. Andrew S. Bajer, University of Oregon
10-7ev2 Dr. Andrew S. Bajer, University of Oregon
10-7f Dr. Andrew S. Bajer, University of Oregon
10-9 T.E. Schroeder/Biological Photo Service
10-14 CNRI/Science Photo Library/Photo Researchers, Inc.
E10-1 Photograph courtesy of The Roslin Institute.

Chapter 11
Opener Rachel Epstein/Stuart Kenter Associates
11-2 Archiv/Photo Researchers, Inc.
11-3a D. Cavagnaro/DRK Photo
11-8 Biophoto Associates/Photo Researchers, Inc.
11-12 Jane Burton/Bruce Coleman Inc.
11-14a P. Marazzi/Science Photo Library/Photo Researchers, Inc.
11-14b K.H. Switak/Photo Researchers, Inc.
11-14c Gary Retherford/Photo Researchers, Inc.
11-15a Dennis Kunkel/Phototake NYC
11-15b Francis Leroy/Biocosmos/Science Photo Library/Photo Researchers, Inc.
11-16a CNRI/Science Photo Library/Photo Researchers, Inc.
11-16b Lawrence Migdale/Pix
E11-1 Nick Kelsh/Peter Arnold, Inc.

Chapter 12
Opener Inga Spence/Visuals Unlimited
12-1aW Stanley N. Cohen/Science Photo Library/Photo Researchers, Inc.
12-4 Courtesy of Orchid Cellmark, Inc., Germantown, Maryland.
12-5 Keith V. Wood/Science VU/Visuals Unlimited
12-6a Norm Thomas/Photo Researchers, Inc.
12-6b Pharmacia Corporation
12-7a Derek Bromhall/Oxford Scientific Films Ltd.
12-7b GenPharm International/Peter Arnold, Inc.
12-7c Choy Hew, National University of Singapore, and Garth Fletcher, Memorial University of Newfoundland, St. John's, Newfoundland.
12-8a AP/Wide World Photos
E12-1 Andrew J. Martinez/Photo Researchers, Inc.

Unit Three
Opener Francois Gohier/Photo Researchers, Inc.

Chapter 13
Opener O. Louis Mazzatenta/NGS Image Collection
13-1 Jim Steinberg/Photo Researchers, Inc.
13-3a John Cancalosi/DRK Photo
13-3b David M. Dennis/Tom Stack & Associates, Inc.
13-3c Chip Clark
13-9a Stephen Dalton/Photo Researchers, Inc.
13-9b Stephen Dalton/Photo Researchers, Inc.
13-9c Douglas T. Cheeseman, Jr./Peter Arnold, Inc.
13-9d Gerard Lacz/Peter Arnold, Inc.
13-10a Photo Lennart Nilsson/Albert Bonniers Forlag AB
13-10b Photo Lennart Nilsson/Albert Bonniers Forlag AB
13-10c Photo Lennart Nilsson/Albert Bonniers Forlag AB
13-11a Stephen J. Krasemann/DRK Photo
13-11b Timothy O'Keefe/Tom Stack & Associates, Inc.
E13-1 CORBIS
E13-2 Frans Lanting/Photo Researchers, Inc.

Chapter 14
Opener David Scharf/Peter Arnold, Inc.
14-3b Luiz C. Marigo/Peter Arnold, Inc.
14-3c Y.R. Tymstra/Valan Photos
14-4a M.P.L. Fogden/Bruce Coleman Inc.

14-4b Tim Davis/Photo Researchers, Inc.
14-5 W. Perry Conway/Tom Stack & Associates, Inc.
14-6 D. Cavagnaro/DRK Photo
14-7a Wayne Lankinen/Valan Photos
14-7b Edgar T. Jones/Bruce Coleman Inc.
14-8a Pat & Tom Leeson/Photo Researchers, Inc.
14-8b Pat & Tom Leeson/Photo Researchers, Inc.
14-9a Mark Smith/Photo Researchers, Inc.
14-9b Thomas Kitchin/Tom Stack & Associates, Inc.
14-9c Mark Smith/Photo Researchers, Inc.
14-10 Tim Laman/NGS Image Collection
14-11 Joy Spurr/Bruce Coleman Inc.
E14-1 Frans Lanting/Minden Pictures

Chapter 15

Opener TSADO/NASA/Tom Stack & Associates, Inc.
15-3 Sidney Fox/Science VU/Visuals Unlimited
15-5 Michael Abbey/Visuals Unlimited
15-6a Milwaukee Public Museum, Photograph Collection
15-6b James L. Amos/Photo Researchers, Inc.
15-6c Carolina Biological Supply Company/Phototake NYC
15-6d Douglas Faulkner/Photo Researchers, Inc.
15-7 llustration by Ludek Pesek/Science Photo Library/Photo Researchers, Inc.
15-8 Terry Whittaker/Photo Researchers, Inc.
15-9 Illustration by Chris Butler/Science Photo Library/Photo Researchers, Inc.
15-12a Tom McHugh/Chicago Zoological Park/Photo Researchers, Inc.
15-12b Frans Lanting/Minden Pictures
15-12c Nancy Adams/Tom Stack & Associates, Inc.
15-13WL Tim D. White/Brill Atlanta
15-15 Neanderthal Museum, Mettmann, Germany
15-16 Jerome Chatin/Getty Images, Inc.

Chapter 16

Opener Stephen Dalton/Photo Researchers, Inc.
16-1a Wayne Lankinen/Bruce Coleman Inc.
16-1b M.C. Chamberlain/DRK Photo
16-1c Maslowski/Photo Researchers, Inc.
16-2a C. Steven Murphree/Biological Photo Service
16-2b Dr. Greg Rouse, Department of Invertebrate Zoology, National Museum of Natural History, Smithsonian Institution
16-2c Dr. Jeremy Burgess/Science Photo Library/Photo Researchers, Inc.
16-3a Hans Gelderblom/Getty Images Inc
16-3b Reprinted by permission of Springer-Verlag from W.J. Jones, J.A. Leigh, F. Mayer, C.R. Woese, and R.S. Wolfe, *Methanococcus jannaschii sp. nov.,* an extremely thermophilic methanogen from a submarine hydrothermal vent. Archives of *Microbiology* 136:254-26
16-5a David M. Phillips/Visuals Unlimited
16-5b Karl O. Stetter, University of Regensburg, Germany
16-5c CNRI/Science Photo Library/Photo Researchers, Inc.
16-6 Karl O. Stetter, University of Regensburg, Germany
16-7 A.B. Dowsett/Science Photo Library/Photo Researchers, Inc.
16-8 CNRI/Science Photo Library/Photo Researchers, Inc.
16-9 Alan L. Detrick/Photo Researchers, Inc.
16-10a Carolina Biological Supply Company/Phototake NYC
16-10b Eric Grave/Science Source/Photo Researchers, Inc.
16-11a P.W. Grace/Science Source/Photo Researchers, Inc.
16-11b Cabisco/Visuals Unlimited
16-12 David M. Phillips/Visuals Unlimited
16-13 Kevin Schafer/Peter Arnold, Inc.
16-14a D.P. Wilson/Eric and David Hosking/Photo Researchers, Inc.
16-14b Lawrence E. Naylor/Photo Researchers, Inc.
16-15 Harry Rogers/Photo Researchers, Inc.
16-16 M.I. Walker/Photo Researchers, Inc.
16-17a Ed Degginger/Color-Pic, Inc.
16-17b Manfred Kage/Peter Arnold, Inc.
16-18 Oliver Meckes & Nicole Ottawa/Eye of Science/Photo Researchers, Inc.
16-19a Robert & Linda Mitchell/Robert & Linda Mitchell Photography
16-19b Elmer Koneman/Visuals Unlimited
16-20 Jeff Lepore/Photo Researchers, Inc.
16-21 David M. Dennis/Tom Stack & Associates, Inc.
16-22 Michael Fogden/DRK Photo
16-23 David Dvorak, Jr.
16-24a John Gerlach/Tom Stack & Associates, Inc.
16-24b John Shaw/Tom Stack & Associates, Inc.
16-25a Dwight R. Kuhn/Dwight R. Kuhn Photography
16-25b Milton Rand/Tom Stack & Associates, Inc.
16-25c Larry Ulrich/DRK Photo
16-26a Maurice Nimmo/A-Z Botanical Collection, Ltd.
16-26b Teresa and Gerald Audesirk
16-27a Dwight R. Kuhn/Dwight R. Kuhn Photography
16-27b David Dare Parker/Auscape International Pty. Ltd.
16-27b inset Matt Jones/Auscape International Pty. Ltd.
16-27c Dwight R. Kuhn/Dwight R. Kuhn Photography
16-27d Teresa and Gerald Audesirk
16-27e Larry West/Photo Researchers, Inc.
16-28a Larry Lipsky/DRK Photo
16-28b Brian Parker/Tom Stack & Associates, Inc.
16-28c Charles Seaborn/Odyssey Productions, Inc.
16-29a Gregory Ochocki/Photo Researchers, Inc.
16-29b Charles Seaborn/Odyssey Productions, Inc.
16-29c Teresa and Gerald Audesirk
16-29d David B. Fleetham/Innerspace Visions
16-30a Kjell B. Sandved/Butterfly Alphabet, Inc.
16-30b David L. Bull/David Bull Photography
16-30c J.H. Robinson/Photo Researchers, Inc.
16-31a Carolina Biological Supply Company/Phototake NYC
16-31b Peter J. Bryant/Biological Photo Service
16-31c Stephen Dalton/Photo Researchers, Inc.
16-31d Werner H. Muller/Peter Arnold, Inc.
16-31e Stanley Breeden/DRK Photo
16-32a Dwight Kuhn/Dwight R. Kuhn Photography
16-32b Tim Flach/Getty Images Inc.
16-32c Teresa and Gerald Audesirk
16-33a Tom Branch/Photo Researchers, Inc.
16-33b Peter J. Bryant/Biological Photo Service
16-33c Carolina Biological Supply Company/Phototake NYC
16-33d Alex Kerstitch/Estate of Alex Kerstitch
16-34a Ray Coleman/Photo Researchers, Inc.
16-34b Alex Kerstitch/Estate of Alex Kerstitch
16-35a Fred Bavendam/Peter Arnold, Inc.
16-35b Ed Reschke/Peter Arnold, Inc.
16-36a Fred Bavendam/Peter Arnold, Inc.
16-36b Kjell B. Sandved/Photo Researchers, Inc.
16-36c Alex Kerstitch/Estate of Alex Kerstitch
16-37 Tom McHugh/Photo Researchers, Inc.
16-38a Peter David/Getty Images/Planet Earth Pictures Ltd
16-38b Mike Neumann/Photo Researchers, Inc.
16-38c Stephen Frink/Getty Images Inc
16-38d Peter Scoones/Getty Images/Planet Earth Pictures Ltd
16-39a Breck P. Kent/Animals Animals/Earth Scenes
16-39b Joe McDonald/Tom Stack & Associates, Inc.
16-39c Cosmos Blank/National Audubon Society/Photo Researchers, Inc.
16-40a David G. Barker/Tom Stack & Associates, Inc.
16-40b Roger K. Burnard/Biological Photo Service
16-40c Frans Lanting/Minden Pictures
16-41a Walter E. Harvey/Photo Researchers, Inc.
16-41b Carolina Biological Supply Company/Phototake NYC
16-41c Ray Ellis/Photo Researchers, Inc.
16-42a Flip Nicklin/Minden Pictures
16-42b Jonathan Watts/Science Photo Library/Photo Researchers, Inc.
16-42c C. and M. Denis-Huot/Peter Arnold, Inc.
16-42d S.R. Maglione/Photo Researchers, Inc.
E16-1 Stanley Breeden/DRK Photo

Unit Four

Opener John Shaw/Tom Stack & Associates, Inc.

Chapter 17

Opener Tom & Pat Leeson/Photo Researchers, Inc.
17-4a Dr. Jeremy Burgess/Science Photo Library/Photo Researchers, Inc.
17-4b John D. Cunningham/Visuals Unlimited
17-7a Lynwood M. Chace/Photo Researchers, Inc.
17-7b Dwight R. Kuhn/Dwight R. Kuhn Photography
17-8UR Ed Reschke/Peter Arnold, Inc.
17-8WR E.R. Degginger/Animals Animals/Earth Scenes
17-9 Teresa and Gerald Audesirk
17-10 J. Robert Waaland, University of Washington/Biological Photo Service
17-11UR Ed Reschke/Peter Arnold, Inc.
17-14a Kenneth W. Fink/Photo Researchers, Inc.
17-14b Belinda Wright/DRK Photo
17-17 Biophoto Associates/Science Source/Photo Researchers, Inc.
17-18R Hugh Spencer/National Audubon Society/Photo Researchers, Inc.
E17-1 Michael Fogden/DRK Photo

Chapter 18

Opener Maximilian Stock Ltd/Animals Animals/Earth Scenes
18-1 Jeff Foott/Bruce Coleman Inc.
18-3b Dr. James L. Castner
18-4 Teresa and Gerald Audesirk
18-5 Dennis Kunkel/Dennis Kunkel Microscopy, Inc.
18-6L Carolina Biological Supply Company/Phototake NYC
18-10 Teresa and Gerald Audesirk
18-11c Runk/Schoenberger/Grant Heilman Photography, Inc.
18-14 John Cancalosi/DRK Photo
18-15 Biophoto Associates/Photo Researchers, Inc.
18-16 Teresa and Gerald Audesirk
18-17a Thomas Eisner, Cornell University.
18-17b Thomas Eisner, Cornell University.
18-18 This item is reproduced by permission of The Huntington Library, San Marino, California.

18-19a Teresa and Gerald Audesirk
18-19b Carolina Biological Supply Company/Phototake NYC
18-20 Teresa and Gerald Audesirk
18-21 Scott Camazine/Photo Researchers, Inc.

Unit Five
Opener Randy Wells/Getty Images Inc.

Chapter 19
Opener Bryan and Cherry Alexander
19-4 Biophoto Associates/Photo Researchers, Inc.
19-5 Manfred Kage/Peter Arnold, Inc.
19-6a John Kennedy/Biological Photo Service/Getty Images Inc
19-6b Fred Bruemmer/OKAPIA/Photo Researchers, Inc.
19-7 Yorgos Nikas/Science Photo Library/Photo Researchers, Inc.
19-8 Manfred Kage/Peter Arnold, Inc.

Chapter 20
Opener Grant Heilman/Grant Heilman Photography, Inc.
20-5 Dennis Kunkel/Dennis Kunkel Microscopy, Inc.
20-6 Photo Lennart Nilsson/Albert Bonniers Forlag
20-7 Kenneth Kaushansky, M.D., University of Washington Medical Center
20-8a CNRI/Science Photo Library/Photo Researchers, Inc.
20-11a Photo Lennart Nilsson/Albert Bonniers Forlag
E20-3b Lou Lainey/TimePix
E20-4 Bill Travis, M.D./National Cancer Institute
E20-5 O. Auerbach/Visuals Unlimited

Chapter 21
Opener Charles Gupton/Stock Boston
21-1 Charles Krebs/Getty Images Inc
21-3 Tobi Zausner
21-4a E.R. Degginger/Photo Researchers, Inc.
21-9b Photo Lennart Nilsson/Albert Bonniers Forlag
21-11b CNRI/Photo Researchers, Inc.
E21-1 De Keerle-Parker/Getty Images, Inc,

Chapter 22
Opener Oliver Meckes/Photo Researchers, Inc.
22-1a David Scharf/Peter Arnold, Inc.
22-2 Photo Lennart Nilsson/Albert Bonniers Forlag
22-3b National Institute for Biological Standards and Control, U.K./Science Photo Library/Photo Researchers, Inc.
22-3c Photo Lennart Nilsson/Albert Bonniers Forlag
22-8a Secchi-Lecaque/Roussel/CNRI/Science Source/Photo Researchers, Inc.
22-8b CNRI/Science Photo Library/Photo Researchers, Inc.
22-9 Oliver Meckes & Nicole Ottawa/Photo Researchers, Inc.
22-12b National Institute for Biological Standards and Control, U.K./Science Photo Library/Photo Researchers, Inc.
22-12c National Institute for Biological Standards and Control, U.K./Science Photo Library/Photo Researchers, Inc.
E22-1 CNRI/Science Photo Library/Photo Researchers, Inc.
E22-2a Lloyd Cluff/CORBIS
E22-2b Sinclair Stammers/Science Photo Library/Photo Researchers, Inc.

Chapter 23
Opener Howard K. Suzuki/Aquatic Life Sculptures
23-5 Feld Entertainment®
23-7b Biophoto Associates/Photo Researchers, Inc.
23-10 Photo courtesy of The Jackson Laboratory, Bar Harbor, Maine.

Chapter 24
Opener Doug Mindell Photography
24-12a Scanning electron micrographs by Robert S. Preston, courtesy of Professor Joseph E. Hawkins Jr., Kresge Hearing Research Institute, University of Michigan Medical School.
24-20a Charles E. Mohr/Photo Researchers, Inc.
24-20b Brian Parker/Tom Stack & Associates, Inc.
E24-1 Ken Graham/Bruce Coleman Inc.

Chapter 25
Opener Brook Kraft/Corbis/Sygma
25-1 Biophoto Associates/Photo Researchers, Inc.
25-3 Tom McHugh/Photo Researchers, Inc.
25-4 Rudie Kuiter/Innerspace Visions
25-5 Michael Fogden/Oxford Scientific Films/Animals Animals/Earth Scenes
25-6a Stephen J. Krasemann/DRK Photo
25-6b Tom McHugh/Photo Researchers, Inc.
25-6c Frans Lanting/Minden Pictures
25-10b CNRI/Science Photo Library/Photo Researchers, Inc.
25-14 C. Edelmann/La Villette/Photo Researchers, Inc.
25-16a Photo Lennart Nilsson/Albert Bonniers Forlag
25-16b Andrew Syred/Microscopix Photolibrary
25-19a Photo Lennart Nilsson/Albert Bonniers Forlag
25-22a Photo Lennart Nilsson/Albert Bonniers Forlag
25-22b Petit Format/Nestle/Science Source/Photo Researchers, Inc.
25-22c Photo Lennart Nilsson/Albert Bonniers Forlag

Chapter 26
Opener Reuters NewMedia Inc./CORBIS
Opener inset Joe McDonald/DRK Photo
26-1a Eric and David Hosking/Frank Lane Picture Agency Limited
26-1b Eric and David Hosking/Frank Lane Picture Agency Limited
26-2a Boltin Picture Library
26-2b Boltin Picture Library
26-2c Boltin Picture Library
26-2d Boltin Picture Library
26-2e Boltin Picture Library
26-3 Frans Lanting/Minden Pictures
26-5 Thomas McAvoy/TimePix
26-7a Renee Lynn/Photo Researchers, Inc.
26-8 Ken Cole/Animals Animals/Earth Scenes
26-9 Richard K. LaVal/Animals Animals/Earth Scenes
26-10 Nuridsany et Perennou/Photo Researchers, Inc.
26-11 Robert & Linda Mitchell/Robert & Linda Mitchell Photography
26-12a Stephen J. Krasemann/Photo Researchers, Inc.
26-12b Hans Pfletschinger/Peter Arnold, Inc.
26-13a M.P. Kahl/DRK Photo
26-13b Marc Chamberlain
26-14 Ray Dove/Visuals Unlimited
26-15 William Ervin/Natural Imagery
26-16 Michael K. Nichols/National Geographic Image Collection
26-17 John D. Cunningham/Visuals Unlimited
26-19a Norbert Wu/Peter Arnold, Inc.
26-19b Frans Lanting/Minden Pictures
26-20 Anne et Jacques Six
26-22a Dwight R. Kuhn Photography
26-22b Teresa and Gerald Audesirk
26-23 Fred Bruemmer/Peter Arnold, Inc.
26-25 Raymond A. Mendez/Animals Animals/Earth Scenes
26-26 Photo Lennart Nilsson/Albert Bonniers Forlag
26-27 William P. Fifer, New York State Psychiatric Institute, Columbia University
26-28a Frans Lanting/Minden Pictures
26-28b Daniel J. Cox/Getty Images, Inc,
26-28c Michio Hoshino/Minden Pictures

Unit Six
Opener Paradise, Part One of The Trilogy of the Earth by Suzanne Duranceau/Illustratrice, Inc.

Chapter 27
Opener Richard Nowitz/Photo Researchers, Inc.
27-1a Prof. S. Cinti/University of Ancona, Italy/CNRI/Phototake NYC
27-1b Thomas Kitchin/Tom Stack & Associates, Inc.
27-6 Tom McHugh/Photo Researchers, Inc.
27-7a Robert & Linda Mitchell Photography
27-7b Tom and Susan Bean, Inc./DRK Photo
27-7c © 2002 Gary Braasch
27-8 inset NASA/Johnson Space Center
E27-3 Frans Lanting/Minden Pictures
E27-3 inset Andy Holbrooke/Corbis/Stock Market

Chapter 28
Opener Hal Beral/Visuals Unlimited
28-3a Henry Ausloos/Animals Animals/Earth Scenes
28-3b Stephen Dalton/Photo Researchers, Inc.
28-3c Frans Lanting/Minden Pictures
28-4a Art Wolfe/Art Wolfe, Inc.
28-4b Frans Lanting/Minden Pictures
28-5a Marty Cordano/DRK Photo
28-5b Paul A. Zahl/Photo Researchers, Inc.
28-5c Ray Coleman/Photo Researchers, Inc.
28-6a Jean-Paul Ferrero and Jean-Michel Labat/Auscape International Pty. Ltd.
28-6b Charles V. Angelo/Photo Researchers, Inc.
28-7 Dr. James L. Castner
28-8a Carmela Leszczynski/Animals Animals/Earth Scenes
28-8b Breck P. Kent
28-8c Ed Degginger/Color-Pic, Inc.
28-8d Ed Degginger/Color-Pic, Inc.
28-9a Robert P. Carr/Bruce Coleman Inc.
28-9b Monica Mather and Bernard Roitberg, Simon Fraser University
28-10a Zig Leszczynski/Animals Animals/Earth Scenes
28-10b Dr. James L. Castner
28-10c Jeff Lepore/Photo Researchers, Inc.
28-11a Thomas Eisner and Daniel Aneshansley, Cornell University.
28-11b Frans Lanting/Minden Pictures
28-12 Teresa Audesirk
28-13a Jim Zipp/Photo Researchers, Inc.
28-13b Stephen J. Krasemann/Photo Researchers, Inc.
28-14aL Dennis Oda/Getty Images Inc
28-14aR Krafft-Explorer/Photo Researchers, Inc.
28-14bL © 2002 Gary Braasch
28-14bR © 2002 Gary Braasch
28-14cR M.P. Kahl/DRK Photo
28-14L Wendy Shattil/Bob Rozinski/Tom Stack & Associates, Inc.
28-17a Robert & Linda Mitchell Photography
28-17b Robert & Linda Mitchell Photography
28-17c Robert & Linda Mitchell Photography
E28-1b Chuck Pratt/Bruce Coleman Inc.
E28-1c Robert & Linda Mitchell Photography
E28-1d Jack Jeffrey/Photo Resource Hawaii Stock Photography

Chapter 29
Opener Biospheres Ventures/Corbis/Sygma
29-12 Tom Walker/Stock Boston
29-13 William E. Ferguson

29-14 Will McIntyre/Photo Researchers, Inc.
E29-1 Thomas A. Schneider/SchneiderStock Photography

Chapter 30

Opener Martin Wendler/Peter Arnold, Inc.
30-2b NASA/Johnson Space Center
30-6a Teresa and Gerald Audesirk
30-6b Tom McHugh/Photo Researchers, Inc.
30-9a Loren McIntyre
30-9b Schafer and Hill/Peter Arnold, Inc.
30-9c Michael Fogden/Oxford Scientific Films Ltd.
30-9d Michael Fogden/DRK Photo
30-9e Frans Lanting/Minden Pictures
30-10 Jacques Jangoux/Peter Arnold, Inc.
30-11a W. Perry Conway/Aerie Nature Series, Inc.
30-11b Aubry Lang/Valan Photos
30-11c Stephen J. Krasemann/DRK Photo
30-12 James Hancock/Photo Researchers, Inc.
30-13a Don & Pat Valenti/DRK Photo
30-13b Brian Parker/Tom Stack & Associates, Inc.
30-13c Tom McHugh/Photo Researchers, Inc.
30-14 William H. Mullins/Photo Researchers, Inc.
30-15 Brian Parker/Tom Stack & Associates, Inc.
30-16 Harvey Payne
30-17a Tom and Susan Bean, Inc./DRK Photo
30-17b W. Perry Conway/Tom Stack & Associates, Inc.
30-17c Nicholas DeVore III/Network Aspen
30-17d Bob Gurr/Valan Photos
30-17e Jim Brandenburg/Minden Pictures
30-18 Teresa and Gerald Audesirk
30-19a © 1993 Gary Braasch
30-19b Thomas Kitchin/Tom Stack & Associates, Inc.
30-19c Stephen J. Krasemann/DRK Photo
30-19d Joe McDonald/Animals Animals/Earth Scenes
30-20a Susan G. Drinker/Corbis/Stock Market
30-20b Teresa and Gerald Audesirk
30-20c Jeff Foott Productions
30-20d Richard Thom/Tom Stack & Associates, Inc.
30-21a Teresa and Gerald Audesirk
30-21b Marty Stouffer Productions/Animals Animals/Earth Scenes
30-21c Wm. Bacon III/National Audubon Society/Photo Researchers, Inc.
30-22 Gary Braasch/Gary Braasch Photography
30-23a Stephen J. Krasemann/DRK Photo
30-23b Jeff Foott/Tom Stack & Associates, Inc.
30-23c Michael Giannechini/Photo Researchers, Inc.
30-23d James Simon/Photo Researchers, Inc.
30-26a T.P. Dickinson/Comstock Images
30-26b Teresa and Gerald Audesirk
30-26b inset Walter Dawn/National Audubon Society/Photo Researchers, Inc.
30-26c Teresa and Gerald Audesirk
30-26c inset Thomas Kitchin/Tom Stack & Associates, Inc.
30-26d Robert R. Given, Marymount College
30-26d inset Stan Wayman/Photo Researchers, Inc.
30-27a Franklin J. Viola/Viola's Photo Visions, Inc.
30-27b Alex Kerstitch/Estate of Alex Kerstitch
30-27c Charles Seaborn/Odyssey Productions, Inc.
30-27d Alex Kerstitch/Estate of Alex Kerstitch
30-28a Stephen J. Krasemann/DRK Photo
30-28b Paul Humann/Jeffrey L. Rotman Photography
30-28c Fred McConnaughey/Photo Researchers, Inc.
30-28d James D. Watt/Getty Images/Planet Earth Pictures Ltd
30-28e Biophoto Associates/Science Source/Photo Researchers, Inc.
30-28f Robert Arnold/Getty Images/Planet Earth Pictures Ltd
30-29 J. Frederick Grassle, Institute of Marine and Coastal Sciences, Rutgers University
E30-1 NASA Headquarters
E30-2 Roy Morsch/Corbis/Stock Market

Index

Teresa Audesirk, Gerald Audesirk,
Bruce E. Byers
Life on Earth, 3rd Edition
MediaTutor Student CD-ROM
ISBN#: 0-13-065352-7

LICENSE AGREEMENT

YOU SHOULD CAREFULLY READ THE FOLLOWING TERMS AND CONDITIONS BEFORE BREAKING THE SEAL ON THE PACKAGE. AMONG OTHER THINGS, THIS AGREEMENT LICENSES THE ENCLOSED SOFTWARE TO YOU AND CONTAINS WARRANTY AND LIABILITY DISCLAIMERS. BY BREAKING THE SEAL ON THE PACKAGE, YOU ARE ACCEPTING AND AGREEING TO THE TERMS AND CONDITIONS OF THIS AGREEMENT. IF YOU DO NOT AGREE TO THE TERMS OF THIS AGREEMENT, DO NOT BREAK THE SEAL. YOU SHOULD PROMPTLY RETURN THE PACKAGE UNOPENED.

LICENSE.
Subject to the provisions contained herein, Prentice-Hall, Inc. ("PH") hereby grants to you a non-exclusive, non-transferable license to use the object code version of the computer software product ("Software") contained in the package on a single computer of the type identified on the package.

SOFTWARE AND DOCUMENTATION.
PH shall furnish the Software to you on media in machine-readable object code form and may also provide the standard documentation ("Documentation") containing instructions for operation and use of the Software.

LICENSE TERM AND CHARGES.
The term of this license commences upon delivery of the Software to you and is perpetual unless earlier terminated upon default or as otherwise set forth herein.

TITLE.
Title, and ownership right, and intellectual property rights in and to the Software and Documentation shall remain in PH and/or in suppliers to PH of programs contained in the Software. The Software is provided for your own internal use under this license. This license does not include the right to sublicense and is personal to you and therefore may not be assigned (by operation of law or otherwise) or transferred without the prior written consent of PH. You acknowledge that the Software in source code form remains a confidential trade secret of PH and/or its suppliers and therefore you agree not to attempt to decipher or decompile, modify, disassemble, reverse engineer or prepare derivative works of the Software or develop source code for the Software or knowingly allow others to do so. Further, you may not copy the Documentation or other written materials accompanying the Software.

UPDATES.
This license does not grant you any right, license, or interest in and to any improvements, modifications, enhancements, or updates to the Software and Documentation. Updates, if available, may be obtained by you at PH's then current standard pricing, terms, and conditions.

LIMITED WARRANTY AND DISCLAIMER.
PH warrants that the media containing the Software, if provided by PH, is free from defects in material and workmanship under normal use for a period of sixty (60) days from the date you purchased a license to it.

THIS IS A LIMITED WARRANTY AND IT IS THE ONLY WARRANTY MADE BY PH. THE SOFTWARE IS PROVIDED 'AS IS' AND PH SPECIFICALLY DISCLAIMS ALL WARRANTIES OF ANY KIND, EITHER EXPRESS OR IMPLIED, INCLUDING, BUT NOT LIMITED TO, THE IMPLIED WARRANTY OF MERCHANTABILITY AND FITNESS FOR A PARTICULAR PURPOSE. FURTHER, COMPANY DOES NOT WARRANT, GUARANTY OR MAKE ANY REPRESENTATIONS REGARDING THE USE, OR THE RESULTS OF THE USE, OF THE SOFTWARE IN TERMS OF CORRECTNESS, ACCURACY, RELIABILITY, CURRENTNESS, OR OTHERWISE AND DOES NOT WARRANT THAT THE OPERATION OF ANY SOFTWARE WILL BE UNINTERRUPTED OR ERROR FREE. COMPANY EXPRESSLY DISCLAIMS ANY WARRANTIES NOT STATED HEREIN. NO ORAL OR WRITTEN INFORMATION OR ADVICE GIVEN BY PH, OR ANY PH DEALER, AGENT, EMPLOYEE OR OTHERS SHALL CREATE, MODIFY OR EXTEND A WARRANTY OR IN ANY WAY INCREASE THE SCOPE OF THE FOREGOING WARRANTY, AND NEITHER SUBLICENSEE OR PURCHASER MAY RELY ON ANY SUCH INFORMATION OR ADVICE. If the media is subjected to accident, abuse, or improper use; or if you violate the terms of this Agreement, then this warranty shall immediately be terminated. This warranty shall not apply if the Software is used on or in conjunction with hardware or programs other than the unmodified version of hardware and programs with which the Software was designed to be used as described in the Documentation.

LIMITATION OF LIABILITY.
Your sole and exclusive remedies for any damage or loss in any way connected with the Software are set forth below. UNDER NO CIRCUMSTANCES AND UNDER NO LEGAL THEORY, TORT, CONTRACT, OR OTHERWISE, SHALL PH BE LIABLE TO YOU OR ANY OTHER PERSON FOR ANY INDIRECT, SPECIAL, INCIDENTAL, OR CONSEQUENTIAL DAMAGES OF ANY CHARACTER INCLUDING, WITHOUT LIMITATION, DAMAGES FOR LOSS OF GOODWILL, LOSS OF PROFIT, WORK STOPPAGE, COMPUTER FAILURE OR MALFUNCTION, OR ANY AND ALL OTHER COMMERCIAL DAMAGES OR LOSSES, OR FOR ANY OTHER DAMAGES EVEN IF PH SHALL HAVE BEEN INFORMED OF THE POSSIBILITY OF SUCH DAMAGES, OR FOR ANY CLAIM BY ANY OTHER PARTY. PH'S THIRD PARTY PROGRAM SUPPLIERS MAKE NO WARRANTY, AND HAVE NO LIABILITY WHATSOEVER, TO YOU. PH's sole and exclusive obligation and liability and your exclusive remedy shall be: upon PH's election, (i) the replacement of your defective media; or (ii) the repair or correction of your defective media if PH is able, so that it will conform to the above warranty; or (iii) if PH is unable to replace or repair, you may terminate this license by returning the Software. Only if you inform PH of your problem during the applicable warranty period will PH be obligated to honor this warranty. You may contact PH to inform PH of the problem as follows:

SOME STATES OR JURISDICTIONS DO NOT ALLOW THE EXCLUSION OF IMPLIED WARRANTIES OR LIMITATION OR EXCLUSION OF CONSEQUENTIAL DAMAGES, SO THE ABOVE LIMITATIONS OR EXCLUSIONS MAY NOT APPLY TO YOU. THIS WARRANTY GIVES YOU SPECIFIC LEGAL RIGHTS AND YOU MAY ALSO HAVE OTHER RIGHTS WHICH VARY BY STATE OR JURISDICTION.

MISCELLANEOUS.
If any provision of this Agreement is held to be ineffective, unenforceable, or illegal under certain circumstances for any reason, such decision shall not affect the validity or enforceability (i) of such provision under other circumstances or (ii) of the remaining provisions hereof under all circumstances and such provision shall be reformed to and only to the extent necessary to make it effective, enforceable, and legal under such circumstances. All headings are solely for convenience and shall not be considered in interpreting this Agreement. This Agreement shall be governed by and construed under New York law as such law applies to agreements between New York residents entered into and to be performed entirely within New York, except as required by U.S. Government rules and regulations to be governed by Federal law.

YOU ACKNOWLEDGE THAT YOU HAVE READ THIS AGREEMENT, UNDERSTAND IT, AND AGREE TO BE BOUND BY ITS TERMS AND CONDITIONS. YOU FURTHER AGREE THAT IT IS THE COMPLETE AND EXCLUSIVE STATEMENT OF THE AGREEMENT BETWEEN US THAT SUPERSEDES ANY PROPOSAL OR PRIOR AGREEMENT, ORAL OR WRITTEN, AND ANY OTHER COMMUNICATIONS BETWEEN US RELATING TO THE SUBJECT MATTER OF THIS AGREEMENT.

U.S. GOVERNMENT RESTRICTED RIGHTS.
Use, duplication or disclosure by the Government is subject to restrictions set forth in subparagraphs (a) through (d) of the Commercial Computer-Restricted Rights clause at FAR 52.227-19 when applicable, or in subparagraph (c) (1) (ii) of the Rights in Technical Data and Computer Software clause at DFARS 252.227-7013, and in similar clauses in the NASA FAR Supplement.

Minimum System Requirements

PC:
- Windows 98/Me/2000/NT
- Pentium II 233 MHz processor
- 64 Mb RAM
- 800 x 600 pixel screen resolution
- Color monitor running "Thousands of Colors" or higher
- Audio capable system with speakers/headphones is recommended
- 4x CD-ROM drive
- Compatible internet browser and multimedia plug-in/ActiveX control
  - Netscape Navigator/Communicator 4.X
    Netscape Communicator 4.78 (c) 1995-2003 is strongly recommended. The installer is included on this CD-ROM.
    **IMPORTANT: Netscape 6 and higher is NOT supported.
  - Internet Explorer 4.X and higher
  - Macromedia Shockwave 8.50 release 326 plug-in
    (c) 1985-2003. The installer is included on this CD-ROM.

Macintosh:
- Power PC with OS 8.5 or greater
- 64 Mb RAM, 40MB Free RAM, Virtual Memory Enabled
- 800 x 600 pixel screen resolution
- Color monitor running "Thousands of Colors" or higher
- Audio capable system with speakers/headphones is recommended
- 4x CD-ROM drive
- Compatible internet browser and multimedia plug-in/ActiveX control
  - Netscape Navigator/Communicator 4.X
    Netscape Communicator 4.78 (c) 1995-2003 is strongly recommended. The installer is included on this CD-ROM.
    **IMPORTANT: Netscape 6 and higher is NOT supported.
  - Macromedia Shockwave 8.50 release 326 plug-in
    (c) 1985-2003. The installer is included on this CD-ROM.

Getting Started

[1.] Installing Netscape 4.x

In order to optimally view this CD-ROM, we recommend using Netscape Navigator/Communicator 4.x as your browser of choice. Netscape Communicator 4.78 is available on this CD. For instructions to install Netscape 4.78 please view the Readme.txt file on the CD-ROM.

Important: Netscape 6.x browsers are not supported as of the release date for this product.

[2.] Installing Macromedia
Shockwave 8.5

The Macromedia Shockwave 8.5 browser plug-in is required. For instructions to install Macromedia Shockwave 8.5 please view the Readme.txt file on the CD-ROM.

[3.] Starting the Program

Macintosh
1. Insert the CD into the drive.
2. A CD icon labeled "LOE3e" will appear on your desktop. Double-click on it to open.
3. Locate the file "Start.html" on the CD-ROM and double-click on it to run it.

PC:
1. Insert the CD into the drive.
2. Locate the "My Computer" icon on the desktop and double click on it to open.
3. Locate the CD-ROM Drive icon (it should be labeled "LOE3e") and double click on it to open.
4. Locate the file "Start.html" on the CD-ROM and double-click on it to run it.

Compatibility Check

When you run the program for the first time, you will see the "Compatibility Check" page, which will help you determine whether you have the browser and plug-in you need to view the content. If you do not have the browser and plug-in you need, the "Compatibility Check" page will provide instructions on how you can install them. **Note: you will not be able to continue until you have the required browser and plug-in.